Chlorinated Solvent Source Zone Remediation

SERDP and ESTCP Remediation Technology Monograph Series

Series Editor: C. Herb Ward, Rice University

Chlorinated Solvent Source Zone Remediation

Edited by

Bernard H. Kueper

Queen's University, Kingston, ON, Canada

Hans F. Stroo

Stroo Consulting LLC, Ashland, OR, USA

Catherine M. Vogel

Noblis Inc., Atlanta, GA, USA

C. Herb Ward

Rice University, Houston, TX, USA

Authors

Michael D. Annable
Robert C. Borden
Michael C. Brooks
Richard A. Brown
Natalie L. Cápiro
Wilson S. Clayton
Michelle Crimi
Kathy L. Davies
Rula A. Deeb
Ronald W. Falta
Jason I. Gerhard
Mark Harkness
Kirk Hatfield
Elisabeth L. Hawley
James W. Jawitz
Paul C. Johnson
Richard L. Johnson
Michael C. Kavanaugh
Jennifer L. Triplett Kingston
Julie Konzuk
Bernard H. Kueper
Herbert Levine
Gregory V. Lowry
David W. Major
Michael C. Marley
James W. Mercer
Kevin G. Mumford
Charles J. Newell
Jennifer L. Nyman
Robert H. O'Laskey
Kurt D. Pennell
P. Suresh C. Rao
Alicia J. Shepard
Robert L. Siegrist
Brent E. Sleep
Hans F. Stroo
Eric J. Suchomel
Neil R. Thomson
Paul G. Tratnyek
Catherine M. Vogel
Douglas I. Walker
Michael R. West
C. Herb Ward
Dean F. Williamson
John T. Wilson
A. Lynn Wood

Springer

Editors
Bernard H. Kueper
Department of Civil Engineering
Queen's University
Kingston, ON, Canada
kueper@civil.queensu.ca

Catherine M. Vogel
Noblis Inc.
Atlanta, GA, USA
catherine.vogel@noblis.org

Hans F. Stroo
Stroo Consulting LLC
Ashland, OR, USA
hans@strooconsulting.com

C. Herb Ward
Civil and Environmental Engineering
Rice University
Houston, TX, USA
wardch@rice.edu

ISSN 1869-6864 ISSN 1869-6856 (electronic)
ISBN 978-1-4614-6921-6 ISBN 978-1-4614-6922-3 (eBook)
DOI 10.1007/978-1-4614-6922-3
Springer New York Heidelberg Dordrecht London

Library of Congress Control Number: 2014932833

Cover design: Cover layout designed by Kenneth C. Arevalo, Noblis Inc., Falls Church, VA 22042

Printed on acid-free paper

Springer is part of Springer Science+Business Media (www.springer.com)

SERDP and ESTCP Remediation Technology Monograph Series
Series Editor: C. Herb Ward, Rice University

SERDP and ESTCP have joined to facilitate the development of a series of monographs on remediation technology written by leading experts in each subject area. This volume provides a review of the state-of-the-art on chlorinated solvent source zone remediation. Previously published volumes in this series include:

- *In Situ* Bioremediation of Perchlorate in Groundwater
- *In Situ* Remediation of Chlorinated Solvent Plumes
- *In Situ* Chemical Oxidation for Groundwater Remediation
- Delivery and Mixing in the Subsurface: Processes and Design Principles for *In Situ* Remediation
- Bioaugmentation for Groundwater Remediation
- Processes, Assessment and Remediation of Contaminated Sediments

SERDP
DOD • EPA • DOE

U.S. Department of Defense Strategic
Environmental Research &
Development Program (SERDP)
4800 Mark Center Drive, Suite 17D08
Alexandria, VA 22350

ESTCP

U.S. Department of Defense Environmental
Security Technology Certification
Program (ESTCP)
4800 Mark Center Drive, Suite 17D08
Alexandria, VA 22350

Preface

In the late 1970s and early 1980s, North America began to grapple with the legacy of past disposal practices for toxic chemicals. With the passage in 1980 of the Comprehensive Environmental Response, Compensation, and Liability Act (CERCLA) in the United States, commonly known as Superfund, it became the law of the land to remediate these sites. The U.S. Department of Defense (DoD), the nation's largest industrial organization, also recognized that it too had a legacy of contaminated sites. Historic operations at Army, Navy, Air Force, and Marine Corps facilities, ranges, manufacturing sites, shipyards, and depots had resulted in widespread contamination of soil, groundwater, and sediment. While Superfund began in 1980 to focus on remediation of heavily contaminated sites largely abandoned or neglected by the private sector, the DoD had already initiated its Installation Restoration Program in the mid-1970s. In 1984, the DoD began the Defense Environmental Restoration Program (DERP) for contaminated site assessment and remediation. Two years later, the U.S. Congress codified the DERP and directed the Secretary of Defense to carry out a concurrent program of research, development, and demonstration of innovative remediation technologies.

As chronicled in the 1994 National Research Council report, "Ranking Hazardous-Waste Sites for Remedial Action," our early estimates on the cost and suitability of existing technologies for cleaning up contaminated sites were wildly optimistic. Original estimates, in 1980, projected an average Superfund cleanup cost of a mere $3.6 million per site and assumed that only around 400 sites would require remediation. The DoD's early estimates of the cost to clean up its contaminated sites were also optimistic. In 1985, the DoD estimated that the cleanup of its contaminated sites would cost from $5 billion to $10 billion, assuming 400–800 potential sites. A decade later, after an investment of over $12 billion on environmental restoration, the cost-to-complete estimates had grown to over $20 billion and the number of sites had increased to over 20,000. By 2007, after spending over $20 billion in the previous decade, the estimated cost to address the DoD's known liability for traditional cleanup (not including the munitions response program for unexploded ordnance) was still over $13 billion. Why did we underestimate the costs of cleaning up contaminated sites? All of these estimates were made with the tacit assumption that existing, off-the-shelf remedial technology was adequate to accomplish the task, that we had the scientific and engineering knowledge and tools to remediate these sites, and that we knew the full scope of chemicals of concern.

However, it was soon and painfully realized that the technology needed to address the more recalcitrant environmental contamination problems, such as fuels and chlorinated solvents in groundwater and dense nonaqueous phase liquids (DNAPLs) in the subsurface, was seriously lacking. In 1994, in the "Alternatives for Ground Water Cleanup" document, the National Research Council clearly showed that as a nation we had been conducting a failed 15-year experiment to clean up our nation's groundwater and that the default technology, pump-and-treat, was often ineffective at remediating contaminated aquifers. The answer for the DoD was clear. The DoD needed better technologies to clean up its contaminated sites, and better technologies could only arise through a better scientific and engineering understanding of the subsurface and the associated chemical, physical, and biological processes. Two DoD organizations were given responsibility for initiating new research, development, and demonstrations to obtain the technologies needed for cost-effective remediation of facilities across the DoD: the Strategic Environmental Research and Development Program (SERDP) and the Environmental Security Technology Certification Program (ESTCP).

SERDP was established by the Defense Authorization Act of 1991 as a partnership of the DoD, the U.S. Department of Energy, and the U.S. Environmental Protection Agency; its mission is "to address environmental matters of concern to the DoD and the Department of Energy through support of basic and applied research and development of technologies that can enhance the capabilities of the departments to meet their environmental obligations." SERDP was created with a vision of bringing the capabilities and assets of the nation to bear on the environmental challenges faced by the DoD. As such, SERDP is the DoD's environmental research and development program. To address the highest priority issues confronting the Army, Navy, Air Force, and Marine Corps, SERDP focuses on cross-service requirements and pursues high-risk and high-payoff solutions to the DoD's most intractable environmental problems. SERDP's charter permits investment across the broad spectrum of research and development, from basic research through applied research and exploratory development. SERDP invests with a philosophy that all research, whether basic or applied, when focused on the critical technical issues, can impact environmental operations in the near term.

A DoD partner organization, ESTCP, was established in 1995 as the DoD's environmental technology demonstration and validation program. ESTCP's goal is to identify, demonstrate, and transfer technologies that address the DoD's highest priority environmental requirements. The program promotes innovative, cost-effective environmental technologies through demonstrations at DoD facilities and sites. These technologies provide a large return on investment through improved efficiency, reduced liability, and direct cost savings. The current cost and impact on DoD operations of environmental compliance are significant. Innovative technologies are reducing both the cost of environmental remediation and compliance and the impact of DoD operations on the environment while enhancing military readiness. ESTCP's strategy is to select laboratory-proven technologies with potential broad DoD application and use DoD facilities as test-beds. By supporting rigorous test and evaluation of innovative environmental technologies, ESTCP provides validated cost and performance information. Through these tests, new technologies gain end-user and regulatory acceptance.

In the 18–22 years since SERDP and ESTCP were formed, much progress has been made in the development of innovative and more cost-effective environmental remediation technology. Since then, recalcitrant environmental contamination problems for which little or no effective technology had been available are now tractable. However, we understand that newly developed technologies will not be broadly used in government or industry unless the consulting engineering community has the knowledge and experience needed to design, cost, market, and apply them.

To help accomplish the needed technology transfer, SERDP and ESTCP have facilitated the development of a series of monographs on remediation technology written by leading experts in each subject area. Each volume has been designed to provide the background in process design and engineering needed by professionals who have advanced training and five or more years of experience. The first volume in this series, *In Situ Bioremediation of Perchlorate in Groundwater*, met a critical need for state-of-the-technology guidance on perchlorate remediation. The second volume, *In Situ Remediation of Chlorinated Solvent Plumes*, addressed the diverse physical, chemical, and biological technologies currently in use to treat what has become one of the most recalcitrant contamination problems in the developed world. The third volume, *In Situ Chemical Oxidation for Remediation of Contaminated Groundwater*, provided comprehensive, up-to-date descriptions of the principles and practices of *in situ* chemical oxidation (ISCO) for groundwater remediation based on a decade of intensive research, development, and demonstration. The fourth volume, *Delivery and Mixing in the Subsurface: Processes and Design Principles for In Situ Remediation*, described the principles of chemical delivery and mixing systems and their design and implementation for effective *in situ* remediation.

The fifth volume, *Bioaugmentation for Groundwater Remediation*, covered the history, current status and future prospects for deliberately adding bacteria and other agents to treat contaminated groundwater. The sixth volume, *Processes, Assessment and Remediation of Contaminated Sediment*, summarized the scientific and practical aspects of managing contaminated sediment sites.

This final volume in the series, *Chlorinated Solvent Source Zone Remediation*, is intended as a companion to the earlier volume that focused on chlorinated solvent plumes. The development of source zone remediation technologies began later in time than the development of plume remediation technologies but has reached the point where practical guidance based on experience and fundamental research can be provided. Remediation of chlorinated solvent source zones is very difficult, at times controversial and must be based on state-of-the-art knowledge of the behavior (migration, distribution and fate) of DNAPLs in the subsurface as well as site-specific geology, chemistry, biology and hydrogeology.

Engineers and scientists with a background in environmental engineering and sciences will find this book helpful in understanding the key issues involved in DNAPL source zone management and remediation. The volume begins with an overview of the current state of the practice that serves as an introduction to the rest of the book. The second chapter summarizes the challenges involved in source zone remediation, which has been and remains contentious, expensive, and difficult, for a variety of reasons. Following are chapters providing more focused discussions of specific aspects of this overall challenge. These chapters cover the following topics:

- Two chapters on source zone characterization, the first summarizing the current issues and techniques and the second focusing on several innovative diagnostic methods.
- Two chapters on modeling, the first focused on modeling source zone remediation itself and the second focused on the responses of downgradient plumes to source remediation.
- A chapter on the use of mass flux and mass discharge information to improve source zone management and remediation.
- A series of chapters on specific source zone remediation methods, including hydraulic displacement and recovery, ISCO, *in situ* chemical reduction (ISCR), enhanced flushing with cosolvents and surfactants, *in situ* bioremediation, and finally source zone monitored natural attenuation. Each of these chapters include a fundamental description of the technology, summaries of their strengths and limitations, specific case studies of their use, and a review of the lessons learned.
- A chapter on combined remedies, discussing the fundamental issues involved in developing effective combined remedies as well as the experience to date in specific combinations of technologies.
- A chapter on the costs of source zone treatment, using several hypothetical but realistic site scenarios, comparing different technologies on a total and net present cost basis.

The last two chapters consider the future of source zone remediation, beginning with a discussion of alternate management strategies that may be useful for source zones, followed by a summary of the research and development needed to improve the state of the practice.

In a single volume covering an area this broad there are topics that cannot be discussed fully, but it is hoped that the topics that are emphasized represent the state of the practice of DNAPL source zone characterization, remediation, and management and that the volume will be a resource for those wishing to further explore the primary literature in the field. Also, it is hoped that the volume will be useful to the technical practitioner as well as the research scientist and engineer in the field.

SERDP and ESTCP are committed to the development of new and innovative technologies to reduce the cost of remediation of soil, groundwater, and sediment contamination as a result of past operational and industrial practices. We are also firmly committed to the widest dissemination of these technologies to ensure that our investments continue to yield savings for not only the DoD but also the nation. In facilitating this monograph series, we hope to provide the broader remediation community with the most current knowledge and tools available in order to encourage full and effective use of these technologies.

Jeffrey A. Marqusee, PhD, Executive Director, SERDP and ESTCP
Andrea Leeson, PhD, Environmental Restoration Program Manager, SERDP and ESTCP

About the Editors

Bernard H. Kueper

Dr. Kueper is a licensed professional engineer and professor in the Department of Civil Engineering at Queen's University in Kingston, Ontario. He received his BASc in Civil Engineering and PhD in Earth Sciences from the University of Waterloo.

Dr. Kueper has been a faculty member at Queen's for 23 years where his research is focused on the fate, transport, and remediation of organic contaminants in soil and groundwater. Dr. Kueper has carried out field and laboratory experiments and developed numerical models focused on the behavior and remediation of chlorinated solvents in both porous and fractured media. He has published over 200 papers, abstracts, and book chapters on the topic of contaminant hydrogeology including guidance for the U.S. Environmental Protection Agency and other organizations.

Dr. Kueper is a former Associate Editor for the Journal of Contaminant Hydrology, Groundwater and the Canadian Geotechnical Journal. He has been a member of several expert panels, review committees, and conference organization committees. Dr. Kueper has taught over 8,000 people outside of the university environment in professional short courses and training seminars on the topics of soil and groundwater contamination, groundwater hydraulics, site investigation techniques, and subsurface remediation. This training has been provided to various groups, regulatory agencies, and licensing bodies in Australia, Brazil, Canada, Denmark, England, Hungary, Switzerland, and the United States.

Hans F. Stroo

Dr. Stroo is an environmental scientist and owner of Stroo Consulting, LLC. He provides technical support on remediation projects for private- and public-sector clients and has served as a technical advisor to the Strategic Environmental Research and Development Program (SERDP) and the Environmental Security Technology Certification Program (ESTCP) for 15 years.

Dr. Stroo received BS degrees in Biology and Soil Science from Oregon State University, an MS in Soil Science from West Virginia University, and a PhD in Soil Science (soil microbiology) from Cornell University.

He was formerly a Principal with Remediation Technologies, Inc. (later RETEC) and with HydroGeoLogic, Inc. (HGL). He has over 25 years of experience in the assessment and remediation of contaminated soil and groundwater, particularly in the development and use of *in situ* bioremediation.

Dr. Stroo has served on several expert review panels for SERDP, other government agencies and private companies. Recently, he served as co-chair of the SERDP workshops on Remediation of Chlorinated Solvents in Groundwater and Remediation of Dense Nonaqueous Phase Liquid (DNAPL) Source Zones. He is coeditor of the SERDP-facilitated monographs *In Situ Bioremediation of Perchlorate in Groundwater*, *Bioaugmentation for Groundwater Remediation*, and *In Situ Remediation of Chlorinated Solvent Plumes*.

Catherine M. Vogel

Ms. Vogel is an Environmental Engineer with Noblis, Inc. She received her BS (Civil/Environmental Engineering) from the Michigan Technological University, MS (Environmental Engineering) from the University of Arizona, and a JD from the University of Houston. Prior to joining Noblis, she served as the program manager for the Air Force Research Laboratory's Biotechnology Program where she managed a large research and development program to develop innovative environmental restoration technologies to address Air Force-contaminated groundwater issues. She also has served as the program manager for the SERDP/ESTCP Environmental Restoration program area. She currently provides advisory services to federal agencies on emerging environmental issues and contaminants, regulatory assessments, strategic program planning and evaluation, and technical consultation support.

C. Herb Ward

Dr. Ward holds the A. J. Foyt Family Chair of Engineering in the George R. Brown School of Engineering at Rice University. He is also Professor of Civil and Environmental Engineering and Ecology and Evolutionary Biology.

Dr. Ward has undergraduate (BS) and graduate (MS, PhD, MPH) degrees from New Mexico State University, Cornell University, and the University of Texas School of Public Health, respectively. He is a registered professional engineer in Texas and a Board-Certified Environmental Engineer by the American Academy of Environmental Engineers.

He has been a faculty member at Rice University for 48 years where he has served as Chair of the Department of Environmental Science and Engineering and the Department of Civil and Environmental Engineering and as the founding Director of the University's Energy and Environmental Systems Institute. He has also served as Director of the U.S. Environmental Protection Agency (USEPA)-sponsored National Center for Ground Water Research and the U.S. Department of Defense (DoD)-sponsored Advanced Applied (Environmental) Technology Development Facility (AATDF).

Dr. Ward has been a member of the USEPA Science Advisory Board and served as Chair of the SERDP Scientific Advisory Board. He is the founding Editor-in-Chief of the international scientific journal Environmental Toxicology and Chemistry.

Dr. Ward received the Frederick George Pohland Medal for Outstanding Contributions to Bridging Environmental Research, Education, and Practice and the Brown and Caldwell Lifetime Achievement Award for Remediation in 2006, the Water Environment Federation Jack Edward McKee Medal for Achievement in Groundwater Restoration in 2007, and the Society for Industrial Microbiology and Biotechnology Charles Thom Award for bioremediation research in 2011 and was recognized as a Distinguished Alumnus by New Mexico State University in 2013.

ACKNOWLEDGMENTS

The editors of this volume gratefully acknowledge the invaluable editorial assistance of Diana Freeman, Rice University and Christina Gannett, Noblis Inc., and graphics support of Dianna Gimon and Kenneth Arevalo, Noblis Inc., and Jonah Munholland, Andrew Logan, Sean Bryck, Paul Hegele and Cindy Zhao, Queen's University.

About the Authors

Michael D. Annable

Dr. Annable is a Professor in the Department of Environmental Engineering Sciences at the University of Florida. He joined the faculty in 1992 after receiving his PhD in Civil and Environmental Engineering from Michigan State University. Dr. Annable has research interests in physical-chemical processes related to field-scale application of innovative technologies for subsurface characterization and remediation. His research includes use of the passive flux meter approach for groundwater monitoring, which has been used at more than 60 field sites. Interests also include impacts of groundwater management on water quantity and quality of lakes, wetlands, and springs.

Robert C. Borden

Dr. Borden serves as Principal Engineer with Solutions-IES, a women-owned, small business, environmental engineering firm located in Raleigh, North Carolina. Much of his work is focused on the natural and enhanced remediation of petroleum hydrocarbons, fuel oxygenates, ethers, chlorinated solvents, explosives, and propellants, including laboratory studies, fieldwork, and mathematical model development.

Dr. Borden received his BS and ME in Civil and Environmental Engineering from the University of Virginia and a PhD in Environmental Engineering from Rice University. He is the author of over 100 publications, principal investigator on over 40 different projects, and Professor of Civil, Construction, and Environmental Engineering at North Carolina State University.

Michael C. Brooks

Dr. Brooks is an Environmental Engineer in the Ground Water and Ecosystems Restoration Research's Subsurface Remediation Branch at the U.S. Environmental Protection Agency (USEPA), Ada, Oklahoma. He received his PhD from the University of Florida in 2000. He has been engaged in groundwater contaminant characterization and remediation work for two decades in the private, academic, and government sectors. His research has included experimental and theoretical studies of air sparging flow patterns, innovative dense nonaqueous phase liquid (DNAPL) characterization and remediation techniques, and energetic contaminant remediation. While a significant portion of this research has been field based, it has also included the use of both deterministic and stochastic modeling approaches to help evaluate field results.

Richard A. Brown

Dr. Brown joined ERM in 1999 as the Director of Technology Development. His responsibilities at ERM include development and implementation of remediation technologies such as bioremediation, chemical oxidation, ozonation, *in situ* metal fixation, *in situ* chemical reduction (ISCR), and evaluation of new technologies for soil and groundwater treatment. He has extensive experience with organic and inorganic contaminants. In 2010, Dr. Brown was made a Technical Fellow at ERM.

Dr. Brown received his BA (Chemistry) from Harvard and an MS (Inorganic and Analytical Chemistry) and PhD (Organometallic Chemistry) from Cornell University. Before joining ERM, Dr. Brown held positions as Senior Technical Consultant for IT Corporation, Vice President for Remediation Technology for Fluor Daniel GTI, Director of Business

Development for Cambridge Analytical Associate's Bioremediation Systems, and Technology Manager for FMC Corporation's Aquifer Remediation Systems. He has worked on new technologies for the investigation and treatment of complex contaminated sites and currently holds 20 U.S. patents.

Natalie L. Cápiro

Dr. Cápiro is a Research Assistant Professor in Civil and Environmental Engineering at Tufts University. She earned her BS in Biological and Environmental Engineering from Cornell University and her MS and PhD in Civil and Environmental Engineering from Rice University. She also completed 3 years of postdoctoral research in Environmental Engineering at the Georgia Institute of Technology. Dr. Cápiro's research interests include environmental biotechnology and bioremediation, fate and transport of persistent organic contaminants, and development and testing of innovative *in situ* remediation technologies.

Wilson S. Clayton

Dr. Clayton is Technical Services Business Unit Manager at Trihydro Corporation. He earned his PhD at Colorado School of Mines (CSM) (Geological Engineering) and BS and MS degrees in Geology from Clemson University and the University of Connecticut, respectively. Dr. Clayton has over 28 years of *in situ* remediation experience spanning the full spectrum of technologies, including biological and chemical technologies as well as more traditional technologies such as air sparging, soil venting, and pump and treat. While being broadly involved in all aspects of remediation implementation, Dr. Clayton's special area of technical expertise is in the application of subsurface reactive transport for *in situ* remediation.

Dr. Clayton previously founded Aquifer Solutions, Inc., and in prior years, served as a corporate-level technology director at Groundwater Technology, Fluor Daniel GTI, and IT Corporation, where he managed research, development, and commercialization of *in situ* oxidation technology.

Michelle Crimi

Dr. Crimi is an Associate Professor in the Institute for a Sustainable Environment at Clarkson University. She received her BS degree in Industrial Hygiene and Environmental Toxicology from Clarkson University, an MS in Environmental Health from Colorado State University, and a PhD in Environmental Science and Engineering from the CSM.

Dr. Crimi's research is focused on *in situ* remediation of contaminated soil and groundwater, chemical oxidation and degradation of organic compounds, impacts of *in situ* remediation on aquifer quality, and human health risk assessment. Dr. Crimi has published extensively in these areas and has delivered invited presentations at numerous workshops and conferences in the United States and abroad. Dr. Crimi is coauthor of the first reference text on ISCO: *Principles and Practices of In Situ Chemical Oxidation Using Permanganate* (2001) and co-edited a more recent book in the SERDP/ESTCP Remediation Technology Monograph Series: *In Situ Chemical Oxidation for Groundwater Remediation.*

Kathy L. Davies

Ms. Davies is a Senior Hydrogeologist with the USEPA Region III office in Philadelphia, Pennsylvania. She received her BS (Geology/Geochemistry) from the University of Michigan and her MS (Geology/Geochemistry) from Arizona State University. She has more than 25 years of experience in the characterization and remediation of contaminated groundwater at hazardous waste sites, with particular interests in contaminant plumes associated with DNAPLs and present in fractured rock environments.

Rula A. Deeb

Dr. Deeb is a Principal at Geosyntec in Oakland, CA. Dr. Deeb's technical expertise includes the cross-media fate and transport of contaminants and the remediation of complex soil and groundwater sites impacted by nonaqueous phase liquids. She received her PhD from the University of California at Berkeley in Civil and Environmental Engineering. Following the completion of her graduate work, she taught environmental engineering principles at Stanford University. As a postdoctoral fellow at UC Berkeley, she developed and implemented research programs in collaboration with scientists and engineers at other universities, consulting firms, and the U.S. Air Force on the remediation of sites impacted with contaminant mixtures. She joined Malcolm Pirnie, Inc. in 2000 where she managed commercial, federal, and municipal projects and clients and served as a technical specialist on key projects company-wide. Following the acquisition of Malcolm Pirnie by ARCADIS in July 2009, Rula served as Vice President and Technical Director for External Outreach in ARCADIS' Environment Division.

Dr. Deeb is heavily engaged in the National Academy of Engineering Frontiers of Engineering program which brings together emerging engineering leaders from industry, academia, and government to discuss pioneering technical work and leading-edge research in various engineering fields and industry sectors. She is the recipient of the 2008 Berkeley Engineering Innovation Young Outstanding Leader Award and is a Board-Certified Environmental Engineering Member of the American Association of Environmental Engineers.

Ronald W. Falta

Dr. Falta is a Professor of Environmental Engineering and Earth Sciences at Clemson University in South Carolina. He received his BS and MS degrees in Civil Engineering from Auburn University and his PhD degree in Mineral Engineering from the University of California, Berkeley.

Dr. Falta's research is primarily in the areas of contaminant transport and remediation, including laboratory- and field-scale studies, as well as mathematical model development. Much of his work has focused on source remediation at sites contaminated by DNAPLs and on assessing the impact of source remediation on plume evolution.

Jason I. Gerhard

Dr. Gerhard is an Associate Professor at the University of Western Ontario. He has BA, BSc, MSc, and PhD degrees from Queen's University (Canada) and was an Assistant Professor at the University of Edinburgh until 2007 when he joined Western as the Canada Research Chair in Geoenvironmental Remediation.

Dr. Gerhard has over 15 years of experience leading a research program that illuminates contaminant behavior with experiments from the pore to the field scales. He has also led the development of several numerical models to assist the design and optimization of remediation systems. As co-director of Research for Subsurface Transport and Remediation (RESTORE), he has generated new understanding of DNAPLs in soil and fractured rock and developed a new remediation technique for coal tars and petroleum hydrocarbons. In addition, he has acted as an expert consultant and an Associate Editor for *Water Resources Research*.

Mark Harkness

Mr. Harkness is a Remediation Engineer at GE Global Research in Niskayuna, New York, where he is part of a multidisciplinary team providing consulting support to GE project managers who wish to apply innovative remedial solutions to soil and groundwater issues. Mr. Harkness received his BS and MS degrees in Chemical Engineering from Rensselaer Polytechnic Institute.

In 23 years at GE, his work has focused on the development of novel remedial solutions for polychlorinated biphenyls (PCBs), petroleum hydrocarbons, and chlorinated solvents. His current specialty is the design of passive bioremediation systems for chlorinated solvents in groundwater. He has served as the GE representative on the steering committee of the Remediation Technologies Development Forum Bioremediation Consortium, and more recently Project SABRE, and is a frequent contributor to journal articles and book chapters in the field of bioremediation.

Kirk Hatfield

Dr. Hatfield is Director of the Engineering School for Sustainable Infrastructure and Environment (ESSIE) at the University of Florida and has been a Professor in the Department of Civil and Coastal Engineering since 1987. He received his MS in Environmental and Civil Engineering from the University of Iowa and a PhD in Civil Engineering from the University of Massachusetts. Dr. Hatfield specializes in groundwater flow and contaminant transport modeling, water resource allocation modeling, surface water and groundwater monitoring, and groundwater remediation. A recent focus has been on application of passive techniques for characterizing mass flux in complex flow systems such as fractured media.

Elisabeth L. Hawley

Ms. Hawley is a Senior Environmental Engineer at ARCADIS, U.S. in Emeryville, California. She earned a BS in Environmental Engineering Science from the University of California at Berkeley in 2000 and an MS in Civil and Environmental Engineering from the University of California at Berkeley in 2001. As a consultant, she has worked on a variety of environmental restoration projects involving site characterization, remediation, fate and transport, sustainability analyses, and emerging contaminants. Ms. Hawley has conducted research and prepared guidance materials on technical impracticability assessments and alternative groundwater remedial strategies for the Air Force, Army, and ESTCP. She is a registered professional engineer in California.

James W. Jawitz

Dr. Jawitz is a Professor in the Soil and Water Science Department at the University of Florida. He received his PhD from the University of Florida in 1999. Dr. Jawitz has research interests that include conflicts between expanding demand for freshwater resources to provide drinking water for the growing population and irrigation water for agriculture and simultaneous pressures to prevent pollution and leave enough water for natural ecosystem functions. Research includes quantifying the benefits of partial removal of aquifer contaminant source zones and evaluating the effects of land use change on water resources.

Paul C. Johnson

Dr. Johnson is the Dean of the Ira A. Fulton Schools of Engineering at Arizona State University and a Professor in the School of Sustainable Engineering and the Built Environment. Prior to that he was a Senior Research Engineer at the Shell Oil/Shell Chemical Westhollow Technology Center in Houston, Texas. He received BS and PhD degrees in chemical engineering from the University of California at Davis and Princeton University, respectively.

Dr. Johnson's research, teaching, and other professional activities include modeling and monitoring to support risk assessment at contaminated soil and groundwater sites as well as the selection, design, monitoring, and optimization of remediation technologies for managing those sites. Dr. Johnson was the Editor in Chief for the National Ground Water Association's journal *Ground Water Monitoring and Remediation* from 2003 to 2012 and serves as a peer reviewer and consultant to the USEPA, state regulatory agencies, DoD, and industry.

Richard L. Johnson

Dr. Richard L. Johnson is Professor in the Division of Environmental and Biomolecular Systems (EBS), Institute of Environmental Health (IEH) at the Oregon Health & Science University (OHSU). His research concerns the transport processes that control the fate and effects of environmental substances, including metals (for remediation), organics (as contaminants), and nanoparticles (for remediation and as contaminants).

He received his PhD in Environmental Science and Engineering from the Oregon Graduate Institute in 1985. He was a Postdoctoral Fellow at the University of Waterloo (Waterloo, Ontario, Canada) from 1985 to 1986 before joining the faculty at OGI and then OHSU.

Michael C. Kavanaugh

Dr. Kavanaugh is a Senior Principal with Geosyntec Consultants, Inc. He is a registered professional engineer and a Board-Certified Environmental Engineer (BCEE) and has over 39 years of consulting experience on water quality, water and wastewater treatment, and groundwater restoration issues. In addition to his consulting practice, Dr. Kavanaugh has completed several assignments with the National Research Council (NRC) including chairing the Water Science and Technology Board and the Board on Radioactive Waste Management. He also chaired the NRC committee on alternatives for groundwater cleanup (1994) and recently chaired an NRC study on the future of subsurface remediation efforts in the United States.

Dr. Kavanaugh was elected into the National Academy of Engineering (NAE) in 1998. He has a BS and MS degrees in Chemical Engineering from Stanford and the University of California, Berkeley, respectively, and a PhD in Civil/Environmental Engineering from the University of California, Berkeley.

Jennifer L. Triplett Kingston

Dr. Kingston is a Technical Specialist with Haley & Aldrich, Inc. in Overland Park, Kansas, where she directs remediation projects and provides technical evaluations for remediation technologies. She has also served as an adjunct professor in the Geological Engineering Department at the Missouri University of Science and Technology.

Dr. Kingston received BS and MS degrees in Geological Engineering and a PhD degree in Civil and Environmental Engineering from the University of Missouri – Rolla and Arizona State University, respectively. Her experience and research include evaluation, selection, and optimization of remediation technologies along with project implementation. She has also assisted in the development of an innovative technology and its eventual design and construction.

Julie Konzuk

Dr. Konzuk is an Associate Engineer with Geosyntec Consultants and has over a decade of environmental consulting experience in the United States, Canada, and Australia. She has BS and PhD degrees in Civil Engineering from Queen's University in Ontario, Canada.

At Geosyntec, Dr. Konzuk has implemented remediation programs in subsurface environments ranging from fractured bedrock to sand and silt/clay environments as well as cleanup of soil and groundwater contaminated with chlorinated solvents, DNAPLs, unexploded ordnance and munitions constituents, heavy metals, nitrated and chlorinated benzenes, and chlorofluorinated compounds. Prior to joining Geosyntec Consultants, she spent several years researching DNAPL and groundwater transport behavior in fractured bedrock and has continued with the applied research during her consulting career. Most recently, Dr. Konzuk was part of a project team that completed an evaluation of DNAPL remediation technology performance and the factors that impact performance, and her team at Geosyntec Consultants developed a software tool (DNAPL TEST) that provides a screening-level analysis of potential DNAPL remediation technology performance given user-inputted site conditions.

Herbert Levine

Mr. Levine is a Regional Hydrogeologist for Superfund, with the USEPA, Region IX in San Francisco, California. He serves as a technical expert for USEPA project managers on hydrogeology issues related to characterization and cleanup of Superfund sites.

Mr. Levine received his BS and MS degrees in Geology from the University of Illinois, Chicago. Over the past 23 years he has provided technical support to Superfund sites throughout the multiple phases of site characterization through remedy selection, implementation, and optimization. He is a coauthor of the USEPA guidance document *Performance Monitoring of MNA Remedies for VOCs in Ground Water*.

Gregory V. Lowry

Dr. Lowry is a Professor in the Department of Civil and Environmental Engineering at Carnegie Mellon University. He teaches Environmental Engineering, Water Quality Engineering, Environmental Fate and Transport of Organic Compounds in Aquatic Systems, Environmental Nanotechnology, and Environmental Sampling and Sample Characterization. Dr. Lowry received his BS (Engineering) from UC Davis and MS and PhD in Civil–Environmental engineering from the University of Wisconsin and Stanford University, respectively.

Dr. Lowry's research interests include mineral-organic macromolecule–water interfacial processes, and transport and reaction in porous media, with a focus on the fundamental physical/geochemical processes affecting the fate of engineered nanomaterials and organic contaminants in the environment. He is also investigating the processes affecting the permanence of CO_2 injected underground for carbon sequestration. He is an experimentalist working on a variety of fundamental and application-oriented research projects developing novel environmental technologies for restoring contaminated sediments and groundwater. His current projects include *in situ* sediment management using innovative sediment caps, DNAPL source zone remediation through delivery of reactive nanoparticles to the NAPL–water interface, and CO_2 capture, sequestration, and monitoring.

David W. Major

Dr. Major is a Principal of Geosyntec Consultants Inc., Associate Editor of Ground Water Monitoring and Remediation, and an Adjunct Professor in the Department of Chemical Engineering and Applied Chemistry, University of Toronto. He received his BSc (1981), MSc (1984), and PhD (1987) in Biology from the University of Waterloo (UW). Dr. Major has over 23 years of experience working with clients, researchers, and regulators to develop practical biological, chemical, and thermal based solutions to remediate contaminated sites. He was inducted into the Space Hall of Fame®, received a UW Science's Alumni of Honor Award, and is a Board-Certified Environmental Scientist of the American Academy of Environmental Engineers and Scientists. Dr. Major has served on several national scientific and regulatory advisory boards.

Michael C. Marley

Mr. Marley is president and co-founder of XDD, LLC in Stratham, New Hampshire, and has over 29 years of experience in environmental and civil engineering. He earned a BS in Civil Engineering from Queen's University in Belfast, Northern Ireland, performed doctoral research and received his MS in Civil and Environmental Engineering from the University of Connecticut, and is a Licensed Environmental Professional (LEP) in the state of Connecticut.

Mr. Marley's expertise focuses on strategies for site closure, including the development and application of innovative remediation technologies for contaminated soils and groundwater.

He has been at the forefront of developing design and application protocols for soil vapor extraction, air sparging, and most recently *in situ* chemical oxidation and reduction technologies. As XDD's primary technical expert Mr. Marley directs all applied research and development efforts and serves as the firm's quality assurance officer. Mr. Marley lectures widely on the design and application of vapor extraction, bioventing, sparging/biosparging, chemical oxidation, and innovative remediation technologies for volatile and semivolatile organic compounds and inorganic compounds. He has been responsible for the modeling support, review, or design of several hundred pilot- and full-scale remediation systems. More recently, Mr. Marley was the principal author of the Electric Power Research Institute-funded study on field applicability of persulfate for chemical oxidation of manufactured gas plant residuals. In addition, Mr. Marley has served as a peer reviewer for Strategic Environmental Research and Development Program (SERDP)/Environmental Security Technology Certification Program (ESTCP) research on *in situ* technologies.

James W. Mercer

Dr. Mercer is a hydrogeologist with Tetra Tech GEO, Inc. He received a BS (Geology) from Florida State University and MS and PhD degrees in Geology from the University of Illinois. Dr. Mercer spent 8 years with the U.S. Geological Survey in the Northeastern Research Group working on contaminant and heat transport issues, including multiphase flow. He cofounded GeoTrans (now Tetra Tech GEO) in 1979 and in 1980 began working on DNAPL issues at Love Canal.

In 1985, Dr. Mercer received the Wesley W. Horner Award of the American Society of Civil Engineers for the work performed at Love Canal. Dr. Mercer continued to work on DNAPL issues and coauthored a book on *DNAPL Site Evaluation* in 1993. In 1994, he received the American Institutes of Hydrology's Thesis Award for contributions to groundwater hydrology. Dr. Mercer has served on the National Research Council's Water Science and Technology Board and was a member of the USEPA Science Advisory Board. He is currently on the Scientific Advisory Board of SERDP.

Kevin G. Mumford

Dr. Mumford is an Assistant Professor in the Department of Civil Engineering at Queen's University, whose research focuses on groundwater contamination and *in situ* remediation. He received his BS and MS degrees from the University of Waterloo and his PhD from McMaster University and held a postdoctoral fellowship at the University of Western Ontario before joining Queen's University in 2010. Dr. Mumford's expertise is in the study of multiphase flow in porous media and mass transfer related to both NAPLs and gases. His research interests include mass transfer and gas dynamics in NAPL source zones, under natural conditions, during remediation applications, and at elevated temperatures during *in situ* thermal treatment.

Charles J. Newell

Dr. Newell is a Vice President of GSI Environmental Inc. and an Adjunct Professor of Civil and Environmental Engineering at Rice University. He received a BS degree in Chemical Engineering and MS and PhD degrees in Environmental Engineering from Rice University.

Dr. Newell has served as a Principal Investigator or Co-PI for five SERDP projects and ten ESTCP projects in the environmental restoration/technologies field. He has been awarded the Hanson Excellence of Presentation Award by the American Association of Petroleum Geologists, the Outstanding Presentation Award by the American Institute of Chemical Engineers, and the 2001 Wesley W. Horner Award by the American Society of Civil Engineers. In 2008 he was cited as Rice University's Outstanding Engineering Alumni.

Jennifer L. Nyman

Dr. Nyman is a Senior Environmental Engineer in the Oakland, California, office of Geosyntec, where she manages large groundwater investigation and remediation projects. She has previously served as technical lead or expert for remediation projects for industrial and government clients. She received her BS degree in Chemical Engineering from Montana State University and her MS and PhD degrees in Environmental Engineering and Science from Stanford University.

Dr. Nyman has specialized expertise in the areas of geochemistry, remedial design, microbiology, bioremediation, and advanced diagnostic tools. She serves on and is a trainer for the Interstate Technology & Regulatory Council team on Attenuation Processes for Metals and Radionuclides.

Robert H. O'Laskey

Mr. O'Laskey is a Senior Geologist at the Emeryville, California, office of ARCADIS U.S., Inc. where he serves as the technical lead for environmental and water resource investigations. He received his BS degree in Geology from State University College of New York at Brockport and his MS degree in Geology from the University of Massachusetts at Amherst. Specific experience includes the design or the rehabilitation of groundwater extraction and water supply wells, planning and implementation of remedial programs for chlorinated solvent contamination sites, development of groundwater monitoring programs at solid waste landfills, and hydrogeologic analysis in support of numerical groundwater fate and transport modeling.

Kurt D. Pennell

Dr. Kurt Pennell is Professor and Chair of the Department of Civil and Environmental Engineering at Tufts University. Dr. Pennell received his BS from the University of Maine, MS from North Carolina State University, and PhD from the University of Florida and completed a postdoctoral fellowship at the University of Michigan.

Prior to moving to Tufts in June 2009, Dr. Pennell was a Professor in the School of Civil and Environmental Engineering at the Georgia Institute of Technology and holds an adjunct appointment in the Department of Neurology at the Emory University School of Medicine. His current research focuses on the environmental fate and neurotoxicity of engineered nanomaterials, groundwater remediation technologies, and role of persistent organic pollutants in Parkinson's disease. Dr. Pennell serves as an Associate Editor for the Journal of Contaminant Hydrology and is a registered professional engineer (PE) and a BCEE.

P. Suresh C. Rao

Dr. Rao is the Lee A. Rieth Distinguished Professor of Civil Engineering and Agronomy at Purdue University. He received his PhD from the University of Hawaii and spent 25 years at the University of Florida before joining Purdue University. Dr. Rao has research interests in multi-scale modeling and analysis of landscape hydrologic and biogeochemical process linkages across human-impact gradients, including intensively managed croplands, urban catchments, and wetlandscapes, with pristine catchments serving as the reference. This work includes resilience-based analysis of coupled natural and engineered complex systems, including food–bioenergy–water sustainability issues and use of models for research and for decision making. Applications include catchment-scale water-quality analyses and aquifer-scale studies on industrial land uses, assessments of groundwater vulnerability, contaminated site remediation, and watershed management.

Alicia J. Shepard

Ms. Shepard is an environmental scientist with HydroGeoLogic, Inc., located in Reston, Virginia. She received her BA (Biology) degree from Macalester College. Ms. Shepard has more than 11 years of experience in project management, communication and outreach activities, and conference/workshop planning relevant to the Department of Defense (DoD).

Ms. Shepard has managed numerous technical workshops that convene groups of experts from multiple disciplines and organizations to review the state of the science, identify gaps, and prioritize research, development, technology demonstration, and technology transfer needs. She has directed the development of strategic plans based on the results of these workshops, including SERDP/ESTCP workshops on coastal and estuarine ecosystems, contaminated sediments, bioavailability, and range sustainability.

Robert L. Siegrist

Dr. Robert L. Siegrist is a University Professor Emeritus of Environmental Science and Engineering at the CSM. From 2001 to 2010, he served as Director of the Environmental Science and Engineering Division at CSM, and during a transitional retirement period, he is sustaining his affiliation with CSM as a Professor in the Civil and Environmental Engineering Department. Dr. Siegrist earned his BS (High Honors) and MS in Civil Engineering and his PhD in Environmental Engineering at the University of Wisconsin. He is a registered Professional Engineer and an AAEES BCEE.

Before joining CSM in 1995, Dr. Siegrist held academic and research positions with the University of Wisconsin, Norwegian Institute for Georesources and Pollution Research, and Oak Ridge National Laboratory. He also served as Founding Partner of RSE, Inc. and a Senior Project Manager with Ayres Associates, Inc. Dr. Siegrist is an internationally recognized expert in *in situ* remediation of contaminated soil and groundwater. He has published 300 technical papers and two reference books and holds two U.S. patents. He was lead author of the first reference text concerned with *in situ* chemical oxidation (ISCO): *Principles and Practices of In Situ Chemical Oxidation Using Permanganate* (2001). He was also lead editor on another text, *In Situ* Chemical Oxidation for Groundwater Remediation (2011). Dr. Siegrist has given invited lectures at more than 100 workshops and conferences in more than 30 countries worldwide. He also has served as a science and engineering advisor to agencies and organizations in the United States and abroad, including the USEPA, Department of Energy, DoD, National Research Council, and the U.S. Government Accountability Office. He also served as a Fellow with the NATO Committee for Challenges to Modern Society.

Brent E. Sleep

Dr. Sleep is currently Chair of the Department of Civil Engineering at the University of Toronto. His research group conducts laboratory and computer modeling studies of bioremediation, thermal remediation, and applications of chemical oxidants and nanoscale zerovalent iron for subsurface remediation. His group is also investigating the transport of pathogens in fractured rock aquifers. Dr. Sleep is an Editor in Chief of the *Journal of Contaminant Hydrology* and an Associate Editor of *Advances in Water Resources*.

Eric J. Suchomel

Dr. Suchomel is a Project Engineer with Geosyntec Consultants, Inc. in San Francisco, California. He received his BSE degree in Chemical Engineering from the University of Iowa and his MSEnvE and PhD degrees in Civil and Environmental Engineering from the Georgia

Institute of Technology. His doctoral research focused on evaluating reductions in mass flux and discharge downgradient of DNAPL source zones following partial source zone mass removal and assessing *in situ* biological reductive dechlorination of residual DNAPL source zones.

Since accepting a position with Geosyntec, Dr. Suchomel has managed remediation projects for government and private clients throughout California. His current work is focused on the development and demonstration of innovative technologies for the remediation of emerging contaminants such as 1,2,3-trichloropropane. He is a registered professional engineer in the State of California.

Neil R. Thomson

Dr. Thomson is a Professor of Civil and Environmental Engineering and a member of the Environmental Modeling and Analysis Group and the Water Institute at the University of Waterloo, Ontario, Canada. At the University of Waterloo he teaches undergraduate and graduate courses in areas such as environmental chemistry, organic contaminants, contaminant transport, and remediation technology design. He received his PhD in Environmental Engineering from the University of Waterloo.

Dr. Thomson provides expert technical assistance on topics that include conceptual groundwater model development, groundwater flow and fate analysis, and remedial alternative selection and system design. He also is a featured lecturer in short courses in North America, the United Kingdom, Brazil, and Mexico. He has over 25 years of research experience in the use of field investigations, laboratory experiments, and numerical models to explore subsurface contaminant fate and remediation issues. His research interests are focused on the environmental fate of contaminants in subsurface systems including immiscible liquids, vapors and pathogens, development and application of simulation tools, and development and assessment of soil and groundwater remediation technologies. He is currently the Editor in Chief of *Groundwater Monitoring & Remediation*.

Paul G. Tratnyek

Dr. Tratnyek is Professor and Associate Head of the Division of EBS, Institute of Environmental Health (IEH) at the Oregon Health & Science University (OHSU). He received his PhD in Applied Chemistry from the CSM in 1987. He served as a National Research Council Postdoctoral Fellow at the USEPA Laboratory in Athens, GA (ERD-Athens), during 1988 and as a Research Associate at the Swiss Federal Institute for Water Resources and Water Pollution Control (EAWAG) from 1989 to 1991. He then served as a National Research Council Postdoctoral Fellow at the USEPA Laboratory in Athens, GA, during 1988 and as a Research Associate at the Swiss Federal Institute for Water Resources and Water Pollution Control (EAWAG) from 1989 to 1991.

Dr. Tratnyek's research concerns the physicochemical processes that control the fate and effects of environmental substances, including minerals, metals (for remediation), organics (as contaminants), and nanoparticles (for remediation, as contaminants, and in biomedical applications). Since 1992, Dr. Tratnyek has led research on the chemistry of permeable reactive barriers containing zerovalent iron and other chemical reductants.

Douglas I. Walker

Mr. Walker is a PhD candidate in the Department of Civil and Environmental Engineering at Tufts University. He received his BS in Civil Engineering from the University of Massachusetts at Dartmouth. Mr. Walker's research involves the fate and transport organic contaminants and engineered nanomaterials in the environment and the role of environmental contaminants in neurodegenerative diseases. His PhD research focuses on the sequestration and metabolism of organic contaminants in cellular components, providing insight into the role of environmental exposures on the onset and progression of Parkinson's disease.

Michael R. West

Dr. West is a licensed professional engineer and holds BSc and PhD degrees from the Department of Civil Engineering at Queen's University in Kingston, Ontario. His PhD dissertation focused on the development and application of three-dimensional numerical models to simulate enhanced *in situ* biodegradation, ISCO, and surfactant flushing. Dr. West has published several articles on the topics of chlorinated solvent source zone remediation, matrix diffusion in fractured rock, and solute transport in fractured media. He is employed at B. Kueper & Associates Ltd. where his work focuses on the analysis and design of groundwater remediation systems, development of geologic models, litigation support, and delineation of NAPL source zones.

Dean F. Williamson

Mr. Williamson is a Principal Technologist with CH2M HILL, Inc., where he has worked for 25 years. As a project manager and senior consultant, he works on some of the most challenging remediation projects across the United States and internationally. He earned a BS at the University of Florida in 1978 and MS at Stanford University in 1986, both degrees in Environmental Engineering. He has over 33 years of experience in the field of soil and groundwater remediation and is a registered Professional Engineer in several states. Mr. Williamson has been involved in the design and implementation of remediation systems with a combined value of over $200 million and covering a wide range of technologies including *in situ* bioremediation, ISCO, *in situ* thermal treatment, *ex situ* thermal desorption, *in situ* air sparging, biosparging, ISCR, SVE, soil mixing, groundwater and leachate extraction and treatment, permeable reactive barriers, and MNA.

John T. Wilson

Dr. Wilson is a research microbiologist for USEPA. He has a BS in Biology from Baylor University, an MA in Microbiology from the University of California at Berkeley, and a PhD in Microbiology from Cornell University.

Dr. Wilson has worked at the Robert S. Kerr Environmental Research Center in Ada, Oklahoma, since 1978. Currently, he conducts research on *in situ* bioremediation of chlorinated solvents and on natural attenuation of BTEX compounds, fuel additives, and chlorinated solvents. In addition to his research activities, he provides training and technical assistance to the USEPA regions and to state agencies on natural attenuation of chlorinated solvents, fuel additives, and BTEX compounds in groundwater.

A. Lynn Wood

Dr. Wood is a Soil Scientist in Ground Water and Ecosystems Restoration Research's Subsurface Remediation Branch, USEPA, Ada, Oklahoma. He holds BS and MS degrees in Agronomy from Oklahoma State University and a PhD in Environmental Science from the University of Oklahoma. He has worked for EPA's Office of Research and Development since 1975. Dr. Wood has conducted extensive research into the transport of organic contaminants in both aqueous solutions and complex mixtures. He led a multidisciplinary work group consisting of scientists from EPA and academia in developing conceptual and mathematical models to describe the influence of organic solvents on the sorption and mobility of nonpolar organic chemicals in porous media. Numerous publications were produced from this work, including EPA reports, journal articles, and dissertations.

External Reviewers

Robert C. Borden
Civil, Construction and Environmental Engineering
North Carolina State University
Raleigh, NC 27695 USA

Richard A. Brown
Environmental Resources Management (ERM)
Ewing, NJ 08618 USA

Mark L. Brusseau
Soil, Water and Environmental Science
The University of Arizona
Tucson, AZ 85721 USA

Elizabeth C. Butler
School of Civil Engineering and Environmental Science
University of Oklahoma
Norman, OK 73019 USA

John Christ
U.S. Air Force Academy
Colorado Springs, CO 80840 USA

Jim Cummings
OSWER Technology Assessment Branch, USEPA
Washington, DC 20460 USA

Thomas Early
Oak Ridge National Laboratory (retired)
Oak Ridge, TN 37831 USA

Murray Einarson
Haley & Aldrich
Walnut Creek, CA 94596 USA

Linda Fiedler
Office of Superfund Remediation and Technology Innovation USEPA
Washington, DC 20460 USA

John Fountain
Marine Earth and Atmospheric Sciences
North Carolina State University
Raleigh, NC 27695 USA

Michael Gefell
ARCADIS U.S., Inc.
Lakewood, CO 80401 USA

Gorm Heron
TerraTherm, Inc.
Keene, CA 93531 USA

George J. Hirasaki
Department of Chemical & Biomolecular Engineering, Rice University
Houston, TX 77005 USA

David Huntley
Geological Sciences
San Diego State University
San Diego, CA 92182 USA

Richard Jackson
Geofirma Engineering Ltd.
Heidelberg, ON Canada N0B 2M1

James W. Jawitz
Environmental Hydrology
University of Florida
Gainesville, FL 32611 USA

Stephen S. Koenigsberg
Brown and Caldwell
Irvine, CA 92612 USA

Fritz Krembs
Trihydro, Inc.
Laramie, WY 82070 USA

Paul D. Lundegard
Lundergard USA
Fullerton, CA 92835 USA

Kira Lynch
Superfund Technology Liaison (STL)
Region 10 ORD, USEPA,
Seattle, WA 98101 USA

Tamzen W. Macbeth
CDM
Helena, MT 59601 USA

Pat McLoughlin
Microseeps, Inc.
Pittsburgh, PA 15238 USA

James W. Mercer
GeoTrans, Inc. (a Tetra Tech Company)
Sterling, VA 20164 USA

John Molson
Département de géologie et de
génie géologique
Université Laval
Québec, QC, Canada G1V 0A6

Robert D. Norris
Brown and Caldwell (retired)
Longmont, CO 80501 USA

Matthew A. Petersen
GE Global Research Center
Niskayuna, NY 12309 USA

Albert Ricciardelli
GZA GeoEnvironmental, Inc.
Norwood, MA 02062 USA

Hanadi S. Rifai
Department of Civil and Environmental
Engineering
University of Houston
Houston, TX 77204 USA

Thomas J. Simpkin
CH2M HILL
Englewood, CO 80112 USA

Brent Sleep
Department of Civil Engineering
University of Toronto
Toronto, ON Canada M5S 1A4

Amy L. Teel
Department of Civil and Environmental
Engineering
Washington State University
Pullman, WA 99164 USA

René Therrien
Department of Geology
and Geological Engineering
Université Laval
Québec, QC, Canada G1V 0A6

Richard J. Watts
Department of Civil and Environmental
Engineering
Washington State University
Pullman, WA 99164 USA

Todd H. Wiedemeier
Wiedemeier & Associates
Evergreen, CO 80439 USA

A. Lynn Wood
Robert S. Kerr Environmental
Research Center, USEPA
Ada, OK 74820 USA

Contents

List of Figures

List of Tables

CHAPTER 1

SOURCE ZONE REMEDIATION: THE STATE OF THE PRACTICE

Bernard H. Kueper,[1] Hans F. Stroo,[2] Catherine M. Vogel[3] and C. Herb Ward[4]

[1]Queen's University, Kingston, ON, Canada K7L 3N6; [2]Stroo Consulting LLC, Ashland, OR 97520, USA; [3]Noblis Inc., Atlanta, GA 30318, USA; [4]Rice University, Houston, TX 77005, USA

1.1 INTRODUCTION

Chlorinated solvents such as perchloroethene (also called tetrachloroethene or PCE), trichloroethene (TCE), 1,1,1-trichloroethane (1,1,1-TCA) and carbon tetrachloride (CT) are found at numerous sites worldwide. These chemicals have a wide variety of uses in industry such as degreasing agents, chemical intermediates, heat-transfer fluids and dry cleaning fluids. Because of their relatively stable nature, desirable chemical properties and low cost to produce, chlorinated solvents have been employed by a wide variety of industries on a frequent basis, particularly since the 1950s coincident with the rise in population and associated industrial activity in many countries.

The United States (U.S.) production of PCE, TCE, 1,1,1-TCA and CT peaked at several hundred million pounds (hundreds of millions of kilograms) per year. Generally, the U.S. production of these particular chlorinated solvents peaked in the 1970s and declined thereafter as a result of environmental concerns (McCarty, 2010). Chlorinated solvents have been found at approximately 80% of all U.S. Superfund sites with groundwater contamination and at more than 3,000 U.S. Department of Defense sites (USEPA, 2004). Chlorinated solvents represent some of the most persistent contaminants in soil and groundwater worldwide.

Table 1.1 lists examples of chlorinated solvents commonly encountered at sites along with their physical and chemical properties. Chlorinated solvents typically enter the subsurface in the form of a dense nonaqueous phase liquid (DNAPL). A DNAPL is an immiscible liquid denser than water and only slightly soluble in water. Consequently, DNAPLs have the potential to migrate as a separate liquid phase to significant depths below the water table where they will slowly dissolve into flowing groundwater, giving rise to aqueous phase plumes of contamination. Although chlorinated solvent DNAPLs are only slightly soluble in water, their solubility is typically orders of magnitude greater than typical drinking water standards. Consequently, even small volumes of DNAPL released to the subsurface can result in contamination of orders of magnitude higher volumes of groundwater. Most chlorinated solvents also have sufficient vapor pressure to allow for vaporization into soil gas, giving rise to vapor phase plumes of contamination.

Although the industry's usage of chlorinated solvents increased throughout the mid-to-late 1900s, recognition of their importance as soil and groundwater contaminants by North American regulators, industry and researchers generally did not occur until the early to mid-1980s. From a global perspective, pioneering work focusing on the fate and transport of chlorinated solvent DNAPLs was initiated by F. Schwille and colleagues in West Germany in 1971 (Schwille, 1981, 1988). Following the discovery of Schwille's work in the early 1980s, coincident with the emerging recognition of the significance of chlorinated solvent DNAPLs in soil

B.H. Kueper et al. (eds.), *Chlorinated Solvent Source Zone Remediation*,
doi: 10.1007/978-1-4614-6922-3_1, © Springer Science+Business Media New York 2014

Table 1.1. Selected Physical and Chemical Properties of Commonly Encountered Chlorinated Solvents at 25°C[a]

Chlorinated Solvent	Density (g/cm^3)	Solubility in Water (mg/L)	Vapor Pressure (torr)	Absolute Viscosity (cP)	U.S. Drinking Water Standard (mg/L)[b]
Perchloroethene (tetrachloroethene)	1.63	200	18.9	0.9	0.005
Trichloroethene	1.46	1,100	75	0.57	0.005
1,1,1-Trichloroethane	1.35	1,300	125	0.84	0.200
Carbon tetrachloride	1.59	825	109	0.97	0.005
1,2-Dichloroethane	1.25	8,500	82.1	0.84	0.005
Dichloromethane	1.33	20,000	415	0.44	0.005
Water[c]	0.997[d]	N/A	23.8[d]	0.894[d]	N/A

Note: °C – degrees Celsius; cP – centipoise; g/cm^3 – grams per cubic centimeter; mg/L – milligrams per liter
[a]Source of data unless specified otherwise: Pankow and Cherry, 1996
[b]Source of data: USEPA, 2013
[c]Properties of water added for reference
[d]Source of data: Streeter and Wylie, 1981

and groundwater in North America, research programs were initiated in Canada and the United States in the mid-1980s with further initiatives launched in Germany and other countries. Recognizing that pump-and-treat systems would need to operate indefinitely given the persistence of chlorinated solvents in the subsurface, research in the 1990s quickly focused on developing and testing *in situ* technologies to either remove or destroy contaminant mass in the source zone.

As a result of approximately 25 years of remediation technology development and implementation at sites, considerable progress has been made in removing contaminant mass and lowering concentrations in soil and groundwater. However, removing sufficient contaminant mass to achieve typical concentration-based cleanup goals in soil and groundwater remains technically challenging at many sites (Stroo et al., 2012; Kavanaugh et al., 2003; NRC, 2005, 2013). This technical challenge has led to the development and implementation of containment technologies and alternative site closure strategies in parallel with continued development and implementation of mass removal technologies. The inability to achieve concentration-based cleanup goals at many sites often poses a dilemma, in that regulators and responsible parties must evaluate the costs and benefits of technology implementation. This dilemma is complicated by the fact that each site is unique in terms of the nature and extent of contamination; it is therefore difficult to predict the performance of any particular technology at a specific site.

A recent National Research Council (NRC) review (NRC, 2013) has estimated that there are over 100,000 contaminated sites where unlimited use and unrestricted exposure will not be possible in a reasonable time frame, and many of these sites contain chlorinated solvents. The report recommends more research specifically aimed at remediation technology development, improved monitoring and modeling, and on formal transition assessments designed to move sites more rapidly into closure or long-term passive or active management. A protocol for these transition assessments should be developed, and they will likely require a more detailed understanding of site conditions than is typically available today, as well as more cost-efficient monitoring and management strategies.

The experience gained in developing and implementing remediation technologies at chlorinated solvent sites over the past 25 years has yielded several insights and lessons that are important for stakeholders to understand. This chapter summarizes these insights and lessons and provides an overview of the entire volume. Following short summaries of the nature of chlorinated solvent source zones and the history of efforts to remediate sites impacted by chlorinated solvents, this chapter provides overviews of various remediation technologies and related topics that are discussed in greater detail in subsequent chapters.

1.2 CHLORINATED SOLVENT SOURCE ZONES

A chlorinated solvent source zone is defined here as that region of the subsurface that contains, or formerly contained, chlorinated solvents in DNAPL form and contains a sufficient mass of chlorinated solvents to sustain concentrations in soil or groundwater above regulatory limits. This definition recognizes that DNAPL may not be present throughout the entire volume of the subsurface where it was once present and that it may be completely depleted as a result of natural processes or the application of engineered remediation technologies.

1.2.1 Evolution of Understanding

Understanding the nature and extent of chlorinated solvents in the subsurface, and what exactly constitutes a source zone, has evolved since widespread recognition in the 1980s that these chemicals can represent significant contaminants in soil and groundwater. For example, early conceptual models of DNAPL migration in the subsurface depicted DNAPL migrating vertically downwards directly beneath the release point and spreading laterally on horizontally continuous major low-permeability units such as clay aquitards (Figure 1.1a). As a result of field experiments performed in the late 1980s and early 1990s, however, it was recognized that even subtle variations in permeability and porous media structure can result in lateral DNAPL migration (Poulsen and Kueper, 1992; Kueper et al., 1993; Brewster et al., 1995). It was also recognized that significant lateral migration can take place as a result of the cumulative effect of encountering lower permeable lenses and laminations, even if they are not each laterally extensive.

Conceptual models progressed such that by the mid-1990s it was commonly known that chlorinated solvent DNAPLs distribute themselves in a tortuous, complex and heterogeneous fashion in the subsurface and that lateral DNAPL migration can occur in all directions including opposite the direction of groundwater flow (Figure 1.1b, d). It was also recognized by the mid-1990s that locating DNAPL in the subsurface is technically challenging and that accurate delineation of the DNAPL source zone in either porous or fractured media may not be feasible at some sites. This reality still exists today and is a major factor limiting the success of remediation efforts at such sites.

In addition to the evolution in understanding the manner in which DNAPL migrates and distributes itself in porous and fractured media, there have been significant advances in understanding the biogeochemistry within source zones and the effects of natural processes on DNAPL depletion. During the last decade it has been recognized that DNAPL may no longer be present throughout the entire volume of the subsurface where it was originally present and that DNAPL may be completely depleted at some sites even though contamination continues to persist. The persistence of contamination at sites where DNAPL has been depleted is attributed to desorption and back diffusion of contaminants from low-permeability zones. It is now common practice to consider the life cycle of a chlorinated solvent source zone and determine for a particular site where the source zone is in terms of its life cycle.

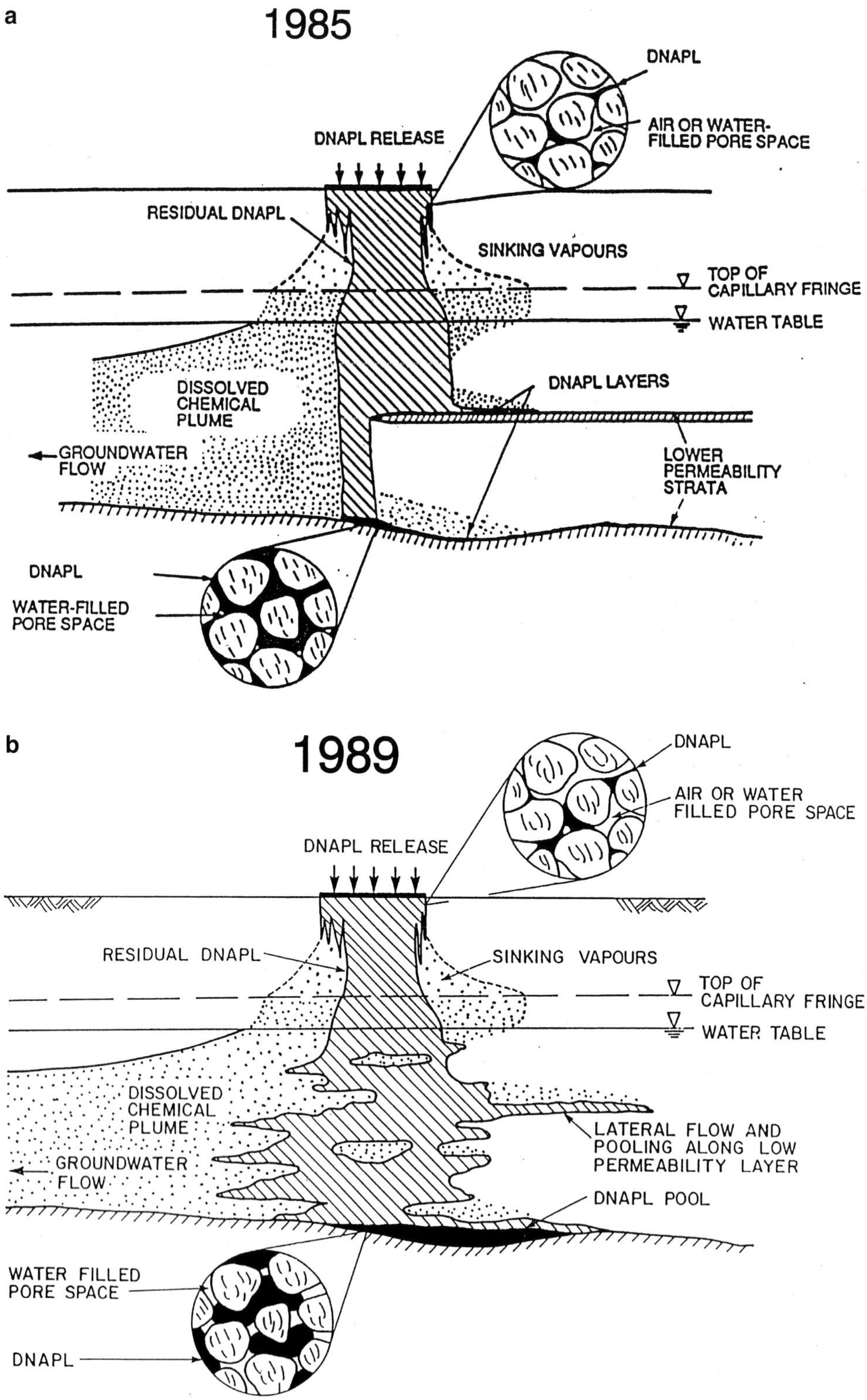

Figure 1.1. (a) Schematic distribution of DNAPL in porous media representative of conceptual models in the mid-1980s (WCGR, 1992). (b) Schematic distribution of DNAPL in porous media representative of conceptual models in the late 1980s (WCGR, 1992). (c) Schematic distribution of DNAPL in porous media representative of conceptual models in the early 1990s (WCGR, 1992). (d) Schematic distribution of DNAPL in porous media representative of conceptual models in the mid-to-late 1990s (Kueper, 1996).

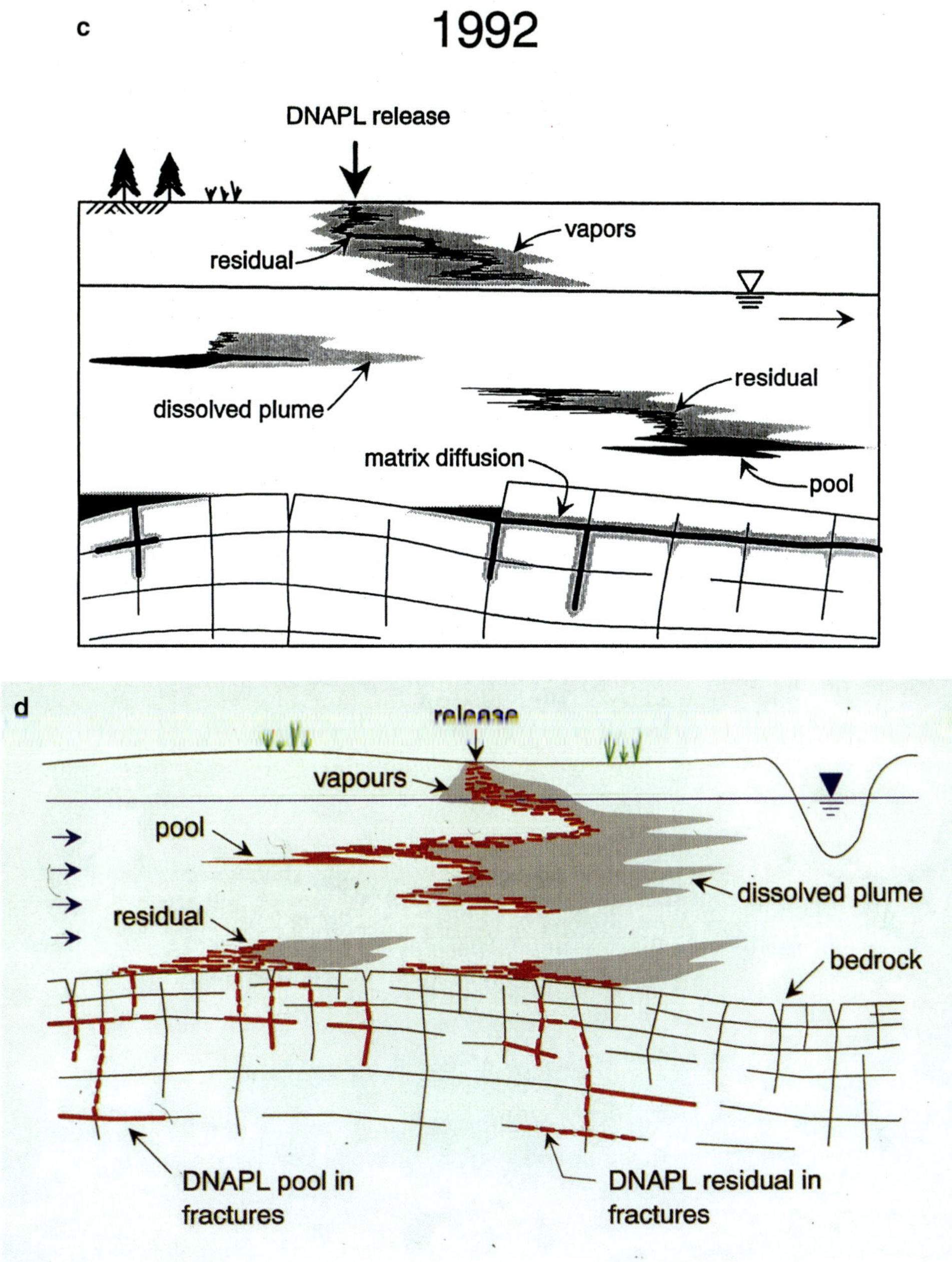

Figure 1.1. (continued)

1.2.2 Life Cycle of a Chlorinated Solvent Source Zone

The life cycle (Figure 1.2) of a chlorinated solvent source zone progresses through five stages beginning with an initial release of DNAPL to the subsurface and culminating in complete DNAPL depletion with contamination persisting as a result of back diffusion from low-permeability zones and desorption from solids. In theory, a sixth life cycle stage exists

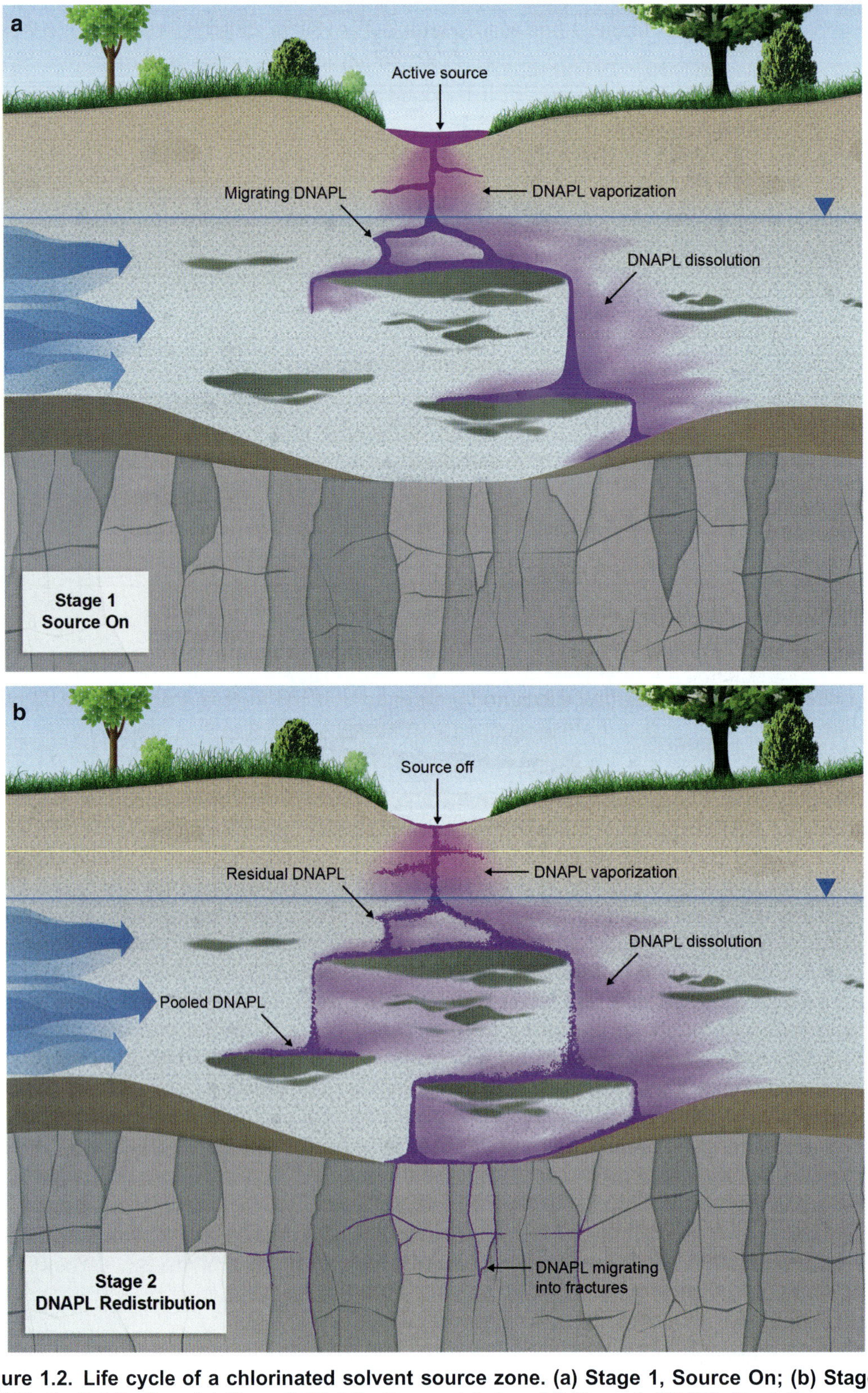

Figure 1.2. Life cycle of a chlorinated solvent source zone. (a) Stage 1, Source On; (b) Stage 2, DNAPL Redistribution; (c) Stage 3, Distribution and Aging; (d) Stage 4, DNAPL Depletion; and (e) Stage 5: Back Diffusion and Desorption.

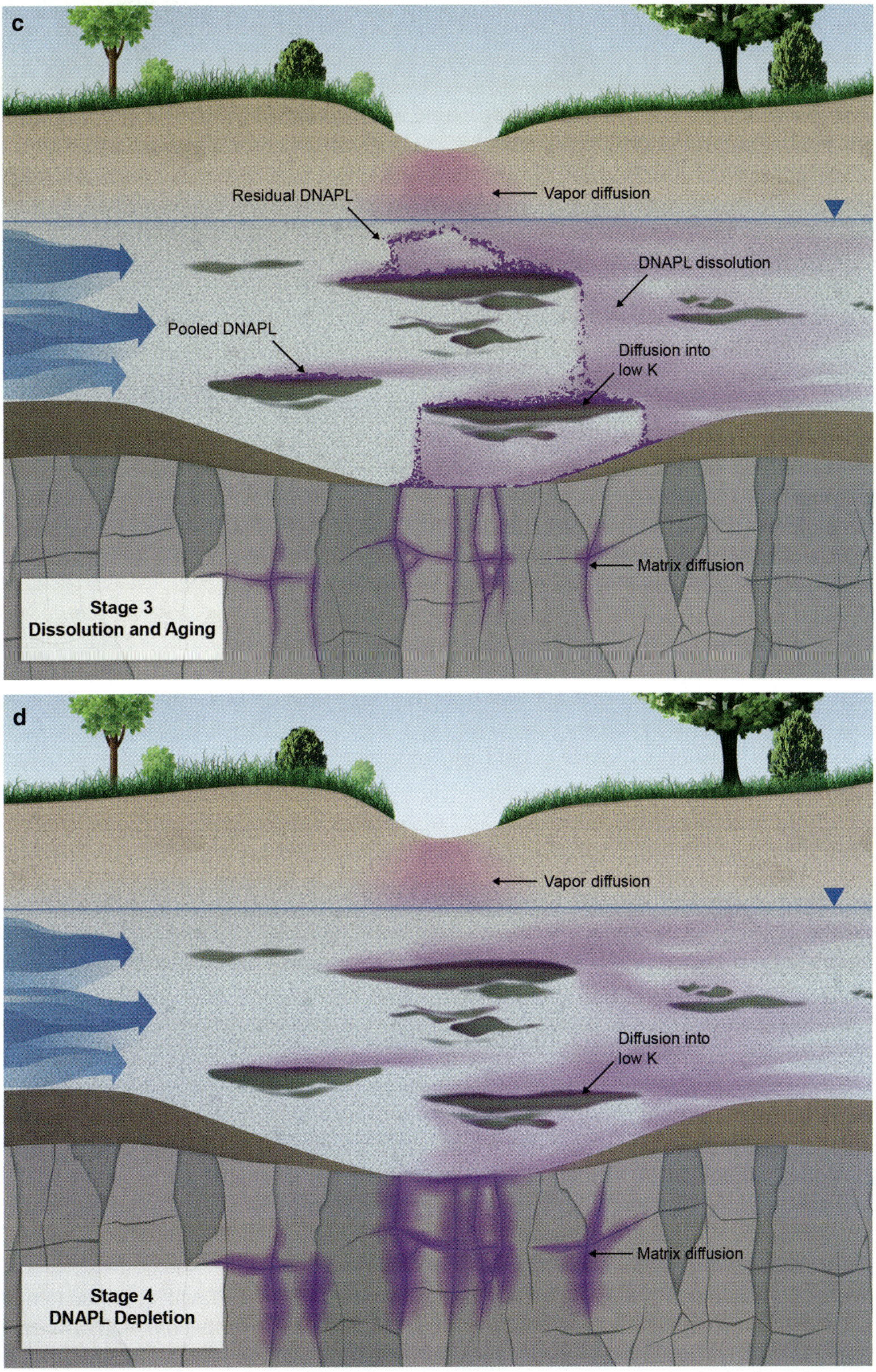

Figure 1.2. (continued)

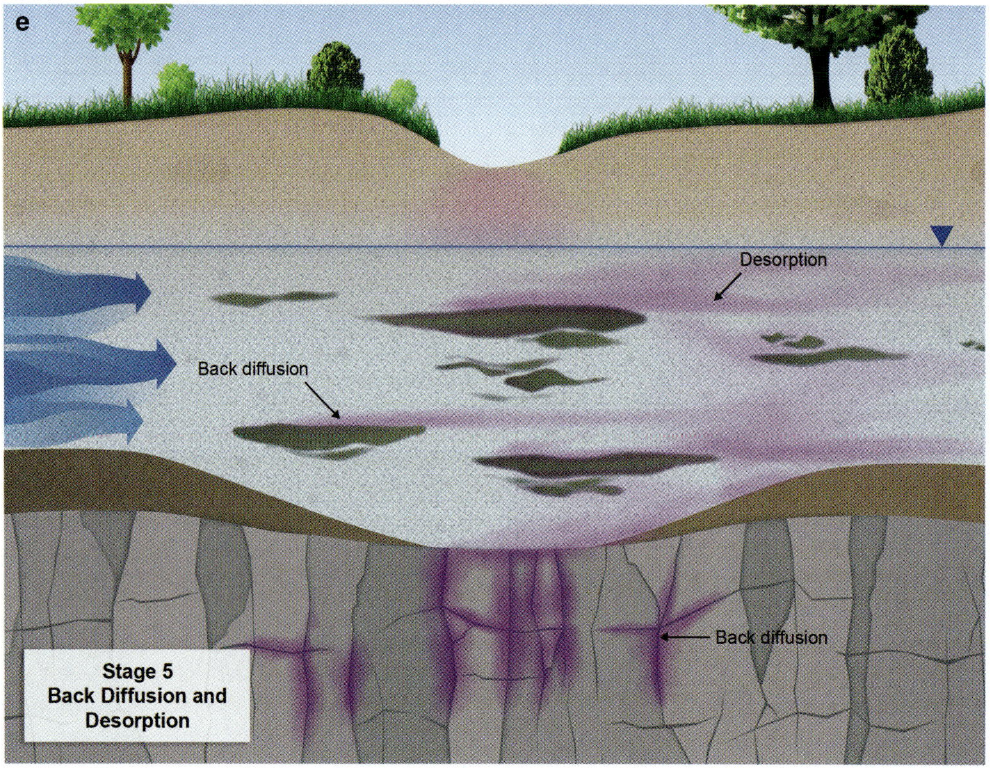

Figure 1.2. (continued)

where all former contaminants are non-detectable in soil, groundwater, and vapor. To our knowledge, there are no documented cases of sites reaching Stage 6.

Stage 1: Initial DNAPL Release

The evolution of a chlorinated solvent source zone begins with an initial release of DNAPL to the subsurface (Figure 1.2a). Releases can be short in duration, such as catastrophic failure of a storage vessel, or longer term in nature, such as persistent leaking from pipes, sumps and disposal ponds operating for many years. The nature, volume and duration of the release influence the subsurface migration pathways of the DNAPL. For example, a slow release at low capillary pressure will result in more lateral spreading of the DNAPL and a sparser distribution of residual and pools compared to a fast, high capillary pressure release (Poulsen and Kueper, 1992; Kueper and Gerhard, 1995). Some sites experience multiple releases of DNAPL through time, with varying DNAPL composition and release locations. Understanding the nature, volume and duration of DNAPL releases is an important component of conceptual site model (CSM) development and the development of site characterization plans.

Stage 2: DNAPL Redistribution

When the release of DNAPL to the subsurface has ceased, DNAPL will continue to migrate until it comes to rest in a static distribution of residual DNAPL (disconnected blobs and ganglia) and DNAPL in accumulations referred to as pools (Figure 1.2b). The particular configuration of residual and pooled DNAPL that arises is referred to as the "DNAPL architecture." The DNAPL architecture is influenced by a variety of factors including the DNAPL viscosity, the DNAPL-water interfacial tension, wettability, the DNAPL density, geologic structure and permeability, the volume of the release and the nature and duration of the release. The DNAPL redistribution stage continues for a period of time following cessation of DNAPL release to the subsurface.

The length of time required for DNAPL redistribution can vary from months (Kueper et al., 1993) to decades (Gerhard et al., 2007) depending on the factors listed above. For example, a low volume release of low viscosity, high interfacial tension DNAPL in a highly permeable setting will redistribute for only a short period (perhaps months), while a large volume release of high viscosity DNAPL in a lower-permeability setting may take much longer to redistribute (perhaps decades or longer). During the DNAPL redistribution period, DNAPL vaporization takes place above the water table with associated vapor phase migration, and DNAPL dissolution into groundwater takes place below the water table giving rise to aqueous phase plumes.

Stage 3: Continued DNAPL Dissolution and Aging

As continued dissolution and vaporization of DNAPL take place, vapor and dissolved phase plumes continue to develop (Figure 1.2c). This ongoing mass discharge results in a continuous reduction of DNAPL mass both above and below the water table with time. The flux of contaminants in groundwater (plume loading) likely peaks during late Stage 2 and early Stage 3, as does the flux of contaminants in vapor form.

For a multicomponent DNAPL, the aqueous solubility of each component will be a function of its molar fraction within the DNAPL and its single component solubility as described by Raoult's Law. As the DNAPL dissolves into groundwater, the higher effective solubility components are depleted faster than the lower effective solubility components, thereby enriching the DNAPL in the lower effective solubility components. This preferential depletion of certain components not only results in changes to the DNAPL composition, but results in changes to the composition of the dissolved phase plume downgradient of the source zone and a lowering of total concentrations within the plume with time. In addition to changes in aqueous phase concentrations stemming from changes in DNAPL composition, total concentrations will be reduced as a result of complete depletion of DNAPL from certain areas of the source zone. In general, DNAPL in higher-permeability pathways will deplete faster than DNAPL in lower-permeability pathways.

Similar aging effects occur for multicomponent DNAPLs in the unsaturated zone as the higher effective vapor pressure components are depleted faster than the lower effective vapor pressure components. This aging also results in changes to the composition of the vapor phase plume and a reduction of total concentrations in soil gas with time. In general, DNAPL in low moisture content regions of the unsaturated zone will deplete faster than DNAPL in high moisture content regions.

During this stage of the chlorinated solvent source zone life cycle, significant quantities of dissolved phase mass can diffuse into low-permeability zones such as silts and clays and the bedrock matrix. If DNAPL is pooled above a capillary barrier, the forward diffusion process can transfer mass directly from the DNAPL into that zone. DNAPL mass is therefore depleted by this mechanism in addition to dissolution into flowing groundwater and vaporization into soil gas.

Stage 4: Complete DNAPL Depletion

Eventually, all DNAPL mass will be depleted as a result of vaporization, dissolution and diffusion mechanisms (Figure 1.2d). The length of time required for complete depletion to occur below the water table will be a function of many factors including the volume of DNAPL originally released to the subsurface, the DNAPL solubility, the groundwater velocity and the degree of biodegradation. For small volume DNAPL releases into high-permeability media, the life span of the DNAPL may be as short as a few years. For large volume DNAPL releases into low to moderate permeability media, the life span could be on the order of hundreds of years. In the unsaturated zone, the length of time required for complete depletion to occur will be a function of many factors including the volume of DNAPL released, the DNAPL vapor pressure, the temperature and the moisture content. For relatively small volumes of high vapor pressure DNAPLs released to warm, dry sediments, the life span may be as short as a few months. Conversely, large volumes of DNAPL released to cold, wet sediments can persist for many decades or longer.

Stage 5: Desorption and Back Diffusion

Following complete depletion of DNAPL, all mass below the water table will be present as contaminants sorbed to surfaces of solids and those present in the aqueous phase (Figure 1.2e). Concentrations in the permeable zones will start to decrease as clean groundwater enters the source zone from upgradient. This lowering of concentrations in the permeable zones triggers back diffusion of contaminant mass from the low-permeability zones. This transfer of mass from low to high-permeability zones continues for a longer period of time than the original forward diffusion process.

How relevant back diffusion is with respect to meeting cleanup goals is largely dependent on the ability of groundwater flow in the permeable zones to dilute the mass diffusing out from the less permeable regions. In general, back diffusion is expected to sustain plumes above cleanup standards for long periods of time in geologic settings where the proportion of low-permeability media exceeds that of the high-permeability media. Examples include fractured clay, fractured sedimentary bedrock and interbedded sand/clay sequences characterized by relatively thin occurrences of sand. Large amounts of organic carbon, low groundwater velocity and little to no biodegradation also serve to prolong Stage 5.

Considerable discussion in the groundwater community has centered on the percentage of current sites in any one of the above five stages. Given better disposal, handling and storage practices, it would be reasonable to conclude that relatively few sites are currently in Stage 1. However, the length of time that a site will remain in Stages 2 and 3 can vary considerably from site to site given the variety of factors that dictate DNAPL redistribution, dissolution and aging time scales. It is also important to recognize that not all regions within a source zone will reach each life cycle stage at the same time.

Some individuals contend that from a management perspective, understanding how much DNAPL still remains within the source zone is unimportant and that expending significant characterization funds to determine this mass is wasted effort (Chapter 13). To others, determining how much DNAPL is present in a source zone is critical from a remediation perspective. Attempting to remediate a source zone with a technology specifically designed for DNAPL removal would be inappropriate, for example, if little or no DNAPL was present, while the design for some remediation technologies such as chemical oxidation or reduction requires a reasonable estimate of the initial mass present.

The fact that source zones may go through a prolonged exhausted stage, where concentrations are low but still exceed cleanup standards, has important implications for site management (Chapter 13). Notably, managers should expect that continued application of certain remediation technologies (treatment trains) may have little benefit in such source zones, while others may have a longer lasting effect. For example, injection of relatively short-lived oxidation reagents that only remove contaminants from transmissive zones without any other secondary effects will not have a long-term beneficial effect on concentrations at sites where transmissive zone concentrations are sustained by back diffusion.

1.3 HISTORICAL REMEDIATION TRENDS

The initial focus of remediation efforts at chlorinated solvent sites in the 1970s and 1980s was to install pump-and-treat systems, largely because this technology was available, relatively straightforward to design, and reliable if implemented correctly. Pump-and-treat systems were typically installed to intercept and cut off aqueous phase plumes of contamination, although there was often an expectation that sources would be depleted in relatively short time frames (years).

The expectation that pump-and-treat systems could deplete sources at many sites in relatively short time frames stemmed from failing to recognize the presence of DNAPL within the source zone. The limitations of pump-and-treat systems soon became widely realized (Travis and Doty, 1990; Pankow and Cherry, 1996) as results demonstrated that contamination persisted for many years following initiation of pump-and-treat activity. Because of the persistent nature of contamination at pump-and-treat sites, the cost per unit mass of contaminant removed became very high. The limited impact on contaminant longevity meant that estimated long-term management costs were high. From the late 1980s onwards, practitioners and researchers began focusing on more aggressive means of remediating sites impacted by chlorinated solvents than implementation of pump-and-treat systems. A variety of technologies have been developed, including the following:

***In Situ* Thermal Treatment (ISTT):** ISTT was originally developed in the petroleum industry for the enhancement of product recovery. ISTT can take various forms including thermal conductive heating (TCH), electrical resistance heating (ERH), steam enhanced extraction (SEE) and radio frequency heating (RFH).

***In Situ* Chemical Oxidation (ISCO):** ISCO was originally developed in the water treatment industry as an *ex situ* treatment technology. For subsurface application at sites impacted by chlorinated solvents, a variety of oxidants have been investigated including permanganate, peroxide, and persulfate.

Surfactant and Cosolvent Flushing: *In situ* flushing with surfactants and cosolvents (primarily alcohols) was originally developed in the petroleum industry for the enhancement of product recovery. A wide variety of surfactants and cosolvents are available for environmental applications including food-grade additives.

***In Situ* Chemical Reduction (ISCR):** The first form of ISCR developed for application at chlorinated solvent sites was the use of a permeable, reactive zero-valent iron (ZVI) barrier to intercept plumes in groundwater. ISCR has since been developed for a wider variety of applications, including injection of reducing agents directly into the chlorinated solvent source zone.

Air Sparging and Soil Vapor Extraction: Air sparging has been applied at chlorinated solvent sites for both plume and source zone treatment below the water table. Soil vapor extraction (SVE) has been widely applied for the removal of volatile organic compounds from the unsaturated zone.

Physical Containment: Physical containment technologies, including the use of sheet piling, injection grouting, and soil-bentonite walls, were originally developed for geotechnical and structural applications within the civil engineering community. These technologies have been applied at chlorinated solvent sites to contain source zones and can be enhanced using groundwater pumping and specialized configurations.

***In Situ* Bioremediation (ISB):** ISB techniques developed after recognition that many chlorinated solvents in groundwater degrade naturally as a result of microbiological activity. ISB techniques first focused on application in contaminant plumes, with subsequent focus on use in chlorinated solvent source zones.

Excavation: Excavation of the source zone can be considered as a remedial technique at some sites. Variations on excavation include *in situ* soil mixing where amendments such as oxidants and reductants are added using large-diameter augurs.

Figure 1.3 illustrates various technologies applied to treat chlorinated solvent source zones based on a survey of 118 sites (NAVFAC and Geosyntec, 2011). Figure 1.3 demonstrates that there is not a dominant technology with universal application at all sites. The particular technology to be implemented at a site depends on a number of factors including the contaminant type, geology, hydrogeology, geochemistry, remedial objectives, budget, and life cycle stage of the source zone.

The relationship between the amount of contaminant mass remaining in a source zone and the resulting concentration of contaminants in groundwater will vary from site to site. Figure 1.4 illustrates some examples of this relationship (Falta et al., 2005). Similar figures are presented and discussed in various chapters in this volume (Chapters 2, 6 and 13). The particular relationship between contaminant mass remaining and contaminant concentration in groundwater is dependent on a number of factors including geology, DNAPL architecture, and choice of technology.

In addition to arriving at a better understanding of source zone remediation technologies, much effort has been put toward understanding the response of the downgradient plume to source zone mass removal. As illustrated conceptually in Figure 1.5, plume response times vary widely from site to site depending on a number of factors such as the presence of low-permeability media that act as storage reservoirs for diffused mass, the sorption characteristics of the contaminant, and groundwater velocity. In general, the length of time required for a plume to dissipate following complete isolation of the source zone, or complete removal of mass from the source zone, will exceed the current age of the plume.

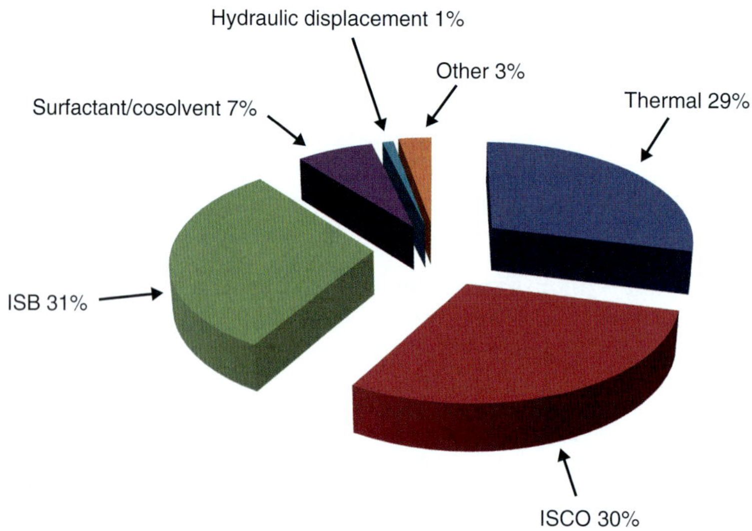

Figure 1.3. Results of survey of field-scale experiences treating chlorinated ethene source zones (from NAVFAC and Geosyntec, 2011). Figure shows the percentages of sites using different *in situ* treatment technologies, from a total of 118 sites.

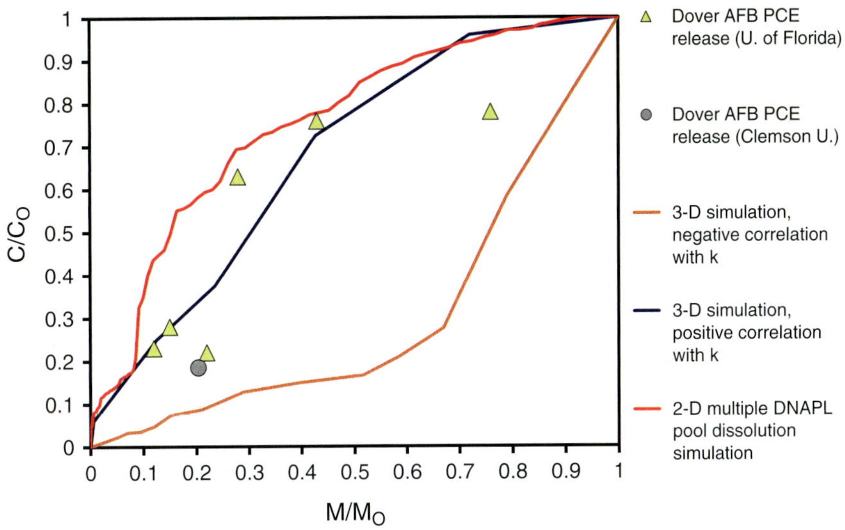

Figure 1.4. Relationship between fraction of mass remaining in a chlorinated solvent source zone (M/M_o) and normalized concentration in groundwater (C/C_o). From Falta et al., 2005.

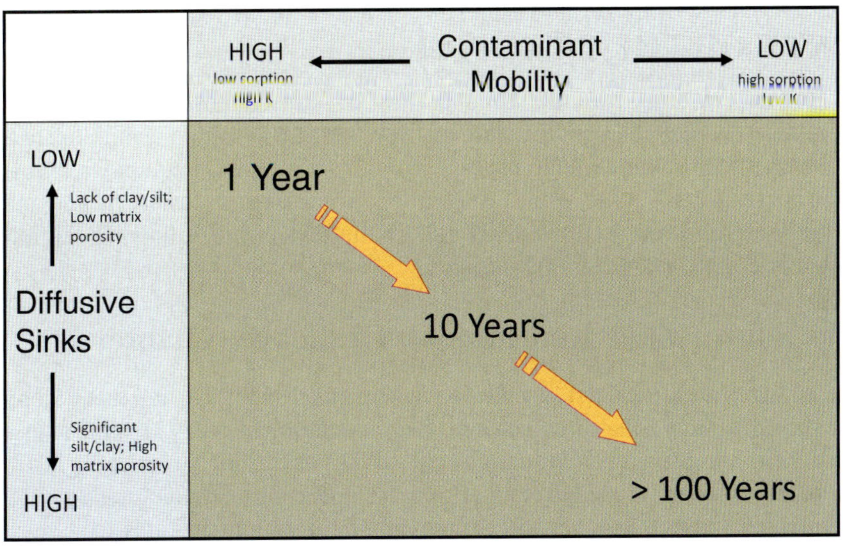

Figure 1.5. Schematic illustrating the influence of various site hydrogeologic factors and chemical properties on plume response times following complete removal of mass from the source zone.

As a result of much experience gained in developing and applying technologies to remediate chlorinated solvent source zones during the past three decades, it is now understood that complete remediation to typical concentration-based cleanup goals in a short period of time is rare, and as a result, integrated strategies are often needed to manage many chlorinated solvent sites. These integrated strategies may require a combination of multiple treatments and/or application of more than one technology, adoption of alternate end points defining success, and long-term management after active remediation. Due to the progress that has been made in developing and applying source zone remediation technologies and the associated experience gained, stakeholders should be able to develop realistic expectations associated with technology deployment and develop cost-effective strategies for managing sites impacted by chlorinated solvents.

1.3.1 Source Containment

This volume focuses on technologies that destroy or remove chlorinated solvents from source zones, with little discussion of source containment. Source containment has been used frequently as part of the overall remedy for sites. Containment can be achieved hydraulically via pump-and-treat or through hydraulic isolation by diverting water around the source (Ankeny and Forbes, 2000). Physical containment can be achieved through installation of sheet piling, soil-bentonite slurry walls and injection grouting. Physical containment often requires some amount of groundwater extraction from within the barrier to manage water levels.

The lack of coverage of containment technologies in this volume does not imply that managers should not consider some form of source containment as part of an overall remedy for a site. Containing a source zone can allow downgradient treatment and/or natural attenuation to remediate the plume more rapidly and may even lead to faster closure of the site overall. With little or no groundwater flux to flush out reagents or add electron acceptors, chemical or biological treatments can continue to degrade contaminants within the contained source zone over several years.

1.4 STATE OF THE PRACTICE

The following sections provide an overview of subsequent chapters of this volume. Collectively, these chapters discuss the state of the practice of various technologies and address important issues such as source zone characterization, advanced diagnostics, numerical modeling, plume response to source zone remediation, flux-based assessment and management, combined remedies and research needs. Much of the text in the following sections has been adopted directly from the chapters of interest.

1.4.1 The Source Zone Remediation Challenge (Chapter 2)

Several studies over the past decade have concluded that partial mass depletion from chlorinated solvent source zones has been a viable remediation strategy at some sites and is likely to provide benefits at other sites (Kavanaugh et al., 2003; NRC, 2005; Stroo et al., 2012). However, barriers to more widespread use of source zone remediation technologies persist. Additional theoretical analysis and assessment tools, improved monitoring techniques for site characterization and performance assessment, and especially field-scale demonstrations that elucidate effective site characterization and technology implementation, as well as the benefits of partial source depletion, are needed to provide a more informed basis for decision making. This information will also provide a basis to estimate the fraction of sites impacted by chlorinated solvents that would be candidates for implementation of source-depletion technologies.

At some chlorinated solvent sites, containment may be the only viable remedial option, and at such sites, containment may be considered a “presumptive remedy” eliminating the need for costly additional studies. However, appropriate guidance for defining the conditions under which chlorinated solvent source zone remediation would be a viable option for site cleanup compared to a containment-only option has not yet been developed. Current strategies of source zone containment have generally proven reliable for limiting routes of human and ecological exposure, provided that the containment system (pump-and-treat, permeable barriers, physical barriers) has been properly designed and maintained. This strategy transfers the burden of site management to future generations and requires long-term financial stability of the responsible parties.

Research is needed to address key knowledge gaps that still exist regarding the effectiveness and costs of chlorinated solvent source zone remediation technologies in a wide range of hydrogeologic settings. Research is needed on the following topics:

- Development, verification and comparison of alternative technologies for measuring mass flux and mass discharge from source zones before and after source depletion.
- Improved predictive tools to estimate the benefits and adverse effects of partial source depletion for a range of treatment technologies and contaminant distribution and geologic scenarios.
- Continued field testing of source-depletion technologies incorporating more than one technology.
- Guidance on the conditions in which source depletion is not likely to be an effective strategy and containment must be relied upon.
- Assessment of the long-term water quality impacts of source-depletion technologies.
- Development of quantitative decision analysis tools that will permit an accounting of all potential costs, benefits and adverse impacts of partial source depletion.

A major challenge is to identify the degree of characterization and post-remediation monitoring needed for effective application and assessment of each of the *in situ* source-depletion technologies.

1.4.2 DNAPL Source Zone Characterization and Delineation (Chapter 3)

Proper site conceptual model development, remediation technology selection and remediation design depend highly on assessing whether DNAPL is present in a source zone, and if so, how much contaminant mass is present as DNAPL and in other phases (aqueous, sorbed and vapor). A converging-lines-of-evidence approach can be used to assess DNAPL presence in both porous and fractured media (USEPA, 2009). The various lines of evidence include visual observation, exceeding saturation and partitioning thresholds for soil concentrations, the use of hydrophobic dye testing, site use and site history, and various aspects of groundwater quality data. Delineation of the spatial extent of chlorinated solvent source zones can be performed by plotting the various lines of evidence (indicators) on plan view and cross section figures. This approach facilitates delineation of both confirmed and potential source zones. At sites where DNAPL has been depleted as a result of either natural processes or remediation efforts, the approach facilitates delineating areas where other processes, such as back diffusion and desorption, may be sustaining plume migration.

The successful implementation of many remediation technologies very much depends on accurate estimates of contaminant mass, both total mass and mass in different phases and locations. Mass estimates can be arrived at through integration of subsurface data using either simple spreadsheet methods or more sophisticated numerical algorithms such as kriging. In practice, there is uncertainty in mass estimates, given the limitations of the tools and resources available, as well as the challenges inherent in locating DNAPLs within the subsurface.

There has been much advancement in source zone characterization and delineation in the past three decades. However, further advancements are required to support optimal selection and implementation of remediation technologies. It is likely that more resources should have been directed toward source zone characterization and delineation at many chlorinated solvent sites where full-scale remediation has been attempted.

1.4.3 Advanced Diagnostic Tools (Chapter 4)

A wide range of innovative diagnostic tools are available for characterization and remedial performance assessment at chlorinated solvent sites. Diagnostic tools are field sampling and laboratory measurement techniques that permit high-resolution description of contaminant distributions or of the *in situ* processes affecting the fate and transport of contaminants. Diagnostic tools are used to focus remedial efforts at locations and for timeframes in which the particular technology deployed is best suited.

The five groups of diagnostic tools evaluated in Chapter 4 include:

1. Multilevel monitoring systems for the delineation of the vertical distribution of contaminants in the dissolved and adsorbed phase
2. Rock matrix characterization for the assessment of chlorinated solvent distribution in consolidated media and performance assessment of *in situ* technologies
3. Mass flux/mass discharge measurement technologies for site characterization and process and performance assessment
4. Compound-specific isotope analysis for confirming *in situ* chemical or biological transformations of chemicals of concern
5. Molecular biological tools for optimizing process performance of *in situ* bioremediation

Collectively, important potential advantages of these diagnostic tools include:

- A more accurate and detailed CSM
- More accurate performance assessment in real time
- Assessment of the feasibility of achieving certain endpoints, such as background concentrations or maximum contaminant levels (MCLs)
- Confirmation of *in situ* processes that result in transformation of the chemicals of concern to nontoxic or less toxic byproducts
- Alternative metrics for performance assessment, such as mass flux and mass discharge

The uncertainty faced by practitioners making decisions regarding remediation at chlorinated solvent sites is a significant challenge, as is the decision of whether to expend additional resources to use one or more new diagnostic tools to decrease that uncertainty. These tools are considered innovative in that they are not routinely used for site characterization or performance assessment in the remediation industry, although their use is increasing. Only a small number of ASTM International standards have been developed to date for applying these diagnostic tools.

1.4.4 Modeling Source Zone Remediation (Chapter 5)

Capabilities to mathematically model chlorinated solvent source zone remediation have improved significantly in the past two decades. Mathematical models can be valuable in assisting with remediation technology selection, design, and application. Advanced numerical methods and growing computing power are making three-dimensional simulations with realistic heterogeneity and complex geochemical reactions a reality for desktop computers. A wide variety of models exist and can be applied by someone skilled in modeling or can be applied in collaboration with the model developers, who are able to make site- or technology-specific modifications. There is ample scope for new models to be developed and existing models to be extended and validated against evermore realistic data sets.

Modeling is often criticized because it inevitably adopts simplifications of reality, generates enormous amounts of data, and creates output that can be incorrectly interpreted as the current or future reality. However, these pitfalls can be avoided by a practitioner trained in both contaminant hydrogeology and modeling. The model must be matched to the project objectives and site conceptual model, attention must be paid to model verification and validation for the key scenario processes, the results need to be interpreted with a sound understanding of the assumptions employed, and uncertainty should be quantified via multiple simulations.

1.4.5 Modeling Plume Response to Source Treatment (Chapter 6)

It is now recognized that plume response to source zone remediation may vary widely from site to site. Plumes present in low to moderate permeability media and plumes present in media containing significant diffusive sinks may persist for decades to hundreds of years following complete source zone remediation. At some sites (late Stage 3), this implies that the lifespan of the plume is currently longer than the lifespan of any remaining DNAPL in the source zone. Because the time scales of interest are long, empirical data sets do not yet exist and mathematical modeling needs to be relied upon to arrive at quantitative estimates of the time scale of plume response to source zone remediation.

The plume response to source remediation is a complex function of many variables, including the fraction of mass remaining in the source compared to the plume, the source concentration compared to regulatory limits, the ratio of plume decay rates to groundwater velocity, the relationship between source mass removal and source discharge and local diffusive effects in the plume. Source remediation typically leads to an eventual reduction of contaminant mass in the plume. At sites where the source concentration is relatively close to regulatory limits, source remediation can be expected to lead to a significant reduction of plume length. However, if the source concentration is many orders of magnitude higher than regulatory limits, it is likely that plume treatment will be needed in addition to source remediation. Source remediation is most effective when it is done before the plume develops, and the bulk of the contaminant is in the source zone rather than the plume. At sites with significant matrix diffusion effects, such as certain fractured rock settings, the plume response to source remediation is expected to be relatively slow.

1.4.6 Flux-Based Site Assessment and Management (Chapter 7)

Contaminant mass flux measurements provide a valuable method to assess conditions at chlorinated solvent sites, including information needed to conduct mass balance calculations and to evaluate natural attenuation rates. The data can be used to assess remedial performance and to support risk assessment based on discharge to receptors. Measurements of mass flux and discharge immediately downgradient of a source zone during source zone remediation provide an assessment of remedial performance and an indication of the likely plume response.

At many sites it may be more appropriate to place a greater emphasis on first defining the source strength function (mass flux leaving the source zone) and then estimate source mass as a secondary effort. This approach provides information that is more directly linked to the plume behavior at the site. Information on flux distribution and mass discharge can lead to more effective application of remedial efforts aimed at reducing the risk associated with the dissolved contaminant plume. Challenges still exist in determining the resolution needed when defining the source mass flux distribution. Larger scale, average measures may lead to lower costs while sacrificing knowledge of local high-flux zones. The link with remedial goals is again critical to defining the level of resolution needed.

1.4.7 Hydraulic Displacement of Dense Nonaqueous Phase Liquids (Chapter 8)

Hydraulic displacement, also referred to as waterflooding and dual phase extraction, is a means of recovering pooled DNAPL from chlorinated solvent source zones. Hydraulic displacement can be implemented in a variety of ways, ranging from pumping DNAPL only from a single extraction well or drain to pumping both DNAPL and water from numerous wells or drains in conjunction with upgradient groundwater injection to accelerate the rate of DNAPL recovery. Hydraulic displacement is capable of recovering significant quantities of pooled DNAPL from a source zone but is not capable of recovering residual DNAPL. Hydraulic displacement may need to be performed for several years depending on well spacing, DNAPL properties, the volume of DNAPL present and the hydraulic conductivity of the porous or fractured medium.

In addition to performing as a DNAPL mass removal technology, hydraulic displacement can be used to stabilize a source zone. Hydraulic displacement can remove pooled and potentially mobile DNAPL, leaving behind a more stable source zone characterized by a significantly increased volume fraction as residual DNAPL. Hydraulic displacement results in an increase in DNAPL-water surface area available for mass transfer. It also results in a greater permeability to water through the DNAPL source zone, increasing water flux for a given hydraulic gradient. All of these effects can be beneficial for the subsequent application of mass transfer- and flushing-based technologies such as surfactant flushing, cosolvent flushing, *in situ* chemical oxidation, *in situ* chemical reduction and enhanced *in situ* bioremediation. In this respect, hydraulic displacement can act as a preconditioning technology as part of a combined remedy approach. This strategy may be particularly advantageous because it is usually less costly to remove DNAPL pools using hydraulic displacement than with thermal, chemical, or biological remediation approaches.

Hydraulic displacement should only be considered for sites exhibiting significant quantities of pooled DNAPL. The existence of pooled DNAPL can be established on the basis of visual observation in cores, visual observation in monitoring wells, and the magnitude of soil and sediment concentrations. A variety of screening calculations can be performed to estimate DNAPL pool heights, DNAPL pool volumes, and the hydraulic gradient required to mobilize a pool.

1.4.8 *In Situ* Chemical Oxidation (Chapter 9)

The principles and practices associated with using *in situ* chemical oxidation (ISCO) to treat dissolved and sorbed phase contamination are generally well established and documented in design manuals and reference texts. The application of ISCO to treat chlorinated solvent source zones containing DNAPL is not as well established and field applications for this purpose have had mixed results. ISCO application to chlorinated solvent source zones typically involves the injection of an oxidant solution into vertical wells to promote distribution of the oxidant through the source zone. The oxidation reaction occurs in the subsurface and fluid extraction is typically not required.

While there is a theoretical basis for the degradation of residual and pooled DNAPL by ISCO, cost-effective remediation generally depends on (1) susceptibility of DNAPL organics to oxidative destruction, (2) rate and extent of dissolution of the DNAPL into the groundwater where the oxidant resides, (3) delivery of an adequate amount of oxidant that contacts DNAPL in the source zone, (4) effects of subsurface conditions on ISCO reactions and (5) ISCO effects on subsurface permeability and biogeochemistry.

Successful implementation of ISCO in chlorinated solvent source zones may require coupling the technology with other site management approaches. Appropriate goals and performance metrics should be established to set realistic expectations for the technology. Monitoring during application of ISCO will allow adjustment of oxidant injection rates and oxidant dosage to optimize technology application. Multiple delivery events may be appropriate for some sites and rebound of contaminant concentrations in groundwater should be expected following cessation of ISCO application.

1.4.9 *In Situ* Chemical Reduction for Source Remediation (Chapter 10)

The remediation technologies grouped under *in situ* chemical reduction (ISCR) are diverse in many respects but can generally be classified according to reductant strength and treatment volume. Other classifications consider factors such as expected reductant longevity in the subsurface, compatibility with other treatment technologies and track record of successful full-scale implementations. Specific ISCR technologies can be applied at a wide range of scales and to a variety of chlorinated solvents. Most ISCR technologies are relatively immature, implying that the full potential value of ISCR applications probably has not yet been realized.

ISCR treatment processes range from mild to very strong in terms of reactivity, so a wide range of contaminants are potentially treatable. The rates and products of these processes can be favorable, but results may vary. In many cases, successful treatment can be achieved by influencing the natural biogeochemical conditions to favor contaminant reduction, and this approach is usually straightforward to deploy. In cases where deployment of strong, artificial reductants is justified, the natural reductant demand should also be considered.

Effective delivery of reactants as fluids to source zones is challenging because of aquifer heterogeneity and risk of DNAPL displacement and must be considered a limitation in the context of ISCR implementation. There has also been considerable interest in delivery of nanoscale zero-valent iron with surface treatments designed to favor partitioning into NAPL. However, delivery of sufficient quantities of those materials through aquifers to source zones will require delivery strategies beyond well injections primarily used today.

Other processes may be stimulated or inhibited by ISCR treatment. In particular, biodegradation is an important component of many ISCR activities (as either the primary or a secondary mode of action) and should be considered in the design of most ISCR systems. Chemical reductants added to the subsurface may have a relatively short lifetime during which the contaminant treatment goal must be achieved. Of the various ISCR approaches, those involving granular ZVI are the only ones that have been applied at a sufficient number of sites to be considered proven. ZVI permeable reactive barriers have been shown to remain active for many years in most cases.

1.4.10 Surfactant and Cosolvent Flushing (Chapter 11)

Over the past two decades, considerable resources have been directed toward the development and demonstration of surfactant and cosolvent flushing for chlorinated solvent source zone remediation. Although these technologies provide distinct advantages and limitations when compared to traditional pump-and-treat approaches, considerable uncertainty remains regarding the effectiveness and long-term benefits of these aggressive source zone treatment methods.

The uncertainty results mainly from the difficulties in both delivering fluids effectively in the subsurface and subsequently recovering and treating the contaminant mass and injected solution.

At the bench scale these technologies have shown promise for reducing the overall remediation time while greatly improving contaminant mass recovery compared to dissolution alone (pump-and-treat). To date, the field-scale results have been less encouraging, in large part because the success of *in situ* technologies in the field is governed by subsurface hydrogeology, in particular the heterogeneity of the aquifer materials and the complexity of contaminant mass distribution. A number of early field trials yielded mixed results and failed to reach desired cleanup goals. However, as the limitations were recognized and refinements were implemented, improved results were achieved.

Despite substantial advances in the technology and successful field applications, surfactant and cosolvent flushing has received little attention as a source zone remediation technology during the past 10 years. Several factors may have contributed to this outcome: (1) relatively high active ingredient costs; (2) the complexity of implementation compared to alternative technologies; (3) reduced performance in low-permeability, heterogeneous media; and (4) the costs and difficulties involved in treating the effluent waste stream.

1.4.11 *In Situ* Bioremediation of Chlorinated Ethene Source Zones (Chapter 12)

Initially, *in situ* bioremediation (ISB) was not considered a feasible technology for treating chlorinated solvent source zones containing DNAPL and still faces some skepticism. However, the experience of the last decade suggests that it can be a viable technology for treating some source zones. Experience with ISB shows that realistic objectives must be set, and complete cleanup over a few years is unlikely. Reducing mass discharge from a source by 90–99% and achieving similar reductions in contaminant concentrations in the permeable portions of the source zone are reasonable expectations, although lower reductions have been observed in some field-scale treatments.

Typically, uncertainty exists in characterization of the source zone and site-specific issues limit performance (stagnant areas, unknown contaminated areas). ISB, like any injection-based technology, is best applied in stages, learning along the way and optimizing the design and operations over time. Some rebound in contaminant concentrations may occur after treatment is completed and the aquifer returns to pretreatment conditions. Although ISB may remove some of the source zone mass and reduce the short-term mass discharge, model simulations indicate that long-term monitoring in combination with natural attenuation may be required at many sites. Recent research suggests that build-up of bacterial biomass and reduced minerals during ISB may sustain treatment after electron donor addition ends. As a result, contaminant concentrations may not rebound for several years after treatment, providing site managers with time to detect rebound and implement contingency actions, if required.

Successful implementation of ISB will require monitoring and implementation of measures to address potential excessive biofouling of injection wells and the treatment zone. The distribution and relative concentrations of the different chloroethenes, along with the geochemical parameters that exist prior to treatment, can be useful in designing an ISB system. For example, substantial preexisting dechlorination suggests that less electron donor will be needed to achieve reducing conditions and may demonstrate that a complete dechlorination pathway already exists. Similarly, a lack of vinyl chloride or ethene given otherwise favorable conditions can suggest a need for bioaugmentation.

Practitioners considering ISB for a source zone should be aware of several potential difficulties. The electron donor demand can be so great that it is not feasible to supply enough donor and/or pH buffer, or it may cause other problems such as methane production or clogging of the subsurface. Long-term performance of ISB is not clear, and treated sources may still sustain a plume that requires ongoing monitoring or even treatment. These challenges require careful design, monitoring, and continuous optimization throughout treatment.

1.4.12 Natural Attenuation of Chlorinated Solvent Source Zones (Chapter 13)

The natural attenuation of a chlorinated solvent source zone (also referred to as source zone natural attenuation, or SZNA) is defined by the authors as "A variety of physical, chemical or biological processes that act without human intervention to reduce the mass, toxicity, mobility, volume, mass flux, mass discharge or concentration of contaminants in soil or groundwater within that volume of the subsurface currently or previously occupied by chlorinated solvents in the form of nonaqueous phase liquids." When SZNA is relied upon as a remedy or remedy component, the authors refer to monitored natural attenuation (MNA) of a chlorinated solvent source zone as "The reliance upon a variety of physical, chemical or biological processes that act without human intervention to achieve site-specific remedial objectives, as verified through ongoing monitoring and data analysis, within and downgradient of that volume of the subsurface currently or previously occupied by chlorinated solvents in the form of nonaqueous phase liquids."

Chlorinated solvent SZNA results from the processes of advection, dispersion, dissolution, sorption/desorption, matrix diffusion, biodegradation, abiotic degradation and vaporization/volatilization. Source zones currently in late Stage 3 or beyond are likely candidates for source zone MNA. Selecting source zone MNA as a remedy or remedy component will likely require a good track record of aqueous phase concentrations to document declining concentration trends with time. Source zone MNA may only be attractive when long-term management over decades to centuries is possible, even following partial source removal. However, because most active technologies will only remove part of the contaminant mass, decades-to-centuries timeframes may also apply to many active remediation projects.

1.4.13 *In Situ* Thermal Treatment of Chlorinated Solvent Source Zones (Chapter 14)

In situ thermal treatment (ISTT) has been applied to chlorinated solvent source zones for approximately 20 years. In that time, it has proven to be capable of rapidly removing the majority (>99%) of the mass of chlorinated solvents from the targeted volume at some sites and reducing the associated groundwater concentrations by two to four orders of magnitude. Given a thorough understanding of contaminant distribution and reasonable access to the contaminants, thermal treatment can be effective for chlorinated solvent mass removal. It may even achieve closure criteria, although in most cases studied, some residual contamination remains, and the groundwater concentrations remain above the MCL values.

Thermal treatment is applied either (1) to increase vapor recovery by inducing boiling and/or increasing the vaporization and volatilization of water and contaminants or (2) to increase liquid recovery by increasing DNAPL mobility by reducing the viscosity and interfacial tension. Given the variety of strategies for implementing the technology and the variety of processes occurring during any form of thermal treatment, it is important to fully understand these processes when selecting or designing a remediation system.

Thermal technology has gone through a rapid evolution. Several different commercially successful approaches have been developed. The approaches most used at the present time are electrical resistance heating (ERH) and thermal conductive heating (TCH), although radio frequency heating, steam enhanced extraction (SEE) and heating while mixing the soil with large augurs are being used commercially as well. The technology evolution also has led to combinations of different thermal techniques with each other, or with other remediation technologies. All thermal technologies require vapor extraction and aboveground engineered systems to capture and treat recovered contaminants.

The state of the practice regarding ISTT has evolved rapidly, to the point that some guidance and standardized protocols are available. A recently released summary of the state of the practice reviewed the design parameters and performance at 182 field sites (Triplett Kingston et al., 2010). Based largely on this review, several practical recommendations have been presented: (1) define reasonable remediation objectives and link the system shutdown criteria to these objectives, (2) adequately characterize the source area since much of the contamination remaining after thermal treatment in some applications is due to insufficient delineation, (3) collect sufficient thermal data in and around the treatment zone, and (4) monitor long enough after treatment to allow groundwater to flow entirely through the treatment zone and allow the treatment zone to cool down to near-ambient conditions.

1.4.14 Combined Remedies (Chapter 15)

The increasing emphasis on deliberately combining remedies for chlorinated solvent source zone remediation has resulted from an increased understanding of both the limitations of individual remediation technologies and the economic and performance benefits of combining remediation technologies. A variety of technology combinations have been used to treat DNAPL source zones, including sequential combinations, concurrent combinations and coupled process combinations. However, little systematic research into optimal strategies for combining technologies has been carried out to date. Because several technical, regulatory and business-related factors may impact remediation planning and technology selection, developing an overall remediation strategy applicable to all DNAPL sites may not be feasible.

Of the available sequential process combinations, ISCO followed by ISB has been perhaps the most widely evaluated combination, both in the laboratory and in the field. ISB is the most frequently suggested active follow-up technology after implementing a primary source treatment method, such as ISTT, ISCO, or surfactant flushing. A key aspect to consider when planning any sequential combination is the post-implementation impacts of the primary treatment technology on the biological and geochemical conditions for the subsequent technology.

Of the available coupled process combinations, the surfactant-ISCO coupled process has received the greatest amount of research interest. Significant progress has been made in understanding the general compatibility of many commonly available surfactants with several commonly used oxidants. However, little information on the field effectiveness or cost-effectiveness of these technologies compared to ISCO alone is yet available. Another coupled process that has been successfully demonstrated is the ISCR-ISB combination. However, little research has been published on the cost competitiveness of these combined technologies compared to ISB or ISCR alone.

Research also has focused on the coupled process combinations of ISTT with other technologies, such as ISCO, ISCR and ISB. The results to date indicate that while some important reaction rates can be accelerated through the application of heat, each technology combined with thermal treatment may have a unique temperature range within which optimal process enhancements occur.

Additional research addressing the following questions would be helpful to advance the state of the practice related to source zone remediation using combined remedies:

- What type of performance can be expected when combining technologies?
- What site conditions favor applying specific technology combinations?
- What potential technology combinations have the greatest synergistic benefits and are the most cost-effective?

1.4.15 Cost Analyses for Remedial Options (Chapter 16)

Chapter 16 presents a general approach to performing cost analyses for a number of remedial options typically applied to chlorinated solvent source zones. The approach employs many assumptions, which may not apply for all sites. The approach employs a net present value calculation to allow the total cost of various remedial options to be compared on a total lifespan basis and significant cost drivers for each option identified. In this analysis, two template sites are used to demonstrate the approach for costing of DNAPL source treatment and identification of cost drivers.

The discount rate employed in the NPV analysis impacts the importance of present versus future costs. Specifically, using a higher discount rate will diminish the importance of future operations and maintenance (O&M) and monitoring costs relative to present capital costs. The evaluation uses a lower discount rate and thus presents a scenario in which the future O&M and monitoring costs have a larger influence on total lifespan costs.

1.4.16 Groundwater Remediation, Management, and the Use of Alternative Endpoints at Highly Complex Sites (Chapter 17)

Complex sites face a number of uncertainties when using today's technology to meet long-term cleanup requirements. A variety of approaches have been used at complex chlorinated solvent sites, including active remediation guided by near-term goals and objectives, the use of MNA over long timeframes, long-term management/containment, or some combination of these approaches. Complex sites have also used a variety of different alternative endpoints, which either waive or substitute for long-term cleanup standards.

Integrated strategies often are essential to manage complex sites efficiently. Case studies show that progress has been made not by choosing between remediation and management, or between remediation and alternative endpoints, but by combining these approaches into a strategy that works for each site. Case studies illustrate how different regulatory agencies and programs have addressed this issue through policy and regulation. Experience has also shown that at highly complex sites, technical challenges may frequently lead to involved discussions of policy, management, stakeholder priorities, stakeholder perceptions, and regulatory interpretation. These factors have the potential to dominate technology selection at complex sites.

1.4.17 Future Directions and Research Needs for Source Zone Remediation (Chapter 18)

Society, industry and regulatory agencies likely will continue to strive to remediate chlorinated solvent source zones for many years to come. Research and development has played a vital role in improving site remediation efforts over the past 25 years and it will continue to be essential in the future. It is apparent, however, that the research community will need to change in response to decreases in funding. Specifically, there is a clear need to develop and participate in research partnerships to leverage funds and capabilities. Research in this area is highly interdisciplinary and requires expertise in several subdisciplines including mathematics, chemistry, geology, hydrogeology, hydrology, geochemistry, microbiology and engineering.

In developing future research partnerships, it will be important to remember that one of the goals of research is to train highly qualified personnel who will be needed to work in the field of environmental protection. It will also be important to consider how the results of research will be communicated. The traditional model is to rely on publication in journal articles and in conference presentations. These avenues will remain essential, but the research community must also realize that these may not be the best ways to disseminate the results of research to important audiences, including practitioners and concerned citizens.

Within the area of effective and efficient characterization of source zones, it is now recognized that the characterization of source zones is often inadequate given the difficulties inherent in working in the subsurface and the complex distribution of contaminants between the DNAPL, aqueous, vapor and sorbed phases. In many cases, the success of source zone remediation is directly related to the quality and thoroughness of site characterization. Specific research needs within the general area of effective and efficient characterization of source zones include (1) source zone delineation and mass estimation, (2) source zone architecture and depletion and (3) increased resolution and fine-scale mapping of geologic properties and contaminant distribution.

Within the area of remediation technologies and strategies, it is recognized that many technologies have been developed to remove mass from chlorinated solvent source zones, yet extensive research has clearly shown technical limits to their success. With this understanding, efforts in recent years have shifted to improving treatment through better delivery techniques, improved formulations or deliberate combinations of technologies, and to improving models and predictions of plume responses to source treatment. Specific research needs within the general area of remediation technologies and strategies include (1) understanding the long-term impacts of applying source zone remediation technologies, (2) development of improved remedial fluid delivery techniques, (3) evaluating the use of combined remedies, (4) development of diagnostic and performance assessment tools, (5) development of improved containment and *ex situ* treatment technologies, (6) development of improved economic analyses, (7) development of improved technology performance models, (8) understanding impacts of various management strategies, and (9) development of technologies for challenging site conditions such as fractured bedrock.

With regard to technology transfer needs, the goals of technology transfer are to improve the state of the practice and to foster the adoption of promising technologies. In this context, the technology users include practitioners, regulators, managers, and the public. Communication needs to occur along a number of avenues, particularly communication and open discussions among stakeholders on a frequent basis. Technology transfer should emphasize development of web-based tools that allow users to quickly find focused information specific to their issues. The need for highly qualified personnel is paramount, especially within the field of contaminant hydrogeology which lacks a universal standard for qualification.

1.5 SUMMARY

Remediating chlorinated solvent source zones remains technically challenging, costly, and contentious in terms of the expectations of various stakeholders associated with any given site. The fundamental properties of chlorinated solvents such as PCE, TCE, 1,1,1-TCA, CT, and others that make them useful for industrial applications also make them difficult to locate and treat in the subsurface. Chlorinated solvent DNAPLs migrate and distribute themselves in a tortuous and complex manner and persist for decades to centuries in many subsurface environments. Reaching typical concentration-based cleanup goals is possible in a short time frame (years) at some sites, but such performance has been relatively rare in practice.

Over the last 25 years, several remediation technologies have been developed and applied at pilot and full scale at various sites. The range of technologies currently available include chemical treatment (ISCO, ISCR, surfactants and cosolvents), thermal treatment (ERH, TCH, SEE), hydraulic displacement, physical containment, SVE in the unsaturated zone, excavation and *in situ* soil mixing, and ISB. Each technology has relative advantages and disadvantages. It may therefore be beneficial to develop combined remedies, using different approaches for different areas or times within an overall remediation strategy. Each chlorinated solvent source zone is unique and requires site-specific evaluation of the appropriate remediation strategy and long-term site management approach.

Experience has shown that making progress often requires stakeholders to develop realistic expectations regarding the ability to remediate chlorinated solvent source zones. Experience also has shown that an adaptive management strategy often leads to the most efficient treatment. These lessons will be valuable as the focus increasingly shifts to remediation of the most complex sites, which include source zones in fractured bedrock and contaminants stored in low-permeability media.

Remediation technology application and assessment has not progressed from simple sites to complicated sites. Rather, remediation technologies have been applied at a wide range of sites exhibiting various levels of complexity, resulting in ambiguous interpretations of technology performance. It may be beneficial for future deployments of chlorinated solvent source zone remediation technologies to progress from simple to complicated sites such that a proper understanding of the effectiveness of a given technology can be obtained. Such an evaluation should include monitoring of downgradient concentrations for several years following technology deployment to properly assess the effects of back diffusion and remaining DNAPL mass on contaminant concentrations. The focus of this monitoring should be on concentrations within the source zone, as well as within the plume downgradient of the source zone.

Remediation of chlorinated solvent source zones is technically challenging and requires the services of highly qualified personnel with appropriate education, knowledge, and experience. There is currently no industry-wide accepted level of qualification for individuals working in the field of chlorinated solvent source zone remediation. Future efforts should be directed at developing an industry-wide level of qualification and developing improved methods of distributing the results of research and past field experiences to all stakeholders.

REFERENCES

Ankeny MD, Forbes JR. 2000. Constant Hydraulic Head Moat and Method for Controlling Regional Ground Water Flow. U.S. Patent No. 6,139,221A. http://www.google.com/pat ents/US6139221?dq=%22ankeny+moat%22. Accessed January 20, 2014.

Brewster ML, Annan AP, Greenhouse JP, Kueper BH, Olhoeft GR, Redman JD, Sander KA. 1995. Observed migration of a controlled DNAPL release by geophysical methods. Ground Water 33:977–987.

Falta RW, Rao PSC, Basu N. 2005. Assessing the impacts of partial mass depletion in DNAPL source zones: I. Analytical modeling of source strength functions and plume response. J Contam Hydrol 78:259–280.

Gerhard JI, Pang T, Kueper BH. 2007. Time scales of DNAPL migration in sandy aquifers examined via numerical simulation. Ground Water 45:147–157.

Kavanaugh MC, Rao PSC, Abriola L, Cherry J, Newell C, Sale T, Destouni G, Falta R, Shoemaker S, Siegrist R, Major D, Mercer J, Teusch G, Udell K. 2003. The DNAPL Remediation Challenge: Is There a Case For Source Depletion? EPA/600/R-03/143. USEPA National Risk Management Research Laboratory, Cincinnati, OH, USA. 129 p.

Kueper BH. 1996. Unpublished. Queen's University, Kingston, ON, Canada.

Kueper BH, Gerhard JI. 1995. Variability of point source infiltration rates for two-phase flow in heterogeneous porous media. Water Resour Res 31:2971–2980.

Kueper BH, Redman JD, Starr RC, Reitsma S, Mah M. 1993. A field experiment to study the behavior of tetrachloroethylene below the water table: Spatial distribution of residual and pooled DNAPL. Ground Water 31:756–766.

McCarty PL. 2010. Groundwater contamination by chlorinated solvents: History, remediation technologies and strategies. In Stroo HF, Ward CH, eds, *In Situ* Remediation of Chlorinated Solvent Plumes. Springer, New York, NY, USA, pp 1–28.

NAVFAC (Naval Facilities Engineering Command), Geosyntec. 2011. DNAPL Technology Evaluation Screening Tool (DNAPL TEST). http://projects.geosyntec.com/DNAPL/dnaplt-est.aspx. Accessed January 20, 2014.

NRC (National Research Council). 2005. Contaminants in the Subsurface: Source Zone Assessment and Remediation. National Academies Press, Washington, DC, USA. 333 p.

NRC. 2013. Alternatives for Managing the Nation's Complex Contaminated Groundwater Sites. Committee on Future Options for Management in the Nation's Subsurface Remediation Effort; Water Science and Technology Board (WSTB); Division on Earth and Life Studies (DELS); National Research Council of the Academies. The National Academies Press, Washington, DC, USA. 422 p.

Pankow JF, Cherry JA, eds. 1996. Dense Chlorinated Solvent and Other DNAPLs in Groundwater. Waterloo Press, Portland, OR, USA. 522 p.

Poulsen M, Kueper BH. 1992. A field experiment to study the behavior of tetrachloroethylene in unsaturated porous media. Environ Sci Technol 26:889–895.

Schwille F. 1981. Groundwater pollution in porous media by fluids immiscible with water. Sci Total Environ 21:173–185.

Schwille F. 1988. Dense Chlorinated Solvents in Porous and Fractured Media: Model Experiments. Translated from the German by J.F. Pankow. Lewis Publishers, Boca Raton, FL, USA. 146 p.

Streeter VL, Wylie EB. 1981. Fluid Mechanics, First SI Metric Edition. McGraw-Hill Ryerson, Whitby, ON, Canada. 562 p.

Stroo HF, Lesson A, Marqusee JA, Johnson PC, Ward CH, Kavanaugh MC, Sale TC, Newell CJ, Pennell KD, Lebrón CA, Unger M. 2012. Chlorinated ethene source remediation: Lessons learned. Environ Sci Technol 46:6438–6447.

Travis CC, Doty CB. 1990. Can contaminated aquifers at Superfund sites be remediated? Environ Sci Technol 24:1464–1466.

Triplett Kingston JL, Dahlen PR, Johnson PC. 2010. State-of-the-practice review of *in situ* thermal technologies. Ground Water Monit Remediat 30:64–72

USEPA (U.S. Environmental Protection Agency). 2004. Cleaning Up the Nation's Waste Sites: Markets and Technology Trends. EPA 542-R-04-015. USEPA Office of Solid Waste and Emergency Response. http://www.clu-in.org/download/market/2004market.pdf. Accessed January 20, 2014.

USEPA. 2009. Assessment and Delineation of DNAPL Source Zones at Hazardous Waste Sites. EPA/600/R-09/119. USEPA National Risk Management Research Laboratory, Cincinnati, OH, USA. 20 p.

USEPA. 2013. Basic Information about Regulated Drinking Water Contaminants and Indicators. http://water.epa.gov/drink/contaminants/basicinformation. Accessed January 20, 2014.

WCGR (Waterloo Centre for Groundwater Research). 1992. Lecture 2 (B. Kueper), Immiscible Liquid Behaviour (Porous Media). Professional Short Course, Site Characterization and Remediation of Dense, Immiscible Phase Liquid Contaminants (DNAPLs) in Porous and Fractured Media. November 2–5, Chicago, IL, USA.

CHAPTER 2

THE SOURCE ZONE REMEDIATION CHALLENGE

Eric J. Suchomel,[1] Michael C. Kavanaugh,[2] James W. Mercer[3] and Paul C. Johnson[4]

[1]Geosyntec Consultants, Inc., San Francisco, CA 94105, USA; [2]Geosyntec Consultants, Inc., Oakland, CA 94607, USA; [3]Tetra Tech GEO, Sterling, VA 20164, USA; [4]Arizona State University, Tempe, AZ 85287, USA

2.1 INTRODUCTION

It is well recognized by stakeholders that restoration of soil and groundwater contaminated by organic chemicals in the form of dense nonaqueous phase liquids (DNAPLs) poses major technical, economic and institutional challenges. As of 2012, few case studies of restoration (defined as achieving concentrations at or below maximum contaminant levels [MCLs], thus allowing for unrestricted use and unlimited exposure at the site) have been reported for source zones where DNAPL releases resulted in significant contaminant plumes. While the consequences of DNAPL releases on groundwater quality are easily observed and have been widely documented, the characterization and remediation of DNAPLs in groundwater remain problematic. At many sites, characterization of the location, distribution and amount of DNAPL causing continued groundwater contamination is difficult and often inaccurate (Mercer et al., 2010). Removal or *in situ* destruction of DNAPLs, even when reasonably well characterized, has proven difficult in saturated zones with any significant degree of heterogeneity (Pankow and Cherry, 1996; Kavanaugh et al., 2003; NRC, 2005).

DNAPL sites continue to represent a significant fraction of the sites impacted by chlorinated solvents and other chemicals characterized as DNAPLs. The U.S. Environmental Protection Agency (USEPA, 1993a) evaluated the likelihood of DNAPL presence at 712 National Priorities List (NPL) sites (roughly 55% of all NPL sites as of 1991). At 44 of these 712 sites, DNAPL was observed directly in the subsurface. The likelihood of DNAPL occurrence at the remainder of the sites was estimated based on a detailed analysis of a subset of 310 sites. Extrapolation of the survey results to the universe of NPL sites indicates that approximately 60% of NPL sites (approximately 775 sites) exhibited a high-to-medium likelihood of having DNAPL present as a source of subsurface contamination. Applying the 60% estimate of DNAPL sites to the current (2012) total of approximately 1,700 NPL sites yields 768 DNAPL sites in the current Superfund program. A similar study of 77 dry cleaner sites in Modesto, California, revealed 9 sites (11.7% of sites considered) with a likelihood of DNAPL occurrence based on elevated groundwater and/or soil gas concentrations (S. Carlton, GeoTrans, Inc., Rancho Cordova, California, personal communication, 2013). Considering recent estimates of approximately 36,000 active dry cleaning facilities in the United States, the potential exists for more than 2,800 active or inactive dry cleaner sites in the United States to have DNAPL present in the subsurface.

Since the early 1980s, the USEPA, Department of Energy, Department of Defense, and private industry have recognized that DNAPL site remediation entails significant technical challenges. For example, through early fundamental and applied research directed by the Robert S. Kerr Environmental Research Center, in Ada, Oklahoma, the USEPA prepared

B.H. Kueper et al. (eds.), *Chlorinated Solvent Source Zone Remediation*,
doi: 10.1007/978-1-4614-6922-3_2, © Springer Science+Business Media New York 2014

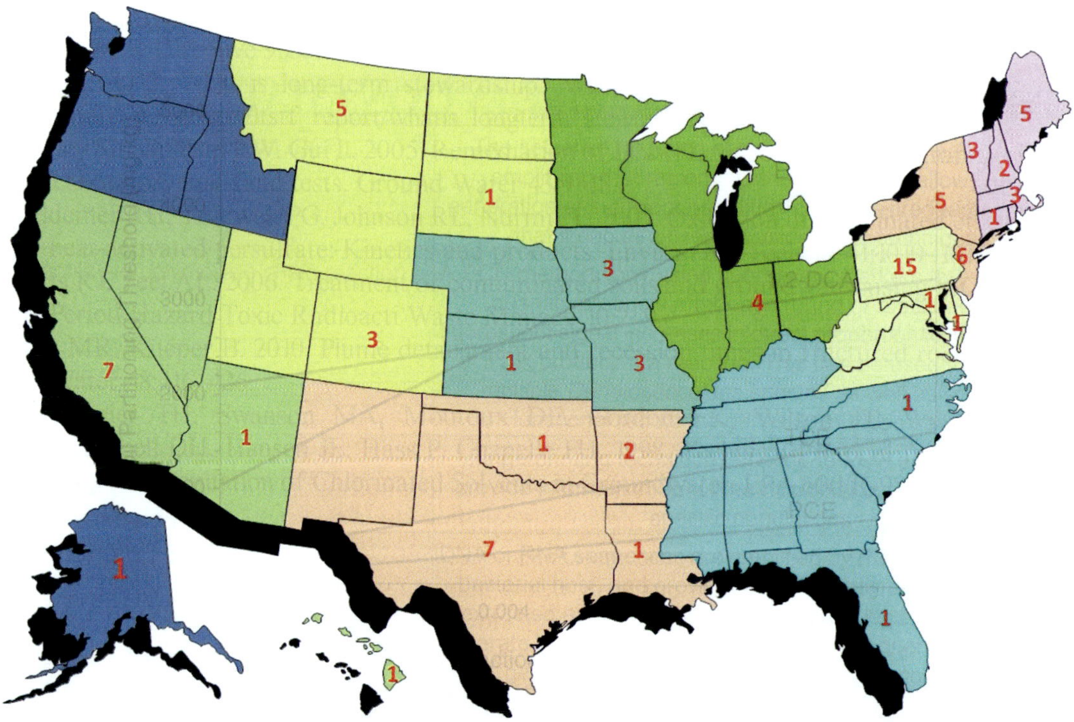

Figure 2.1. Locations of groundwater technical impracticability (TI) waivers. Total is approximately 85 sites as of November 2011. From Charsky, 2012.

numerous technical guidance documents to advise stakeholders on characterization and remediation of DNAPL sites (USEPA, 1994). In addition to relevant guidance documents, USEPA also developed policy guidance in recognition of the technical challenges confronting cleanup of DNAPL sites. The applicable or the relevant and appropriate requirement (ARAR) technical impracticability (TI) waiver guidance document published in 1993 provided USEPA's recommended approach for assessing the feasibility of meeting established performance goals at Superfund and Resource Conservation and Recovery Act (RCRA) sites where remediation was considered impracticable from "an engineering perspective, taking cost into consideration" (USEPA, 1993b). In 2011, USEPA released a clarifying memorandum stating that regulators "should not consider the mere presence of DNAPL alone but should provide a sufficient, science-based justification for invoking a TI waiver" (USEPA, 2011c). This clarification is consistent with the infrequent use of this policy instrument since 1993, with only 85 TI waivers (Figure 2.1) addressing groundwater-contaminated sites granted from a universe of approximately 1,500 Superfund sites as of November 2011 (Environmental Security Technology Certification Program, ESTCP, 2011; Charsky, 2012).

In 1994, a committee established by the National Research Council published *Alternatives for Groundwater Cleanup* (NRC, 1994), summarizing the results of research and practical experiences through 1992 on the use of pump-and-treat technology, the dominant approach at that time for remediation of groundwater contaminated by DNAPLs. Groundwater sites impacted by DNAPLs were considered to be the most difficult sites to remediate (Category 4 in a scale of 1–4). At that time, none of the known or suspected DNAPL sites had achieved required cleanup levels, usually MCLs. While significant technical advances in DNAPL site characterization and remediation technologies have been achieved since 1994 (Stroo et al., 2012), it is difficult to predict the impact of any of the suite of *in situ* technologies now available to remediate DNAPL source zones on the time frames needed to achieve groundwater remediation performance objectives.

More recent reviews of DNAPL source removal case studies also have shown that significant mass removal can be achieved when DNAPL sources are well defined, but complete restoration to concentration levels that allow for unrestricted use and unlimited exposure in these source areas is difficult (Kavanaugh et al., 2003; NRC, 2005; ITRC, 2011). A recent summary of case studies of DNAPL sites that had met regulatory closure criteria indicated that MCLs were achieved at 4 of the 13 sites (USEPA, 2009). However, even in this study, remedial action objectives (RAOs) other than MCLs were the basis for most decisions on final disposition of the site (no further action or long-term monitoring (LTM)) (Table 2.1). Of the four sites where MCLs were achieved, two were impacted by chlorinated solvents and two were impacted by petroleum hydrocarbons and/or polycyclic aromatic hydrocarbons.

Due in part to the number of sites potentially impacted by DNAPL and the limited number of documented DNAPL sites where the final remedial objectives have been achieved, there is still a lack of consensus regarding the ability of proven or innovative technologies to achieve groundwater remediation performance objectives in source zones at DNAPL sites within a reasonable time frame, and at a cost commensurate with perceived benefits or risk reduction (NRC, 2005). In particular there continues to be a lack of consensus regarding the appropriateness of applying intensive and often costly remediation technologies for DNAPL extraction or destruction in the source zone, if such partial mass removal will not have a quantifiable and substantial impact on the time required to meet RAOs and on the life cycle costs of a containment remedy, such as pump-and-treat.

This chapter summarizes the challenges facing practitioners responsible for selecting, designing and operating source remediation technologies. The chapter starts by briefly describing the general management options and the difficulties associated with predicting the performance of source remediation efforts. The chapter then summarizes the state of the art for source characterization and source remediation technologies and then discusses both the benefits and the risks of source depletion. The chapter concludes with a discussion of the potential impacts of source depletion on the life cycle costs for site management.

2.2 SOURCE MANAGEMENT OPTIONS

While the option of long-term containment at DNAPL-impacted sites has disadvantages, this option may represent the optimal use of financial resources, when considering the time value of money. For many DNAPL sites, given the uncertainties associated with the application of demonstrated or innovative technologies and the risk of insufficient performance, combined with the difficulties of predicting the benefits of partial DNAPL source depletion, potentially responsible parties in both the private sector (industry) and the public sector (government) often have been reluctant to undertake intensive source zone remediation. The current practical consequence of this lack of consensus is that for many DNAPL groundwater-contaminated sites, site remediation strategies are dominated by containment technologies, coupled with LTM. For example, pump-and-treat and monitored natural attenuation (MNA), both of which can contain contaminated sites, were used as groundwater remedies, either alone or in combination, at 71% of 164 Superfund sites between 2005 and 2008. During this same time period, institutional controls were included more often than any other groundwater remedy in site decision documents (USEPA, 2010).

This strategy has been effective at limiting the spread of contaminants at these sites and significantly reducing or even eliminating the risk of human and ecological exposures to these chemicals outside of the source zones. However, failure to remove the DNAPL source from the

Table 2.1. Summary of 2009 USEPA DNAPL Case Studies Reportedly Achieving Maximum Contaminant Levels (MCLs)

Site Name, Location	NAPL Type	Regulatory Oversight Program	Site COCs and RAOs for Groundwater	Technologies Applied	Reported Results	MCLs Achieved?	ICs in Place?
Camelot Cleaners, West Fargo, North Dakota	Chlorinated solvents	USEPA, Region 8 Superfund	*COCs:* PCE (up to 89,000 μg/L) *RAOs:* Reduce total VOCs to less than 1,000 μg/L in groundwater	Electrical resistive heating	5,188 pounds of VOC mass removed; groundwater concentrations reduced by 99.98%; all groundwater samples under 1,000 μg/L	No	No
Visalia Pole Yard, Visalia, California	Creosote, diesel with PCP	USEPA Superfund NPL with state lead	*COCs and RAOs* PCP, RAO: 1 μg/L Benzo(a)pyrene, RAO: 0.2 μg/L TCDD equivalent, RAO: 30 pg/L	Grout wall, P&T, *in situ* thermal desorption, excavation	1,330,000 pounds of organics recovered; risk-based concentration goals and MCLs met for groundwater	Yes	Yes – Land use restricted
Former Southern California Edison MGP Site, Long Beach, California	Polyaromatic hydrocarbon, petroleum hydrocarbons	Voluntary Action under DTSC Expedited Remedial Action Program	*COCs:* TPH (912 μg/L) Benzene (4.82 μg/L) Naphthalene (20 μg/L) Benzo(a)pyrene (0.34 μg/L) *RAOs:* No RAOs specified for groundwater	ISCO (ozonation), excavation	Concentrations of benzo (a)pyrene reduced to below MCL (0.2 μg/L)	Yes	Yes – Land use restricted
Former Cowboy Cleaners Site, Broomfield, Colorado	Chlorinated solvents	Colorado Voluntary Cleanup Program	*COCs:* PCE (1,900 μg/L) *RAOs:* No numerical RAO specified	ISCO (potassium permanganate)	PCE concentrations in source area reduced to 48 μg/L	No	Yes – Land use restricted
Dry Clean USA No. 11502, Orlando, Florida	Chlorinated solvents	Florida Department of Environmental Protection Dry Cleaning Program	*COCs:* PCE (27,300 μg/L) *RAOs:* Reduce PCE in groundwater to 3 μg/L (Florida MCL)	ISCO (hydrogen peroxide), P&T, SVE	PCE concentrations in groundwater reported to range from ND to 3 μg/L	Yes	No

(*continued*)

Table 2.1. (continued)

Site Name, Location	NAPL Type	Regulatory Oversight Program	Site COCs and RAOs for Groundwater	Technologies Applied	Reported Results	MCLs Achieved?	ICs in Place?
Naval Submarine Base Kings Bay Site 11, Camden County, Georgia	Chlorinated solvents	RCRA Corrective Action	*COCs:* PCE (8,500 μg/L) TCE (550 μg/L) *cis*-DCE (1,300 μg/L) Vinyl chloride (4,500 μg/L) *RAOs:* Reduce groundwater concentration to below Georgia MCLs	P&T, ISCO (Fenton's Reagent), biostimulation, MNA	Concentrations of chlorinated hydrocarbons decreased to 13.9 μg/L; MNA implemented as a final remedy	No	N/A – Awaiting status of MNA
Former Sta-Lube Site, Rancho Dominguez, California	Chlorinated solvents	California Regional Water Quality Control Board, Los Angeles Region	Methylene chloride (2,600,000 μg/L) RAO: Reduce methylene chloride concentration to 50 μg/L	SVE, P&T, excavation, ISCO (catalyzed hydrogen peroxide to activate sodium persulfate)	Methylene chloride levels reduced by 94 to 97% within 4 months following treatment and below the 50 μg/L cleanup goal within 5 months	Yes[a]	No
Pasley Solvents and Chemicals, Inc., Hempstead, New York	Chlorinated solvents	USEPA Superfund	*COCs:* 1,1-DCA, 1,1-DCE, *trans*-DCE, chloroform, TCE, toluene, chlorobenzene, ethylbenzene, and xylenes (37,000 μg/L total VOCs 320 μg/L TCE) *RAOs:* Reduce concentrations to residential risk levels and MCLs	SVE, air sparging	Concentrations of total VOCs in monitoring wells ranged from ND to 4 μg/L, below MCLs	Yes	No

Notes: *μg/L* microgram(s) per liter, *1,1-DCA* 1,1-dichloroethane, *1,1-DCE* 1,1-dichloroethene, cis-DCE *cis*-dichloroethene, *COC* compound of concern, *DTSC* Department for Toxic Substances Control, *IC* institutional control, *ISCO in situ* chemical oxidation, *MGP* manufactured gas plant, *MNA* monitored natural attenuation, *ND* non-detect, *P&T* pump-and-treat, *PCE* perchloroethene, *PCP* pentachlorophenol, *pg/L* picogram(s) per liter, *RAOs* remedial action objectives, *RCRA* Resource Conservation and Recovery Act, *SVE* soil vapor extraction, *TCDD* 2,3,7,8-tetrachlorodibenzo-p-dioxin, *TCE* trichloroethene, *TPH* total petroleum hydrocarbon, trans-DCE *trans*-dichloroethene, *VOC* volatile organic compound

[a]Although described as achieving MCLs in the report, methylene chloride concentrations were above the MCL when groundwater monitoring was discontinued

groundwater does not restore the aquifer, limits reuse of the site, and requires that both long-term institutional controls and assurances of sufficient financial resources to maintain these controls must be in place. Stakeholder concerns over the long-term stability of these engineering and/or institutional controls, coupled with issues related to natural resource damage claims when drinking water sources are contaminated, and litigation risks due to property damage or trespass, continue to provide a powerful incentive to responsible parties for improving the capabilities of technologies to remove or destroy DNAPLs in source zones.

Several obstacles have prevented widespread application of source zone depletion technologies. These include:

- Setting functional RAOs (such as achieving MCLs in the source zone) that are likely to be technically impractical at a substantial number of sites (NRC, 2005)
- Uncertainty regarding the long-term effectiveness and cost of source depletion options and the length of time required to achieve RAOs
- Limited number of well-documented case studies that could reduce the uncertainties regarding the likely effectiveness of source depletion technologies

An additional obstacle is the current limited availability of cost cap insurance policies to protect contractors from financial losses in the event of cost overruns for source removal remedies contracted under fixed price arrangements.

An additional uncertainty at many sites is the fraction of the contaminant mass in the DNAPL source zone (and downgradient) that may be present in diffusion-controlled, low-permeability zones inaccessible to fluid flushing technologies (Parker et al., 1996; Liu and Ball, 2002; AFCEE, 2007). In some geological settings, DNAPL constituents will diffuse into these zones relatively quickly due to very high concentration gradients, but these same constituents will diffuse more slowly out of these zones after the DNAPL has been removed as the concentration gradients will be significantly reduced. If the metric for successful remediation is achievement of MCLs, the source depletion goal must include depletion of the dissolved- and sorbed-phase mass in addition to the DNAPL mass, which may be significant depending on the amount, type and age of the DNAPL release in the source area as well as matrix characteristics such as effective porosity. *In situ* technologies for source zone depletion are generally limited in their ability to remove contaminant mass from these low-permeability zones; however, thermal technologies may overcome this limitation at some sites. Conversely, it is likely that continued release of contaminants from these low-permeability zones will be at mass discharge rates substantially lower than those prior to source depletion (Sale et al., 2008a). Whether this reduction in source zone mass discharge will be sufficient to warrant implementation of a source depletion technology is not currently predictable and remains an important research topic.

2.3 PERFORMANCE PREDICTION LIMITATIONS

Modeling codes and decision tools are available to (1) predict the performance of DNAPL source zone remediation technologies, (2) predict the beneficial and adverse impacts after remediation has been attempted and (3) guide the decision process for selecting technologies or to achieve desired end points. However, the reliability of the output from these codes and tools is largely a function of the quality of the input. Given the difficulty in accurately characterizing the geology, hydrogeology and contaminant distribution at a particular site, it follows that predictions of remediation technology performance are typically associated with significant uncertainty. Consequently, DNAPL models primarily are used in a conceptual mode. Modeling codes and decision tools are discussed in Chapters 5 and 6 of this book.

Much of the difficulty in quantifying treatment benefit results from the inherent uncertainty in determining the magnitude and distribution of the DNAPL source zone mass prior to remediation. Uncertainties in predicting remedial performance, life cycle costs, and benefits further confound both economic and technical analyses and comparisons of technical options for DNAPL source zone depletion. For example, the use of traditional economic analyses to compare alternatives that require different time frames to achieve remedial goals is meaningless when the time of remediation exceeds about 50 years using standard Office of Management and Budget (OMB)-defined net discount rates. Typical net discount rates used in such comparisons range from 3 to 7%. Unless the discount rate is set to zero (no discounting of future expenditures), net present value of costs in the future (in particular greater than 50 years from the present) will approach zero.

One likely strategy for achieving benefits through partial source mass depletion is to reduce contaminant mass discharge to a level less than the natural attenuation capacity of an aquifer with respect to the dissolved plume. Under such conditions, the contaminant mass discharge for the DNAPL constituents becomes less than the rate of contaminant degradation in the plume, and as a result the plume gradually shrinks until a smaller steady-state plume is achieved. Such a strategy is most likely to be beneficial for small DNAPL source zones at sites that are inactive. However, at many chlorinated solvent sites, natural attenuation by abiotic or microbial degradation may be of insufficient magnitude because of inadequate microbial densities and non-optimum geochemical conditions in the plume. Modifications of these conditions will often be necessary to achieve acceptable degradation rates to be protective of potential receptors.

The following section summarizes the current state of the practice with respect to management of DNAPL sites, focusing on site characterization and remediation.

2.4 SOURCE CHARACTERIZATION AND MONITORING

The challenges involved in characterizing DNAPL sources are daunting. DNAPL migration pathways are governed by even slight differences in permeability and often occur over relatively large areas compared to the original release points. Thin accumulations can be difficult to detect, and contaminants that later sorb and/or diffuse into the aquifer materials are difficult to find. To help the reader appreciate the issues involved, the following sections provide a brief overview of DNAPL sites and their evolution over time, a summary of the source characterization challenges, and a description of the conventional and more innovative tools that can be useful in dealing with these challenges.

2.4.1 Overview of a DNAPL Site

DNAPLs represent continuing, long-term sources of dissolved constituents to groundwater. Although DNAPL compounds generally have low aqueous solubilities (less than 10,000 milligrams per liter [mg/L]), they are orders of magnitude greater than the relevant health-based cleanup levels, which are in the low parts per billion (ppb) range for chlorinated solvents (Pankow and Cherry, 1996). Consequently, it is common for a dissolved or an aqueous plume at concentrations above regulatory limits to develop downgradient of the DNAPL. Therefore, a DNAPL-impacted site consists of a DNAPL zone and a plume emanating from the source area. DNAPL-impacted sites can vary significantly in size, from small (dry cleaners) to very large (the Motorola 52nd Street Superfund site, including additional downgradient sources and similar sites with plumes extending for several miles downgradient). The challenges associated with characterizing and remediating DNAPL-impacted sites will vary with the size and complexity of the sites, along with evaluating the potential benefits of corrective remedial actions.

The DNAPL zone can include one or more of the following long-term, continuing sources of constituents of concern to the aqueous plume:

- Residual DNAPL (ganglia)
- Pooled DNAPL
- Sorbed constituents
- Dissolved constituents in the pore water or diffused into fine-grained media

Residual DNAPL comprises isolated blobs and ganglia of organic liquid trapped by capillary pressure. Residual DNAPL is formed at the trailing end of a migrating DNAPL body and typically occurs at saturations of no more than approximately 25% of pore space (Kueper et al., 1993). Pooled DNAPL represents a continuous accumulation of organic liquid above a capillary barrier such as a layer of silt, fine sand, or clay. DNAPL saturations within pools are typically between approximately 25 and 80% of pore space. Depending on the stratigraphy, the source zone may contain a complex distribution of residual and pooled DNAPL that is difficult to locate and characterize. This was demonstrated by Kueper and co-workers (Kueper et al., 1993; Poulsen and Kueper, 1992) at a field site located at Canadian Forces Base Borden, where perchloroethene (PCE) DNAPL was released into a sand aquifer both above and below the water table. The resulting pathways and complex distribution of residual and pooled DNAPL were found to be governed by small-scale porous media structure at the scale of millimeters.

Both residual and pooled DNAPL will dissolve into flowing groundwater giving rise to aqueous-phase plumes of dissolved contamination. Pooled DNAPL has less contact with flowing groundwater than does residual DNAPL given the relative differences in the DNAPL-specific surface area exposed to the groundwater. In addition, pools contain DNAPL at higher saturations than zones of residual DNAPL. Consequently, the rate of dissolution is low in pools compared to residual DNAPL, and the DNAPL lifetime in pools will generally be much longer. At most sites, the DNAPL was released decades ago, and in many of these older sources, only pools will remain. For DNAPL sites without pools, it is possible that residual DNAPL has been eliminated by dissolution and no longer represents a source; in fact, the aqueous plume may be detached from the source zone (Sale et al., 2008a).

At older sites, significant transfer of dissolved constituents into fine-grained media via diffusion may have occurred, depending on the site geology. At sites where sufficient sorption and diffusion have occurred, back diffusion of constituents is a continuing source to the aqueous plume and can occur in both the DNAPL zone and within the aqueous plume where fine-grained media exists (Parker et al., 1996; Liu and Ball, 2002; AFCEE, 2007). Finally, desorption of constituents is another phenomenon contributing DNAPL constituents to the aqueous plume.

A life cycle refers to all the stages of a process or a site from beginning to end. Once the components of a life cycle of a particular process are determined, a life cycle analysis (LCA) can be conducted to examine all major steps along the way in terms of environmental, societal, and economic concerns. Ideally, alternatives at each step are evaluated in terms of risks, energy and material usage, costs and benefits, and an optimal approach is determined. The focus of the following section is the major components of the "life cycle of a DNAPL site investigation and remedy." This differs from the "life cycle of a source zone" as discussed and illustrated in Chapters 1 and 13 of this book, which focus on the aging of DNAPL source zones and changes in transfer of mass with time.

2.4.2 Life Cycle of a DNAPL Site Investigation and Remedy

The life cycle of a DNAPL site investigation and remedy can include the following: (1) contamination discovery and reporting, (2) regulatory decision (resource allocation), (3) decision on investigation strategy based on the initial Conceptual Site Model (CSM), (4) dissolved plume characterization (often performed in phases), (5) determining location of the source or the sources, (6) characterizing the source or the sources (often performed in phases), (7) determining that one or more sources is a DNAPL, (8) characterizing DNAPL sources if possible, (9) performing health and environmental risk assessment, (10) establishing remedial action objectives, performance measures and points of compliance, (11) defining an acceptable time frame for restoration, (12) performing a remedy assessment and selecting the remedy, (13) designing the remedy which could include multiple components, (14) constructing the remedy, (15) performance assessment – monitoring remedy impacts and tracking performance measures, (16) optimizing and enhancing the remedy if necessary, (17) LTM, (18) pursuing a TI wavier or other alternative endpoint consideration, and (19) closure. During each of these steps, regulatory approval is generally required, adding to the life cycle time frame. In addition, supplemental characterization, updates to the CSM, remedy optimization and remedy enhancements and modifications can occur throughout remedial design and implementation phases. Thus, life cycle evaluation for a DNAPL site is rarely a linear process.

In general, for Comprehensive Environmental Response, Compensation, and Liability Act (CERCLA) NPL sites, the remedy selection criteria that must be considered are outlined in the National Contingency Plan (NCP): overall protection of human health and the environment, compliance with ARARs, long term effectiveness and permanence, reduction of toxicity, mobility or volume, short-term effectiveness, implementability, cost, state acceptance, and community acceptance. Other potential factors in remedy selection include residuals and by-product formation/control, process and mechanical reliability, ease of operation/control/implementation, and permitting and sustainability metrics (remedy carbon footprint).

At CERCLA and other sites, when groundwater is considered a potential source of drinking water, ARARs will include MCLs for any chemical regulated under the Safe Drinking Water Act as the RAOs throughout the impacted aquifer. As discussed previously, MCLs can be difficult to achieve at sites containing DNAPL. If at some point in the life cycle of a DNAPL site it is determined that ARARs cannot be met in a reasonable time frame, the NCP allows for ARAR waivers (technical impracticability waivers under CERCLA and RCRA) if it can be demonstrated that the standard cannot be met for a specified area or zone of the site. Under this scenario, a zone of the site must be established within which the ARAR will be waived. At non-Superfund sites, other options that have been considered to address this problem include alternative points of compliance and implicit acceptance of long time frames (greater than 100 years) to reach remedy objectives (ESTCP, 2011). The USEPA-recommended process for restoring contaminated groundwater at Superfund sites is summarized in Figure 2.2.

2.4.3 Characterization and Monitoring Challenges

Locating DNAPL source zones can be difficult, especially if the DNAPL is distributed in complex patterns due to multiple release points and subsurface heterogeneities (USEPA, 1992). Under these conditions, direct observation of DNAPL may not be possible. Further, heterogeneities, including fractures, can produce changes in soil/rock concentrations that vary by orders of magnitude over short distances. Kueper and Davies (2009) present a converging-lines-of-evidence approach to assessing the likelihood of DNAPL presence at a site as well as an approach to delineating the spatial extent of the DNAPL source zone.

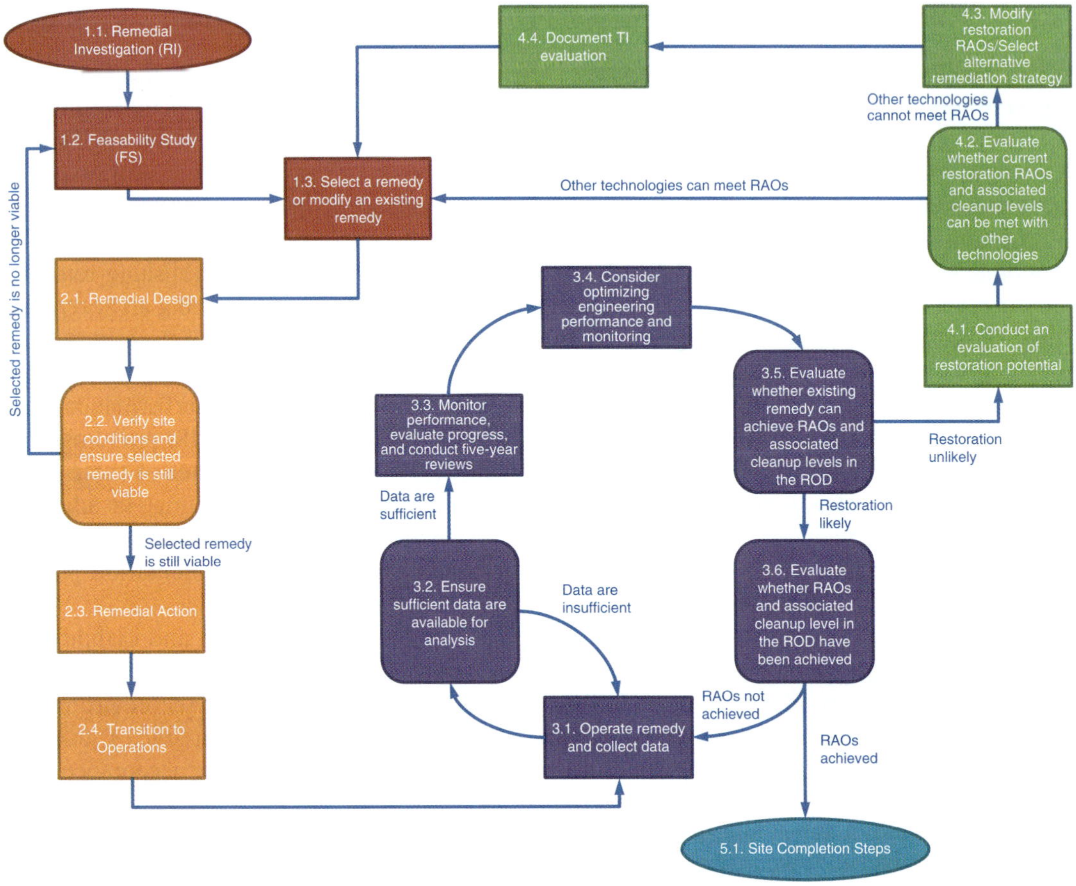

Figure 2.2. USEPA-recommended process for restoring contaminated groundwater at Superfund sites (from USEPA, 2011b). Note: ROD - Record of Decision.

Even if the general area of the DNAPL source zone can be estimated, detailed characterization of the DNAPL extent and distribution is often challenging and inadequate (USEPA, 1992; Cohen and Mercer, 1993). Another difficulty in DNAPL source zone characterization is minimization of the potential for causing cross contamination, especially to deeper stratigraphic layers. The potential for cross contamination is low for smaller, older releases due to matrix diffusion and depletion of the DNAPL caused by dissolution into the aqueous phase as discussed earlier. For more recent and/or larger releases of chlorinated solvents, the potential for cross contamination is high.

To minimize cross contamination, USEPA (1992) suggested an outside-in approach to DNAPL source zone characterization. Even with improved characterization tools such as aquifer profiling using multiple depth-discrete sampling, the outside-in approach is generally preferable. This approach, in which data are first collected downgradient of the source zone, allows better definition of stratigraphy and back tracking to depth-discrete, pooled and residual DNAPL. This information often can be used to improve the later detailed source zone characterization (Mercer et al., 2010).

Another challenge associated with characterizing a DNAPL source zone is measuring the impacts of mass removal on the flux of contaminants from the source. Typically, site-specific cleanup standards consist of chemical-specific concentrations, often MCLs. These concentrations are often in the very low ppb and are rarely, if ever, achieved in and immediately downgradient of a DNAPL source zone. Consequently, as a practical matter (and not necessarily a regulatory opinion), mass flux measurements and mass discharge estimates have been suggested as alternative performance measures, requiring nontraditional characterization techniques to measure mass flux and discharge before and after remediation (Kavanaugh et al., 2003; ITRC, 2010).

Since some DNAPL is likely to remain in place following remedy implementation, post-remedy site characterization also typically involves LTM, which is used to evaluate the degree to which the remedial measure achieves its objectives (USEPA, 2012). LTM is a significant, persistent and growing cost in remediation at sites relying on containment. Consequently, efforts are being made to optimize LTM and reduce long-term costs. This optimization may take the form of reduced samples in time and space based on statistical evaluation of the number and location of wells necessary to monitor a plume (AFCEE, 2006) and/or more efficient sampling (sensor systems or diffusion samplers).

2.4.4 Conventional Methods

Given the complexity and cost of DNAPL site characterization, it is critical to use a dynamic and flexible approach based on a CSM that optimizes characterization over time (Mercer et al., 2010; USEPA, 2011a). The CSM forms the basis for designing a data collection program, which is phased using an outside-in approach and may initially rely on noninvasive and minimally invasive methods such as soil, gas, and surface sampling prior to drilling or subsurface probing. Stratigraphic information can help identify preferential pathways and traps for DNAPL. Dissolved concentrations in groundwater, which is an integrator of contamination and can be used to more easily find potential DNAPL source contamination than point soil samples, can be used along with site history information to trace back to likely DNAPL source zones.

Conventional methods relied largely on groundwater monitoring wells. For deeper contamination and for stratigraphy (consolidated layers) where direct-push technologies cannot be utilized, monitoring wells are still required. Primary limitations associated with monitoring wells include the possibility of drilling through a DNAPL zone and not identifying it and a general lack of vertical contamination distribution delineation. Vertical delineation is aided by the use of well clusters screened at different depths and use of multi-level sampling systems within a single well (Einarson and Cherry, 2002). Other conventional methods include soil gas surveys, surface geophysics and soil/rock core sampling. These techniques may be helpful in locating potential DNAPL source areas and/or stratigraphy that may serve as DNAPL traps, but the information they provide may not allow determination of the structure and quantity of DNAPL. Figure 2.3 illustrates an example estimated DNAPL distribution for Launch Complex 34 Solid Waste Management Unit (SWMU) Number CC054 based on interpolation of 1,200 soil samples collected from 90 discrete boring locations.

2.4.5 Innovative Methods

Although sampling technologies for soil characterization are mature, innovations for soil characterization include the use of equilibrium partitioning calculation and soil chemistry combined with graphical methods to determine DNAPL source structure and quantity. Innovative techniques for DNAPL characterization in groundwater include improvements in direct-push technology and sensors. Kram et al. (2002) compared several DNAPL characterization methods and concluded that direct-push sensors (cone penetrometer testing, laser-induced fluorescence, and membrane interface probes) and the down-hole NAPL FLUTe™ liner methods are particularly cost-effective components of integrated DNAPL site characterization.

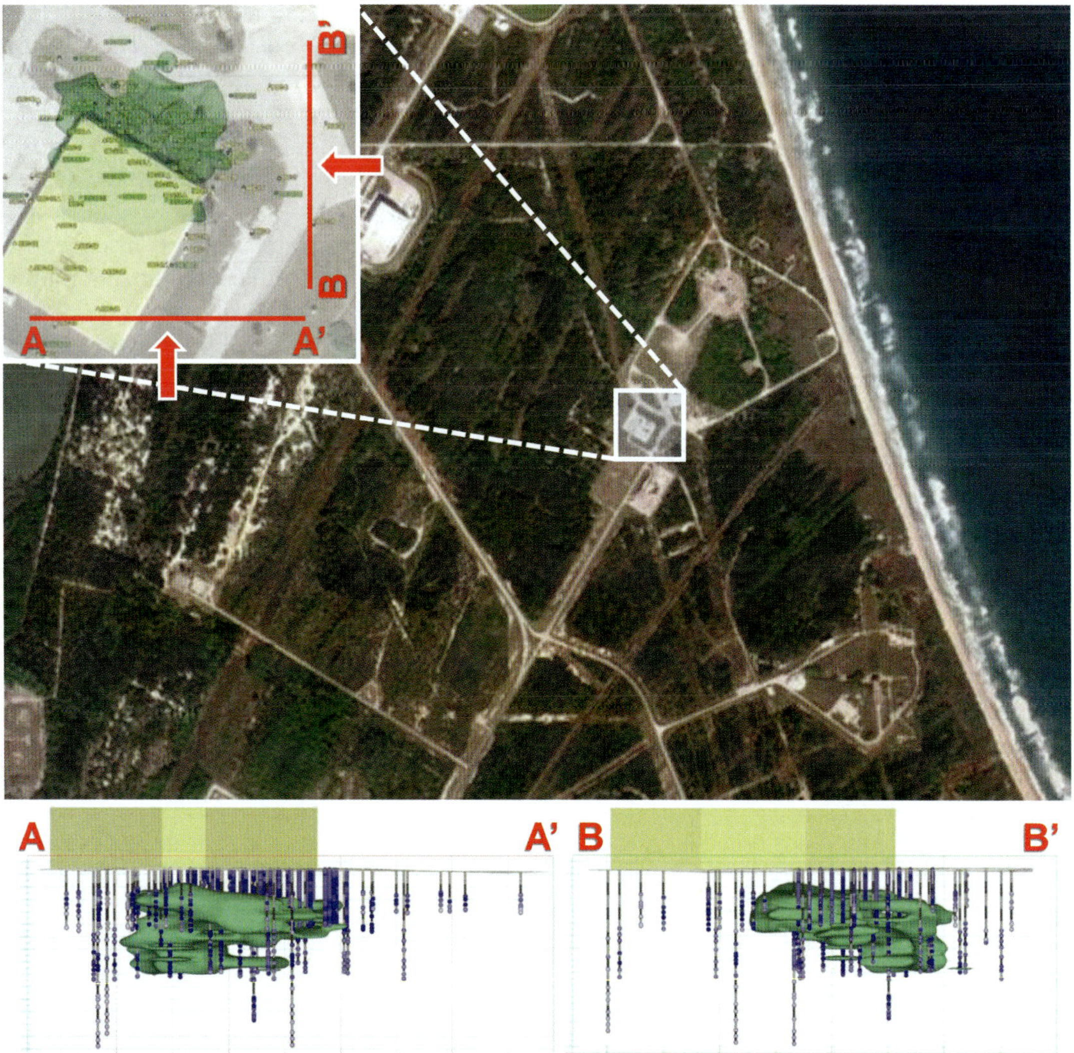

Figure 2.3. Estimated DNAPL distribution at Launch Complex 34 SWMU No. CC054 based on interpolation of 1,200 soil samples collected from 90 discrete boring locations (from Geosyntec and Tetra Tech 2007).

These methods are useful in shallow (less than 100 ft [30 m] below ground surface) subsurface environments that are accessible using direct-push technologies.

As mentioned, mass flux and discharge measurements are used to assess DNAPL impacts to groundwater. They also provide estimates of DNAPL source strength and mass loading. Mass flux and discharge can also be used as metrics to assess the impact of source remediation and are a key input to evaluate monitored natural attenuation. Methods to measure mass flux include using (Mercer et al., 2010) (1) water quality data from transects (multiple locations and depths) and groundwater velocity, (2) sorptive permeable media in downgradient wells to intercept contaminated groundwater and release resident tracers (the passive flux meter method), and (3) plume capture by pumping (the integral pumping method). An advantage of using the transect method is that data may be helpful in backtracking up gradient, following the location of the highest concentrations to identify depth-specific DNAPL source locations. The strengths and limitations of the flux measurement methods are evaluated in a recent ITRC (2010) report.

2.5 REMEDIATION TECHNOLOGIES

A number of technologies have been or are currently being applied to address DNAPL in the subsurface. The performance and limitations of many of these remedial technologies are discussed by Stroo et al. (2012) and are summarized below. The most effective approach to remediating DNAPL sites may be strategic combinations of two or more of these techniques, often starting with high energy approaches and polishing with more passive methods.

Remediation technologies have been described in detail in various documents (ITRC, 2004, 2005a, b; USEPA, 2004; 2006), and individual technologies are discussed in various chapters of this book. All technologies currently considered viable for DNAPL source treatment have specific strengths and limitations. The following presents a brief synopsis of the current state of the practice understanding of the technologies most applicable to DNAPL source depletion.

2.5.1 Hydraulic Displacement

Hydraulic displacement involves the recovery of pooled DNAPL by pumping from either vertical wells or horizontal drains. The rate of DNAPL recovery can be enhanced by water injection, which produces a favorable alteration of capillary pressures (Gerhard et al., 2001, 1998; Alexandra et al., 2012). This technology is best suited to sites containing appreciable amounts of pooled DNAPL. The technology does not remove residual DNAPL and therefore removal is limited by the amount of DNAPL mass present in pools. Hydraulic displacement can be viewed as a source zone stabilization technology in that pooled DNAPL is removed and it is often used as an initial means of DNAPL mass removal, followed by a mass transfer-based technology.

2.5.2 *In Situ* Chemical Oxidation

In situ chemical oxidation (ISCO) can be applied to treat DNAPL sources, particularly if the contaminants of concern are chlorinated ethenes. Technology practice guidance documents for ISCO have been developed and are widely available (Siegrist et al., 2011). ISCO is an attractive technology because it can destroy contaminants rapidly *in situ*, at moderate cost. However, the technology has a number of significant limitations related to oxidant delivery and concentration rebound following treatment (McGuire et al., 2006; Krembs et al., 2010). The most commonly used oxidants include catalyzed hydrogen peroxide (CHP) and permanganate (Krembs et al., 2010), although work to develop improved ISCO formulations is ongoing. In particular, use of activated persulfate is increasing because it is relatively stable in the subsurface compared to CHP and is capable of treating a wide variety of contaminants (Watts and Teel, 2006; Waldemer et al., 2007).

Other research has focused on improving hydrogen peroxide application through the use of stabilizing agents (Watts and Teel, 2006) and permanganate application through the use of agents to control formation of manganese dioxide (MnO_2) particles (Smith et al., 2008). Work to evaluate ISCO in combination with other source depletion technologies suggests that ISCO can be compatible with other technologies such as *in situ* bioremediation (ISB) (ESTCP, 2010) and thermal treatment (Pennell et al., 2009), provided that the secondary effects of ISCO such as oxidation of natural organic material and metals are managed.

2.5.3 *In Situ* Chemical Reduction

In situ chemical reduction (ISCR) is a proven technology for treating groundwater plumes, typically as zero-valent iron (ZVI)-permeable reactive barriers (Gillham and Vogan, 2010).

However, effective delivery of micron-sized ZVI within source areas is difficult, and recent work has focused on alternative delivery mechanisms such as large-auger subsurface mixing (Wadley et al., 2005) and injection of nanoscale ZVI (Gavaskar et al., 2004). Research to improve nanoscale ZVI surface properties and thereby improve delivery within the subsurface is ongoing. The major problems include particle agglomeration and inadequate penetration of the nanoscale ZVI into the formation (Saleh et al., 2007; Berge and Ramsburg, 2009). ISCR can be used in conjunction with other source depletion technologies, particularly ISB, where the addition of ZVI may make the subsurface amenable to subsequent anaerobic biodegradation (Gavaskar et al., 2004), and materials that combine ZVI with electron donors are commercially available. In spite of advances in material formulation and delivery, however, ZVI appears to be more appropriate as a source containment technology (Taghavy et al., 2010).

2.5.4 Surfactant and Cosolvent Flushing

In recent years little work has focused on enhanced flushing with surfactants and/or cosolvents. These technologies can be effective, although, as with any flushing technology, contaminants are preferentially removed from transmissive zones but can be unaffected when present in hydraulically inaccessible zones (Childs et al., 2006; Ramsburg et al., 2005). In addition, chemical costs for some surfactant formulations can be high relative to other technologies (Fountain et al., 1996). Most recent work has instead focused on coupling surfactants and cosolvents with other source treatment technologies such as ISCO to improve distribution during delivery (Dugan et al., 2010).

2.5.5 *In Situ* Bioremediation

In the past several years, ISB has been recognized as a potentially effective treatment of DNAPL sources (ITRC, 2005b). While not appropriate for all DNAPL sources (ITRC, 2008), case study reviews have indicated that in some cases ISB can be as effective as other source remediation technologies (McGuire et al., 2006). In particular, ISB may enhance dissolution rates from residual DNAPL during treatment, with dissolution enhancement factors (compared to natural dissolution) between two and five reported (Sleep et al., 2006; Amos et al., 2008). ISB treatment may be relatively long lasting, with sustained treatment for several years, and can be optimized through the use of bioaugmentation cultures to reduce treatment lag times and cost (ESTCP, 2005; NAVFAC, 2007).

ISB also may be effective for treating hydraulically inaccessible zones through diffusion of slow-release electron donors such as vegetable oil into less permeable materials, although work to demonstrate degradation in these less permeable materials is ongoing as part of ESTCP project ER-1740. Some secondary effects, notably pH decreases, methane production, and increases in dissolved iron and arsenic concentrations have been observed during implementation of ISB source zone remedies, but such impacts are typically localized to the treatment area (ITRC, 2008). Finally, ISB is also attractive as a potential combined remedy (or polishing treatment) for other source treatment technologies such as ISCO (ESTCP, 2010), surfactant flushing (Ramsburg et al., 2004) and *in situ* thermal treatment (Costanza et al., 2009).

2.5.6 Natural Attenuation of Sources

Natural attenuation is defined as any combination of “physical, chemical, or biological processes that, under favorable conditions, act without human intervention to reduce the mass, toxicity, mobility, volume, or concentration of contaminants in soil or groundwater”

(Wiedemeier et al., 1998). These processes include advection, dispersion, adsorption, volatilization, and chemical or biological stabilization, transformation, or destruction of contaminants. Monitoring to evaluate whether these natural attenuation processes are ongoing as expected is referred to as MNA. For sites where an integrated approach to source management is applied (source mass depletion combined with less aggressive approaches for managing the dissolved plume), natural attenuation may be a viable approach, even in the presence of DNAPL (Rao et al., 2001; ITRC, 2002). However, this approach requires that an MNA remedy will achieve remedial objectives within a reasonable time frame and in some cases may require removal or isolation of a DNAPL source (USEPA, 1998). Further discussion of source zone natural attenuation and use of source zone monitored natural attenuation as a remedy component is provided in Chapter 13 of this book.

2.5.7 *In Situ* Thermal Treatment

Thermal treatment can be highly effective with respect to source zone mass depletion with mass removal in excess of 99% possible even in heavily contaminated sources (Johnson et al., 2009). In addition, it may be particularly attractive for source zones characterized by significant amounts of low-permeability media, where flushing technologies may be less effective. However, mass removal may be incomplete within the low-permeability zones, and effectiveness may be limited in zones where groundwater flow is sufficiently fast to reduce the effect of subsurface heating (Costanza and Pennell, 2007). Recent research indicates that thermal treatment may be compatible with ISB if low-temperature treatment is applied (Costanza et al., 2009) or if sufficient time is allowed for microbial species to recolonize the subsurface after treatment (Beyke and Powell, 2005). Work to evaluate thermal treatment as a combined remedy with ISCR is also ongoing (Siegrist et al., 2011).

The most significant limitation of thermal treatment is generally a higher unit cost compared to other technologies, with reported costs for treatment ranging from $88 to $300 per cubic yard (McDade et al., 2005). Other challenges include maintaining hydraulic control during treatment and keeping adequate heating within subsurface saturated zones with higher water fluxes entering the heating zone. It should also be noted that thermal remedies are subject to the same limitations of all source zone remedial strategies if the DNAPL source zone is incompletely defined. However, when life cycle costs are considered, the cost of thermal treatment may compare well with other remedial strategies. In particular, the impact of aggressive source depletion using thermal technologies on post-remedy mass discharge from source zones should be carefully evaluated.

2.6 SOURCE REMEDIATION ISSUES

2.6.1 DNAPL Remediation Effectiveness

As discussed here and in previous documents (Kavanaugh et al., 2003), and despite the advances in source treatment technologies described above, remediation is unlikely to achieve cleanup to MCLs in the source zone at a significant fraction of complex sites impacted with DNAPLs. However, DNAPL source remediation can significantly reduce groundwater concentrations of the DNAPL constituents and can remove substantial amounts of DNAPL mass from the subsurface (source depletion). DNAPL mass removal is considered substantial when it would result in a meaningful reduction in mass discharge from the source zone and/or a reduction in the length of time to reach site closure or other final management stages of a site, such as LTM. The amount of DNAPL mass removal required for meaningful reductions in mass discharge and time to closure will vary between sites and depends on numerous factors such as site hydrogeology, DNAPL release history and distribution in the subsurface, uncertainties in estimating DNAPL mass present in the subsurface, and potential problems related to remediation of dissolved-phase plumes downgradient of a DNAPL source due to diffusional phenomena.

Select pilot- and full-scale implementations of DNAPL source remediation were evaluated by Kavanaugh et al. (2003). The case studies indicated that substantial (greater than 50%) DNAPL mass removal could be achieved, with corresponding significant decreases in contaminant concentration in groundwater. It should be noted that none of the implementations evaluated achieved MCLs at the time of the study. Other case study evaluations have reported similar remedial performance, with estimated DNAPL mass removals exceeding 90% in some cases (NRC, 2005; ITRC, 2011).

McGuire et al. (2006) evaluated the performance of four DNAPL source depletion technologies (ISCO, ISB, thermal treatment and surfactant/cosolvent flushing) at a total of 59 sites impacted by chlorinated volatile organic compounds. Performance was evaluated on the basis of volatile organic compound (VOC) concentration reduction and rebound, with limited evaluation of alternative performance metrics such as DNAPL mass removal and mass discharge reduction, probably due to limitations in the available data. Evaluation of remedial performance with respect to differences in site hydrogeology was also limited. Median concentration reductions for the four technologies exceeded 80%; however, the concentration reductions varied significantly. For 11 of the 59 sites where DNAPL mass removal could be estimated, the relationship between aqueous concentration reduction and mass removal was within 30% of a 1:1 relationship (50% mass removal corresponds to a 50% concentration reduction).

A similar performance evaluation was conducted by Krembs et al. (2010), focusing on the performance of ISCO for sites with and without DNAPL present. Of the 242 sites evaluated, 104 reportedly contained DNAPL. The performance evaluation incorporated site hydrogeology in sorting the case studies following the approach developed by the NRC (2005) along with alternative performance metrics (cleanup levels other than MCLs, mass reduction goals, etc.). No clear differences in ISCO performance as a function of site hydrogeology were observed, although practitioners may vary the implementation approaches for sites with different geological conditions. For sites where DNAPL mass removal data were available, the estimated amount of relative mass removal was greater than the relative concentration reduction (the relationship did not follow a 1:1 ratio). The study also indicated that, while application of ISCO at a DNAPL site is unlikely to achieve MCLs in the source zones, it can achieve alternative remedial objectives such as significant mass removal if realistic expectations of remedy performance are specified.

Johnson et al. (2009) presented a state of the practice overview of *in situ* thermal treatment of DNAPL source zones. Unlike many remedial performance reviews, the overview incorporated information on the hydrogeological settings and where treatment was applied, as well as a variety of performance metrics such as concentration reduction, estimated mass removal and mass discharge. In general, thermal applications were limited in more complex hydrogeologic settings, likely reflecting the risks and uncertainties related to remediation in those environments. Only a small subset of the evaluated applications had sufficient data to allow for an evaluation of mass discharge reductions, reflecting the fact that this important information is seldom collected as part of performance monitoring. Of the applications with sufficient data to perform an evaluation, the mass discharge reduction from the DNAPL source zone ranged over several orders of magnitude, from reductions of less than a factor of 10 to reductions of greater than a factor of 10,000. Only one application in a complex hydrogeology (fractured bedrock) had sufficient data to perform an evaluation of mass discharge reduction, and in this case the reduction was limited (less than a factor of 10).

When assessing the performance of the remedial technologies used for DNAPL source depletion, both the inherent limitations of a technology and any limitations related to design and operational decisions made by the practitioners implementing the technology will have an impact. For example, a key conclusion of the overview of *in situ* thermal treatment presented by Johnson et al. (2009) is that many treatment systems were designed to remediate a DNAPL source that was smaller than the actual source zone, limiting the effectiveness of the remedy. In these cases, additional characterization work could have resulted in a more complete understanding of the extent of the DNAPL source, thus improving remedy performance. In addition, the operational period of many *in situ* thermal remedies appears to be somewhat arbitrary, with little regard for system optimization or monitoring. Other DNAPL source remedies often have similar practitioner-related limitations, such as determining when to cease amendment injections for ISCO or ISB. The ability to discriminate between technology- and practitioner-related limitations has not been resolved to date. However, in our experience surprises are common and implementation is often less than ideal. As a result, source treatments should be carefully designed and operated to minimize performance issues resulting from the implementation, as opposed to any inherent limitations of the remedial technology.

2.6.2 Technology Applicability

Attempts have been made to evaluate DNAPL mass removal performance in the context of site conditions (ITRC, 2002; Pankow and Cherry, 1996). Kavanaugh et al. (2003) presented a conceptual evaluation of the potential applicability of different technologies for depleting DNAPL mass under two different hydrogeological settings: hydraulically accessible, where permeable zones are connected to wells, and hydraulically inaccessible, where impermeable zones or permeable zones are not connected to wells. The evaluation further considered two different DNAPL loading scenarios: pooled DNAPL in significant mass and DNAPL distributed as residuals with a relatively small total mass. For hydraulically accessible zones, many technologies were considered to be potentially effective for DNAPL mass depletion, particularly for residual DNAPLs with small mass. In contrast, most technologies were not expected to be practical or effective for depletion of DNAPL mass in hydraulically inaccessible zones.

Similarly, a committee organized by the NRC (2005) has developed a multidimensional screening matrix for evaluating potential technology performance based on factors including remedial objectives, remedial technology, and site hydrogeological setting. Five relatively generic hydrogeological setting types are considered in the screening matrix:

Type I – Granular media with mild heterogeneity and moderate-to-high permeability
Type II – Granular media with mild heterogeneity and low permeability
Type III – Granular media with moderate-to-high heterogeneity
Type IV – Fractured media with low matrix porosity
Type V – Fractured media with high matrix porosity

For a specific site, the screening matrix can be used to compare the expected efficacy of different DNAPL source depletion technologies. Technologies are compared on the basis of achieving functional goals, such as plume containment, mass or concentration reductions, and risk management rather than absolute objectives such as protection of human health and the environment.

In deciding whether DNAPL source depletion is appropriate for a given site, the key question is how much mass removal can reasonably be achieved and whether this mass removal will be sufficient to meet remedial objectives within the source zone within some reasonable time frame. In addition to the applicability of source depletion technologies, a number of site-specific factors are important when considering whether containment or source depletion may be appropriate. Some key factors include the distribution of contaminants within the source zone, source hydrogeological setting, age of the source, and whether the source area has been sufficiently characterized for the remedial approach selected. It should also be noted that this discussion is focused primarily on remediation and/or containment of the DNAPL source. However, many large DNAPL sites currently have downgradient groundwater plumes that can extend for thousands of feet, and containment in those cases may also refer to remedial efforts to prevent further spread of impacts downgradient of a DNAPL source.

2.6.3 Contaminant Distribution

Contaminants within a DNAPL source zone may occur in four different phases:

1. DNAPL (pooled or residual)
2. Gas in soil vapor
3. Dissolved in groundwater
4. Sorbed on aquifer material

The processes governing distribution of each of these phases within the subsurface have been extensively described in the literature (Feenstra et al., 1996; Falta et al., 1989; Freeze and Cheery, 1979; Sale et al., 2008a; and Schwarzenbach et al., 1993) and are not repeated here. The total mass of contaminant present within a source zone will be equal to the sum of the mass within each of these four phases, with the distribution of mass within the phases varying based on site-specific characterizations. In addition, each of the four phases may occur in hydraulically accessible or inaccessible zones, alternatively described as mobile and immobile pore space (Payne et al., 2008). The distribution of contaminants between these four phases, and particularly the distribution within the immobile and immobile pore space, is a critical factor affecting source treatment performance of any *in situ* technology.

Back diffusion is a process that has been recognized for many years (Parker et al., 1996), but interest has increased in recent years. Back diffusion occurs when the concentration of dissolved-phase contaminants in groundwater within the mobile pore space is greater than the concentration within the immobile pore space, resulting in diffusion of the contaminants into the immobile pore space. Once concentrations within the mobile pore space decrease, contaminants will begin to diffuse from the immobile zones into the mobile pore space. This process is known as back diffusion or matrix diffusion and can result in persistent groundwater plumes, even following depletion of DNAPL that may have been present in the source (Parker et al., 1996; Liu and Ball, 2002; AFCEE, 2007; Sale et al., 2008a; West and Kueper, 2010). Depending on site geology and DNAPL release history, the amount of mass present within the immobile pore space may be significant and represent a long-term secondary source of contamination to the groundwater. Figure 2.4 illustrates the distribution of PCE in the lower aquifer and aquitard at Dover Air Force Base, Delaware, prior to and following groundwater extraction and treatment (Liu and Ball, 2002). The data show that pump-and-treat provided significant concentration reductions in the aquifer but that diffused mass persisted in the aquitard.

Contaminant distribution within each of the four phases described above for a given source is related to the hydrogeological setting. Using the framework of the five geological type

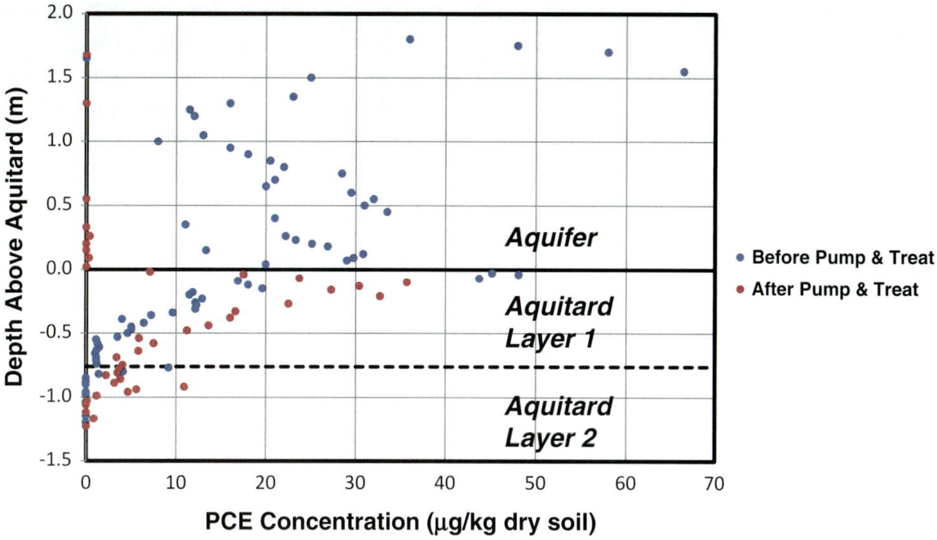

Figure 2.4. PCE in the lower aquifer and aquitard at Dover Air Force Base, Delaware, prior to and following groundwater extraction and treatment (from Liu and Ball, 2002).

settings developed by NRC (2005), contaminant distribution as it relates to hydrogeological setting can be described as follows (Sale and Newell, 2010):

- Type I settings are characterized by granular media with mild heterogeneity and moderate-to-high permeability, resulting in limited contaminant storage within immobile pore space. In this setting, the dominant contaminant storage mechanisms are dissolution from DNAPL and sorption to the aquifer material.
- Type II settings are uncommon but are characterized by granular media with mild heterogeneity and low permeability, resulting in a source that essentially comprises entirely immobile pore space. In this setting the dominant contaminant storage and transport mechanism is diffusion.
- Type III settings are characterized by granular media with moderate-to-high heterogeneity. In these settings, DNAPL constituents will diffuse from the mobile pore space into the immobile pore space. After the DNAPL is depleted, the contaminants present in the immobile pore space will act as a persistent source to groundwater.
- Type IV settings are characterized by fractured media with low matrix porosity. In these settings the primary contaminant storage mechanism is dissolution from DNAPL, with limited mass storage within the low-permeability zones. When the bulk of the DNAPL has dissolved, the plume may persist due to the presence of residual DNAPL in low-flow zones such as dead-end fractures.
- Type V settings are characterized by fractured media with high matrix porosity. In these settings DNAPL constituents will diffuse from the fractures into the porous aquifer material. Once the bulk of DNAPL in the fractures has been depleted, back diffusion from the aquifer material can sustain dissolved contaminant concentrations within the fractures.

In addition to changes in contaminant distribution based on hydrogeology, the age of a source zone will also influence the distribution. Immediately following a release, the contaminants may be present largely as DNAPL, with little or no migration of contaminants into the

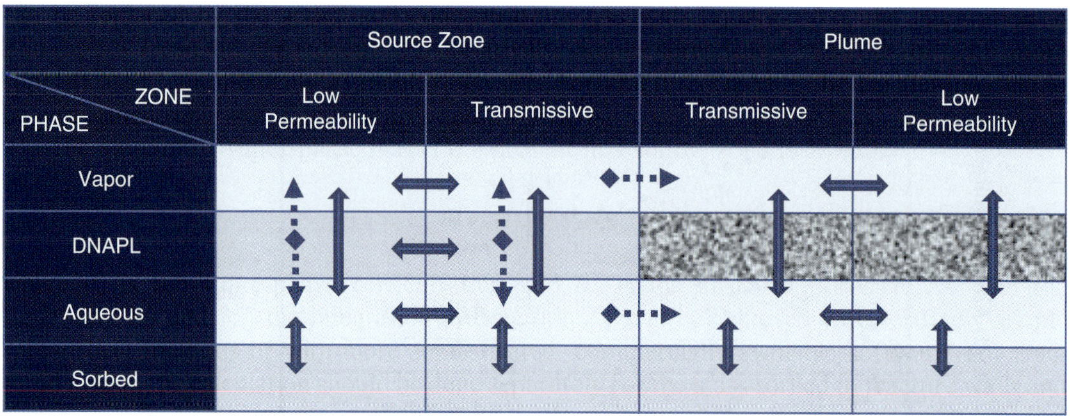

Figure 2.5. Lattice of 17 potentially relevant contaminant fluxes illustrated by the 14-compartment model (from Sale and Newell, 2010).

immobile pore space (life cycle Stages 1 and 2, Chapter 1). As a source zone ages, more of the DNAPL will move into the aqueous, vapor, and sorbed phases and contaminants will be present in the immobile pore space (life cycle Stage 3, Chapter 1). Eventually the source zone will reach the weathered stage where DNAPL has largely been depleted from the more transmissive zones and the groundwater plume is sustained by back diffusion from low-permeability zones located both in the source zone and in the downgradient plume (life cycle Stages 4 and 5, Chapter 1). As discussed, the impact of the back diffusion on groundwater concentrations will vary.

A key ongoing challenge is describing contaminant distribution within a DNAPL source. One approach that has been advanced is an eight-compartment conceptual model to describe the distribution of contaminants within a source zone (Sale et al., 2008a; Sale and Newell, 2010). This model, which is also applied to the plume zone and thus has 14 total compartments, identifies each of the four phases in which organic contaminants can exist and includes separate compartments for mobile and immobile pore spaces. The model incorporates the conceptual mass transfer links between the different compartments. While this model is helpful conceptually, it is difficult to use quantitatively because of the common lack of information on contaminant distribution in all eight compartments, which potentially can be present in more than one hydrostratigraphic zone. Furthermore, the complexity of the model can make it difficult to understand intuitively, particularly for non-practitioners, and to date this approach has not been widely incorporated into decision analysis for source depletion. If data at a particular site are sufficient to quantify the distribution of contaminants using this model, it may become part of a more accurate site-specific CSM, which remains the most effective method for conveying information regarding contaminant distribution within a DNAPL source and downgradient plume. Figure 2.5 illustrates a lattice of 17 potentially relevant contaminant fluxes illustrated by the 14-compartment model (Sale and Newell, 2010).

2.7 BENEFITS OF SOURCE DEPLETION

While some sources may be restored entirely, allowing for unrestricted use and unlimited exposure, in many cases the reasonable expectation is that well-implemented source treatment will reduce mass and concentrations by perhaps 1–2 orders of magnitude, and significant continuing management will be required for many years following active remediation (NRC, 2005; ITRC, 2011). As a result, it often is important to quantify the benefits of partial remediation. To determine how much remediation is needed to allow a transition to passive treatment or to meet remedial objectives regarding the size or the longevity of the plume, nontraditional metrics such as mass discharge are needed. The following sections discuss the impacts of source depletion on mass discharge and plume behavior to illustrate the opportunities and difficulties associated with considering such alternative criteria for assessing the performance of remedial actions.

2.7.1 Source Depletion and Mass Discharge

The decision to implement a source depletion or containment technology will be influenced by an understanding of where the majority of the contaminant mass in a source zone is located, as governed by source hydrogeology, age, and properties of the DNAPL (Mercer and Cohen, 1993). The complexity of contaminant distribution within a source zone makes *a priori* prediction of reductions in mass discharge or downgradient plume concentrations from remediation efforts difficult. Recent work has focused on evaluating how contaminant distribution within a source impacts remedial performance.

Source architecture is a key feature governing performance of remedial technologies, both in terms of containment mass removal and subsequent impacts to the groundwater plume (Saenton et al., 2002; Fure et al., 2006). The term *source architecture* refers to the distribution of DNAPLs between residual ganglia and pools as well as the distribution of contaminants between mobile and immobile pore space within the source zone (Sale and McWhorter, 2001; Lemke et al., 2004). Source architecture will be a function of the type of DNAPL, DNAPL release history, DNAPL properties, source zone hydrogeology and quantity of DNAPL released cumulatively.

Extensive modeling and laboratory studies have evaluated source zone architecture and associated remedy performance. Sale and McWhorter (2001) modeled dissolution of a heterogeneous DNAPL architecture in a uniform porous media (Type I hydrogeological setting) and observed that contaminant dissolution occurs at the upgradient edge of a DNAPL, implying that contaminant loading to downgradient plumes will continue as long as DNAPL persists. From this observation, Sale and McWhorter (2001) concluded that nearly complete contaminant removal would be required to achieve "significant near-term improvements in groundwater quality," where a significant improvement was considered to be the several orders-of-magnitude reduction in aqueous concentrations required to achieve risk-based MCLs.

In contrast, Rao and Jawitz (2003) modeled dissolution of a uniform DNAPL within a heterogeneous porous media (Type III hydrogeological setting). Results of their analysis showed that DNAPL mass depletion in the range of 70–90% could achieve 70–98% reductions in contaminant loading to downgradient plumes. Although contaminant reductions of this magnitude may not result in achieving MCLs within a source zone, they may allow for water quality objectives to be met in the downgradient plumes if the reduction in mass discharge results in a stable or a shrinking plume.

Christ et al. (2005) used numerical modeling to evaluate the potential benefits of ISB and surfactant flushing on reducing the longevity of DNAPL sources with varying architecture. The modeled results showed reductions in source longevity regardless of architecture when ISB was applied. However, for sources in which the architecture was dominated by pooled DNAPL the reduction in source longevity was limited, with DNAPL pools present post-remediation sustaining the groundwater plume as long as centuries. For sources with architecture dominated by residual DNAPL (ganglia), ISB and surfactant flushing reduced source longevity to a decade or less theoretically.

In a series of two-dimensional flushing experiments, Suchomel and colleagues (Suchomel and Pennell, 2006; Suchomel et al., 2007) evaluated reductions in contaminant mass discharge

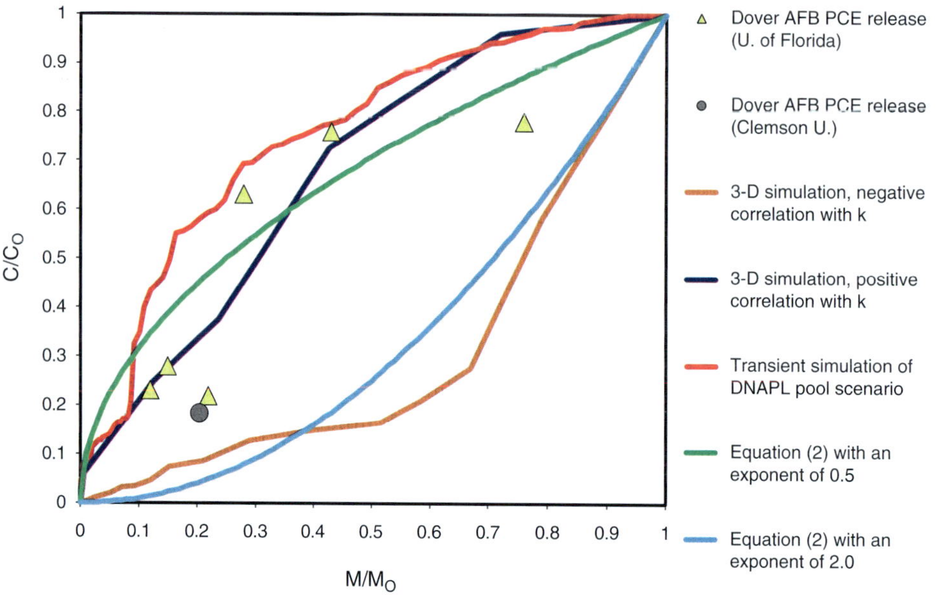

Figure 2.6. Modeled (*lines*) and observed (*points*) normalized DNAPL source zone concentrations as a function of normalized source zone mass removal (from Falta et al., 2005a).

following partial source depletion. The experiments were conducted in a Type I hydrogeological setting with DNAPL architectures ranging from pool dominated to ganglia dominated. Results of the experiments indicated that mass discharge can be substantially reduced (40–90%) following partial source depletion (53–85%), consistent with the findings of Rao and Jawitz (2003). Pool-dominated source architectures had the most significant reductions in mass discharge but also had the lowest fraction of total contaminant mass removal and diminishing returns with respect to mass removal efficiency (the mass of contaminant removed per pore volume flushed consistently declined over time). This result is consistent with preferential dissolution of DNAPL ganglia relative to pools and illustrates the trade-off between reducing source longevity and reducing contaminant mass discharge.

Based on these modeling and limited laboratory evaluations, a number of functions have been advanced to describe source architecture and relate source zone mass depletion to downgradient groundwater quality. Among these are the gamma function (Falta et al., 2005a; 2005b), beta function (Park and Parker, 2008), and ganglia-to-pool ratio, also quantified as the pool fraction (Christ et al., 2005). These functions are useful analytical tools, but determining a representative value for a field site has proven difficult. For example, there has been limited success in locating and quantifying pooled DNAPL with current characterization technologies. Basu et al. (2008) suggested the use of a stream tube model along with partitioning tracer testing to estimate source architecture and predict future source depletion, although attempts to validate this approach in the field have been limited to date. Figure 2.6 illustrates an example of normalized DNAPL source zone concentrations as a function of normalized source zone mass removal (Falta et al., 2005a).

Given the limited ability to estimate source architecture with current characterization technologies, research and fieldwork have focused on relating source architecture and remedial performance to upscaled mass transfer approaches. Park and Parker (2008) developed a model to evaluate the effects of remediation on source mass depletion, mass discharge from a source and downgradient plume attenuation for source architectures dominated by pooled and residual DNAPL. The model results were consistent with approaches that do not incorporate

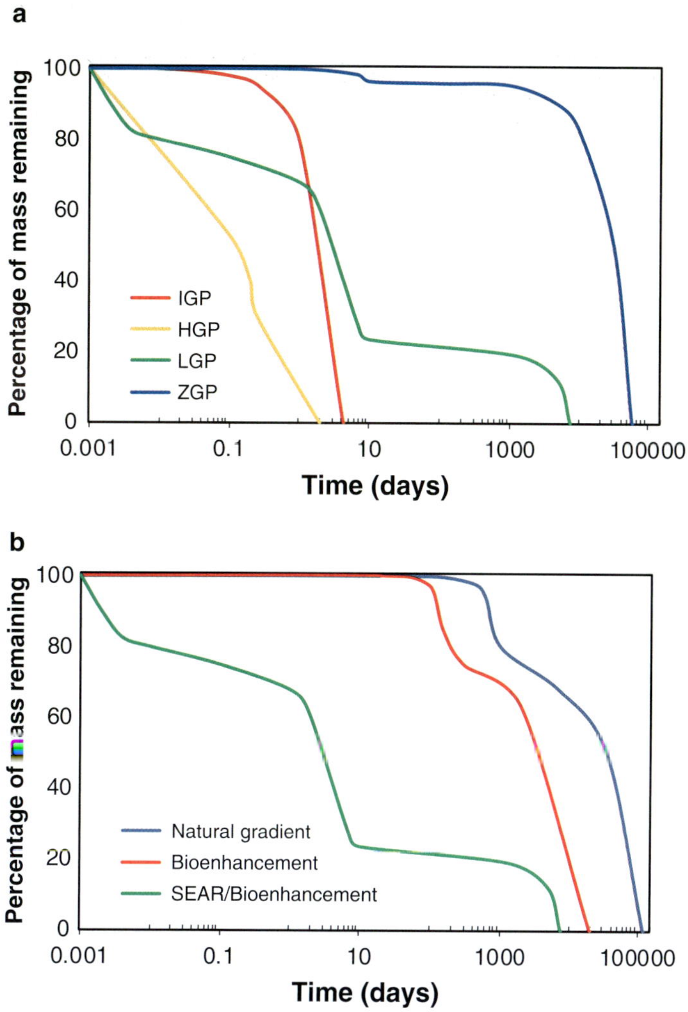

Figure 2.7. Modeled percentage of DNAPL mass remaining as a function of time for (a) surfactant flushing followed by bioremediation for DNAPL distributions ranging from only residual DNAPL (IGP) to only pooled DNAPL (ZGP) and (b) three alternative remediation strategies. The results illustrate the two-stage DNAPL mass recovery that has been observed in the field. From Christ et al. (2005).

upscaled mass transfer coefficients, indicating that partial source depletion reduced source longevity in all cases but that mass discharge reductions were more significant for sources comprising DNAPL ganglia and less significant for pooled DNAPL sources. However, the upscaled mass transfer relationship used by the model to describe dissolution from the DNAPL source was developed by fitting several parameters to time-series mass discharge data, limiting its use as an *a priori* predictive tool.

Christ et al. (2010) refined the approach developed by Park and Parker (2008) to allow estimation of the mass transfer coefficient based only on concentrations at the downgradient boundary of the source and incorporating a function relating beta to the ganglia-to-pool ratio. This refined approach to upscaled mass transfer is useful in that it can simulate the two-stage mass recovery that has been observed in the field (Brusseau et al., 2007) and may occur in sources that comprise spatially distinct regions of DNAPL ganglia with high contaminant mass discharge and DNAPL pools with lower, but sustained, mass discharge (Figure 2.7).

Research on estimating field-scale upscaled (bulk) mass transfer coefficients based on site-specific data has been conducted as part of ESTCP Project ER-200833. This approach involves relating contaminant mass extraction rates observed during source zone pump-and-treat to the bulk mass transfer rate from DNAPL to groundwater in the source. Bulk mass transfer rates were estimated using this method at a field site impacted by a multicomponent NAPL (jet fuel; benzene, toluene, ethylbenzene, and total xylenes [BTEX]; and naphthalene) before and following a pilot-scale test of thermal enhanced extraction. Observed benzene concentrations in site monitoring wells correlated well to concentrations modeled with the field-scale estimate of the bulk mass transfer rate, indicating that this may be a promising approach for assessing the impact of partial source removal on mass discharge.

In reviewing the work completed to date relating to source depletion, source mass discharge, and source architecture, it must be stressed that this has focused mostly on bench-scale laboratory studies and modeling experiments. The ability to translate this work to the field scale (practicably characterize the gamma function or the pool fraction) has not been demonstrated, and given this limitation, some researchers have questioned the utility of these source characterization metrics (Basu et al., 2008). As a result, current research on source architecture has focused on better predicting plume response to source depletion through groundwater modeling and innovative field characterization technologies such as multi-tracer push–pull testing and measurements of contaminant mass flux (Stroo et al., 2012).

2.7.2 Source Depletion and Plume Behavior

While the current state of the practice suggests that source remediation is unlikely to achieve MCLs in the source zone, benefits to downgradient groundwater quality may nevertheless be realized. Many DNAPL constituents will attenuate naturally in groundwater systems, either through biotic process (Wiedemeier et al., 1998) or abiotic processes (Danielson and Hayes, 2004). As noted earlier, if the rates of these natural attenuation processes are sufficiently rapid, they can result in groundwater plumes that are either stable or shrinking. They may also be sufficient to control contaminants that are back diffusing out of immobile pore space within the groundwater plume.

If the above natural attenuation processes are ongoing at a site, partial source depletion may offer the following benefits:

- For a stable or a shrinking contaminant plume – accelerating the rate at which the plume shrinks and/or decreasing the overall longevity of the plume
- For an expanding contaminant plume – reducing the mass loading to the plume so that the plume may stabilize or begin to shrink

The first of these was illustrated by Chapelle et al. (2004) based on a review of data from a field site in Georgia where source treatment via ISCO was implemented. Following ISCO treatment, concentrations of PCE in the source zone decreased from up to 5,000 micrograms per liter (μg/L) to <100 μg/L, and a resulting order-of-magnitude concentration reduction was observed in the downgradient groundwater plume. However, natural attenuation of DNAPL constituents at some sites may occur only at very slow rates or not at all. At these sites, relatively little benefit to downgradient groundwater quality may arise from source depletion.

Another potential limitation to the benefit of source depletion on downgradient groundwater quality is the potential presence of contaminants in the immobile pore space within a groundwater plume. Reductions in mass loading due to source depletion/containment may result in back diffusion of contaminants that may essentially act as a secondary contaminant source within the groundwater plume and sustain the plume for a significant duration, even with source depletion. The impact of back diffusion will vary from site to site and within the same site. Back diffusion sustaining a downgradient plume has been shown at field sites following implementation of source containment technologies (AFCEE, 2007; Sale et al., 2008a).

This phenomenon was explored in detail at a large site in Northern California impacted by chlorinated solvents (Newell et al., 2011). Source areas at the site have been contained either through installation of slurry walls or by source zone pump-and-treat. Mass discharge rates for plume containment pump-and-treat wells at the site were compiled and compared to estimates developed using two analytical models: a simplified advection–dispersion–retardation model and a model incorporating matrix diffusion within the downgradient plume (AFCEE, 2007). In general, the matrix diffusion model better approximated the observed mass discharge at the site, with a normalized root mean square error of 9% for the matrix diffusion model compared to 34% for the advection–dispersion–retardation model. However, inspection of the data indicated that the matrix diffusion model tended to overpredict systematically the amount of mass discharge at a given well, potentially resulting in an overly pessimistic view of plume cleanup.

While the previously described study did not investigate the cause for the overprediction of mass discharge by the matrix diffusion model, the model used was highly simplified. Sale et al. (2008b) have developed an analytical solution to describe matrix diffusion that incorporates other phenomena such as contaminant degradation in both the mobile and immobile pore space and fully models the contaminant distribution within the transmissive zone. This solution may more accurately describe changes in mass discharge due to matrix diffusion, but the highly complex nature of the solution has limited its use to date, although efforts to incorporate it into a practitioner-friendly tool for evaluating matrix diffusion are ongoing.

Given the above examples it is apparent that benefits to downgradient plume concentrations can be realized by source mass depletion, but the extent of the benefits will be highly dependent on the distribution of contaminant mass within the subsurface. The benefits of mass depletion may be related to changes in the natural plume behavior (improved natural attenuation performance) or reductions in plume size due to coupling of active plume remediation with source depletion. A number of researchers are developing screening-level models to assess the potential benefits of source depletion. These include the REMChlor model developed by Falta (2008), an analytical solute transport model, and mass balance-based models developed by Newell and Adamson (2005). Examples of how these models may be used to estimate plume responses to source depletion are discussed in Chapter 6 of this book. However, the currently available screening-level models are limited in some respects, and such limitations need to be considered when using the models for decision-making purposes.

2.8 RISKS OF IMPLEMENTING SOURCE DEPLETION TECHNOLOGIES

The previous sections have discussed the potential benefits related to source depletion, including decreased mass discharge, reductions in downgradient plume concentrations and reductions in remedial time frames. However, negative impacts related to source depletion are also possible. These risks should be considered when making a decision to implement source depletion, with the objective of doing what can be done to improve subsurface conditions weighed against the objective of doing no harm to the subsurface. Specific risks are discussed in detail elsewhere (Kavanaugh et al., 2003) and are summarized below.

- Expansion of a source may occur, particularly when using surfactant flushing or thermal treatment for source depletion (the LC34, Cape Canaveral Air Station, Florida, case study presented in Kavanaugh et al., 2003), with potential contaminant migration into low-permeability media or downward migration into deeper permeable zones.
- Changes to source architecture that reduce the potential for source depletion or increased contaminant mass transfer from DNAPL to groundwater. Such changes could include mobilization of DNAPL ganglia resulting in formation of new DNAPL pools, reducing the efficacy of mass removal technologies. Alternatively, changes to the subsurface permeability distribution due to DNAPL removal could produce transient increases in contaminant flux if the changes to the groundwater flow field result in access to previously inaccessible portions of the DNAPL source.
- Certain source depletion technologies can change the physical characteristics of the source, through formation of mineral precipitates such as manganese dioxide during ISCO, excessive growth of biomass during ISB, or changes to subsurface conditions (localized soil consolidation and/or surface slumping at some sites) related to heating during *in situ* thermal remediation. In many cases, these changes can result in reductions in source permeability, reducing future access to any source material remaining following treatment.
- Similarly, source depletion technologies that rely on injection of chemicals, particularly ISCO and ISB, can also alter geochemical properties of the source, notably pH, oxidation reduction potential (ORP) and salinity. These changes to geochemical properties can limit biological growth and potentially promote the transport of metals with ORP-dependent solubility such as chromium and arsenic.
- Some source depletion technologies (ISCO, *in situ* thermal treatment) may temporarily reduce subsurface biomass, biodiversity and functionality. These impacts appear to be transient in nature, however, and recent research indicates that biological activity can recover following implementation of these technologies.

In addition to the potential risks related to source depletion technologies, other potential negative impacts include an inability to develop or use the site both during and potentially following remediation, financial and/or legal consequences if the target remedial endpoints are not met, and stakeholder concerns if target remedial endpoints are not met and a remedial action is deemed to have "failed."

2.9 ESTIMATING IMPACTS OF SOURCE DEPLETION ON LIFE CYCLE COSTS

Along with benefits related to mass discharge and concentration reductions, that is, risk management benefits, the potential value of source depletion in the view of the site owner may be in reducing the life cycle cost to achieve cleanup objectives. Annual and life cycle costs for cleanup of sites impacted with DNAPL can be very high. For example, the USEPA 28 Site Study (USEPA, 1999) estimated a median annual cost of $180,000 (range $30,000–$4,000,000) for operation and maintenance (O&M) of pump-and-treat systems for DNAPL source containment. Since many of these containment systems are operating without concurrent implementation of source depletion technologies and may need to operate for centuries to achieve remedial objectives, the life cycle costs are clearly very high in constant dollars.

It is difficult to predict how source depletion will impact life cycle costs given the uncertainties associated with estimating the impacts of partial mass removal on source lifetime and mass discharge; however, potential economic benefits of complete or partial removal may include the following:

- Earlier site closure when regulatory requirements are met
- Lower annual and life cycle costs due to either reduced O&M costs or creating conditions that allow for implementation of more passive remedial technologies (MNA)
- Removal of liabilities and accrued reserves for remediation
- Enhanced land value due to reductions or removal of residual contamination
- The ability to conduct land transactions that would not be otherwise possible due to land-use restrictions or the need to retain liability and access associated with remedies requiring long-term O&M

While each of the economic benefits listed above could provide cost savings, cost–benefit analyses of potential source depletion remedies are complicated by a number of factors, including differences in accounting practices, discount rates used in analysis and tax consequences of using different remedial technologies (NRC, 1997), and are thus highly site specific. In addition, regulatory drivers specifying restoration goals continue to be the primary reason for remediating DNAPL source zones. Partial source depletion is generally not capable of meeting ARARs, thus limiting the utility of cost–benefit analyses.

The decision to implement source zone depletion technologies at a DNAPL site is based on highly site-specific conditions and criteria, and numerous regulatory, technology and stakeholder factors must be considered. The current decision process, as practiced in the United States, generally has resulted in selection of containment over source depletion. Pump and treat and/or MNA, both potential containment technologies, were used as groundwater remedies, either alone or in combination, at 71% of 164 Superfund sites where Records of Decision were issued between 2005 and 2008. During this same time period, institutional controls were included more often than any other groundwater remedy in site decision documents (USEPA, 2010).

2.10 SUMMARY AND FUTURE NEEDS

Several studies over the past decade have concluded that partial mass depletion from DNAPL source zones has been a viable remediation strategy at certain sites and is likely to provide benefits at a number of additional sites (Kavanaugh et al., 2003; NRC, 2005; Stroo et al., 2012). However, barriers to more widespread use of DNAPL source zone technologies persist. Additional theoretical analysis and assessment tools (performance prediction tools, cost–benefit assessment tools, technology failure analysis, reliability of long-term management), improved monitoring techniques for site characterization and performance assessment, and especially field-scale demonstrations that elucidate effective site characterization and technology implementation as well as the benefits of partial source depletion are needed to provide a more informed basis for decision making on whether to undertake DNAPL source zone depletion at both sites with a containment remedy in place and at new DNAPL sites.

This information will also provide a basis to estimate the fraction of DNAPL-impacted sites that would be candidates for implementation of source depletion technologies. At some DNAPL sites, containment may be the only viable remedial option, and at such sites, containment may be considered a "presumptive remedy" eliminating the need for costly additional studies. However, appropriate guidance for defining the conditions under which DNAPL source remediation would be a viable option for site cleanup compared to a containment-only option has not yet been reported. The current strategy of source zone containment has generally proven reliable for limiting routes of human and ecological exposure to chemical contaminants emanating from DNAPL-impacted sites, provided that the containment system (pump-and-treat or permeable barriers) has been properly designed and maintained. However, this strategy poses long-term risks, transfers the burden of site management to future generations, and requires long-term financial stability of the responsible parties. Furthermore, these long-term risks are generally difficult or impossible to quantify accurately. It is thus imperative that sufficient resources be devoted to resolving the many uncertainties in DNAPL source zone characterization and depletion technologies to ensure that source depletion at DNAPL sites is implemented to the maximum extent practicable.

In particular, research is needed to address key knowledge gaps that still exist regarding the effectiveness and cost of these technologies for DNAPL source removal in a wide range of hydrogeologic settings. Research is needed on the following topics: (1) development, verification and comparison of alternative technologies for measuring mass flux and mass discharge from DNAPL source areas before and after source depletion; (2) improved predictive tools to estimate the benefits and adverse effects of partial source depletion for a range of DNAPL treatment technologies and DNAPL distribution and geologic scenarios; (3) continued field testing of DNAPL source depletion technologies incorporating more than one technology (e.g., thermal, *in situ* flushing or *in situ* chemical oxidation combined with biodegradation); (4) guidance on the conditions in which source depletion is not likely to be an effective strategy and containment must be relied upon; (5) assessment of the long-term water quality impacts of source depletion technologies; and (6) development of quantitative decision analysis tools that will permit an accounting of all potential costs, benefits, and adverse impacts of partial DNAPL source depletion. A major challenge is to identify the degree of characterization and post-remediation monitoring needed for effective application of each of the *in situ* source depletion technologies.

REFERENCES

AFCEE (Air Force Center for Engineering and the Environment). 2006. Monitoring and Remediation Optimization System (MAROS), Software version 2.2, User's guide. http://www.gsi-net.com/es/software/software-gratis/maros.html. Accessed September 20, 2013.

AFCEE. 2007. AFCEE Source Zone Initiative Final Report. http://www.clu-in.org/download/contaminantfocus/dnapl/Chemistry_and_Behavior/AFCEE-szi-2007a.pdf. Accessed September 20, 2013.

Alexandra R, Gerhard JI, Kueper BH. 2012. Hydraulic displacement of dense nonaqueous phase liquids for source zone stabilization. Ground Water 50:765–774.

Amos BK, Suchomel EJ, Pennell KD, Löffler FE. 2008. Microbial activity and distribution during enhanced contaminant dissolution from a NAPL source zone. Water Res 42:2963–2974.

Basu NB, Fure AD, Jawitz JW. 2008. Predicting dense nonaqueous phase liquid dissolution using a simplified source depletion model parameterized with partitioning tracers. Water Resour Res 44:W07414. doi 10.1029/2007WR006008.

Berge ND, Ramsburg CA. 2009. Oil-in-water emulsions for encapsulated delivery of reactive iron particles. Environ Sci Technol 43:5060–5066.

Beyke G, Powell T. 2005. Heat enhanced bioremediation of chlorinated solvents using electrical resistance heating. In Abstracts of the 8th International *In Situ* and Onsite Bioremediation Symposium, Baltimore, MD, USA, June 6–9.

Brusseau ML, Nelson NT, Zhang Z, Blue JE, Rohrer J, Allen T. 2007. Source-zone characterization of a chlorinated-solvent contaminated Superfund site in Tucson, AZ. J Contam Hydrol 90:21–40

Chapelle FH, Bradley PM, Casey CC. 2004. Accelerated cleanup follows Fenton's ISCO and substrate addition. USEPA Technology News and Trends. December. http://clu-in.org/products/newsltrs/tnandt/view.cfm?issue=1204.cfm#3. Accessed September 20, 2013.

Charsky M. 2012. Joint technical presentation and panel discussion: Technical impracticability (TI) waivers. In Abstracts of the 2012 meeting of the Technical Support Project, Oklahoma City, OK, USA, May 1–3.

Childs J, Acosta E, Annable MD, Brooks MC, Enfield CG, Harwell JH, Hasegawa M, Knox RC, Rao PSC, Sabatini DA, Shiau B, Szekeres E, Wood AL. 2006. Field demonstration of surfactant-enhanced solubilization of DNAPL at Dover Air Force Base, Delaware. J Contam Hydrol 82:1–22.

Christ JA, Ramsburg CA, Abriola LM, Pennell KD, Löffler FE. 2005. Coupling aggressive mass removal with microbial reductive dechlorination for remediation of DNAPL source zones: A review and assessment. Environ Health Perspect 113:465–477.

Christ JA, Ramsburg CA, Pennell KD, Abriola LM. 2010. Predicting DNAPL mass discharge from pool-dominated source zones. J Contam Hydrol 114:18–34.

Cohen RM, Mercer JW. 1993. DNAPL Site Evaluation. CRC Press, Boca Raton, FL, USA. 384 p.

Costanza J, Pennell KD. 2007. Distribution and abiotic degradation of chlorinated solvents in heated field samples. Environ Sci Technol 41:1729–1734.

Costanza J, Fletcher KE, Löffler FE, Pennell KD. 2009. Fate of TCE in heated Fort Lewis soil. Environ Sci Technol 43:909–914.

Danielsen KM, Hayes KF. 2004. pH dependence of carbon tetrachloride reductive dechlorination by magnetite. Environ Sci Technol 38:4745–4752.

Dugan PJ, Siegrist RL, Crimi M. 2010. Coupling surfactants/cosolvents with oxidants for enhanced DNAPL removal: A review. Remediat J 20:27–49.

Einarson MD, Cherry JA. 2002. A new multilevel ground water monitoring system using multichannel tubing. Ground Water Monit Remediat 22:52–65.

ESTCP (Environmental Security Technology Certification Program). 2005. Bioaugmentation for Remediation of Chlorinated Solvents: Technology Development, Status, and Research Needs. http://www.serdp.org/. Accessed January 23, 2014.

ESTCP. 2010. Cost and Performance Report: Remediation of DNAPL Through Sequential *In Situ* Chemical Oxidation and Bioaugmentation. ESTCP Project 200116. http://www.serdp.org/. Accessed January 23, 2014.

ESTCP. 2011. Alternative Endpoints and Approaches Selected for the Remediation of Contaminated Groundwater. ESTCP Project ER-200832. Final Report. http://www.serdp.org/. Accessed January 23, 2014.

ESTCP. 2012. Improved Field Evaluation of NAPL Dissolution and Source Longevity. ESTCP Project ER-200833. http://www.serdp.org/. Accessed January 23, 2014.

ESTCP. 2013. Basic research addressing contaminants in low permeability zones. ESTCP Project ER-1740. http://www.serdp.org/. Accessed January 23, 2014.

Falta RW. 2008. Methodology for comparing source and plume remediation alternatives. Ground Water 46:272–285.

Falta RW, Javandel I, Pruess K, Witherspoon P. 1989. Density driven flow of gas in the unsaturated zone due to evaporation of volatile organic compounds. Water Resour Res 25:2159–2169.

Falta RW, Rao PSC, Basu N. 2005a. Assessing the impacts of partial mass depletion in DNAPL source zones. 1. Analytical modeling of source strength functions and plume response. J Contam Hydrol 78:259–280.

Falta RW, Basu N, Rao PSC. 2005b. Assessing impacts of partial mass depletion in DNAPL source zones. 2. Coupling source strength functions to plume evolution. J Contam Hydrol 79:45–66.

Feenstra S, Cherry JA, Parker BL. 1996. Conceptual models for the behavior of dense nonaqueous phase liquids (DNAPLs) in the subsurface. In Pankow JF, Cherry JA, eds, Dense Chlorinated Solvents in Groundwater. Waterloo Press, Portland, OR, USA, pp 53–88.

Fountain JC, Starr RC, Middleton T, Beikirch M, Taylor C, Hodge D. 1996. A controlled field test of surfactant-enhanced aquifer remediation. Ground Water 34:910–916.

Freeze RA, Cherry JA. 1979. Groundwater. Prentice Hall, Englewood Cliffs, NJ, USA. 604 p.

Fure AD, Jawitz JW, Annable MD. 2006. DNAPL source depletion: Linking architecture and flux response. J Contam Hydrol 85:118–140.

Gavaskar A, Tatar L, Condit W. 2004. Cost and performance report: Nanoscale zero-valent iron technologies for source remediation. NAVFAC Contract Report No. CR-05-007-ENV. NAVFAC, Port Hueneme, CA, USA. http://www.cluin.org/download/remed/cr-05-007-env.pdf. Accessed January 23, 2014.

Geosyntec, LFR, Tetra Tech NUS. 2007. Launch Complex 34 SWMU No. CC054 corrective measures study report, Cape Canaveral Air Force Station, Florida, USA, February.

Gerhard JI, Kueper BH, Hecox G. 1998. Waterflooding for the removal of pooled DNAPL. Ground Water 36:283–292.

Gerhard JI, Kueper BH, Hecox G, Schwarz R. 2001. Site-specific waterflood design for the recovery and stabilization of pooled DNAPL. Ground Water Monit Remediat 21:71–88.

Gillham R, Vogan J. 2010. Iron barrier walls for chlorinated solvent remediation. In Stroo HF, Ward CH, eds, *In Situ* Remediation of Chlorinated Solvent Plumes. Springer, New York, NY, USA, pp 537–571.

ITRC (Interstate Technology and Regulatory Council). 2002. DNAPL Source Reduction: Facing the Challenge. Washington, DC, USA. http://www.itrcweb.org. Accessed September 20, 2013.

ITRC. 2004. Strategies for Monitoring the Performance of DNAPL Source Zone Remedies. Washington, DC, USA. http://www.itrcweb.org. Accessed September 20, 2013.

ITRC. 2005a. Technical and Regulatory Guidance for *In Situ* Chemical Oxidation of Contaminated Soil and Groundwater. Washington, DC, USA. http://www.itrcweb.org. Accessed September 20, 2013.

ITRC. 2005b. Overview of *In Situ* Bioremediation of Chlorinated Ethene DNAPL Source Zones. BioDNAPL-1. Washington, DC, USA. http://www.itrcweb.org. Accessed September 20, 2013.

ITRC. 2008. Technical and Regulatory Guidance: *In Situ* Bioremediation of Chlorinated Ethene DNAPL Source Zones. BioDNAPL-2. Washington, DC, USA. http://www.itrcweb.org. Accessed September 20, 2013.

ITRC. 2010. Technology Overview: Use and Measurement of Mass Flux and Mass Discharge at Contaminated Sites. Washington, DC, USA. http://www.itrcweb.org. Accessed September 20, 2013.

ITRC. 2011. Technical and Regulatory Guidance: Integrated Strategies for Chlorinated Solvent Sites. Washington, DC, USA. http://www.itrcweb.org. Accessed September 20, 2013.

Johnson P, Dahlen P, Triplett Kingston J, Foote E, Williams S. 2009. State of Practice Overview: Critical Evaluation of State-of-the-Art *In Situ* Thermal Treatment Technologies for DNAPL Source Zone Treatment. http://www.serdp.org/. Accessed January 23, 2014.

Kavanaugh MC, Rao PSC, Abriola L, Cherry J, Newell C, Sale T, Destouni G, Falta R, Shoemaker S, Siegrist R, Major D, Mercer J, Teusch G, Udell K. 2003. The DNAPL Remediation Challenge: Is There a Case for Source Depletion? EPA/600/R-03/143. U.S. Environmental Protection Agency, Washington, DC, USA.

Kram ML, Keller AA, Rossabi J, Everett LG. 2002. DNAPL characterization methods and approaches, Part 2: Cost comparisons. Ground Water Monit Remediat 22:46–61.

Krembs FJ, Siegrist RL, Crimi M, Furrer RF, Petri BG. 2010. ISCO for groundwater remediation: Analysis of field applications and performance. Ground Water Monit Remediat 30:42-53.

Kueper BH, Davies KL. 2009. Assessment and delineation of DNAPL source zones at hazardous waste sites. EPA/600/R-09/119. USEPA, Cincinnati, OH, USA.

Kueper BH, Redman D, Starr RC, Reitsma S, Mah M. 1993. A field experiment to study the behavior of tetrachloroethylene below the water table: Spatial distribution of residual and pooled DNAPL. Ground Water 31:756–766.

Lemke LD, Abriola LM, Lang JR. 2004. Influence of hydraulic property correlation on predicted dense nonaqueous phase liquid source zone architecture, mass recovery and contaminant flux. Water Resour Res 40:W12417/doi 10.1029/2004WR003061.

Liu C, Ball WP. 2002. Back diffusion of chlorinated solvents from a natural aquitard to a remediated aquifer under well-controlled field conditions: Predictions and measurements. Ground Water 40:175–184.

McDade JM, McGuire TM, Newell CJ. 2005. Analysis of DNAPL source depletion costs at 36 field sites. Remediat J 15:9–18.

McGuire TM, McDade JM, Newell CJ. 2006. Performance of DNAPL source depletion technologies at 59 chlorinated solvent-impact sites. Ground Water Monit Remediat 26:73–84.

Mercer JW, Cohen RM. 1993. DNAPL Site Evaluation. CRC Press, Boca Raton, FL, USA.

Mercer JW, Cohen RM, Noel MR. 2010. DNAPL site characterization issues at chlorinated solvent sites. In Stroo HF, Ward CH, eds, *In Situ* Remediation of Chlorinated Solvent Plumes. Springer, New York, NY, USA. 805 p.

NAVFAC (Naval Facilities Engineering Command). 2007. Lessons Learned on Bioaugmentation of DNAPL Source Zone Areas. ESTCP Project ER-0008. http://www.clu-in.org/download/techfocus/biochlor/ER-0008-Less-Learned.pdf. Accessed January 23, 2014.

Newell CJ, Adamson DT. 2005. Planning-level source decay models to evaluate impact of source depletion on remediation timeframe. Remediat 15:27–47.

Newell CJ, McDade J, Seyedabbasi A, Gandhi D, Gallinatti J, Cocianni V, Ferguson DJ. 2011. Potential impact of matrix diffusion on a large pump-and-treat system. In International Symposium on Bioremediation and Sustainable Environmental Technologies. Reno, NV, USA, June 27–30.

NRC (National Research Council). 1994. Alternatives for Ground Water Cleanup. National Academies Press, Washington, DC, USA.

NRC. 1997. Valuing Groundwater. Economic Concepts and Approaches. National Academies Press, Washington, DC, USA.

NRC. 2005. Contaminants in the Subsurface: Source Zone Assessment and Remediation. National Academies Press, Washington, DC, USA.

Pankow JF, Cherry JA, eds. 1996. Dense Chlorinated Solvents and Other DNAPLs in Groundwater. Waterloo Press, Portland, OR, USA.

Park E, Parker JC. 2008. Effects of mass reduction, flow reduction and enhanced biodecay of DNAPL source zones. Transport Porous Media 73:95–108.

Parker BL, Cherry JA, Gillham RW. 1996. Effects of molecular diffusion on organic chemical flow and transport in fractured geologic porous media. In Pankow JF, Cherry JA, eds, Dense Chlorinated Solvents and Other DNAPLs in Groundwater. Waterloo Press, Portland, OR, USA. 522 p.

Payne F, Quinnan JA, Potter ST. 2008. Remediation Hydraulics. CRC Press, Boca Raton, FL, USA.

Pennell KD, Löffler FE, Costanza J, Fletcher KE, Ramaswamy NS, Otaño G, Callaghan J. 2009. Investigation of Chemical Reactivity, Mass Recovery and Biological Activity during Thermal Treatment of DNAPL Source Zones. SERDP Project 1419 Final Report. ESTCP, Washington, DC, USA.

Poulsen M, Kueper BH. 1992. A field experiment to study the behavior of tetrachloroethylene in unsaturated porous media. Environ Sci Technol 26:889–895.

Ramsburg CA, Abriola LM, Pennell KD, Löffler FE, Gamache M, Amos BK, Petrovskis EA. 2004. Stimulated microbial reductive dechlorination following surfactant treatment at the Bachman Road site. Environ Sci Technol 38:5902–5914.

Ramsburg CA, Pennell KD, Abriola LM, Daniels G, Drummond CD, Gamache M, Hsu HL, Petrovskis EA, Rathfelder KM, Ryder JL, Yavaraski TP. 2005. A pilot-scale demonstration of surfactant-enhanced PCE solubilization at the Bachman Road site. 2. System operation and evaluation. Environ Sci Technol 39:1791–1801.

Rao PSC, Jawitz JW. 2003. Comment on "Steady-state mass transfer from single-component dense non-aqueous phase liquids in uniform flow fields" by Sale TC and McWhorter DB. Water Resour Res 39:1069.doi:10.1029/2002WR001423.

Rao PSC, Jawitz JW, Enfield CG, Falta RW, Annable MD, Wood AL. 2001. Technology integration for contaminated site remediation: Cleanup goals and performance criteria. In Thornton S, Oswald S, eds, Groundwater Quality: Natural and Enhanced Restoration of Groundwater Pollution. Publication No. 275. International Association of Hydrological Sciences, Wallingford, UK, pp 571–578.

Saenton S, Illangasekare TH, Soga K, Saba TA. 2002. Effects of source zone heterogeneity on surfactant-enhanced NAPL dissolution and resulting remediation end-points. J Contam Hydrol 59:27–44.

Sale TC, McWhorter DB. 2001. Steady state mass transfer from single-component DNAPLs in uniform flow fields, Water Resour Res 37:393–404.

Sale TC, Newell CJ. 2010. Impacts of source management on chlorinated solvent plumes. In Stroo HF, Ward CH, eds, *In Situ* Remediation of Chlorinated Solvent Plumes. Springer, New York, NY, USA, pp 185–216.

Sale T, Newell C, Stroo H, Hinchee R, Johnson P. 2008a. Frequently Asked Questions Regarding Management of DNAPL Sites. ESTCP, Arlington, VA, USA. http://www.serdp.org/. Accessed January 23, 2014.

Sale TC, Zimbron J, Dandy D. 2008b. Effects of reduced contaminant loading on downgradient water quality in an idealized two layer system. J Contam Hydrol 102:72–85.

Saleh N, Sirk K, Liu Y, Phenrat T, Dufour B, Matyjaszewski K, Tilton R, Lowry GV. 2007. Surface modifications enhance nanoiron transport and DNAPL targeting in saturated porous media. Environ Eng Sci 24:45–57.

Schwarzenbach R, Gschwend PM, Imboden DM. 1993. Environmental Organic Chemistry. John Wiley and Sons, New York, NY, USA. 681 p.

Siegrist RL, Crimi, M, Brown RA. 2011. *In situ* chemical oxidation: Technology description and status. In Siegrist RL, Crimi M, Simpkin TJ, eds, *In Situ* Chemical Oxidation for Groundwater Remediation. Springer, New York, NY, USA. 545 p.

Sleep BE, Seeperstad DJ, Mo K, Heidorn CM, Hrapovic L, Morrill PL, McMaster ML, Hood ED, Lebrón C, Lollar BS, Major DW, Edwards EA. 2006. Biological enhancement of tetrachloroethene dissolution and associated microbial community changes. Environ Sci Technol 40:3623–3633.

Smith MM, Silva JAK, Munakata-Marr J, McCray JE. 2008. Compatibility of polymers and chemical oxidants for enhanced groundwater remediation. Environ Sci Technol 42:9296–9301.

Stroo HF, Leeson A, Marqusee JA, Johnson PC, Ward CH, Kavanaugh MC, Sale TC, Newell CJ, Pennel KD, Lebrón CA, Unger M. 2012. Chlorinated ethene source remediation: Lessons learned. Environ Sci Technol 46:6438–6447.

Suchomel EJ, Pennell KD. 2006. Reductions in contaminant mass discharge following partial mass removal from DNAPL source zones. Environ Sci Technol 40:6110–6116.

Suchomel EJ, Ramsburg CA, Pennell KD. 2007. Evaluation of trichloroethene recovery processes in heterogeneous aquifer cells flushed with biodegradable surfactants. J Contam Hydrol 94:195–214.

Taghavy A, Costanza J, Pennell KD, Abriola LM. 2010. Effectiveness of nanoscale zero-valent iron for treatment of a PCE-DNAPL source zone. J Contam Hydrol 118:143–151.

USEPA (U.S. Environmental Protection Agency). 1992. Estimating potential for occurrence of DNAPL at superfund sites. OSWER Publication 9355.4-07FS. Office of Solid Waste and Emergency Response, Washington, DC, USA.

USEPA. 1993a. Evaluation of the likelihood of DNAPL presence at NPL sites – national results. EPA 540R-93-073. Washington, DC, USA.

USEPA. 1993b. Guidance for evaluating the technical impracticability of groundwater restoration. OSWER Directive 9234.2-25. Office of Solid Waste and Emergency Response, Washington, DC, USA.

USEPA. 1994. DNAPL site characterization quick reference fact sheet. EPA 540/F-94/049. Office of Solid Waste and Emergency Response, Washington, DC, USA.

USEPA. 1998. Evaluation of subsurface engineered barriers at waste sites. EPA 542/R-98/005. Office of Solid Waste and Emergency Response, Washington, DC, USA.

USEPA. 1999. Groundwater cleanup: Overview of operating experience at 28 sites. EPA 542/R-99/006. Washington, DC, USA.

USEPA. 2004. In-situ thermal treatment of chlorinated solvents: Fundamentals and field applications. EPA 542/R-04/010. Washington, DC, USA.

USEPA. 2006. Engineering forum issue paper: *In situ* treatment technologies for contaminated soil. EPA 542/F-06/013. Washington, DC, USA.

USEPA. 2009. DNAPL remediation: Selected projects where regulatory closure goals have been achieved. EPA 542/R-09/008. Washington, DC, USA.

USEPA. 2010. Superfund remedy report, 13th ed. EPA 542/R/10/004. Washington, DC, USA.

USEPA. 2011a. Environmental cleanup best management practices: Effective use of the project life cycle conceptual site model. EPA 542 F/11/011. Washington, DC, USA.

USEPA. 2011b. Groundwater road map. Recommended process for restoring contaminated groundwater at Superfund sites. OSWER directive 9283.1-34. Washington, DC, USA.

USEPA. 2011c. Clarification of OSWER's 1995 technical impracticability waiver policy, OSWER Directive 9355.5-32. Washington, DC, USA.

USEPA. 2012. What is long-term stewardship? Washington DC, USA. http://www.epa.gov/landrevitalization/ltstf_report/whatis_longterm_stewardship.htm. Accessed January 23, 2014.

Wadley SL, Gillham RW, Gui L. 2005. Remediation of DNAPL source zones with granular iron: Laboratory and field tests. Ground Water 43:9–18.

Waldemer RH, Tratnyek PG, Johnson RL, Nurmi JT. 2007. Oxidation of chlorinated ethenes by heat-activated persulfate: Kinetics and products. Environ Sci Technol 41:1010–1015.

Watts RJ, Teel AL. 2006. Treatment of contaminated soils and groundwater using ISCO. Pract Period Hazard Toxic Radioacti Waste Manage 10:2–9.

West MR, Kueper B. 2010. Plume detachment and recession times in fractured rock. Ground Water 48:416–426.

Wiedemeier TH, Swanson MA, Moutoux DE, Gordon EK, Wilson JT, Wilson BWH, Kampbell DH, Hansen JE, Hass P, Chapelle FH. 1998. Technical Protocol for Evaluating Natural Attenuation of Chlorinated Solvents in Groundwater. EPA 600/R-98/128. Washington, DC, USA.

CHAPTER 3

DNAPL SOURCE ZONE CHARACTERIZATION AND DELINEATION

Bernard H. Kueper[1] and Kathy L. Davies[2]

[1]Queen's University, Kingston, ON, Canada K7L 3N6; [2]U.S. Environmental Protection Agency, Region III, Philadelphia, PA 19103, USA

3.1 INTRODUCTION

The ability to remediate a chlorinated solvent source zone depends greatly on the adequacy of source zone characterization. Source zone characterization refers to defining the nature and extent of contamination within a geologic and hydrogeologic framework, which acts as a source of contributing contamination to the vapor or the aqueous phase. In general, chlorinated solvent contaminants can be present in the vapor phase, sorbed phase, aqueous phase and in the form of dense nonaqueous phase liquids (DNAPLs). The applicability of a particular remediation technology depends on how much contamination is present in each of these four forms, and their relative distribution within varying permeability media. Young source zones (Stage 1, 2 or early Stage 3; Chapter 1) may still contain significant amounts of contaminant mass present as DNAPL in moderate to high-permeability media, whereas in older source zones (late Stage 3, Stage 4 or 5; Chapter 1) the majority of the remaining contaminant mass may be in low-permeability media where diffusion, not dissolution of DNAPL, is the primary transport mechanism.

For source zones where DNAPL is present, the architecture of the DNAPL distribution will be of interest. The DNAPL architecture refers to the range of DNAPL saturations and the lateral continuity and thickness of DNAPL pools and zones of residual DNAPL. Specific physical and chemical properties of interest include the DNAPL density, DNAPL—water interfacial tension, DNAPL—air interfacial tension, DNAPL composition, DNAPL component effective solubility, and DNAPL viscosity and wettability. Determining the potential for DNAPL migration is also of fundamental importance. The time scales of DNAPL migration following an initial release to the subsurface depend on a number of factors including the volume released, physical properties of the DNAPL, geologic structure, and permeability (Gerhard et al., 2007; Richards et al., 2012; Reynolds and Kueper, 2004). At some chlorinated solvent sites, the length of time required for the DNAPL to cease migration (reach the end of Stage 2; Chapter 1) following the initial release to the subsurface may be a few years to a decade or longer.

In addition to the form and distribution of contaminant mass, it is important to understand the geologic framework of the source zone. This aspect of source zone characterization refers to defining the spatial extent and structure of major lithologic units with attention given to both large- and small-scale features. Properties of interest in unconsolidated deposits include the magnitude of hydraulic conductivity at both the small and large scales, anisotropy in hydraulic conductivity, porosity, bedding thickness and orientation, soil fraction organic carbon, grain size distribution, entry pressure, wettability and soil dry bulk density. Properties of interest in fractured media include the magnitude of bulk rock hydraulic conductivity, orientation of

B.H. Kueper et al. (eds.), *Chlorinated Solvent Source Zone Remediation*,
doi: 10.1007/978-1-4614-6922-3_3,

fracture sets, fracture spacing, fracture aperture and properties of the rock matrix including the porosity, fraction organic carbon, dry bulk density and pore throat distribution.

Source zone characterization also includes defining the water table depth, lateral and vertical components of the hydraulic gradient, groundwater flow directions and groundwater velocities. Seasonal variations in the rates and directions of groundwater flow need to be defined as well as any transience due to tidal or similar fluctuations and any groundwater—surface water interactions that may be occurring. In addition to groundwater velocity, aqueous-phase contaminant velocity typically needs to be defined, including rates of diffusion in low-permeability media. Above the water table, the rate of vapor-phase diffusion may be of interest, in which case soil moisture content and soil organic carbon content should be assessed.

This chapter focuses on characterizing the source zone with respect to the nature and extent of contamination. As pointed out above, source zone characterization also includes defining the hydrogeologic and geologic characteristics relevant for remediation. These topics are beyond the scope of this chapter and can be found in other sources such as the many books that address geology and hydrogeology. Also of relevance to source zone characterization are Chapter 7 in this book, *Flux-Based Site Assessment and Management*; Chapter 4, *Advanced Diagnostics*; Chapter 17, *Alternate Management Strategies*; and Chapter 8, *Hydraulic Displacement of Dense Nonaqueous Phase Liquids* that includes a section on capillary pressure, DNAPL saturation and relative permeability concepts. DNAPL site characterization methods are also described in Mercer et al. (2010).

3.2 ASSESSING DNAPL PRESENCE

Assessing the presence of DNAPL is a fundamental component of source zone characterization. At many DNAPL sites, however, the DNAPL is not directly observed or encountered. This failure to detect DNAPL stems largely from the fact that the primary means of data collection at DNAPL sites is through the use of drilling and direct push methods. Drilling and direct push methods provide an opportunity to observe cores, to obtain samples for laboratory analysis, and to construct a monitoring well or piezometer. The volume of media that is removed and/or contacted using drilling and direct push methods, however, is small compared to the size of the source zone. Given the tortuous and sparse nature of DNAPL migration, it is not surprising that DNAPL is not directly encountered or observed at some DNAPL sites. As a result, it is important that a converging-lines-of-evidence approach be employed in assessing DNAPL presence.

The various lines of evidence that can be relied upon to assess the presence of DNAPL are summarized in Figure 3.1. Lines of evidence A and B can be relied upon to conclude that DNAPL is present at a site, regardless of whether any of lines of evidence C through H are satisfied. However, the lack of A or B does not imply that DNAPL is absent at a site. Rather, some or all of the other lines of evidence (C through H) will need to be relied upon to interpret whether DNAPL is present or absent. The various lines of evidence in Figure 3.1 are not ranked in order of importance, nor are they all required to arrive at an interpretation of DNAPL presence or absence. The types of data collected and relied upon to assess DNAPL presence can vary from site to site depending on the geological conditions, type of contaminant, age of the site, availability of technology and budget. In general, the lines of evidence in Figure 3.1 are applicable to both porous and fractured media. Lines of evidence B and C, however, are not applicable to fractured rock. At most DNAPL sites, some of the lines of evidence A through H (see Figure 3.1) will be negative.

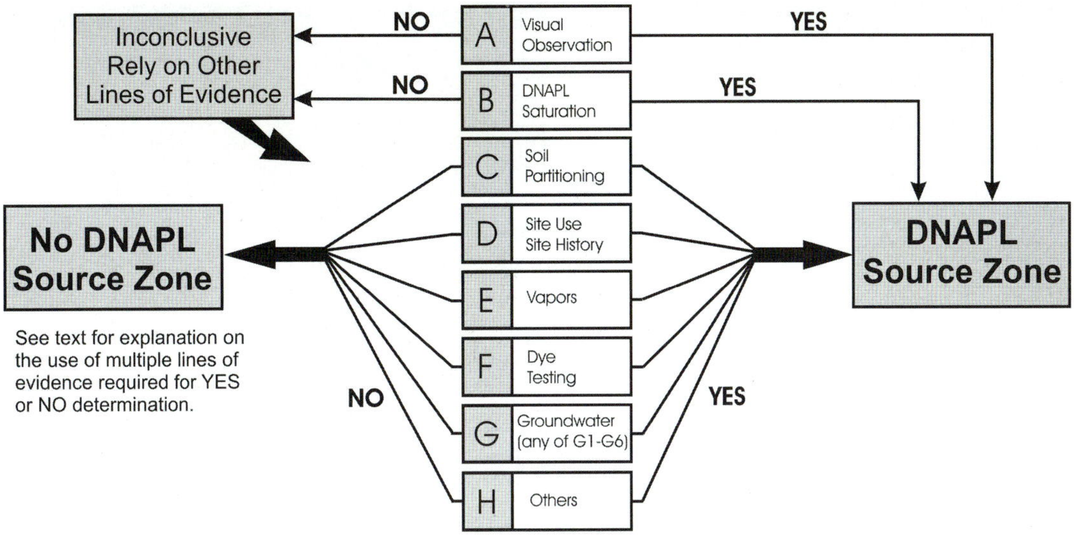

Figure 3.1. Converging-lines-of-evidence approach for assessing DNAPL presence (USEPA, 2009).

3.2.1 Visual Observation (Line of Evidence A)

The visual observation of DNAPL in soil cores, in monitoring wells, in excavations, and in drill return water is conclusive evidence of DNAPL presence. If sufficient volume of the DNAPL can be recovered, it is customary to have the sample analyzed for component composition, DNAPL—water interfacial tension, viscosity and density. It is recommended that the viscosity be measured at the subsurface temperature of interest. If a nonaqueous phase liquid (NAPL)—water interface probe indicates the presence of DNAPL at the bottom of a monitoring well, extraction well or piezometer, it is prudent to obtain a sample using a bailer or a pump to confirm that the interface probe response correctly identified the presence of DNAPL. Similarly, if NAPL, staining or a sheen is visually observed in a core, it is recommended to submit a soil sample to the laboratory for quantitative analysis of volatile organic compounds (VOC) and semivolatile organic compounds (SVOC). The types of compounds detected can allow light nonaqueous phase liquids (LNAPLs) to be distinguished from DNAPL, which may not be obvious upon visual observation of NAPL in core. In addition, the magnitude of soil concentrations can be used to estimate the saturation of the NAPL and provide a determination of whether the NAPL exists at residual or pooled saturations (see below).

Figure 3.2 provides photographs of DNAPL observed in porous media. Figure 3.2a illustrates pooled DNAPL that was encountered during excavation activities at a site located in the United States. The DNAPL is dark brown in color and is relatively easy to distinguish from the native gray- and light brown-colored soil. Figure 3.2b illustrates pooled DNAPL (black and brown in color) present in a fine sand layer between two layers of clay in a soil core obtained from below the water table using a sonic drilling rig. Figure 3.2c illustrates residual DNAPL present in horizontally bedded fine- to medium-grained sands at Canadian Forces Base Borden (Ontario, Canada). The perchloroethene (PCE) DNAPL was dyed red using Sudan IV prior to release to the subsurface (Kueper et al., 1993). Without the presence of the red dye, the colorless DNAPL would not be visually identifiable. Figure 3.2d illustrates pooled PCE DNAPL from the same field experiment. The DNAPL pool was perched on a 1 centimeter (cm) (0.4 inch[in]) thick silt lens.

Figure 3.2. Photographs of DNAPL encountered in porous media. (a) Dark brown pooled DNAPL encountered during excavation activities above the water table (Kueper, 2005). (b) Pooled DNAPL observed in sand layer between clay layers in soil core obtained below the water table with sonic drilling rig at an industrial facility (Kueper, 2011). (c) Residual DNAPL in horizontally bedded sands at Canadian Forces Base Borden (Kueper et al., 1993). (d) Pooled DNAPL in horizontally bedded sands at Canadian Forces Base Borden (Kueper et al., 1993).

Visual observation of DNAPL at fractured rock sites is typically limited to accumulations in monitoring wells or observations on ribbon samplers impregnated with hydrophobic dye that are placed into an open hole prior to well completion. Because of the aggressive nature of drilling in rock, it is rare for DNAPL to persist in fractures and be observed in retrieved rock cores. The presence of sheens on drill return water, however, is consistent with DNAPL presence.

3.2.2 Soil Concentration

Most site investigations performed in unconsolidated deposits involve obtaining soil samples for laboratory analysis from both above and below the water table. These soil samples typically are obtained at the time that borings are drilled for the purposes of piezometer installation, monitoring well installation, extraction well installation, and characterization of geology and contaminant distribution. In some instances, soil samples are obtained in association with trenching or excavation activities. The soil chemical concentrations reported by the laboratory can be used to assess whether DNAPL is present in the soil sample at saturations representing pooled DNAPL, at saturations representing residual DNAPL, or not present at all.

Soil samples are typically taken from a core or an exposed soil surface relatively quickly after extraction of the core from the subsurface or exposure of the soil surface so that losses due to vaporization and volatilization are small. Soil samples are typically taken from areas exhibiting obvious staining, visual DNAPL presence, positive dye test results or high organic vapor monitoring (OVM) readings. In some cases, soil samples are taken from directly above a capillary barrier regardless of any obvious indication of impact since this is most likely where DNAPL is or may have been present. In other cases, soil samples are taken at elevations where nearby membrane interface probe (MIP) results or other types of vertical profiling data indicate the presence of impacts.

Although traditional soil sampling programs often specified fixed depth intervals from which to obtain samples from borings, this practice has been largely replaced by targeted sampling based upon the observations and indications of impact noted above. This is important given the irregular and somewhat random nature of DNAPL occurrence in soil cores. Also of note is that the drilling and subsequent soil sampling methods can potentially alter the distribution of DNAPL. It is common to specify both the minimum and the maximum number of samples to obtain from a boring and have a sampling plan outlining the decision criteria used to satisfy that requirement. Field personnel should consult any relevant state or federal guidance regarding sampling techniques (methanol preservation, sample volume), shipping requirements, duplicates and holding times associated with soil sampling.

3.2.2.1 Saturation Threshold (Line of Evidence B)

The soil concentration (C_D) corresponding to a prescribed NAPL saturation is calculated as

$$C_D = \frac{S_r \phi \rho_N 10^6}{\rho_b} + C^T \quad \text{(Eq. 3.1)}$$

where C_D is the soil concentration (milligrams per kilograms [mg/kg]) of interest, S_r is the prescribed NAPL saturation (portion or pore space), ϕ is the total porosity (portion of bulk soil volume), ρ_N is the NAPL density (grams per cubic centimeters [g/cm^3]), ρ_b is the dry soil bulk density (g/cm^3), and C^T is the maximum amount of contaminant (in mg/kg) that can be stored in the soil sample in the aqueous, vapor, and sorbed phases expressed on a dry weight basis. C^T can be calculated using Equation 3.2 as discussed below. The use of Equation 3.1 requires the site-specific assessment and/or estimation of the total porosity, dry soil bulk density, NAPL density, and C^T. Equation 3.1 is valid for both DNAPLs and LNAPLs above or below the water table. In the event that the laboratory-reported soil concentration is expressed on a wet weight basis, adjustments need to be made for moisture content. In the event that more than one chemical is detected in the soil sample, the sum of all detected compounds (total soil concentration) should be used to compare to C_D.

For practical purposes, a NAPL saturation of 0.25 (25% of pore space) can be used as the threshold saturation representing pooled NAPL. The precise value of saturation demarking pooled from residual NAPL is site specific and depends on a number of factors including grain size distribution, NAPL—water interfacial tension, and NAPL saturation history. A NAPL saturation of between 0.05 and 0.10 (5—10%) should be used to arrive at a reliable conclusion of NAPL presence.

Example Calculation: Figure 3.3 presents a plot of soil concentration versus DNAPL saturation for various DNAPLs in sandy media characterized by a total porosity of 0.40 (40%) and a dry bulk density of 1.63 g/cm^3. The C_T term is approximated as 500 mg/kg for each DNAPL for the purposes of this example. Adopting an approximate threshold DNAPL saturation of 0.05 (5%) to

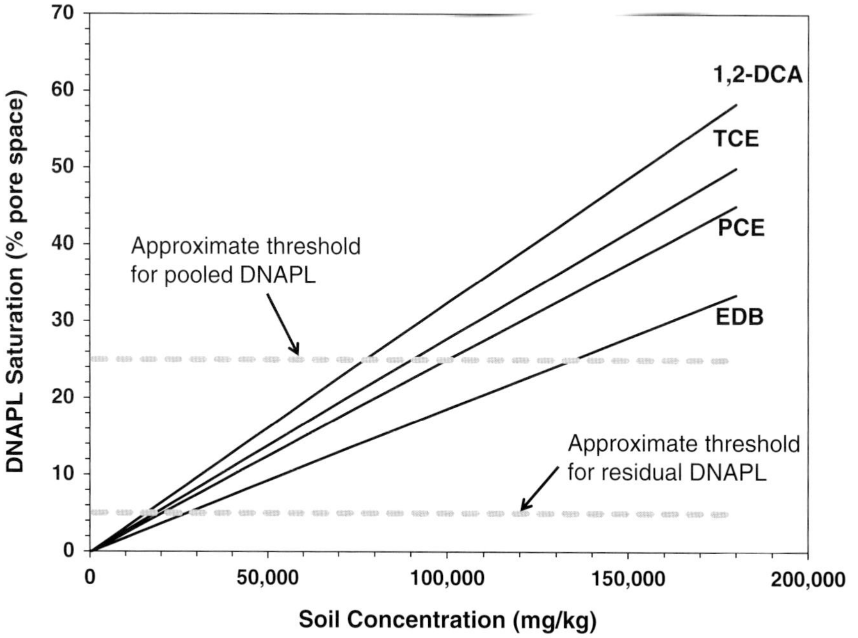

Figure 3.3. Soil concentration versus DNAPL saturation in sandy media for various chlorinated solvents. 1,2-DCA = 1,2-dichloroethane, TCE = trichloroethene, PCE = perchloroethene, EDB = ethylene dibromide.

represent residual DNAPL presence, Figure 3.3 indicates that soil concentrations in excess of between approximately 16,000 and 27,500 mg/kg would be indicative of residual DNAPL depending on the specific DNAPL type. Adopting an approximate threshold DNAPL saturation of 0.25 (25% of pore space) to represent pooled DNAPL presence, Figure 3.3 indicates that soil concentrations in excess of between approximately 77,000 and 135,000 mg/kg would be indicative of pooled DNAPL depending on the specific DNAPL type. This example assumes that each DNAPL of interest is a single-component DNAPL and that only the compound comprising the DNAPL was detected in the soil sample.

3.2.2.2 Partitioning Threshold (Line of Evidence C)

The soil concentration (C^T) corresponding to the theoretical maximum combined total amount of sorbed-, aqueous-, and vapor-phase chemical present in a soil sample absent the presence of NAPL is given by (Feenstra et al., 1991)

$$C_i^T = \frac{C_i}{\rho_b}\left(K_d\rho_b + \theta_w + H'\theta_a\right) \qquad \text{(Eq. 3.2)}$$

where C_i^T is the soil concentration (mg/kg) threshold for component i, C_i is the effective solubility (milligrams per liter [mg/L]) of component i, ρ_b is the dry soil bulk density (g/cm^3), K_d is the soil—water partition coefficient [milliliters per gram (mL/g)] approximated using $K_d = K_{oc}f_{oc}$, θ_w is the water-filled porosity (dimensionless), H' is the unitless Henry's Law constant (dimensionless), θ_a is the air-filled porosity (dimensionless), K_{oc} is the organic carbon—water partition coefficient (mL/g) and f_{oc} is the soil fraction organic carbon content (dimensionless). The water-filled porosity and the air-filled porosity sum to the total porosity (not the effective porosity).

The dry soil bulk density, soil fraction organic carbon, water-filled porosity and air-filled porosity are site-specific values. It is not typical to evaluate these parameters using the same soil sample submitted for concentration analysis. These parameters typically are evaluated using unimpacted soil samples from the same geologic unit as the impacted samples. For some chlorinated solvents of interest, the first term in parentheses in Equation 3.2 dominates the value of C_i^T. This term represents the sorbed phase fraction and depends on the magnitude of the soil fraction organic carbon. Consequently, some site investigations focus on obtaining several representative samples of the soil fraction organic carbon, and fewer samples for the other soil properties. In some cases, these other soil properties are estimated using literature values.

The use of Equation 3.2 is straightforward for the case of a single-component DNAPL or the case of a multicomponent DNAPL where the composition is known. In the event that a multicomponent DNAPL is encountered and the component composition is not known, the following can be used:

$$\sum_{i=1}^{n} \frac{C_{obs,i}^T}{C_{S,i}^T} \geq 1 \qquad \text{(Eq. 3.3)}$$

where $C_{obs,i}^T$ is the reported concentration (mg/kg) of component i, $C_{S,i}^T$ is the single-component soil partitioning concentration (mg/kg) of component i calculated using Equation 3.2 with the single-component solubility as input, and n is the number of components in the soil sample. If DNAPL is present in the soil sample, the summation in Equation 3.3 will exceed unity.

Both Equations 3.2 and 3.3 are subject to uncertainty given the natural variability of soil properties and the fact that soil concentration analyses may not detect all components present in the soil sample. Consequently, Equations 3.2 and 3.3 are typically considered to be indirect lines of evidence of DNAPL presence that should be evaluated along with other lines of evidence. Marginally exceeding the thresholds in Equations 3.2 and 3.3 could be the result of using inaccurate values of soil properties. There may also be spatial variability of contaminant mass within the obtained soil sample that could make the laboratory-reported value appear low compared to what is locally present within the sample.

Example Calculation: Figure 3.4 presents the partitioning threshold calculated using Equation 3.2 as a function of soil fraction organic carbon for the same four single-component DNAPLs considered in Figure 3.3 (1,2-DCA, TCE, PCE, and EDB). The air-filled porosity is assumed to be zero, the water-filled porosity is set to 40%, and the dry soil bulk density is set to 1.63 g/cm^3. The single-component solubilities for 1,2-DCA, TCE, PCE, and EDB are set to 8,500, 1,100, 200, and 4,200 mg/L, respectively. The organic carbon—water partition coefficients for 1,2 DCA, TCE, PCE, and EDB are set to 14, 126, 364, and 92 mL/g, respectively. Figure 3.4 illustrates that of the four compounds considered, 1,2 DCA is least sensitive to the value of soil fraction organic carbon because of its relatively low organic carbon—water partitioning coefficient. However, 1,2 DCA exhibits the highest partitioning threshold at very low values of soil fraction organic carbon because it has the highest aqueous-phase solubility of the four compounds considered.

3.2.3 Site Use/Site History (Line of Evidence D)

The presence of DNAPL in the subsurface is often associated with certain storage and handling activities, industries and manufacturing processes. Chlorinated solvents have a wide range of uses including degreasing metal, decaffeinating coffee, dry cleaning, intermediates in plastic production, as heat transfer fluids and as components of both pesticides and herbicides.

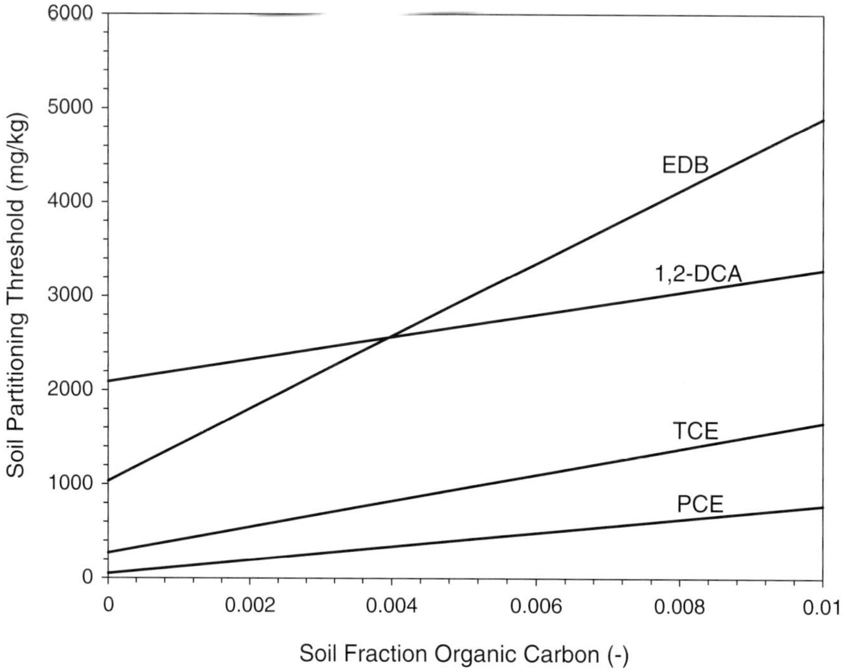

Figure 3.4. Soil partitioning threshold versus soil fraction organic carbon in sandy media for various chlorinated solvents.

Information regarding past DNAPL use and handling can be obtained from employee interviews, purchase and sale records, engineering and process handling documents and building plans. Common release points for DNAPLs to the subsurface include unlined drum storage areas, unlined ponds and burn pits, leaky underground storage tanks, septic leach fields and floor drains.

3.2.4 Vapor Concentrations (Line of Evidence E)

The presence of vapors in the unsaturated zone may be consistent with current or past DNAPL presence. However, it should be noted that solvent vapors can arise due to off-gassing from groundwater or soil moisture, independent of DNAPL presence. Furthermore, off-gassing from groundwater or soil moisture can produce soil vapor concentrations of the same magnitude as vaporization directly from DNAPL. For this reason, there is no vapor concentration threshold that can be used to distinguish vapors due to off-gassing from vapors due to vaporization of DNAPL. The presence of vapors can be useful, however, in deciding where to collect additional data to assess DNAPL presence.

Many chlorinated solvent DNAPLs will be depleted in the unsaturated zone due to vaporization much faster than they will be depleted in the saturated zone due to dissolution into groundwater. For this reason, it would be incorrect to conclude that DNAPL is not present below the water table solely on the basis of DNAPL absence above the water table. Fast rates of DNAPL depletion above the water table are promoted by high temperature, high vapor pressure and low soil moisture content. Once DNAPL has depleted itself above the water table, vapors can persist for long periods of time and will continue to migrate laterally and vertically due to diffusion.

3.2.5 Hydrophobic Dye Testing (Line of Evidence F)

The use of hydrophobic dyes such as Oil-Red-O and Sudan IV can be very useful in identifying DNAPL presence in soil core and soil samples. These dyes partition into an organic fluid such as a chlorinated solvent DNAPL but do not partition into water or onto soil. Even dark brown and black DNAPLs appear red in color following contact with the dye. Hydrophobic dye testing is particularly useful in identifying colorless DNAPL, DNAPL at low saturation, and DNAPL that has a color similar to the soil. Hydrophobic dye testing is typically carried out on sections of soil core exhibiting an abrupt increase in OVM readings, visible staining and any visible sign of DNAPL presence.

Hydrophobic dye testing can be carried out by filling a small glass sample vial one-third full with the soil being tested. Water is then added to the vial to the two-thirds full mark, and a small amount of the powdered dye is added. The vial is capped and shaken for several seconds. If DNAPL is present in the soil, red liquid in the form of either a film or droplets will appear against the glass in the upper one-third of the vial. If large amounts of DNAPL are present, red liquid may also appear within the soil in the bottom one-third of the vial.

Figure 3.5a illustrates hydrophobic dye standards that were prepared at the Solvents Recovery Service of New England (SRSNE) Superfund Site in Southington, Connecticut, USA (BBL, 2003). These standards were prepared using soil and DNAPL obtained from the site so that field personnel could distinguish residual from pooled DNAPL, and a detection limit could be established. The standards were prepared using DNAPL saturations ranging from 0 to 30% of pore space. Saturations of 2% of pore space or less were difficult to detect, demonstrating that the method is subject to false negatives at low DNAPL saturations. As seen in Figure 3.5a, the sample corresponding to pooled DNAPL (30% of pore space) exhibits a red DNAPL film on the inside surface of the upper part of the vial as well as a layer of red DNAPL at the very bottom of the vial. Excess dye appears as black particles in some vials. Figure 3.5b shows positive results for a sample from the same field site.

In addition to the hydrophobic dye jar shake test described above, paper impregnated with powdered hydrophobic dye can be obtained and used to assess DNAPL presence in soil. In this technique, the soil sample is pressed onto the paper and subsequently removed. If sufficient DNAPL was present in the soil, a visible reaction with the dye-impregnated paper will be apparent. Other variations include the use of downhole flexible hydrophobic membranes (USEPA, 2004), sample vials that can be purchased with hydrophobic dye present inside, and hydrophobic dye that can be sprayed directly onto a soil core.

3.2.6 Groundwater Concentrations (Line of Evidence G)

It is customary to sample groundwater on a routine basis and to evaluate concentrations as a means of inferring DNAPL presence. Groundwater concentrations can be evaluated a number of ways, and it is typical to perform the evaluation after each round of sampling.

3.2.6.1 Concentrations Above 1% Effective Solubility in Groundwater

The 1% rule of thumb has been in place since the early 1990s when investigators observed that chlorinated solvent concentrations in groundwater at DNAPL sites are often well below textbook solubility values. Prior to that time, some investigators concluded that DNAPL was absent unless concentrations in groundwater were near solubility values. The 1% rule of thumb states that sampled groundwater concentrations in excess of 1% effective solubility indicate that the sampled groundwater may have come in contact with DNAPL at some point along its

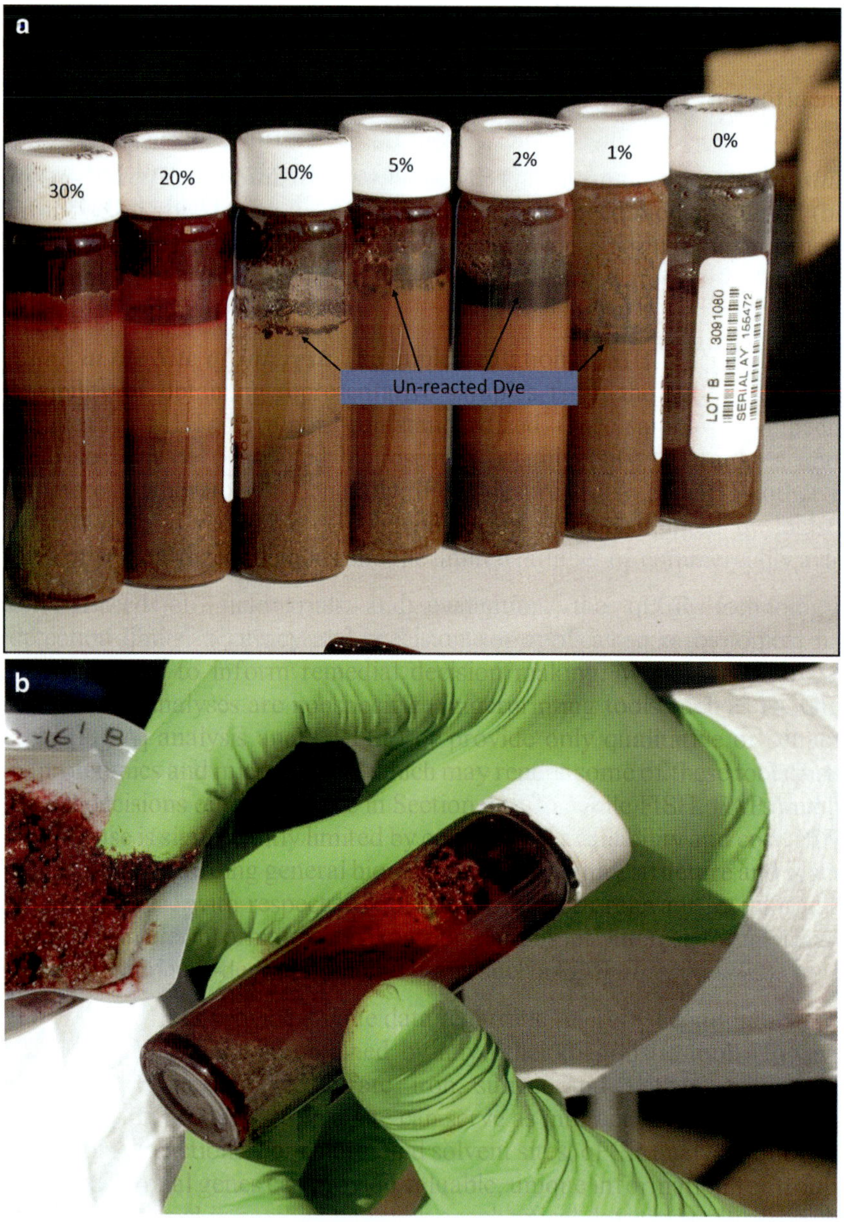

Figure 3.5. Hydrophobic dye testing using Sudan IV at the SRSNE Superfund Site, Southington, Connecticut. (a) Hydrophobic dye testing standards prepared using DNAPL and soil from the site. The *percent number* shown on the cap of the vial indicates the DNAPL saturation as a percentage of pore space. (b) Positive hydrophobic dye testing results at SRSNE in both a glass sample vial and a plastic dish (*left side* of the figure).

flow path. This rule of thumb does not imply that DNAPL is necessarily present at the monitoring location where 1% effective solubility is exceeded, and thus it cannot be assumed that the spatial extent of concentrations in groundwater exceeding 1% effective solubility is coincident with the spatial extent of DNAPL occurrence.

Concentrations above 1% effective solubility in groundwater can occur inside the DNAPL source zone, outside of the DNAPL source zone, at sites with totally depleted DNAPL and at sites where DNAPL was never present below the water table. Many investigators argue that the 1% effective solubility rule of thumb is no longer needed, while others argue that it serves as a useful trigger for gathering other types of data to evaluate DNAPL presence. In some cases, other percentages of effective solubility are taken as the threshold value to indicate that sampled groundwater may have come in contact with DNAPL at some point along its flow path.

The reasons that sampled groundwater concentrations downgradient of a DNAPL source zone can be much less than the effective solubility include the dilution effects of hydrodynamic dispersion, possible dilution effects associated with the groundwater sampling technique, nonoptimal monitoring well or sample collection location, and degradation processes. Because it can be difficult to accurately quantify some of these effects, it is difficult to determine the distance between the DNAPL and the monitoring location of interest.

The effective solubility of a compound in groundwater is calculated as

$$C_i = m_i S_i \quad \text{(Eq. 3.4)}$$

where C_i is the effective solubility, m_i is the mole fraction of component i in the DNAPL, and S_i is the single-component (textbook) solubility of component i. When evaluating a data set where it is thought that significant degradation of the parent compound is occurring, the daughter product concentrations can be converted to equivalent parent product concentrations and then added to the parent product concentration.

If the composition of the DNAPL is not known, it will not be possible to calculate the mole fractions required in Equation 3.4. In such cases, the following equation can be used:

$$\sum_{i=1}^{n} \frac{C_i^{obs}}{S_i} = \alpha \quad \text{(Eq. 3.5)}$$

where C_i^{obs} is the sampled groundwater concentration (mg/L) of component i, S_i is the single-component solubility of component i, α is the cumulative mole fraction of the sample and n is the number of components in the groundwater sample. Equation 3.5 assumes that the degree of borehole dilution, dispersion, and degradation are identical for each component of interest. To set Equation 3.5 equivalent to the 1% rule of thumb, α would need to exceed 0.01. Similarly, α would need to exceed 0.1 in order for Equation 3.5 to be equivalent to exceeding 10% effective solubility. The use of Equation 3.5 can be prone to false negatives if the groundwater sample was not analyzed for some of the components present in the DNAPL.

3.2.6.2 Persistent Plume

The presence of a persistent plume extending from a suspected DNAPL release location is consistent with DNAPL presence. Given an estimate of groundwater velocity and accounting for sorption, the rate of advance of an aqueous-phase plume not sustained by DNAPL dissolution can be calculated. If site monitoring data indicate that the plume has detached from the suspected release location and is migrating at the calculated rate, then DNAPL is likely not present at the suspected release location. On the other hand, if the upgradient extent of the plume remains stationary, particularly in moderate- to high-permeability media, then it can be concluded that DNAPL may be present. However, care must be taken to rule out a plume being sustained by back diffusion from low-permeability media when concluding that DNAPL is present. This caveat is particularly relevant in fractured sedimentary rock where the majority of contaminant mass may be present in the rock matrix, not the open fractures.

Related to the presence of a persistent plume sustained by DNAPL dissolution is the persistent presence of low molecular weight alcohols such as ethanol and methanol in groundwater. Low molecular weight alcohols, having entered the subsurface not in association with a DNAPL, would likely migrate quickly (low sorption) and biodegrade readily. The continued presence of low molecular weight alcohols at a monitoring location, however, could be attributable to the continued partitioning of the alcohol from a DNAPL to groundwater. This behavior has been observed at sites where chlorinated solvents, such as TCE and PCE, have been disposed of with alcohols.

3.2.6.3 Presence of Contamination in Anomalous Locations

Aqueous-phase contamination in areas not directly downgradient of known release locations can be consistent with DNAPL presence. A simple example would be the presence of dissolved-phase impacts at depth in a system experiencing upward groundwater flow. In this case, the dissolved-phase impacts may be attributable to the presence of DNAPL beneath the monitoring point in question. A second example would be the presence of dissolved-phase impacts at depth in a system experiencing lateral groundwater flow. Since vertical transverse dispersion is typically a weak mixing mechanism, the presence of contamination at depth in a horizontal flow system is difficult to explain without the presence of a source (DNAPL) at depth. A third example is the presence of dissolved-phase impacts hydraulically upgradient or significantly side gradient of a known DNAPL release location. In these cases, the dissolved-phase impacts may be attributable to the dissolution of DNAPL that has migrated laterally according to geologic structure.

3.2.6.4 Groundwater Concentration Trends with Depth

Abrupt reversals in dissolved-phase concentrations in groundwater are consistent with DNAPL presence. In this case, the spatial distribution of dissolved-phase concentrations is a reflection of the spatial distribution of upgradient DNAPL occurrences. Care must be taken to distinguish between abrupt reversals in dissolved-phase concentrations attributable to DNAPL presence from those attributable to solute transport in highly heterogeneous porous media. Vertical profiling of groundwater concentrations can be performed using small-scale sampling devices and push techniques such as a MIP in unconsolidated material.

3.2.6.5 Groundwater Concentration Trends with Time

Groundwater data characterized by steadily decreasing concentrations of high-solubility compounds and steadily increasing or stable concentrations of low-solubility compounds can be an indication of the presence of a multicomponent DNAPL. As DNAPL dissolution takes place, the DNAPL becomes depleted in the higher solubility compounds and enriched in the lower solubility compounds (referred to as weathering). The dissolution of multicomponent DNAPLs and the implications for source zone natural attenuation are discussed in Chapter 13.

At sites where a single-component DNAPL is present, steadily decreasing concentrations are not necessarily indicative of a lack of DNAPL presence. Concentrations downgradient of a DNAPL source zone will decline as DNAPL mass is depleted. Unfortunately, it is difficult to identify the point in time at which DNAPL has been completely depleted (Stage 4; Chapter 1) on the basis of dissolved-phase concentrations alone at sites where desorption and back diffusion are significant processes sustaining plume presence.

3.3 ASSESSING THE PRESENCE OF DIFFUSED MASS

At sites having low-permeability media such as silt, clay and fractured sedimentary rock, substantial contaminant mass may have diffused into these media over time. Determining the amount of mass present in low-permeability media and associated concentrations is often an important part of developing a site conceptual model and selecting a remediation technology.

In porous media, the primary means of assessing contamination present in low-permeability units is through the use of traditional soil sampling. Laboratory analysis of soil samples provides a quantitative measure of concentration, from which mass can be calculated (see Section 3.5). An alternative to traditional soil sampling is the use of a MIP, which provides a relative measure of concentration, not a direct quantitative measurement. The advantage of a MIP over traditional soil sampling, however, is that a continuous profile of relative concentration can be obtained in a relatively quick manner without waiting for laboratory turnaround. A MIP is often deployed on the same string as a cone penetrometer. Used in this fashion, both geology and relative concentration can be assessed at the same location. At many sites, MIP data are obtained early in an investigation and are relied upon to target specific locations and depth intervals for subsequent traditional soil sampling. If pooled DNAPL is present, MIP technology should be used with caution to avoid leaving vertical pathways through capillary barriers.

Figure 3.6 illustrates a MIP profile obtained at a site impacted by chlorinated solvents. The profile on the left represents geology from cone penetrometer results. The profile on the right represents the relative concentration (volts) of chlorinated volatile organic compounds (CVOCs). The highest concentrations of CVOCs are seen to be present in a 4 meter (m) (13 foot [ft]) thick low-permeability layer. The peak concentration occurs in the middle of the low-permeability layer, illustrating a concentration gradient resulting in back diffusion from the low-permeability layer into the overlying and underlying higher permeability layers. Chlorinated solvents are known to have been used at this site as early as the 1960s.

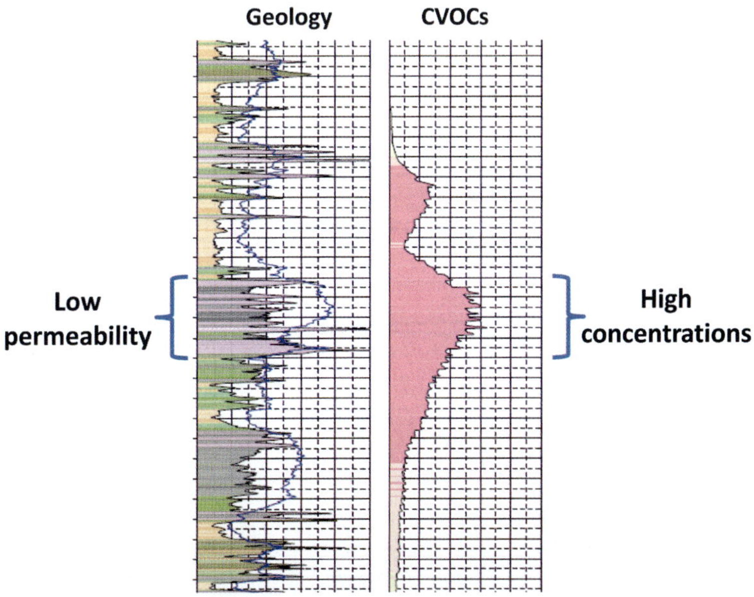

Figure 3.6. Geology and CVOC profiles with depth. Highest concentrations are present in low-permeability media. Entire profile is 27 m (88.6 ft) thick.

As discussed in Chapter 13, a profile of soil concentrations through a low-permeability layer can be fit with mathematical models to estimate the age of the contamination and the magnitude of diffusive flux. A profile of soil concentrations can provide conclusive evidence of whether back diffusion is occurring from the low-permeability layer into the underlying and overlying layers (such as illustrated in Figure 3.6) or whether forward diffusion into the low-permeability layer is occurring.

In fractured sedimentary rock, the majority of the contaminant mass may be stored in the rock matrix compared to the open fractures. At sites where the DNAPL release occurred several decades ago, it may be possible that little DNAPL remains in open fractures and that the plume is being sustained primarily by back diffusion from the rock matrix. The presence or the absence of contaminant mass in the rock matrix can be assessed through laboratory measurements on closely spaced samples of rock core. The samples are typically crushed in the field and preserved in methanol prior to submission to the laboratory for quantitative analysis.

Figure 3.7 illustrates a profile of TCE concentrations in bedrock core from the former Naval Air Warfare Center (NAWC) site located in West Trenton, New Jersey (ESTCP, 2012). Trichloroethene was utilized at the site as a heat transfer fluid. Bedrock consists of steeply dipping mudstone units characterized by low matrix porosity and low matrix permeability. The concentration profile presented in Figure 3.7 indicates that forward diffusion of TCE is occurring into the rock matrix at this particular location at the site; the high concentration peaks are associated with the presence of fractures. The diffusion profiles between fractures are not fully developed, with relatively low concentrations of TCE within many depth intervals.

The relatively undeveloped matrix diffusion profile from the sampling at the NAWC site and illustrated in Figure 3.7 can be explained by the low matrix permeability and moderate-to-high rock matrix fraction organic carbon. Figure 3.8 illustrates the pore throat distribution for a sample of core collected within a few meters of the core for which TCE concentration data are illustrated in Figure 3.7. The maximum pore throat radius is 0.012 microns (μ), consistent with

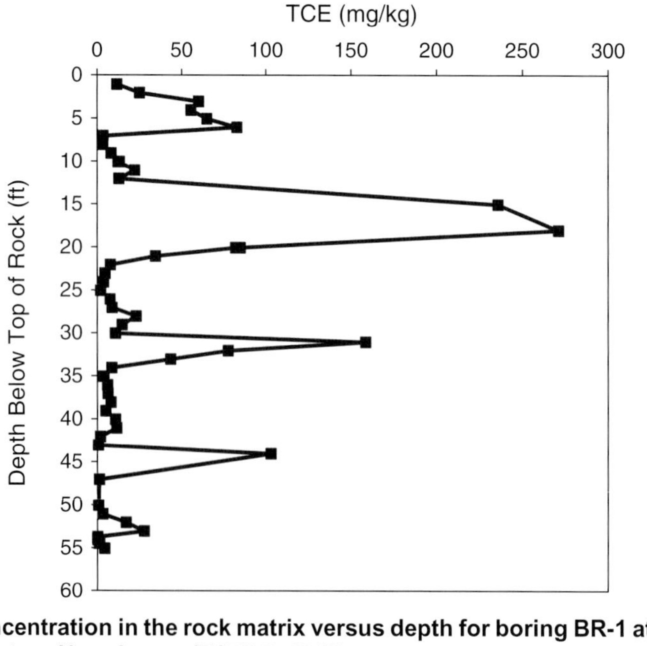

Figure 3.7. TCE concentration in the rock matrix versus depth for boring BR-1 at the former NAWC located in West Trenton, New Jersey (ESTCP, 2012).

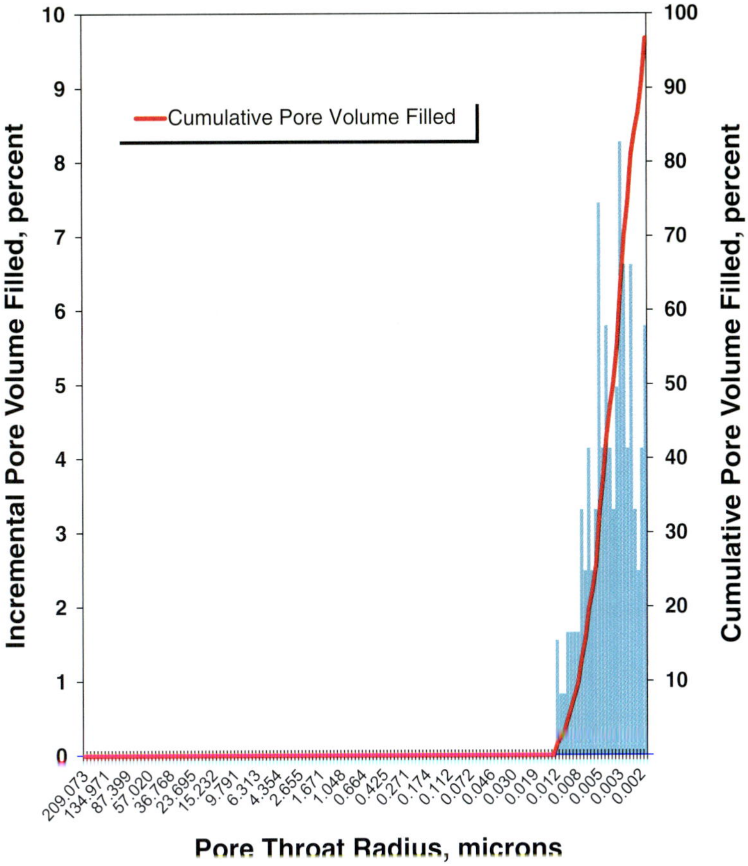

Figure 3.8. Pore throat radius distribution for mudstone sample obtained 30 ft below ground surface at the former NAWC site located in West Trenton, New Jersey (ESTCP, 2012). Mercury porosimetry test method ASTM D4404, API RP40. Total matrix porosity is 3.1%.

the low matrix permeability of the rock (measured on the order of 10^{-14} to 10^{-13} square centimeter [cm^2], which is equivalent to hydraulic conductivity between 10^{-9} and 10^{-8} centimeter per second [cm/s]). The measured rock matrix fraction organic carbon ranges from 0.36 to 1.2%. Chapter 14 discusses the use of thermal conductive heating to remove TCE from the rock matrix at the NAWC site.

3.4 SOURCE ZONE DELINEATION

Following assessment of DNAPL presence, it is customary to delineate the spatial extent of the source zone using the DNAPL lines of evidence (DNAPL indicators) discussed in Section 3.2. If it has been determined that DNAPL is present at a site, delineation of the source zone is synonymous with delineation of the spatial extent of DNAPL.

When delineating the spatial extent of DNAPL, it is recommended that the various DNAPL indicators be plotted in plan view and in cross section. To address uncertainty, both a confirmed and a potential DNAPL source zone can be delineated. The confirmed DNAPL source zone is that volume of the subsurface within which compelling and multiple lines of evidence indicate that DNAPL is present. It is possible, however, to have negative lines of evidence within the confirmed DNAPL source zone because of the complex and irregular nature of DNAPL distribution in the subsurface. The potential DNAPL source zone is always larger than the confirmed DNAPL source zone and is defined as that volume of the subsurface within which some lines of evidence indicate that DNAPL is present, but the lines of evidence are not as numerous, consistent, or compelling as those within the confirmed DNAPL source zone. Some positive lines of evidence may fall outside of both the confirmed and potential DNAPL source zones. Exceeding 1% effective solubility in groundwater or marginally exceeding the soil partitioning threshold, for example, can occur within the plume downgradient of the DNAPL.

The accuracy to which a DNAPL source zone can be delineated depends on the quantity and types of data that have been collected as well as the experience of the investigator. The amount of acceptable uncertainty in delineating the source zone will vary from site to site depending on a number of factors including the available budget and the remediation objectives. The decision to employ a remediation technology and/or strategy specifically targeting DNAPL removal may not be justified at a poorly characterized site.

One site at which both confirmed and potential DNAPL source zones were delineated is the SRSNE Superfund Site in Southington, Connecticut. Figure 3.9 presents a plan view depicting both the confirmed (referred to as probable in the figure) and potential DNAPL source zones in overburden. The symbols plotted correspond to various DNAPL indicators. The primary reason for extending the potential source zone east of the Quinnipiac River was the presence of site-related constituents in groundwater on the other side of the flow divide beneath the river (presence of contamination in anomalous locations). One possible explanation for the presence

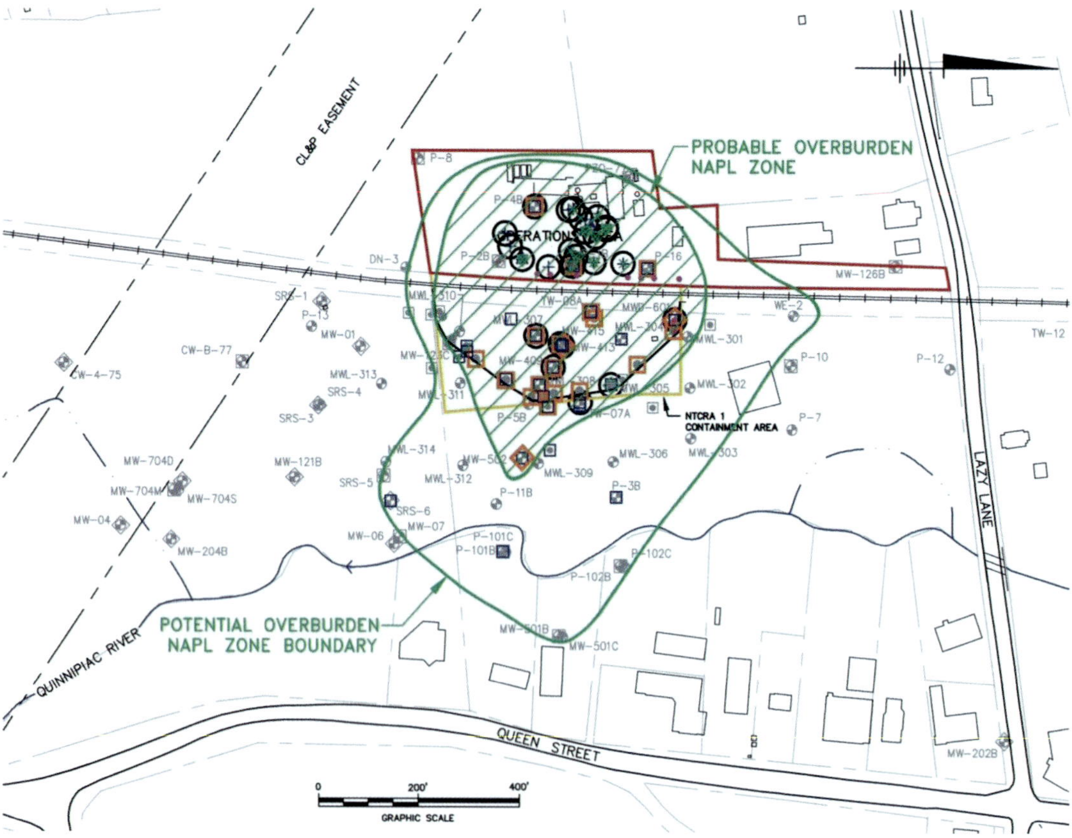

Figure 3.9. Probable and potential DNAPL source zones in overburden at the SRSNE Site, Southington, Connecticut (BBL, 1998).

of these constituents is the migration of DNAPLs beneath the river to the east. A second possible explanation, not dependent on the presence of DNAPL in overburden east of the river, is the upward discharge of impacted bedrock groundwater in response to the upward component of the hydraulic gradient that is known to exist in that area.

3.5 MASS ESTIMATES

Estimating the amount of contaminant mass present at a site is an important part of the site conceptual model development and the remediation technology selection and design process. In the general case, contaminant mass can be present as vapor phase, sorbed phase, aqueous phase and in the form of DNAPL. In addition to estimating how much mass is present in each of the four forms listed above, it can be important to understand the distribution of each of the four forms of mass above the water table, below the water table, in low-permeability media and in relatively higher permeability media.

The potential effectiveness of a remediation technology depends on the form, amount and/or location of contaminant mass. For example, the effectiveness of reaction-based technologies, such as *in situ* chemical oxidation, *in situ* chemical reduction, surfactant flushing and *in situ* bioremediation, depends on the amount of contaminant mass because the amount of reactant/substrate to inject must be stoichiometrically balanced against the contaminant mass. These technologies are also very sensitive to delivery constraints such as permeability and fracture continuity as the reactant/substrate is typically delivered to the contaminant via injection wells.

3.5.1 Calculation Procedures for Estimating Mass

Although the following is given as an example of how to estimate mass in the various phases within soil and the aquifer for the consideration and design of the remediation, adequate site characterization and monitoring of the remedial technology during and after implementation are critical to the proper evaluation of the effectiveness of the technology.

The amount of aqueous-phase mass (M_A) present in a source zone is given by

$$M_A = \int c_w \theta_w dV \qquad \text{(Eq. 3.6)}$$

where c_W is the aqueous-phase concentration of the compound of interest, θ_w is the water-filled porosity and V is the overall volume of the source zone. In practice, the integration in Equation 3.6 can be carried out in either a simplistic fashion using spreadsheet approximations or a more sophisticated manner using commercial software based on kriging or other types of estimation and interpolation techniques. In fractured media, the calculation should be carried out separately for fractures and the matrix.

The amount of sorbed-phase mass (M_S) present in a source zone is given by

$$M_S = \int c_w K_d \rho_b dV \qquad \text{(Eq. 3.7)}$$

where K_d is the soil—water distribution coefficient and ρ_b is the dry soil bulk density. The soil—water distribution coefficient is typically estimated as $K_d = K_{oc} f_{oc}$ where K_{oc} is the organic carbon—water partitioning coefficient and f_{oc} is the soil fraction organic carbon. In some cases, a laboratory-measured sorption isotherm using soil and groundwater from the site is used as the basis for arriving at K_d. Equation 3.7 is cast in terms of the aqueous-phase concentration, c_w. In the event that sufficient soil data are available, the $c_w K_d$ grouping of terms can be replaced by the actual soil concentrations. As with Equation 3.6, the integration in Equation 3.7 can be carried out with relatively simple spreadsheet methods or with more sophisticated, commercially available software. For fractured media, the calculation should be performed separately for fractures and the matrix.

The amount of vapor-phase mass (M_v) present in a source zone is given by

$$M_V = \int c_v \theta_a dV \qquad \text{(Eq. 3.8)}$$

where c_v is the vapor-phase concentration and θ_a is the air-filled porosity. As with the Equations 3.6 and 3.7, the integration in Equation 3.8 can be achieved with relatively simple spreadsheet methods or with more sophisticated, commercially available software. For fractured media, the calculation should be done separately for the mass sorbed to fracture walls and the mass sorbed within the matrix.

The amount of DNAPL mass (M_N) present in a source zone is given by

$$M_N = \int \theta S_n \rho_n dV \qquad \text{(Eq. 3.9)}$$

where θ is the total porosity, S_n is the DNAPL saturation and ρ_n is the DNAPL density. In porous media, the values of S_n for use in Equation 3.9 can be calculated on the basis of soil concentrations (Equation 3.1).

A practical alternative to the use of Equation 3.9 is to assign a DNAPL bulk retention capacity to the source zone. The bulk retention capacity is defined as the total volume of DNAPL that has been retained as residual and pooled DNAPL in a unit volume of the subsurface (USEPA, 2009). The bulk retention capacity accounts for the presence of residual and pooled DNAPL in certain lenses, laminations, and fractures as well as for the absence of DNAPL in other lenses, laminations, and fractures. In unconsolidated deposits, the bulk retention capacity may range from 0.005 to 0.032 (Kueper et al., 2003). The precise value is site specific and depends on a number of factors including geologic properties, DNAPL properties, and volume and rate of the DNAPL release. In fractured rock, the bulk retention capacity will be much lower than in unconsolidated deposits because the DNAPL is typically restricted to open fractures, which account for a very small portion of the overall rock mass. In fractured media, the bulk retention capacity can be in the range of 0.0002—0.002 (Kueper et al., 2003). If a bulk retention capacity approach is used, the volume of DNAPL can be converted to a mass of DNAPL knowing the DNAPL density.

***Example Calculation – Porous Media*:** Consider a porous media DNAPL source zone where the DNAPL of interest is TCE with a solubility of 1,100 mg/L, a density of 1,460 kg/m^3, and an organic carbon—water partition coefficient of 126 mL/g. A 3-m (10-ft)-thick unsaturated zone is present having an average water-filled porosity of 0.20 and an average air-filled porosity of 0.20. Trichloroethene impacts occur from ground surface to a depth of 20 m (approximately 66 ft). The geology consists of horizontally bedded fine- to medium-grained sands having an average dry soil fraction organic carbon content of 0.002, a porosity of 0.40, and a dry soil bulk density of 1.63 g/cm^3. The source zone is 2,000 m^2 (0.5 acre) in area, which results in a total volume of 40,000 m^3 (52,000 cubic yard [cy]), with 6,000 m^3 (approximately 8,000 cy) above the water table and 34,000 m^3 (approximately 44,000 cy) below the water table. The volume-averaged vapor-phase concentration is 47,100 micrograms per cubic meter (μg/m^3), the volume-averaged aqueous-phase concentration is 3 mg/L, and the DNAPL bulk retention capacity is 0.001. Employing Equations 3.6 through 3.8, the masses of aqueous-phase, sorbed-phase, and vapor-phase TCE in the unsaturated zone are 3.6 kg (7.9 pounds [lb]), 7.4 kg (16.3 lb), and

0.06 kg (0.13 lb), respectively. The masses of aqueous-phase and sorbed-phase TCE in the saturated zone are 40.8 kg (89.9 lb) and 41.8 kg (92.2 lb), respectively.

The ratio of sorbed-phase to aqueous-phase TCE mass is greater in the unsaturated zone than the saturated zone because of the difference in water-filled porosity. For fully saturated media, the fraction of mass dissolved in groundwater compared to that sorbed is given by $1/R$ where R is the retardation factor defined as

$$R = 1 + \frac{\rho_b}{\theta} K_d \qquad \text{(Eq. 3.10)}$$

For this example, $R = 2.02$ which is consistent with the nearly identical mass of TCE in the aqueous and sorbed phases in the saturated zone. Using a DNAPL bulk retention capacity of 0.001, the masses of DNAPL in the unsaturated and saturated zones are 8,760 kg (29 drums) and 49,640 kg (163 drums), respectively.

Example Calculation – Fractured Rock: Consider a TCE source zone below the water table in fractured sandstone characterized by a matrix porosity of 0.05 and a fracture porosity of 6E-4. The fracture porosity corresponds to three orthogonal sets of 200-μ aperture fractures spaced 1 m (3 ft) apart. The matrix is characterized by a fraction organic carbon of 0.002 and a dry bulk density of 2.57 g/cm^3. The source zone is 2,000 m^2 (0.5 acre) in area and is 20 m (65.6 ft) thick, which results in a total volume of 40,000 m^3 (same volume as in the above porous media example). It is assumed that all DNAPL has been depleted from the fractures (Stage 5 source zone; Chapter 1). The volume-averaged concentration of TCE in the aqueous phase is 3 mg/L throughout both the fractures and the matrix. Using the above calculation procedures, the mass of aqueous-phase TCE in the open fractures and the rock matrix is 0.07 and 6 kg, respectively. The mass of sorbed-phase TCE in the open fractures is assumed to be zero. The mass of sorbed-phase TCE in the matrix is calculated to be 78 kg. In this example, the majority of the estimated contaminant mass exists in the rock matrix primarily as a result of sorption.

3.6 SUMMARY

Proper site conceptual model development, remediation technology selection and remediation design depend highly on assessing whether DNAPL is present in a source zone and, if so, how much contaminant mass is present as DNAPL and in other phases (aqueous, sorbed, and vapor). This chapter presents a converging-lines-of-evidence approach to assessing DNAPL presence in both porous and fractured media. The various lines of evidence include visual observation, exceeding saturation and partitioning thresholds for soil concentrations, use of hydrophobic dye testing, site use and site history, and various aspects of groundwater quality data. Delineation of the spatial extent of DNAPL source zones can be performed by plotting the various lines of evidence (DNAPL indicators) on plan view and cross-section figures. This approach facilitates the delineation of both confirmed and potential DNAPL source zones. At sites where DNAPL has been depleted as a result of either natural processes or remediation efforts, the approach facilitates delineating areas where other processes, such as back diffusion and desorption, may be sustaining plume migration.

The successful implementation of many remediation technologies very much depends on accurate estimates of contaminant mass, both total mass and mass in different phases and locations. Mass estimates can be arrived at through integration of subsurface data using either simple spreadsheet methods or more sophisticated numerical algorithms such as kriging. In practice, there is uncertainty in mass estimates, given the limitations of the tools and resources available as well as the challenges inherent in locating DNAPLs within the subsurface.

ACKNOWLEDGEMENTS

Portions of this chapter were originally published in a peer-reviewed U.S. Environmental Protection Agency (USEPA) document. However, the views expressed in this chapter are those of the coauthors and do not necessarily reflect the official views of the USEPA. Mention of trade names does not reflect endorsement by the USEPA.

REFERENCES

BBL (Blasland, Bouck & Lee). 1998. Remedial Investigation Report, Solvents Recovery Service of New England, Superfund Site, Southington, Connecticut. U.S. Environmental Protection Agency (USEPA) Region 1, Boston, MA, USA.

BBL. 2003. NAPL Delineation Pilot Study, SRSNE Site, Southington, Connecticut. Technical Memorandum. USEPA Region 1, Boston, MA, USA.

ESTCP (Environmental Security Technology Certification Program). 2012. Dense Non Aqueous Phase Liquid (DNAPL) Removal from Fractured Rock Using Thermal Conductive Heating (TCH). Project ER200715, Final Report. Prepared by Lebrón CA, Phelan D, Heron G, LaChance J, Nielsen SG, Kueper B, Rodriguez D, Wemp A, Baston D, Lacombe P, Chapelle FH for ESTCP, Alexandria, VA, USA. http://www.serdp.org/. Accessed January 23, 2014.

Feenstra S, Mackay DM, Cherry JA. 1991. Presence of residual NAPL based on organic chemical concentrations in soil samples. Ground Water Monit Remediat 11:128—136.

Gerhard JI, Pang T, Kueper BH. 2007. Time scales of DNAPL migration in sandy aquifers examined via numerical simulation. Ground Water 45:147—157.

Kueper BH. 2005. Photograph Courtesy of B.H. Kueper, Queen's University, Kingston, Ontario, Canada.

Kueper BH. 2011. Photograph Courtesy of B.H. Kueper, Queen's University, Kingston, Ontario, Canada.

Kueper BH, Redman D, Starr RC, Reitsma S, Mah M. 1993. A field experiment to study the behavior of tetrachloroethylene below the watertable: Spatial distribution of residual and pooled DNAPL. Ground Water 31:756—766.

Kueper BH, Wealthall GP, Smith JWN, Leharne SA, Lerner DN. 2003. An Illustrated Handbook of DNAPL Transport and Fate in the Subsurface. Prepared for the United Kingdom Environment Agency. R&D Publication 133. ISBN 1844320669.

Mercer JW, Cohen RM, Noel MR. 2010. DNAPL Site Characterization Issues at Chlorinated Solvent Sites. In Stroo HF, Ward CH, eds, *In Situ* Remediation of Chlorinated Solvent Plumes. Springer, New York, NY, USA, pp 217—280.

Reynolds DA, Kueper BH. 2004. Multiphase flow and transport through fractured heterogeneous porous media. J Contam Hydrol 71:89—110.

Richards A, Gerhard JI, Kueper BH. 2012. Hydraulic displacement of dense, nonaqueous phase liquids. Ground Water 50:765—774.

USEPA (U.S. Environmental Protection Agency). 2004. Site Characterization Technologies for DNAPL Investigations. EPA/542/R-04/017. USEPA Office of Superfund Remediation and Technology Innovation (OSRTI), Washington, DC, USA. 165 p. http://www.clu-in.org/download/char/542r04017.pdf. Accessed January 23, 2014.

USEPA. 2009. Assessment and Delineation of DNAPL Source Zones at Hazardous Waste Sites. EPA/600/R-09/119. USEPA National Risk Management Research Laboratory, Cincinnati, OH, USA. 20 p.

CHAPTER 4

ADVANCED DIAGNOSTIC TOOLS

Rula A. Deeb,[1] Jennifer L. Nyman,[1] Elisabeth L. Hawley,[2] Michael C. Kavanaugh[1] and Robert H. O'Laskey[2]

[1]Geosyntec, Oakland, CA 94607, USA; [2]ARCADIS, Emeryville, CA 94608, USA

4.1 INTRODUCTION

4.1.1 What Are Diagnostic Tools?

Despite more than 40 years of experience, groundwater industry professionals worldwide still face many technical, economic and regulatory challenges for cleanup of contaminated soil and groundwater. In the past decade, the advent of innovative diagnostic tools has improved site assessment and remediation practices and has piqued interest as a basis for making decisions about remediation strategies, site management and site closure. Diagnostic tools hold promise for improving and optimizing remedial efforts with respect to duration and spatial extent.

Generally speaking, diagnostic tools utilize a variety of advances in microbiology, chemistry, geophysics and other disciplines to obtain a more accurate evaluation of complex environmental processes. These include both field and laboratory measurement techniques. Diagnostic tools typically enable high-resolution mapping of contaminant distribution or *in situ* processes affecting contaminant fate and transport. Diagnostic tools can be used with a wide range of technologies, including pump-and-treat technologies and *in situ* technologies used for source and plume treatment (USEPA, 2003; NRC, 2005; ITRC, 2004; Kavanaugh and Kresic, 2008). Regardless of the type of treatment (chemical oxidation and reduction technologies, biologically mediated oxidation or reduction technologies, surfactant flooding, alcohol flooding, hydraulic displacement, multiphase extraction and thermal technologies), the appropriate application of diagnostic tools can lead to significant improvements and drive greater efficiencies during remedy operation to achieve remedial action objectives (RAOs).

4.1.2 Why Are Diagnostic Tools Needed?

Historically, site characterization and performance assessment have relied on two-dimensional measurements and representations of dissolved contaminant concentrations obtained from monitoring wells in aquifers, such as changes in maximum concentrations and plume extent over time. Such an approach often has limitations that influence remedial decision making and impact the evaluation of technology effectiveness at these sites. For example, locating and delineating dense nonaqueous phase liquids (DNAPLs) at chlorinated solvent sites represents a difficult remediation challenge, as has been well documented by the USEPA (2003), NRC (2005), and ITRC (2004). Diagnostic tools provide distinct advantages over conventional tools for site characterization and remedy performance monitoring.

In situ technologies are often applied in complex and highly heterogeneous subsurface environments. This heterogeneity complicates monitoring strategies because of the potential for inaccurate results if the monitoring systems are placed in locations that do not reflect the

B.H. Kueper et al. (eds.), *Chlorinated Solvent Source Zone Remediation*,
doi: 10.1007/978-1-4614-6922-3_4, © Springer Science+Business Media New York 2014

controlling processes in the subsurface. It has been established that in both unconsolidated and consolidated media, preferential flow paths are the norm and not the exception (Payne et al., 2008; Sale et al., 2008). In addition to hydrogeologic complexities in the subsurface, monitoring *in situ* remedies is complicated further by the diversity of microbial populations in the subsurface, spatial variability of geochemical properties (pH, oxidation-reduction potential, temperature, chemical speciation), and complex distribution of any NAPL (NAPL architecture). This complexity results in the need to develop diagnostic tools that address site-specific features.

During site characterization, the development of an accurate conceptual site model (CSM) depends on identifying the major controlling features of the site such as the preferential flow paths, presence and extent of NAPLs, major continuing sources of contaminant release to the groundwater, and likely receptors that could be impacted. The use of diagnostic tools that provide a more accurate and detailed CSM can encourage the selection and better design of *in situ* remedies.

Some diagnostic tools, such as molecular biological tools, can provide confirmation of *in situ* processes that result in transformation of the chemicals of concern to nontoxic by-products and provide estimates of the rate of transformation. These tools can be used to help resolve key issues before making critical site strategic decisions including the efficacy of monitored natural attenuation (MNA) and the ability of remedial strategies to target high-concentration zones including preferential contaminant migration pathways or DNAPL source zones. Acceptance of MNA by site stakeholders often depends on using several lines of evidence to confirm that risks of further migration and receptor impacts are acceptably low (USEPA, 1999). Demonstrating that the chemicals of concern are being depleted by biotic or abiotic processes is essential for remedy acceptance. Thus, the development of diagnostic tools that provide reliable, accurate, and transparent data is essential for reducing the time and cost of achieving either site closure or long-term management.

4.1.3 Overview of Tools Discussed in This Chapter

This chapter summarizes the attributes and utility of five innovative diagnostic tools that are important for source zone management because they are capable of evaluating the following site characterization and remedial performance assessment issues:

1. Delineation of the vertical distribution of contaminants in the dissolved and adsorbed phase through the use of multilevel monitoring systems (MLM systems)
2. Assessment of chlorinated solvent distribution and performance of *in situ* technologies in consolidated media with rock crushing and other rock matrix characterization techniques
3. Use of mass flux/mass discharge measurement technologies for site characterization and process and performance assessment
4. Confirmation of *in situ* chemical or biological transformations of chemicals of concern via compound-specific isotope analysis (CSIA)
5. Use of molecular biological tools (MBTs) to improve *in situ* bioremediation

This chapter is based on lessons learned from field demonstration of these five diagnostic tools at three U.S. Department of Defense (DoD) sites: Watervliet Arsenal (New York), Vandenberg Air Force Base (California), and Fort Lewis Logistics Center (Washington). The field demonstrations and related reports were completed under Environmental Security Technology Certification Program (ESTCP) Project ER-200318 (Malcolm Pirnie, Inc., 2011a, b). These five diagnostic tools are a subset of many tools currently in use or being developed that may provide unique information for remedy selection, optimization or performance assessment.

4.1.4 Utility of Diagnostic Tools for Optimizing Remedial Strategies

The use of innovative diagnostic tools, such as those discussed in this report, may lead to increased costs of site investigation or monitoring activities. Cost analyses of each of the tools evaluated in this chapter are summarized in later sections. However the increased costs for investigation, process monitoring to improve remedial design or operation, or performance monitoring to assess progress toward achieving RAOs can result in many important outcomes. Examples include a better understanding of the CSM that leads to appropriate implementation of *in situ* remedies to achieve greater mass removal rates, shorter remedial time frames and lower overall life cycle costs. A decision framework applicable to the use of innovative diagnostic tools is discussed in the next section.

4.2 VALUE ADDED THROUGH APPLICATION OF DIAGNOSTIC TOOLS

4.2.1 Value of Information Analysis

One of the most significant challenges in the cleanup of contaminated sites is determining whether the degree of uncertainty in the values of relevant physical, chemical and/or biological parameters is small enough so that decision makers can be reasonably confident of making remediation and/or site closure decisions based on data. During most phases of the remedial process, decision making under uncertainty is a dominant theme. A rich and detailed body of literature exists on this topic, a review of which is beyond the scope of this chapter. Nonetheless, determining whether additional information is needed to enhance the quality of site decisions is a primary function of site managers. In simple terms, one must decide whether the value gained by using alternative diagnostic tools is sufficient to warrant their cost. This is a classic *Value Of Information* (VOI) problem within the context of decision making under conditions of uncertainty.

A VOI analysis is relevant to many societal issues other than subsurface hydrogeology and remediation. A review of the VOI literature (Yokota and Thompson, 2004) indicates that this perspective has been applied to a wide variety of disciplines including general medical care, clinical trials, general environmental health, toxicology and risk assessment. According to the authors of this review paper, the Value Of Information analysis "extends decision analysis to evaluate the benefits of collecting additional information to reduce or eliminate uncertainty in a specific decision-making context" (Yokota and Thompson, 2004).

Conceptually, VOI analysis is described in economic terms, requiring an analysis of the impact of additional information on the expected value of the decision. The decision maker(s) must compare the expected value of a decision made with the imperfect (uncertain) information at hand to the expected value of the decision with the new information to be gathered. Over the past few decades, numerous publications have attempted to develop algorithms that would provide a basis for conducting a quantitative VOI analysis regarding hydrogeologic data as it applies to risk-based remedial action decisions (Massman and Freeze, 1987; Reichard and Evans, 1989; Freeze et al., 1992; James and Freeze, 1993; Back et al., 2007; James and Gorelick, 1994; Russell and Rabideau, 2000; Cardiff et al., 2010). These algorithms are data intensive and usually require estimates of the probability distribution function for all parameters in the algorithm.

These types of analyses, while conceptually appealing, are limited in the context of site remediation because of the extent of uncertainty in multiple attributes of the problem.

As thought experiments, they provide value in defining which uncertainties are most critical to the decision process. However, for practical purposes, the decision to obtain additional site information to reduce uncertainties is usually based on professional judgment or essentially a qualitative VOI analysis.

4.2.2 Basis for Selection of Diagnostic Tools

A qualitative VOI analysis can be applied to selection of innovative diagnostic tools relying on various specific or relative attributes of the tools themselves and their applicability to a site-specific issue. Some of these attributes include the following:

- **Maturity of the Tool:** A diagnostic tool is similar to any new technology that must pass through a maturation process including proof of concept, field testing and finally commercialization. Some of the innovative tools discussed in this chapter are generally commercially available and have had varying degrees of field testing and evaluation but are not yet widely or routinely used.
- **Applicability to Site Characteristics:** Some diagnostic tools are only suitable for certain geologic site conditions. For example, rock matrix characterization is only appropriate for consolidated media.
- **Applicability to Specific *In Situ* Technology:** Certain diagnostic tools are only applicable to a specific technology. For example, molecular biological tools are only appropriate for *in situ* processes relying on microbial transformations, such as enhanced reductive dechlorination (ERD).
- **Implementation at the Site of Interest:** The ease of implementing a diagnostic tool at a particular site is also a relevant criterion for selection. Physical constraints at a site, such as aboveground structures, may limit the applicability of a given tool. Complex operating requirements and associated components of a tool may also limit its usefulness.
- **Data Quality of the Tool:** Sufficient field data should be available to determine if the diagnostic tool provides *detection limits* relevant to the chemicals of concern and that the reported values of the data produced through the use of the tool are of sufficient *accuracy* and *precision* to improve decision making.
- **Uniqueness of Data Gathered by the Tool:** If a diagnostic tool provides unique data that cannot be obtained using other methods, that tool has essentially a competitive advantage compared to other techniques. In this case, the value of the information must be considered in the context of the remedial process decision.
- **Cost Relative to Similar Methods:** A final criterion for selecting these tools is the relative cost of application of the diagnostic tool compared to alternative or conventional techniques or other tools that can provide equivalent information. This criterion is only applicable if there are competing methods for obtaining the same data. Detailed cost analyses are not provided in this chapter because the state of the practice is evolving rapidly but some current cost information for many of these tools is provided in ITRC (2011) and Malcolm Pirnie reports (2011a, b).

In the remainder of this chapter, the applicability of each of the five diagnostic tools to characterize and assess performance of *in situ* technologies at chlorinated solvent sites is evaluated qualitatively using the above criteria to assess the VOI obtained by their use.

4.3 DEPTH DISCRETE SAMPLING

Depth discrete sampling has the ability to characterize vertical variability in contaminant concentrations. This information can assist in evaluating the architecture of DNAPL source zones and allow for targeting high mass areas during remedial design. The optimization process can result in greater mass removal rates, shorter remedial time frames, and lower life cycle remediation costs. Diagnostic tools for depth discrete characterization include several multilevel groundwater monitoring systems and, for consolidated media, rock matrix characterization.

This section addresses tools for the collection of laboratory samples. It should be noted that diagnostic tools that interrogate the subsurface using surrogates for contaminant presence, such as membrane interface probes and laser-induced fluorescence probes, are also available and often can provide important information that complements the tools discussed below.

4.3.1 Multilevel Monitoring Systems

4.3.1.1 Description

MLM systems are designed to collect depth discrete samples or measurements over a single vertical profile of the subsurface, in contrast to conventional groundwater monitoring which generates data that represents a vertically averaged measurement from each well (ITRC, 2004; Einarson, 2006). MLM systems provide a better understanding of the vertical distribution of contaminants as well as the changes in concentration with depth and over time within the contaminant plume compared to conventional monitoring.

Figure 4.1 presents results from a multilevel investigation demonstrating the heterogeneity of PCE mass flux immediately downgradient of a source area in an unconfined sandy aquifer in Milford, New Hampshire, USA, studied by Guilbeault et al. (2005). The results reflect both multiple releases within the cross-sectional area as well as multiple DNAPL layers vertically through the profile. It should be readily apparent from this study that such information can greatly improve the design of source remediation systems.

The general-purpose term *MLM system* refers to an engineered assembly of various components installed in a single borehole to achieve the following:

- Obtain depth discrete measurements of water pressure (or hydraulic head)
- Acquire groundwater samples at specific depths for analysis
- Conduct tests to measure the hydraulic characteristics of the monitored interval
- Allow repeated sampling from the same location

For the purposes of this evaluation, an MLM system is defined as a single-cased (or *single-tube*) entity capable of monitoring at least two discrete intervals within a borehole. Several MLM systems fitting this definition are described in the literature, and some are listed in Table 4.1. Einarson (2006) provides an overview encompassing all types of MLM systems and well nests and clusters used in North America.

4.3.1.2 Comparative Costs

Comparisons of costs between MLM/nested well systems are complex and require specifications to narrow the monitoring purpose and scope. Cost components to consider when evaluating an MLM system include the following: purchase and shipment to the site; installation labor, equipment and materials; on-site vendor assistance; labor costs associated with

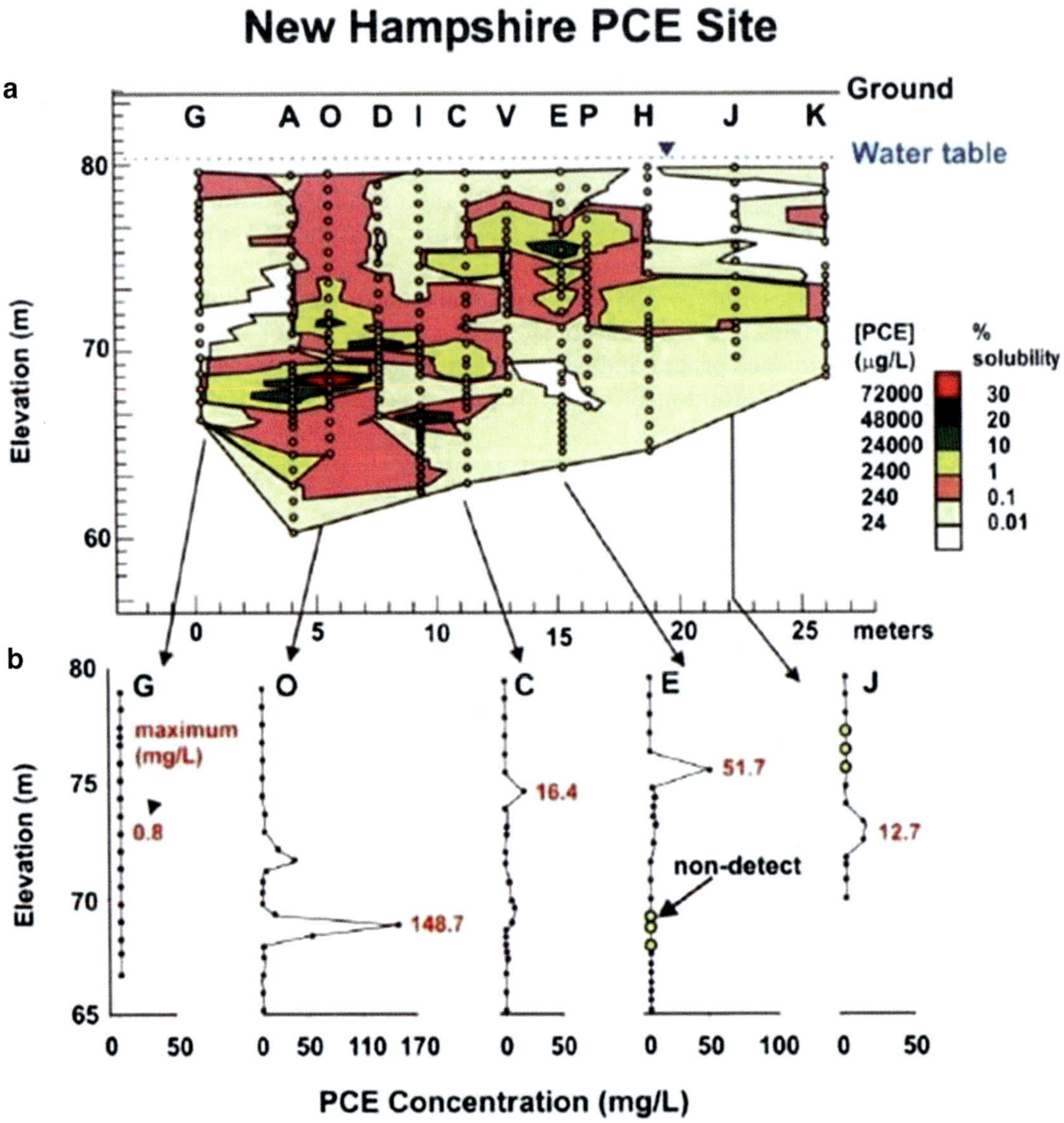

Figure 4.1. Results from characterization of a PCE site showing the complex three-dimensional structure of the plume downgradient from a source. (a) Shows the PCE distribution, expressed as percent of DNAPL solubility (240 milligrams per liter [mg/L]), projected onto the cross section. (b) Presents the profiles of PCE concentration (mg/L) versus depth. Such multilevel monitoring results can help target treatment efforts on the areas with the highest contaminant concentrations and mass flux. From Guilbeault et al. (2005).

MLM system sampling; analytical costs for sample analysis; and purchase, operation, and maintenance costs for downhole equipment (Westbay systems).

In addition to the cost of the hardware, there are associated costs for drilling and sampling and any other operational costs such as electricity, purged water disposal, and decommissioning. Each system, however, has inherent capabilities and limitations so that selection based only on cost considerations is not practical, with much more consideration required on the intended monitoring purpose and scope in the context of many of the factors listed above.

Table 4.1. Example Multilevel Monitoring Systems

MLM System	Number of Monitoring Intervals	Description and Key References
Solinst CMT® (Continuous Multichannel Tubing)	2–7	Continuous length of three- or seven-chamber polyethylene tubing. Sampling ports are constructed at specified depths, and the remainder of the chamber below each port is sealed off. The system is installed in boreholes with sand or bentonite in the annular space. See Einarson and Cherry (2002), van Dijk (2005).
Solinst Waterloo system	2–24	A modular system where each monitoring module comprises a screened interval, a sample entry port, and a tubing extending to the surface. A system is constructed by coupling together monitoring modules and blank casing. Monitoring modules are hydraulically isolated by bentonite backfill in the annular space in unconsolidated formations or by packers in bedrock. See www.solinst.com.
Westbay system	Not limited[a]	A modular system where each module comprises an upper and lower packer, a pumping port, and a monitoring port. A system is constructed by coupling together modules and blank casing. Development and monitoring are performed by lowering specialized tools down the interior of the casing to the port depth. See Meyer et al. (2008) and www.slb.com.
Groundwater FLUTe™[b]	1–20	A polyurethane-coated nylon liner everted into a borehole using interior water pressure exceeding formation pressure. Sampling ports are predesigned and built into the liner system. Discrete monitoring zones are separated by exterior spacers. The system requires a stable borehole, such as core holes in fractured rock. See Cherry et al. (2007).
ZIST™ (Zone Isolation Sampling Technology by BESST, Inc.	1–2[c]	The system consists of a standard polyvinyl chloride (PVC) well construction in which a pump and sensor/data logger dock into a well screen receptacle coupled between the well screen and riser pipe. The monitoring zone is isolated from water in the riser pipe, allowing for continuous monitoring of groundwater quality and head conditions between sampling events and greatly reduced purge volumes during sampling. See www.besstinc.com.

[a]Limited only by the number of modular systems that can be fit into the borehole length
[b]A trademark name for Flexible Liner Underground Technologies
[c]Can be used in conjunction with nested wells installed in a single borehole; however, the number of monitoring zones is limited by the number of nested wells that can be constructed in a borehole. Thus, the ZIST™ system is not considered to be an MLM system in the same category as the other systems

4.3.1.3 Advantages/Disadvantages

The primary advantage of the use of MLM systems and nested wells is the capability to develop improved conceptual site models by providing the following:

- Ability to assess vertical hydrogeologic characteristics including hydraulic gradients
- Identification of high-permeability zones and areas of predominant contaminant flux or reactant injection
- Better definition of exposure risks based on the improved understanding of contaminant distribution

The primary disadvantage of the use of MLM systems is the requirement to collect more samples and incur additional costs for system procurement and installation, increasing both analytical and investigation costs. However, each zone may not require sampling during all monitoring events. This should be evaluated based on initial data from all zones in the system collected during the characterization phase. An additional disadvantage may be the physical size limitation of any sampling or access ports, which may limit the ability to lower certain data loggers, transducers, bailers and pumps down the hole.

4.3.1.4 Attributes for Selection

Maturity of the Tool: MLM systems are mature tools that are commercially available and have been used widely. Each MLM system evaluated in this chapter has been used in numerous investigations of sites distributed across North America, and some of the systems have been used on other continents. The manufacturers provide detailed information about their MLM systems on their websites. The Westbay system was the first MLM system to enter the marketplace (in the late 1970s). Descriptions of the system are provided by Black et al. (1986) and more recently by Meyer et al. (2008). It was followed by the Waterloo system which was commercialized in the late 1990s. Cherry and Johnson (1982) described an early preproduction version of the Waterloo system, and Parker et al. (2006) described a recent version of the Waterloo system installed in rotasonic holes. The CMT system described by Einarson and Cherry (2002) and the FLUTe system described by Cherry et al. (2007) entered the marketplace in the late 1990s. The ZIST™ nested well system also entered the marketplace in the late 1990s. Widespread reporting on the uses of MLM systems can be found in site characterization reports and conference proceedings.

Applicability to Site Characteristics and Specific In Situ Technology: MLM systems can be applied to both consolidated and unconsolidated geologies but are most beneficial when used in settings with a high degree of heterogeneity, where contaminant distribution cannot be predicted using noninvasive methods (modeling). Due to the materials used in their construction, some MLM systems may be incompatible with certain remedial technologies, such as those involving chemical oxidants or high temperatures.

Implementation at the Site of Interest: MLM systems are generally implementable but do often require specialized knowledge during installation and initial rounds of sampling. For example, the use of a trained, onsite technician is recommended for the installation of the Westbay, FLUTe and ZIST systems. Installation difficulties can arise because of borehole irregularities, MLM system construction requirements or bridging of well backfill materials. Less than ideal borehole conditions may prevent the MLM system from reaching the bottom of the borehole.

Data Quality: MLM systems can provide more accurate and precise data than conventional long-screen groundwater monitoring devices by minimizing elements of sample bias. Sample bias refers to influences on contaminant concentrations during sampling and analysis that cause the measured concentrations to differ from true concentrations in the formation immediately beyond the sampling interval. Many factors can cause sample bias, such as reactions with MLM system materials (sorption or leaching), volatilization losses during sampling, and formation mixing due to screen length. Purging the system storage volume (water contained in the system between the intake and the discharge) can be labor intensive and may lead to sample bias by introducing air or increasing turbidity. Each MLM system is designed to reduce system storage volume and target discrete intervals of the formation.

Of the five MLM systems identified in this chapter, the Westbay system has the lowest system storage volume, because the groundwater sample is collected downhole in a canister made of glass or steel without headspace. However, subsequent transfer from the canister to sampling bottles at the surface may create bias similar to that created by other MLM systems. The four other MLM systems convey water to the surface through small-diameter tubing that must be purged before sample collection but minimize system storage volume.

***Uniqueness of Data Gathered by the Tool*:** MLM systems provide unique data regarding discrete values of hydrogeologic properties (hydraulic head and hydraulic conductivity) and changes in contaminant distribution over time that cannot be obtained by conventional long-screen sampling methods.

***Cost Relative to Similar Methods*:** The MLM systems will increase cost compared to conventional groundwater monitoring systems, but they provide an understanding of contaminant distribution that cannot be obtained conventionally. Optimization of injection strategies during *in situ* treatment to target vertical intervals not being fully treated and source zone areas with the greatest mass can result in greater mass removal rates, shorter remedial time frames and lower overall life cycle costs than would be achieved without the use of MLM systems. Optimization of extraction strategies with targeted well placement can result in lower pumping rates and increased mass removal.

4.3.2 Rock Matrix Characterization

4.3.2.1 Description

Rock matrix characterization, which measures chlorinated solvent mass in rock samples as well as various physical properties of the rock matrix, is a diagnostic tool used in consolidated geologic environments. Rock matrix characterization emphasizes revealing the contaminant distribution in the rock matrix where groundwater is nearly immobile due to low permeability. The method protocol can include collecting samples at high density (approximately one sample per foot of core) including at fractures (one of the fracture faces) and bedding planes, at lithologic changes, and from matrix blocks between fractures. The technique generally involves collecting small rock core samples at various depths, crushing the rock and then extracting chlorinated solvents from the rock matrix with methanol for subsequent laboratory measurement. Laboratory results are presented in terms of mass of contaminant per mass of rock, which can be converted to an equivalent pore water concentration using simple partitioning relationships. Figure 4.2 illustrates a stainless steel rock crusher into which methanol is placed prior to crushing the rock sample. Discrete rock matrix samples can also be obtained for purposes of measuring the matrix dry bulk density, matrix fraction organic carbon, matrix entry pressure, matrix permeability, and matrix porosity. Method protocols are discussed by Lebron et al. (2012) and Rodriquez (2012).

The data obtained through rock matrix characterization is useful for a number of reasons including the following:

- Assessment of the mass of contaminants that has diffused into the rock matrix
- Estimation of the distance to which contaminants have diffused into the rock matrix, including an assessment of whether forward diffusion is still continuing at the location of interest and the attenuating effect of forward diffusion on plume velocity in fractures
- Estimation of the time scale and relevance of back diffusion that may occur following removal of contaminants from open fractures

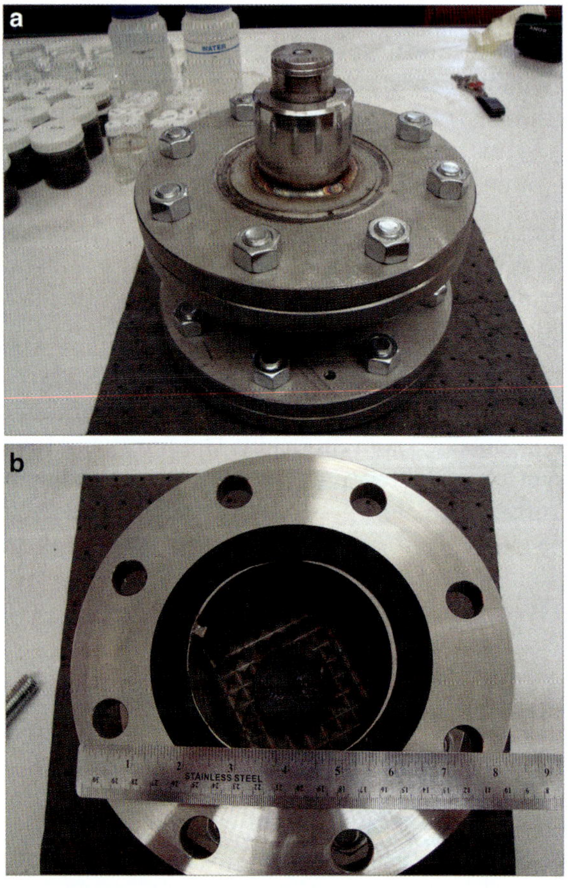

Figure 4.2. Rock crusher developed by Queen's University under ESTCP project ER-0715. Methanol is placed into interior cavity prior to crushing (Lebron et al., 2012).

- Determination of whether the contaminant has entered the rock matrix as an NAPL
- Evaluation of the use of remedial technologies such as chemical oxidation, enhanced *in situ* biodegradation, and thermal conductive heating, which have the potential to address contaminant mass that has diffused into the rock matrix.

4.3.2.2 Technology Status

Investigators have been obtaining discrete rock matrix samples for contaminant and physical property characterization for many years (Parker, 2007). Early uses were largely restricted to research purposes, but the technique is gaining acceptance as part of fractured bedrock investigations at contaminated sites.

4.3.2.3 Advantages/Disadvantages

Rock matrix characterization provides a better understanding of contaminant distribution in fractured rock systems compared to conventional groundwater monitoring tools. One major reason so little is known about contaminant migration and fate in fractured sedimentary rock is that traditional monitoring approaches involve only sampling water from the fractures. Several uncertainties are introduced by focusing solely on the water phase. Field studies using the rock core volatile organic compound (VOC) analysis method show that contaminant mass storage is dominated by the rock matrix rather than the fractures and the contaminant concentrations in the fractures and the matrix may not be in equilibrium (Hurley and Parker, 2002; Sterling et al., 2005). Therefore, sampling only the groundwater from the fractures cannot provide the overall mass distribution. Furthermore, when conventional boreholes are drilled, the water from a fracture in one section of the borehole migrates to another section of the borehole due to differences in hydraulic head between the two sections. This creates an unnatural flow and contaminant transport condition within the system known as borehole cross-connection. This condition will also persist across the screened interval of a conventional monitoring well, and as a consequence, results from sampling the well do not reflect the natural system (Price and Williams, 1993; Sterling et al., 2005). In general, the distribution of chlorinated solvents in the water phase has strong spatial variability due to heterogeneity in source zone contaminant mass distributions, fracture network, and matrix characteristics, accompanied by temporal variability in groundwater flow. Contaminant distribution in the rock matrix changes slowly over time.

4.3.2.4 Attributes for Selection

***Maturity of the Tool*:** The rock matrix characterization technique is mature and can be performed provided that rock core samples are obtained during drilling. The means of subdividing the rock core for subsequent laboratory analysis varies depending on the type of measurement required (intact rock for measurement of matrix porosity, permeability, or entry pressure; crushed rock for VOC analysis). Applications of rock matrix characterization are presented by Sterling et al. (2005), Parker (2007), and Rodriguez (2012).

***Applicability to Site Characteristics*:** The rock matrix characterization technology is applicable to any site where bedrock groundwater has become contaminated with organic compounds, particularly chlorinated volatile organic compounds (CVOCs). However, it has been shown to be most valuable in geologic settings consisting of sedimentary rock, such as sandstone, shale, limestone, dolostone and siltstone, where matrix porosity may be appreciable (1–20% range).

***Applicability to Specific In Situ Technology*:** Rock matrix characterization is not specific to a particular remedial technology. The technology is applicable for evaluation of the efficacy of *in situ* remedial technologies with the potential to remove contaminants from the rock matrix, such as chemical oxidation, bioremediation and thermal conductive heating.

***Implementation at the Site of Interest*:** This technique can be implemented but involves significant time and effort because it requires drilling rock cores and detailed sampling and analyses of the core samples.

***Data Quality*:** The detection limits, accuracy, and precision of rock matrix characterization have been found sufficient for the analysis of chlorinated solvents in fractured rock and for the assessment of rock matrix physical properties.

***Uniqueness of Data*:** Rock matrix characterization provides unique data that cannot be obtained by other methods.

***Cost*:** The criterion of comparative cost is not applicable because there are no competing methods for obtaining the same data. While the cost of any particular laboratory measurement may not be excessive, the cost of drilling using a coring method in comparison to a non-coring method must be considered.

4.4 MASS FLUX/MASS DISCHARGE MEASUREMENT TOOLS

4.4.1 Uses

Mass flux is a measurement of the mass of contaminants passing through a unit cross-sectional area over time. Mass discharge is the area-integrated mass flux across an entire cross-sectional area of interest, such as a plume. As discussed in Chapter 7, mass flux/mass discharge measurements are being used increasingly to develop more robust and more quantitative conceptual site models to guide the site characterization process and the remedial response. Most commonly, mass discharge measurement techniques are performed to assess remedial performance and as a decision-making tool for transitioning between two technologies. As an integrated measurement of source zone or plume strength, mass discharge can be a key performance metric for evaluating MNA or *in situ* treatment. Local mass flux measurements can also guide remedial progress, ensuring that sufficient amendments are delivered to high-flux zones. However, other applications are under increasing consideration. For example, mass discharge can be used to assess the potential risks of site contaminants to downgradient receptors such as water supply wells or surface water bodies and perhaps prioritize source areas for remediation. This section presents an assessment of the current status of mass flux/mass discharge measurement tools.

4.4.2 Description of Techniques

Several different methods are available for measuring mass flux/mass discharge including synoptic sampling (transect method), time-integrated sampling (passive flux meters or PFMs) and spatially integrated sampling (steady-state pumping, recirculation flux measurement (RFM), integrated pumping tests (IPTs) and modified IPTs). Many of these methods were developed under the umbrella of the Strategic Environmental Research and Development Program (SERDP) DNAPL Source Zone Initiative (SERDP, 2004; 2006). Brief descriptions of each method and key references to obtain further information to implement the method are listed in Table 4.2.

4.4.3 Advantages/Disadvantages

Over the last decade, there has been increased recognition that mass discharge is a key indicator of the severity or the strength of a contaminant release, particularly as it relates to potential risks to downgradient receptors (Feenstra et al., 1996; Einarson and Mackay, 2001; Rao et al., 2001; Newell et al., 2011). Many scientific panels (USEPA, 2003; SERDP, 2004, 2006; NRC, 2005) and guidance development teams (ITRC, 2004; 2010b) have recommended that *in situ* remediation efforts should focus on reducing contaminant mass discharge in conjunction with reducing contaminant concentrations to numerical standards. In heterogeneous geologic media, focusing remediation efforts on high-flux zones can often result in significant reductions in contaminant mass discharge and be more efficient at remediation in terms of mass removed per dollars spent or gallons treated. In addition, mass discharge measurements can provide a different perspective on the magnitude and average impact of site contamination. For example, high concentrations in one well may be recognized to be of relatively minor significance if contaminant mass discharge from the site is low overall (that is, only a small total mass of contaminant per unit time actually migrates with the groundwater).

Table 4.2. Mass Flux/Mass Discharge Measurement Methods

Measurement Technique	Typical Pumping Required	Description Key References
Synoptic sampling along transects	Minutes	Synoptic point measurements of concentration along a well transect with independently derived flow estimates Einarson and Mackay, 2001; Farhat et al., 2005; Kubert and Finkel, 2006
Passive flux meter (PFM)	None	Time-integrated point measurements along a well transect using compound-specific sorbants Annable et al., 2005; Campbell et al., 2006; de Jonge and Rothenberg, 2006; Hatfield et al., 2004; Hatfield et al., 2002
Steady-state pumping (SSP)	Days to weeks	Spatially integrated value derived from steady-state pumping and sampling of a transect of extraction wells extending across the entire plume Einarson and Mackay, 2001; Buscheck, 2002
Recirculation flux measurements (RFM)	Days to weeks	Spatially integrated value derived from steady-state pumping and sampling of extraction and injection well pairs installed across the plume Goltz et al., 2009; Huang et al., 2004
Integral pumping tests (IPTs)	Days to weeks	Spatially integrated value derived from transient pumping and sampling of extraction wells and use of an inversion algorithm to estimate depth-weighted average concentration Bauer et al., 2004; Bayer-Raich et al., 2004; Bockelmann et al., 2003; Bockelmann et al., 2001
Modified integral pumping tests (MIPTs)	Days to weeks	Spatially integrated value derived from transient pumping and sampling of extraction wells and use of an analytical method to estimate Darcy flux Brooks et al., 2008

The primary disadvantage is the increased cost for mass flux and discharge analyses. Cost assessments should consider the possible need for more detailed site characterization. Useful mass flux and discharge measurements often require careful delineation and hydraulic characterization of the primary advective pathways which can increase the cost of site investigation but reduce life cycle costs of the site cleanup.

4.4.4 Status and Attributes for Selection

Maturity of the Tool: Mass flux/mass discharge measurement is maturing as a diagnostic tool. Formal attention from state regulators and other ITRC members combined with implementation at a significant number of sites (61 case studies cited in ITRC, 2010b) indicates that these methods are gaining more widespread acceptance. The most common methods being used include synoptic point sampling along a transect (33 sites), PFMs (7 sites) and IPT tests (6 sites). In addition, mass flux has been designated as a performance metric in a Record of Decision (ROD) for at least one site (USEPA, 2009). Confidence and acceptance in mass flux/mass discharge measurement methods will likely grow among the environmental remediation community as other studies are conducted and regulatory acceptance grows.

Applicability to Site Characteristics: When considering the value of mass flux/mass discharge measurement at a site, one must take into account site characteristics including the geologic setting, depth to groundwater, stability of groundwater flow, and previous site characterization data. Aquifers with low transmissivities and fractured rock environments likely have limited applicability for spatially integrated mass discharge measurement techniques relying on groundwater extraction systems. Sites with a shallow depth to groundwater are easier to characterize with synoptic sampling or time-integrated measurements along transects. Sites with narrow low-yield aquifers may be dewatered if pumping techniques are used for mass flux measurement. Optimal measurements are obtained where groundwater flow is fairly stable in magnitude and direction so that flow lines are perpendicular to mass flux transects. Previous site characterization efforts (knowledge of the plume extent, flow pathways, detailed understanding of site geology, geochemistry) facilitate the use of mass flux measurement methods and decrease the cost of implementation.

***Applicability to Specific In Situ Technology*:** Mass flux/mass discharge measurements are not technology specific. They can be used as a performance metric for all types of *in situ* source area remedial technologies as well as pump-and-treat systems. However, these techniques may not provide real-time data with sufficient frequency to enhance decision making, while *in situ* source area remediation is in progress. This is particularly true for short-term but aggressive remedies such as thermal remediation, although it may be less true for longer term source zone bioremediation projects with multiple injections. Mass discharge measurements can add value to baseline site characterization and remedial technology selection by better defining the horizontal and vertical extent of hot spots commonly targeted for *in situ* cleanup. A comparison between pre- and post-remediation mass discharge measurements downgradient of the source area should be considered for evaluating the completeness of cleanup.

***Implementation at the Site of Interest*:** The ease of implementation of mass flux/mass discharge measurement methods depends on site characteristics and the level of prior site characterization. Implementation may be impeded or have an excessive cost due to the great depth or volume of contaminated groundwater, site access constraints and the presence of multiple source areas. The decision to use point measurements along transects (synoptic sampling or PFMs) or spatially integrated sampling from pumping wells may depend upon site-specific hydrogeologic factors that might include the variability in groundwater flow rate and direction and complexity of advective flow paths. Shifts in groundwater flow direction can cause significant errors in mass flux/mass discharge measurements but are less significant with pumping methods. Sites with shifting flow fields will need relatively dense monitoring networks or careful evaluation of plume capture for accurate data collection.

***Data Quality*:** Do the available mass flux/mass discharge measurement methods produce data of sufficient accuracy and precision to improve site decision making? Each mass flux/mass discharge measurement technique has somewhat unique elements of uncertainty. However, many of the limitations to data quality associated with these tools also apply to conventional monitoring strategies. For synoptic point measurements along a transect, factors that improve the accuracy of the calculated mass flux/mass discharge values include having a robust CSM of the subsurface geology and hydrogeology, well-defined hydraulic properties of advective flow paths and focused measurements along several transects positioned along the flow path. For time-integrated sampling along a transect using PFMs, field tests have indicated significant uncertainties in mass flux/mass discharge calculations due to monitoring well construction techniques (Brooks et al., 2008) and due to vertical flow in a well that magnifies the apparent horizontal mass discharge. Spatially integrated mass discharge measurements based on samples collected from pumping wells depend on substantially complete capture of the contaminant plume for accurate results.

***Uniqueness of Data*:** Some types of mass flux/mass discharge can be measured using data typically collected at a site (contaminant concentrations, hydraulic conductivity values, gradient measurements, pumping rates). Measurements of mass flux/mass discharge can provide another set of data or evaluation technique for assessing remedial performance in terms of the effect on the plume downgradient of a source area. In this sense, these measurements are unique diagnostic tools.

***Cost Relative to Similar Methods*:** Absolute and relative costs of mass flux/mass discharge measurements are site specific and depend on factors such as the degree of heterogeneity in geologic characteristics and the nature and extent of groundwater contamination. Depth of the plume has a direct impact on cost. Despite the potential increase in site characterization costs, the information provided with this technology can lead to significant cost savings in the future if the operation time of a remedy can be reduced, thereby decreasing monitoring and life cycle costs.

4.5 COMPOUND-SPECIFIC ISOTOPE ANALYSIS

4.5.1 Description

CSIA is an analytical technique used to generate an isotopic signature or ratio for different compounds. Isotope fractionation makes CSIA a useful technique to distinguish between concentration decreases due to degradative versus nondegradative processes for a wide variety of compounds, which is unique information in the assessment of performance of *in situ* remedies. This section provides an overview of CSIA as a diagnostic tool at chlorinated solvent sites during site characterization, performance assessment during active remediation, process optimization, and monitoring efforts. To date, carbon has been the most common element analyzed for isotope ratios at chlorinated solvent sites.

CSIA can detect isotopic shifts due to abiotic as well as biological reactions. Although CSIA is not technically a MBT, the ITRC (2011) included CSIA and MBTs in its guidance on Environmental Molecular Diagnostic (EMD) tools useful for characterizing and monitoring biological and abiotic degradation or transformation reactions. Figure 4.3 illustrates the changes in $^{13}C/^{12}C$ that typically occur during reductive dechlorination of TCE. The carbon isotope ratios of each successive chloroethene increase (become less negative) over time as the compounds are biodegraded, because the ^{13}C-containing compounds are slightly more difficult to biodegrade than those with only ^{12}C atoms.

Stable isotopes are typically reported as the ratio of two isotopes of an element in a sample relative to the ratio of those two isotopes in an established reference standard. For carbon, these isotopes are ^{12}C and ^{13}C, with ^{13}C being the heavier isotope with one extra neutron. The ratio of stable carbon isotopes in a compound may change due to physical, chemical or biological processes. This change in the ratio of stable isotopes is known as *fractionation*, and it occurs because of differences in the rates of reaction for different molecular species. Heavier isotopes react more slowly because the reaction rate and/or dissociation energy for heavy and light isotopes of a molecule differ (Sueker, 2001). The result is that as the reaction proceeds, the reactant that remains has a progressively higher content of the heavy isotope because the molecules containing the light isotope have reacted more quickly to form the product compared to those containing the heavier isotope.

For chlorinated hydrocarbons, nondegradative processes, such as volatilization, vaporization, dissolution, dispersion, advection and sorption, have been found to be non-fractionating or minimally fractionating under equilibrium conditions (the isotope ratio has been found to remain unchanged or minimally altered via CSIA) (USEPA, 2008). Therefore, isotope

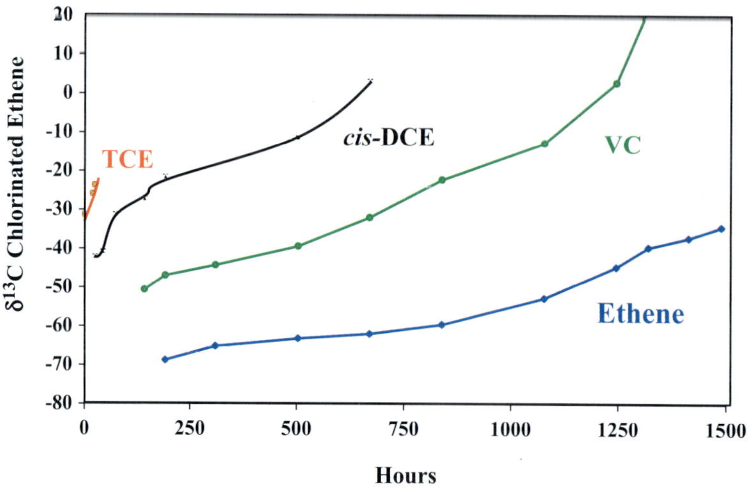

Figure 4.3. Changes in ^{13}C enrichment of TCE and daughter products during reductive dechlorination in laboratory microcosms. Modified from Slater et al. (2001).

fractionation makes CSIA a useful technique to distinguish between concentration decreases caused by degradative versus nondegradative processes and can therefore serve as a useful diagnostic tool during both characterization and remediation of chlorinated solvent sites.

In general, contaminant degradation is confirmed when enrichment in ^{13}C is measured in the compound undergoing degradation. For example, if degradation processes are active, three trends can be observed. First, chlorinated compounds in the source area should become enriched in ^{13}C over time relative to the original source material. Second, contaminant samples collected along a flow path should become enriched in ^{13}C in the downgradient direction. Finally, carbon isotope ratios in contaminants subjected to chemical or biological remediation should respond by becoming enriched in ^{13}C. CSIA has also been used to distinguish between multiple chlorinated solvent sources that originated from solvents produced in different lots or in some cases by different manufacturers (Hunkeler et al., 2004). Table 4.3 lists potential uses of CSIA during site characterization, active remediation and long-term monitoring for chlorinated solvent sites. Chlorinated compound degradation products cannot be more enriched in ^{13}C than the original compound; however, the daughter product can be enriched relative to the parent compound by further degradation (TCE $\rightarrow$ DCE $\rightarrow$ VC).

4.5.2 Regulatory Acceptance

As the use of CSIA becomes more prevalent for chlorinated solvent sites, the regulatory community has become aware of its significance, and the use of CSIA data has increased the regulatory acceptance of remedies. However, a generally limited understanding of use and interpretation of CSIA data still exists within the regulatory community. To address this need, ITRC started the EMD team in 2010 and recently published detailed factsheets on several EMDs including CSIA (ITRC, 2011). Regulatory community awareness and/or acceptance are expected to increase along with the use of CSIA in field applications.

4.5.3 Attributes for Selection

Maturity of the Tool: CSIA applications are gaining acceptance for use at chlorinated solvent sites to complement traditional site investigation and remediation performance monitoring

Table 4.3. Compound-Specific Isotope Analysis Applications at Chlorinated Solvent Sites

	Site Characterization	Active Remediation	Long-Term Monitoring
Goal	**Improve conceptual site model (CSM)**	**Improve remedy design and application**	**Improve performance assessment and accelerate site closure**
CSIA Applications	• Baseline CSIA measurements prior to active remediation and/or long-term monitoring • Forensics application for source differentiation • Qualitative and/or quantitative evidence for biodegradation including MNA to guide decisions on selection of remediation strategy • Contaminant fate and transport numerical modeling	• Qualitative and/or quantitative evidence for biodegradation during remedy application including evidence for MNA • Qualitative and/or quantitative evidence for abiotic degradation during remedy application (e.g., degradation versus dilution; Liang et al., 2007) • Mechanism of biological degradation of chlorinated solvents • Quantification of the rate of degradation (Morrill et al., 2005)	• Monitoring progress of biological or abiotic degradation • Measurement of potential rebound effects after completion of an active remedy

techniques. To date, CSIA has been applied most frequently to carbon isotopes, so CSIA for carbon isotopes can be considered a mature technology. For other atoms of interest at chlorinated solvent sites, such as hydrogen, oxygen and chlorine, CSIA has not been performed to the same extent as for carbon. Use of CSIA for these other atoms remains a topic of active research and shows promise for future applications at chlorinated solvent sites (USEPA, 2008; Abe et al., 2009). CSIA is commercially available although the number of CSIA vendors is somewhat limited.

***Applicability to Site Characteristics*:** Application of CSIA is generally not influenced by site geology or hydrologic conditions. It has been used at sites with consolidated (Watervliet Arsenal; see Malcolm Pirnie, 2011d) and unconsolidated (Fort Lewis; see Malcolm Pirnie, 2011c) media and found applicable to both. The most important site condition that influences the use of CSIA is the nature of the contaminants present. CSIA has been applied most widely for chlorinated solvents, benzene, toluene, ethylbenzene and total xylenes (BTEX) and methyl tertiary butyl ether (MTBE). A few applications of CSIA for PAHs and PCBs have also been reported (Sueker, 2001; USEPA, 2008). With current technology, the heaviest compounds that can be analyzed for shifts in carbon isotope ratios contain 12–13 carbon atoms (USEPA, 2008). Application of CSIA is generally not influenced by site geology or hydrologic conditions.

***Applicability to Specific In Situ Technology*:** CSIA is applicable to many remedial technologies including *in situ* bioremediation, abiotic *in situ* technologies, and MNA (Lollar et al., 2001; Morrill et al., 2005). In addition, CSIA is useful in providing information regarding the mechanisms for degradation which can then be used to ascertain whether a remedy is performing as designed. To date, most of the applications have involved biodegradation. However, several *in situ* chemical oxidation (ISCO) applications have been reported including ISCO with Fenton's reagent and persulfate (Ahad and Slater, 2008; Marchesi et al., 2009) as well as ISCO with permanganate which was demonstrated at Watervliet Arsenal (Malcolm Pirnie, 2011d) and also at other sites (Poulson and Naraoka, 2002; Hunkeler et al., 2003).

Theoretically, CSIA can be applied at any site where reactive processes in groundwater produce a change in the ratio of stable isotopes.

Implementation at the Site of Interest: CSIA requires the collection and submission of samples from the site. It does not have complex requirements and is therefore implementable. The correct interpretation of CSIA data requires knowledge of site geology and geochemistry, so these elements must be accurately characterized before implementation of CSIA.

Data Quality: To obtain an accurate isotopic reading with acceptable detection limits for reductive daughter products present in lower concentrations, a large volume of groundwater may need to be collected. When analyzing for low concentrations, tedious purge-and-trap methods may be necessary prior to analysis to concentrate the sample. Overall, CSIA data and gas chromatography data should complement each other. Currently CSIA has no standard analytical methods. Therefore, methods and results can be highly variable among laboratories conducting this work. It is important to use the same methods and laboratories on a given project so that results are comparable. Guidelines to achieve acceptable data quality are provided by USEPA (2008).

Uniqueness of Data: CSIA are unique in that they can provide evidence that chemicals of concern are being transformed by *in situ* biotic or abiotic processes as opposed to concentration changes due to dilution or other physical processes.

Cost Relative to Similar Methods: CSIA offers unique data; thus, no similar methods exist to which to compare its cost.

4.6 MOLECULAR BIOLOGICAL TOOLS

4.6.1 Overview

MBTs include a suite of assays targeting biomolecules such as nucleic acids (deoxyribonucleic acid [DNA] and ribonucleic acid [RNA]), lipids and proteins to provide evidence regarding the composition and/or activity of microbial communities. These assays can have utility in remediation applications that rely on biological degradation mechanisms to attenuate contaminants including bioremediation and MNA (SERDP, 2005; Stroo et al., 2006).

4.6.1.1 Principles of MBT Methods

Generally, MBTs using gene-based approaches target DNA, the genetic template from which all biological molecules are derived. DNA is transcribed into RNA when the microorganisms are metabolically active. RNA is then translated during synthesis of proteins (for example, enzymes) which carry out metabolic processes for the microorganisms, such as production of lipids. Molecular biological tools can be used for qualitative descriptions and, in some cases, quantitative estimates of DNA, RNA, proteins or lipids. These targets dictate whether the assays are focused on identifying the presence of a capability or on assessing activity within a system. The closer the MBT is to assessing the functions of interest, the more can be said about whether the desired function is actually occurring in the environment. Figure 4.4 illustrates the relationship between genetic and enzymatic MBTs and activity.

MBTs target either a population of interest (16S ribosomal RNA (rRNA) gene specific to *Dehalococcoides* spp.) or a biological function of interest (a functional gene such as the vinyl chloride reductase gene [*vcrA*] from *Dehalococcoides* spp.) and can show correlations with bioremediation parameters (pH, total organic carbon [TOC]). Table 4.4 provides a

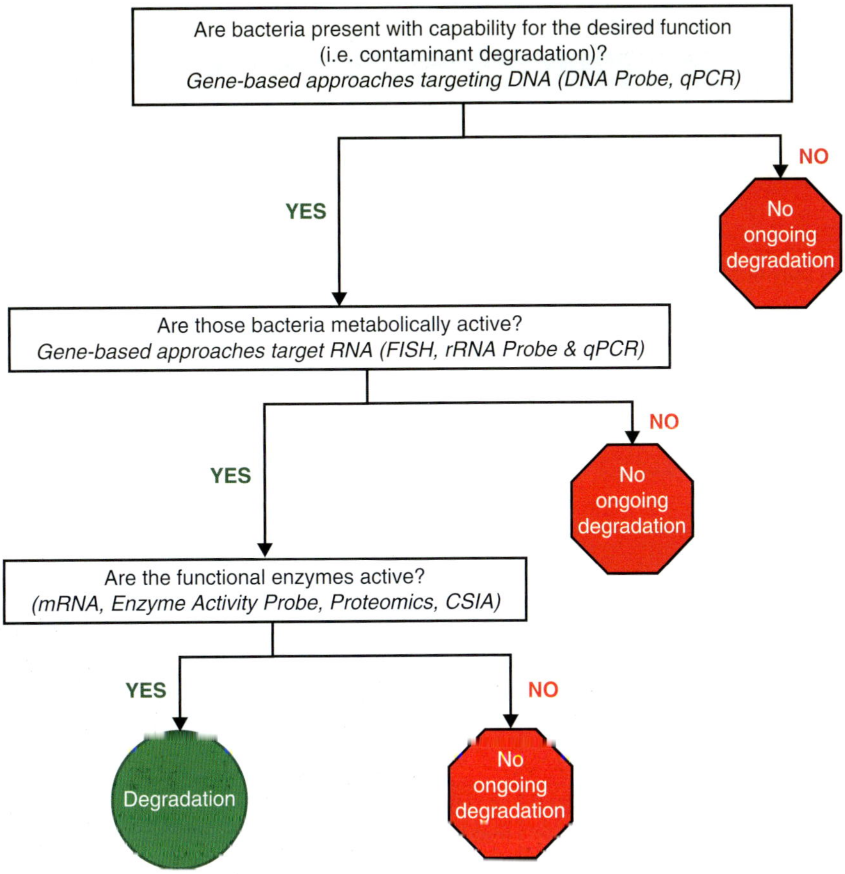

Figure 4.4. Hierarchy of relationship between molecular biological tools and the activity of interest, biological degradation. Courtesy ME Watwood, from a presentation by Howard et al. (2005).

non-exhaustive list of MBTs that have been applied to evaluate contaminated field sites in the context of bioremediation.

4.6.1.2 Context of Use of MBTs for Bioremediation

Biological processes that directly or indirectly impact contaminant degradation are complex. MBTs can be used to provide evidence that desired or undesired biological processes are occurring, when natural attenuation as a stand-alone remedy is sufficient or whether enhancements such as chemical amendments or bioaugmentation are necessary (SERDP, 2005). Ultimately, the remedial decisions supported by data generated from MBTs include the following:

- Is biodegradation occurring at the site, and is the rate of attenuation sufficient to meet RAOs? In general, several MBTs can indicate biodegradation activity; however, MBTs cannot directly measure biological attenuation rates.
- Are native contaminant-degrading populations present at the site, or is bioaugmentation necessary? Quantitative polymerase chain reaction (qPCR) provides direct evidence quickly and relatively cheaply regarding the presence of contaminant-degrading populations including *Dehalococcoides* species and demonstrates growth during biostimulation. qPCR can be used to determine the need for bioaugmentation during an initial design of bioremediation.

Table 4.4. Molecular Biological Tools

MBT	Description
Gene-Based MBTs	
Polymerase chain reaction (PCR)	The selective targeting of a region of DNA or RNA and amplifying that region exponentially. This allows for generation of sufficient quantities of the desired target such that it can be observed using other techniques
Quantitative polymerase chain reaction (qPCR)	A PCR methodology using fluorescent markers to detect and quantify products after each round of reaction. The fluorescent signal increases in direct proportion to the amount of PCR product in a reaction. This can be used to assess multiple *Dehalococcoides* species as well as specific functional reductase genes from those species
T-RFLP and DGGE	Fingerprinting techniques applied to PCR products that provide information about the microbial community and/or specific populations
Clone Library	DNA or RNA sequences are amplified by PCR, inserted into a bacterial host, and grown until sufficiently abundant to be extracted and sequenced. The technique can be used to generate libraries of closely related gene sequences (i.e., 16S rRNA gene sequences from different *Dehalococcoides* spp.)
Microarray	A technology used to detect and quantify the presence of a very large number of targeted genes (DNA) and/or their expression (RNA) in a single assay
Fluorescence *in situ* hybridization (FISH)	A whole-cell assay that uses molecular fluorescent probes that bind, or hybridize, to either the DNA chromosome or the RNA present within microbial cells. FISH probes have been developed that target chlorinated solvent-degrading populations, such as *Dehalococcoides*
Lipid or enzyme-based MBTs	
Phospholipid fatty acid (PLFA) analysis	The extraction, separation, and analysis of fatty acids to gauge viable biomass, composition, and metabolic status of a microbial community. It is generally nonspecific and does not provide information on key populations of interest (i.e., contaminant-degrading populations)
Enzyme activity probes (EAPs)	Substrates that can bind to specific enzymes of interest and are subsequently transformed by those enzymes into fluorescent products if the enzyme is both present and active. EAPs have been developed for a suite of enzymes that cometabolically degrade chlorinated ethenes
Proteomics	Methods to identify proteins expressed by a microbial cell and to determine their role in the physiology of that cell. Proteomics can provide direct evidence of any one targeted protein or all proteins such as biodegrading enzymes being expressed by microbial populations

Notes: DGGE – denaturing gradient gel electrophoresis; T-RFLP – terminal restriction fragment length polymorphism

- Are contaminant-degrading populations active at the site, or do they need to be stimulated? Enzyme activity probes (EAP) provide the most direct measure of activity, but limited methods are available; only aerobic cometabolic degradation pathways have been demonstrated. RNA-based qPCR provides direct evidence regarding the activity of *Dehalococcoides* populations. MBTs are particularly useful if degradation by-products are short-lived or difficult to measure using standard sampling and analytical techniques.
- Are processes or conditions limiting biodegradation rates that could be optimized to facilitate achieving RAOs? The ability to evaluate *Dehalococcoides* populations with qPCR in response to bioremediation operations provides data to troubleshoot any conditions that may be adversely impacting Dehalococcoides growth and activity (pH, redox conditions).
- What other biological processes may be contributing to undesirable secondary water quality impacts, and can they be mitigated? Fluorescent *in situ* hybridization (FISH) and qPCR can evaluate the presence of methanogenic populations. However, MBT data are generally unnecessary to evaluate generation of methane, which can be easily determined by direct analysis. qPCR methods are available for other populations that might cause secondary water quality impacts, such as metal reducers; however, their application to this issue has been limited at best.

4.6.2 Regulatory Acceptance

As the use of MBTs becomes more prevalent, the regulatory community has become aware of their significance and MBT-based data have been used to gain regulatory acceptance of remedies. However, a generally limited understanding of the use and interpretation of MBT data still exists within the regulatory community. To address this need, ITRC's EMD team was begun in 2010 to advance the use of molecular biological and chemical diagnostic techniques or tools that can identify and quantify key microorganisms (taxonomy) and their genes (function) (ITRC, 2010a). It is the goal of the EMD team to summarize the fundamental background and current status of available EMDs and provide objective guidance on the best practices for using them and evaluating, applying, and interpreting the results of EMDs. Technical and regulatory guidance should lead to greater use and confidence in these diagnostics and help site managers faced with major decisions about site design, management, and resolution. The ITRC EMD team recently published a series of ten factsheets (ITRC, 2011) describing the use of MBTs in site characterization and remedial evaluation.

4.6.3 Attributes for Selection

Maturity of the Tool: MBTs have varying degrees of technology maturity. While many MBTs are not commercially available at the present time, qPCR for the 16S ribosomal RNA (rRNA) gene is commercially available for a few key organisms such as *Dehalococcoides* spp. and methanogens. qPCR for functional genes is commercially available for a few key genes (reductive dehalogenase genes, sulfate reductase), and denaturing gradient gel electrophoresis terminal restriction fragment length polymorphism (DGGE/T-RFLP) community analyses are also commercially available. For *Dehalococcoides* spp. qPCR has been applied to quantify concentrations of these organisms in environmental samples and correlate with observed dehalogenation activity (Lu et al., 2006; Ritalahti et al., 2006). Ongoing projects funded through

SERDP and ESTCP (ER-1683, ER-1587, and ER-200708) are researching protocols for use of these and other MBTs at chlorinated solvent-contaminated sites.

Applicability to Site Characteristics: Application of MBTs is generally not influenced by site geology or hydrologic conditions.

Applicability to Specific In Situ Technology: MBTs are applicable for *in situ* remedial processes involving microbial transformations, such as enhanced biodegradation (bioremediation) and intrinsic biodegradation (MNA), and can provide information on other biological processes that may be contributing to undesirable secondary water quality impacts and the extent to which they can be mitigated.

Implementation at the Site of Interest: The implementation of MBTs depends partly upon the ease and feasibility of obtaining representative samples from the subsurface. Methods for obtaining samples for MBTs have not been standardized. Once samples have been obtained, molecular analyses can be conducted by research or university laboratories and, in some cases, by commercial laboratories. Generally, the widespread use of MBTs other than qPCR for *Dehalococcoides* and reductive dehalogenase genes can be limited by logistical issues such as the lack of standardized methods and the limited number of commercially available labs.

Data Quality: Based on field trials and evaluation, the qPCR technology provides sufficient detection limits, accuracy and precision for application to biological processes at chlorinated solvent sites to inform remedial decision making. While MBTs are, in general, accurate, many of the analyses are nonspecific (fingerprinting tools such as DGGE, phospholipid fatty acid [PLFA] analysis and T-RFLP) or provide only qualitative or semiquantitative information (proteomics and microarrays), which may render some of these tools insufficient to address remedial decisions outlined above in Section 4.6.1.2. While FISH, EAPs and proteomics are accurate, their use is significantly limited by commercial availability and cost. PLFA analysis can provide indications regarding general biomass and community structure and is accurate, but it is generally nonspecific with respect to biological processes relevant to chlorinated solvent sites. It does not provide information on key populations of interest to chlorinated solvent biodegradation (contaminant-degrading populations such as *Dehalococcoides*).

Uniqueness of Data: MBTs provide unique data regarding microbial community structure and potential functions that cannot be obtained by any other diagnostic tool. However, routine geochemical data, such as dissolved nitrate, iron, sulfate and methane concentrations, and analyses of degradation products from a site can serve as indirect but reliable and informative evidence for microbial reactions. For chlorinated solvent sites, qPCR analyses of organisms (16S rRNA gene) and functional genes can provide valuable, unique information for troubleshooting ERD, but they are best used in combination with geochemical data and only when geochemical data are insufficient to address all performance assessment issues. In addition, MBTs have only limited ability to quantify contaminant attenuation rates relative to standard approaches, which limits their decision-making impact. Geochemical data and CSIA can be coupled with hydraulic modeling to better evaluate attenuation rates controlled by biodegradation.

Cost Relative to Similar Methods: MBTs offer unique data, and therefore no similar methods exist to which to compare their cost. Commercial availability is currently limited to research or university laboratories for most MBTs, although a few commercial laboratories are providing some MBT analyses. The costs and availability of MBT assays are changing rapidly and are dependent on the type of MBT required. The additional expenditure required for MBTs should be weighed against the usefulness of the information they could potentially provide for decision making at sites.

4.7 VALUE OF INFORMATION ANALYSIS SUMMARY

4.7.1 Summary of Diagnostic Tools

This chapter has discussed a wide range of innovative diagnostic tools for characterization and remedial performance assessment at chlorinated solvent-contaminated sites. The diagnostic tools evaluated in this chapter are field sampling and laboratory measurement techniques that permit high-resolution description of contaminant distributions or of *in situ* processes affecting the fate and transport of contaminants. Diagnostic tools are used to focus remedial efforts at locations and for time frames in which the remedy will do the most good. The five groups of diagnostic tools evaluated in this chapter include the following:

1. MLM systems for the delineation of the vertical distribution of contaminants in the dissolved and adsorbed phase
2. Rock matrix characterization for the assessment of chlorinated solvent distribution in consolidated media and performance assessment of *in situ* technologies
3. Mass flux/mass discharge measurement technologies for site characterization and process and performance assessment
4. CSIA for confirming *in situ* chemical or biological transformations of chemicals of concern
5. MBTs for optimizing process performance of *in situ* bioremediation

The advantages of each of the diagnostic tools have been summarized in the previous sections for each tool individually. These tools are designed to provide information on performance of *in situ* technologies that would otherwise not be available from conventional techniques. Collectively, important potential advantages of these diagnostic tools include (1) a more accurate and detailed CSM; (2) more accurate performance assessment in real time; (3) assessment of the feasibility of achieving certain endpoints, such as background concentrations or maximum contaminant levels (MCLs); (4) confirmation of *in situ* processes that result in transformation of the chemicals of concern to nontoxic or less toxic by-products; and (5) alternative metrics for performance assessment, such as mass flux/mass discharge.

Innovative diagnostic tools provide decision makers addressing cleanup of chlorinated solvent sites with information that may improve the timeliness and accuracy of regulatory action. The use of MBTs can establish new lines of evidence that can, for example, help establish the validity for choosing MNA as a remedy or for transitioning to MNA from active remediation.

4.7.2 Summary of Attributes for Selection

The uncertainty faced by practitioners while making decisions regarding remediation at such sites is a significant challenge, as is the decision of whether to expend additional resources to use one or more new diagnostic tools to decrease that uncertainty. These tools are considered innovative in that they are not routinely used for site characterization or performance assessment in the remediation industry, although their use is increasing. Only a small number of ASTM (formerly the American Society for Testing and Materials, now known as ASTM International) standards have been developed to date for applying these diagnostic tools. Qualitative guidelines regarding the evaluation of attributes of the tools relevant to a value of information analysis are provided in Table 4.5.

Table 4.5. Summary of Evaluation of Value of Information Attributes

Attribute	Multilevel Monitoring Systems	Rock Matrix Characterization	MF/MD	CSIA	MBTs
Maturity of the tool	Mature; commercially available	Mature; commercially available	Maturing; some tools commercially available	Mature; commercially available	Variable among tools; some tools commercially available
Applicability to site characteristics	Applicable	Applicable to consolidated media	Consider site characteristics carefully	Applicable	Applicable
Applicability to specific *in situ* technology	May be incompatible with certain oxidants or high temperatures	Applicable	Applicable	Applicable to bioremediation, abiotic *in situ* treatment (e.g., chemical oxidation), MNA	Applicable to processes involving biological transformations
Ease of implementation	Generally implementable	Involves significant time and effort	Depends on site and level of prior characterization	Implementable	Some tools limited by logistical issues (e.g., lack of standardized methods)
Detection limits, accuracy and precision	More precise than conventional monitoring; several variables impact accuracy	Sufficient for chlorinated solvents in fractured rock	Depends on accuracy of prior characterization; mixed results regarding PFMs	Some variability; important to follow guidelines to achieve acceptable data quality	Sufficient for some tools (e.g., PCR); other tools are qualitative
Uniqueness of data	Unique	Unique	Unique	Unique	Unique
Cost	Short-term costs likely to result in long-term savings	NA; provides unique data	Site specific	NA; provides unique data	NA; provides unique data

4.7.3 Recommendations for Use

A summary of recommendations for each of the tools discussed in this chapter is provided in Table 4.6. When implemented according to these recommendations, the tools discussed in this chapter can provide sufficient value of information for decision making to justify the additional investment beyond conventional characterization and performance assessment.

Table 4.6. Recommendations for Application of Innovative Diagnostic Tools at Chlorinated Solvent Contaminated Sites

Diagnostic Tool	Recommendations
Multilevel monitoring	1. Use for vertical delineation of hydrogeologic properties and contaminant concentrations, particularly at highly homogeneous sites 2. Balance relevant criteria for the selection of the most appropriate MLM system for a given site
Rock matrix characterization	3. Consider rock matrix characterization as a characterization tool at consolidated sites, but carefully weigh the potential value of information collected from the technique against its cost
Mass flux/discharge measurement	4. Calculate mass flux/discharge at all contaminated sites, if possible, because it can be used to improve remedial decisions made at various stages of the cleanup process 5. Consider the site hydrogeologic setting when selecting the mass flux/mass discharge measurement method 6. Follow best practices during field implementation to increase the accuracy, usefulness, and cost-effectiveness of mass flux/mass discharge measurement methods
Compound-specific isotope analysis	7. Use CSIA for multiple purposes throughout site characterization and remediation 8. Conduct baseline CSIA measurements and analyses to confirm that the required detection limits are achievable 9. Use CSIA data to complement conventionally generated analytical data and vice versa, overall
Molecular biological tools	10. Use qPCR at chlorinated solvent sites to (1) troubleshoot engineered bioremediation or monitored natural attenuation or (2) provide a supporting line of evidence for biodegradation 11. Evaluate standard geochemical parameters in groundwater before using MBTs to characterize and troubleshoot many operational issues related to biodegradation of chlorinated solvents 12. Avoid conducting routine molecular evaluation of methanogenic populations unless site-specific conditions require detailed evaluation of these populations

ACKNOWLEDGEMENTS

The work summarized in this chapter was funded by ESTCP under project ER-0318. The authors would like to acknowledge project team members for their significant contribution to the project's success. These include, in alphabetical order, Lisa Alvarez-Cohen (University of California, Berkeley), Michael Annable (University of Florida), John Cherry (University of Guelph), Murray Einarson (Haley and Aldrich), Ken Goldstein (The Louis Berger Group), Mark Goltz (Air Force Institute of Technology), Kirk Hatfield (University of Florida), Douglas Mackay (University of California, Davis), Tamzen Macbeth (CDM Smith), Daria Navon (ARCADIS), Beth Parker (University of Guelph), Suresh Rao (Purdue University), Kent Sorenson (CDM Smith) and Andrew Vitolins (ARCADIS).

REFERENCES

Abe Y, Aravena R, Zopfi J, Shoukar-Stash O, Cox E, Roberts JD, Hunkeler D. 2009. Carbon and chlorine isotope fractionation during aerobic oxidation and reductive dechlorination of vinyl chloride and *cis*-1,2-dichloroethene. Environ Sci Technol 43:101–107.

Ahad J, Slater G. 2008. Differentiating biotic from abiotic (Fenton's reaction) degradation of toluene: A test of compound-specific stable carbon isotopes. Sci Total Environ 401:194–198.

Annable MD, Hatfield K, Cho J, Klammler H, Parker BL, Cherry JA, Rao PSC. 2005. Field-scale evaluation of the passive flux meter for simultaneous measurement of groundwater and contaminant fluxes. Environ Sci Technol 39:7194–7201.

Back PE, Rosén L, Norberg T. 2007. Value of information analysis in remedial investigations. Ambio 36:486–493.

Bauer S, Bayer-Raich M, Holder T, Kolesar C, Muller D, Ptak T. 2004. Quantification of groundwater contamination in an urban area using integral pumping tests. J Contam Hydrol 75:183–213.

Bayer-Raich M, Jarsjo J, Liedl R, Ptak T, Teutsch G. 2004. Average contaminant concentration and mass flux in aquifers from time-dependent pumping well data - Analytical framework. Water Resour Res 40:W08303.

Black WH, Smith HR, Patton FD. 1986. Multiple-level ground water monitoring with the MP system. Proceedings of the National Water Well Association's Conference on Surface and Borehole Geophysical Methods and Ground Water Instrumentation. Denver, CO, USA, October 15–17, pp 41–61.

Bockelmann A, Ptak T, Teutsch G. 2001. An analytical quantification of mass fluxes and natural attenuation rate constants at a former gasworks site. J Contam Hydrol 53:429–153.

Bockelmann A, Zamfirescu D, Ptak T, Grathwohl P, Teutsch G. 2003. Quantification of mass fluxes and natural attenuation rates at an industrial site with a limited monitoring network: A case study. J Contam Hydrol 60:97–121.

Brooks MC, Wood LA, Annable MD, Hatfield K, Cho J, Holbert C, Rao PSC, Enfield CG, Lynch K, Smith RE. 2008. Changes in contaminant mass discharge from DNAPL source mass depletion: Evaluation at two field sites. J Contam Hydrol 102:140–153.

Buscheck TE. 2002. Mass Flux Estimates to Assist Decision-Making. Technical Bulletin. ChevronTexaco, Houston, TX, USA.

Campbell T, Hatfield K, Klammler H, Annable MD, Rao PS. 2006. Magnitude and directional measures of water and Cr(VI) fluxes by passive flux meter. Environ Sci Technol 40:6392–6397.

Cardiff M, Liu X, Kitanidis PK, Parker J, Kim U. 2010. Cost optimization of DNAPL source and plume remediation under uncertainty using a semi-analytic model. J Contam Hydrol 113:25–43.

Cherry JA, Johnson PE. 1982. A multi-level device for monitoring in fractured rock. Ground Water Monit Rev 2:41–44.

Cherry JA, Parker BL, Keller C. 2007. A new depth discrete multilevel monitoring approach for fractured rock. Ground Water Monit Remediat 27:57–70.

de Jonge H, Rothenberg G. 2006. New device and method for flux-proportional sampling of mobile solutes in soil and groundwater. Environ Sci Technol 39:274–282.

Einarson M. 2006. Multi-level Groundwater Monitoring. In DM Nielsen, ed, Practical Handbook of Environmental Site Characterization and Ground-Water Monitoring, 2nd ed. CRC Press, Boca Raton, FL, USA, pp 808–845.

Einarson MD, Cherry JA. 2002. A new multi-level ground-water monitoring system utilizing multichannel tubing. Ground Water Monit Remediat 22:52–65.

Einarson MD, Mackay DM. 2001. Predicting impacts of ground water contamination. Environ Sci Technol 35:66A–73A.

Farhat SK, Newell CJ, Nichols EM. 2005. Mass Flux Toolkit. Groundwater Services, Houston, TX, USA. http://www.gsi-net.com/en/software/free-software/mass-flux-toolkit.html. Accessed February 26, 2013.

Feenstra S, Cherry J, Parker B. 1996. Conceptual Models for the Behavior of DNAPLs in the Subsurface. In Pankow J, Cherry J, eds, Dense Chlorinated Solvents and Other DNAPLs in Groundwater: History, Behavior and Remediation. Waterloo Press, Waterloo, ON, Canada, pp 53–88.

Freeze RA, Bruce J, Massman, J, Sperling T, Smith L. 1992. Hydrogeological decision analysis: 4. The concept of data worth and its use in the development of site investigation strategies. Ground Water 30:574–588.

Goltz MN, Close ME, Yoon H, Huang J, Flintoft MJ, Kim S, Enfield C. 2009. Validation of two innovative methods to measure contaminant mass flux in groundwater. J Contam Hydrol 106:51–61.

Guilbeault MA, Parker BL, Cherry JA. 2005. Mass and flux distributions from DNAPL zones in sandy aquifers. Ground Water 43:70–86.

Hatfield K, Annable MD, Kuhn S, Rao PS, Campbell T. 2002. A new method for quantifying contaminant flux at hazardous waste sites. In Thornton SF, Oswald SE, eds, Groundwater Quality: Natural and Enhanced Restoration of Groundwater Pollution. Internat Assoc Hydrological Sci Pub. No. 275, pp 25–32.

Hatfield K, Annable M, Cho JH, Rao PSC, Klammler H. 2004. A direct passive method for measuring water and contaminant fluxes in porous media. J Contam Hydrol 75:155–181.

Howard MH, Clingenpeel SR, Leiser OP, Rothermell JS, Watwood ME. 2005. Molecular and physiological characterization of aerobic TCE degradation potential. Proceedings, 8th In Situ and On-Site Bioremediation Symposium, Baltimore, MD, USA, Paper No. G-31.

Huang J, Close ME, Pang L, Goltz MN. 2004. Innovative method to measure flux of dissolved contaminants in groundwater. Proceedings, Fourth International Conference on Remediation of Chlorinated and Recalcitrant Compounds. Monterey, CA, USA, May 24–27. Battelle Press. Abstract No. 1D-08.

Hunkeler D, Aravena R, Parker BL, Cherry JA, Diao X. 2003. Monitoring oxidation of chlorinated ethenes by permanganate in groundwater using stable isotopes: Laboratory and field studies. Environ Sci Technol 37:798–804.

Hunkeler D, Chollet N, Pittet X, Aravena R, Cherry JA, Parker BL. 2004. Effect of source variability and transport processes on carbon isotope ratios of TCE and PCE in two sandy aquifers. J Contam Hydrol 74:265–282.

Hurley JC, Parker BL. 2002. Rock core investigation of DNAPL penetration and TCE mobility in fractured sandstone. In Stolle D, Piggott AR, Crowder JJ, eds, Proceedings of the 55th Canadian Geotechnical and 3rd Joint IAH-CNC and CGS Groundwater Specialty Conferences, Ground and Water: Theory to Practice. Canadian Geotechnical Society. Niagara Falls, Ontario, Canada, October 20–23, 2001, pp 473–480.

ITRC (Interstate Technology & Regulatory Council). 2004. Strategies for Monitoring the Performance of DNAPL Source Zone Remedies. http://www.itrcweb.org/Documents/DNAPLs-5.pdf. Accessed February 26, 2013.

ITRC. 2010a. ITRC Technical Project Teams. http://www.itrcweb.org/Documents/2010Team-Descriptions.pdf. Accessed February 26, 2013.

ITRC. 2010b. Use and Measurement of Mass Flux and Mass Discharge. Int-DNAPL-1. www.itrcweb.org/Guidance/GetDocument?documentID=49. Accessed February 26, 2013.

ITRC. 2011. Environmental Molecular Diagnostics Fact Sheets. EMD-1. Washington, DC: Interstate Technology & Regulatory Council, Environmental Molecular Diagnostics Team. www.itrcweb.org/GuidanceDocuments/EMD1.pdf. Accessed February 26, 2013.

James BR, Freeze RA. 1993. The worth of data in predicting aquitard continuity in hydrogeological design. Water Resour Res 29:2049–2065.

James BR, Gorelick SM. 1994. When enough is enough: The worth of monitoring data in aquifer remediation design. Water Resour Res 30:3499–3513.

Kavanaugh M, Kresic N. 2008. Large Urban Groundwater Basins: Water Quality Threats and Aquifer Restoration. In Dimkic M, Brauch HJ, Kavanaugh M, eds, Groundwater Management in Large River Basins. IWA Publishing, London, UK, pp 520–600.

Kubert M, Finkel M. 2006. Contaminant mass discharge estimation in groundwater based on multi-level point measurements: A numerical evaluation of expected errors. J Contam Hydrol 84:55–80.

Lebron CA, Phelan D, Heron G, Lachance J, Nielsen SG, Kueper BH, Rodriguez D, Wemp A, Baston D, Lacombe P, Chapelle FH. 2012. Dense Non Aqueous Phase Liquid (DNAPL) Removal from Fractured Rock Using Thermal Conductive Heating (TCH). Final Report ESTCP Project ER-200715. Prepared for ESTCP, Arlington, VA, USA.

Liang X, Dong Y, Kuder T, Krumholz LR, Philp RP, Butler EC. 2007. Distinguishing abiotic and biotic transformation of tetrachloroethylene and trichloroethylene by stable carbon isotope fractionation. Environ Sci Technol 41:7094–7100.

Lollar BS, Slater GF, Sleep B, Witt M, Klecka GM, Harkness M, Spivack J. 2001. Stable carbon isotope evidence for intrinsic bioremediation of tetrachloroethene and trichloroethene at Area 6, Dover Air Force Base. Environ Sci Technol 35:261–269.

Lu X, Wilson JT, Kampbell DH. 2006. Relationship between *Dehalococcoides* DNA in ground water and rates of reductive dechlorination at field scale. Water Res 40:3131–3140.

Malcolm Pirnie, Inc. 2011a. Guidance Report: Diagnostic Tools for Performance Evaluation of Innovative *In Situ* Remediation Technologies at Chlorinated Solvent-Contaminated Sites. ESTCP Project Number ER-200318. Prepared for ESTCP, Arlington, VA, USA.

Malcolm Pirnie, Inc. 2011b. Final Report: Vandenberg Air Force Base, Diagnostic Tools for Performance Evaluation of Innovative *In Situ* Remediation Technologies at Chlorinated Solvent-Contaminated Sites. ESTCP Project Number ER-200318. Prepared for ESTCP, Arlington, VA, USA.

Malcolm Pirnie, Inc. 2011c. Final Report: Fort Lewis, Diagnostic Tools for Performance Evaluation of Innovative *In Situ* Remediation Technologies at Chlorinated Solvent-Contaminated Sites. ESTCP Project Number ER-200318. Prepared for ESTCP, Arlington, Alexandria, VA, USA.

Malcolm Pirnie, Inc. 2011d. Final Report: Watervliet Arsenal, Diagnostic Tools for Performance Evaluation of Innovative *In Situ* Remediation Technologies at Chlorinated Solvent-Contaminated Sites. ESTCP Project Number ER-200318. Prepared for ESTCP, Arlington, Alexandria, VA, USA.

Marchesi M, Aravena R, Otero N, Soler A, Gil I, Sra KS, Thomson NR, Mancini S. 2009. Assessment of in-situ chemical oxidation (ISCO) performance for chlorinated solvents contaminated groundwater using stable carbon isotope at laboratory and field scale. Geophysical Research Abstracts 11. EGU2009-10201-1. http://adsabs.harvard.edu/abs/2009EGUGA..1110201M. Accessed February 26, 2013.

Massman J, Freeze RA. 1987. Groundwater contamination from waste management sites – the interaction between risk-based engineering design and regulatory policy. Water Resour 23:368–380.

Meyer JR, Parker, BL, Cherry JA. 2008. Detailed hydraulic head profiles as essential data for defining hydrogeologic units in layered fractured sedimentary rock. Environ Geol 56:27–44.

Morrill PL, Lacrampe-Couloume G, Slater GF, Sleep BE, Edwards EA, McMaster ML, Major DW, Sherwood Lollar B. 2005. Quantifying chlorinated ethene degradation during reductive dechlorination at Kelly AFB using stable carbon isotopes. J Contam Hydrol 76:279–293.

Newell CJ, Farhat SK, Adamson DT, Looney BB. 2011. Contaminant plume classification system based on mass discharge. Ground Water 49:914–919.

NRC (National Research Council). 2005. Contaminants in the Subsurface: Source Zone Assessment and Remediation. National Academy Press, Washington, DC, USA.

Parker BL. 2007. Investigating contaminated sites on fractured rock using the DFN approach. Proceedings of the 2007 USEPA/NGWA Fractured Rock Conference: State of the Science and Measuring Success in Remediation. Portland, ME, USA, September 24–26, pp 150–168.

Parker BL, Cherry JA, Swanson BJ. 2006. A multilevel system for high-resolution monitoring in rotasonic boreholes. Ground Water Monit Remediat 26:57–73.

Payne FC, Quinnan JA, Potter ST. 2008. Remediation Hydraulics. CRC Press, Boca Raton, FL, USA.

Poulson SR, Naraoka H. 2002. Carbon isotope fractionation during permanganate oxidation of chlorinated ethylenes (cDCE, TCE, PCE). Environ Sci Technol 36:3270–3274.

Price M, Williams AT. 1993. The influence of unlined boreholes on groundwater chemistry: A comparative study using pore-water extraction and packer sampling. J Instit Water Environ Manage 7:651–659.

Rao PSC, Jawitz JW, Enfield CG, Falta RW, Annable MD, Wood AL. 2001. Technology integration for contaminated site remediation: Clean-up goals and performance criteria. In Thornton SF, Oswald SE, eds, Groundwater Quality: Natural and Enhanced Restoration of Groundwater Pollution. Internat Assoc Hydrological Sci Pub. No. 275, pp 571–578.

Reichard EG, Evans JS. 1989. Assessing the value of hydrogeologic information for risk-based remedial action decisions. Water Resour Res 25:1451–1460.

Ritalahti KM, Amos BK, Sung Y, Wu Q, Koenigsberg SS, Löffler FE. 2006. Quantitative PCR targeting 16S rRNA and reductive dehalogenase genes simultaneously monitors multiple *Dehalococcoides* strains. Appl Environ Microbiol 72:2765–2774.

Rodriquez DJ. 2012. Assessment of Thermal Heating for the Removal of Chlorinated Solvents from Fractured Bedrock. PhD Thesis. Queens University, Kingston, Ontario, Canada. http://hdl.handle.net/1974/7515. Accessed February 26, 2013.

Russell KT, Rabideau AJ. 2000. Decision analysis for pump-and-treat design. Ground Water Monit Remediat 20:159–168.

Sale T, Newell C, Stroo H, Hinchee R, Johnson P. 2008. Frequently Asked Questions Regarding Management of Chlorinated Solvents in Soils and Groundwater. ESTCP. Project ER-0530. Prepared for ESTCP, Arlington, VA, USA. http://www.estcp.org/Technology/upload/ER-0530-FAQ.pdf. Accessed February 28, 2013.

SERDP (Strategic Environmental Research & Development Program). 2004. Annual Report: DNAPL Source Zone Initiative. ESTCP, Arlington, VA, USA.

SERDP. 2005. Expert Panel Workshop Report: Research and Development Needs for the Environmental Remediation Application of Molecular Biological Tools. Retrieved from http://www.clu-in.org/products/tins/tinsone.cfm?num=68982420. Accessed February 28, 2013.

SERDP. 2006. Expert Panel Workshop Report: Reducing the Uncertainty of DNAPL Source Zone Remediation. ESTCP, Arlington, VA, USA.

Slater GF, Sherwood-Lollar B, Sleep BE, Edwards EA. 2001. Variability in carbon isotopic fractionation during biodegradation: Implications for field applications. Environ Sci Technol 35:901–907.

Sterling SN, Parker BL, Cherry JA, Williams JH, Lane Jr JW, Haeni FP. 2005. Vertical cross contamination of trichloroethylene in a borehole in fractured sandstone. Ground Water 43:557–573.

Stroo HF, Leeson A, Shepard AJ, Koenigsberg SS, Casey CC. 2006. Monitored natural attenuation forum: Environmental remediation applications of molecular biological tools. Remediat J 16:125–137.

Sueker JK. 2001. Isotope applications in environmental investigations: Theory and use in chlorinated solvent and petroleum hydrocarbon studies. Remediat J 12:5–24.

USEPA (U.S. Environmental Protection Agency). 1999. Use of Monitored Natural Attenuation at Superfund, RCRA Corrective Action, and Underground Storage Tank Sites. Final OSWER Directive. EPA/540/R-99/009. USEPA, Washington, DC, USA. http://www.epa.gov/swerust1/directiv/d9200417.pdf. Accessed July 22, 2013.

USEPA. 2003. The DNAPL Remediation Challenge: Is There a Case for Source Depletion. EPA/600/R-03/143. http://www.clu-in.org/download/remed/600r03143.pdf. Accessed July 22, 2013.

USEPA. 2008. A Guide for Assessing Biodegradation and Source Identification of Organic Ground Water Contaminants using Compound Specific Isotope Analysis (CSIA). EPA 600/R-08/148. http://www.epa.gov/nrmrl/pubs/600r08148.html Accessed May 9, 2013.

USEPA. 2009. Amendment to the Record of Decision for the Commencement Bay – South Tacoma Channel Superfund Site, Operable Unit 1, Well 12A, Tacoma, Washington. USEPA Region 10, Seattle, WA, USA.

van Dijk G. 2005. Bentonite usage hits new ground. GeoDrilling International, November, pp 36–39.

Yokota F, Thompson K. 2004. Value of information literature analysis: A review of applications in health risk assessment. Med Decision Making 24:287–298.

CHAPTER 5

MODELING SOURCE ZONE REMEDIATION

Jason I. Gerhard,[1] Bernard H. Kueper[2] and Brent E. Sleep[3]

[1]University of Western Ontario, London, Ontario, Canada N6A 5B9; [2]Queen's University, Kingston, Ontario, Canada K7L 3N6; [3]University of Toronto, Toronto, Ontario, Canada M5S 1A4

5.1 INTRODUCTION

The modeling of multiphase flow phenomena has a long history in petroleum engineering in the form of reservoir simulators. The principles developed in that field were adapted to contaminant hydrogeology beginning in the early 1980s as awareness of the historical legacy of light nonaqueous phase liquid (LNAPL) and dense nonaqueous phase liquid (DNAPL) releases emerged (Pinder and Abriola, 1986). Challenges at that time included both adequately incorporating the physics of DNAPL movement and addressing the computational limits that hindered their application to field-scale scenarios. In the last 25 years, profound advances in the understanding of the physics, geochemistry, and microbiology of DNAPL source zones combined with the development of sophisticated numerical algorithms that integrate with groundwater flow and contaminant transport codes have produced remediation modeling tools of real practical value.

These models have a number of uses in the context of remediating DNAPL source zones. Models provide cost-effective tools to explore "what if" scenarios related to understanding the development and architecture of DNAPL source zones and their ultimate fate. Numerical modeling results may aid in choosing between remediation alternatives, designing both pilot-scale and full-scale remediation applications, optimizing the performance of a chosen remediation technology, or exploring the potential consequences of inaction. One may explore these questions with respect to hypothetical site conditions or incorporate detailed site-specific data (geology, hydrogeology, DNAPL configuration or DNAPL release history, geochemistry, etc.). The complexity of the modeling project can range from simple spreadsheet calculations to a suite of numerical simulations completed on a supercomputer cluster. The degree of model sophistication, the data requirements, and the simulations undertaken depend on the modeling objectives, the available data, and the complexity of the site and budget.

A site-specific, calibrated numerical model can be a tool for not only choosing a remediation approach but also designing the remediation system and then optimizing that system prior to, or in conjunction with, field implementation. Chapter 8, *Hydraulic Displacement*, details an example in which modeling was employed to design and optimize a dual-phase (DNAPL and water) extraction system for more than 7.5 million liters (2 million gallons) of pooled DNAPL below a chemical manufacturing facility in the southern United States. Optimization involved a suite of simulations exploring how to maximize a reduction in DNAPL mass and pool height while minimizing capital costs (number of wells, orientation of wells, total length of wells) and operation and maintenance costs (pumping duration, volume of groundwater produced). The modeling achieved substantial savings in both capital and operation and maintenance (O&M) costs by eliminating the wells predicted to be least productive and demonstrating that a pulsed pumping strategy could minimize the extracted volume of contaminated water without reducing the total DNAPL volume recovered (Gerhard et al., 2001).

B.H. Kueper et al. (eds.), *Chlorinated Solvent Source Zone Remediation*,
doi: 10.1007/978-1-4614-6922-3_5,

In addition to site-specific modeling, another common application is to further understanding of DNAPL remediation performance at the field scale in a manner that is more widely applicable than at a single site. Sensitivity studies permit examining the influence of a single variable (mean hydraulic conductivity, variance of hydraulic conductivity) or set of source zone characteristics (DNAPL composition, DNAPL mass, presence of low permeability zones) on remediation performance. Such studies typically employ a base case scenario and then explore how results are affected by varying one or more scenario conditions. A few examples include (1) examining the benefits of partial DNAPL mass removal on aqueous phase mass flux leaving the source zone (Lemke and Abriola, 2006; Falta et al., 2005; Christ et al., 2006); (2) understanding the influence of back diffusion on plume detachment and recession times following source removal in fractured rock for various aged sources (West and Kueper, 2010); (3) examining the influence of chemical parameters on the success of injecting micellar solutions composed of surfactants and alcohols for LNAPL recovery (Bernardez et al., 2009); (4) understanding how variations in porous media structure and DNAPL type influence the amount of DNAPL that can be recovered using dual-phase extraction (Alexandra et al., 2012); (5) exploring the influence of injected humic acid concentration on the remediation of LNAPL below the water table (Molson et al., 2002); (6) understanding how the degree of concentration rebound following application of chemical oxidation in fractured clay varies with matrix porosity, matrix organic carbon content, fracture aperture, oxidant dosage, and hydraulic gradient (Mundle et al., 2007); (7) investigating the influence of nonaqueous phase liquid (NAPL) type and source zone architecture on NAPL recovery via hot-water flooding (O'Carroll and Sleep, 2009); and (8) understanding how the length of time to achieve boiling in fractured rock using thermal technologies varies with fracture aperture, rock matrix properties and the rate of energy application (Baston et al., 2010; Baston and Kueper, 2009).

An important aspect of modeling is an appreciation of the uncertainty in model predictions. Uncertainty in model input parameters leads to uncertainty in model predictions. For *in situ* chemical oxidation, for example, uncertainty in site-specific model results can stem from incomplete knowledge of the DNAPL mass present, the DNAPL distribution, the distribution of natural oxidant demand, and site heterogeneity. Uncertainty of this type can be evaluated and even quantified using sensitivity analyses and can often be reduced by using those results to guide the collection of more field data. Uncertainty also arises from model error associated with accuracy of the conceptual model and the associated physical, chemical, and biological processes included in the model. For example, does a bioremediation model assume first-order decay or a more robust representation of degradation kinetics including bacteria growth and competition? Does the model allow for three-dimensional groundwater flow, or only one- or two-dimensional flow? Model error can be evaluated by comparing the results of different models, but more complex models require more detailed input datasets, and adequate site-specific measurements are needed.

This chapter presents an introductory overview of DNAPL source zone remediation modeling. First, the broad types of models available are discussed. Next, the components of a remediation model are presented, providing insight into the relationships between site data, simulated processes, and modeling results. The chapter then discusses the modeling process, highlighting the major steps in conducting a remediation modeling study. Finally, several examples of DNAPL remediation modeling are presented to illustrate how the concepts discussed can be applied in practice.

5.2 ANALYTICAL VERSUS NUMERICAL MODELS

An *analytical model* is a model in which the solution to the system's governing equation can be written explicitly without the need for numerical approximations. Analytical models are typically only available for relatively simple geometries and processes. One exists, for example, to the advection–dispersion equation describing solute transport including linear sorption for one-dimensional geometry and a continuous and constant source (Van Genuchten and Alves, 1982). In the context of multiphase systems, an analytical model exists for one-dimensional DNAPL injection from a continuous source into a homogeneous, one-dimensional column including the effects of capillarity (McWhorter and Sunada, 1990). The main advantage of an analytical model is the speed of computing the solution (typically on the order of seconds) and thus the large number of simulations that can be completed. For example, Lemke and Bahrou (2009) used the analytical solution to the advection–dispersion equation to investigate the sensitivity of increased cancer risk to 22 different variables associated with perchloroethene (PCE) transport in groundwater via 20,000 simulations. The main disadvantage of an analytical model is the restrictive conditions that need to be employed. For example, in the Lemke and Bahrou (2009) study it was necessary to assume a simplified geology (homogeneous, isotropic aquifer) with a simplified flow field (one-dimensional flow, constant hydraulic gradient) and a continuous input of PCE at its aqueous solubility in a single location, with uniform and basic aquifer conditions (linear sorption and first-order decay throughout). Such models are valuable for learning about the relationship between key system variables and for performing screening-level evaluations of remediation performance. Chapter 6, for example, makes extensive use of analytical solutions to illustrate the response of a plume to mass removal from the DNAPL source zone in both porous media and fractured rock.

A *numerical model* is a model in which the system's governing equations are solved using numerical techniques. Numerical techniques allow solving complex equations that cannot be solved analytically. A variety of numerical methods exist for solving the equations efficiently and in a manner that balances the accuracy of the final solution with the speed of computation. Advantages of numerical models include the complexity of the scenarios that can be investigated, including few limits on the geometry or heterogeneity of the subsurface, the complexity of the DNAPL source zone, the time-varying nature of conditions such as hydraulic gradients or recharge, and the complexity of the natural and engineered processes being considered. Disadvantages of numerical models include the amount of time required to conduct simulations (from minutes/hours for straightforward scenarios to days/weeks for sophisticated problems). Also, the use of numerical methods to solve equations introduces approximation error, the magnitude of which is a function of the spatial discretization and time step size chosen: error is reduced by increasing resolution, which increases computational time. Furthermore, increased sophistication often corresponds to increased data requirements and an increased number of model parameters to be specified. Section 5.5 presents three examples of numerical modeling for evaluating DNAPL source zone remediation.

Choosing between an analytical and a numerical model is a critical point in the modeling process. Several analytical and semi-analytical models have evolved into screening-level tools. These employ simplified scenarios to give order-of-magnitude estimates, such as for predicting the dissolved-phase mass flux leaving the source zone (e.g., Parker and Park, 2004; Jawitz et al., 2005; Christ et al., 2006). Analytical models and the related screening tools typically focus on the dissolved-phase plume and its response to conditions in the aquifer or source zone. These models are listed and discussed at length in other chapters in this series, including Chapter 6, *Modeling Plume Response to Source Treatment*, and Chapter 13, *Monitored Natural Attenuation of Chlorinated Solvent Source Zones*, of this volume as well as Chapter 6, *Modeling*

Remediation of Chlorinated Solvent Plumes, of *In Situ Remediation of Chlorinated Solvent Plumes* (Stroo and Ward, 2010). These models can be quick to run and easy to use but are limited in the simplicity of the scenarios they consider relative to the complexity of actual sites. They may be useful for considering a single aspect of source zone behavior, such as gross dissolution rate, but provide little information on other aspects (mass distribution, influence of remediation). Moreover, users of the results must pay careful attention to the assumptions built into the model and the implications for interpreting the results. As mentioned in Section 5.1, model simplification is one source of uncertainty in model results and one that is particularly difficult to quantify.

Considering DNAPL subsurface migration and source zone evolution following a DNAPL release typically requires the use of a numerical model for simulation. Numerical models are also required for considering the time-dependent dissolution of DNAPL sources in forming dissolved solvent plumes and simulation of most remediation technologies. Numerical models are also essential for modeling complex, site-specific data (e.g., three-dimensional DNAPL distribution, hydraulic head distribution, heterogeneity of permeability). Numerical models are advantageous for considering multiple implications of a remediation technology; for example, how surfactant flooding will not only affect mass flux but also DNAPL distribution, potential downward remobilization, and surfactant distribution. The decision to use an analytical or numerical model is, therefore, dependent on the objectives of the project. Using a numerical model may require significantly more investment of time and resources, in particular for providing the requisite input data. More input parameters, each associated with uncertainty, introduce the second source of model uncertainty discussed in Section 5.1. Thus, when moving from an analytical/screening model to a numerical model, the user largely replaces the first type of uncertainty (simplification) with the second type (input data uncertainty). The advantage is that this second type of uncertainty is much more readily evaluated and quantified by running sensitivity simulations that examine the range of outputs for a reasonable range of inputs; this valuable procedure is discussed further in Section 5.4. It must be pointed out that numerical models also employ simplifications of reality, and it is important to know the assumptions built into any model when considering it for use and interpreting its results. However, numerical models are better suited than analytical models for evaluating some key simplifications: some processes in the model can be turned on and off to examine the influence of model complexity. For example, a study of bioremediation in fractured rock can compare cases in which reductive dechlorination is assumed to occur in both fractures and matrix versus occurring only in the fractures. Because the complexity of DNAPL source zone remediation is most commonly addressed with numerical models, and analytical/screening models are dealt with elsewhere in this series, the rest of this chapter deals primarily with numerical modeling. However, it is acknowledged that many of the steps described below with respect to the modeling process apply to the use of both types of models.

5.3 THE COMPONENTS OF A REMEDIATION MODEL

Figure 5.1 illustrates the major components of a DNAPL remediation model and how those components interact. Central to the model is the *multiphase flow* code, which contains equations that describe the flow of each fluid phase (DNAPL, groundwater) through the subsurface. Typically, they incorporate a Darcy-type equation relating the flux to the gradient in each phase. These equations are linked since they all depend on the volume fraction each fluid occupies in the pore space (fluid saturations) and the fluid pressures (water pressure, DNAPL pressure, and capillary pressure). Solving these equations simultaneously determines the flow of water and DNAPL as a function of time. Models that permit inclusion of the unsaturated zone will

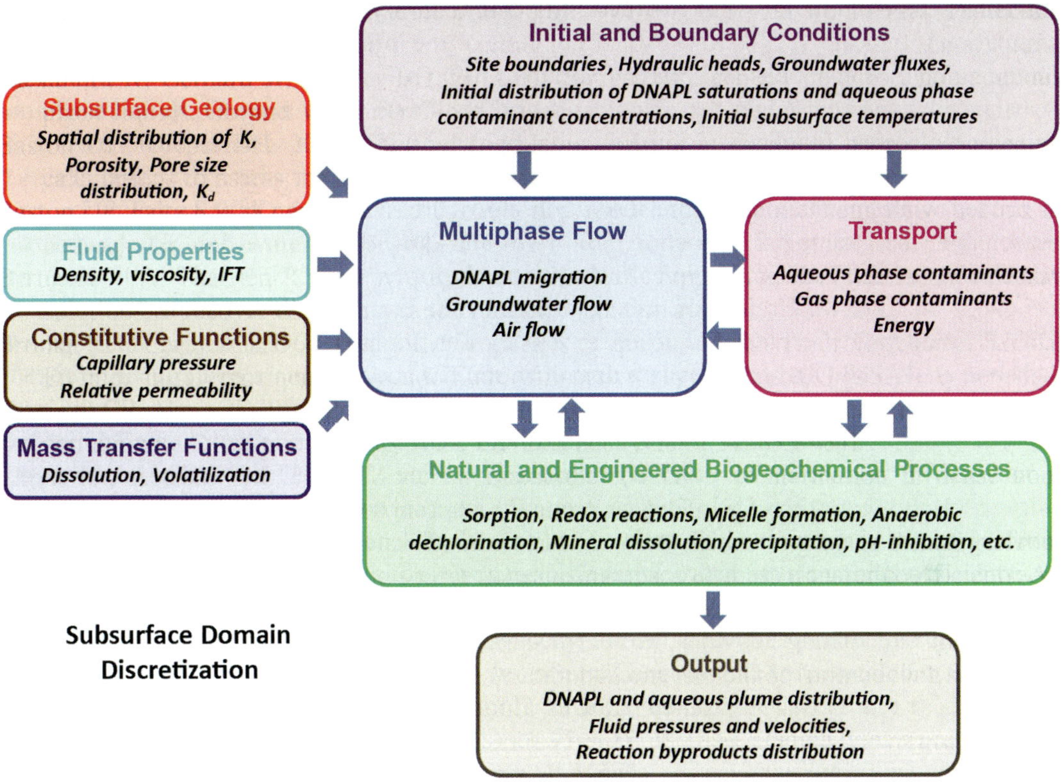

Figure 5.1. Primary components of a DNAPL source zone remediation numerical model. ***Boxes*** **represent key modules and** ***arrows*** **represent the exchange of information between modules. Each module is discussed in the text.**

account for the air phase as well, either as a passive (ambient pressure, static) fluid or as an active phase that can flow in response to pressure gradients (e.g., vacuum extraction). The number and type of governing equations in the numerical model will define the type of scenarios that can be simulated.

Early in the development of a model, a domain is defined by the number of dimensions to be considered in the problem. One-dimensional simulations have limited applicability to DNAPL scenarios apart from laboratory column experiments. Field scenarios are typically described by two- or three-dimensional (2-D, 3-D) domains. Two-dimensional horizontal domains have limited value for exploring DNAPL scenarios because of the importance of variability with depth due to capillary and gravity forces, heterogeneity, and the vertical resolution of monitoring wells. Therefore, two-dimensional representations of a source zone are typically performed in vertical cross sections oriented parallel to the dominant groundwater flow direction. Three-dimensional domains have recently become more commonplace with the availability of increased computing power.

Linked to the decision of the number of dimensions to employ is the overarching decision of grid resolution and, therefore, the total number of elements in the domain. Increased grid resolution (smaller elements) is usually beneficial as it provides higher resolution and a more accurate solution to the governing equations. It may also permit incorporating more realistic subsurface heterogeneity. DNAPL migration is sensitive to variations in capillary properties on scales from meters to centimeters, and the delivery of injected remedial fluids is also highly sensitive to the dominant flow pathways, themselves impacted by the distribution of soils and the architecture of the DNAPL. However, the core memory and time required to complete a simulation is typically related to the square or cube of the total number of elements, depending on the number of equations and primary variables being solved and the number of dimensions. A balance is typically sought that maximizes grid resolution while maintaining run times that fit with the project timelines. Grid refinement or mesh adaptation schemes are employed by some models to increase grid resolution in regions of the domain where increased accuracy is needed while maintaining coarse resolution elsewhere in order to keep computation costs reasonable (see Example 2 in Section 5.5). Two- and three-dimensional DNAPL source zone simulations at the field scale typically adopt on the order of 100,000–1,000,000 elements. Guidance on model discretization can be found in the literature, for example, on grids for DNAPL modeling (Gerhard and Kueper, 2003c), for adequately discretizing heterogeneity (Ababou et al., 1989) and for adequate discretization for aqueous phase contaminant transport (Zheng and Bennett, 2002).

Once the computational grid is defined and oriented, a remediation model requires a number of key inputs in order to adequately define the scenario under consideration (Figure 5.1). *Boundary conditions* define, along the external boundaries of the scenario domain, how the modeled subsurface interacts with the external environment. This may include, for example, a confining clay or bedrock layer along the base of the domain that may be considered effectively impermeable. Wells, whether for municipal/industrial water production or associated with remediation activities, are defined via point sources or sinks applied within the domain at the location of the screened interval(s). Boundary conditions for the DNAPL may include those that define the source, such as a point source with constant DNAPL flow rate representing a cracked pipe, or a patch source representing a broad lagoon from which DNAPL leaks. Boundary conditions can be changed over time; so, for example, the release may be terminated after a defined time or defined volume has entered the subsurface. Boundary conditions can also be employed for simulating certain DNAPL removal approaches, such as passive interceptor trenches and horizontal or vertical dual-phase extraction wells.

Initial conditions are also required, specifying the initial value of every major model variable throughout the domain. For fluid phases, this includes the initial distribution of fluid saturations and pressures throughout the domain. If a DNAPL release into a pristine aquifer is being simulated, then the initial conditions usually comprise a subsurface completely saturated by groundwater below the water table and groundwater pressures set to hydrostatic equilibrium. Instead, if a known DNAPL source zone is being examined, then the initial distribution of DNAPL must be defined. As well, the concentration distribution of each species under consideration must be defined at the start of the simulation. With respect to natural and engineered reactions, initial conditions may include the distribution of the fraction of organic carbon (f_{OC}) for sorption, initial microbial community populations for bioremediation, initial mineral concentrations for pH buffering processes, and initial subsurface temperatures for thermal processes.

Another key input for DNAPL remediation models is the *subsurface geology*. DNAPL migration, as well as the flow of injected remedial fluids and transport of dissolved-phase constituents, is highly sensitive to the subsurface distribution of intrinsic permeability. For this reason, the homogeneous domains sometimes employed for groundwater flow modeling are usually inappropriate for DNAPL source zone modeling. The strength of numerical modeling is the ability to consider in detail the impact of site heterogeneity on DNAPL source zones and their remediation. Figure 5.2, for example, illustrates a heterogeneous permeability field and the distribution of residual (blue) and pooled (green) DNAPL resulting from the release of 2,000 gallons of trichloroethene (TCE) at the water table into a fine- to medium-grained sand aquifer. The distribution of DNAPL illustrated in Figure 5.2 was utilized as the initial condition in

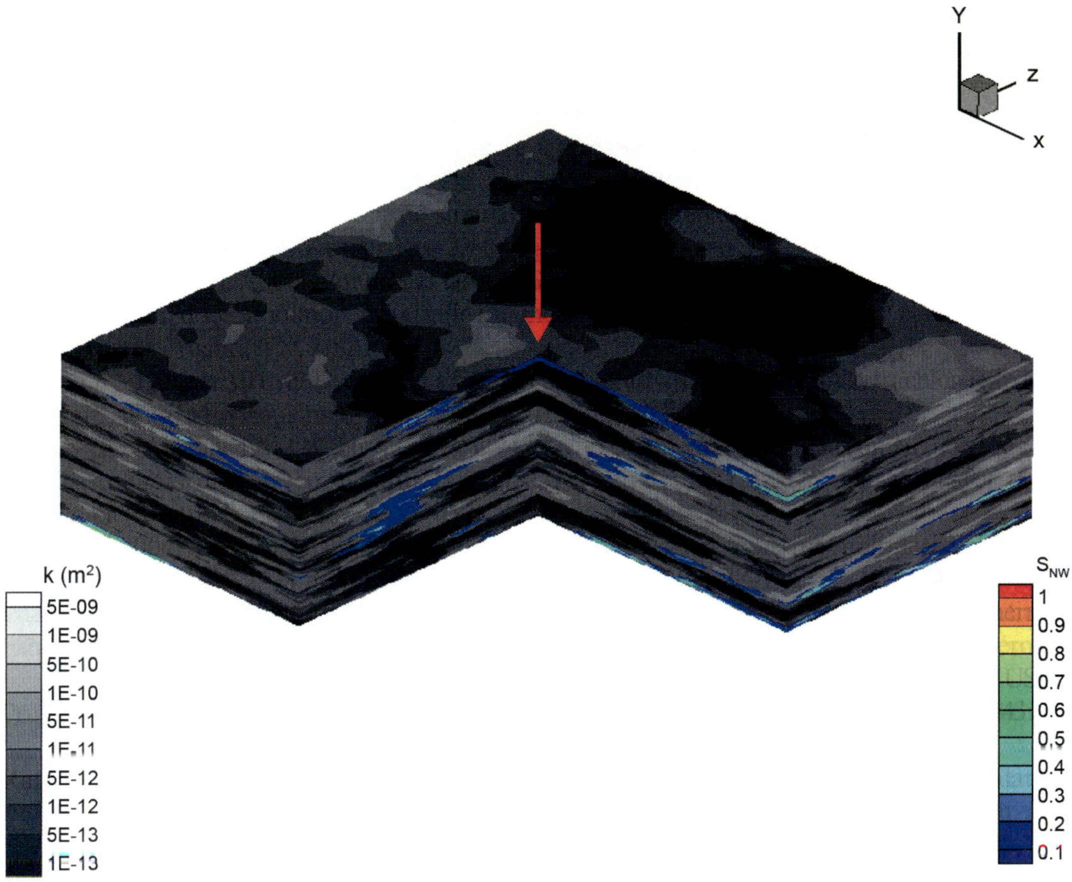

Figure 5.2. Distribution of permeability (*grey scale*) and DNAPL saturation (*color scale*) in a fine- to medium-grained, heterogeneous sand aquifer following the release of 2,000 gallons of TCE (Alexandra et al., 2012). Domain is 20 meters [m] × 20 m × 5 m high; figure employs a 10 m × 10 m × 5 m cutout to reveal the interior.

performing numerical simulations investigating the use of dual-phase extraction (water flooding) on the recovery of pooled DNAPL (Alexandra et al., 2012). The spatial variability of residual and pooled DNAPL illustrated in Figure 5.2 is a direct reflection of the structure and scale of heterogeneity incorporated into the underlying permeability field.

The subsurface geology required to set up a remediation model is typically based on site investigation results such as borehole logs, geophysical surveys, and cone penetrometer results. In a simplified approach, each overall lithological unit may be assumed to have homogeneous properties. However, such an approach must be employed with caution, given the significant impact of aquifer heterogeneity on the development of DNAPL source zones (Figure 5.2). Field experiments have demonstrated that even relatively mild heterogeneity leads to complex and tortuous DNAPL migration pathways (Poulsen and Kueper, 1992; Kueper et al., 1993), corresponding to increased horizontal spreading and increased depth of vertical movement for a given volume of DNAPL released, relative to that expected in homogeneous environments.

There are two primary means of incorporating heterogeneous permeability into a remediation model. One involves including explicit representation of site-specific conditions based on collected data. This can be as simple as incorporating the major lithological units present at a site based on geologic interpretation or incorporating spatial variability of geology using numerical interpolation algorithms such as kriging that honors data at boring locations. A second means of incorporating heterogeneous geology is generating a representation of the permeability field using site-wide statistical properties, with or without conditioning to data at boring locations (Figure 5.2). However, when employing a heterogeneous permeability field based on site-wide statistical properties alone, it is important to acknowledge that each randomly generated field represents only one possible realization of the soil structure associated with the set of prescribed statistical properties. Multiple simulations can be run, each with a different realization of the heterogeneous unit. This approach, known as Monte Carlo simulation, can provide the average values of metrics of interest as well as an estimate of the variance in the metrics. However, to get meaningful statistics, a substantial number of runs are often required (e.g., at least 25 but often upwards of 100), with more needed when significant uncertainty exists in the parameters and correlation between parameters.

It is not only critical to define the permeability values at each location in the model domain, but also to define the associated entry pressure. Entry pressure for a lens is the minimum capillary pressure that must exist at the base of an accumulated DNAPL pool before it will penetrate that underlying lens (capillary barrier). Entry pressure can be measured in the laboratory with site soil, groundwater, and DNAPL, or it can be estimated based on scaling relationships to permeability, porosity, and DNAPL/water interfacial tension (Leverett and Lewis, 1941). The concepts of capillary and entry pressure are described more fully in Section 8.2.1, Fundamentals of Multiphase Flow, in Chapter 8. The distribution of entry pressures in the subsurface has an important impact on DNAPL migration in source zones (Gerhard and Kueper, 2003c). Other parameters that may need to be defined in the context of subsurface geology but that often vary less widely than permeability or entry pressure include porosity, pore size distribution and f_{OC}.

As shown in Figure 5.1, remediation models also require definition of *fluid properties*. These include the density and viscosity of all fluid phases being considered (groundwater, steam, DNAPL, emulsified vegetable oil), as well as the interfacial tension between them. Depending on the complexity of the model's design, other fluid properties that may be required include the compressibility of the fluids, the dependence of fluid density and viscosity on chemical composition, and the aqueous solubility of the DNAPL compounds. Field values of the fluid properties for chlorinated solvent DNAPLs are often considerably different from textbook values for laboratory grade solvents. The employed values may need to be chosen carefully because DNAPL migration and recovery rates are sensitive to density and viscosity, and the extent to which capillary barriers influence DNAPL movement depends on DNAPL/water interfacial tension. While these values are often taken as constants, some processes or technologies may alter fluid properties in time and space. For example, preferential dissolution of the high solubility components of a multicomponent DNAPL may alter the density and viscosity of the remaining nonaqueous phase, and surfactant flooding may cause an increase in chlorinated solvent solubility and a decrease in DNAPL/water interfacial tension locally where the surfactant solution contacts the DNAPL body. Fluid properties are often determined from measurements of site samples, particularly if DNAPL can be recovered from monitoring locations or test pits.

Defining the model scenario also requires *constitutive functions* (Figure 5.1). These are relationships that describe the local scale interaction of the immiscible fluids (air, DNAPL, water) sharing the pore space. They include the capillary pressure–saturation relationship (Pc-S) and the relative permeability–saturation relationship (kr-S). Brief details are provided here to convey their importance to source zone remediation modeling; a more complete description of constitutive functions for multiphase flow in porous media is provided in Section 8.2.1. The Pc-S function defines the relationship between the capillary pressure (the pressure difference between the fluids) and the fraction of porosity occupied by each fluid. All Pc-S functions are hysteretic: the relationship between capillary pressure and saturation is different depending on if DNAPL is invading (and displacing water) or receding (and water is displacing DNAPL). In addition, the Pc-S functions control the amount of residual DNAPL remaining in the soil in the vadose zone and below the water table. Caution is necessary when choosing the mathematical form and the defining parameters for the constitutive functions in a modeling project. A well-known Pc-S relationship is the Van Genuchten (1980) function, which was first developed for water infiltration into soil but has frequently been adopted for DNAPL codes. However, the standard form of this function does not include a distinct entry pressure, which can lead to unrealistic predictions of DNAPL behavior (Gerhard and Kueper, 2003c). A second well-known Pc-S relationship is the Brooks–Corey (Brooks and Corey, 1964) function; its inclusion of a distinct entry pressure ensures DNAPL predictions exhibit the required sensitivity to capillary barriers (Grant et al., 2007). The Pc-S function parameters can be measured in a laboratory with site-specific soil and fluids. Alternatively, they may be assumed based upon more easily measured parameters such as mean grain size and soil type or may be estimated based upon permeability, grain size distribution, and interfacial tension.

Relative permeability–saturation functions are also important components of a constitutive model. One relationship is necessary for each fluid phase; thus, in a groundwater/DNAPL system, there will be a groundwater (or wetting phase) relative permeability–saturation function (krw-S) as well as one for the organic (or nonwetting phase) (krn-S). These functions describe the effective permeability of the soil locally for each fluid as a function of saturation, since the fluids impede each other in the pore space. These functions play an important role in controlling the timescale of DNAPL and groundwater movement in numerical model predictions, thereby directly impacting such results as the rate of DNAPL recovery by dual-phase recovery and the rate of DNAPL dissolution and mass flux in groundwater out of the source zone. The kr-S functions, like Pc-S functions, exhibit hysteresis, and their dependence on the direction of saturation change and saturation history should be accounted for (Gerhard and Kueper, 2003b). The kr-S functions are particularly difficult to measure, and determining them for site-specific soils and fluids is usually possible only in research laboratories. Typically, kr-S functions are parameterized via associated Pc-S parameters (pore-size distribution index) and with expert judgment. Sensitivity simulations can help address uncertainty in constitutive functions and their parameters.

With the model components described thus far in this chapter, a scenario involving the movement of DNAPL could be simulated, such as a near-surface release and subsequent physical recovery via induced hydraulic gradients. However, to simulate DNAPL dissolution and transport in groundwater, additional model components are required. For this purpose, a multiphase flow model must be coupled with a transport model (Figure 5.1). Such models solve the advection–dispersion equation for mass transport of each dissolved-phase chemical species in groundwater. Provided with an initial spatial distribution of dissolved chlorinated solvent and equipped with the appropriate boundary conditions, many of which are common to those required for the multiphase flow code, this module predicts the distribution of dissolved mass at future times. MT3D (Zheng, 1990), for example, is a freely available code for this purpose. The modeling of chlorinated solvent plumes is discussed in detail in Stroo and Ward (2010, Chapter 6). Some models also incorporate an energy transport equation in order to solve for the movement of heat, as required in steam flushing applications. Depending on the domain discretization and the velocity of the groundwater, the numerical algorithms in the model component responsible for solving for mass transport can be significantly more costly in terms of time to complete a simulation than those for multiphase flow.

The coupling of multiphase flow and contaminant transport models requires defining the *mass transfer function(s)* (Figure 5.1). These govern the local rate of DNAPL dissolution into groundwater and the volatilization rate of DNAPL and dissolved compounds to air.

While applied locally to each element, the choice of the mass transfer functions can be highly influential on overall model outcomes such as the predicted lifetime of the DNAPL and the predicted risk at downgradient receptors. The simplest treatment of DNAPL dissolution is to assume an equilibrium process in which the concentration of solvent in groundwater adjacent to DNAPL is always equal to the compound's effective solubility; this typically ensures the DNAPL dissolves at the maximum possible rate. However, this likely is only applicable under a narrow range of conditions. DNAPL dissolution is known, more generally, to be a kinetic (or rate-limiting) process where the rate of mass transfer is a function of a number of local factors such as the groundwater velocity, local contaminant concentration in groundwater, and the interfacial area (IFA) between the DNAPL and groundwater. Since the IFA is usually not known, a number of empirical mass transfer functions have been derived from laboratory experiments (Powers et al., 1994; Saba and Illangasekare, 2000). Caution is required when employing an empirical mass transfer relationship because their validity for field-scale modeling is unknown and it has been shown that simulations employing different functions predict very different rates of DNAPL dissolution for otherwise identical conditions (Kokkinaki et al., 2013). Recently, IFA has become directly measurable via synchrotron imaging experiments. As a consequence, it has been possible to validate numerical routines that predict DNAPL/water IFA and to use these routines as the basis for mass transfer functions that avoid empirical correlations (Grant and Gerhard, 2007a). A DNAPL remediation model using an IFA-based mass transfer function was shown to successfully predict the complete dissolution of a complex DNAPL spill in a heterogeneous, two-dimensional sand box experiment (Grant and Gerhard, 2007b).

Essential components of a remediation model are the natural and engineered biogeochemical processes that correspond to specific remediation technologies (Figure 5.1). These modules take into account the specific impacts of remedial approaches on the properties of the fluid phases, the dissolved mass, the soil, and the interactions between these processes and compartments. This model component also handles the sorption of dissolved compounds onto soil particles and the retardation this imparts to the groundwater plume. A variety of sorption isotherms may be available, and some models permit assigning a heterogeneous distribution of f_{OC} to the subsurface. The interactions between subsurface processes are critical as there are often important feedback loops between model components (Figure 5.1). For example, DNAPL saturation impacts groundwater velocity, which impacts the DNAPL mass transfer rate, all of which are affected by interactions with surfactant micelles. Moreover, there is significant scope for variability in the complexity of remediation modules, and it is necessary to understand the details of the processes included and excluded in order to understand the significance of the modeling results.

For example, consider a reaction module that accounts for chemical oxidation by potassium permanganate. At its most simple, it could be designed to simply eliminate any DNAPL that contacts permanganate. If designed to be more comprehensive, it could take into account (1) the stoichiometry of the reaction of the oxidant and dissolved chlorinated solvent, eliminating the appropriate amounts of each, (2) the kinetic rates of the reactions and their dependence on the local concentrations of both the solvent compounds and the oxidant, (3) the production of manganese dioxide as a by product of the reaction and its tendency to clog the pore space and encapsulate DNAPL, (4) the consumption of oxidant by organic aquifer material and (5) the influence of organic carbon on sorption and the reduction in the local sorptive capacity of the aquifer as organic aquifer material (OAM) is destroyed.

As another example, bioremediation reaction modules can also vary greatly in their complexity. Most straightforward is first-order, sequential anaerobic dechlorination. From that starting point, there exist a wide range of models with increasing complexity (for a list, see the table provided in Kouznetsova et al., 2010). Advanced reaction modules can include (1) multiple bacterial communities (fermentors, sulfate reducers, various dechlorinating consortia) whose populations can grow, decay, and migrate; (2) complex organic substrates that can transport and partition into DNAPL; (3) advanced kinetic reaction rate equations with multiple dependencies (on hydrogen concentration, solvent concentration, bacteria concentration); (4) bioclogging of pore space; (5) pH sensitivity of bacteria and a wide array of aqueous redox and mineral geochemical reactions that influence pH; and (6) chlorinated solvent toxicity factors and competitive interferences. Similar long lists of possible reactions and processes can be catalogued for the majority of remediation technologies.

The set of reactions and processes that are appropriate to include in a numerical model for each remediation technology is an open question. There is a trade-off to be considered: on one hand, a more complex model is a model that more fully represents reality and therefore has the potential to provide more accurate simulations with less reliance on simplifying assumptions. On the other hand, the associated increase in model complexity corresponds to increased run time, additional model variables that require measurement or estimation and often less extensive model validation (less confidence in the model). Assumptions are inherent in any modeling exercise and are important for reducing complexity and thus making the problem manageable. However, those assumptions should be made explicit so that care can be taken to ensure that the results are considered within the bounds of those assumptions. For example, if the DNAPL body is assumed to be immobile – an implicit assumption for source zone modeling that does not include a multiphase flow model component – it must be recognized that the modeling results cannot address the question of whether remediation activities might remobilize the DNAPL.

The output generated by a model is the ultimate product of the modeling process and has the potential to provide a comprehensive picture of the state of all of the variables and processes under consideration at the end of the simulation (Figure 5.1). For the multiphase flow model, this typically includes images that reveal the spatial distribution of DNAPL, and the pressure and velocities in both groundwater and DNAPL phases over time. With respect to the transport and reaction modules, images are produced of the dissolved species plumes and for all reactants (e.g., potassium permanganate) and some products (e.g., manganese dioxide). Snapshots of these as a function of time are typically available, permitting the generation of animated sequences that demonstrate the evolution of key processes of interest with time. Quantitative data obtained as a function of time include mass balances on DNAPL and dissolved-phase chemical species, mass fluxes crossing boundaries, concentrations at wells, average DNAPL saturation, average capillary pressure, pool–ganglia ratios, etc. Obtaining the most valuable model output is often contingent upon carefully defining the problem and the information required to provide insight into its solution in the early stages of the modeling process.

5.4 THE MODELING PROCESS

There is a process, or sequence of steps, that is common to virtually all modeling exercises; these are illustrated in Figure 5.3. The first step is to *define the objectives* of the project. The objectives should be defined explicitly and in detail in order to be a strategic guide to the type of model chosen, the degree of modeling complexity undertaken, the number of simulations, and the type of output generated and analyzed. Examples of modeling objectives include the following:

- Examine the influence of DNAPL fluid properties on the rate and extent of its migration following release.
- Recreate a historical DNAPL release and consider the amount of DNAPL that may have crossed the property boundaries by a specified time.

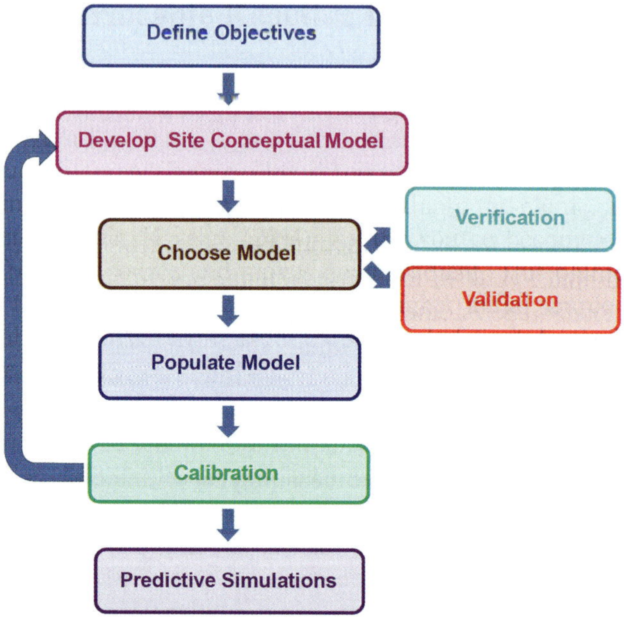

Figure 5.3. Outline of the modeling process. Details of each step in the process are presented in the text.

- Optimize the number, location, and pumping rates for dual-phase extraction wells at a DNAPL site.
- Compare enhanced *in situ* bioremediation to *in situ* chemical oxidation strategies for a specific site.
- Examine the influence of fracture density and orientation and forward- and back-diffusion processes on the potential success of DNAPL bioremediation in fractured bedrock source zones.
- Examine the sensitivity of the success of thermal treatment at a site to key site characteristics and engineering design parameters.

The second step in the modeling process is to *develop a conceptual site model* that appropriately represents the scenario under consideration (Figure 5.3). The conceptual site model is similar to that which underpins the development of a site investigation plan and list of site remedial options. Information is required to generate the initial and boundary conditions required for the model (as described more fully in Section 5.2). This includes determining the distribution and variability of hydraulic conductivity, estimating the distribution of DNAPL and aqueous phase mass and verifying groundwater flow characteristics. An important component of defining the conceptual site model is defining the processes of significance for the modeling scenario, the site, and the technologies under consideration. It is through this stage that literature, laboratory, and field data are collected with which appropriate values can be assigned to all model parameters. Furthermore, parameters with significant uncertainty may be designated for investigation with sensitivity simulations.

The third step involves *choosing the model* for the study (Figure 5.3). It is necessary to ensure that the model has the processes and attributes appropriate to the developed conceptual site model and therefore able to address the project objectives. It is also important to ensure that the model has been verified and validated for the processes simulated. *Verification* refers to the procedure of ensuring that the coded equations are correctly solved; this is usually accomplished by comparing numerical results to those of analytical solutions. For this reason, verification is usually only possible for simple cases (see Section 5.2) and typically not for many remediation-specific processes. Other tests are therefore valuable to provide confidence that the remediation reactions and interactions are coded and solved correctly. *Validation* refers to the procedure of ensuring that the model accurately predicts real-world system behavior. This is usually accomplished by comparing model predictions to detailed laboratory experiments for each modeled process. It is worthwhile considering the validation of each separate model component (Figure 5.1), such as DNAPL migration, DNAPL dissolution, and energy transport. Validation for remediation processes is the most challenging because of the complexity of running and comprehensively measuring *in situ* behavior for many advanced remediation processes. Despite the challenge, model validation is an important step in providing confidence that the model is reliable for the intended purpose and at the scale of interest.

A wide variety of models are available. As mentioned above, analytical or semi-analytical models are typically employed for simple and rapid screening level calculations. Table 5.1 provides examples of existing models, focusing specifically on those numerical multiphase flow codes that include remediation modules. If an existing model is not appropriate for a particular project, then an existing model can often be modified, either by the user if the code is open source or in collaboration with the developers. Of course, developing a new code is also an option and the references provided in Table 5.1 provide details on model development options.

Once a model has been chosen, the next task is to populate the model with representative site parameters. Typically, the data requirements of the model increase with the degree of sophistication of the conceptual site model and the complexity of the processes simulated. Analytical tests and/or microcosm, batch, or column laboratory experiments with site soil, groundwater, and DNAPL may be beneficial in providing some of the more critical model parameters, such as partitioning coefficients, microbial concentrations, reaction rates, constitutive function parameters, and DNAPL fluid properties. Often a project can benefit from an iterative process of data gathering and preliminary simulations to further identify key data that would reduce uncertainty in the results.

The next step is to calibrate the model to demonstrate that it can adequately simulate measured, site-specific results. Typically, this entails utilizing the constructed model to simulate one or more field pilot tests (a single injection well bioremediation pilot, a localized dual-phase extraction test, a conservative tracer, and/or partitioning tracer test). As part of this exercise, it is not uncommon to tune the model, often varying a key unknown (mean hydraulic conductivity, first-order biodegradation rate), to ensure the best match to the results. The objective is to develop confidence that the model has adequately incorporated the conceptual model and site data so as to provide meaningful results at the full scale of the problem of interest. As illustrated in Figure 5.3, if the model cannot be successfully calibrated, this may indicate that some revision of the conceptual model, chosen model, and/or site data may be warranted, for example, including some subsurface process originally neglected and/or conducting additional site investigation. For this reason, there can be significant benefit to planning for a dynamic interaction between phased site investigation and modeling activities.

The final step in the modeling process is conducting predictive simulations. This may include deterministic simulations of specific conditions, such as a set of runs examining the influence of injection strategies for optimizing a flushing-based remediation technology. In addition, this may include a set of sensitivity simulations, which explore the variability of the predicted results to key aspects of the system design or key uncertainties in the model design or in the site data. This may also involve comparative simulations between different technologies applied to a specific site or the behavior of a particular technology across many types of sites.

Table 5.1. Examples of DNAPL Source Zone Remediation Numerical Models

Model	Others in Family	Remediation Technologies	Other Features	Selected References
DNAPL3D	DNAPL3D-MT DNAPL3D-RX DNAPL3D-FRAC	ISCO Bioremediation SEAR STAR	– Multicomponent – Fractured clay/rock – IFA-based DNAPL dissolution – Extensive validation	Kueper and Frind, 1991 Gerhard and Kueper, 2003a, b, c Grant and Gerhard, 2007a, b West et al., 2008 Pang, 2010 MacPhee et al., 2012
COMPSIM		Bioremediation Steam flushing ERH	– Multicomponent – Temperature dependence – Gas phase transport – Energy transport	Sleep and Sykes, 1993 McClure and Sleep, 1996 O'Carroll and Sleep, 2007
UTCHEM		Bioremediation SEAR Cosolvent/polymer flooding	– Multicomponent – Microemulsions – Geochemical reactions – Freely available	Delshad et al., 1996 Pope and Delshad, 2000
TMVOC	T2VOC TOUGH2 TOUGH2-MP TOUGHREACT	Bioremediation SVE Steam flushing	– Two-region continua – Massively parallel version – Geochemical reactions	Falta et al., 1995 Pruess et al., 1999 Pruess, 2004 Pruess and Battistelli, 2002 Xu et al., 2008
MVALOR	VALOR MISER	SVE Bioventing SEAR Bioremediation	– Multicomponent – Density-dependent flow	Abriola et al. 1992, 1997 Rathfelder et al., 2000 Christ and Abriola, 2007
STOMP		SVE Steam flushing		White and Oostrom, 2003 White et al., 2004
CompFlow	CompFlow Bio	SVE Bioremediation (aerobic + anaerobic)	– Multicomponent – Fracture networks	Unger et al., 1995, 1996 Slough et al., 1999
MOFAT	MOTRANS	SVE		Kaluarachchi and Parker, 1988, 1990
NUFT	NUFT-C	ERH	– Multicomponent – Energy transport – Massively parallel version	Nitao, 1998 Carrigan and Nitao, 2000 Glassley et al., 2001

ERH electrical resistance heating, *ISCO in situ* chemical oxidation, *SEAR* surfactant-enhanced aquifer remediation, *STAR* self-sustaining treatment for active remediation, *SVE* soil vapor extraction

The complexity of DNAPL source zones, subsurface heterogeneity, and remediation reactions suggests caution in pursuing the objective of predicting the actual outcome of applying a technology at a specific site. Modeling may be more confidently applied for comparing and contrasting scenarios of interest to provide insight into the most likely historical behavior, most promising courses of action or worst-case scenario. In addition, models have a key role to play in quantifying and evaluating uncertainty (due to uncertainty in field data) and in reducing that uncertainty (conducting a supplemental field investigation to quantify parameters revealed to be associated with the highest uncertainty). Such modeling, properly conducted, has substantial power to improve site-specific remedial outcomes and reduce capital and operation and maintenance costs for site investigation and long-term management. The examples provided below illustrate this while underscoring the concepts described in the preceding sections.

5.5 EXAMPLES

5.5.1 Numerical Simulation of DNAPL Remediation in Unconsolidated Media Using Chemical Oxidation

In this study (West and Kueper, 2012), the objective was to develop a better understanding of the influence of key site characteristics on the success of *in situ* chemical oxidation (ISCO). Modeling was an obvious approach due to the lack of field remediation data that covers the range of site characteristics, and numerical modeling was chosen due to the complexity of processes involved in ISCO as well as the key role of subsurface heterogeneity. The model used was West et al. (2008), which is a three-dimensional DNAPL remediation model (DNAPL3D-RX) designed for simulating ISCO of TCE or PCE using potassium permanganate. The model has been extensively verified (e.g., Gerhard et al., 2001; West et al., 2008) and also validated against laboratory experiments for DNAPL migration in one (Gerhard and Kueper, 2003a, b) and two dimensions (Grant et al., 2007), DNAPL dissolution of pools and residual in two-dimensional heterogeneous porous media (Grant and Gerhard 2007b), and ISCO in one-dimensional columns (West et al., 2008). The model was applied to examine the influence of geologic heterogeneity and DNAPL mass on the remediation of TCE and PCE in a three-dimensional, heterogeneous sandy aquifer. The model incorporates a suite of processes associated with ISCO, such as a linear relationship between soil permeability and the concentration of manganese dioxide precipitate (rind); this allows the model to account for local permeability reductions due to precipitate formation. It also incorporates second-order reactions between the oxidant and the target contaminant and between the oxidant and naturally occurring OAM (natural oxidant demand). The model also incorporates some simplifying assumptions, such as an equilibrium relationship for dissolution of DNAPL into flowing groundwater. The model was validated against column experiments of ISCO that included the mass transfer limitations and permeability reductions incurred due to rind formation (West et al., 2008).

The study employed a suite of sensitivity simulations in which a base case was established and then each site characteristic was varied individually to observe its influence on the results. Figure 5.4 illustrates the model configuration for the base case simulation. A three-dimensional domain was chosen due to the importance of heterogeneity on bypassing and contact time (and rind formation) between the injected remedial fluid and the DNAPL. The generated heterogeneous permeability field is characterized by a mean of 3×10^{-12} m^2, a permeability variance of 1.74 (for ln k), a horizontal correlation length (exponential model) of 3 m (isotropic in the horizontal plane), and a vertical correlation length of 0.2 m. The domain discretization (0.4 m by 0.4 m by 0.05 m blocks, total of 125,000 nodes) was chosen to ensure sufficient resolution of

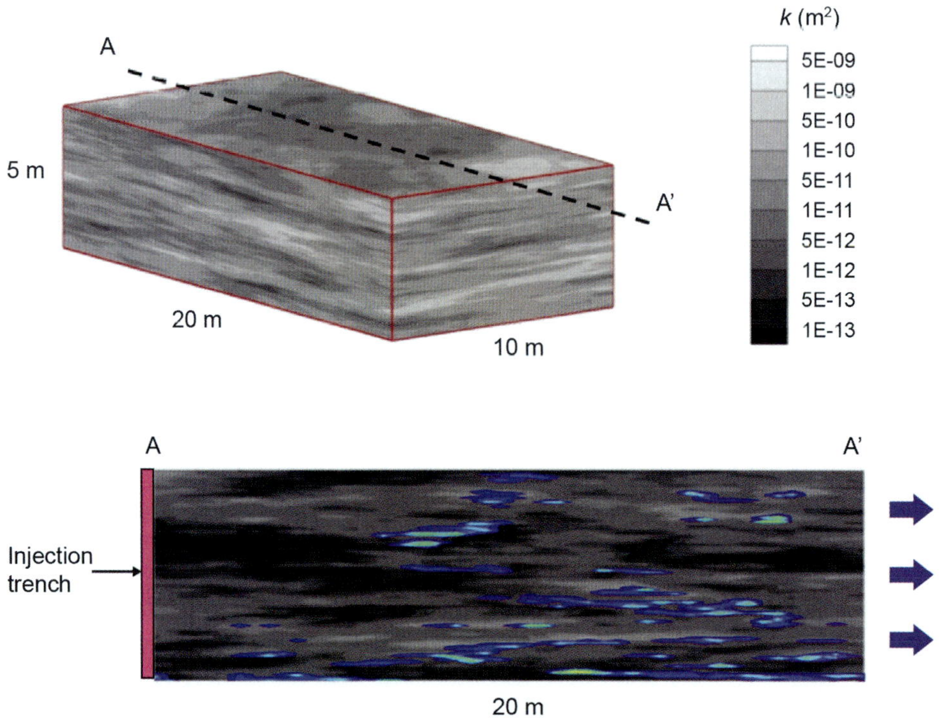

Figure 5.4. Distribution of permeability in three-dimensional heterogeneous sandy aquifer employed in the base case ISCO example simulation (*upper portion of figure*). Cross-section A–A' illustrates the initial distribution of DNAPL residual and pools along the centerline of the three-dimensional aquifer (*lower portion of figure*). The color scale for the DNAPL saturation distribution is presented in Figure 5.2. Oxidant is injected into a trench along the upgradient end of the domain with subsequent flow to the right subject to a hydraulic gradient of 0.05. From West and Kueper (2012).

heterogeneity while ensuring reasonable run times for the model. The model domain incorporates a spatially variable distribution of naturally occurring organic carbon that is negatively correlated to permeability with a mean value of 0.003 and a variance of 0.24 (ln f_{OC}). These parameters correspond to a moderately heterogeneous sandy aquifer with a moderate degree of naturally occurring organic carbon. Potassium permanganate is injected into a trench along the upgradient face of the domain and allowed to migrate subject to a hydraulic gradient of 0.05.

Table 5.2 summarizes the suite of simulations performed in this example, with the single parameter altered from the base case highlighted in bold type (West, 2009). Simulation 1 represents the base case, with simulations 2 through 5 representing sensitivity to the permeability distribution. Simulations 6a and 6b evaluate sensitivity to whether the initial DNAPL distribution was subject to hydraulic displacement (removal of pooled DNAPL), simulation 7 represents a larger volume DNAPL release, and simulation 8 incorporates PCE DNAPL rather than TCE DNAPL.

For each run, the model produced data on all of the aqueous species concentrations as well as DNAPL saturations (at each node, as domain averages, and at flux fences such as the downgradient boundary) as a function of time. There is only room here to present a few select results. Figure 5.5 illustrates the distribution of DNAPL saturation, aqueous phase TCE concentration, and oxidant concentration at $t = 1$ month and $t = 2.33$ years of oxidant injection for the base case. Oxidant is injected at a concentration of 2.5 grams per liter (g/L) continuously for 2.33 years in the base case. Figure 5.5 clearly illustrates the influence of heterogeneity in

Table 5.2. Numerical Simulations Investigating ISCO Effectiveness on DNAPL in Unconsolidated Media (the Single Parameter Altered from the Base Case Highlighted in Bold Type) (West and Kueper, 2012)

Template Site	Simulation Number	DNAPL	Initial Mass (kg)	Initial Volume (m^3)	Mean *k* (m^2)	Mean ln *k* (ln (m^2))	Variance ln *k* (ln (m^2))^2
Base case	**1**	**TCE**	**3,520**	**2.41**	**3.03×10^{-12}**	**−27.36**	**1.74**
High mean *k*	2	TCE	3,496	2.39	**3.02×10^{-11}**	−25.06	1.74
Low mean *k*	3	TCE	3,535	2.42	**3.04×10^{-13}**	−29.66	1.74
Low heterogeneity	4	TCE	3,355	2.30	1.87×10^{-12}	−27.43	**0.87**
High heterogeneity	5	TCE	3,186	2.18	7.41×10^{-12}	−27.26	**3.48**
Small DNAPL volume (post-HD)	6a	TCE	785	**0.54**	3.03×10^{-12}	−27.36	1.74
Small DNAPL volume (pre-HD)	6b	TCE	803	**0.55**	3.03×10^{-12}	−27.36	1.74
Large DNAPL volume	7	TCE	7,343	**5.03**	3.03×10^{-12}	−27.36	1.74
High-density DNAPL	8	**PCE**	3,871	2.37	3.03×10^{-12}	−27.36	1.74

HD hydraulic displacement, *kg* kilogram(s)

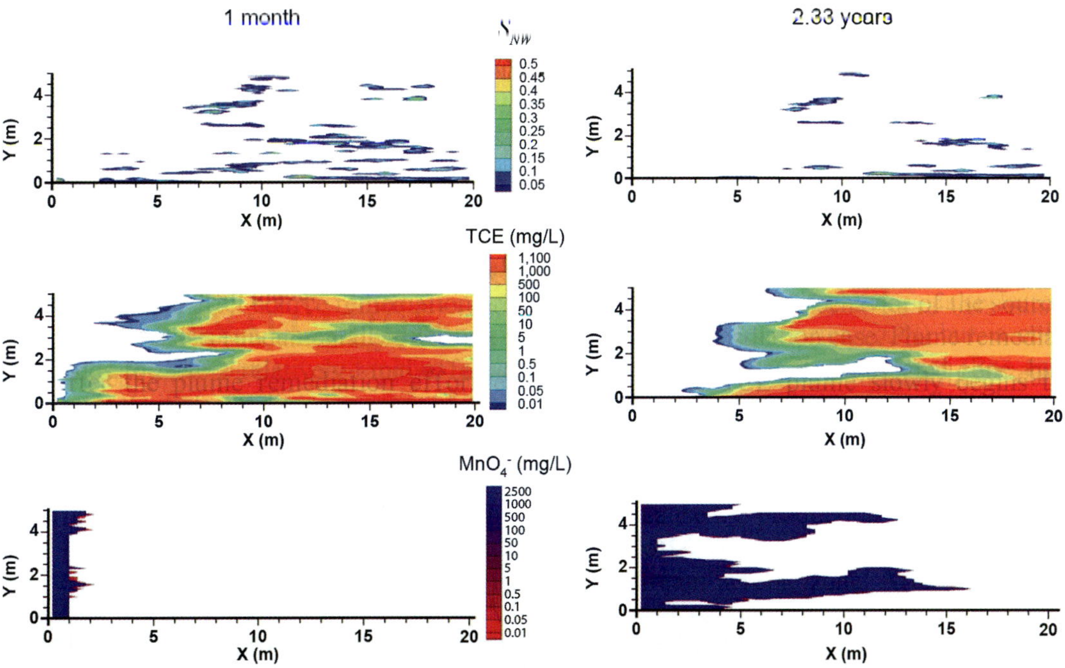

Figure 5.5. Distribution of DNAPL saturation (*upper panel*), aqueous phase TCE concentration (*middle panel*), and oxidant concentration (*lower panel*) for the base case simulation at $t = 1$ month (*left column of panels*) and $t = 2.33$ years (*right column of panels*) along cross-section A–A' (West, 2009). *Note*: mg/L – milligram(s) per liter.

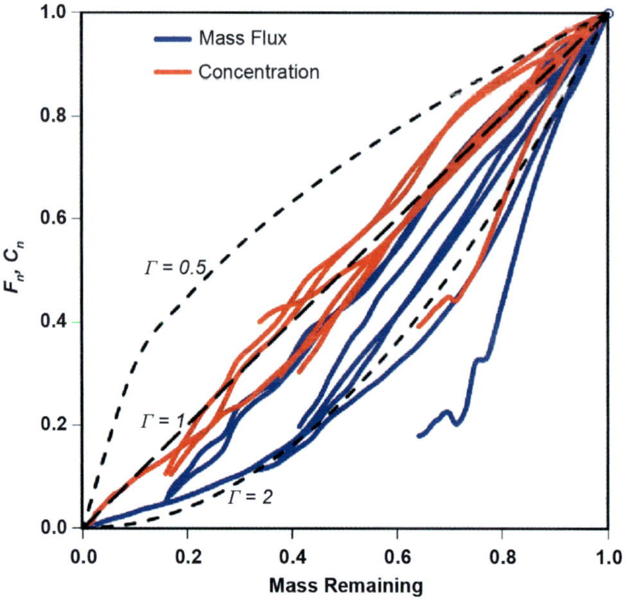

Figure 5.6. Normalized concentration (C_n) and normalized mass flux (F_n) at the downgradient domain boundary versus mass of DNAPL remaining in the source zone for the suite of simulations outlined in Table 5.2. The parameter Γ is intended to account for source zone architecture and heterogeneity in the relationship $F_n = \textit{Mass Remaining}^{\Gamma}$. From West and Kueper (2012).

permeability on the migration pathways of oxidant (lower panel) and the corresponding reduction in DNAPL saturation (upper panel) and reduction in aqueous phase TCE concentrations (middle panel). Preferential pathways for oxidant migration (higher permeability zones) are seen towards the top and bottom of the section within which the majority of contaminant removal takes place.

One of the main benefits of a sensitivity study is a quantitative comparison between key metrics from all of the simulations. Figure 5.6 summarizes the results of the suite of simulations in terms of normalized concentration (C_{n}) and normalized mass flux (F_{n}) at the downgradient domain boundary versus mass of DNAPL remaining in the source zone. Concentration and mass flux are normalized with respect to values at $t = 0$ when remediation begins. A mass remaining of 1.0 represents the initial mass of DNAPL in the source zone (Table 5.2), while a mass remaining of 0.0 corresponds to complete removal of all DNAPL. Figure 5.6 illustrates that mass flux decreases more quickly than concentration in all cases. This stems from the fact that the oxidant solution migrates more readily in higher permeability zones than lower permeability zones, which in turn have the greatest influence on mass flux leaving the source zone. The most beneficial result is represented by simulation 7, which assumed the largest DNAPL volume released.

5.5.2 Application of *In Situ* Chemical Oxidation for DNAPL Remediation in Fractured Rock

This study focused on the performance of ISCO in fractured rock and the key factors affecting success. The model chosen was DNAPL 3D-RX-FRAC which permits simulations of source zone remediation in fractured clay and rock by employing an adaptive grid refinement routine; this achieves high spatial accuracy in the vicinity of fractures, thereby ensuring that DNAPL migration through fractures as well as forward and backward matrix diffusion is

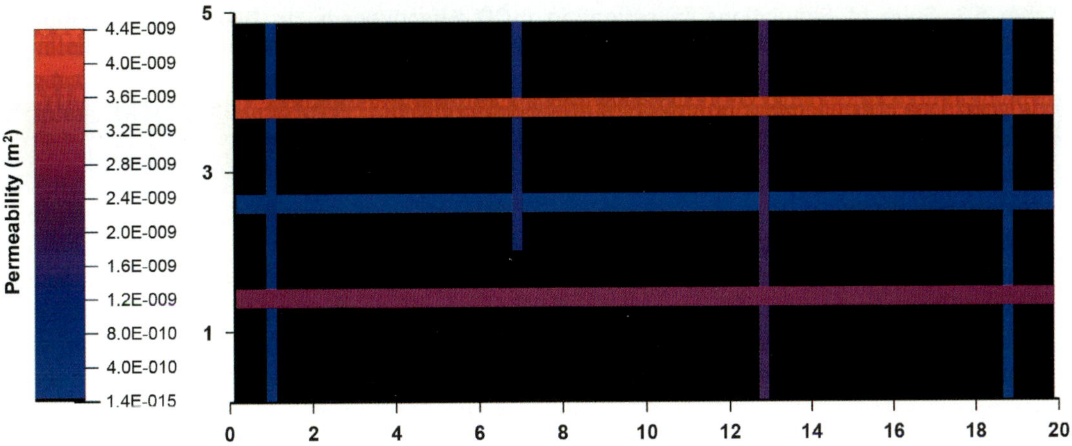

Figure 5.7. Distribution of intrinsic permeability for the base case sandstone domain for two-dimensional ISCO simulations in fractured rock. The orthogonal fractures are constant aperture within each fracture, but apertures vary between fractures. The horizontal and vertical distances are in meters. Note that, in this and subsequent figures, the domain is vertically exaggerated, and the fracture apertures are enhanced, to assist visualization (since the actual width of each fracture is equal to its aperture, they would not otherwise be visible at this scale) (Pang, 2010).

accurately simulated (Pang, 2010). The model was utilized to explore the influence of site characteristics (fractured rock type, DNAPL type) and engineering techniques (oxidant concentration, continuous versus pulsed injection) on ISCO remedial performance. This work was conducted in two dimensions (cross section) because of the computational expense of multiphase flow and transport modeling of fractured rock domains. As a consequence, while manganese dioxide formation was included, it was not permitted to clog the pore space since that was considered to have unrealistic consequences on a two-dimensional flow field. In all other respects, the chemical reaction module of West et al. (2008) was employed (see previous example 5.5.1). Since validation experiments were not available apart from those already cited for sand columns, confidence in the model was built by comparison against analytical solutions for fractured transport and matrix diffusion for simplified (single fracture intersection) scenarios (Pang, 2010).

The domain was designed to be large enough to represent a reasonable portion of a DNAPL source zone in fractured rock, while limiting computational expense. Representative sandstone and fluid parameters were obtained from field and laboratory studies. The 5 m (16 feet [ft]) deep by 20 m (66 ft) long cross-sectional domain was characterized by a distribution of fracture apertures (25–230 micrometers [μm]) and spacing of 6 m (20 ft) for vertical and 1 m (3.3 ft) for horizontal fractures (Figure 5.7). Each fracture was assigned a single fracture aperture, but fracture apertures were varied across the domain. The DNAPL entry pressures varied as a function of fracture aperture, meaning that DNAPL would preferentially enter the largest fractures. The matrix permeability was five orders of magnitude lower than the smallest fracture, and its entry pressure was sufficiently high that DNAPL never entered the matrix porosity. The matrix was assigned a porosity = 7.7% and f_{OC} = 0.05%, while the fractures were assigned porosity = 1 and f_{OC} = 0. The domain was discretized into 3,072 nodes, with matrix nodes far from fractures exhibiting dimensions 0.25 m × 0.25 m and resolution increasing exponentially (e.g., 0.00025 m × 0.00076 m adjacent to vertical fractures), and the fracture nodes themselves were assigned a width equal to their aperture (e.g., 0.000125 m). The adaptive grid refinement scheme permitted the simulation time for one run to be reduced from several months to several weeks on a desktop personal computer.

In each of the simulations conducted, a DNAPL release was first simulated across the top of the domain and then redistribution allowed until the DNAPL stopped moving. The boundary

Table 5.3 Suite of Simulations Investigating ISCO Effectiveness on DNAPL Remediation in Fractured Rock (the Single Variable Altered from the Base Case Identified in Bold Type) (Pang, 2010)

Template Site	Simulation No.	Bedrock	DNAPL	$KMnO_4$ Concentration (mg/L)	$KMnO_4$ Delivery Strategy
Base case	**1**	**Sandstone**	**TCE**	**2,500**	**Continuous injection for 2 years**
Bedrock 2	2	**Shale**	TCE	2,500	Continuous injection for 2 years
Bedrock 3	3	**Granite**	TCE	2,500	Continuous injection for 2 years
High DNAPL density	4	Sandstone	**PCE**	2,500	Continuous injection for 2 years
High $KMnO_4$ concentration	5	Sandstone	TCE	**5,000**	Continuous injection for 1 year
Low $KMnO_4$ concentration	6	Sandstone	TCE	**500**	Continuous injection for 10 years
Short $KMnO_4$ pulses	7	Sandstone	TCE	2,500	**3 months on/3 months off for 4 years**
Medium $KMnO_4$ pulses	8	Sandstone	TCE	2,500	**6 months on/6 months off for 4 years**
Long $KMnO_4$ pulses	9	Sandstone	TCE	2,500	**12 months on/12 months off for 4 years**

conditions permitted DNAPL to exit the domain through the sides and bottom, such that this represented a portion of the source zone near the release. This was followed by a 20-year site-aging period during which an ambient, horizontal hydraulic gradient of 0.005 from left to right was applied. This resulted in DNAPL dissolution, aqueous phase contaminant transport in the fractures, forward matrix diffusion, and solvent sorption onto the OAM in the matrix. Next, a 2-year treatment period was simulated during which potassium permanganate ($KMnO_4$) was injected at a constant concentration along the left boundary. During this period, the hydraulic gradient across the domain was increased to 0.025 to represent active amendment conditions.

In all simulations, the total amount of oxidant injected substantially exceeded the theoretical amount required to destroy all of the contaminant mass in the domain. Finally, a posttreatment stage was simulated where the ambient hydraulic gradient of 0.005 was reapplied for a period of 5 years. This provided an opportunity to consider the longer-term influence of the oxidant as well as quantify concentration rebound associated with back diffusion.

Table 5.3 presents the suite of nine simulations conducted in this study, with the single variable altered from the base case identified in bold type (Pang, 2010). The base case involved the release of TCE DNAPL into fractured sandstone and ISCO applied via a 2-year continuous injection of $KMnO_4$ at 2,500 mg/L. Simulations 2 and 3 explored the same scenario in shale and granite, each exhibiting a distribution of fracture apertures, and matrix characteristics, representative of those rock types. Simulation 4 examined the difference associated with PCE DNAPL in the sandstone domain. Simulations 5 and 6 explored the influence of the $KMnO_4$ concentration injected, while simulations 7–9 explored whether benefit was observed for pulsed injection of the oxidant, while keeping constant the total mass injected.

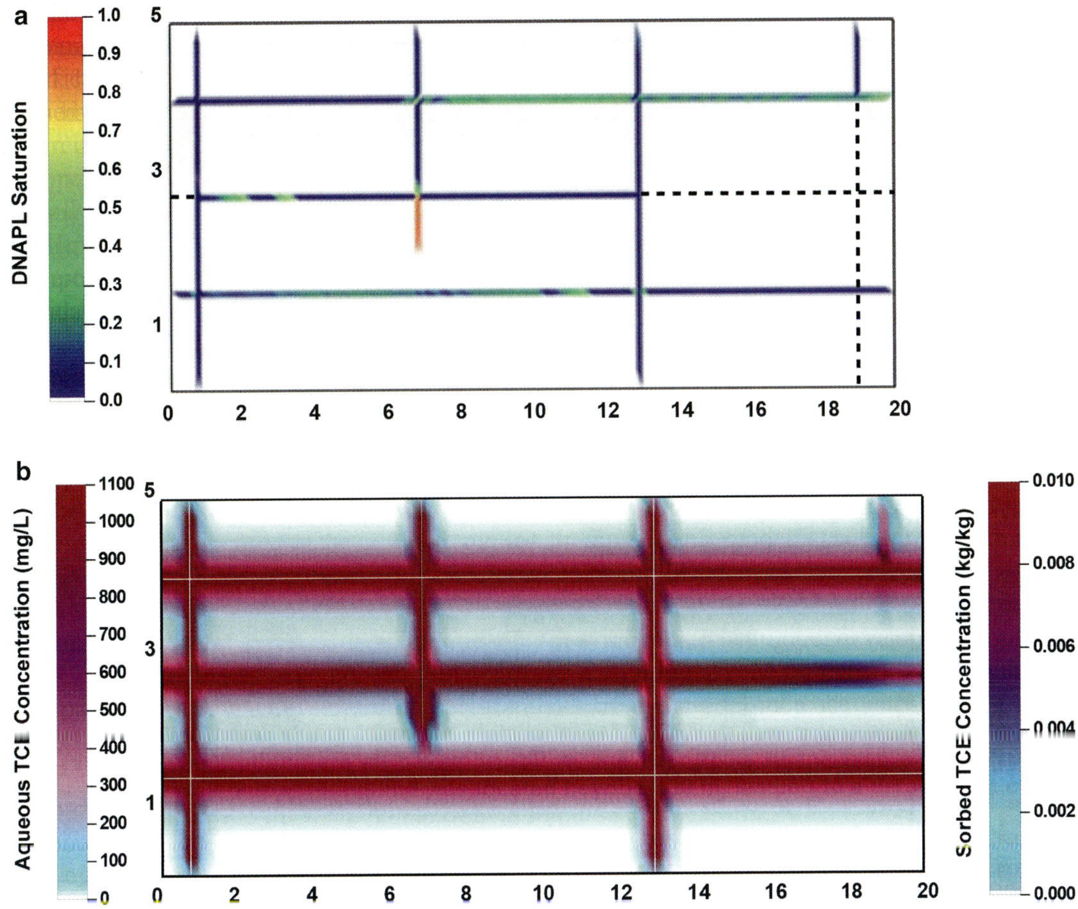

Figure 5.8. (a) DNAPL distribution in the sandstone base case simulation at the end of the DNAPL release period, after all fluids have achieved hydrostatic equilibrium. The DNAPL volume is 0.0036 m^3 (5.3 kg of TCE) and the pool to residual ratio is 71%:29%. DNAPL capillary pressures were not sufficient to invade some fractures, as denoted by the *dashed lines*. Note the highest DNAPL saturations in the vertical dead-end fracture. DNAPL dissolution and diffusion processes were ignored during the DNAPL release period. (b) The distribution of TCE mass after the 20-year aging period under a 0.005 hydraulic gradient causing groundwater flow from left to right. At this time, no DNAPL remains and 99% of the TCE mass resided in the matrix, 97% sorbed, and 2% in the aqueous phase. The significant porosity and fraction of organic carbon of the sandstone matrix combine to make it a substantial sink for TCE (Pang, 2010).

Like the previous example, it is useful with modeling to examine both the qualitative and the quantitative results. Figure 5.8a illustrates the base case scenario at the end of the DNAPL release period, revealing a distribution of connected-phase pools and immobile residual DNAPL associated with the fracture distribution and connectivity within the sandstone. Figure 5.8b illustrates the sandstone base case at the end of the 20-year site-aging period, revealing that all of the TCE DNAPL has disappeared by this time and the majority of the mass (approximately 98%) resided as sorbed phase in the matrix; this is a result of the significant matrix porosity of the sandstone and the relatively high f_{OC}, which provides an important sink for the forward diffusing TCE. Figure 5.9a presents the base case after the 2-year application of potassium permanganate. It is observed that some TCE mass reduction has occurred, primarily in the most permeable horizontal fractures; this is corroborated by the distribution of $KMnO_4$ at that time presented in Figure 5.9b.

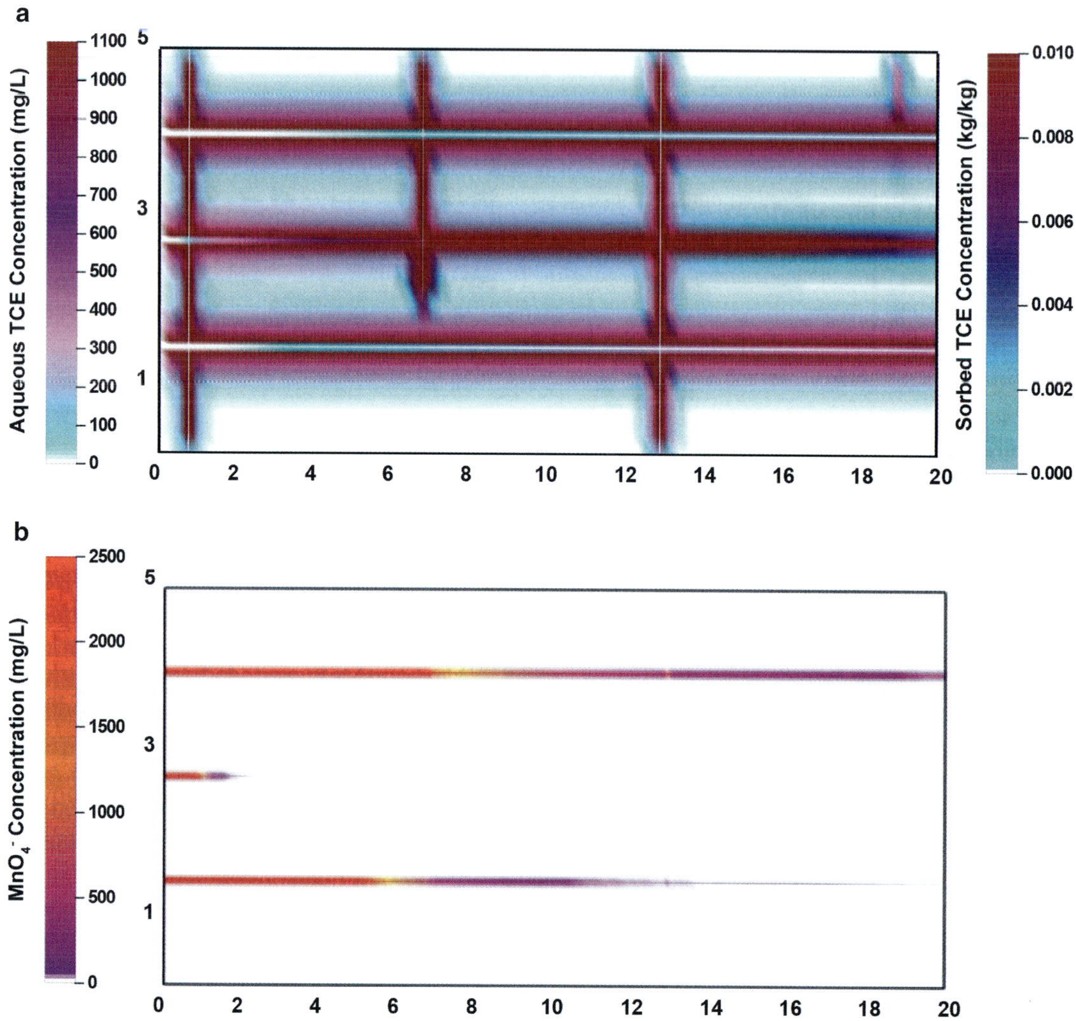

Figure 5.9. (a) The distribution of TCE for the base case at year 22, immediately after the 2-year ISCO treatment period under an engineered 0.025 hydraulic gradient. TCE mass reduction, relative to Figure 5.8b, is observed in the most permeable horizontal fractures due to both oxidation and groundwater flushing. (b) The distribution of MnO_4 at the same time, which reveals the limited penetration of oxidant into the matrix, due to its reaction with TCE in the fractures and its diffusion and subsequent destruction by OAM in the matrix (Pang, 2010).

These figures indicate that, as expected, the majority of the treatment and groundwater flushing associated with the increased gradient has occurred within the fractures and the sorbed and aqueous mass within the matrix is relatively unaffected. As expected, the greatest degree of treatment is associated with the most permeable horizontal fractures (see Figure 5.7). This is confirmed by Figure 5.10, which presents a plot of aqueous TCE concentration in a 1 m (3.3 ft) cross section perpendicular to the uppermost horizontal fracture. Figure 5.10 illustrates how the oxidant flush dramatically reduces the TCE concentration in the fracture during the treatment period, but has almost no effect on the TCE concentration in the adjacent matrix: two years of $KMnO_4$ diffusion has a minor effect on mass loading associated with 20 years of TCE diffusion. Figure 5.10 further illustrates how the 5 year post-treatment period results in a slight decrease of TCE concentration in the matrix in conjunction with a concentration rebound in the fracture due to back diffusion.

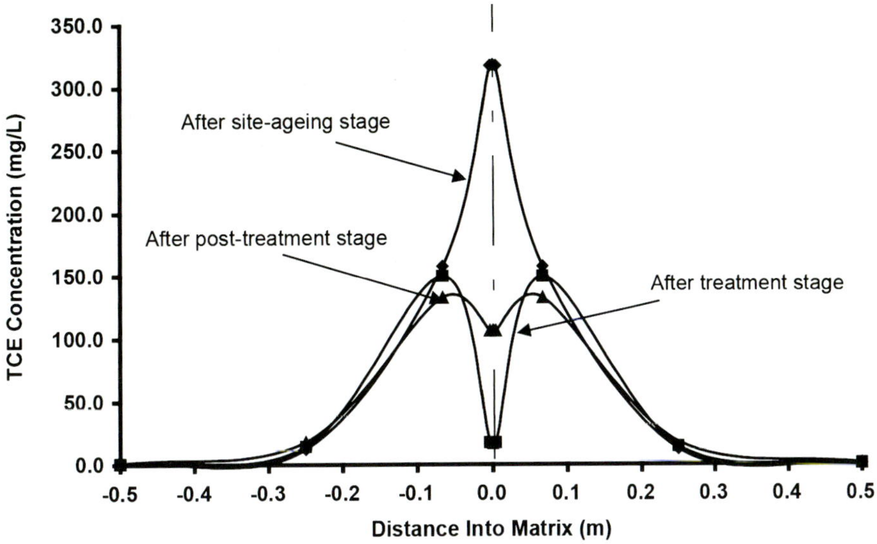

Figure 5.10. Concentration profiles of aqueous TCE perpendicular to the uppermost horizontal fracture during the base case simulation. The fracture is denoted by the *dashed line*. The limited influence of ISCO on TCE in the matrix after the 2-year treatment period is apparent, as is the concentration rebound in the fracture after the subsequent 5 year post-treatment period (Pang, 2010).

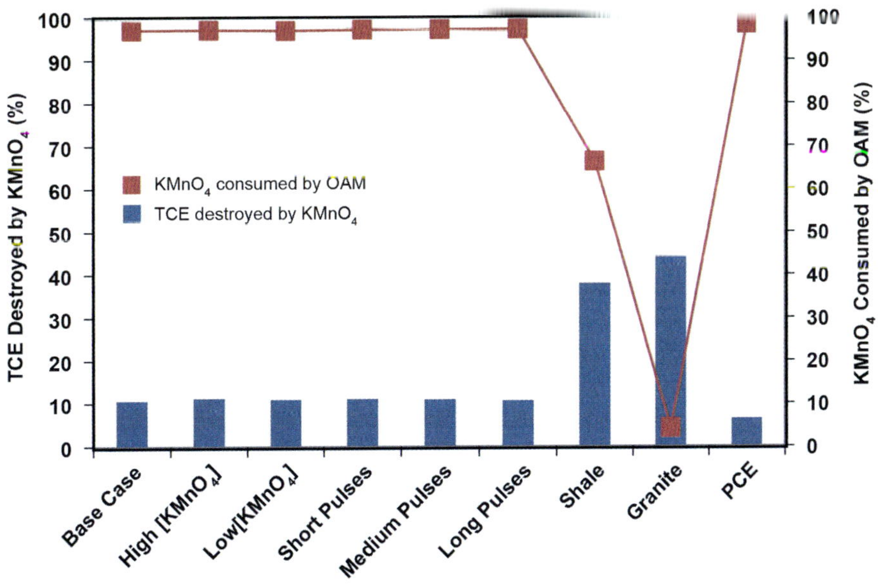

Figure 5.11. Destruction efficiency for the suite of simulations in Table 5.3. Fraction of TCE (DNAPL, aqueous and sorbed) destroyed by oxidant is plotted as *blue bars* using the left-hand axis. Fraction of injected oxidant consumed by organic aquifer material is plotted as *red points* using the right-hand axis (data from Pang, 2010).

Figure 5.11 reveals that across all of the sensitivity simulations, on average less than 10% of the chlorinated solvent mass present in the domain at the start of treatment was destroyed by the injected oxidant. The exceptions are the shale and granite rock types; in these a larger fraction of the TCE was destroyed because a number of factors (larger fractures, reduced effective diffusion coefficients and reduced f_{OC}) meant that matrix loading during the site-aging period was much reduced and thus much more of the TCE mass was retained in the fractures. Overall, these simulations suggest that the efficiency of ISCO injection into fractured rock may be very poor. In shale and sandstone, this occurs because most of the oxidant is likely to be consumed by the OAM in the matrix, while in granite, where OAM is minimal, most of the injected oxidant is likely to flow through the source zone's dominant fractures and thus quickly exit the source zone; in these cases, contaminant mass in the matrix is not effectively treated.

5.5.3 Application of Steam Flushing for DNAPL Remediation

In this study, the goal was to visualize the effects of steam flushing on a DNAPL pool perched on a low permeability lens. Since this is a remediation approach with numerous interacting processes and the objective was to observe all of the associated implications (DNAPL mass reductions, aqueous mass flux, DNAPL redistribution including potential downward migration), a numerical model was required.

The Compsim model (Sleep and Sykes, 1993; O'Carroll and Sleep, 2007) simulates three-dimensional, three-phase flow with DNAPL dissolution and volatilization and energy transport. Fluid and chemical properties (viscosities, densities, solubilities, vapor pressures, Henry's law constants, heat capacities, diffusion coefficients) are temperature dependent. Temperature dependencies of these properties are based on published empirical relationships (e.g., Sleep and Ma, 1997; O'Carroll and Sleep, 2009). Capillary pressure relationships are also temperature dependent, following She and Sleep (1998). Compsim has been applied to simulate hot-water flushing (O'Carroll and Sleep, 2007, 2009) and steam flushing (She, 1997) with validation against two-dimensional laboratory experiments. In addition to the standard fluid and chemical properties required for simulation of multiphase flow and transport, in simulating energy transport, the following are also required: heat capacities of soils, water and DNAPL components; latent heats of vaporization of water and DNAPL components; thermal conductivities of soils and water and DNAPL components in each of the water, gas and DNAPL phases; and critical and boiling point temperatures of water and DNAPL components.

Compsim is applied here to demonstrate the use of steam flushing for removal of PCE DNAPL perched on a low permeability layer. The two-dimensional simulation domain is shown in Figure 5.12. A total of 3,600 grid blocks (60 of equal size in each direction) were used in the simulation. The upper and lower more permeable layers (Layers 1 and 3) have horizontal and vertical permeabilities of 1×10^{-12} m^2 and 2×10^{-13} m^2, respectively. The low permeability layer (layer 2) in the middle of the domain has horizontal and vertical permeabilities of 1×10^{-15} m^2 and 2×10^{-16} m^2, respectively. A total of 1730 L of PCE DNAPL was injected

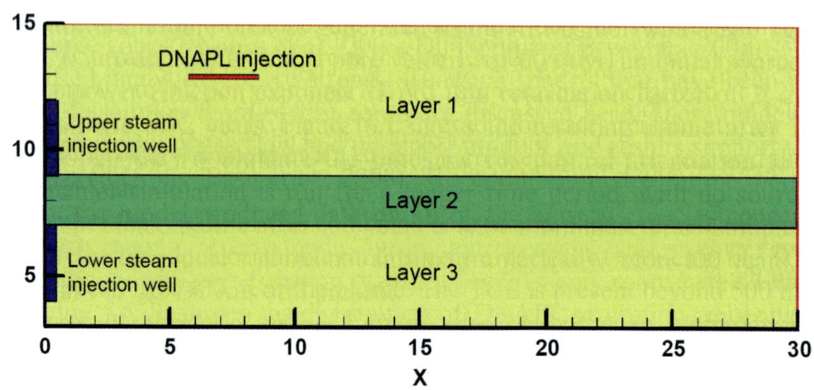

Figure 5.12. Simulation domain for steam flushing example.

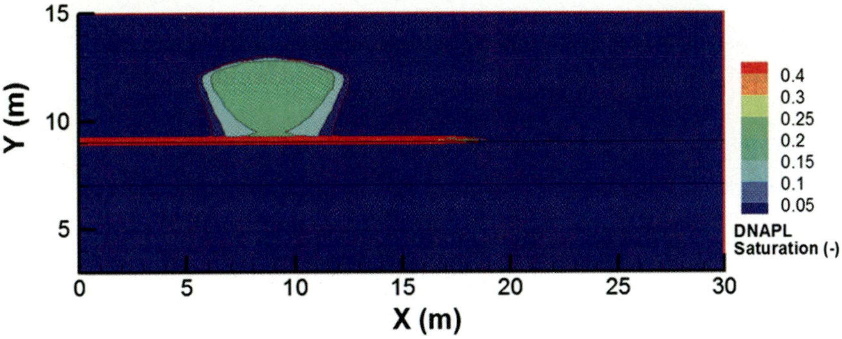

Figure 5.13. DNAPL saturations (percent of pore space) after 1,000 days of DNAPL redistribution.

into the system at the location shown in Figure 5.12 over 4 days with water outflow allowed on the right end of the system via a constant head boundary. In this simulation, all other boundaries are considered to be impermeable. After the injection period, the DNAPL was allowed to redistribute for 1,000 days, with the resulting DNAPL distribution shown in Figure 5.13. The DNAPL is pooled on the low permeability layer with a zone of DNAPL at residual saturation below the injection point. No DNAPL has breached the low permeability layer due to the high entry pressure of the layer and the limited pool height of DNAPL above the layer.

Following the DNAPL redistribution period, steam is injected at the left side of the system, above and below the low permeability layer. She and Sleep (1999) demonstrated in lab experiments that downward movement of DNAPL following desaturation of capillary barriers due to heating and water vaporization is a concern with steam flushing. To ensure that DNAPL mobilized downwards is volatilized and captured by the extraction system, in this simulation steam is injected (12 kilograms per hour per meter [kg/hr/m] at 127 degrees Celsius [°C], 100% steam quality) both above and below the low permeability layer (injection wells shown in Figure 5.12). Figure 5.14a d shows conditions in the system after 24 days. Due to the reduced height of the lower layer compared to the upper layer, the steam front (as indicated by the temperature profile in Figure 5.14a) has moved farther in the lower layer compared to the upper layer. The heating of the low permeability layer occurs only from heat conduction, so the temperature front lags behind the temperature front in the upper and lower layers as a result. In addition, the distribution of gas phase saturations (Figure 5.14b) indicates that the middle layer has only started to desaturate at the top and bottom, as movement of water vapor out of this layer is limited by the low permeability of the layer.

The injection of steam and formation of the high-temperature steam zone has led to volatilization of DNAPL from this zone (Figure 5.14c). The co-boiling temperature of PCE and water at 1 atmosphere (atm) pressure is approximately 90°C, so water and PCE volatilization may begin at temperatures below the temperature at which water alone would boil for a given *in situ* pressure. Although DNAPL has been removed from the steam zone, some dissolved PCE remains in the water at the top of the low permeability layer in the steam zone (Figure 5.14d).

The temperatures, DNAPL saturations, and normalized aqueous phase PCE concentrations after 48 days of steam flushing are shown in Figure 5.15. By 48 days, the steam zone (Figure 5.15a) has moved 20 m in the zone above the low permeability layer and 22 m below the layer, while in the low permeability layer, the steam zone has only moved about 10 m from the injection point. The expansion of the steam zone has resulted in the DNAPL

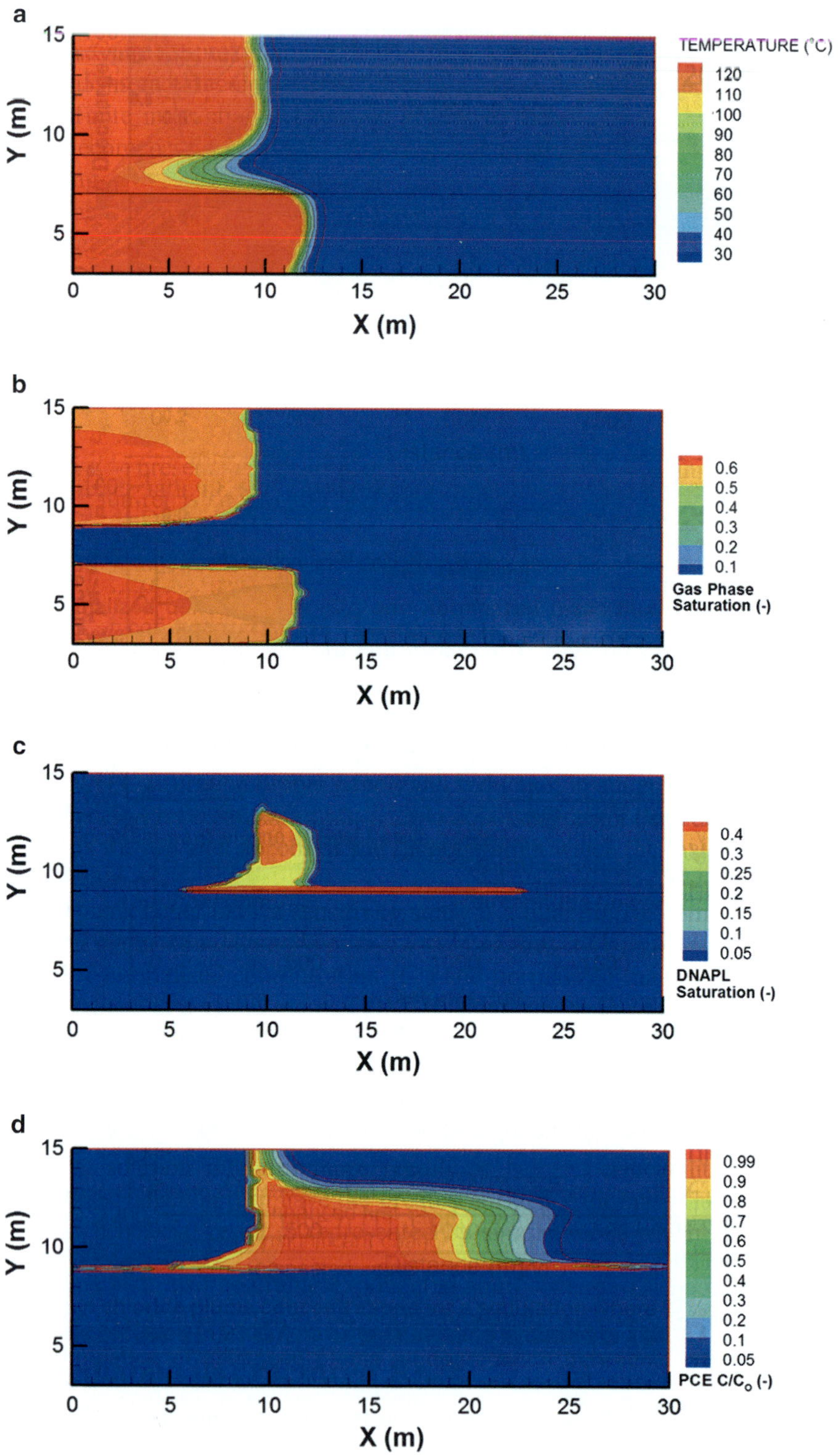

Figure 5.14. (a) Temperatures after 24 days of steam flushing. (b) Gas phase saturations after 24 days of steam flushing. (c) DNAPL saturations after 24 days of steam flushing. (d) Normalized PCE concentrations (to a PCE solubility of 150 mg/L) after 24 days of steam flushing.

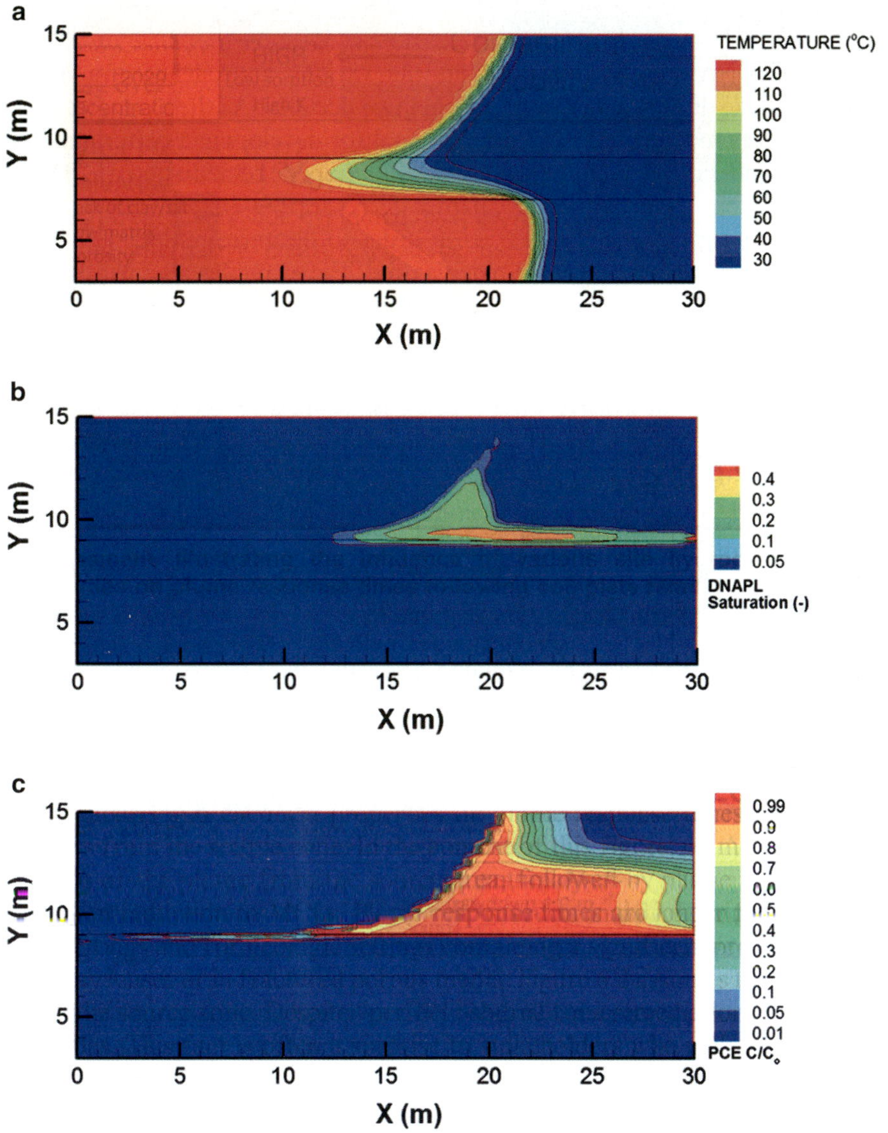

Figure 5.15. (a) Temperatures after 48 days of steam flushing. (b) DNAPL saturations after 48 days of steam flushing. (c) Normalized PCE concentrations (to a PCE solubility of 150 mg/L) after 48 days of steam flushing.

reaching the right boundary (Figure 5.15b) where it would be removed by an extraction well. The low permeability layer is still mostly water saturated across the domain, and as a consequence, a small amount of dissolved PCE (Figure 5.15c) remains at the top of the low permeability layer below the original location of the DNAPL. A computational time of 72 minutes was required for simulating the 48 days of steam flushing with the Compsim model (compiled under Red Hat Linux with the Portland Group Fortran compiler with full optimization) on a 3.2 GHz Intel I7 processor.

5.6 SUMMARY

Modeling of DNAPL source zones and the various approaches employed for their remediation has made dramatic strides in the past two decades. As the examples in this chapter illustrate, numerical modeling can be a valuable tool for developing a better understanding of the *in situ* and engineering characteristics that assist or impede remediation of a DNAPL source zone. As illustrated in the example in Chapter 8, *Hydraulic Displacement*, it can also be valuable for comparing "what if" scenarios in order to optimize remedial performance and minimize capital and O&M costs for a particular site. The examples all illustrate the wealth of qualitative and quantitative outputs that can be generated, which can provide understanding of complex processes to stakeholders and provide justification for decisions in the field. The examples illustrate how uncertainty in key input parameters can influence model results and suggest how numerical models provide a means of assessing the importance of that uncertainty for a given scenario. Coupling sensitivity modeling to an interactive field investigation program can be a very effective way of reducing this kind of uncertainty in the remediation program.

Advanced numerical methods and growing computing power are making three-dimensional simulations with realistic heterogeneity and complex reactions a reality for desktop computers. A wide variety of models exist and can be applied by someone skilled in modeling or can be applied in collaboration with the model developers, who are able to make site- or technology-specific modifications. Moreover, there is an ample scope for new models to be developed and existing models to be extended and validated against ever-more realistic data sets. Modeling is often criticized because it inevitably uses simplifications of reality, generates enormous amounts of data, and creates output that can be incorrectly interpreted as the current or future reality. However, these pitfalls can be avoided by a practitioner trained in both contaminant hydrogeology and modeling. As outlined in this chapter, the model must be matched to the project objectives and site conceptual model, attention must be paid to model verification and validation for the key scenario processes, site characteristics (input data) must be well characterized, the results need to be interpreted with a sound understanding of the assumptions employed, and uncertainty – defining an envelope of possible outcomes – should be quantified via multiple simulations. As illustrated elsewhere in this book, analytical and semi-analytical screening models can be useful tools, particularly with respect to the mass flux and plume development emanating from source zones (Chapter 6). However, for examining scenarios within the source zone itself, simulating the biogeochemical aspects of remediation, examining DNAPL migration and (re)distribution as pools and residual, and better understanding the important feedback mechanisms between *in situ* processes, numerical models are particularly valuable.

REFERENCES

Ababou RD, McLaughlin D, Gelhar LW, Tompson AFB. 1989. Numerical simulation of three-dimensional saturated flow in randomly heterogeneous porous media. Transp Porous Media 4:549–565.

Abriola LM, Rathfelder KM, Yadav S, Maiza M. 1992. VALOR Code Version 1.0: A PC Code for Simulating Subsurface Immiscible Contaminant Port. EPRI TR-101018. Electric Power Research Institute, Palo Alto, CA, USA.

Abriola LM, Lang J, Rathfelder K. 1997. Michigan Soil-Vapor Extraction Remediation (MISER) Model—A Computer Program to Model Bioventing of Organic Chemicals in

Unsaturated Geological Material. EPA/600R-97/099. U.S. Environmental Protection Agency, Washington, DC, USA.

Alexandra R, Gerhard JI, Kueper BH. 2012. Hydraulic displacement of dense, non-aqueous phase liquids. Ground Water 50:1745–6584.

Baston DP, Kueper BH. 2009. Thermal conductive heating in fractured bedrock: Screening calculations to assess the effect of groundwater influx. Adv Water Resour 32:231–238.

Baston DP, Falta RW, Kueper BH. 2010. Numerical modeling of thermal conductive heating in bedrock. Ground Water 48:836–843.

Bernardez LA, Therrien R, Lefebvre R, Martel R. 2009. Simulating the injection of micellar solutions to recover diesel in a sand column. J Contam Hydrol 103:99–108.

Brooks RH, Corey AT. 1964. Hydraulic Properties of Porous Media. Hydrology Paper 3. Civil Engineering Department, Colorado State University, Fort Collins, CO, USA.

Carrigan CR, Nitao JJ. 2000. Predictive and diagnostic simulation of *in situ* electrical heating in contaminated, low-permeability soils. Environ Sci Technol 34:4835–4841.

Christ JA, Abriola LM. 2007. Modeling metabolic reductive dechlorination in dense non-aqueous phase liquid source-zones. Adv Water Resour 30:1547–1561.

Christ JA, Ramsburg CA, Pennell KD, Abriola LM. 2006. Estimating mass discharge from dense nonaqueous phase liquid source zones using upscaled mass transfer coefficients: An evaluation using multiphase numerical simulations. Water Resour Res 42:W11420, doi:10.1029/2006WR004886.

Delshad M, Pope GA, Sepehrnoori K. 1996. A compositional simulator for modeling surfactant enhanced aquifer remediation. J Contam Hydrol 23:303–327.

Falta RW, Pruess K, Finsterle S, Battistelli A. 1995. T2VOC User's Guide, Lawrence Berkeley Laboratory Report LBL-36400. Berkeley, CA, USA.

Falta RW, Rao PS, Basu N. 2005. Assessing the impacts of partial mass depletion in DNAPL source zones: I. Analytical modeling of source strength functions and plume response. J Contam Hydrol 78:259–280.

Gerhard J, Kueper BH. 2003a. Capillary pressure characteristics necessary for simulating DNAPL infiltration, redistribution, and immobilization in saturated porous media. Water Resour Res 39:1212, doi:10.1029/2002WR001270.

Gerhard JI, Kueper BH. 2003b. Relative permeability characteristics necessary for simulating DNAPL infiltration, redistribution, and immobilization in saturated porous media. Water Resour Res 39:1213, doi:10.1029/2002WR001490.

Gerhard JI, Kueper BH. 2003c. The influence of constitutive model parameters on the predicted migration of DNAPL in heterogeneous porous media. Water Resour Res 39:1279, doi:10.1029/2002WR001570.

Gerhard JI, Kueper BH, Hecox GR, Schwarz R. 2001. Site-specific design for dual phase recovery and stabilization of pooled DNAPL. Ground Water Monit Remediat Spring:71–88.

Glassley WE, Nitao JJ, Grant CW, Boulos TN, Gokoffski MO, Johnson JW, Kercher JR, Levatin JA, Steefel CI. 2001. Performance Prediction for Large-Scale Nuclear Waste Repositories. Final Report, Lawrence Livermore National Laboratory, Livermore, CA, USA. 42 p.

Grant GP, Gerhard JI. 2007a. Simulating the dissolution of a complex dense nonaqueous phase liquid source zone: 1. Model to predict interfacial area. Water Resour Res 43:W12410, doi:10.1029/2007WR006038.

Grant GP, Gerhard JI. 2007b. Simulating the dissolution of a complex dense nonaqueous phase liquid source zone: 2. Experimental validation of an interfacial area-based mass transfer model. Water Resour Res 43:W12409, doi:10.1029/2007WR006039.

Grant GP, Gerhard JI, Kueper BH. 2007. Multidimensional validation of a numerical model for simulating a DNAPL release in heterogeneous porous media. J Contam Hydrol 92:109–128.

Jawitz JT, Fure AD, Demmy GG, Berglund S, Rao PSC. 2005. Groundwater contaminant flux reduction resulting from nonaqueous phase liquid mass reduction. Water Resour Res 41: W10408, doi.org/10.1029/2004WR003825 doi:10.1029/2004WR003825#doilink .

Kaluarachchi JJ, Parker JC. 1988. User's Guide to MOFAT-2D, version 1.2, A Two-Dimensional Vertical Section Finite Element Code for Flow and Multispecies Transport in Three Fluid Phase Porous Media Systems. Virginia Polytechnic Institute and State University, Blacksburg, VA, USA. (with unlimited use rights to the U.S. Environmental Protection Agency).

Kaluarachchi JJ, Parker JC. 1990. Modeling multicomponent organic chemical transport in three-fluid-phase porous media. J Contam Hydrol 5:349–374.

Kokkinaki A, O'Carroll DM, Werth CJ, Sleep BE. 2013. An evaluation of Sherwood–Gilland models for NAPL dissolution and their relationship to soil properties. J Contam Hydrol 155:87–98.

Kouznetsova L, Mao X, Robinson C, Barry DA, Gerhard JI, McCarty PL. 2010. Biological reduction of chlorinated solvents: Batch-scale geochemical modeling. Adv Water Resour 33:969–986.

Kueper BH, Frind EO. 1991. Two-phase flow in heterogeneous porous media: 1. Model development. Water Resour Res 27:1049–1057.

Kueper BH, Redman D, Starr RC, Reitsma S, Mah M. 1993. A field experiment to study the behavior of tetrachloroethylene below the water table: Spatial distribution of residual and pooled DNAPL. Ground Water 31:756–766.

Lemke LD, Abriola LM. 2006. Modeling DNAPL mass removal in nonuniform formations: Linking source zone architecture and system response. Geosphere 2:74–82.

Lemke L, Bahrou A. 2009. Partitioned multiobjective risk modeling of carcinogenic compounds in groundwater. Stoch Environ Res Risk Assess 23:27–39.

Leverett MC, Lewis WB. 1941. Steady flow of gas-oil-water mixtures through unconsolidated sands. Trans AIME Pet Eng Div 142:107–116.

MacPhee SL, Gerhard JI, Rein G. 2012. A novel method for simulating smoldering propagation and its application to STAR (Self-sustaining treatment for active remediation). Environ Model Softw 31:84–98.

McClure PD, Sleep BE. 1996. Simulation of bioventing for soil and ground-water remediation. J Environ Eng 122:1003–1012.

McWhorter D, Sunada D. 1990. Exact integral solutions for two-phase flow. Water Resour Res 26:399–413.

Molson JW, Frind EO, Van Stempvoort DR, Lesage S. 2002. Humic acid enhanced remediation of an emplaced diesel source in groundwater. 2. Numerical model development and application, J Contam Hydrol 54:277–305.

Mundle K, Reynolds DA, West M, Kueper BH. 2007. Concentration rebound following in-situ chemical oxidation in fractured clay. Ground Water 45:692–700.

Nitao JJ. 1998. User's Manual for the USNT Module of the NUFT Code, Version 2.0. Report UCRL-MA-130653. Lawrence Livermore National Laboratory, Livermore, CA, USA. 65 p.

O'Carroll DM, Sleep BE. 2007. Hot water flushing for immiscible displacement of a viscous NAPL. J Contam Hydrol 91:247–266.

O'Carroll DM, Sleep BE. 2009. Role of NAPL thermal properties in the effectiveness of hot water flooding. Transp Porous Media 79:393–405.

Parker JC, Park E. 2004. Modeling field-scale dense nonaqueous phase liquid dissolution kinetics in heterogeneous aquifers. Water Resour Res 40:W05109, doi.org/10.1029/2003WR002807 doi:10.1029/2003WR002807#doilink.

Pang TW. 2010. DNAPL Remediation of Fractured Rock Evaluated Via Numerical Simulation. PhD Thesis, Institute for Infrastructure and Environment, University of Edinburgh, Edinburgh, United Kingdom.

Pinder GF, Abriola LM. 1986. On the simulation of nonaqueous phase organic compounds in the subsurface. Water Resour Res 22:109S–119S.

Pope G, Delshad M. 2000. Volume I: User's Guide for UTCHEM-9.0, A Three-Dimensional Chemical Flood Simulator. Reservoir Engineering Research Program, University of Texas at Austin, Austin, TX, USA. 153 p.

Poulsen M, Kueper BH. 1992. A field experiment to study the behavior of tetrachloroethylene in unsaturated porous media. Environ Sci Technol 26:889–895.

Powers SE, Abriola LM, Weber WJ Jr. 1994. An experimental investigation of nonaqueous phase liquid dissolution in saturated subsurface systems: Transient mass transfer rates. Water Resour Res 30:321–332.

Pruess K. 2004. The TOUGH Codes—A family of simulation tools for multiphase flow and transport processes in permeable media. Vadose Zone J 3:738–746.

Pruess K, Battistelli A. 2002. TMVOC, A Numerical Simulator for Three-Phase Non-isothermal Flows of Multicomponent Hydrocarbon Mixtures in Saturated-Unsaturated Heterogeneous Media. LBNL-49375. Lawrence Berkeley National Laboratory, Berkeley, CA, USA.

Pruess K, Oldenburgh C, Moridis G. 1999. TOUGH2 User's Guide, Version 2.0. LBNL-43134. Lawrence Berkeley National Laboratory, Berkeley, CA, USA. 198 p.

Rathfelder KM, Lang JR, Abriola LM. 2000. A numerical model (MISER) for the simulation of coupled physical, chemical, and biological processes in soil vapor extraction and bioventing systems. J Contam Hydrol 43:239–270.

Saba T, Illangasekare TH. 2000. Effect of groundwater flow dimensionality on mass transfer from entrapped nonaqueous phase liquid contaminants, Water Resour Res 36:971–979.

She HY. 1997. Laboratory Evaluation of *In Situ* Steam Flushing for NAPL Removal from Soil. PhD Thesis, University of Toronto, Toronto, Ontario, Canada.

She HY, Sleep BE. 1998. The effect of temperature on capillary pressure-saturation relationships for air-water and perchloroethylene-water systems. Water Resour Res 34:2587–2597.

She HY, Sleep BE. 1999. Removal of PCE from soil by steam flushing in a two-dimensional laboratory cell. Ground Water Monit Remediat 19:70–77.

Sleep BE, Ma Y. 1997. Thermal variation of organic fluid properties and impact on thermal remediation feasibility. J Soil Contam 6:281–306.

Sleep BE, Sykes JF. 1993. Compositional simulation of groundwater contamination by organic compounds: 1. Model development and verification. Water Resour Res 29:1697–1708.

Slough KJ, Sudicky EA, Forsyth PA. 1999. Numerical simulation of multiphase flow and phase partitioning in discretely fractured geologic media. J Contam Hydrol 40:107–136.

Stroo HF, Ward CH, eds. 2010. *In Situ* Remediation of Chlorinated Solvent Plumes. SERDP/ESTCP Environmental Remediation Technology Monograph Series. Springer, New York, NY, USA. 802 p.

Unger AJA, Sudicky EA, Forsyth PA. 1995. Mechanisms controlling vacuum extraction coupled with air sparging for remediation of heterogeneous formations contaminated by dense nonaqueous phase liquids. Water Resour Res 31:1913–1925.

Unger AJA, Forsyth PA, Sudicky EA. 1996. Variable spatial and temporal weighting schemes for use in multi-phase compositional problems. Adv Water Resour 19:1–27.

Van Genuchten MTh. 1980. A closed-form equation for predicting the hydraulic conductivity of unsaturated soils. Soil Sci Am J 44:892–898

Van Genuchten MTh, Alves WJ. 1982. Analytical Solutions of the One-Dimensional Convective-Dispersive Solute Transport Equation. USDA Technical Bulletin 1661. U.S. Government Printing Office, Washington, DC, USA. 149 p.

West MR. 2009. Mathematical Modeling of Source Zone Remediation. PhD Thesis, Queen's University, Kingston, Ontario, Canada.

West MR, Kueper BH. 2010. Plume detachment and recession times in fractured rock. J Ground Water 48:416–426.

West MR, Kueper BH. 2012. Numerical simulation of DNAPL source zone remediation with *in situ* chemical oxidation (ISCO). Adv Water Resour 44:126–139.

West MR, Grant GP, Gerhard JI, Kueper BH. 2008. The influence of precipitate formation on the chemical oxidation of TCE DNAPL with potassium permanganate. Adv Water Resour 31:324–338.

White MD, Oostrom M. 2003. STOMP, Subsurface Transport Over Multiple Phases. Version 3.0. User's Guide. PNNL-14286. Pacific Northwest National Laboratory, Richland, WA, USA.

White MD, Oostrom M, Lenhard RJ. 2004. A practical model for mobile, residual, and entrapped NAPL in porous media. Ground Water 42:734–746.

Xu T, Sonnenthal E, Spycher N, Pruess K. 2008. TOUGHREACT User's Guide: A Simulation Program for Non-isothermal Multiphase Reactive Geochemical Transport in Variably Saturated Geologic Media, V1.2.1. LBNL-55460-2008. Lawrence Berkeley National Laboratory, Berkeley, CA, USA. 195 p.

Zheng C. 1990. MT3D, A Modular Three-Dimensional Transport Model for Simulation of Advection, Dispersion and Chemical Reactions of Contaminants in Groundwater Systems. USEPA, Washington, DC, USA. 170 p.

Zheng C, Bennett GD. 2002. Applied Contaminant Transport Modeling, 2nd ed. John Wiley and Sons, Inc., New York, NY, USA, 621 p.

CHAPTER 6

MODELING PLUME RESPONSES TO SOURCE TREATMENT

Ronald W. Falta[1] and Bernard H. Kueper[2]

[1]Clemson University, Clemson, SC 29634, USA; [2]Queen's University, Kingston, ON, Canada K7L 3N6

6.1 INTRODUCTION

Source zone remediation at a site may be undertaken for a number of reasons. In many instances, it may be necessary to remove contaminant mass to reduce risks of direct exposure to people living or working in the immediate vicinity of the source zone. At contaminated industrial sites being redeveloped for other purposes such as housing, shopping centers, and schools, the direct reduction of source zone contaminant exposure that results from source zone remediation is probably the key consideration. The reduction in potential contaminant exposure resulting from source remediation would typically be modeled with either a simplified analytical solution or one of the more sophisticated source remediation models described in the previous chapter.

When the risk of direct exposure to source zone contamination is removed, attention usually focuses on the dissolved contaminant plume. Chlorinated solvent source zones can create large dissolved plumes, which in some cases may be several miles in length and persist for many decades. The development of methods for treating these plumes has received intense effort over the past two decades (e.g., Stroo and Ward, 2010). The benefit of plume remediation is often fairly obvious: dissolved concentrations in the treated zone are reduced, at least temporarily. However, the long-term effect of plume remediation may be less certain, particularly if the plume remediation is not permanent and if the chlorinated solvent source is not removed.

Considerable debate has focused on the effectiveness of source remediation for reducing or eliminating dissolved plumes (USEPA, 2003). Dense nonaqueous phase liquid (DNAPL) source zones often contain thousands of pounds of chlorinated solvents, yet regulatory drinking water standards (maximum contaminant levels, or MCLs) may be only 5 micrograms per liter (μg/L) or less. If MCLs are applied to groundwater plumes near the source, then it may be very difficult to achieve plume remediation through source remediation alone (Sale and McWhorter, 2001; USEPA, 2003). However, it has been shown that source remediation usually leads to a reduction in contaminant discharge into the plume (e.g., Rao and Jawitz, 2003; Falta, 2003; USEPA, 2003; Falta et al., 2005a; Park and Parker, 2005; Wood et al., 2005; Basu et al., 2006; Christ et al., 2006; Fure et al., 2006; Basu et al., 2008; DiFilippo and Brusseau, 2008; Kaye et al., 2008; Christ et al., 2010). Depending on the site conditions and history, this reduced discharge into the plume may lead to reductions in the plume size, concentration, mass, or longevity. The reduced discharge into a plume that results from source remediation may also increase the effectiveness of plume remediation efforts. The relationship between the plume dynamics and source behavior is often complex, varying with time, and it depends on the site

B.H. Kueper et al. (eds.), *Chlorinated Solvent Source Zone Remediation*,
doi: 10.1007/978-1-4614-6922-3_6, © Springer Science+Business Media New York 2014

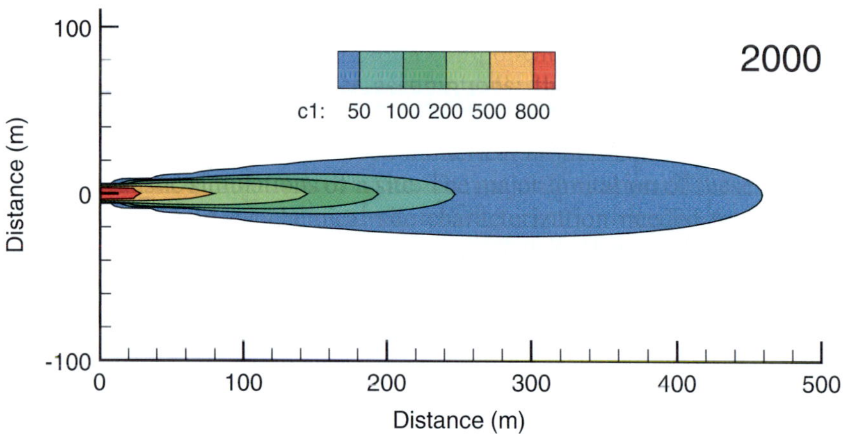

Figure 6.1. Steady-state dissolved 1,2-DCA plume resulting from a hypothetical 1980 release calculated using the REMChlor (Falta, 2008) transport model. Concentrations range from 1,000 μg/L at the source to 5 μg/L at the leading edge. The length units are in meters (m).

characteristics and remediation history. These complexities can hinder source and dissolved plume remediation decisions, and they may be difficult to explain to site stakeholders.

Consider a simplified example, consisting of a hypothetical 1980 release of 200 kilograms (kg) (440 pounds) of 1,2-dichloroethane (1,2-DCA) into an aquifer with a groundwater velocity of 60 meters per year (m/yr) (200 ft/yr). If the initial (1980) source zone concentration was 1 milligram per liter (mg/L), the retardation factor was 2, and the 1,2-DCA plume biodegradation half-life was 2 years, then simulations show that the plume would stabilize at a length of about 450 m (1,475 ft) by the year 2000 (Figure 6.1). Now suppose that it is necessary to shrink this plume so that it does not exceed a length of 250 m (820 ft) over the time period of 2010–2020. It is not obvious what magnitude of source remediation in terms of mass removal or plume remediation in terms of increased decay rates would be needed to achieve this goal.

Understanding the future impact of source remediation on a plume is greatly aided by using a model that couples source zone remediation to the plume response and also considers possible plume remediation. These types of coupled source/plume models are the focus of this chapter.

6.2 KEY ISSUES TO CONSIDER IN MODELING PLUME RESPONSES

Plume response to source remediation depends on the site hydrogeology and geochemistry, source contaminant mass and distribution, history of the initial source contaminant release, and timing and effectiveness of the source remediation. In addition, because source remediation is often coupled with plume remediation, these should be considered together when assessing the overall plume response.

The overall nature of the site hydrogeology is perhaps the most important factor in plume response to source remediation. For example, a highly permeable sand and gravel aquifer will respond much differently than a low permeability silt or clay formation due to the large difference in groundwater velocity (v) and Darcy flux (V_d). Systems that contain large contrasts in permeability (e.g., clay lenses in a sandy aquifer) will respond differently than more homogeneous systems due to diffusion of contaminants into and out of the low permeability material. Similarly, fractured porous materials such as fractured sandstones or limestones

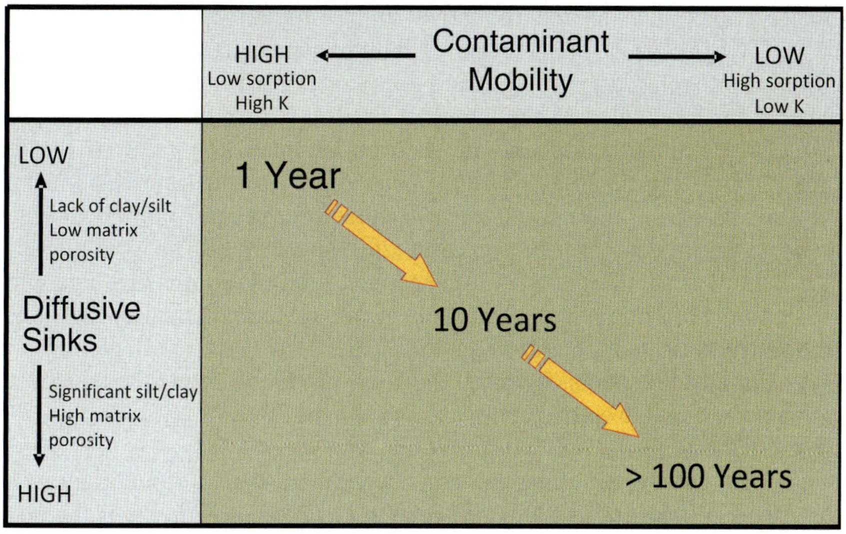

Figure 6.2. Schematic illustrating the influence of various site hydrogeologic factors and chemical properties on plume response times following complete removal of DNAPL mass from the source zone.

will respond differently than fractured crystalline rocks because of matrix diffusion effects (West and Kueper, 2010).

Figure 6.2 presents a simple schematic diagram illustrating the influence of various site hydrogeologic factors and chemical properties on plume response times following complete removal of mass from the source zone. In the context of this figure, plume response is defined as a detachment of the plume from the source area, followed by plume recession (shrinking) and concentration reduction to MCLs. Plume response times are longer for low permeability, high sorption settings and for geologic settings containing a significant proportion of diffusive sinks such as clay lenses or in fractured porous media. Figure 6.2 assumes that all mass has been removed from the source zone. Despite "perfect" source zone remediation, plumes may persist for many decades. This fact is often a surprise to stakeholders who make decisions regarding source zone remediation without prior consideration of expected plume response times. Sale and Newell (2010) provide further discussion on the impact of the hydrogeological setting on plume response to source management.

Site geochemistry also plays a key role in plume response because it controls plume decay rates and plume retardation. It is well known that dissolved chlorinated solvents can be destroyed by biodegradation processes such as reductive dechlorination, anaerobic oxidation, aerobic oxidation, and aerobic cometabolism (Stroo and Ward, 2010; Wiedemeier et al., 1999; NRC, 2000; Alvarez and Illman, 2006). Many forms of plume remediation involve manipulation of the geochemistry to destroy contaminants by biotic or abiotic means. A comprehensive review of *in situ* plume remediation methods is found in Stroo and Ward (2010).

Lumping the geochemical effects into bulk transport parameters, the most important parameters are the contaminant retardation factor R and the contaminant first-order plume decay rate, λ_p. Using a retardation factor requires the assumption of instantaneous reversible linear adsorption; this simple model of adsorption is commonly used in practice. The retardation factor directly affects the rate at which a plume advances and recedes. The use of a first-order decay assumption is also very common in contaminant transport models, and the first-order decay rate is the key parameter controlling the rate of contaminant destruction.

Some plume reactions such as the reductive dechlorination of perchloroethene (PCE) or trichloroethene (TCE) result in the creation of daughter products. Models of these reactions must therefore include retardation factors and decay rates for the daughter products as well as yield coefficients for production of the daughter products from the parent compounds.

The source contaminant mass and its distribution play key roles in the contaminant discharge and flux distribution leaving the source zone. To illustrate relationships between contaminant flux/discharge and source zone architecture, several numerical simulations of source zone behavior are described below. Falta (2003) performed fine-grid numerical simulations of DNAPL pool dissolution in a homogeneous porous medium, using a geometry initially studied by Sale and McWhorter (2001) in their steady-state analytical model. The simulated dissolved concentrations at different stages of TCE DNAPL source depletion are shown in Figure 6.3. As the DNAPL pools dissolve, the total discharge from the source zone slowly declines, with a more rapid drop occurring as some of the DNAPL pools disappear at approximately 16 years (Figure 6.4).

A more complex geological setting was considered in numerical simulations of source zone depletion described in Basu et al. (2008) and Falta et al. (2005a) (Figure 6.5). These simulations considered emplaced DNAPL pools and ganglia, with DNAPL saturations that were either positively or negatively correlated with the intrinsic permeability. The dissolved chlorinated solvent discharge when the DNAPL is negatively correlated with permeability shows a very rapid decline at early times, followed by a gradual tailing that lasts for hundreds of years (Figure 6.6). This occurs because only a small amount of the DNAPL is located in the high-permeability zones; as this material is removed, the discharge drops very quickly. The long-term tailing of the discharge is due to the remaining contaminant slowly dissolving and diffusing from the lower permeability materials.

The case with a positive correlation between DNAPL saturation and permeability shows a more gradual decline of discharge at early times, with the discharge dropping to zero after about 33 years (Figure 6.7). The shorter DNAPL source life in this case is due to the fact that most of the DNAPL is present in high permeability zones, where it can be readily dissolved by the flowing groundwater.

These simulations show that the dissolved chlorinated solvent discharge from a source zone is a complex function of the source zone architecture, that is, the three-dimensional (3D) distribution of DNAPL relative to the 3D distribution of permeability and capillary entry pressure. Despite the complexities and uncertainties present in the source zone, the dissolved contaminant discharge with time can be characterized by several features: (1) an early maximum discharge, (2) a monotonically decreasing discharge as the source mass is depleted, and (3) a discharge that slowly goes to zero as the source mass disappears. These observations have led several researchers to suggest that the source discharge could be approximated as a power function of the source mass (Rao et al., 2001; Rao and Jawitz, 2003; Parker and Park, 2004; Zhu and Sykes, 2004).

Using averaged properties of the source zone, the net water flow through the source zone is the product of the Darcy flux and the cross-sectional area perpendicular to the flow, A. Then the mass discharge leaving the source zone is

$$\dot{m}_s = V_d A C_s(t) \qquad \text{(Eq. 6.1)}$$

where C_s is the flow-averaged concentration of dissolved chlorinated solvent leaving the source zone. With this notation, the power function describing the source strength can be written as follows:

$$\frac{C_s(t)}{C_0} = \left(\frac{M(t)}{M_0}\right)^{\Gamma} \qquad \text{(Eq. 6.2)}$$

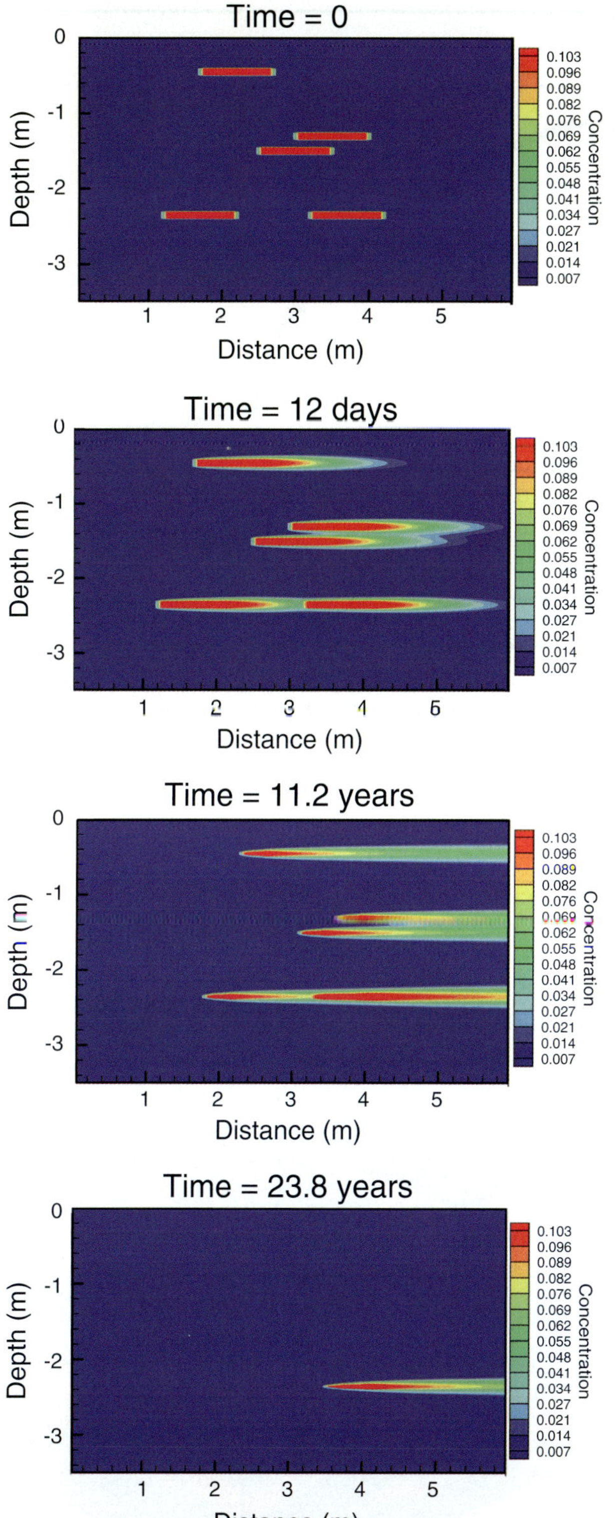

Figure 6.3. Numerical simulations of TCE DNAPL pool dissolution in a homogeneous medium, adapted from Falta (2003). The units on the color scale are g/L.

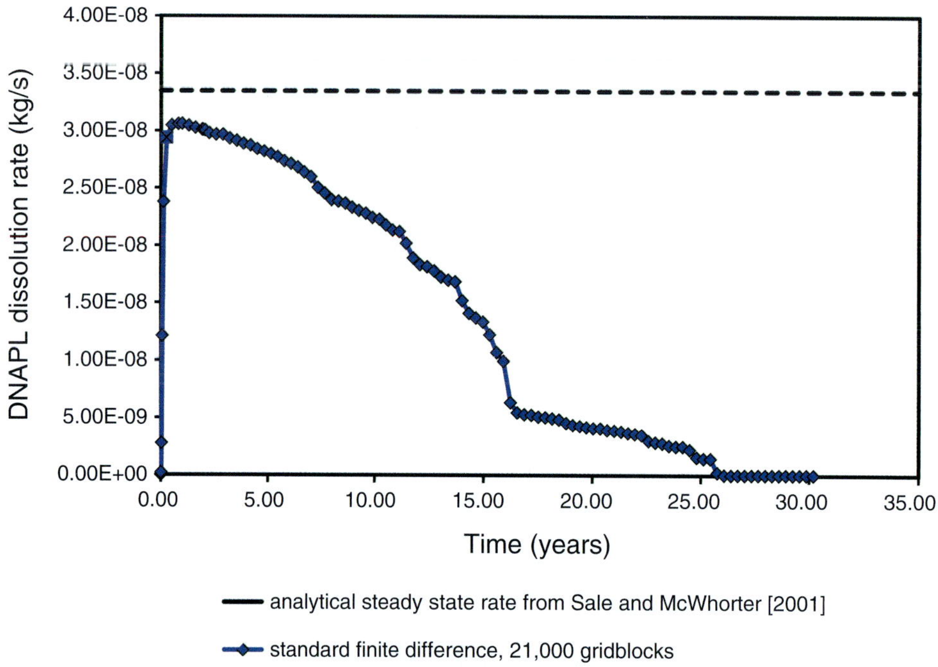

Figure 6.4. Simulated dissolved chlorinated solvent discharge from homogeneous pool dissolution case (*blue line*; adapted from Falta, 2003). The *dashed black line* is the dissolution rate from Sale and McWhorter (2001).

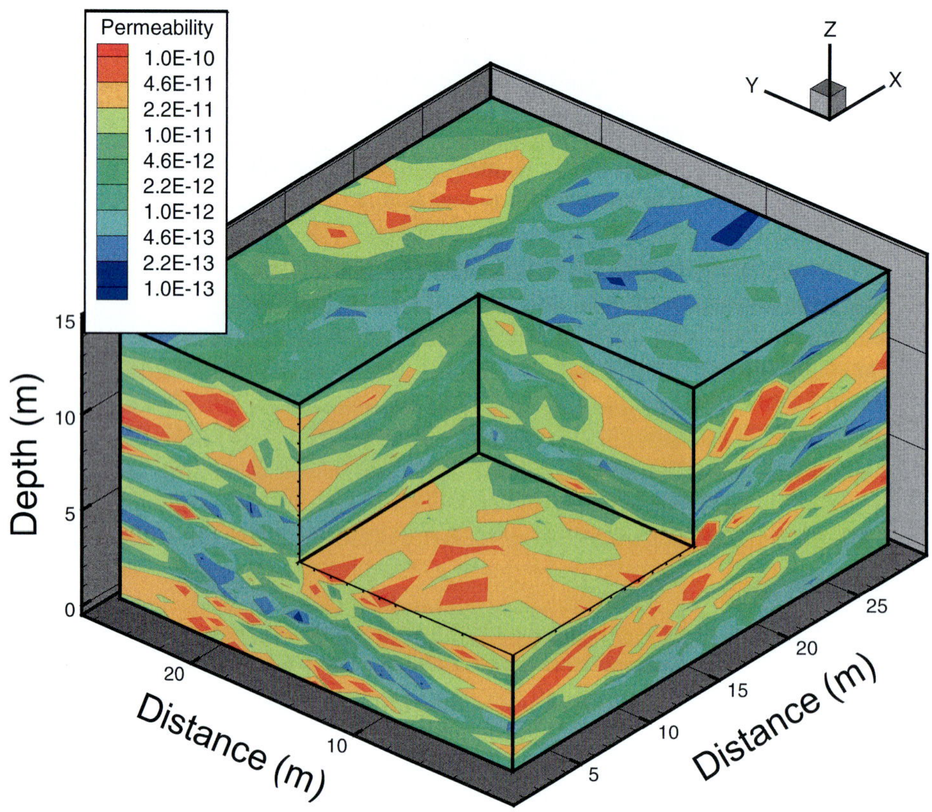

Figure 6.5. Permeability field used in numerical DNAPL source zone depletion simulation, adapted from Basu et al. (2008). Permeability is assigned units of m^2.

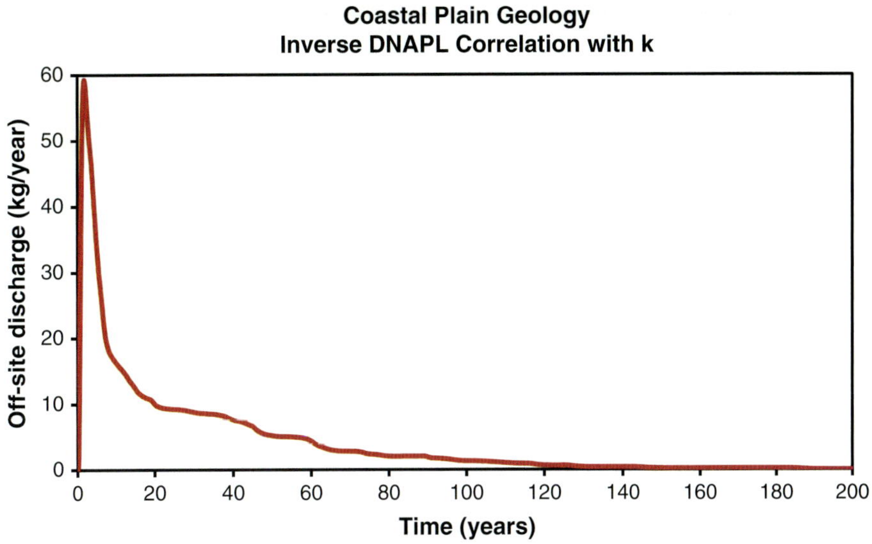

Figure 6.6. Calculated TCE discharge from source zone for case where most of the DNAPL is located in low permeability zones.

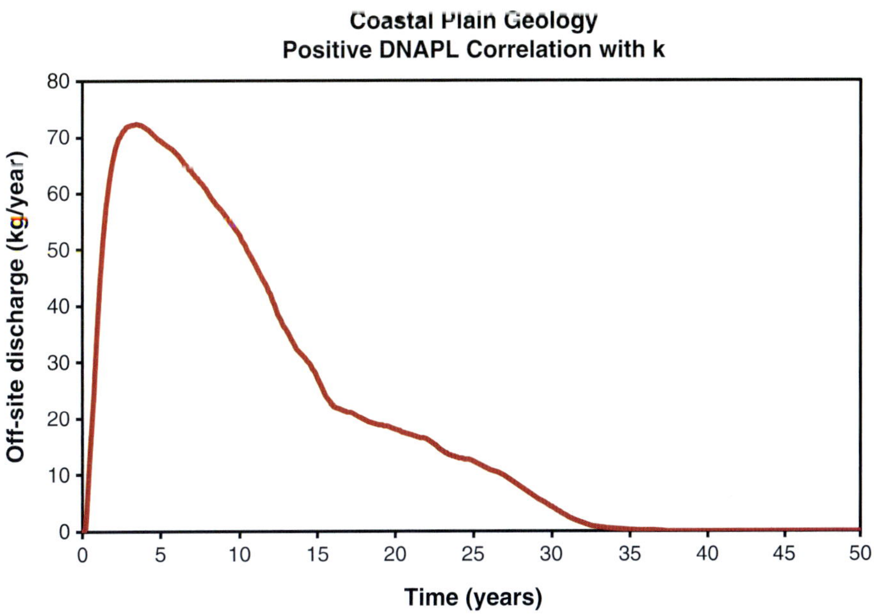

Figure 6.7. Calculated TCE discharge from source zone for case where most of the DNAPL is located in high permeability zones. Time scale is different than in Figure 6.6.

where C_0 is the averaged source concentration that corresponds to a source mass of M_0, and $M(t)$ is the time-dependent source mass. The key fitting parameter in this model is the exponent Γ, which determines the shape of the power function curve. The source zone discharge data from Figures 6.4, 6.6, and 6.7 are plotted as a function of source mass in Figure 6.8. This figure also shows data from field DNAPL remediation experiments funded by SERDP, conducted at

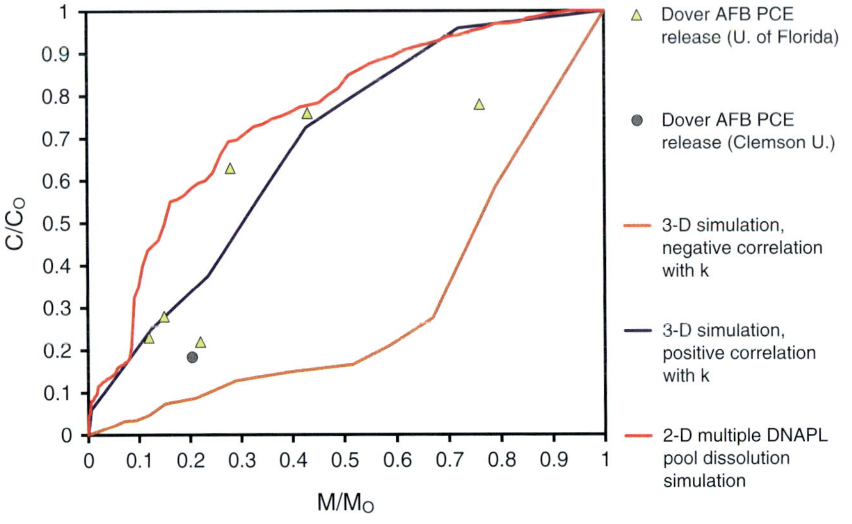

Figure 6.8. Source zone dissolved concentrations as a function of source zone mass (adapted from Falta et al., 2005a). The *red line* corresponds to the case illustrated in Figure 6.4, the *orange line* corresponds to Figure 6.6 and the *blue line* corresponds to Figure 6.7.

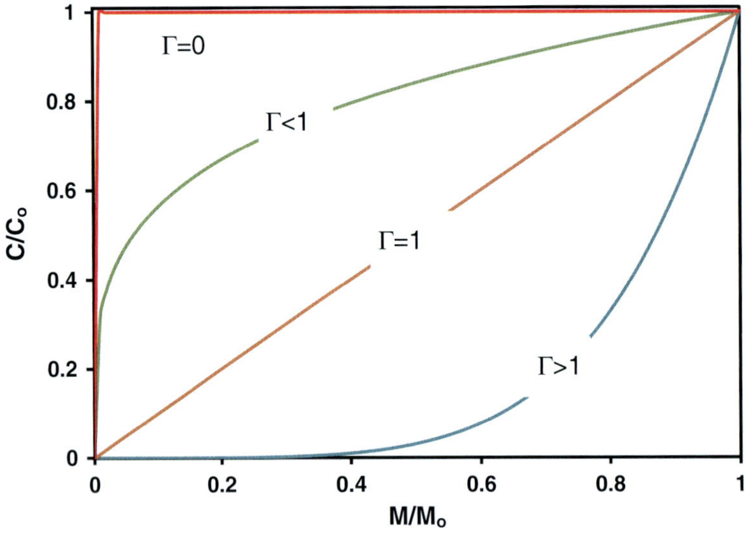

Figure 6.9. Power function representation of source mass/source discharge relationship (Equation 6.2).

Dover Air Force Base (AFB) (Brooks et al., 2002, 2004; Wood and Falta, 2003). The scaled power function produces similar curves, as illustrated in Figure 6.9. When Γ is zero, the source concentration is constant, while a Γ value of one leads to a linear dependency between the source mass and discharge. Larger exponents result in rapid initial drops in discharge, followed by extensive tailing. Similar figures are presented in other chapters illustrating normalized concentration and normalized mass flux versus mass remaining, at the downgradient boundary of a DNAPL source zone subject to remediation by chemical oxidation (Chapter 5) and *in situ* bioremediation (Chapter 12).

The timing and effectiveness of source remediation obviously play a large role in the plume response. If the source is nearly depleted by dissolution before source remediation is attempted, then most of the contaminant mass is already in the plume, and source remediation would probably have little effect. Similarly, an ineffective source remediation effort would not be expected to have a beneficial impact on plume response.

Many chlorinated solvent source remediation projects attempt to either destroy or remove as much source mass as is practically possible. These remediation efforts nearly always occur some time, often many years, after the initial chlorinated solvent release. If it is assumed that the power function model (Equation 6.2) is valid for remediation efforts, then the effect of remediation on the source discharge can be estimated by rescaling Equation 6.2 (Falta et al., 2005a). Starting with a source mass of M_1 at the start of remediation (t_1), we can define a remediation effectiveness, X, as the fraction of M_1 that is removed by the remediation. Then the mass remaining in the source following remediation (M_2) is calculated as

$$M_2 = (1 - X)M_1 \tag{Eq. 6.3}$$

where the value of M_1 can be calculated from M_0 and C_0 by solving a source zone mass balance equation that incorporates the power function relationship, which is described later in this chapter. The resulting discharge concentration from the source immediately following remediation is then

$$C_2 = C_0\left(\frac{(1 - X)M_1}{M_0}\right)^{\Gamma} \tag{Eq. 6.4}$$

and at later times, the source concentration is calculated using Equation 6.2:

$$C_s(t) = C_2\left(\frac{M(t)}{M_2}\right)^{\Gamma} \tag{Eq. 6.5}$$

where M can be calculated by solving a mass balance equation on the source zone that starts at a time of t_2 with a mass of M_2.

6.3 MODELING LIMITATIONS

Mathematical models are idealized representations of the real field system. They are developed by applying mass balance concepts to control volumes, incorporating various phenomenological relationships to describe physical, biological, and chemical processes. Generally speaking, there are two main types of models: analytical and numerical. While these are described in more detail in the next section, analytical models are typically thought of as simplified or screening models because their development requires many simplifying assumptions.

Most analytical models for chemical transport require the assumption of constant and uniform groundwater flow, and they use a concentration gradient-driven dispersion term to simulate additional chemical spreading that is actually due to small-scale velocity variations. Analytical models normally assume linear instantaneous reversible adsorption and zero- or first-order chemical decay reactions. The limitations of analytical models are due to these assumptions; they cannot, for example, simulate complex 3D flow fields in heterogeneous systems with nonlinear coupled reactions. Despite these limitations, analytical models are popular because they require few input parameters and can be run quickly and efficiently with a modest level of training. At sites where many key parameters are uncertain, analytical models can be run with a range of parameter values to screen or bracket the expected site behavior.

Numerical models require far fewer assumptions; they can be applied to complex 3D heterogeneous porous and fractured media with transient multiphase flows and various coupled nonlinear reactions. These advanced numerical models can in principle provide much more detailed and realistic simulations of a site. The major limitation of these models is that the level of hydrogeological and geochemical site characterization needed to fully parameterize the models is not always available. Furthermore, the exact contaminant source mass, architecture, and history are not always known at field sites. Therefore, while the advanced numerical models can be used to simulate complex processes under finely controlled and characterized experimental conditions, they may be subject to uncertainties when applied to field conditions unless parameter sensitivity is studied. These models may also require substantial investment in training and in time to set up simulations. The most advanced models can strain computer resources and require large parallel computers in order to complete simulations in a reasonable time.

Even in the absence of detailed site characterization data, the advanced numerical models can be useful for developing conceptual models of site behavior. This is done by running simulations that are generically realistic even though they may not precisely represent the site characteristics. The insights gained from these numerical simulations of complex systems often aid in the development of simpler models, such as the source strength model described in the previous section (Equations 6.2, 6.3, 6.4, and 6.5).

6.4 AVAILABLE MODELS

Four general types of models are available for simulating plume response to source treatment. These include 3D numerical multiphase flow and transport models, 3D numerical groundwater transport models, analytical mass balance models, and probabilistic models.

A large number of 3D multiphase flow models are available that can simulate DNAPL transport, remediation, and subsequent plume behavior. These models generally use methods developed in the fields of petroleum and geothermal reservoir engineering with modifications to make them more suitable for modeling contaminant transport. Most of these models can simulate the simultaneous movement of a gas, aqueous, and NAPL phase in 3D heterogeneous porous or fractured media, with full phase interference effects and mass transfer between the phases. Examples of these types of codes include UTCHEM (Delshad et al., 1996), STOMP (White et al., 1995), NUFT (Nitao, 1996), MUFTE (Helmig et al., 1994), MAGNAS (Panday et al., 1994), T2VOC (Falta et al., 1995), CompFlow (Unger et al., 1996), MISER (Abriola et al., 1997), and TMVOC (Pruess and Battistelli, 2002).

All of these codes can simulate DNAPL dissolution and subsequent plume development. Some of the codes (STOMP, NUFT, MUFTE, T2VOC, TMVOC) can simulate thermal remediation of DNAPL source zones, while others can model surfactant flooding (UTCHEM, MISER). While these types of models are comprehensive in their process descriptions, they tend to be computationally demanding, because they solve strongly coupled nonlinear governing equations that control multiphase flow. This computational demand limits the number of grid blocks that can be used in a simulation run on a single processor. In practice, it is difficult to design a simulation grid that has high resolution in the source zone with sufficient spatial extent to model the full plume formation without resorting to a large number of grid blocks. For this reason, these models mainly have been used to simulate DNAPL source zone emplacement and remediation. Newer parallel versions of these codes (e.g., TMVOC-MP by

Zhang et al., 2007) allow simulations to be run on tens or hundreds of processors, but these parallel versions have not yet been widely used for modeling plume formation.

A more practical approach to modeling plume response to source remediation may be to couple a 3D multiphase DNAPL migration code with a 3D solute transport code such as MT3D (Zheng and Wang, 1999) or RT3D (Clement, 1997; Clement et al., 1998). In this approach, a multiphase flow module is used to model the DNAPL migration, while a solute transport module is used to model aqueous-phase transport and associated reactions. Examples of studies employing this strategy include Parker and Park (2004) who coupled a DNAPL percolation model with MT3D, Grant and Gerhard (2007) who coupled the DNAPL3D multiphase flow code (Gerhard et al., 1998) with MT3D, Christ et al. (2010) who coupled UTCHEM with MT3D, and West et al. (2008) who coupled DNAPL3D with RT3D to create DNAPL3D-RX, which has been used to simulate source zone remediation using chemical oxidation (West and Kueper, 2012) and enhanced *in situ* bioremediation.

Other source/plume models assume an immobile, dissolving DNAPL source with subsequent plume transport of the dissolved contaminant. ISCO3D (Zhang and Schwartz, 2000) models single-phase transport of dissolved chlorinated solvents from an immobile DNAPL source that dissolves according to first-order mass transfer kinetics. This model includes chemical oxidation of the dissolved contaminant.

The SEAM3D model (Waddill and Widdowson, 2000; Widdowson, 2003) is an extension of the MT3D groundwater transport code that allows for chlorinated solvent biodegradation by reductive dechlorination, oxidation, and cometabolism; it includes a NAPL dissolution package that can simulate dissolution of multicomponent NAPLs. The Natural Attenuation Software (NAS) package (Chapelle et al., 2003; Widdowson et al., 2004) combines SEAM3D with other analytical and numerical models to calculate cleanup times and levels of source reduction needed for plume reduction or stabilization. Although the NAS model uses numerical transport codes, it is not designed for simulating complex 3D flows. Instead, it provides a convenient platform for running screening-type simulations with simplified flow fields but with more complex biogeochemical reactions than can be considered by purely analytical models.

Analytical plume models are widely used to evaluate natural attenuation of plumes at chlorinated solvent sites. A study of 45 chlorinated solvent sites by McGuire et al. (2004) found that mathematical models were used at 60% of the sites and that the analytical model BIOCHLOR (Aziz et al., 2000) was the most commonly used model. BIOCHLOR simulates advection and dispersion of a solute with sequential first-order decay reactions. The source is modeled as either a constant concentration or as an exponentially declining source concentration. It can be shown that an exponentially declining source concentration is equivalent to the use of Equation 6.2 with an exponent (Γ) equal to one (Newell et al., 1996; Parker and Park, 2004; Zhu and Sykes, 2004; Falta et al., 2005a). BIOCHLOR can simulate two different reaction zones, provided that steady-state concentrations are present in the upgradient zone. BIOCHLOR does not directly model source or plume remediation. BIOCHLOR provides an approximate solution to the equation governing solute transport, as discussed by West et al. (2007).

Falta et al. (2005b) developed an analytical transport model that couples source remediation to plume evolution. This model uses the simplified source depletion/remediation model described by Equations 6.2, 6.3, 6.4, and 6.5 to solve a source zone mass balance equation. The resulting source solution, which is described in the next section, then provides a time-dependent source concentration boundary condition for the analytical plume model. The analytical transport model described in Falta et al. (2005b) was improved by Falta (2008) with a new plume model that incorporates three reaction zones in space, with time-varying reaction rates in each spatial zone to simulate transient plume remediation activities. The Falta (2008) model, called REMChlor (Remediation Evaluation Model for Chlorinated Solvents), was further developed to add a graphical user interface and user's guide (Falta et al., 2007). The model was released publicly by the EPA in late 2007 and is available from the EPA website http://www.epa.gov/ada/csmos/models/remchlor.html (accessed November 19, 2012). The REMChlor model is more fully described later in this chapter.

A probabilistic version of REMChlor called PREMChlor was recently developed by Liang (2009) and Liang et al. (2010a, b). The PREMChlor model couples the deterministic REMChlor source/plume remediation model with the GoldSim probabilistic model (http://www.goldsim.com). PREMChlor allows all of the input parameters in REMChlor to be sampled from probability density functions during Monte Carlo simulations where REMChlor is run hundreds or thousands of times. This allows the user to incorporate uncertainty into the modeling process. The PREMChlor model is a GoldSim Player application that runs on the GoldSim Player program (version 9.60 SP4; GSP9604.exe). GoldSim Player may be downloaded free of charge from the GoldSim website (http://www.goldsim.com; accessed November 19, 2012). The PREMChlor GoldSim Player graphical user interface, a complete user's guide, and the FORTRAN dynamic link library are available from the first author (faltar@clemson.edu).

6.5 REMCHLOR DESCRIPTION

The REMChlor model consists of a FORTRAN program that is linked to a Visual Basic graphical user interface (GUI). As with many groundwater models, the GUI is used to prepare the input file used by the FORTRAN program, and the GUI produces plots from the output files from the FORTRAN program. The mathematical method used in REMChlor consists of an analytical source remediation mass balance model coupled to an analytical plume remediation model. The source mass balance model lumps the entire source into a single compartment. Assuming that the source is removed by water flowing through the source zone (Figure 6.10), and allowing for additional first-order decay of the source, the mass balance equation is as follows:

$$\frac{dM(t)}{dt} = -Q(t)C_s(t) - \lambda_s M(t) \quad \text{(Eq. 6.6)}$$

where $Q = V_d A$, and λ_s is the rate of first-order source decay by processes other than flushing. Equation 6.6 assumes that any first-order decay of the source zone is proportional to the source mass; Parker and Falta (2008) give analytical solutions for the case where the first-order source decay is proportional to the source mass present in the aqueous phase only. Following Parker and Park (2004) and Zhu and Sykes (2004), Falta et al. (2005a) used Equation 6.2 to relate the

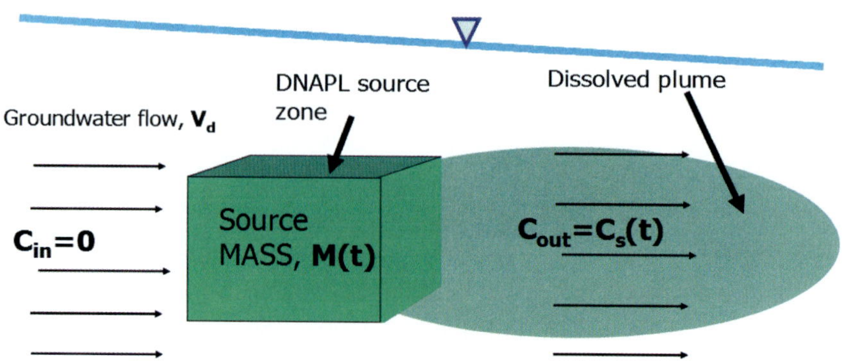

Figure 6.10. Conceptual model of source zone with time-dependent discharge into plume.

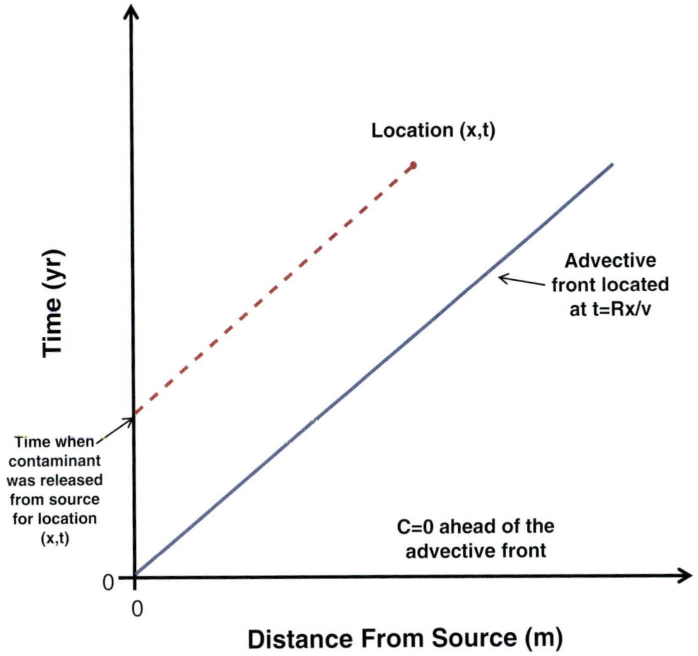

Figure 6.11. Distance–time plot for advective transport (from Falta, 2008).

source concentration to the source mass in Equation 6.6. Solving this equation for a constant flushing rate (Q), the source mass is

$$M(t) = \left\{ \frac{-QC_0}{\lambda_s M_0^{\Gamma}} + \left(M_0^{1-\Gamma} + \frac{QC_0}{\lambda_s M_0^{\Gamma}} \right) e^{(\Gamma-1)\lambda_s t} \right\}^{\frac{1}{1-\Gamma}} \quad \text{(Eq. 6.7)}$$

and the source concentration is as follows:

$$C_s(t) = \frac{C_0}{M_0^{\Gamma}} \left\{ \frac{-QC_0}{\lambda_s M_0^{\Gamma}} + \left(M_0^{1-\Gamma} + \frac{QC_0}{\lambda_s M_0^{\Gamma}} \right) e^{(\Gamma-1)\lambda_s t} \right\}^{\frac{1}{1-\Gamma}} \quad \text{(Eq. 6.8)}$$

Equivalent expressions were developed by Parker and Park (2004) and Zhu and Sykes (2004) for the case without first-order source decay. Equation 6.8 is used to describe the source concentration in REMChlor prior to any source remediation activities.

Source remediation is simulated by removing a specified fraction (X) of the mass remaining in the source zone at a specified time. Following remediation, Equations 6.7 and 6.8 are simply recalculated, using Equations 6.3, 6.4, and 6.5 to account for the source removal effect on source mass and source concentration. This approach is not technology specific, and it allows for a mass conservative assessment of the effects of source remediation on source longevity and discharge.

The plume model in REMChlor is based on a one-dimensional (1D) streamtube that allows reactions to be space and time dependent. Longitudinal dispersion is added by considering a bundle of streamtubes with a normally distributed velocity field. This results in linearly scale-dependent longitudinal dispersion. Transverse and vertical dispersion are added using Domenico's (1987) approximation, which is valid for small to moderate dispersivities (Falta, 2008).

The REMChlor streamtube model assumes a constant velocity and retardation factor, and it includes only advection and reactions. The advective transport of a solute in a streamtube can be shown graphically on a distance–time plot (Figure 6.11), where the x-axis represents

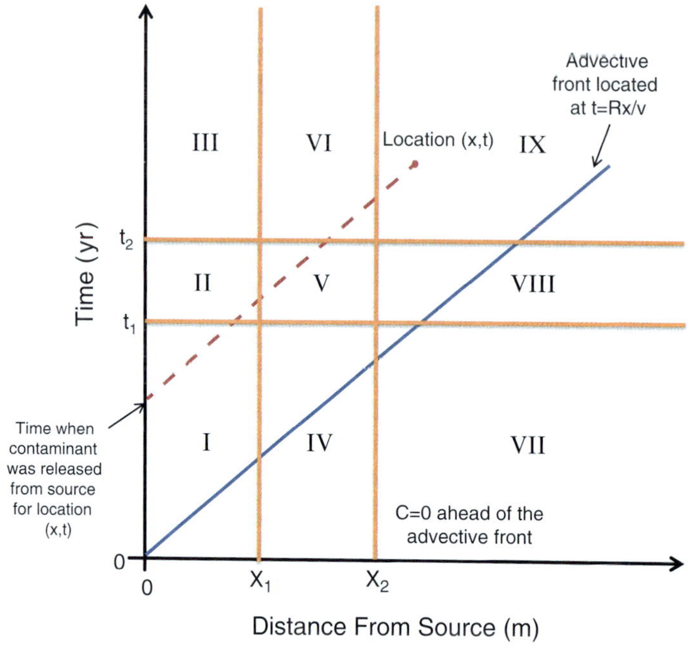

Figure 6.12. Distance–time plot for advective transport with multiple sets of reaction rates (from Falta, 2008).

downgradient distance from the source, and the *y*-axis represents time since the initial source release.

At any downgradient location (x) at any time (t), it can quickly be determined whether or not the advective front has passed that location. If the advective front has passed the x, t location, then the time that parcel of solute was released from the source is

$$t_{release} = t - Rx/v \qquad \text{(Eq. 6.9)}$$

In the absence of any reactions, the concentration at that x,t location would be equal to the source concentration at a time of $t_{release}$. The solute residence (travel) time in the streamtube is needed to account for reactions in the streamtube. For the case shown in Figure 6.11, the solute residence time would be equal to $t - t_{release}$ or Rx/v, so the solute has undergone a reaction for this length of time. This residence–time reaction model can be generalized to the case of spatially and temporally varying reaction rates by dividing the distance–time plot into reaction zones (Figure 6.12).

Calculating the solute residence time in each reaction zone in Figure 6.12 (see Falta, 2008 for details) is relatively straightforward. Then, the cumulative effect of the solute passing through the multiple reaction zones is computed by solving a sequential set of batch reaction differential equations (Falta, 2008). As the solute enters each reaction zone, the concentration is known from the reactions in the previously encountered reaction zone. The REMChlor model assumes that a parent compound undergoes first-order decay with parent–daughter reactions in each reaction zone, with a maximum of three daughter products. Therefore, the parent and daughter compound concentrations entering and leaving each reaction zone are calculated until the desired position (x,t) in the plume is reached. With one parent compound and three daughter compounds, the solute decay reactions are characterized by four first-order reaction rates, and three yield coefficients in each reaction zone in Figure 6.12.

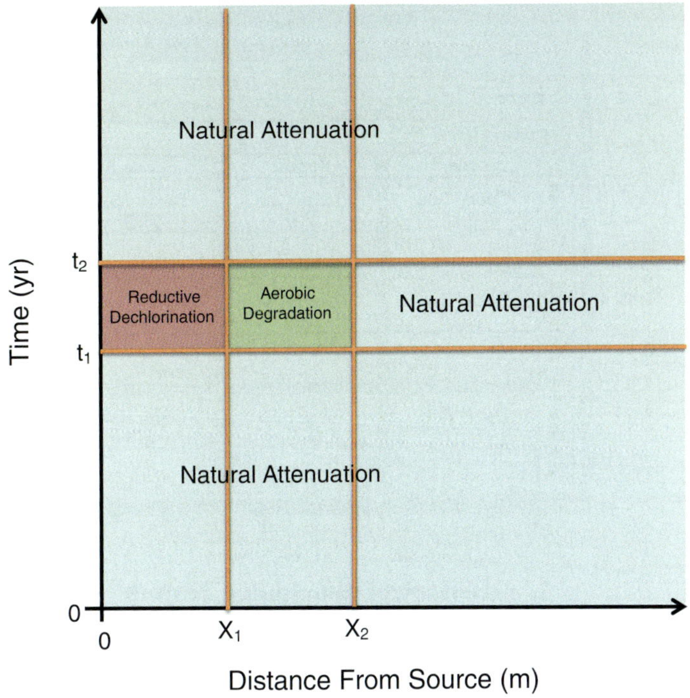

Figure 6.13. Possible design of an enhanced plume remediation scheme with an enhanced reductive dechlorination zone for the destruction of PCE and TCE and an enhanced aerobic degradation zone for destruction of DCE and VC. The other zones are set to natural background conditions to represent pre- and post-remediation conditions in the plume.

The reason for having all of these reaction zones in the plume model is so that enhanced plume remediation can be simulated. Figure 6.13 shows an example of how the reaction zones might be used to simulate plume remediation of PCE and its daughter products TCE, *cis*-1,2-dichloroethene (DCE), and vinyl chloride (VC) using a combination of enhanced reductive dechlorination and enhanced aerobic degradation.

For this example, it was assumed that PCE was released into the source zone at time zero, with plume remediation beginning at time t_1 and extending for a period of $(t_2 - t_1)$ years. The plume remediation scheme is assumed to consist of two spatial zones. The first zone, which is immediately downgradient of the source, uses an electron donor to increase the rate of reductive dechlorination of PCE and TCE. This zone extends for a distance of x_1. In the second zone, which starts at x_1, aerobic conditions are created to enhance the destruction of DCE and VC. Downgradient of this zone, beyond x_2, conditions revert to background or natural attenuation conditions. Figure 6.14 shows where the reaction rates corresponding to this type of example would be entered on the REMChlor input screen. Each of the decay rates here corresponds to one of the distance–time zones in Figure 6.13, and the rates are entered independently for each of the four plume compounds (in this case, PCE, TCE, DCE, and VC).

Although the plume reaction zones were designed to simulate enhanced bioremediation, they can be used to approximate any solute destruction process that can be modeled by a first-order decay reaction. For example, Liang et al. (2010b) use a very narrow reaction zone (5 inches) (0.16 m) with very high decay rates to simulate a zero valent iron reactive barrier.

The output from the REMChlor program consists of plume concentrations of the parent chlorinated solvent compound, along with up to three daughter compound concentrations at

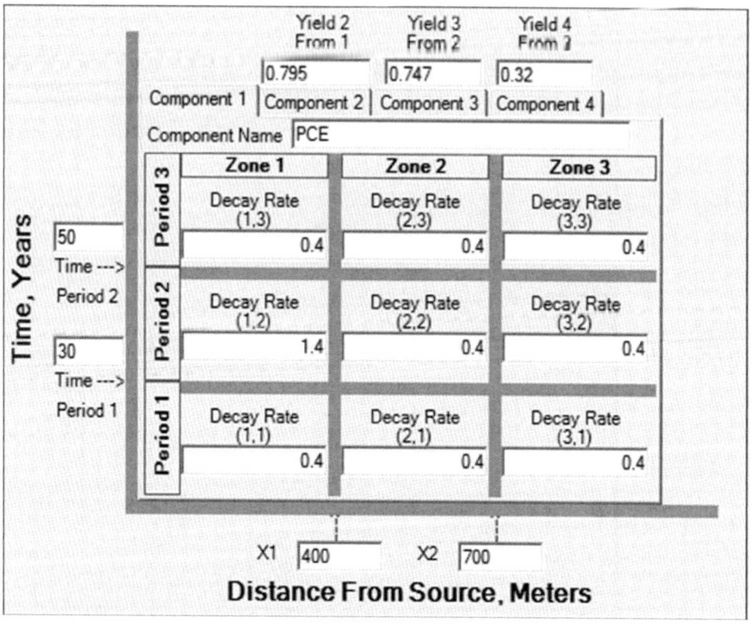

Figure 6.14. Part of REMChlor input screen where plume reaction rates are entered. Rates are entered separately for each plume species.

specified locations in space and time. The program includes a simple x–y plotting routine for making plume center line plots, and it produces formatted files that can be used to make contour plots. REMChlor also calculates the mass discharge (kg/yr) (lbs/yr) of each dissolved component crossing control planes at specified x locations and times and the lifetime excess cancer risks for contaminants considered to be carcinogenic. Cancer risk from exposure to carcinogens is quantified using the chronic daily intake (*CDI*) of the carcinogen (mg/kg-day) and a cancer risk slope factor (*SF*) that has units of risk per mg/kg-day. Further details of the cancer risk calculations are found in Falta et al. (2007).

6.6 ILLUSTRATIVE EXAMPLES OF REMCHLOR SIMULATIONS

6.6.1 Comparing Source and Plume Remediation Effects

The plume shown in Figure 6.1 was generated using REMChlor with a 200 kg (42 gallon) release of 1,2-DCA into an aquifer with a pore velocity of 60 m/yr, an initial source concentration of 1 mg/L, a power function exponent (Γ) of 1, a retardation factor of 2, and a plume biodegradation half-life of 2 years. Figure 6.1 shows the resulting plume after 20 years of development (1980–2000). To evaluate the potential for natural attenuation to reduce the plume, the REMChlor simulation is run for a longer time period, with no source or plume remediation. The simulated plume after 100 years is shown in Figure 6.15. Comparing the two figures, it is clear that the plume still has not shrunk dramatically. After 100 years, 74% of the original source mass of 1,2-DCA is still present.

Recalling the earlier scenario, suppose it was necessary to shrink this plume during the period of 2010–2020, so that the plume does not exceed a length of 250 m after 2020. REMChlor can be used to simulate source and/or plume remediation to evaluate what

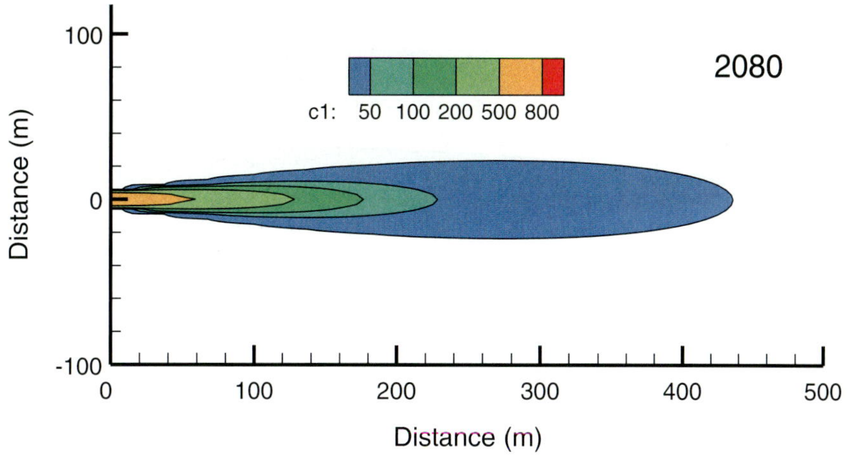

Figure 6.15. Simulated 1,2-DCA plume 100 years after initial release (this is the same case shown in Figure 6.1). The concentration units are μg/L and the dimensions are in meters.

magnitude of remediation (in terms of source removal or enhanced plume decay rates) would be needed to meet this cleanup goal.

A simulation of enhanced plume biodegradation is developed by simply modifying the plume decay rate matrix to include the enhanced decay rate over the desired distance and time. Applying an enhanced decay rate of 1.34 per year (1,2-DCA half-life of 0.5 year) for a 10-year period (2010–2020) produces the results illustrated in Figure 6.16 (middle panel). However, while plume remediation is very effective in reducing the length of the plume and meeting the 2020 cleanup goal, the effect is only temporary, and the plume rebounds once the enhanced biodegradation effort stops (Figure 6.16, bottom panel). This suggests that if enhanced biodegradation were chosen as a remediation option for this site, the process would need to be continued indefinitely.

The long life of the source in this case leads to a very persistent plume. Under these conditions, source remediation might be an attractive option. It is assumed here that source remediation may remove 90% of the remaining source mass. This is simulated in REMChlor by specifying the removal fraction (0.9 for 90% removal) and the time that the remediation is performed (2010–2011, or 30–31 years after the 1980 release). The resulting plume concentrations before and after source remediation are shown in Figure 6.17. The behavior of the source remediation case is fundamentally different from the plume remediation case. Plume remediation tends to have a relatively rapid effect on plume concentrations in the treated zone, but these effects may not be permanent if the source is long lived and not controlled. Source remediation, on the other hand, does not immediately affect the plume concentrations. In Figure 6.17, the middle panel shows the simulated plume 3–4 years after source remediation. Over the first 120 m, the source concentrations are reduced substantially (by a factor of 10); however, beyond this distance there is no effect on the plume. The positive effects of source remediation on the plume take time to "wash" through the plume zone. The time required depends on the groundwater pore velocity and the contaminant retardation factor. Eventually the plume stabilizes to a smaller size that corresponds to the reduced contaminant discharge leaving the source after remediation. From the bottom panel in Figure 6.17, it is apparent that the groundwater plume following source remediation is much smaller than before, and in this case, it has met the remediation objectives.

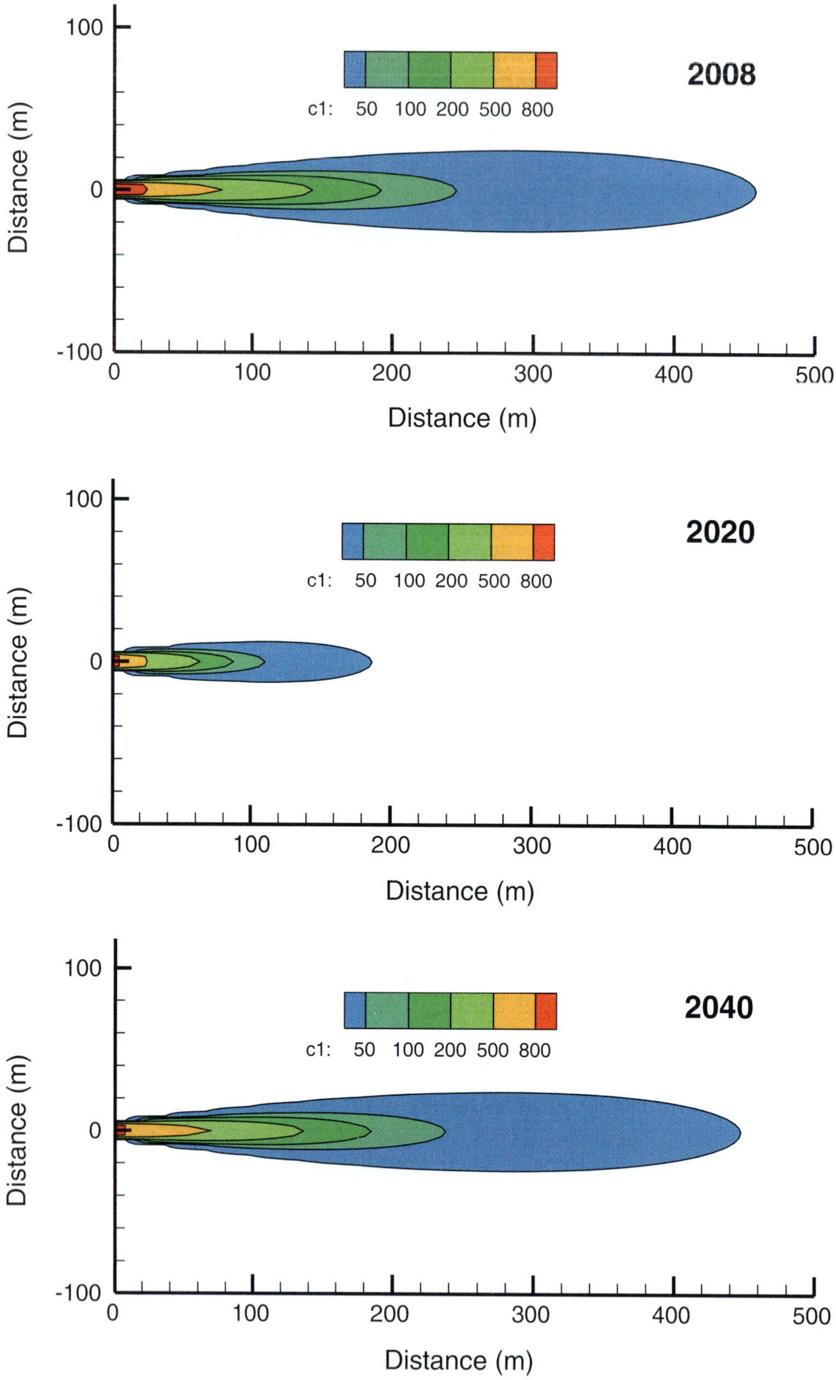

Figure 6.16. **Simulation of enhanced plume biodegradation between 2010 and 2020 over first 300 m of plume for the 1,2-DCA case. The *top panel* shows the plume prior to remediation, the *middle panel* shows the plume at the end of the remediation period and the *bottom panel* shows the plume 20 years after remediation ends.**

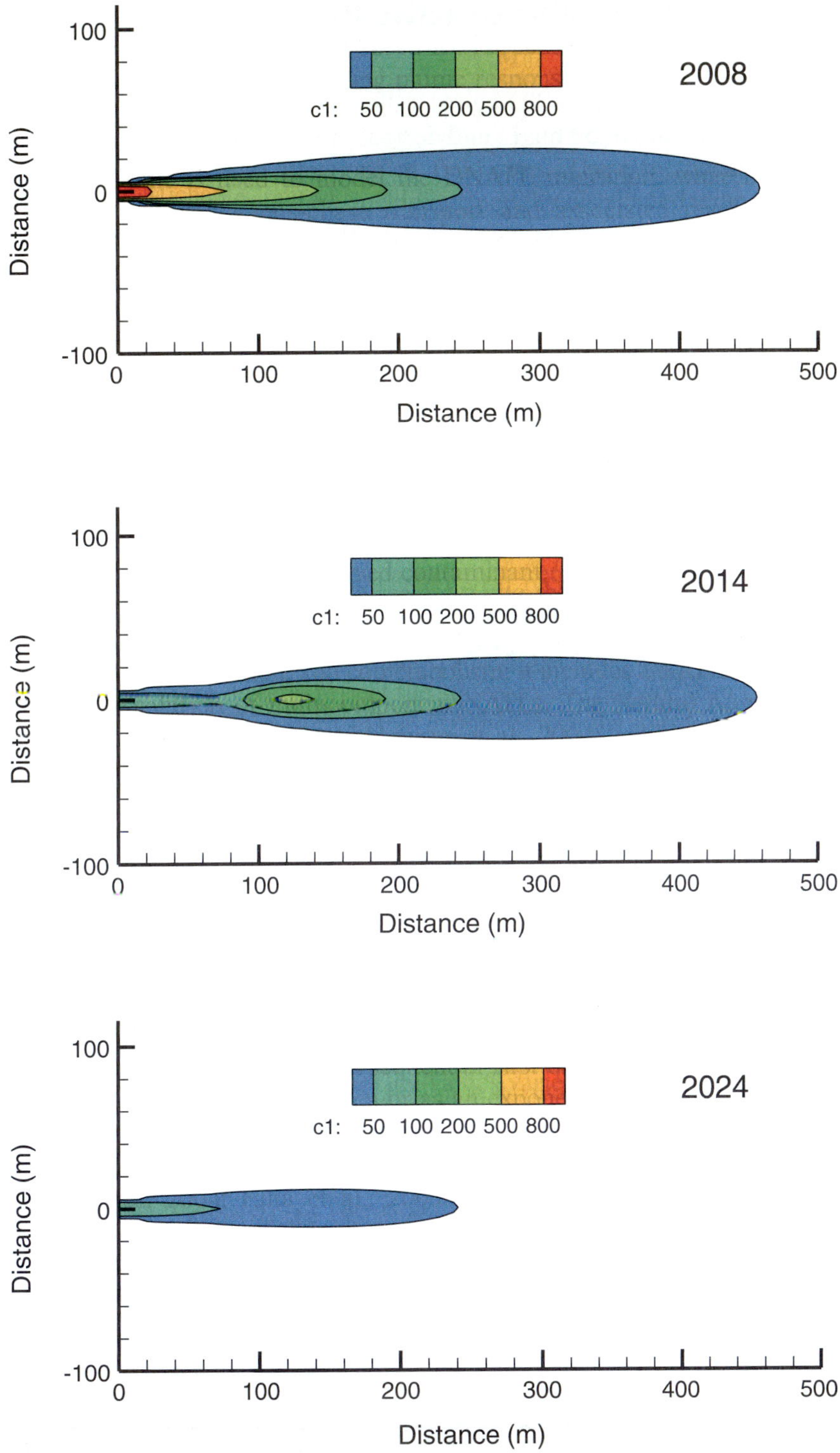

Figure 6.17. **Simulation of removal of 90% of the source between 2010 and 2011 for the 1,2-DCA case. The *top panel* shows the plume prior to source remediation, the *middle panel* shows the plume 3–4 years after source remediation and the *bottom panel* shows the plume 13 years after source remediation.**

6.6.2 Modeling a Permeable Reactive Barrier

The source and plume behavior depend very strongly on the characteristics of the source. If we take the previous example of a 200 kg (42 gallons) release of 1,2-DCA, increase the initial source concentration from 1 to 50 mg/L, and decrease the Γ value from 1.0 to 0.5, the resulting plume behavior is much different (Figure 6.18). If this release occurred in the year 2000, by 2010 most of the contaminant mass would have been released into the plume, and the source would be completely depleted within 14 years (2014). As the source becomes depleted, the plume naturally detaches from the source and moves downgradient. This plume initially has higher concentrations than the last example because of the higher initial source concentration. As a result, the plume is able to travel farther from the source before it decays below the regulatory limit (5 μg/L in these figures).

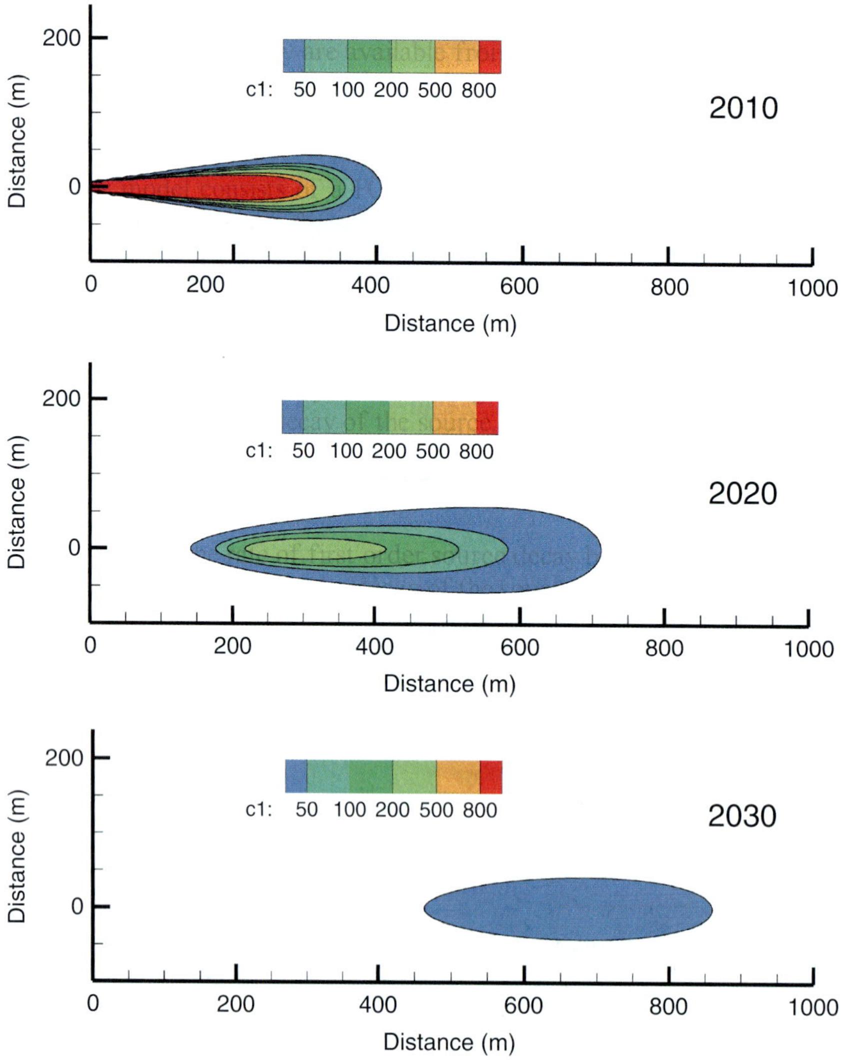

Figure 6.18. Simulation of a 200 kg (42 gallons) release of 1,2-DCA in the year 2000 with an initial source concentration of 50 mg/L and Γ = 0.5. This is a natural attenuation simulation with no remediation. The concentration units are μg/L and the dimensions are in meters.

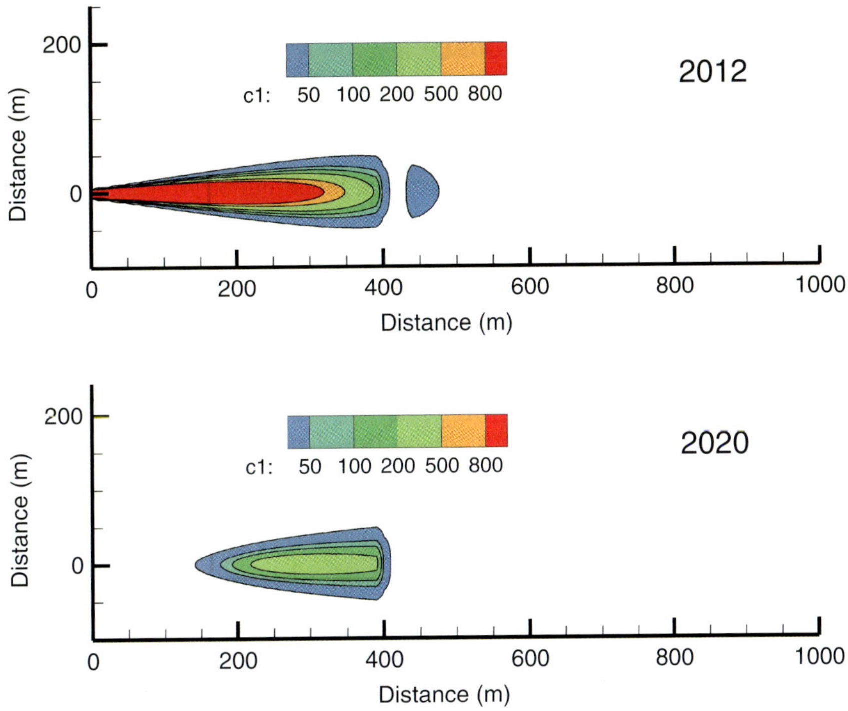

Figure 6.19. Simulated plume after the installation of a PRB 400 m from the source in 2011. The plume concentrations are all below 5 μg/L by the year 2030. At the time of the PRB wall installation a small part of the plume was located beyond 400 m, leading to a detachment of that part of the plume from the main plume (*top figure*).

Remediation of this plume requires a much different strategy than in the previous example. If we anticipate starting remediation in the year 2011, it is clear from the simulation that little would be gained from source remediation because the source is completely exhausted within a few years without any action, and most of the contaminant mass is in the plume.

REMChlor can also be used to simulate a permeable reactive barrier (PRB) in the plume zone. This is done by simply creating a very narrow reaction zone, with a high destruction rate. If data on the destruction efficiency of the wall are available, the value of the equivalent first-order decay rate for the wall can be calculated from the following:

$$\lambda = \frac{-V_d}{L\phi} \ln\left(\frac{C_{\text{down}}}{C_{\text{up}}}\right) \qquad \text{(Eq. 6.10)}$$

where L is the thickness of the PRB in the direction of flow, C_{down} is the plume concentration immediately downgradient of the wall, ϕ is the porosity, and C_{up} is the plume concentration immediately upgradient of the wall.

For the example shown in Figure 6.19, we can consider the effect of a 6 inch (0.152 m) thick PRB, located 400 m (1,310 ft) downgradient of the source, installed in the year 2011. If we have data that indicate that the wall would be capable of a 100-fold reduction of dissolved concentration, the appropriate value of λ from Equation 6.10 would be 1,818 per year. Figure 6.19 shows the simulated plume 1 year after installation of the PRB (2012) and 9 years after the installation (2020).

6.6.3 Modeling Remediation of a High-Strength Persistent Source

The next simulation, which is described in Falta (2008), represents a more difficult site with a long-lived source and a high source concentration; the plume decay rates are low, and the groundwater velocity is high. The scenario in this case is a 1,620 kg (264 gallon) release of PCE DNAPL in 1979. The initial source concentration is 100 mg/L, which is approximately one-half the solubility of pure PCE. The source concentration and mass are assumed to be linearly related ($\Gamma = 1$), the Darcy flux is 10 m/yr (33 ft/yr), and the porosity is 0.33. The source zone area perpendicular to flow is 3 m deep and 10 m wide, so the water flow through the source zone is 300 m^3/yr. This results in an initial PCE mass discharge into the plume of 30 kg/yr. This high rate of PCE discharge is sufficient to generate a substantial plume, but it is low enough compared to the initial source mass that the source will persist for many years.

In the plume, the PCE is assumed to undergo reductive dechlorination to TCE, then *cis*-1,2-DCE, and then VC. It is assumed, however, that only half of the DCE is converted to VC by the reaction; the other half is destroyed. The retardation factor for all components (PCE, TCE, DCE, VC) is equal to 2. The decay rates of each component in the plume are low, corresponding to the low range of values observed in the field (Wiedemeier et al., 1999; Aziz et al., 2002): PCE, 0.4 per year; TCE, 0.15 per year; DCE, 0.1 per year; and VC, 0.2 per year. The MCLs for PCE, TCE, DCE, and VC are 5, 5, 70, and 2 μg/L, respectively.

Figure 6.20 shows the four plume components 30 years after the release (2009). At this time, the plume of the parent compound (PCE) has nearly stabilized at a length of about 500 m, but the daughter plumes (TCE, DCE, VC) are still growing, as evidenced by Figures 6.21 and 6.22, which show the plumes in 2039 and 2079. In 2039, the plumes extend more than 1,000 m from the source, and about 57% of the original PCE mass remains in the source zone. Even after 100 years (2079), the plumes of DCE and VC are still growing, and about 15% of the original PCE mass still remains in the source zone.

Source remediation might be considered at a field site having the characteristics of this example, because the source is long lived and has a high discharge into the plume. In the next example, a source remediation effort that removes 90% of the source remaining is conducted in 2009. Because $\Gamma = 1$, this results in an immediate reduction of the PCE discharge into the plume of 90%. However, it takes many years for this action to have an effect on the plume, as illustrated by Figure 6.23. In this figure, only the VC plumes are shown, but similar behavior is seen in the other plumes.

By the time source remediation took place, 30 years after the initial release, a large amount of PCE mass has already been released into the plume, 43% of the original spill mass. Although the PCE in the plume has been degrading during this period by reductive dechlorination, this decay produces large amounts of TCE, which in turn leads to DCE and then to VC, so that there is a large accumulation of all of these components in the plume by 30 years. After source remediation, the concentrations of the plume components are all reduced by a factor of 10 near the source, and this zone of lower concentration moves downgradient at the retarded groundwater velocity (15 m/yr) (49 ft/yr).

6.6.4 Combining Source and Plume Remediation

Plume remediation would probably be needed at this site in addition to the source remediation. In the next simulation, we can consider both source and plume remediation, using a two-zone plume remediation scheme based on enhanced biodegradation. The motivation for choosing a two-zone biodegradation approach is based on the observation that PCE and TCE degradation is readily enhanced through the addition of electron donors such as hydrogen,

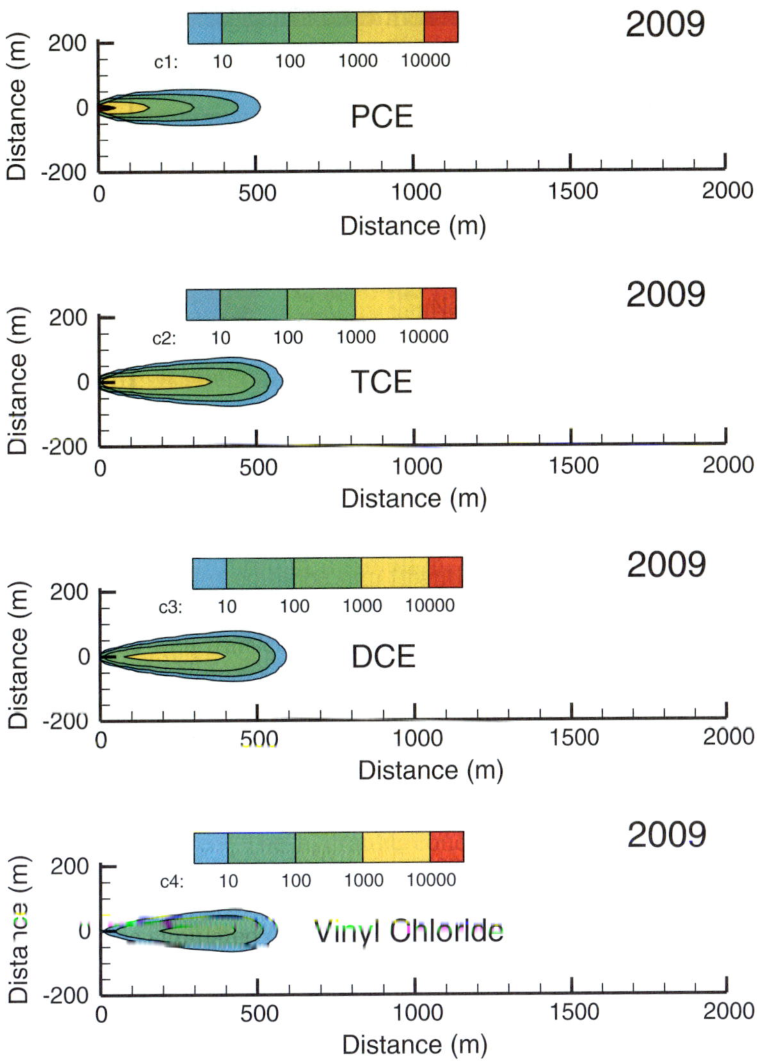

Figure 6.20. Plume concentrations in 2009 following a 1979 release of 1,620 kg of PCE. No remediation has taken place and outer contours of the plumes represent 1 μg/L. The units of the plume contours are μg/L and the dimensions are in meters.

lactate, or molasses (Wiedemeier et al., 1999; Alvarez and Illman, 2006; Stroo and Ward, 2010). These additions act to increase the rate of reductive dechlorination of PCE and TCE, but in many cases the effect is limited on the rates of DCE and VC degradation. The rate of DCE and VC biodegradation can be enhanced by creating a separate aerobic treatment zone downgradient from the reductive dechlorination zone (NRC, 2000; Wiedemeier et al., 1999; Alvarez and Illman, 2006). This aerobic zone, however, would probably have little effect on any PCE and TCE that traveled beyond the anaerobic zone.

The two-zone plume remediation scheme is illustrated on a time–distance diagram in Figure 6.24. The reductive dechlorination zone extends 400 m (1,312 ft) from the source, followed by a 300 m (984 ft) aerobic degradation zone. Both zones are created in 2009 and are maintained for 20 years (2029), when conditions revert to background (natural attenuation) conditions.

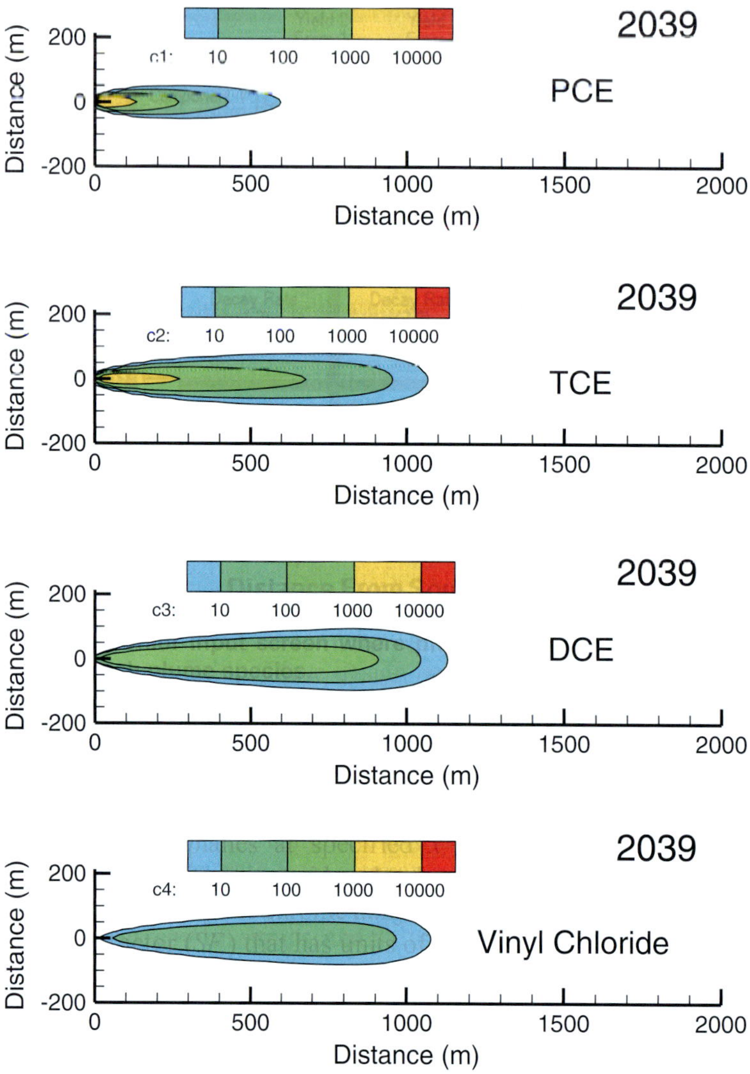

Figure 6.21. Plume concentrations in 2039 following a 1979 release of 1,620 kg of PCE. No remediation has taken place.

In the reductive dechlorination zone, the PCE degradation rate is increased from 0.4 to 1.4 per year and the TCE rate is increased from 0.15 to 1.5 per year. The DCE rate in this zone is only increased from 0.1 to 0.2 per year, and the VC rate remains unchanged at 0.2 per year. In the aerobic degradation zone, the PCE and TCE degradation rates remain at their background levels (0.4 and 0.15 per year), while the DCE rate is increased to 3.5 per year and the VC rate is increased to 3.6 per year.

Figure 6.25 shows the results of this combined source and plume remediation effort on the TCE plume. The reductive dechlorination zone is very effective at reducing the TCE and PCE concentrations over the first 400 m of the plume. The TCE is present beyond 500 m in the year 2029 because some of the PCE and TCE were present in the anaerobic zone when the remediation effort began; these compounds are not affected by the aerobic treatment in this simulation.

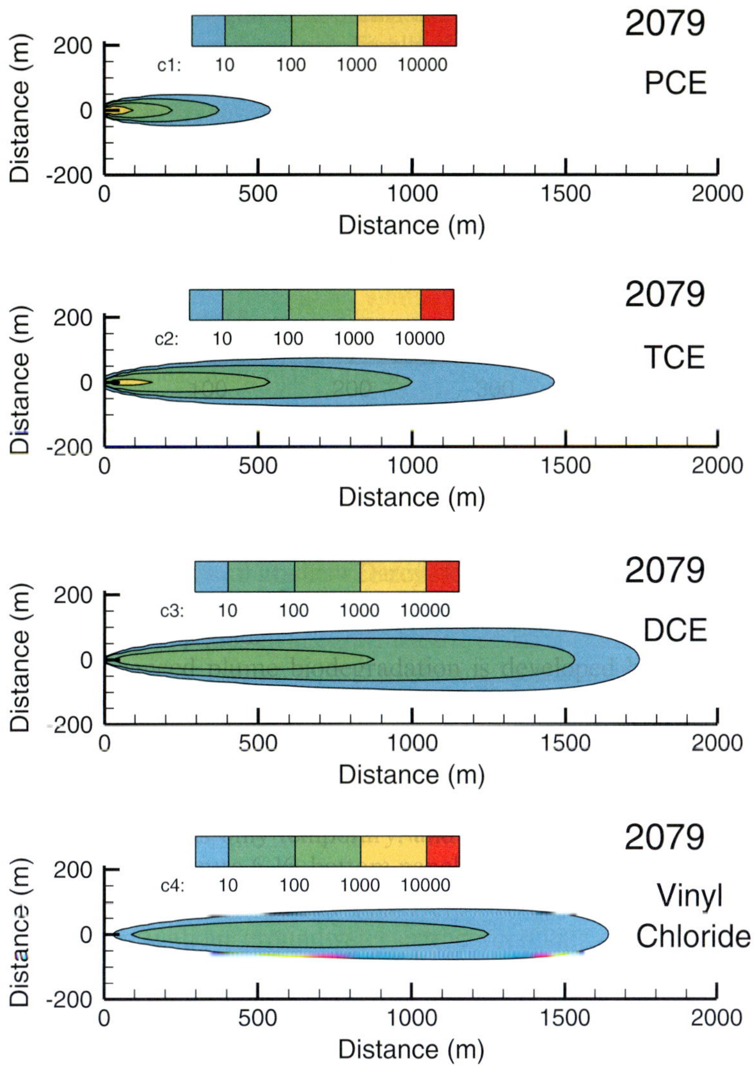

Figure 6.22. Plume concentrations in 2079 following a 1979 release of 1,620 kg of PCE. No remediation has taken place.

After the plume remediation effort ends in 2029, the TCE plume slowly begins to rebound, but because of the source remediation, the reformed plume is much smaller and has a lower concentration than without remediation (Figure 6.22). The VC plume is shown in Figure 6.26.

The VC behavior is much different from the TCE because VC production in the anaerobic zone is actually increased because of the increased dechlorination of PCE and TCE. Nevertheless, the VC is destroyed very effectively in the downstream aerobic zone. After the plume remediation effort is stopped in 2029, the VC plume rebounds somewhat but less than in the natural attenuation case (Figure 6.22) due to the greatly reduced discharge of the ultimate parent compound (PCE) from the source zone. The VC plume at later times is also significantly smaller than it was when only source remediation was performed (Figure 6.23), showing the benefit of adding the plume remediation to the source remediation.

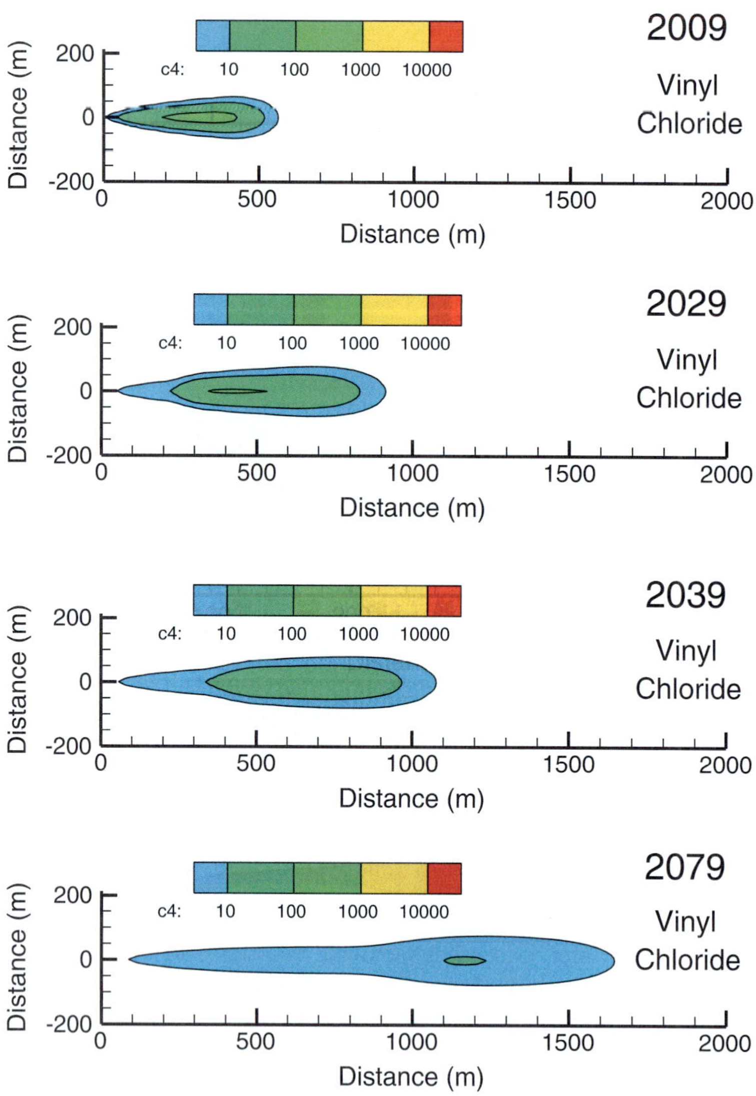

Figure 6.23. Vinyl chloride plume concentrations for a simulation where 90% of the remaining PCE source mass was removed in 2009. The original release occurred in 1979.

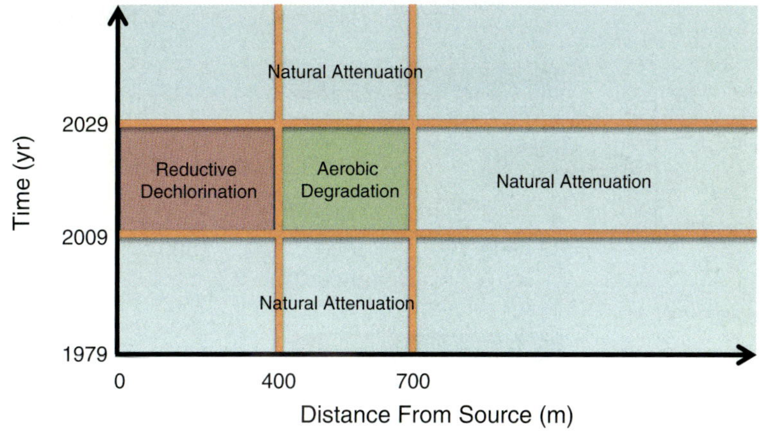

Figure 6.24. Time–distance diagram showing the operation of the enhanced biodegradation remediation in the plume.

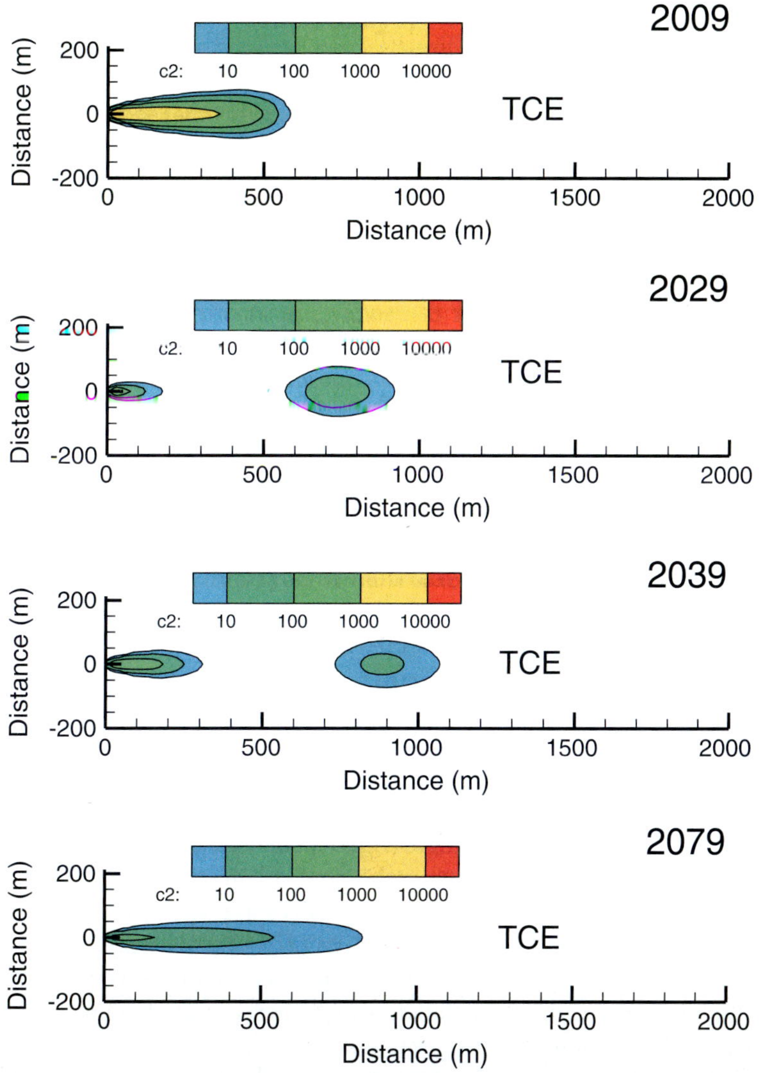

Figure 6.25. TCE plume concentrations for a simulation that considers both source and plume remediation.

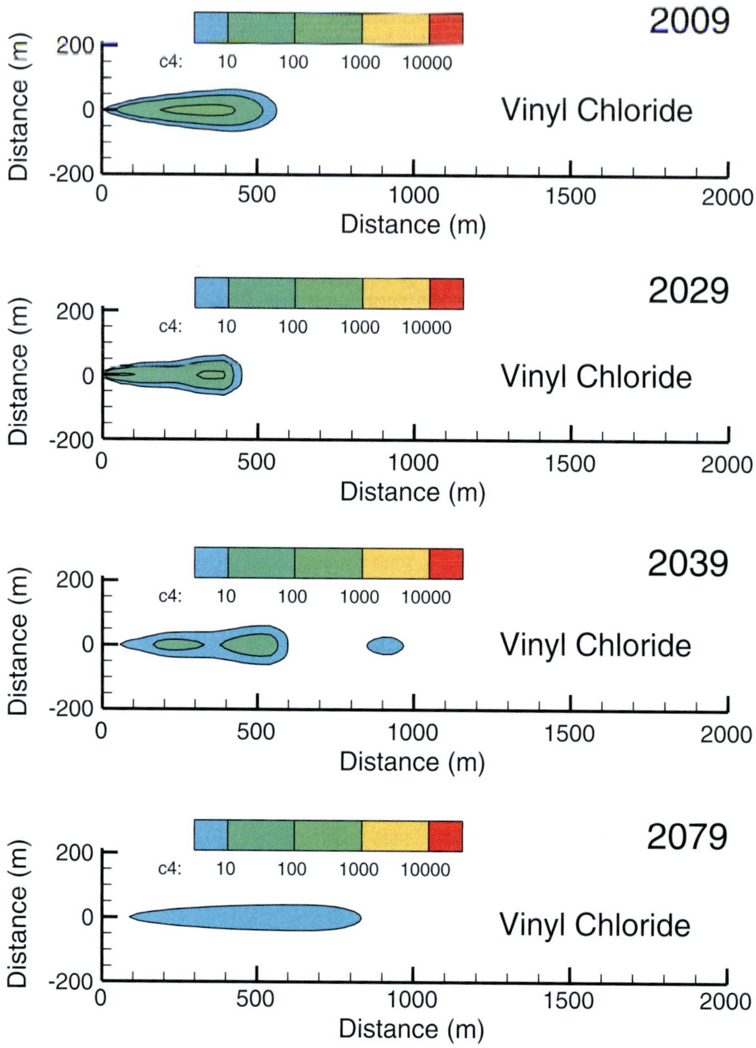

Figure 6.26. VC plume concentrations for a simulation that considers both source and plume remediation.

6.7 MODELING PLUME RESPONSES AT CHALLENGING SITES

6.7.1 Layered Sand–Clay Sequences

The previous examples illustrated the effects of various remediation strategies on plume response times in unconsolidated porous media void of diffusive sinks. In geologic settings with a significant proportion of low permeability diffusive sinks, however, plume response times will generally be longer because of back-diffusion. Back-diffusion within the source zone can sustain a source for many years after all DNAPL has been removed, while back-diffusion within the plume will delay plume detachment times, increase the longevity of the plume, and slow the rate of plume migration.

Figure 6.27 illustrates two examples of layered sand/clay sequences in which back-diffusion may be expected to dominate plume response times following source zone remediation.

Figure 6.27. Examples of clay/sand sequences. *Upper panel* illustrates sharp contact between clay and medium-grained sand. *Lower panel* illustrates small-scale clay layers within fine- to medium-grained sands. Photos by B.H. Kueper, Queen's University, Kingston, ON, Canada.

The amount of contaminant mass that will initially forward-diffuse from the transmissive sands into the clay layers is dependent primarily on the length of time that the layers are exposed to contamination and the storage capacity (aqueous and sorbed phase) of the clay. Once the source of contamination is eliminated, back-diffusion will occur, transferring contaminants from the clay back into the sands. The length of time required to back-diffuse a given amount of contaminant mass far exceeds the initial amount of time required to forward-diffuse that same amount of mass (Reynolds and Kueper, 2002).

With respect to relevance, the degree to which back-diffusion will influence sampled contaminant concentrations in the transmissive sands is dependent on a number of factors including the thickness of the sand layers relative to the thickness of the clay layers, the length and position of the monitoring well screen relative to the location of the sand/clay contact, and the magnitude of groundwater flux in the sand layers relative to the diffusive flux exiting the clay layers. In general, the slower the rate of groundwater flow in the sands, the greater the influence of back-diffusion on sampled concentrations.

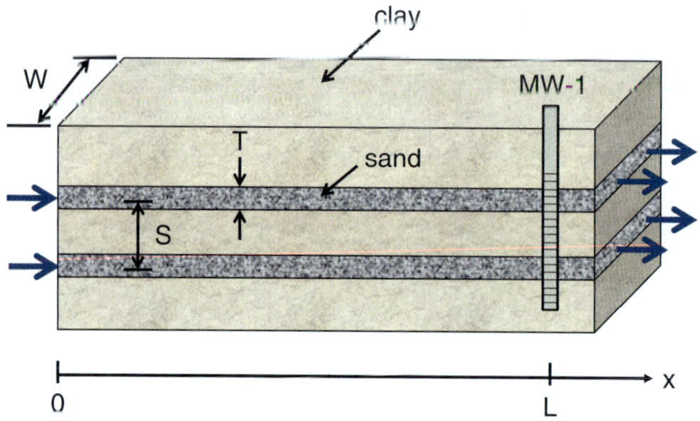

Figure 6.28. Conceptual model of layered sand/clay sequence employed in example simulations. *T* is the sand thickness, *S* is the spacing of the sand layers and *L* is the distance from the downgradient edge of the source zone to the monitoring well.

Table 6.1. Summary of Base Case Model Input Parameters for Layered Sand/Clay Simulations

Parameter	Value
Sand lens thickness	0.15 m (6 inches)
Distance to MW-1 (*L*)	30 m (100 ft)
Clay porosity	0.40
Width of source (*W*)	25 m (80 ft)
Decay of solute	None
Clay fraction organic carbon	0.002
Hydraulic conductivity of sand	1.0E-03 cm/s
Hydraulic gradient	0.01
Spacing of sand lenses	0.5 m (20 inches)
Contaminant	TCE
Source concentration	1,100 mg/L

Figure 6.28 presents a simplified conceptual model of a layered sand/clay sequence. The sequence is dominated by clay containing equally spaced sand layers subject to steady-state horizontal groundwater flow. The left edge of the domain represents the downgradient edge of a DNAPL source zone from which all contaminant mass has been removed at time t^* (age of source). Monitoring well MW-1 samples groundwater from the sand layers at distance L from the downgradient edge of the source zone. A variety of model runs are presented here using the analytical model of West et al. (2004) adapted for the conceptual model presented in Figure 6.28. The model assumes complete mixing of solute concentrations across the thickness of the sand layers, a finite width source (W), two-dimensional solute transport within the plane of the sand layers subject to advection, longitudinal and transverse dispersion, sorption and decay, and one-dimensional diffusion within the clay layers subject to sorption and decay.

Table 6.1 summarizes the model base case input parameters. The contaminant of interest is TCE, which is assumed to provide a source concentration to the sand layers equal to 1,100 mg/L from $t = 0$ up until the time at which the source is instantaneously removed (t^*). From t^*

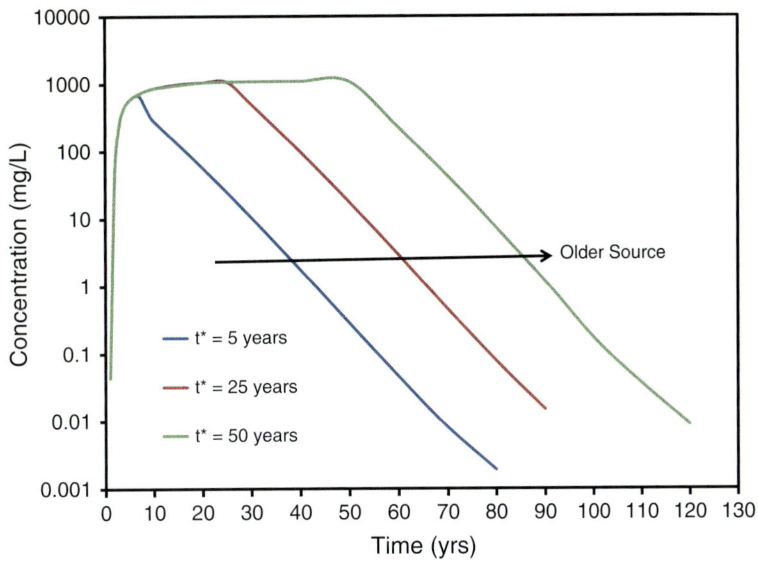

Figure 6.29. Base case TCE concentration in monitoring well MW-1 located 30 m from the source as a function of time for various values of t^*.

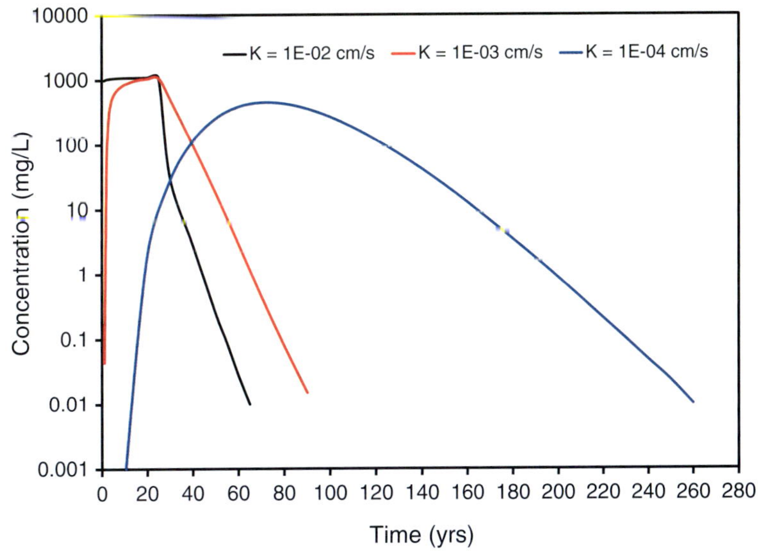

Figure 6.30. Concentration of TCE in monitoring well MW-1 located 30 m from the source as a function of time for various values of the sand layer hydraulic conductivity. The source is completely remediated at $t^* = 25$ years.

onwards, clean groundwater ($C = 0$) enters the sand layers at the left side of the domain. Figure 6.29 illustrates the base case concentration of TCE in monitoring well MW-1 located 30 m from the source as a function of time for various values of t^*. The effect of completely eliminating the source at time t^* quickly leads to declining concentrations at MW-1 but is subject to a long period of tailing as a result of back-diffusion from the clay layers.

Figure 6.30 illustrates the concentration of TCE in monitoring well MW-1 located 30 m (98 ft) from the source as a function of time for various values of the sand layer hydraulic

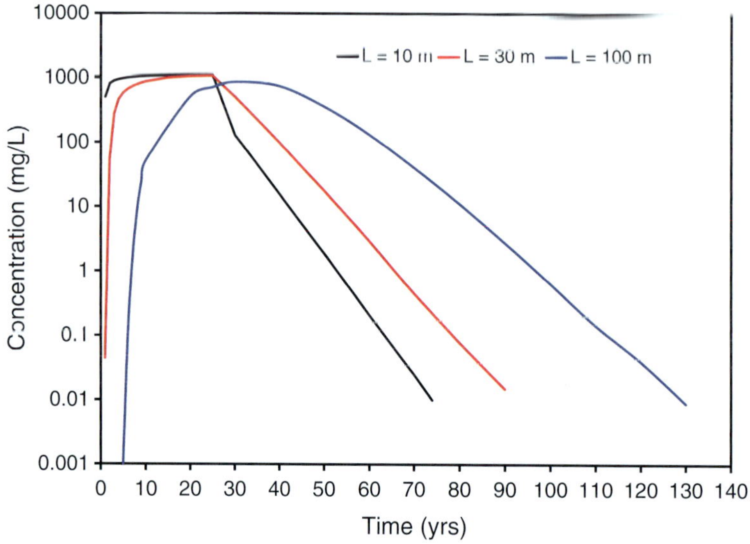

Figure 6.31. Concentration of TCE in monitoring wells located 10, 30, and 100 m from the downgradient edge of the source zone. The source is completely remediated at $t^* = 25$ years. Sand hydraulic conductivity is 1E-03 cm/s.

conductivity. It is assumed that the source is completely remediated at $t^* = 25$ years. The highest considered value of sand hydraulic conductivity (1E-02 cm/s) is associated with the fastest decline in concentrations at MW-1 because of the beneficial effect of clean water flushing on diluting the back-diffusing concentrations. However, the lowest considered value of sand hydraulic conductivity (1E-04 cm/s) is associated with a very different plume response. In this case, concentrations at MW-1 continue to increase until approximately $t = 70$ years even though the source was completely removed at $t = 25$ years. Concentrations start to tail downwards after 70 years to values of approximately 10 µg/L at $t = 260$ years. The latter case is an example of plume response dominated by back-diffusion effects.

Figure 6.31 illustrates the concentration of TCE versus time in monitoring wells located 10 m (33 ft), 30 m (98 ft), and 100 m (328 ft) from the downgradient edge of the source zone. The source is completely remediated at $t^* = 25$ years, and the sand layers are assigned a hydraulic conductivity of 1.0E-03 cm/s. Figure 6.31 demonstrates that the benefits of source remediation manifest themselves earlier at locations closest to the source but that slow tailing of concentrations is observed at all of the considered distances as a result of back-diffusion.

6.7.2 Fractured Bedrock Sites

The layered sand/clay examples in the previous section considered a geologic setting with a moderate proportion of diffusive sinks. In contrast, fractured clay and sedimentary rock represent geologic settings that are often completely dominated by diffusive effects. This stems from the fact that advective flow and transport are limited to the open fractures, which represent only a very small volume of the overall clay or rock mass. While the vast majority of groundwater flow takes place in the open fractures, the vast majority of contaminant mass is stored in the porous matrix. The influence of forward-diffusion on attenuating the rate of downgradient plume advance in fractured rock and clay has been studied by several authors including Sudicky and Frind (1982), Reynolds and Kueper (2001), and Lipson et al. (2005). The example simulations presented here focus on the influence of back-diffusion following source treatment and are taken from West and Kueper (2010).

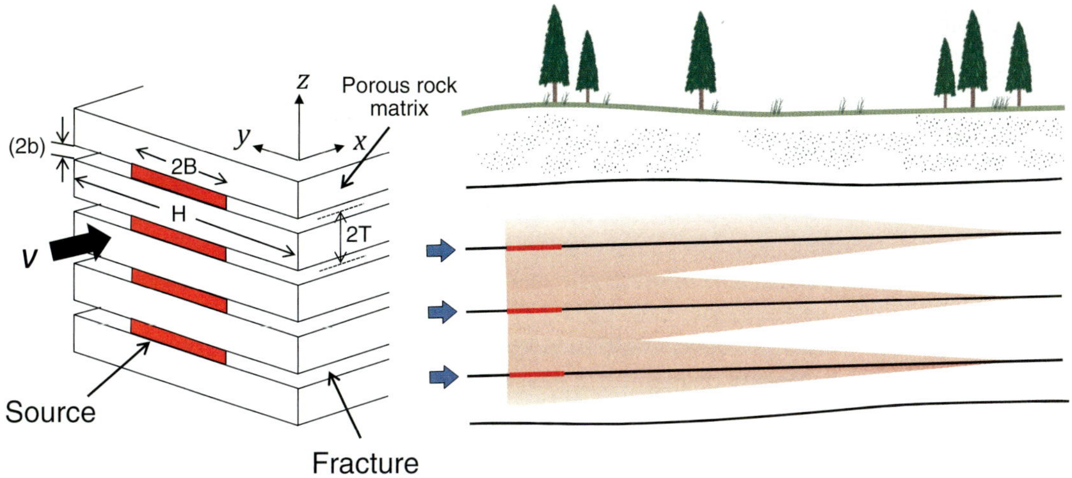

Figure 6.32. Conceptual model adopted for fractured rock plume response simulations (West and Kueper, 2010).

Figure 6.32 illustrates the conceptual model upon which the example simulations are based. Bedrock contains equally spaced, parallel, constant aperture fractures with a finite width source at the inflow to the system. Solute transport in the fractures is subject to advection, dispersion, sorption, decay, and diffusion into the matrix. Diffusive transport in the matrix is subject to sorption and decay. Figure 6.33 illustrates the three types of sources considered. A Type A source provides a constant concentration (C_s) with time at the inflow to the system. An aqueous-phase plume evolves in the downgradient direction and achieves a steady-state length ($L_p = L_{max}$) at time $t = t_s$. The plume is defined here as all concentrations exceeding 5 μg/L. A Type B source provides a constant concentration (C_s) until time t^*, beyond which concentrations throughout the source are maintained at $C = 0$ (complete remediation). As a result of the complete removal of the source, the plume detaches itself and eventually begins receding. A Type C source provides a constant concentration (C_s) until time t^*, beyond which concentrations at the source exhibit an exponential decay with time. The exponential decay in concentration is presumed to result from either natural source depletion processes or implementation of a mass removal technology in the source zone. The rate of source concentration decay is characterized by a half-life, independent of the plume solute biodegradation half-life. The plume will eventually detach from the source location and begin to recede after reaching its maximum length. Table 6.2 summarizes the base case input parameters to the model. The contaminant of interest is TCE, and the bedrock properties are characteristic of fractured sandstone.

The time at which source zone remediation begins to manifest itself in the form of a beneficial plume response is of primary interest at many sites. Figure 6.34 illustrates the length of the TCE plume (L_p) as a function of time for a Type B source for three different values of t^* (the time at which source concentrations are lowered to zero). The maximum length of the plume (L_{max}) occurs at approximately 100 years for $t^* = 25$ years and increases to approximately 150 years for $t^* = 75$ years, with the corresponding L_{max} ranging between 380 m (1,246 ft) and 400 m (1,312 ft). The time at which L_{max} occurs is the time to recession. The plume requires an additional 100 years to diminish below 5 ppb throughout the domain. The Type B simulation is plotted in plan view in Figure 6.35 to further illustrate the characteristics of the plume for $t^* = 25$ years.

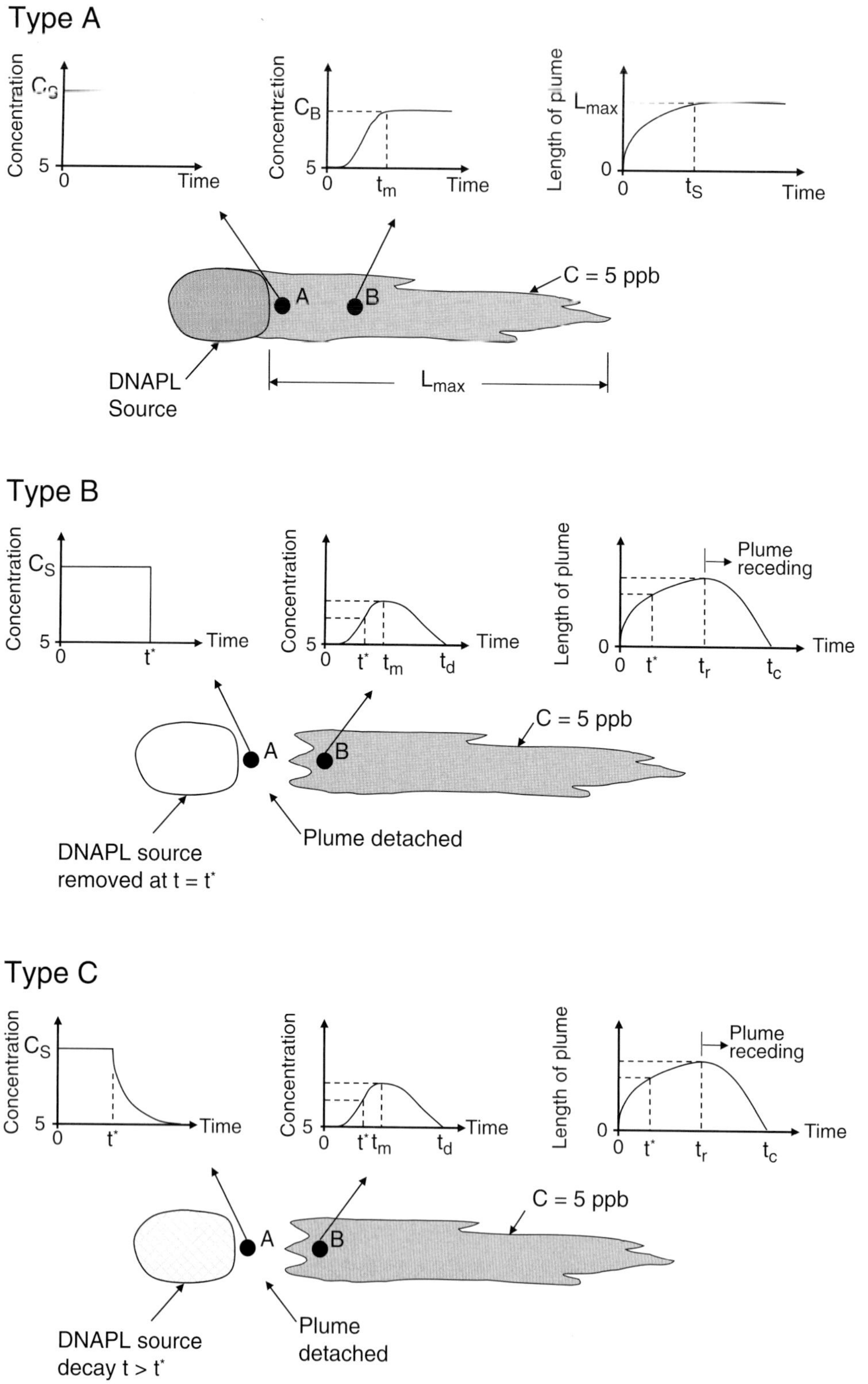

Figure 6.33. Three types of sources considered in fractured rock plume response simulations (West and Kueper, 2007).

Table 6.2. Base Case Input Parameters for Fractured Rock Simulations

Parameter	Symbol and Unit	Value
Initial source concentration	C_s (mg/L)	789
Source zone width	$2B$ (m)	20
Hydraulic gradient	∇h	0.005
Fracture aperture	$2b$ (μm)	140
Fracture spacing	$2T$ (m)	1.42
Fracture aqueous-phase decay half-life	$T(\lambda)_{1/2}$ (yrs)	15
Fracture retardation factor	R	1.0
Fracture longitudinal dispersivity	α_L (m)	1.0
Fracture transverse dispersivity	α_T (m)	0.1
Matrix aqueous-phase decay half-life	$T(\lambda')_{1/2}$ (yrs)	15
Matrix porosity	θ'	0.077
Matrix fraction organic carbon	f_{oc}	0.0036
Matrix dry bulk density	ρ_b (g/cm^3)	2.49
Matrix tortuosity	τ	0.2
Free-water diffusion coefficient[a]	D^o (m^2/s)	1.0×10^{-9}
Octanol–carbon partition coefficient[b]	K_{oc} (mL/g)	126

Parameters taken from Lipson et al. (2005) unless otherwise noted
[a]Pankow and Cherry, 1996
[b]Pankow and Cherry, 1996

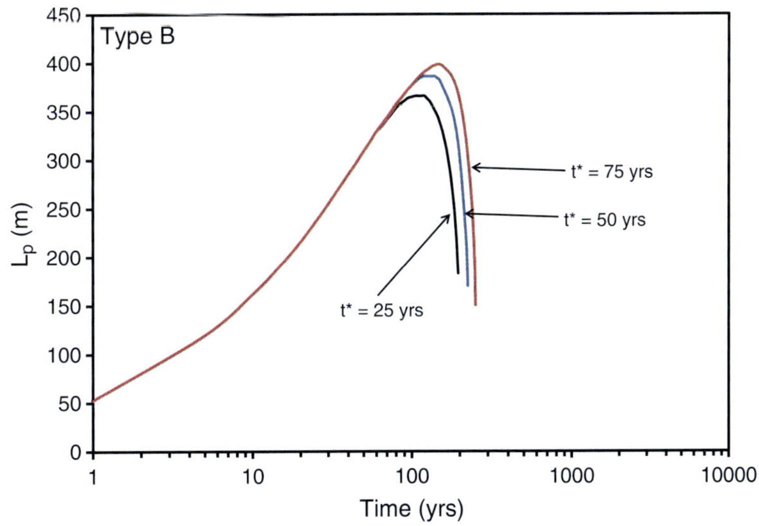

Figure 6.34. Influence of t^* on Type B source zone treatment in fractured sandstone example (West and Kueper, 2007).

Figures 6.34 and 6.35 clearly illustrate that back-diffusion can sustain the growth and persistence of a plume in the fractures for significant periods of time after complete source removal is implemented. From Figure 6.35, the time to recession of the leading edge occurs at approximately 105 years. Thereafter, the leading edge gradually recedes until compliance is achieved.

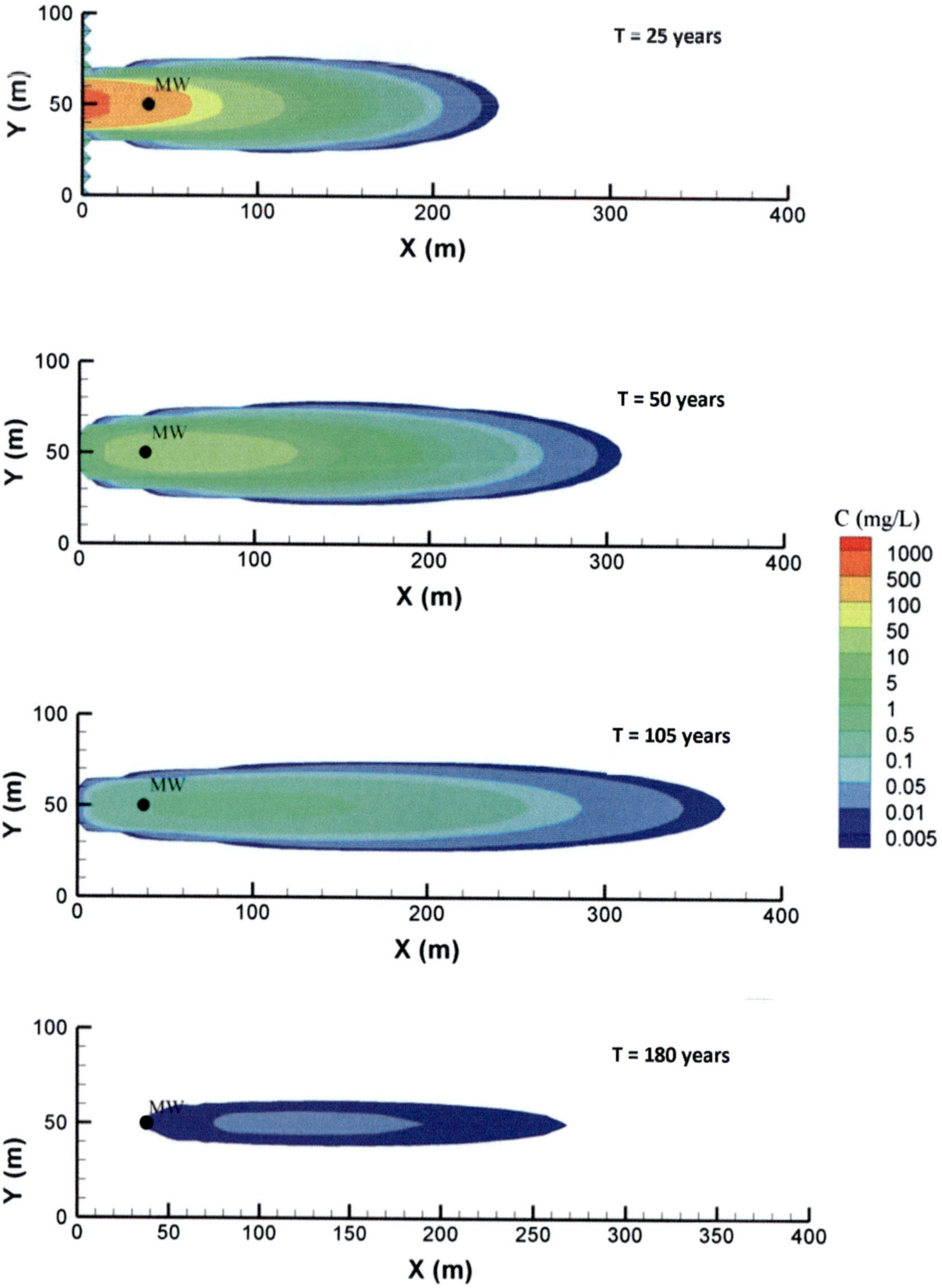

Figure 6.35. Distribution of TCE concentrations within the facture plane at $t = 25$, 50, 105 and 180 years for a Type B source with complete source remediation at $t^* = 25$ years (West and Kueper, 2007).

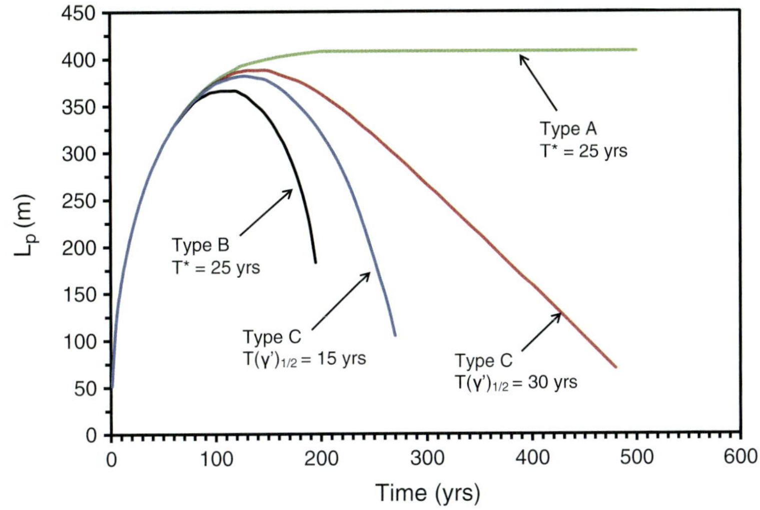

Figure 6.36. Comparison of Type *A*, *B*, and *C* source zone functions on maximum plume length achieved and the time at which plume recession begins (West and Kueper, 2007).

It can also be observed that the leading edge of the plume recedes faster than the back edge of the plume moves forward (detaches). This stems from the fact that back-diffusion from the rock matrix near the source contributes more mass to the fractures than back-diffusion at the leading edge of the plume because of the larger amount of mass sequestered in the matrix near source.

The influence of Type *A*, *B*, and *C* source zone functions on the length of the plume (L_p) versus time is presented in Figure 6.36. For Type *A*, steady state is achieved after 200 years; this curve is the benchmark for subsequent evaluation. When considering Type *B* and *C* source zone functions, differences in L_p versus time can be observed for each source decay rate with a t^* of 25 years. When examining all curves, it can be observed that some treatment or concentration decay ($T(\gamma)_{1/2} = 30$ or 15 years) of the source reduces the time to recession (t_r) approximately the same degree that complete source removal does, and decreasing $T(\gamma)_{1/2}$ from 30 to 15 years reduces the time to compliance ($C \leq 0.005$ mg/L) by approximately twofold. The difference between $T(\gamma)_{1/2} = 15$ years and the instantaneous Type *B* treatment is marginal with respect to the recession time (t_r); the difference is approximately 20 years. The differences become more significant when considering the time to compliance.

6.8 CONCLUSIONS

A wide range of modeling approaches are possible for simulating plume response to source treatment. These range from advanced 3D multiphase numerical models that can simulate the source treatment effort itself to 3D groundwater reactive transport models that model the plume response to a specified source discharge function to simple analytical models that couple idealized source discharge models to idealized reactive plume models. The analytical models are most appropriate for use at small sites or at sites with fairly straightforward hydrogeology. They are also useful for making screening-level calculations at more complex sites, where they can help the user develop a better general understanding of how the site will respond to various forms of remediation. The advanced 3D models are used less often but are justified at sites with complex hydrogeology or contaminant transport.

The plume response to source remediation is a complex function of many variables, including the fraction of mass remaining in the source compared to the plume, source concentration compared to regulatory limits, ratio of plume decay rates to groundwater velocity, relationship between source mass removal and source discharge, and local diffusive effects in the plume.

Source remediation typically leads to an eventual reduction of contaminant mass in the plume. At sites where the source concentration is relatively close to regulatory limits, source remediation can be expected to lead to a significant reduction of plume length. However, if the source concentration is many orders of magnitude higher than regulatory limits, it is likely that plume treatment will be needed in addition to source remediation. Source remediation is most effective when it is done before the plume develops, and the bulk of the contaminant is in the source zone rather than the plume. At sites with significant matrix diffusion effects, the plume response to source remediation is expected to be relatively slow.

Models are tools that allow practitioners to quantify these generalizations about plume response to source remediation. They can be used to help decide if source remediation alone will be sufficient to address a contaminated site or whether plume remediation will also be needed. They can evaluate what would likely happen under natural attenuation conditions and how the plume would respond to temporary or permanent plume treatment efforts. These insights, gained from modeling, then allow the practitioner to establish reasonable goals and expectations for what can be achieved through source remediation.

REFERENCES

Abriola LM, Jang J, Rathfelder K. 1997. Michigan Soil-Vapor Extraction Remediation (MISER) Model – A computer program to model bioventing of organic chemicals in unsaturated geological material. EPA/600R-97/099. U.S. Environmental Protection Agency, Washington, DC, USA.

Alvarez PJJ, Illman WA. 2006. Bioremediation and Natural Attenuation. Wiley-Interscience, Hoboken, NJ, USA.

Aziz CE, Newell CJ, Gonzales JR, Hass P, Clement TP, Sun Y. 2000. BIOCHLOR Natural Attenuation Decision Support System User's Manual Version 1.0. EPA/600/R-00/008. Office of Research and Development, U.S Environmental Protection Agency, Washington, DC, USA.

Aziz CE, Newell CJ, Gonzales JR. 2002. BIOCHLOR Natural Attenuation Decision Support System Version 2.2 User's Manual Addendum. http://www.epa.gov/ada/download/models/biochlor22.pdf. Accessed November 20, 2012.

Basu NB, Rao PSC, Poyer JC, Annable MD, Hatfield K. 2006. Flux – Based assessment at a manufacturing site contaminated with trichloroethylene. J Contam Hydrol 86:105–127.

Basu NB, Rao PSC, Falta RW, Annable MD, Jawitz JW, Hatfield K. 2008. Temporal evolution of DNAPL source and contaminant flux distribution: Impacts of source mass depletion. J Contam Hydrol 95:93–109.

Brooks MC, Annable MD, Rao PSC, Hatfield K, Jawitz JW, Wise WR, Wood AL, Enfield CG. 2002. Controlled release, blind tests of DNAPL characterization using partitioning tracers. J Contam Hydrol 59:287–201.

Brooks MC, Annable MD, Rao PSC, Hatfield K, Jawitz JW, Wise WR, Wood AL, Enfield CG. 2004. Controlled release, blind tests of DNAPL remediation by ethanol flushing. J Contam Hydrol 69:281–297.

Chapelle FH, Widdowson MA, Brauner JS, Mendez E, Casey C. 2003. Methodology for Estimating Times of Remediation Associated with Monitored Natural Attenuation. Water-Resources Investigations Report 03-4057. U.S. Geological Survey, Reston, VA, USA.

Christ JA, Ramsburg CA, Pennell KD, Abriola LM. 2006. Estimating mass discharge from dense nonaqueous phase liquid source zones using upscaled mass transfer coefficients: An evaluation using multiphase numerical simulations. Water Resour Res 4:W11420.

Christ JA, Ramsburg CA, Pennell KD, Abriola LM. 2010. Predicting DNAPL mass discharge from pool-dominated source zones. J Contam Hydrol 114:18–34.

Clement TP. 1997. RT3D – A modular computer code for simulating reactive multi-species transport in 3-dimensional groundwater aquifers. PNNL-11720. Pacific Northwest National Laboratory, Richland, WA, USA.

Clement TP, Sun Y, Hooker BS, Petersen JN. 1998. Modeling multispecies reactive transport in ground water. Ground Water Monit Remed 18(2):79–92.

Delshad M, Pope GA, Sepehrnoori K. 1996. A compositional simulator for modeling surfactant enhanced aquifer remediation. J Contam Hydrol 23:303–327.

DiFillippo EL, Brusseau ML. 2008. Relationship between mass flux reduction and source zone mass removal: Analysis of field data. J Contam Hydrol 98:22–35.

Domenico PA. 1987. An analytical model for multidimensional transport of a decaying contaminant species. J Contam Hydrol 91:49–58

Falta RW. 2003. Modeling sub-grid block scale DNAPL pool dissolution using a dual domain approach. Water Resour Res 39:1360–1368.

Falta RW. 2008. Methodology for comparing source and plume remediation alternatives. Ground Water 46:272–285.

Falta RW, Pruess K, Finsterle S, Battistelli A. 1995. T2VOC User's Guide, Report LBL-36400. Lawrence Berkeley National Laboratory, Berkeley, CA, USA.

Falta RW, Rao PSC, Basu N. 2005a. Assessing the impacts of partial mass depletion in DNAPL source zones: I. Analytical modeling of source strength functions and plume response. J Contam Hydrol 78:259–280.

Falta RW, Basu N, Rao PSC. 2005b. Assessing the impacts of partial mass depletion in DNAPL source zones: II. Coupling source strength functions to plume evolution. J Contam Hydrol 79:45–66.

Falta RW, Stacy MB, Ahsanuzzman ANM, Wang M, Earle RC. 2007. REMChlor Remediation Evaluation Model for Chlorinated Solvents User's Manual Version 1.0. U.S. Environmental Protection Agency. http://www.epa.gov/ada/csmos/models/remchlor.html. Accessed November 20, 2012.

Fure AD, Jawitz JJ, Annable MD. 2006. DNAPL source depletion: Linking architecture and flux response. J Contam Hydrol 85:118–140.

Gerhard J, Kueper B, Hecox G. 1998. The influence of waterflood design on the recovery of mobile DNAPLs. Ground Water 36:283–292.

Grant GP, Gerhard JI. 2007. Simulating the dissolution of a complex dense nonaqueous phase liquid source zone: 2. Experimental validation of an interfacial area-based mass transfer model. Water Resour Res 43:W12409.

Helmig R, Braun C, Emmert M. 1994, MUFTE – A numerical model for simulation of multiphase flow processes in porous and fractured porous media, program documentation (HGK-208). Technical Report 94/3. Institut fur Wasserbau, Universitat Stuttgart, Stuttgart, Germany.

Kaye AJ, Cho J, Basu NB, Chen X, Annable MD, Jawitz JW. 2008. Laboratory investigation of flux reduction from dense non-aqueous phase liquid (DNAPL) partial source zone remediation by enhanced dissolution. J Contam Hydrol 102:17–28.

Liang H. 2009. Probabilistic Remediation Evaluation Model for Chlorinated Solvents Considering Uncertainty. PhD thesis. Clemson University, Clemson, SC, USA.

Liang H, Falta RW, Newell CJ, Farhat SK, Rao PSC, Basu N. 2010a. Decision & Management Tools for DNAPL Sites: Optimization of chlorinated solvent source and plume remediation considering uncertainty, environmental security testing and certification program. ESTCP Project Final Report ER-0704. Alexandria, VA, USA.

Liang H, Falta RW, Henderson JK, Shoemaker S. 2010b. Probabilistic simulation of source and plume remediation for a site contaminated by trichloroethylene. Ground Water Monit Remediat 32:131–141.

Lipson D, Kueper BH, Gefell MJ. 2005. Matrix diffusion-derived plume attenuation in fractured bedrock. Ground Water 43:30–39.

McGuire TM, Newell CJ, Looney BB, Vangelas KM, Sink CH. 2004. Historical analysis of monitored natural attenuation: A survey of 191 chlorinated solvent sites and 45 solvent plumes. Remediat J 15:99–112.

Newell CJ, McLeod RK, Gonzales JR. 1996. BIOSCREEN Natural Attenuation Decision Support System User's Manual, Version 1.3. EPA/600/R-96/087. USEPA National Risk Management Research Laboratory, Washington, DC, USA.

Nitao J. 1996. Reference Manual for the NUFT Flow and Transport Code. Report UCRL-ID-113520. Lawrence Livermore National Laboratory, Livermore, CA, USA.

NRC (National Research Council). 2000. Natural Attenuation for Groundwater Remediation, National Academy Press, Washington, DC, USA.

Panday S, Wu YS, Huyakorn PS, Springer EP. 1994. A three-dimensional multiphase flow model for assessing NAPL contamination in porous and fractured media, porous medium simulation examples. J Contam Hydrol 16:131–156.

Pankow JF, Cherry JA. 1996. Dense Chlorinated Solvents and Other DNAPLs in Groundwater, Waterloo Press, Waterloo, ON, Canada.

Park E, Parker JC. 2005. Evaluation of an upscaled model for DNAPL dissolution kinetics in heterogeneous aquifers. Adv Water Resour 20:1280–1291.

Parker JC, Falta RW. 2008. Comparison of alternative upscaled model formulations for simulating DNAPL source dissolution and biodecay. Adv Water Resour 31:1325–1332.

Parker JC, Park E. 2004. Modeling field-scale dense nonaqueous phase liquid dissolution kinetics in heterogeneous aquifers. Water Resour Res 40:W05109.

Pruess K, Battistelli A. 2002. TMVOC, A Numerical Simulator for Three-Phase Non-isothermal Flows of Multicomponent Hydrocarbon Mixtures in Saturated-Unsaturated Heterogeneous Media. Report LBNL-49375. Lawrence Berkeley National Laboratory, Berkeley, CA, USA.

Rao PSC, Jawitz JW. 2003. Comment on "Steady state mass transfer from single-component dense non-aqueous phase liquids in uniform flow fields" by Sale TC, McWhorter DB. Water Resour Res 39:1068.

Rao PSC, Jawitz JW, Enfield CG, Falta R, Annabel MD, Wood AL. 2001. Technology integration for contaminated site remediation: Cleanup goals and performance metrics. Ground Water Quality, Sheffield, UK. pp 410–412.

Reynolds DA, Kueper BH. 2001. Multiphase flow and transport in fractured clay/sand sequences. J Contam Hydrol 51:41–62.

Reynolds DA, Kueper BH. 2002. Numerical examination of the factors controlling DNAPL migration through a single fracture. Ground Water 40:368–377.

Sale TC, McWhorter DB. 2001. Steady-state mass transfer from single-component dense non-aqueous phase liquids in uniform flow fields. Water Resour Res 37:393–404.

Sale TC, Newell CJ. 2010. Impacts of Source Management on Chlorinated Solvent Plumes. In Stroo HF, Ward CH, eds, *In Situ* Remediation of Chlorinated Solvent Plumes. Springer, New York, NY, USA, pp 185–216.

Stroo HF, Ward CH, eds. 2010. *In Situ* Remediation of Chlorinated Solvent Plumes. Springer, New York, NY, USA.

Sudicky EA, Frind EO. 1982. Contaminant transport in fractured porous media: Analytical solutions for a system of parallel fractures. Water Resour Res 18:1634–1642.

Unger AJA, Forsyth PA, Sudicky EA. 1996. Variable spatial and temporal weighting schemes for use in multi-phase compositional problems. Adv Water Resour 19:1–27.

USEPA (U.S. Environmental Protection Agency). 2003. The DNAPL Remediation Challenge: Is there a case for source depletion? EPA/600/R-03/143. EPA Office of Research and Development, Washington, DC, USA.

Waddill DW, Widdowson MA. 2000. SEAM3D – A numerical model for three dimensional solute transport and sequential electron acceptor-based bioremediation in groundwater. ERDC/EL TR-00-18. U.S. Army Engineer Research and Development Center, Vicksburg, MS, USA.

West M, Kueper BH. 2007. Plume detachment and recession times following source treatment in bedded fractured rock. Proceedings, USEPA/NGWA Fractured Rock Conference: State of the Science and Measuring Success in Remediation. Portland, ME, USA, September 24–26, pp 343–356. http://www.clu-in.org/products/siteprof/2007fracrock/000tableofcontents-2_.pdf. Accessed January 30, 2014.

West MR, Kueper BH. 2010. Plume detachment and recession times in fractured rock. Ground Water 48:416–426.

West MR, Kueper BH. 2012. Numerical simulation of DNAPL source zone remediation with in situ chemical oxidation (ISCO). Adv Water Resour 44:126–139.

West M, Kueper BH, Novakowski KS. 2004. Semi-analytical solutions for solute transport in fractured rock using a strip source of finite width. Adv Water Resour 27:1045–1059.

West MR, Kueper BH, Ungs M. 2007. On the use and error of approximation in the Domenico (1987) solution. Ground Water 45:126–135.

West MR, Grant GP, Gerhard JI, Kueper BH. 2008. The influence of precipitate formation on the chemical oxidation of TCE DNAPL with potassium permanganate. Adv Water Resour 31:324–338.

White MD, Oostrom M, Lenhard RJ. 1995. Modeling fluid flow and transport in variable saturated porous media with the STOMP simulator. Adv Water Resour 18:353–364.

Widdowson MA. 2003. SEAM3D: A numerical model for three-dimensional solute transport and sequential electron acceptor-based biodegradation in ground water. Final Report to the U.S. Army Engineer Research and Development Center, Vicksburg, MS, USA.

Widdowson MA, Mendez E, Chapelle F, Casey CC. 2004. Natural Attenuation Software User's Manual, Version 2. http://www.nas.cee.vt.edu/index.php. Accessed November 20, 2012.

Wiedemeier TD, Rifai HS, Newell CJ, Wilson JT. 1999. Natural Attenuation of Fuels and Chlorinated Solvents in the Subsurface. John Wiley & Sons, Inc., New York, NY, USA.

Wood AL, Falta RW. 2003. SERDP and NRMRL Sponsor Field Test of Cosolvent-Enhanced DNAPL Removal. EPA 542-N-03-003. Technology News and Trends, Issue 6. USEPA, Washington, DC, USA.

Wood AL, Enfield CG, Espinoza FP, Annable M, Brooks MC, Rao PSC, Sabatini D, Knox R. 2005. Design of aquifer remediation systems: (2) Estimating site-specific performance and benefits of partial source removal. J Contam Hydrol 81:148–166.

Zhang H, Schwartz FW. 2000. Simulating the in situ oxidative treatment of chlorinated compounds by potassium permanganate. Water Resour Res 36:3031–3042.

Zhang K, Yamamoto H, Pruess K. 2007. TMVOC-MP: A parallel numerical simulator for three-phase non-isothermal flows of multicomponent hydrocarbon mixtures in porous/fractured media. Report LBNL-63827. Lawrence Berkeley National Laboratory, Berkeley, CA, USA.

Zheng C, Wang PP. 1999. MT3DMS: A modular three-dimensional multispecies transport model for simulation of advection, dispersion and chemical reactions of contaminants in groundwater systems; documentation and user's guide. Contract Report SERDP-99-1. U.S. Army Engineer Research and Development Center, Vicksburg, MS, USA.

Zhu J, Sykes JF. 2004. Simple screening models of NAPL dissolution in the subsurface. J Contam Hydrol 72:245–258.

CHAPTER 7

FLUX-BASED SITE ASSESSMENT AND MANAGEMENT

Michael D. Annable,[1] Michael C. Brooks,[2] James W. Jawitz,[1] Kirk Hatfield,[1]
P. Suresh C. Rao[3] and A. Lynn Wood[2]

[1]University of Florida, Gainesville, FL 32611, USA; [2]National Risk Management Research Laboratory, U.S. Environmental Protection Agency, Ada, OK 74820, USA; [3]Purdue University, West Lafayette, IN 47907, USA

7.1 INTRODUCTION

Characterization of dense nonaqueous phase liquid (DNAPL) contaminated sites poses significant challenges, but continual advances have led to improvements in both site characterization methods and decision making in recent years. In the area of site risk/liability characterization there has been increased interest in shifting from a focus on concentration measurements to quantifying flux of both contaminants and groundwater and estimation of source mass (Einarson and Mackay, 2001; API, 2002, 2003; ITRC, 2003; Kavanaugh and Rao, 2003; NRC, 2004). As a result of this shift, new approaches have been developed for measuring both flux and mass discharge at DNAPL-contaminated sites. Flux-based data can be used to determine site mass discharge, conduct plume, and source mass balances and track changes in mass over time. This information can support reassessment of the site (risk/liability) conceptual model and allows quantitative evaluation of the importance of source and plume management approaches and associated benefits (risk/liability reduction).

The primary technical challenge associated with DNAPL contaminated sites is a result of the density contrast between the immiscible phase and water, which contributes to DNAPL migrating under gravitational forces and/or becoming trapped on and within layers of geologic media with contrasting oil entry (and wettability) properties. This process results in complex distributions of DNAPL in the subsurface that challenge our ability to locate the DNAPL and to characterize in detail the spatial distribution (Kueper et al., 1993; Illangasekare et al., 1995). This difficulty, in turn, leads to significant challenges in designing, implementing and evaluating remedial technologies.

In the early stages of a site assessment, the focus is generally on determining the type and extent of the contamination present. This is generally achieved by measuring the concentration of contaminants in water samples and obtaining some understanding of the spatial extent of the contaminated zone. Such efforts usually start with an assessment of the dissolved plume, usually viewed in relation to a regulatory maximum contaminant level (MCL) established for each constituent. If the contaminant is suspected to be the result of a nonaqueous phase liquid (NAPL) spill then the concentrations observed may be compared to the solubility limit of the NAPL mixture present in the aquifer. At this stage of an investigation, attention might turn to identifying the presence and extent of a source zone that is defined by the NAPL-contaminated region.

In this framework, the site is divided conceptually into a source zone and a plume (Figure 7.1). At a more advanced stage of site assessment, decisions are made on (1) further characterization of the source and plume zones, (2) evaluation of the technologies for

B.H. Kueper et al. (eds.), *Chlorinated Solvent Source Zone Remediation*,
doi: 10.1007/978-1-4614-6922-3_7, © Springer Science+Business Media New York 2014

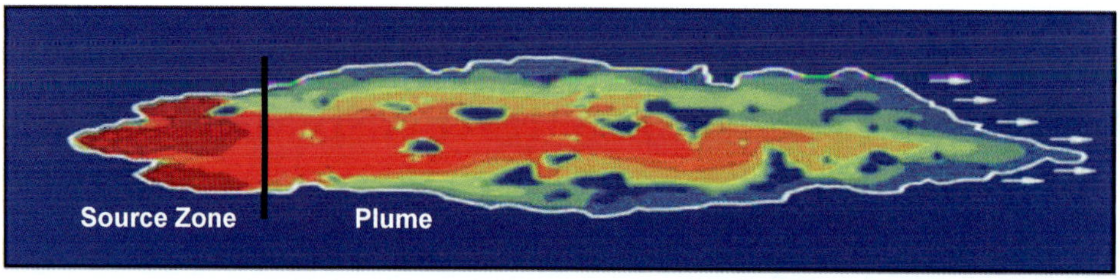

Figure 7.1. Conceptual site with source zone and plume.

remediation of either or both zones to meet regulatory compliance and risk/liability reduction and (3) optimized implementation of cleanup technologies to achieve remediation goals in a cost- and time-effective fashion. The final stages involve legacy stewardship, i.e., long-term management of residual levels of contamination.

The discussion here assumes that an exact distinction between the source zone and the plume can be made, but it is recognized that zones of contamination that result from diffusion of dissolved contaminant mass into low-conductivity units both in and near the source zone may exist. These might be described as secondary source zones and are distinguished from the original source zone (i.e., NAPL-contaminated zone). Back diffusion of mass from these zones has been identified as a mechanism that extends plume longevity (Parker et al., 2008).

In recent years, site characterization technologies have been developed to characterize contaminant mass flux and integrated mass discharge at contaminated sites. Here we define mass flux as the contaminant mass moving across a unit area of porous media (aquifer) perpendicular to the groundwater flow direction. This measure has units of mass/area time. A typical unit might be grams/square meter day (g/m^2 day). In contrast, integrated mass flux comprises the total mass crossing some area of interest (the control plane defining the total cross section of the source zone) over time. In contaminant plumes, the area of interest is generally taken to be large enough to contain the entire plume, and thus the magnitude becomes the mass discharge of the plume at some distance along the plume axis and at the time of measurement. The units are thus mass/time with a typical unit of grams/day (g/day) or kilograms/year (kg/yr). This mass flux/mass discharge information can be valuable during all phases of site management, including assessing risk, quantifying natural and enhanced attenuation processes, assessing remedial performance and measuring loads to receiving bodies (ITRC, 2010).

The methods currently available for quantifying both local mass flux and integrated mass discharge can be categorized by spatial and temporal scales of measurement. Point-scale techniques include multilevel samplers and passive flux meters (PFMs) within screened wells. Both methods collect data to determine the local mass flux [mass per unit area per time] that can be spatially integrated to determine mass discharge [mass per time]. Alternatively integral methods, such as integral pump tests (IPT), collect data averaged over fully screened wells when large volumes of water are pumped from aquifers. This information, while lacking spatial resolution, has advantages in that local plume cores are incorporated. The utility of each scale of site flux characterization is explored in terms of site management decisions including plume or source management and remedial design improvements.

In response to the challenges presented by DNAPL sites, advances have been made in evaluating the use of flux measurements for contaminated site assessment and management (Basu et al., 2006; Falta et al., 2005a, b). A major component of this advance has focused on establishing relationships between the mass present in an NAPL source zone and the contaminant mass flux (or source strength) leaving the source zone and forming a dissolved contaminant plume. Recent research has focused on establishing these relationships (Fure et al., 2006; Parker and Park, 2004; Falta et al., 2005a; Jawitz et al., 2005; Lemke et al., 2004). This work has established the fundamental knowledge needed to advance site management strategies based on mass flux, mass discharge and mass balance. Site evaluations may also benefit from considering the age of the site, defined as the relative mass remaining in the source area compared to the original mass estimate, and some estimate of plume longevity and risk given various remedial options.

It must be acknowledged that because of the inherent complexities at DNAPL sites, uncertainties prevail in all aspects of site assessment and management. Thus, technology advances are focused on reducing such uncertainties and (1) include better, faster and cheaper methods for measuring contaminant spatial distribution, usually by sampling at multiple points (locations) at a site and (2) as an alternative to distributed point measurements focus on approaches that provide spatially and/or temporally integrated metrics. The first approach has the disadvantage that a large number of points in space will have to be sampled to decrease interpolation errors, while the second approach has the limitation that the metrics may lack sufficient spatial or temporal resolution. The choice between these two groups of technologies is determined by cost constraints and utility of the data (the level of detail required to make a certain decision, with some acceptable level of uncertainty). Thus, site assessment represents a trade-off between cost effectiveness of assessment and robustness and utility of the information generated to help make reliable decisions.

Management decisions at contaminated field sites are typically based on the available data compiled on subsurface conditions related to both hydrogeology and geochemistry. Final decisions are usually linked to risk drivers, regulatory requirements and future plans for site utilization. Critical to any decision on remedial strategies and long-term stewardship is site characterization data quantity and quality.

The utility of local mass flux and/or integrated mass discharge measurements has been slow to develop, likely because of the nearly complete reliance on concentration as a regulatory measure. When reducing concentration to a target level is set as the remedial objective, directing additional site resources to characterizing mass flux is seen as unwarranted. The key to changing this model is recognizing the link between mass flux and plume response (Soga et al., 2004; Christ et al., 2006; DiFilippo and Brusseau, 2008; Basu et al., 2009; ITRC, 2010). Understanding this link quantitatively can provide those making technical decisions on site management a means to make sound decisions on site remedial efforts and the associate benefits in terms of plume response.

In addition, better understanding of the flux distribution from source zones and the complexity of those distributions can lead to cost savings on remedial design where systems are targeted to zones that provide the most significant load to the contaminant plume. In some cases site remedial efforts may be misdirected toward regions of high concentrations that may not be contributing the largest load to the site plume (low-conductivity zones with high concentrations). An understanding of flux characteristics at a site can also lead to better prioritization of limited site resources given that at most sites all the mass cannot be removed (particularly at DNAPL sites) and efforts should target those regions responsible for plume behavior.

The ITRC DNAPL team identified six areas where mass flux and/or mass discharge can improve site management (ITRC, 2010):

1. Site conceptual model and site characterization
2. Threat and exposure evaluation
3. Remediation selection and design

4. Performance monitoring and optimization
5. Compliance and long-term monitoring
6. Site prioritization

This list provides objectives for collection and use of flux data. At many sites multiple uses may be realized, though often the most critical among these is simply enhancing the understanding of how the site functions to sustain plume conditions.

This chapter provides an overview of the utility of contaminant flux data collection for improving understanding of DNAPL site conditions, with the objective of improving site management decisions. Flux data collection can improve evaluation of risk/liability projections and selection and design of remedial activities. Combined with traditional site data collection and historical site monitoring data that provide estimates of mass in plumes or sources, the flux data can be used to estimate mass balances useful for decisions on source versus plume remediation.

7.2 MASS DISCHARGE AND MASS FLUX AS MEASURES IN SITE CHARACTERIZATION AND MANAGEMENT

7.2.1 Fundamental Background

In the context of this chapter, the terms mass flux and mass discharge describe the rate of contaminant mass transport from the source to the plume and within the plume. In a more general sense, the terms flux and discharge describe the transport of quantities such as heat, mass or volumes across an area of interest. These terms are usually linked in a mass or heat balance framework and are incorporated in a system mass balance in an effort to more rigorously define the system conditions. Considering our more narrowly defined problem involving a contaminant source from which mass is removed under natural gradient groundwater flow generating a much larger dissolved contaminant plume, we define mass flux, J [M/L^{2}T], as the product of the Darcy flux, q [L/T], and the local concentration of contaminant, C [M/L^3], in the aqueous phase:

$$J = qC \qquad \text{(Eq. 7.1)}$$

The Darcy flux can be calculated by applying Darcy's Law along the direction of the plume axis at the location of interest:

$$q = -K\frac{dh}{dx} \qquad \text{(Eq. 7.2)}$$

where K is the hydraulic conductivity [L/T] and h is the head measurement at the location of interest. For this calculation the x-direction is selected as the maximum gradient direction. The Darcy and mass flux terms are vector quantities and thus should be specified with an associated direction. In the case of contaminated sites with a well-developed plume structure, the mean vectors for flux are typically associated with the plume growth direction. It is noted that the groundwater flow direction is dynamic and that the contaminant mass flux changes direction with groundwater flow (Rein et al., 2009); however, this discussion is focused on longer time scale plume dynamics in which case the average Darcy flux direction may be considered.

The contaminant mass flux is a local quantity that can be measured or calculated across a transect orthogonal to the plume axis. This transect, or control plane, A, then becomes the spatial domain over which the local mass flux values are integrated over the plume extent (width and height within the control plane) to quantify mass discharge, M_D:

$$M_D = \int_A J dA \quad \text{(Eq. 7.3)}$$

The mass discharge is a scalar quantity in which a positive value reflects contaminant mass transport in the plume growth direction. Mass discharge is likely to change with position along the plume axis (typically in the x direction) and time. Mass discharge measurements quantified near the downgradient edge of the source zone are defined as the source zone mass discharge. How this mass discharge changes with natural dissolution of the source defines the source strength function, $M_D(0,t)$. Proposed formulas for source strength functions will be discussed later. The link between the source zone and plume can be used to evaluate the source/plume mass balance. While often difficult to quantify accurately, even approximate estimates of mass present in both the source and plume can be used to evaluate site conditions.

One measure of the site status has been referred to as the site *age* (Jawitz et al., 2005). A simple definition of site age is the fraction of the initial source zone mass that has been removed from the source zone through dissolution, volatilization or degradation. Given an estimate of the initial mass and the current mass, the age is simply the ratio of the mass lost to the initial mass present. A site in which 90% of the mass was removed from the source zone would be considered an aged site, while those with a small fraction, for example 10%, would be considered a young site. Obviously many factors collectively determine how fast a site ages. Some major factors include the solubility limit of the contaminant, groundwater flow velocity, size of the source zone in the flow direction and average NAPL saturation along the flow path. Basic calculations of mass removal based on equilibrium zero-order removal can be used to provide preliminary estimate of the dissolution process. Sites with a low-solubility DNAPL, such as perchloroethene (PCE) (solubility 150 milligrams per liter [mg/L]), will age more slowly than a comparable site with trichloroethene (TCE) (solubility 1,100 mg/L).

The mass flux at a specified location within a plume can be averaged over different fractions of the cross section. This may be useful for measurements taken over larger scales such as an IPT with multiple wells forming a transect or a traditional sampling from the wells. Average mass flux over portions of the transect represented by area B may also be useful for calculating the flux-averaged concentration perhaps representing what would be expected at an impacted pumping well that has a capture zone of the same area. Thus the average mass flux can be calculated as follows:

$$\overline{J} = \frac{\int_B J\, dB}{B} \quad \text{(Eq. 7.4)}$$

A flux-averaged concentration can be calculated for a well having a capture area B simply using $C = J/q$. This simple approach allows one to relate flux values to target concentration on a flux average basis. Through these basic relationships targets based on concentration and flux can be linked. If degradation processes are active in the contaminant plume these can be considered and used to lower target remedial objectives for source flux and or mass discharge.

When considering strategies for short- and long-term site management, the distribution of the contaminant mass between phases (as an immiscible liquid, sorbed on aquifer solids or dissolved in groundwater) at the site and projections of mass discharge history are important to consider. For example, a trend of increasing mass discharge over time near the source zone would suggest early stages of source depletion, and projections of future trends would depend on source mass and release dynamics. However, a declining trend of mass discharge would imply that much of the DNAPL mass has been depleted and might be present in the dissolved plume if degradation losses do not play a major role.

It is important then to recognize that while contaminant release dynamics within the source zone contribute to source mass discharge trends, the growth and attenuation of the dissolved plume are controlled by hydrodynamic characteristics of the aquifer (advection–dispersion, diffusive mass transfer with low-conductivity zones or volatile mass transfer with the unsaturated zone) and biogeochemical conditions (aerobic or anaerobic). When the contaminant mass discharge rate at the edge of the source zone is equal to the overall mass attenuation rate in the dissolved plume, steady-state conditions prevail and the plume becomes static (no growth). If the attenuation rate is greater than the source release rate, one can expect shrinking plumes, and plumes continue to grow when the release rate exceeds the attenuation rate. Both source mass discharge rates and in-plume attenuation rates change with time, which contributes to temporal instability of plume shape and size.

The collection of measurements of mass discharge from the source zone to the plume may allow for estimates of projections of mass discharge over the life of the DNAPL source area. To support these projections some knowledge of current or past mass present in the source area and/or plume is desirable. Often sites lack sufficient information to rigorously estimate the mass discharge history or source strength function. In this case, historical concentration data may be used to support these estimates. If the historical concentration trends are linked to recent flux measurements then an estimated mass discharge history can be determined.

7.2.2 Relationship of Mass Depletion and Mass Discharge and Flux

Several modeling approaches have been used to evaluate how mass discharge or flux changes as a result of mass removal from DNAPL source zones (Sale and McWhorter, 2001; Rao et al., 2002; Rao and Jawitz, 2003; Lemke et al., 2004; Parker and Park, 2004; Enfield et al., 2005; Jawitz et al., 2005; Wood et al., 2005). The impacts of reductions in mass discharge resulting from source treatment on the dissolved plume have also been examined in a modeling analysis through development of models such as REMCHLOR (Falta et al., 2005a, b; Basu et al., 2008; Falta, 2008). Results from these models suggest that a wide range of flux behavior may occur as a function of hydrogeological conditions and DNAPL distributions. Initial field measurements in hydraulically isolated test cells suggested that significant reductions in contaminant mass discharge are indeed observed after removal of NAPL source mass (Brooks et al., 2004). However, it could be argued that tests performed in isolated flow domains (test cells) do not represent actual field situations at NAPL sites. More recent measurements at field sites have demonstrated that significant flux reduction can be achieved through mass removal from DNAPL source zones (Brooks et al., 2008; DiFilippo and Brusseau, 2008).

Three simplified analytical source strength function models that have been proposed to simulate DNAPL dissolution from source zones are discussed here. These models have been compared to complex multiphase flow and transport models, and results support their use for simple long-term predictions (Basu et al., 2008). The equilibrium streamtube model (ESM) is based on a Lagrangian approach where the DNAPL source zone is conceptualized as a collection of non-interacting streamtubes, with hydrodynamic and DNAPL heterogeneity represented by various travel times and DNAPL saturations within streamtubes (Jawitz et al., 2003, 2005):

$$\frac{C_f(T)}{f_c C_s} = \frac{1}{2} - \frac{1}{2} erf\left(\frac{\ln T - \mu_{\ln\tau}}{\sigma_{\ln\tau}\sqrt{2}}\right) \quad \text{(Eq. 7.5)}$$

Here C_f [ML^{-3}] is the flux-averaged concentration exiting the source zone; T is the number of pore volumes of natural flowing groundwater that have moved through the source zone [-]; f_c is the fraction of the streamtubes initially containing DNAPL [-]; C_s is the solubility limit [ML^{-3}]; τ is defined as reactive travel time that combines the travel time (t) and the trajectory-averaged DNAPL content $\hat{S}$ ($\hat{S} = \frac{\hat{S}_N \eta}{\theta_w}$), where $\hat{S}_N$ is the trajectory-averaged DNAPL saturation along a streamline [L^3L^{-3}], η is the porosity [L^3L^{-3}], and θ_w is the water content [L^3L^{-3}]; $\mu_{\ln\tau}$ and $\sigma_{\ln\tau}$ are the mean and standard deviation of the lognormal variable $\ln\tau$. The travel time and the DNAPL content $\hat{S}$ are both assumed to be lognormal variables, so the reactive travel time obeys the lognormal distribution.

The Power Law source depletion model (PLM) initially proposed by Rao et al. (2002) has been incorporated in the model REMCHLOR by Falta et al. (2005a). The power function relates the flux-averaged concentration $C_f(T)$ to the source mass through an empirical DNAPL heterogeneity measure Γ, a function of both media and DNAPL variability:

$$\frac{C_f(T)}{f_c C_s} = \left[\frac{M(T)}{M_0}\right]^{\Gamma} \quad \text{(Eq. 7.6)}$$

where $M(T)$ is the mass remaining in the source zone [M] and M_0 is the initial source mass [M]. Falta et al. (2005a) derived an analytical solution for dissolution under the PLM:

$$\frac{C_f(T)}{f_c C_s} = \frac{1}{M_0^{\Gamma}}\left[\frac{(\Gamma - 1)M_{D,0}}{M_0^{\Gamma}} t + M_0^{1-\Gamma}\right]^{\frac{\Gamma}{1-\Gamma}} \quad \text{(Eq. 7.7)}$$

where $M_{D,0}$ is the initial mass discharge (M/T).

Parker and Park (2004) presented a simplified model for estimating DNAPL source zone mass depletion using an effective Damkohler number, Da:

$$\frac{C_f(T)}{C_s} = 1 - \exp[-Da] \quad \text{(Eq. 7.8)}$$

Here, $Da = k_{eff} L_s / \bar{q}$ where k_{eff} [T^{-1}] is the field-scale effective mass transfer coefficient, L_s is the source zone length in the mean flow direction [L], and $\bar{q}$ [LT^{-1}] is the average Darcy flux for the source zone. The effective mass transfer coefficient (k_{eff}) is related to changes in the DNAPL mass and the average groundwater velocity (Parker and Park, 2004).

The first two models, ESM and PLM, are plotted in Figure 7.2 to provide a general shape of the dissolution curves. The most simplified case is a constant mass discharge as reflected by $\Gamma = \sigma_{\ln\tau} = 0$. This was the model proposed by Sale and McWhorter (2001) but is only representative of a domain with a homogeneous flow field and DNAPL distribution (Rao and Jawitz, 2003). The other extreme case is for very high values of Γ or $\sigma_{\ln\tau}$, in which case the mass discharge declines rapidly but has extreme tailing behavior (Figure 7.2). This might be representative of very heterogeneous sites with long tailing from limited DNAPL pools or release from back-diffusion zones.

7.2.3 Site Mass Balance Considerations

One of the main benefits of considering site conditions in terms of mass flux and mass discharge at control planes is linking this information to a site-wide mass balance. The mass in the source zone is linked to the mass in the plume through the historical source strength function. While measuring mass quantities in the source is quite challenging using direct measurements, using a mass balance approach can lead to useful estimates for the site.

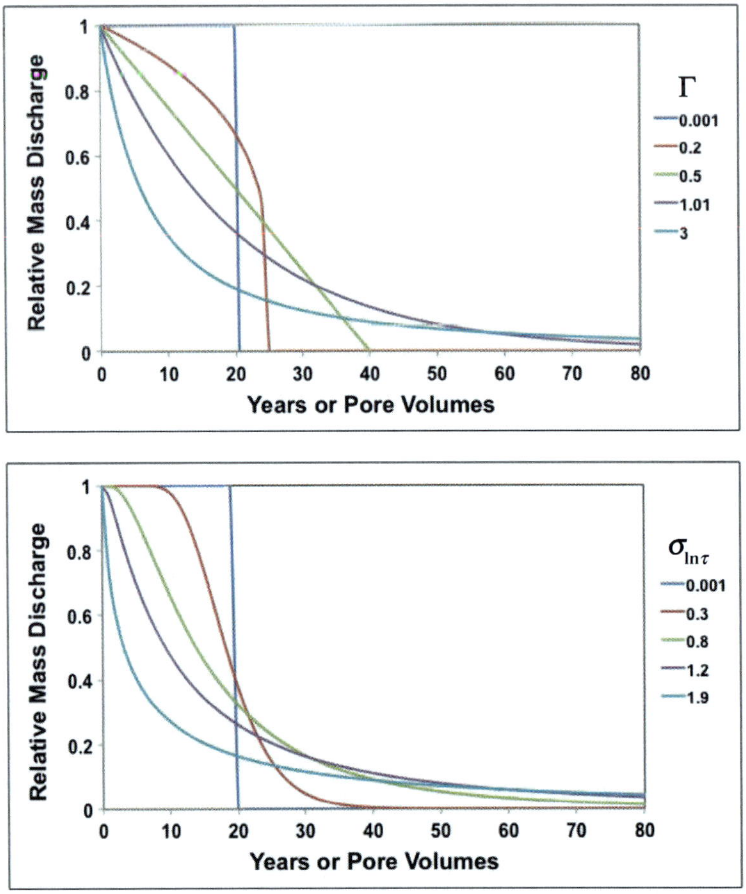

Figure 7.2. Power Law (*upper* graph) and equilibrium streamtube (*lower* graph) models for source zone dissolution.

One approach for source mass estimation, under the assumption of a simple exponential source decay model, makes use of the decay rates (k) estimated from monitoring wells ideally located near the source:

$$M_{t2} = M_{t1}\ \exp(-(t_2 - t_1)k) = \frac{qAC_{t1}}{k}\exp(-(t_2 - t_1)k) \qquad \text{(Eq. 7.9)}$$

where M_{t1} and M_{t2} are the source masses [M] at two different times, and C_{t1} is the concentration [ML^{-3}] in the monitoring well at time t_1 [T]. Using Equation 7.9, the present DNAPL source mass (M_{t2}) can be estimated.

A second method, also assuming exponential source decay, for source mass estimation assumes (1) zero degradation rates in the plume, or alternatively uses the molar sum of all by-products; (2) a measure of the source strength at some time $t2$ ($M_{D,t2}$) and (3) an estimated plume mass $M_{P,t2}$ at $t2$ to calculate the initial source mass and source discharge:

$$M_{P,t2} = \int_0^{t2} M_{D,0}\ \exp\left(\frac{M_{D,0}}{M_0}t\right)dt = M_0\left(1 - \exp\left(\frac{M_{D,0}}{M_0}\right)t_{t2}\right) \qquad \text{(Eq. 7.10)}$$

$$M_{D,t2} = M_{D,0}\ \exp\left(\frac{M_{D,0}}{M_0} t_{t2}\right) \quad \text{(Eq. 7.11)}$$

Given that there are a number of approaches in which mass flux can be approximated or the source functions analyzed using historical data, we define five different methods that can be used to analyze source strength functions at field sites. These methods attempt to utilize commonly recorded historical field data, including spatially distributed temporal concentration histories, well transects and extraction well concentration histories. The methods are briefly discussed in the next two sections.

7.2.4 Mass Discharge and Flux During Plume Evolution

At field sites that have undergone years of investigation, generally there are a number of well locations at varying depths throughout the plume and source zones. To take advantage of these data to characterize source zone behavior, a plume matching technique can be employed by coupling an advection–dispersion model with a source strength function as a concentration boundary condition. Ideally, this method would employ an advanced, calibrated site groundwater flow model.

As solute migrates away from the source zone into the contaminant plume, the contaminant concentration can be affected by a number of processes other than source zone behavior (biodegradation, sorption, diffusion and volatilization). For this reason, wells near the source zone are considered to more directly reflect the source zone behavior than those farther from the source zone. Given this, a maximum spatial distance from the source zone might be used to select wells for estimating a site source strength function. Removing data in this fashion should only be done if the number of measurements can remain relatively high so that model optimization is not hampered by a lack of data.

7.2.5 Estimating Source Strength Functions

The mass discharge from DNAPL sources can vary over several orders of magnitude between different sites (ITRC, 2010). In addition, the mass flux within a given site can vary widely over time and space. For example, a recent study using multilevel samplers at four sites (Guilbeault et al., 2005) found a range of integrated mass load estimates ranging from 41 to 85 g/day for TCE, but the individual mass flux values varied enormously over short distances. The fundamental basis for quantifying different source strength functions is discussed in the sections that follow.

7.2.5.1 Use of Time Series Data

Two approaches can be used to fit an unknown source strength function to concentration time series field data. The first is a global fit to all data trying to optimize the function parameters, such as Γ or $\sigma_{\ln\tau}$. While ideally the entire plume response (all available data) would be fit to a single set of source strength function parameters, in reality a distribution of parameters that best describes the plume response at a field site is probable. This limitation is likely to hold for field sites that are heavily aged or have particularly large contaminant plumes.

An alternative approach to the global optimization is the use of individual wells to fit source strength function parameters. Source strength function parameters can be obtained for individual wells by fitting to the limited data set for each well. While the uncertainty is much higher for each individual well fit, the distribution of parameter values can be used to estimate a range of likely values, with some outliers that are poorly linked to the source behavior and perhaps responding to other processes taking place in the plume. Because optimization at each individual well will be for a limited number of parameters, an exhaustive search can be effective and less computationally intensive for this analysis.

7.2.5.2 Use of Transects

Transect methods allow one to follow mass discharge history either temporally or spatially. In addition, transect methods tend to reduce the effects of noise in concentration data but under ideal conditions require that wells are arranged perpendicular to the flow field down-gradient of the source zone. Mass discharge across transects can be calculated using a number of methods, but in general it can be summarized as

$$M_D = \sum_{i=1}^{n} q_i C_i A_i \qquad \text{(Eq. 7.12)}$$

Transect mass fluxes can be measured directly using passive flux meters (Annable et al., 2005; Basu et al., 2009), multilevel samplers, stochastic methods (Schwede and Cirpka, 2009), and combinations of these techniques in addition to others. To obtain mass discharge, methods are required to interpolate flux data over the transect. A point concentration with a representative area, A_i, coupled with groundwater flux, q_i, is used as reflected in Equation 7.12. Kübert and Finkel (2006) argue that there are merits to other ways to calculate mass discharge via multipoint samplers. A more rigorous stochastic analysis of the transects used may be necessary to improve accuracy and quantify uncertainty (Klammler et al., 2012).

Temporal Transects

Because well transects can be used to couple flux-averaged concentration with groundwater flow to produce mass flux and integrated mass discharge values, observing the change in these flux values in time can produce a meaningful mass discharge time series that can be directly fit to source strength functions. In the application of this approach, it is assumed that the transect used to quantify mass discharge is close enough to the source zone to reduce error introduced by other processes occurring in the field (change in flow field direction).

Spatial Transects

Similar to transect data collected in time, transect data in space involves having multiple well transects spaced a significant distance apart down the plume centerline. This allows for translation of concentrations in time without the need of significantly long time series data. This translation from space to time is done via travel time using assumed retardation and groundwater velocity between transects. Careful consideration must be made to quantify any losses, such as complete degradation of the product, changes in hydro-stratigraphy between transects, and remediation.

7.2.5.3 Use of Extraction Wells

The final method discussed here is the evaluation of source strength through changes in extraction well data, namely, changes in concentrations in time in extracted contaminated groundwater. This method allows prediction of removal efficiencies and analysis of source behavior changes over time during remedial operations. While fitting a source strength function to the cumulative mass removal of these data is possible, the source strength functions that describe this pumping may not accurately reflect the source strength function for natural groundwater gradients.

For sites with concentration time series data from extraction wells, that data can be fit to assumed source strength function characteristics. As an example, the simplest model to consider is the exponential decay model. In this model the mass flux is proportional to the remaining mass:

$$\frac{M_A}{M_{A,0}} = \frac{J}{J_0} = \frac{Cq}{J_0} \quad \text{(Eq. 7.13)}$$

Here M_A is the mass of contaminant per unit area perpendicular to the groundwater Darcy flux q, J is the mass flux, C is contaminant concentration, and the 0 subscripts denote initial values. Thus writing an equation for mass loss

$$\frac{dM_A}{dt} = -Cq \quad \text{(Eq. 7.14)}$$

The solution for the exponential decay model can be written as:

$$C = C_0 \exp\left(-\frac{J_0}{M_{A,0}}t\right) \quad \text{(Eq. 7.15)}$$

This function can be easily fit to a concentration time series data to determine the unknown values C_0 and $(J_0/M_{A,0})$.

The more general power source function model can also be applied to $C(t)$ data. The solution used to fit to $C(t)$ data employs the following relationship between source zone mass and mass flux:

$$\frac{J}{J_0} = \left(\frac{M_A}{M_{A,0}}\right)^{\Gamma} \quad \text{(Eq. 7.16)}$$

This can be solved, yielding:

$$C = C_0\left[1 + \frac{J_0(\Gamma - 1)}{M_{A,0}}\right]^{\frac{\Gamma}{1-\Gamma}} \quad \text{(Eq. 7.17)}$$

This function with three fitting parameters C_0, Γ, and $J_0/M_{A,0}$ can also be fit to the field data sets.

7.2.6 Remedial Considerations

The underlying goal of source zone or DNAPL treatment is to reduce the risks posed by the contamination to humans and the environment. Aqueous contaminant concentrations can be viewed as surrogate measures of risk and are often the primary metric for regulatory decisions. Under this traditional framework, the goal for NAPL source treatment technologies is therefore to achieve acceptable aqueous concentrations such as drinking water standards by either removing or destroying NAPL mass in the source zone. Unfortunately, current NAPL remedial technologies are unable to completely eliminate NAPL from source areas at all sites, and partial mass removal from NAPL source zones is unlikely to be sufficient to meet drinking water standards at all locations within a site. Other potential benefits may, however, be achieved as a result of source treatment, and one proposed benefit is a reduction in contaminant mass discharge (M/T), from the NAPL source area (Rao et al., 2002; Kavanaugh and Rao, 2003; Stroo et al., 2003; ITRC, 2003; NRC, 2004). The magnitude of the benefit realized needs to be quantified by measuring mass flux and discharge before and after remedial efforts are implemented.

Flux measurement should be conducted using the same protocol before and after remedial treatment, and differences in hydrologic conditions should be considered when interpreting the data. Also critical is the need to provide adequate time between the remedial activity and the post-remedial flux measurement to ensure that conditions are representative of natural groundwater flow and contaminant dissolution from the original source area. Too often there is a rush to measure the effect of the remedial action, and the measurements are only indicative of modifications still significantly influenced by remedial actions such as elevated temperatures or residual flushing agents. A general rule may be to require at least two pore volumes of natural gradient water flow to pass through the source area extending to the monitoring location prior to post-remedial monitoring of mass flux. For many sites this could mean 2–5 years following remediation. Often site managers may find this length of time unacceptable. Partly for this reason very few studies have been completed to date with high-quality measurements of pre- and post-remediation flux measurements (Brooks et al., 2008).

7.3 METHODS FOR MASS DISCHARGE AND MASS FLUX MEASUREMENT

The methods currently available for field measurement of mass flux and discharge are reviewed in the following sections. The methods for quantifying both local contaminant mass flux, $J(x,y,z)$, and integrated mass discharge (M_D) can be categorized by spatial and temporal scales of measurement (Figure 7.3). Point-scale techniques include multilevel samplers and passive flux meters (Hatfield et al., 2004) within screened wells. Both methods collect data to determine the local mass flux [mass per unit area per time] that can be spatially integrated to determine mass discharge [mass per time]. Alternatively integral methods, such as integral pump tests (Bockelmann et al., 2001; Goltz et al., 2008), collect data averaged over fully screened wells and large volumes of water pumped from aquifers. This information, while lacking spatial resolution, has advantages in that local plume cores are incorporated.

Local measures of mass flux, $J(x,y,z)$, can be calculated using the local flux average concentration, $C(x,y,z)$; the local hydraulic conductivity, $K(x,y,z)$; and the local head gradient, dh/dx, applied in the maximum flow direction (implied horizontal flow). All of these measures can involve significant effort to obtain at the local scale. Thus, average values are often applied

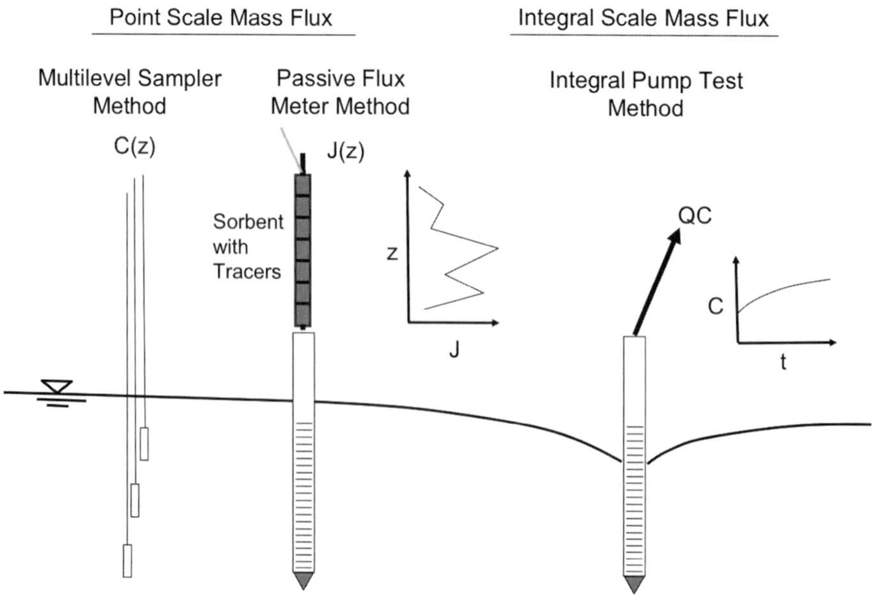

Figure 7.3. Point and integral mass flux methods in well screens typically 1.5–9 m (5–30 ft) in length and well spacing 3–15 m (10–50 ft).

for *dh*/*dx* and *K*. The flux-averaged concentration distribution is obtained using multilevel samplers (Guilbeault et al., 2005). In this approach, spatial variability is determined for local mass flux that is simply a function of the concentration distribution. Any correlation between the Darcy flux and concentration is not considered.

7.3.1 Transect Method

The most traditional method for calculating contaminant discharge from a NAPL source area or at some location within a plume uses the product of the Darcy flux, cross-sectional area within a transect of wells and contaminant concentration in the well (the Transect Method as described in API, 2003). The Darcy flux can be calculated as the product of independent measures of the hydraulic gradient and hydraulic conductivity. To calculate the contaminant mass flux, groundwater samples are collected from the fully screened wells forming the transect. The contaminant flux is calculated as the product of the Darcy flux for each well and the flux-averaged concentration collected from each well. In this method, simplification of the flux is acknowledged because average values are used for the hydraulic conductivity, hydraulic gradient, and flux-averaged concentration. Each of these values could have variability in space, and local mass flux could be significantly different from the average values. It is assumed that the integral values used are appropriate for calculating average fluxes and integrated fluxes over the well transect. This assumption has not been verified for field sites and forms a need for future research.

The transect method also can apply to a line of multilevel samplers downgradient of the source area or across a plume at a location of interest. In this case the calculation is made for each sampling location to determine the local mass flux using the local contaminant concentration. The local Darcy flux can only be calculated if the local hydraulic conductivity and hydraulic gradient are measured. The hydraulic conductivity may be available from locally conducted slug tests or borehole flow meter analysis in adjacent screened wells. Typically this information is not available, and the site average hydraulic conductivity is used. The local hydraulic gradient, while possible to measure, is rarely determined, and average values from monitoring wells generally are used instead.

The primary advantage of transect methods is that well-established methods are used to collect the data. Further, these methods often can incorporate infrastructure already present at the site. These methods also generate some spatial information on mass flux distributions, and finally, transect sampling generally produces minimal wastes requiring disposal or handling.

The primary disadvantages of transect methods include the necessity to acquire independent measurements of the Darcy flux and the potential errors associated with sampling a relatively small volume of the groundwater. The independent measures of flux typically include measurements for hydraulic gradient, which for low-gradient systems can introduce errors (Devlin and McElwee, 2007). The methods interrogate a small volume of the aquifer and are instantaneous measures. Thus the data must be spatially integrated to obtain contaminant mass loads. Minimizing the errors involved can require large numbers of samples. For example, Li et al. (2007) evaluated the measurement densities needed to have a reasonable certainty and concluded that sampling approximately 1–4% of the domain would be required. Errors associated with the transect method are difficult to determine. However, a study by Kübert and Finkel (2006) indicated that errors were typically in the range of 20–50%. This assessment was based on a modeling approach where the plume covers a relatively large portion of the control plane. A recent study concluded that the error involved in field sampling using multilevel transects was significantly greater than the error in an integrated pump test at the same site and that a relatively large number of samples would be needed to reduce the uncertainty to a similar level (Béland-Pelletier et al., 2011).

7.3.2 Passive Flux Meters

An alternate approach to measuring groundwater and contaminant fluxes involves deployment of a permeable, sorbent pack (passive flux meter [PFM]) in the wells located along a transect and screened across the vertical extent of the contaminant plume (Hatfield et al., 2002, 2004; Annable et al., 2005). The PFM sorbent material is selected to capture the dissolved contaminants in the groundwater as it flows through the flux meter during the designated period of exposure under natural gradient groundwater flow conditions. Thus, flux measurements using this approach are referred to as passive in contrast to methods that require pumping water.

The PFMs placed in wells are pre-saturated with resident tracers that are desorbed and depleted as groundwater flows through the well. Groundwater fluxes are calculated from the depletion of tracer mass, while the mass of contaminants captured on the sorbent is used to estimate contaminant fluxes. By analyzing the sorbent for the residual tracers and the captured contaminants in vertical segments of the PFM, the depth distribution of groundwater and contaminant fluxes along the well screen interval can be determined. Hatfield et al. (2004) presented the theoretical development for PFM, and Annable et al. (2005) reported on field-scale validation efforts, while Basu et al. (2006) presented a field-scale PFM application for site characterization (Figure 7.4).

The use of PFMs can be optimized by selecting tracers with appropriate retardation on the sorbent used and by installing the PFMs for a time period such that tracer loss and contaminant mass accumulation are quantifiable. These design parameters are based on existing estimates of the groundwater flux and contaminant concentrations at the site. More advanced interpretation of PFM data is provided by Klammler et al. (2007a,b).

The primary advantage of the PFM approach is that spatial information on cumulative water and contaminant mass flux are obtained. Methods that exist to estimate uncertainty indicate that the PFM approach produces low errors in integrated mass load (Kübert and Finkel, 2006). The PFM method can be used to generate local estimates of horizontal aquifer conductivity given measured hydraulic gradients. The method produces minimal solid waste volumes. The passive nature allows application in remote settings, and the method does integrate information over time.

The primary disadvantage is similar to that of the multilevel sampling, i.e., the PFM interrogates small volumes of the aquifer, so the errors involved may be large. Also, the data must be spatially integrated to obtain contaminant mass discharge and total water discharge.

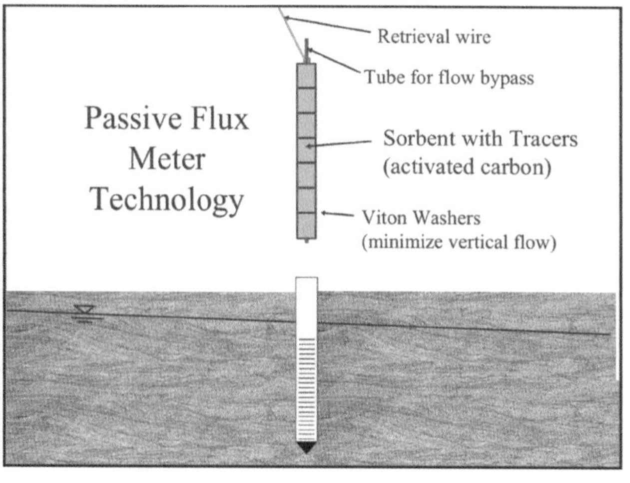

Figure 7.4. Installation of the passive flux meter in a screened well.

Tracers, typically alcohols, are required to obtain Darcy flux measurements, and these tracers must be approved for use in the subsurface. Finally, PFMs are not yet demonstrated for use in fractured media environments, although modifications of the PFM have been developed and are currently being tested in fractured bedrock.

To date the method has been deployed at more than 40 sites, primarily government facilities. The largest deployment to date was at Fort Lewis, Washington, USA, where a PFM network totaling 700 linear feet was deployed in three 10-well transects. The deepest application to date has been 67 m (220 feet [ft]) at a landfill near Perth, Australia.

7.3.3 Integral Pump Tests

The IPT technique for measuring mass flux or discharge is based primarily on measuring the contaminant concentration time series in the effluent of multiple pumping wells aligned perpendicular to the prevailing direction of groundwater flow. The use of concentration time series information to estimate contaminant flux or mass discharge was initially described by Teutsch et al. (2000) and Ptak et al. (2000), and applications of the technique were described by Bockelmann et al. (2001, 2003). The measured concentration of contaminants is used with independent estimates of the natural gradient Darcy flux. In this respect the method is similar to the transect method with the difference being the significant volume of water that is interrogated during the test. This method provides an integral measure of the contaminant concentration across the well transect and avoids the concern related to missing local plume cores within the transect.

The primary advantage of the IPT is the integral nature of the flux measurements. The method generates contaminant mass flow estimates and does supply some limited information on the spatial distribution of mass flux across the well transect. The method can be applied to deep aquifers and have no limitations based on well design.

The primary disadvantage of the IPT is the volume of waste that may be generated. These secondary wastes can be significant and potentially costly depending on site infrastructure. In an effort to overcome this challenge, methods have been developed and tested in which tandem circulation wells are used without groundwater extraction (Goltz et al., 2008, 2009). Integral pump-based methods can require a lengthy deployment time depending on aquifer characteristics. Finally, the method requires independent measures of the Darcy flux, although these measurements can be incorporated into the pump test (Brooks et al., 2008).

7.4 FORT LEWIS DNAPL CASE STUDY FOR MASS FLUX EVALUATION

To demonstrate the utility of flux assessments for improved site management, data collected at Fort Lewis Army Base (now Joint Base Lewis-McChord) Washington, USA is considered. Fort Lewis offers a unique opportunity because it is a large DNAPL contaminated site with extensive characterization data and a major source zone remedial effort that included collection of flux data as a component of the remedial performance assessment. These data will be used to look beyond remedial performance assessment to consider how the data would benefit remedial design and overall site management.

The Fort Lewis TCE contaminant plume, which is approximately 3 kilometers (1.4 miles) long, is the result of a large DNAPL source area at the East Gate Disposal Yard (EGDY) (Figure 7.5). Extensive source zone characterization was undertaken resulting in the delineation of three primary DNAPL contaminated regions identified as NAPL Areas 1, 2, and 3 (Figure 7.6). These areas were evaluated for a variety of remedial technologies, and resistive heating was selected as the remedial option. The footprint for each of the three areas was

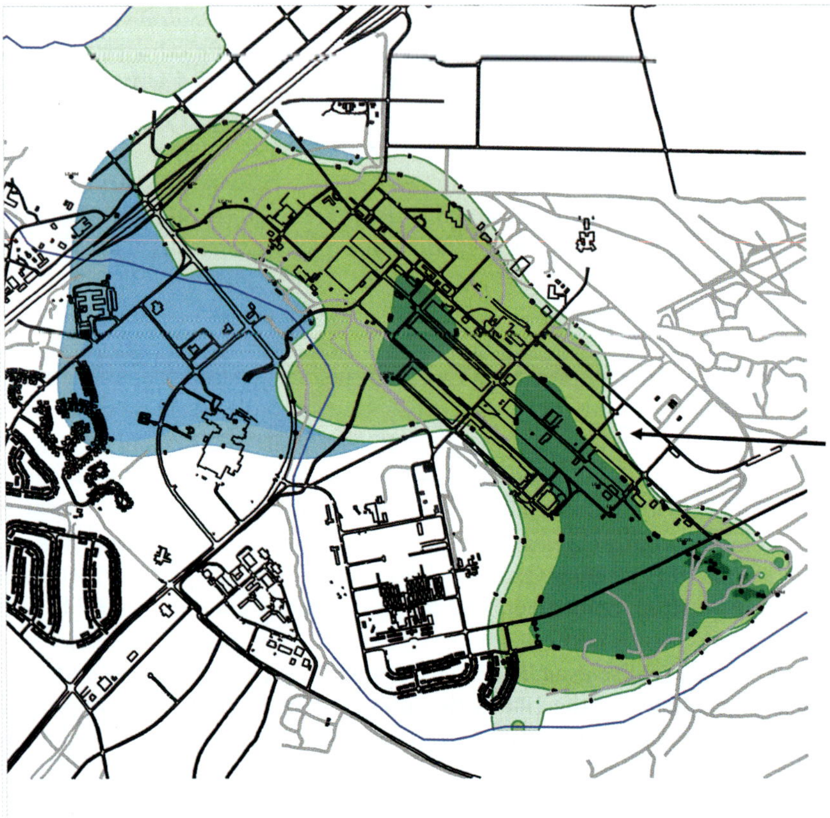

Figure 7.5. TCE contaminant plume at Fort Lewis c. 2003.

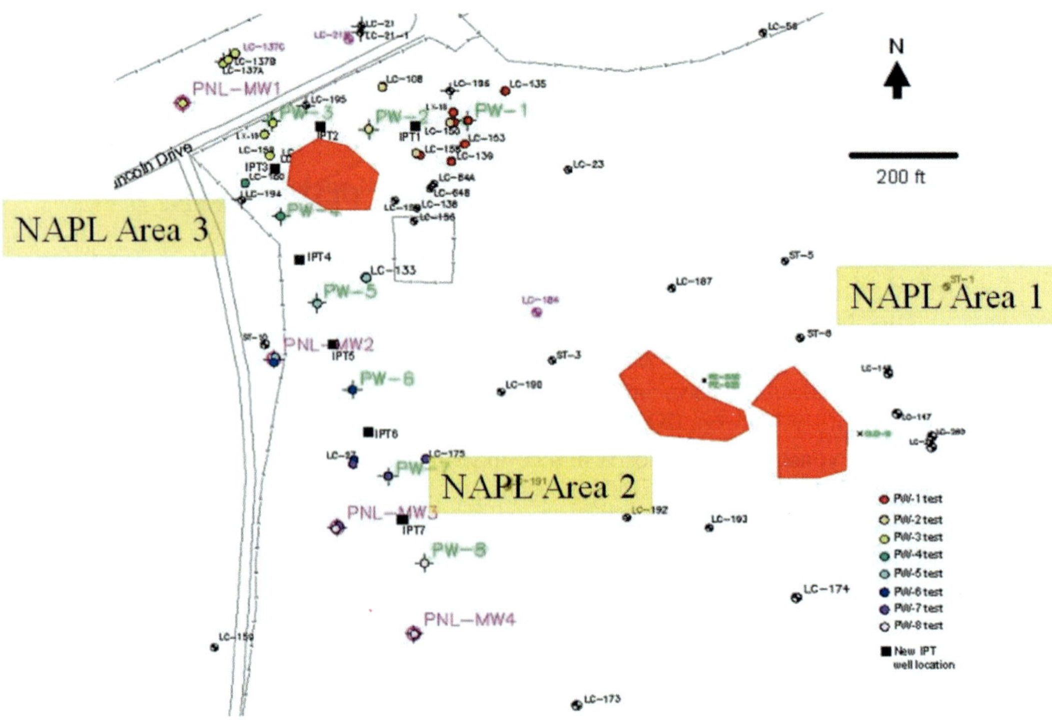

Figure 7.6. Fort Lewis NAPL treatment areas.

based on site characterization efforts including cores, groundwater samples, historical site activities and drum removal. These treatment areas were sequentially addressed with electrical resistance heating over a period of approximately 3 years (2005–2007). Details on the remedial performance of the resistive heating of NAPL Area 1 have been reported by Brooks et al. (2008). NAPL Area 1 was treated first and was the most upgradient source and thus was ideal for remedial evaluation. The focus of remedial assessment using flux data was on NAPL Areas 1 and 3. NAPL Area 2 was not evaluated by pre- and post-flux measurements; however, information from alternative measures can be used to provide mass balance evaluation. Thus, all three source areas are discussed here for both remedial design considerations and source mass balance assessments.

The EGDY source areas 1 and 3 were evaluated by constructing transects of wells screened across the contaminated surficial aquifer to quantify mass flux distributions and integral mass discharge. For NAPL Area 1 a single transect of ten wells was located immediately downgradient of the treatment zone (Figure 7.6). For NAPL Area 3, since mass was entering the treatment zone from smaller undefined upgradient sources, a transect of eight wells was constructed immediately upgradient of the treatment area. On the downgradient side of NAPL Area 3 the saturated thickness of the surficial aquifer was approximately 15 m (50 ft), and therefore two transects of ten wells each were constructed, forming a shallow and a deep downgradient transect. These installations comprise the efforts toward collecting local source flux measurements using PFMs.

The mass discharge leaving the entire EGDY source area was also of interest to assess the input to the entire plume system. To evaluate this aspect of the site, a site-wide IPT was conducted along the downgradient region of the EGDY source area. This assessment comprised 15 wells fully screened over the surficial aquifer. This measurement was conducted shortly after conductive heating of NAPL Area 2. Thus this measurement represents an intermediate mass discharge value, having been impacted by remedial efforts in Areas 1 and 2. These remediation impacts will be considered in subsequent analyses.

The mass flux assessments at the three well transects installed near the treatment areas were conducted using PFMs followed by IPT. The initial samples from the IPT were assumed to represent values obtained from low-flow sampling of the fully screen wells and thus were used in calculations of mass discharge based on the traditional transect method. Thus for each flux measurement event there are three approaches for quantifying mass discharge. Each approach reports values for individual wells located in the transect, and these are summed to quantify the transect mass discharge.

7.4.1 Remedial Design Considerations

While the design of the three resistive heating systems was not based on the flux data collected as part of the performance assessment, the data were evaluated for potential design value. The two areas in which local flux measurements were made are NAPL Areas 1 and 3. The transects of wells were evaluated using PFMs segmented in vertical intervals of approximately 30 centimeters (cm) or 12 inches (in). Thus for the wells forming the downgradient transect at NAPL Area 1, each well had a vertical flux profile consisting of about 25 samples. These data along the entire transect of ten wells were used to produce a flux distribution plot using Surfer, a contouring software package (Figure 7.7).

The flux distribution highlights the variable nature of the TCE mass flux emanating from the DNAPL source area. This variability is likely the result of individual pools of DNAPL located upgradient of the flux control plane. The flux distribution demonstrated that the majority of the mass flux is localized within the control plane. This observation was made by

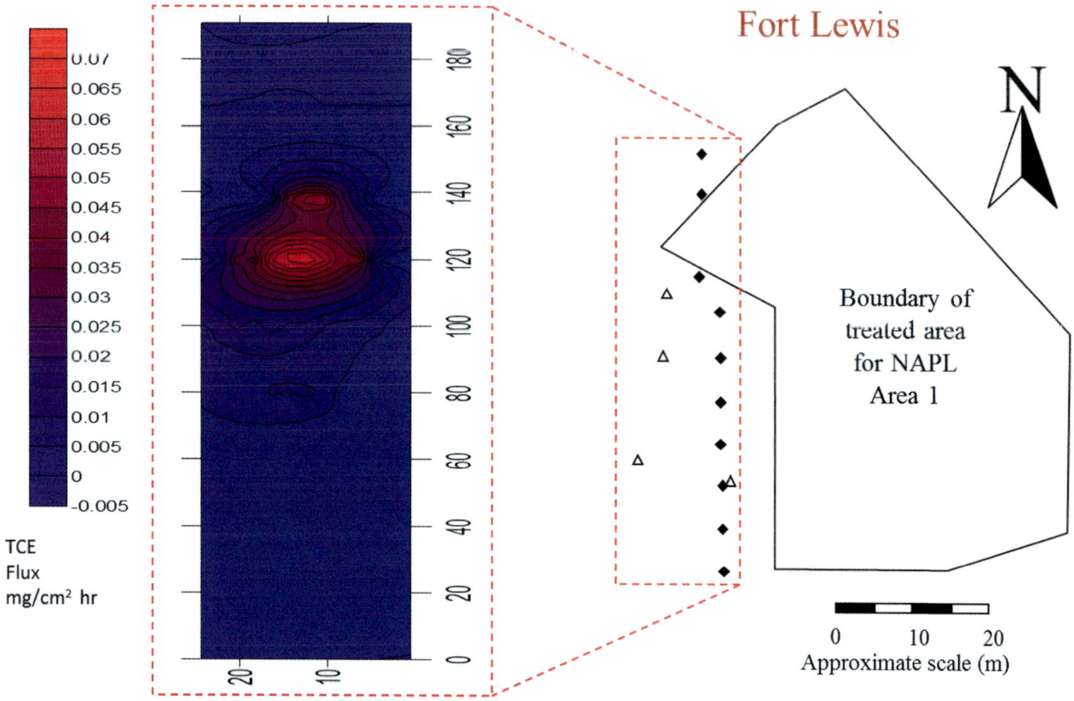

Figure 7.7. Fort Lewis NAPL treatment Area 1.

others investigating DNAPL sites (Guilbeault et al., 2005). In the data observed for NAPL Area 1, quantitatively about 90% of the mass flux is contained with about 33% of the control plane cross section. This mass originates from DNAPL sources located upgradient of the flux observed at the control plane. This information is useful for reducing the size of the remedial treatment zone by focusing on these upgradient regions.

To consider focusing the treatment design based on control plane flux data, the predominant groundwater flow direction must be employed. The control plane flux data set can be considered an instantaneous measurement that is the result of groundwater flow imposed on the DNAPL source area for the travel time required between source and control plane. For NAPL Area 1 this travel time is approximately the travel time through the treatment area given that groundwater sampling has established that contaminant concentrations are very low upgradient of the conductive heating treatment area. The travel distance is approximately 50 m or roughly 160 ft (Figure 7.7). Given an average Darcy velocity based on the PFMs of 0.25 m/day (0.82 ft/day) and assuming a porosity of 0.33 the travel time is approximately 70 days. Thus the predominant gradient over the previous 70 days should be considered to locate upgradient source areas. Recognizing that gradient shifts in direction and magnitude are evident at this and many other sites, the mean gradient magnitude-weighted direction should be calculated. For many sites the gradient direction and magnitude shift between wet and dry seasons. This information should be considered when selecting appropriate dates to perform instantaneous flux measurements at the site for purposes of design considerations.

Flux distributions were also conducted at NAPL Area 3 in both upgradient and downgradient well transects. This data was also contoured providing TCE and *cis*-dichloroethene (*cis*-DCE) flux distributions (Figure 7.8). The upgradient flux distribution was less localized than that observed at NAPL Area 1. This difference was likely due to the much longer travel distance between the source of the mass flux and the control plane. The relatively long time and

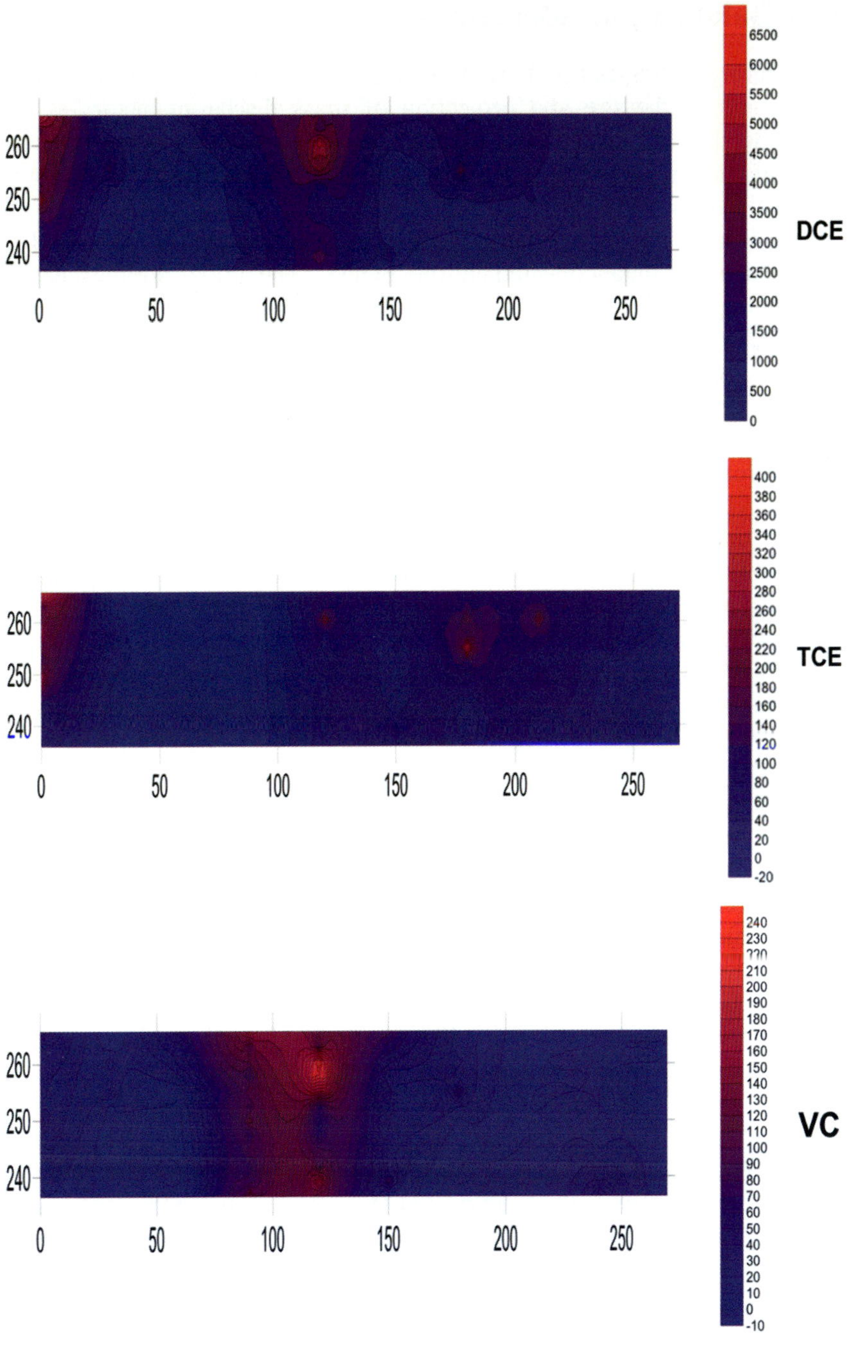

Figure 7.8. Shallow downgradient mass flux distribution NAPL Area 3. Flux values are in units of mg/m^2 hr.

travel distance allowed greater mixing and dispersion to occur. This result also demonstrates that for design purposes the control plane should be as close to the source area as possible.

The downgradient control plane at NAPL Area 3 was about 15 m (50 ft) in saturated thickness and was much thicker than the control plane at NAPL Area 1, because the till unit at the base of the surficial aquifer declines rapidly in the treatment zone. The flux distribution along this control plane is again more localized providing evidence of DNAPL pools in NAPL Area 3.

7.4.2 Mass Balance Considerations

When considering strategies for short- and long-term site management some knowledge of distribution of mass at the site and projection of mass discharge history is important to consider. The collection of mass discharge measurements from the source area to the plume allows for estimates of projections of mass discharge over the life of the NAPL source area. To support these projections some knowledge of current or past mass present in the source area and/or in the plume is desirable. Estimates of mass in the source area are inherently difficult as was demonstrated using data collected at the EGDY. The initial source characterizations conducted resulted in estimates of mass presented in each of the treatment areas. The initial mass estimates were 11,800, 13,400, and 43,100 kg (25,960, 29,480, and 94,820 pounds [lb]) TCE for the three areas, respectively. The resistive heating events removed 2,580, 1,090, and 840 kg (5,676, 2,398, and 1,848 lb), respectively (Table 7.1).

From this analysis it was concluded that the initial mass estimates were grossly high for all three treatment areas. This observation was based on the fact that significantly diminishing returns were observed for the mass recovery, suggesting that high fractions of the original mass were removed. Of much more value are the mass removals during resistive heating as they are considered to be much more reliable. From this data one can conclude that NAPL Area 1 was the most significant in terms of mass present and NAPL Areas 2 and 3 were lower.

The mass discharges measured at NAPL Areas 1 and 3 were 6,700/7,300 millimoles per day (mmol/day) (IPT/PFM) and 2,300/8,800 mmol/day (Table 7.2). This information can be used to estimate mass discharge histories for both treatment areas. In order to perform the calculations, some assumptions are required. A model must be assumed for the mass discharge–mass remaining relationship (Falta et al., 2005a; Jawitz et al., 2005). The simplest model to assume is an exponential decay model. Using pre- and post-mass discharge measurements, the mass at the beginning of the treatment was 2,600 kg (5,720 lb) and the mass after treatment was 19 kg (42 lb). Thus a mass removal of 99.3% is estimated. In addition, if a date is assumed for the initial spill of DNAPL (for this analysis 1950 was assumed), an initial mass in NAPL Area 1 of 5,200 kg (11,400 lb) is estimated. Applying the same analysis to NAPL Area 3 provides an initial mass estimate of 2,800 kg (6,200 lb). While NAPL Area 2 did not have local pre- and post-treatment mass flux measurements, based on the mass removed, we can estimate an additional source area mass of 3,100 kg (6,800 lb). Using all of the estimates, the total initial TCE mass estimate for the EGDY was 10,000 kg (22,000 lb). This estimate can be compared to mass estimates for the TCE currently present in the dissolved plume resulting from the EGDY of approximately 5,000 kg (11,000 lb).

The site-wide IPT conducted at EGDY also provides insight into the site-wide mass balance (Table 7.3). While the measurement was impacted by treatment of NAPL Areas 1 and 2, since it was conducted 6 months after heating was terminated in Area 1, it still provided valuable data for comparison. In the northern and central sections the results can be compared with the mass

Table 7.1. Summary of Estimated NAPL Mass Removed *Ex Situ* by Constituent and Area

Treatment Zone	TCE (kg)	*cis*-DCE (kg)	TPH (kg)
Area 1	2,580	410	40,170
Area 2	1,090	250	11,340
Area 3	840	280	530

Notes: Numbers provided in table are rounded. NA2 configuration changed; all estimates based on revised configuration. Total mass removal estimates based on stoichiometric means. Thermal treatment objective to maximize chlorinated volatile organic compound (TCE, *cis*-DCE) removal, not total petroleum hydrocarbon (TPH) removal.

Table 7.2. Mass Flux Estimates for Different Techniques

Test	Mass Flow (mmol/day)	
	Integrated Pump Test	Passive Flux Meter
NA1 Pretreatment	6,700	7,300
NA1 Post-treatment	17	85
NA3 Upgradient pretreatment	708	1,280
NA3 Downgradient deep pretreatment	644	3,500
NA3 Downgradient shallow pretreatment (first test)	1,704	5,300
NA3 Downgradient shallow pretreatment (second test)	1,371	10,800
EGDY (North end only – IPT 2 and 3)	3,475	NA

Table 7.3. Mass Flow and Mass Flux Values for Each Well Section Southern Test Interval (IPT1)

Well	Percentage of Control Plane Cross-Sectional Area	Mass Flow for the Well Section (mmol/day)	Mass Flux for the Well Section (μmol/day m^2)
Southern Test Interval (IPT1)			
FX5	25.21	29.4	42.2
PW6	16.81	2.9	6.3
FX6	16.81	7.8	16.9
PW7	15.13	4.0	9.6
FX7	13.45	4.8	13.0
PW8	12.61	10.4	29.9
Middle Test Interval (IPT2)			
PW3	14.33	463	1,157.9
FX3	14.33	2,160	5,403.5
PW4	14.33	231	577.1
FX4	16.00	138	310.4
PW5	16.00	159	356.0
FX5	25.00	78	112.2
Northern Test Interval (IPT3)			
PW1	19.77	34.1	105.1
FX1	18.64	8.9	28.8
PW2	15.82	32.1	123.5
FX2	21.47	121	341.9
PW3	24.29	52.1	130.5

Table 7.4. Mass Flow Across the Control Plane for Each Test Interval and Total

Test Interval	Mass Flow			
	mmol/day	mol/yr	kg-TCE[1]/day	kg-TCE[1]/yr
Southern (IPT1)	59.4	21.7	0.008	2.85
Middle (IPT2)	3,227	1,178	0.42	155
Northern (IPT3)	248	90.5	0.033	11.9
TOTAL	3,535	1,290	0.46	170

[1]Conversion from moles to mass of TCE assumes only TCE present and does not consider that the mass flow comprised TCE, DCE isomers, and VC. Thus, the mass flow values in these columns are approximate numbers for generalized purposes only in terms of assessing the amount of originally disposed TCE being carried downgradient from the source area

discharge from NAPL Area 3. The site-wide based IPT mass discharge of 3,450 mmol/day compares favorably with the IPT measurement of 3,200 mmol/day, while the PFM was higher at 8,800 mmol/day (Table 7.4).

In order to estimate the site-wide mass discharge prior to treatment the mass discharge from NAPL Area 1 must be summed with an unknown contribution from NAPL Area 2. The complication arises because NAPL Area 2 is downgradient of NAPL Area 1 and thus mass discharge from this zone may be reduced due to the presence of TCE in solution from NAPL Area 1. Total estimates of mass discharge from EGDY can be conducted under two extreme cases, NAPL Area 2 contributes no additional mass, and NAPL Area 2 contributes a scaled mass input based on results from other zones. This analysis resulted in an estimated mass discharge of 11,000–14,000 mmol/day.

7.5 BORDEN COAL TAR EMPLACEMENT EXPERIMENT CASE STUDY FOR MASS FLUX EVALUATION

Most work conducted to date on mass flux evaluation has been directed toward chlorinated solvent DNAPL sites. However, the basic concepts are equally applicable to manufactured gas plant (MGP) DNAPL contaminated sites as well as more generally to any contaminated site that has a source and a plume. Decisions need to be made about how best to manage the site with limited resources. Even in the context of monitored natural attenuation of polycyclic aromatic hydrocarbons (PAHs), concepts associated with a flux-based approach are important because of the relative balance between the source strength and the rate of degradation in the plume, which determines whether a plume at a given site is advancing, is receding, or has reached steady state conditions.

Important differences exist between a chlorinated solvent DNAPL site and a coal tar DNAPL site, in particular, the behavior of the source function. As noted earlier, the source strength function has been investigated by a number of researchers in recent years, and one description that has been used is the PLM (Equation 7.6) Power Law model (Rao et al., 2002; Parker and Park, 2004; Zhu and Sykes, 2004; Falta et al., 2005a, b; Falta, 2008). The PLM is a convenient screening-level mathematical description of source function behavior (Figure 7.2). The exponent, Γ, is inherently a function of the DNAPL spatial distribution, hydraulic conductivity spatial distribution, and correlation between the two. In general, however, the PLM does not fully account for the dissolution behavior associated with a multicomponent NAPL. The Borden coal tar emplacement experiment provides an illustration of this point.

King and Barker (1999) and King et al. (1999) described results from a field experiment in which 74 kg of coal tar creosote was emplaced in a shallow aquifer at the Canadian Forces Base in Borden, Canada, in August 1991. The resulting source and plume have been monitored since the emplacement, and more recent results from this long-term study were described by Fraser et al. (2008) and Thomson et al. (2008). The emplacement method consisted of mixing creosote above grade with ~5,800 kg of coarse sand, resulting in a DNAPL bulk saturation of ~7%. The mixed material was then placed below the water table in two ~2 m × 2 m × 0.5 m (6 ft × 6 ft × 1.5 ft) blocks approximately 1 m (3 ft) apart (King and Barker, 1999), resulting in a total cross-sectional flow area of ~8 m^2 (ca. 75 ft^2), with a source thickness relative to the flow direction of 0.5 m (1.6 ft).

Given the emplacement method, a reasonable assumption is that the DNAPL is distributed uniformly within the coarse sand. Theoretically, a uniformly distributed single component DNAPL within a matrix of homogeneous hydraulic conductivity would produce a source function in which the vast majority of the DNAPL would have to be depleted before reductions in source mass discharge occurred. This corresponds to a very low PLM exponent (see Figure 7.2). While the overall dissolution behavior of a multicomponent DNAPL as a whole may be similar to the single component DNAPL, the dissolution behavior of individual components in a multi-component DNAPL is very different.

One relationship used to describe dissolution from a multi-component DNAPL is Raoult's Law (Banerjee, 1984), which states that dissolution of the *i*th species is a function of the aqueous solubility and DNAPL mole fraction:

$$C_i = \chi_i \gamma_i S_i \quad \text{(Eq. 7.18)}$$

where C_i is the aqueous concentration [ML^{-3}] of the *i*th solute resulting from equilibrium partitioning from the DNAPL; χ_i and γ_i are the mole fraction [-] and the activity coefficient [-] of the *i*th solute in the organic phase, respectively; and S_i is the pure-phase aqueous solubility limit [ML^{-3}] of the *i*th solute. Often γ_i is assumed to be 1, which is the value in an ideal organic fluid. Applications of Raoult's Law to predict solute partitioning between coal tar and water have been described by Lee et al. (1992), King and Barker (1999), D'Affonseca et al. (2008) and Fraser et al. (2008).

Reasonable agreements between measurements and estimates based on Raoult's Law have been reported, with the difference between the two ranging from factors of 2 to 4 (Lee et al., 1992; King and Barker, 1999). Others however have noted order-of-magnitude differences (Lane and Loehr, 1992) or have noted kinetic limitations in mass transfer between water and coal tar and thus questioned the appropriateness of an equilibrium model such as Raoult's Law (Mahjoub et al., 2000). Alternatively, polyparameter linear free energy relationships, which predict partitioning based on a more mechanistic description of the intermolecular interactions, also have been used to predict solute partitioning between water and coal tars (Endo and Schmidt, 2006).

Application of Raoult's Law to the Borden emplacement study is shown in Figure 7.9, where the source strength function based on data provided in King and Barker (1999) is shown in Figure 7.9a and the source function based on the same data is shown in Figure 7.9b. For the more soluble compounds, the source function can be approximated by using a PLM with an exponent of 1 (Henderson et al., 2009). However, for the less soluble compounds, the early time data deviates from PLM behavior, resulting from the fact that the mole fraction of the less soluble compounds is actually increasing with time as the more soluble compounds leave the DNAPL. After some duration, however, any *i*th species that was initially a less soluble compound becomes the most soluble compound, and its dissolution behavior at that point appears to follow a PLM with an exponent of unity (Figure 7.9b).

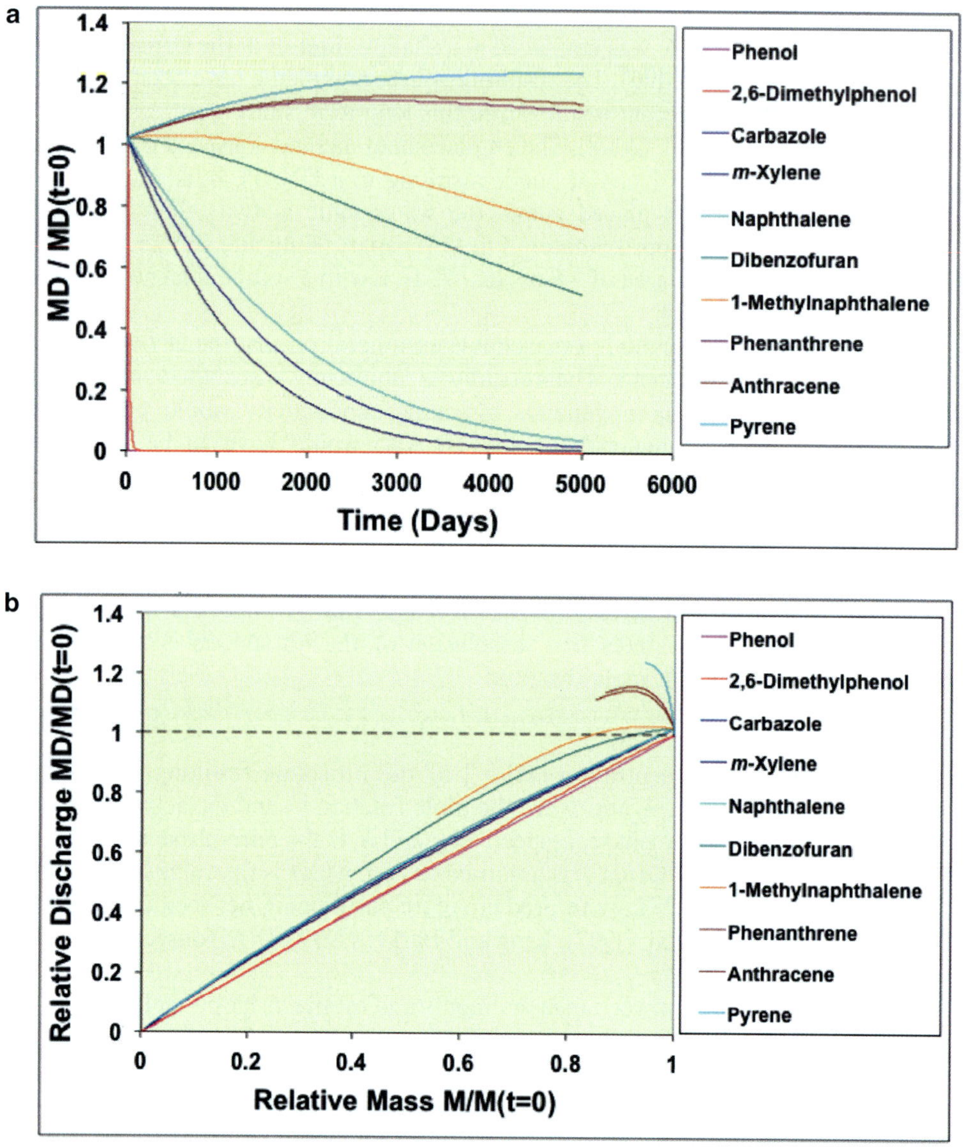

Figure 7.9. (a) Source strength function (relative mass discharge versus time) for select PAH compounds based on Raoult's Law and data from the Borden coal tar emplacement experiment. (b) Source function (relative mass discharge versus relative mass) for the same select PAH compounds based on Raoult's Law and data from the Borden coal tar emplacement experiment. Data reported by King and Barker (1999).

7.5.1 Source Strength Predictions

Shown in Figure 7.10 is mass discharge (M_D) as a function of time for ten compounds as reported by Fraser et al. (2008); six measurements were completed between 278 and 514 days after emplacement. This data provides a useful opportunity to test how well simple models of source strength functions predict future behavior using intermediate-scale pilot tests. It was assumed that the first three measurements collected between 278 and 1,357 days after

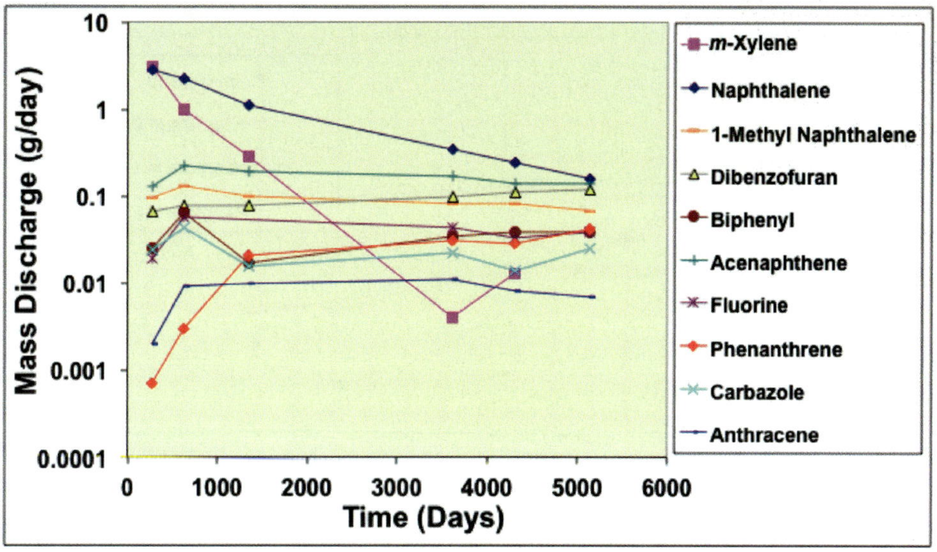

Figure 7.10. Source mass discharge (g/day) as a function of time (days) for ten monitored compounds. Data reported by Fraser et al. (2008).

emplacement were part of site monitoring activities and therefore were used to make predictions of mass discharge. These predictions in turn were compared to measurements completed for $3{,}619 \leq t \leq 5{,}140$ days.

If the first three measurements are viewed in isolation, it appears that monotonically decreasing trends in M_D occurred for *m*-xylene and naphthalene, while monotonically increasing trends in M_D occurred for phenanthrene and anthracene. With the exception of fluorine, mass discharge for the remaining compounds increased over the first two measurements but decreased with the third. Results for fluorine were not reported for $t = 1{,}357$ days, but M_D increased over the first two measurements. Conceptually, this behavior can be explained using Raoult's Law: as the most soluble compounds leave the NAPL, the mole ratio of the remaining components increases, resulting in a higher effective solubility. However, it should be kept in mind during the assessment of this type of data that the response at the source zone control plane is an integrated measurement that reflects not only dissolution processes but also sorption/desorption processes between the NAPL and the control plane as well as potential mass loss due to diffusion or volatilization.

Power Law models can be fit to *m*-xylene and naphthalene because they show decreasing trends. Model fits were completed for $\Gamma = \{0.5, 1.0, 2.0\}$ using linear regression and are shown in Figure 7.11. For $\Gamma = 1.0$, a linear transformation of the mass discharge data is required prior to linear regression using the natural logarithm (i.e., $y = \ln[M_D(t)]$, where y is the transformed data upon which linear regression is completed). Likewise, for $\Gamma = 2.0$, a linear transformation of the mass discharge data is needed using the inverse of the square root (i.e., $y = M_D(t)^{-1/2}$, where y is the transformed data upon which linear regression is completed). As shown in Figure 7.11, the mass discharge measurements completed for $3{,}619 \leq t \leq 5{,}140$ days fall in between model fits for $\Gamma = \{1.0, 2.0\}$. The model fit using $\Gamma = 0.5$ results in the largest discrepancy between measured data and model predictions. However, the rapid decline in the source strength as predicted by a PLM with $\Gamma = 0.5$ is generally considered unlikely.

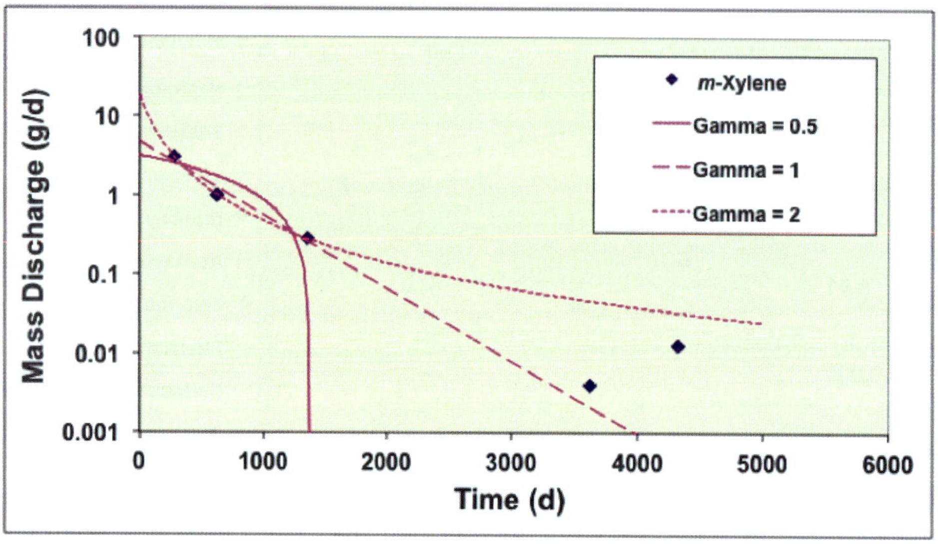

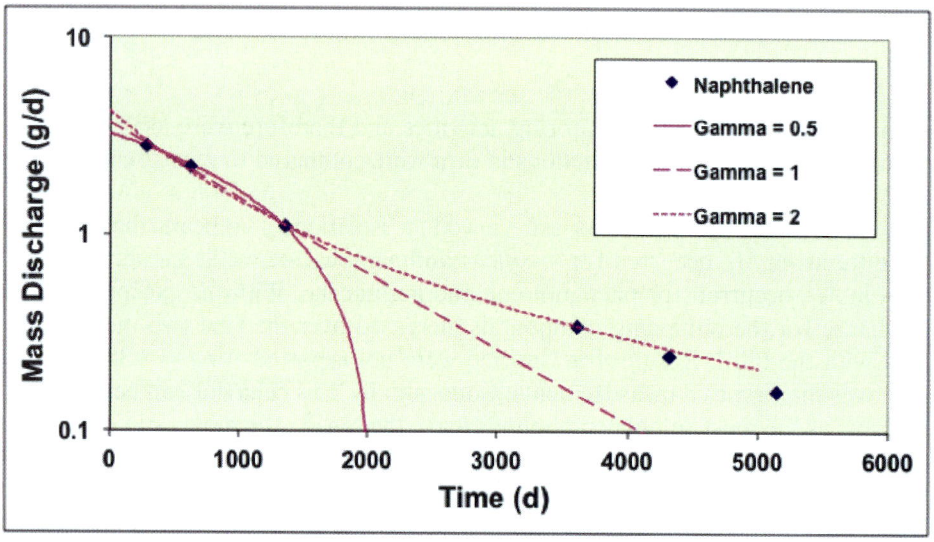

Figure 7.11. PLM fits to measured data for *m*-xylene and naphthalene.

7.5.2 Source Mass Estimation

Estimates of DNAPL source mass typically have high uncertainty due primarily to spatial heterogeneity in both subsurface characteristics and contaminants relative to sparse sampling densities. Methods that have been used to estimate source mass include the integration of core data, partitioning tracer tests and geophysical methods.

Once a PLM has been fit to the data, it can be used to estimate the initial source mass, M_0 (i.e., Equation 7.7). For *m*-xylene, estimates of M_0 were 2,170, 2,204 and 3,441 g for $\Gamma = 0.5$, 1 and 2, respectively, while the emplaced mass of *m*-xylene was 2,850 g. Estimates of the initial mass were therefore within a factor of 2 of the true mass. For naphthalene, estimates of M_0 were 3,298, 4,176 and 6,076 g for $\Gamma = 0.5$, 1 and 2, respectively, while the emplaced mass of naphthalene was 7,110 g. Estimates of initial mass were therefore within a factor of 2.2 of the true mass.

For any field-based method of DNAPL mass estimate, these results have to be considered as very good. However, the controlled nature of the study may be responsible for the agreement between the estimates and the actual values, and applications of this approach to more complex field sites need further evaluation.

7.5.3 Borden Coal Tar Plume Behavior

Another important feature of the Borden coal tar emplacement study with respect to flux-based site management was the evaluation of the plume behavior with respect to the source strength and the attenuation capacity within the plume. Plume behavior for the first 4 years of the study was reported by King and Barker (1999) and King et al. (1999) for seven compounds: phenol, *m*-xylene, naphthalene, phenanthrene, dibenzofuran, carbazole and 1-methylnaphthalene. Phenol (reported solubility of 82,000 mg/L) was observed to completely dissolve from the source and migrate downgradient as a discrete slug and was completely degraded after 2 years. For *m*-xylene and carbazole, the plume was observed to migrate downgradient and then recede as the mass flux from the source decreased below the natural attenuation capacity within the plume. The spatial extent of dibenzofuran reached steady state, indicating a balance between the mass discharge from the source and the rate of plume degradation. Naphthalene and 1-methylnaphthalene continued to migrate downgradient over the entire study period, indicating that source mass discharge exceeded degradation. The phenanthrene plume was also observed to expand and recede; however, the degree of sorption associated with this compound hampered assessment over the study duration.

Fraser et al. (2008) extended the analysis presented by King and Barker (1999) and King et al. (1999) to a duration of 14 years and included additional compounds (anthracene, acenaphthene, biphenyl and fluorene) in their analysis. The compounds were placed into one of the three classes based on their plume behavior. The first class consisted of those compounds that were significantly attenuated over the 14-year study duration (phenol, *m*-xylene and naphthalene) and represents the expected result when the natural attenuation capacity exceeds the source strength. The second class included the compounds for which the plume continued to expand over the entire 14-year study (phenanthrene and dibenzofuran) and represents the expected result when the source strength exceeds the downgradient natural attenuation capacity. The final class comprised those compounds that appeared to reach a steady plume extent (1-methylnaphthalene, carbazole, biphenyl, fluorene, anthracene and acenaphthene) and represents the expected result when the natural attenuation capacity and source strength are balanced.

From the perspective of flux-based site management, and its application to future MGP sites, the Borden emplacement study highlights the dependence of the plume and its spatial extent on the source mass, source age, source strength and natural attenuation capacity. Manipulation of source mass and strength through source remediation or plume degradation through plume remediation are the options that can be used to optimize the management of the contaminated site if complete DNAPL source depletion is not feasible.

7.6 CONCLUSIONS

Contaminant mass flux measurements provide a valuable method to assess conditions at DNAPL contaminated sites, including information needed to conduct mass balance calculations for a site and to evaluate natural attenuation rates and site age. The data can be used for a number of purposes, including remedial performance, risk assessment and site prioritization based on discharge from a source zone or to a receiving body. Measurements of mass flux and discharge immediately downgradient of a source area provide an early assessment of remedial performance and an indication of the likely plume response.

Existing technologies make it possible to define the spatial extent and mass present at DNAPL-contaminated source and plume zones. The challenge is managing the cost associated with obtaining this information, with the need for data resolution dependent on site management objectives. Some costs may be reduced by collecting data at integral or coarse scales; however, loss of detailed spatial information could lead to less efficient applications of remedial efforts. Thus the resolution of the data collected for defining source characteristics is inherently tied to site management including remedial strategies and risk reduction. For quantitative information on site characterization cost estimates for DNAPL sites see Kram et al. (2002).

We conclude that at many sites it may be more appropriate to place a greater emphasis on first defining the source strength function (mass flux leaving the source zone) and then estimate source mass as a secondary effort. This approach provides information that is more directly linked to the plume behavior at the site. Information on flux distribution and mass discharge can lead to more effective application of remedial efforts aimed at reducing the risk associated with the dissolved contaminant plume. Challenges still exist in determining the resolution needed when defining the source mass flux distribution. Again, integral measures may lead to lower costs while sacrificing knowledge of local high-flux zones. The link with remedial goals is again critical to defining the level of resolution needed.

ACKNOWLEDGEMENTS

The work upon which this chapter is based was supported by the U.S. Environmental Protection Agency (USEPA) through its Office of Research and Development with funding provided by the Strategic Environmental Research and Development Program (SERDP), a collaborative effort involving the USEPA, the U.S. Department of Energy (DOE), and the U.S. Department of Defense (DoD). It has not been subjected to Agency review and, therefore, does not necessarily reflect the views of the Agency and no official endorsement should be inferred.

REFERENCES

Annable MD, Hatfield K, Cho J, Klammler H, Parker BL, Cherry JA, Rao PSC. 2005. Field-scale evaluation of the passive flux meter for simultaneous measurement of groundwater and contaminant fluxes. Environ Sci Technol 39:7194–7201.

API (American Petroleum Institute). 2002. Estimating Mass Flux for Decision-Making: An Expert Workshop. American Petroleum Institute, Washington DC.

API (American Petroleum Institute). 2003. Groundwater Remediation Strategies Tool. Regulatory Analysis and Scientific Affairs Department. Publication 4730. American Petroleum Institute, Washington DC.

Banerjee S. 1984. Solubility of organic mixtures in water. Environ Sci Technol 18:587–591.

Basu NB, Rao PSC, Poyer IC, Annable MD, Hatfield K. 2006. Flux-based assessment at a manufacturing site contaminated with trichloroethylene. J Contam Hydrol 86:105–127.

Basu NB, Fure AD, Jawitz JW. 2008. Simplified contaminant source depletion models as analogs of multiphase simulators. J Contam Hydrol 97:87–99.

Basu NB, Rao PSC, Poyer IC, Nandy S, Mallavarapu M, Naidu R, Davis GB, Patterson BM, Annable MD, Hatfield K. 2009. Integration of traditional and innovative characterization techniques for flux-based assessment of dense non-aqueous phase liquid (DNAPL) sites. J Contam Hydrol 105:161–172.

Béland-Pelletier C, Fraser M, Barker J, Ptak T. 2011. Estimating contaminant mass discharge: A field comparison of the multilevel point measurement and the integral pumping investigation approaches and their uncertainties. J Contam Hydrol 122:63–75.

Bockelmann A, Ptak T, Teutsch G. 2001. An analytical quantification of mass fluxes and natural attenuation rate constants at a former gasworks site. J Contam Hydrol 53:429–453.

Bockelmann A, Zamfirescu D, Ptak T, Grathwohl P, Teutsch G. 2003. Quantification of mass fluxes and natural attenuation rates at an industrial site with a limited monitoring network: A case study. J Contam Hydrol 60:97–121.

Brooks MC, Annable MD, Rao PSC, Hatfield K, Jawitz JW, Wise WR, Wood AL, Enfield CG. 2004. Controlled release, blind test of DNAPL remediation by ethanol flushing. J Contam Hydrol 69:281–297.

Brooks MC, Wood AL, Annable MD, Hatfield K, Cho J, Holbert C, Rao PSC, Enfield CG, Lynch K, Smith RE. 2008. Changes in contaminant mass discharge from DNAPL source mass depletion: Evaluation at two field sites. J Contam Hydrol 102:140–153.

Christ JA, Ramsburg CA, Pennell KD, Abriola LM. 2006. Estimating mass discharge from dense nonaqueous phase liquid source zones using upscaled mass transfer coefficients: An evaluation using multiphase numerical simulations. Water Resour Res 42:W11420, doi:10.1029/2006WR004886.

D'Affonseca FM, Blum P, Finkel M, Melzer R, Grathwohl P. 2008. Field scale characterization and modeling of contaminant release from a coal tar source zone. J Contam Hydrol 102:120–139.

Devlin JF, McElwee CD. 2007. Effects of measurement error on horizontal hydraulic gradient estimates. Ground Water 45:62–73.

DiFilippo EL, Brusseau ML. 2008. Relationship between mass-flux reduction and source-zone mass removal: Analysis of field data. J Contam Hydrol 98:22–35.

Einarson MD, Mackay DM. 2001. Predicting impacts of groundwater contamination. Environ Sci Technol 35:66A–73A.

Endo S, Schmidt TC. 2006. Prediction of partitioning between complex organic mixtures and water: Application of polyparameter linear free energy relationships. Environ Sci Technol 40:536–545.

Enfield CG, Wood AL, Brooks MC, Annable MD, Rao PSC. 2005. Design of aquifer remediation extraction systems: (1) Describing hydraulic structure and NAPL architecture using tracers. J Contam Hydrol 81:125–147.

Falta RW. 2008. Methodology for comparing source and plume remediation alternatives. Ground Water 46:272–285.

Falta RW, Rao PSC, Basu N. 2005a. Assessing the impacts of partial mass depletion in DNAPL source zones: I. Analytical modeling of source strength functions and plume response. J Contam Hydrol 78:259–280.

Falta RW, Basu N, Rao PSC. 2005b. Assessing the impacts of partial mass depletion in DNAPL source zones: II. Coupling source strength functions to plume evolution. J Contam Hydrol 79:45–66.

Fraser M, Barker JF, Butler B, Blaine F, Joseph S, Cooke C. 2008. Natural attenuation of a plume from an emplaced coal tar creosote source over 14 years. J Contam Hydrol 100:101–115.

Fure AD, Jawitz JW, Annable MD. 2006. DNAPL source depletion: Linking architecture and flux response. J Contam Hydrol 85:118–140.

Goltz MN, Huang J, Close ME, Flintoft MJ, Pang L. 2008. Use of tandem circulation wells to measure hydraulic conductivity without groundwater extraction. J Contam Hydrol 100:127–136.

Goltz MN, Close ME, Yoon H, Huang J, Flintoft MJ, Kim S, Enfield C. 2009. Validation of two innovative methods to measure contaminant mass flux in groundwater. J Contam Hydrol 106:51–61.

Guilbeault MA, Parker BL, Cherry JA. 2005. Mass and flux distributions from DNAPL zones in sandy aquifers. Ground Water 43:70–86.

Hatfield K, Rao PSC, Annable MD, Campbell T. 2002. Device and Method for Measuring Fluid and Solute Fluxes in Flow Systems. U.S. Patent No. 6,402,547 B1. U.S. Patent Office, Washington DC, USA.

Hatfield K, Annable MD, Cho J, Rao PSC, Klammler H. 2004. A direct passive method for measuring water and contaminant fluxes in porous media. J Contam Hydrol 75:155–181.

Henderson JK, Falta RW, Freedman DL. 2009. Simulation of the effect of remediation on EDB and 1,2-DCA plumes at sites contaminated by leaded gasoline. J Contam Hydrol 108:29–45.

Illangasekare TH, Ramsey J, Jensen KH, Butts BM. 1995. Experimental study of movement and distribution of dense organic contaminants in heterogeneous aquifers. J Contam Hydrol 20:1–25.

ITRC (Interstate Technology & Regulatory Council). 2003. Assessing the Performance of DNAPL Source Reduction Remedies. Dense Nonaqueous Phase Liquids Team, Washington, DC, USA.

ITRC. 2010. Use and Measurement of Mass Flux and Mass Discharge. MASSFLUX-1. Integrated DNAPL Site Strategy Team, Washington, DC, USA.

Jawitz JW, Annable MD, Demmy GG, Rao PSC. 2003. Estimating non-aqueous phase liquid spatial variability using partitioning tracer higher temporal moments. Water Resour Res 39:1192, doi:10.1029/2002WR001309.

Jawitz JW, Fure AD, Demmy GG, Berglund S, Rao PSC. 2005. Groundwater contaminant flux reduction resulting from nonaqueous phase liquid mass reduction. Water Resour Res 41: W10408.

Kavanaugh MC, Rao PSC, eds. 2003. The DNAPL Remediation Challenge: Is There a Case for Source Depletion? EPA/600/R-03/143. National Risk Management Research Laboratory, Office of Research and Development, U.S. Environmental Protection Agency, Cincinnati, OH, USA.

King MWG, Barker JF. 1999. Migration and natural fate of a coal tar creosote plume 1. Overview and plume development. J Contam Hydrol 39:249–279.

King MWG, Barker JF, Devlin JF, Butler BJ. 1999. Migration and natural fate of a coal tar creosote plume 2. Mass balance and biodegradation indicators. J Contam Hydrol 39:281–307.

Klammler H, Hatfield K, Annable MD. 2007a. Concepts for measuring horizontal groundwater flow directions using the passive flux meter. Adv Water Resour 30:984–997.

Klammler H, Hatfield K, Annable MD, Agyei E, Parker BL, Cherry JA, Rao PSC. 2007b. General analytical treatment of the flow field relevant to the interpretation of passive fluxmeter measurements. Water Resour Res 43:W04407, doi:10.1029/2005WR004718.

Klammler H, Hatfield K, Luz JAG, Annable MD, Newman M, Cho J, Peacock A, Stucker V, Ranville J, Cabaniss S, Rao PS. 2012. Contaminant discharge and uncertainty estimates from passive flux meter measurements. Water Resour Res 48:W02512, doi:10.1029/2011WR010535.

Kram ML, Keller AA, Rossabi J, Everett LG. 2002. DNAPL characterization methods and approaches, part 2: Cost comparison. Ground Water Monit Remed 22:46–61.

Kübert M, Finkel M. 2006. Contaminant mass discharge estimation in groundwater based on multi-level point measurements: A numerical evaluation of expected errors. J Contam Hydrol 84:55–80.

Kueper BH, Redman D, Starr RC, Reitsma S, Mah M. 1993. A field experiment to study the behavior of tetrachloroethylene below the water table: Spatial distribution of residual and pooled DNAPL. Ground Water 31:756–766.

Lane WF, Loehr RC. 1992. Estimating the equilibrium aqueous concentrations of polynuclear aromatic hydrocarbons in complex mixtures. Environ Sci Technol 26:983–990.

Lee LS, Rao PSC, Okuda I. 1992. Equilibrium partitioning of polycyclic aromatic hydrocarbons from coal tar into water. Environ Sci Technol 26:2110–2115.

Lemke LD, Abriola LM, Lang JR. 2004. Influence of hydraulic property correlation on predicted dense nonaqueous phase liquid source zone architecture, mass recovery and contaminant flux. Water Resour Res 40:W12417.

Li KB, Goovaerts P, Abriola LM. 2007. A geostatistical approach for quantification of contaminant mass discharge uncertainty using multilevel sampler measurements. Water Resour Res 43:W06436.

Mahjoub B, Jayr E, Bayard R, Gourdon R. 2000. Phase partition of organic pollutants between coal tar and water under variable experimental conditions. Water Res 34:3551–3560.

NRC (National Research Council). 2004. Contaminants in the Subsurface: Source Zone Assessment and Remediation. The National Academies Press, Washington DC, USA.

Parker JC, Park E. 2004. Modeling field-scale dense nonaqueous phase liquid dissolution kinetics in heterogeneous aquifers. Water Resour Res 40:W05109, doi: 10.1029/2003WR002807.

Parker BL, Chapman SW, Guilbeault MA. 2008. Plume persistence caused by back diffusion from thin clay layers in a sand aquifer following TCE source-zone hydraulic isolation. J Contam Hydrol 102:86–104.

Ptak T, Schirmer M, Teutsch G. 2000. Development and performance of a new multilevel groundwater sampling system. In Wickramanayake GB, Gavaskar AR, Kelley ME, Nehring KW, eds, Risk, Regulatory and Monitoring Considerations: Remediation of Chlorinated and Recalcitrant Compounds. Battelle Press, Columbus, OH, USA, pp 95–102.

Rao PSC, Jawitz JW. 2003. Comment on "Steady state mass transfer from single-component dense nonaqueous phase liquids in uniform flow fields" by Sale TC, McWhorter DB. Water Resour Res 39:1068.

Rao PSC, Jawitz J, Enfield C, Falta R, Annable M, Wood A. 2002. Technology integration for contaminated site remediation: Cleanup goals and performance criteria. In Thornton S, Oswald S, eds, Groundwater Quality 2001 Proceedings, Oxford: IAHS Publication 275, pp 571–578.

Rein A, Bauer S, Dietrich P, Beyer C. 2009. Influence of temporally variable groundwater flow conditions on point measurements and contaminant mass flux estimations. J Contam Hydrol 108:118–133.

Sale TC, McWhorter DB. 2001. Steady-state mass transfer from single-component dense nonaqueous phase liquids in uniform flow fields. Water Resour Res 37:393–404.

Schwede RL, Cirpka OA. 2009. Use of steady-state concentration measurements in geostatistical inversion. Adv Water Resour 32:607–619.

Soga K, Page JWE, Illangasekare TH. 2004. A review of NAPL source zone remediation efficiency and the mass flux approach. J Hazard Mater 110:13–27.

Stroo HF, Unger M, Ward CH, Kavanaugh MC, Vogel C, Leeson A, Marqusee JA, Smith BP. 2003. Remediating chlorinated solvent source zones. Environ Sci Technol 37:224A–230A.

Teutsch G, Ptak T, Schwarz R, Holder T. 2000. Ein neus integrales verfahren zur quantifizierung der grundwasserinnission: I. Theoretische grundlagen. Grundwasser 4:170–175.

Thomson NR, Fraser MJ, Lamarche C, Barker JF, Forsey SP. 2008. Rebound of a coal tar creosote plume following partial source zone treatment with permanganate. J Contam Hydrol 102:154–171.

Wood AL, Enfield CG, Annable MD, Brooks MC, Rao PSC, Sabatini D, Knox R. 2005. Design of aquifer remediation extraction systems: (2) Estimating site-specific performance and benefits of partial source removal. J Contam Hydrol 81:148–166.

Zhu J, Sykes JF. 2004. Simple screening models of NAPL dissolution in the subsurface. J Contam Hydrol 72:245–258.

CHAPTER 8

HYDRAULIC DISPLACEMENT OF DENSE NONAQUEOUS PHASE LIQUIDS

Bernard H. Kueper[1] and Jason I. Gerhard[2]

[1]Queen's University, Kingston, ON, Canada K7L 3N6; [2]University of Western Ontario, London, ON, Canada N6A 5B9

8.1 INTRODUCTION

Hydraulic displacement is a mass removal technology that involves recovering dense nonaqueous phase liquids (DNAPLs) from either vertical wells or horizontal drains. Hydraulic displacement is capable of removing significant quantities of pooled DNAPL from the subsurface in either porous or fractured media, but it is not suitable for the removal of residual DNAPL. Hydraulic displacement can stabilize a source zone: DNAPL pools are removed, thereby reducing the potential for future undesirable mobilization of DNAPL during drilling, hydraulic testing and construction of subsurface barriers.

In addition to mass removal and source zone stabilization, hydraulic displacement is a means of preconditioning a source zone for implementing mass transfer based remediation techniques. Technologies such as enhanced *in situ* bioremediation (EISB), surfactant flushing, cosolvent flushing, and *in situ* chemical oxidation (ISCO) rely upon mass transfer of components from the DNAPL to groundwater. The rate of mass transfer from pooled DNAPL is small compared to that from residual DNAPL. This stems from both the lower specific DNAPL surface area for pools compared to the residual and from the lower relative permeability to water within a pool compared to within a zone of residual DNAPL. Preconditioning a source zone with hydraulic displacement will allow for more efficient and cost effective application of subsequent, mass transfer based technologies.

Hydraulic displacement is also referred to as waterflooding (Craig, 1971; Willhite, 1986) and dual phase extraction (Gerhard et al., 2001). The technology was first developed in the petroleum industry as a means of recovering oil from reservoirs. At DNAPL sites, hydraulic displacement can be implemented in several ways. The simplest implementation is to pump DNAPL from any well in which it has accumulated. This will lead to a gradual reduction of DNAPL saturations within the formation adjacent to the well. The rate of DNAPL recovery can be enhanced by simultaneously extracting water and DNAPL from the well (the so-called dual phase or multiphase extraction). Water extraction results in an increased hydraulic gradient across the DNAPL pool in the formation, which in turn results in a favorable capillary pressure gradient facilitating faster rates of DNAPL migration to the well. The hydraulic gradient across DNAPL pools in the formation can be further increased by injecting water into wells around the perimeter of the source zone. The advantages of water injection and withdrawal include faster rates of DNAPL recovery and ability to mobilize pools that were not initially in contact with recovery wells or drains. A disadvantage of water injection and withdrawal is that any extracted water may need to be treated prior to either reinjection or disposal.

B.H. Kueper et al. (eds.), *Chlorinated Solvent Source Zone Remediation*,
doi: 10.1007/978-1-4614-6922-3_8, © Springer Science+Business Media New York 2014

Hydraulic displacement is a candidate technology at sites with appreciable quantities of pooled DNAPL. DNAPL pooling is promoted by the presence of horizontal capillary barriers, large volume DNAPL releases, high DNAPL–water interfacial tension and low DNAPL density. Pooled DNAPL can be detected through visual observation in monitoring and extraction wells, visual observation in soil cores, and soil analytical results that exceed a prescribed saturation threshold. If these methods of detection do not indicate the presence of pooled DNAPL, then the source zone is probably not a candidate for hydraulic displacement.

This chapter focuses primarily on the application of hydraulic displacement at sites impacted by chlorinated solvent DNAPLs, although some discussion of other types of DNAPLs also is included. The chapter begins with screening calculations including examples (Section 8.2), followed by a discussion of the state of the practice including numerical simulation examples (Section 8.3), followed by a case history (Section 8.4) and concludes with a summary. Section 8.2 also includes definitions of several multiphase flow terms pertinent to DNAPL behavior and hydraulic displacement that may not be familiar to some practitioners in the field of contaminant hydrogeology.

8.2 SCREENING CALCULATIONS

A variety of screening calculations are presented here to (1) estimate the height of DNAPL pools that may be present in a source zone given prescribed geologic and fluid properties, (2) estimate the volume of DNAPL present in pools, (3) estimate the saturation of DNAPL from a soil analytical result and (4) assess the magnitude of hydraulic gradient required to mobilize DNAPL pools. These screening calculations are presented to illustrate several relevant principles and to assist investigators in deciding whether hydraulic displacement is a candidate technology for the site of interest. Although a number of the input parameters to the screening calculations may be atypical compared to many hydrogeologic investigations, these can be measured in commercial laboratories with a history of servicing the petroleum industry. In addition, some input parameters can be estimated on the basis of simple correlations and scaling expressions.

8.2.1 Fundamentals of Multiphase Flow

Hydraulic displacement involves manipulating fluid pressures so that DNAPL will migrate towards recovery wells. This is only possible for DNAPL that exists as a connected phase through the pore space (pooled DNAPL). If the DNAPL content is reduced to such a degree that it becomes disconnected, then it can no longer flow in response to fluid pressure gradients and is rendered immobile (residual DNAPL). These two fundamental goals of hydraulic displacement, DNAPL recovery and DNAPL stabilization, are best understood through consideration of capillary pressure and relative permeability relationships.

Figure 8.1a illustrates a typical capillary pressure–saturation (*Pc–S*) curve for a porous medium. Although the primary focus of this chapter is porous media, the concepts are equally applicable to fractures in clay and rock (Kueper and McWhorter, 1991; Reitsma and Kueper, 1994; Longino and Kueper 1999a, b).

The capillary pressure (P_c) is defined as the difference between the nonwetting- and the wetting-phase pressures ($P_c = P_N - P_W$). In this chapter, which addresses the hydraulic displacement of DNAPL below the water table, the nonwetting phase is deemed to be the DNAPL and the wetting phase to be the groundwater (water). Although exceptions to this convention can occur, the DNAPL is typically the nonwetting phase in most groundwater systems. The water saturation (S_W) is defined as the percentage of the pore space occupied

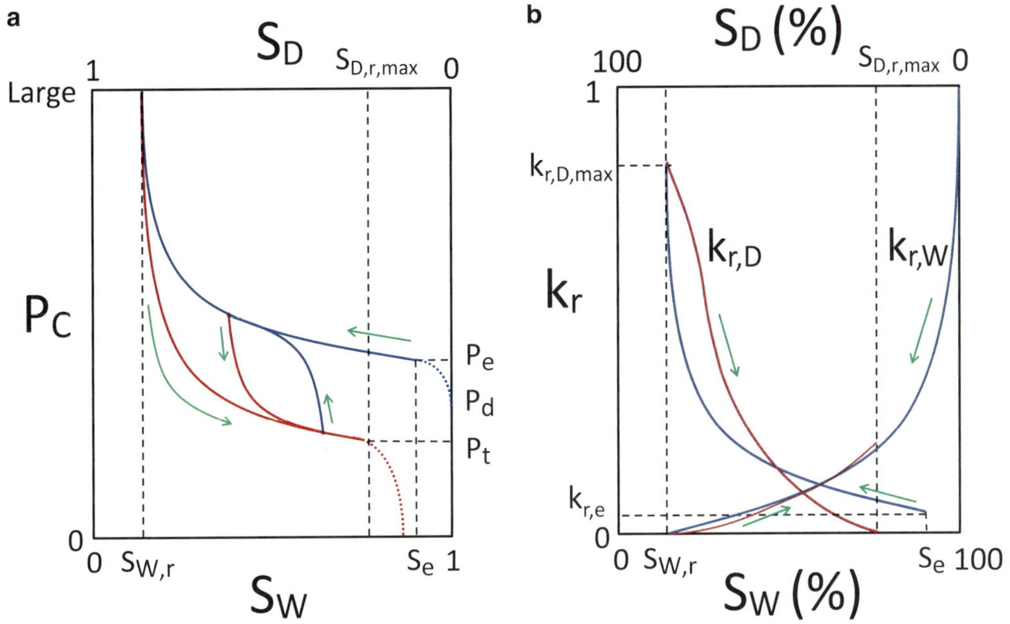

Figure 8.1. (a) Typical capillary pressure–saturation curve shown; (b) general relationship between relative permeability and saturation.

by water, while the DNAPL saturation (S_D) is defined as the percentage of the pore space occupied by DNAPL. The sum of the water and DNAPL saturations is always 100% in a DNAPL–water two-phase system ($S_W + S_D = 100$).

Figure 8.1a illustrates that higher DNAPL saturations are associated with higher capillary pressures. The horizontal axes represent the water saturation (S_W) as a percentage of pore space (lower axis) and the DNAPL saturation (S_D) as a percentage of pore space (upper axis). The displacement pressure (P_d) is defined as the threshold capillary pressure required for the initial entry of DNAPL into a water-saturated porous medium. At this point, some water has been displaced but there is no connected pathway of DNAPL through the unit of interest. The entry pressure (P_e) is defined as the threshold capillary pressure required for a connected pathway of DNAPL to have formed through the unit of interest. The unit of interest may be, for example, a layer of fine sand, a layer of silt, or a layer of clay. The saturation corresponding to the entry pressure is referred to as the emergence saturation (S_e).

If the capillary pressure increases beyond the entry pressure, progressively smaller pore throats are invaded and the DNAPL saturation increases. This is termed a drainage process in that the water is being drained from the porous medium as a result of the DNAPL invasion. However, not all of the water can be drained; a certain amount of irreducible water saturation remains, referred to as the residual water saturation ($S_{W,r}$), even at high capillary pressures. The formation of DNAPL pools accumulated on low permeability strata typically occurs as a result of a drainage process; thus, the relationship between saturation and capillary pressure in such a pool is best represented by the drainage P_c–S curve.

After the capillary pressure starts to decrease, DNAPL exits the unit of interest and the water saturation increases. This is known as a wetting or an imbibition process. The wetting and drainage P_c–S relationships do not overlap; this hysteresis in P_c–S behavior occurs because pore drainage and pore wetting processes are distinct. As capillary pressure decreases and the wetting process proceeds, the DNAPL saturation decreases as some of the DNAPL is displaced from the unit. At the same time some of the DNAPL is snapped off to form residual globules or bypassed to form residual ganglia; these are disconnected from the connected phase DNAPL, are trapped by capillary forces, and are difficult to displace. The highest capillary pressure on the wetting curve at which no connected-phase DNAPL remains, and thus only residual globules and ganglia of DNAPL exist, is referred to as the terminal pressure (P_t). Within the drainage and wetting limbs of the capillary pressure–saturation relationship, additional hysteresis loops can occur along scanning curves initiated when drainage switches to wetting (or vice versa), for example, when hydraulic displacement of a DNAPL pool is initiated.

The amount of residual DNAPL left behind a migrating DNAPL body depends on the maximum capillary pressure attained along the drainage path. If very high capillary pressures were attained, then the maximum amount of residual DNAPL saturation ($S_{D,r,\max}$) is left behind. If only lower capillary pressures were attained along drainage, then a lower residual DNAPL saturation is left behind. This dependence of residual DNAPL saturation on the maximum capillary pressure attained along drainage is referred to as the initial-residual relationship (Land, 1968). Kueper et al. (1993) present the results of a field experiment involving the release of 231 liters (L) (61 gallons [gal]) of perchloroethene (PCE) DNAPL into a fine- to medium-grained sand aquifer. The measured residual DNAPL saturations ranged from 1 to 29% of pore space. Further discussion of capillary pressure–saturation relationships is provided by Brooks and Corey (1966) and Gerhard and Kueper (2003a).

In addition to the relationship between capillary pressure and saturation, it is also important to consider the relationship between relative permeability and saturation. Relative permeability is a dimensionless measure that accounts for the fact that the effective permeability available to a particular fluid phase in a two-phase flow system is dependent on the saturation of that fluid phase. If the saturation of the phase is low, it will have a low relative permeability because it only occupies a low fraction of the pore space. If the saturation of the phase is high, it will have a high relative permeability, reflecting the fact that it occupies a large fraction of the pore space. Relative permeability ranges from 0 to 1. Effective permeability, which governs the ability of a phase to flow (towards a recovery well), is the product of relative permeability and the porous medium's intrinsic permeability of the pore spaces.

Figure 8.1b presents the general relationship between relative permeability and fluid saturation for a two-phase DNAPL–water system. The horizontal axes are identical to Figure 8.1a, but here the vertical axis represents relative permeability. The right-hand blue curve represents the relative permeability to the wetting phase ($k_{r,W}$) under drainage conditions, which has a value of 1 prior to any DNAPL entry. Once DNAPL starts to enter the porous medium, $k_{r,W}$ decreases and if DNAPL saturations become high enough, $k_{r,W}$ is reduced to zero (effectively no flow of water, corresponding to residual wetting-phase saturation $S_{w,r}$). If water subsequently displaces DNAPL (wetting), $k_{r,W}$ will increase, but not to its original value of 1. The maximum relative permeability to water on wetting is less than 1 because of the presence of residual DNAPL ($S_{D,r,\max}$).

The left-hand blue curve in Figure 8.1b represents the relative permeability to DNAPL ($k_{r,D}$) under drainage conditions. For DNAPL entering an initially water-saturated porous medium, the first relative permeability to DNAPL ($k_{r,e}$) occurs at the emergence saturation (S_e). As DNAPL continues to enter the porous medium and increase in saturation, $k_{r,D}$ increases and can reach a maximum value ($k_{r,D,\max}$) at the residual water saturation ($S_{W,r}$). The maximum relative permeability to DNAPL is less than one because some of the pore space is occupied by the residual water. If water subsequently begins to displace DNAPL from the porous medium, $k_{r,D}$ will decrease along the left-hand red curve in Figure 8.1b, eventually reaching a value of zero corresponding to the elimination of all connected-phase DNAPL leaving only residual DNAPL ($S_{D,r,\max}$).

As with the capillary pressure–saturation curves, the drainage relative permeability curves are relevant in increasing DNAPL saturations such as during the formation or the advance of a DNAPL pool, while the wetting curves are relevant in decreasing DNAPL saturations such as during hydraulic displacement activities. As illustrated, the values of relative permeability can be significantly less than one, which has implications for DNAPL recovery by hydraulic displacement. For example, as saturations within a DNAPL pool are being depleted as a result of hydraulic displacement, the ratio of DNAPL to water in the recovered fluid stream decreases with time because of relative permeability effects. As with the P_c–S relationship, hysteresis scanning loops associated with drainage/wetting reversals are present between the main drainage and main wetting relative permeability curves. Additional details on relative permeability curves are provided by Demond and Roberts (1987) and Gerhard and Kueper (2003b).

8.2.2 Height of DNAPL Pools

The maximum vertical height of pooled DNAPL that can accumulate above a capillary barrier (Figure 8.2) without allowing the DNAPL to penetrate the barrier is given by (Kueper et al., 1993):

$$H = \frac{P_c'' - P_c'}{\Delta \rho g} \quad \text{(Eq. 8.1)}$$

where H is the maximum height of DNAPL so that the capillary barrier is not penetrated, P_c'' is the entry pressure of the capillary barrier, P_c' is the capillary pressure at the top of the pool, $\Delta\rho$ is the difference in density between the DNAPL and water, and g is acceleration attributable to gravity. Equation 8.1 assumes a one-dimensional hydrostatic system in which neither the DNAPL nor the groundwater is moving. The analysis presented here also assumes that the DNAPL is nonwetting with respect to water. As illustrated in Figure 8.2, it is assumed that the DNAPL pool is located above a single capillary barrier, as opposed to being sandwiched between two capillary barriers.

The capillary pressure at the top of the DNAPL pool (P_c') is dependent on the saturation history associated with pool placement. The maximum value of P_c' will be the entry pressure of the porous medium within which the pool resides. This assumes that the top of the pool exists under drainage conditions. However, if the top of the pool exists under wetting

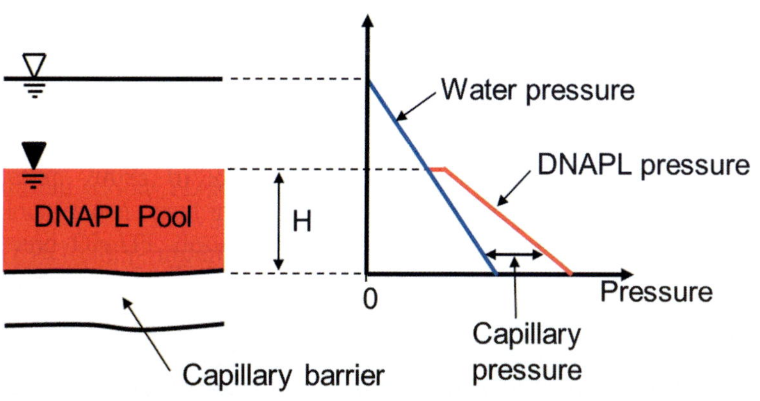

Figure 8.2. Distribution of water pressure, DNAPL pressure, and capillary pressure in a DNAPL pool under hydrostatic conditions.

conditions, P_c' may be as low as the terminal pressure. As discussed by Gerhard and Kueper (2003a), the terminal pressure can be reasonably approximated as half the entry pressure. Also of note is that the pool height predicted by Equation 8.1 is independent of the depth below the water table.

The entry pressure can be measured by commercial laboratories using an undisturbed sample of porous media, collected, for example, using a Shelby tube. It is also convenient to estimate the entry pressure of a particular porous medium given a value of permeability and the entry pressure measured for a different porous medium. The Leverett function can be used to correlate entry pressure with permeability as follows (Leverett, 1941):

$$P_e^{\dim} = \frac{P_e}{\sigma}\left[\frac{k}{\phi}\right]^{\alpha} \qquad \text{(Eq. 8.2)}$$

where $P_e^{\dim}$ is a dimensionless entry pressure, P_e is a measured entry pressure, σ is the interfacial tension between the two fluids used to measure P_e, k is the intrinsic permeability of the sample used to measure P_e, ϕ is the porosity of the sample, and α is an exponent often set to 0.5. Kueper and Frind (1991) found α to be 0.65 and $P_e^{\dim}$ to be 0.186 for a set of PCE–water capillary pressure curves measured using samples of sand with hydraulic conductivity varying from 4.3×10^{-3} centimeters/second (cm/s) to 1.2×10^{-2} cm/s.

Equations 8.1 and 8.2 illustrate that higher DNAPL pool heights are promoted by lower DNAPL density, higher DNAPL–water interfacial tension, lower permeability capillary barriers, and larger differences in permeability between the capillary barrier and the host material in which the pool resides. Although not illustrated by Equations 8.1 and 8.2, higher DNAPL pool heights will occur in a flow system experiencing an upward component of groundwater flow, and a shorter pool height will occur in response to a downward component of groundwater flow (Kueper et al., 2008). This dependence of pool height on the vertical component of groundwater flow is why hydraulic testing activities such as pumping groundwater from beneath zones of pooled DNAPL run the risk of vertical pool mobilization.

Example Calculation: Consider a medium-grained sand aquifer characterized by a hydraulic conductivity of 1×10^{-2} cm/s and a porosity of 40%. The aquifer contains horizontal lenses of fine sand having a hydraulic conductivity of 1×10^{-3} cm/s and a porosity of 40%. The air–water entry pressure of the fine sand was measured to be 5,000 Pascals (Pa). The DNAPL of interest is dichloroethane (1,2-DCA) having a density of 1.25 grams (g) per cm^3 and an interfacial tension with water of 0.025 Newtons/meter (N/m).

Equation 8.2 provides a dimensionless entry pressure of 0.11 using the fine sand properties above, an air–water surface tension of 0.072 N/m, and $\alpha = 0.5$. Equation 8.2 is then used to calculate the DNAPL entry pressure of the fine sand to be 1,736 Pa. Similarly, Equation 8.2 is used to calculate the DNAPL entry pressure of the medium sand to be 549 Pa. Setting P_c'' to 1,736 Pa and P_c' to 274.5 Pa (terminal pressure of the medium sand using the ratio of $P_t/P_e = 0.5$, assuming wetting conditions at top of the DNAPL pool), the maximum stable DNAPL pool height is calculated to be 0.59 m (1.96 feet [ft]).

Figure 8.3 illustrates the relationship between the maximum stable DNAPL pool height calculated using Equation 8.1 versus the hydraulic conductivity of the capillary barrier for a variety of chlorinated solvent DNAPLs, assuming that the pool resides in a medium-grained sand having a hydraulic conductivity of 1×10^{-2} cm/s and a porosity of 40%. Figure 8.3 adopts a dimensionless entry pressure of 0.11 and assumes that the top of the pool exists under wetting conditions. Each DNAPL is assumed to have an interfacial tension with water of 0.025 N/m. Figure 8.3 illustrates that lower density DNAPLs and lower hydraulic conductivity capillary barriers promote larger DNAPL pool heights.

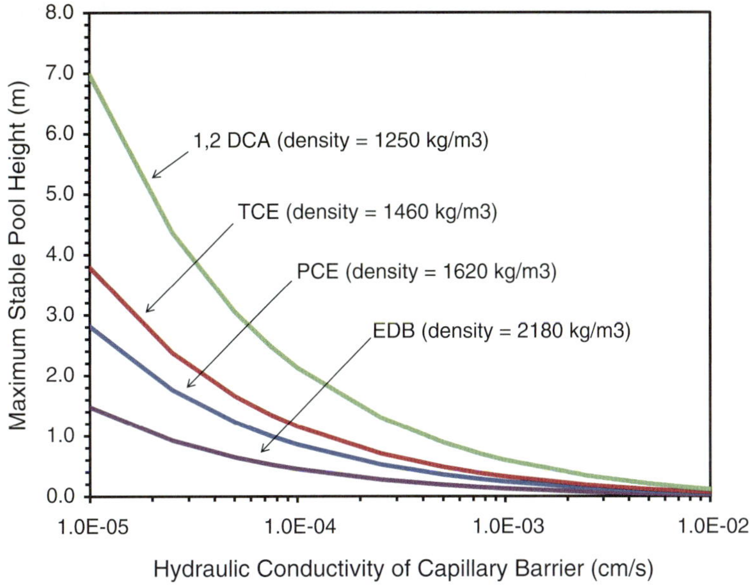

Figure 8.3. Maximum stable DNAPL pool height versus hydraulic conductivity of capillary barrier.

8.2.3 Volume of DNAPL in a Pool

The total volume of DNAPL in a pool can be calculated as:

$$V_D = A\phi \int_0^H S_D dz \qquad \text{(Eq. 8.3)}$$

where V_D is the DNAPL volume of interest, A is the area (footprint in plan view) of the pool, H is the height of the pool, S_D is the DNAPL saturation, and z is the vertical Cartesian coordinate. As illustrated in Figure 8.4, Equation 8.3 assumes that the DNAPL pool is of uniform thickness, H, at all locations. Equation 8.3 requires knowledge of how the DNAPL saturation varies with height in the pool. This can be calculated on the basis of the relationship between capillary pressure and fluid saturation using the Brooks and Corey (1966) capillary pressure–saturation relationship:

$$S_e = \left[\frac{P_c}{P_d}\right]^{-\lambda} \qquad \text{(Eq. 8.4)}$$

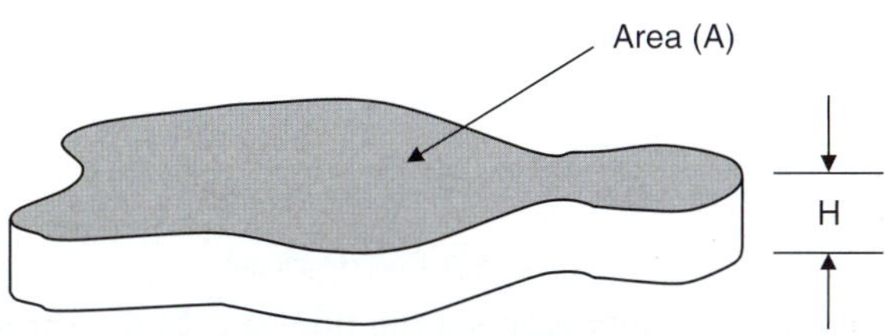

Figure 8.4. Schematic illustration of uniform-thickness DNAPL pool.

where P_c is the capillary pressure between the DNAPL and water, P_d is the porous media displacement pressure (approximately equal to the entry pressure), λ is the pore size distribution index, and S_e is an effective wetting-phase (water) saturation given by:

$$S_e = \frac{S_W - S_r}{1 - S_r} \qquad \text{(Eq. 8.5)}$$

where S_W is the wetting-phase (water) saturation expressed as a fraction of pore space and S_r is the residual wetting-phase saturation ($S_{W,r}$). Note that the wetting-phase (water) and nonwetting-phase (DNAPL) saturations sum to unity: $S_W + S_D = 1.0$. This allows Equations 8.4 and 8.5 to be combined, and the DNAPL saturation can be expressed as a function of the capillary pressure:

$$S_D = 1 - \left(\frac{P_C}{P_d}\right)^{-\lambda}(1 - S_r) - S_r \qquad \text{(Eq. 8.6)}$$

Evaluation of Equation 8.3 following substitution of Equation 8.6 provides the following relationship between DNAPL volume and pool height (Longino and Kueper, 1995):

$$V_D = A\phi\left(H - S_rH - \frac{P_d^{\lambda}(1 - S_r)}{(1 - \lambda)(\Delta\rho g)}\left[(P_d + \Delta\rho gH)^{1-\lambda} - P_d^{1-\lambda}\right]\right) \qquad \text{(Eq. 8.7)}$$

where $\Delta\rho$ is the difference in density between the DNAPL and water (i.e., $\Delta\rho = \rho_D - \rho_W$) and g is the acceleration due to gravity. Equation 8.7 assumes that the pool of interest is under drainage conditions. When the pool is under wetting conditions, Equation 8.7 should be viewed as an approximation.

Example Calculation: Figure 8.5 illustrates the relationship between DNAPL volume per m^2 of pool area versus pool height for a variety of hydraulic conductivity values. The hydraulic conductivity values correspond to the porous medium in which the pool resides, not the

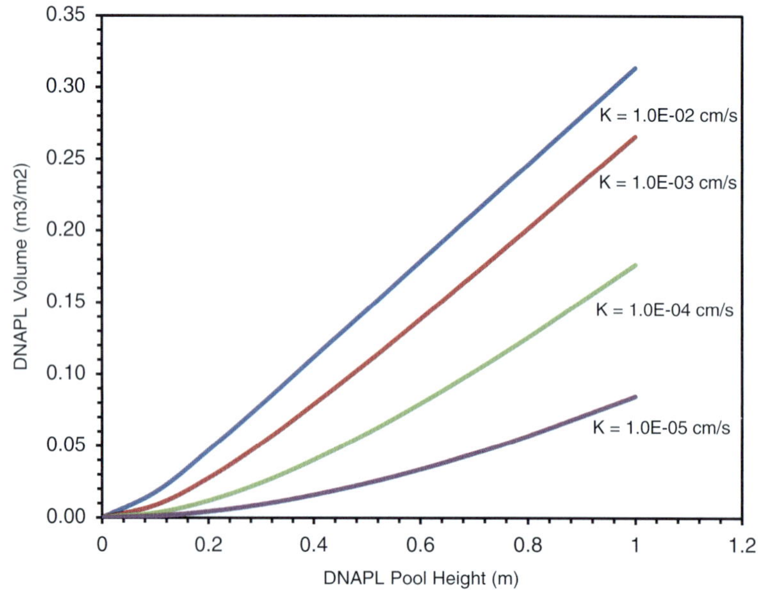

Figure 8.5. DNAPL volume per m^2 pool area versus pool height for various values of hydraulic conductivity.

capillary barrier upon which the pool is perched. For this calculation, it is assumed that a sufficient capillary barrier exists to support the DNAPL pool. Figure 8.5 assumes that the dimensionless displacement pressure is 0.11 as in Section 8.2.2, that the DNAPL of interest is TCE with a DNAPL–water interfacial tension of 0.025 N/m and a density of 1,460 kilograms per cubic meter (kg/m^3), that the porosity is 0.4, that the residual wetting-phase saturation is 0.15, and that the pore size distribution index is 2.5. Figure 8.5 illustrates that the DNAPL volume per area of pool increases with pool height and that this volume is greater in higher hydraulic conductivity media for a given pool height. This demonstrates that pools most amenable to hydraulic displacement exist in coarser-grained media and that estimates of both pool height and capillary properties are required to estimate DNAPL volume. Although not demonstrated by Figure 8.5, well-sorted porous media, low DNAPL–water interfacial tension, and high DNAPL density all promote greater DNAPL pool volumes. If these parameters are not well defined, a sensitivity analysis should be performed and decisions be made accordingly.

8.2.4 DNAPL Saturation Corresponding to Measured Soil Concentration

Soil sampling is a technique commonly employed to determine the presence of DNAPL at a site. The magnitude of the soil concentration can be used to establish whether the sample was obtained from a DNAPL pool, a zone of residual DNAPL, or a zone absent of DNAPL. Soil sampling can therefore be a primary means of assessing whether hydraulic displacement should be considered to remove DNAPL from a particular site. The reported soil concentration can be converted to an estimate of DNAPL saturation using the following:

$$S_D = \frac{(C_D - C_T)\rho_b}{\phi \rho_D 10^6} \qquad \text{(Eq. 8.8)}$$

where S_D is the DNAPL saturation (unitless), C_D is the soil concentration (mg/kg) reported by the laboratory, ϕ is the porosity (unitless), ρ_D is the DNAPL density (g/cm^3), ρ_b is the dry soil bulk density (g/cm^3), and C_T is the theoretical maximum possible amount of chemical (mg/kg) present in the soil sample in the aqueous, vapor, and sorbed phases.

Example Calculation: Figure 8.6 presents a plot of soil concentration versus DNAPL saturation for various DNAPLs in sandy media characterized by an effective porosity of 0.30 and a dry bulk density of 1.63 g/ cm^3. The C_T term is approximated as 500 mg/kg for each DNAPL for the purposes of this example. Chapter 3 discusses the C_T term in detail, including example calculations. Adopting an approximate threshold DNAPL saturation of 25% to represent pooled DNAPL, Figure 8.6 indicates that soil concentrations in excess of between 60,000 and 100,000 mg/kg would be indicative of pooled DNAPL.

8.2.5 Hydraulic Gradient Required to Mobilize Pooled DNAPL

The capillary pressure (P_c) between DNAPL and water, assuming that DNAPL is the nonwetting liquid and water is the wetting liquid, is given by:

$$P_c = P_D - P_W \qquad \text{(Eq. 8.9)}$$

where P_D is the DNAPL pressure and P_W is the water pressure. A static (not moving) DNAPL pool perched upon a horizontal capillary barrier (finer-grained layer) exhibits an increasing capillary pressure with depth in the pool (Figure 8.2) and the same capillary pressure laterally from end to end within the pool assuming that there is no groundwater flow. Under conditions

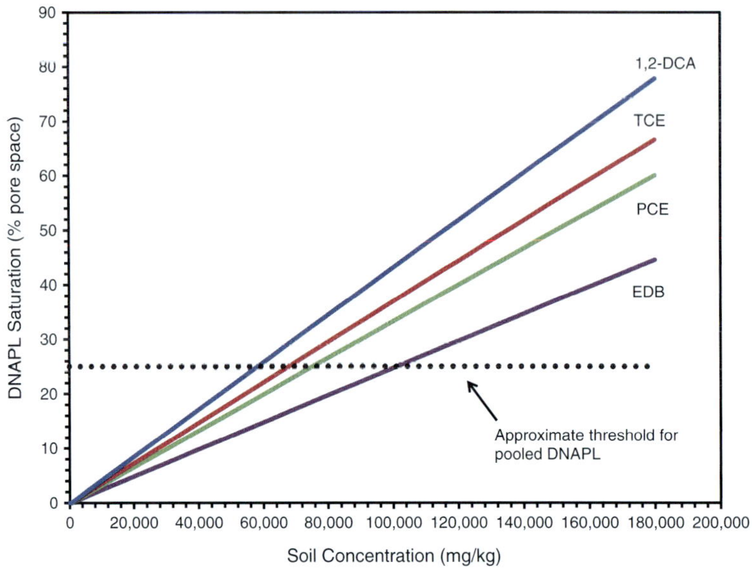

Figure 8.6. Soil concentration versus DNAPL saturation in sandy media for various chlorinated solvents. Note: *1,2-DCA* = 1,2 dichloroethane, *TCE* = trichloroethene, PCE = perchloroethene, *EDB* = ethylene dibromide.

of groundwater flow, however, a hydraulic gradient exists across the pool and the capillary pressure distribution is altered. The hydraulic gradient results in a higher capillary pressure at the downgradient end of the pool compared to the upgradient end of the pool. Without a sufficient capillary barrier against the downgradient end of the pool, the capillary pressure imbalance caused by the hydraulic gradient results in pool mobilization. This, in addition to gravity drainage, is the fundamental mechanism by which hydraulic displacement brings about DNAPL recovery.

Consider the DNAPL pool located in an inclined sand layer between two clay layers in Figure 8.7. The sand layer consists of the host sand within which the pool resides and a lower permeability (higher displacement pressure) sand adjacent to the down dip end of the pool that acts as a capillary barrier. The simplest expression relating the stable pool length (L) to the hydraulic gradient across the pool is given by:

$$\frac{\Delta\rho}{\rho_W} L \sin\alpha + \Delta h > \frac{P_c(L) - P_c(0)}{\rho_W g} \qquad \text{(Eq. 8.10)}$$

where $\Delta\rho$ is the difference between the DNAPL density (ρ_D) and water density (ρ_W), α is the dip of the bedding below horizontal, g is acceleration attributable to gravity, Δh is the difference in hydraulic head between the up dip end of the pool and the down dip end of the pool ($h(0) - h(L)$), $P_c(L)$ is the capillary pressure at the down dip end of the pool, and $P_c(0)$ is the capillary pressure at the up dip end of the pool. If the left-hand side of Equation 8.10 exceeds the right-hand side, the pool will move through the capillary barrier in the sand layer at location L. For a vertical system (dip = 90°) in the absence of groundwater flow ($\Delta h = 0$), Equation 8.10 reduces to Equation 8.1.

Example Calculation: Consider a DNAPL pool in a horizontal sand layer (dip = 0°) where the sand in which the DNAPL resides is characterized by a hydraulic conductivity of 1.0E-02 cm/s, an entry pressure of 549 Pa, and a terminal pressure of 274.5 Pa. Figure 8.8 presents a plot of the maximum stable pool length (L) versus the hydraulic gradient across the pool for two example

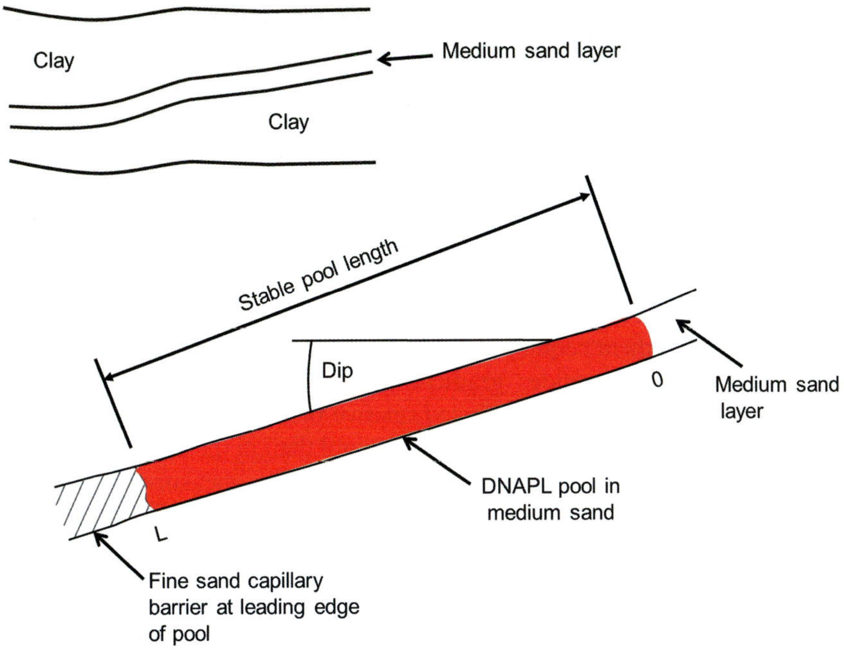

Figure 8.7. DNAPL pool located in dipping sand layer adjacent to capillary barrier.

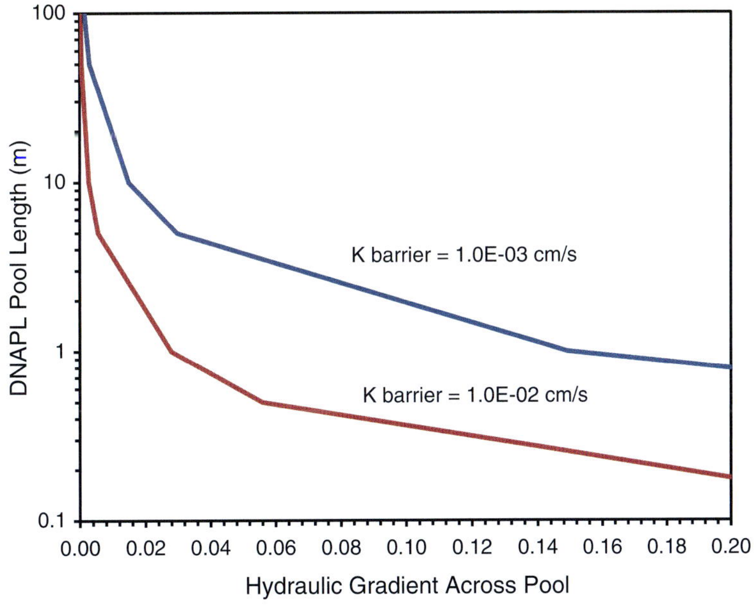

Figure 8.8. DNAPL pool length versus hydraulic gradient across pool in horizontally bedded porous media.

capillary barriers adjacent to the pool at location L. In this calculation, $P_c(L)$ is set to the entry pressure of the capillary barrier, and $P_c(0)$ is set to the terminal pressure. The lower curve in Figure 8.8 is for the case of a homogeneous sand layer where the capillary barrier has the same physical properties as the host sand. In this case, a hydraulic gradient of 0.05 would result in a pool length (L) of approximately 0.5 m. The upper curve assumes that the capillary barrier has an entry pressure of 1,736 Pa. In both cases, it is clear that small increases in the hydraulic gradient result in substantially smaller pool lengths for moderate-to-large initial pool lengths. The shape of the curves in Figure 8.8 demonstrates that it requires increasingly higher hydraulic gradients to mobilize shorter pools. Hydraulic displacement is therefore best suited for the removal of moderate- to large-size pools, not very short pools.

Equation 8.10 illustrates that the hydraulic gradient required to mobilize pooled DNAPL in horizontally bedded media is independent of the density and viscosity of the DNAPL. In dipping media, the DNAPL density is a factor. Equation 8.10 can also be used to calculate the hydraulic gradient required to arrest vertical or down slope DNAPL migration and to calculate the hydraulic gradient required to create up slope movement of DNAPL.

Equation 8.10 does not consider gradient magnification within the DNAPL pool. As discussed by Kueper et al. (2008), the reduced relative permeability to water within a DNAPL pool leads to steeper hydraulic gradients within the pool than in the surrounding media outside of the pool. As a result a relatively weak hydraulic gradient, as measured between piezometers located a certain distance away from the ends of the pool, actually represents a steeper hydraulic gradient within the pool. This effect is beneficial for the hydraulic displacement of DNAPL pools.

Of note is that Equation 8.10 and Figure 8.8 are intended to represent the hydraulic gradient required to mobilize a DNAPL pool that is not connected to a recovery well. In the case of a DNAPL pool that extends to a recovery well, gravity drainage will be a primary recovery mechanism. Any hydraulic gradient applied across such a pool will accelerate the rate of DNAPL recovery.

8.3 STATE OF THE PRACTICE

8.3.1 Evaluation and Implementation Process

Hydraulic displacement is routinely considered as a remedy component at sites having significant quantities of pooled DNAPL. The general steps taken to evaluate the feasibility of a hydraulic displacement remedy are as follows:

- Significant quantities of pooled DNAPL are established to be present at the site. Pooled DNAPL presence is determined by visual observation in soil cores, visual observation in monitoring wells, and soil concentrations exceeding a specified saturation threshold (Equation 8.8). The magnitude of groundwater concentrations, however, is not a good indicator of the presence of DNAPL at residual versus pooled saturations.
- The geology at the site includes sufficient capillary barriers to support DNAPL pools, consistent with seeing pooled DNAPL on the basis of visual observation and soil concentrations.
- The site history is consistent with the release of significant quantities of DNAPL to the subsurface.
- Screening calculations are performed to estimate DNAPL volume given site specific data.
- The remedial action objectives (RAOs) include one or more of the following needs: (1) recover DNAPL, (2) stabilize the source zone by reducing pooled DNAPL saturations to residual levels, and (3) increase the rate of mass transfer from DNAPL to groundwater.

- A hydraulic displacement pilot test is implemented. The pilot test may be designed using screening calculations and/or three-dimensional multiphase flow modeling. Screening calculations and multiphase flow modeling will require certain site-specific input parameters such as hydraulic conductivity, lateral extent of capillary barriers, displacement pressures (which can be correlated with hydraulic conductivity using Equation 8.2), and DNAPL properties such as viscosity, interfacial tension, and density.
- Full-scale design is performed using the results of the pilot test, site specific charac terization data, screening calculations, and three-dimensional numerical modeling.

An important decision in full-scale application of hydraulic displacement is whether or not active groundwater injection will take place. This will increase hydraulic gradients and thus intensify the ability to mobilize isolated pools and increase the rate of DNAPL recovery. If recovered groundwater can be reinjected into upgradient wells without treatment, then the cost of groundwater treatment will be limited, and active groundwater injection may be warranted. However, if the cost of groundwater treatment is significant, a passive approach involving the recovery of DNAPL only or the recovery of DNAPL and groundwater only (without injection) may be in order.

The primary design issues surrounding hydraulic displacement include the number and spacing of recovery and injection wells, whether to utilize vertical wells or horizontal drains, hydraulic gradient to be generated if injection wells or drains are included as part of the design, pump selection, and produced fluid treatment. It is typical to recover both DNAPL and water from recovery wells with a general decline in the oil to water ratio with time as the pools are depleted. Recovery wells can be equipped with separate pumps for the DNAPL and water phases. Performance monitoring typically includes measuring the following: water levels so that hydraulic gradients can be calculated, cumulative DNAPL recovery on a well-by-well basis, and groundwater injection and recovery rates.

8.3.2 Three-Dimensional Numerical Modeling of Hydraulic Displacement

8.3.2.1 Study Overview

Alexandra et al. (2012) present a comprehensive evaluation of hydraulic displacement in a representative sandy aquifer using three-dimensional multiphase flow modeling. This study examined the sensitivity of source zone mass removal and DNAPL distribution to (1) DNAPL type, (2) degree of porous media heterogeneity, (3) release volume, and (4) hydraulic gradient. The study utilized the three-dimensional two-phase flow reactive transport model DNAPL3D-RX (West et al., 2008). Other models are also available to simulate hydraulic displacement as discussed in Chapter 5 of this book.

DNAPL releases into a heterogeneous porous media domain were simulated prior to performing hydraulic displacement simulations. The model domain spanned 20.0 m (65.6 ft) in both lateral dimensions and 5.0 m (16.4 ft) in the vertical dimension. A nodal spacing of 0.4 m (1.3 ft) in the x- and z-directions and 0.05 m (0.16 ft) in the y-direction discretized the domain into 250,000 nodes. Each node was assigned a permeability value using a spatially correlated random field generator (FGEN) (Robin et al., 1993), thereby creating a heterogeneous, horizontally stratified deposit. The base case permeability field was log-normally distributed and characterized by a mean log-normal permeability (ln-k) (m^2) of -27.6 and a variance of ln-k of 2.0 (moderate heterogeneity). Correlation lengths of 3.0 m in the horizontal plane and 0.2 m for the vertical dimension were used to create horizontally stratified deposits with grain sizes ranging from silt to medium sand. The base case simulation release involved activating the source boundary condition until 7.57 m^3 (2,000 gal) of TCE DNAPL had entered the domain. The source boundary condition was then eliminated, and the infiltrated DNAPL was allowed to redistribute until migration effectively ceased. This distribution of DNAPL was then utilized as the initial condition for the hydraulic displacement simulation.

Hydraulic displacement was simulated employing a total of six fully penetrating, vertical wells: three for water injection in the upgradient portion of the domain and three for the extraction of both DNAPL and water in the downgradient portion of the domain. Table 8.1 lists the base case (Run 1) and 12 sensitivity simulations, each one modifying a single parameter from the base case. Table 8.2 presents relevant fluid properties. For each simulation, the parameters evaluated with time included (1) the volume of DNAPL recovered by the extraction wells, (2) the number of nodes invaded by DNAPL and the fractions exhibiting pooled and residual saturations (the ganglia to pool ratio; see Glossary), and (3) the distribution of DNAPL saturations.

Table 8.1. Schedule of Simulations (Alexandra et al., 2012)

Run #	DNAPL	Volume (m^3)[a]	Mean K (m/s)	Variance	Well Spacing (m)	Drawdown (m)
1	TCE	7.57	1×10^{-5}	2.00	6.4	2.0
2	TCE	1.89	1×10^{-5}	2.00	6.4	2.0
3	TCE	18.9	1×10^{-5}	2.00	6.4	2.0
4	TCA	7.57	1×10^{-5}	2.00	6.4	2.0
5	PCE	7.57	1×10^{-5}	2.00	6.4	2.0
6	TCE	7.57	1×10^{-4}	2.00	6.4	2.0
7	TCE	7.57	1×10^{-6}	2.00	6.4	2.0
8	TCE	7.57	1×10^{-5}	1.00	6.4	2.0
9	TCE	7.57	1×10^{-5}	4.00	6.4	2.0
10	TCE	7.57	1×10^{-5}	2.00	6.4	0.5
11	TCE	7.57	1×10^{-5}	2.00	6.4	1.0
12	TCE	7.57	1×10^{-5}	2.00	8.0	2.0
13	TCE	7.57	1×10^{-5}	2.00	3.6	2.0

[a]7.57 m^3 = 2,000 U.S. gallons; 1.89 m^3 = 500 U.S. gallons; 18.9 m^3 = 5,000 U.S. gallons
Notes: PCE – perchloroethene; TCA – 1,1,1 trichloroethane; TCE – trichloroethene

Table 8.2. Fluid Properties (Alexandra et al., 2012)

Compound	Density[a] (kg/m^3)	Aqueous Solubility[a] (mg/L)	Effective Diffusion Coefficient[b] (m^2/s)	Viscosity[a] (cP)
1,1,1-Trichloroethane	1,350	1,300	6.4E-10	0.84
Trichloroethene	1,460	1,100	6.6E-10	0.57
Perchloroethene	1,630	200	6.0E-10	0.89

[a]Pankow and Cherry (1996)
[b]Wilke and Change (1955), Lyman et al. (1982).

8.3.2.2 Results and Discussion

Figures 8.9 and 8.10 superimpose on the permeability field the spatial distribution of DNAPL saturations before and after hydraulic displacement, respectively, for the base case simulation. Visual inspection indicates that a substantial number of discrete DNAPL pools with moderate-to-high saturations (S_{NW} ranging from 0.3 to 0.7) are present in the subsurface prior to hydraulic displacement. In contrast, Figure 8.10 reveals that, with a few exceptions, the saturation of pools generally decreases following hydraulic displacement and that the remaining DNAPL is spread through a larger portion of the domain.

The DNAPL recovery results for all 13 simulations are presented in Figures 8.11, 8.12, and 8.13, with each curve providing the cumulative DNAPL volume recovered from all extraction wells. Characteristic of oil and DNAPL extraction data (Sale and Applegate, 1997), all curves exhibit an initial rapid recovery rate that decreases and eventually plateaus over time. This is because of the high relative permeability to DNAPL that exists at the initially high DNAPL saturations and the rapid decrease in DNAPL relative permeability as saturations decrease, respectively. Table 8.3 provides the initial DNAPL volume, volume recovered by hydraulic displacement, and percentage recovered for each simulation.

Figure 8.11 illustrates that DNAPL recovery rate and final extracted volume are highly sensitive to the original amount released. Moreover, the percentage of DNAPL recovered increases with the volume released. Only 9% of the 1.89 m^3 (500 gal) TCE release was recovered, compared to 31% of the 7.57 m^3 (2,000 gal) TCE release and 45% of the 18.9 m^3 (5,000 gal) TCE release; the 45% is an underestimate of the potential total because 200 days of hydraulic displacement was insufficient to reach the recovery plateau in this simulation. The final recovered volume of the lower density DNAPL (TCA) is approximately twice that for

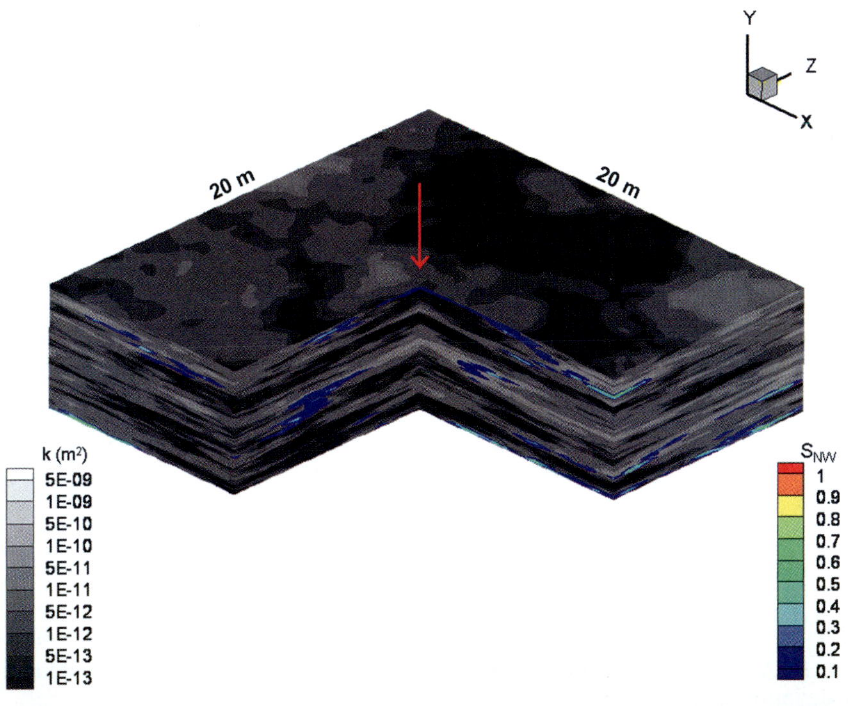

Figure 8.9. Base case DNAPL distribution before hydraulic displacement. *Grey scale* represents hydraulic conductivity field, with *darker shades* representing lower permeabilities. *Color scale* represents DNAPL saturation expressed as fraction of pore space (Alexandra et al., 2012).

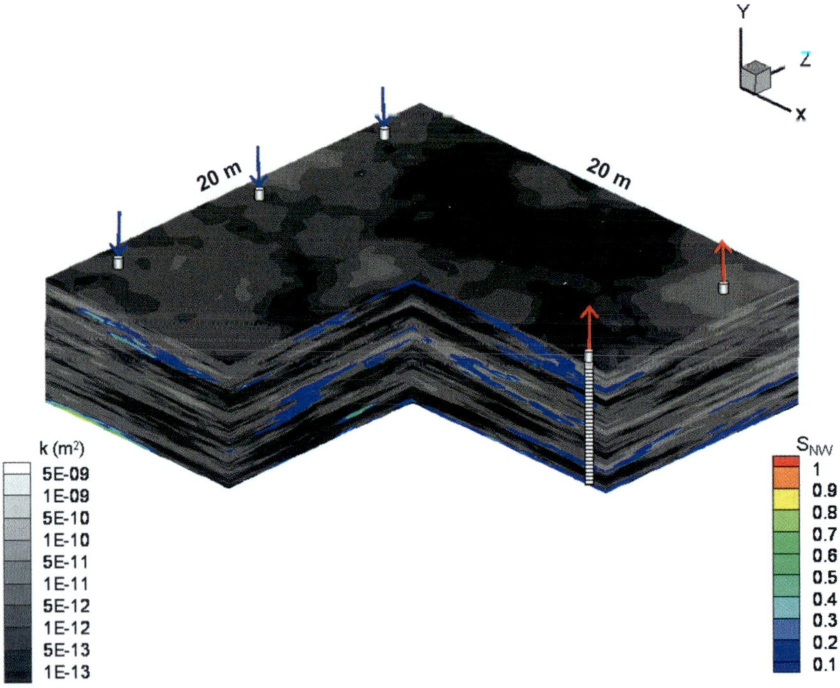

Figure 8.10. Base case DNAPL distribution after hydraulic displacement. *Grey scale* represents hydraulic conductivity field, with *darker shades* representing lower permeabilities. *Color scale* represents DNAPL saturation expressed as fraction of pore space (Alexandra et al., 2012).

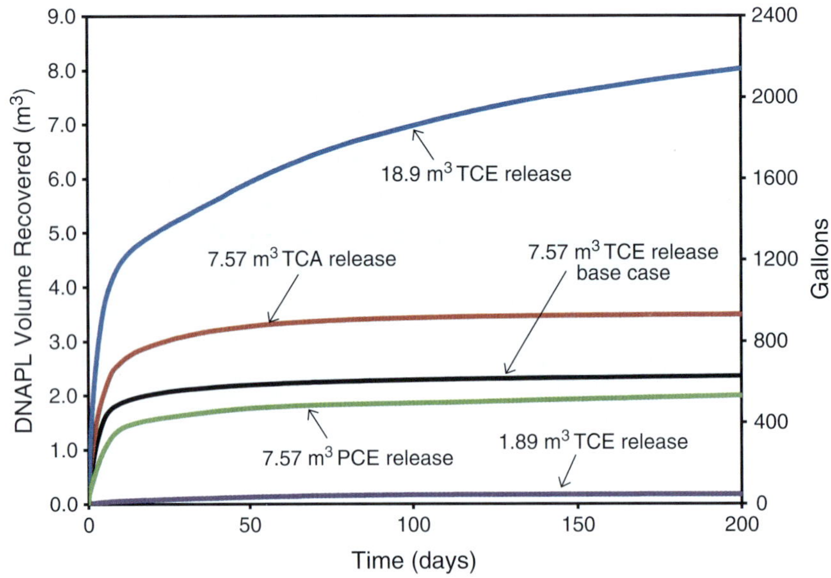

Figure 8.11. DNAPL volume recovery comparisons for DNAPL type and initial volume of DNAPL released (Alexandra et al., 2012).

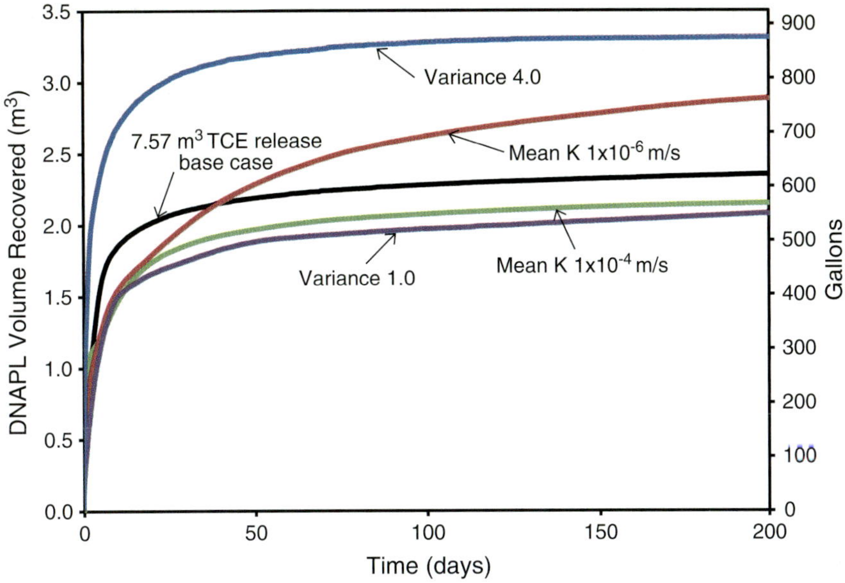

Figure 8.12. DNAPL volume recovery comparisons for different mean hydraulic conductivity and degree of heterogeneity (variance) (Alexandra et al., 2012).

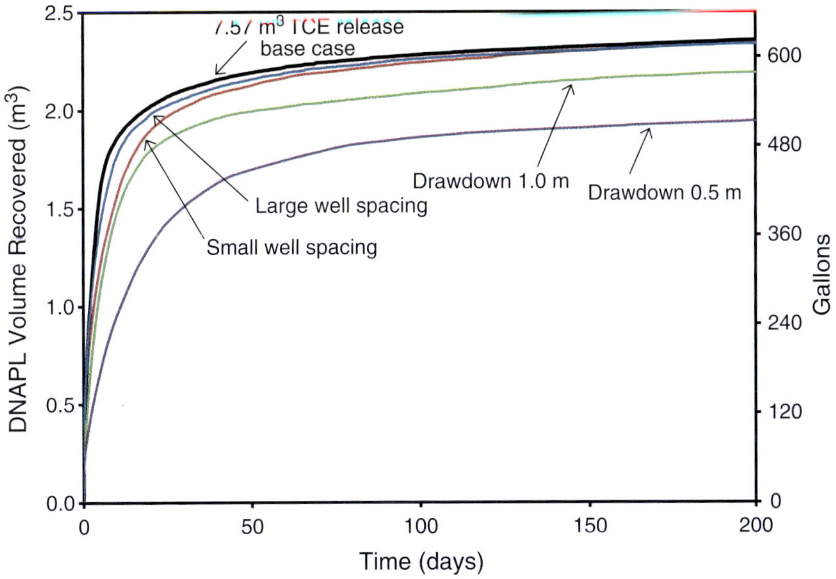

Figure 8.13. DNAPL volume recovery comparisons for varying drawdown in recovery wells and well spacing (Alexandra et al., 2012).

the same released volume of the higher density DNAPL (PCE), which can be attributed to the greater ability of lower density DNAPL to form pools above capillary barriers.

Figure 8.12 reveals that DNAPL recovery is sensitive to the mean and variance of hydraulic conductivity of the aquifer. Both lower mean hydraulic conductivity and higher variance of permeability correspond to increased final amounts of DNAPL recovered. These results contrast with general expectations that overall higher hydraulic conductivity and smaller variations in hydraulic conductivity (less heterogeneity) typically are considered beneficial for remedial flushing technologies (because of improved sweep efficiency). In the case of hydraulic displacement and the capture of DNAPL, however, the technology benefits from overall lower hydraulic conductivity and greater variations in hydraulic conductivity because these conditions provide the capillary barriers capable of supporting pooled DNAPL (lower pool fraction). A larger proportion of capillary barriers provides for more lateral DNAPL spreading, more formation of pools, accumulation of higher saturations within pools, and less vertical plunging of the infiltrating DNAPL prior to application of hydraulic displacement.

Figure 8.13 demonstrates that DNAPL recovery for the base case release scenario is less sensitive to injection/extraction well spacing than to the other parameters investigated. While some early sensitivity to well spacing is demonstrated, the final recovered volumes are not significantly different. This conclusion may be limited to this particular study, however, since it is expected that, in general, fewer wells would ultimately result in less DNAPL recovery. Figure 8.13 and Table 8.3 demonstrate, as expected, that decreases in extraction well drawdown resulted in reductions in the DNAPL volume recovered because of the weaker hydraulic gradients that were achieved.

Figure 8.14 presents the pre- and post-hydraulic displacement GTP (ganglia to pool) ratios for all simulations. In all cases, hydraulic displacement preferentially removes mass from the pools, resulting in substantial increases in the GTP ratio (reduction in the pool fraction). Since the objective of hydraulic displacement is primarily source stabilization, an increased GTP ratio is one measure of the treatment's success, because it indicates that DNAPL saturations are shifting from pools (potentially mobile) to residual (immobile). The greatest increases (GTP ratio between 0.50 and 0.70) are associated with those simulations exhibiting the highest initial DNAPL saturations and largest DNAPL recovery volumes. This is not surprising; those are the simulations in which DNAPL exhibits the highest mobility. As discussed by Alexandra et al. (2012), a strict definition of residual is employed; hence, Figure 8.14 may report GTP ratios that are low compared to other studies employing this metric.

Table 8.3. DNAPL Recovery by Hydraulic Displacement (Alexandra et al., 2012)

Run #	Case	Initial Volume (m^3)	Volume Recovered (m^3)	Fraction Recovered (%)
1	2,000 gal. TCE (base case)	7.63	2.36	31.0
2	500 gal. TCE	1.89	0.18	9.4
3	5,000 gal. TCE	18.83	8.50	45.2
4	2,000 gal. TCA	7.73	3.49	45.1
5	2,000 gal. PCE	7.69	2.02	26.3
6	High mean (k) TCE	7.57	2.15	28.4
7	Low mean (k) TCE	7.57	2.89	38.1
8	Low var (k) TCE	7.57	2.08	27.5
9	High var (k) TCE	7.57	3.31	43.7
10	Very low drawdown TCE	7.57	1.94	25.6
11	Low drawdown TCE	7.57	2.19	28.9
12	Large well spacing TCE	7.57	2.33	30.8
13	Small well spacing TCE	7.57	2.34	30.9

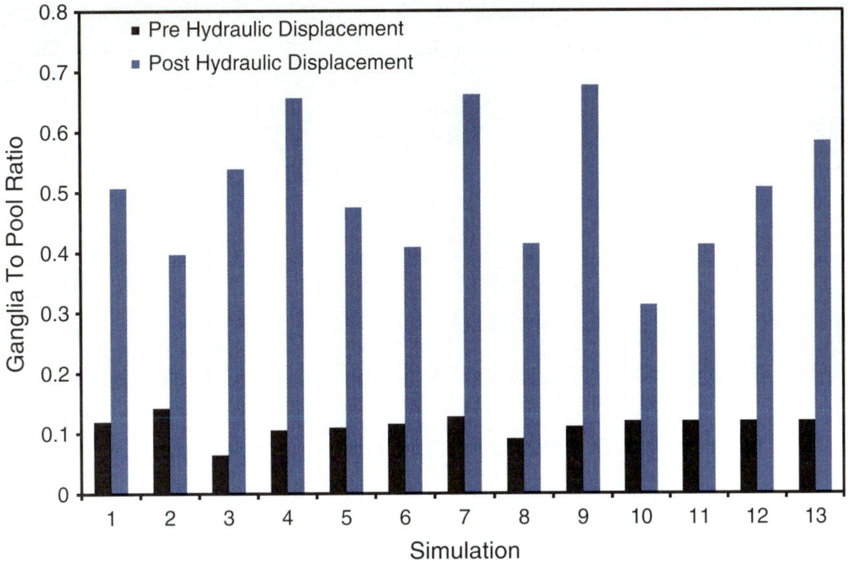

Figure 8.14. Ganglia to pool ratio before and after hydraulic displacement (Alexandra et al., 2012).

8.4 CASE HISTORIES

8.4.1 Chlorinated Solvent Hydraulic Displacement Remedy Design

Hydraulic displacement was chosen as the primary remediation approach for a chemical manufacturing facility in the southern United States. Site investigation revealed a laterally extensive pool of chlorinated solvent DNAPL perched upon a clay layer situated, on average, 5 m (16 ft) below the water table. The DNAPL pool, residing in a relatively homogeneous silty sand fill deposit, exhibited a variable thickness of up to 4.2 m (13.8 ft) and comprised a total volume of approximately 7,500 m^3 (~47,000 drums). The primary objectives of the proposed hydraulic displacement remedy were to (1) rapidly recover significant volumes of DNAPL and (2) reduce the effective pool height to stabilize the pool and thereby reduce the potential for subsequent pool mobilization. This latter objective was important in this instance because a main drinking water aquifer for the region was situated below the clay aquitard. Numerical modeling was used to design and optimize the horizontal drain dual phase recovery system prior to implementation. The objectives of the optimization process included the following: (1) maximize the DNAPL volume recovered per unit length of installed drain; (2) minimize the volume of water recovered from the subsurface, which required treatment; and (3) reduce operation and maintenance costs of a system expected to operate for years up to a decade. Gerhard et al. (2001) provide a thorough account of the methodology and results of this study, and the key details and results are highlighted here.

A detailed site investigation provided information on the topography of the upper surface of the clay aquitard and the varying thickness of the DNAPL pool perched above it. A three-dimensional numerical simulation domain was created that encompassed the entire DNAPL pool in both lateral and vertical extents. The domain was 224 m wide, 301 m long, and 5 m high (730 × 990 × 16 ft). The domain was discretized into 110,080 nodes, each 3.5 m (11.5 ft) in both lateral directions and 0.25 m (0.8 ft) vertically. A total of 47 soil cores were obtained on site, each providing a measure of the depth to the fill/clay interface, and these results were interpolated to provide a continuous mapping of the undulating aquitard surface that

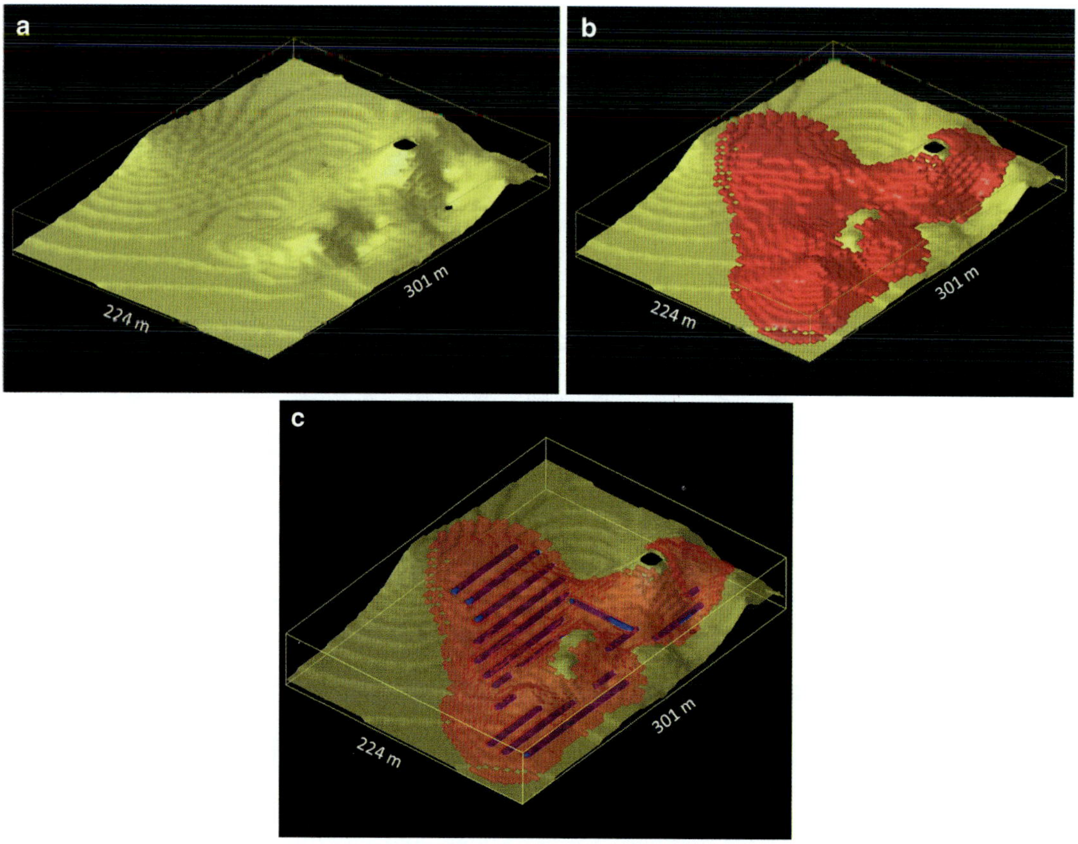

Figure 8.15. (a) Top surface of the clay aquitard in the model domain. (b) DNAPL pool's outer surface (*red*) overlying the top surface of the clay aquitard (*yellow*) in the model domain (Gerhard et al., 2001). (c) Location of the screened portions of the horizontal drains (*blue*) for the base case hydraulic displacement system. Also pictured is the DNAPL pool's outer surface (*transparent red*) overlying the top surface of the clay aquitard (*yellow*) in the model domain. Drain 15, oriented perpendicular to the rest, represents the pilot test drain installed at the site.

supported the DNAPL pool. Figure 8.15a illustrates that the clay surface contained numerous hills and valleys, varying up to 6.6 m (22 ft) vertically. Measurements of the DNAPL/water interface in 57 monitoring wells provided estimates of DNAPL pool thickness across the site. The interpolated pool provided as the starting condition in the numerical domain is presented in Figure 8.15b, illustrating significant variability in thickness ranging from 0 to 4.2 m (13.8 ft). The model was able to provide an estimate of DNAPL saturations and DNAPL pool volume assuming that the pool was at hydrostatic equilibrium. Of the estimated 7,500 m^3 (1.9 million gallons) DNAPL in the pool, 6,150 m^3 (1.6 million gallons) (82%) was predicted to be potentially mobile (not residual) and thus the upper limit of what could potentially be recovered by hydraulic displacement.

The base case system design incorporated 15 horizontal drains, the screened portions of which are plotted in Figure 8.15c so as to be visible within the DNAPL pool. A relatively tight drain spacing of 14 m (46 ft) was used, and the drains were located within the fill immediately above the clay to maximize gravity drainage of DNAPL during recovery. This design decision was based on a preliminary sensitivity modeling study that investigated the influence of drain spacing and water injection in a simplified model of the site conditions. This study revealed that, for the conditions at this site (including a relatively dense DNAPL), maximizing DNAPL recovery and minimizing water intake resulted from minimizing drain spacing and using all drains for withdrawal instead of using some for water injection (Gerhard et al., 2001). The base case drain layout aimed to ensure adequate coverage of the entire DNAPL pool and also to maximize pool height above the drains while accounting for the undulating topography of the clay surface. Therefore, each drain was located to balance pool depth against uninterrupted screened length in the fill while not extending screens beyond the pool extremities. Figure 8.16 presents the base case drain configuration (and the number identifier assigned to each drain) plotted relative to the DNAPL pool thickness and the topography of the clay surface.

Drain 15 (shown in Figure 8.15c trending perpendicular to the rest of the drains) was installed on site as a pilot test drain. This horizontal drain with a 55 m (180.4 ft) long screen was employed in a two-stage pilot test in which it was actively pumped for days 0 to15 and then again from days 29 to 42. The water level observed in the drain is plotted in Figure 8.17. Figure 8.18 presents the cumulative DNAPL recovery versus time, showing that approximately 23 m^3 of chlorinated solvent DNAPL was recovered during the pilot test. An almost equal volume of contaminated water was simultaneously removed during the pilot test.

The numerical model used in design was DNAPL3D (Gerhard et al., 1998), a version of the code used in the previous example and discussed in more detail in Chapter 5. The pilot test data was used to calibrate the numerical model for site specific conditions. In particular, it was used to establish an appropriate, homogeneous fill permeability value to use in the design simulations.

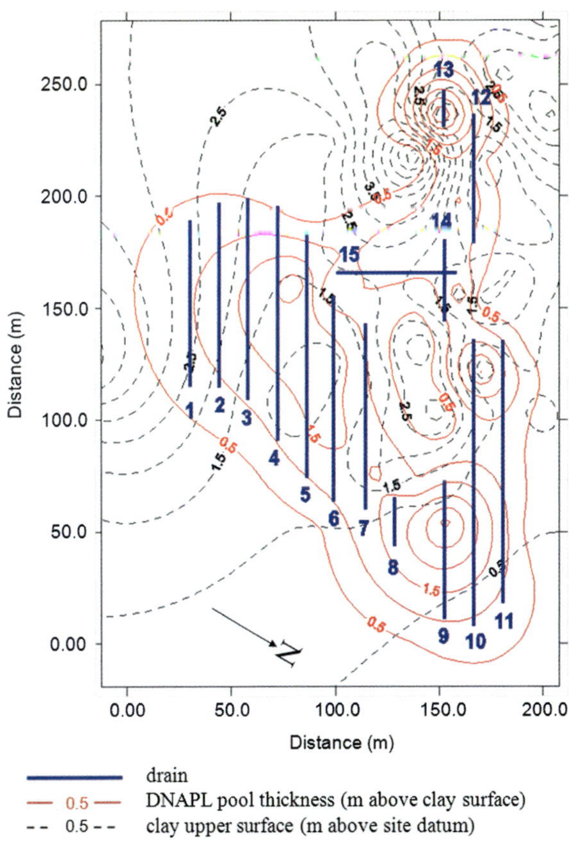

Figure 8.16. Base case drain locations and drain identification numbers (*blue*) overlying contours representing the elevation of the clay aquitard surface (*dashed*) and the DNAPL pool thickness (*red*) (Gerhard et al., 2001).

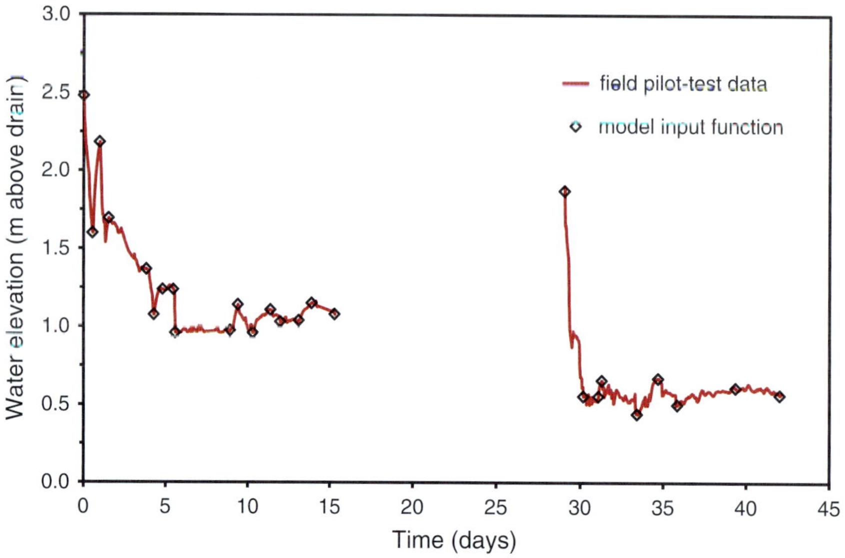

Figure 8.17. Observed water level at recovery drain during single-well pilot test (*red*) and straight line function employed as corresponding input to the numerical model simulation of the pilot test (*blue*) (Gerhard et al., 2001).

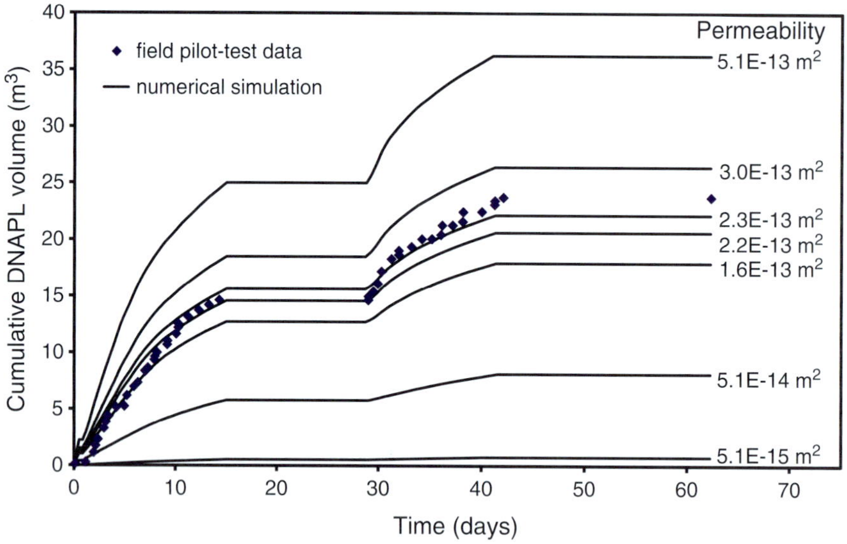

Figure 8.18. Cumulative DNAPL volume recovered for seven simulations of the pilot test, each employing a different value of intrinsic permeability for the fill unit (*black*) and the observed results from the actual pilot test (*blue*) (Gerhard et al., 2001).

Figure 8.17 provides the function of water elevation versus time that was interpolated from the observed data; this function was used as the prescribed drawdown for the single recovery drain in the simulated pilot test. Table 8.4 provides the key fluid and soil property values used in all of the simulations. The DNAPL was assigned fluid properties determined from laboratory measurements on six samples obtained from the site. Gerhard et al. (2001) provide full details of the model calibration simulations.

Figure 8.18 presents the pilot test modeling results, illustrating that the recovered DNAPL volume over the 28 days of pumping is predicted to be in the range of 1 to 37 m^3 when the fill

Table 8.4. Soil and Fluid Parameters Used in Modeling

Description	Value
Porosity (θ)	0.30
Water density (ρ_W)	1,000 kg/m^3
Water viscosity	0.001 Pa · s
DNAPL density (ρ_D)	1,460 kg/m^3
DNAPL viscosity	0.00628 Pa · s
DNAPL–water interfacial tension	0.0221 N/m
Pore size distribution index (λ)	2.5
Residual water saturation ($S_{w,r}$)	0.15
Maximum DNAPL residual ($S_{D,r,max}$)	0.18
Terminal pressure (P_T)	10 Pa

permeability brackets the range 5.1×10^{-15} to 5.1×10^{-13} m^2 (hydraulic conductivity of 0.004 to 0.4 m/day). The figure demonstrates that employing an intrinsic permeability value of 2.3×10^{-13} m^2 provided a good fit to the actual DNAPL recovery pilot test data. This result provided confidence that the model was able to adequately predict site specific DNAPL recovery with time. The calibrated intrinsic permeability value was adopted for the design simulations. Additional sensitivity analyses demonstrated that uncertainty in other key model parameters (residual water saturation, correlated displacement pressure, pore size distribution, and porosity) had minor impact on the predicted DNAPL recovery volume (Gerhard et al., 2001). Hence, calibrating the model via intrinsic permeability was a valuable way to minimize uncertainty in the predicted performance of the full-scale system.

Table 8.5 presents the suite of six full-scale design simulations conducted to optimize the remedial action objectives. The base case (Run 1) employed all 15 horizontal drains shown in Figures 8.15c and 8.16, totaling 1,106.5 m (3,630 ft) of screened length. The drains were simulated as equivalent to a free exit boundary to both water and DNAPL that was residing within a 0.5 m (1.6 ft) wide coarse gravel pack. When in operation, the drains were simulated as maintaining a constant water pressure equivalent to 1.125 m (3.7 ft) drawdown relative to hydrostatic conditions. All simulations were conducted for 3,000 days (8.2 years).

Figure 8.19 provides the cumulative DNAPL volume recovered versus time for each of the 15 drains in Run 1 (base case). As expected, each drain recovers DNAPL at a high rate initially and the rate decreases with time; this is similar to what is observed in primary extraction curves for petroleum reservoirs. Here this is due to decreasing DNAPL pool thickness and corresponding decreasing relative permeability to DNAPL flow as DNAPL saturations near the drains decrease. The figure reveals that the initial DNAPL recovery rate and the total volume recovered after eight years can vary significantly between drains. Drain position within the pool is the key factor, with those beneath the largest DNAPL accumulations for a substantial screened length (drains 5, 9, and 10) providing the most DNAPL. This occurs because DNAPL recovery by gravity drainage dominates at this site. In addition, the deceleration of the recovery rate is not uniform, with some drains (8, 11, 12, 13, and 14) not yet approaching a horizontal asymptote even after 8 years of pumping. These drains are not flanked by nearby drains which compete for DNAPL, allowing the overlying pool to deplete more slowly and thus to sustain higher recovery rates.

Table 8.5. Hydraulic Displacement System Drain Configurations (Gerhard et al., 2001)

System	Number of Drains	Drains Included (#)	Total Drain Length (m)	Pulsed Pumping	Notes
Run 1	15	1–15	1, 106.5	No	Base case
Run 2	13	3–15	951.0	No	Remove drains 1, 2
Run 3	12	3–14	896.0	No	Also remove drain 15
Run 4	10	3–6, 8–13	780.0	No	Also remove drains 7, 14
Run 5A	9	3–6, 8–10, 12, 13	664.0	No	Also remove drain 11
Run 5B	9	3–6, 8–10, 12, 13	664.0	Yes	Even/odd pulsed 180-day cycle

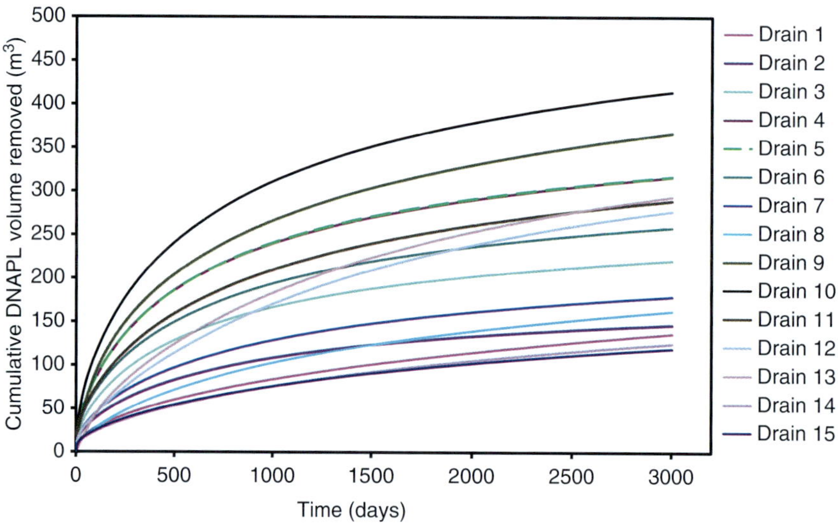

Figure 8.19. Cumulative DNAPL volume recovered for each drain, Run 1 (Gerhard et al., 2001).

Figure 8.20 illustrates that, for the system as a whole, 59% of the initial mobile pool volume (3,600 m^3) was predicted to be recovered in Run 1 after 8 years of continuous operation. The cumulative DNAPL recovery curve for the system exhibits the same shape as the cumulative DNAPL recovery curve for each well, and, although it indicates that the recovery rate is decreasing continually, it has not yet reached a plateau after 8 years. In operating the real system, cost/benefit analysis would have to be employed to determine the optimal time to terminate pumping.

Since installing each additional drain was a dominant capital cost, it was advantageous to consider eliminating drains and minimizing total drain length to make the system more cost effective. However, this needed to be accomplished in accordance with the overall objectives of maximizing DNAPL recovery and pool height reduction. Runs 2, 3, 4, and 5A represent systematic reductions in total system drain length by sequentially eliminating drains (Table 8.5) with the poorest recovery efficiency. Figure 8.20 plots the DNAPL recovery efficiency for each drain, where recovery efficiency is defined as the cumulative DNAPL recovered after 3,000 days in a drain divided by its screened length. For example, Run 2 used the same design as Run 1 except that drains 1 and 2 were eliminated because they exhibited the lowest recovery efficiency values. This is not surprising because they are long drains located at the extremity of the DNAPL pool

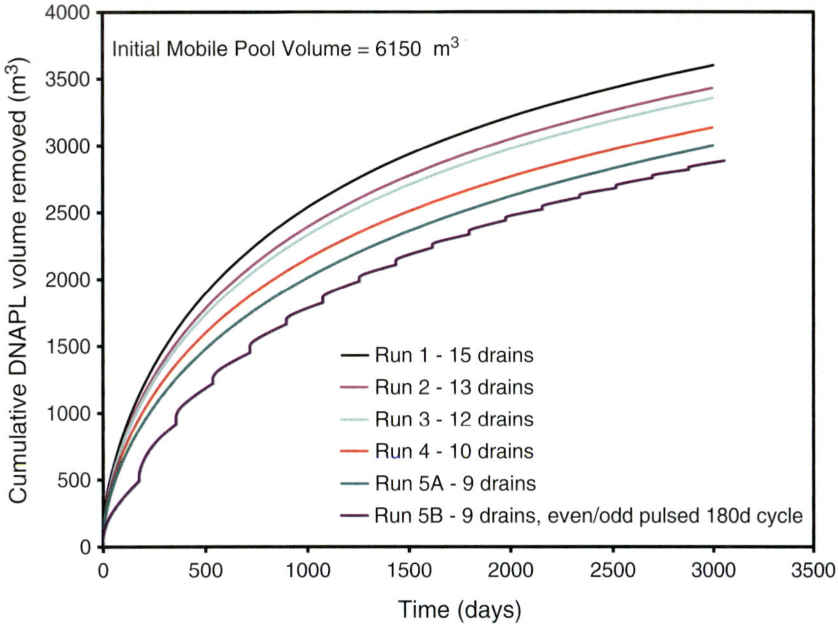

Figure 8.20. Cumulative DNAPL volume recovered versus time for the entire system, all runs (Gerhard et al., 2001).

on a sloping face of the underlying clay (Figures 8.15c and 8.16). Figure 8.20 illustrates that, as expected, Run 2 recovers less DNAPL overall relative to Run 1 (3,430 m^3 versus 3,600 m^3). However, this is only a 5% DNAPL reduction while obtaining the benefit of reducing total drain length by 15%. The explanation, as observed in Figure 8.21, is that drain 3 increases its recovery efficiency by 46% over Run 1 due to the increased amount of DNAPL available when its neighboring drains were eliminated.

Figure 8.22 plots the total amount of DNAPL recovered and the overall system recovery efficiency for each of the simulations. System recovery efficiency is defined as the total volume recovered by all the drains divided by the total system drain length. The figure demonstrates that the system continuously increases in overall DNAPL recovery efficiency as the least efficient drains are removed from Run 1 through Run 5A. This occurs because both the least efficient drains are removed and the remaining drains adjacent to the abandoned locations increase their efficiency, thereby reducing the negative impact on DNAPL recovery of removing those drains. In Run 5A the system has been reduced to the nine most efficient drains for DNAPL recovery, exhibiting a decrease in total drain length of 40% relative to the base case at the cost of decreasing DNAPL recovery by only 20% after 8 years.

Minimizing water recovery was also identified as one of the main objectives of the system design, since all water brought to the surface would require treatment. Figure 8.23 presents the cumulative amount of water pumped from each drain for the base case simulation. It indicates that water recovery curves exhibit a distinct shape to the DNAPL recovery curves (compare to Figure 8.19). The rate of water recovery in each drain begins at a small value and increases with time as the relative permeability to water increases around each drain because of decreasing DNAPL saturations and decreasing pool height. The amount of water recovered varies dramatically between drains and is not strongly correlated to the amount of DNAPL recovered. Drain 11 is predicted to remove the greatest amount of water, a result associated with its length (116 m) [380 ft] and its proximity to the edge of the DNAPL pool along its entire screen. In Run 4, Drain 11 exhibits the lowest DNAPL recovery efficiency; it is also responsible for 40%

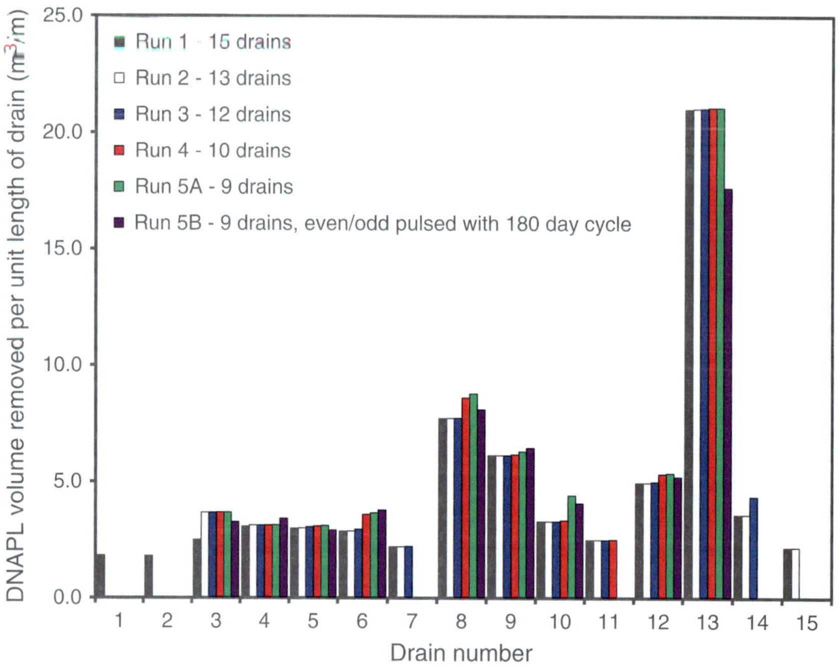

Figure 8.21. Drain recovery efficiency for individual drains, all runs after 8.2 years (Gerhard et al., 2001).

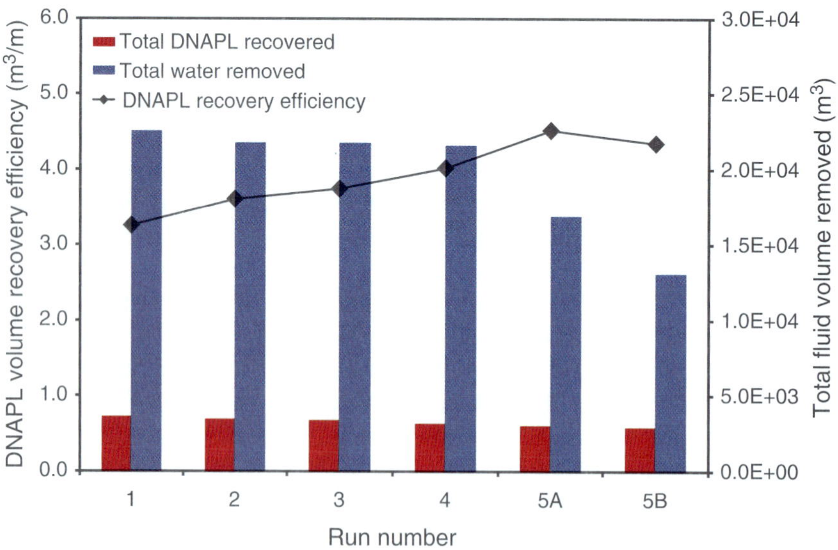

Figure 8.22. System performance after 8.2 years of hydraulic displacement for all runs. System DNAPL recovery efficiency (*black*; *left vertical axis*), total DNAPL volume recovered (*blue*; *right vertical axis*), and total water volume removed (*blue*; *right vertical axis*) (Gerhard et al., 2001).

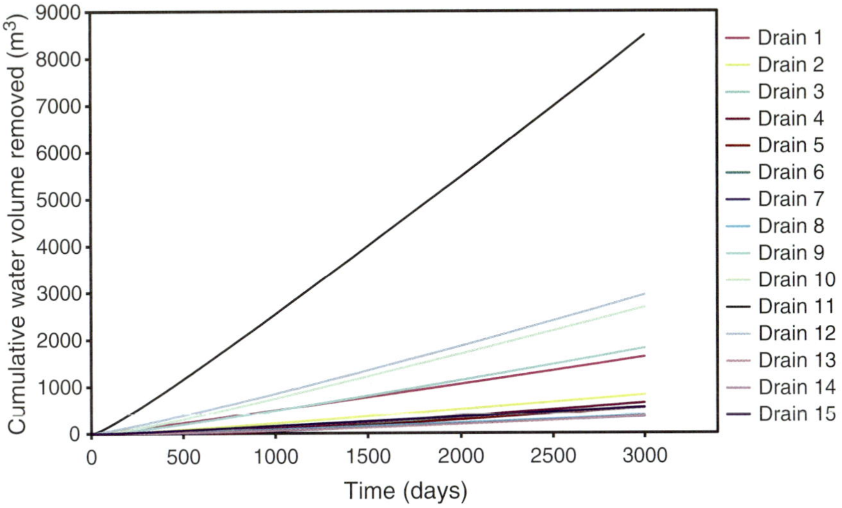

Figure 8.23. Cumulative water volume removed by each drain for Run 1 (Gerhard et al., 2001).

of the total water removed by the ten-drain system, and thus it is removed for Run 5A. Figure 8.22 reveals that this not only results in a significant increase in overall system efficiency with respect to DNAPL recovery, but also that the total amount of water removed is predicted to drop from 21,560 to 16,850 m^3.

A second strategy considered for improving system optimization was operating the drains in a pulsed pumping mode. Run 5B used the same nine-drain configuration as Run 5A but operated all drains in a cycle: even-numbered drains for 180 days followed by odd-numbered drains for 180 days. One advantage of this approach is that each drain is operated for only half of the 8-year treatment period, which would translate into significant operation and maintenance cost savings. Figure 8.20 illustrates that the cumulative DNAPL recovery curve is significantly less for Run 5B than for 5A at the beginning but approaches the 5A total near the end of the treatment. This is because the system experiences an increased rate of DNAPL recovery each time a set of drains is activated, corresponding to removal of the DNAPL that has accumulated in the quiescent drain plus the increased DNAPL flow rates that result from DNAPL saturations increasing during each pump-off period. This has a dramatic effect; despite only operating the system for half of the 3,000 days, the amount of DNAPL removed is reduced by only 4% after 8 years (from 3,000 to 2,890 m^3). Figure 8.22 shows that this reduces the system DNAPL recovery efficiency by only a small amount (4.5 to 4.3 m^3/m). Figure 8.24, presenting the cumulative DNAPL and water recovery histories for Runs 5A and 5B, illustrates that pulsed pumping has a dominant influence on the amount of water recovered by the system, with a reduction after 8 years of 33% (16,850 to 13,060 m^3).

Apart from these considerations of system efficiency and optimization with respect to cost/benefit, it is appropriate to ask whether the system design achieves its primary objectives of reducing DNAPL saturations and mobile pool height so that the potential for remobilization is minimized. Figure 8.25 presents the distribution of DNAPL saturations for the pool prior to hydraulic displacement, revealing that it is dominated by the high saturations associated with significant pool heights (highest fraction of locations exhibit 70 to 80% DNAPL). The figure illustrates that, after 8 years, hydraulic displacement is predicted to result in a dramatic shift towards low-to-residual saturations (highest fraction of locations exhibit 10 to 20% DNAPL).

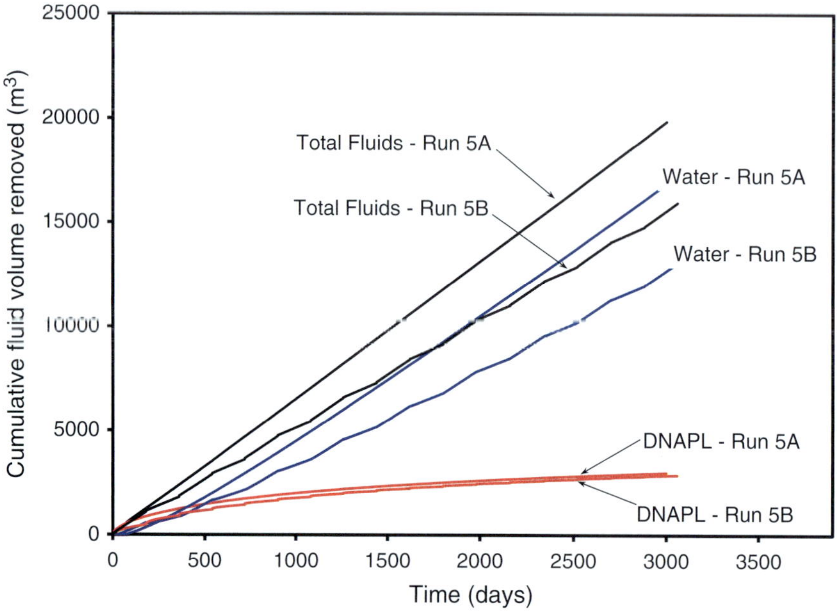

Figure 8.24. Effect of pulsed pumping on the cumulative volumes removed of DNAPL (*red*), water (*blue*) and their sum (*black*) (Gerhard et al., 2001).

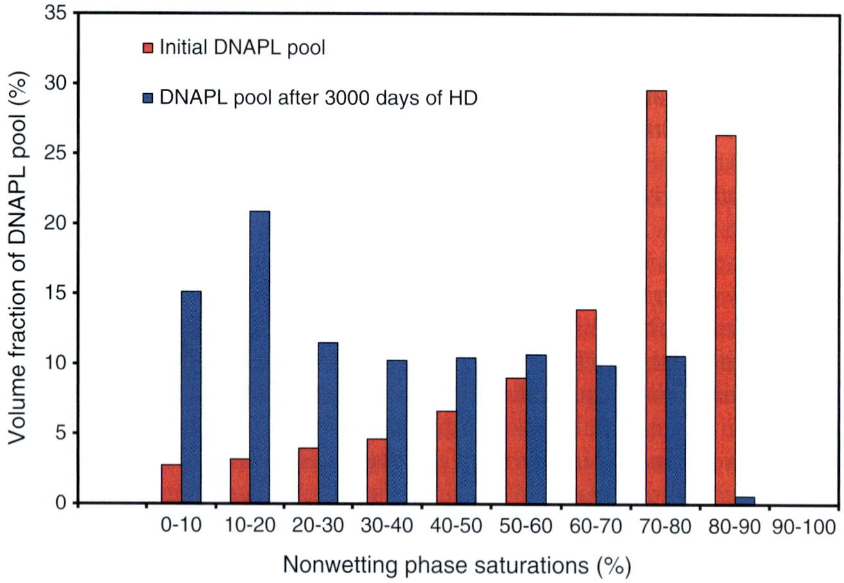

Figure 8.25. Distribution of DNAPL saturations before (*red*) and after (*blue*) 8 years of hydraulic displacement for the final design scenario, Run 5B (Gerhard et al., 2001).

Figure 8.26a plots the change in the distribution of DNAPL saturations above 70%, revealing a dramatic decrease after hydraulic displacement in areas where the DNAPL pool was initially most thick. Figure 8.26b plots the change in the distribution of DNAPL saturations below 20%, showing a corresponding increase in the same areas. All six of the system configurations assessed in this study produced similar results in these respects; in all cases,

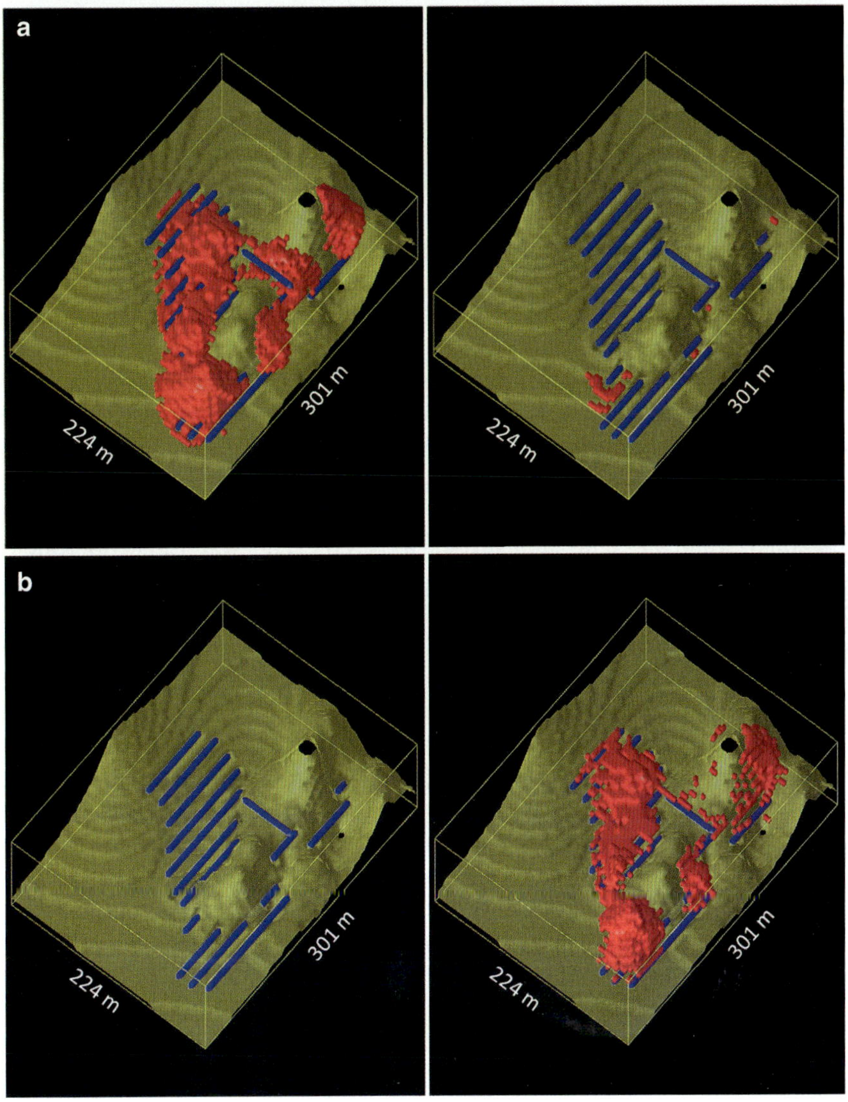

Figure 8.26. (a) Distribution of DNAPL at saturations exceeding 70% prior to hydraulic displacement (*left*) and after 8 years of hydraulic displacement (*right*), (b) distribution of DNAPL at saturations below 20% prior to hydraulic displacement (*left*) and after 8 years of hydraulic displacement (*right*).

the effective pool heights are substantially reduced, with high saturations mostly eliminated in the lower half of the pool and low-to-residual saturations generated in the upper half of the pool.

8.4.2 Creosote DNAPL Hydraulic Displacement Performance

The following case history is taken from Sale and Applegate (1997). Although the DNAPL of interest was creosote, not a chlorinated solvent, the performance data illustrate many fundamental aspects of expected DNAPL recovery during implementation of a hydraulic displacement remedy. The hydraulic displacement remedy discussed here was implemented at a former creosote-based wood treating facility located in Laramie, Wyoming. The facility treated railroad ties from the mid-1930s through to the early 1980s. The creosote was characterized with a density of 1,030 kg/m^3 and a viscosity of 54 centipoise (cP) at 20°C. The geology at the site consists of braided stream deposits between 8 (2.4 m) and 10 ft (3.0 m) thick that generally coarsen downwards, underlain by bedrock. The hydraulic conductivity at the base of the overburden is approximately 1.8×10^{-1} cm/s (500 ft/day). The pooled creosote was located primarily in the bottom 0.6 m (2 ft) of the overburden deposit over an area of approximately 80 acres.

The chosen design consisted of using vertically stacked horizontal drain lines (horizontal wells) to recover separately DNAPL from the lower drains and water from the upper drains (Figure 8.27). By pumping water separately from the DNAPL, the hydraulic gradient imposed across the DNAPL pools can be controlled independent of the rate of DNAPL recovery. The hydraulic gradient was further controlled by injecting water into lower drain lines located between the DNAPL recovery drains. DNAPL recovery and water injection drains were spaced approximately 200 ft (60.9 m) apart. The overall 80 acre treatment area was subdivided into several units so that the hydraulic displacement remedy could be applied incrementally.

Figure 8.28 presents a plot of cumulative DNAPL recovery and the rate of water injection as a function of time for Unit 7. The left-hand vertical axis represents cumulative DNAPL recovery (gallons), while the right-hand vertical axis represents water injection rate (gallons per minute [gpm]). The plot of cumulative DNAPL recovered versus time shows a typical response to hydraulic displacement with the highest rates of recovery (slope of the cumulative DNAPL volume recovered versus time plot) in the beginning followed by a continual decrease in the rate of DNAPL recovery. DNAPL recovery rates are the highest at the beginning because of high relative permeability to DNAPL stemming from high DNAPL saturations. As the DNAPL saturations within the pools are decreased, the DNAPL relative permeability decreases and the rate of DNAPL recovery drops. Figure 8.28 illustrates that approximately 50,000 gallons (189,270 L) of DNAPL were recovered after 150 days, after which very little additional DNAPL was recovered.

Figure 8.28 illustrates that the water injection rate was reduced to zero after approximately 310 days of operation. As discussed above, very little additional DNAPL was recovered after about 150 days of operation. Reducing the water injection rate to zero (no water injection) is referred to as a shutdown test. As shown in Figure 8.28, water injection was shut down about day 310 and resumed on approximately day 370. The purpose of the shutdown test is to allow any remaining potentially recoverable DNAPL to accumulate around DNAPL recovery drains by gravity drainage without the expense of operating the water injection system.

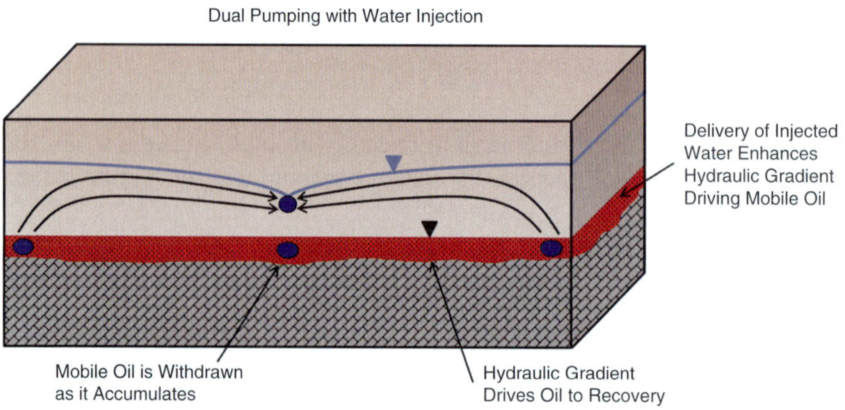

Figure 8.27. Schematic illustration of vertically stacked recovery drains and lower water injection drains (after Sale and Applegate, 1997).

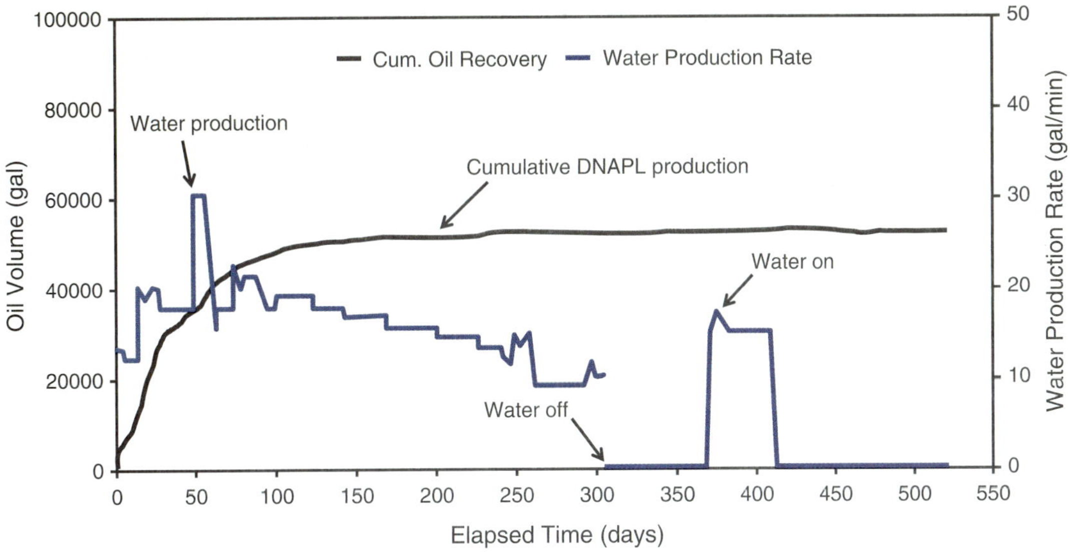

Figure 8.28. Cumulative DNAPL and water injection rate versus time for Unit 7 (after Sale and Applegate, 1997).

Figure 8.28 illustrates that very little additional DNAPL was recovered after resuming water injection on day 370, indicating that DNAPL saturations were successfully reduced to residual saturations.

Also of interest in Figure 8.28 is the change in water injection rate that occurred on day 50 of operation. Prior to day 50, the water injection rate was approximately 20 gpm (76 L/min). On day 50 the water injection rate was increased by 50% to 30 gpm (114 L/min), resulting in an increased rate of DNAPL recovery. The increased rate of DNAPL recovery is demonstrated by the change in slope of the cumulative DNAPL recovery versus time curve at day 50. This illustrates the fundamental relationship between DNAPL recovery rate and groundwater hydraulic gradient. Higher groundwater hydraulic gradients lead to greater manipulation of capillary pressures and a resulting greater driving force for DNAPL recovery. As discussed earlier in this chapter, water injection provides a more aggressive hydraulic displacement remedy, but the cost of water injection (and the treatment of additional produced water) must be considered.

8.5 SUMMARY

Hydraulic displacement, also referred to as waterflooding and dual phase extraction, is a means of recovering pooled DNAPL from the subsurface. Hydraulic displacement can be implemented in a variety of ways, ranging from pumping DNAPL only from a single extraction well or drain to pumping both DNAPL and water from numerous wells or drains in conjunction with upgradient groundwater injection to accelerate the rate of DNAPL recovery. Hydraulic displacement is capable of recovering significant quantities of pooled DNAPL from a source zone but is not capable of recovering residual DNAPL. As illustrated and discussed in this chapter, hydraulic displacement may need to be performed for several years depending on well spacing, DNAPL properties, volume of DNAPL present, and hydraulic conductivity of the porous medium.

In addition to performing as a DNAPL mass removal technology, hydraulic displacement can be used to stabilize a source zone. Hydraulic displacement can remove pooled and potentially mobile DNAPL, leaving behind a more stable source zone characterized by a significantly increased volume fraction as residual DNAPL. Hydraulic displacement results in an increase in DNAPL–water interfacial area available for mass transfer. It also results in a greater permeability to water through the DNAPL source zone, increasing water flux for a given hydraulic gradient. All of these effects can be beneficial for the subsequent application of mass transfer- and flushing-based technologies such as surfactant flushing, cosolvent flushing, ISCO, and enhanced *in situ* bioremediation. In this respect, hydraulic displacement can act as a preconditioning technology as part of a combined remedy approach. This strategy may be particularly advantageous because it is usually less costly to remove DNAPL pools using hydraulic displacement than with thermal, chemical, or biological remediation approaches.

Hydraulic displacement should only be considered for sites exhibiting significant quantities of pooled DNAPL. The existence of pooled DNAPL can be established on the basis of visual observation in core, visual observation in monitoring wells, and magnitude of soil concentrations. A variety of screening calculations can be performed to estimate DNAPL pool heights, DNAPL pool volumes, and hydraulic gradient required to mobilize a pool.

The evaluation and design of a hydraulic displacement remedy include a number of steps: (1) establishing that significant quantities of pooled DNAPL are present at the site, (2) determining that the geology at the site includes sufficient capillary barriers to support DNAPL pools, (3) verifying that the site history is consistent with the release of significant quantities of DNAPL to the subsurface, (4) performing screening calculations to estimate the DNAPL volume present, (5) verifying that the remedial action objectives (RAOs) for the site will benefit from the removal of pooled DNAPL, (6) executing a hydraulic displacement pilot test, and (7) implementing a full-scale design using the results of the pilot test, site specific characterization data, screening calculations, and three-dimensional numerical modeling.

REFERENCES

Alexandra R, Gerhard JI, Kueper BH. 2012. Hydraulic displacement of dense, non-aqueous phase liquids. Ground Water 50:765–774.

Brooks RH, Corey AT. 1966. Properties of porous media affecting fluid flow. Journal of the Irrigation and Drainage Division: Proceedings, American Society of Civil Engineers 92 (IR2):61–88.

Craig FF. 1971. The Reservoir Engineering Aspects of Waterflooding. Society of Petroleum Engineers of American Institute of Mining Engineers. New York, NY, USA.

Demond AH, Roberts PV. 1987. An examination of relative permeability relations for two-phase flow in porous media. Water Resour Bull 23:617–628.

Gerhard JI, Kueper BH. 2003a. Capillary pressure characteristics necessary for simulating DNAPL infiltration, redistribution, and immobilization in saturated porous media. Water Resour Res 39:1212, doi:10.1029/2002WR001270.

Gerhard JI, Kueper BH. 2003b. Relative permeability characteristics necessary for simulating DNAPL infiltration, redistribution, and immobilization in saturated porous media. Water Resour Res 39:1213, doi:10.1029/2002WR001490.

Gerhard JI, Kueper BH, Hecox G, Schwarz R. 2001. Site-specific waterflood design for the recovery and stabilization of pooled DNAPL. Ground Water Monit Remediat 21:71-88.

Gerhard JI, Kueper BH, Hecox G. 1998. Waterflooding for the removal of pooled DNAPL. Ground Water 36:283-292.

Kueper BH, Frind EO. 1991. Two-phase flow in heterogeneous porous media. 2. Model application. Water Resour Res 27:1059-1070, doi:10.1029/91WR00267.

Kueper BH, McWhorter DB. 1991. The behavior of dense, non-aqueous phase liquids in fractured clay and rock. Ground Water 29:716–728.

Kueper BH, Redman JD, Starr RC, Reitsma S, Mah M. 1993. A field experiment to study the behavior of tetrachloroethylene below the watertable: Spatial distribution of residual and pooled DNAPL. Ground Water 31:756–766.

Kueper BH, Shi J, Gefell MJ. 2008. Hydraulic gradient magnification in NAPL pools: Implications for MGP NAPL vertical mobility. Proceedings of the Sixth International Conference, Remediation of Chlorinated and Recalcitrant Compounds, Monterey, CA, USA. Battelle Press, Columbus, OH, USA, Paper K-004.

Land CS. 1968. Calculation of imbibition relative permeability for two and three phase flow from rock properties. Soc Petrol Eng J 8:149–156.

Leverett MC. 1941. Capillary behaviour in porous solids. Trans AIME 142:159–172.

Longino BL, Kueper BH. 1995. The use of upward gradients to arrest downward DNAPL migration in the presence of solubilizing surfactants. Can Geotech J 32:296–308.

Longino BL, Kueper BH. 1999a. Non-wetting phase retention and mobilization in rock fractures. Water Resour Res 35:2085–2093.

Longino BL, Kueper BH. 1999b. Effects of capillary pressure and viscosity ratio on DNAPL retention and mobilization in a rough-walled fracture. Environ Sci Technol 33:2447–2455.

Lyman WJ, Reehl WF, Rosenblatt DH. 1982. Handbook of Chemical Property Estimation Methods. Environmental Behavior of Organic Chemicals. McGraw-Hill, New York, NY, USA.

Reitsma S, Kueper BH. 1994. Laboratory measurement of capillary pressure saturation curves in a natural rock fracture. Water Resour Res 30:865–878.

Pankow JF, Cherry JA, eds. 1996. Dense Chlorinated Solvents and Other DNAPLs in Groundwater. Waterloo Press, Portland, OR, USA.

Sale T, Applegate D. 1997. Mobile NAPL recovery: Conceptual, field, and mathematical considerations. Ground Water 35:418–426.

West MR, Grant GP, Gerhard JI, Kueper BH. 2008. The influence of precipitate formation on the chemical oxidation of TCE DNAPL with potassium permanganate. Adv Water Res 31:324–338.

Wilke CR, Chang P. 1955. Correlation of diffusion coefficients in dilute solutions. AIChE J 1:264–270.

Willhite GP. 1986. Waterflooding. SPE Textbook Series. The Society of Petroleum Engineers, Richardson, TX, USA.

CHAPTER 9

IN SITU CHEMICAL OXIDATION

Robert L. Siegrist,[1] Michelle Crimi,[2] Neil R. Thomson,[3] Wilson S. Clayton[4] and Michael C. Marley[5]

[1]Colorado School of Mines, 80401 Golden, CO, USA; [2]Clarkson University, 13699 Potsdam, NY, USA; [3]University of Waterloo, Waterloo, ON, Canada, N2L 3G1; [4]TriHydro Corporation 80439 Evergreen, CO, USA; [5]XDD, 03885 Stratham, NH, USA

9.1 TECHNOLOGY DESCRIPTION

This section discusses the topic of *in situ* chemical oxidation (ISCO) including highlights of the overall principles and practices of ISCO for groundwater remediation. Much of the material in this section has been adapted from the recent book *In Situ Chemical Oxidation for Groundwater Remediation* (Siegrist et al., 2011).

9.1.1 Overview of ISCO

In situ chemical oxidation is one of the technologies that have the potential for cost effective remediation of soil and groundwater contaminated by organic chemicals (ITRC, 2005; Huling and Pivetz, 2006; Krembs et al., 2010; Tsitonaki et al., 2010; Siegrist et al., 2011). Many of the most prevalent organic contaminants of concern (COCs) at sites can be destroyed using a chemical oxidant such as catalyzed hydrogen peroxide (H_2O_2), potassium permanganate ($KMnO_4$), activated sodium persulfate ($Na_2S_2O_8$), ozone (O_3), or combinations of oxidants such as ozone and hydrogen peroxide or hydrogen peroxide and sodium persulfate. Under the right conditions, oxidants can transform and often mineralize many COCs including solvents (such as perchloroethene [PCE], trichloroethene [TCE], 1,1,1-trichloroethane [1,1,1-TCA], and 1,4-dioxane), fuels (such as benzene, toluene, methyl tertiary butyl ether), phenols (pentachlorophenol), polycyclic aromatic hydrocarbons (PAHs) (such as naphthalene, phenanthrene), explosives (trinitrotoluene) and pesticides (lindane). Degradation reactions tend to involve electron transfer or free radical processes with simple to complex pathways. Oxidation reactions typically follow second-order kinetics. The need for activation to generate reactive species and the sensitivity to matrix conditions, such as temperature, pH and salinity, vary with the different oxidants and specific contaminants.

An oxidant and amendments (if needed) can be delivered into a target treatment zone (TTZ) within the subsurface at varied concentrations and mass loading rates in liquid, gas or solid phases. Delivery has most commonly been accomplished through permeation by vertical direct-push injection probes or flushing by vertical groundwater wells (Figure 9.1) (Krembs et al., 2010). Other delivery approaches have included horizontal wells, infiltration galleries, soil mixing and hydraulic or pneumatic fracturing.

B.H. Kueper et al. (eds.), *Chlorinated Solvent Source Zone Remediation*,
doi: 10.1007/978-1-4614-6922-3_9, © Springer Science+Business Media New York 2014

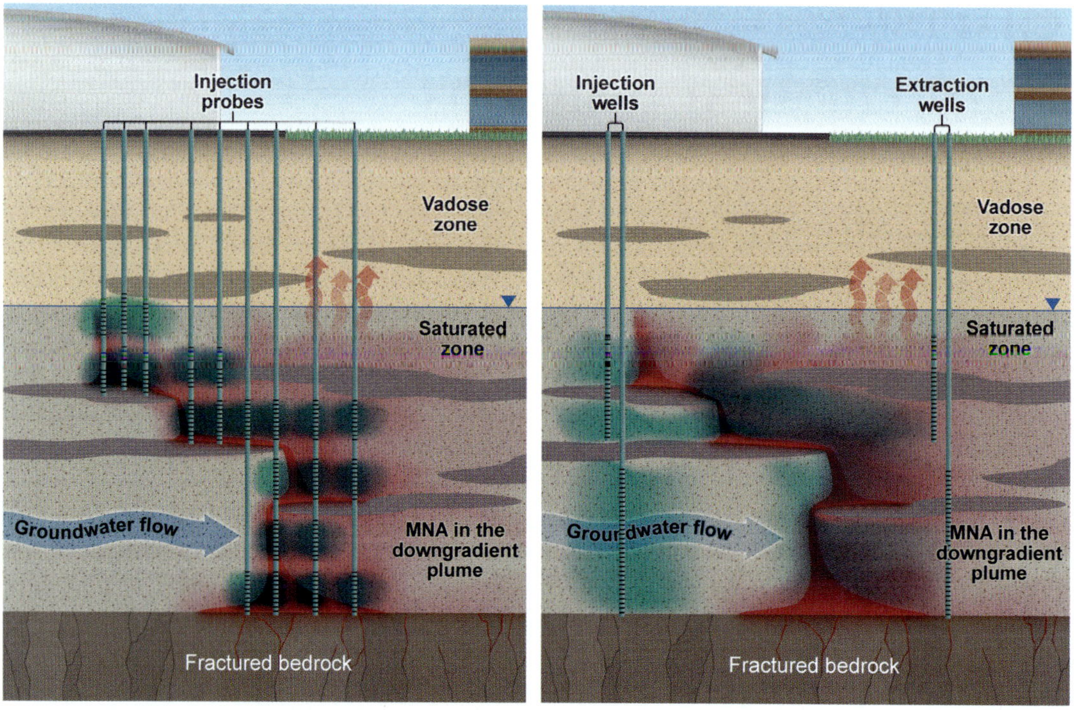

Figure 9.1. ISCO using (*left*) direct-push injection probes or (*right*) well-to-well flushing to deliver oxidants (shown in *blue*) into a TTZ of groundwater contaminated by DNAPL compounds (shown in *red*) (Siegrist et al., 2011).

9.1.2 Historical Evolution of ISCO

There is a long history of using chemical oxidation to destroy organic contaminants in water within the municipal and industrial water and waste treatment industry. The first step in the evolution of ISCO involved research and development (R&D) to adapt use of chemical oxidants like hydrogen peroxide and ozone to treat organic COCs in groundwater that was pumped to the surface and containerized in tank-based reactors (i.e., *ex situ* treatment) (Barbeni et al., 1987; Glaze and Kang, 1988; Bowers et al., 1989; Watts and Smith, 1991; Venkatadri and Peters, 1993). While a concept underlying ISCO was patented in 1986 (Brown and Norris, 1986), the first commercial *in situ* application of hydrogen peroxide occurred in 1984 to treat groundwater contaminated with formaldehyde (Brown et al., 1986).

Beginning in 1990, researchers began to explore hydrogen peroxide and modified Fenton's reagent oxidation as applied in soil and groundwater environments (Watts et al., 1990; Watts and Smith, 1991; Watts et al., 1991; Tyre et al., 1991; Ravikumur and Gurol, 1994; Gates and Siegrist, 1993, 1995; Watts et al., 1997). Research also was initiated with alternative oxidants such as ozone (Bellamy et al., 1991; Nelson and Brown, 1994; Marvin et al., 1998) and potassium permanganate (Vella et al., 1990; Vella and Veronda, 1994; Gates et al., 1995; Schnarr et al., 1998; West et al., 1997; Siegrist et al., 1998a, b, 1999; Yan and Schwartz, 1998, 1999; Tratnyek et al., 1998; Urynowicz and Siegrist, 2000). The development of ISCO continues to expand as evidenced by more recent research with newer oxidants like sodium persulfate (Brown et al., 2001; Block et al., 2004; Liang et al., 2004a, b; Crimi and Taylor, 2007).

In the late 1990s and early 2000s, case study reports became available (USEPA, 1998; ESTCP, 1999), followed by the publication of the first reference book (Siegrist et al., 2001) and the first technical and regulatory reference manual (ITRC, 2001). These documents provided valuable insight into principles and practices, field experiences and regulatory requirements. However, they did not provide the state of the science knowledge and engineering know-how needed for a standard of practice to ensure effective, timely and cost effective site-specific application of ISCO alone or in combination with other remedial options. As a result, the implementation of ISCO was hampered by uncertain and variable design and application practices. This caused ISCO performance to be unpredictable for some site conditions and remediation applications.

To advance the science and engineering of ISCO and resolve questions regarding its design and performance, ISCO research and development efforts were escalated during the early 2000s. This research was catalyzed in a large part by the promising potential of ISCO and growing interest in its use, notably at Department of Defense (DoD) sites across the country. Around 2002, a major ISCO research program was launched within DoD's Strategic Environmental Research and Development Program (SERDP) and Environmental Security Technology Certification Program (ESTCP), as part of its *Dense Nonaqueous Phase Liquid Source Zone Cleanup focus area.*[1]

A portfolio of ISCO projects and activities, both within the DoD ISCO initiative and through other sponsored research programs, has increased the understanding of ISCO. Advancements have included improved fundamental understanding of the following:

- COC oxidation chemistry and treatment (Gates-Anderson et al., 2001; Jung et al., 2004; Qiu et al., 2004; Smith et al., 2004; Watts et al., 2005a; Forsey et al., 2010)
- Oxidant interactions with subsurface media (Siegrist et al., 2002; Crimi and Siegrist, 2003, 2004a, b; Anipsitakis and Dionysiou, 2004; Shin et al., 2004; Jung et al., 2005; Monahan et al., 2005; Mumford et al., 2005; Bissey et al., 2006; Jones, 2007; Teel et al., 2007; Sun and Yan, 2007; Sirguey et al., 2008; Urynowicz et al., 2008; Woods, 2008; Xu and Thomson, 2008, 2009)
- Destruction of DNAPLs (Crimi and Siegrist, 2005; Heiderscheidt, 2005; Kim and Gurol, 2005; Urynowicz and Siegrist, 2005; Watts et al., 2005b; Siegrist et al., 2006; Smith et al., 2006; Heiderscheidt et al., 2008a; Petri et al., 2008)
- Examination of newer oxidants (Liang et al., 2004a, b; Crimi and Taylor, 2007; Waldemer et al., 2007; Liang and Lee, 2008)
- Oxidant transport processes and deliverability (Choi et al., 2002; Lowe et al., 2002; Struse et al., 2002; Lee et al., 2003; Tunnicliffe and Thomson, 2004; Heiderscheidt, 2005; Ross et al., 2005; Zhang et al., 2005; Petri, 2006; Hønning et al., 2007; Heiderscheidt et al., 2008a; Petri et al., 2008; Smith et al., 2008; Christiansen, 2010, 2011; Christiansen et al., 2010; Silva, 2011)
- Combining ISCO with other remedies (Sahl, 2005; Dugan, 2006; Sahl and Munakata-Marr, 2006; Sahl et al., 2007; Dugan et al., 2010)
- Treatability test methods (Haselow et al., 2003; Mumford et al., 2004; ASTM, 2007)
- Formulation of mathematical models and decision support tools (Kim and Choi, 2002; Heiderscheidt, 2005; Heiderscheidt et al., 2008b)

Research efforts have revealed that the successful use of ISCO for remediation of contaminated soil and groundwater fundamentally depends on the reaction chemistry of the oxidant used and its ability to degrade the COCs. Success also depends on effective delivery of

---

[1] http://www.serdp.org/Featured-Initiatives/Cleanup-Initiatives/DNAPL-Source-Zones, January 24, 2014

oxidants into the subsurface, which is determined by the oxidant reactive transport under the hydrogeological and geochemical conditions present. Oxidant reaction chemistry (both with the COCs and with natural organic matter [NOM] and minerals) and subsurface transport affect ISCO application, COC destruction, and cost effectiveness. In addition, the application of ISCO can be affected by, and can also effect changes in, the pre-ISCO subsurface conditions (such as groundwater flow direction and velocity, redox potential [Eh], pH, temperature, dissolved organic carbon), which should be considered and accounted for during ISCO system design and implementation.

9.1.3 Retrospective Analysis of ISCO Performance and Costs

Field applications of ISCO have continued to grow over the past decade. These field experiences have enabled a retrospective analysis of ISCO system design features and site conditions and the factors affecting performance and cost. Of 242 ISCO projects examined by Krembs et al. (2010), PCE or TCE were the targeted COCs at 70% of the sites, the subsurface conditions were characterized as permeable at 75% of the sites and oxidants were delivered using permanent or temporary injection wells at 70% of the sites. For 99 full-scale ISCO projects that attempted to meet a specific goal and reported success, the results were as follows: 21% of 28 projects attempting to achieve drinking water maximum contaminant limits (MCLs) met this goal (all of these projects were characterized by lower concentrations without reported nonaqueous phase liquids [NAPLs]); 44% of 25 projects attempting to achieve alternative concentration limits (ACLs) met this goal; 33% of six projects attempting to reduce the COC mass by a certain percentage met this goal; 82% of 34 projects attempting to reduce the COC mass and/or time to cleanup met this goal; and 100% of six projects attempting to evaluate effectiveness and optimize future injections met this goal.

Krembs (2008) reported the median total cost for 55 ISCO projects to be $220,000; the median unit cost was $94 per cubic yard (cy) ($123 per cubic meter) treated based on 33 projects with unit cost data. McDade et al. (2005) reported median and unit costs of $230,000 and $125/cy ($163 per cubic meter), respectively, for 13 ISCO projects. It is important to recognize that the cost of an ISCO project can vary by an order of magnitude or more depending on various factors. For example, sites with fuel hydrocarbons and permeable subsurface conditions typically cost less than those with DNAPLs or complex subsurface conditions. High unit costs can also result where ISCO has been used to treat relatively smaller source zones.

The effectiveness of ISCO varies: at some sites, ISCO has been applied and the destruction of the target COCs has occurred, enabling cleanup goals to be met in a cost effective and timely manner, whereas at other sites, ISCO applications have had uncertain or unsatisfactory treatment performance. Poor ISCO performance often has been attributed to application of an inadequate oxidant dose or volume, inadequate oxidant delivery caused by low permeability zones and formation heterogeneity, and excessive oxidant loss due to interaction with natural subsurface materials or the presence of large masses of DNAPLs (Siegrist et al., 2001, 2006, 2008a). In some applications, concerns have arisen over secondary effects, such as mobilization of metals, loss of well screen and formation permeability, and gas evolution and fugitive emissions, as well as health and safety practices (Siegrist et al., 2001; Crimi and Siegrist, 2003; Krembs, 2008; Krembs et al., 2010).

Rebound is a condition where post-treatment concentrations of target COCs in groundwater within a TTZ return to levels near or even above those present prior to treatment. Rebound is a relatively common occurrence with ISCO. It may be a negative condition or could reflect an inherent shortcoming of ISCO or a site-specific performance deficiency. The rebound observed at an ISCO treated site can be beneficial if it is used in an observational approach to refine the conceptual site model (CSM) and refocus subsequent treatment. The use of ISCO can be viewed as an ongoing, iterative process that will take advantage of contaminant rebound rather than view it as an indication that the technology was inappropriate for a site or was applied improperly. It is noted that rebound also can occur with other approaches and technologies that depend on injection of remedial amendments.

9.1.4 ISCO System Selection, Design, and Implementation

In situ chemical oxidation system selection, design and implementation practices should rely on a clear understanding of ISCO and its applicability to a given set of contaminant and site conditions to achieve site-specific remediation objectives. A number of key issues may be relevant and need to be addressed regardless of the oxidant and delivery system being employed, including: (1) amenability of the target COCs to degradation by oxidants; (2) effectiveness of the oxidant for NAPL destruction; (3) requirements for effective implementation (oxidant dose, concentration, volume delivered, method of delivery) for a given TTZ in a given subsurface setting; (4) nonproductive oxidant loss due to interactions with NOM, reduced inorganic species, and some mineral phases in the subsurface; (5) nonproductive oxidant consumption due to auto-decomposition reactions and free radical scavenging reactions; (6) potential adverse effects (mobilizing metals such as chromium, forming toxic byproducts, reducing formation permeability, generating off-gases and heat) and (7) potential to combine ISCO with other remediation technologies and approaches (Siegrist et al., 2011).

The selection, design and implementation of a remedial action are generally accomplished within a phased project approach. During the feasibility study for a given site, consideration of ISCO as a viable remedial option is often based on the general benefits that ISCO can offer. These include rapid and extensive reactions with various COCs, applicability to many subsurface environments, ability to tailor ISCO to a site and rapid implementation that can support property transfers and site redevelopment projects. Potential limitations for ISCO are also considered during decision making, including the resistance of some COCs to complete chemical oxidation, the level of nonproductive loss exerted by the subsurface, the stability of the oxidant in the subsurface, the constraints on effective oxidant distribution, the possible fugitive gas emissions, the potential for contaminant rebound and the effects of chemical oxidant addition on water quality.

If ISCO is selected as a viable alternative for a particular site and a site-specific design must be accomplished, many choices and decisions have to be made. For example, choices must be made between oxidant type (hydrogen peroxide, persulfate, permanganate, ozone), delivery method (direct push probes, injection wells, air sparging wells), process control and performance monitoring. These choices should be made carefully to improve the likelihood that the ISCO system will yield a sufficient concentration of a suitable oxidant in contact with the target COCs under amenable conditions over a sufficient period of time for the COCs to be destroyed.

The ISCO systems that can be, and have been, applied in the field are highly varied in their features. Different oxidants and additives (such as stabilizers or activators) have been used. Concentrations and injection flow rates can vary widely, and a variety of subsurface delivery methods can be employed. In addition, overall site remediation goals and regulatory constraints may influence the remediation objectives established for the ISCO technology. To improve the confidence level in the choices and decisions made, treatability studies and field-scale pilot tests are frequently necessary. If ISCO is selected for a site, remedial design and system construction must be accomplished. Finally, ISCO operation and performance monitoring ensues. Given the properties of chemical oxidants, the use of ISCO requires diligent attention to safety and waste management issues (Siegrist et al., 2011). Eventually site closure can be achieved through ISCO alone or in conjunction with another remediation technology or approach.

9.2 KEY CONCEPTS OF ISCO AND DNAPL SOURCE ZONES

While the previous section highlighted the general principles and practices of ISCO for remediation of soil and groundwater, there has been great interest in the potential of ISCO to remediate groundwater where contamination includes DNAPL source zones present in a variety of subsurface geological media. The theoretical basis for the degradation of DNAPL residuals by ISCO is well established, *but* its cost effective realization generally depends on several factors: (1) susceptibility of DNAPL organics to destruction using chemical oxidants, (2) rate and extent of interphase mass transfer of the DNAPL, (3) ability to deliver and transport the oxidant in the subsurface, (4) effects of subsurface conditions on ISCO reactions and (5) ISCO effects on subsurface permeability and biogeochemistry. These issues are discussed in detail in the following sections.

9.2.1 Chemically Reactive Zones and Mass Transfer

In situ chemical oxidation involves the creation of a chemically reactive subsurface environment. This reactive environment is one where a chemical oxidant is introduced with the sole purpose of oxidizing the identified target COCs. By definition, the reaction process results in the oxidation state of a target COC increasing. For example, the oxidation of TCE involves a half reaction where the valence state of carbon is increased from +I to +IV as shown by Equation 9.1 and the associated oxidant is reduced (lowering of oxidation state) as noted below in Section 9.2.2:

$$C_2Cl_3H + 4H_2O \rightarrow 2CO_2 + 3Cl^- + 9H^+ + 6e^- \qquad \text{(Eq. 9.1)}$$

Chemical oxidation can involve more than electron transfer (hydrogen transfer, hydride transfer, oxygen donation) and thus includes all reactions that convert a compound to a higher oxidation state (Stewart, 1964). *In situ* chemical oxidation is an *in situ* mass destructive technology where the objective is to create reaction conditions so that when the COCs are either contacted or intercepted, they are immobilized or degraded into nontoxic end products. The underlying reaction mechanisms and pathways can be very complex and in many cases are unknown. Complete mineralization to CO_2 and H_2O may not occur. Reaction intermediates, which form temporally before they are oxidized to end products, are difficult to identify. Forsey et al. (2010) indicate that ketones and carboxylic acids are potential oxidation reaction end products that may form during the oxidation of coal tars with permanganate. In some cases these end products are nontoxic and can be used as a carbon source for indigenous microbial communities.

A thermodynamic analysis can be used to determine the feasibility of a reaction to take place, but the results provide no information relating to the oxidation reaction rate or kinetics (Schwarzenbach et al., 1993). In some cases the reaction may be deemed feasible, but the kinetics will be far too slow to be useful. In general, the reaction between a chemical oxidant and target COC can be described by a second-order kinetic expression as given by Equation 9.2 (Siegrist et al., 2011):

$$\frac{dC_{COC}}{dt} = -k_{ox}[C_{COC}][C_{ox}] \qquad \text{(Eq. 9.2)}$$

where C_{COC} and C_{ox} are the concentration of the COC and the oxidant, respectively, and k_{ox} is the second-order rate coefficient with respect to the organic compound. As shown by Equation 9.2, the rate of destruction of the COC depends not only on the reaction rate coefficient k_{ox} but also on the product of the concentration of the COC and the oxidant.

To maximize the chemical oxidation reaction, the concentration of the oxidant needs to remain elevated. This can be controlled, in part, by system design. However, as the concentration of the target COC decreases, so will the overall oxidation rate. Second-order rate coefficients for most environmentally relevant DNAPL compounds dissolved in water have been experimentally determined (see IscoKin database at http://cgr.ebs.ogi.edu/iscokin/; accessed January 24, 2014). Reaction rate coefficients are sensitive to temperature variations and pH conditions (Schwarzenbach et al., 1993).

While the *in situ* destruction of DNAPL contaminants (such as PCE and TCE) is one key attribute offered by ISCO, the second attribute is that this technology has the ability to increase mass transfer from DNAPL contaminated regions. There is potential for gas–NAPL or gas–solid reactions, but nearly all reactions of interest that occur between a chemical oxidant and a target organic compound in groundwater take place in the aqueous phase and not in the DNAPL phase. Thus, DNAPL dissolution or mass transfer from the DNAPL phase to the aqueous phase is extremely important since it will, in large part, control the degree of mass destruction and hence source zone treatment effectiveness. The rate at which constituents dissolve from the DNAPL determines the dissolved phase plume concentrations and its longevity. The presence of a DNAPL results in dissolved phase DNAPL constituents that create a concentration gradient across a stagnant boundary layer between the DNAPL and the bulk aqueous solution (Figure 9.2). The larger and more sustained the concentration gradient, the greater the mass of DNAPL depleted.

The mass transfer of a single component DNAPL into the aqueous phase is generally expressed by a macroscopic variation of the stagnant film model given by Equation 9.3 (Schwarzenbach et al., 1993):

$$\frac{dC_{aq}}{dt} = -k_{diss}\left(C^{*} - C_{aq}\right) \qquad \text{(Eq. 9.3)}$$

where C_{aq} is the aqueous concentration of the COC, C^{*} is the aqueous solubility limit and k_{diss} is the lumped mass transfer or dissolution rate coefficient. In a porous medium, the lumped mass transfer rate coefficient cannot be determined from first principles, and hence, various methods have been developed to estimate it from system parameters such as pore size distribution metrics, DNAPL saturation, molecular diffusion coefficient and groundwater velocity (Siegrist et al., 2011). As shown by Equation 9.3, the rate at which DNAPL mass is depleted is a product of the dissolution rate coefficient and the concentration difference or gradient between C^{*} and C_{aq}. Oxidation reactions in the aqueous phase can decrease the concentration of the DNAPL constituents in the bulk solution and, thus, increase or steepen the concentration gradient, which in turn will increase the overall rate of DNAPL mass removal from the system. Results from laboratory studies have shown that a 6 to 10 times increase in the

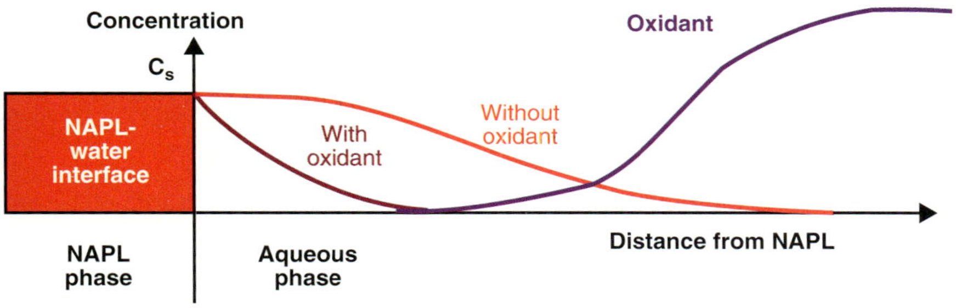

Figure 9.2. Schematic of the dissolved phase concentration gradients near a DNAPL/aqueous phase interface with and without an oxidant present in close proximity.

mass transfer rate is possible to achieve when a DNAPL residual or pool is treated using permanganate (Schnarr et al., 1998; MacKinnon and Thomson, 2002). Research with catalyzed hydrogen peroxide (CHP) has similarly demonstrated increases in mass transfer rates (Watts et al., 2005b; Smith et al., 2006).

In regions of the DNAPL source zone where preferential flow pathways and non-advective locations are present, diffusive dominated transport is the principal mechanism for the migration of DNAPL compounds out of these locations and for the migration of oxidants into these locations. Persistent chemical oxidants, those that are stable in the subsurface for an extended period of time, are particularly advantageous in these situations. When delivered into the treatment zone by way of the preferential flow pathways, these oxidants can diffuse from a preferential pathway into more non-advective regions driven by an oxidant concentration gradient (Figure 9.3). This is in the opposite direction of the gradient associated with the aqueous DNAPL constituents. This counter-diffusion, or two-way diffusion process, increases diffusive mass transfer by decreasing the distance between the DNAPL/aqueous phase interface and the zone where the concentrations of dissolved DNAPL compounds have been reduced as a result of oxidation. Favorable mass transfer rate enhancements have been estimated for these situations, assuming that the chemical oxidant is persistent.

9.2.2 Oxidant Properties

The current chemical oxidants in widespread use are hydrogen peroxide (H_2O_2), potassium or sodium permanganate ($KMnO_4$, $NaMnO_4$), sodium persulfate ($Na_2S_2O_8$), and ozone (O_3). Siegrist et al. (2011) provide a comprehensive overview of these oxidants including chemistry principles and contaminant treatability. The following tables provide insights into some of the key properties of these oxidants. Table 9.1 provides some characteristics of the chemical oxidants in use, while Table 9.2 lists reactive species along with their respective standard electrode potential.

9.2.2.1 Hydrogen Peroxide

Hydrogen peroxide is a strong oxidant that has a high standard reduction potential and is delivered into a source zone as a reagent solution (Table 9.1). Hydrogen peroxide has the potential for direct oxidation of many organic compounds; however, the reaction kinetics are too slow for use (Watts and Teel, 2005). When catalyzed, H_2O_2 can generate a wide range of

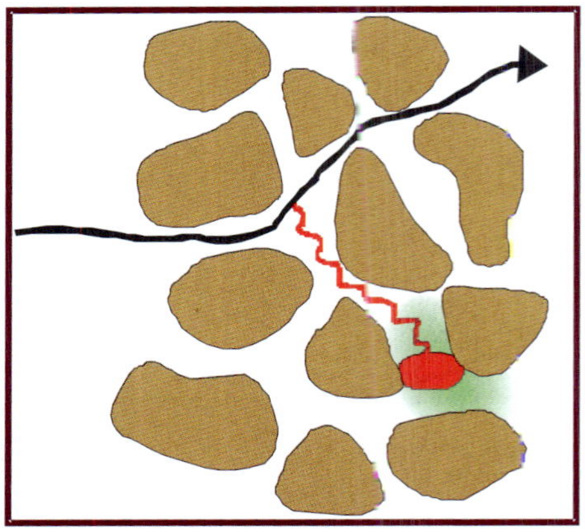

Figure 9.3. Illustration of the two-way diffusion process occurring between a preferential flow pathway (*black arrow*) and an isolated DNAPL residual (*red*).

Table 9.1. Characteristics of Chemical Oxidants Used for Destruction of Organic Contaminants (adapted from Huling and Pivetz, 2006, as presented in Siegrist et al., 2011)

Oxidant[1]	Oxidant Chemical	Commercial Form	Activator	Reactive Species
Hydrogen peroxide	H_2O_2	Liquid	None, Fe(II), Fe(III)	$OH^\bullet$, $O_2^{\bullet-}$, $HO_2^\bullet$, HO_2^-
Permanganate	$KMnO_4$ or $NaMnO_4$	Powder, liquid	None	MnO_4^-
Persulfate	$Na_2S_2O_8$	Powder	None, Fe(II), Fe(III), heat, H_2O_2, high pH	SO_4^{2-}, $SO_4^{\bullet-}$
Ozone	O_3 (in air)	Gas	None	O_3, $OH^\bullet$
Peroxone	H_2O_2 plus O_3 (in air)	Liquid, gas	O_3	O_3, $OH^\bullet$
Percarbonate	$Na_2CO_3{\cdot}1.5H_2O_2$	Powder	Fe(II)	$OH^\bullet$
Calcium peroxide	CaO_2	Powder	None	H_2O_2, HO_2^-

Table 9.2. Reactive Species and Their Electrode Potentials (adapted from Huling and Pivetz, 2006, as presented in Siegrist et al., 2011)

Reactive Species Name	Formula	Electrode Potential (Eh), volts (V)
Hydroxyl radical	$OH^\bullet$	+2.8 V
Sulfate radical	$SO_4^{\bullet-}$	+2.6 V
Ozone	O_3	+2.1 V
Persulfate anion	SO_4^{2-}	+2.1 V
Hydrogen peroxide	H_2O_2	+1.77 V
Permanganate anion	MnO_4^-	+1.7 V
Perhydroxyl radical	$HO_2^\bullet$	+1.7 V
Oxygen	O_2	+1.23 V
Hydroperoxide anion	HO_2^-	−0.88 V
Superoxide radical	$O_2^{\bullet-}$	−2.4 V

free radicals and other reactive species that can attack the target organic compound. Hence, hydrogen peroxide is usually applied so that it is catalyzed to generate free radicals such as $OH^\bullet$ (Table 9.1). This can be accomplished by coinjection of a catalyst (such as $FeSO_4$) or through H_2O_2 interactions with naturally occurring minerals. The reactive species formed in a CHP system include oxidants and in some cases reductants. A CHP system should not be confused with the classic Fenton's reagent (Walling, 1975) where dilute hydrogen peroxide and ferrous iron are combined under acidic pH conditions to produce hydroxyl radicals as given by Equation 9.4:

$$H_2O_2 + Fe^{+2} \rightarrow Fe^{+3} + OH^- + OH^\bullet \qquad \text{(Eq. 9.4)}$$

Research has shown that when hydrogen peroxide and ferrous iron are applied *in situ* using a significantly higher H_2O_2 concentration, a much more complex set of reactions occurs (Watts and Teel, 2005). A CHP system involves numerous reactive species and mechanisms that result in significant contaminant transformation or degradation. Notable reactive species are the hydroxyl radical, superoxide anion, perhydroxyl radical and hydroperoxide anion

(Siegrist et al., 2011). Radicals have an unpaired electron and are highly reactive and unstable. Thus, their *in situ* transport distances are short. Therefore, for a CHP system to treat a DNAPL source zone effectively, it must be applied directly in the source zone given that limited transport of the reactive species will occur.

Many stabilizers (chelating agents) have been considered as a means to enhance the transport of hydrogen peroxide and ferrous iron *in situ* (Watts et al., 1999, 2007; Kakarla and Watts, 1997). Ethylenediaminetetraacetic acid (EDTA) is one of the most popular chelating agents used to reduce the decomposition of hydrogen peroxide by suppressing the catalytic activity of naturally occurring transition metals (Jones and Williams, 2002; Ramo, 2003). Sun and Pignatello (1992) assessed the effectiveness of 50 chelating agents and determined that nitrilotriacetic acid (NTA) and hydroxyethyliminodiacetic acid (HEIDA) were the most effective chelating agents for iron. While currently used or proposed chelating agents are poorly biodegradable (e.g., EDTA), associated with health issues (e.g., NTA is a carcinogen), or not very effective (e.g., citrate), Xu and Thomson (2007) explored a newly available and environmentally friendly chelating agent, ethylenediamine disuccinate (EDDS), to stabilize hydrogen peroxide. They found that EDDS was able to reduce hydrogen peroxide decomposition rates in the presence of various aquifer materials by close to 40% in column trials. Schmidt et al. (2011) reported that phytate (25 millimolars [mM]) was effective as a stabilizer for transport of H_2O_2 through iron-coated sand, resulting in H_2O_2 concentrations two orders of magnitude greater than without phytate.

Rapid destruction of DNAPLs, such as TCE, PCE and carbon tetrachloride, has been demonstrated with CHP systems (Watts and Teel, 2005; Yeh et al., 2003). Smith et al. (2004, 2006) demonstrated that the superoxide and hydroperoxide anions are the reactive species responsible for the degradation of some recalcitrant organic contaminants.

Watts et al. (1999) reported that some reactive species formed by CHP may enhance desorption of sorbed contaminants. Corbin et al. (2007) isolated the reactive species in CHP reactions that are responsible for enhanced contaminant desorption – the superoxide species. Therefore, some of the reaction oxygen intermediates formed during CHP oxidation, including superoxide, may act as surfactants or have other properties that enhance solubilization of hydrophobic contaminants. Organic intermediates produced by contaminant degradation reactions may contribute as well. Ndjou'ou and Cassidy (2006) reported that the organic intermediates generated from CHP oxidation of some complex organic wastes, such as NAPL-phase fuel hydrocarbons, may produce organic surfactants as intermediates during the reaction, and these can even become concentrated enough in solution that they exceed the critical micelle concentration. Thus, enhanced solubilization of the remaining contaminant may occur, increasing its availability for aqueous phase degradation reactions.

9.2.2.2 Potassium Permanganate

Potassium permanganate, which is the most commonly used reagent in ISCO applications, is a crystalline solid that is derived from mined potassium ores. Neutral permanganate solutions prepared with water are relatively stable because water is the only solvent that reduces permanganate very slowly in the presence of manganese dioxide (or dust). Manganese species have potential valence states ranging from +1 to +7, of which the +2, +4 and +7 states are the only ones that are stable over a wide range of acidity. Manganese in permanganate has the highest oxidation state. Permanganate is typically delivered into a source zone as a reagent solution, but other applications have used the solid form (see Siegrist et al., 1999).

Because permanganate is highly reactive with compounds containing C=C bonds, the most common use of permanganate-based ISCO is to remediate groundwater contaminated by PCE, TCE, dichloroethene (DCE), and vinyl chloride (VC) (Siegrist et al., 2001). In addition to chlorinated ethenes, recent studies have shown that permanganate is able to oxidize other contaminants such as hexahydro-1,3,5-trinitro-1,3,5-triazine (RDX), some PAHs and pesticides (Tollefsrud and Schreier, 2002; Adam et al., 2004; Waldemer and Tratnyek, 2006; Forsey et al., 2010). Oxidation of target COCs by permanganate occurs by electron transfer and no radicals are involved. Thus, the reaction rate is slower and the range of reactivity is limited.

The most common permanganate reaction employed in environmental engineering is the complete reduction of permanganate (Mn^{7+}) to manganese dioxide (MnO_2) (Mn^{4+}) (Schnarr et al., 1998; Siegrist et al., 2001; Crimi and Siegrist, 2004b), which is a three-equivalent reaction as shown in Equation 9.5. This reaction proceeds differently as a function of pH as shown in Equations 9.5–9.7 (Siegrist et al., 2011):

$$MnO_4^- + 2H_2O + 3e^- \rightarrow MnO_2(s) + 4OH^- \quad 3.5 > pH < 12 \qquad \text{(Eq. 9.5)}$$

$$MnO_4^- + 8H^+ + 5e^- \rightarrow Mn^{2+} + 4H_2O \quad pH < 3.5 \qquad \text{(Eq. 9.6)}$$

$$MnO_4^- + 1e^- \rightarrow MnO_4^{2-} \quad pH > 12 \qquad \text{(Eq. 9.7)}$$

An important product of the reduction of permanganate is manganese oxide, which forms at the point of reaction. Manganese oxide is a water insoluble solid that is highly polar and tends to coagulate by aggregation to form hydrated colloids with high water contents (Perez-Benito and Arias, 1991). The degree to which the colloids will aggregate depends on the presence of other ions as well as the system Eh and pH (Morgan and Stumm, 1963). Laboratory research has demonstrated that the precipitation of manganese oxides and the formation of CO_2, another reaction product, decreased the hydraulic conductivity between 50 and 90% (Schroth et al., 2001) in sand-packed columns, caused flow divergence around regions of higher DNAPL saturations in a two-dimensional physical model (Li and Schwartz, 2004), and resulted in pore plugging and the formation of a distinct manganese oxide layer in the vicinity of the NAPL that reduced the post-treatment mass transfer (Mackinnon and Thomson, 2002; Conrad et al., 2002; Urynowicz and Siegrist, 2005). The longevity of the trapped CO_2 is temporary, but the manganese oxide deposition is believed to be stable as long as the geochemical conditions are conducive (pH >3; Appelo and Postma, 1999). As a result, some have hypothesized that treatment of a DNAPL source zone by permanganate will initially result in some degree of mass destruction until manganese oxide deposition has reached a level to impact hydraulic conductivity and mass transfer. When this point is reached, the manganese oxides will adversely affect the hydraulic performance of the delivery system and/or the rate of mass destruction will decrease.

It has been observed, however, that the deposition of MnO_2 is dependent on the architecture of the DNAPL source zone, the concentration of permanganate delivered and the velocity of the oxidant-amended groundwater across the DNAPL–water interface (Petri et al., 2008; Heiderscheidt et al., 2008a, b). Research findings revealed lower oxidant concentrations and higher flow velocities can result in more dispersed deposition of MnO_2 and limited impacts on mass transfer and DNAPL destruction. Other research has explored the use of additives (hexametaphosphate) to stabilize the MnO_2 solids and limit their deposition at DNAPL–water interfaces (Crimi et al., 2009).

9.2.2.3 Sodium Persulfate

Sodium persulfate is the most common and effective form of persulfate salt for ISCO applications and is delivered into a source zone as a reagent solution. The solubility of sodium persulfate that controls the amount of oxidant mass that can be delivered is high (730 grams per liter [g/L]), although density-driven transport is a concern at higher oxidant concentrations.

The persulfate anion is not affected by sorption (Liang et al., 2008), and it is more stable in the subsurface as compared to peroxide and ozone (Brown and Robinson, 2004; Johnson et al., 2008; Sra et al., 2010).

The chemistry of persulfate is quite complex and not fully understood. Oxidation by persulfate at ambient conditions may occur either by direct oxidation (Equation 9.8) or through the activation of persulfate to sulfate radicals using ferrous iron (Equation 9.9), heat (Equation 9.10), hydrogen peroxide (Equation 9.11), or hydroxide (i.e., base) (Equation 9.12) (House, 1962; Huang et al., 2002):

$$S_2O_8^{2-} + 2e^- \rightarrow 2SO_4^{2-} \quad \text{(Eq. 9.8)}$$

$$S_2O_8^{2-} + 1e^- \rightarrow SO_4^{2-} + SO_4^{\bullet -} \quad \text{(Eq. 9.9)}$$

$$S_2O_8^{2-} \rightarrow 2SO_4^{\bullet -} \quad \text{(Eq. 9.10)}$$

$$S_2O_8^{2-} + OH^{\bullet} \rightarrow SO_4^{2-} + SO_4^{\bullet -} + \frac{1}{2}O_2 + H^+ \quad \text{(Eq. 9.11)}$$

$$SO_4^{\bullet -} + OH^- \rightarrow SO_4^{2-} + OH^{\bullet} \quad \text{(Eq. 9.12)}$$

Direct persulfate oxidation is not able to destroy some organic contaminants such as saturated chlorinated solvents and normal chain alkanes (Liang et al., 2008). A higher reaction activation energy is required for persulfate to react with these organic compounds. Compared to unactivated persulfate, activated persulfate reacts with a wider range of organic compounds through production of sulfate and hydroxyl free radicals. As shown in Equations 9.9–9.11, sulfate radicals are produced as a result of persulfate activation and initiate a chain of reactions that may form other free radicals and other highly oxidative species such as the hydroxyl radical, superoxide anion, perhydroxyl radical and hydroperoxide anion (Furman et al., 2010; Siegrist et al., 2011). These types of reactive species are very similar to the suite of reactive species generated in a CHP system. However, for activated persulfate, the reactive species generated can vary with the activation approach employed. A free radical reaction, in general, involves a chain of initiation, propagation and termination reactions. Kolthoff and Miller (1951), House (1962) and Liang et al. (2004a, b; 2007) have studied this three-step chain reaction. As a result of persulfate activation and initiation of radical-based chemical reactions, the reaction rate of organic compound degradation can be significantly increased until all reactions are terminated (Liang et al., 2007).

Persulfate has been shown to be an effective oxidant for a variety of DNAPL compounds including chloroethenes (PCE, TCE, DCE, VC). The reactivity of persulfate with halogenated aliphatics is variable and appears to depend on the nature of the carbon bonding, the degree of halogenation and the persulfate activation approach. Literature suggests that persulfate is able to degrade polychlorinated biphenyls (PCBs) and some PAHs (Yukselen-Aksoy et al., 2010).

9.2.2.4 Ozone

Ozone is a gas that is delivered to the TTZ as a mixture with ambient air or pure oxygen. In the saturated zone, the gas distribution is controlled by the same processes that occur during *in situ* air sparging (IAS) (Thomson and Johnson, 2000). Oxidation of the target COC can occur in the gas channel after the contaminant has volatized (gas phase oxidation) or in the aqueous phase once the ozone has dissolved (aqueous phase oxidation). In either case the controlling process is mass transfer between the gas and aqueous phase, which is limited to very near the gas phase channels (Braida and Ong, 2001). Thus, for successful treatment of a DNAPL source zone, closely spaced gas channels are required within the TTZ, and ozone gas delivery must normally occur for an extended period of time (several months).

Oxidation can occur by direct reaction with ozone or by the hydroxyl radical that forms from ozone during the following decomposition reaction (Siegrist et al., 2011):

$$O_3 + H_2O \rightarrow O_2 + 2OH^{\bullet} \qquad \text{(Eq. 9.13)}$$

When hydrogen peroxide is present in the water or coinjected into the source zone along with ozone (as used in a peroxone ISCO system), then a more aggressive treatment system is possible (Siegrist et al., 2011). Ozone is a short-lived oxidant with a relatively low solubility and high reactivity with a range of nontarget compounds. With minimal transport in the aqueous phase, ozone has to be injected directly into a DNAPL source zone to enable effective treatment.

9.2.3 Oxidant Reactions in the Subsurface

The oxidants in common use for ISCO can react with DNAPL compounds that are prevalent in DNAPL source zones (such as PCE, TCE, DCE). In addition, naturally occurring reductants and catalysts at many contaminated sites can be reactive and, thus, influence oxidant persistence. Typically, the role of the aquifer solids overshadows that of the dissolved groundwater species. Inorganic species containing iron (Fe), manganese (Mn), sulfur (S) and the NOM associated with the aquifer solids are usually of concern. The possibility of multiple inorganic species, as well as a range of NOM, creates an extremely heterogeneous environment in which reactions may occur (Figure 9.4).

The result of the interaction between the selected oxidant and aquifer material leads to either an increase in the consumption of the oxidant by the aquifer solids or an enhancement in the oxidant decomposition rate. When an oxidant is consumed, the reactive species associated

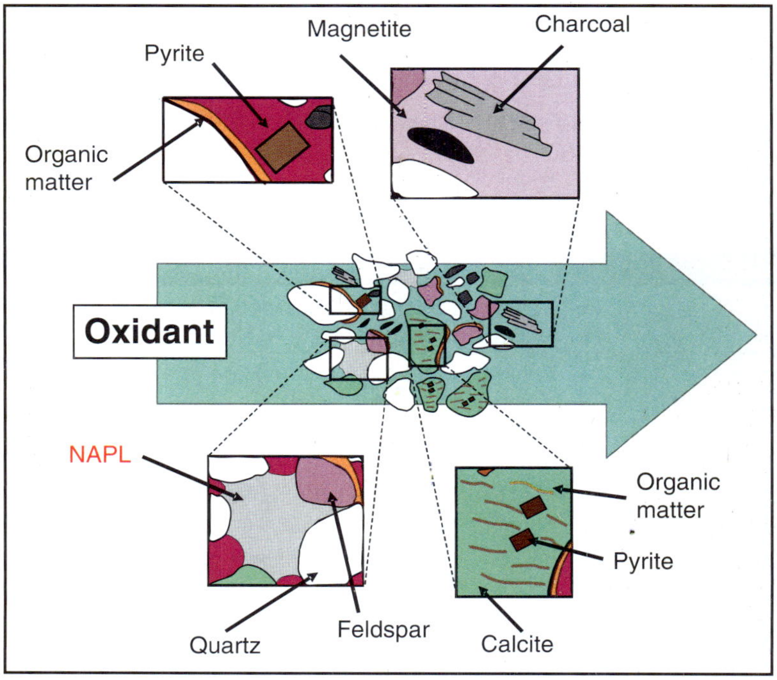

Figure 9.4. Pore-scale conceptual model for natural oxidant interactions showing the possibility of reaction with reduced aquifer solid species, reaction with dissolved DNAPL species, and transport of unreacted oxidant (adapted from Mumford et al., 2005).

with the aquifer solids are finite, and hence there exists a finite consumption or natural oxidant demand (NOD). Once the maximum NOD is satisfied, there is minimal additional oxidant and aquifer material interaction. Conversely, an enhancement in the oxidant decomposition rate implies that infinite interaction capacity is available, that is, all oxidant will eventually be consumed by decomposition reactions. To capture these behavioral differences and the associated underlying processes for all oxidant behavior, the term *nonproductive oxidant depletion* can be used rather than *natural oxidant demand*. Regardless of the way in which nonproductive depletion may manifest itself, it can result in (1) a decrease in the transport of the oxidant within the subsurface, (2) a decrease in the reaction rate with the target COCs and (3) a reduced mass of oxidant available. All of these can result in an inefficient ISCO system. Quantification of the nonproductive oxidant depletion is a requirement for site-specific assessment and the design of cost effective ISCO treatment systems. Table 9.3 describes the observed nonproductive behavior of the four widely used oxidants in the presence of uncontaminated aquifer materials.

To illustrate the inherent differences between the persistence of peroxide, permanganate and persulfate in the presence of aquifer solids, consider the following injection scenario: (1) an uncontaminated aquifer is subject to the injection of an oxidant (peroxide, permanganate or persulfate) in two sequential episodes spaced 30 days apart; (2) following injection the oxidant solution remains immobile as it reacts with the aquifer material; (3) the controlling *in situ* kinetic parameters are taken from bench-scale efforts performed on the same aquifer material and (4) the injection concentration is adjusted so that the oxidation strength is identical for each oxidant. Figure 9.5 depicts simulated oxidant concentration profiles over 60 days, illustrating the following features:

- Peroxide concentration is rapidly reduced.
- The peroxide profiles following injection 1 and injection 2 are identical.
- The permanganate concentration profile following injection 1 decreases quickly and then slows down as it approaches an asymptote.
- The decrease in the permanganate concentration profile following injection 2 is much less than after injection 1, reflecting the consumption of much of the fast reacting NOM.
- The persulfate concentration profile is nearly identical following injection 1 and injection 2.

Table 9.3. Characteristics of the Nonproductive Behavior of Common Chemical Oxidants

Oxidant	Nonproductive Behavior
Ozone	• Enhanced first-order degradation rate • Infinite interaction, infinite NOD
Hydrogen peroxide	• Enhanced first-order degradation rate • Insignificant change in NOM • Repeated decomposition behavior • Infinite interaction, infinite NOD
Permanganate	• Fast initial and slow later consumption rates • Significant change in NOM • MnO_2 catalyzed decomposition • Finite interaction but with increased persistence with each injection • Definitive NOD
Persulfate	• Enhanced first-order degradation rate • Slight to moderate decrease in NOM • Repeated decomposition behavior • Infinite interaction, infinite NOD

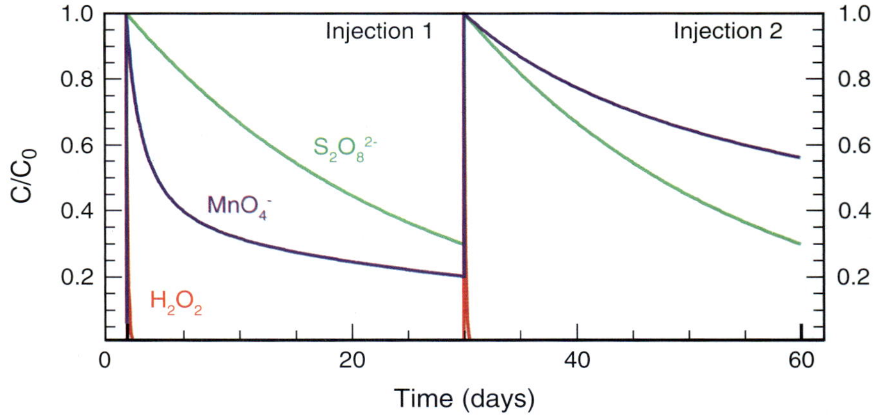

Figure 9.5. Concentration profiles following two sequential injection episodes into a synthetic aquifer for peroxide (*red*), permanganate (*purple*), and persulfate (*green*).

- The persulfate reaction rate is substantially less than peroxide; thus, persulfate is more persistent in this system than peroxide.

Aside from the important nonproductive reactions just noted, other competing and nonproductive reactions occur in ISCO systems. For oxidants involving free radical generation, radical scavenging reactions can decrease the concentration of free radicals available to degrade targeted contaminants. Therefore, scavenging reactions can also result in system treatment inefficiency. Common groundwater anions such as nitrate, chloride, carbonate and bicarbonate are thought to be the most important radical scavengers. Relevant reactions with the bicarbonate ion and the hydroxyl radical or sulfate radical are given below (Buxton et al., 1988; Huie et al., 1991):

$$OH^{\bullet} + HCO_3^- \rightarrow H_2O + CO_3^{\bullet -} \quad \text{(Eq. 9.14)}$$

$$SO_4^{-\bullet} + HCO_3^- \rightarrow SO_4^{2-} + CO_3^{\bullet -} + H^+ \quad \text{(Eq. 9.15)}$$

The scavenging effect of carbonate or bicarbonate anions on the efficiency of persulfate reactivity has been observed in experiments (Huang et al., 2002; Liang et al., 2006; Waldemer et al., 2007). Waldemer et al. (2007) used PCE as a model compound with three different concentrations of carbonate to investigate changes in the pseudo first-order kinetic rates. The results showed, as expected, that an increase in the carbonate concentration decreases the rate of PCE oxidation with persulfate.

9.3 CURRENT PRACTICES FOR DNAPL SOURCE ZONE TREATMENT USING ISCO

In this section, the current state of practice for ISCO applied to sites with DNAPL source zones is highlighted. This section conveys what practitioners working at DNAPL sites have normally done, even though this may not represent best practices based on recent research and field experiences. In Section 9.4, remedial design issues are presented with alternative and recommended practices that ideally should be followed to help achieve remedial objectives in a cost effective manner for DNAPL source zones.

To reveal the current state of practice for ISCO applied to DNAPL source zones, it is insightful to examine case studies of field projects. The site conditions and ISCO performance data for ISCO projects at sites with DNAPL source zones are summarized in Example 1. These case studies reveal the general types of approaches that have been and are being used to apply ISCO to achieve varied treatment goals at sites with contrasting site conditions and contamination characteristics.

Example 1. Examples of ISCO Applications to Sites with DNAPL Source Zones

Naval Submarine Base Kings Bay, St. Marys, Georgia. Modified Fenton's reagent was used to treat a chlorinated solvent DNAPL source zone beneath a landfill (reported in NAVFAC, 2000, and Chapelle et al., 2005). After the first application, monitoring data indicated that the source zone had not been completely delineated prior to the design of the first ISCO injection event. Additional modified Fenton's reagent was applied in three additional delivery events. A total of 0.23 pore volumes were applied throughout the course of the project. The cumulative dose of oxidant applied was 24 grams (g) oxidant/kilogram (kg) of impacted media. COC concentrations in the source zone were reduced to below MCLs. However, the offsite plume that was not targeted by ISCO remediation remained and has not attenuated to below MCLs over an approximately 10-year monitoring period.

Sun Belt Precision Products. Four applications of permanganate were applied to a source zone at a site with a known DNAPL release (reported in USEPA, 2003, and ITRC, 2005). The relatively homogeneous sand treatment zone was particularly amenable to ISCO treatment. Only 0.16 pore volumes were delivered over the course of the project, which the design team indicated was intentional to minimize physical displacement. A design radius of influence of 7 feet (ft) (2.1 meters [m]) was used for injections. While NOD was not reported, the sand stratigraphy suggests that it was likely low. Aqueous phase groundwater concentrations were reduced by three to four orders of magnitude throughout the treatment zone. ACLs were met in the treatment zone, allowing the site to obtain closure.

U.S. Department of Energy Portsmouth Gaseous Diffusion Plant. The X-701B site has been treated with permanganate delivered through horizontal wells (West et al., 1997; 1998) and CHP delivered with direct push points (Cross et al., 2006; Cross and Baird, 2008; Wymore et al., 2010). The DNAPL source zone is located in a heterogeneous formation directly overlying low permeability shale. The permanganate recirculation delivered approximately 0.7 pore volumes. While transient COC reductions in groundwater were observed, significant rebound occurred throughout the treatment zone. Catalyzed hydrogen peroxide delivered in multiple injection events was able to reduce aqueous and soil phase COC concentrations in the majority of the treatment zone, but was unable to adequately contact the DNAPL located directly above the shale at the base of the treatment zone. The source zone is currently being remediated by excavation.

U.S. Army Corps of Engineers, Eastland Woolen Mill Superfund Site (Corinna, Maine). Laboratory treatability tests, field pilot tests and full-scale implementation of ISCO were conducted for the site and completed in 2007 (Osgerby et al., 2006). The COCs at the site include 1,2,3-trichlorobenzene; 1,2,4-trichlorobenzene; 1,2-dichlorobenzene; 1,3-dichlorobenzene; 1,4-dichlorobenzene; benzene; and chlorobenzene. The target area is a portion of the site composed of basal till and weathered/fractured bedrock that could not be excavated and contained residual DNAPL. Bench-scale testing showed activated persulfate and CHP to be potentially applicable; however, field pilot testing demonstrated activated persulfate to be a superior ISCO process for the site due to the reactive nature of site soils

(continued)

Example 1. (continued)

that severely limited distribution of the CHP. Full-scale application of activated persulfate was performed between 2005 and 2007 in four phases for treatment of overburden basal till (Phases I and II), weathered bedrock (Phase III), and fractured bedrock (Phase IV). This final phase demonstrated that distribution of the oxidant and treatment of the bedrock would be feasible, though some fracture bedding planes would require high-pressure injections. Tracer tests also indicated that oxidant mass flux into secondary porosity features within the fracture matrix would be an important mechanism to treat the contaminant mass within the bedrock. Nearly 200,000 pounds (lbs) (91,000 kg) of activated persulfate reagents were applied during the four phases of treatment. Over the four phases of injection, an approximate 60% reduction in soil contaminant mass and 90% reduction in groundwater contaminant mass (ISCO objective was an 85% reduction in groundwater COC concentrations) have been achieved.

Industrial Site in New Hampshire, Multiple Source Areas. Remedial investigations identified the contamination as a mixed DNAPL including chlorinated ethanes and chlorinated ethenes. The primary COCs were identified as 1,1,1-TCA and TCE. The contamination was found in silty fine sands; the silt content increased progressively silty with depth. ISCO was selected as the preferred remedial technology. The objective of the remediation was to treat the source and allow the plume to attenuate naturally. As the soils at the site became increasingly silty with depth, the potential existed for preferential flow into the more shallow portions of the injection area that could cause failure of the ISCO to meet its goal. Therefore, modeling of the vertical placement of the injection wells was performed to optimize the distribution of reagents in the subsurface. Bench-scale testing was performed for CHP, which was the ISCO technology selected; the work was completed in 2005–2010 (Smith et al., 2012). During the bench-scale tests, it was determined that H_2O_2 was unstable in the presence of the site soils, which would result in inadequate distribution of the H_2O_2. A variety of CHP stabilization agents were tested to develop a protocol that would allow for better reagent distribution. Direct injection, via installed polyvinyl chloride (PVC) wells, of the stabilized H_2O_2 was selected for the remedial design. The four identified source areas were treated with 76,000 lbs (34,500 kg), 142,000 lbs (64,500 kg), 110,000 lbs (50,000 kg), and 76,000 lbs (34,500 kg) of stabilized H_2O_2, respectively, within 645,200 gallons (gal) of solution. The ISCO application (in most source areas, one or two applications were required) achieved a reduction of >80% of dissolved phase chlorinated compounds, which exceeded the ISCO application objective for the site.

Active Manufacturing Facility (New England). The COCs at this facility were 1,1,1-TCA (historical levels over 100 miligrams per liter [mg/L] detected), PCE, and 1, 4-dioxane. The ISCO goal was to reduce COCs concentrations to <1 mg/L to allow natural attenuation of the residual plume prior to reaching downgradient receptors. As the DNAPL source area was under an active manufacturing facility, safety concerns eliminated the application of CHP and alkaline activated persulfate (AAP) was selected for the ISCO application. Alkaline activated persulfate required adjustment of the aquifer pH to alkaline conditions (typically >pH 10.5). Bench-scale testing demonstrated that the site-specific soils had both a high buffering capacity and nontarget oxidant demand, but that AAP could be successful in achieving destruction of the COCs to below target levels; the work was completed in 2007–2008 (Crawford et al., 2009). The full-scale application required approximately 55,000 gal (208,000 L) of oxidant solution over two separate field events to treat a 30 ft × 60 ft (9 m by 18 m) source zone over a targeted saturated zone thickness of approximately 15 ft (4.6 m), which is equivalent to approximately 0.45 pore volumes of the treatment zone for each event. Two distinct geological

(continued)

Example 1. (continued)

units were identified within the TTZ: an upper sandy silt zone overlying a deeper, lower permeability glacial till zone. This required installation of two separate injection well systems at two discrete depths to optimize oxidant distribution into each soil layer. The post-treatment results showed that the groundwater goals were met in all performance monitoring locations. A year of post-ISCO sampling was completed, and sampling confirmed that a >99% reduction of the target contaminants was achieved, with final concentrations approaching MCLs in many monitoring wells.

To obtain more quantitative insight into ISCO design and performance, databases of project applications have been compiled. Krembs developed a database of ISCO projects that were carried out at 242 sites (Siegrist et al., 2010; Krembs et al., 2010; Krembs, 2008). A subset of these sites was categorized as DNAPL sites. Based on critical review of those sites categorized as having DNAPL contamination, the following general trends were noted for treatment of DNAPL sites using ISCO:

- *In situ* chemical oxidation practitioners use longer delivery events, greater injection volumes and higher oxidant doses for DNAPL sites. The average duration of delivery was longer for DNAPL sites as well.
- A median cumulative injection volume of 0.13 pore volumes was delivered at DNAPL sites. While this number is low, it is more than twice the median volume delivered at sites where DNAPL was not believed to be present (median of 0.056 pore volumes).
- At the 82 DNAPL sites studied, it was more common to use ISCO with other technologies (79%) than to apply ISCO alone (62%), especially post-ISCO (56% vs. 40% of 59 sites).
- Enhanced bioremediation and monitored natural attenuation (MNA) were the most frequently used post-ISCO technologies applied at DNAPL sites (33 and 21% for 34 sites, respectively). In this analysis, MNA was considered as a coupled technology when source documents specifically stated that it would be used, which likely results in an underestimation of the frequency of the use of MNA after ISCO.
- Practitioners attempted to meet MCLs immediately after ISCO at only 14% of 110 DNAPL sites reviewed.
- No DNAPL sites met MCLs following ISCO.
- DNAPL sites were able to attain ACLs in 39% of cases.
- Rebound in COC concentrations in groundwater was observed in at least one monitoring location at 82% of DNAPL sites.

Based on anecdotal reports of technology developers and practitioners, commonly cited reasons for not meeting the desired ISCO performance goals at sites with DNAPL source zones include (1) a lack of contact between the COCs to be treated and the injected reagents (oxidants, catalysts or activators), (2) back diffusion from low permeability and heterogeneous materials, (3) inadequate desorption of sorbed COCs, (4) NAPL solubility and mass transfer limitations and (5) incomplete characterization of the extent or mass distribution of the contaminants (Siegrist et al., 2008b).

9.4 REMEDIAL DESIGN ISSUES AND APPROACHES

9.4.1 DNAPL Source Zones and Viability of ISCO

Dense nonaqueous phase liquid source zones challenge all treatment technologies, particularly where the treatment goals are unrealistic for the site-specific conditions. For example, to date there are no documented cases where DNAPL source zones have been treated and maintained at MCL concentrations by any remediation technology. *In situ* chemical oxidation is applicable to source zones, however, only when treatment goals are realistic and site conditions are amenable; ISCO alone is not recommended for source zones if significant treatment of DNAPL pools is required. Furthermore, ISCO alone is not recommended where >99% concentration, mass, or mass flux reduction is the treatment goal, unless the site is relatively homogeneous (hydraulic conductivity varies by less than a factor of 1,000) and permeable ($>10^{-4}$ centimeters per second [cm/s]) (Siegrist et al., 2011). Concentration, mass and mass flux reductions of 50–90% are achievable and have been well documented in source zones using ISCO; treatment to 99% reduction of concentration, mass or mass flux is feasible but less likely in source zones (Siegrist et al., 2011). The probability of meeting treatment goals, regardless of their stringency, decreases with increasing complexity (such as increasing heterogeneity and/or decreasing permeability) (Krembs et al., 2010; Siegrist et al., 2011). Site complexity reduces the ability to establish contact between oxidants and contaminant – the more challenging the contact of any remediation amendment with contaminants, the less likely successful treatment will be achieved.

Table 9.4 provides an overview of potential oxidant types, delivery methods, potential for rebound and performance goals for various DNAPL source zones (single compound or mixtures), and associated geological subsurface conditions. A discussion of remedial design considerations and approaches is given in the following sections.

9.4.2 Remedial Design Issues

A number of specific issues should be addressed to successfully implement ISCO at DNAPL sites. These issues can be categorized as conceptual site model, site treatment goals, site-specific limitations, effects of source zone treatment, and contaminant rebound (Table 9.5). Further discussion of each of these remedial design issues is given below.

9.4.2.1 Comprehensive Conceptual Site Model

It is well established that geological heterogeneities have a significant impact on source zone architecture (Schwille, 1988; Kueper et al., 1993) (Figure 9.6). Dense nonaqueous phase liquid migration, in particular, has been studied quite extensively as a function of site geology (Pinder and Abriola, 1986; Kueper et al., 1989; Mackay and Cherry, 1989; Kueper and Frind, 1991a, b; Johnson and Pankow, 1992; Poulsen and Kueper, 1992; Held and Illangasekare, 1995a, b; Illangasekare et al., 1995a, b; Ball et al., 1997; Liu and Ball, 2002; Sale et al., 2007). These studies emphasize the complex behavior of DNAPL within source zones, including lateral spread over finer grained media layers as it sinks within the aquifer, entry into and storage within these lower permeability layers, and DNAPL pooling in coarser layers. When ISCO is applied to a DNAPL source zone, for example the DNAPL site depicted in Figure 9.6, it is both efficient and cost effective to deliver the oxidant to regions with high contaminant mass densities. When a site is poorly characterized and complexities are misunderstood or ignored, ISCO systems can bypass zones of high contaminant mass densities. While no site characterization effort will result in a complete understanding of the subsurface conditions, it is important to have a comprehensive CSM. The more geologically complex the site, the higher the degree of characterization that is required.

Table 9.4. Characteristics of DNAPL Source Zones and General Amenability to Successful Application of ISCO

DNAPL Type Subsurface Conditions[a]	Simple Mixtures[b] (Solvents [PCE, TCE, DCE,. . .])		Complex Mixtures[b] (e.g., Solvents w/Fuels, Coal Tars)	
	Globs and ganglia (higher GTP ratio)	Pools (lower GTP ratio)	Globs and ganglia (higher GTP ratio)	Pools (lower GTP ratio)
A. Permeable and homogeneous (21%)	Oxidant type: A Delivery: P, I Rebound: LM Goal: ACL or MF Combined: N	Oxidant type: S Delivery: P Rebound: LM Goal: ACL or MF Combined: Y	Oxidant type: S Delivery: P, I Rebound: LM Goal: ACL or MF Combined: Y	Oxidant type: S Delivery: P, I Rebound: HD Goal: ACL or MF Combined: Y
C. Permeable and heterogeneous (47%)	Oxidant type: A Delivery: P Rebound: LM Goal: MCL, ACL, MF Combined: N	Oxidant type: S Delivery: P, M, O Rebound: HD Goal: MCL, ACL, MF Combined: Y	Oxidant type: A Delivery: P Rebound: LM Goal: MCL, ACL, MF Combined: N	Oxidant type: S Delivery: P, M, O Rebound: HD Goal: MCL, ACL, MF Combined: Y
B. Impermeable and homogeneous (3%)	Oxidant type: S Delivery: M, O Rebound: U Goal: ACL Combined: Y	Conditions not likely to occur	Oxidant type: S Delivery: M, O Rebound: U Goal: ACL Combined: Y	Conditions not likely to occur
D. Impermeable and heterogeneous (15%)	Oxidant type: S Delivery: M, O Rebound: U Goal: ACL Combined: Y	Conditions not likely to occur	Oxidant type: S Delivery: M, O Rebound: U Goal: ACL Combined: Y	Conditions not likely to occur
E. Consolidated material with low matrix porosity (7%)	Oxidant type: A Delivery: I, O Rebound: HD Goal: MF Combined: Y	Conditions not likely to occur	Oxidant type: S Delivery: I, O Rebound: HD Goal: MF Combined: Y	Conditions not likely to occur
F. Consolidated material with high matrix porosity (7%)	Oxidant type: A Delivery: I, O Rebound: HD Goal: MF Combined: Y	Conditions not likely to occur	Oxidant type: S Delivery: I, O Rebound: HD Goal: MF Combined: Y	Conditions not likely to occur

Note: *GTP* ganglia to pool ratio
Oxidant type: A = All oxidants of equal applicability; S = Certain oxidants likely more effective
Delivery method: P = Probes surgically emplaced; I = Injection wells; M = Physical mixing; O = Other enhancements (e.g., well-to-well recirculation)
Rebound potential: *LM* limited and can normally be managed with subsequent targeted injections, *HD* high likelihood and difficult to manage with subsequent injections
Goal achievement: *MCL* maximum contaminant levels, *ACL* alternative concentration limits, *MF* mass flux reduction. Note that for DNAPL source zones, setting a goal of MCLs will almost always require implementation of a remediation approach or technology following ISCO
Combined remedy needed: Y = almost always needed; N = not always necessary; U = uncertain
[a]The classification of subsurface conditions was developed by Krembs et al. (2010) and the percentage listed in parentheses represents the percentage of sites ($n = 209$ sites) compiled by Krembs et al. (2010) where ISCO was applied (includes all types of sites, not just those with DNAPL source zones)
[b]In the opinion of the authors, the general attributes given are associated with more successful applications of ISCO

Table 9.5. Conditions That Challenge Successful ISCO Treatment of DNAPL Source Zones

Category	Challenging Conditions
Conceptual site model	Site can be poorly characterized with respect to contaminant mass density (pools vs. residual) and distribution, total contaminant mass present, and degree of media heterogeneity
	Site geology can be challenging/complex, limiting contact of oxidant and contaminant
	Significant contaminant mass can have moved into lower permeability media via diffusion and/or mass transport
	Prediction of oxidant and/or oxidant amendment distribution at the site can be poor
Site treatment goal	Treatment goal can be unrealistically stringent for a source zone site
Site-specific limitations	Amount of oxidant necessary to treat the mass of contaminant present can be cost prohibitive
	Volume of oxidant necessary to treat the mass of contaminant present at the site can be prohibitive in terms of time onsite (cost) or the ability of the formation to accept the volume
	Project schedule, budget, and timeframe cannot accommodate multiple oxidant delivery events
Effects of source zone treatment	Byproducts of ISCO can be incompatible with aquifer physical or chemical characteristics and/or contribute to implementation/contact challenges
Rebound	Rebound is more prevalent at source zone sites

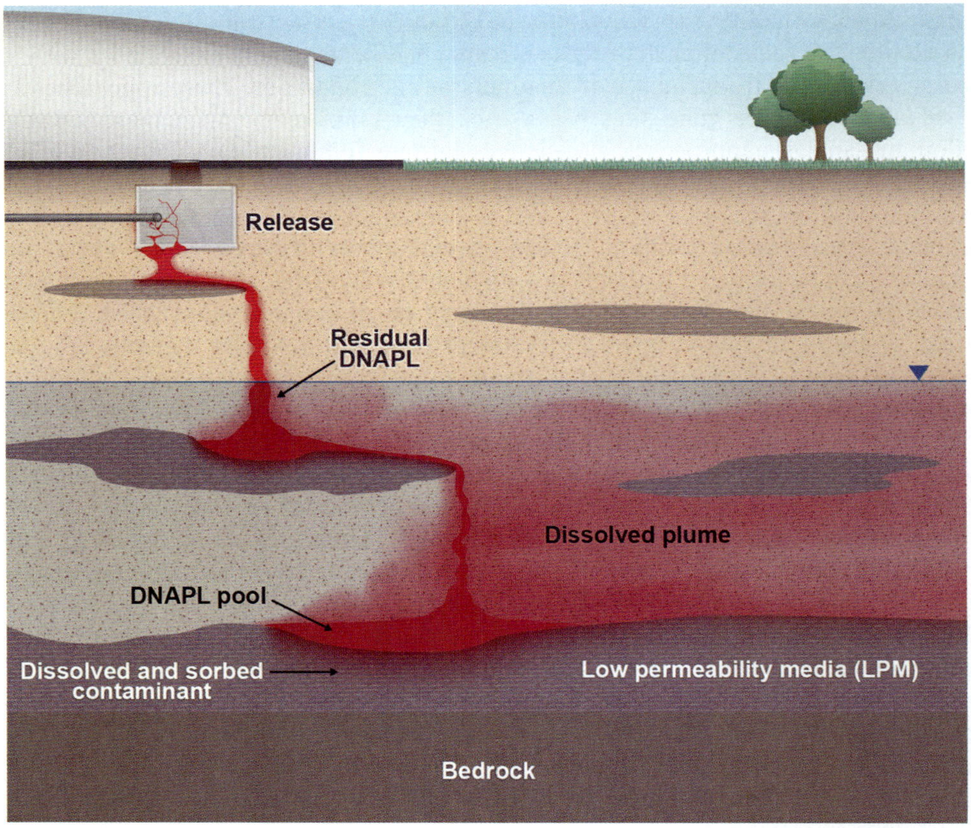

Figure 9.6. Generic contaminated site CSM for a DNAPL source (DNAPL contamination is shown in *red*) (Siegrist et al., 2011).

For source zone treatment, it is critical to understand the contaminant phases, mass, architecture and age, in addition to typical hydrogeology, geochemistry and other site data. At some sites it may be cost prohibitive to attempt to treat all of the mass in a source zone. However, because of the DNAPL architecture, it is often viable to treat enough mass to cost effectively reduce COC mass flux (expressed as mass/time/area) and mass discharge (flux integrated across a cross section of the plume perpendicular to flow) (mass/time). Treatment objectives and monitoring should be established as a function of the phase in which contaminants are present. For example, it is unrealistic and inappropriate to monitor only groundwater concentration within source zones that contain high DNAPL and sorbed phases of contaminant. This is true both for pre-ISCO site characterization and post-ISCO monitoring. Contaminant mass will dictate the chemical oxidant needs for treatment. Dense nonaqueous phase liquid architecture will dictate design features such as the oxidant delivery approach and delivery point spacing. For example, localized DNAPL pools call for closely spaced delivery of oxidants and direct-push injection (multiple events) may be appropriate, whereas residual DNAPL spread relatively uniformly across a site may call for wells installed at a wider spacing. Where contamination has migrated into lower permeability layers, more aggressive delivery approaches, including soil mixing, may be considered. A comprehensive CSM will not only guide ISCO design, it will also serve to guide post-treatment expectations. When it is known that the contaminant is located within lower permeability zones, longer-term back diffusion and site recontamination could be anticipated, and follow-on injections could be planned accordingly.

Current site characterization efforts use systematic planning, dynamic work strategies and real-time field measurements to reduce CSM uncertainty and guide the treatment effort. This dynamic approach is termed the accelerated site characterization process by the American Society for Testing and Materials (ASTM) and the Triad approach by the U.S. Environmental Protection Agency (USEPA) (ITRC, 2003, 2007; ASTM, 2004; Crumbling et al., 2001, 2003; USEPA, 2008). The focus of each of these is managing the uncertainty associated with both the site characteristics and the influence of treatment on site conditions. These approaches are even more valuable to source zone sites because of their complexity. With these iterative data collection and analysis approaches, decisions regarding follow-on data collection efforts can be made near real time to fill data gaps based on information about the site as data is collected. For example, in a source zone where NAPL is suspected, grab sampling can identify a high contaminant concentration location. Follow-on and more refined efforts, where time and budget allow, can be localized in that area with high density data collection rather than moving forward with a grid sampling approach that may be of limited value when a hot spot is identified. During and following ISCO treatment of a site, a similar dynamic approach can and should be used to assess delivery and treatment effectiveness.

Modeling is one tool that can help improve understanding of subsurface conditions and the viability of remediation using ISCO. For example, by simulating possible processes and matching observed data, potential source zone locations and architectures can be estimated. These simulations can help to select appropriate oxidant and delivery approaches for the likely source architecture and hydrogeology. By performing simulations using expected upper and lower limits for site parameters, the effect of uncertainty can also be examined. Examples of ISCO-specific modeling tools include CDISCO and CORT3D.

The conceptual design for ISCO (CDISCO) tool was developed with support from the ESTCP under Projects ER-0626 (Borden et al., 2012) and ER-0623 (Siegrist et al., 2010) and is intended to assist with the conceptual design of injection systems for ISCO using permanganate. The conceptual design for ISCO tool allows the user to evaluate permanganate delivery as a function of aquifer thickness, permeability, total NOD, NOD kinetics, permanganate concentration, injection flow rate and injection duration. It does not, however, model contaminant concentrations and in fact assumes an equal distribution across the treatment zone to which the model is applied.

While CDISCO is an invaluable tool to help design a permanganate delivery system, often models applied to source zones need to consider greater complexity than CDISCO captures. The three dimensional chemical oxidation reactive transport model (CORT3D) was developed with support from SERDP under Projects ER-1290 (Siegrist et al., 2006) and ER-1294 (Illangasekare et al., 2006). CORT3D is based on a modified form of reactive transport in three dimensions (RT3D) version 2.5 (Clement, 1997, 2001; Clement et al., 1998, 2000; Clement and Johnson, 2002; Johnson et al., 2006) and MODFLOW, 2000 (McDonald and Harbaugh, 1988; Harbaugh and McDonald, 1996a, b; Harbaugh et al., 2000). It allows for simulation of three-dimensional (3-D) ISCO reactive transport in groundwater (Heiderscheidt, 2005). CORT3D includes NAPL dissolution, equilibrium or rate-limited sorption, second-order kinetic contaminant oxidation, kinetic oxidation of NOD – both a fast kinetic portion and a slow kinetic portion – and different diffusion coefficients for each aqueous species. The code tracks three aqueous mobile components (contaminant, aqueous chloride and aqueous oxidant) and five immobile components (NAPL, sorbed contaminant, manganese oxide, fast NOD and slow NOD). The code was developed for permanganate, but it may be applicable to other oxidants with further investigation and appropriate oxidant-specific assumptions.

The best predictive tools are only as reliable as the data that is entered into them. Therefore, sophisticated modeling tools are only worth the expense and effort of using them if comprehensive CSM and monitoring plans exist that can be implemented during and after ISCO.

9.4.2.2 Treatment Goals

As discussed in Section 9.3, application of ISCO to DNAPL sites has not been successful in achieving MCLs, though it is capable of meeting less stringent remedial action objectives (such as ACLs). Generally, it is more reasonable to deploy ISCO in the source zone with the intent to reduce the source zone mass and perhaps decrease the mass discharge so that a combined remedy such as ISCO followed by MNA can be effective.

9.4.2.3 Oxidant Type and Concentration

Some oxidants are more viable for treatment of source zones than others based on their stability (longevity) in the subsurface and their ability to be distributed throughout a TTZ. Because of their persistence, permanganate and persulfate, under some activation approaches, tend to be more viable for source zone treatment compared to CHP or ozone (Siegrist et al., 2011). On the other hand, where contaminants such as coal tar or chlorinated solvents are trapped at high saturations, the more aggressive ISCO oxidants, such as CHP or ozone, might facilitate movement of the contaminant, which could increase accessibility and make the DNAPL more amenable to ISCO and combined treatment technologies and approaches.

Where contaminants are present at high concentrations, such as in source zones, it may seem intuitive to deliver high oxidant concentrations in order to rapidly and extensively destroy contaminants. Several factors, in addition to contaminant mass, should be considered when selecting an appropriate oxidant concentration. Extensive nonproductive oxidant depletion occurs due to interaction with the natural constituents of the subsurface, regardless of the oxidant selected. The higher the concentration of oxidant that is delivered, the faster this nonproductive oxidant depletion occurs. Oxidants react more rapidly with most contaminants than with subsurface media; therefore, high concentrations are not problematic where the full contaminant mass is present in the aqueous phase. This is rarely the case, particularly in source zones. It is only efficient to provide oxidant in excess of what is consumed by contaminant reactions and rapid reactions with the media so it can persist long enough to allow for the transfer of remaining mass from the nonaqueous or sorbed phases and to allow for transport into lower permeability areas via diffusion or downgradient via advection and dispersion. In addition to nonproductive oxidant consumption, using a concentration higher than necessary can exacerbate the generation of byproducts, such as gas or acidity/alkalinity, with effects that may be undesirable.

Another issue associated with the need for a large mass of oxidant is that when the large mass is delivered at high concentration, the oxidant solution can experience density-driven flow downward through the aquifer due to gravitational forces. Schincariol and Schwartz (1990) reported that this could happen in subsurface systems when the density difference between fluids is as little as 0.0008 g/cm^3, which is equal to the density difference caused by a concentration of 1,000 mg/L of sodium chloride. This effect has been noted with permanganate by several investigators (Nelson, 1999; Stewart, 2002; Siegrist et al., 2006). While this density effect can be advantageous if expected, treatment can be ineffective in cases where the contaminated zone is completely bypassed.

9.4.2.4 Oxidant Delivery

Certain delivery approaches are more amenable to source zone treatment as well. Typically treatment of source zones requires longer delivery of a greater volume of remediation fluids; therefore, it is advantageous to use a delivery approach that will accommodate this effort. It may ultimately be cost effective to install delivery wells (as opposed to direct push probes) or recirculation systems to accommodate longer delivery periods. Also, soil mixing and hydraulic fracture-emplaced ISCO have the potential to be effective approaches for source zone treatment, particularly where these approaches are closely matched to site conditions. Soil mixing can help overcome heterogeneities through the physical mixing process. Fracturing might be used to facilitate delivery in low permeability formations, although closely spaced fractures will likely be needed to overcome diffusion-limited mass transport. The key point is that more expensive and aggressive delivery approaches improve cost effectiveness as site complexity (hydrogeology and contaminant mass distribution) increases.

Delivery into complex source zones can be enhanced by improved hydraulic delivery and distribution, and improved oxidant stability and reactivity with the COCs. Improved hydraulic delivery can improve contact between the oxidant and COC to increase treatment efficiency and effectiveness. Improved oxidant stability can lead to increased contact duration, which is particularly important where contaminant oxidation is controlled by mass transfer limitations. Oxidant stability can be improved by the addition of amendments to alter its chemical or physical properties. Table 9.6 summarizes methods that are being developed for enhanced ISCO delivery.

Table 9.6. Emerging Approaches for Enhanced ISCO Delivery (after Siegrist et al., 2011)

Approach	Description and Examples
Stabilized oxidants	Phytate, citrate, and malonate have been shown to increase the half-life of H_2O_2 during CHP reactions (Watts et al., 2007)
Encapsulated oxidants	Potassium permanganate in solid microcapsules (Ross et al., 2005) or within a paraffin matrix (Dugan et al., 2009) to achieve controlled release into groundwater (Lee and Schwartz, 2007; Lee et al., 2009)
Polymers and delivery aids	Use of polymers like xanthan to overcome ISCO delivery problems caused by zones of low permeability, site heterogeneity, and excessive NOD (Smith et al., 2008; Crimi et al., 2011)
Density-driven delivery	Density-driven flow was used to deliver permanganate to NAPL located in groundwater on top of an aquitard and a model (MIN3P-D) could simulate the observed behavior (Henderson et al., 2009)
Mechanical mixing	Mechanical mixing of oxidant solutions into shallow soils and subsurface porous media using augers for deep soil mixing or tilling and excavator mixing in shallow soils

In situ chemical oxidation system design must accommodate the treatment limitations of sorbed and DNAPL phase contaminants in source zones as dictated by the rate and extent of mass transfer into the aqueous phase. In addition to controlling the volume, duration, concentration and approach for oxidant delivery, the number of planned delivery events is another design parameter that can be controlled to accommodate the challenges of treating source zones. In fact, for sites with low rates of mass transfer from the sorbed and DNAPL phases, or from lower permeability media, it may be more cost effective to plan multiple delivery events over time with a lag time between them to accommodate contaminant re-equilibration with the aqueous phase than to plan single, longer duration events where oxidant may be used less efficiently.

9.4.2.5 Cost Considerations

Cost is a key consideration that can influence the viability of successful ISCO for source zone sites. Between the extensive amount of oxidant that can be consumed in DNAPL source zones with high contaminant mass densities and oxidant depletion due to reactions with the media within that treatment zone, chemical costs can be very high for source zone treatment (see Example 2).

Example 2. Oxidant Requirements Using a Mass Balance Approach

Site X contains high concentrations of TCE – up to 100 mg/L in groundwater and 1,000 mg/kg in porous media. Presence of DNAPL is suspected. Permanganate injection via wells has been deemed most viable for the site and multiple injections are planned. The contamination is between 10 ft (3.05 m) and 30 ft (9.1 m) bgs over an area 100 ft (30.5 m) in length and 100 ft (30.5 m) wide.

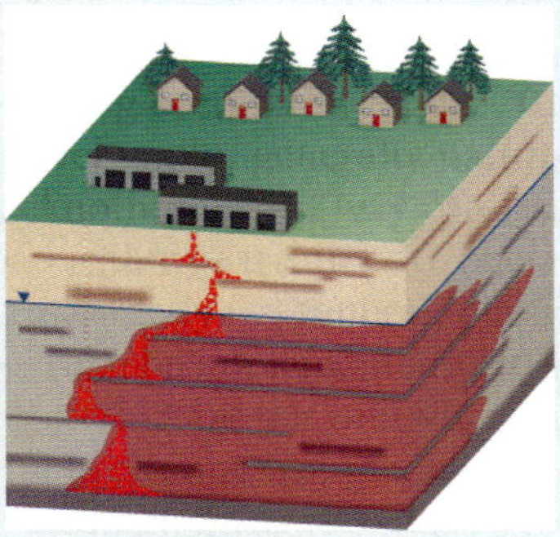

Assuming a porosity of 0.30 and a bulk density of 1.6 g/cm^3, the total mass of TCE estimated in the treatment zone is approximately 9,250 kg (~20,400 lbs). With a reaction stoichiometry of 1.8 g of permanganate (as MnO_4^-) per g of TCE, approximately 16,650 kg (~36,600 lbs) of permanganate is needed to oxidize the TCE. This mass does not consider the natural demand of the media for permanganate, which would need to be added in excess. Using a simple mass balance approach and an assumption of an NOD of 2 g of permanganate per kg of media, an additional (approximately) 18,160 kg (~40,000 lbs) of permanganate is necessary. If potassium permanganate is used, this translates to 76,600 lbs of $KMnO_4$. At a current cost of \$2.25 per pound of $KMnO_4$, the total cost of permanganate alone is \$172,350. This is \$23/cy of media treated accounting for only the chemical costs. Major assumptions include the following: (1) all of the permanganate is consumed within the target area with no bypass through or around, (2) this mass of oxidant can be delivered uniformly at a reasonable concentration and (3) all contaminant is present in an available aqueous phase within the treatment timeframe. The first assumption presumes a fairly homogeneous formation. Any degree of heterogeneity will cause some extent of bypass. The second assumption depends on numerous design variables. If, for example, a concentration of 1 wt% $KMnO_4$ is used, 1.09 million L (2.9 million gal) is required. This would equal not quite 2 pore volumes of fluid. An underlying concern is whether the formation and budget will accept this volume and associated time onsite. A recirculation approach may be one means to facilitate delivery of this volume/mass of oxidant.

In situ chemical oxidation projects at DNAPL sites have, on average, a higher total cost of treatment relative to other sites (Krembs et al., 2010), which can be attributed to the longer duration of oxidant delivery, greater volume of reagents, greater mass of oxidant and a higher number of delivery events necessary for effective treatment. Sites with large zones of very high contaminant mass density, such as DNAPL pools, are unrealistic to treat with ISCO alone. However, partial mass removal using ISCO followed by alternative treatment approaches that are more cost effective for residual mass may be appropriate. Partial mass removal using ISCO, in fact, may provide enough treatment to reduce mass flux to an acceptable limit, depending on the contaminant architecture. Alternatively, ISCO may be appropriate for residual mass following excavation or other source removal approaches. As discussed earlier, ISCO is best considered as part of a comprehensive approach for source zone remediation.

9.4.2.6 Impacts of ISCO Source Zone Treatment on Aquifer Conditions

Table 9.7 provides an indication of the major subsurface interactions that may occur during ISCO treatment of DNAPL source zone, while Table 9.8 presents some oxidant specific interactions. As discussed below, these interactions can affect aquifer conditions with beneficial or adverse effects.

In situ chemical oxidation can alter subsurface geochemistry (oxidation–reduction potential [ORP], pH) and cause changes to the mobility of metals. Metals of concern include antimony, arsenic, barium, cadmium, hexavalent chromium, copper, iron, lead, mercury, nickel and selenium. Typically, changes are temporary and conditions return to pre-ISCO conditions after the aquifer has re-equilibrated. In source zones containing DNAPL where high oxidant concentrations, oxidant mass, or treatment duration occurs, it is possible for these effects to be amplified over long distances and/or durations.

Gas evolution during ISCO can fill pore spaces and reduce permeability of the formation. For example, gas evolution (CO_2 byproduct) during permanganate ISCO of DNAPL has been reported where groundwater flow has been altered, bypassing zones where significant gas is produced such as the NAPL interface (Reitsma and Marshall, 2000; Lee et al., 2003; Li and Schwartz, 2003). Also, oxygen generation during CHP reactions can be extensive (Watts et al., 1999). While no controlled studies have evaluated the impact in transport systems, it is reasonable to infer that localized O_2 production during CHP reactions can impact permeability much like CO_2. In fact, O_2 generation during CHP reactions is much faster and more aggressive than CO_2 generation during permanganate reactions, so its impact is likely even more pronounced.

Chloride is a byproduct of the reaction of oxidants with many common DNAPLs including TCE and PCE. When oxidation is occurring within source zones, extensive amounts of chloride can be generated local to the point of reaction. For example, Petri (2006) measured concentrations in excess of 3,000 mg/L when treating nonaqueous phase chlorinated solvents with permanganate near the point of reaction. In permanganate systems, chloride does not create challenges to treatment; however, in free radical oxidation systems, such as persulfate and CHP, chloride can act as a free radical scavenger, limiting productive oxidation of contaminants. Furthermore, chloride can form complexes with amendments used to catalyze or facilitate catalysis of free radical oxidation reactions such as iron, impacting oxidation kinetics (Sung and Morgan, 1980). Finally, when present in high concentrations (>3,500 mg/L), chloride can react with contaminants and natural organic carbon to form halogenated intermediates that are potentially more toxic than the parent contaminant (Aiken, 1992; Waldemer et al., 2007).

As discussed in Section 9.2.2.2, manganese dioxide solids (denoted MnO_2, although actually the Mn is in a mixed oxidation state) are a byproduct of ISCO unique to permanganate. Manganese dioxide solids have the potential to alter subsurface hydraulic conductivity and

Table 9.7. Major Subsurface Interactions that Can Potentially Occur During ISCO

Interaction	Type	Comments
Altered behavior of COCs	Organic COCs	Formation of films and rinds due to chemical oxidation at NAPL–water interfaces is possible; this can reduce mass transfer into the aqueous phase where oxidation reactions typically occur
	Metal co-COCs	Oxidation reactions can change the redox state of certain metals and change their toxicity and mobility in groundwater (e.g., Cr^{+3} to Cr^{+6}, more toxic and more mobile; As^{+3} to As^{+6}, less toxic and less mobile)
Permeability loss in injectors and porous media	Particle generation	Formation of reaction products can result in generation of particles, which can clog injectors, well screens, and boreholes (e.g., oxidation using MnO_4^- can generate MnO_2 particles)
	Gas evolution	Oxidation reactions can generate gas (O_2 and CO_2), which can become entrapped in soil and groundwater and reduce effective permeability for flow
Biogeochemical changes	Chemical impurities	Oxidant formulations can include impurities, which can affect groundwater quality (certain permanganate formulations contain metal impurities)
	pH	Oxidation of organics can yield acidity and depress subsurface pH in zones of intense reaction (at NAPL–water interfaces)
	Ion behavior	Permanganate ISCO delivers ions into the subsurface, which can affect sorption and cation exchange sites (e.g., Na^+ can cause dispersion of expandable clays; K^+ and Na^+ can displace sorbed cations)
	Natural organic matter (NOM)	Oxidation of NOM can reduce the f_{oc} and K_d of the subsurface within a TTZ
	Dissolved organic carbon	Oxidation of NOM and hydrophobic organics can generate dissolved organic carbon, which can serve as a substrate for microbial reactions
	Microbiology	Oxidation within a subsurface TTZ can perturb ambient microbial ecology initially depressing biomass, diversity and activity but subsequently stimulating biomass and activity levels to pre-ISCO levels or higher

permeability near the point of oxidant injection and within a TTZ, affecting transport of permanganate in the subsurface. Crimi and Ko (2009) and Crimi et al. (2009) determined that MnO_2 particles in groundwater could be controlled, inhibiting deposition using the amendment sodium hexametaphosphate (SHMP). When SHMP is included in a permanganate oxidation solution, the resulting MnO_2 particles are smaller in size and remain suspended in solution (mobile). During one-dimensional (1-D) transport evaluations of MnO_2 mobility both with and without SHMP in various porous media (organic matter, clay, mineralogy), SHMP enhanced MnO_2 stability in groundwater and resulted in significantly less MnO_2 deposition, particularly within the contaminant source zone. Currently, Crimi et al. (2011) are conducting a field demonstration to evaluate the use of SHMP with permanganate to enhance delivery and

Table 9.8. **Subsurface Interactions for Contrasting Oxidants Used for ISCO**

Feature	Catalyzed H_2O_2	$KMnO_4$ or $NaMnO_4$	Persulfate	O_3
Subsurface effects on potential rate and extent of oxidation of COCs				
Effect of nonproductive reactions	Can lead to high mass delivery of oxidant solutions	Can lead to high mass delivery of oxidant solutions	Can lead to high mass delivery of oxidant solutions	Can lead to high mass delivery of oxidant solutions
Effect of pH	Optimum <4	Effective at pH 3–12	Optimum >11 for alkaline activation; other activation approaches are less pH dependent	Optimum pH <7
Effect of total dissolved solids	Little to no effect	Little to no effect	Little to no effect	Little to no effect
Effect of CO_3^{-2}	Radical scavenger	No effect	Radical scavenger	Radical scavenger
Effect of temperature	Affects rate of reaction	Affects rate of reaction	Increased temperature increases rate of activation and reaction	Affects rate of reaction
Potential for oxidation to alter pre-ISCO subsurface conditions				
pH in target treatment zone (TTZ)	Can be lowered to acidic conditions depending on buffering capacity	Can be lowered to acidic conditions depending on buffering capacity	High pH can persist following reaction using alkaline pH activation; pH can be lowered to acidic conditions in other activation approaches depending on buffering capacity	Can be lowered to acidic conditions depending on buffering capacity
pH at DNAPL interfaces	Can be lowered toward pH 2	Can be lowered toward pH 2	Can be lowered toward 2	Can be lowered toward pH 2
Temperature	Appreciable increases possible depending on oxidant concentration used	Minor increases possible under normal applications	Minor increases possible under unactivated or iron activated conditions; appreciable increases can result from H_2O_2 or heat activation	Moderate to appreciable increases possible depending on oxidant concentration used
Metal mobilization	Limited potential	Potential for mobilization of redox metals, notably Cr^{+3}	Potential for mobilization of redox metals, notably Cr^{+3}	Potential for mobilization of redox metals, notably Cr^{+3}
Permeability loss	Potential for loss at injection wells and in formation due to gas evolved	Potential for substantial losses at injection wells due to MnO_2 formation	Dependent on activation approach; H_2O_2 activation can lead to permeability losses due to gas evolved; alkaline pH activation could potentially lead to formation of precipitates in/near injection	Potential for loss at injection wells and in formation due to gas evolved
Microbiology	Potential to disrupt biomass levels and activity within TTZ but effects are short term	Potential to disrupt biomass levels and activity within TTZ but effects are short term	Potential to disrupt biomass levels and activity within TTZ but effects are short term	Potential to disrupt biomass levels and activity within TTZ but effects are short term

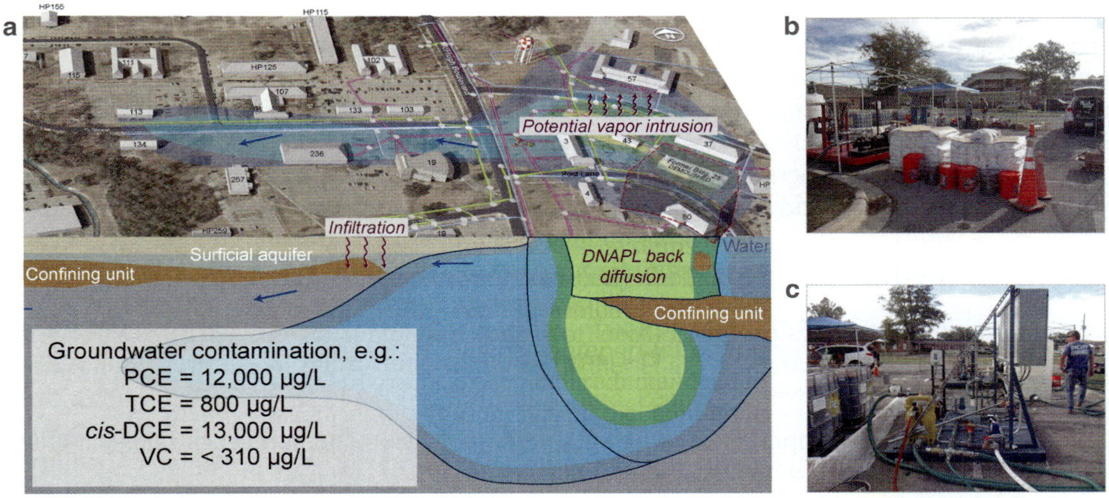

Figure 9.7. Field demonstration Marine Corps Base Camp Lejeune, Jacksonville, North Carolina, of ISCO for remediation of chlorinated solvent contaminated groundwater using coinjection of oxidant and amendments to enhance delivery (from Crimi et al., 2011). The ISCO system includes coinjection of $KMnO_4$ (5 g/L), SHMP (5 g/L), and xanthan gum (0.5 g/L) during a single-well injection event to treat a 10 meter (m) (33 foot) diameter zone of sand and silty sand in the depth interval of 11 to 16 m below ground surface (bgs).

distribution. The demonstration is ongoing at the Marine Corps Base Camp Lejeune in Jacksonville, North Carolina, USA (Figure 9.7). The test area is located within the site of the former base dry cleaning facility, which operated from the 1940s until 2004 when the building was demolished. The subsurface in the test area is comprised of sands of varying permeability with some interbedded silt layers. Perchloroethene contamination has been detected at high concentrations in soil and groundwater samples.

9.4.2.7 Rebound of DNAPL Compound Concentrations

A general rule of thumb used during groundwater monitoring is that a change of $<10\%$ in the field parameters and contaminant concentrations between three consecutive monitoring events can be used to demonstrate re-equilibration or rebound. Re-equilibration may also be presumed to have occurred when geochemical parameter concentrations measured at certain monitoring locations are within the 95% confidence interval for the average baseline condition. A substantial length of time may be required for complete re-equilibration to occur depending on rates of advection versus contaminant desorption, dissolution and back diffusion.

Rebound after an initial ISCO DNAPL source zone treatment event is common. The data presented by Krembs et al. (2010) indicated that rebound (based on an increase of 25% or more) was observed in at least one well following ISCO application at 78% of DNAPL sites and at 56% of the sites where DNAPL was not suspected to be present. While a number of factors may contribute to rebound conditions, the primary reason for rebound at DNAPL sites is because oxidation reactions occur in the aqueous phase (see Section 9.2.1). Dense nonaqueous phase liquid and sorbed contaminants must first dissolve or desorb to be accessible for destruction by oxidants. Once oxidants are depleted due to their reaction with contaminants and/or natural media constituents, any remaining contaminant will continue to partition into the aqueous phase.

While rebound has a negative connotation when associated with performance assessment at treated sites, there are indeed positive attributes of rebound. Rebound marks the transfer of contaminants into a more treatable aqueous phase when identified appropriately as rebound and is not attributable to migration of contaminant into the treated zone from upgradient areas. Rebound itself is an indication that there is untreated mass at the site and additional delivery events are required. Furthermore, rebound data can be used to help locate specific source areas where effort, time and budget can be focused for follow-on treatment.

9.4.3 Remedial Design and Implementation

Implementation of ISCO for treatment of DNAPL source zones should follow the same overall considerations that apply to the general application of ISCO (Siegrist et al., 2011), including:

- Selection of an appropriate approach for delivery that is compatible with site conditions and is consistent with a realistic CSM
- Safe work planning, procedures and equipment
- Provision of appropriate equipment and experienced personnel for chemical handling, mixing and injection operations
- Clear implementation targets for oxidant dose and volume based on a rational design
- An appropriate monitoring program that includes process and performance monitoring
- An adaptive approach that allows dynamic work planning and facilitates use of real-time data in the field

However, since the treatment of DNAPL source zones is inherently challenging, the tolerance for inappropriate application of a technology like ISCO is minimal.

The primary challenge for successful treatment of DNAPL source zones using ISCO is to achieve effective contact between the oxidant and DNAPL. For this reason, careful selection of an implementation approach that is consistent with site geologic conditions is especially critical when treating a source zone. Hence, a realistic CSM is critical because it provides the framework for identifying the specific target locations that require treatment.

Developing an appropriate ISCO implementation approach involves professional judgment as well as technical analysis. Often, several potentially viable implementation approaches exist, and the most favorable approach might consider nontechnical factors related to accessibility of treatment areas, cost, etc. The primary elements of an ISCO implementation approach include the following:

- Consideration of combinations of ISCO and other treatment technologies or use of ISCO as part of a treatment train (such as integrating ISCO with bioremediation, thermal, or MNA)
- Clear treatment objectives defining the TTZ and the treatment endpoints
- Identification of site-specific conditions that may impact the oxidant delivery method or the design oxidant dose and volume
- Selection of an appropriate oxidant and delivery method that overcomes the identified limitations to the maximum extent possible

9.4.3.1 Combining ISCO with Other Technologies and Approaches

Combined remedies often are required for treating DNAPL sources (see Chapter 15). Using ISCO as a component of a combined remedy can be accomplished simultaneously or through various temporal or spatial combinations with other approaches (Siegrist et al., 2011). Enhanced *in situ* bioremediation (EISB) and MNA are most frequently combined with ISCO. ISCO can be an effective pretreatment step to bioremediation by providing partial treatment and making COCs more bioavailable. ISCO has been shown to temporarily decrease microbial abundance and diversity, but these typically rebound with time. In some cases, bioremediation and MNA may actually be enhanced by ISCO as the oxidant may add terminal electron acceptors or partially oxidize contaminants making them more amenable to biodegradation.

In situ chemical oxidation is one component of a comprehensive approach for source zone remediation. A case study review by Krembs et al. (2010) found that sites with NAPL present were more likely to have ISCO coupled with other technologies, especially post-ISCO technologies such as EISB and MNA. Currently, remediation of source zones is increasingly viewed as best accomplished by combining remedies simultaneously or sequentially for different zones of contamination (NRC, 2005) (Figure 9.8). *In situ* chemical oxidation can be appropriately

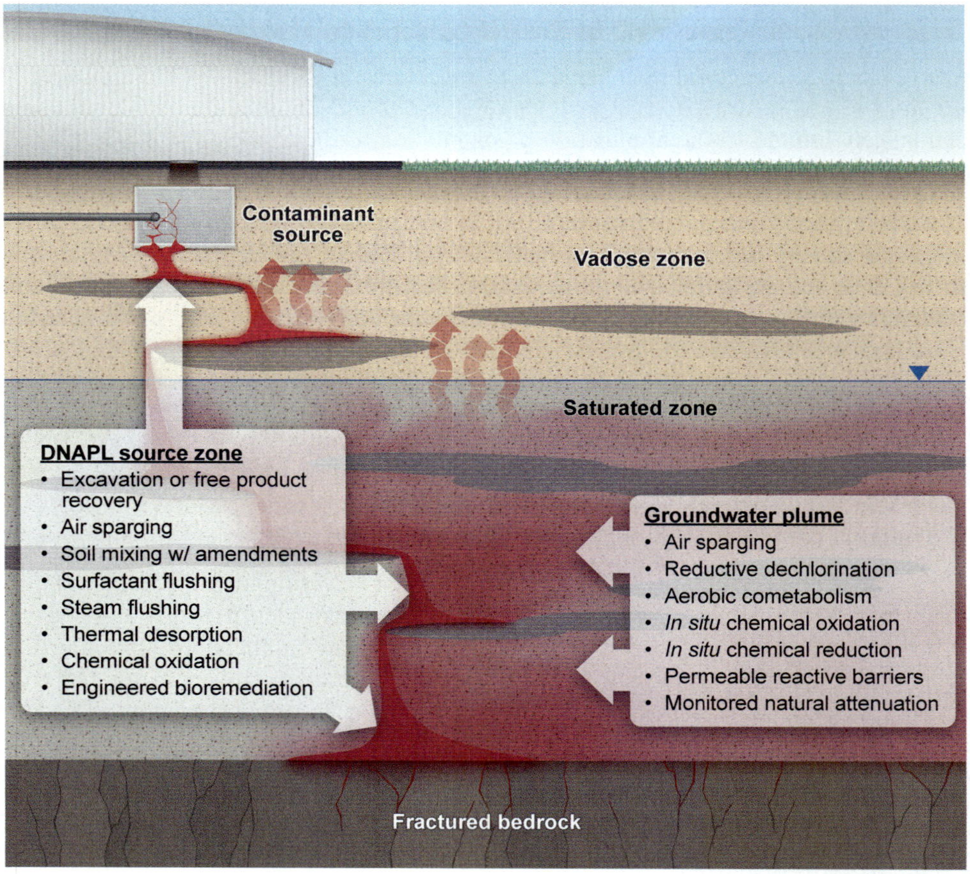

Figure 9.8. Illustration of *in situ* technologies that might be viable for a combined remedy for a site with a DNAPL source zone and associated groundwater plume (DNAPL contamination is shown in *red*) (Siegrist et al., 2011).

coupled with pre-ISCO excavation, air sparging, soil mixing with amendments, surfactant flushing, steam flushing, thermal desorption or engineered bioremediation. On the other end of the treatment train, ISCO can be coupled appropriately with air sparging, reductive dechlorination, aerobic biodegradation, *in situ* chemical reduction (ISCR), permeable reactive barriers or MNA.

Coupling ISCO with different remediation technologies can improve the probability of meeting remediation objectives, especially at sites with unique challenges or aggressive cleanup goals. To optimize the performance of ISCO and the combined remedy, careful consideration of potential interactions is recommended prior to implementation. Siegrist et al. (2011) describe principles of combining ISCO with other approaches, along with key considerations and guidance.

9.4.3.2 ISCO Treatment Objectives

For DNAPL source zone treatment, typically ISCO treatment objectives are related to COC mass reduction although in some situations a numerical concentration-based treatment objective may be adopted. In some cases, an appropriate ISCO treatment objective may stem from an evaluation of the impacts of source zone treatment upon other portions of the site and related exposure pathways. The relationships between source treatment and plume mass discharge or concentration are often an important consideration in developing source area treatment goals (see Chapter 6).

In developing site-specific ISCO treatment objectives, it is critical to differentiate between the vadose zone and saturated zone portions of the source. Vadose zone treatment using ISCO has been largely limited to the application of ozone due to the challenges of applying aqueous oxidant solutions to the vadose zone (Siegrist et al., 2011).

9.4.3.3 Overcoming Site-Specific Limitations on ISCO Application

For ISCO to be successful in treating a DNAPL source zone, several potential limitations must be overcome:

- Difficulty in delivering an adequate mass of oxidant to meet stoichiometric demands uniformly across the TTZ
- Limited persistence of some oxidants and the resulting difficulty of maintaining sufficient oxidant concentrations to sustain elevated COC reaction rates and allow diffusive transport to occur
- Potential for adverse secondary geochemical effects

These limitations form an important framework for selecting an appropriate ISCO implementation approach. When dealt with in a site-specific context, identification and consideration of the potential limitations for ISCO application can lead to the development of a more effective implementation approach. Low permeability media and geologic heterogeneity are two primary potential limitations for ISCO application, particularly within DNAPL source areas. Low permeability conditions in some cases restrict the volume of oxidant that can be delivered, which may prevent an adequate oxidant dose from being applied. Low permeability conditions also require higher injection well density, which increases costs.

When oxidants are injected into the subsurface in a heterogeneous formation, the amendments tend to be transported primarily in the more transmissive zones. Therefore, oxidant delivery into the lower permeability zones occurs more gradually over time by advection or diffusion transport processes. For ISCO, the relative rates of oxidant transport into the lower permeability zones and oxidant consumption are critical implementation considerations. Oxidants that are more persistent will be more amenable to delivery into lower permeability zones in heterogeneous environments.

9.4.3.4 Oxidant Selection and Delivery Methods

The primary oxidant delivery methods are batch injection, continuous injection (constant pressure systems or metered flow systems) and continuous recirculation or mechanical mixing. These are discussed individually below. Although in theory each method can be used with any oxidant, selection of the appropriate oxidant and delivery method should be considered together since in some cases oxidant characteristics may limit the available delivery alternatives. For example, oxidant formulations including percarbonate are typically a solid slurry and will have a limited radius of influence compared to aqueous phase oxidants. Therefore, percarbonate applications may require a higher injection well density or may rely on mechanical soil mixing or other means of delivery such as hydraulic or pneumatic fracturing.

Batch oxidant injection is the most prevalent ISCO implementation approach. Batch injection involves injecting a specified volume of oxidant solution into the subsurface in one or more discrete injection events. For DNAPL source zone treatment, multiple batch injection events should be anticipated to achieve effective COC treatment. Batch injection can be performed using either permanent injection wells or temporary injection points. Direct push drilling of temporary injection points is a common approach, which may facilitate targeting of specific vertical treatment intervals. Proper abandonment of temporary injection points using tremie grouting or other techniques is critical to ensure that the penetrations do not serve as conduits for daylighting or preferential flow of injected fluids.

Continuous oxidant injection systems involve automated delivery systems that maintain a steady delivery of oxidant over a sustained period of time. Ozone gas systems are by default continuous or pulsed injection systems. Similarly, aqueous oxidant solutions can be delivered continuously over extended time periods using either flow metering or constant pressure automated systems. Continuous oxidant delivery may be advantageous for treatment of source zones, where elevated COC concentrations lead to large stoichiometric oxidant dose requirements. Also, continuous oxidant delivery may be advantageous to overcome limitations associated with low permeability and heterogeneous environments.

Oxidant recirculation systems take advantage of groundwater extraction to provide hydraulic control, to increase gradients for oxidant transport through the treatment zone, and to minimize the net addition of water volume to the TTZ. Recirculation systems are generally most practicable in moderate to high permeability environments, although they can still be used in lower permeability environments. Oxidant recirculation systems may require relatively sophisticated fluid handling to mitigate potential impacts of precipitant formation and other secondary geochemical effects related to pH, oxidant presence, potential off-gassing, etc.

Mechanical soil mixing can be accomplished using large diameter augers, rotary drums or other configurations. Mechanical soil mixing has the primary advantage of being able to overcome the delivery limitations in heterogeneous and low permeability settings. However, mechanical soil mixing requires unobstructed access and is limited to a relatively shallow TTZ. Mechanical soil mixing of ISCO reagents involves adding either aqueous oxidant solutions or solid-phase oxidant directly to the soil during the mixing process. Mechanical soil mixing may impact the geotechnical properties of soils, and a geotechnical engineering analysis should be performed to assess the suitability for post-treatment construction, if applicable.

9.5 PROJECT MONITORING AND SYSTEM OPTIMIZATION

9.5.1 Program Components and Elements

In situ chemical oxidation monitoring programs are typically designed to meet several objectives under complex, uncertain and difficult-to-predict conditions. A typical monitoring program includes baseline, delivery and treatment performance monitoring activities, which are conducted to achieve the following goals: (1) collect the necessary data to measure and document achievement of the ISCO treatment operational objectives and treatment milestones, (2) monitor the ISCO process to continually confirm that the design is performing as intended and (3) provide engineering performance parameters for subsequent applications and/or post-ISCO technologies (Siegrist et al., 2011). The elements that are typically addressed in a monitoring plan include:

- Clearly defined data needs and data quality objectives
- Frequency of observations and sampling
- Duration of the monitoring period
- Number and location of monitoring points
- Environmental media to be sampled
- Field and laboratory analytical methods
- Quality assurance/quality control procedures
- Data analysis, evaluation, and documentation

While the types of data collected (contaminant, oxidant, intermediates/byproducts, geochemical changes) for sites where ISCO is being used to treat DNAPL source zones may not differ relative to sites with dilute contaminant plumes, the data collection locations, frequency and duration should be greater for sites with DNAPL source zones. Greater spatial density of data collection will help to ensure that oxidant is transported throughout the TTZ and makes contact with the DNAPL for an adequate time to achieve the expected mass depletion. Also, frequent monitoring will help guide modifications to the ISCO system, if necessary, to achieve target concentrations and oxidant longevity. Long-term monitoring will not only help to assess if treatment objectives have been met, it will also help to establish whether rebound (re-equilibration in the case of source zones) is occurring or likely to occur.

Monitoring only groundwater may underestimate the extent of treatment. Source zones can have contaminant present in multiple phases, including NAPL, sorbed, aqueous and vapor phases. In many cases it is important that monitoring includes each of these phases, although it is common to monitor only the aqueous phase via monitoring wells. However, significant DNAPL mass depletion can occur at a site without significant changes to post-treatment groundwater concentrations after sites have re-equilibrated. Furthermore, monitoring only groundwater provides a limited understanding of potential exposure pathways and associated risk, such as vapor phase concentrations. An integrated monitoring approach will often be required for ISCO applied to DNAPL source zones. Such an approach would include sampling and analysis of COC concentrations in groundwater along with porous media phases complemented by diagnostic techniques such as vertical profiling (using membrane interface probes) and other chemical analyses (chloride concentrations and compound-specific isotopes). Samples collected for analyses need to be carefully handled and preserved to avoid errors and artifacts.

An important foundation for a monitoring program associated with an ISCO project involves the establishment of operational objectives. These are defined as a set of specific delivery and treatment performance-related objectives that, if met, should result in the achievement of the treatment goals for the ISCO remedy. Selecting operational objectives is a project-specific decision. In most cases, they are beneficial for use as a standard against which to compare performance measurements and as a basis for follow-up optimization activities to improve ISCO treatment efficiency. For example, an oxidant recirculation project may include a minimum runtime operational objective to ensure that the treatment system infrastructure is capable of meeting oxidant delivery requirements. If the system runtime does not meet the operational objectives, then optimization activities are needed.

Care should be taken to ensure that the operational objectives are realistic and achievable and, if achieved, lead to successful ISCO treatment. Operational objectives may be established based on bench or field pilot testing, modeling, literature review, previous experience or engineering judgment. Consultation with prospective or contracted ISCO treatment contractors may be beneficial to help define these objectives. Some ISCO contractors have developed unique and proprietary knowledge of ISCO treatments under certain hydrogeochemical and contaminant settings. Typical operational objectives focus on ensuring that the oxidant makes contact with the contaminants within the TTZ for an adequate amount of time to achieve the desired contaminant mass/concentration reduction. Operational objectives that may be considered for an example activated persulfate ISCO application include the following:

- Co-deliver a minimum persulfate concentration of *X* mg/L and activator concentration of *Y* mg/L to the zones of highest groundwater contamination within the TTZ.
- Maintain a minimum of *Z* days of persulfate residence time within the TTZ for the oxidation reaction to occur.
- Minimize loss of oxidant from the TTZ during the delivery process due to daylighting or short-circuiting.
- Maintain hydraulic control and mitigate spread of contamination outside the TTZ.
- Utilize a real-time measurement and data analysis routine to optimize/adapt the injection/monitoring program, ensure cost effective treatment and protect sensitive receptors in the area of treatment.
- Experience zero health and safety incidents.

9.5.2 Monitoring Phases

9.5.2.1 Baseline Monitoring

Baseline monitoring is performed to establish pre-existing conditions and, at a minimum, must include the analysis of environmental samples from impacted media (groundwater, aquifer solids, etc.) at the site for target contaminant(s) and general water quality parameters (dissolved oxygen [DO], ORP, temperature, conductivity) (Siegrist et al., 2011). These baseline analytes are selected because they may affect oxidant performance and may be useful in understanding the ISCO delivery needed; parameters such as ORP, DO and conductivity may be used as surrogate tracers for oxidant migration and distribution. Based on detailed site understanding, the baseline testing may need to be expanded to other parameters, such as potential biological or abiotic interferences, reduced or redox-sensitive metals (if present), products of ISCO treatment (chloride, manganese oxides, acetone, sulfate), fraction organic carbon (f_{OC}) of the sediments and biological sampling to evaluate pretreatment biodiversity.

In general, the baseline monitoring program focuses on characterizing the conditions that affect the injection and monitoring program design, and it is critical to provide a basis for oxidant delivery and contaminant treatment performance assessment. Therefore, much of the baseline sampling focuses on hydrogeological parameters, water chemistry and contaminant phase distribution. Baseline sampling generally should include (1) locations upgradient of the contaminated zone to establish ambient conditions, (2) locations within the TTZ to establish ambient baseline conditions and (3) locations downgradient of the TTZ for treatment performance monitoring (e.g., if downgradient migration of oxidants is expected or if mass flux evaluations are conducted as a treatment cessation criterion). *In situ* chemical oxidation treatment zones are often placed within the core of the contaminated zone and are often subject to some degree of influx or rebound.

Baseline monitoring is typically performed to achieve the following specific objectives:

- Determine pre-existing contaminant concentrations in the different phases.
- Estimate pre-existing contaminant mass.
- Estimate pre-existing mass flux emanating from the contaminant source.
- Estimate ambient variability in water quality concentrations due to natural conditions such as precipitation, seasonal or tidal influence and wet/drought cycles if historical data is available or multiple rounds of baseline monitoring are possible.
- Perform field test for pre-existing geochemical conditions.
- Quantify background metal concentration if metals are at levels of concern.
- Quantify background vadose zone contaminant vapor concentrations if vapor migration is of concern.
- Estimate background biological degradation processes.

Baseline monitoring can be accomplished in many different ways. The most common is to supplement an existing well network and historic database of contaminant concentration data with data collected from a few new borings and wells within the TTZ. Often rapid assessment tools such as a membrane interface probe can assist in developing good data on the contaminant mass and phase distribution. In this way, valuable historical data can be used to aid in assessing the ambient variability of contaminant concentrations, and a one-time analysis of a few key analytes including geochemistry (if missing from historic data set), metals or microbiology can be used to meet the remaining baseline data objectives. Future performance monitoring data can be compared to the historic database to assess treatment efficiency outside the range of historic fluctuation, for example.

Sites seeking to expedite schedules by performing ISCO infrastructure construction and baseline sampling in the same mobilization should be cautious. Well drilling, construction and development activities can cause significant disturbance to the surrounding groundwater. Care must be taken to comply with state requirements, and avoid sampling wells too soon after well development to ensure that baseline samples are representative of the equilibrated formation. Multiple rounds of baseline monitoring can help prevent inaccurate baseline assessment.

Mass flux measurement can be complex and generally requires a detailed understanding of the source and associated dissolved plume geometry and hydrogeological setting. Mass flux measurement requires an understanding of groundwater hydraulics and the dissolved COC distribution. Groundwater hydraulics can be measured using standard slug or pump testing techniques, tracer dilution testing or passive flux meters. The COC distribution can be measured using single or multiple transects of multilevel samplers. Careful consideration should be given to the mass flux measurement methods because method applicability depends upon site conditions.

Baseline monitoring data must be adequate prior to proceeding with the ISCO treatment; ISCO delivery and treatment performance can only be judged after an adequate understanding of the pre-existing conditions.

9.5.2.2 Delivery Performance Monitoring

Delivery performance monitoring is conducted to measure and evaluate the nature and extent of oxidant delivery and hydraulic effects on the water table and includes physical measurements and analysis of environmental samples from the site for parameters indicative of oxidant presence. It does not typically include analysis of target COCs since oxidation reactions are occurring.

In general, the delivery monitoring focuses on characterization of the nature and extent of oxidant delivery for the purposes of evaluating achievement of the design conditions, deciding the scope of continued oxidant injection activities and/or triggering optimization actions. One delivery performance monitoring approach is the observational method (Terzaghi and Peck, 1948). The observational method is an adaptive ISCO implementation method for sites with uncertainty. If the observational method is applied, then delivery performance monitoring data are used to evaluate and optimize the ISCO injection process. The data are compared to trigger levels specified in a site-specific developed implementation or operations plan, and the results are used to determine the next phase of work, whether it be for additional injections, lower pressure application, or delivery cessation because the delivery objectives have been achieved. As such, plans are highly site specific. Delivery performance monitoring is typically performed during and shortly after oxidant delivery for the following purposes:

- Monitor oxidant and activator injection flow rates and volume.
- Monitor water levels and hydraulic controls.
- Monitor injection pressures to ensure aquifers are not plugging or excessive pressures are not applied that could cause well damage, formation deformation or contaminant spreading outside the TTZ.
- Monitor for the appearance of the oxidant to determine if adequate subsurface distribution of the oxidant and activators is occurring; it is very important to recognize the potential for preferential flow and mobile zones in a formation given the pore volumes of oxidant solution delivered and the spacing of observation wells.
- Test changes in ORP, conductivity, pH, DO and temperature to indirectly monitor oxidation reactions and distribution.
- Monitor byproducts to help determine effective radius and the mass of contamination treated.
- Monitor and manage aquifer conditions that affect oxidant chemistry such as pH.
- Monitor nearby utilities for potential breakout (sewer and storm drainage lines).
- Monitor migration/displacement of VOCs in vapor or groundwater during injection.
- Maintain protection of sensitive receptors, human health and the environment.

The timeframe and measurement parameters depend on the oxidant applied, the type of contamination being treated, and the site-specific oxidant persistence. For some oxidants (CHP and ozone), delivery performance monitoring also includes the measurement of oxidant impact on groundwater temperature and off-gas generation in the vadose zone and wellhead space. As such, delivery performance monitoring is also an important component of a comprehensive ISCO health and safety program. Many of the parameters to be monitored are measured using field instrumentation (a YSI 600XL [or similar] water quality meter equipped with a flow-through cell to collect water samples utilizing standard low-flow sampling techniques) or through field kit methods (hydrogen peroxide or persulfate using a CHEMetrics CHEMets® Kit [or similar]).

9.5.2.3 Treatment Performance Monitoring

In situ chemical oxidation, as with any *in situ* technology, is subject to the relatively slow pace of subsurface hydrogeological and contaminant phase distribution equilibrium and mass transfer processes. As such, despite the rapid rates of ISCO contaminant destruction reactions, the re-equilibrated results of an ISCO application are typically not immediately observable. Therefore, most ISCO applications implement treatment performance monitoring. The treatment performance monitoring is typically initiated after oxidant delivery objectives are achieved and the project focus moves toward concentrations of target COCs. Treatment performance monitoring is intended to determine whether the specific treatment goals and milestones established by stakeholders during the design phase are met.

Treatment performance is generally monitored to achieve the following objectives:

- Measure progress toward, and achievement or failure to achieve, the ISCO treatment goals (the degree of source removal or source mass flux reduction)
- Evaluate the extent and uniformity of COC treatment within the TTZ
- Evaluate changes in aquifer solids and aquifer conditions due to the treatment process
- Evaluate effects of recontamination and/or rebound
- Monitor conditions for system optimization through additional applications of ISCO or polishing using another technology within a treatment train approach

Common ISCO treatment goals, as discussed previously, typically include the following:

- Reducing the contaminant mass in the source zone and/or mass discharge
- Numerical standard in a specific phase (soil gas, groundwater or aquifer solids)
- Removal of visible or mobile NAPL
- Achieving a risk-based contamination target

Treatment performance monitoring is often designed to include evaluation of the impact to the contaminant mass whether it is a specific ISCO treatment goal or not because ISCO is most commonly used as a mass reduction technology and evaluation of the impacts to mass is typically the most direct measurement of the effectiveness of the treatment technology. If mass flux reduction is the ISCO treatment goal, it may become important to understand the relationship between mass reduction and mass flux. At many sites, the relationship is not linear, and a large mass reduction may be needed to reduce the mass flux and meet the ISCO treatment goal. Different approaches can be used to estimate contaminant concentrations and mass levels in a target treatment zone and locations downgradient from it. For performance assessment related to ISCO, it is important to pay careful attention to (1) general issues and concerns with sampling and analysis of volatile organic compounds (e.g., West et al., 1995; Oesterreich and Siegrist, 2009) and (2) the effects of ISCO on NOM and associated changes in partitioning behavior (e.g., Woods et al., 2012).

Monitoring dissolved contaminant concentrations in the presence of oxidant requires appropriate sample preservation to prevent further oxidation reactions that may degrade contaminants within the sample container and introduce a negative bias to laboratory results (Huling et al., 2011). Several emerging and innovative tools are available to support an evaluation of ISCO effectiveness. Compound-specific isotope analysis is a useful tool to evaluate whether contaminant destruction has occurred; it may be especially useful to document contaminant mass destruction in source areas where dissolved concentration rebound through dissolution of NAPL or sorbed phase mass into groundwater masks treatment effectiveness. In addition, measurement of chloride concentrations can be used as an indicator of chlorinated solvent mineralization.

Depending upon site-specific concerns of the stakeholders and others involved in the project, an objective of the treatment monitoring plan may also include monitoring the impact of ISCO on subsurface conditions. This may be a concern in site-specific cases as ISCO has the potential to result in changes to microbial populations, geochemical conditions, nontarget organic compounds and aqueous concentration of metals (Siegrist et al., 2011).

In addition, ISCO is often used as part of a treatment train, and other remediation technologies, such as EISB or MNA, may follow ISCO treatment. In this situation, one of the objectives of the treatment performance monitoring may be to establish a baseline condition for follow-up remedy. Therefore, additional parameters beyond those discussed here, such as microbiological analysis, may need to be added to the ISCO treatment performance monitoring.

9.6 STATUS AND AREAS FOR ADVANCEMENTS

9.6.1 Status

In situ chemical oxidation has been developed and deployed for *in situ* remediation of organically contaminated soil and groundwater. The principles and practices associated with ISCO applied to treat dissolved and sorbed phase contamination are generally well established and documented in design manuals and reference texts (ITRC, 2005; Huling and Pivetz, 2006; Siegrist et al., 2011). The application of ISCO to treat DNAPL source zones is not as well established, and field applications for this purpose have had mixed results. While there is a theoretical basis for the degradation of DNAPL residuals by ISCO, cost effective remediation generally depends on the following: (1) susceptibility of DNAPL organics to oxidative destruction, (2) rate and extent of interphase mass transfer of the DNAPL, (3) delivery of an adequate amount of oxidant that contacts DNAPL in the TTZ, (4) effects of subsurface conditions on ISCO reactions and (5) ISCO effects on subsurface permeability and biogeochemistry. These issues have been discussed in the previous sections.

The theoretical considerations related to the application of ISCO to treat DNAPL source zones can be summarized as follows:

- DNAPL source zones tend to be present in highly heterogeneous subsurface settings, both physical and chemical, and delivery of oxidants and amendments is challenging to accomplish and difficult to predict.
- Chemical oxidants, by nature, are nonselective and thus reactions are competitive; in some cases nontarget reactions (nonproductive reactions) can dominate the overall oxidant use.
- A large mass of oxidant is necessary to treat a DNAPL source zone; because use of high concentrations of oxidant can result in greater nonproductive consumption due to NOD and faster decomposition, high fluid volumes are generally necessary.
- Higher oxidant concentrations can give rise to density-driven transport.
- Delivery must overcome physical heterogeneities.
- DNAPL/oxidant contact will maximize treatment. Due to mass transfer limitations of DNAPL, this requires a longer duration of oxidant delivery and/or persistence than a typical aqueous phase application.

- Preferential pathways will cause non-ideal conditions and increase diffusion mass transfer limitations. DNAPL sites will often have significant mass trapped in lower-permeability zones. Over time, DNAPL compounds can back-diffuse into more transmissive zones, and thus, greater oxidant persistence is advantageous.
- Dense nonaqueous phase liquid architecture must be suitable for reagent delivery.

The key challenges to the successful application of ISCO for treatment of DNAPL source zones include:

- Inappropriate goals, schedule, and budget for the site
- Lack of a comprehensive CSM
- Inadequate understanding of site-specific limitations related to contaminant and hydrogeological characteristics
- Limited availability of appropriate predictive tools
- Limited understanding of the effects of ISCO reactions on aquifer physical and biogeochemical processes
- Incomplete and/or inappropriate monitoring (such as groundwater concentrations only) to determine mass depletion
- Contaminant rebound

Successful implementation of ISCO depends on understanding and anticipating these challenges. Approaches for managing these challenges include:

- Coupling ISCO with other site management approaches
- Setting appropriate goals and performance metrics
- Using dynamic characterization and assessment approaches and striving to effectively monitor the right locations and phases to assess DNAPL destruction
- Applying appropriate predictive tools
- Applying enough oxidant solution (volume) at an appropriate concentration for long enough to reflect contaminant mass transfer limitations
- Enhanced delivery approaches
- Multiple delivery events
- Use of rebound data to plan future injections

9.6.2 Areas for Advancement

9.6.2.1 Emerging Approaches and Technologies

Areas for advancement of ISCO, including effective application for DNAPL source zones, have been summarized by Siegrist et al. (2011) and include emerging approaches and technologies in areas such as:

- Using ISCO as part of an integrated approach coupled with other approaches or technologies for improved treatment efficiency and effectiveness
- Enhanced delivery methods for ISCO
- Improved ISCO monitoring and assessment

Interest in using ISCO combined with other technologies is increasing. Using ISCO as a component of a combined remedy can be accomplished simultaneously or through temporal or spatial implementation approaches. *In situ* chemical oxidation and other technologies can be applied sequentially to the same zone, such as ISCO followed by bioremediation, or ISCO can be used to treat a portion of a site, with different technologies applied to other portions. For example, ISCO may be used to treat the source zone, and ISCR may then be used to treat the associated groundwater plume. Potential benefits of combining technologies include improved effectiveness, shorter treatment times, less rebound, and lower overall treatment costs. Examples of emerging approaches and technologies coupled with ISCO include the following: (1) ISCO via multiple oxidants, (2) ISCO with EISB, (3) ISCO with MNA, (4) ISCO with ISCR, (5) ISCO with surfactants and (6) ISCO with thermally enhanced recovery. Issues to consider when selecting and designing a coupled remedy include compatibility of reactants, particularly if used simultaneously (e.g., surfactants and oxidants), and the implication of the byproducts of one technology serving as the inputs for follow-on approaches (such as ISCO followed by MNA – ISCO byproducts will serve as inputs for bioprocesses and other geochemical interactions).

Improving and advancing oxidant delivery approaches is of particular interest because delivery effectiveness is strongly affected by certain hydrogeological and geochemical conditions. Oxidant delivery can be enhanced by improving hydraulic delivery and distribution of the solution containing the oxidant and/or improving oxidant stability and targeted reactivity toward contaminants. An array of approaches are under development, including investigation into oxidant stabilization and/or encapsulation, the use of polymers and other delivery aids to enhance contact between oxidant and contaminant, density-driven delivery and mechanical mixing approaches.

Successful ISCO application also requires appropriate monitoring and assessment of oxidant delivery and treatment effectiveness. The integrated use of field methods and real-time analysis in an expedited approach (Triad) can enhance ISCO system design and implementation. The development and deployment of *in situ* sensors and continuous data logging also can aid in understanding how an ISCO application is progressing.

9.6.2.2 Research Needs and Breakthrough Areas

Research needs and breakthrough areas to enhance ISCO development and deployment include ISCO process chemistry, ISCO delivery, ISCO system design and ISCO process control and assessment (Siegrist et al., 2011). While extensive knowledge about ISCO-related reaction and transport processes is available to guide design and implementation, knowledge gaps need additional attention. Research is needed to more effectively and predictably apply ISCO to a broader range of sites and achieve equal or better outcomes in less time and expense. Additional understanding of reaction mechanisms and kinetics for complex contaminant conditions (pesticides) and/or complex mixtures (manufactured gas plant residuals) is needed. The understanding of oxidant interactions with subsurface constituents needs to improve, along with approaches to minimize nonproductive oxidant consumption. The applicability of ISCO to large, dilute plumes can be improved through such development. The long-term impacts of ISCO on water quality and on specific subsurface constituents (NOM), including influences on post-ISCO natural attenuation, need further attention. Innovative methods for delivering oxidants that will improve movement into or control contaminant migration out of lower permeability zones are necessary. Advances in delivery need to be coupled with tools to understand and predict delivery and treatment effectiveness. Finally, the development of robust sensors to understand oxidant delivery, contaminant treatment, contaminant migration and biogeochemical impacts (which can be deployed *in situ* and transmit data for real-time decision making) would have a beneficial impact on ISCO success.

REFERENCES

Adam ML, Comfort SD, Morley MC, Snow DD. 2004. Remediating RDX-contaminated ground water with permanganate: Laboratory investigations for the Pantex perched aquifer. J Environ Qual 33:2165–2173.

Aiken G. 1992. Chloride interference in the analysis of dissolved organic carbon by the wet oxidation method. Environ Sci Technol 26:2435–2439.

Anipsitakis GP, Dionysiou DD. 2004. Radical generation by the interaction of transition metals with common oxidants. Environ Sci Technol 38:3705–3712.

Appelo CAJ, Postma D. 1999. Geochemistry, Groundwater and Pollution. A.A. Balkema Publishers, Leiden, The Netherlands.

ASTM (American Society for Testing Materials). 2004. Standard Guide for Accelerated Site Characterization for Confirmed or Suspected Petroleum Releases. E1912-98. ASTM International, West Conshohocken, PA, USA.

ASTM. 2007. Standard Test Method for Estimating the Permanganate Natural Oxidant Demand of Soil and Aquifer Solids. D7262-07. ASTM International, West Conshohocken, PA, USA. 5 p.

Ball WP, Liu C, Xia G, Young DF. 1997. A diffusion-based interpretation of tetrachloroethene and trichloroethene concentration profiles in a groundwater aquitard. Water Resour Res 33:2741–2757.

Barbeni M, Nfinero C, Pelizzetti E, Borgarello E, Serpon N. 1987. Chemical degradation of chlorophenols with Fenton's reagent. Chemosphere 16:2225–2237.

Bellamy WD, Hickman PA, Ziemba N. 1991. Treatment of VOC-contaminated groundwater by hydrogen peroxide and ozone oxidation. J Water Pollut Control Fed 63:120–128.

Bissey LL, Smith JL, Watts RJ. 2006. Soil organic matter-hydrogen peroxide dynamics in the treatment of contaminated soils and groundwater using catalyzed H_2O_2 propagations (modified Fenton's reagent). Water Res 40:2477–2484.

Block PA, Brown RA, Robinson D. 2004. Novel activation technologies for sodium persulfate *in situ* chemical oxidation. Proceedings, Fourth International Conference on the Remediation of Chlorinated and Recalcitrant Compounds, Monterey, CA, USA, May 24–27, Paper 2A-05.

Borden RC, Cha KY, Simpkin T, Lieberman MT. 2012. Development of a Design Tool for Planning Aqueous Amendment Injection Systems Soluble Substrate Design Tool. Final Report and User's Guide, ESTCP Project ER-200626. June. 40 p. http://www.serdp.org/. Accessed January 25, 2014.

Bowers AR, Gaddipati P, Eckenfelder WW, Monsen RM. 1989. Treatment of toxic or refractory wastewaters with hydrogen peroxide. Water Sci Technol 21:477–486.

Braida WJ, Ong SK. 2001. Air sparging effectiveness: Laboratory characteristics of air-channel mass transfer zone for VOC volatilization. J Hazard Mater 87:241–258.

Brown RA, Norris RD. 1986. Method for Decontaminating a Permeable Subterranean Formation. U.S. Patent 4,591,443. U.S. Patent Office, Washington, DC, USA.

Brown RA, Robinson D. 2004. Response to naturally occurring organic material: Permanganate versus persulfate. In Gavaskar AR, Chen ASC, eds, Remediation of Chlorinated and Recalcitrant Compounds, Proceedings, Fourth International Conference, Monterey, USA, Battelle Press, Columbus, OH, USA, May 24–27.

Brown RA, Norris RD, Westray M. 1986. *In situ* treatment of groundwater. Presented at Haz Pro '86, Baltimore, MD, USA, April 1–3.

Brown RA, Skaladany G, Robinson D, Fiacco RJ. 2001. Comparing permanganate and persulfate treatment effectiveness for various organic contaminants. Proceedings, The First International Conference on Oxidation and Reduction Technologies for In-Situ Treatment of Soil and Groundwater, Niagara Falls, Ontario, Canada, June 25–29.

Buxton GV, Greenstock CL, Helman WP, Ross AB. 1988. Critical review of rate constants for reactions of hydrated electrons, hydrogen atoms and hydroxyl radicals (OH·/O·−) in aqueous solution. J Phys Chem Ref Data 17:513–886.

Chapelle FH, Bradley PM, Casey CC. 2005. Behavior of a chlorinated ethene plume following source-area treatment with Fenton's reagent. Ground Water Monit Remediat 25:131–141.

Choi H, Lim H-N, Hwang T-M, Kang J-W. 2002. Transport characteristics of gas phase ozone in unsaturated porous media for in-situ chemical oxidation. J Contam Hydrol 57:81–98.

Christiansen CM. 2010. Methods for Enhanced Delivery of *In Situ* Remediation Amendments in Contaminated Clay Till. PhD Thesis. Technical University of Denmark, Kgs. Lyngby, Denmark.

Christensen KE. 2011. Evaluating Dense Non-aqueous Phase Liquid Dissolution Kinetics and Chemical Oxidation in a Three-dimensional Bench-scale Fracture Network. PhD Thesis. Colorado School of Mines, Golden, CO, USA.

Christiansen CM, Damgaard I, Broholm MM, Kessler T, Klint KE, Nilsson B, Bjerg PL. 2010. Comparison of delivery methods for enhanced *in situ* remediation in clay till. Ground Water Monit Remediat 30:107–122.

Clement TP. 1997. A Modular Computer Code for Simulating Reactive Multi-Species Transport in 3-D Groundwater Systems. PNNL-11720. Pacific Northwest National Laboratory, Richland, WA, USA, 59 pp.

Clement TP. 2001. Generalized solution to multispecies transport equations coupled with a first-order reaction network. Water Resour Res 37:157–163.

Clement TP, Johnson CD. 2002. What's New in RT3D Version 2.5. Pacific Northwest National Laboratory, Richland, WA. 20 p. http://www.eng.auburn.edu/~clemept/Computer_codes/RT3Dv25_Update.pdf. Accessed January 25, 2014.

Clement TP, Sun Y, Hooker BS, Petersen JN. 1998. Modeling multispecies reactive transport in ground water. Ground Water Monit Remediat 18:79–82.

Clement TP, Johnson CD, Sun Y, Klecka GM, Bartlett C. 2000. Natural attenuation of chlorinated solvent compounds: Model development and field-scale application. J Contam Hydrol 42:113–140.

Conrad SH, Glass RJ, Peplinski WJ. 2002. Bench-scale visualization of DNAPL remediation processes in analog heterogeneous aquifers: Surfactant floods and *in situ* oxidation using permanganate. J Contam Hydrol 58:13–49.

Corbin JF III, Teel AL, Allen-King RM, Watts RJ. 2007. Reactive oxygen species responsible for the enhanced desorption of dodecane in modified Fenton's systems. Water Environ Res 79:37–42.

Crawford S, Smith BA, O'Shaughnessy K, Hagelin N, Jacobson R. 2009. Application of alkaline activated persulfate and evaluation of treatment residuals. Conference on Design and Construction Issues at Hazardous Waste Sites, Philadelphia, PA, USA.

Crimi M, Ko S. 2009. Control of manganese dioxide particles resulting from *in situ* chemical oxidation using permanganate. Chemosphere 74:847–853.

Crimi ML, Siegrist RL. 2003. Geochemical effects associated with permanganate oxidation of DNAPLs. Ground Water 41:458–469.

Crimi ML, Siegrist RL. 2004a. Association of cadmium with MnO_2 particles generated during permanganate oxidation. Water Res 38:887–894.

Crimi ML, Siegrist RL. 2004b. Impact of reaction conditions on MnO_2 genesis during permanganate oxidation. J Environ Eng 130:562–572.

Crimi ML, Siegrist RL. 2005. Factors affecting effectiveness and efficiency of DNAPL destruction using potassium permanganate and catalyzed hydrogen peroxide. J Environ Eng 131:1716–1723.

Crimi ML, Taylor J. 2007. Experimental evaluation of catalyzed hydrogen peroxide and sodium persulfate for destruction of BTEX contaminants. Soil Sediment Contam 16:29–45.

Crimi M, Quickel M, Ko S. 2009. Enhanced permanganate *in situ* chemical oxidation through MnO_2 particle stabilization: Evaluation in 1-D transport systems. J Contam Hydrol 105:69–79.

Crimi M, Silva JAK, Palaia T. 2011. Cooperative Technology Demonstration: Polymer-Enhanced Subsurface Delivery and Distribution of Permanganate. ESTCP Project ER-0912. http://www.serdp.org/. Accessed November 2, 2012.

Cross PE, Baird D. 2008. Initial Results of ISCO for a Large TCE DNAPL Source Area. Presentation at the Sixth International Conference on Remediation of Chlorinated and Recalcitrant Compounds (Monterey, CA; May 2008). In Sass BM (Conference Chair), Remediation of Chlorinated and Recalcitrant Compounds – 2008. ISBN 1-57477-163-9, Battelle Press, Columbus, OH, USA, Abstract C-018.

Cross PE, Baird D, Wymore R. 2006. Modified Fenton's Reagent Remediation of a Large TCE Source Zone. Presentation at the Fifth International Conference on Remediation of Chlorinated and Recalcitrant Compounds (Monterey, CA; May 2006). In Sass BM (Conference Chair), Remediation of Chlorinated and Recalcitrant Compounds – 2006. ISBN 1-57477-157-4, Battelle Press, Columbus, OH, USA, Abstract D-68.

Crumbling DM, Groenjes C, Lesnik B, Lynch K, Shockley J, VanEe J, Howe R, Keith L, McKenna G. 2001. Applying the concept of effective data to contaminated sites could reduce costs and improve cleanups. Environ Sci Technol 35:405A-409A.

Crumbling DM, Griffith J, Powell DM. 2003. Improving decision quality: Making the case for adopting next-generation site characterization practices. Remediat 13:91–111.

Dugan P. 2006. Coupling *In Situ* Technologies for DNAPL Remediation and Viability of the PITT for Post-Remediation Performance Assessment. PhD Thesis. Colorado School of Mines, Golden, CO, USA.

Dugan PJ, Vlastnik E, Ivy S, Swearingen L, Swearingen J. 2009. Micro-encapsulated oxidant technology: Enhancing *in situ* chemical oxidation (ISCO) with selective oxidation using controlled-release permanganate. Proceedings, Annual Conference on Soils, Sediments, Water, and Energy, Amherst, MA, USA, October 19–22. http://www.umasssoils.com/posters2009/chem_ox.htm. Accessed November 2, 2012.

Dugan PJ, Siegrist RL, Crimi M. 2010. Coupling surfactants/cosolvents with oxidants for enhanced DNAPL removal: A review. Remediat J 20:27–50.

ESTCP (Environmental Security Technology Certification Program). 1999. Technology Status Review: *In Situ* Oxidation. ESTCP, Arlington, VA, USA. 50 p. http://www.serdp.org/. Accessed June 25, 2012.

Forsey SP, Thomson NR, Barker JF. 2010. Oxidation kinetics of polycyclic aromatic hydrocarbons by permanganate. Chemosphere 79:628–636.

Furman O, Teel AL, Watts RJ. 2010. Mechanism of base activation of persulfate. Environ Sci Technol 44:6423–6428.

Gates DD, Siegrist RL. 1993. Laboratory Evaluation of Chemical Oxidation Using Hydrogen Peroxide. Report from The X-231B Project for *In Situ* Treatment by Physicochemical Processes Coupled with Soil Mixing, ORNL/TM-12259. Oak Ridge National Laboratory, Oak Ridge, TN, USA.

Gates DD, Siegrist RL. 1995. *In situ* chemical oxidation of trichloroethylene using hydrogen peroxide. J Environ Eng 121:639–644.

Gates DD, Siegrist RL, Cline SR. 1995. Chemical Oxidation of Contaminants in Clay or Sandy Soil. Proceedings, American Society of Civil Engineering (ASCE) National Conference on Environmental Engineering, Pittsburgh, PA, USA, July 1995.

Gates-Anderson DD, Siegrist RL, Cline SR. 2001. Comparison of potassium permanganate and hydrogen peroxide as chemical oxidants for organically contaminated soils. J Environ Eng 127:337–347.

Glaze WH, Kang JW. 1988. Advanced oxidation processes for treating groundwater contaminated with TCE and PCE: Laboratory studies. J Am Water Works Assoc 5:57–63.

Harbaugh AW, McDonald MG. 1996a. User's Documentation for MODFLOW-96: An Update to the U.S. Geological Survey Modular Finite-Difference Ground-Water Flow Model. Open-File Report 96-485. U.S. Geological Survey, Reston, VA, USA. 56 p.

Harbaugh AW, McDonald MG. 1996b. Programmer's Documentation for MODFLOW-96: An Update to the U.S. Geological Survey Modular Finite-Difference Ground-Water Flow Model. Open-File Report 96-485. U.S. Geological Survey, Reston, VA, USA, 56 p.

Harbaugh AW, Banta ER, Hill MC, McDonald MG. 2000. MODFLOW-2000: The U.S. Geological Survey Modular Ground-Water Model – User Guide to Modularization Concepts and the Ground-Water Flow Process Model. Open-File Report 00-92. U.S. Geological Survey, Reston, VA, USA, 130 p.

Haselow JS, Siegrist RL, Crimi ML, Jarosch T. 2003. Estimating the total oxidant demand for *in situ* chemical oxidation design. Remediat J 13:5–15.

Heiderscheidt JL. 2005. DNAPL Source Zone Depletion During *In Situ* Chemical Oxidation (ISCO): Experimental and Modeling Studies. PhD Thesis. Colorado School of Mines, Golden, CO, USA.

Heiderscheidt JL, Siegrist RL, Illangasekare TH. 2008a. Intermediate-scale 2D experimental investigation of *in situ* chemical oxidation using potassium permanganate for remediation of complex DNAPL source zones. J Contam Hydrol 102:3–16.

Heiderscheidt JL, Crimi ML, Siegrist RL, Singletary M. 2008b. Optimization of full-scale permanganate ISCO system operation: Laboratory and numerical studies. Ground Water Monit Remediat 28:72–84.

Held RJ, Illangasekare TH. 1995a. Fingering of dense non-aqueous phase liquids in porous media: 1. Experimental investigation. Water Resour Res 31:1213–1222.

Held RJ, Illangasekare TH. 1995b. Fingering of dense non-aqueous phase liquids in porous media: 2. Analysis and classification. Water Resour Res 31:1223–1231.

Henderson TH, Mayer KU, Parker BL, Al TA. 2009. Three-dimensional density-dependent flow and multicomponent reactive transport modeling of chlorinated solvent oxidation by potassium permanganate. J Contam Hydrol 106:195–211.

Hønning J, Broholm M, Bjerg P. 2007. Role of diffusion in chemical oxidation of PCE in a dual permeability system. Environ Sci Technol 41:8426–8432.

House DA. 1962. Kinetic and mechanisms of oxidations by peroxydisulfate. Chem Rev 62:185–203.

Huang KC, Couttenye RA, Hoag GE. 2002. Kinetics of heat-assisted persulfate oxidation of methyl tert-butyl ether (MTBE). Chemosphere 49:413–420.

Huie RE, Clifton CL, Neta P. 1991. Electron transfer reaction rates and equilibria of the carbonate and sulfate radical anions. International Journal of Radiation Applications and Instrumentation, Part C: Radiat Phys Chem 38:477–481.

Huling SG, Pivetz BE. 2006. Engineering issue paper: In-situ chemical oxidation. EPA 600-R-06-072. U.S. Environmental Protection Agency, National Risk Management Research Laboratory, Cincinnati, OH, USA. 60 p. http://www.epa.gov/tio/tsp/issue.htm#EF. Accessed November 3, 2012.

Huling SG, Ko S, Pivetz B. 2011. Groundwater sampling at ISCO sites: Binary mixtures of volatile organic compounds and persulfate. Ground Water Monit Remediat 31:72–79.

Illangasekare TH, Armbruster EJ III, Yates DN. 1995a. Non-aqueous-phase fluids in heterogeneous aquifer: Experimental study. J Environ Eng 121:571–579.

Illangasekare TH, Ramsey JL, Jensen KH, Butts M. 1995b. Experimental study of movement and distribution of dense organic contaminants in heterogeneous aquifers. J Contam Hydrol 20:1–25.

Illangasekare TH, Munakata Marr J, Siegrist RL, Soga K, Glover KC, Moreno-Barbero E, Heiderscheidt JL, Saenton S, Matthew M, Kaplan AR, Kim Y, Dai D, Gago JL, Page JWE. 2006. Mass Transfer from Entrapped DNAPL Sources Undergoing Remediation: Characterization Methods and Prediction Tools. CU-1294. ER 1294. Final Report. Strategic Environmental Research and Development Program (SERDP), Arlington, VA, USA. 435 p. http://www.serdp.org/. Accessed November 2, 2012.

ITRC (Interstate Technology & Regulatory Council). 2001. Technical and Regulatory Guidance for *In Situ* Chemical Oxidation of Contaminated Soil and Groundwater (ISCO-1). Washington, DC, USA. http://www.itrcweb.org/guidancedocument.asp?TID=13. Accessed November 2, 2012.

ITRC. 2003. Technology and Regulatory Guidance for the Triad Approach: A New Paradigm for Environmental Project Management. SCM-1, December. ITRC, Washington, DC, USA. http://www.itrcweb.org/guidance/getdocument?documentid=90. Accessed June 19, 2013.

ITRC. 2005. Technical and Regulatory Guidance for *In Situ* Chemical Oxidation of Contaminated Soil and Groundwater, 2nd ed (ISCO-2). Prepared by the ITRC *In Situ* Chemical Oxidation Team. http://www.itrcweb.org/guidancedocument.asp?TID=13. Accessed December 6, 2009.

ITRC. 2007. Triad Implementation Guide. SCM-3. ITRC Sampling, Characterization, and Monitoring Team, Washington, DC, USA. http://www.itrcweb.org. Accessed November 2, 2012.

Johnson CD, Truex MJ, Clement TP. 2006. Natural and Enhanced Attenuation of Chlorinated Solvents Using RT3D. PNNL-15937. Pacific Northwest National Laboratory, Richland, WA, USA.

Johnson RL, Pankow JF. 1992. Dissolution of dense chlorinated solvents into groundwater. 2. Source functions for pools of solvent. Environ Sci Technol 26:896–901.

Johnson RL, Tratnyek PG, O'Brien Johnson R. 2008. Persulfate persistent under thermal activation conditions. Environ Sci Technol 42:9350–9356.

Jones LJ. 2007. The Impact of NOD Reaction Kinetics on Treatment Efficiency. MS Thesis. University of Waterloo, Waterloo, Ontario, Canada.

Jones PW, Williams DR. 2002. Chemical speciation simulation used to assess the efficiency of environment-friendly EDTA alternative for use in pulp and paper industry. Inorg Chim Acta 339:41–50.

Jung H, Kim J, Choi H. 2004. Reaction kinetics of ozone in variably saturated porous media. J Environ Eng 130:432–441.

Jung H, Ahn Y, Choi H, Kim IS. 2005. Effects of in-situ ozonation on indigenous microorganisms in diesel contaminated soil: Survival and regrowth. Chemosphere 61:923–932.

Kakarla PKC, Watts RJ. 1997. Depth of Fenton-like oxidation in remediation of surface soils. J Environ Eng 123:11–17.

Kim J, Choi H. 2002. Modeling *in situ* ozonation for the remediation of nonvolatile PAH-contaminated unsaturated soils. J Contam Hydrol 55:261–285.

Kim K, Gurol MD. 2005. Reaction of nonaqueous phase TCE with permanganate. Environ Sci Technol 39:9303–9308.

Kolthoff IM, Miller IK. 1951. The chemistry of persulfate. I. The kinetics and mechanism of the decomposition of the persulfate ion in aqueous medium. J Am Chem Soc 73:3055–3059.

Krembs FJ. 2008. Critical Analysis of the Field Scale Application of *In Situ* Chemical Oxidation for the Remediation of Contaminated Groundwater. MS Thesis. Colorado School of Mines, Golden, CO, USA.

Krembs FJ, Siegrist RL, Crimi M, Furrer RF, Petri BG. 2010. ISCO for groundwater remediation: Analysis of field applications and performance. Ground Water Monit Remediat 30:42–53.

Kueper BH, Frind EO. 1991a. Two-phase flow in heterogeneous porous media. 1. Model development. Water Resour Res 27:1049–1057.

Kueper BH, Frind EO. 1991b. Two-phase flow in heterogeneous porous media. 2. Model application. Water Resour Res 27:1059–1070.

Kueper BH, Abbot W, Farquhar G. 1989. Experimental observations of multiphase flow in heterogeneous porous media. J Contam Hydrol 5:83–95.

Kueper BH, Redman D, Starr RC, Reitsma S, Mah M. 1993. A field experiment to study the behavior of tetrachloroethylene below the water table: Spatial distribution of residual and pooled DNAPL. Ground Water 31:756–766.

Lee BS, Kim JH, Lee KC, Kim YB, Schwartz FW, Lee ES, Woo NC, Lee MK. 2009. Efficacy of controlled-release KMnO4 (CRP) for controlling dissolved TCE plume in groundwater: A large flow-tank study. Chemosphere 74:745–750.

Lee ES, Schwartz FW. 2007. Characteristics and applications of controlled-release $KMnO_4$ for groundwater remediation. Chemosphere 66:2058–2066.

Lee ES, Seol Y, Fang YC, Schwartz FW. 2003. Destruction efficiencies and dynamics of reaction fronts associated with the permanganate oxidation of trichloroethylene. Environ Sci Technol 37:2540–2546.

Li XD, Schwartz FW. 2003. Permanganate Oxidation Schemes for the Remediation of Source Zone DNAPLs and Dissolved Contaminant Plumes. In Henry SM, Warner SD, eds, Chlorinated Solvent and DNAPL Remediation. American Chemical Society, Washington, DC, USA, pp 73–85.

Li XD, Schwartz FW. 2004. DNAPL mass transfer and permeability reduction during *in situ* chemical oxidation with permanganate. Geophys Res Lett 31:L06504, doi:10.1029/2003GL019218.

Liang C, Lee IL. 2008. *In situ* iron activated persulfate oxidative fluid sparging treatment of TCE contamination: A proof of concept study. J Contam Hydrol 100:91–100.

Liang C, Bruell CJ, Marley MC, Sperry KL. 2004a. Persulfate oxidation for *in situ* remediation of TCE. I. Activated by ferrous ion with and without a persulfate-thiosulfate redox couple. Chemosphere 55:1213–1223.

Liang C, Bruell CJ, Marley MC, Sperry KL. 2004b. Persulfate oxidation for *in situ* remediation of TCE. II. Activated by chelated ferrous ion. Chemosphere 55:1225–1233.

Liang C, Wang Z S, Mohanty N. 2006. Influences of carbonate and chloride ions on persulfate oxidation of trichloroethylene at 20 C. Sci Total Environ 370:271–277.

Liang C, Huang CF, Mohanty N, Lu CJ, Kurakalva RM. 2007. Hydroxypropyl-β-cyclodextrin-mediated iron-activated persulfate oxidation of trichloroethylene and tetrachloroethylene. Ind Eng Res 46:6466–6479.

Liang C, Lee I-L, Hsu I-Y, Liang C-P, Lin Y-L. 2008. Persulfate oxidation of trichloroethylene with and without iron activation in porous media. Chemosphere 70:426–435.

Liu C, Ball WP. 2002. Back diffusion of chlorinated solvent contaminants from a natural aquitard to a remediated aquifer under well-controlled field conditions: Predictions and measurements. Ground Water 40:175–184.

Lowe KS, Gardner FG, Siegrist RL. 2002. Field pilot test of *in situ* chemical oxidation through recirculation using vertical wells. Ground Water Monit Remediat 22:106–115.

Mackay DM, Cherry JA. 1989. Ground water contamination: Limits of pump-and-treat remediation. Environ Sci Technol 23:630–636.

MacKinnon LK, Thomson NR. 2002. Laboratory-scale *in situ* chemical oxidation of a perchloroethylene pool using permanganate. J Contam Hydrol 56:49–74.

Marvin BK, Nelson CH, Clayton W, Sullivan KM, Skladany G. 1998. *In Situ* Chemical Oxidation of Pentachlorophenol and Polycyclic Aromatic Hydrocarbons: From Laboratory Tests to Field Demonstration. In Wickramanayake GB, Hinchee RE, eds, Physical, Chemical, and Thermal Technologies: Remediation of Chlorinated and Recalcitrant Compounds. Battelle Press, Columbus, OH, USA, pp 383–388.

McDade JM, McGuire TM, Newell CJ. 2005. Analysis of DNAPL source-depletion costs at 36 field sites. Remediat J 15:9–18.

McDonald MG, Harbaugh AW. 1988. A Modular Three-Dimensional Finite-Difference Ground-Water Flow Model. Techniques of Water-Resources Investigations, Book 6, Chapter A1. U.S. Geological Survey, Reston, VA, USA, 586 p.

Monahan MJ, Teel AL, Watts RJ. 2005. Displacement of five metals sorbed on kaolinite during treatment with modified Fenton's reagent. Water Res 39:2955–2963.

Morgan JJ, Stumm W. 1963. Colloid-chemical properties of manganese dioxide. J Coll Sci 19:347–359.

Mumford KG, Lamarche CS, Thomson NR. 2004. Natural oxidant demand of aquifer materials using the push-pull technique. J Environ Eng 130:1139–1146.

Mumford KG, Thomson NR, Allen-King RM. 2005. Bench-scale investigation of permanganate natural oxidant demand kinetics. Environ Sci Technol 39:2835–2840.

NAVFAC (Naval Facilities Engineering Service Center). 2000. Site 11, Old Camden County Landfill Remedial Action Operation: NSB Kings Bay, GA. Remedial Action Operation Optimization Summary Report. NFESC, Port Hueneme, CA, USA, 8 p.

Ndjou'ou A-C, Cassidy D. 2006. Surfactant production accompanying the modified Fenton oxidation of hydrocarbons in soil. Chemosphere 65:1610–1615.

Nelson MD. 1999. The Geochemical Reactions and Density Effects Resulting from the Injection of $KMnO_4$ for PCE DNAPL Oxidation in a Sandy Aquifer. MSc Thesis. University of Waterloo, Waterloo, ON, Canada.

Nelson CH, Brown RA. 1994. Adapting ozonation for soil and ground water cleanup. Chem Eng 11:EE18-EE22.

NRC. 2005. Contaminants in the Subsurface: Source Zone Assessment and Remediation. National Academies Press, Washington, DC, USA. 372 p.

Osgerby IT, Smith BA, Crawford SC, Sperry K, Boeckeler AJ, Getchell SA, McGrath D, Acone SE, Hathaway E. 2006. Case Study: History of a Successful ISCO Project: Eastland Woolen Mills Superfund Site (Corinna, Maine). Proceedings, The 5th International Conference on Remediation of Chlorinated and Recalcitrant Compounds, Monterey, CA, USA.

Oesterreich RC, Siegrist RL. 2009. Quantifying volatile organic compounds in porous media: Effects of sampling method attributes, contaminant characteristics and environmental conditions. Env Sci Technol 43:2891–2898.

Perez-Benito JF, Arias C. 1991. Occurrence of colloidal manganese dioxide in permanganate reactions. J Colloid Interface Sci 152:70–84.

Petri BG. 2006. Impacts of Subsurface Permanganate Delivery Parameters on Dense Nonaqueous Phase Liquid Mass Depletion Rates. MS Thesis, Environmental Science and Engineering Division, Colorado School of Mines, Golden, CO, USA.

Petri B, Siegrist RL, Crimi ML. 2008. Effects of groundwater velocity and permanganate concentration on DNAPL mass depletion rates during *in situ* oxidation. J Environ Eng 134:1–13.

Pinder GF, Abriola LM, 1986. On the simulation of non-aqueous phase organic compounds in the subsurface. Water Resour Res 22:109S-119S.

Poulsen MM, Kueper BH. 1992. A field experiment to study the behavior of tetrachloroethylene in unsaturated porous media. Environ Sci Technol 26:889–895.

Qiu Y, Kuo CH, Zappi ME, Fleming EC. 2004. Ozonation of 2,6- 3,4- and 3,5-dichlorophenol isomers within aqueous solutions. J Environ Eng 130:408–416.

Ramo J. 2003. Hydrogen Peroxide-Metals-Chelating Agents: Interactions and Analytical Techniques. Oulu University Press, Oulu, Finland.

Ravikumur JX, Gurol M. 1994. Chemical oxidation of chlorinated organics by hydrogen peroxide in the presence of sand. Environ Sci Technol 28:394–400.

Reitsma S, Marshall M. 2000. Experimental Study of Oxidation of Pooled NAPL. In Wickramanayake GB, Gavaskar AR, Chen ASC, eds, Chemical Oxidation and Reactive Barriers: Remediation of Chlorinated and Recalcitrant Compounds. Battelle Press, Columbus, OH, USA, pp 25–32.

Ross C, Murdoch LC, Freedman DL, Siegrist RL. 2005. Characteristics of potassium permanganate encapsulated in polymer. J Environ Eng 131:1203–1211.

Sahl J. 2005. Coupling *In Situ* Chemical Oxidation (ISCO) with Bioremediation Processes in the Treatment of Dense Non-Aqueous Phase Liquids (DNAPLs). MS Thesis, Environmental Science and Engineering Division, Colorado School of Mines, Golden, CO, USA.

Sahl J, Munakata-Marr J. 2006. The effects of *in situ* chemical oxidation on microbial processes: A review. Remediat J 16:57–70.

Sahl JW, Munakata-Marr J, Crimi ML, Siegrist RL. 2007. Coupling permanganate oxidation with microbial dechlorination of tetrachloroethene. Water Environ Res 79:5–12.

Sale T, Illangasekare T, Zimbron J, Rodriguez D, Wilking B, Marinelli F. 2007. AFCEE Source Zone Initiative. Final Report. Submitted to Air Force Center for Engineering and the Environment, Brooks City-Base, TX, USA.

Schincariol RA, Schwartz FW. 1990. An experimental investigation of variable density flow and mixing in homogeneous and heterogeneous media. Water Resour Res 26:2317–2329.

Schmidt JT, Ahmad M, Teel AL, Watts RJ. 2011. Hydrogen peroxide stabilization in one-dimensional flow columns. J Contam Hydrol 126:1–7.

Schnarr MJ, Truax CL, Farquhar GJ, Hood ED, Gonullu T, Stickney B. 1998. Laboratory and controlled field experiments using potassium permanganate to remediate trichloroethylene and perchloroethylene DNAPLs in porous media. J Contam Hydrol 29:205–224.

Schroth MH, Oostrom M. Wietsma TW, Istok JD. 2001. In-situ oxidation of trichloroethene by permanganate: Effects on porous medium hydraulic properties. J Contam Hydrol 44:185–201.

Schwarzenbach RP, Gschwend PM, Imboden DM. 1993. Environmental Organic Chemistry. John Wiley & Sons, Inc., New York, NY, USA.

Schwille F. 1988. Dense Chlorinated Solvents in Porous and Fractured Media. Translated by JF Pankow. Lewis Publishers, Chelsea, MI, USA. 146 p.

Shin W-T, Garanzuay X, Yiacoumi S, Tsouris C, Gu B, Mahinthakumar G. 2004. Kinetics of soil ozonation: An experimental and numerical investigation. J Contam Hydrol 72:227–243.

Siegrist RL, Lowe KS, Murdoch LD, Slack WW, Houk TC. 1998a. X-231A Demonstration of *In Situ* Remediation of DNAPL Compounds in Low Permeability Media by Soil Fracturing with Thermally Enhanced Mass Recovery or Reactive Barrier Destruction. Oak Ridge National Laboratory Report ORNL/TM-13534. U.S. Department of Energy Office of Technology Development, Washington, DC, USA. 407 p.

Siegrist RL, Lowe KS, Murdoch LC, Case TL, Pickering DA, Houk TC. 1998b. Horizontal Treatment Barriers of Fracture-Emplaced Iron and Permanganate Particles. In North Atlantic Treaty Organization (NATO)/Committee on the Challenges for Modern Society (CCMS) Pilot Study Special Session on Treatment Walls and Permeable Reactive Barriers. EPA 542-R-98-003. Washington, DC, USA, pp 77–82.

Siegrist RL, Lowe KS, Murdoch LC, Case TL, Pickering DL. 1999. *In situ* oxidation by fracture emplaced reactive solids. J Environ Eng 125:429–440.

Siegrist RL, Urynowicz MA, West OR, Crimi ML, Lowe KS. 2001. Principles and Practices of *In Situ* Chemical Oxidation Using Permanganate. Battelle Press, Columbus, OH, USA. 336 p.

Siegrist RL, Urynowicz MA, Crimi ML, Lowe KS. 2002. Genesis and effects of particles produced during *in situ* chemical oxidation using permanganate. J Environ Eng 128:1068:1079.

Siegrist RL, Crimi ML, Munakata-Marr J, Illangasekare T, Lowe KS, Van Cuyk S, Dugan P, Heiderscheidt J, Jackson S, Petri B, Sahl J, Seitz S. 2006. Reaction and Transport Processes Controlling *In Situ* Chemical Oxidation of DNAPLs. ER-1290 Final Report. DoD Strategic Environmental Research and Development Program (SERDP), Washington, DC, USA. 235 p. http://www.serdp.org/. Accessed November 2, 2012.

Siegrist RL, Crimi ML, Munakata-Marr J, Illangasekare T, Dugan P, Heiderscheidt J, Petri B, Sahl J. 2008a. Chemical Oxidation for Clean Up of Contaminated Ground Water. In Annable MD, Teodorescu M, Hlavinek P, Diels L, eds, Methods and Techniques for Cleaning-Up Contaminated Sites. NATO Science for Peace and Security Series. Springer Publishing, Dordrecht, The Netherlands, pp 45–58.

Siegrist RL, Petri B, Krembs F, Crimi ML, Ko S, Simpkin T, Palaia T. 2008b. *In Situ* Chemical Oxidation for Remediation of Contaminated Ground Water. Proceedings, ISCO Technology Practices Workshop. ESTCP ER-0623. Golden, CO, USA, March 7–8, 2007. 77 p. http:// www.serdp.org/. Accessed November 2, 2012.

Siegrist RL, Crimi ML, Petri B, Simpkin T, Palaia T, Krembs FJ, Munakata-Marr J, Illangasekare T, Ng G, Singletary M, Ruiz N. 2010. *In Situ* Chemical Oxidation for Groundwater Remediation: Site Specific Engineering and Technology Application. ER-0623. Version PRv1.01, October 29, 2010). Final Report. Version PRv1.01, October 29, 2010). ESTCP, Arlington, VA, USA. http://www.serdp.org/. Accessed January 25, 2014.

Siegrist RL, Crimi M, Simpkin TJ, eds. 2011. *In Situ* Chemical Oxidation for Groundwater Remediation. Springer, New York, NY, USA. 678 p.

Silva JAK. 2011. The Utility of Polymer Amendment for Enhancing *In Situ* Remediation Effectiveness. PhD Thesis. Colorado School of Mines, Golden, CO, USA.

Sirguey C, de Souza e Silva PT, Schwartz C, Simonnot M. 2008. Impact of chemical oxidation on soil quality. Chemosphere 72:282–289.

Smith BA, Teel AL, Watts RJ. 2004. Identification of the reactive oxygen species responsible for carbon tetrachloride degradation in modified Fenton's systems. Environ Sci Technol 38:5465–5469.

Smith BA, Teel AL, Watts RJ. 2006. Mechanism for the destruction of carbon tetrachloride and chloroform DNAPLs by modified Fenton's reagent. J Contam Hydrol 85:229–246.

Smith BA, Marley MC, O'Shaughnessy K, Maclean DA, Santos DA. 2012. In-Situ Chemical Oxidation (ISCO) Using Catalyzed Hydrogen Peroxide to Remediate a DNAPL Source Area. Proceedings, 8th International Conference on Remediation of Chlorinated and Recalcitrant Compounds, Monterey, CA, May 19–22.

Smith MM, Silva JAK, Munakata-Marr J, McCray JE. 2008. Compatibility of polymers and chemical oxidants for enhanced groundwater remediation. Environ Sci Technol 42:9296–9301.

Sra KS, Thomson NR, Barker JF. 2010. Persistence of persulfate in uncontaminated aquifer materials. Environ Sci Technol 44:3098–3104.

Stewart CL. 2002. Density-Driven Permanganate Solution Delivery and Chemical Oxidation of a Thin Trichloroethene DNAPL Pool in a Sandy Aquifer. MSc Thesis. University of Waterloo, Waterloo, Ontario, Canada.

Stewart R. 1964. Oxidation Mechanisms. W.A. Benjamin, New York, NY, USA.

Struse AM, Siegrist RL, Dawson HE, Urynowicz MA. 2002. Diffusive transport of permanganate during *in situ* oxidation. J Environ Eng 128:327–334.

Sun H, Yan Q. 2007. Influence of Fenton oxidation on soil organic matter and its sorption and desorption of pyrene. J Hazard Waste Hazard Mater 144:164–170.

Sun Y, Pignatello JJ. 1992. Chemical treatment of pesticide wastes. Evaluation of Fe(III) chelates for catalytic hydrogen peroxide oxidation of 2,4-D at circumneutral pH. J Agric Food Chem 40:332–337.

Sung W, Morgan JJ. 1980. Kinetics and product of ferrous iron oxygenation in aqueous systems. Environ Sci Technol 14:561–567.

Teel AL, Finn DD, Schmidt JT, Cutler LM, Watts RJ. 2007. Rates of trace mineral-catalyzed decomposition of hydrogen peroxide. J Environ Eng 133:853–858.

Terzaghi K, Peck RB. 1948. Soil Mechanics in Engineering Practice. John Wiley & Sons, New York, NY, USA.

Thomson NR, Johnson RL. 2000. Air distribution during *in situ* air sparging: An overview of mathematical modeling. J Hazard Mater 72:265–282.

Tollefsrud E, Schreier CG. 2002. Effectiveness of chemical oxidation to remove organochlorine pesticides from soil. Proceedings, Third International Conference on Remediation of Chlorinated and Recalcitrant Compounds, Monterey, CA, USA. Battelle Press, Columbus, OH, USA, Paper 2C-16.

Tratnyek PG, Johnson TL, Warner SD, Clarke HS, Baker JA. 1998. *In Situ* Treatment of Organics by Sequential Reduction and Oxidation. In Wickramanayake GB, Hinchee RE, eds, Physical, Chemical, and Thermal Technologies: Remediation of Chlorinated and Recalcitrant Compounds. Battelle Press, Columbus, OH, USA, pp 371–376.

Tsitonaki A, Petri B, Crimi M, Mosbæk H, Siegrist RL, Bjerg PL. 2010. *In situ* chemical oxidation of contaminated soil and groundwater using persulfate: A review. Crit Rev Environ Sci Technol 40:55–91.

Tunnicliffe BS, Thomson NR. 2004. Mass removal of chlorinated ethenes from rough-walled fractures using permanganate. J Contam Hydrol 75:91–114.

Tyre BW, Watts RJ, Miller GC. 1991. Treatment of four biorefractory contaminants in soils using catalyzed hydrogen peroxide. J Environ Qual 20:832–838.

Urynowicz MA, Siegrist RL. 2000. Chemical Degradation of TCE DNAPL by Permanganate. In Wickramanayake GB, Gavaskar AR, Chen ASC, eds, Chemical Oxidation and Reactive Barriers: Remediation of Chlorinated and Recalcitrant Compounds Series C2-6. Battelle Press, Columbus, OH, USA, pp 75–82.

Urynowicz MA, Siegrist RL. 2005. Interphase mass transfer during chemical oxidation of TCE DNAPL in an aqueous system. J Contam Hydrol 80:93–106.

Urynowicz MA, Balu B, Udayasankar U. 2008. Kinetics of natural oxidant demand by permanganate in aquifer solids. J Contam Hydrol 96:87–194.

USEPA (U.S. Environmental Protection Agency). 1998. Field Applications of *In Situ* Remediation Technologies: Chemical Oxidation. EPA 542-R-98-008. USEPA OSWER, Washington, DC, USA. http://www.epa.gov/swertio1/download/remed/chemox.pdf. Accessed November 2, 2012.

USEPA. 2003. The DNAPL Remediation Challenge: Is There a Case for Source Depletion?. 600-R-03-143. USEPA Office of Research and Development, Washington, DC, USA. December.

USEPA. 2008. Triad Central Website. http://www.triadcentral.org. Accessed January 25, 2014.

Vella PA, Veronda B. 1994. Oxidation of Trichloroethylene: A Comparison of Potassium Permanganate and Fenton's Reagent. In *In Situ* Chemical Oxidation for the Nineties, Vol 3. Technomic Publishing, Inc., Lancaster, PA, USA, pp 62–73.

Vella PA, Deshinsky G, Boll JE, Munder J, Joyce WM. 1990. Treatment of low-level phenols with potassium permanganate. Res J Water Pollut Control Fed 62:907–914.

Venkatadri R, Peters RW. 1993. Chemical oxidation technologies: Ultraviolet light/hydrogen peroxide, Fenton's reagent, and titanium dioxide assisted photocatalysis. J Hazard Waste Hazard Mater 10:107–149.

Waldemer RH, Tratnyek PG. 2006. Kinetics of contaminant degradation by permanganate. Environ Sci Technol 40:1055–1061.

Waldemer RH, Tratnyek PG, Johnson RL, Nurmi JT. 2007. Oxidation of chlorinated ethenes by heat-activated persulfate: Kinetics and products. Environ Sci Technol 41:1010–1015.

Walling C. 1975. Fenton's reagent revisited. Account Chem Res 8:125–131.

Watts RJ, Smith BR. 1991. Catalyzed hydrogen peroxide treatment of octachlorodibenzo-pdioxin (OCCD) in surface soils. Chemosphere 23:949–955.

Watts RJ, Teel AL. 2005. Chemistry of modified Fenton's reagent (catalyzed H2O2 propagation-CHP) for *in situ* soil and groundwater remediation. J Environ Eng 131:612–622.

Watts RJ, Rausch RA, Leung SW, Udell MD. 1990. Treatment of pentachlorophenol contaminated soils using Fenton's reagent. J Hazard Waste Hazard Mater 7:335–345.

Watts RJ, Leung SW, Udell MD. 1991. Treatment of contaminated soils using catalyzed hydrogen peroxide. Proceedings, First International Symposium on Chemical Oxidation. Technomic, Nashville, TN, USA. February 20–22.

Watts RJ, Jones AP, Chen P, Kenny A. 1997. Mineral-catalyzed Fenton-like oxidation of sorbed chlorobenzenes. Water Environ Res 69:269–275.

Watts RJ, Foget MK, Kong SH, Teel AL. 1999. Hydrogen peroxide decomposition in model subsurface systems. J Hazard Mater B69:229–243.

Watts RJ, Sarasa J, Loge FJ, Teel AL. 2005a. Oxidative and reductive pathways in manganese-catalyzed Fenton's reactions. J Environ Eng 131:158–164.

Watts RJ, Howsawkeng J, Teel AL. 2005b. Destruction of a carbon tetrachloride dense nonaqueous phase liquid by modified Fenton's reagent. J Environ Eng 131:1114–1119.

Watts RJ, Finn DD, Cutler LM, Schmidt JT, Teel AL. 2007. Enhanced stability of hydrogen peroxide in the presence of subsurface solids. J Contam Hydrol 91:312–326.

West OR, Siegrist RL, Mitchell TJ, Jenkins RA. 1995. Measurement error and spatial variability effects on characterization of volatile organics in the subsurface. Env Sci Technol 29:647–656.

West OR, Cline SR, Holden WL, Gardner FG, Schlosser BM, Thate JE, Pickering DA, Houk TC. 1997. A Full-Scale Field Demonstration of *In Situ* Chemical Oxidation through Recirculation at the X-701B Site. ORNL/TM-13556. Oak Ridge National Laboratory, Oak Ridge, TX, USA. 114 p.

West OR, Cline SR, Siegrist RL, Houk TC, Holden WL, Gardner FG, Schlosser RM. 1998. A Field-Scale Test of *In Situ* Chemical Oxidation through Recirculation. Proceedings, Spectrum '98 International Conference on Nuclear and Hazardous Waste Management, Denver, CO, USA, September 13–18, pp 1051–1057.

Woods LM. 2008. *In Situ* Remediation Induced Changes in Subsurface Properties and Trichloroethene Partitioning Behavior. MSc Thesis. Environmental Science and Engineering Division, Colorado School of Mines, Golden, CO, USA.

Woods PL, Siegrist RL, Crimi M. 2012. Effects of *in situ* remediation using oxidants or surfactants on subsurface organic matter and sorption of trichloroethene. J Ground Water Monit Remediat 32:96–105.

Wymore R, Cross PE, Baird D. 2010. Case Study: CHP Treatment of a TCE DNAPL Plume. Presentation at the Seventh International Conference on Remediation of Chlorinated and Recalcitrant Compounds (Monterey, CA; May 2010). In Fields KA, Wickramanayake GB (Chairs), Remediation of Chlorinated and Recalcitrant Compounds – 2010. ISBN 978-0-9819730-2-9. Battelle Memorial Institute, Columbus, OH, USA, Abstract D-005.

Xu X, Thomson NR. 2007. Stability of hydrogen peroxide in the presence of aquifer solids using a green chelating reagent, Chemosphere 69:755–762.

Xu X, Thomson NR. 2008. Estimation of the maximum consumption of permanganate by aquifer solids using a modified chemical oxygen demand test. ASCE J Environ Eng134:353–361.

Xu X, Thomson NR. 2009. A long-term bench-scale investigation of permanganate consumption by aquifer materials. J Contam Hydrol 110:73–86.

Yan YE, Schwartz FW. 1998. Oxidation of Chlorinated Solvents by Permanganate. In Wickramanayake GB, Hinchee RE, eds, Physical, Chemical, and Thermal Technologies: Remediation of Chlorinated and Recalcitrant Compounds. Battelle Press, Columbus, OH, USA, pp 403–408.

Yan YE, Schwartz FW. 1999. Oxidative degradation and kinetics of chlorinated ethylenes by potassium permanganate. J Contam Hydrol 37:343–365.

Yeh CK-J, Wu H-M, Chen T-C. 2003. Chemical oxidation of chlorinated non-aqueous phase liquid by hydrogen peroxide in natural sand systems. J Hazard Mater 96:29–51.

Yukselen-Aksoy Y, Khodadoust AP, Reddy KR. 2010. Destruction of PCB 44 in spiked subsurface soils using activated persulfate oxidation. Water Air Soil Pollut 209:419–427.

Zhang H, Ji L, Wu F, Tan J. 2005. *In situ* ozonation of anthracene in unsaturated porous media. J Hazard Mater 120:143–148.

CHAPTER 10

IN SITU CHEMICAL REDUCTION FOR SOURCE REMEDIATION

Paul G. Tratnyek,[1] Richard L. Johnson,[1] Gregory V. Lowry[2] and Richard A. Brown[3]

[1]Oregon Health & Science University, Portland, OR 97239, USA; [2]Carnegie Mellon University, Pittsburgh, PA 15213, USA; [3]Environmental Resources Management, Ewing, NJ 08618, USA

10.1 INTRODUCTION

In situ chemical reduction (ISCR) has been defined in various ways since the term first began appearing in the late 1990s (Szecsody et al., 2004; Brown et al., 2006; Brown, 2008; Dolfing et al., 2008; Brown, 2010; Tratnyek, 2010). In general, and for the purposes of this chapter, ISCR refers to the category of *in situ* groundwater remediation technologies in which treatment occurs primarily by chemical reduction of contaminants. The emphasis of ISCR is on abiotic processes, but contaminant reduction by biogenic minerals is included if the role of microbial activity in the contaminant reduction is indirect. The reducing conditions necessary for ISCR can arise from natural *intrinsic* biogeochemical processes, can be generated by stimulating *in situ* microbial activity, or can be created by adding chemical reductants.

The chemical reductants that contribute to ISCR include reduced metal species such as Fe^{2+}, Fe_3O_4, and Fe^0, reduced sulfur species (HS^-, $S_2O_4^{2-}$, FeS, FeS_2) and other labile, electron-donating substances (natural organic matter [NOM]). The branch of ISCR that relies on intrinsic reductants is limited to minerals that contain Fe^{II}, S^{-II} and/or S^{-I}, and NOM, whereas the more highly engineered ISCR technologies may employ stronger reductants, such as dithionite ($S_2O_4^{2-}$) or zero-valent metals. More details on the reductants used for ISCR are given in Section 10.2.2.

The reduction processes responsible for contaminant treatment by ISCR include degradation of organics and sequestration of inorganics. Contaminants that are subject to reduction under ISCR conditions include (1) organic compounds with chloro-, nitro- or other readily reducible functional groups; (2) metal oxyanions that become less mobile upon reduction; and (3) a few nonmetal inorganics such as nitrate and perchlorate. More detail on the contaminants that are treatable by ISCR is given in Section 10.2.1, but the scope of this chapter is limited primarily to chlorinated solvents.

The application of ISCR for chlorinated solvent source remediation is a more recent development than its use for plume containment or treatment (Brown, 2010). This chapter describes the technical basis, engineering aspects, past experiences and future prospects for ISCR treatment of chlorinated solvent source zones. The discussion begins with a summary of the fundamentals, including the chemistry of contaminant reduction reactions and the geotechnical aspects of implementation. The chapter presents case studies for each of the major variations of ISCR in current use and discusses lessons learned from this experience. Finally, the chapter summarizes the overall status of this approach to source zone remediation, with a forecast of likely near-term developments and an assessment of barriers to future progress.

B.H. Kueper et al. (eds.), *Chlorinated Solvent Source Zone Remediation*,
doi: 10.1007/978-1-4614-6922-3_10,

10.1.1 Technology Development

The fact that groundwater and sediment contaminants can be reduced by abiotic pathways (i.e., pathways that do not directly involve microorganisms) has been well documented in the research literature for more than 30 years. Most of the early work on these processes has been summarized in several reviews (Macalady et al., 1986; Tsukano, 1986; Wolfe and Macalady, 1992). More recently, there have been many academic studies of organic contaminant degradation reactions using model systems designed to represent the natural reductants that are most likely to be responsible for abiotic reduction reactions in soils, sediments and groundwaters.

Three types of naturally occurring, abiotic reductants have been studied most thoroughly:

1. Minerals or their amorphous analogs that derive reducing properties from Fe^{II}. These include magnetite (Lee and Batchelor, 2002a; Gorski et al., 2010), green rust (Lee and Batchelor, 2002b; Erbs et al., 1999; O'Loughlin and Burris, 2004; Ayala-Luis et al., 2012), ferruginous clays (Cervini-Silva et al., 2002; Elsner et al., 2004), goethite with adsorbed Fe^{II} (Amonette et al., 2000; Kenneke and Weber, 2003; Elsner et al., 2004; Butler et al., 2011; Zhang et al., 2011), and possibly minerals commonly associated with basalt (Ingram et al., 2000).
2. Minerals deriving their reducing properties from S^{-II} (or S^{-I}) as well as Fe^{II}. Mackinawite (Butler and Hayes, 1999; Butler and Hayes, 2001; Butler et al., 2011) and pyrite (Kriegman-King and Reinhard, 1994; Lee and Batchelor 2002a, b) are the most studied of these minerals, but other Fe^{II} and $S^{-II/-I}$ phases of possible significance include greigite, marcasite and amorphous FeS.
3. Redox-active moieties associated with natural organic matter. These are mainly quinones (Tratnyek and Macalady, 1989; Schwarzenbach et al., 1990; Uchimiya and Stone, 2009) but could also include thiol groups and/or complexed metals (Xia et al., 1998; Struyk and Sposito, 2001; Szulczewski et al., 2001).

Few documents have attempted a comprehensive review of the peer-reviewed literature on these intrinsic ISCR reductants; the most thorough and comprehensive review to date is He et al. (2009). However, there are numerous detailed reviews of the recent literature on specific classes of abiotic contaminant reduction pathways (Haderlein and Schwarzenbach, 1995; Haderlein et al., 2000; Tratnyek and Macalady, 2000; Amonette, 2002; Totten and Assaf-Anid, 2003; Adriaens et al., 2004; Elsner and Hofstetter, 2011; O'Loughlin et al., 2011; Strathmann, 2011), and these reactions are attributed to the same putative reductants as listed above.

Despite this extensive background, most in the remediation engineering community have long assumed that *in situ* abiotic reduction of contaminants in the environment is insignificant as a component of natural attenuation or as a basis for remediation technologies. Recently, however, a growing number of field studies have characterized sites where a significant portion of the overall contaminant degradation appears to be due to direct reactions with reducing mineral phases (Ferrey et al., 2004; Darlington et al., 2008). This trend has created interest in the role these phases play in natural attenuation and in the prospects for manipulating such systems to generate more or better reductants *in situ* (Becvar et al., 2008; He et al., 2009).

Along with these developments, there has been a rapid growth of remediation technologies that rely on zero-valent iron (ZVI) to serve as the chemical reductant of contaminants (Tratnyek et al., 2003a; Henderson and Demond, 2007; Gillham et al., 2010). This field has become fairly mature in recent years, and considerable competition exists among vendors of ZVI for remediation applications. This competition has led to efforts to engineer better forms of ZVI or to develop other enhancements that will improve ZVI reactivity, longevity or delivery. One major strategy for enhancing ZVI performance involves bimetallic combinations of ZVI and catalytic metals such as Pd or Ni (Grittini et al., 1995; Gui and Gillham, 2003). Another strategy involves nanoscale zero-valent iron (nZVI), where the nano size of these particles provides greater reactive surface area and may impart other nano-size effects that could enhance remediation performance (Li et al., 2006; Tratnyek and Johnson, 2006; Lowry, 2007; Geiger and Carvalho-Knighton, 2009; Comba et al., 2011; Scott et al., 2011; Crane and Scott, 2012; Mueller et al., 2012). Other approaches to engineering better reductants range from manipulation of metallurgical properties (Landis et al., 2001) to replacement of Fe^0 with more strongly reducing metals such as Zn^0 (Li and Klabunde, 1998; Arnold et al., 1999; Tratnyek et al., 2010).

Even more recently, it has been recognized that all of the above chemical reductants form the basis for a coherent family of remediation technologies, which has come to be known as *in situ* chemical reduction or ISCR (Brown et al., 2006; Brown, 2008; Brown, 2010). This recognition may allow researchers and practitioners an opportunity to advance all forms of ISCR technologies by utilizing some of the general properties of reductants that determine the strengths, weaknesses, similarities and differences among the options. Furthering this goal is a major objective of this chapter.

10.1.2 State of the Practice

In situ chemical reduction is now widely recognized as a major category of remedial options, encompassing a range of technologies including some that are well established, others that are emerging and a few that are still under development. No consensus has been reached on any one scheme for classification of ISCR technologies, but two approaches seem promising.

The first classification approach distinguishes between technologies based on the relatively natural, intrinsic reductants (including minerals of geological origin or minerals formed by *in situ* stimulation of otherwise natural biogeochemical processes) and technologies using addition of chemical reductants that do not otherwise occur in nature, such as zero-valent metals. This distinction was made previously in the summary of ISCR reductants in the preceding section. In Figure 10.1, it is represented as a continuum of *effective strength*, from relatively mild, generally natural, to stronger, typically engineered, reductants.

Another useful way of classifying ISCR technologies distinguishes between those used to create reactive treatment zones that intercept contaminant plumes and those that target source zones directly. In general, plume treatment technologies are designed to be longer lasting, and often can be implemented with relatively less invasive methods. Source zone treatment requires more reductant per unit volume treated, and this is usually accomplished by emplacement into the source zone using comparatively disruptive methods.

Both of these classification schemes are used in Figure 10.1 to provide a summary map of ISCR technologies that currently are established or emerging. The ISCR technologies included in the map are briefly described below, and case studies of each are given in Section 10.3.2. Combined remediation and other variations on these technologies are addressed in Section 10.2.5. A broader perspective on combined remedies is given in Chapter 15 of this volume.

Abiotic Monitored Natural Attenuation. Contaminant degradation during monitored natural attenuation (MNA) is usually dominated by biodegradation, but recently it has become recognized that abiotic degradation pathways can be significant (Ferrey et al., 2004). In these cases, referred to here as *abiotic MNA* (Brown et al., 2007; Brown, 2010), significant contaminant degradation occurs by direct reaction with mild reductants, primarily ferrous iron and iron sulfide minerals, that are generated from natural biogeochemical processes. Abiotic MNA applies only to contaminants that are relatively labile to reduction and concentrations of contaminants that are relatively small, e.g., part per billion [ppb] levels of trichloroethene.

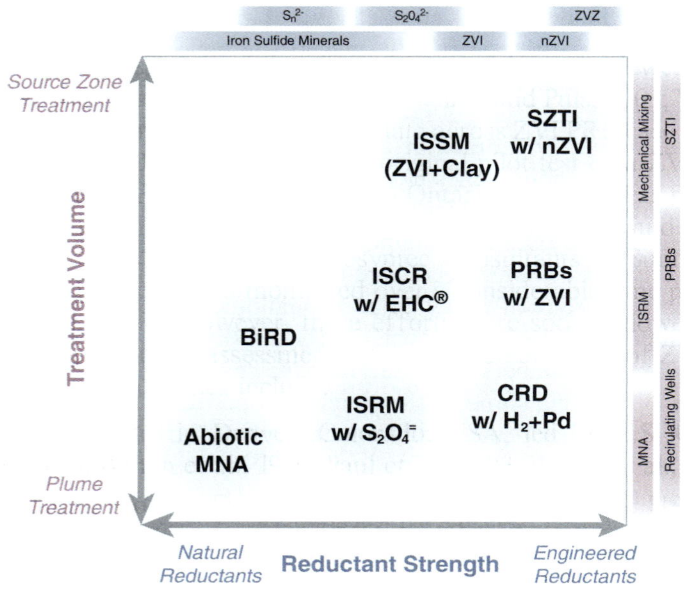

Figure 10.1. Map of ISCR technologies in current practice. The *horizontal dimension* represents the continuum from naturally occurring and/or mild reductants (Fe^{II} and $S^{II-/I-}$ containing minerals) to the generally strong chemical reductants used in fully engineered systems (Fe^0 and Zn^0). The *vertical dimension* represents the various modes of application, from emplacement of reductants to intercept plumes (PRBs) to injection of reductants that target the source zone (nZVI). Acronyms used to identify the specific technologies are defined in the following subsections.

Biogeochemical Reductive Dechlorination. The term *biogeochemical reductive dechlorination* (BiRD) was coined by Kennedy et al. (2006a, b) to describe the process of stimulating abiotic reduction of chlorinated solvents by forming reactive iron sulfides. In this scenario, iron sulfides are created by stimulating microbial sulfate reduction in the presence of iron. The application of BiRD requires the presence of sufficient carbon source, sulfate and iron. Carbon and sulfate generally must be added, and iron may be added although naturally present iron minerals are often sufficient. Biogeochemical reductive dechlorination is primarily employed as a barrier technology, and to date these usually have been biowalls of mulch amended with gypsum and goethite.

In Situ Redox Manipulation. In soil matrices with significant iron (>1 wt%), applications of moderately strong chemical reductants such as sodium dithionite or calcium polysulfide cause reduction of the ferric iron associated with the mineral matrix. The resulting Fe^{II}-bearing minerals can then serve as the reducing agent to effect reductive transformation of contaminants. An example of such a process is the technology known as *in situ* redox manipulation (ISRM) (Fruchter et al., 2000; Szecsody et al., 2004), where dithionite, a soluble chemical reductant, is injected into the subsurface to reduce native ferric iron to adsorbed and structural ferrous iron, which can in turn reduce contaminants such as chromate, carbon tetrachloride, trichloroethene and some munition compounds.

Permeable Reactive Barriers. Technologies that mitigate contaminant plumes by *in situ* placement of permeable, reactive material transverse to groundwater flow are known as reactive treatment zones (RTZs) or more commonly, permeable reactive barriers (PRBs).

A wide range of materials (leaf litter, fish bones, activated carbon, etc.) can be used in PRBs to effect a variety of contaminant removal processes, but the most significant is granular ZVI in various forms (Scherer et al., 2000; Tratnyek et al., 2003a; Comba et al., 2011). Permeable reactive barriers may be placed near a source zone or downgradient, such as before a receptor, depending on site-specific considerations. Emplacement of the early ZVI PRBs was done by trenching, but now they often are constructed by hydraulic or pneumatic fracturing, soil mixing, or direct injection of micron- or nano-sized ZVI. Several other ISCR technologies (ISRM and source zone targeted injection [SZTI]) can have operational characteristics that overlap with PRBs.

Source Zone Targeted Injection. The injection of chemical reductants to directly target source zone contamination is not yet widely utilized, but a growing range of applications is being considered. The reductants that have been studied the most include ZVI (both micron- and nano-sized iron) and polysulfide foam, although other forms of chemical reductants are feasible. In general, the materials are either particulate, such as ZVI types, or liquids and foams. Particulate forms of reductant are attractive because they may remain resident and reactive in the source area for longer times than do liquids/foams and therefore provide residual treatment capacity. Recently, SZTI with nanoscale zero-valent iron (nZVI) has attracted a great deal of attention for these reasons, but the fine particulate nature of this reductant raises other challenges related to longevity and emplacement, as discussed below in Section 10.2.4.6. Calcium polysulfide has been used for the *in situ* treatment of hexavalent chromium source zones. In this treatment method, calcium polysulfide foam is injected throughout the source zone to reduce Cr^{VI} to the less mobile Cr^{III} form (Graham et al., 2006). It has been employed at many locations in the United States and elsewhere (Fruchter, 2002).

In Situ Soil Mixing. Relatively shallow contaminated sites can be treated by mixing with a variety of treatment agents using large-diameter augurs. The most prominent example of *in situ* soil mixing (ISSM) for ISCR involves both ZVI and clay (Shackelford et al., 2005; Wadley et al., 2005). Typically, a mixture of clay (5–10%) and microscale ZVI (0.5–2%) is mixed into a soil matrix using large-diameter (4–8 feet (ft); 1.2–2.4 meters (m) diameter) augurs or soil mixers (e.g., Lang Tool). The clay has the potential to disperse dense nonaqueous phase liquid (DNAPL) as a Pickering emulsion (Roy-Perreault et al., 2005) and can also inhibit the movement of contaminated groundwater by decreasing overall aquifer permeability. The mixing ensures uniform contact between the emulsion and the ZVI. The main application of this technology thus far has been DNAPL zones (Water Science and Technology Board, 2004), including soils at depths as great as 50 ft (15.2 m).

Catalytic Reductive Dechlorination. Dechlorination by noble metal-catalyzed hydrogenolysis has been adapted for *in situ* remediation of contaminated groundwater, and this process is being called catalytic reductive dechlorination (CRD). Although CRD has performed well in bench-scale tests (Davie et al., 2008) and at least one extended pilot test (McNab et al., 2000), deactivation of the Pd catalyst occurs, especially when the groundwater contains sulfide (Reinhard et al., 2006; Munakata and Reinhard, 2007). This deactivation is reversible upon treatment with a suitable oxidant such as sodium hypochlorite (Lowry and Reinhard, 2000) or air-saturated water (Munakata and Reinhard, 2007). Improved catalyst formulations, for example, incorporating Au clusters on Pd or using zeolite supports to separate the Pd from constituents in the water that deactivate the catalyst, can improve resistance to deactivation and increase the time needed before regeneration (Schüth et al., 2000; Heck et al., 2009). However, implementation of CRD using in-well recirculating reactors remains promising. The reduction of other common groundwater contaminants including perchlorate and nitrate by Pd-based catalysts has also been studied (Wang et al., 2009; Choe et al., 2010).

Combined ISCR and In Situ Bioremediation. Amendments that combine chemical reductants (especially ZVI) with materials that stimulate microbial activity (organic carbon in various forms) are available as commercial products. The products include EHC® and Daramend® (FMC Environmental Solutions), ABC® + (Redox Tech, LLC), and emulsified zero-valent iron (EZVI) (National Aeronautics and Space Administration). The relative significance of abiotic versus biotic contaminant degradation by these amendments is usually not known, but they are included among the case studies described below. Additional discussion of these technologies is presented in Chapter 12 (*In Situ* Bioremediation of Sources).

10.2 TECHNICAL BACKGROUND

This section provides an overview of the general aspects of ISCR, including the processes that lead to contaminant removal (Section 10.2.1), conditions and agents that cause reduction (Section 10.2.2), strategies for accomplishing source zone remediation (Section 10.2.3), engineering aspects of implementation (Section 10.3.3) and considerations regarding the compatibility of ISCR in combination with other remediation technologies (Section 10.3.2).

10.2.1 Chlorinated Solvent Degradation Under Reducing Conditions

Dehalogenation can occur by several reductive pathways (Elsner and Hofstetter, 2011). The simplest results in replacement of a C-bonded halogen atom with a hydrogen and is known as *hydrogenolysis* or *reductive dehalogenation*. For a generic chlorinated aliphatic compound, RCl, hydrogenolysis is described by the half reaction:

$$RCl + H^+ + e^- \rightarrow RH + Cl^- \qquad \text{(Eq. 10.1)}$$

Polychlorinated compounds can undergo sequential hydrogenolysis, resulting in a characteristic sequence of partially dechlorinated products. For polychlorinated methanes, sequential hydrogenolysis via chloroform, dichloromethane, and others is represented by the solid lines in Figure 10.2a. For polychlorinated ethenes, sequential hydrogenolysis results in the dichloroethenes and vinyl chloride, which is shown with solid arrows in Figure 10.2b.

In both cases, each step in the sequence is much less favorable, thermodynamically and kinetically, than the one before; therefore, sequential hydrogenolysis tends to result in partially dechlorinated products that are persistent and therefore problematic. This is a well-known problem with anaerobic biodegradation of carbon tetrachloride (CT), perchloroethene (PCE) and trichloroethene (TCE). In the case of CT, degradation appears to stall at chloroform and methylene chloride, while degradation of PCE/TCE stalls at the 1,2-dichloroethenes and vinyl chloride. Under conditions where abiotic reduction of contaminants proceeds mainly by sequential hydrogenolysis, as is often observed when CT reacts with ZVI (Matheson and Tratnyek, 1994; Támara and Butler, 2004), the problem of accumulated intermediate dechlorination products may also arise in ISCR.

In contrast to hydrogenolysis, the other major dehalogenation pathway involves eliminating two halogens, leaving behind a pair of electrons that usually forms a carbon–carbon double bond. Where the pathway involves halogens on adjacent carbons, it is known as vicinal dehalogenation or reductive β-elimination, and where both chlorines are eliminated from the same carbon, it is known as gem- or α-elimination. This pathway tends to produce characteristic

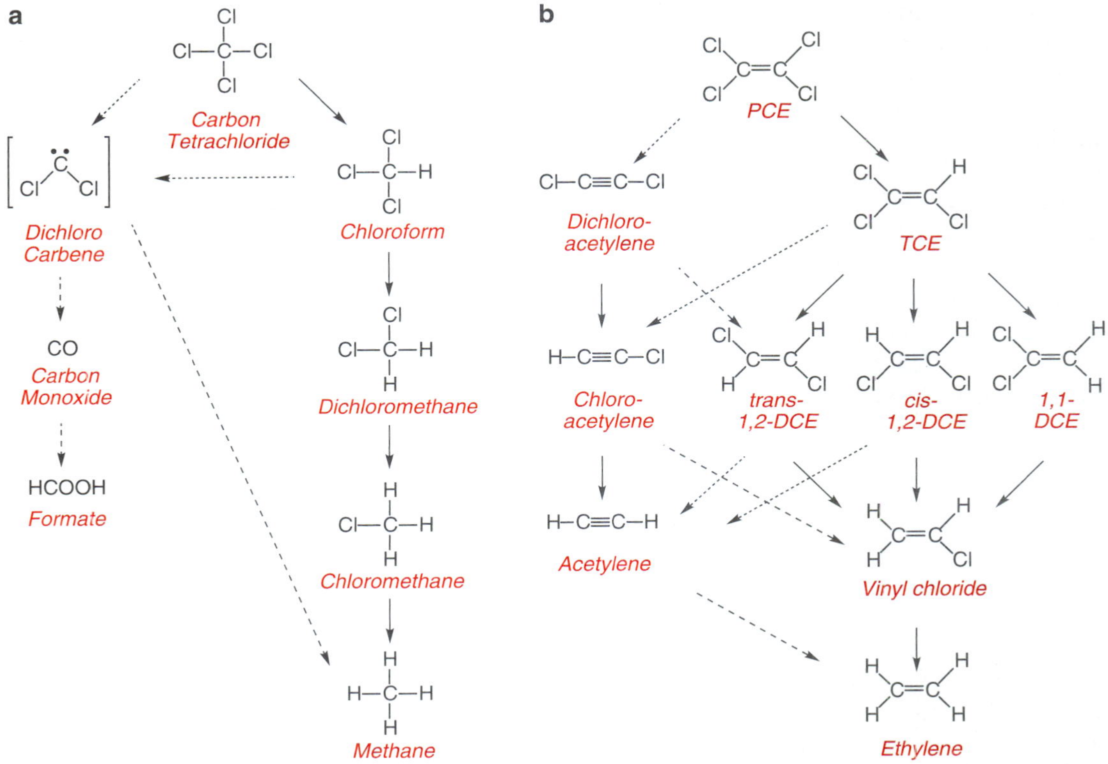

Figure 10.2. Reaction schemes for reductive transformations of (a) carbon tetrachloride and other chloromethanes and (b) perchloroethene and other chloroethenes. *Solid arrows* represent hydrogenolysis; *finely dashed arrows* are for β- and α-reductive elimination; and *coarse dashed arrows* show hydrogenation, hydrolysis and other steps. The scheme for chloromethanes (a) is synthesized from several sources (McCormick and Adriaens, 2004; Tratnyek, 2010); the scheme for chloroethenes (b) is reproduced from Tratnyek et al. (2003a), which in turn was adapted from Arnold and Roberts (2000).

products: alkenes from vicinal dihaloalkanes (Equation 10.2), alkynes from vicinal dihaloalkenes (Equation 10.3), and fully dehalogenated products from dihalomethanes (Equation 10.4).

$$R_1{-}CCl_2{-}CCl_2{-}R_2 + 2\,e^- \longrightarrow (Cl)(R_1)C{=}C(Cl)(R_2) + 2\,Cl^- \qquad \text{(Eq. 10.2)}$$

$$(Cl)(R_1)C{=}C(Cl)(R_2) + 2\,e^- \longrightarrow R_1{-}C{\equiv}C{-}R_2 + 2\,Cl^- \qquad \text{(Eq. 10.3)}$$

$$R_1{-}CCl_2{-}R_2 + 2\,e^- \longrightarrow R_1{-}C({=}O){-}R_2 + CO,\ CH_4,\ \text{etc.} \qquad \text{(Eq. 10.4)}$$

After reductive elimination, hydrogenolysis of polychlorinated contaminants can occur, resulting in another series of partially dechlorinated intermediates. On the left side of Figure 10.2b, this process is shown for PCE.

In addition to the two major reductive pathways for dechlorination, two additional reactions must be considered: *hydrogenation*, which involves addition of hydrogens across a C–C double or triple bond (Equation 10.5), and *dehydrohalogenation*, which involves elimination of H^+ and X^- to give a new C–C double bond (Equation 10.6). Hydrogenation has been invoked to explain the distribution of products observed in several studies involving chlorinated alkenes and Fe^0 (Arnold and Roberts, 2000) and is particularly important where a noble metal like Pd is present to act as a catalyst for activation of H_2. Note that H_2(surf) in Equation 10.5 represents all of the various forms of surface-activated hydrogen (e.g., $H^\bullet$) and is not meant to imply that the reaction necessarily involves adsorbed diatomic molecular hydrogen. Dehydrohalogenation (Equation 10.6) has received relatively little attention as a reaction that might contribute to degradation of chlorinated ethenes by Fe^0, even though it can be base-catalyzed (Roberts et al., 1993), which might make it favored under alkaline conditions, such as those created by corrosion of zero-valent metals.

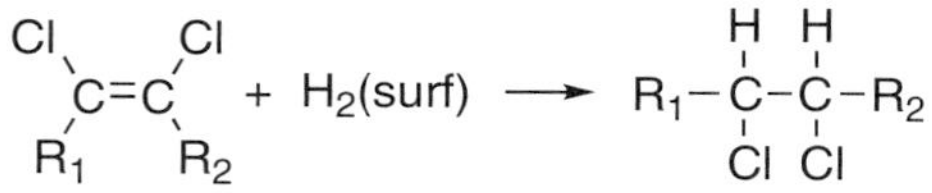

(Eq. 10.5)

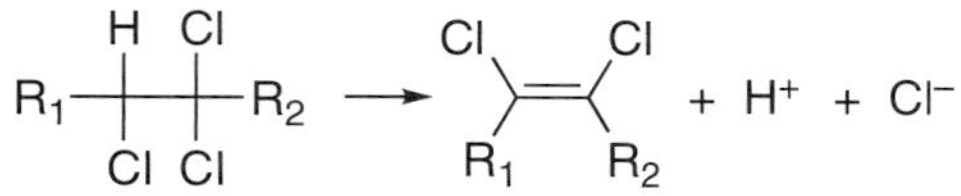

(Eq. 10.6)

The relative significance of hydrogenolysis, reductive elimination, and dehydrohalogenation in the degradation of chlorinated solvents during remediation depends on the contaminant structure, properties of the reductant, and environmental conditions. In general, hydrogenolysis tends to be more important for contaminants that are more nearly per-halogenated (all positions available for substitution are occupied by chlorines) and reductants that are relatively mild and therefore favor 1-electron reduction processes (Equation 10.1). Reductive elimination tends to be most important for contaminants with few chlorines per carbon center and strong reductants that favor two-electron processes (Equations 10.2, 10.3 and 10.4). Dehydrohalogenation is not a reduction reaction and therefore does not require a reductant, but this reaction is base-catalyzed so it is favored at higher pHs.

To improve on the qualitative rules of thumb above, some research has concentrated on developing quantitative models for predicting the relative rates of reduction reactions for the various chlorinated solvents (Tratnyek et al., 2003b). Most of this work has focused on dechlorination by zero-valent iron, and for this reductant—for the limited range of conditions addressed in existing studies—several quantitative structure–activity relationships (QSARs) have been reported (Tratnyek et al., 2003b; Onanong et al., 2007; Cwiertny et al., 2010). However, these relationships do not take into account the reductant type (Fe^0 and Fe^{II} minerals are likely to produce different relative rates of dechlorination) or the reductant concentrations; thus, they predict only relative rates not absolute rates under specific environmental conditions. In principle, this restriction can be overcome by cross-correlation of reduction rates between two systems (Kenneke and Weber, 2003) or incorporation of site conditions into the QSAR model (Peijnenburg et al., 1991), but neither approach has been applied at the field scale, to date.

10.2.2 Reductants Contributing to ISCR

The dechlorination reactions responsible for most chlorinated solvent degradation during ISCR are written generically as reduction half reactions in Equations 10.1, 10.2, 10.3 and 10.4. To complete these reactions, they must be balanced with oxidation half reactions involving the reducing agents (reductants) that are relevant under ISCR conditions. Identifying the relevant reductants and their corresponding oxidation half reactions is complicated by a number of factors, including equilibrium speciation, kinetically controlled metastable phases, and spatial heterogeneity. This complexity precludes a complete understanding of which reductants are responsible for contaminant degradation during ISCR in any particular field site, but the general principles presented below provide a framework for assessing, predicting, and optimizing the performance of ISCR over all types of sites.

The reductants that play the largest overall role in natural and engineered ISCR processes are the reduced Fe and S mineral phases that were summarized in Section 10.1.1. The relationships between the most stable forms of these phases are shown as redox potential (Eh-pH) diagrams in Figure 10.3a (for Fe minerals) and Figure 10.3c (for Fe/S minerals). From Figure 10.3a, for example, it is evident that the half reaction for contaminant reduction by Fe^0 at pH <6 results in $Fe^{2+}_{(aq)}$ but at pH >7 the thermodynamically preferred oxidation product is Fe_3O_4 (magnetite). At slightly higher (more positive) potentials, reduction by Fe_3O_4 favors Fe_2O_3 (hematite) as the oxidation product. In the presence of S, FeS_2 (pyrite) largely replaces Fe_3O_4 as the most stable intermediate valence phase (Figure 10.3c).

While the thermodynamically stable phases are favored over the long run, biogeochemical cycling of iron and sulfur or purposeful manipulations of geochemistry for remediation tend to produce metastable phases, which are inclined to be more reactive and therefore can be the dominant reductants of contaminants. To represent these conditions, some of the equilibrium phases in Figure 10.3a, c have been suppressed, resulting in Figure 10.3b, d. In the absence of S, the most important metastable reducing solid phase is probably $Fe_3(OH)_8$ (green rust), and in the presence of S, the phase of greatest interest is amorphous FeS.

Where they occur naturally in subsurface environments, the reducing phases represented in Figure 10.3 ultimately arise from and are sustained by microbial activity. This process is mainly due to the metabolism of iron-reducing and sulfate-reducing bacteria, which produce the Fe^{II} and S^{II} that go on to form Fe^{II} oxides and sulfides as diagramed in Figure 10.4. The four scenarios represented in Figure 10.4 are well supported by laboratory studies of dechlorination in biogeochemical model systems (Lee and Batchelor 2002a, b; Elsner et al., 2004; McCormick and Adriaens, 2004; Butler et al., 2011; Jeong et al., 2011) and are consistent with recent reports of abiotic MNA in the field (Ferrey et al., 2004; Elsner et al., 2010).

However, all of these reducing phases can also be generated *in situ* through purely abiotic processes by introducing a strong chemical reductant such as dithionite or ZVI. For example, green rust has been detected in and downgradient from ZVI PRBs, and it has been proposed that this phase provides additional opportunity for reduction of contaminants or intermediates of contaminant reduction (Johnson and Tratnyek, 1994; Kamolpornwijit et al., 2004). Recently, it has been shown that green rusts can be synthesized with improved characteristics for remediation (Ayala-Luis et al., 2012; Larese-Casanova et al., 2010), and this might represent a new class of solid-phase reductants engineered for source zone treatment by ISCR.

The reductants emphasized in Figures 10.3 and 10.4 are all solid phases, which implies that contaminant degradation occurs only after direct contact between these reactants. This requires mass transport of the contaminant to the reductant surface and formation of a precursor complex between the two species, either of which can limit the rate of contaminant degradation. In contrast, the analogous steps tend not to be significant barriers to reaction of contaminants with solution-phase reductants; thus, other factors being equal, solution-phase reductants can produce faster degradation rates. Solution-phase reductants that are relevant to ISCR include soluble complexes of Fe^{II} (Fe^{II}-citrate, Fe^{II}-porphyrins), lower-valent forms of S (disulfite, dithionite, thiosulfate, polysulfides), redox-active moieties associated with NOM (quinones) and other redox-active biomolecules such as cobalamins and others.

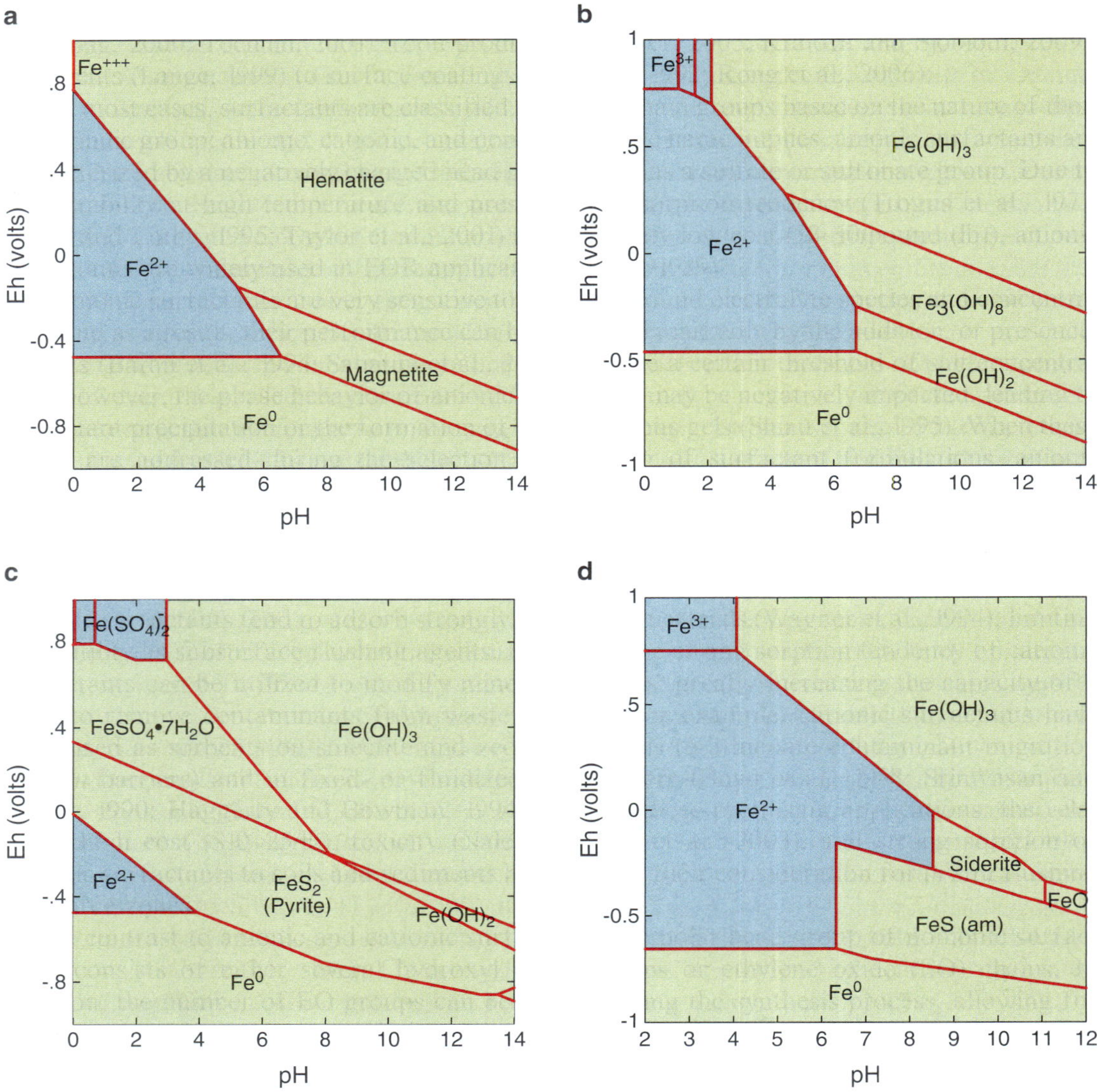

Figure 10.3. Eh-pH diagrams for iron–water (a, b) and iron–sulfur–water (c, d) systems. The *left pair* (a, c) shows the most stable species, and the *right pair* (b, d) shows metastable phases. *Blue* shading represents solution-phase species and *yellow* represents solid phases. Calculations were performed using the Geochemist's Workbench software suite and its default database of thermodynamic stability constants. *Red lines* represent the conditions with equimolar concentrations of reactants and products.

A selection of these redox couples is summarized on the right side of Figure 10.5 along a redox ladder that represents their standard reduction potentials at pH 7. The figure also shows ranges of potentials of two redox couples involving solid phases (Fe^{0}_{nano}/Fe^{III} and Fe^{II}/Fe^{III} oxides), and comparing these with the data for soluble reductants shows that the two groups overlap in their strength as reductants. Adding the iron sulfide phases and soluble forms of reduced sulfur to Figure 10.5 would not expand the overall range of potentials covered. All of the above are sufficiently strong reductants to make dechlorination thermodynamically

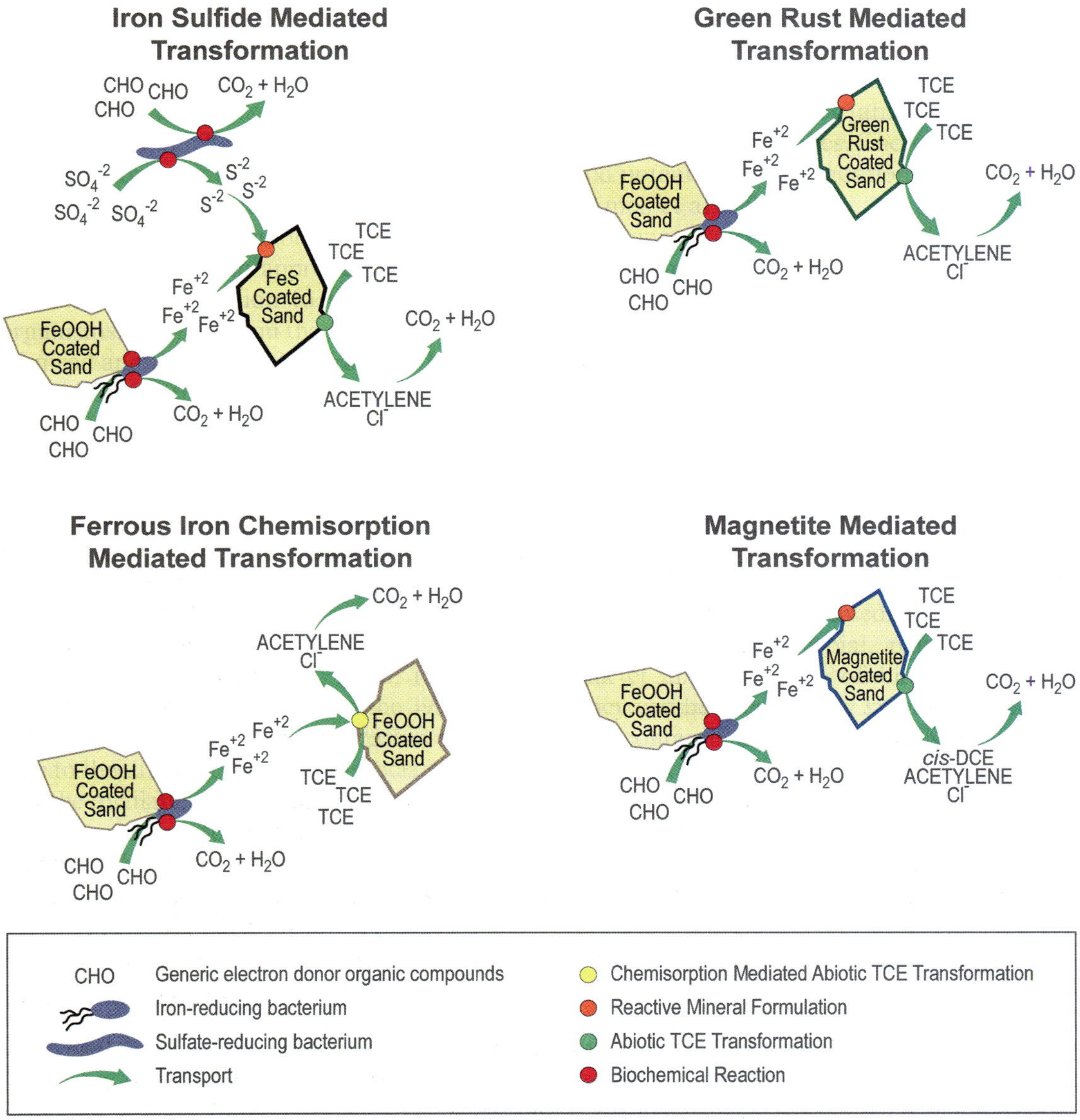

Figure 10.4. Conceptual models for *in situ* chemical reduction of TCE initiated by microbial metabolism that generates solid-phase reductants: FeS, green rust, FeII-doped goethite, and magnetite. Figure from Stroo and Ward (2010), which was adapted from Becvar et al. (2008), used with permission.

favorable (Haas and Shock, 1999; Dolfing, 2003); however, this analysis alone does not ensure that these reactions will be kinetically facile.

The link between reductant strength and kinetics of contaminant dechlorination is made in Figure 10.5 by plotting (left of the axis) the one-electron reduction potentials (E^1) for dissociative electron attachment. This step is often the rate-limiting step in dechlorination reactions, and therefore, E^1 is usually a good predictor of dechlorination rates (Tratnyek et al., 2003b; Bylaska et al., 2011). Comparison of the E^1 for dechlorination of various chlorinated solvents to the

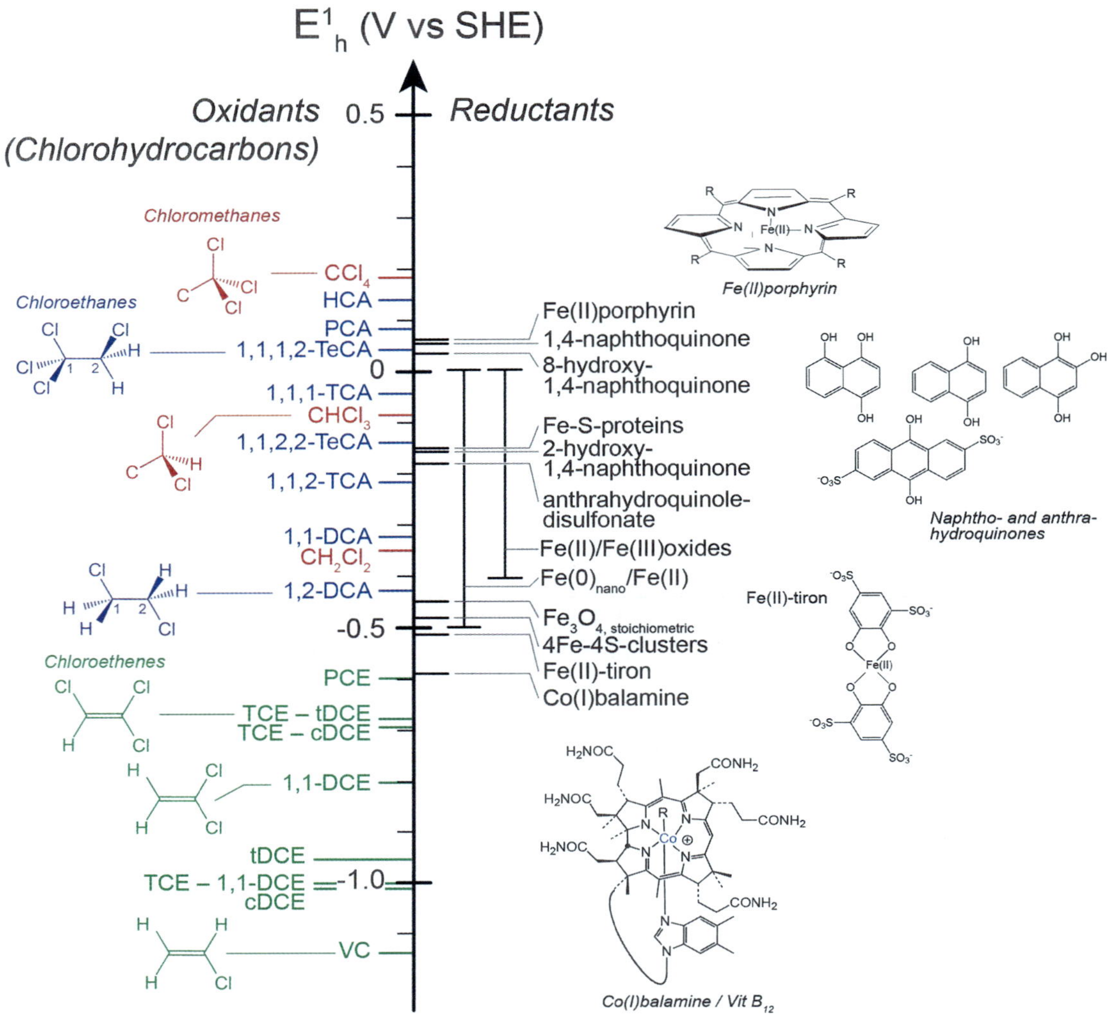

Figure 10.5. Redox ladder of one-electron reduction potentials for dechlorination (*left*) and standard reduction potentials putative reductants involved in ISCR (*right*). All potentials are for pH 7. Figure from Elsner and Hofstetter (2011), used with permission.

standard potential for various reductants shows that the initial electron attachment step is only favorable for the more highly chlorinated methanes and ethanes. Even for the less chlorinated alkanes and alkenes, the overall dechlorination reaction (Equations 10.1, 10.2, 10.3 and 10.4) is favorable under most conditions.

Another aspect of contaminant reduction implied by Figures 10.4 and 10.5 is the scenario by which a redox couple of intermediate potential serves to mediate or "shuttle" electrons from the original electron donor to a terminal electron acceptor. This effect can serve to bypass a kinetic barrier to direct reduction of the terminal acceptor by the original donor, if the steps involving the mediator are more facile. It has been shown that the reduction of contaminants can be significantly accelerated by electron transfer mediation involving aqueous complexes of iron, iron oxides, natural organic matter, and model compounds for the quinone moieties associated with natural organic matter (Van der Zee and Cervantes, 2009; Sposito, 2011; Zhang and Weber, 2009). These effects are likely to play a key role in the sustained performance of natural ISCR processes, but they are hard to quantify under field conditions. Electron transfer mediation can also be utilized to achieve enhanced remediation performance in engineered ISCR technologies, although this aspect of ISCR is not yet well developed (Johnson and Tratnyek, 1994).

Throughout the above discussion of contaminant reduction pathways (Section 10.2.1) and the reductants responsible for ISCR processes (this section), half reactions were written that explicitly involve electron transfer (Equations 10.1, 10.2, 10.3 and 10.4), and redox potentials were used to represent relative strength of prospective electron donors and acceptors (Figures 10.3 and 10.5). Most contaminant dechlorination reactions do occur by electron transfer mechanisms, but exceptions exist (e.g., see Equation 10.5 and Smolen et al., 1999), and the potential difference between the electron donor and acceptor does generally correlate with redox reaction rates (Miehr et al., 2004). Furthermore, it is true that faster reduction rates are generally observed in systems that exhibit lower, more negative, overall oxidation–reduction potential (ORP) of the system. However, the latter relationship is weak and indirect, and therefore not a reliable predictor of field scale performance of ISCR, as discussed further in Section 10.3.1.

A full explanation of why the system ORP (as measured by immersing a Pt electrode in a sample of site water or suspension of subsurface sediments) is often not predictive of dechlorination rates and why ORP, per se, should not be interpreted as the *cause* of contaminant reduction is beyond the scope of this chapter. However, many of the fundamental aspects of this issue have recently been described as they apply to SZTI with nZVI (Shi et al., 2011), and earlier analyses of the general issue of ORP interpretation are cited therein. In the case of SZTI with nZVI, measured ORP has a very complex relationship to the availability of nZVI as an electron donor to effect dechlorination of chlorinated solvents. In ISCR systems dominated by reduced sulfur species, the relationship between ORP and contaminant degradation rates is likely to be even more distant because the mechanism of contaminant reduction by iron sulfides is more complex.

10.2.3 Treatment of Source Zones

As with most other *in situ* restoration techniques, delivery of reductants to the target treatment zone is critical to the effectiveness of ISCR. In the context of source zone treatment, three types of commonly encountered sources of contaminants are considered here: aqueous plumes (Section 10.2.3.1), entrapped nonaqueous phase liquid (NAPL) sources (Section 10.2.3.2), and diffusional sources (e.g., slow back diffusion of contaminants entrapped in clay lenses) (Section 10.2.3.3). Treating each type of source presents different challenges associated with delivery of materials and will have different requirements in terms of reactivity and reactive lifetime.

10.2.3.1 Treatment of Aqueous Plumes

Although the focus of this book is on treatment of source zones, rather than dissolved groundwater plumes, there are a number of reasons to discuss *in situ* chemical reduction of aqueous groundwater plumes. Perhaps the most important reason is that many NAPL and diffusion source zones are difficult to access directly because they are located under structures in active use or because the location of the source has not been or cannot be sufficiently characterized to allow for targeted treatment of the source. In these cases, cutting off the source zone with, for example, a permeable reactive barrier may represent the best long-term solution. This approach is practical with ZVI PRBs because they can have very long reactive lifetimes and may require little monitoring, which results in overall favorable life cycle costs when compared with alternatives (Mak and Lo, 2011). Permeable reactive barriers in difficult-to-reach environments (under buildings and in deep groundwater aquifers) may eventually be possible using nZVI. However, challenges with their emplacement (Section 10.2.4.6) will have to be addressed, as will issues related to the relatively short reactive lifetime of nZVI.

10.2.3.2 Treatment of Nonaqueous Phase Liquid Zones

Assuming that the location of the NAPL sources can be determined with sufficient accuracy, delivering reactants to NAPL zones still presents a number of challenges that must be considered if ISCR is to be successful. These challenges are of two general types: hydraulic and partitioning.

Hydraulic challenges arise because ISCR reactive materials are typically delivered using water as the delivery vehicle. Nonaqueous phase liquid zones, and in particular NAPL pools, may have relatively limited contact with flowing groundwater. This situation is common because many sources have been in place for decades, and areas with high NAPL–water contact will have already dissolved away. This limited water–NAPL contact can have significant consequences on the effect of added ISCR reagents. For example, in the nZVI case, the nZVI may not be delivered directly to the NAPL–water interface, but instead will form a reactive barrier in the vicinity of the entrapped NAPL. In this scenario (Figure 10.6), the reaction will be limited by the rate of mass transfer of contaminants from the NAPL phase (present in the layer or in the media or in the intermediate K-layer) into the aqueous phase, where the dissolved contaminants are then transported to the reactive nZVI surfaces (Fagerlund et al., 2012).

Partitioning challenges arise from the hydrophobic nature of NAPLs, so that water-soluble reagents may move with the groundwater, but will tend to stay in the water phase. A significant amount of research has focused on designing reductants such as nZVI that partition to the NAPL–water interface (Figure 10.7) (Saleh et al., 2005) or completely into the NAPL phase

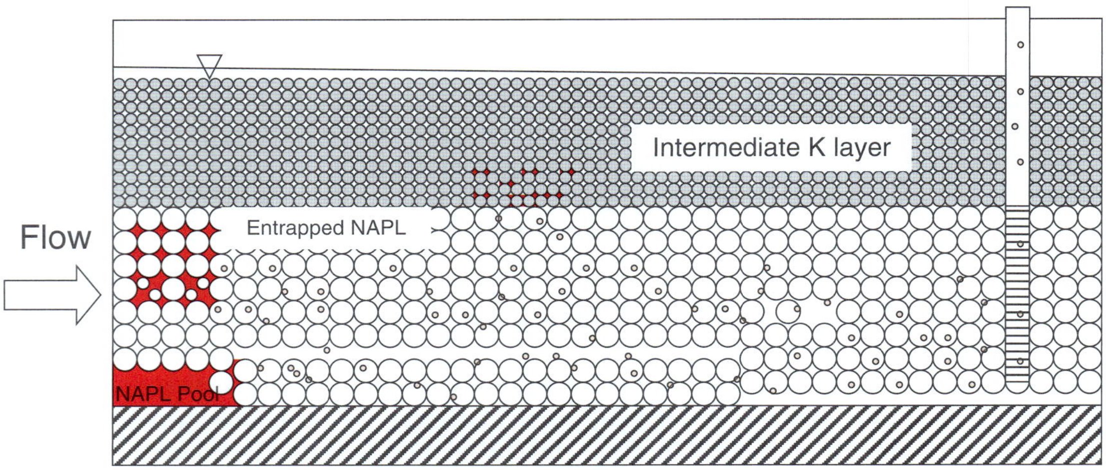

Figure 10.6. Schematic illustrating an entrapped NAPL source of contaminants to an aquifer. Slow dissolution of the entrapped NAPL serves as a long-term source of contamination to the aquifer. Emplacement of nZVI or other particulate reductant in the vicinity of the NAPL or into the NAPL can decrease the source mass, thereby decreasing the mass emission from the source and its lifetime. Challenges lie in providing sufficient reductant to eliminate a significant mass of the entrapped NAPL.

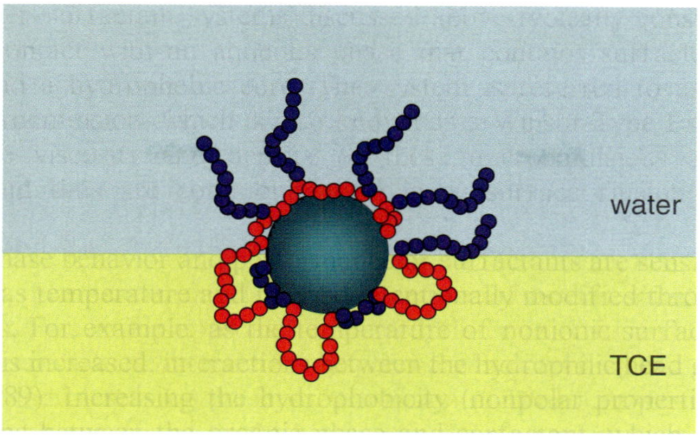

Figure 10.7. Hydrophobic segments (*red*) of the polymer coating swell in TCE (NAPL) while hydrophilic segments (*blue*) swell in water. This modification of nZVI could provide thermodynamic affinity of the particles to the NAPL–water interface, causing them to partition to where they can effect maximum NAPL degradation.

(Ramsburg and Pennell, 2001; Ramsburg and Pennell Kurt, 2002; Ramsburg et al., 2004; Ramsburg et al., 2005). Promoting contact between nZVI and the NAPL will, other factors being equal, increase the rate of reaction because the reaction is proportional to the concentration of dissolved TCE at the nZVI surface (Liu et al., 2007).

However, the approach illustrated in Figure 10.7 still requires that reductants reach the NAPL interface, which can be difficult to achieve because the NAPL may be located in poorly accessible (low conductivity) regions of the aquifer. Furthermore, even if the nZVI can be concentrated at the NAPL–water interface, this will result in rapid oxidation of the nZVI (Fagerlund et al., 2012), the products of which then remain at the NAPL–water interface, thereby preventing attachment of fresh reactive particles (Phenrat et al., 2011). Overall, the concept of targeted delivery of particulate reagents to the NAPL–water interface *in situ* has proven to be conceptually and practically challenging and has not yet been demonstrated to be a feasible technology for field applications.

10.2.3.3 Treatment of Diffusion Source Zones

Diffusion source zones are created when high-concentration sources such as NAPL remain in contact with lower-permeability media for an extended period of time (years). Such sources can occur at the surfaces of the low-permeability media (beneath pools of NAPL) or deeper within media because of NAPL movement down fractures, and subsequent diffusion. Back diffusion of contaminants from the lower-permeability media such as clay lenses can result in a very long-lived, low concentration sources (Figure 10.8).

To treat source zones such as those represented in Figure 10.8, the delivered reductant must have sufficient longevity and reactive capacity to sustain adequate degradation rates of contaminants that arrive by diffusion at an interface where groundwater is moving (the surface of the diffusion source). It is necessary in this case to maintain high concentrations of reactant at these interfaces, requiring emplacement of significant quantities of reductant or replenishment of the reductant by repeated or sustained injection via recirculation, for example.

Nanoparticle-sized reagents may ultimately be designed with the desired traits for treating diffusion sources (Khandelwal et al., 1998; Rabideau et al., 1999). For example, particulate reductants could be emplaced into the higher conductivity regions surrounding a clay lens

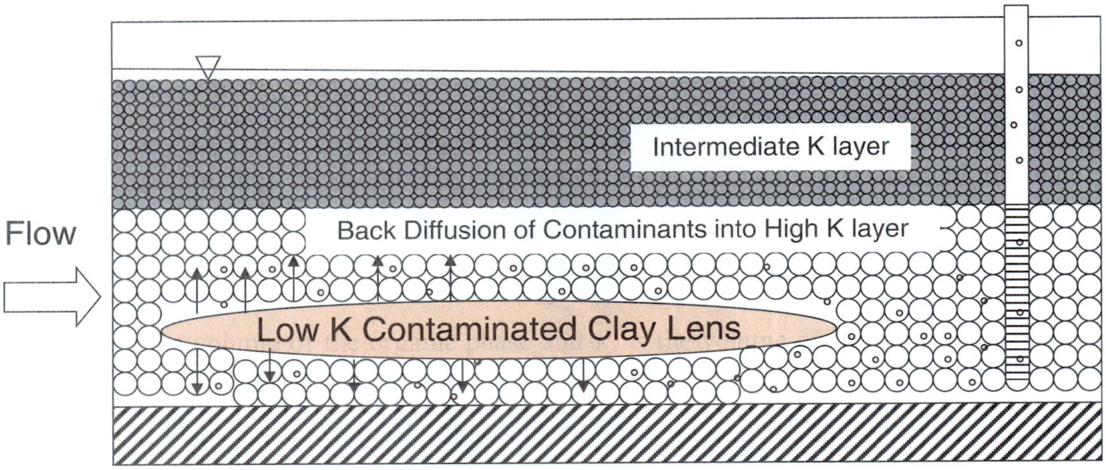

Figure 10.8. Schematic illustrating a diffusional source of contaminants to an aquifer. Slow diffusion of contaminants from clay lenses serves as a long-term source of contamination to the aquifer. nZVI or other particulate reductants can be emplaced in the high K region surrounding the lenses in order to degrade contaminants as they released.

serving as the source (Figure 10.8). The relatively low mobility of these materials (Section 10.2.4.6) could allow them to remain in the region surrounding the lenses and serve as a reactive barrier that degrades contaminants as they are released. This remediation strategy requires reductants that have sufficient capacity and reactivity to produce the required decrease in contaminant concentration, but also have a relatively long reactive lifetime so that they do not need to be replaced frequently.

The process described above primarily treats the mass leaving the diffusion source and has relatively little impact on mass within the diffusion source. If the latter is to be addressed, then the reductant to be delivered must have a number of additional characteristics. For example, it must have a sufficiently large diffusion coefficient that it can penetrate into the diffusion source. Also, it must be sufficiently inert to persist as it diffuses, but also sufficiently reactive to degrade the contaminants when the two come together. Once again, it must have been delivered to the interface with the diffusion source in appropriately high concentrations and for an extended period to control the long-term flux of contaminants from the diffusion source.

This is a challenging set of criteria to meet. Fortunately, the overall process of diffusion and reaction can probably be predicted with reasonable accuracy using numerical modeling (Fagerlund et al., 2012). Thus, if delivery of the reductant to the interface can be accomplished, the effectiveness of the ISCR likely can be assessed.

10.2.4 Strategies for Delivery

Reagents for ISCR are diverse in their chemical form, and some forms involve unique considerations for effective delivery in the subsurface. *In situ* chemical reduction reagents can include dissolved aqueous species (sodium dithionite), NAPLs (edible oils), emulsions and foams (calcium polysulfide), gases (H_2), and particles (ZVI and nZVI). Approaches and challenges for delivery of aqueous solutions of dissolved aqueous species are similar to those for delivery of oxidants for ISCO (Simpkin et al., 2011) (Chapter 9 of this volume). However, delivery of the NAPLs, emulsions, foams, gases, and particulate reagents used in ISCR can present unique challenges. A range of delivery strategies have been utilized for *in situ* delivery of these various forms of reductants, and they are summarized briefly below.

10.2.4.1 Delivery of Aqueous Solutions of Dissolved Reductants

Perhaps the most common approach for delivery of reductants is by injection of an aqueous solution of a dissolved species via wells into the subsurface. This approach is applicable to moderately permeable media, where the reductants can be emplaced into the formation over distances of meters in time frames of a few days. In high-permeability media, such as gravels, groundwater velocities can be sufficiently large that the residence time of the reductant within the treatment zone may be too small to achieve the required degradation of contaminants or to provide any residual capacity for reduction. In this case, alternate pumping strategies, for example, recirculation and/or injection–extraction ("push–pull"), may be more effective.

In low-permeability materials (clays, tills), solution injection rates can be too low to provide reasonable zones of influence from a given injection well, so hydraulic or pneumatic fracturing has become a common approach to enhance distribution. At least two strategies are used in this type of fracturing. The first is to use the fractures to deliver reductants directly to the source. The second is to use the fractures to focus the groundwater flow through the fractures filled with reductant such that treatment occurs within the fractures. With respect to chlorinated solvent source zones, there has been little evaluation of the relative importance of these mechanisms at the field scale.

10.2.4.2 Improved Delivery of Injected Dissolved Reagents

The delivery of aqueous solutions of reductants is challenging because of the presence of heterogeneity in the porous media and subsequent flow field. In heterogeneous media, targeted delivery of dissolved reductants to source zones is especially difficult if higher flow zones exist that lead the reductant away from the source zone. Several strategies have been developed to improve contact between the reductant and the source. Two of them, shear-thinning fluids and density-enhancing agents, are briefly discussed here.

Shear-thinning fluids have the characteristic that their viscosity decreases as shear velocity increases (Kaplan et al., 1994; Kaplan et al., 1996; Cantrell et al., 1997; Truex et al., 2011a, b). As a result, under high flow rate (injection) conditions, a shear-thinning fluid may be delivered to a high hydraulic conductivity zone, but under lower flow conditions (normal groundwater flow), that fluid will be sufficiently viscous that, once emplaced, flow through that zone will be decreased and as a result increased flow through lower conductivity zones (where the source might be located) will occur. Significant increases in viscosity can be achieved easily with relatively low concentrations (0.1–1 gram per liter [g/L]) of food-grade biopolymers such as xanthan and guar gum. Shear-thinning polymers are commercially available (Truex et al., 2011b). These materials have an added advantage, in the context of ISCR, because they provide readily biodegradable carbon sources that help drive aquifer systems to more strongly reducing conditions.

Shear-thinning fluids have been widely used in the petroleum industry to enhance oil recovery, but they have received less consideration in reductant delivery. One reason may be the compatibility of the viscosity agents and the reductants. For example, the presence of these biopolymers may decrease the reactivity of ZVI or nZVI that is injected along with the shear-thinning fluid (Phenrat et al., 2009a; Tratnyek et al., 2011; Truex et al., 2011b).

Density-driven flow also can be used to deliver aqueous phase dissolved reductants (Henderson et al., 2009; Heiderscheidt et al., 2011). The addition of almost any solute to water increases the density of the resulting solution. For highly soluble solutes, such as inorganic salts, resulting densities can be quite large (>1.2 grams per milliliter (g/mL)). If these solutions are injected into the subsurface, significant downward movement due to gravity can occur. If managed correctly, this enhanced downward movement can be utilized to increase contact between the injected solution and source zones. Several commonly used chemical reductants (sodium dithionite and calcium polysulfide) are very soluble and can be prepared readily at concentrations well suited for density-driven injection.

10.2.4.3 Delivery of Nonaqueous Phase Liquids, Emulsions, and Foams

Nonaqueous phase liquids can be used to emplace reactive materials in the subsurface. The NAPL can be the reactive material itself, or it can be a carrier for other materials. For example, nonaqueous, edible oils have been combined with reductants (especially ZVI) to create biogeochemical zones favorable for ISCR (Quinn et al., 2005). The NAPLs have an important advantage over aqueous liquids because they can become immobilized in the subsurface and as a result can provide a practical approach for maintaining high concentrations of reductants at or near source zones. Furthermore, the low solubility of NAPLs allows them to remain in place and serve as a long-term sources of reductant.

However, it is challenging to deliver nonaqueous fluids in the subsurface in a predictable manner. At least two strategies can improve deliverability of NAPLs: emulsions and fracturing. Emulsions of oil and micron-sized ZVI have been examined in laboratory and field studies (Quinn et al., 2005). The reactivity and transport of oil-in-water emulsions containing nZVI have also been demonstrated in the laboratory (Berge and Ramsburg, 2009). The emulsion may provide multiple benefits:

- Enhanced transport of the ZVI through the subsurface and to the NAPL interface, thereby requiring less dissolution of NAPL constituents for them to be reduced by the ZVI.
- Protection of the ZVI from the processes that contribute to natural reductant demand, thereby increasing the ZVI lifetime and possibly its efficiency (proportion of its capacity that goes to reduction of contaminants rather than water and other geochemical oxidants).
- Provision of organic carbon that can stimulate the biogeochemical conditions that are favorable to microbial dechlorination process.

In practice, the movement of emulsions through the subsurface is problematic for three primary reasons: (1) distribution of emulsions is highly influenced by physical and chemical heterogeneities within the aquifer, (2) emulsions tend to be unstable in porous media (porous media have historically been used to break emulsions created during chemical synthesis) and (3) the emulsions typically used (water–oil–water emulsions) are very viscous (up to 100s of centipoise (cP)) and difficult to inject in porous media, even in relatively high conductivity sand aquifers (Quinn et al., 2005).

Foams are liquid–gas emulsions that have been studied and used to remediate DNAPL and heavy metals from unsaturated soils (Wang and Mulligan, 2004; Shen et al., 2011). The foam is created from mixing a surfactant solution and nonwetting gas. They are very low density and the stability of the foam (time to collapse) can be designed based on surfactant choice to allow for delivery in the subsurface. Foams have been used in at least one field demonstration (Hirasaki et al., 1997) to remove TCE DNAPL. Foams have historically been developed to solubilize and mobilize the contaminants. However, the appropriate choice of surfactants and gas could potentially make foams a viable delivery tool for ISCR reactants in the unsaturated zone.

Movement of NAPL phases, including emulsions and foams, can be enhanced by fracturing. However, precise delivery of these fluids to specific locations (source zones) has proven difficult, even in relatively homogeneous media because it can be difficult to control propagation of the fractures. As discussed above, fracturing can be used to focus groundwater flow and treatment within the fractures, and this may provide the best opportunity for utilizing NAPLs to increase ISCR.

10.2.4.4 Delivery of Gases

Reducing gases, particularly hydrogen, have been used to enhance *in situ* reduction both in the unsaturated zone (Evans et al., 2011) and in the saturated zone (Newell et al., 1997; Newell et al., 1998; Newell et al., 2000). Delivery of gases in water-saturated media is typically accomplished using *in situ* sparging, although the use of hollow-fiber membranes also has been investigated (Fang et al., 2002). However, hydrogen solubility in water is only 1–2 mg/L, equivalent to ~800 micromolar (μM) at ambient pressure and groundwater temperature, so delivering significant quantities of dissolved H_2 by sparging groundwater is not practical. As with other types of specialized gas sparging, relatively low flow rates and pulsed operations are commonly employed.

In most aquifer materials, especially those with grain sizes less than ~1 millimeter (mm), gases travel almost exclusively in channels, rather than as individual bubbles (Johnson et al., 1993; Clayton, 1998). This limits the contact between water and gas to the flow of water around the gas channel. However, during pulsed operation, significant amounts of gas can remain trapped within the medium, and this trapped gas can provide a long-term low-level source of reductant between pulses. This trapped gas can persist from days to months, depending on the rate of groundwater flow.

One of the important limitations of sparging in source zone treatment is that it can again be difficult to deliver the reactants to the source zone, particularly if those sources are associated with lower-permeability layers (pools on top of those layers or diffusion sources within those layers.) As a result, sparging may be better suited to cutting off plumes from those source zones, rather than in trying to treat the zones directly (Johnson and Johnson, 2012).

10.2.4.5 Emplacement of Solids

Direct *in situ* mixing of solid-phase reductants and aquifer materials (ISSM) is an attractive approach to ISCR of source zones if conditions allow it (aquifer materials are amenable to auguring and the depth is not prohibitive). If excellent contact between the reductants and the source materials can be achieved, this approach can be quite successful (Olson et al., 2012). However, it does necessitate a significant input of resources. For example, a dense network of augured holes, including significant overlap between the holes, will likely to be necessary if remediation is to be complete. This approach has potential disadvantages, in addition to cost. It may result in mobilization of DNAPLs as part of the auguring process. This mobilization could result in additional downward movement of the DNAPLs, increasing their environmental impact.

As discussed above, in the context of source zone treatment, emplacement of solid reductants as PRBs will probably be useful only when cutoff of the resultant groundwater plume represents the best treatment strategy. In this context, reductant PRBs have been used in conjunction with impermeable barriers (sheet pile, slurry walls) in funnel and gate configurations (Gillham et al., 2010). Fractures propped open with reactants have been proposed for treatment of groundwater plumes (as mini-PRBs) as well as for direct treatment of sources.

As discussed above, the focus of source treatment can be either to deliver directly reactants to the source or to capture groundwater flow from the source and treat it within the fractures. In the context of ISCR, emplacement of granular metals or metals mixed with sand, within the fractures, can be an effective approach.

10.2.4.6 Injection of nZVI Suspensions

Of all the *in situ* approaches discussed here, injection of suspensions containing nZVI has received the most attention from researchers. Nanoscale zero-valent iron has a number of desirable characteristics that have contributed to this interest:

- It is highly reactive toward a number of important contaminants, including many of the chlorinated solvents.
- In principle, nZVI or micron-sized ZVI particles can be made mobile in a range of porous media types.
- The surface properties of these materials can be modified to increase their mobility in the subsurface and to provide NAPL targeting, although the benefits of such targeting are yet to be proven, as discussed above.
- With sufficient mobility, injection using existing deep wells allows for emplacement of reductants in the deep subsurface, or in other difficult to reach areas, such as under buildings.
- If the selectivity of nZVI toward the target contaminants can be improved, the emplacement of particulate materials provides excess reductant that will remain in place for extended times (to treat back diffusion sources, for example).

In practice, delivery of nZVI to treat source zones has proven to be more difficult than anticipated. A number of important challenges remain:

- Magnetic effects of the commonly used nZVI materials are sufficiently large that aggregation (Phenrat et al., 2007) and subsequent straining become an important process limiting nZVI mobility except at very low nZVI concentrations ($<<1$ g/L) (Phenrat et al., 2009a).
- Magnetic properties also may contribute to deposition within the subsurface because nZVI particles have a tendency to deposit on previously deposited nZVI particles.
- Three-dimensional flow pathways tend to cause nZVI transport directions to shift easily as plugging of the formation occurs, particularly in the immediate vicinity of injection wells.
- Reliance on water as the injection fluid for the dispersion limits the amount of nZVI that can be delivered directly to entrapped DNAPL (Figure 10.9).

The combination of these challenges makes it difficult to deliver nZVI in porous media and to deliver the quantities of nZVI necessary to significantly reduce the contaminant masses characteristic of source zones.

As a rule of thumb, travel distances of approximately 1.5 m (5 ft) are required to make nZVI injection an attractive remediation alternative. This spacing allows for an injection gallery with wells approximately 3 m (10 ft) apart. Lower mobility will require too many wells/injection points to distribute the material in the subsurface to create a PRB for plume treatment or to deliver sufficient material to entrapped NAPL. While some laboratory and field data suggest that nZVI transport distances of >2 m (6.5 ft) should be possible (Bennett et al., 2010), there are few field examples of where significant quantities of nZVI have been directly observed at distances of more than 1.5 m (5 ft) from an injection point (Johnson et al., 2013). Within those

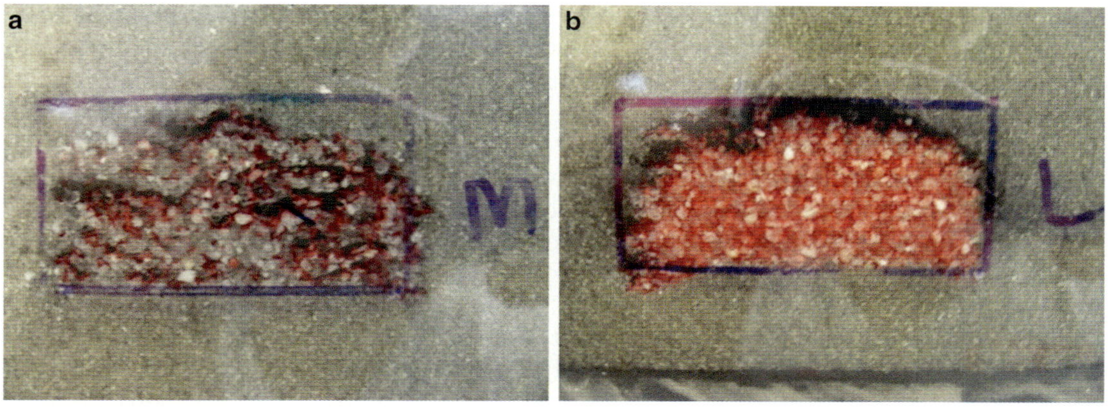

Figure 10.9. Images of polymer-coated nZVI targeting entrapped NAPL *in situ*. A 3 g/L dispersion of nZVI was flowed past the entrapped NAPL, and a fraction of the nZVI remained in the source. (a) A source with low NAPL saturation allows some nZVI to access the entire source. (b) A source with high NAPL saturation diverts water and nZVI above the source and only allows for attachment at the upper surface. Reprinted with permission from Phenrat et al. (2011). Copyright 2011, American Chemical Society.

few examples is evidence that flow was primarily along preferential flow pathways, in agreement with expectations and with laboratory studies (Phenrat et al., 2010a).

It should be noted that travel distance is operationally defined. Sometimes travel distance is considered the distance where nZVI is detected; however, this distance may represent only a small fraction (<1%) of the concentration of nZVI in the injection solution (Johnson et al., 2013). In other cases, it may be defined as the distance where a sizeable fraction (e.g., 25%) of the injected concentration is measured. Inconsistent use of the term travel distance has led to confusion over what is truly achievable in terms of travel distance for nZVI.

These field scale results suggest that it will be difficult to deliver significant quantities of nZVI to NAPL source zones when relying on water as a carrier for the nZVI. These data do, however, suggest that emplacement of nZVI into higher conductivity regions lying above diffusion sources may be possible. This remediation strategy can provide a reactive barrier to degrade contaminants as they diffuse from the low conductivity region to the higher conductivity regions. Progress continues to be made with regard to minimizing aggregation and increasing mobility through the use of surface modification, and the current state of research is summarized below.

A very important limitation of nZVI transport in the subsurface is its aggregation and subsequent filtration by the aquifer materials (Johnson et al., 2013). This leads to pore plugging, changes in flow patterns, and an inability to deliver sufficient material to the source (Phenrat et al., 2010b). The transport of polymer-coated nZVI in porous media can be good at low concentrations (100s of mg/L), but delivery of concentrated nZVI suspensions (~10 g/L) is required to make nZVI a useful remediation alternative, regardless of whether the goal is to emplace a PRB for plume treatment, to target NAPL directly, or to emplace nZVI in high conductivity regions to prevent back diffusion of contaminants. Effective delivery of concentrated solutions may be achieved through the use of surface modifications of nZVI using polymers (polypropylene glycol or carboxymethylcellulose) to prevent aggregation and deposition. It may also be achievable through the use of excess polymer in the injection solution, which can modify the surfaces of porous media as the solution carrying the nZVI moves through. Modification of the porous media further decreases deposition of nZVI, which vastly improves transport (Kim et al., 2012). Shear-thinning fluids for delivery of nZVI may also prove useful, as this method has been used to deliver 2-micron-sized ZVI particles in porous media (Truex et al., 2011a).

Two approaches may be used to modify the surfaces of nZVI. The first is to physically adsorb (physisorb) polymers to the nZVI surface (Phenrat et al., 2008). Polymers readily adsorb to nZVI, and modification is achieved simply by dispersing the nZVI in the presence of the dissolved polymer. This simple procedure allows one to select from a variety of polymer types available. It also allows for the selection of polymers compatible with the site's geochemical conditions, and polymers that can potentially serve as a biostimulant for microbial degradation of contaminants if free polymer is present in the injection solution. The presence of poly-aspartate coating on nZVI increased microbial biomass in an organic carbon limited marine soil by several orders of magnitude (Kirschling et al., 2010). Polymers with average molecular weights greater than ~1,000 g/mol will typically adsorb to nZVI with densities ranging from a 0.25 to 2 mg/m^2 (Golas et al., 2010). Adsorption of large molecular weight polymers (e.g., >2,000 g/mole) occurs at multiple points on the nZVI surface and is therefore effectively irreversible over time scales of interest, for example, months to years (Kim et al., 2009).

The second modification approach is to synthesize the nZVI in the presence of the polymeric modifier (carboxymethylcellulose) (He and Zhao, 2007; Liu et al., 2008). In this scenario the polymer is also used to help control the size distribution of the nZVI being synthesized. This process seems to provide surface-modified nZVI that is quite stable against aggregation and has improved transport characteristics over other types of nZVI (Bennett et al., 2010; He et al., 2010).

It should be noted that the adsorption of the polymeric coating onto nZVI will also alter the reactivity of the nZVI (Tratnyek et al., 2011). The coating generally decreases reactivity by a factor of 4 to 10 (Saleh et al., 2007), due to a combination of site blocking and an effective decrease in the surface concentration of the contaminant at the nZVI–water interface (Phenrat et al., 2009b). Typically this reduction in nZVI reactivity will not decrease reactivity to levels that make it impractical, but the reactivity of surface-modified material should be used in design calculations, especially for a PRB scenario where the length of the reactive zone will depend on the reaction rate of the target contaminant with the nZVI.

10.2.5 Combined Remedies

Some of the most significant improvements in remediation efficiency over the last decade have come through combined remedies in which two or more approaches are combined to take advantages of both. Two strategies for enhancing ISCR are the addition of heat and the arrangement of ISCR and other treatment processes into combinations of spatial zones. Further discussion of combined remedies is provided in Chapter 15 of this volume.

10.2.5.1 Heat

Heat represents a straightforward approach for increasing the reactivity of ISCR agents. Since the rates of many abiotic processes increase in a predictable manner with increasing temperature (they show Arrhenius-type behavior), even modest increases in temperature can have a significant effect. Thus, a small amount of electrical resistance heating that raises the temperature 10 or 20°C (50 or 68°F) can result in a several-fold increase in rates. Of course, rates of reactions that compete with the target reaction may also be increased, and as a result the persistence of the reactants may be reduced. However, a number of scenarios have been proposed that are designed to take advantage of increased reaction rates for both oxidants and reductants (La Mori et al., 2010; Truex et al., 2011a, b).

10.2.5.2 Sequential Zones

The deliberate combination of treatment methods into sequential zones has been proposed fairly frequently. Three early efforts of this type were the systematic study of various sequenced reactive barriers for treatment of chlorinated solvents (Fiorenza et al., 2000), a sequential reactive treatment zone proposed for reductive and then oxidative treatment of explosives contaminated groundwater (Tratnyek et al., 2001), and pretreatment zones for protecting ZVI PRBs from premature failure (Kenneke and McCutcheon, 2003). Recently, this approach has been further developed and referred to as *multibarrier* design (Bastiaens et al., 2005). There are, of course, many possible combinations of treatment zones in a multibarrier design, and it will be some time before the most beneficial combinations are identified and recognized as proven.

Most multibarrier concepts entail combination of a chemical (ISCR) treatment process with some form of enhanced bioremediation. In some cases, the intent is clearly to form distinct zones that isolate treatment processes over a scale of several meters of more (Becvar et al., 2008), whereas other approaches attempt to stimulate abiotic and biotic contaminant degradation in closer proximity (millimeters to centimeters). Examples of the latter (involving products such as EHC®) were summarized in Section 10.1.2. All these scenarios still have uncertainty about whether the abiotic and biotic processes are synergistic or antagonistic: the conditions within a ZVI PRB would seem to be unfavorable for most forms of microbial activity, yet recent work has generally found that the overall effect on remediation performance is positive (Cullen et al., 2011; Kirschling et al., 2010; Van Nooten et al., 2008).

10.3 IMPLEMENTATION

As with any remedy for contaminated groundwater, the implementation of ISCR involves three stages: (1) site delineation and characterization, (2) technology assessment and selection and (3) technology design and implementation. Each of these stages is discussed in the following subsections, with emphasis on aspects unique to ISCR and relevant to source zone remediation.

10.3.1 Site Delineation and Characterization

In addition to the site delineation and characterization requirements that are prerequisite to any sort of remedial action (type and extent of contamination, lithology and hydrology of the contaminated zone, proximity to receptors), several considerations require special attention if ISCR is to be a candidate for source zone treatment.

The first of these requirements is whether the types of contaminants at the site are subject to degradation by reduction. As described in Section 10.2.1, chlorinated solvents, explosives, and pesticides containing nitro groups and a few other types of organic contaminants can be reduced under ISCR conditions. Organic compounds without these functional groups—such as aliphatic and aromatic hydrocarbons (surfactants, most fuel components, polycyclic hydrocarbons [PAHs])—generally will be unaffected by ISCR. Furthermore, with respect to dechlorination, relatively highly chlorinated aliphatic compounds, such as CT and PCE, will be much more labile to reduction than less highly chlorinated aromatic compounds. These trends in susceptibility to ISCR are the inverse of what is expected, and observed, under conditions of ISCO.

The second type of characterization requirement that has aspects unique to ISCR concerns the site geochemistry: specifically, the type, quantity and availability of redox-active materials that will act as sinks for reductants in the treatment zone. The most rigorous approach to this issue would be to inventory all the potentially important materials (reducing minerals, natural organic matter, reducible contaminants, dissolved oxygen) and develop a biogeochemical reactive-transport model for the system based on principles summarized in Section 10.2.2. This, however, is rarely feasible, so a more practical alternative approach is needed. One approach that probably offers the best compromise between rigor and practicality involves standardized assays of reductant demand that are analogous to the assays for oxidant demand that are now widely used to determine the applicability of ISCO to particular sites (Haselow et al., 2003; ASTM, 2007).

The analogy between reductant demand and oxidant demand extends to a range of issues, from the operationally defined nature of these parameters to the variety of qualifiers needed to refine the terminology (natural, total, aquifer, and others). All of these issues have been discussed extensively for oxidant demand (Lee and Batchelor, 2003; Urynowicz et al., 2008; Xu and Thomson, 2008; Xu and Thomson, 2009), but the concept of reductant demand is much less well developed. Few previous publications have explicitly addressed this subject (Ford, 2002; Brown and Robinson, 2004; Johnson et al., 2013), but to be consistent with that work, the term used in this chapter is natural reductant demand (NRD). While the main significance of NRD is to characterize the capacity of the system to consume reductant, any assay for NRD will depend on the reductant used and the time of exposure to the reductant used in the assay (just as NOD depends on the exposure time to a particular oxidant) (Borda et al., 2009).

Currently, there are no validated or widely used protocols for characterizing NRD of an aquifer system. Instead, current practice is to rely on oxidation–reduction potential (ORP) measurements made with an inert working electrode (usually Pt). As a general rule, negative ORPs (vs. the normal hydrogen electrode, NHE) suggest reducing conditions, which means low NRD to be overcome with added reductant, and together these things favor ISCR. Furthermore, treatability by ISCR generally will be more favorable at more negative ORPs.

However, the use of ORP in this way has important caveats. First, ORP measurements generally do not reflect the contribution of solid phases, such as minerals and natural organic matter; these can be significant sinks for reductants. In addition, the ORP response of dissolved-phase redox-active species varies greatly in sensitivity (Fe^{II} is high, but ammonia is low), reversibility (electrode reactions involving N and S are irreversible), and electrode material and condition (H_2 is electrode active on Pt but not on carbon). Due to these limitations, ORP should be regarded only as a qualitative indicator of the prospects for source zone treatment by ISCR.

Beyond NRD and ORP, other biogeochemical conditions that should be considered in assessing the applicability of ISCR to any particular site are the background concentrations of dissolved species indicative of the major microbial terminal electron-accepting processes (TEAPs). These include O_2, iron, nitrate, sulfate, sulfide, methane and hydrogen. Carbonate concentrations (because it can cause precipitation) and pH should also be measured.

10.3.2 Technology Assessment and Selection

Several broad classes of technology are typically considered for chlorinated volatile organic compound (CVOC) sites in addition to ISCR, including physical removal (excavation), containment, extraction (aqueous and volatile; with and without applied heat), chemical destruction and biological treatment. Even within the category of ISCR, the range of technologies (summarized in Section 10.1.2) is rather diverse. To select from among all of these options, four general criteria need to be assessed:

1. **Initial Site Conditions:** (1) degree of chlorination of the contaminants (number of chlorine atoms per carbon atom), (2) oxidation–reduction state, (3) pH, (4) site permeability and heterogeneity (in lower-permeability and more heterogeneous media, technology selection may be directed toward aggressive emplacement methods, such as mixing or fracturing) and (5) size of treatment zone in relation to the remediation technology (very large sites may be better managed using containment strategies rather than aggressive source treatment).
2. **Performance Goals:** (1) the treatment goal, (2) the time needed to treat or to achieve the treatment goal and (3) the length of time the agent remains active to address mass transfer issues associated with posttreatment rebound.
3. **Implementability:** (1) site lithology (permeability, heterogeneity) that can control the distribution of reactants, (2) site conditions/accessibility (e.g., if the CVOCs are located under buildings, in areas of subsurface pipes, one or more technologies may be eliminated from consideration) and (3) depth of contamination may also preclude the use of some technologies.
4. **Cost Criteria:** (1) the chemical agent(s) to be applied, (2) the application methods, (3) the number of applications and (4) monitoring requirements.

Among these criteria are several aspects of particular importance for ISCR. With regard to site conditions, the aquifer should, in general, be in a reduced state prior to ISCR to minimize reductant demand and mineral precipitation. Ideally, the pH should be slightly acidic ($4 < pH < 7$) and ISCR approaches will, in general, be more effective if the contaminants are highly chlorinated (at least one carbon atom with >50% maximum chlorination).

Since ISCR can be applied in a number of ways (recirculation of reactants, containment using PRBs, direct mixing in source zones), it is difficult to evaluate the strengths and weaknesses of ISCR in a general way. Of the ISCR approaches currently in use, application of granular zero-valent metals (ZVMs) represents a technology that fills a niche without many direct competitors. This is due in part to the comparatively long useful lifetime of granular ZVMs in the subsurface as PRBs. Since transport of ZVMs, including nano- and micro-sized materials, is difficult, the emplacement techniques should be considered carefully. These can include trenching (PRB construction), mechanical mixing, and movement in induced fractures. At the same time, ZVMs are not well suited for distribution via groundwater wells, and application of ISCR in this context utilizes divalent metal such as Fe^{II} species and/or reductant anions ($S_2O_4^{2-}$).

10.3.3 Technology Design and Implementation

For many ISCR applications, design and implementation are intimately coupled to the technology itself. This is particularly the case for ZVM applications, where the mode of delivery will often be chosen simultaneously with the reductant. The situation is less defined for dissolved species delivered by some form of aqueous injection such as recirculation wells. In this case, tradeoffs between several candidate reductants and delivery options may need to be evaluated.

The objective of the design process is to provide detailed engineering parameters for implementing the technology. The design process may employ both bench scale and pilot scale testing in order to identify these aspects:

- The dosage of reactive agent required to reach the treatment goal. The beneficial and nonbeneficial (reductant demand) consumption of the reactive agents.
- An estimate of the number of sequential treatments necessary to achieve cleanup goals.

- The construction and spacing of application points, if used, including permanent points, temporary points, *in situ* mixing design (if used) and monitoring system.
- Design of the monitoring program including frequency, duration, and density of monitoring points necessary to characterize the performance of the technology. In the context of ISCR, this program should include both a strategy to monitor the delivery and persistence of the reactants as well as to monitor the destruction of the CVOCs.

Of particular design relevance in the context of ISCR are reductant demand, persistence of the reductant in the medium due to competitive reactions, efficacy of the degradation process under site-specific conditions, and the impacts of the reductants on biohydrogeochemical conditions in the aquifer (precipitate formation, pH, dissolved metals concentration).

In situ chemical reduction technologies, like ISCO, come with some inherent level of risk because large quantities of reactive materials are being handled and/or injected under pressure. As a consequence, health and safety aspects of technology implementation are important.

The most common concern of ISCR implementation almost certainly deals with the mobility of nano- and micrometer-sized ZVM. As discussed above, historical interpretation of ZVM transport has, in general, been overly optimistic. Design and implementation of ZVM injection systems, excluding fracturing, should be undertaken with the idea that transport distances will likely be less than 1 m (3.28 ft), except in exceptional circumstances, and as a result, target areas must be small or the number of injection points must be large.

10.3.4 Case Studies

Well documented and characterized sites provide case studies for most of the ISCR remediation technologies. Most ISCR technologies (abiotic MNA, BiRD or ISRM) have only a few such sites, so selection of which case studies to feature below was straightforward. However, for the ZVI-based approaches, a plethora of informative case studies exist. In these cases, the sites with the longest assessment history have been featured to provide evidence regarding longevity.

10.3.4.1 Abiotic Monitored Natural Attenuation Case Study

In 2003, at the Air Force Center for Environmental Excellence (AFCEE) Technology Transfer Workshop, John T. Wilson presented a paper titled "Abiotic reactions may be the most important mechanism in natural attenuation of chlorinated solvents" (Wilson, 2003). This presentation was one of the first public discussions of the role of reduced minerals in the attenuation of chlorinated solvents. This observation was based on the study of a *cis*-1,2- dichloroethene (DCE) plume at the Twin Cities Army Ammunition Plant (TCAAP) in Arden Hills, Minnesota, USA. It is of note that the TCAAP site was studied as a biologically mediated MNA site (Wilson et al., 2001).

Ferrey and co-workers discussed the analysis of abiotic MNA of the site (Ferrey et al., 2004), comparing laboratory microcosm results to 13 years of monitoring data. They concluded that the reductions in measured groundwater concentrations at the site were consistent with dechlorination rates measured in the microcosms. The rates of dechlorination for autoclaved and live microcosms were observed to be similar, and they attributed the abiotic degradation to magnetite, which was detected in aquifer solids from the site. This result was consistent with their modeling of dechlorination by magnetite using surface-area normalized rate constants derived sample data and literature sources (Lee and Batchelor, 2002a, b).

10.3.4.2 BiRD Case Study

At Dover Air Force Base, Delaware, USA, a portion of a solvent plume was treated with Epson salt ($MgSO_4 \bullet 7H_2O$) and sodium lactate (Kennedy et al., 2006a). The solution was injected into a contaminated groundwater plume in five wells positioned 3 m (10 ft) apart and perpendicular to groundwater flow. Sediment was sampled prior to and 8 months postinjection. Significant iron sulfide minerals were developed in the sandy aquifer matrix. Trichloroethene reduction was observed a few weeks after injection, and 95% reductions in PCE, TCE and *cis*-DCE concentrations were achieved in less than 1 year.

10.3.4.3 ISRM Case Study

In Fort Lewis, Washington, USA, an extensive treatability study was performed to determine optimal conditions for ISRM (Szecsody et al., 2000), and then a buffered solution of sodium dithionite ($Na_2S_2O_4$) was injected at the site to reduce the iron in the phyllosilicate clays of the aquifer sediments (Vermeul et al., 2000). After four injections (Figure 10.10), more than 40% of the ferric iron was reduced to ferrous iron, and TCE degradation was observed. Downgradient from the treated, reduced zone, TCE concentrations were decreased by as much as 92%.

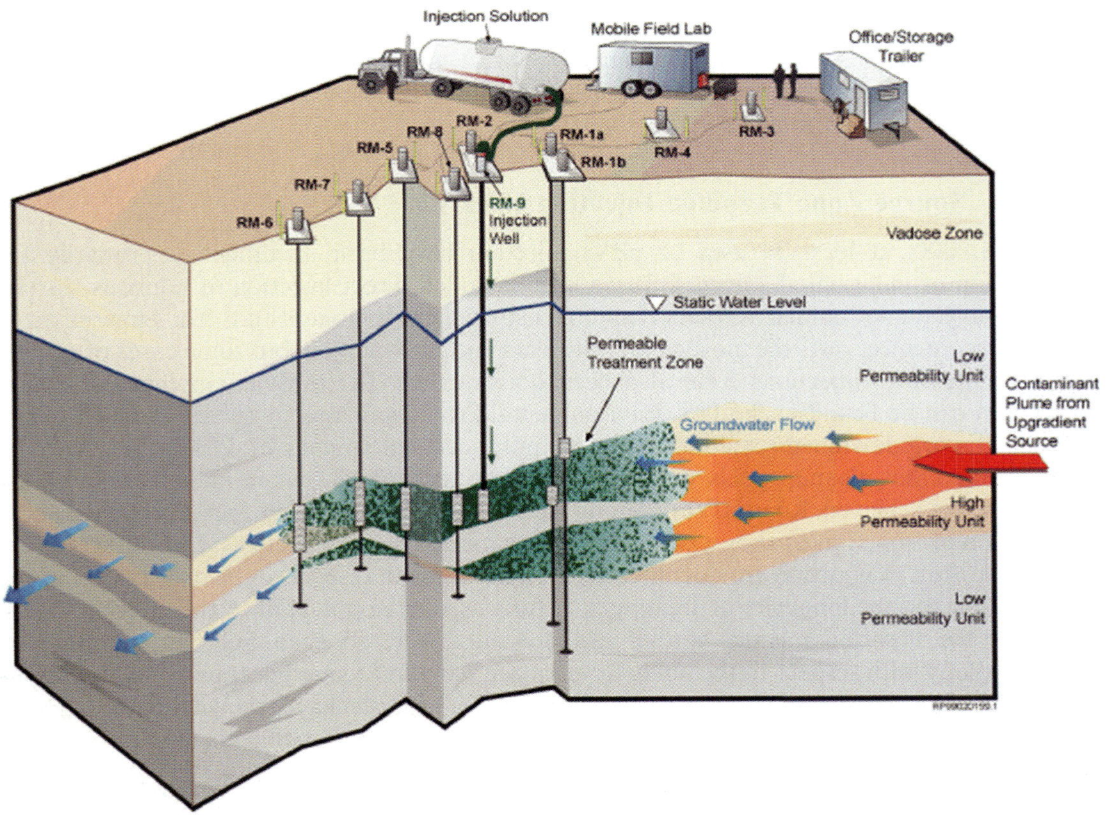

Figure 10.10. Conceptual model and treatment plan for ISRM proof-of-principle test site at the Fort Lewis Logistics Center, WA. From Vermeul et al. (2000), used with permission.

10.3.4.4 Conventional ZVI PRB

Case studies of ZVI PRBs are so numerous that they have been reviewed many times (U.S. Environmental Protection Agency, 1999; Vidic, 2001; Wilkin and Puls, 2003; Wilkin et al., 2003) and there even have been several meta-analyses of data across ZVI PRB case study sites (Powell et al., 2002; Henderson and Demond, 2007). The original pilot test of a ZVI PRB by Gillham and co-workers at Canadian Forces Base Borden in Ontario, Canada, was studied extensively and documented (O'Hannesin and Gillham, 1998; Wadley et al., 2005), and the first full-scale commercial application of a ZVI PRB by Geosyntec Consultants in Sunnyvale, California, USA, was also described in detail and monitored over a considerable time period (Sorel et al., 2003; Warner and Sorel, 2003). However, these efforts were soon followed by several even more thorough and comprehensive assessments of *in situ* performance of ZVI PRBs. Three of the most impactful early case studies include:

1. Denver Federal Center in Denver, Colorado, USA, led by U.S. Geological Survey investigators (McMahon et al., 1999; Paul et al., 2003; Wilkin et al., 2003)
2. Naval Air Station Moffett Field in Mountain View, California, USA, led by Battelle investigators (Gavaskar et al., 1998; ESTCP, 1999; Gavaskar, 1999; Yabusaki et al., 2001; Jeen et al., 2011)
3. Elizabeth City, New Jersey, USA, led by USEPA investigators (Puls et al., 1996; Blowes et al., 1999a, b; Blowes and Mayer, 1999; Puls et al., 1999a, b; Mayer et al., 2001; Furukawa et al., 2002; Paul et al., 2003; Wilkin et al., 2005; Lee and Wilkin, 2010)

Conventional ZVI PRBs are not applied directly to source zones, although they are frequently used for source zone containment. Direct application of ZVI to source zones is mainly through targeted injection or mixing into source zones, and case studies for each of these are presented in the two sections that follow.

10.3.4.5 Source Zone Targeted Injection of nZVI

Since 2000, at least 34 uses of nZVI injection have been documented, primarily for remediation of chlorinated solvents. These have included direct injection of aqueous slurries of nZVI, injection of emulsions (EZVI) and the use of shear-thinning fluids (for 1 micron-sized ZVI). Pressure injection is the most commonly used injection method, but some cases of gravity fed (low pressure) injections have also been documented. The most comprehensive list of these sites can be found at the U.S. Environmental Protection Agency (USEPA) CLU-IN site (http://clu-in.org/download/remed/nano-site-list.pdf; accessed January 27, 2014).

In some cases cleanup goals were met using the technology, but in many cases cleanup goals were not achieved. Reasons for missing cleanup targets vary by site, but it is likely that the distribution of iron at most sites was insufficient to meet cleanup objectives. Unlike PRBs, very few cases of nZVI injection are documented sufficiently with respect to the distribution of the injected iron and the longevity of treatment to fully assess the potential of these technologies. Some of the best-documented studies are presented here to shed light on the state of the technology with respect to the ability to emplace the nZVI using injection wells. Injections of either nZVI emulsions (EZVI) or stabilized nZVI slurries have shown limited success for emplacing the nZVI uniformly in the subsurface at reasonable distances from the injection wells.

In the most prominent case study of EZVI emplacement (Krug et al., 2010), pneumatic fracturing and direct-push injection was used at a site that consisted of fine to medium sand. With coring they found evidence of EZVI emplaced 1.8 and 0.8 m (6 and 2.6 ft) from the injection well for pneumatic fracturing and direct-push injection, respectively. They also observed significant variability in the vertical distribution of the EZVI in those soil cores. A second example used a recirculating well system to inject an aqueous dispersion of nZVI; Henn and Waddill (2006) showed evidence of nZVI at one monitoring well approximately 1.2 m (4 ft) from an injection well. He et al. (2010) conducted a field test using nZVI that was prepared in the field by borohydride reduction of $FeSO_4$ in carboxymethylcellulose (CMC). With this material, they achieved transport distances of 1.5 m (5 ft); however, breakthrough at that distance occurred more rapidly than was expected based on a simple geometric calculation, probably because of preferential flow in a high hydraulic conductivity layer present at the site (F. He, Oak Ridge National Laboratory, personal communication, 2011). As a consequence, the nZVI was likely delivered into only a small fraction of the vertical injection interval. All of these documented cases suggest that transport of significant quantities of nZVI in the subsurface has been limited using existing injection methods and available nZVI formulations.

Emplacement of injected ZVI (~2 ± 1 micron-sized particles) in a TCE source zone, and the subsequent effect on TCE concentrations in monitoring wells, was documented by Truex et al. (2011a). They injected approximately 190 kg (420 lbs) of these mZVI particles at Joint Base Lewis-McChord (formerly Fort Lewis) near Tacoma, Washington, USA, using a shear-thinning polymer (SlurryPro™) present at 0.019 wt%. They injected a total volume of 13,660 L (3,610 gallons) at approximately 83 L/minute (22 gallons per minute) over a 2.5-hour period. Unlike previous reports for injections of nZVI, this injection method in this very high-permeability formation did not cause undue increases in pressure at the injection well (about 1.5 m head). The presence of ZVI was measured at a monitoring well as far as 4 m (13 ft) away from the injection well; however, the maximum concentration of ZVI found at any well away from the injection well was only 26% of the injection concentration at a distance of 1.2 m (4 ft). Most measurements were less than 10% of the injected ZVI concentration, and only 2.6% of the injected ZVI concentration was found at a distance of 4 m (13 ft). The lack of a trend in measured concentration of ZVI versus distance from the injection point suggests that heterogeneity likely influenced the distribution of ZVI in the subsurface.

Despite the nonuniform distribution of nZVI and an initial increase in TCE concentrations in monitoring wells due to the displaced water during injection, TCE concentrations decreased over a 44-day period after injection in all monitoring wells to levels below those observed before treatment, and the formation of reduction products (ethane, acetylene and 1,2-DCE) was observed. Electrical resistance heating of the site to approximately 50°C further enhanced TCE degradation and mass removal. The ZVI kept aqueous phase concentrations of TCE low during the heating which minimized migration of TCE in the vapor phase (Truex et al., 2011a). Overall, this demonstrates the potential to emplace particulate phase reductants for ISCR in a source zone, but highlights the issues surrounding their use, notably the difficulty in obtaining uniform emplacement at significant distances away from an injection well.

10.3.4.6 *In Situ* Soil Mixing Case Study

A number of *in situ* soil mixing demonstrations have been presented in the literature (Shackelford et al., 2005; Fjordboge et al., 2012a, b; Olson et al., 2012). The results from Camp Lejeune, North Carolina, USA, are representative of those observed elsewhere. A total of 22,900 m^3 (30,000 cubic yards) of soils were treated to an average depth of 7.6 m (25 ft) with 2% ZVI and 3% sodium bentonite (dry-weight basis, Olsen et al., 2012). After one year the total concentrations of CVOCs in soil samples decreased by site-wide average and median values of 97% and >99%, respectively. Because of the addition of the bentonite, pre- and post-mixing average hydraulic conductivity values were 1.7×10^{-5} and 5.2×10^{-8} meters per second (m/s),

Figure 10.11. Apparatus used for ISSM at Camp Lejeune, NC. (a) Crane-mounted rotary table and hollow kelly bar, (b) 3.0 m diameter augers. From Olsen et al. (2012), used with permission.

respectively, indicating a reduction of about 2.5 orders of magnitude. To control volatile organic compound (VOC) volatilization during mixing, a 4-m diameter "hood" was used (Figure 10.11).

10.4 SYNTHESIS AND RECOMMENDATIONS

10.4.1 Advantages and Limitations: Lessons Learned

The remediation technologies grouped under ISCR, as defined here, are diverse in many respects, including the two classification criteria emphasized in Figure 10.1 (treatment volume and reductant strength) and other considerations such as expected longevity, compatibility with other treatment technologies, and track record of successful full-scale implementations. Diversity with respect to the first two criteria is an advantage of ISCR, in that specific ISCR technologies can be applied to a wide range of scales and contaminants. However, diversity with respect to the latter considerations reflects the relative immaturity of ISCR as a family of technologies, and this means the full potential value of ISCR applications probably has not yet been realized.

The background and experience currently available on remediation of contaminated groundwater with ISCR, as summarized in this chapter, lead to the following specific conclusions and recommendations:

- **Treatability of Contaminants:** *In situ* chemical reduction treatment processes range from mild to very strong, so a wide range of contaminants are potentially treatable from CT with dissolved forms of Fe^{II} to chlorinated aromatics with H_2/Pd. The rates and products of these processes can be favorable, but results may vary.
- **Deployment:** In many cases, successful treatment can be achieved by influencing the natural biogeochemical conditions to favor contaminant reduction, and this approach is usually more straightforward to deploy. In cases where deployment of strong, artificial reductants is justified, the natural reductant demand should also be considered.
- **Delivery:** Effective delivering of reactants as fluids to source zones is challenging because of aquifer heterogeneity and risk of NAPL displacement and must be considered a limitation in the context of ISCR implementation. In some cases delivery may be

improved using amendments such as viscous fluids. There has also been considerable interest in delivery of nZVI with surface treatments designed to favor partitioning into NAPL. However, delivery of sufficient quantities of those materials through aquifers to source zones will require delivery strategies beyond well injections primarily used today.

- **Complementary Processes:** Other processes may be stimulated or inhibited by ISCR treatment. In particular, biodegradation is an important component of many ISCR activities (as either the primary or a secondary mode of action) and should be considered in the design of most ISCR systems. The addition of organic and/or inorganic reductants to directly or indirectly facilitate biodegradation represents an important component of long-term treatment.
- **Longevity:** The reductant lifetime is an important consideration. As with other *in situ* chemical treatment approaches, chemical reductants added to the subsurface may have a relatively short lifetime during which the contaminant treatment goal must be achieved, or sufficiently reducing conditions must be created so that contaminant degradation will continue after the added reductant is exhausted.

Of the ISCR approaches discussed above, those involving granular zero-valent iron are the only ones that have been applied at a sufficient number of sites to be considered *proven*. Zero-valent iron PRBs have been shown to remain active for many years in most cases. In the context of source zone treatment, *in situ* soil mixing with granular ZVI has been shown to be reliably effective.

10.5 FUTURE PROSPECTS AND NEEDS

10.5.1 Near-Term Prospects Without Additional Research

In situ chemical reduction has developed over the past decade from a plume containment technology to a technique that is being used to treat and/or contain source zones. Despite some limitations, it is a promising technology that can help practitioners manage source zones more effectively and efficiently in the future (Stroo et al., 2012). It is likely that use of ISCR will continue to grow (just as with ISCO some years ago) based on pooling of available technology and information alone, even without or before any new technology development.

Increased adoption of ISCR techniques for source zone treatment is expected as an outgrowth of increased awareness and acceptance of ISCR. The technology can provide rapid mass destruction *in situ*, potentially with continued long-term treatment of any residual contamination remaining after treatment. *In situ* chemical reduction also offers remarkable flexibility, with different formulations and products for different objectives and site conditions. Practitioners have learned how best to deploy the different reductants and how to combine them with other processes such as biodegradation, so increased use of existing products and development of improved materials is likely.

10.5.2 Longer-Term Research Needs

Despite the increased use and attention, ISCR still has several important limitations as a source zone treatment technology. Research and development in this area are necessary to overcome or mitigate these limitations and to expand the use of the ISCR technologies. Key areas for future research include the following:

- Better understanding of the interactions between chemical and biological processes. The emphasis should be on the *interactions*, because the processes are linked.

For example, a more refined distinction between biotic and abiotic is not necessarily that useful, but a more quantitative understanding of the biotic–abiotic interactions should lead to more efficient use of the combined processes.

- Better methods to characterize *in situ* reducing conditions. Effective diagnosis and design requires tools that can distinguish *intensity* from *capacity*. The current methods are not adequate for characterizing field sites to select or design and ISCR technique.
- A method to predict dechlorination rates from a readily measurable or controllable system property. This has been the stated goal of many research studies over the last decade, but a variety of factors have prevented the emergence of a practical method of quantitatively assessing the performance of ISCR at any particular field site based on general characteristics of the site and contaminants.
- More emphasis on improving the overall efficiency of treatment rather than alternative reductants. Most of the most viable candidate reductants are already known, but there are many ways to improve their performance after deployment (catalysts or other activation strategies, innovative delivery techniques or longevity enhancements).
- Methods to deliver reductants *in situ* in a controlled manner and with a more uniform distribution are needed.

ACKNOWLEDGEMENTS

The authors would like to recognize the helpful and substantive input from James E. Szecsody (Pacific Northwest National Laboratory) and permission to reproduce figures from Erica Becvar (AFCEE), Patrick Evans (CDM), and Thomas Hofstetter (EAWAG).

Portions of this work were funded by the U.S. Department of Defense, Strategic Environmental Research and Development Program (SERDP), Department of Energy, Subsurface Biogeochemical Research (SBR) Program, and the Environmental Protection Agency and National Science Foundation (Center for Environmental Implications of Nanotechnology (CEINT)). This document has not been reviewed by any of the sponsoring agencies and therefore it does not necessarily reflect their views and no official endorsement should be inferred.

REFERENCES

Adriaens P, Gruden C, McCormick ML. 2004. Biogeochemistry of halogenated hydrocarbons. Treatise Geochem 9:511–539.

Amonette JE. 2002. Iron redox chemistry of clays and oxides: Environmental applications. In Fitch A, ed, Electrochemistry of Clays. Clay Minerals Society, Aurora, CO, USA. CMS Work Lect 10:89–147.

Amonette JE, Workman DJ, Kennedy DW, Fruchter JS, Gorby YA. 2000. Dechlorination of carbon tetrachloride by Fe(II) associated with goethite. Environ Sci Technol 34:4606–4613.

Arnold WA, Roberts AL. 2000. Pathways and kinetics of chlorinated ethylene and chlorinated acetylene reaction with Fe(0) particles. Environ Sci Technol 34:1794–1805.

Arnold WA, Ball WP, Roberts AL. 1999. Polychlorinated ethane reaction with zero-valent zinc: Pathways and rate control. J Contam Hydrol 40:183–200.

ASTM (American Society for Testing and Materials). 2007. Standard Test Method for Estimating the Permanganate Natural Oxidant Demand of Soil and Aquifer Solids. D7262-07. ASTM International, West Conshohocken, PA, USA. Vol 04.09.

Ayala-Luis KB, Cooper NGA, Koch CB, Hansen HCB. 2012. Efficient dechlorination of carbon tetrachloride by hydrophobic green rust intercalated with dodecanoate anions. Environ Sci Technol 46:3390–3397.

Bastiaens L, Dries J, Vos J, Simons Q, De Smet M, Diels L. 2005. Comparison of different multibarrier concepts designed for treatment of groundwater containing mixed pollutants. IAHS Publ 298:45–51.

Becvar E, Evans P, Lebron C, Stroo H, Wilson JT, Wymore R. 2008. Workshop on In Situ Biogeochemical Transformation of Chlorinated Solvents. AFD-080429–058. Brooks City Base, TX, USA. 65 p.

Bennett P, He F, Zhao D, Aiken B, Feldman L. 2010. In situ testing of metallic iron nanoparticle mobility and reactivity in a shallow granular aquifer. J Contam Hydrol 116:35–46.

Berge ND, Ramsburg CA. 2009. Oil-in-water emulsions for encapsulated delivery of reactive iron particles. Environ Sci Technol 43:5060–5066.

Blowes DW, Mayer KU. 1999. An in situ permeable reactive barrier for the treatment of hexavalent chromium and trichloroethylene in ground water, Volume 3-Multicomponent Reactive Transport Modeling. EPA/600/R-99/095c. U.S. Environmental Protection Agency, Ada, OK, USA. 39 p.

Blowes DW, Gillham RW, Ptacek CJ, Puls RW, Bennett TA, O'Hannesin SF, Hanton-Fong CJ, Bain JG. 1999a. An in situ permeable reactive barrier for the treatment of hexavalent chromium and trichloroethylene in ground water, Volume 1-Design and Installation. EPA/600/R-99/095a. U.S. Environmental Protection Agency, Ada, OK, USA. 111 p.

Blowes DW, Puls RW, Gillham RW, Ptacek CJ, Bennett TA, Bain JG, Hanton-Fong CJ, Paul CJ. 1999b. An in situ permeable reactive barrier for the treatment of hexavalent chromium and trichloroethylene in ground water, Volume 2-Performance Monitoring. EPA/600/R-99/095b. U.S. Environmental Protection Agency, Ada, OK, USA. 207 p.

Borda MJ, Venkatakrishnan R, Gheorghiu F. 2009. Status of nZVI technology: Lessons learned from North American and international implementations. In Geiger Cherie L, Carvalho-Knighton KM, eds, Environmental Applications of Nanoscale and Microscale Reactive Metal Particles. Symposium Series 1027. American Chemical Society, Washington, DC, USA, pp 219–232.

Brown RA. 2008. Developments in in situ chemical reduction (ISCR) technology. Proceedings, 6th International Conference on Remediation of Chlorinated and Recalcitrant Compounds, Monterey, CA, USA, May 19–22. Battelle Press, Columbus, OH, USA, Paper C-066.

Brown RA. 2010. Chemical oxidation and reduction for chlorinated solvent remediation. In Stroo HF, Ward CH, eds, *In Situ* Remediation of Chlorinated Solvent Plumes. Springer, New York, NY, USA, pp 481–535.

Brown RA, Robinson D. 2004. Response to naturally occurring organic material: Permanganate versus persulfate. Proceedings of the 4th International Conference on Remediation of Chlorinated and Recalcitrant Compounds, Monterey, CA, USA, May 24–27. Battelle Press, Columbus, OH, USA. 2A.06/1-2A.06/8.

Brown RA, Lewis RL, Fiacco RJ, Jr, Leahy MC. 2006. The technical basis for in situ chemical reduction (ISCR). Proceedings, 5th International Conference on Remediation of Chlorinated and Recalcitrant Compounds, Monterey, CA, USA, May 22–25. Battelle Press, Columbus, OH, USA. d 04 ppr/1-d 04 ppr/8.

Brown RA, Wilson JT, Ferrey M. 2007. Monitored natural attenuation forum: The case for abiotic MNA. Remediat J 17:127–137.

Butler EC, Hayes KF. 1999. Kinetics of the transformation of trichloroethylene and tetrachloroethylene by iron sulfide. Environ Sci Technol 33:2021–2027.

Butler EC, Hayes KF. 2001. Factors influencing rates and products in the transformation of trichloroethylene by iron sulfide and iron metal. Environ Sci Technol 35:3884–3891.

Butler EC, Dong Y, Krumholz LR, Liang X, Shao H, Tan Y. 2011. Rate controlling processes in the transformation of tetrachloroethylene and carbon tetrachloride under iron reducing and sulfate reducing conditions. In Tratnyek PG, Grundl TJ, Haderlein SB, eds, Aquatic Redox Chemistry. American Chemical Society, Washington, DC, USA, ACS Symp Ser 1071:519–538.

Bylaska EJ, Salter-Blanc AJ, Tratnyek PG. 2011. One-electron reduction potentials from chemical structure theory calculations. In Tratnyek PG, Grundl TJ, Haderlein SB, eds, Aquatic Redox Chemistry. American Chemical Society, Washington, DC, USA. ACS Symp Ser 1071:37–64.

Cantrell KJ, Kaplan DI, Gilmore TJ. 1997. Injection of colloidal Fe0 particles in sand with shear-thinning fluids. J Environ Eng 123:786–791.

Cervini-Silva J, Larson RA, Wu J, Stucki JW. 2002. Dechlorination of pentachloroethane by commercial Fe and ferruginous smectite. Chemosphere 47:971–976.

Choe JK, Shapley JR, Strathmann TJ, Werth CJ. 2010. Influence of rhenium speciation on the stability and activity of Re/Pd bimetal catalysts used for perchlorate reduction. Environ Sci Technol 44:4716–4721.

Clayton WS. 1998. A field and laboratory investigation of air fingering during air sparging. Ground Water Monit Remediat 18:134–145.

Comba S, Molfetta A, Sethi R. 2011. A comparison between field applications of nano-, micro-, and millimetric zero-valent iron for the remediation of contaminated aquifers. Water Air Soil Pollut 215:595–607.

Crane RA, Scott TB. 2012. Nanoscale zero-valent iron: Future prospects for an emerging water treatment technology. J Hazard Mater 211–212:112–125.

Cullen LG, Tilston EL, Mitchell GR, Collins CD, Shaw LJ. 2011. Assessing the impact of nano- and micro-scale zerovalent iron particles on soil microbial activities: Particle reactivity interferes with assay conditions and interpretation of genuine microbial effects. Chemosphere 82:1675–1682.

Cwiertny DM, Arnold WA, Kohn T, Rodenburg LA, Roberts AL. 2010. Reactivity of alkyl polyhalides toward granular iron: Development of QSARs and reactivity cross correlations for reductive dehalogenation. Environ Sci Technol 44:7928–7936.

Darlington R, Lehmicke L, Andrachek RG, Freedman DL. 2008. Biotic and abiotic anaerobic transformations of trichloroethene and *cis*-1,2-dichloroethene in fractured sandstone. Environ Sci Technol 42:4323–4330.

Davie MG, Cheng H, Hopkins GD, Lebron CA, Reinhard M. 2008. Implementing heterogeneous catalytic dechlorination technology for remediating TCE-contaminated groundwater. Environ Sci Technol 42:8908–8915.

Dolfing J. 2003. Thermodynamic considerations for dehalogenation. In Häggblom MM, Bossert ID, Bossert ID, eds, Dehalogenation: Microbial Processes and Environmental Applications. Kluwer, Boston, MA, USA, pp 89–114.

Dolfing J, van Eekert M, Seech A, Vogan J, Mueller J. 2008. In situ chemical reduction (ISCR) technologies: Significance of low Eh reactions. Soil Sediment Contam 17:63–74.

Elsner M, Hofstetter TB. 2011. Current perspectives on the mechanisms of chlorohydrocarbon degradation in subsurface environments: Insight from kinetics, product formation, probe molecules and isotope fractionation. In Tratnyek PG, Grundl TJ, Haderlein SB, eds, Aquatic Redox Chemistry. American Chemical Society, Washington, DC, USA, ACS Symp Ser 1071:407–439.

Elsner M, Schwarzenbach RP, Haderlein SB. 2004. Reactivity of Fe(II)-bearing minerals toward reductive transformation of organic contaminants. Environ Sci Technol 38:799–807.

Elsner M, Couloume GL, Mancini S, Burns L, Lollar BS. 2010. Carbon isotope analysis to evaluate nanoscale Fe(0) treatment at a chlorohydrocarbon contaminated site. Ground Water Monit Remediat 30:79–95.

Erbs M, Hansen HCB, Olsen CE. 1999. Reductive dechlorination of carbon tetrachloride using iron(II) iron(III) hydroxide sulfate (green rust). Environ Sci Technol 33:307–311.

ESTCP (Environmental Security Technology Certification Program). 1999. Permeable Reactive Wall Remediation of Chlorinated Hydrocarbons in Groundwater. Cost Performance Report for CU-9604. Alexandria, VA, USA. 52 p.

Evans PJ, Fricke RA, Hopfensperger K, Titus T. 2011. In situ destruction of perchlorate and nitrate using gaseous electron donor injection technology. Ground Water Monit Remediat 31:103–112.

Fagerlund F, Illangasekare TH, Phenrat T, Kim HJ, Lowry GV. 2012. PCE dissolution and simultaneous dechlorination by nanoscale zero-valent iron particles in a DNAPL source zone. J Contam Hydrol 131:9–28.

Fang Y, Hozalski RM, Clapp LW, Novak PJ, Semmens MJ. 2002. Passive dissolution of hydrogen gas into groundwater using hollow-fiber membranes. Water Res 36:3533–3542.

Ferrey ML, Wilkin RT, Ford RG, Wilson JT. 2004. Nonbiological removal of cis-dichloroethylene and 1,1-dichloroethylene in aquifer sediment containing magnetite. Environ Sci Technol 38:1746–1752.

Fiorenza S, Oubre CL, Ward CH. 2000. Sequenced Reactive Barriers for Groundwater Remediation. CRC Press, Boca Raton, FL, USA. 730 p.

Fjordboge AS, Lange IV, Bjerg PL, Binning PJ, Riis C, Kjeldsen P. 2012a. ZVI-Clay remediation of a chlorinated solvent source zone, Skuldelev, Denmark: 2. Groundwater contaminant mass discharge reduction. J Contam Hydrol 140–141:67–79.

Fjordboge AS, Riis C, Christensen AG, Kjeldsen P. 2012b. ZVI-Clay remediation of a chlorinated solvent source zone, Skuldelev, Denmark: 1. Site description and contaminant source mass reduction. J Contam Hydrol 140–141:56–66.

Ford RG. 2002. Solid phase redox characterization. In Wilkin RT, Ludwig RD, Ford RG, eds, Workshop on Monitoring Oxidation-Reduction Processes for Ground-Water Restoration. Workshop Summary, Dallas, TX, USA. April 25–27. EPA/600/R-02/002. U.S. Environmental Protection Agency, Cincinnati, OH, USA, pp 105–113.

Fruchter J. 2002. In-situ treatment of chromium-contaminated groundwater. Environ Sci Technol 36:464A–472A.

Fruchter JS, Cole CR, Williams MD, Vermeul VR, Amonette JE, Szecsody JE, Istok JD, Humphrey MD. 2000. Creation of a subsurface permeable treatment zone for aqueous chromate contamination using in situ redox manipulation. Ground Water Monit Remediat 20:66–77.

Furukawa Y, Kim J-W, Watkins J, Wilkin RT. 2002. Formation of ferrihydrite and associated iron corrosion products in permeable reactive barriers of zero-valent iron. Environ Sci Technol 36:5469–5475.

Gavaskar A. 1999. Design and construction techniques for permeable reactive barriers. J Hazard Mater 68:41–71.

Gavaskar A, Sass B, Gupta N, Hicks J, Yoon S, Fox T, Sminchak JR. 1998. Performance Evaluation of a Pilot-Scale Permeable Reactive Barrier at Former Naval Air Station Moffett Field, Mountain View, California. Final Report for ESTCP CU-9604. Battelle Press, Columbus, OH, USA. 173 p.

Geiger CL, Carvalho-Knighton KM, eds. 2009. Environmental applications of nanoscale and microscale reactive metal particles. ACS Symp Ser 1027. 306 p.

Gillham RW, Vogan J, Gui L, Duchene M, Son J. 2010. Iron barrier walls for chlorinated solvent remediation. In Stroo HF, Ward CH, eds, *In Situ* Remediation of Chlorinated Solvent Plumes. Springer, New York, NY, USA, pp 537–571.

Golas PL, Lowry GV, Tilton RD, Matyjaszewski K. 2010. Designing polymer stabilizers by ATRP for iron nanoparticles used in groundwater remediation. ACS National Meeting, March 21–25, San Francisco, CA, USA. American Chemical Society, Washington, DC, USA. COLL-304.

Gorski CA, Nurmi JT, Tratnyek PG, Hofstetter TB, Scherer MM. 2010. Redox behavior of magnetite: Implications for contaminant reduction. Environ Sci Technol 44:55–60.

Graham MC, Farmer JG, Anderson P, Paterson E, Hillier S, Lumsdon DG, Bewley RJF. 2006. Calcium polysulfide remediation of hexavalent chromium contamination from chromite ore processing residue. Sci Total Environ 364:32–44.

Grittini C, Malcomson M, Fernando Q, Korte N. 1995. Rapid dechlorination of polychlorinated biphenyls on the surface of a Pd/Fe bimetallic system. Environ Sci Technol 29:2898–2900.

Gui L, Gillham RW. 2003. Preparation and regeneration of nickel-iron for reduction of organic contaminants. In Henry SM, Warner SD, eds, Chlorinated Solvent and DNAPL Remediation: Innovative Strategies for Subsurface Cleanup. American Chemical Society, Washington, DC, USA. ACS Symp Ser 837:206–216.

Haas JR, Shock EL. 1999. Halocarbons in the environment: Estimates of thermodynamic properties for aqueous chloroethylene species and their stabilities in natural settings. Geochim Cosmochim Acta 63:3429-3441.

Haderlein SB, Schwarzenbach RP. 1995. Environmental processes influencing the rate of abiotic reduction of nitroaromatic compounds in the subsurface. In Spain JC, ed, Biodegradation of Nitroaromatic Compounds. Plenum, New York, NY, USA, pp 199–225.

Haderlein SB, Hofstetter TB, Schwarzenbach RP. 2000. Subsurface chemistry of nitroaromatic compounds. In Spain JC, Hughes JB, Knackmus HJ, eds, Biodegradation of Nitroaromatic Compounds and Explosives. Lewis, Boca Raton, FL, USA, pp 311–356.

Haselow JS, Siegrist RL, Crimi M, Jarosch T. 2003. Estimating the total oxidant demand for in situ chemical oxidation design. Remediat J 13:5–16.

He F, Zhao D. 2007. Manipulating the size and dispersibility of zerovalent iron nanoparticles by use of carboxymethyl cellulose stabilizers. Environ Sci Technol 41:6216–6221.

He F, Zhao D, Paul C. 2010. Field assessment of carboxymethyl cellulose stabilized iron nanoparticles for in situ destruction of chlorinated solvents in source zones. Water Res 44:2360–2370.

He Y, Su C, Wilson J, Wilkin R, Adair C, Lee TR, Bradley P, Ferrey M. 2009. Identification and characterization methods for reactive minerals responsible for natural attenuation of chlorinated organic compounds in ground water. EPA/600/R-09/115. U.S. Environmental Protection Agency, Ada, OK, USA. 150 p.

Heck KN, Nutt MO, Alvarez P, Wong MS. 2009. Deactivation resistance of Pd/Au nanoparticle catalysts for water-phase hydrodechlorination. J Catal 267:97–104.

Heiderscheidt JL, Illangasekare TH, Borden RC, Thomson NR. 2011. Principles of ISCO related subsurface transport and modeling. In Siegrist RL, Crimi M, Simpkin TJ, eds, *In Situ*

Chemical Oxidation for Groundwater Remediation. Springer, New York, NY, USA, pp 233–284.

Henderson AD, Demond AH. 2007. Long-term performance of zero-valent iron permeable reactive barriers. A critical review. Environ Eng Sci 24:401–423.

Henderson TH, Mayer KU, Parker BL, Al TA. 2009. Three-dimensional density-dependent flow and multicomponent reactive transport modeling of chlorinated solvent oxidation by potassium permanganate. J Contam Hydrol 106:195–211.

Henn KW, Waddill DW. 2006. Utilization of nanoscale zero-valent iron for source remediation – A case study. Remediat J 16:57–77.

Hirasaki GJ, Miller CA, Szafranski R, Tanzil D, Lawson JB, Meinardus HW, Jin M, Londergan J, Jackson RE, Pope GA, Wade WH. 1997. Field demonstration of the surfactant/foam process for aquifer remediation. SPE Annual Technical Conference and Exhibition, Society of Petroleum Engineers, San Antonio, TX, USA, October 5–8, Paper EPE 39292:1–16.

Ingram JC, Cortez MM, Bates DL, McCurry MO. 2000. Reductive dechlorination of trichloroethylene and carbon tetrachloride at iron oxides and basalt minerals. In Eller PG, Heineman W, eds, Nuclear Site Remediation. American Chemical Society, Washington, DC, USA, ACS Symp Ser 778:267–281.

Jeen SW, Gillham RW, Przepiora A. 2011. Predictions of long-term performance of granular iron permeable reactive barriers: Field-scale evaluation. J Contam Hydrol 123:50–64.

Jeong HY, Anantharaman K, Han Y-S, Hayes KF. 2011. Abiotic reductive dechlorination of cis-dichloroethylene by Fe species formed during iron- or sulfate-reduction. Environ Sci Technol 45:5186–5194.

Johnson RL, Johnson PC. 2012. *In situ* sparging for delivery of gases in the subsurface. In Kitanidis PK, McCarty PL, eds, Delivery and Mixing in the Subsurface. Springer, New York, NY, USA, pp 193–216.

Johnson RL, Johnson PC, McWhorter DB, Hinchee RE, Goodman I. 1993. An overview of in situ air sparging. Ground Water Monit Remediat 13:127–135.

Johnson RL, Nurmi JT, O'Brien G, Fan D, Shi Z, Salter-Blanc Alexandra J, Tratnyek, PG, Lowry, GV. 2013. Field scale transport and transformation of carboxymethylcellulose stabilized nano zero valent iron. Env Sci Tech 47:1573–1580.

Johnson TL, Tratnyek PG. 1994. A column study of carbon tetrachloride dehalogenation by iron metal. 33rd Hanford Symposium on Health and the Environment. Battelle Pacific Northwest Laboratories, Richland, WA, USA, Vol 2, pp 931–947.

Kamolpornwijit W, Liang L, Moline GR, Hart T, West OR. 2004. Identification and quantification of mineral precipitation in Fe0 filings from a column study. Environ Sci Technol 38:5757–5765.

Kaplan DI, Cantrell KJ, Wietsma TW. 1994. Formation of a barrier to groundwater contaminants by the injection of zero-valent iron colloids: suspension properties. 33rd Hanford Symposium on Health and the Environment, Vol 2. Battelle Pacific Northwest Laboratories, Richland, WA, USA, pp 821–837.

Kaplan DI, Cantrell KJ, Wietsma TW, Potter MA. 1996. Retention of zero-valent iron colloids by sand columns: Application to chemical barrier formation. J Environ Qual 25:1086–1094.

Kennedy LG, Everett JW, Becvar E, DeFeo D. 2006a. Field-scale demonstration of induced biogeochemical reductive dechlorination at Dover Air Force Base, Dover, Delaware. J Contam Hydrol 88:119–136.

Kennedy LG, Everett JW, Gonzales J. 2006b. Assessment of biogeochemical natural attenuation and treatment of chlorinated solvents, Altus Air Force Base, Altus, Oklahoma. J Contam Hydrol 83:221–236.

Kenneke JF, McCutcheon SC. 2003. Use of pretreatment zones and zero-valent iron for the remediation of chloroalkenes in an oxic aquifer. Environ Sci Technol 37:2829–2835.

Kenneke JF, Weber EJ. 2003. Reductive dehalogenation of halomethanes in iron- and sulfate-reducing sediments. 1. Reactivity pattern analysis. Environ Sci Technol 37:713–720.

Khandelwal A, Rabideau AJ, Shen P. 1998. Analysis of diffusion and sorption of organic solutes in soil-bentonite barrier materials. Environ Sci Technol 32:1333–1339.

Kim H-J, Phenrat T, Tilton RD, Lowry GV. 2009. Fe0 nanoparticles remain mobile in porous media after aging due to slow desorption of polymeric surface modifiers. Environ Sci Technol 43:3824–3830.

Kim H-J, Phenrat T, Tilton RD, Lowry GV. 2012. Effect of kaolinite, silica fines and pH on transport of polymer-modified zero valent iron nano-particles in heterogeneous porous media. J Colloid Interf Sci 370:1–10.

Kirschling TL, Gregory KB, Minkley EG, Jr, Lowry GV, Tilton RD. 2010. Impact of nanoscale zero valent iron on geochemistry and microbial populations in trichloroethylene contaminated aquifer materials. Environ Sci Technol 44:3474–3480.

Kriegman-King MR, Reinhard M. 1994. Transformation of carbon tetrachloride by pyrite in aqueous solution. Environ Sci Technol 28:692–700.

Krug T, O'Hara S, Watling M, Quinn J. 2010. Emulsified zero-valent nano-scale iron treatment of chlorinated solvent DNAPL source areas. Final Report. ESTCP ER-0431. Environmental Security Technology Certification Program, Alexandria, VA, USA, 763 p.

La Mori P, Kirkland E, Faircloth H, Bogert R, Kershner M. 2010. Combined thermal and zero-valent iron in situ soil mixing remediation technology. Remediat J 20:9–25.

Landis RL, Gillham RW, Reardon EJ, Fagan R, Focht RM, Vogan JL. 2001. An examination of zero-valent iron sources used in permeable reactive barriers. International Containment and Remediation Technology Conference and Exhibition, Orlando, FL, USA, June 10–13. Abstract 420.

Larese-Casanova P, Cwiertny DM, Scherer MM. 2010. Nanogoethite formation from oxidation of Fe(II) sorbed on aluminum oxide: Implications for contaminant reduction. Environ Sci Technol 44:3765–3771.

Lee TR, Wilkin RT. 2010. Iron hydroxy carbonate formation in zerovalent iron permeable reactive barriers: Characterization and evaluation of phase stability. J Contam Hydrol 116:47–57.

Lee W, Batchelor B. 2002. Abiotic reductive dechlorination of chlorinated ethylenes by iron-bearing soil minerals. 1. Pyrite and magnetite. Environ Sci Technol 36:5147–5154.

Lee W, Batchelor B. 2002. Abiotic reductive dechlorination of chlorinated ethylenes by iron-bearing soil minerals. 2. Green rust. Environ Sci Technol 36:5348–5354.

Lee W, Batchelor B. 2003. Reductive capacity of natural reductants. Environ Sci Technol 37:535–541.

Li WF, Klabunde KJ. 1998. Ultrafine zinc and nickel, palladium, silver coated zinc particles used for reductive dehalogenation of chlorinated ethylenes in aqueous solution. Croat Chem Acta 71:853–872.

Li X-Q, Elliott DW, Zhang W-X. 2006. Zero-valent iron nanoparticles for abatement of environmental pollutants: materials and engineering aspects. Crit Rev Solid State Mater Sci 31:111–122.

Liu J, He F, Durham E, Zhao D, Roberts CB. 2008. Polysugar-stabilized Pd nanoparticles exhibiting high catalytic activities for hydrodechlorination of environmentally deleterious trichloroethylene. Langmuir 24:328–336.

Liu Y, Phenrat T, Lowry GV. 2007. Effect of TCE concentration and dissolved groundwater solutes on NZVI-promoted TCE dechlorination and H_2 evolution. Environ Sci Technol 41:7881–7887.

Lowry GV. 2007. Nanomaterials for groundwater remediation. In Wiesner MR, Bottero J-Y, eds, Environmental Nanotechnology. McGraw Hill, New York, NY, USA, pp 297–336.

Lowry GV, Reinhard M. 2000. Pd-catalyzed TCE dechlorination in groundwater: solute effects, biological control, and oxidative catalyst regeneration. Environ Sci Technol 34:3217–3223.

Macalady DL, Tratnyek PG, Grundl TJ. 1986. Abiotic reduction reactions of anthropogenic organic chemicals in anaerobic systems. J Contam Hydrol 1:1–28.

Mak MSH, Lo IMC. 2011. Environmental life cycle assessment of permeable reactive barriers: Effects of construction methods, reactive materials and groundwater constituents. Environ Sci Technol 45:10148–10154.

Matheson LJ, Tratnyek PG. 1994. Reductive dehalogenation of chlorinated methanes by iron metal. Environ Sci Technol 28:2045–2053.

Mayer KU, Blowes DW, Frind EO. 2001. Reactive transport modeling of an in situ reactive barrier for the treatment of hexavalent chromium and trichloroethylene in groundwater. Water Resour Res 37:3091–3104.

McCormick ML, Adriaens P. 2004. Carbon tetrachloride transformation on the surface of nanoscale biogenic magnetite particles. Environ Sci Technol 38:1045–1053.

McMahon PB, Dennehy KF, Sandstrom MW. 1999. Hydraulic geochemical performance of a permeable reactive barrier containing zero-valent iron, Denver Federal Center. Ground Water 37:396–404.

McNab WW, Jr, Ruiz R, Reinhard M. 2000. In-situ destruction of chlorinated hydrocarbons in groundwater using catalytic reductive dehalogenation in a reactive well: Testing and operational experiences. Environ Sci Technol 34:149–153.

Miehr R, Tratnyek PG, Bandstra JZ, Scherer MM, Alowitz M, Bylaska EJ. 2004. The diversity of contaminant reduction reactions by zero-valent iron: Role of the reductate. Environ Sci Technol 38:139–147.

Mueller NC, Braun J, Bruns J, Cernik M, Rissing P, Rickerby D, Nowack B. 2012. Application of nanoscale zero valent iron (NZVI) for groundwater remediation in Europe. Environ Sci Pollut Res 19:550–558.

Munakata N, Reinhard M. 2007. Palladium-catalyzed aqueous hydrodehalogenation in column reactors: Modeling of deactivation kinetics with sulfide and comparison of regenerants. Appl Catal B Environ 75:1–10.

Newell CJ, Fisher RT, Hughes JB. 1997. Direct hydrogen addition for the in-situ biodegradation of chlorinated solvents. Proceedings, Petroleum Hydrocarbons and Organic Chemicals in Ground Water: Prevention, Detection, and Remediation Conference, Houston, TX, USA, November 12–14, pp 791–800.

Newell CJ, Hughes JB, Fisher RT, Haas PE. 1998. Subsurface hydrogen addition for the in-situ bioremediation of chlorinated solvents. Proceedings, 1st International Conference on Remediation of Chlorinated and Recalcitrant Compounds, Monterey, CA, USA, May 21–28. Battelle Press, Columbus, OH, USA, pp 47–52.

Newell CJ, Haas PE, Hughes JB, Khan TA. 2000. Results from two direct hydrogen delivery field tests for enhanced dechlorination. Proceedings, 2nd International Conference on Remediation of Chlorinated and Recalcitrant Compounds, Monterey, CA, USA, May 22–25. Battelle Press, Columbus, OH, USA, pp 21–38.

O'Hannesin SF, Gillham RW. 1998. Long-term performance of an in situ "iron wall" for remediation of VOCs. Ground Water 36:164–170.

O'Loughlin EJ, Burris DR. 2004. Reduction of halogenated ethanes by green rust. Environ Toxicol Chem 23:41–48.

O'Loughlin EJ, Boyanov MI, Antonopoulos DA, Kemner KM. 2011. Redox processes affecting the speciation of technetium, uranium, neptunium, and plutonium in aquatic and terrestrial environments. In Tratnyek PG, Grundl TJ, Haderlein SB, eds, Aquatic Redox Chemistry. American Chemical Society, Washington, DC, USA, ACS Symp Ser 1071: 477–517.

Olson MR, Sale TC, Shackelford CD, Bozzini C, Skeean J. 2012. Chlorinated solvent source-zone remediation via ZVI-clay soil mixing: 1-year results. Ground Water Monit Remediat 32:63–74.

Onanong S, Comfort SD, Burrow PD, Shea PJ. 2007. Using gas-phase molecular descriptors to predict dechlorination rates of chloroalkanes by zerovalent iron. Environ Sci Technol 41:1200–1205.

Paul CJ, McNeil MS, Beck Jr FP, Clark PJ, Wilkin RT, Puls RW. 2003. Capstone report on the application, monitoring, and performance of permeable reactive barriers for ground-water remediation: Vol 2 – Long-Term Monitoring of PRBs: Soil and Ground Water Sampling. EPA/600/R-03/045b. U.S. Environmental Protection Agency, Washington, DC, USA.

Peijnenburg WJGM, 't Hart MJ, den Hollander HA, van de Meent D, Verboom HH, Wolfe NL. 1991. QSARs for predicting biotic and abiotic reductive transformation rate constants of halogenated hydrocarbons in anoxic sediment systems. Sci Total Environ 109/110:283–300.

Phenrat T, Saleh N, Sirk K, Tilton RD, Lowry GV. 2007. Aggregation and sedimentation of aqueous nanoscale zerovalent iron dispersions. Environ Sci Technol 41:284–290.

Phenrat T, Saleh N, Sirk K, Kim H-J, Tilton RD, Lowry GV. 2008. Stabilization of aqueous nanoscale zerovalent iron dispersions by anionic polyelectrolytes: Adsorbed anionic polyelectrolyte layer properties and their effect on aggregation and sedimentation. J Nanopart Res 10:795–814.

Phenrat T, Kim H-J, Fagerlund F, Illangasekare T, Tilton RD, Lowry GV. 2009a. Particle size distribution, concentration, and magnetic attraction affect transport of polymer-modified Fe0 nanoparticles in sand columns. Environ Sci Technol 43:5079–5085.

Phenrat T, Liu Y, Tilton RD, Lowry GV. 2009b. Adsorbed polyelectrolyte coatings decrease Fe nanoparticle reactivity with TCE in water: Conceptual model and mechanisms. Environ Sci Technol 43:1507–1514.

Phenrat T, Song JE, Cisneros CM, Schoenfelder DP, Tilton RD, Lowry GV. 2010. Estimating attachment of nano- and submicrometer-particles coated with organic macromolecules in porous media: Development of an empirical model. Environ Sci Technol 44:4531–4538.

Phenrat T, Cihan A, Kim H-J, Mital M, Illangasekare T, Lowry GV. 2010. Transport and deposition of polymer-modified Fe^0 nanoparticles in 2-D heterogeneous porous media: Effects of particle concentration, Fe^0 content, and coatings. Environ Sci Technol 44:9086–9093.

Phenrat T, Fagerlund F, Illangasekare T, Lowry GV, Tilton RD. 2011. Polymer-modified Fe^0 nanoparticles target entrapped NAPL in two dimensional porous media: Effect of particle concentration, NAPL saturation, and injection strategy. Environ Sci Technol 45:6102–6109.

Powell RM, Powell PD, Puls DW. 2002. Economic analysis of the implementation of permeable reactive barriers for remediation of contaminated groundwater. EPA/600/R-02/034. U.S. Environmental Protection Agency, Ada, OK, USA. 31 p.

Puls RW, Paul CJ, Powell RM. 1996. Remediation of Chromate-Contaminated Groundwater Using Zero-Valent Iron: Field Test at USCG Support Center, Elizabeth City, North Carolina. Final Report. EPA/600/R-02/034. U.S. Environmental Protection Agency, Washington, DC, USA. 19 p.

Puls RW, Blowes DW, Gillham RW. 1999a. Long-term performance monitoring for a permeable reactive barrier at the U.S. Coast Guard Support Center, Elizabeth City, North Carolina. J Hazard Mater 68:109–124.

Puls RW, Paul CJ, Powell RM. 1999b. The application of in situ permeable reactive (zero-valent iron) barrier technology for the remediation of chromate-contaminated groundwater: a field test. Appl Geochem 14:989–1000.

Quinn J, Geiger C, Clausen C, Brooks K, Coon C, O'Hara S, Krug T, Major D, Yoon W-S, Gavaskar A, Holdsworth T. 2005. Field demonstration of DNAPL dehalogenation using emulsified zero-valent iron. Environ Sci Technol 39:1309–1318.

Rabideau AJ, Shen P, Khandelwal A. 1999. Feasibility of amending slurry walls with zero-valent iron. J Geotech Geoenviron Eng 125:330–333.

Ramsburg CA, Pennell KD. 2001. Experimental and economic assessment of two surfactant formulations for source zone remediation at a former dry cleaning facility. Ground Water Monit Remediat 21:68–82.

Ramsburg CA, Pennell KD. 2002. Density-modified displacement for DNAPL source zone remediation: density conversion and recovery in heterogeneous aquifer cells. Environ Sci Technol 36:3176–3187.

Ramsburg CA, Pennell KD, Kibbey TC, Hayes KF. 2004. Refinement of the density-modified displacement method for efficient treatment of tetrachloroethene source zones. J Contam Hydrol 74:105–131.

Ramsburg CA, Pennell KD, Abriola LM, Daniels G, Drummond CD, Gamache M, Hsu H-L, Petrovskis EA, Rathfelder KM, Ryder JL, Yavaraski TP. 2005. Pilot-scale demonstration of surfactant-enhanced PCE solubilization at the Bachman Road site. 2. System operation and evaluation. Environ Sci Technol 39:1791–801.

Reinhard M, Hopkins GD, Cunningham J, Lebron CA. 2006. From laboratory study to full-scale application: Treating groundwater for TCE removal using catalyzed reductive dechlorination. Proceedings, 232nd ACS National Meeting, September 10–14, San Francisco, CA, USA. American Chemical Society. ENVR-68.

Roberts AL, Jeffers PM, Wolfe NL, Gschwend PM. 1993. Structure-reactivity relationships in dehydrohalogenation reactions of polychlorinated and polybrominated alkanes. Crit Rev Environ Sci Technol 23:1–39.

Roy-Perreault A, Kueper BH, Rawson J. 2005. Formation and stability of polychlorinated biphenyl pickering emulsions. J Contam Hydrol 77:17–39.

Saleh N, Phenrat T, Sirk K, Dufour B, Ok J, Sarbu T, Matyjaszewski K, Tilton RD, Lowry GV. 2005. Adsorbed triblock copolymers deliver reactive iron nanoparticles to the oil/water interface. Nano Lett 5:2489–2494.

Saleh N, Sirk K, Liu Y, Phenrat T, Dufour B, Matyjaszewski K, Tilton RD, Lowry GV. 2007. Surface modifications enhance nanoiron transport and NAPL targeting in saturated porous media. Environ Eng Sci 24:45–57.

Scherer MM, Richter S, Valentine RL, Alvarez PJJ. 2000. Chemistry and microbiology of permeable reactive barriers for in situ groundwater clean up. Crit Rev Environ Sci Technol 30:363–411.

Schüth C, Disser S, Schüth F, Reinhard M. 2000. Tailoring catalysts for hydrodechlorinating chlorinated hydrocarbon contaminants in groundwater. Appl Catal B Environ 28:147–152.

Schwarzenbach RP, Stierli R, Lanz K, Zeyer J. 1990. Quinone and iron porphyrin mediated reduction of nitroaromatic compounds in homogeneous aqueous solution. Environ Sci Technol 24:1566–1574.

Scott TB, Popescu IC, Crane RA, Noubactep C. 2011. Nano-scale metallic iron for the treatment of solutions containing multiple inorganic contaminants. J Hazard Mater 186:280–287.

Shackelford CD, Sale TC, Liberati MR. 2005. In-situ remediation of chlorinated solvents using zero valent iron and clay mixtures: A case history. Proceedings, 18th GRI Conference held in conjunction with Geo-Frontiers 2005, January 24–26, Austin, TX, USA. Geo-Institute of the American Society of Civil Engineers, Reston, VA, USA, pp 3699–3707.

Shen X, Zhao L, Ding Y, Liu B, Zeng H, Zhong L, Li X. 2011. Foam, a promising vehicle to deliver nanoparticles for vadose zone remediation. J Hazard Mater 186:1773–1780.

Shi Z, Nurmi JT, Tratnyek PG. 2011. Effects of nano zero-valent iron (nZVI) on oxidation-reduction potential (ORP). Environ Sci Technol 45:1586–1592.

Simpkin TJ, Palaia T, Petri BG, Smith BA. 2011. Oxidant delivery approaches and contingency planning. In Siegrist RL, Crimi M, Simpkin TJ, eds, *In Situ* Chemical Oxidation for Groundwater Remediation. Springer, New York, NY, USA, pp 449–480.

Smolen JM, Weber EJ, Tratnyek PG. 1999. Molecular probe techniques for the identification of reductants in sediments: Evidence for reduction of 2-chloroacetophenone by hydride transfer. Environ Sci Technol 33:440–445.

Sorel D, Warner SY, Longino BL, Honniball JH, Hamilton LA. 2003. Performance monitoring and dissolved hydrogen measurements at a permeable zero valent iron reactive barrier. In Henry SM, Warner SD, eds, Chlorinated Solvent and DNAPL Remediation. American Chemical Society, Washington, DC, USA, ACS Symp Ser 837:278–285.

Sposito G. 2011. Electron shuttling by natural organic matter: Twenty years after. In Tratnyek PG, Grundl TJ, Haderlein SB, eds, Aquatic Redox Chemistry. American Chemical Society, Washington, DC, USA, ACS Symp Ser 1071:113–127.

Strathmann TJ. 2011. Redox reactivity of organically complexed iron(II) species with aquatic contaminants. In Tratnyek PG, Grundl TJ, Haderlein SB, eds, Aquatic Redox Chemistry. American Chemical Society, Washington, DC, USA, ACS Symp Ser 1071:283–313.

Stroo HF, Leeson A, Marqusee J, Johnson PC, Ward CH, Kavanaugh MC, Sale TC, Newell CJ, Pennell KD, Lebron CA, Unger M. 2012. Chlorinated ethene source remediation: Lessons learned. Environ Sci Technol 46:6438–6447.

Stroo HF, Ward CH. 2010. Future directions and research needs for chlorinated solvent plumes. In Stroo HF, Ward CH, eds, *In Situ* Remediation of Chlorinated Solvent Plumes. Springer, New York, NY, USA, pp 699–725.

Struyk Z, Sposito G. 2001. Redox properties of standard humic acids. Geoderma 102:329–346.

Szecsody JE, Fruchter JS, Sklarew DS, Evans JC. 2000. In Situ Redox Manipulation of Subsurface Sediments from Fort Lewis, Washington: Iron reduction and TCE Dechlorination Mechanisms. Report PNNL-13178. Pacific Northwest National Laboratory, Richland, WA, USA. 89 p.

Szecsody JE, Fruchter JS, Williams MD, Vermeul VR, Sklarew D. 2004. In situ chemical reduction of aquifer sediments: enhancement of reactive iron phases and TCE dechlorination. Environ Sci Technol 38:4656–4663.

Szulczewski MD, Helmke PA, Bleam WF. 2001. XANES spectroscopy studies of Cr(VI) reduction by thiols in organosulfur compounds and humic substances. Environ Sci Technol 35:1134–1141.

Támara M, Butler EC. 2004. Effects of iron purity and groundwater characteristics on rates and products in the degradation of carbon tetrachloride by iron metal. Environ Sci Technol 38:1866–1876.

Totten LA, Assaf-Anid NM. 2003. Abiotic dehalogenation by metals. In Häggblom MM, Bossert ID, Bossert ID, eds, Dehalogenation: Microbial Processes and Environmental Applications. Kluwer, Boston, MA, USA, pp 261–290.

Tratnyek PG. 2010. Chemical reductants for ISCR: The potential for improvement. Proceedings, 7th International Conference on Remediation of Chlorinated and Recalcitrant Compounds, Monterey, CA, USA, May 24–27. Battelle Press, Columbus, OH, USA, D-054.

Tratnyek PG, Johnson RL. 2006. Nanotechnologies for environmental cleanup. NanoToday 1:44–48.

Tratnyek PG, Macalady DL. 1989. Abiotic reduction of nitro aromatic pesticides in anaerobic laboratory systems. J Agric Food Chem 37:248–254.

Tratnyek PG, Macalady DL. 2000. Oxidation-reduction reactions in the aquatic environment. In Mackay D, Boethling RS, eds, Handbook of Property Estimation Methods for Chemicals: Environmental and Health Sciences. Lewis, Boca Raton, FL, USA, pp 383–415.

Tratnyek PG, Johnson RL, Johnson TL, Miehr R. 2001. In Situ Remediation of Explosives Contaminated Ground-Water with Sequential Reactive Treatment Zones. SERDP Final Report. Project CU-1176. Strategic Environmental Research and Development Program, Alexandria, VA, USA. 32 p.

Tratnyek PG, Scherer MM, Johnson TJ, Matheson LJ. 2003a. Permeable reactive barriers of iron and other zero-valent metals. In Tarr MA, ed, Chemical Degradation Methods for Wastes and Pollutants: Environmental and Industrial Applications. Marcel Dekker, New York, NY, USA, pp 371–421.

Tratnyek PG, Weber EJ, Schwarzenbach RP. 2003b. Quantitative structure-activity relationships for chemical reductions of organic contaminants. Environ Toxicol Chem 22:1733–1742.

Tratnyek PG, Salter AJ, Nurmi JT, Sarathy V. 2010. Environmental applications of zerovalent metals: Iron vs. zinc. In Erickson LE, Koodali RT, Richards RM, eds, Nanoscale Materials in Chemistry: Environmental Applications. American Chemical Society, Washington, DC, USA, ACS Symp Ser 1045:165–178.

Tratnyek PG, Salter-Blanc AJ, Nurmi JT, Amonette JE, Liu J, Wang C, Dohnalkova A, Baer DR. 2011. Reactivity of zerovalent metals in aquatic media: Effects of organic surface coatings. In Tratnyek PG, Grundl TJ, Haderlein SB, eds, Aquatic Redox Chemistry. American Chemical Society, Washington, DC, USA, ACS Symp Ser 1071:381–406.

Truex MJ, Macbeth TW, Vermeul VR, Fritz BG, Mendoza DP, Mackley RD, Wietsma TW, Sandberg G, Powell T, Powers J, Pitre E, Michalsen M, Ballock-Dixon SJ, Zhong L, Oostrom M. 2011a. Demonstration of combined zero-valent iron and electrical resistance heating for in situ trichloroethene remediation. Environ Sci Technol 45:5346–5351.

Truex MJ, Vermeul VR, Mendoza DP, Fritz BG, Mackley RD, Oostrom M, Wietsma TW, Macbeth TW. 2011b. Injection of zero-valent iron into an unconfined aquifer using shear-thinning fluids. Ground Water Monit Remediat 31:50–58.

Tsukano Y. 1986. Transformations of selected pesticides in flooded rice-field soils: A review. J Contam Hydrol 1:47–63.

USEPA (U.S. Environmental Protection Agency). 1999. Field Applications of In Situ Remediation Technologies: Permeable Reactive Barriers. Report EPA 542-R-99-002. Washington, DC, USA. 114 p.

Uchimiya M, Stone AT. 2009. Reversible redox chemistry of quinones: Impact on biogeochemical cycles. Chemosphere 77: 451–458.

Urynowicz MA, Balu B, Udayasankar U. 2008. Kinetics of natural oxidant demand by permanganate in aquifer solids. J Contam Hydrol 96:187–194.

Van der Zee FP, Cervantes FJ. 2009. Impact and application of electron shuttles on the redox (bio)transformation of contaminants: A review. Biotechnol Adv 27:256–277.

Van Nooten T, Springael D, Bastiaens L. 2008. Positive impact of microorganisms on the performance of laboratory-scale permeable reactive iron barriers. Environ Sci Technol 42:1680–1686.

Vermeul VR, Williams MD, Evans JC, Szecsody JE, Bjornstad BN, Liikala TL. 2000. In Situ Redox Manipulation Proof-of-Principle Test at the Fort Lewis Logistics Center. Final Report PNNL-13357. Pacific Northwest National Laboratory, Richland, WA, USA. 330 p.

Vidic RD. 2001. Permeable reactive barriers: Case study review. GWRTAC E-Series Technology Evaluation Report TE-01-01. Ground Water Remediation Technologies Analysis Center, Pittsburgh, PA, USA. 49 p.

Wadley SLS, Gillham RW, Gui L. 2005. Remediation of DNAPL source zones with granular iron: Laboratory and field tests. Ground Water 43:9–18.

Wang S, Mulligan CN. 2004. An evaluation of surfactant foam technology in remediation of contaminated soil. Chemosphere 57:1079–1089.

Wang Y, Sakamoto Y, Kamiya Y. 2009. Remediation of actual groundwater polluted with nitrate by the catalytic reduction over copper-palladium supported on active carbon. Appl Catal A: General 361:123–129.

Warner SD, Sorel D. 2003. Ten years of permeable reactive barriers: Lessons learned and future expectations. In Henry SM, Warner SD, eds, Chlorinated Solvent and DNAPL Remediation: Innovative Strategies for Subsurface Cleanup. American Chemical Society, Washington, DC, USA, ACS Symp Ser 837:36–50.

Water Science and Technology Board. 2004. Contaminants in the Subsurface: Source Zone Assessment and Remediation. U.S. National Academies Press, Washington, DC, USA.

Wilkin RT, Puls RW. 2003. Capstone Report on the application, monitoring, and performance of permeable reactive barriers for ground-water remediation: Vol 1 – Performance evaluations at two sites. EPA 600-R-03–045a. U.S. Environmental Protection Agency, Cincinnati, OH, USA.

Wilkin RT, Puls RW, Sewell GW. 2003. Long-term performance of permeable reactive barriers using zero-valent iron: geochemical and microbiological effects. Ground Water 22:165–168.

Wilkin RT, Su CM, Ford RG, Paul CJ. 2005. Chromium-removal processes during groundwater remediation by a zerovalent iron permeable reactive barrier. Environ Sci Technol 39:4599–4605.

Wilson JT. 2003. Abiotic Reactions May be the Most Important Mechanism in Natural Attenuation of Chlorinated Solvents. AFCEE Technology Transfer Workshop, Brooks AFB, San Antonio, TX, USA, February 24–27.

Wilson JT, Kampbell DH, Ferrey M, Estuesta P. 2001. Evaluation of the protocol for the natural attenuation of chlorinated solvents: Case study at the Twin Cities Army Ammunition Plant. Office of Research and Development NRMRL, U.S. Environmental Protection Agency, Ada, OK, USA.

Wolfe NL, Macalady DL. 1992. New perspectives in aquatic redox chemistry: Abiotic transformations of pollutants in groundwater and sediments. J Contam Hydrol 9:17–34.

Xia K, Weesner F, Bleam WF, Bloom PR, Skyllberg UL, Helmke PA. 1998. XANES studies of oxidation states of sulfur in aquatic and soil humic substances. Soil Sci Soc Am J 62:1240–1246.

Xu X, Thomson NR. 2008. Estimation of the maximum consumption of permanganate by aquifer solids using a modified chemical oxygen demand test. J Environ Eng 134:353–361.

Xu X, Thomson NR. 2009. A long-term bench-scale investigation of permanganate consumption by aquifer materials. J Contam Hydrol 110:73–86.

Yabusaki S, Cantrell K, Sass B, Steefel C. 2001. Multicomponent reactive transport in an in situ zero-valent iron cell. Environ Sci Technol 35:1493–1503.

Zhang H, Weber EJ. 2009. Elucidating the role of electron shuttles in reductive transformations in anaerobic sediments. Environ Sci Technol 43:1042–1048.

Zhang H, Colón D, Kenneke JF, Weber EJ. 2011. The use of chemical probes for the characterization of the predominant abiotic reductants in anaerobic sediments. In Tratnyek PG, Grundl TJ, Haderlein SB, eds, Aquatic Redox Chemistry. American Chemical Society, Washington, DC, USA, ACS Symp Ser 1071:539–557.

CHAPTER 11

SURFACTANT AND COSOLVENT FLUSHING

Kurt D. Pennell, Natalie L. Cápiro and Douglas I. Walker

Department of Civil and Environmental Engineering, Tufts University, Medford, MA 02155, USA

11.1 INTRODUCTION

Surfactant and cosolvent flushing gained attention in the early 1990s as a potential *in situ* remediation technology that could greatly enhance the extraction and recovery of chlorinated solvents from the subsurface. In the hypothetical unconfined aquifer remediation scenario shown in Figure 11.1, a solution containing active ingredients (for example, 5% surfactant or 90% ethanol) is injected upgradient of a highly contaminated source area containing dense nonaqueous phase liquid (DNAPL). The treatment solution enhances removal of DNAPL and sorbed contaminants as it flows through the targeted treatment zone. The treatment solution and contaminants are then extracted from a downgradient well for aboveground treatment, reuse, and/or disposal.

When implemented appropriately, surfactant and/or cosolvent flushing technologies can be highly effective for removing large amounts of contaminant mass from the subsurface over short timeframes – weeks to months – depending on aquifer permeability. As with most *in situ* technologies that involve subsurface delivery and recovery of active ingredients (for example, steam flushing or chemical oxidation), surfactant and cosolvent flushing are most amenable to relatively homogeneous subsurface systems with sufficient permeability to allow the injected fluid to be delivered and recovered efficiently. In addition, successful application of surfactant and cosolvent flushing technologies typically requires detailed site characterization, laboratory treatability tests, and mathematical modeling. These efforts are necessary to ensure that an accurate conceptual site model (CSM) is developed, and contaminant mass recovery as well as fluid injection and extraction are optimized to achieve remediation goals.

Surfactant and cosolvent flushing technologies are based in large part on enhanced oil recovery (EOR) research undertaken in the 1960s and 1970s. The goal of EOR was to develop efficient down-hole methods to recover residual oil from formations following secondary oil production (Holm and Csaszar, 1962; Taber, 1981; Lake, 1989). In EOR applications, surfactant formulations may be augmented with cosolvents, polymers, and/or salt (brine) to improve the recovery of residual crude oil from petroleum reservoirs following primary and/or secondary production (Nelson, 1989); similar strategies can be implemented for DNAPL source zone remediation.

While many lessons can be learned from EOR studies conducted at both the laboratory- and pilot-scale (Lake, 1989; Pope and Wade, 1995), these approaches often are not directly transferable to DNAPL source zone remediation. A number of important differences exist between EOR and DNAPL source zone remediation, including subsurface conditions, contaminant

B.H. Kueper et al. (eds.), *Chlorinated Solvent Source Zone Remediation*,
doi: 10.1007/978-1-4614-6922-3_11, © Springer Science+Business Media New York 2014

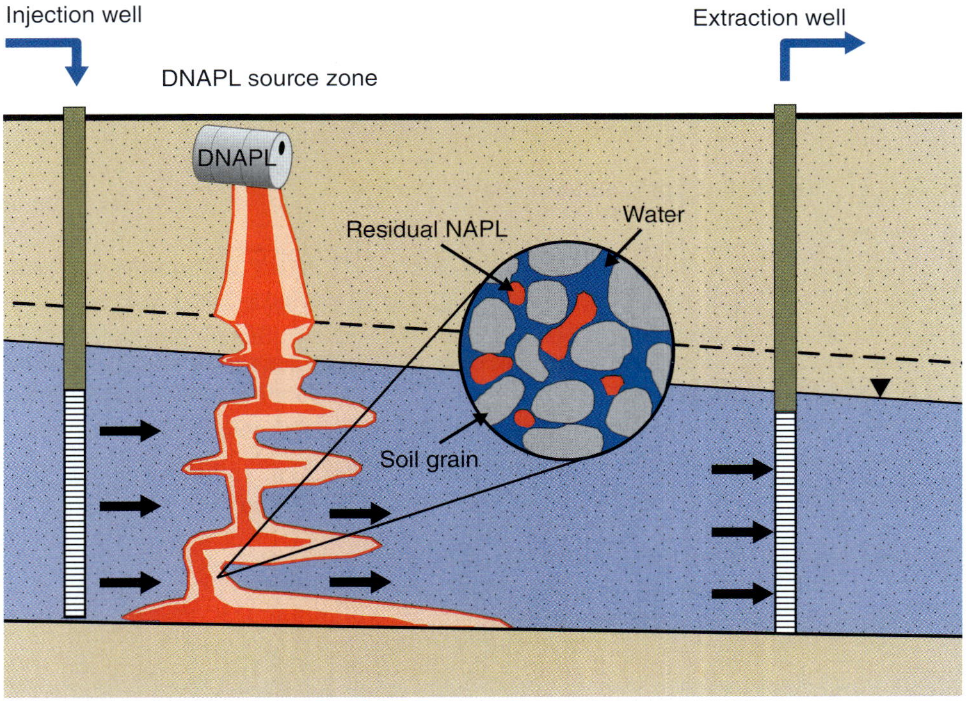

Figure 11.1. Conceptual model of an *in situ* flushing system for treatment of a DNAPL source zone in an unconfined aquifer.

Table 11.1. Summary of Common Differences Between Enhanced Oil Recovery and DNAPL Source Zone Remediation

Property or Parameter	Enhanced Oil Recovery (EOR)	DNAPL Source Zone Remediation
Porous media	Typically consolidated (e.g., sandstone), fully confined formations located well below the surface (e.g., 4,000–6,000 ft bgs), ranging from relatively homogeneous or heterogeneous	Typically unconsolidated (e.g., aquifer), unconfined formations located near the surface (<100 ft bgs), ranging from relatively homogeneous to highly heterogeneous
Physical and chemical conditions	Often high pressure, high temperature, and high salinity (% levels, brine)	Typically low pressure, low temperature (10–20°C), low salinity (ppm levels)
Organic liquid-contaminant properties	Crude oil with densities less than or similar to water, viscosities often >1 cP	Organic liquids with densities less (LNAPL) or greater (DNAPL) than water, viscosities often <1 cP
Recovery/remediation goals	Primary + secondary oil recoveries range from 10% to 35%, EOR may achieve 50–60% recovery, leaving 40–50% as residual oil	Drinking water MCL, often in the ppb range, risk-based solid-phase concentration, combined remedies may be utilized to achieve goals
Active ingredient toxicity and biodegradability	Generally not considered, may select active ingredients that are recalcitrant to minimize potential losses and prolong shelf life	Human and ecotoxicity are important considerations, regulatory approval required to inject into aquifer formations, biodegradation often desirable

Note: °C – degrees Celsius; bgs – below ground surface; cP – centipoise; ft – feet; LNAPL – light nonaqueous phase liquid; MCL – maximum concentration limit; ppb – part(s) per billion properties, and regulatory criteria, which are summarized in Table 11.1. Failure to recognize these differences, particularly fluid density contrasts and subsurface heterogeneity, can negatively impact recovery efficiencies as well as the overall cost and performance of surfactant and cosolvent flushing.

Careful consideration of site constraints during the selection of active ingredients and design of flushing strategies is essential in order to achieve effective DNAPL recovery at a reasonable cost. In fact, a number of pilot- and field-scale tests of surfactant and cosolvent flushing conducted in the 1990s yielded marginal DNAPL recovery efficiencies and were expensive when compared to competing technologies (Mulligan et al., 2001; McDade et al., 2005; McGuire et al., 2006). As a result, use of surfactant and cosolvent flushing for DNAPL remediation waned during most of the 2000s, but has received renewed attention as an additive for surfactant enhanced *in situ* chemical oxidation (S-ISCO) and for treatment of light nonaqueous phase liquid (LNAPL) source zones. In addition, the coupling of surfactant or cosolvent flushing with bioremediation in a treatment train has been documented at several chlorinated solvent sites (Mravik et al., 2003; Ramsburg et al., 2004a).

In the sections that follow, the basic principles of surfactant and cosolvent flushing are presented, followed by practical considerations for the proper design and implementation of these technologies at the field scale. Particular attention is directed toward the limitations and potential pitfalls of surfactant and cosolvent flushing, with the intent of guiding researchers and practitioners toward cost effective solutions based on the lessons learned over the past two decades. The potential to couple surfactant and cosolvent flushing with other technologies is also discussed because much of the current interest in these agents involves combined remedies or treatment trains.

11.1.1 Surfactants

Surfactant flushing typically involves injection of an aqueous solution containing between 1% and 10% active ingredient, which consists primarily of the surfactant, but may also include salts and organic compounds to improve stability and performance. The unique properties of surfactants arise from the fact that they possess both nonpolar (hydrophobic) and polar (hydrophilic) groups, and are therefore, referred to as amphiphilic compounds. The hydrophobic group or "tail" of a surfactant typically consists of a long-chain hydrocarbon, while the hydrophilic group consists of a polar "head" group, such as sulfate or ethylene glycol (Rosen, 1989).

The unique chemical structure of surfactants results in their strong tendency to accumulate at the interface between two phases and thereby alter interfacial properties. For example, in air–water systems the hydrophobic tail is oriented along the surface of the water, while the hydrophilic head group resides in the aqueous phase. Such accumulation or *surface excess* of surfactant at the air–water interface typically results in a surface tension reduction from approximately 72 dyne/centimeter (dyne/cm) (pure water) to 30–40 dyne/cm depending upon the properties of the surfactant (Rosen, 1989; Hiemenz and Rajagopalan, 1997). Similar behavior is observed in organic liquid–water systems and solid–water systems, where the hydrophobic group associates with the organic liquid or solid phase and the hydrophilic group extends into the aqueous phase (Valsaraj et al., 1988; Graciaa et al., 1993). For most nonaqueous phase liquid (NAPL)–water systems, the interfacial tension (IFT) can be reduced from approximately 30–40 dyne/cm to 5 dyne/cm or lower depending on the particular surfactant formulation. In EOR applications, salts and cosolvents are frequently added to the surfactant solution to achieve ultra-low IFTs ($<$0.01 dyne/cm) (Lake, 1989).

The accumulation at, and modification of, interfaces gave rise to the term surfactant, which is a contraction of the phrase *surf*ace-*act*ive ag*ent*. A large number of surfactants are produced commercially, with applications ranging from pharmaceuticals (Lawrence, 1994; Drummond and Fong, 2000; Tochilin, 2001), food products (Stauffer, 2005; Kralova and Sjoblom, 2009), detergents (Lange, 1999) to surface coatings (Claesson, 1992; Rong et al., 2006).

In most cases, surfactants are classified into three main groups based on the nature of their hydrophilic group: anionic, cationic, and nonionic. As the name implies, anionic surfactants are characterized by a negatively charged head group, such as a sulfate or sulfonate group. Due to their stability at high temperature and pressure, low sorption tendency (Trogus et al., 1977; Adeel and Luthy, 1995; Taylor et al., 2001) and relatively low cost ($3–10/pound (lb)), anionic surfactants are widely used in EOR applications (Lake, 1989).

Anionic surfactants are very sensitive to the background electrolyte species and concentration, and as a result, their performance can be improved markedly by the addition, or presence, of salts (Baran et al., 1994; Sabatini et al., 2000). Above a certain threshold of salt concentration, however, the phase behavior of anionic surfactant may be negatively impacted, leading to surfactant precipitation or the formation of highly viscous gels (Shiau et al., 1995). When these issues are addressed during the selection and testing of surfactant formulations, anionic surfactant floods can be extremely effective for treating contaminated source zones (Londergan et al., 2001).

Cationic surfactants contain a positively charged head group, which typically consists of an amine or a quaternary ammonium group. Since most solids possess a net negative charge, cationic surfactants tend to adsorb strongly to aquifer materials (Wagner et al., 1994), limiting their utility as subsurface flushing agents. However, the strong sorption tendency of cationic surfactants can be utilized to modify mineral surfaces, greatly increasing the capacity of a solid to remove contaminants from waste streams. For example, cationic surfactants have been used as sorbents on smectite and zeolite minerals to attenuate contaminant migration (*in situ* barriers) and in fixed- or fluidized-bed reactors (Boyd et al., 1988; Srinivasan and Fogler, 1990; Haggerty and Bowman, 1994). Despite these promising applications, the relatively high cost ($10–25/lb), toxicity (Nalecz-Jawecki et al., 2003), and strong sorption of cationic surfactants to soils and sediments have limited their consideration for *in situ* flushing of source zones.

In contrast to anionic and cationic surfactants, the polar head group of nonionic surfactants consists of either several hydroxyl (OH) groups or ethylene oxide (EO) chains. In addition, the number of EO groups can be varied during the synthesis process, allowing for adjustment of the polarity or hydrophilic properties of the surfactant molecule. For this reason, nonionic surfactants are often characterized by the ratio of hydrophilic (polar) to hydrophobic (nonpolar) groups, represented by the hydrophile–lipophile balance (HLB). The HLB was defined by Rosen (1989):

$$\mathrm{HLB} = 20 \times \frac{MW_H}{MW_H + MW_L} \qquad \text{(Eq. 11.1)}$$

where MW_H and MW_L are the molecular weights of the hydrophilic and lipophilic portions of the nonionic surfactant, respectively.

For subsurface remediation applications, nonionic surfactants with HLBs between 12 and 15 (classified as detergents) are typically selected because they readily dissolve in water and do not strongly partition into organic liquids. Although nonionic surfactants are relatively insensitive to pH and salt levels, temperature can strongly influence their behavior. When the temperature is raised beyond a certain point, termed the *cloud point*, the solution can become turbid (milky-white) due to the formation of surfactant-rich and surfactant-depleted phases (Shinoda and Saito, 1968; Kunieda and Shinoda, 1982). In most subsurface scenarios, cloud point is not an issue due to the relatively low temperature of most groundwater (10–15°C), but it can be important during aboveground treatment of effluent waste streams and surfactant separation processes (reuse of the surfactant).

Nonionic surfactants frequently have been considered for *ex situ* soil washing (Deshpande et al., 1999) and *in situ* flushing remediation technologies (Ramsburg and Pennell, 2001; Ramsburg et al., 2005) because they possess several potential advantages. These include their relatively low cost ($2–5/lb), low toxicity, and limited potential for degradation under aerobic and anaerobic conditions (Ramsburg et al., 2004a; Amos et al., 2007).

Although most nonionic surfactants are relatively innocuous and many such as sorbitan ethoxylates have been approved for food-grade applications, nonylphenyl ethoxylates (NPEs) are a notable exception to this general rule. NPE surfactants are acutely toxic to aquatic organisms, can persist in the environment, and are linked to disruption of the endocrine system (Nimrod and Benson, 1996; Gray and Metcalfe, 1997). With limited exceptions, the use of NPEs is prohibited within the European Union (EU), while the United States Environmental Protection Agency (USEPA) recently (USEPA, 2012) released a final report on NPE alternatives through the Design for the Environment Alternative Assessment Program. As a result of these actions, NPEs should be avoided as potential candidates for environmental remediation applications.

In addition to the three surfactant classes discussed above, two additional classes of surfactants exist: amphoteric surfactants and biosurfactants. Amphoteric surfactants, which may function as cationic, anionic, or nonionic depending upon the pH, are widely used in cosmetics and eye drops. However, they are prohibitively expensive for subsurface remediation uses.

Biosurfactants are compounds produced or derived from microbial, plant or animal sources (Muthusamy et al., 2008; Pacwa-Pociniczak et al., 2011). Use of biosurfactants in remediation gained attention in the 1990s because they can be produced *in situ* and have been shown to enhance the biodegradation of hydrocarbon contaminants (Zhang and Miller, 1992). Due to the difficulty and cost associated with producing sufficient quantities of biosurfactants, and the availability of commercial surfactants with similar structures, biosurfactants have not been considered for use in pilot- or field-scale remediation efforts. Nevertheless, naturally occurring biosurfactants could play an important role in desorption and biodegradation of groundwater contaminants, particularly at sites undergoing monitored natural attenuation (Kuyukina et al., 2005; Mulligan, 2005).

11.1.2 Cosolvents

Cosolvent flushing is based on the injection of concentrated organic solutions, typically consisting of greater than 70% active ingredient (cosolvent) and water, through the contaminant source zone. In the late 1980s and early 1990s, cosolvents were investigated as a means to alter the sorption–desorption behavior of hydrophobic organic compounds (HOCs) in soils (Morris et al., 1988; Brusseau et al., 1991; Rao et al., 1991). In the presence of miscible or partially miscible cosolvents, this work established a log-linear relationship demonstrating enhancement of HOC solubility with increasing addition of cosolvent. This enhanced solubility was accompanied by reductions in the HOC sorption capacity and the sorption–desorption rate limitations. A range of cosolvents was investigated in these early studies, with particular focus on low molecular-weight alcohols.

Relevant physical–chemical properties of a homologous series of low molecular-weight alcohols are shown in Table 11.3. These data illustrate two key properties of alcohol cosolvents. First, low molecular-weight alcohols are less dense than water, which has important implications for subsurface delivery and recovery, especially when the resident fluids (groundwater or DNAPL) possess a greater density than the injected cosolvent solution. Second, the lowest molecular-weight alcohols, such as ethanol, are completely miscible in water, which facilitates mixing and allows for delivery of a single fluid phase at high cosolvent concentrations (e.g., 90% ethanol + 10% water, that is, water dissolved in ethanol). In pilot- and field-scale applications of cosolvent flushing (described in Section 11.4), the most commonly employed alcohol is ethanol due to its abundant supply as a fuel additive and relatively low cost (~$1–2/gallon (gal); $0.26–$0.53/L), assuming exemption from federal and state taxes.

11.2 SURFACTANT AND COSOLVENT RECOVERY MECHANISMS

Although surfactants and cosolvents are often categorized together in most of the remediation literature, several important distinctions must be recognized to ensure proper selection, design, and implementation of *in situ* flushing technologies. As can be seen from the chemical formulas shown in Tables 11.2 and 11.3, both surfactants and cosolvents possess hydrophilic and hydrophobic moieties. However, the hydrophilic portion of a cosolvent is typically limited to a single alcohol group (–OH), while surfactants contain hydrophilic moieties that may include

Table 11.2. Properties of Representative Anionic, Cationic, and Ionic Surfactants used in Subsurface Remediation

Surfactant Name	Chemical Structure	Class	Mol. Wt. (g/mol)	HLB	CMC (mg/L)
Sodium dioctyl sulfosuccinate[a]	$C_{20}H_{37}O_7NaS$	Anionic	444	na	1,700
Sodium dodecyl sulfate[b]	$C_{12}H_{25}O_4NaS$	Anionic	288	na	2,300
Cetyltrimethylammonium bromide[a]	$C_{19}H_{42}NBr$	Cationic	364	na	335
Polyoxyethylene (n = 20) sorbitan monooleate[c]	$C_{24}H_{44}O_6(CH_2CH_2O)_{20}$	Nonionic	1,310	15.0	35
Lauryl alcohol ethoxylate[d] (n = 9)	$C_{12}H_{26}O(CH_2CH_2O)_9$	Nonionic	590	13.4	54

Note: CMC – critical micelle concentration; g – gram(s); HLB – hydrophile–lipophile balance; mg/L – milligram(s) per liter; mol-mole(s); na – not applicable
[a]Rosen (1989)
[b]Shiau et al. (1994)
[c]Taylor et al. (2001)
[d]Zhou and Rhue (2000)

Table 11.3. Selected Physical–Chemical Properties of a Homologous Series of Low Molecular-Weight Alcohols (Mackay et al. 2006)

Cosolvent	Chemical Formula	Mol. Wt. (g/mol)	Density 20 °C (g/mL)	Aqu. Sol. 20 °C (mg/L)	Vapor Pressure 20 °C (mm Hg)
Methanol	CH_3OH	32.04	0.79	Miscible	97.9
Ethanol	C_2H_5OH	46.07	0.79	Miscible	44.7
Propanol	C_3H_7OH	60.10	0.80	Miscible	15.0
Butanol	C_4H_9OH	74.12	0.81	770,000	4.4
Pentanol	$C_5H_{11}OH$	88.15	0.81	19,940	1.6
Hexanol	$C_6H_{13}OH$	102.17	0.81	5,900	0.5

Note: g/mL – gram(s) per milliliter; mm Hg – millimeter(s) mercury sulfate (anionic surfactants), amine (cationic surfactants), and ethylene oxide groups (nonionic surfactants).

This difference in chemical structure allows both the hydrophobic and hydrophilic groups of a surfactant to be varied, and typically results in a much greater affinity or tendency for surfactants to accumulate at interfaces compared to typical alcohol cosolvents. Thus, on a molar or mass basis, surfactants are usually far more effective than alcohols for contaminant recovery, and therefore, are typically delivered with active ingredient concentrations of 5% weight (wt) or less. In contrast, alcohol cosolvents are applied at concentrations ranging from 70% to 95% active ingredient. The sizable difference in the applied concentration of surfactants versus cosolvents has important practical implications with respect to active ingredient cost, fluid delivery and recovery, and safety issues, all of which will be discussed in subsequent sections of this chapter.

The sections below focus on the two primary contaminant recovery mechanisms that occur during surfactant and cosolvent flushing: (1) enhanced solubility or solubilization, which may impact both sorbed phase and nonaqueous liquid phase (NAPL) contaminants, and (2) displacement or mobilization of NAPLs existing as either residual or pooled DNAPL. Residual DNAPL refers to the entrapped droplets and ganglia at residual saturation (typically 10–30% of the pore volume) left after DNAPL and water flow through the subsurface, and pools refer to DNAPL that has accumulated in higher-saturation zones, often perched above layers or lenses of low-permeability media (see Chapter 1).

11.2.1 Enhanced Solubility (Solubilization)

In addition to their tendency to accumulate at interfaces, surfactants have the ability to self-aggregate, to form micelles above a specific concentration, referred to as the critical micelle concentration (CMC). The CMC of nonionic surfactants in aqueous solutions is typically on the order of 10–100 mg/L, while the CMC of anionic surfactants is usually greater than 1,000–2,000 mg/L, and can be influenced by the presence of salts. At concentrations less than the CMC, surfactants exist as individual molecules or monomers and have minimal effects on the solubility of hydrophobic compounds, except in the case of extremely hydrophobic compounds where slight solubility enhancements are observed below the CMC (Kile and Chiou, 1989; Edwards et al., 1991; Pennell et al., 1997).

When the surfactant concentration approaches the CMC, surfactant monomers begin to aggregate to form micelles, consisting of a hydrophobic core surrounded by a hydrophilic shell or mantle (Figure 11.2). Here, the mantle stabilizes the surfactant micelle in the aqueous solution, providing a nonpolar core into which hydrophobic (nonpolar) organic compounds can readily partition. When the surfactant concentration is increased above the CMC, the number of micelles in solution continues to increase, while the number of monomers remains constant. As a result, the total volume of micelle cores increases with increasing surfactant concentration, yielding a corresponding linear increase in the overall (or apparent) solubility of the hydrophobic compound (Figure 11.2). The change in the NAPL–water IFT with increasing surfactant concentration is also shown in Figure 11.2 on the right-hand axis. Here, it is important to note that above the CMC, the addition of more surfactant to the system does not result in further reductions in IFT because the NAPL–water interface is already saturated with surfactant.

When the solubility of an organic species increases linearly above the CMC, the capacity of a surfactant to solubilize an organic compound is often represented by the weight solubilization ratio (WSR):

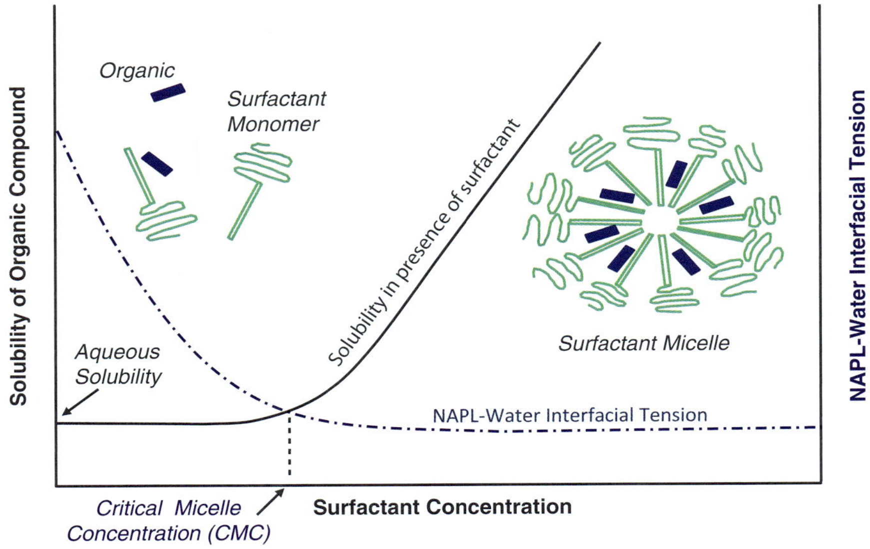

Figure 11.2. Influence of surfactant concentration on the aqueous solubility of hydrophobic organic compounds (*solid line*) and NAPL–water interfacial tension (*dashed line*). At surfactant concentrations greater than the CMC, a linear increase in the aqueous solubility is typically observed.

$$\mathrm{WSR} = \frac{C_w^o - C_{w,sol}^o}{C_w^{surf} - C_{w,CMC}^{surf}} \qquad \text{(Eq. 11.2)}$$

where C_w^o is the mass concentration of organic species in water, $C_{w,sol}^o$ is the solubility of the organic species in water (surfactant-free), C_w^{surf} is the mass concentration of surfactant in water, and $C_{w,CMC}^{surf}$ is CMC of the surfactant in water.

Measured WSR values for relevant surfactant-organic species typically range from 0.1 to 2.0. For example, the WSR of trichloroethene (TCE)-DNAPL in 4% Tween® 80 [polyoxyethylene (20) sorbitan monooeleate] has been reported to be 1.74, which represents a TCE solubility of 63,000 mg/L, compared to 1,100 mg/L in water at room temperature (Suchomel et al., 2007). Similarly, the solubility of dodecane was shown to increase by six orders of magnitude (from 3.7 micrograms per liter (μg/L) to 3,500 mg/L) in the presence of 4% Tween® 80 (Pennell et al., 1993).

The WSR can be expressed as the molar solubilization ratio (MSR), which can then be used to compute a micelle-water partition coefficient (K_{mw}) (Edwards et al., 1991):

$$K_{mw} = \frac{X_m}{X_w} \qquad \text{(Eq. 11.3)}$$

Here, X_m is the mole fraction of organic in the micellar phase, defined as $X_m = \mathrm{MSR}/(1 + \mathrm{MSR})$ and X_w is the mole fraction of organic in free water, defined as $X_w = C_{w,sol}^o \times V_w$, where V_w is the molar volume of water (0.01805 L/mol). Several correlations have been developed to estimate the value of K_{mw} based on the octanol–water partition coefficient (K_{ow}) of the organic species and properties of the surfactant such as the number of carbon atoms in the hydrophobic group (Jafvert et al., 1994; Pennell and Abriola, 1997).

The NAPL–water–surfactant systems discussed above typically consist of excess NAPL (free product) in contact with an aqueous phase that contains surfactant micelles with a hydrophilic shell and a hydrophobic core. This system is referred to as an oil-in-water or NAPL-in-water microemulsion, which is also known as a Winsor Type I system (Figure 11.3). In most cases, the viscosity and density of these microemulsions are similar to that of groundwater, and thus are commonly used in subsurface flushing and soil washing technologies.

However, the phase behavior and performance of surfactants are sensitive to environmental conditions such as temperature and may be intentionally modified through the addition of cosolvents and salts. For example, as the temperature of nonionic surfactants or salinity of anionic surfactants is increased, interactions between the hydrophilic head group and water are reduced (Rosen, 1989). Increasing the hydrophobicity (nonpolar properties) of a surfactant increases interactions between the organic phase and surfactant, which further reduces the NAPL–water IFT and increases the solubilization capacity (Figure 11.3).

With sufficient increases in the salinity (anionic surfactants) or temperature (nonionic surfactants), a separate middle phase, enriched in surfactant and organic, may form between the oil and aqueous phases; this is known as a Winsor Type III system. If the salt concentration or temperature is increased even further, the surfactant can become so hydrophobic that it partitions into the oil phase, forming reverse micelles consisting of a hydrophilic core surrounded by a hydrophobic mantle (Winsor Type II system). In practice, it is usually easier to control the salt concentration than the temperature, and salinity control has been used to develop commercial anionic surfactants having low IFTs and high solubilization capacities.

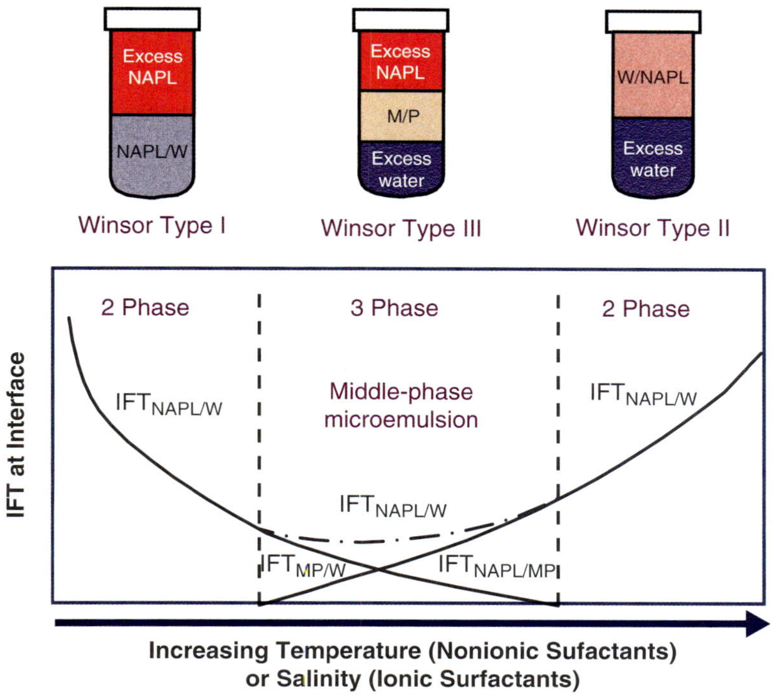

Figure 11.3. Effect of temperature and salinity on the phase behavior and interfacial tension of NAPL–water–surfactant systems.

In EOR applications, anionic surfactant formulations that yield middle-phase microemulsions are often desired because of the ultra-low IFTs that can be achieved.

To evaluate experimentally the phase behavior of surfactant/oil systems, equal volumes of NAPL (oil) and aqueous solution are equilibrated in graduated pipettes and changes in phase volumes are monitored as a function of salt concentration and species (salinity scan). Assuming ideal mixing, the solubilization ratio (S_r) can be defined as (Roshanfekr and Johns, 2011; Kanan et al., 2012):

$$S_r = \frac{V_o}{V_s} \qquad \text{(Eq. 11.4)}$$

where V_o is the volume of the organic phase and V_s is the volume of the aqueous surfactant phase. The solubilization ratio then can be related to the IFT between the organic phase and surfactant solution using the correlation of Huh (1979):

$$\mathrm{IFT}_{os} = \frac{C}{S_r^{\ 2}} \qquad \text{(Eq. 11.5)}$$

where c is an empirical constant equal to 0.3 dyne/cm. Using this approach, salinity scans can be performed rapidly to indentify optimal mixtures of surfactant, salt, and cosolvent for a particle oil or NAPL.

In contrast to surfactants, solubility enhancements resulting from cosolvent addition are based on changes in the properties of the fluid phases, as opposed to the formation of micelles or microemulsions. In general, water-miscible or partially miscible linear alcohols are effective solubilizing agents for low molecular-weight chlorinated hydrocarbons such as perchloroethene (PCE) and TCE, whereas higher molecular-weight cosolvents are required for more strongly hydrophobic compounds.

Much of the early work related to cosolvent effects on hydrocarbon solubility was established in the drug delivery literature, where the cosolvency power (σ_{cs}) can be defined as the ratio of the solubility of target hydrocarbon in the cosolvent and in water (Li and Yalkowsky, 1998):

$$\sigma_{cs} = \mathrm{Log}\left(\frac{C^{o}_{cs,sol}}{C^{o}_{w,sol}}\right) \qquad \text{(Eq. 11.6)}$$

where $C^{o}_{cs,sol}$ is the solubility of the organic species in the cosolvent. The cosolvency power can be related to the octanol–water partition coefficient (K_{OW}) through the following relationship (Yalkowsky and Roseman, 1981):

$$\sigma_{cs} = M\ \mathrm{Log}\ K_{ow} + N \qquad \text{(Eq. 11.7)}$$

where M and N are empirical fitting parameters obtained from experimental data. In the case of ethanol, the values of M and N are reported to range from 0.3 to 0.95 for organic solutes with Log K_{OW} values ranging from −4.90 to 8.23 (Li and Yalkowsky, 1998). An example of the solubility enhancement of toluene as a function of increasing volume fraction of either methanol or ethanol is shown in Figure 11.4.

The liquid–liquid phase behavior and composition of systems containing an organic liquid, cosolvent, and water can be further described through the use of phase diagrams. A representative phase diagram for a ternary system consisting of PCE-DNAPL, *n*-butanol (cosolvent), and water, expressed on a mole fraction basis, is shown in Figure 11.5. This system exhibits Type II phase behavior arising from the presence of two partially miscible pairs (PCE/water and

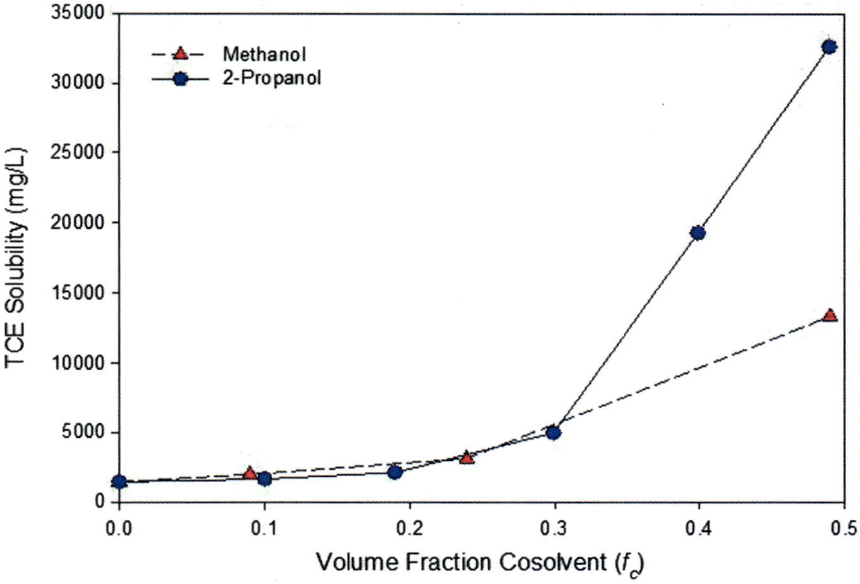

Figure 11.4. Solubility enhancement of TCE in water–isopropanol and water–methanol mixtures. Adapted with permission from Pinal et al. (1990). Copyright (1996) American Chemical Society.

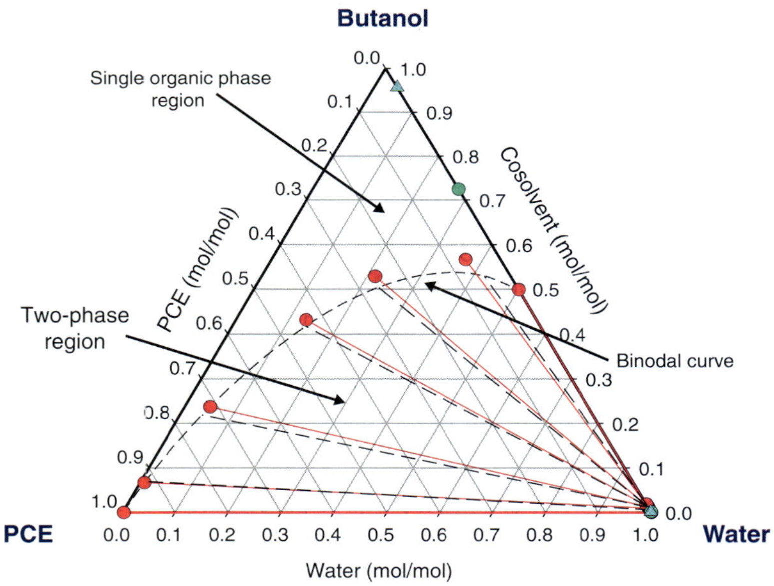

Figure 11.5. Ternary phase diagram for a system consisting of PCE, water and *n*-butanol (cosolvent) showing single- and two-phase regions and tie lines of constant phase composition. Adapted from Ramsburg et al. (2004b). Copyright (2004), with permission from Elsevier.

butanol/water) and one completely miscible pair (PCE/butanol). Type II systems are characterized by a single-phase region located above a two-phase region. These two regions are separated by a binodal curve, represented by the dashed line and the series of composition end points shown in Figure 11.5. The single-phase region consists of water and PCE dissolved in butanol, while the two-phase region consists of mixtures of PCE and water. Although not visible in Figure 11.5, a very small single-phase range exists in the lower right-hand corner of the phase diagram that consists of PCE and butanol dissolved in water. This single-phase region corresponds to the portion of the phase diagram that would typically represent a water flood or partitioning interwell tracer test (PITT).

Within the two-phase region below the binodal curve, the equilibrium partitioning of a cosolvent (butanol) between the aqueous and organic phases is defined by the slope of the tie line. These tie lines connect the binodal curve to one of the lower corners of the phase diagram and represent a line of constant phase composition. Predictions of liquid–liquid equilibrium (LLE) composition of such three phase systems can be obtained from the universal quasi-chemical theory (UNIQUAC), developed by Abrams and Prausnitz (1975). As shown in Figure 11.5, UNIQUAC predictions (dashed lines), accurately capture the phase transition (binodal curve) and the measured tie lines (composition of the two-phase region). For the example case shown in Figure 11.5, the tie lines slope downward from the organic phase (PCE) side of the diagram (left-hand side) toward the lower right-hand corner of the diagram (pure water). This trend indicates that butanol preferentially partitions into organic phase (PCE) relative to the aqueous phase. Such cosolvent partitioning behavior can result in swelling of the organic liquid phase and alteration of the organic phase density.

Thus, partitioning alcohols can be used to modify the density of organic liquids. For example, Ramsburg and coworkers (Ramsburg et al., 2003, 2004a, b) demonstrated *in situ* density conversion of TCE and PCE from DNAPLs to LNAPLs by flushing a contaminated source with aqueous solutions and emulsions containing *n*-butanol to improve source zone remediation (detailed in Section 11.2.2). Alternatively, if the cosolvent preferentially dissolves or partitions into the aqueous phase relative to the organic liquid, the resulting tie lines would slope downward from the water side of the phase diagram (right-hand side) toward the lower left-hand corner of the diagram (pure PCE). Such non-partitioning cosolvents, such as ethanol, are often selected for *in situ* flushing applications to minimize losses and swelling of NAPL present in the treatment zone.

11.2.2 NAPL Displacement ("Mobilization")

From the perspective of contaminant mass recovery, the displacement or mobilization of free product (NAPL) is far more efficient than solubility enhancement alone and is typically the goal of EOR applications involving surfactant and cosolvent floods. NAPL mobilization occurs when the reduction in IFT between the aqueous and organic liquid phases, coupled with viscous and gravitational forces, is sufficient to overcome the capillary forces holding the NAPL within the pore space. As noted previously, surfactants tend to accumulate at interfaces, thereby reducing interfacial forces, including the surface tension of water (air–water interface) and the IFT between water and NAPLs (liquid–liquid interface) (Figure 11.2). At concentrations above the CMC, surfactants typically reduce the IFT between water and organic liquid by at least one order of magnitude.

For example, Pennell et al. (1996) reported that the IFT between water and PCE was reduced from 47.8 dyne/cm to 5.0 dyne/cm in the presence of 4% Witconol® 2722 [polyoxyethylene (20) sorbitan monooleate] and to 0.09 dyne/cm with a 4% solution containing a 1:1 mixture of Aerosol® AY [sodium diamyl sulfosuccinate] and Aerosol® OT [sodium dioctyl sulfosuccinate] and 500 mg/L $CaCl_2$. After injection, approximately 10 pore volumes (PVs) of the 4% Witconol® 2722 surfactant solution were passed through a soil column containing residual PCE, and more than 95% of the initial mass was recovered via micellar solubilization, with only minimal free product mobilization. In contrast, when less than two PVs of 4% Aerosol® AY/OT were introduced, approximately 80% of the PCE mass was displaced from the column as free product and the remaining 19% was solubilized. The findings clearly demonstrate that ultra-low IFTs are not necessary to achieve substantial NAPL displacement in unconsolidated porous media and that surfactant formulations can be selected to either minimize or promote mobilization of free product.

Similarly, when sufficient cosolvent is added to a ternary system (water, cosolvent, and NAPL), the IFT will decrease and the two phases become completely miscible (single-phase region located above the binodal curve). To illustrate this point, a hypothetical ternary phase diagram for a system consisting of DNAPL, water, and a non-partitioning cosolvent such as ethanol is shown in Figure 11.6. Here, the tie lines slope downward from the cosolvent side of the phase diagram toward the pure DNAPL apex, indicating that the cosolvent preferentially partitions into water rather than the DNAPL.

As discussed by Falta (1998), a critical tie line can be drawn tangent to the binodal curve (plait point), which separates the cosolvent axis into two regions, indicating the minimum amount of cosolvent necessary to achieve a completely miscible single phase. When the injected cosolvent concentration is greater than this critical value, a mobilized bank of free product is likely to form ahead of the cosolvent front. Conversely, when the injected cosolvent concentration falls below the critical value, partial mobilization is likely to occur, resulting in a smaller mobilized bank of free product and enhanced dissolution within the cosolvent pulse (Pope, 1980; Falta, 1998). Although ternary phase diagrams can be used to evaluate the phase behavior and performance of surfactant formulations, the amount of surfactant used in source zone flushing is relatively small (<5% wt) and is typically dissolved in the aqueous phase.

The above discussion clearly demonstrates the ability of surfactants and cosolvents to reduce the IFT of NAPL–water systems. However, to utilize these data to inform practical decisions, one must be able to couple the measured IFT reductions with processes governing

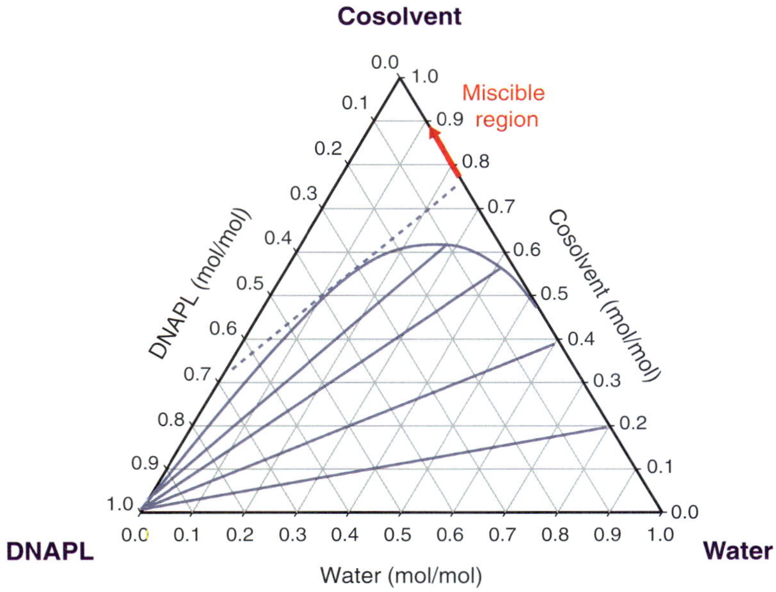

Figure 11.6. Hypothetical ternary phase diagram for a system consisting of DNAPL, water, and cosolvent, illustrating the method used to identify the amount of cosolvent required to promote mobilization of free product during flushing.

NAPL retention in porous media. For this purpose, Pennell et al. (1996) developed an approach to estimate, *a priori*, the potential for mobilization of entrapped NAPLs based on the relative contributions of gravitational (buoyancy), viscous and capillary forces. The theoretical development of this concept results in the combination of two dimensionless numbers, the capillary number (N_{Ca}) and the Bond number (N_B), to yield the total trapping number (N_T).

First, the capillary number (N_{Ca}) accounts for ratio of the viscous forces to the capillary forces:

$$N_{Ca} = \frac{q_w \mu_w}{\gamma_{ow} \cos\theta} \quad \text{(Eq. 11.8)}$$

where q_w is the Darcy velocity of the aqueous phase (positive in the upward direction), μ_w is the aqueous phase viscosity, γ_{ow} is the IFT between the organic and aqueous phases, and θ is the contact angle.

Second, the Bond number (N_B) accounts for the ratio of the buoyancy forces to capillary forces:

$$N_B = \frac{\Delta\rho g k_i k_{rw}}{\gamma_{ow} \cos\theta} \quad \text{(Eq. 11.9)}$$

where $\Delta\rho$ is the difference between the aqueous and organic phase densities (i.e., $\rho_w - \rho_o$), g is the gravitational constant, k_i is the intrinsic permeability of the porous media, and k_{rw} is the relative permeability of the aqueous phase.

When expressed in this form, the Bond and capillary numbers can be directly combined to yield the total trapping number (N_T):

$$N_T = \sqrt{N_{Ca}^2 + 2N_{Ca}N_B \sin\alpha + N_B^2} \quad \text{(Eq. 11.10)}$$

where α is the angle the system flow makes with the horizontal axis. A more detailed discussion of the conceptual and mathematical development of these equations is provided by Pennell et al. (1996).

In most *in situ* flushing scenarios, the flow field can be approximated as horizontal (α equal to 0°); thus, Equation 11.10 reduces to the following form:

$$N_T = \sqrt{N_{Ca}^2 + N_B^2} \quad \text{(Eq. 11.11)}$$

In the case of vertical flushing systems flowing in the direction of the buoyancy force (α is equal to 90°), Equation 11.10 becomes

$$N_T = |N_{Ca} + N_B| \quad \text{(Eq. 11.12)}$$

With respect to sign convention, it is important to note that the vertical upward direction is considered positive (+), while the vertical downward direction is considered to be negative (−). A similar mathematical analysis for the mobilization of residual NAPL in rock fractures is presented by Longino and Kueper (1999).

To determine empirically the values of the total trapping number corresponding to the initiation or onset of NAPL mobilization, as well as complete NAPL displacement, Pennell et al. (1996) conducted a series of one-dimensional column studies. These experiments were performed in glass columns packed with several different size fractions of quartz sand and operated with different surfactant formulations and flow rates to provide a range of initial residual saturations and trapping number values. In all cases, PCE-DNAPL mobilization was observed once the total trapping number exceeded 2×10^{-5}, with nearly complete PCE displacement when the value of N_T exceeded 1×10^{-4} (Figure 11.7).

Based on the saturation curves shown in Figure 11.7, when the value of N_T is less than approximately 2×10^{-5}, mobilization of residual NAPL is not expected to occur. In contrast, if the value of N_T exceeds 1×10^{-4}, substantial or complete mobilization of entrapped NAPL is expected. All parameters needed to compute the value of N_T can be measured independently or estimated from reference data, allowing one to assess the potential for mobilization of entrapped NAPL for a given soil–NAPL–surfactant or NAPL–cosolvent system.

In addition to the total trapping number, it is also important to consider the angle the mobilized DNAPL travels relative to the horizontal x axis, which can be defined as (Pennell et al., 1996; Jin et al., 2007):

$$\gamma = \arctan\left(\frac{N_B}{N_{Ca}}\right) \quad \text{(Eq. 11.13)}$$

Equation 11.13 shows that the angle of downward flow by mobilized DNAPL is represented by the ratio of the gravitational to viscous forces. In a practical sense, this means that it is advantageous for the horizontal viscous force to be large enough, and the vertical (downward) buoyancy force to be small enough, so that the γ angle is relatively small (<45°) so that any mobilized DNAPL can be captured by multiphase extraction wells. Based on a detailed analysis of previous laboratory- and field-scale surfactant floods of DNAPL source zone, Jin et al. (2007) demonstrated that when both the total trapping number (N_T) and the angle of DNAPL flow (γ) are considered, the flushing systems can be designed to minimize the risk of downward DNAPL mobilization.

The total trapping number (N_T) concept has been incorporated into several multidimensional, multiphase flow, and transport models, including UTCHEM (Delshad et al., 2000; Jin et al., 2007) and MISER (Abriola et al., 1997), to simulate NAPL mobilization during surfactant

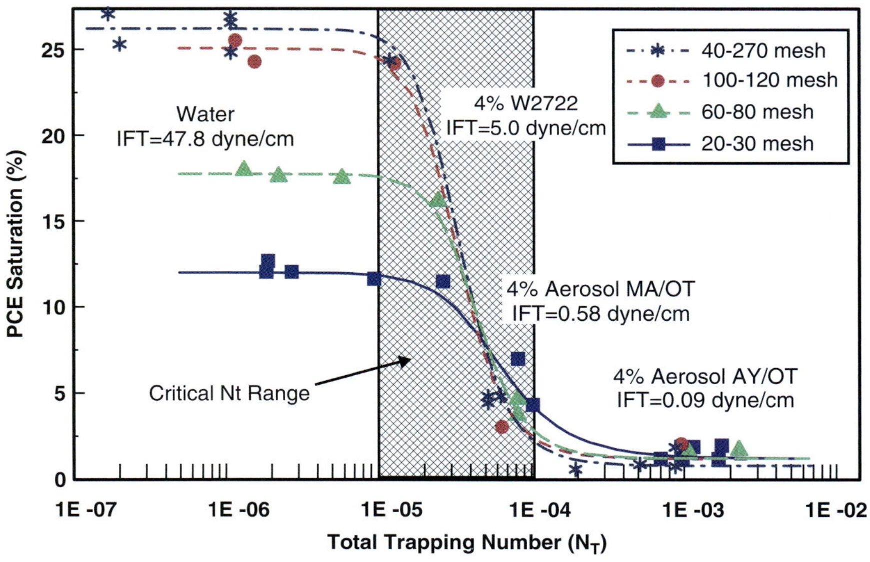

Figure 11.7. Displacement of PCE-DNAPL as a function of total trapping number. The *shaded area* corresponds to the critical trapping number region, below which DNAPL displacement does not occur, and above which nearly complete NAPL displacement is observed. Adapted with permission from Pennell et al. (1996), Copyright (1996) American Chemical Society.

and/or cosolvent flushing. In the MISER model, the capillary pressure–saturation relationship for the aqueous and organic liquid phase system is represented by the van Genuchten functional form (van Genuchten, 1980), and the influence of IFT on the capillary pressure–saturation relationship is incorporated using Leverett scaling (Cinar and Orr, 2005).

To validate the resulting numerical model in a multidimensional system, data obtained from rectangular columns containing residual PCE-DNAPL in Wurtsmith Air Force Base (AFB) aquifer material were simulated by Li et al. (2007). As shown in Figure 11.8, the model accurately captured the onset and extent of PCE-DNAPL mobilization, as well as the migration of mobilized PCE-DNAPL, which formed a bank with an angle of approximately 60°. The numerical model was also capable of accurately predicting the results of similar experiments conducted with 20–30 mesh Ottawa sand. These experiments were conducted at two different flow velocities that yielded bank angles of 78° and 80° and preferential flow of the surfactant solution along the top of the PCE-DNAPL bank in the region where downward migration had depleted residual PCE-DNAPL (Li et al., 2007). It is important to note that the numerical models described here account for the dependence of the capillary pressure–saturation relationships on IFT. Thus, their application is not limited to the mobilization of residual (entrapped) NAPL, and such models also account for the mobilization of NAPLs from higher-saturation pools as well as the migration of free product banks.

Clearly, NAPL mobilization has the potential to yield far more efficient recovery of contaminant mass when compared to enhanced solubility, and if implemented properly, could dramatically reduce the volume of solution delivered, the active ingredient costs and

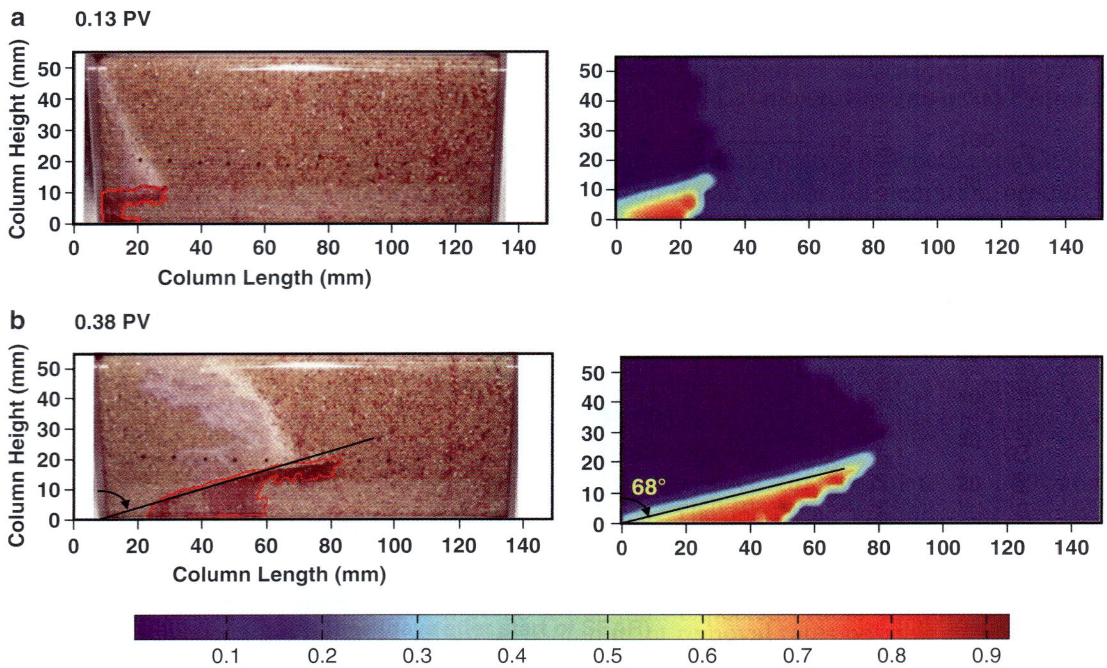

Figure 11.8. Comparison of observed and simulated PCE saturation profiles during low-IFT flushing (0.09 dyne/cm) of a fully saturated rectangular column initially containing PCE-DNAPL at a residual saturation 13%. The PCE-DNAPL is dyed red to facilitate observation and data are presented after injecting 0.13 and 0.38 PVs of a 4% 1:1 mixture of Aerosol® AY/OT and 500 mg/L $CaCl_2$. Adapted with permission from Li et al. (2007). Copyright (2007) American Chemical Society.

the treatment time needed for *in situ* flushing. For source zones contaminated with LNAPLs, such as gasoline, mobilization is typically desirable because the displaced LNAPL will tend to migrate upward toward the water table. Indeed, many of the recent *in situ* flushing applications conducted by commercial vendors have focused on low concentration (<1% active ingredient), low-IFT formulations of surfactant, cosolvent and salt to treat gasoline, and other petroleum hydrocarbon source zones.

However, for DNAPL-contaminated sites, mobilization remains problematic because of the potential for uncontrolled and/or downward migration of DNAPL in response to IFT reductions (Longino and Kueper, 1995). Chapter 8, for example, discusses the fact that the maximum stable pool height that can accumulate above a capillary barrier is proportional to the DNAPL–water IFT. To address the potential mobilization of DNAPL, upward alcohol flushing (Lunn and Kueper, 1997), the use of dense brine solutions (Miller et al., 2000), and *in situ* modification of NAPL density (Roeder et al., 2001; Ramsburg and Pennell, 2002) have been proposed. In addition, careful design of the flushing process, including selection of the active ingredient formulation (solubilization capacity), optimization of flow conditions (hydraulic gradient) and additives such as polymers (viscosity) can potentially mitigate or eliminate the downward migration of mobilized DNAPL during flushing (Jin et al., 2007).

Lunn and Kueper (1999a) performed bench-scale experiments in which the density of PCE-DNAPL was manipulated using a multistep cosolvent flooding approach. An aqueous partitioning pre-flood solution of 10% by volume isobutanol reduced the NAPL density to approximately 1.0 g/mL due to swelling of the NAPL. A composite alcohol main flood, containing 65% by volume ethylene glycol and 35% by volume 1-propanol, maintained miscibility while achieving neutral buoyancy and near stable displacement of the NAPL. Lunn and Kueper (1999b) demonstrated similar density modification using a 2-butanol pre-flood followed by a 1-propanol main flood. Roeder et al. (2001) also developed a cosolvent flushing strategy to convert DNAPL to LNAPL in a horizontal flow regime. In this approach, a low-density alcohol (*tert*-butanol) preferentially partitions into the DNAPL, thereby lowering the density, and then a high-density solution containing glycerol and sucrose as dense additives was used to the displace the mobile DNAPL.

Ramsburg and Pennell (2002) developed a density-modified displacement (DMD) method, in which an aqueous alcohol solution (6% butanol) or emulsion (15% butanol) is flushed through the source zone, resulting in butanol partitioning into the DNAPL. The partitioning process causes the organic liquid to swell and ultimately results in the conversion from a DNAPL to an LNAPL. The density conversion step is then followed by a low-IFT surfactant flood that displaces the newly converted LNAPL upward to a recovery well (Ramsburg et al., 2003, 2004b). This approach provides for efficient recovery of DNAPL, while mitigating downward migration of both the dissolved plume and mobilized free product. Thus, the approach can be implemented in unconfined systems using a conventional horizontal (line-drive) flow configuration. In addition to the above approach, Shook et al. (1998) and Kostarelos et al. (1998) developed a *neutral buoyancy* surfactant flushing approach to address the density of the solubilized DNAPL plume. Here, a low molecular-weight alcohol, such as ethanol or isopropanol, is added to the injected surfactant solution at a concentration that will result in a plume density similar to the resident solution. Polymers can also be added to increase the viscosity of the injected solution to optimize viscous and gravity forces to achieve efficient recovery of the solubilized contaminant (Kostarelos et al., 1998; Jayanti et al., 2002).

11.3 SURFACTANT FLUSHING CASE STUDIES

While surfactant-based source zone treatment has been highly effective under controlled laboratory conditions (Ramsburg and Pennell, 2001; Taylor et al., 2001; Suchomel and Pennell, 2006; Suchomel et al., 2007), field-scale applications initially yielded mixed results. A number of early field demonstrations were conducted at sites contaminated by petroleum hydrocarbons, polychlorinated biphenyl (PCB) compounds and wood-treating creosote NAPLs. A few of these early field demonstrations showed success with LNAPL mobilization (Advanced Applied Technology Demonstration Facility [AATDF], 1997), enhanced recovery of sorbed-phase hydrophobic compounds (Abdul et al., 1992; Abdul and Ang, 1994) and removal of residual DNAPL (Fountain et al., 1996; AATDF, 1997) in subsurface systems that were relatively homogeneous and underlain by a confining layer.

However, a significant number of early studies faced difficulties caused by insufficient site characterization and limited laboratory (treatability studies) evaluations using site contaminants and soils (Roote, 1998). Technical challenges included contaminants perched on clay lenses and low-permeability zones, poorly screened surfactant solutions, reductions in aquifer permeability due to surfactant-induced dispersion of fines, and formation of highly viscous emulsions, all of which can contribute to low contaminant mass recoveries (Fountain et al., 1996; AATDF, 1997; Roote, 1998).

Based on the lessons learned from these laboratory and pilot-scale studies, subsequent field trials showed substantial improvement. For example, a number of well-controlled, field-scale tests of surfactant flushing indicated that DNAPL recoveries in the range of 60–70% can be expected (Rao et al., 1997; Holzmer et al., 2000; Jawitz et al., 2000; Brooks et al., 2004; Soga et al., 2004) and that mass recoveries of greater than 90% are achievable (Heath et al., 1994; Londergan et al., 2001; Abriola et al., 2005; Ramsburg et al., 2005). Detailed descriptions of two field sites, Hill AFB operable unit (OU) 2 and the Bachman Road site, where surfactant flushing was successfully applied to yield 90% + DNAPL mass recoveries using anionic and nonionic surfactants, respectively, are presented below.

A summary of these and other example surfactant flushing field demonstrations, including the surfactant formulation and estimated mass recovery, are provided in Table 11.4. It should be noted that assessments of remedial performance are typically accomplished through some measure of mass recovery, which often involves considerable uncertainty with respect to the initial mass of NAPL present in the treated source zone. Thus, the mass recovery values reported in Table 11.4 should be viewed as estimated values.

11.3.1 Hill Air Force Base Operable Unit 2

The recovery of more than 98% of the NAPL mass using an anionic surfactant was demonstrated at the Hill AFB Operable Unit 2 (OU2) in northern Utah, USA (Londergan et al., 2001). The contaminated alluvial aquifer consisted sand and gravel beds of variable thickness overlaying a silty-clay deposit with hydraulic conductivities ranging from 13 to 35 meters (m)/day. From 1967 to 1975, the site was utilized to dispose of spent degreasing solvents into two unlined trenches (Radian, 1992), which resulted in an estimated 45,000 gallons of DNAPL released into the aquifer (USAF, 1999; Meinardus et al., 2000). DNAPL collected from the site was composed primarily of chlorinated solvents, including 60% TCE, 10% trichloroethane (TCA), and 5% PCE, with the remainder predominantly solubilized oil and grease (Jackson and Dwarakanath, 1999). The collected DNAPL had an IFT, density and viscosity of approximately 9 dyne/cm, 1.38 g/mL, and 0.78 centipoise (cP), respectively (Oolman et al., 1995).

Table 11.4. Representative Examples of Surfactant Flushing Field Demonstrations

Field Site	Surfactant Formulation	NAPL	Amount Recovered (Estimated Recovery)[a]	References
Dover AFB, DE	3.3% Aerosol® MA + 3.3% isopropanol + 0.4% $CaCl_2$	PCE	46 L (68%)	Childs et al., 2006
Bachman Road Oscoda, MI	**6% Tween® 80**	**PCE**	**19 L (90%+)**	**Abriola et al., 2005; Ramsburg et al., 2005**
Alameda Point, CA	5% Dowfax® 8390 + 2% Aerosol® MA + 3% NaCl + 1% $CaCl_2$	TCA, TCE, DCA, DCE	325 kg (97%)	Hasegawa et al., 2000
Camp Lejeune, Marine Corps Base, NC	4% Alfoterra® 145-4PO sulfate + 16% propanol + 0.2% $CaCl_2$	PCE	288 L (72%)	Delshad et al., 2000; Holzmer et al., 2000
Hill AFB OU1, UT	3% Brij® 97 + 2.5% pentanol	Jet fuel, chlorinated solvents	396 L (72%)	Jawitz et al., 2001
Hill AFB OU1 (Cell 5), UT	2.2% Aerosol® OT + 2.1% Tween® 80 + 0.4% $CaCl_2$	Jet fuel, chlorinated solvents	14.4 kg (42, 97%)	Knox et al., 1997
Hill AFB OU1 (Cell 6), UT	4.3% Dowfax® 8390	Jet fuel, chlorinated solvents	1.5 kg (85-95%)	Knox et al., 1997
Hill AFB OU2, UT	3.5% Aerosol® MA + 1.0% NaCl + air injection (foam)	TCE, TCE, PCE, CT	68.8 L (93%)	Hirasaki et al., 1997; Szafranski et al., 1998; Meinardus et al., 2002
Hill AFB OU2, UT	**7.6% Aerosol® MA + 4.5% isopropanol + 0.7% NaCl**	**TCE, TCE, PCE, CT**	**363 L (98.5%)**	**Londergan et al., 2001**
Coast Guard Station, Traverse City, MI	3.6% Dowfax® 8390	PCE, Jet Fuel	3.3 g PCE + 47 kg TH[b]	Knox et al., 1997
Thouin Sand Quarry, Quebec, Canada	9.2% butanol + 9.2% Hostapur® SAS 60 + 13.2% toluene + 13.2% d-limonene	TCE, PCE, waste oil	532 kg (86%)	Martel et al., 1998
Canadian Forces Base Borden, Ontario, Canada	2% 1:1 Rexophos® 25/97 + Alkasurf® NP10	PCE	67 L (69%)	Fountain et al., 1996

Case studies in bold font are discussed in more detail below

Note: CT – carbon tetrachloride; DCA – dichloroethane; DCE – dichloroethene; TCA – trichloroethane

[a]Mass recovery values are based on estimates of the initial mass present, except when the DNAPL was released intentionally (e.g., Dover AFB)

[b]Recovery not reported, mass of PCE and total hydrocarbons (TH) recovered

Prior to surfactant flushing, a pump-and-treat source recovery system was operated from 1992 to 1998, which recovered more than 37,000 gallons of DNAPL (Betts et al., 1998). Production of free-phase DNAPL from the recovery well was observed to decrease significantly prior to surfactant flushing, indicating that the majority of remaining DNAPL existed as discrete, immobile, low-saturation ganglia entrapped within soil pores. The initial source zone characterization performed using a PITT indicated an average DNAPL saturation of 2.7% prior to surfactant flushing.

Surfactant flushing was performed in two phases, a pilot-scale Phase I and a full-scale implementation Phase II. The pilot phase was conducted to ensure that the flushing scheme provided adequate hydraulic control of the swept zone, which was accomplished through a series of injection and extraction wells installed in a 3 × 3 line-drive pattern (Brown et al., 1999). Unlike efforts at Hill AFB OU1, this injection and extraction scheme did not require installation of physical barriers to maintain hydraulic control (McCray et al., 2011). The surfactant formulations used for both stages of the project were selected based on extensive laboratory studies, which included batch and column experiments and hydrogeological characterization of the aquifer formation (Dwarakanath et al., 1999). A steam stripper was used to reduce contaminant levels in the effluent waste stream to below an approved limit of 16 mg/L prior to discharge to the industrial wastewater treatment plant. To minimize foam formation during the steam stripper treatment, 100 mg/L of silicone antifoaming agent was added to the effluent waste stream.

Phase I of the surfactant flood consisted of a 0.6-day injection of 7.7 wt% sodium dihexyl sulfosuccinate (Aerosol® MA) at two gallons per minute (gpm) through the central injection well, followed by 8.5 days of water flooding. During surfactant flushing, total volatile organic compound (VOC) concentrations in the extraction wells increased from approximately 750 mg/L to 15,000 mg/L, with an estimated 78 gallons of DNAPL removed due to solubilization. Approximately 107 gallons of DNAPL-free product were also recovered, resulting in a total of 185 gallons of DNAPL removed during Phase I flushing. High surfactant and tracer recovery, coupled with drawdown evident in the surrounding injection wells indicated hydraulic control of the swept area, consistent with model simulations.

Following the successful completion of Phase I, the remaining DNAPL present at the site was characterized by a second PITT prior to Phase II surfactant flooding, which was then followed by flushing with 7,000 mg/L NaCl for 1 day. The salt pre-flood was completed to ensure that the aquifer salinity was raised to the optimum value for the formation of the microemulsion to achieve maximum contaminant recovery. Following NaCl flooding, a mixture containing 7.55 wt% Aerosol® MA, 4.47% isopropanol, and 7,000 mg/L NaCl was injected at 2.5 gpm through all three injection wells for a period of 3.4 days. This surfactant formulation was chosen because of its ability to reduce the DNAPL-microemulsion IFT to 0.02 dyne/cm and to increase the solubility of the three principal chlorinated constituents to 620,000 mg/L.

During Phase II surfactant flushing, total VOC concentrations increased from approximately 750 mg/L to a maximum of 39,000 mg/L, although very little free product was recovered. Flux-averaged effluent concentrations of TCA, TCE, and PCE from the monitoring well located in the center of the injection and extraction wells are given in Figure 11.9.

The higher VOC concentrations observed during surfactant flooding, combined with the minimal mass of free DNAPL recovered, indicated that the primary mechanism for enhanced DNAPL removal was enhanced solubility due to the formation of an oil-in-water microemulsion. After injecting with 2.4 PVs of surfactant over the course of 3.4 days, the treatment zone was flushed with water for an additional 11 days. Comparison of the pre- and post-PITT

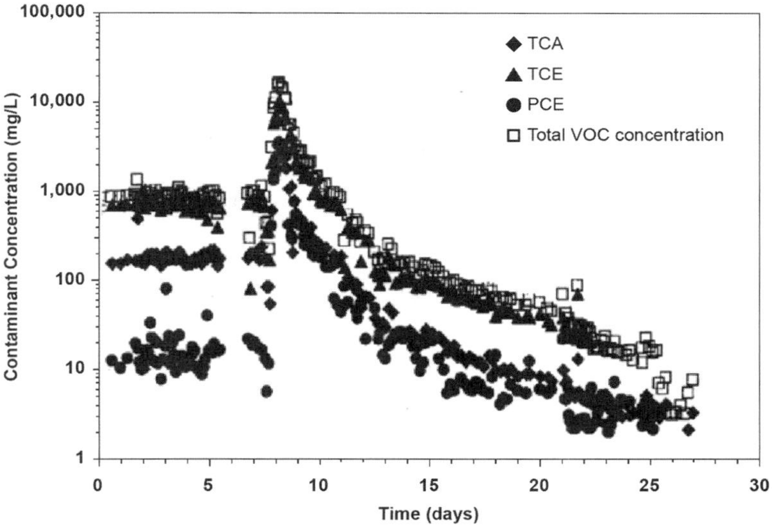

Figure 11.9. Measured VOC concentrations at monitoring well (MW) 1 during Phase II. Adapted with permission from Londergan et al. (2001).

indicated that 143 gallons of DNAPL were removed from the aquifer during Phase II, thereby reducing the final averaged DNAPL saturation to less than 1%.

Based on the contaminant source zone characterization pre- and post-remediation, integrated effluent concentrations and mass recovered in the effluent steam stripper, between 341 and 363 gallons of DNAPL were removed during Phase I and II surfactant flooding. Comparison of the pre- and post-remediation PITTs indicated that the averaged DNAPL saturation was reduced from 2.7% to 0.035%, indicating that approximately five gallons (19 L) of DNAPL remained in the source zone area. Total VOC concentrations measured during post-remediation monitoring were less than 10 mg/L, two orders of magnitude less than the 1,000 mg/L level measured during the initial water flooding. Based on the simplified dissolution model of Johnson (1992) and Johnson and Pankow (1992), the estimated total time for complete dissolution of the remaining DNAPL at an induced groundwater velocity of 8 m/day (26 ft/day) would be less than 1 year. By incorporating the laboratory treatability tests for surfactant selection, extensive site characterization, and proper remediation design and implementation, this study demonstrated that surfactant flushing can effectively remove entrapped DNAPL from a heterogeneous alluvial aquifer formation.

Additional pilot and larger-scale applications of surfactant flushing were conducted within the Hill AFB OU2 area. In particular, surfactant foams were used to improve mobility control, which resulting in greater sweep efficiency of the treated zones (Hirasaki, et al., 1997; Szafranski et al., 1998). Here, a surfactant formulation consisting of 3.5% Aerosol® MA + 1.0% NaCl was introduced followed by air injection to create a foam bank. Measurements of mass discharge were conducted pre- and post-treatment of the OU2 area, using both passive flux meters and integral pumping tests. These studies provided independent confirmation that surfactant-based source zone treatment resulted in substantial reductions in contaminant discharge from the site, which was found to decrease from approximately 76 g/day to 5 g/day (TCE) 1 year after treatment.

11.3.2 Bachman Road Site, Oscoda, Michigan

A pilot-scale field demonstration was performed at the Bachman Road site in Oscoda, Michigan, USA to evaluate the ability of surfactant flushing to recover PCE-DNAPL beneath a former dry cleaning facility (Ramsburg et al., 2004a, 2005; Abriola et al., 2005). This site is both representative of small DNAPL source zones associated with dry cleaners and unique among previously documented surfactant flushing trials due to the unknown source release history and location in an active commercial and residential area.

The Michigan Department of Public Health first detected VOCs in water samples from private supply wells along Bachman Road in late 1979, and 14 monitoring wells were subsequently installed and sampled between 1985 and 1989 during a remedial investigation initiated by the Michigan Department of Environmental Quality (MDEQ). Data from the monitoring wells were used to delineate four distinct plumes of groundwater contamination in the area (Figure 11.10). The southern chlorinated ethene plume (Plume B, Figure 11.10) was relatively narrow (170 m at its widest point for the 5 μg/L contour) and appeared to emanate from the vicinity of a former dry cleaning facility, discharging into Lake Huron approximately 230 m downgradient from the source zone. The contaminated formation was composed of glacial outwash sand deposits with relatively low organic carbon content (0.02% wt) and a confining clay layer approximately 7.6 m below ground surface (bgs). Depth to groundwater varied seasonally between 2.4 m and 3.0 m, with groundwater flow generally west to east at an estimated rate of 0.13 m/d.

Injection of a bromide tracer (8,175 L, 1,000 mg/L) yielded a high mass recovery (85%) that agreed well with model predictions (~92%) and indicated that the mathematical model of the site effectively captured solute transport and hydraulic control within the treatment zone (Abriola et al., 2005). Although free-phase PCE was not detected, pretreatment aqueous PCE concentrations in excess of 80 mg/L and soil core concentrations in excess of 100 milligrams per kilogram (mg/kg) were measured at locations downgradient from the former dry cleaning facility. The spatial distribution DNAPL saturations were assessed through a PITT consisting of 790 mg/L 2-propanol, 855 mg/L 2,4-dimethyl-3-pentanol, 475 mg/L 2-octantol, and 210 mg/L 3,5,5-trimethyl-1-hexanol, injected at a constant rate of 1.9 L/min. Effluent breakthrough curves (BTCs) were analyzed at each multilevel sampling location within the swept area; however, a substantial number of samples were below analytical quantification limits and required extrapolation by fitting (nonlinear regression) an exponential function to the distal portion of the BTCs (Jin et al., 1995; Rao et al., 1997; Annable et al., 1998; Sillan, 1999; Cain et al., 2000). Average DNAPL saturations based on the exponential extrapolation technique ranged from 0.15 ± 0.02% to 0.74 ± 0.20% along the measured flow paths (Ramsburg et al., 2005).

Laboratory batch studies using site materials quantified the sorptive capacity of the silty and clay materials and confirmed the presence of a DNAPL source based on equilibrium-phase distribution calculations, with DNAPL saturations of up to 0.7% (Abriola et al., 2005). Throughout the initial investigation, which also included drive point sampling, data collected from soil and water samples indicated an absence of substantial concentrations of PCE transformation products (TCE, *cis*-dichloroethene [*cis*-DCE] and vinyl chloride [VC]). The favorable hydrogeology (relatively homogeneous, shallow, sandy formation underlain by clay) and low suspected DNAPL saturations made this site an ideal candidate for the application of surfactant flushing using micellar solubilization as the recovery mechanism.

The location of the test site beneath an occupied building presented a significant challenge for hydraulic capture design because the existing structure prohibited the use of a traditional line-drive flood in the direction of groundwater flow. To address these unique conditions, a suite of mathematical models was employed to design a surfactant delivery and extraction

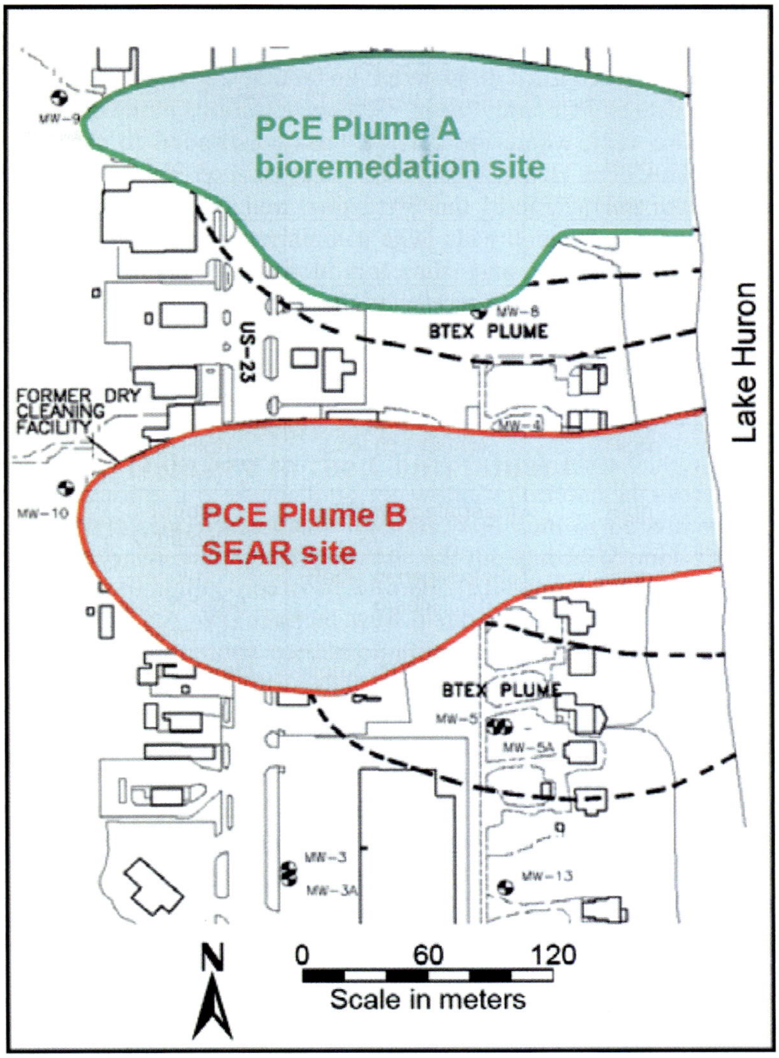

Figure 11.10. Estimated plume areas at the Bachman Road Site based on 1994 monitoring well data. Adapted with permission from Abriola et al. (2005). Copyright (2005) American Chemical Society.

system to flush a region (approximately 4.3 m × 6.7 m; 14 ft × 22 ft) beneath the northwest corner of the building (Abriola et al., 2005). The remediation monitoring systems incorporated five bundles of multilevel samplers (MLSs) that were installed beneath the building using angled boring methods for sampling the treatment zone. On the basis of laboratory treatability tests and cost analyses (Ramsburg and Pennell, 2001), the commercially available nonionic, food-grade surfactant Tween® 80 was selected for flushing (Abriola et al., 2005).

A solution of 45.8 millimolar (mM) (6 wt%) Tween® 80 was introduced over the entire aquifer thickness for 5 days using three laboratory size piston pumps at 0.19 L/min (Ramsburg et al., 2005). Surfactant injection continued for a second 5-day period through wells screened over the top and bottom 1.2 m of saturated depth. This design was used to target regions suspected to be highly contaminated, while reducing overall surfactant usage and cost

(Abriola et al., 2005). Following post-treatment water flushing (2 days), the extraction system continued to produce 19.7 L/minute (min) until day 31, at which time the flow rate was increased to 28.4 L/min to maximize capture of injected surfactant and solubilized PCE until day 47. Analytical measurements of PCE and Tween® 80 concentrations in the extraction well effluent are presented in Figure 11.11, where the vertical lines correspond to changes in the influent solution and/or flow rate.

Upon detecting the surfactant in the extraction well, the PCE concentration increased to a value of 44 mg/L, consistent with PCE solubilization. Concentrations of PCE in the effluent quickly became transient, suggesting a reduction in source mass. However, a second peak was evident near day 10 and corresponded to the targeted injection implemented from day 5 until day 10, before the decrease in PCE concentration observed during post-treatment water flooding (steps 3 and 4, Figure 11.11). Approximately 19 L of PCE were extracted during pilot-scale operations, which was separated from the effluent stream using two low-profile air strippers (Kibbey et al., 2001). The clean water entering the extraction well via its upstream radius of influence, and the remediation design, which used the injection of clean water to produce surfactant sweep across the natural groundwater gradient, accounts for the dilution of influent values (~60 g/L) compared to measured effluent surfactant concentrations (up to ~10 g/L). Successful hydraulic control throughout the test facilitated the recovery of 95% of the injected surfactant from the extraction well. Surfactant recovery was comparable to previous surfactant flushing demonstrations at Hill AFB OU2 (~94% surfactant recovery reported by Londergan et al., 2001) and consistent with pre-test modeling efforts (~95% reported by Abriola et al., 2005).

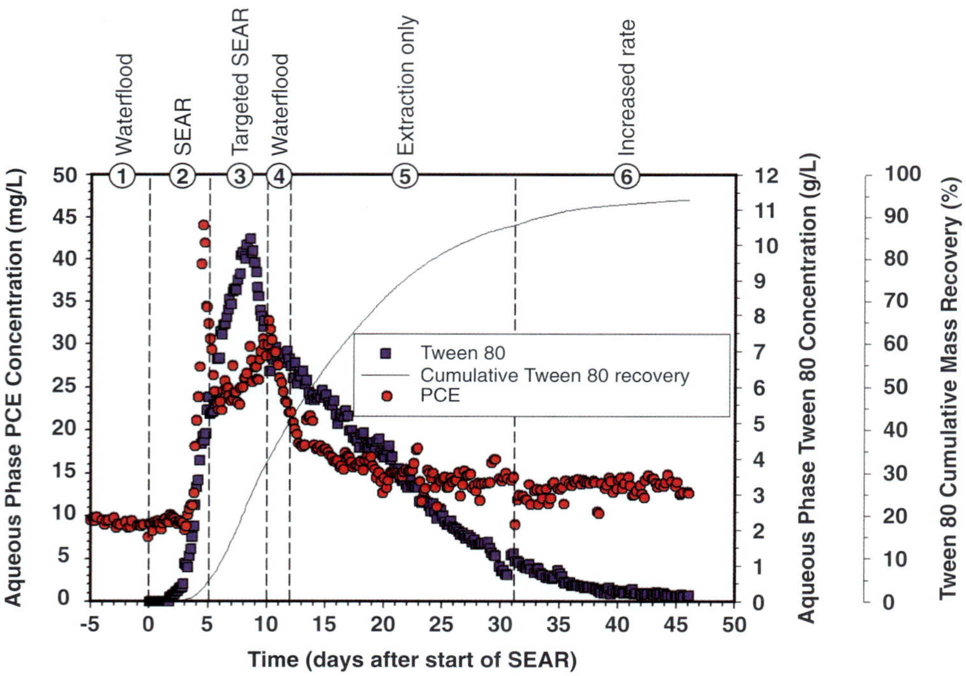

Figure 11.11. Concentrations of PCE and Tween® 80 measured in the extraction well. ***Vertical lines*** **represent a change to the influent solution or flow rate: 1. water flood; 2. injection of Tween 80 over entire aquifer thickness; 3. Tween 80 injection targeting suspected DNAPL; 4. water flood; 5. extraction only; 6. increased rate of extraction before pumping was terminated at day 47. Adapted with permission from Ramsburg et al. (2005). Copyright (2005) American Chemical Society. Note: SEAR– surfactant enhanced aquifer remediation.**

Post-treatment monitoring revealed that PCE concentrations were reduced by up to two orders of magnitude within the source zone, with no evidence of rebound after more than 450 days following the cessation of surfactant flushing. Elevated concentrations (more than 2 orders of magnitude greater than pretreatment values) of PCE dechlorination products (TCE and *cis*-DCE), as well as elevated levels of organic acids (acetate), were observed in monitoring wells 450 days following surfactant flushing (Ramsburg et al., 2004a). Fermentation of residual Tween® 80 (detected at concentrations of 50–2,750 mg/L 450 days after surfactant flushing) provided suitable electron donors(s) that stimulated native microbial dechlorination activity in the oligotrophic aquifer (Ramsburg et al., 2004a). These data supported implementation of a combined remediation strategy coupling aggressive mass removal by surfactant flushing and microbial reductive dechlorination to treat any contaminants remaining after flushing.

11.3.3 Surfactant Flushing Potential and Limitations

Although surfactant-based remediation strategies have shown great promise as an effective strategy to rapidly treat NAPL source zones, the technology is not without limitations. Perhaps the most critical issue, which impacts all technologies that involve *in situ* flushing of active ingredients, is the ability to deliver the surfactant solution throughout the treatment zone, and more importantly, to extract the mobilized and/or solubilized contaminant and surfactant. Another important limitation is that surfactants tend to disperse fine particles as a result of their emulsifying properties, which can lead to particle mobilization and pore clogging (Liu and Roy, 1995; Renshaw et al., 1997; Rao et al., 2006).

A considerable body of literature exists related to the sorption of surfactants on solids, which was expanded to include soils and aquifer materials during the 1990s. Since most natural minerals possess a net negative charge, interactions between a hydrophilic head group of a surfactant and mineral surfaces generally decrease for the different classes of surfactants in following order: cationic >> nonionic > anionic. At low surface coverage, surfactant monomers are thought to lie parallel to the mineral surface, and as the surfactant concentration increases, the monomers orient perpendicular to the surface to form bilayers (Rosen, 1989; Pennell et al., 1996). In general, however, surfactants exhibit a limiting or maximum sorption capacity for mineral surfaces, and the resulting isotherms can be described using a Langmuir sorption model (Adeel and Luthy, 1995; Pennell et al., 1996).

The salient point is that a limiting or maximum sorption capacity exists in predominantly mineral solids, after which no additional net loss of surfactant to the solid phase will occur. As an example, consider a 125 m^3 volume treatment zone that will be flushed with five PVs of a 4% surfactant solution. Assuming a porosity of 0.35 and a soil bulk density of 1.70 g/cm^3, the pore volume would be approximately 44 m^3 or 44,000 L, and the mass of solids would be 212,500 kg. If the surfactant cost of $3/lb and the Langmuir maximum sorption capacity (S_{max}) was 0.3 mg/g, approximately 64 kg of surfactant would be adsorbed at a cost of $420. However, the cost of surfactant (9,450 kg) for flushing the treatment zone would be approximately $62,000; thus, the estimated cost of losses due to sorption represents less than 1% of the total surfactant costs and would decrease if more surfactant was applied.

In addition to sorption on the solid phase, surfactants may partition strongly into NAPLs. Such partitioning has been shown to result in the formation of viscous emulsions that can reduce soil conductivity (Zimmerman et al., 1999; Jain and Demond, 2002). In practice, potential losses of surfactant due to sorption and partitioning should be characterized through laboratory and/or pilot-scale treatability tests conducted prior to full-scale field implementation and serve to highlight the need for careful selection of a surfactant formulation that is both effective for contaminant recovery and compatible with the hydrogeology of the site.

As with all aggressive source zone remediation technologies, surfactant flushing entails significant upfront costs, which are largely driven by the cost of the surfactant and the effluent treatment system. To minimize surfactant costs, it is advisable to consider surfactants that are used in other commercial applications, such as food products, detergents, and pharmaceuticals. Two such types of surfactants are the nonionic sorbitan ethoxylates such as Tween® and the anionic sulfosuccinates (Aerosol®), which typically cost on the order of $2–4/lb (Ramsburg and Pennell, 2001) and are widely used in whipped toppings and other food products. These surfactants have the added benefit of relatively low toxicity and biodegradability, an important consideration for subsurface applications. In contrast, specialty surfactants can cost from $10–40/lb, and formulations that include alcohols, salts, and other organics can substantially increase material costs.

Treatment of the effluent waste stream is critical to the success of any *in situ* flushing strategy, and thus it is critical to address operational issues and costs associated with aboveground separation, treatment, and reuse of both the surfactant and contaminants (Krebs-Yuill et al., 1995; Ramsburg and Pennell, 2001). As a consequence, it is important to consider surfactant formulation and flushing options that will minimize the injected volume and amount (concentration) of surfactant required to achieve remediation goals. In the future, surfactant flushing is most likely to be applied to sites with (1) well-defined source zones that have minimal heterogeneity and relatively high permeability, (2) LNAPL contaminants that can be effectively treated with several PVs of solution using low-IFT surfactant formulations to achieve mobilization, and (3) in combination with other technologies to improve performance (for example, chemical oxidation or bioremediation).

11.4 COSOLVENT FLUSHING CASE STUDIES

Laboratory studies (Imhoff et al., 1995; Lunn and Kueper, 1997) have demonstrated the effectiveness of cosolvent flushing as an NAPL source zone remediation technology. Lunn and Kueper (1997) reported efficient recovery of more than 90% of the PCE initially present in a two-dimensional aquifer cell during flushing with a 90% ethanol cosolvent formulation. Effluent PCE concentrations of nearly 100,000 mg/L were achieved and DNAPL mobilization was not observed. However, this study also demonstrated that all alcohols do not behave in a similar manner. For example, when using 90% 1-propanol for flushing, PCE recovery remained high (up to 96%), but the primary contaminant recovery mechanism was PCE mobilization.

In the field, several successful pilot- and full-scale demonstrations of cosolvent flushing have been conducted, most notably at Hill AFB, Dover AFB, and the former Sages dry cleaner site (Table 11.5). At Hill AFB OU1, two field demonstrations of cosolvent flushing were conducted, which recovered approximately 80% and 85% of the NAPL mass from separate 3 m × 5 m (10 ft × 16 ft) test cells that were vertically confined by interlocking sheet pile walls (Rao et al., 1997; Falta et al., 1999). In a subsequent field demonstration of cosolvent flushing conducted at the Dover AFB, a known amount of PCE was released into a test cell that was then flushed with a 70% ethanol solution, which recovered approximately 64% of the initial PCE mass (Rao et al., 1997). At the Sages dry cleaner site, which is discussed in detail below, injection of a 95% ethanol solution achieved similar PCE mass recovery (62–65%) without test cell confinement, while downgradient aqueous PCE concentrations were reduced by up to 92% following cosolvent flushing (Jawitz et al., 2000).

Several important technical advances were realized in these studies, most notably the use of gradient injection (increasing the concentration of cosolvent delivered with time) to avoid

Table 11.5. Representative Field-Scale Demonstrations of Cosolvent Flushing. The case study in bold font is discussed in more detail below.

Field Site	Cosolvent	DNAPL	Amount Recovered (Estimated Recovery)[a]	References
Dover AFB Test Cell, DE	70% Ethanol	PCE	53 L (64%)	Rao et al., 1997
Sages Dry Cleaner, Jacksonville, FL	**95% Ethanol**	**PCE**	**30 L (63%)**	**Jawitz et al., 2000**
Hill AFB OU1, UT	80% *t*-Butanol + 15% Hexanol	Jet fuel, chlorinated solvents	99 mg/kg[b] (80%)	Falta et al., 1999
Hill AFB OU1, UT	70% Ethanol + 12% Pentanol	Jet fuel, chlorinated solvents	~ 300 L TH[c] (85%)	Rao et al., 1997

[a]Mass recovery values are based on estimates of the initial mass present, except when the DNAPL was released intentionally (e.g., Dover AFB)
[b]Based on soil concentrations of total of monitored contaminants
[c]Based on total hydrocarbons (TH) monitored

density override effects (Rao et al., 1997), placement of injection wells to promote upward migration of the miscible effluent in unconfined systems (Jawitz et al., 2000), and the potential for enhanced contaminant biodegradation in the downgradient plume (Ramakrishnan et al., 2005). The latter two points are discussed below in a detailed description of the Sages dry cleaning site.

11.4.1 Sages Dry Cleaner Site

A full-scale demonstration of cosolvent flushing was completed at the abandoned PCE-contaminated Sages dry cleaning facility in Jacksonville, Florida, USA that was in operation from 1968 to 1973 and again from 1979 to 1989. The primary goal of this project was to determine whether *in situ* flushing with a cosolvent could be an effective means of remediation. Site geology was comprised of unconsolidated sediments, primarily well-sorted sand/gravel with the depth to water table approximately 3 m (10 ft) bgs. The average saturated hydraulic conductivity, estimated from slug test data, was about 6 m/day in the upper sand unit (0–9 m bgs) and about 3 m/day in the lower sand to silty sand zone (9–10.7 m bgs) (Jawitz et al., 2000). A thin clay layer (0.15–0.3 m thick) was detected at approximately 10.7 m bgs in most soil borings; however, the clay layer was judged to be discontinuous with very fine sand to silty fine sand observed beneath (Levine Fricke, 1998).

The PCE release history at this site was unknown. However, the suspected release location was a former floor-drain sump open directly to the subsurface. Concentrations of PCE in groundwater samples collected from existing monitoring wells during the initial site assessment ranged from 70 mg/L to 150 mg/L. About 15 cm of free-phase PCE were detected at the bottom (9.8 m bgs) of a former supply well. After recovery of approximately 100 mL of PCE from this well, subsequent site visits over a period of several months revealed that additional PCE (typically 10–20 mL) had entered the well. This finding indicated that, at some depth above the bottom of the well, DNAPL existed at high enough saturations to form a continuous, free-flowing phase.

Initial site characterization using direct-push cone penetrometer testing (CPT) technologies identified an oblong PCE source 7.9–9.4 m bgs in an area of approximately 7.3 m by 2.7 m (Figure 11.12) (Jawitz et al., 2000; Mravik et al., 2003). Numerical simulations using a homogeneous domain with a uniform thickness of 1.5 m and a saturated hydraulic conductivity of 6 m/day (equal to that measured in the upper sand unit, Figure 11.12) were performed by Sillan (1999). These simulations were used to select a well configuration consisting of six recovery wells (RWs) encircling three injection wells (IWs). The six recovery wells were located just outside the perimeter of the initial estimated horizontal extent of the PCE source zone. Further, seven MLSs were installed within the NAPL source zone at five sampling depths (8.1, 9.1, 9.4, 9.7, and 9.9 m bgs). Soil core samples collected during well installation indicated PCE concentrations as high as 40,000 mg/kg (Strbak, 2000).

The DNAPL source zone was further characterized before and after alcohol flushing using tracer tests (Jin et al., 1995; Annable et al., 1998; Sillan, 1999; Cain et al., 2000). Pre-pilot PITT and an interfacial tracer tests were conducted to estimate (1) the mass of DNAPL PCE present in the swept volume of the injection and extraction system; (2) the swept volume of the injection and recovery wells; and (3) the arrival time of alcohol to the recovery system. The tracers used in the pre-test tracer study included methanol (2,200 mg/L) as the non-partitioning tracer, and *n*-hexanol (820 mg/L), 2,4-dimethyl-3-pentanol (DMP, 440 mg/L), and 2-ethyl-1-hexanol (*e*-Hex, 440 mg/L) as the partitioning tracers. A tracer pulse of approximately 0.20 PVs (3.8 h) was delivered to the IWs during steady-state water flow. Data from these samples yielded BTCs that

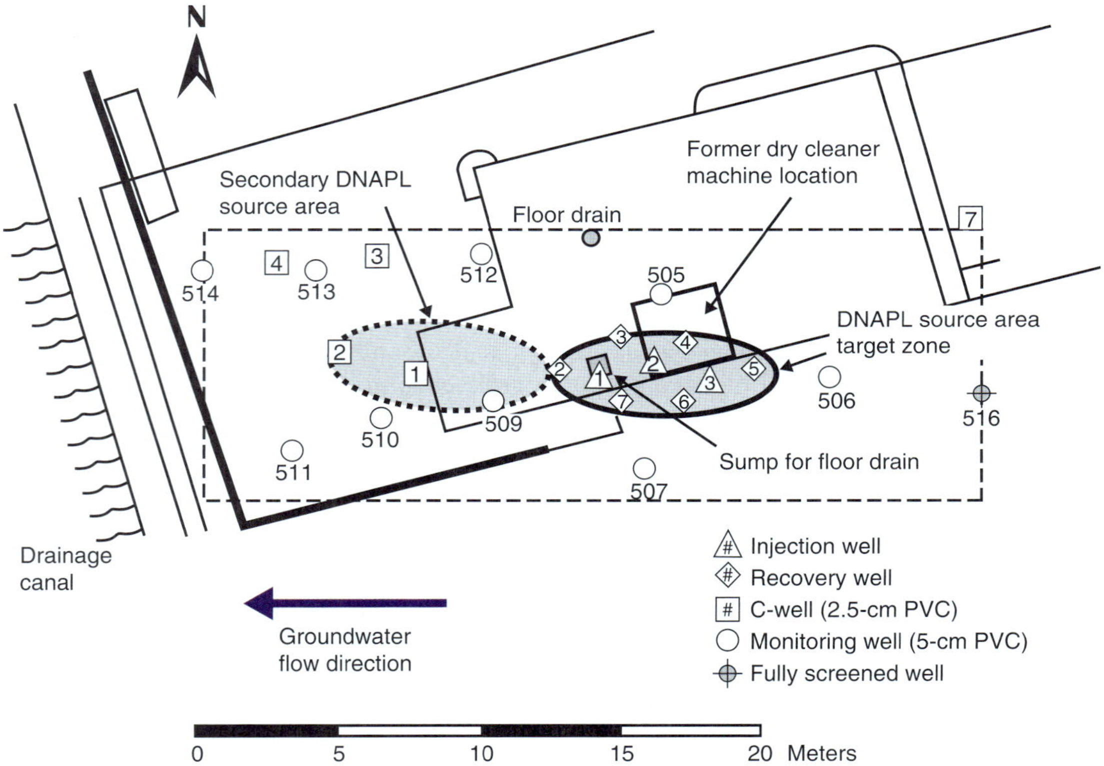

Figure 11.12. Former dry cleaner site map indicating former buildings, suspected areas of contamination, and well locations. (*Dashed line* indicates grid area used for generation of model contour plots.) Reprinted with permission from Mravik et al. (2003). Copyright (2003) American Chemical Society. Note: PVC – polyvinyl chloride.

were exponentially extrapolated when incomplete (Rao et al., 1997; Abriola et al., 2005; Ramsburg et al., 2005). Temporal moment analysis was applied and the average PCE saturation, S_N, within the swept volume of each recovery well was calculated (Sillan, 1999; Jawitz et al., 2000). Pre-test PITT results revealed that the largest quantities of PCE (>50% of the total) were found in the swept volumes of two wells (RW-3 and RW-7, Figure 11.12). These two wells were on opposite sides of the sump that was the suspected point of PCE entry into the subsurface. The total estimated volume of PCE within the swept zone of all wells was approximately 68 L, equivalent to an overall average S_N of 0.004.

After approximately 2 hours of flushing with only water, it was determined that a capture zone sufficient for the pilot test had been established. Over the next 3.5 days (4.5 PV), roughly 34,000 L of the cosolvent, consisting of 95% ethanol and 5% water, were injected to the three IWs. After completing alcohol injection, water flooding was initiated to remove the remaining ethanol and was continued for 4.5 days (i.e., an additional 5.8 PV). The extracted effluent was treated with a macro-porous polymer extraction (MPPE) system, which consisted of a column containing MPPE material into which the PCE will preferentially partition to separate the PCE from the cosolvent. The MPPE columns resulting from this treatment were then regenerated with low-pressure steam stripping, and the vaporized steam was condensed into free-phase PCE, which could then be disposed (Jawitz et al., 2000). Regulatory constraints prevented the reinjection of the ethanol/water mixture during this demonstration.

Ethanol and PCE BTCs from the six RWs were used to estimate enhanced dissolution and mobilization of PCE during cosolvent flushing (Levine Fricke, 1998; Jawitz et al., 2000). Of particular interest, were the two wells (RW-3 and RW-7) where a majority of the PCE mass was measured during the pre-test tracer. Based on the results of the pre-test PITT study, the highest estimated mass of DNAPL PCE in the swept volume was found at RW-7, (approximately 14 L), with PCE concentrations of approximately 15 mg/L measured during the initial water flushing stage. The PCE concentration in RW-7 peaked at approximately 1,300 mg/L (highest of all RWs), and ethanol concentration reached a plateau at approximately 25% of the injected concentration.

Note that dilution resulting from the 2:1 extraction to injection ratio at the RWs was manifested as reduced measured flux-averaged concentrations of ethanol and PCE compared to those in fluids arriving at the MLSs. As was seen at other RWs, it is believed that volume of alcohol flushed through this zone was not sufficient to remove all the accessible PCE-DNAPL. Similarly, prior to ethanol flushing, it was estimated that 11.6 L of PCE was encountered at RW-3 and the aqueous PCE concentration was approximately 40–50 mg/L (Figure 11.13). Ethanol breakthrough occurred after extracting approximately 8,000–10,000 L of groundwater from RW-3 (23–26 h after beginning the test).

After breakthrough, the PCE concentration increased rapidly to 1,000–1,200 mg/L (Figure 11.13). Although the ethanol concentration remained high, the PCE concentration began to decrease after a total volume pumped of 15,000 L. The PCE concentration did not decrease rapidly, indicating that heterogeneities exist in the aquifer. After extracting approximately 40,000 L of groundwater from this recovery well, the PCE concentration returned to the pre-breakthrough concentration of approximately 40–50 mg/L.

A post-test PITT study followed, using the same tracers and volumes described above for the pre-pilot injection. Determined from numerical integration of the BTCs, the total amount of PCE extracted from all wells was approximately 43 L, representing 63% of the initial mass, leaving a residual average S_N of 0.0014 (Jawitz et al., 2000). Substantially more PCE was extracted from RWs 3, 6, and 7 than from RWs 2, 4, and 5 (Figure 11.12), in agreement with the preflushing partitioning tracer estimates of the amount of PCE within the swept volume of each well.

Integration of the ethanol BTCs at all wells indicated that approximately 92% of the injected alcohol had been recovered by the end of the 5.8 PV water flood. Hydraulic

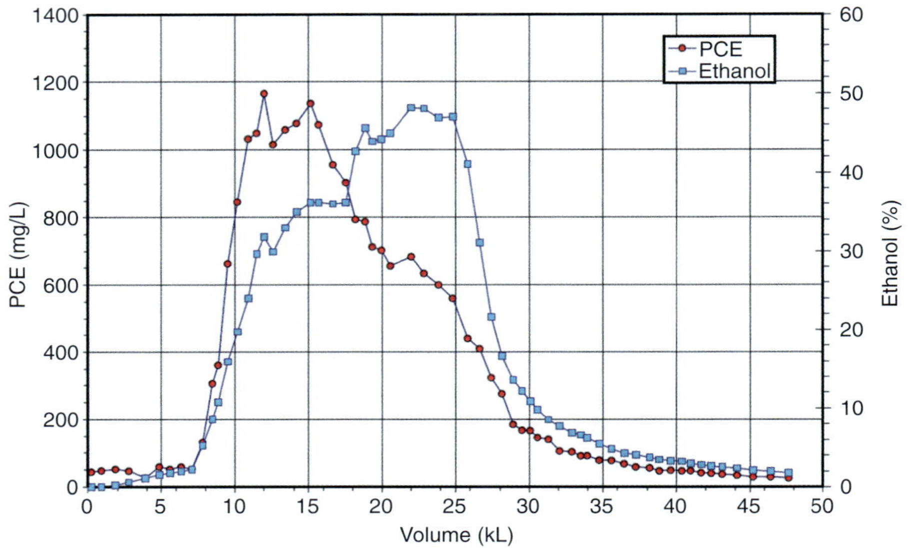

Figure 11.13. Concentrations of PCE and ethanol measured in RW3. Adapted with permission from Jawitz et al. (2000). Copyright (2000) American Chemical Society. Note: kL – killiliter(s).

containment in the source area was discontinued after the measured ethanol concentrations fell below 217 mM (1%), approximately 10 days after the cosolvent flushing, and the system was allowed to return to natural gradient conditions.

During post-treatment monitoring, assessment of residual biotreatment and enhanced dechlorination of PCE was conducted by evaluating groundwater monitoring results. The approximately 2.72 kiloliters (kL) of ethanol remaining in the subsurface was evaluated as a potential electron donor source for enhancement of biological processes in the source zone and downgradient areas during this phase. Initially, groundwater monitoring in IW and RW wells detected ethanol concentrations up to 250 mM (1.1%) through the first 2 years of monitoring. During the first year of sampling, PCE concentrations in all the RWs rebounded to near initial concentrations of 4–350 micromolar (μM) (0.7–58 mg/L) and remained steady at these levels. Recovery wells 3, 4, 6, and 7 were sampled throughout the entire post-flush monitoring; after approximately 1 year, concentrations began decreasing, and after 3 years, the average PCE concentration in the source zone was below 100 μM (16.6 mg/L). The decrease in PCE concentrations coincided with the decrease in ethanol to less than 100 mM (0.5%) (maximum concentration) by the third year. This observation was in agreement with microcosms conducted with site materials demonstrating that concentrations of ethanol must be decrease below 1% for microbial metabolism (Ramakrishnan et al., 2005).

The fermentation of ethanol led to the formation of potential electron donors, acetate and hydrogen, and subsequent enhancement of multiple microbial processes. As acetate and hydrogen increased, methane increased and sulfate concentrations decreased (i.e., sulfate-reduction increased) simultaneously with the decrease in PCE. Quarterly groundwater monitoring over the 3-year period showed decreasing concentrations of PCE in the source zone from initial values of 4 to 350 μM to less than 150 μM during the last sampling event. Initially there was little to no daughter product formation in the source zone, but after 3 years, measured concentrations were 242 μM for *cis*-1,2-dichloroethene, 13 μM for VC, and 0.43 μM for ethene. Primers and probes targeting *Dehalococcoides mccartyi* 16S rRNA genes verified the presence of these obligatory hydrogenotrophic PCE-to-ethene organisms that were supplied electron donor (hydrogen) as a result of microbial fermentation of ethanol (Mravik et al., 2003; Ramakrishnan et al., 2005). The measured changes in geochemistry and the formation of chloroethene daughter products (i.e., removal of PCE parent compound) indicate that *in situ* biotransformations were enhanced in areas exposed to the residual ethanol.

11.4.2 Cosolvent Flushing Implementation and Limitations

Although cosolvent flushing has shown promise as an effective strategy for NAPL source zone remediation, the technology faces several challenges that have limited implementation on a broader scale. The first issue is material cost; the injected cosolvent solution is typically on the order of 80–95% active ingredient, which is necessary to achieve performance on par with that of typical surfactant formulations (5% active ingredient). Even when ethanol is purchased without federal and state alcohol consumption taxes, the cost for cosolvent alone is still approximately $2/gallon and can rise to $5/gallon because of its demand as an alternative fuel.

The second issue is the density of the concentrated cosolvent solution, which is typically 20% less than that of the resident pore fluid (water). This large density contrast can result in marked *density override* effects, in which the injected solution flows over the resident pore water and treatment zone. The reverse effect occurs when attempting to recover post-flood alcohol, in which the injected water flows under the cosolvent solution which can become stagnated near the top of the water table and in the capillary fringe. To address this issue, a number of approaches have been implemented, including gradient injection (increasing alcohol levels over time), selective well screening and aggressive extraction pumping (Rao et al., 1997; Jawitz et al., 2000).

In addition to these implementation issues, the use of highly concentrated cosolvent solutions (>90% cosolvent) raises important safety concerns about the flammability of the injected cosolvent solution and the extracted effluent stream. Thus, despite some documented successes, the complexity, cost, and safety concerns related to cosolvent flushing have limited adoption of this technology for source zone treatment. In the future, cosolvents are likely to receive the most attention in the remediation field as low-concentration additives to enhance the effectiveness and performance of other technologies (surfactant flushing, chemical oxidation, and as electron donors for bioremediation).

11.5 COMBINED REMEDIATION STRATEGIES FOR *IN SITU* FLUSHING

While the field studies discussed in Sections 11.3 and 11.4 demonstrate that reductions in source zone mass can approach 90%, it is now recognized that physical–chemical remediation technologies, including surfactant and cosolvent flushing, will not completely remove all of the contaminant mass from DNAPL source zones (Freeze and McWhorter, 1997; Stroo et al., 2003, 2012). In fact, source zone treatments with *in situ* flushing are likely to increase the mobility and distribution of the residual mass, which may lead to increased aqueous phase concentrations in the short term. Following *in situ* flushing, not only will residual mass remain, but active ingredients employed in flushing solutions that are not completely extracted from the swept volume may also persist in lower permeability zones and sorbed by the solid phase (Abriola et al., 2005). Thus, post-treatment source zones are likely to contain substantial amounts of residual DNAPL contamination as well as flushing solution constituents.

Due to the inefficiencies of the stand-alone technologies, attention has shifted toward developing complementary *in situ* remediation approaches that can be combined at the same time (in parallel) or sequentially (in series) to more efficiently treat contaminant source zones (Ramsburg et al., 2004a; Christ et al., 2005; Friis et al., 2006; Amos et al., 2007;

Costanza et al., 2009). In a sequential or "*treatment train*" approach, an aggressive *in situ* treatment technology, such as cosolvent or surfactant flushing, is used to remove significant contaminant mass in a relatively short time-frame, while a second "polishing" technology (often microbial reductive dechlorination) is then applied to remove or detoxify the remaining contaminant mass. Such sequential remediation strategies have the potential to take advantage of efficient mass removal achieved by aggressive treatment technologies, while addressing limitations associated an individual technology (e.g., flow bypassing) that lead to incomplete mass removal.

Although the potential impacts of aggressive *in situ* treatment technologies on the effectiveness of combined remedies, and long-term plume behavior have not been sufficiently explored (Stroo et al., 2003), the pilot-scale studies conducted at the Bachman Road (Section 11.3.2) and Sages (Section 11.4.1) dry cleaning sites provide evidence that surfactant flushing and cosolvent flushing can be successfully combined with reductive dechlorination to treat residual contaminant mass (Ramsburg et al., 2004a; Ramakrishnan et al., 2005). In both studies, residual flushing solution, ethanol or Tween® 80, was fermented by native microbial communities to form potential electron donors to stimulate microbial reductive dechlorination (at minimum detoxification to *cis*-DCE). The Bachman Road data were supported by laboratory-based studies that further defined the potential benefits of Tween® 80 on reductive dechlorination (Amos et al., 2007). However, selection of compatible surfactants and appropriate concentrations is critical due to potential surfactant toxicity or inhibition toward of microbial populations relevant to desired degradation pathway (Yeh et al., 1999; McGuire and Hughes, 2003; Ramsburg et al., 2004a).

The potential benefits of surfactant flushing coupled with subsequent bioremediation were evaluated by Christ et al. (2005). Results of these model predictions clearly demonstrate that bioremediation alone will provide only minimal benefits when compared to natural gradient dissolution, while aggressive treatment of the source zone with surfactant enhanced aquifer remediation (SEAR) followed by bioremediation can dramatically increase the rate of mass removal and reduce source longevity by several orders of magnitude (Figure 11.14).

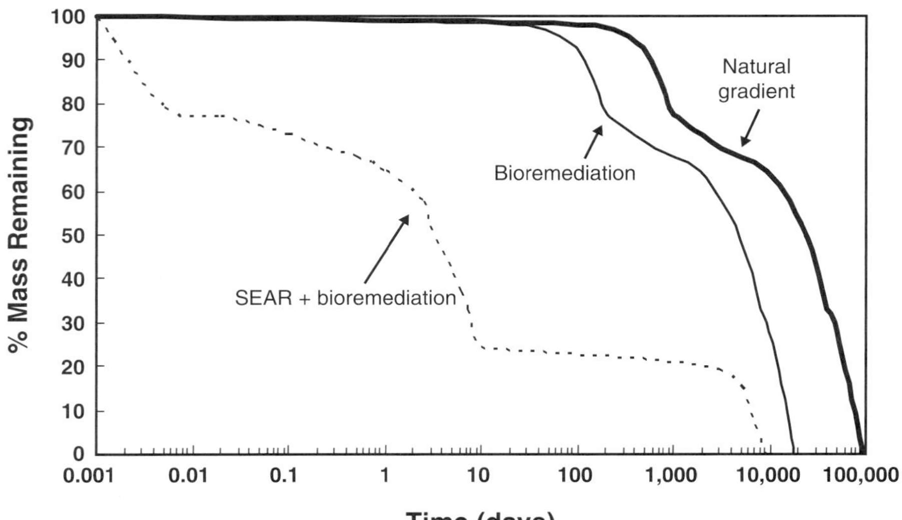

Figure 11.14. Percent DNAPL mass remaining as a function of time for three alternative remediation strategies; natural gradient dissolution alone; bioremediation alone; and surfactant flushing (4% Tween® 80) followed by bioremediation. Reproduced with permission from Environmental Health Perspectives (Christ et al., 2005).

Although surfactant and cosolvent flushing have shown utility for removing contaminant mass from the subsurface, both technologies require aboveground treatment of the effluent waste stream. For this reason, it is advantageous to couple recovery-based flushing technologies with those that promote *in situ* destruction of the contaminant. In addition to bioremediation, which was discussed above primarily in the treatment train or sequential treatment scenarios, combining surfactants or cosolvents directly with *in situ* chemical oxidation (ISCO) has received considerable research attention in the past few years (Dugan et al., 2010). This interest stems from the fact that ISCO is most effective for dissolved-phase contaminants and both surfactants and cosolvents have the potential to dramatically increase contaminant mass transfer from the DNAPL and sorbed phase to the aqueous phase. To date, at least four U.S. patents have been issued related to the injection of coupled ISCO technologies (Cowdery et al., 2004; Shiau, 2008; Dugan et al., 2009; Hoag and Collins, 2011). A number of laboratory-scale batch reactor and column studies have been performed to test surfactant or cosolvents with chemical oxidants, using either sequential or combined treatment strategies (Zhai et al., 2006; Li and Hanlie, 2008; Tsai et al., 2009). In general, these studies demonstrated a modest enhancement in oxidation of dissolved-phase chlorinated solvents in the presence of surfactant or cosolvent. However, additional laboratory and field-based research studies are needed to further understand and optimize ISCO when combined with surfactants (S-ISCO) or cosolvents (C-ISCO).

11.6 SUMMARY AND CONCLUSIONS

Over the past two decades, considerable resources have been directed toward the development and demonstration of NAPL source zone remediation technologies, including surfactant and cosolvent flushing. Although these technologies provide distinct advantages and limitations when compared to traditional pump-and-treat approaches, considerable uncertainty remains regarding the effectiveness and long-term benefits of these aggressive source zone treatment methods. The uncertainty results mainly from the difficulties in both delivering fluids effectively in the subsurface and subsequently recovering and treating the contaminant mass and injected solution.

At the bench scale, these technologies have shown promise for reducing the overall remediation time while greatly improving contaminant mass recovery compared to dissolution alone (pump-and-treat). To date, the field-scale results have been less encouraging, in large part because the success of *in situ* technologies in the field is governed by subsurface hydrogeology, in particular the heterogeneity of the aquifer materials and the complexity of contaminant mass distribution. A number of early field trials yielded mixed results and failed to reach desired clean up goals (Fountain et al., 1996; AATDF, 1997; Smith et al., 1997). However, as the limitations were recognized and refinements were implemented, improved results were achieved (AATDF, 1997; Jawitz et al., 1998). For example, the benefits of utilizing surfactant foams to improve flushing effectiveness in heterogeneous media were demonstrated at Hill AFB OU2 (Hirasaki et al., 2000; Meinardus et al., 2002).

Under ideal subsurface and operating conditions, *in situ* surfactant and cosolvent flushing have been shown to consistently recover greater than 80–90% of the contaminant mass. As the extent and complexity of a contaminated source zone increases, however, mass recovery rates on the order of 50–70% are more realistic. Nevertheless, such reductions in source zone contaminant mass are likely to reduce substantially source longevity and may greatly reduce contaminant mass flux emanating from the treated source zone, thereby decreasing the risks of environmental and health effects (USEPA, 2003). In addition, recent efforts to combine aggressive, short-term flushing technologies with low impact, long-term strategies (bioremediation) hold particular promise as a means to reduce treatment cost and source longevity and to achieve remedial goals.

Despite substantial advances in the technology and successful field applications, surfactant and cosolvent flushing have received minimal attention as source zone remediation technology during the past 10 years. Several factors may have contributed to this outcome: (1) relatively high active ingredient costs; (2) the complexity of implementation compared to alternative technologies; (3) reduced performance in low permeability, heterogeneous media; and (4) the costs and difficulties involved in treating the effluent waste stream. However, it is likely that the use of surfactants and cosolvents for source zone remediation will increase in the future, particularly at LNAPL sites, and as additives to improve delivery of other reagents (oxidants) or as one step in a treatment train (flushing + bioremediation) designed to enhance the overall remedy performance and achieve cleanup targets.

REFERENCES

Advanced Applied Technology Demonstration Facility (AATDF). 1997. Technology Practices Manual for Surfactants and Cosolvents. Houston, TX, USA. http://www.clu-in.org/PRODUCTS/AATDF/toc.htm. Accessed September 16, 2013.

Abdul AS, Ang CC. 1994. In situ surfactant washing of polychlorinated biphenyls and oils from a contaminated field site: Phase II pilot study. Ground Water 32:727–734.

Abdul AS, Gibson TL, Ang CC, Smith JC, Sobczynski RE. 1992. In situ surfactant washing of polychlorinated biphenyls and oils from a contaminated site. Ground Water 30:219–231.

Abrams DS, Prausnitz JM. 1975. Statistical thermodynamics of liquid mixtures: A new expression for the excess gibbs energy of partly or completely miscible systems. AIChE J 21:116–128.

Abriola LM, Lang JR, Rathfelder K. 1997. Michigan soil-vapor extraction remediation (miser) model - a computer program to model bioventing of organic chemicals in unsaturated geological material. EPA/600R-97/099. U.S. Environmental Protection Agency, Ada, OK, USA.

Abriola LM, Drummond CD, Hahn EJ, Hayes KF, Kibbey TCG, Lemke LD, Pennell KD, Petrovskis EA, Ramsburg CA, Rathfelder KM. 2005. Pilot-scale demonstration of surfactant-enhanced pce solubilization at the bachman road site. 1. Site characterization and test design. Environ Sci Technol 39:1778–1790.

Adeel Z, Luthy RG. 1995. Sorption and transport kinetics of a nonionic surfactant through an aquifer sediment. Environ Sci Technol 29:1032–1042.

Amos BK, Daprato RC, Hughes JB, Pennell KD, Löffler FE. 2007. Effects of the nonionic surfactant tween 80 on microbial reductive dechlorination of chlorinated ethenes. Environ Sci Technol 41:1710–1716.

Annable MD, Rao PSC, Hatfield K, Graham WD, Wood AL, Enfield CG. 1998. Partitioning tracers for measuring residual NAPL: Field-scale test results. J Environ Eng 124:498–503.

Baran JR, Pope GA, Wade WH, Weerasooriya V. 1994. Phase behavior of water/perchloroethylene/anionic surfactant systems. Langmuir 10:1146–1150.

Betts PC, Aurelius M, Ashburn A, Curtis M. 1998. Design, construction and initial operation of the source recovery system. Proceedings, 1st International Conference on Remediation of Chlorinated and Recalcitrant Compounds. In Wickramanayake GB, Hinchee RE, Nonaqueous Phase Liquids: Remediation of Chlorinate and Recalcitrant Compounds. Battelle Press, Columbus, OH, USA, Vol C1-2, pp 187–192.

Boyd SA, Mortland MM, Chiou CT. 1988. Sorption characteristics of organic compounds on hexadecyltrimethylammonium-smectite. Soil Sci Soc Am J 52:652–657.

Brooks MC, Annable MD, Rao PSC, Hatfield K, Jawitz JW, Wise WR, Wood AL, Enfield CG. 2004. Controlled release, blind test of DNAPL remediation by ethanol flushing. J Contam Hydrol 69:281–297.

Brown CL, Delshad M, Dwarakanath V, Jackson RE, Londergan JT, Meinardus HW, McKinney DC, Oolman T, Pope GA, Wade WH. 1999. Demonstration of surfactant flooding of an alluvial aquifer contaminated with dense nonaqueous phase liquids. In Brusseau ML, Sabatini DA, Gierke JS, Annable ME, eds, Field Testing of Innovative Subsurface Remediation Technologies. ACS Symposium Series. American Chemical Society, Washington, DC, USA, Chapter 6, pp 64–85.

Brusseau ML, Wood AL, Rao PSC. 1991. Influence of organic solvents on the sorption kinetics of hydrophobic organic contaminants. Environ Sci Technol 25:903–910.

Cain RB, Johnson GR, McCray JE, Blanford WJ, Brusseau ML. 2000. Partitioning tracer tests for evaluating remediation performance. Ground Water 38:752–761.

Childs J, Acosta E, Annable MD, Brooks MC, Enfield CG, Harwell JH, Hasegawa M, Knox RC, Rao PSC, Sabatini DA, Shiau B, Szekeres E, Wood AL. 2006. Field demonstration of surfactant-enhanced solubilization of DNAPL at Dover Air Force Base, Delaware. J Contam Hydrol 82:1–22.

Christ JA, Ramsburg CA, Loeffler FE, Pennell KD, Abriola LM. 2005. Coupling aggressive mass removal with microbial reductive dechlorination for remediation of DNAPL source zones - a review and assessment. Environ Health Perspect 113:465–477.

Cinar Y, Orr FM. 2005. Measurement of three-phase relative permeability with IFT variation. SPE Res Eval Eng 8:33–43.

Claesson P. 1992. Poly(ethylene oxide) surface coatings: Relations between intermolecular forces, layer structure and protein repellency. Colloid Surface A 77:109–118.

Costanza J, Flethcher KE, Löffler FE, Pennell KD. 2009. Fate of tce in heated Fort Lewis soil. Environ Sci Technol 43 909–914.

Cowdery RC, Applegate JL Warner KM. 2004. Co-oxidation method and co-oxidation reagent for decontamination of soil and groundwater. U.S. Patent 6869535.

Delshad M, Pope GA, Yeh L, Holzmer FJ. 2000. Design of the surfactant flood at Camp Lejeune. In Wickramanayake GB, Gavaskar AR, Gupta N, eds, Treating Dense Nonaqueous-Phase Liquids (DNAPLs): Remediation of Chlorinated and Recalcitrant Compounds. Battelle Press, Columbus, OH, USA, pp 203–210.

Deshpande S, Shiau BJ, Wade D, Sabatini DA, Harwell JH. 1999. Surfactant selection for enhancing ex situ soil washing. Water Res 33:351–360.

Drummond CJ, Fong C. 2000. Surfactant self-assembly objects as novel drug delivery vehicles. Curr Opin Colloid Interface Sci 4:449–456.

Dugan PJ, Siegrest RL, Crimi ML. 2009. Method and compositions for treatment of subsurface contaminants. U.S. Patent 7553105.

Dugan PJ, Siegrist RL, Crimi ML. 2010. Coupling surfactants/cosolvents with oxidants for enhanced DNAPL removal: A review. Remediat J 20:27–49.

Dwarakanath V, Kostarelos K, Pope GA, Shotts D, Wade WH. 1999. Anionic surfactant remediation of soil columns contaminated by nonaqueous phase liquids. J Contam Hydrol 38:465–488.

Edwards DA, Luthy RG, Liu Z. 1991. Solubilization of polycyclic aromatic hydrocarbons in micellar nonionic surfactant solutions. Environ Sci Technol 25:127–133.

Falta RW. 1998. Using phase diagrams to redict the performance of cosolvent floods for NAPL remediation. Ground Water Monit Remediat 18:94–102.

Falta RW, Lee CM, Brame SE, Roeder E, Coates JT, Wright C, Wood AL, Enfield CG. 1999. Field test of high molecular weight alcohol flushing for subsurface nonaqueous phase liquid remediation. Water Resour Res 35:2095–2108.

Fountain JC, Starr RC, Middleton T, Beikirch M, Taylor C, Hodge D. 1996. A controlled field test of surfactant-enhanced aquifer remediation. Ground Water 34:910–916.

Freeze RA, McWhorter DB. 1997. A framework for assessing risk reduction due to DNAPL mass removal from low-permeability soils. Ground Water 35:111–123.

Friis AK, Albrechtsen HJ, Cox E, Bjerg PL. 2006. The need for bioaugmentation after thermal treatment of a tce-contaminated aquifer: Laboratory experiments. J Contam Hydrol 88:235–248.

Graciaa A, Lachaise J, Cucuphat C, Bourrell M, Salager JL. 1993. Interfacial segregation of an ethyl oleate/hexadecane oil mixture in microemulsion systems. Langmuir 9:1473–1478.

Gray MA, Metcalfe CD. 1997. Induction of testis-ova in japanese medaka (*oryzias latipes*) exposed to *p*-nonylphenol. Environ Toxicol Chem 16:1082–1086.

Haggerty GM, Bowman RS. 1994. Sorption of chromate and other inorganic anions by organozeolite. Environ Sci Technol 28:452–458.

Hasegawa MH, Shau BJ, Sabatini DA, Knox RC, Harwell JH, Lago R, Yeh L. 2000. Surfactant-enhanced subsurface remediation of DNAPLs at the former Naval Air Station Alameda, California. In Wickramanayake GB, Gavaskar AR, Gupta N, eds, Treating Dense Nonaqueous-phase Liquids (DNAPLs): Remediation of Chlorinated and Recalcitrant Compounds. Battelle Press, Columbus, OH, USA, pp 219–226.

Heath WO, Richardson RL, Goheen SC. 1994. Heating solid earthen material, measuring moisture and resistivity. U.S. Patent 5330291.

Hiemenz PC, Rajagopalan R. 1997. Principles of Colloid and Surface Chemistry. CRC Press, Boca Raton, FL, USA.

Hirasaki GJ, Miller CA, Szafranski R, Tanzil D, Lawson JB, Meinardus H, Jin M, Londergan JT, Jackson RE. 1997. Field demonstration of the surfactant/foam process for aquifer remediation. Proceedings, SPE Annual Technical Conference and Exhibition, October 5–8. San Antonio, TX, USA, 39292-MS.

Hirasaki GJ, Jackson RE, Jin M, Lawson JB, Londergan J, Meinardus H, Miller CA, Pope GA, Szafranski R, Tanzil D. 2000. Field demonstration of the surfactant/foam process for remediation of a heterogeneous aquifer contaminated with DNAPL. In Fiorenza S, Oubre CL, Ward CH, eds, NAPL Removal: Surfactants, Foams, and Microemulsions. Lewis Publishers, Boca Raton, FL, USA, pp 1–166.

Hoag GE, Collins J. 2011. Soil remediation method and composition. U.S. Patent 7976241.

Holm LW, Csaszar AK. 1962. Oil recovery by solvents mutually soluble in oil and water. Soc Petrol Eng J 2:129–144.

Holzmer FJ, Pope GA, Yeh L. 2000. Surfactant-enhanced aquifer remediation of PCE DNAPL in low-permeability sand. In Wickramanayake GB, Gavaskar AR, Gupta N, eds, Treating Dense Nonaqueous-Phase Liquids (DNAPLs): Remediation of Chlorinated and Recalcitrant Compounds. Battelle Press, Columbus, OH, USA, pp 211–218.

Huh C. 1979. Interfacial-tensions and solubilizing ability of a microemulsion phase that coexists with oil and brine. J Colloid Interf Sci 71:408–426.

Imhoff PT, Gleyzer SN, McBride JF, Vancho LA, Okuda I, Miller CT. 1995. Cosolvent-enhanced remediation of residual dense nonaqueous phase liquids: Experimental investigation. Environ Sci Technol 29:1966–1976.

Jackson RE, Dwarakanath V. 1999. Chlorinated decreasing solvents: Physical-chemical properties affecting aquifer contamination and remediation. Ground Water Monit Remediat 19:102–110.

Jafvert CT, Hoof PLV, Chu W. 1994. Solubilization of non-polar compounds by non-ionic surfactant micelles. Water Res 28:1009–1017.

Jain V, Demond AH. 2002. Conductivity reduction due to emulsification during surfactant-aquifer remediation. 2. Formation of emulsion *in situ*. Environ Sci Technol 36:5434–5440.

Jawitz JW, Annable MD, Rao PSC, Rhue RD. 1998. Field implementation of a winsor Type I surfactant/alcohol mixture for in situ solubilization of a complex LNAPL as a single phase microemulsion. Environ Sci Technol 32:523–530.

Jawitz JW, Sillan RK, Annable MD, Rao PSC, Warner K. 2000. In situ alcohol flushing of a DNAPL source zone at a dry cleaner site. Environ Sci Technol 34:3722–3729.

Jawitz JW, Annable MD, Rao PSC, Rhue RD. 2001. Evaluation of remediation performance and cost for field-scale single-phase microemulsion (SPME) flushing. J Environ Sci Health A 36:1437–1450.

Jayanti S, Britton LN, Dwarakanath V, Pope GA. 2002. Laboratory evaluation of custom-designed surfactants to remediate NAPL source zones. Environ Sci Technol 36:5491–5497.

Jin M, Delshad M, Dwarakanath V, McKinney DC, Pope GA, Sepehrnoori K, Tilburg CE, Jackson RE. 1995. Partitioning tracer test for detection, estimation, and remediation performance assessment of subsurface nonaqueous phase liquids. Water Resour Res 31:1201–1211.

Jin M, Hirasaki GJ, Jackson RE, Kostarelos K, Pope GA. 2007. Control of downward migration of dense nonaqueous phase liquid during surfactant flooding by design simulations. Water Resour Res 43:W01412.

Johnson RL. 1992. The Dissolution of Dense Immiscible Solvents into Groundwater: Implications for Site Characterization and Remediation. Merkel Dekker, New York, NY, USA.

Johnson RL, Pankow JE. 1992. Dissolution of dense chlorinated solvents into groundwater: Source functions for pools of solvent. Environ Sci Technol 26:896–901.

Kanan K, Yousef H, Kayali I. 2012. Nanostructured microemulsion phase behavior using aot or extended surfactant combined with a cationic hydrotrope. J Surf Eng Mater Adv Technol 2:53–60.

Kibbey TC, Pennell KD, Hayes KF. 2001. Application of sieve-tray air strippers to the treatment of surfactant-containing wastewaters. AIChE J 47:1461–1470.

Kile DE, Chiou CT. 1989. Water solubility enhancements of DDT and trichlorobenzene by some surfactants below and above the critical micelle concentration. Environ Sci Technol 23:823–838.

Knox RC, Sabatini DA, Harwell JH, Brown RE, West CC, Blaha F, Griffin C. 1997. Surfactant remediation field demonstration using a vertical circulation well. Ground Water 35:948–953.

Kostarelos K, Pope GA, Rouse BA, Shook GM. 1998. A new concept: The use of neutrally-buoyant microemulsions for DNAPL remediation. J Contam Hydrol 34:383–397.

Kralova I, Sjoblom J. 2009. Surfactants used in food industry: A review. J Disper Sci Technol 30:1363–1383.

Krebs-Yuill B, Harwell JH, Sabatini DA, Knox RC. 1995. Economic considerations in surfactant-enhanced pump-and-treat remediation. In Sabatini DA, Know RC, Harwell JH, eds, Surfactant-Enhanced Subsurface Remediation. ACS Symposium Series 594. American Chemical Society, Washington, DC, USA, pp 265–278.

Kunieda H, Shinoda K. 1982. Phase behavior in systems of nonionic surfactant/water/oil around the hydrophile-lipophile-balance-temperature (HLB-temperature). J Disper Sci Technol 3:233–244.

Kuyukina MS, Ivshina IB, Makarov SO, Litvinenko LV, Cunningham CJ, Philp JC. 2005. Effect of biosurfactants on crude oil desorption and mobilization in a soil system. Environ Int 31:155–161.

Lake LW. 1989. Enhanced Oil Recovery. Prentice Hall, Old Tappan, NJ, USA.

Lange KR, ed. 1999. Surfactants: A Practical Handbook. Hanser Gardner Publications, Cincinnati, OH, USA.
Levine Fricke L. 1998. Cosolvent flushing pilot test report: Former sages dry cleaner. Florida Department of Environmental Protection, Jacksonville, FL, USA.
Lawrence MJ. 1994. Surfactant systems: Their use in drug delivery. Chem Soc Rev 23:417–424.
Li A, Yalkowsky SH. 1998. Predicting cosolvency. 1. Solubility ratio and solute log kow. Ind Eng Chem Res 37:4470–4475.
Li Z, Hanlie H. 2008. Combination of surfactant solubilization with permanganate oxidation for DNAPL remediation. Water Res 42:605–614.
Li Y, Abriola LM, Phelan TJ, Ramsburg CA, Pennell KD. 2007. Experimental and numerical validation of the total trapping number for prediction of DNAPL mobilization. Environ Sci Technol 41:8135–8141.
Liu MW, Roy D. 1995. Surfactant induced interactions and hydraulic conductivity changes in soil. Waste Manag 15:463–470.
Londergan JT, Meinardus HW, Mariner PE, Jackson RE, Brown CL, Dwarakanath V, Pope GA, Ginn JS, Taffinder S. 2001. DNAPL removal from a heterogeneous alluvial aquifer by surfactant-enhanced aquifer remediation. Ground Water Monit Remediat 21:57–67.
Longino BL, Kueper BH. 1995. The use of upward gradients to arrest downward dense, nonaqueous phase liquid (DNAPL) migration in the presence of solubilizing surfactants. National Research Council of Canada, Ottawa, ON, Canada.
Longino BL, Kueper BH. 1999. Nonwetting phase retention and mobilization in rock fractures. Water Resour Res 35:2085–2093.
Lunn SRD, Kueper BH. 1997. Removal of pooled dense, nonaqueous phase liquid from saturated porous media using upward gradient alcohol floods. Water Resour Res 33:2207–2219.
Lunn SRD, Kueper BH. 1999a. Manipulation of density and viscosity for the optimization of DNAPL recovery by alcohol flooding. J Contam Hydrol 38:427–445.
Lunn SRD, Kueper BH. 1999b. Risk reduction during chemical flooding: Preconditioning DNAPL density in situ prior to recovery by miscible displacement. Environ Sci Technol 33:1703–1708.
Mackay D, Shiu WT, Ma KC, Lee SC. 2006. Handbook of Physical-Chemical Properties and Environmental Fate for Organic Chemicals. CRC Press, Boca Raton, FL, USA.
Martel R, Gelinas PJ, Saumure L. 1998. Aquifer washing by micellar solutions: 3. Field test at the thouin sand pit (L'Assomption, Quebec, Canada). J Contam Hydrol 30:33–48.
McCray JE, Tick GR, Jawitz JW, Gierke JS, Brusseau ML. 2011. Remediation of NAPL source zones: Lessons learned from field studies at Hill and Dover AFB. Ground Water 49:727–744.
McDade JM, McGuire TM, Newell CJ. 2005. Analysis of DNAPL source depletion costs at 36 field sites. Remediat 15:9–18.
McGuire T, Hughes JB. 2003. Effects of surfactants on the dechlorination of chlorinated ethenes. Environ Toxicol Chem 22:2630–2638.
McGuire TM, McDade JM, Newell CJ. 2006. Performance of DNAPL source depletion technologies at 59 chlorinated solvent impacted sites. Ground Water Monit Remediat 26:73–84.
Meinardus HW, Dwarakanath V, Jackson RE, Jin M, Ginn JS, Stotler GC. 2000. Full-scale characterization of a DNAPL source zone with pitts. In Wickramanayake GB, Gavaskar AR, Gupta N, eds, Treating Dense Nonaqueous Phase Liquids (DNAPLs): Remediation of Chlorinated and Recaleitrant Compounds. Battelle Press, Columbus, OH, USA, Vol C2-2, pp 17–24.
Meinardus HW, Dwarakanath V, Ewing J, Hirasaki GJ, Jackson RE, Jin M, Ginn JS, Londergan JT, Miller CA, Pope GA. 2002. Performance assessment of NAPL remediation in heterogeneous alluvium. J Contam Hydrol 54:173–193.

Miller R, Fainerman VB, Makievski AV, Krägel J, Grigorieva DO, Kazakov VN, Sinyachenko OV. 2000. Dynamics of protein and mixed protein/surfactant adsorption layers at the water/fluid interface. Adv Colloid Interf Sci 86:39–82.

Morris KR, Abramowitz R, Pinal R, Davis P, Yalkowsky SH. 1988. Solubility of aromatic pollutants in mixed solvents. Chemosphere 17:285–298.

Mravik SC, Sillan RK, Wood AL, Sewell GW. 2003. Field evaluation of the solvent extraction residual biotreatment technology. Environ Sci Technol 37:5040–5049.

Mulligan CN. 2005. Environmental applications for biosurfactants. Environ Pollut 133:183–198.

Mulligan CN, Yong RN, Gibbs BF. 2001. Surfactant-enhanced remediation of contaminated soil: A review. Eng Geol 60:371–380.

Muthusamy K, Gopalakrishnan S, Ravi TK, Sivachidambaram P. 2008. Biosurfactants: Properties, commercial production and application. Curr Sci 94:736–747.

Nalecz-Jawecki GA, Grabińska-Sota EB, Narkiewicz PA. 2003. The toxicity of cationic surfactants in four bioassays. Ecotoxicol Environ Saf 54:87–91.

Nelson R. 1989. Chemically enhanced oil recovery: The state of the art. Chem Eng Prog 85:50–57.

Nimrod AC, Benson WH. 1996. Estrogenic responses to xenobiotics in channel catfish (*Ictalurus punctatus*). Mar Environ Res 42:155–160.

Oolman T, Godard ST, Jin M, Pope GA, Kirchner K. 1995. DNAPL flow behavior in a contaminated aquifer: Evaluation of field data. Ground Water Monit Remediat 15:125–137.

Pacwa-Pociniczak M, Paza G, Piotrowska-Seget Z, Cameotra S. 2011. Environmental applications of biosurfactants: Recent advances. Int J Mol Sci 12:633–654.

Pennell KD, Abriola LM. 1997. Surfactant enhanced aquifer remediation: Fundamental processes and practical application. In Sikdar SK, Irvine RL, eds, Bioremediation: Principles and Practice. Technomic Publication, Lancaster, PA, USA, pp 693–750.

Pennell KD, Abriola LM Jr, Weber WJ. 1993. Surfactant-enhanced solubilization of residual dodecane in soil columns. 1. Experimental investigation. Environ Sci Technol 27:2332–2340.

Pennell KD, Pope GA, Abriola LM. 1996. Influence of viscous and buoyancy forces on the mobilization of residual tetrachloroethylene during surfactant flushing. Environ Sci Technol 30:1328–1335.

Pennell KD, Adinolfi AM, Abriola LM, Diallo MS. 1997. Solubilization of dodecane, tetrachloroethylene, and 1,2-dichlorobenzene in micellar solutions of ethoxylated nonionic surfactants. Environ Sci Technol 31:1382–1389.

Pinal R, Rao PSC, Lee LS, Cline PV, Yalkowsky SH. 1990. Cosolvency of partially miscible organic solvents on the solubility of hydrophobic organic chemicals. Environ Sci Technol 4:639–647.

Pope GA. 1980. The application of fractional flow theory to enhanced oil recovery. Soc Petrol Eng J 20:191–205.

Pope GA, Wade WH. 1995. Lessons from enhanced oil recovery research for surfactant-enhanced aquifer remediation. In Surfactant-Enhanced Subsurface Remediation. ACS Symposium Series, Vol 594. American Chemical Society, Washington, DC, USA, pp 142–160.

Radian Corporation. 1992. Final Remedial Investigation for Operable Unit 2, Sites WP07 and SS21. Hill Air Force Base, Utah, USA.

Ramakrishnan V, Ogram AV, Lindner AS. 2005. Impacts of co-solvent flushing on microbial populations capable of degrading trichloroethylene. Environ Health Perspect 113:55–61.

Ramsburg CA, Pennell KD. 2001. Experimental and economic assessment of two surfactant formulations for source zone remediation at a former dry cleaning facility. Ground Water Monit Remediat 21:68–82.

Ramsburg CA, Pennell KD. 2002. Density-modified displacement for dense nonaqueous-phase liquid source-zone remediation: Density conversion using a partitioning alcohol. Environ Sci Technol 36:2082–2087.

Ramsburg CA, Pennell KD, Kibbey TCG, Hayes KF. 2003. Use of a surfactant stabilized emulsion to deliver *n*-butanol for density modified displacement of trichloroethene-NAPL. Environ Sci Technol 37:4246–4253.

Ramsburg CA, Abriola LM, Pennell KD, Loffler FE, Gamache M, Amos BK, Petrovskis EA. 2004a. Stimulated microbial reductive dechlorination following surfactant treatment at the Bachman Road site. Environ Sci Technol 38:5902–5914.

Ramsburg CA, Pennell KD, Kibbey TCG, Hayes KF. 2004b. Refinement of the density-modified displacement method for efficient treatment of tetrachloroethene source zones. J Contam Hydrol 74:105–131.

Ramsburg CA, Pennell KD, Abriola LM, Daniels G, Drummond CD, Gamache M, Hsu H-l, Petrovskis EA, Rathfelder KM, Ryder JL, Yavaraski TP. 2005. Pilot-scale demonstration of surfactant-enhanced pce solubilization at the bachman road site. 2. System operation and evaluation. Environ Sci Technol 39:1791–1801.

Rao PSC, Lee LS, Pinal R. 1991. Solubility, Sorption and Transport of Hydrophobic Organic Chemicals in Complex Mixtures. Environmental Research Brief. U.S. Environmental Protection Agency, Washington, DC, USA.

Rao PSC, Annable MD, Sillan RK, Dai D, Hatfield K, Graham WD, Lynn Wood A, Enfield CG. 1997. Field-scale evaluation of in situ cosolvent flushing for enhanced aquifer remediation. Water Resour Res 33:2673–2686.

Rao P-H, He M, Yang X, Zhang Y-C, Sun S-Q, Wang J-S. 2006. Effect of an anionic surfactant on hydraulic conductivities of sodium- and calcium-saturated soils. Pedosphere 16:673–680.

Renshaw CE, Zynda GD, Fountain JC. 1997. Permeability reductions induced by sorption of surfactant. Water Resour Res 33:371–378.

Roeder E, Falta RW, Lee CM, Coates JT. 2001. DNAPL to LNAPL transitions during horizontal cosolvent flooding. Ground Water Monit Remediat 21:77–88.

Rong MZ, Zhang MQ, Ruan WH. 2006. Surface modification of nanoscale fillers for improving properties of polymer nanocomposites: A review. Mater Sci Technol 22:787–796.

Roote DS. 1998. In Situ Flushing. Technology Status Report TS-98-01. Ground-Water Remediation Technologies Analysis Center, Pittsburgh, PA, USA.

Rosen MJ. 1989. Surfactants and Interfacial Phenomena, 2nd ed. John Wiley and Sons, New York, NY, USA.

Roshanfekr M, Johns RT. 2011. Prediction of optimum salinity and solubilization ratio for microemulsion phase behavior with live crude at reservoir pressure. Fluid Phase Equilibr 304:52–60.

Sabatini DA, Knox RC, Harwell JH, Wu B. 2000. Integrated design of surfactant enhanced DNAPL remediation: Efficient supersolubilization and gradient systems. J Contam Hydrol 45:99–121.

Shiau B. 2008. In-situ surfactant and chemical oxidant flushing for complete removal of contaminants and methods of using same. U.S. Patent 7364386.

Shiau BJ, Sabatini DA, Harwell JH. 1994. Solubilization and microemulsification of chlorinated solvents using direct food additive (edible) surfactants. Ground Water 32:561–569.

Shiau BJ, Sabatini DA, Harwell JH. 1995. Properties of food grade (edible) surfactants affecting subsurface remediation of chlorinated solvents. Environ Sci Technol 29:2929–2935.

Shinoda K, Saito H. 1968. The effect of temperature on the phase equilibria and the types of dispersions of the ternary system composed of water, cyclohexane, and nonionicsurfactant. J Colloid Interf Sci 26:70–74.

Shook GM, Pope GA, Kostarelos K. 1998. Prediction and minimization of vertical migration of DNAPLs using surfactant enhanced aquifer remediation at neutral buoyancy. J Contam Hydrol 34:363–382.

Sillan RK. 1999. Field-scale evaluation of in situ co-solvent flushing for enhanced aquifer remediation. University of Florida, Gainesville, FL, USA.

Smith JA, Sahoo D, McLellan HM, Imbrigiotta TE. 1997. Surfactant-enhanced remediation of a trichloroethene-contaminated aquifer. 1. Transport of triton x-100. Environ Sci Technol 31:3565–3572.

Soga K, Page JWE, Illangasekare TH. 2004. A review of NAPL source zone remediation efficiency and the mass flux approach. J Hazard Mater 110:13–27.

Srinivasan KR, Fogler HS. 1990. Use of inorgano-organo-clays in the removal of priority pollutants from industrial wastewaters: Adsorption of benzo(a)pyrene and chlorophenols from aqueous solutions. Clay Clay Miner 38:287–293.

Stauffer CE. 2005. Emulsifiers for the food industry. Volume 4. In Shahidi F, ed, Bailey's Industrial Oil and Fat Products: Products and Applications. John Wiley and Sons Inc., New York, NY, USA, pp. 229–267.

Strbak L. 2000. In situ flushing with surfactants and cosolvents. U.S. Environmental Protection Agency, Washington, DC, USA. http://www.clu-in.org. Accessed December 4, 2012.

Stroo HF, Unger M, Ward CH, Kavanaugh MC, Vogel TM, Leeson A, Marqusee JA, Smith BP. 2003. Remediating chlorinated solvent source zones. Environ Sci Technol 37:224A–230A.

Stroo HF, Leeson A, Marqusee JA, Johnson PC, Ward CH, Kavanaugh MC, Sale TC, Newell CJ, Pennell KD, Lebron CA, Unger M. 2012. Chlorinated ethene source remediation: Lessons learned. Environ Sci Technol 46:6438–6447.

Suchomel EJ, Pennell KD. 2006. Reductions in contaminant mass discharge following partial mass removal from DNAPL source zones. Environ Sci Technol 40:6110–6116.

Suchomel EJ, Ramsburg CA, Pennell KD. 2007. Evaluation of trichloroethene recovery processes in heterogeneous aquifer cells flushed with biodegradable surfactants. J Contam Hydrol 93:195–214.

Szafranski R, Lawson J, Hirasaki G, Miller C, Akiya N, King S, Jackson R, Meinardus H, Londergan J. 1998. Surfactant/foam process for improved efficiency of aquifer remediation. In Rehage H, Peschel G, eds, Structure, Dynamics and Properties of Disperse Colloidal Systems. Vol. 111. Springer Berlin, Germany, pp 162–167.

Taber JJ. 1981. Research on enhanced oil recovery: Past, present and future. In Shah DO, ed, Surface Phenomenon in Enhanced Oil Recovery. Plenum Press, New York, NY, USA, pp 13–52.

Taylor TP, Pennell KD, Abriola LM, Dane JH. 2001. Surfactant enhanced recovery of tetrachloroethylene from a porous medium containing low permeability lenses, 1. Experimental studies. J Contam Hydrol 48:325–350.

Tochilin VP. 2001. Structure and design of polymeric surfactant-based drug delivery systems. J Control Release 73:137–172.

Trogus FJ, Sophany T, Schechter RS, Wade WH. 1977. Static and dynamic adsorption of anionic and nonionic surfactants. SOC Petrol Eng J 17:337–344.

Tsai TT, Kao CM, Yeh TY, Liang SH, Chien HY. 2009. Application of surfactant enhanced permanganate oxidation and bidegradation of trichloroethylene in groundwater. J Hazard Mater 161:111–119.

URS Greiner Woodward Clyde and Duke Engineering and Services. 1999. Final report for the dense nonaqueous phase liquid (DNAPL) source delineation project, OU2. U.S. Air Force, Hill Air Force Base, Utah, USA.

USEPA (U.S. Environmental Protection Agency). 2003. The DNAPL remediation challenge: Is there a case for source depletion? EPA/600/R-03/143. Office of Research and Development, Washington, DC, USA.

USEPA. 2012. DfE Alternatives Assessment for Nonylphenol Ethoxylates. http://www.epa.gov/dfe/pubs/projects/npe/ Accessed September 16, 2013.

Valsaraj K, Gupta A, Thibodeaux L, Harrison D. 1988. Partitioning of chloromethanes between aqueous and surfactant micellar phases. Water Res 22:1173–1183.

van Genuchten MT. 1980. A closed-form equation for predicting the hydraulic conductivity of unsaturated soils. Soil Sci Soc Am J 44:892–898.

Wagner J, Chen H, Brownawell BJ, Westall JC. 1994. Use of cationic surfactants to modify soil surfaces to promote sorption and retard migration of hydrophobic organic compounds. Environ Sci Technol 28:231–237.

Yalkowsky SH, Roseman TJ. 1981. Solubilization of drugs by cosolvents. In Yalkowsky SH, ed, Techniques of Solubilization of Drugs. Marcel Dekker, New York, NY, USA, pp 91–134.

Yeh DH, Pennell KD, Pavlostathis SG. 1999. Effect of tween surfactants on methanogenesis and microbial reductive dechlorination of hexachlorobenzene. Environ Toxicol Chem 18:1408–1416.

Zhai X, Hua I, Rao PSC, Lee LS. 2006. Cosolvent-enhanced chemical oxidation of perchloroethylene by potassium permanganate. J Contam Hydrol 82:61–74.

Zhang Y, Miller RM. 1992. Enhanced octadecane dispersion and biodegradation by a pseudomonas rhamnolipid surfactant (biosurfactant). Appl Environ Microbiol 58:3276–3282.

Zhou M, Rhue RD. 2000. Screening commercial surfactants suitable for remediating DNAPL source zones by solubilization. Environ Sci Technol 34:1985–1990.

Zimmerman JB, Kibbet TCG, Cowell MA, Hayes KF. 1999. Partitioning of ethoxylated nonionic surfactants into nonaqueous-phase organic liquids: Influence on solubilization behavior. Environ Sci Technol 33:169–176.

CHAPTER 12

IN SITU BIOREMEDIATION OF CHLORINATED ETHENE SOURCE ZONES

Hans F. Stroo,[1] Michael R. West,[2] Bernard H. Kueper,[3] Robert C. Borden,[4] David W. Major[5] and C. Herb Ward[6]

[1]Stroo Consulting, LLC, Ashland, OR 97520, USA; [2]B. Kueper & Associates, Ltd, Kingston, ON, Canada K7L 4V1; [3]Queen's University, Kingston, ON, Canada K7L 3N6; [4]North Carolina State University, Raleigh, NC 27695, USA; [5]Geosyntec Consultants, Inc., Guelph, ON, Canada N1G 5G3; [6]Rice University, Houston, TX 77005, USA

12.1 INTRODUCTION

In situ bioremediation (ISB) was quickly adopted during the mid-1990s to treat dissolved chlorinated solvent plumes and to contain source zones by creating permeable reactive barriers (Stroo and Ward, 2010). Initially, ISB was not considered a feasible technology for treating dense nonaqueous phase liquid (DNAPL) source zones, and still faces some skepticism (AFCEE et al., 2004; Borden, 2003; ITRC, 2008). However, the experience of the last decade suggests that it can be a viable technology for treating some source zones. ISB has become one of the most commonly used source remediation technologies because it is relatively inexpensive, sustainable and adaptable to a wide range of site-specific conditions. This chapter summarizes both the promise and the potential pitfalls of using ISB to deplete source zones.

Seagren et al. (1993, 1994) showed that biological activity could enhance the dissolution of nonaqueous phase liquid (NAPL) and thereby shorten the longevity of the source. This work provided an initial theoretical basis for treating chlorinated solvent sources with ISB. However, there was little commercial interest in ISB for source zones for several years, until research demonstrated that dechlorinating bacteria could survive and degrade chlorinated ethenes even at the high concentrations found near trichloroethene (TCE) and perchloroethene (PCE or tetrachloroethene) DNAPL (Nielsen and Keasling, 1999; Harkness et al., 1999; Yang and McCarty, 2000). Following laboratory demonstrations of enhanced dissolution of PCE DNAPL (Carr et al., 2000; Cope and Hughes, 2001), field-scale pilot demonstrations were performed in the early 2000s (Peterson et al., 2000; Battelle, 2004; Hood et al., 2008). Those results have increased the confidence that ISB can be useful for at least some DNAPL source zones (ITRC, 2008).

However, ISB is not appropriate for all chlorinated ethene source zones, particularly those with a significant fraction of DNAPL present as pools (Christ et al., 2005). ISB has been adopted mainly as a primary remediation technology for low-strength DNAPL source zones (Sale et al., 2008) and as a polishing technology for higher-strength sources after more aggressive approaches have been used (ITRC, 2008; Sleep et al., 2006). A recent survey of 118 sites that have used *in situ* source-zone remediation at a field scale (NAVFAC and Geosyntec, 2011) found that ISB had been used at roughly 30% of the sites (Figure 12.1). That survey and others (McDade et al., 2005; McGuire et al., 2006; ITRC, 2008) also have

B.H. Kueper et al. (eds.), *Chlorinated Solvent Source Zone Remediation*,
doi: 10.1007/978-1-4614-6922-3_12,

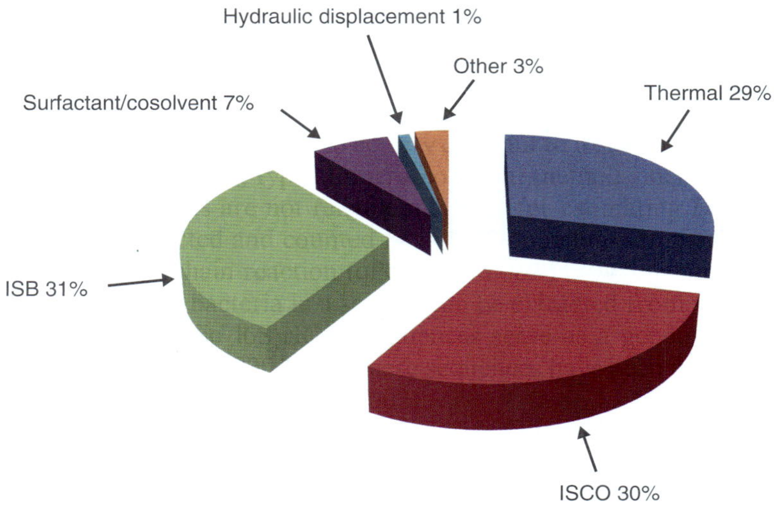

Figure 12.1. Results of survey of field-scale experiences treating chlorinated ethene source zones (from NAVFAC and Geosyntec, 2011). Figure shows the percentages of sites using different *in situ* treatment technologies, from a total of 118 sites. Note: ISCO – *in situ* chemical oxidation.

documented that ISB can achieve reductions in contaminant concentrations that are similar to those measured using other technologies, generally for lower cost.

Despite its rapid adoption it is important to note that, to the authors' knowledge, there are no well-documented case studies of ISB at a heavily contaminated chlorinated solvent site that have demonstrated site-wide reductions in contaminant concentrations to below maximum contaminant levels (MCLs). The technology is slow compared to thermal or chemical treatments, and it is extremely difficult to deliver chemical reagents to all of the contaminated areas in the subsurface; thus, some residual contamination should be expected (ITRC, 2011). Source-zone ISB designs and operations must address several significant challenges that often are not fully appreciated. However, source-zone ISB is a technology that is still developing, and further innovations will occur. Use of ISB for source treatment will continue because it is an economically attractive technology and it also can be a very effective one, given appropriate conditions and remedial objectives.

This chapter summarizes the technical basis for ISB of chlorinated solvent source zones, the advantages and limitations of the technology, the challenges faced when designing and operating an ISB system for a source zone and the options available for addressing site-specific challenges. The chapter also discusses realistic performance expectations and monitoring of these systems and concludes with a summary of the lessons learned to date.

The chapter includes descriptions of four well-documented case studies to provide examples of tank- to field-scale results and the different implementation strategies that have been used. Source-zone ISB is sufficiently innovative that no peer-reviewed report of a full-scale treatment was found in preparing this chapter. However, these case studies provide valuable lessons regarding the appropriate uses of the technology and the performance that can be expected. Finally, the chapter includes a description of a recently developed numerical model of source-zone ISB, with the conclusions drawn from using the model for a range of site conditions.

12.2 TECHNOLOGY DESCRIPTION

12.2.1 Technical Basis

ISB of chlorinated ethenes generally relies on sequential reductive dechlorination, with each step in the process removing one chlorine atom at a time and replacing it with a hydrogen atom (Figure 12.2) (Bradley and Chapelle, 2010). In this sequence PCE is reduced to TCE, which in turn is reduced to dichloroethene (primarily the *cis*-1,2-DCE isomer), which then is reduced to vinyl chloride (VC) and finally to ethene (Vogel and McCarty, 1985; Mohn and Tiedje, 1992). This process has been studied for more than 20 years and has been used to treat chlorinated ethenes in the dissolved phase (i.e., in the plume) for well over a decade (Stroo, 2010). Although several dechlorinating bacteria are capable of the initial steps in dechlorination (PCE to TCE and DCE), to date only bacteria in the genus *Dehalococcoides* (*Dhc*) have been shown to be capable of gaining energy from the complete dechlorination of DCE to VC and eventually to ethene (Maymó-Gatell et al., 2001; He et al., 2003; Löffler et al., 2013b). All known *Dhc* strains are now classified as *Dehalococcoides mccartyi* (Löffler et al., 2013a).

In reductive dechlorination, the chlorinated ethene serves as the electron acceptor, hydrogen serves as the electron donor, and the dechlorinating bacteria generally obtain their carbon from acetate (Bradley and Chapelle, 2010). Stimulating reductive dechlorination typically relies on adding complex organic carbon compounds (electron donor sources such as vegetable oil or lactate) that are fermented, producing acetate and hydrogen. The addition of these electron donor sources into the subsurface stimulates the growth and activity of dechlorinating bacteria by creating a sufficiently anaerobic groundwater treatment zone and generating hydrogen through fermentation reactions (ITRC, 2005). The hydrogen and injected electron donor are referred to as *substrates*, reduced compounds whose oxidation can be linked to reduction of the contaminant compound. The anaerobic treatment zone is created through the consumption of oxygen and other electron acceptors (for example, nitrate and sulfate) during the biodegradation of the electron donor source.

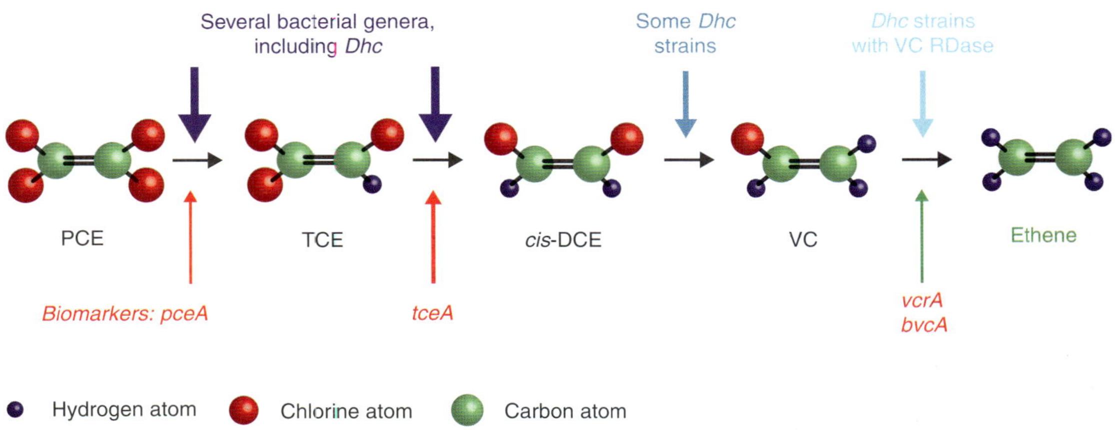

Figure 12.2. Reductive dechlorination of chlorinated ethenes. Bacteria capable of energy-yielding reductive dechlorination of each of the ethenes are noted in *blue text*, and useful biomarkers for particular steps are identified in *red. Dhc* = *Dehalococcoides mccartyi*; RDase = reductive dehalogenase; pceA is a marker for PCE RDase; tceA is a marker for TCE RDase; and vcrA and bvcA are markers for two identified VC RDases.

The hydrogen serves as the ultimate electron donor for anaerobic bacteria that reduce the chlorinated ethenes. Thus the electron donor addition both helps create favorable redox conditions (highly reducing methanogenic or sulfate-reducing conditions, with Eh values <-100 milivolts [mV]) and also fuels the reductive dechlorinating bacteria (notably *Dhc*). This form of ISB is often referred to as enhanced reductive dechlorination (ERD), to distinguish it from aerobic bioremediation techniques.

ISB through reductive dechlorination has proven to be a useful approach to treat chlorinated ethenes because it is capable of treating all chloroethenes, and because it has proven to be relatively easy to implement and control under field conditions, compared to other potential biological approaches such as aerobic cometabolism (described below). However ISB is not suitable for all sites. Key limitations include the presence of significant DNAPL accumulations, moderately acidic pH or low alkalinity, low permeability or highly heterogeneous aquifers or very rapid groundwater velocities (Table 12.1).

12.2.2 Enhanced Dissolution and Source Removal During ISB

Under ideal conditions, aqueous phase chlorinated solvents such as PCE and TCE may be reduced all the way to nontoxic end products (ethene and ethane). In some cases, however, substantial amounts of lesser chlorinated compounds (primarily *cis*-DCE) often are produced (Carr et al., 2000; Yang and McCarty, 2000, 2002; Sleep et al., 2006). This incomplete dechlorination has been attributed to insufficient residence time (Amos et al., 2007b), low pH (Adamson et al., 2003; Eaddy, 2008; McCarty et al., 2007; Robinson et al., 2009) and inhibitory parent compound concentrations (Chu et al., 2003; Yu et al., 2005; Sabalowsky and Semprini, 2010a, b).

During ISB in the presence of DNAPL, the parent compound (e.g., PCE) dissolves into the aqueous phase and is then degraded by bacteria. This reduces the PCE concentration in the aqueous phase, allowing more of the DNAPL to dissolve (Seagren et al., 1993, 1994; Cope and Hughes, 2001). However as PCE is degraded, TCE, DCE and VC are produced. Since TCE, DCE and VC are somewhat hydrophobic, these daughter products can partition back into the DNAPL phase. The simultaneous dissolution, biotransformation and back partitioning can lead to temporary sequestering of the daughter products in the DNAPL phase, complicating data interpretation (Ramsburg et al., 2010). Aqueous phase PCE concentrations may be low, even though a substantial amount of this material is still present in the form of DNAPL. Also, the mass flux of TCE, DCE and VC released to the downgradient plume may be lower than the actual production rate because partitioning back into the DNAPL will reduce aqueous phase concentrations of these compounds.

Cleanup of chlorinated solvent source zones is often limited by the low aqueous solubility of the major contaminants (for example, PCE and TCE) and slow mass transfer from DNAPL to the dissolved phase (Kueper et al., 2003). However, ISB can accelerate DNAPL removal through several different mechanisms. Removal of DNAPL during ISB will be accelerated by the increased DNAPL dissolution rate, although repartitioning of the parent or daughter products (e.g., TCE, DCE and VC) back into DNAPL can occur as well, though it may be downgradient of the original DNAPL accumulation. The relative importance of these two competing processes depends on a variety of factors including the transformation rates, effective aqueous solubility of the various chlorinated ethenes, and rate of groundwater flow. If the DCE degradation rate is very high, DCE will not accumulate substantially and the DNAPL removal rate will be controlled primarily by the PCE and TCE degradation rates and dissolution of these compounds by flowing groundwater. However, it is common for the DCE

Table 12.1. Site Conditions that Affect the Applicability of ISB (Modified from AFCEE et al., 2004)

Factor	Favorable Condition	Less Favorable Condition	Impact on ISB	Ease of Managing
Geochemistry				
pH	5.5 to 8	<4.5 or >9	High	Difficult
Temperature	>10°C	<10°C	Low	Difficult
Alkalinity	High (>100 mg/L $CaCO_3$)	Low	Moderate	Difficult
Competing electron acceptors	Low concentrations or influx	High concentrations or influx	Moderate	Moderate
Hydrogeology				
Depth to water	Moderate	Deep or very shallow	Low to Moderate	Difficult
Hydraulic conductivity	Med to High (>1 ft/day)	Low (<1 ft/day)	High	Moderate
Groundwater velocity	>0.1 ft/day or <3 ft/day	>3 ft/day or stagnant	High	Moderate – if slow Difficult – if fast
Aquifer matrix	Granular; unconsolidated media	Rock; consolidated media	High	Difficult
Heterogeneity	Low to moderate	High (>100 variation in local scale K)	Moderate	Difficult
Source-Zone Characteristics				
Contaminant	Single component	Multicomponent mixtures	Moderate to High	Difficult
Depth of source	Shallow	Deep	Low to Moderate	Moderate
Age of source/ plume maturity	Recent (<10 year)	Mature (>10 year)	High	Difficult
Volume	Small (<500 cy)	Large (>500 cy)	Moderate	Moderate
Microbiology				
Native heterotrophic population	Present in pre-ISB screening	Absent in pre-ISB screening	Moderate to High	Difficult
Dehalorespirators	Present in pre-ISB screening	Absent in pre-ISB screening	Low	Easy
Other Factors				
Location of onsite infrastructure	Vapor intrusion risks are acceptable	Vapor intrusion risk is a concern	Moderate	Moderate
Site Access	Site is easy to access	Site is remote or otherwise difficult to access	Moderate	Moderate

Note: cy – cubic yard

degradation rate to be slower than that for TCE and PCE, and for some accumulation to occur, with resulting partitioning of DCE and perhaps VC back into the NAPL phase.

The enhancement of chloroethene dissolution through biotransformation is illustrated in Figure 12.3 (from Yang and McCarty, 2000). Laboratory columns containing droplets of PCE DNAPL were flushed with a feed solution saturated with PCE and containing 1.7 milimolar (mM)

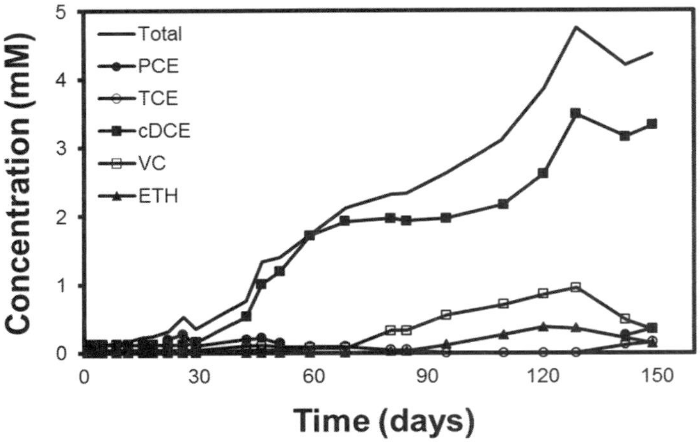

Figure 12.3. Concentration of PCE (*filled circle*), TCE (*open circle*), *cis*-DCE (*filled square*), VC (*open square*), ethene (*filled triangle*) and the total (without marker) in the column effluent with time. Reprinted with permission from Yang and McCarty (2000). Copyright (2000) American Chemical Society.

sodium benzoate and 20 mg/L yeast extract. The columns were then inoculated with an enrichment culture known to degrade PCE to ethene. Throughout the experiment, PCE and TCE concentrations remained low due to rapid transformation of these compounds to DCE, VC, and ethene, even though significant amounts of PCE DNAPL were still present. At the start of the experiment, partitioning of DCE back into the DNAPL droplets could have reduced the flushing rate. However by 45 days, total ethenes (sum of PCE, TCE, DCE, VC and ethene) were greater than the aqueous PCE solubility (~1 mM), demonstrating that ISB was enhancing DNAPL removal. At 130 days, total ethenes were over 4 mM, indicating DNAPL removal was enhanced by more than a factor of 4.

High electron donor concentrations can also impact dissolution rates. Macbeth et al. (2006) reported that high concentrations of dissolved electron donor and/or fermentation products potentially can enhance DNAPL mass transfer rates directly through cosolvency, desorption and/or partitioning from DNAPL to dissolved organic compounds. However, aqueous phase concentrations must be more than 1% (10,000 mg/L) to have a significant impact on TCE solubility (Hood et al., 2007). Low solubility electron donors can also impact chloroethene mobility. Hiortdahl and Borden (2011) reported a four- to five-fold increase in the effective solubility of PCE when flushing columns containing trapped PCE DNAPL with emulsified vegetable oil (EVO). However, partitioning of chloroethenes to oil droplets attached to aquifer material can reduce mobility (ESTCP, 2006).

At some sites, injection and extraction systems are used to recirculate groundwater containing electron donor through the treatment area. The increased groundwater velocity due to recirculation can increase mass transfer rates, potentially accelerating DNAPL removal. In theory, ISB can significantly enhance flushing rates and DNAPL removal. However in practice, the observed enhancement may be lower than expected. Table 12.2 (adapted from Sleep et al., 2006 and Amos et al., 2008) shows the measured enhancements in controlled laboratory experiments. Glover et al. (2007) showed that PCE dissolution could be enhanced by up to a factor of 13 in a 5-centimeter (cm) (2-inch [in]) flow cell. However, the cumulative PCE removal rate was only increased by a factor of 1.7 in 2-dimensional (2-D) aquifer cells containing a nonuniform PCE DNAPL distribution due to a shrinking DNAPL source zone and bioclogging/pore blockage from methane gas generation (Sleep et al., 2006).

Table 12.2. Summary of Selected DNAPL ISB Experiments (Adapted from Sleep et al., 2006 and Amos et al., 2008)

Reference Study	System Description	Major Dechlorination Product(s)	Dissolution Enhancement
Yang and McCarty, 2000	1-D column containing aquifer material and nonuniform residual PCE DNAPL with pentanol as substrate	*cis*-DCE	5 maximum
Carr et al., 2000	Continuous-flow stirred-tank reactor with PCE/tridecane NAPL (0.13 mol/mol PCE/tridecane) with formate as substrate	*cis*-DCE	14 – cumulative based on model comparison
Cope and Hughes, 2001	1-D column containing glass beads with PCE/tridecane NAPL (0.13 mol/mol PCE/tridecane) with pyruvate as substrate	TCE, VC	5.0 to 6.5 cumulative
Yang and McCarty, 2002	1-D column containing aquifer material and PCE DNAPL with different substrates (pentanol, calcium oleate, and olive oil)	*cis*-DCE	Cumulative 1.6 – pentanol 2.3 – oleate 1.9 – olive oil
Adamson et al., 2003	3-D pilot-scale simulated aquifer containing sand and PCE DNAPL with HRC as substrate	*cis*-DCE, VC	Not determined
Sleep et al., 2006	2-D aquifer cell containing aquifer material and PCE DNAPL and methanol–ethanol as substrate	*cis*-DCE, VC	3.3 maximum 1.7 cumulative
Da Silva et al., 2006	3-D pilot-scale simulated aquifer containing sand and PCE DNAPL and HRC and lactate as substrate	*cis*-DCE, VC	1.6
Glover et al., 2007	Custom 5-cm flow cell containing sand mixtures (course- to fine-grained) and PCE DNAPL and methanol as substrate	*cis*-DCE	<1.5 to 13
Amos et al., 2008	1-D sand columns with residual PCE DNAPL and pyruvate as substrate	No significant degradation	1
Amos et al., 2008	1-D sand columns with residual NAPL mixture of 25% PCE/75% hexadecane (mol/mol) and pyruvate as substrate	*cis*-DCE	13.6 maximum 4.6 cumulative
Philips et al., 2011	Three layer diffusion cell with TCE DNAPL and formate as substrate	*cis*-DCE	2.4 maximum

Note: mol – mole(s).

It is important to remember that the enhancement factor represents the improvement of dissolution over flushing the source zone with water only. Compared to ambient conditions, active recirculation can significantly increase water movement, especially through the more transmissive portions of the source, accelerating contaminant dissolution and increasing DNAPL removal by both physical and biological processes. If multiple pore volumes of water are flushed through a source zone, even the relatively low enhancement factors commonly measured (roughly 2–3) could significantly reduce contaminant mass in the source zone, as well as the dissolved concentrations and mass discharge from the source after flushing (ITRC, 2008). Without active flushing, ISB still can reduce source-zone concentrations and mass discharge (McGuire et al., 2006), although mass removal likely will be much slower, and a significant fraction of the contaminant mass may remain after treatment is stopped.

Note also that the enhancement factor generally is calculated as an average enhancement over a relatively large volume, typically the entire source zone. However, the degree of enhancement is likely to differ widely between different regions, based on the spatial differences in the ability to deliver reagents and move water through the subsurface. These variations can be important – for example, lower enhancement in some areas may indicate the need for additional targeted treatment.

12.2.3 Microbiology of Chlorinated Ethene Biodegradation

Several microbiological mechanisms for chlorinated ethene biodegradation exist – direct aerobic oxidation, anaerobic oxidation, aerobic cometabolism, cometabolic reductive dechlorination and direct reductive dechlorination by organohalide respiration. In addition, the microbial activities during ISB can stimulate chemical reduction of chlorinated ethenes, a process termed biogeochemical degradation (Brown et al., 2009). This chapter focuses on anaerobic biodegradation, particularly organohalide respiration, because it is the most important process for source-zone treatment. However, a brief discussion of the other processes is also presented below. More extensive information is available in reviews of the biodegradation of chlorinated ethenes (Bradley and Chapelle, 2010) and of *Dehalococcoides* and reductive dechlorination (Löffler et al., 2013b).

12.2.3.1 Aerobic Oxidation, Anaerobic Oxidation and Cometabolic Biodegradation

Direct aerobic oxidation of chlorinated ethenes is restricted to DCE and VC (Coleman et al., 2002; Hartmans et al., 1985), though VC has the greatest tendency to undergo oxidation. Vinyl chloride oxidation can occur at very low oxygen concentrations that may appear to be anaerobic (Gossett, 2010; Bradley and Chapelle, 2011). Several strains of aerobic bacteria have been found that can grow on VC, and aerobic DCE biodegradation also occurs, although to date only one strain (a *Polaromonas* strain identified as strain JS666) has been isolated and proven to be capable of growth on DCE (Coleman et al., 2002). This process is not effective for the most common parent compounds (PCE and TCE) and is therefore most appropriate for plume treatment and has not been applied to source zones.

Anaerobic oxidation of DCE and VC has been suggested by several studies that have measured mineralization of these daughter products under nominally anoxic conditions (e.g., Bradley and Chapelle, 1996; Bradley et al., 1998). However, it has proven difficult to verify that this mechanism is actually responsible for the observed disappearances, and it has not been used in engineered remediation systems, though it may contribute to losses observed during natural or enhanced biodegradation (Bradley and Chapelle, 2010).

The other important biological process affecting chlorinated solvents is called cometabolism. Cometabolism refers to a situation in which an organism can degrade a contaminant without deriving any benefit, so that for example, it can grow on one compound while fortuitously degrading the contaminant. There are two cometabolic processes affecting chlorinated solvents, termed aerobic and anaerobic cometabolism. These are described briefly in the following paragraphs.

Aerobic cometabolism of TCE has been known for over 20 years (Wilson and Wilson, 1985), and cometabolism of DCE also can occur (McCarty and Semprini, 1994), but no examples of aerobic cometabolism of PCE have been reported. The enzymes responsible for aerobic cometabolism are a variety of oxygenase enzymes that can be expressed by bacteria growing on a range of substrates including alkanes, phenol, toluene, ammonia and VC (Bradley and Chapelle, 2010). This process may be important in mixed-waste plumes where hydrocarbons are present or downgradient of anaerobic bioremediation systems where methane and oxygen mix. However, it has not been applied to treat chlorinated solvent source zones.

Anaerobic cometabolism can occur under reducing conditions, but it is a much less efficient process than the energy-yielding reduction of solvents described in the next section. From a practical perspective, anaerobic cometabolism is largely a side effect of the actions taken to stimulate the *Dehalococcoides* bacteria capable of complete dechlorination to ethene. Cometabolic dechlorinators can be important contributors to the total biodegradation achieved, and they can consume a significant fraction of the total electron donor sources added during biostimulation. Anaerobic cometabolism of PCE and TCE occurs under highly reducing conditions and can be mediated by a wide variety of organisms, including methanogens and other anaerobic bacteria (Bouwer and McCarty, 1983; Fathepure et al., 1987; Vogel and McCarty, 1985). These bacteria contain reduced transition-metal cofactors that fortuitously dechlorinate the solvents (Löffler et al., 2013b). However, the dechlorination rates decrease by an order of magnitude with each chlorine removed, so this process yields very little further reduction of DCE and VC, and little or no ethene is produced (Gantzer and Wackett, 1991). Anaerobic cometabolism can represent simply an inefficient process contributing to the overall removal of the solvents, but it also can pose a problem if it produces a sustained increase in the DCE and especially VC concentrations.

12.2.3.2 Organohalide Respiration and *Dehalococcoides*

The most important microbial process in chlorinated solvent source-zone ISB is direct energy-yielding reductive dechlorination. This process has been referred to by several terms, notably chlororespiration, dechlororespiration, halorespiration and dehalorespiration. The preferred term for the reductive dehalogenation of chlorinated ethenes, as well as similar metabolic processes responsible for the degradation of a wide variety of halogenated compounds, is organohalide respiration (Löffler et al., 2013b). This term refers to the fact that the organisms "breathe" organohalide compounds such as chlorinated ethenes, using them as electron acceptors in the same way that mammals use oxygen (McCarty, 1997).

Organohalide respiration of chlorinated ethenes is restricted to a few genera of bacteria, and respiration of DCE and VC is so far known to be mediated only by strains of *Dehalococcoides mccartyi* (Löffler et al., 2013a). The organisms capable of dechlorinating PCE to *cis*-1,2-DCE include the first such bacterium isolated, *Dehalobacter restrictus* (Holliger et al., 1993; Holliger et al., 1998). Some *Dehalobacter* isolates in fact require PCE or TCE as electron acceptors. Several other PCE-to-*cis*-DCE-dechlorinating bacteria have been identified, including strains of *Desulfuromonas* (Krumholtz et al., 1996; Sung et al., 2003), *Geobacter lovleyi* (Sung et al., 2006), *Sulfurospirillum multivorans* (Luitjen et al., 2003) and *Desulfitobacterium* (Maillard et al., 2005).

The first bacterium known to dechlorinate PCE to VC and ethene was originally named *Dehalococcoides ethenogenes* strain 195 (Maymó-Gatell et al., 1997). Strain 195 grows with PCE, TCE, *cis*-DCE, and 1,1-DCE electron acceptors. However, Strain 195 does not use VC as an electron acceptor and only slowly dechlorinates VC by a cometabolic process (Maymó-Gatell et al., 2001). Later studies have identified other *Dehalococcoides mccartyi* strains that more rapidly dechlorinate PCE to ethene and can use VC directly as an electron acceptor (He et al., 2003; Sung et al., 2006; Müller et al., 2004). In addition, several mixed cultures have been identified that reduce chlorinated ethenes completely to ethene (Duhamel et al., 2002; Richardson et al., 2002).

Dehalococcoides strains have proven to be very difficult to isolate and grow in pure culture (Löffler et al., 2013b). As a result, typical microbiological methods including plate counts or most-probable-number methods are not reliable methods for estimating *Dhc* numbers. *Dehalococcoides* cells can be detected and counted in an environmental sample, however, based on the quantitative polymerase chain reaction (qPCR) method. The 16S subunit of the ribosomal ribonucleic acid (RNA) of the bacteria in a sample can be extracted from water or soil samples and analyzed by qPCR. Specific 16S rRNA sequences serve as a genetic fingerprint of *Dhc* (Löffler et al., 2000; Fennel et al., 2001), and these sequences have been used to monitor *Dhc* in environmental samples (Hendrickson et al., 2002).

The ability of some *Dhc* strains to respire VC depends on an enzyme, vinyl chloride reductive dehalogenase (VC RDase). Gene probes have been developed for deoxyribonucleic acid (DNA) sequences unique to this enzyme, allowing detection and enumeration of two different VC RDase genes – vcrA (Müller et al., 2004) and bvcA (Krajmalnik-Brown et al., 2004). There are other VC reducing genes that are not detected by these assays (Ritalahti et al., 2006; Scheutz et al., 2008). However, the evidence so far suggests that these markers, and particularly vcrA, are useful for characterization and monitoring at most sites (van der Zaan et al., 2010).

Dhc cells have several interesting features that suggest a highly specialized lifestyle. They require hydrogen as an electron donor and also require a reduced organic compound such as acetate as a carbon source (Löffler et al., 2013b). They do not produce their own vitamin B_{12} and must rely on other bacteria in the community to supply it (He et al., 2007). *Dhc* are not tolerant of even moderate acidity, and activity (particularly VC reduction activity) declines rapidly below a pH of about 6.0 (Vainberg et al., 2006; Fogel et al., 2009). *Dhc* cells are very small, which can be beneficial for bacteria that live on compounds that usually are present at very low concentrations, by maximizing the surface area-to-volume ratio (Duhamel et al., 2004).

The *Dhc* genome is also very small, one of the smallest known for a free-living organism, which is consistent with the high degree of specialization (Giovannoni et al., 2005). The *Dhc* strains share most of their core genes on strongly conserved regions of the genome, but also have so-called high plasticity regions that allow rapid transfer of some genes such as those that code for VC RDases (McMurdie et al., 2009). High plasticity regions may allow the native *Dhc* population to adapt to new substrates without carrying copies of rarely used genes in all of its cells.

Dhc strains may not compete effectively with other bacteria for the early steps in reductive dechlorination, but they flourish when DCE and VC reduction are occurring (Becker, 2006). Therefore, high numbers of *Dhc* in groundwater ($>10^6$ cells/L) are generally found only when complete dechlorination is occurring (Lu et al., 2006, 2009; Van der Zaan et al., 2010), although analyses of the VC RDase gene also should be used to confirm complete biodegradation potential (Ritalahti et al., 2010). Other compounds also can inhibit *Dhc* activity, notably other chlorinated ethenes and cocontaminants such as chloroform (CF) or 1,1,1-trichloroethane (1,1,1-TCA) (Maymó-Gatell et al., 2001; Yu et al., 2005).

The actions taken to stimulate biological reductive dechlorination also may cause chemical reduction of chlorinated ethenes, particularly reductive dechlorination by ferrous iron (Fe[II]) minerals (Cwiertny and Scherer, 2010). This abiotic degradation is sometimes referred to as biogeochemical transformation because it is a result of coupled biological and chemical processes (Becvar et al., 2008). Reduced minerals are formed as a result of the fermentation of organic compounds and the creation of highly anaerobic conditions, and some of these minerals, such as ferrous sulfides, can directly reduce chlorinated ethenes (Butler and Hayes, 1999; Lee and Batchelor, 2002). This process can be important for natural attenuation (Ferrey et al., 2004), but it also can be responsible for some of the contaminant removal during ISB treatment of a source and may contribute to the sustained treatment observed after source-zone ISB (Adamson et al., 2011).

12.2.4 Practical Implications

The important points for practitioners to understand regarding the microbiology of reductive dechlorination include:

- Several different anaerobic bacteria can gain energy by rapidly reducing PCE and TCE to *cis*-1,2-DCE.
- Only some strains of *Dehalococcoides mccartyi* are known to rapidly dechlorinate all of the chlorinated ethenes to ethene.
- Efficient complete dechlorination depends on the presence of an enzyme (VC RDase). Some forms of this enzyme can be monitored using genetic probes.
- The presence and abundance of *Dhc* and VC RDases can be monitored effectively, and the numbers can be linked to the rate and extent of dechlorination (Lu et al., 2009).
- *Dhc* strains are specialized bacteria that need other organisms, and function in complex consortia, so an entire anaerobic community must be maintained for effective ISB.
- *Dhc* strains require highly anaerobic conditions (preferably methanogenic, with redox potential (Eh) values at least <-100 mV). They are inhibited by even mildly acidic conditions (pH $<$ 6.0) and some cocontaminants (notably TCA and CF).

12.3 TECHNICAL CHALLENGES

ISB has the potential to remediate highly contaminated source zones containing DNAPL. However, several significant technical challenges are involved. The growth and activity of many dechlorinating microorganisms are inhibited by high chlorinated solvent concentrations and low pH. Effectively delivering electron donors to the contaminant can be challenging due to substrate fermentation to methane in areas with low contaminant concentrations, pore blockage with biomass and gas bubbles, and heterogeneity in contaminant distribution and aquifer permeability.

12.3.1 Toxicity

Yang and McCarty (2000) showed that some dechlorinating bacteria can survive at chlorinated solvent concentrations near the aqueous solubility, raising the potential to greatly enhance removal of the more accessible DNAPL (ganglia) and sorbed contaminants (Adamson et al., 2003). However, Amos et al. (2008) reported minimal dechlorination and no enhancement of PCE dissolution in bioaugmented laboratory columns containing PCE DNAPL and concluded that the organism used in this work (*S. multivorans*) did not tolerate saturating PCE concentrations. However when one part PCE was mixed with three parts hexadecane, the effective PCE solubility was reduced to 300 micromolar (μM) (50 mg/L), allowing extensive dechlorination of the mixed PCE DNAPL producing *cis*-DCE.

A variety of investigators have reported toxic inhibition by PCE (Huang and Becker, 2011; Amos et al., 2007a; Yu et al., 2005; Philips et al., 2011) and daughter products (Sabalowsky and

Semprini, 2010a, b; Cupples et al., 2004, Yu et al., 2005). The upper limit for PCE dechlorination appears to vary for different organisms and mixed cultures from approximately 400–800 μM (Yu et al., 2005; Duhamel et al., 2002; Amos et al., 2007a) with higher concentrations tolerated by mixed cultures that can form biofilms and aggregates and thereby provide some biomass protection (Amos et al., 2007a; Sabalowsky and Semprini, 2010b).

12.3.2 pH

Aquifer pH also can have an important impact on dechlorination. During ISB, complex organic substrates (molasses, whey, vegetable oil) and short chain fatty acids are added to the aquifer as an electron donor. The complex substrates are fermented releasing hydrogen (H_2) and acetic acid that can be used by dechlorinators to reduce more highly chlorinated compounds to less chlorinated ones. In the process, H_2 and acetic acid are consumed releasing carbonic acid (H_2CO_3) and hydrochloric acid (HCl), which can cause a drop in pH. A variety of investigators have reported low pH as a contributing factor to reduced dechlorination efficiency (Adamson et al., 2003; Eaddy, 2008; McCarty et al., 2007; Robinson et al., 2009). Inhibition from high chlorinated ethene concentrations also can be compounded by nonneutral pH conditions (Sabalowsky and Semprini, 2010a).

Many biological processes are sensitive to pH and most microorganisms function efficiently in near neutral conditions (Lowe et al., 1993). Zhuang and Pavlostathis (1995) found that neutral pH was optimum for reductive dechlorination by a methanogenic mixed culture capable of dechlorinating PCE to VC. *Desulfitobacterium dichloroeliminans* strain DCA1 has an optimal pH range of 7.2–7.8 but maintained some activity down to pH ~ 5.4 (Fogel et al., 2007). Rowlands (2004) reported that the KB-1™ bioaugmentation culture has an optimal range of 6.0–8.3 and is completely inhibited below pH ~ 5.0. A pH of 6.0–6.8 is optimum for the dechlorination of PCE by the SDC-9™ bioaugmentation culture (Vainberg et al., 2006). Using a bioaugmentation culture enriched from Savannah River Site aquifer material, Eaddy (2008) found that dechlorination of PCE and TCE slowed at pH of 6.0 with increased accumulation of *cis*-DCE and VC. At pH 5.5, reduction of *cis*-DCE to VC and VC to ethene was completely inhibited.

The pH that microorganisms are exposed to during ISB is controlled by a variety of factors including the background aquifer pH and buffering, acidity produced during ISB and alkaline materials added during ISB to control pH decline. In low rainfall areas, the background pH is often neutral to slightly alkaline. However in humid areas, the background pH may be lower than optimal for ISB because of long-term leaching by acidic rainfall. When present, naturally occurring calcium carbonate ($CaCO_3$) can neutralize some acidity, in the process releasing bicarbonate ion (HCO_3^-) and carbonic acid (H_2CO_3). Above the water table, H_2CO_3 will degas as CO_2 essentially stripping acidity from the water. Below the water table, however, CO_2 may not be able to degas causing a buildup of dissolved carbonate (CO_3^{2-}) and HCO_3^-, stopping $CaCO_3$ dissolution. If $CaCO_3$ does not dissolve, it will not be effective in buffering pH (McCarty et al., 2007). Alkalinity can be produced by the reduction of nitrate (NO_3^-), sulfate (SO_4^{2-}) and iron hydroxides ($Fe(OH)_3$). Ion exchange to clays and iron hydroxides can strongly buffer pH. If the pH is near neutral, sorption of H^+ to clays and iron oxides can limit pH declines. However if the pH is already low, large amounts of base may be required to raise the pH because of the large amount of H^+ and other acidic materials sorbed to the aquifer material.

Large amounts of base may be required to maintain neutral pH during ISB. McCarty et al. (2007) calculated that 800 mg/L of bicarbonate alkalinity would be required to reduce the acidity produced during reductive dechlorination of 1.2 mM of TCE using 0.9 mM glucose as the electron donor. According to this calculation, ISB will require approximately six times more bicarbonate than electron donor. Adding these large amounts of alkalinity can be a logistical challenge and can increase the dissolved solids concentration of the groundwater significantly.

A variety of alkaline materials are available to control pH declines during ISB including hydroxides (NaOH, KOH, $Ca(OH)_2$ and $Mg(OH)_2$) and carbonates (Na_2CO_3, $NaHCO_3$ and $CaCO_3$). Sodium hydroxide and KOH are very soluble, so large amounts of alkalinity can be added to an aquifer. However, concentrated solutions of NaOH and KOH have pH > 13, which is also inhibitory to bacteria. Calcium hydroxide and $Mg(OH)_2$ are much less soluble; thus, water in equilibrium with these materials will have pH values of approximately 12.5 and 10.5, respectively. The lower solubility of these materials, however, makes them more difficult to distribute in the subsurface. As described above, $CaCO_3$ is relatively insoluble, making it a relatively ineffective for controlling pH. Mixtures of Na_2CO_3 and $NaHCO_3$ can be prepared to have a pH between 8.3 and 10, making them useful for ISB. However, carbonates provide relatively little alkalinity per unit mass, so large amounts of these materials are required.

12.3.3 Substrate Delivery to the Contaminant

Effectively delivering the organic substrate/electron donor can be a major challenge and may limit ISB efficiency. For example, Sleep et al. (2006) studied DNAPL ISB in two 2-D boxes packed with homogenized aquifer material with DNAPL source zones containing 10 milliliters (mL) of neat PCE. Each box was flushed with plain groundwater for 112 days followed by electron donor solutions containing methanol, ethanol and/or acetate. One of the boxes was bioaugmented with the KB-1™ culture, while one was not bioaugmented. In the bioaugmented box, the sum of PCE, TCE, *cis*-DCE, VC and ethene reached a maximum of 3.0 mM immediately downgradient from the DNAPL source zone indicating a solubility enhancement of 3.3.

The PCE removal declined after 681 days, even though approximately 35% of the initial PCE DNAPL was still present in the box. The substantial decline in PCE removal corresponded to an increase in methane production, suggesting that electron donor delivery to the DNAPL was being limited by consumption by methanogens or pore blockage by biomass and/or gas bubbles. Numerical simulations by Chu et al. (2003) indicate that pore blockage may divert substrate away from the contaminant, reducing ISB efficiency. In the field, substrate delivery will be further complicated by spatial variations in aquifer permeability. Numerical model simulations presented later in this chapter indicate that spatial variations in permeability, and the resulting potential for much of the subsurface flow to bypass some contaminated regions, may significantly reduce ISB efficiency.

12.4 ADVANTAGES AND LIMITATIONS

Source-zone ISB has several advantages that have made it a popular technology for appropriate sites and remedial objectives. However, it also has important limitations that often are not fully appreciated by practitioners. The advantages and limitations are summarized in the following sections.

12.4.1 Advantages

The advantages of ISB for treating chlorinated solvent source zones include the following:

- *Low cost* relative to certain other technologies, particularly capital costs. ISB has proven to be less costly than other common source treatment technologies, notably ISCO and *in situ* thermal treatment (ISTT) (McDade et al., 2005). Cost is typically site specific, however, and economies of scale can benefit certain technologies.

- *Similar performance to other injection-based technologies*, in terms of reductions in groundwater concentrations and mass discharge. The performance of ISB is typically assessed by monitoring groundwater concentrations of the chlorinated ethenes for some period of time following treatment. By this measure, ISB is expected to yield reductions of roughly one to two orders of magnitude (90–99% reductions) in total concentrations within the source zone at appropriate sites, although the total mass removed by some technologies such as ISTT is generally greater (McGuire et al., 2006). The long-term performance and the reductions in the overall restoration timeframes due to ISB or other source treatment technologies are difficult to evaluate at this time (ITRC, 2008).
- *Flexible design and operations.* ISB is adaptable to a wide variety of site conditions, and the ability to modify or expand the treatment system allows an adaptive management strategy that responds to the inevitable uncertainties while treating source zones. The ISB designs can be adapted to accommodate site infrastructure (buildings and piping) and operations can be adapted to respond to interim results (pulsing or reversing water flow, adjusting electron donor concentrations, adding wells to treat stagnant areas).
- *Ability to treat other cocontaminants.* ISB may treat other contaminants mixed with the chlorinated ethenes, including other solvents (dichloroethane and carbon tetrachloride) as well as other oxidized contaminants of concern that can be found with chlorinated solvents (e.g., hexavalent chromium, perchlorate and some explosives).
- *Ability to combine with other technologies.* As mentioned earlier, ISB may be used in combination with several other treatment technologies as part of a combined remedy. This feature can be particularly valuable for source zones because rarely can any one technology alone result in site closure.
- *Robust treatment.* ISB has proven robust with respect to upsets. Once biomass is built up, the system can handle downtimes and temporary changes in the operating conditions with little effect on the performance, and there is generally ample time to respond to upsets.

12.4.2 Limitations

The key limitations in using ISB to treat chlorinated solvent source zones include the following:

- *Possible increases in VC concentrations.* Temporary or even long-lasting increases in VC concentrations may be a concern at some sites. Such increases raise concerns regarding vapor intrusion into nearby buildings, exposures to workers or the public, and the potential for expansion of plumes. However, temporary increases are expected and generally can be managed successfully.
- *Relatively slow treatment.* Compared to ISTT, which may be completed within a few months of onsite treatment, ISB of a source zone may require many years of active treatment (ITRC, 2007). The time requirement is caused by both the lag time required for acclimation and growth of dechlorinating bacteria before optimal performance can be achieved, and the relatively slow contaminant removal rates possible during ISB.
- *Particularly slow treatment of DNAPL pools.* ISB is probably not appropriate for treating sources with extensive pools of DNAPL because the mass transfer from these pools can be very slow (Glover et al., 2007; ITRC, 2008).
- *Sensitivity to site characteristics.* Site-specific conditions can make ISB difficult or even infeasible. Low permeability or highly heterogeneous sites may be difficult to

treat effectively because of the difficulty in delivering substrates throughout the aquifer. Treatment depths may be limited by the drilling capabilities or by cost. Other site-specific problems might include the presence of competing electron acceptors, inhibitory cocontaminants or geochemical conditions such as unfavorable pH values.

- *Possible microbiological limitations.* Microbial populations capable of complete reductive dechlorination may not be present at a site, or they may be unable to compete with other microbial populations, causing incomplete degradation and accumulations of *cis*-DCE or VC. Bioaugmentation may be needed, or in some cases, aboveground treatment may be required to manage the daughter products.
- *Potential degradation of secondary water quality parameters.* Changes in pH and redox conditions as a result of bioremediation may lead to solubilization of metals such as arsenic or iron or may lead to an increase in the total dissolved solids. Acidification may be a particularly difficult issue that limits biodegradation and requires additions of buffer. The biodegradation process may cause harmful byproducts such as methane or hydrogen sulfide to be produced and remain in the subsurface.
- *Biofouling may cause operational difficulties.* Biofouling of wells may occur, as large concentrations of electron donors cause microbes to proliferate around injection or extraction wells. This biofouling can be controlled in many cases, but often causes increased costs and operational problems (Geosyntec, 2005).
- *Potential inhibition due to acidification.* As discussed earlier, *Dhc* cells are sensitive to pH, especially acidity, and in particular, the final reaction of VC reduction does not seem to occur below a pH of approximately 6.0. The potential for acidification should be considered carefully when selecting an electron donor source and when operating and monitoring the system.
- *Potential for rebound.* Although rebounds in contaminant concentrations within the source zone are not typically observed after ISB treatment, a long post-treatment monitoring record has not been kept, and there is reason to suspect that rebound could eventually occur at some treated sites (Adamson et al., 2011).
- *Potential changes in permeability.* Changes in permeability may occur during ISB for several reasons including (1) biomass clogging, (2) mineral precipitation in pores and (3) mobilization of colloidal particles. Biomass can grow near DNAPL accumulations and divert groundwater around these areas, at least temporarily. Such clogging on a local scale was observed during the ISB demonstration project at Dover Air Force Base (Section 12.10.2) and complicated interpretation of the results (Lebrón et al., 2007). Creation of reducing conditions can cause precipitation of several types of minerals, notably iron sulfides. Colloidal particles could potentially be mobilized by the reduction of iron oxides or changes in interparticle bond strength, but no field evidence indicates that this process represents a significant issue.

12.5 IMPLEMENTATION OPTIONS

ISB is a flexible technology. A wide range of electron donor sources can be used, and they can be delivered in a variety of ways. The technology can be used for different purposes and in different ways, depending on the goals and site characteristics (Figure 12.4 depicts the general design options). For source zones, ISB can be used as the primary technology or as a polishing

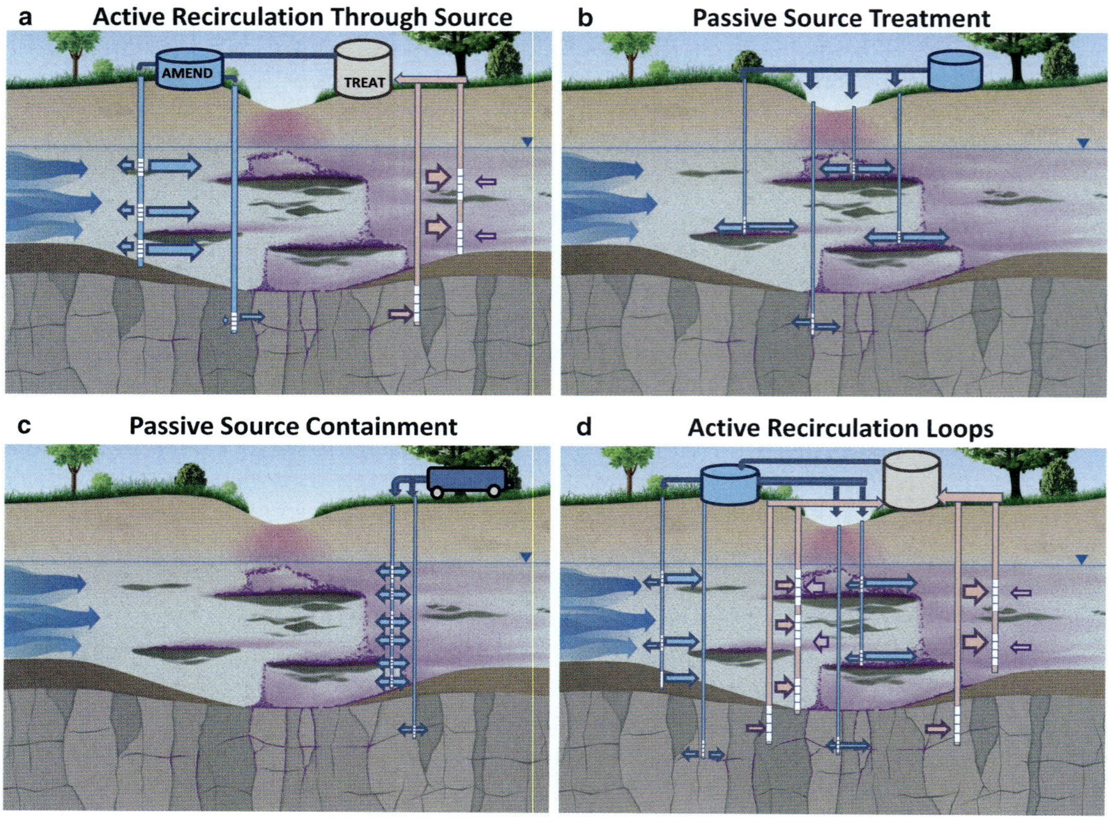

Figure 12.4. Possible designs for ISB systems to treat or contain a source zone. Active systems (a and d) rely on constant or frequent injections of water-soluble rapidly degraded electron donors, generally with some aboveground treatment of extracted water prior to reinjection. Passive systems (b and c) use one-time or infrequent additions of slowly solubilized electron donors.

technology. It can be implemented in a very active manner or in a much more passive approach. Bioaugmentation may or may not be used. Finally, ISB can be combined with other technologies to exploit synergies between biological and non-biological processes. These options are briefly described in the following sections.

12.5.1 Primary or Polishing Technology

ISB may be used as a primary treatment technology for some source zones. It can be very effective for treating chlorinated solvent source zones in which the contamination is present primarily in ganglia or sorbed to the aquifer matrix. However, it may not significantly reduce the cleanup time for sites with a significant fraction of the contamination present as pools (Christ et al., 2005; Table 12.3). Originally, many people assumed that ISB was useful only as a polishing technology following more aggressive technologies such as ISTT. Experience has shown, however, that ISB can be effective even at sites with DNAPL present (ITRC, 2007), and the ability to inject long-lasting electron donors such as vegetable oil can allow the use of ISB to control even high-strength sources and slowly remediate them over several years (ITRC, 2008; Borden et al., 2007).

Table 12.3. Impact of Pool Fraction on Source Longevity Without Treatment, With *In Situ* Bioremediation (ISB) Alone, or With ISB After Surfactant Flushing (Modified from Christ et al., 2005; Reproduced with Permission from *Environmental Health Perspectives*)

Scenario	Source Longevity (years)		
	Natural Gradient Dissolution	Source-Zone Bioremediation	SEAR[a] + Biopolishing
1. No Pools (GTP = ∞, PF = 0)[b]	36	7	0.01
2. Low PF (GTP > 1, PF < 0.5)	54	11	0.01
3. High PF (GTP < 1, PF > 0.5)	245	50	24
4. Pools Only (GTP = 0, PF = 1)	817	163	157

[a]Surfactant Enhanced Aquifer Remediation (4% Tween® for 10 days, assumed to remove 99.9% of the PCE mass
[b]Ganglia to Pool Ratio and Pool Fraction (0–1.0)

12.5.2 Active or Passive Treatment

Implementation of ISB can occur in several ways. The most fundamental distinction between different ISB methods is whether the treatment is considered *active* or *passive* (Stroo and Norris, 2009). Active treatment uses recirculation of fluids through the treatment zone on a continuous or near-continuous basis, with constant or frequent injections of a highly soluble electron donor solution such as lactate. In contrast, passive treatment relies on infrequent injections of long-lasting electron donors such as vegetable oils with little or no recirculation of fluids. More information on the types of electron donors available and the treatment options available is provided in AFCEE et al. (2004) and ITRC (2008).

For treating highly concentrated chlorinated solvent source zones, active treatment appears to be the most common approach (ITRC, 2008). It attempts to maximize the delivery of electron donors (and other amendments) throughout the treatment zone, leading to faster remediation. By continuously recirculating fluids through the target treatment zone, contaminant dissolution, and mixing with electron donor is enhanced. Active treatment is also highly flexible. Some or all of the injection and extraction wells can be reversed to improve delivery, additional wells can be added as needed to target stagnant or high-strength areas, and the concentrations and type of electron donor can be changed to respond to changing site conditions The principal negatives include the relatively high cost for constant operations and maintenance (O&M) and the potential for biofouling of wells.

Passive treatment is a slower process that relies largely on the slow migration of electron donors following injections. Distribution also may be more limited than active treatment, often requiring a larger number of injection points or wells. As a result, passive treatment approaches may be more expensive if drilling costs are high. There is also less opportunity to respond to uncertainties or changing conditions. However, passive treatment largely avoids the problems due to biofouling, and the potential for degradation of secondary water quality parameters may be much lower. Finally, the cost is generally much less than for active treatment, and passive source-zone treatment can be successful under the right conditions and given compatible remedial objectives (Borden et al., 2007).

12.5.3 Mass Removal or Source Containment

The majority of this chapter is concerned with mass removal as the primary purpose of ISB at a source zone. ISB can also be used to contain a source zone, however, by stimulating anaerobic biological activity within and/or immediately downgradient of the DNAPL source zone, to stop or reduce the flux of contaminants leaving the source zone. Containment is typically achieved by establishing a reactive barrier downgradient of the source, through injection of an electron donor into wells or injection points along a transect perpendicular to the groundwater flow path (see Figure 12.4). Alternatively, physical barriers may be established, with biological treatment in defined areas (so-called funnel-and-gate applications). Finally, hydrophobic electron donors such as vegetable oil may be injected into and near the source zone to sequester the DNAPL and degrade DNAPL constituents over time as they solubilize (Henry et al., 2007). Slow-release electron donors injected directly into source zones can degrade contaminants as they diffuse from lower-permeability zones, potentially enhancing cleanup rates. In contrast to mass removal applications, effective containment typically focuses on achieving and sustaining complete reductive dechlorination of all chlorinated ethenes in the aqueous phase.

12.5.4 Biostimulation or Bioaugmentation

Biostimulation relies on stimulating the native microbial population through creating appropriate conditions for their growth and activity (neutral pH, Eh <-100 mV, available hydrogen). In many cases the native population is fully capable of efficient and complete reductive dechlorination, although it may require a lag time of several months for growth and migration *in situ* (Henry, 2010). Managers may prefer to rely on the native microbial population because of the costs associated with adding bacteria to the subsurface (bioaugmentation), and in some cases because of regulatory concerns about introducing "foreign" organisms or genes.

In some cases, however, bioaugmentation may be needed or helpful, especially for chlorinated solvent contamination (Stroo et al., 2010). Several cultures containing *Dhc* strains capable of compete dechlorination have been developed for bioaugmentation of chlorinated solvent sites (ESTCP, 2005). Bioaugmentation can ensure that microorganisms capable of complete dechlorination are present at a site, and can decrease the lag time before complete dechlorination is observed (Ellis et al., 2000; Geosyntec, 2007; Lebrón et al., 2007). Models useful for designing bioaugmentation systems have recently been published (Schaefer et al., 2009a).

Bioaugmentation may be particularly useful for source-zone remediation for two reasons. First, the O&M costs for active ISB treatment of a source zone can be relatively high, and therefore any time lost due to inefficient treatment can be costly. Second, the fact that the contaminant concentrations are much higher than in the plume means that the potential for VC accumulations and exposure of humans to VC is greater, and bioaugmentation can reduce the magnitude and duration of any VC accumulation. Bioaugmentation also may be useful in treating any residual contamination after aggressive source-zone treatment technologies, such as ISTT, that may kill most or all of the *Dhc* in the target treatment zone (Friis et al., 2006). While deciding whether to bioaugment can be a difficult decision in some cases, the development of molecular biological and stable isotope techniques, and the experience from field-scale use, has improved the decision-making process (Stroo et al., 2013).

12.5.5 Combining ISB with Other Technologies

ISB is well suited to combinations with other technologies, either separated in time or space, or even applied together to exploit useful synergies. Perhaps the first example of combining ISB with another technology was the use of residual surfactants or cosolvents as electron donors (Mravik et al., 2003; Ramsburg et al., 2004). Since it is difficult to remove all of these chemicals from the subsurface after their use to enhance flushing of NAPLs, it makes sense to biodegrade these materials *in situ* and to take advantage of the fact that they can be fermented to produce hydrogen and to create a reducing environment and thereby stimulate reductive dechlorination (Amos et al., 2007b). However, these materials have proven to be expensive for direct treatment of DNAPL source zones and can have other limitations, such as the potential for enhancing DNAPL migration (NRC, 2005), so this combination has found little commercial use to date.

ISB also is well suited to use after more aggressive DNAPL recovery or destruction technologies, and such treatment trains may have a useful synergy. Significantly, recovery or destruction of DNAPL as an initial step will reduce the subsequent electron donor demand, making ISB more effective and cost-effective. For example, recovery of DNAPL pools using hydraulic displacement (Chapter 8) can be a valuable first step in that the DNAPL-water surface area available for mass transfer is increased, and electron donor demand is reduced. Other treatment technologies also will remove some of the electron donor demand, but may have other potential interactions worth considering.

In situ chemical oxidation may be compatible with ISB because it does not entirely sterilize the treated zone, and the site can be recolonized naturally, or bioaugmented if necessary (Sahl and Munakata-Marr, 2006). In some cases, partial oxidation of native organic compounds may stimulate subsequent biological activity by increasing the dissolved organic carbon (DOC) in the groundwater (Sahl et al., 2007). Most source-zone treatments with ISCO will require additional treatment (Krembs et al., 2010); therefore, ISB after ISCO may be an attractive combination, offering reductions in contaminant mass, followed by sustained biodegradation of the residual contaminants (Hrapovic et al., 2005; Sutton et al., 2010). However, there may be some concerns for subsequent *Dhc* activity, notably the rapid pH changes that can occur, especially after ISCO treatment in poorly buffered systems (Munakatta-Marr et al., 2011). If permanganate is used as the chemical oxidant, substantial amounts of solid phase MnO_2 may be present, requiring additional electron donor to achieve strongly reducing conditions. It often is prudent to wait for some period (months to years) for the geochemical conditions to re-equilibrate after ISCO.

In situ thermal treatment also may be combined with ISB. Although ISTT can leave the subsurface near-sterile, recolonization or bioaugmentation (after electron donor addition) can successfully reestablish reductive dechlorination (Friis et al., 2005). In fact, ISTT may stimulate biological activity after treatment or downgradient from the heated area by increasing the electron donor availability (Costanza et al., 2009; Fletcher et al., 2011). The temperatures typically achieved during ISTT (between 100°C and 300°C) are lethal to *Dhc*, but mild heating (30°C to as much as 40°C) may increase *Dhc* growth and reductive dechlorination rates (Friis et al., 2007). Note that it may take several months after ISTT for the subsurface to cool sufficiently for reductive dechlorination to proceed.

Some interest also has been shown in combining ISB with electrokinetics to deliver electron donors into lower-permeability materials (Gent et al., 2001). The concept takes advantage of the ability of electrokinetic remediation systems to preferentially move charged compounds into and through low-permeability materials (Acar et al., 1995). This technique is discussed further in Section 12.12.1.

To date the most commercially successful combination of ISB with another technology has been the combination of *in situ* chemical reduction (ISCR) and ISB. Commercially available materials combine a chemical reductant such as microscale zero-valent iron (ZVI) with electron donors. A material has been developed by the U.S. National Aeronautics and Space Administration (NASA) that combines emulsified edible oil and nanoscale ZVI to target DNAPL source zones (EZVI). In fact the material combines three processes – chemical reduction, biodegradation and partitioning of the DNAPL into the emulsified oil (Quinn et al., 2005). Field testing has indicated EZVI can be effective when delivered successfully. However, effectively delivering the EZVI to the contaminant can be challenging (O'Hara et al., 2006).

12.6 DESIGN CONSIDERATIONS

Designing an ISB system for a source zone involves balancing several interrelated factors (Figure 12.5). Good remediation designs require careful consideration of the site conditions and the available alternatives for implementing any technology. The flexible nature of ISB allows for many options. The design should be based on as complete a conceptual site model (CSM) as possible, one that includes an understanding of the local hydrogeology, the source-zone architecture (an estimate of the amounts of DNAPL in pools and ganglia) and the contaminant distribution between sorbed and DNAPL phases and between regions of differing permeability. Given the importance of delivery for ISB success, the CSM should also include an understanding of the local lithology and the degree of heterogeneity, as well as the adsorptive capacity of the subsurface materials.

An understanding of these subsurface characteristics is important when evaluating the materials that can be used as chemical amendments (particularly the electron donors and pH buffers). The donors and other amendments selected in turn strongly impact the injection and delivery strategy (passive or active treatment) and the design parameters (notably the spatial and vertical distributions of the injection and extraction wells and the well screenings). Finally, the microbiological and biogeochemical conditions should be evaluated before and during startup to determine whether bioaugmentation will be beneficial and/or to modify the design prior to full-scale operation. These issues are discussed briefly in the following sections.

12.6.1 Site-Specific Challenges

Each site poses a unique set of challenges for a source-zone ISB system. These include the site features, as well as regulatory, public, infrastructure and characterization challenges (Simpkin and Norris, 2010). The site features that can present major technical challenges can be classified as results of the subsurface hydraulics, lithology, contaminant distribution, geochemistry, cocontaminants or microbiology of the site. These technical challenges are described briefly in the remainder of this section.

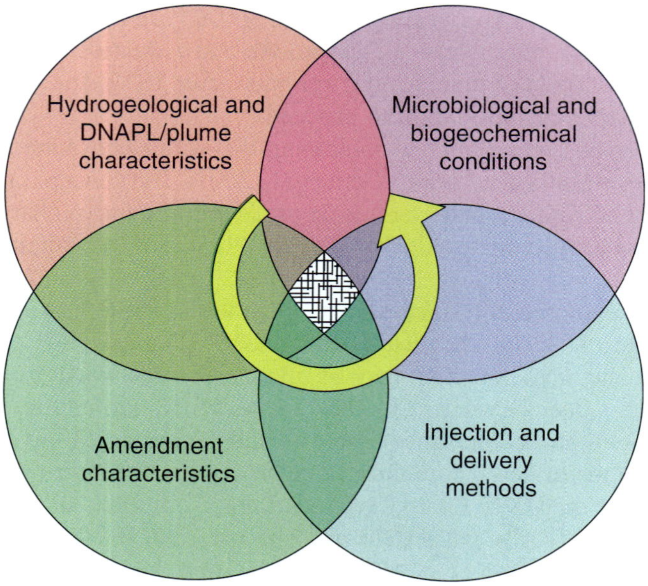

Figure 12.5. Design considerations for implementing ISB of chlorinated solvent source zones.

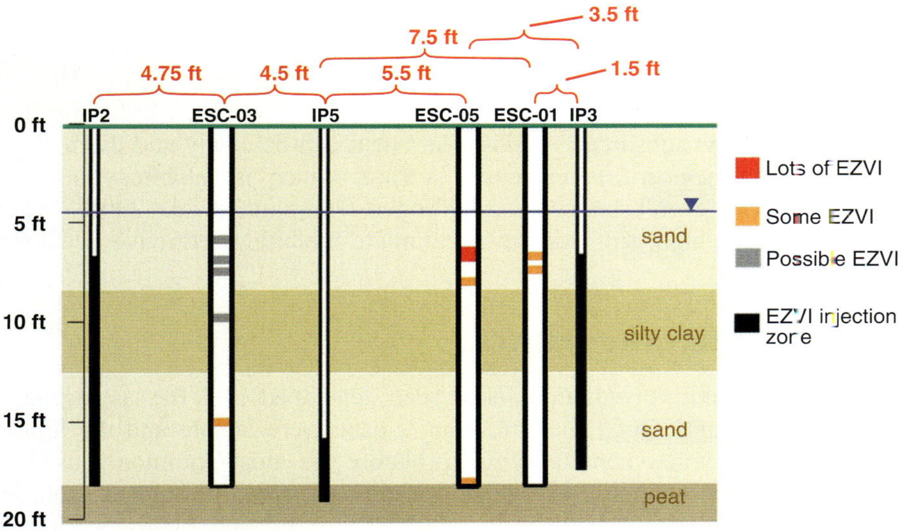

Figure 12.6. Visual observations of the distribution of emulsified oil:ZVI mixture (EZVI) during field demonstration at Cape Canaveral site. Most of the readily apparent black iron particles were found in only a small fraction of the total volume in extracted soil cores (ESC-01, 03, and 05) located near the injection points (IP 1, 2 and 3) within relatively thin sandy layers. Graphic provided courtesy of Geosyntec, Guelph, Ontario, Canada.

Potential hydraulic problems include the presence of preferential pathways and the relatively stagnant zones where little advective flux occurs. Preferential pathways can strongly limit delivery of amendments, even in aquifers in unconsolidated materials lacking clear lenses or fractures that obviously dominate flow. One of the best visual examples of the difficulties faced when injecting electron donors into the subsurface is from the demonstration of EZVI performed at Cape Canaveral in relatively permeable and homogeneous materials (Figure 12.6). In this case, the EZVI material was black, so the injected material could be easily observed in cores taken at varying distances from the injection points. The results from the core observations demonstrate that the vast majority of the material traveled through a very small proportion of the total volume, and any effects are likely to be spatially limited, at least initially (Quinn et al., 2005). The related hydraulic issue is the presence of stagnant zones, where DNAPL treatment is much slower and less effective, because amendments must reach the contaminated areas primarily through diffusion.

Contaminant distribution is almost always a difficult challenge, particularly since most chlorinated solvent sites have been contaminated for more than 20 years, allowing time for slower processes to affect the distribution. Effective treatment is particularly challenging if contaminants have migrated over time into regions with low permeability or poor accessilibity (Feenstra et al., 1996; Chapman and Parker, 2005). In such cases, delivering remedial agents to these zones via injection is difficult, and treatment will be relatively slow and incomplete because it will depend upon the rate of back diffusion. Another aspect of contaminant distribution is the source architecture (Lemke et al., 2004; Fure et al., 2006). Ganglia, particularly those in regions with higher permeability, are much more rapidly depleted than pools, because of the large differences in the surface area:volume relationships. Consequently, even enhanced dissolution by ISB will take a considerable time to remove pooled DNAPL accumulations.

The most common geochemical challenge, beyond overcoming the electron donor demand, is the pH and buffering capacity. Although conditions can be too alkaline for effective reductive dechlorination, acidity is by far the more common issue. Even slightly acidic conditions (pH < 6.0) can strongly inhibit the complete dechlorination to ethene, and partial dechlorination is inhibited below pH 5.5. Poorly buffered sites easily can become too acidic for dechlorination as fermentation of the electron donors proceeds. Modifying the pH of a significant volume of the subsurface is possible, but it can be costly and difficult. The other common geochemical concern is excessive sulfate, which is inhibitory because sulfate reducers compete with dechlorinators for hydrogen (Heimann et al., 2005). When sulfate concentrations are very high, sulfides can accumulate inhibiting reductive dechlorination by *Dhc* (He et al., 2005).

12.6.2 Amendment Alternatives

Several electron donor formulations have been developed over the last decade, and this rapid development is continuing. The first donors used were lactate and the lighter volatile fatty acids such as butyrate. Lactate is still probably the most commonly used donor for source-zone treatment because it is soluble (so it can be delivered through the source zone via recirculated water), it breaks down to provide both acetate and hydrogen, it is relatively inexpensive and the dosage can be adjusted to respond to site conditions and treatment responses. Other sources have been used including vegetable oils, molasses, and other soluble carbohydrate materials, alcohols (ethanol and methanol), whey, chitin and slow-release forms of lactate, notably Hydrogen Release Compound (HRC®). The properties and characteristics of these materials are summarized in Table 12.4. For source-zone ISB, the soluble materials, especially lactate, are probably the most commonly used, followed by the semisoluble vegetable oils and HRC®.

The soluble electron donor sources (alcohols, fatty acids, sugars) are compatible with either recirculation or direct injection techniques. These materials are inexpensive on a per-volume basis, and the ability to circulate them is a key advantage for source-zone treatment in particular. Contaminated regions distant from injection points can be treated more effectively by constant or near-constant recirculation than by injections, particularly if techniques such as pulsing or reversals in direction are used. Significantly, the active distribution and the extended duration of treatment result in sustained high concentration gradients between the more transmissive and the less permeable regions, promoting the migration of electron donors into the less permeable regions and thereby enhancing treatment of the diffused contaminants.

The less soluble sources (HRC®, vegetable oils) are compatible with passive treatment, involving one-time or infrequent injections to place the materials in the source zone. These materials provide long-term treatment, as they can persist for perhaps 2–5 years, depending on the contaminant levels and in the electron acceptor influx (AFCEE et al., 2004). The materials also may migrate to the less permeable areas following injection, and the DNAPL constituents can partition into the oil phase if these materials are used (ESTCP, 2006). However, the semisoluble electron donors will eventually be depleted, and reinjections may be needed at most sites, especially aerobic sites with rapid groundwater velocities. Slow-release electron donors are also commonly used after more aggressive source depletion or as a temporary barrier downgradient of the source-zone treatment.

It should also be noted that product development is ongoing. Recent developments have included products and procedures that combine materials, for example using lactate for initial treatment and oils for sustained treatment, or products like EZVI that seek to combine biological and chemical reduction. The distinctions between commercial electron donor sources may become less clear, and different products may well be used for different times during overall site remediation (initial treatment or reinjections to address residual contamination) or

Table 12.4. Substrates Used for Enhanced Anaerobic Bioremediation (From AFCEE et al., 2004)

Substrate	Typical Delivery Techniques	Form of Application	Frequency of Injection
Soluble Substrates			
Lactate and Butyrate	Injection wells or circulation systems	Acids or salts diluted in water	Continuous to monthly
Methanol and Ethanol	Injection wells or circulation systems	Diluted in water	Continuous to monthly
Sodium Benzoate	Injection wells or circulation systems	Dissolved in water	Continuous to monthly
Molasses, High Fructose Corn Syrup	Injection wells	Dissolved in water	Continuous to monthly
Whey (soluble)	Direct injection or injection wells	Dissolved in water or slurry	Monthly to annually
Slow-Release Substrates			
HRC® or HRC-X™	Direct injection	Straight injection	Annually to biannually for HRC® (typical); Every 3 to 4 years for HRC-X™; potential for one-time application
Vegetable Oils	Direct injection or injection wells	Straight oil injection with water push or high oil/water content (>20% oil) emulsions	One-time application (typical)
Vegetable Oil Emulsions	Direct injection or injection wells	Low oil content (<10%) microemulsions suspended in water	Every 2 to 3 years (typical)
Solid Substrates			
Mulch and Compost	Trenching or excavation	Trenches, excavations, or surface amendments	One-time application (typical)
Chitin (solid)	Trenching or injection of a chitin slurry	Solid or slurry	Annually to biannually; potential for one-time application

for different areas or objectives, such as enhanced source removal using lactate vs. injections of edible oils for source containment.

Other chemical amendments also may be needed for optimal treatment. The most common amendment added is an alkaline material to raise the pH. *Dehalococcoides* are pH sensitive, and reductive dechlorination is markedly inhibited below a pH of about 6.0 (Vainberg et al., 2006; Eaddy, 2008). Given that the groundwater at many sites is poorly buffered and mildly acidic before treatment, and fermentation of the large masses of electron donors that can be needed for source-zone bioremediation can be highly acidifying, pH buffering may be needed at many sites (Robinson et al., 2009). Potential water-soluble buffering materials include calcium carbonates ($CaCO_3$), sodium salts ($NaHCO_3$, Na_2CO_3), caustic soda (NaOH), magnesium carbonates ($MgCO_3$) or magnesium hydroxide ($Mg(OH)_2$). Materials also have been formulated specifically for groundwater buffering applications (AquaBupH™, a suspension of buffer particles in emulsified soybean oil). Vitamin B_{12} and yeast extract have been considered for use because *Dhc* relies on external sources of B_{12}. However, it may not be necessary to add these materials because other bacteria produce sufficient amounts. Similarly, inorganic nutrients may be included in commercial biostimulation products, but research has not demonstrated that they are necessary (AFCEE et al., 2004).

12.6.3 Electron Donor Requirements

The amount of electron donor that needs to be added depends on the amendment type chosen. For many slow-release substrates (EVOs or HRC), the substrate-loading rate is generally calculated on meeting the total demand from both the native electron acceptors (such as oxygen, nitrate and ferric iron) and the contaminant mass, including the continued influx of electron acceptors over the design life (AFCEE, 2007). Soluble substrates can be added frequently, so the loading rate need only achieve and sustain reducing conditions for a few days to weeks.

A spreadsheet tool has been developed to assist practitioners in determining the site-specific electron acceptor demand and the substrate required to meet that demand over the design life of the application (AFCEE et al., 2004). In addition, vendors have developed similar spreadsheets for their own materials. All spreadsheets should be used with some caution, as the uncertainty surrounding the estimates of the masses of contaminants and other electron acceptors within source zones often necessitates using large safety factors to estimate the dosage needed. In fact, it is common to include five- to tenfold increases in the initial electron donor estimates to address this uncertainty (ITRC, 2008).

It is also important to realize that other factors, such as the oil retention capacity of the treatment zone matrix, can determine the amount of electron donor that can be added at any one time (Borden et al., 2008). Lower safety factors may be justified when slow-release substrates are used for long-term control, especially when monitoring during and after a prior technology has yielded accurate estimates of contaminant mass or biodegradation rates. Lower safety factors also may be used for the soluble substrates, because the total demand can be met over numerous injections so the dosage can be adjusted in response to the operational monitoring results.

12.6.4 Injection and Delivery

The options for injection and delivery include direct injection, using permanent wells or direct-push injection points, and recirculation, using either natural gradient flow or forced recirculation. Natural gradient flow involves extracting water from one area, often clean water from upgradient of the contaminated area, amending it with electron donors and other amendments if needed and then injecting it immediately upgradient and/or within the source zone. Forced recirculation involves extraction from downgradient and injection upgradient and/or within the source.

Direct-push injection offers flexibility in the locations and depth intervals treated, but it also offers less control of the treatment area and the ability to deliver amendments. It also is likely to be less costly than establishing permanent wells, though it is important to realize that subsequent reinjections may be needed for adequate treatment. The forced recirculation strategy establishes a closed system that will reduce or eliminate groundwater influx and cause upgradient mounding and flow around the treatment area. Other options worth considering at some sites include the use of horizontal or directional wells to improve delivery, or even use of injection trenches for shallow zones.

When selecting and designing the system, one should ensure that an adequate amendment mass to treat the estimated amount of contaminant and other electron acceptors can be delivered, that the delivery will be as targeted to the contaminant mass as feasible, and that contingency plans are identified. Source-zone characterization is inherently uncertain, and the field-scale performance of ISB can be difficult to predict accurately, so a well-designed monitoring plan is needed, with clear links to contingency actions. Example contingencies include installing additional injection points to treat possible stagnant or bypassed zones, adding amendments into previously treated areas to target residual contamination, or injecting amendments long after treatment if eventual rebound occurs. The injection point spacing, the amendment concentrations and the injected water volumes should be based on calibrated groundwater models and careful mass estimates, but contingency plans will still be needed given the uncertainties in most site characterizations.

12.7 REMEDIAL OBJECTIVES

Careful definition of the remedial action objectives is always critical, including when using ISB for a chlorinated solvent source zone. Complete mass removal and rapid site closure are not likely using any source-zone remediation technology, and ISB is less aggressive and takes more time than other source-zone technologies such as thermal treatment or *in situ* soil mixing (Kavanaugh et al., 2003). Unfortunately, near-complete removal is needed to meet typical cleanup objectives such as MCLs, and ISB is unlikely to meet such criteria (Wright et al., 2010; ITRC, 2008). Therefore, it often is necessary to develop less stringent *functional* objectives for source treatment, such as mass discharge reduction or removal to the extent that monitored natural attenuation (MNA) can be implemented reliably (NRC, 2005). Some of these functional objectives are relatively easy to achieve using ISB, while others are more difficult, so it is important to develop realistic performance expectations when selecting the technology.

Figure 12.7 provides an initial screening-level evaluation of the ability of ISB to meet common remedial objectives for a source zone. The most common goal is probably to meet concentration criteria at a downgradient location after ISB is completed. This objective is difficult to assess in general, as it is highly dependent on site-specific factors. Based on experience to date, it seems reasonable to expect 90–99% reductions in concentrations immediately downgradient from the source. Similar reductions may be expected eventually at points further downgradient, although it may be less pronounced and take longer to measure if the travel time is significant and if there is significant sorbed contamination between the source and the compliance point.

A very common goal of source-zone ISB is *mass removal* and is based on the total contaminant mass removed or destroyed as a fraction of the estimated original mass. Unfortunately, this objective is often hard to measure accurately, given the uncertainties in most starting source mass estimates. Generally it is possible to measure the direct removal in extracted water, but estimates of mass destroyed are less precise, usually relying on total chloride increases and daughter product concentrations. Estimates of mass removal that have been made indicate more than 90% of the original mass can be removed under favorable conditions (McDade et al., 2005; GeoSyntec, 2004). In many cases, the goal is simply to remove mass to the extent practicable, although many would argue that the goal should be to remove sufficient DNAPL from the source zone to make a significant difference in the future site care requirements (for example, to allow MNA after ISB).

Less commonly, the primary goal of ISB is *flux reduction* (i.e., reductions in the mass discharge from the source and/or the concentrations in groundwater leaving the source zone). In terms of the risks to potential downgradient receptors, flux reduction is a more relevant metric than mass or concentration reductions, and interest in this approach is therefore increasing (ITRC, 2010; Cai et al., 2012). The flux reduction may be more or less than the

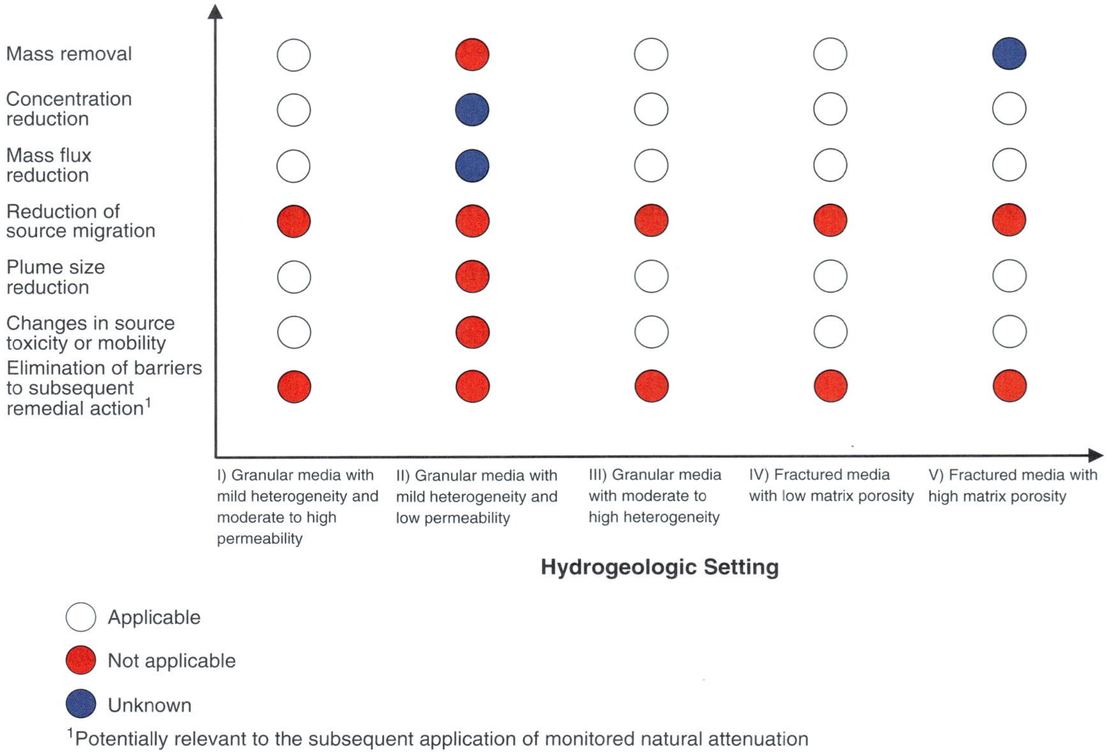

Figure 12.7. Ability of ISB to meet various remedial objectives for source-zone treatment in different hydrogeological settings (modified from ITRC, 2005).

mass removal, depending largely on the DNAPL architecture (Basu et al., 2008). However, if >90% of the mass can be removed, it is likely that a similar reduction in concentrations or discharge should be achieved as well (Sale and McWhorter, 2001; Stroo et al., 2003; Wood et al., 2004).

Plume life reduction is often the assumed goal of source depletion. That is, it is assumed that source depletion will reduce the plume longevity to a more reasonable timeframe (e.g., from centuries to <30 years). However, the lifetime of the dissolved phase plume may not be reduced nearly as much as the mass removal might suggest, because plumes tend to exhibit first-order decreases over time, and storage of contaminants in the plume (sorbed phase, diffused mass) can sustain the plume even after the source zone is completely removed or contained (Sale and Newell, 2010).

A related goal that is often the objective of the responsible parties is *life cycle cost reduction* (generally expressed not as total costs but as net present value). This objective can be difficult to define or estimate, given the limited history of any aggressive source depletion efforts. Plumes will continue to require some form of containment for long periods of time following most source depletion efforts, and the long-term costs for these containment technologies can represent a substantial fraction of the life cycle costs, even for MNA. The clearest opportunities to reduce life cycle costs through treatment are at sites where source depletion will allow practitioners to turn off an existing pump-and-treat system and rely on MNA for post-treatment site management. Of course in some cases the rate of spending may be as important, or even more important, than the total life cycle costs, and these economic drivers can affect the decision to treat the source zone as well as the choice of the technology.

12.8 PREDICTING PERFORMANCE

Although the relative contributions of the various mechanisms resulting in mass removal and concentration/flux reduction at a particular site of interest are not always clear, some level of performance prediction is often required as part of ISB remedial designs and remedy evaluations. Past experiences are of value, but mathematical model predictions are often useful. Models provide guidance when selecting ISB initially, when designing the ISB remedy, when deciding whether and how to modify ISB systems during operations, when establishing performance expectations regarding timeframes for mass removal and concentration/flux reduction, and when designing post-remedy monitoring programs. The remainder of this section provides an overview of modeling of source-zone ISB.

Modeling to predict ISB performance generally can be performed using either a simple screening-level model or a more sophisticated model that incorporates several of the important hydraulic, chemical, and biological processes discussed in this chapter. The level of sophistication appropriate for the model chosen will depend on the site conditions, level of existing information, and the modeling objectives. Screening-level models are appropriate for limited resources and information and are often analytical models based on exact solutions to a series of equations. More complex models, requiring more site-specific information and greater resources to develop and use, are generally numerical models (models using numerical time-stepping procedures that allow more realistic models of greater complexity than can be described by analytical models). Modeling is discussed in greater detail in Chapters 5 and 6.

The screening-level models used for simulating natural or enhanced biodegradation are typically based on the concept of first-order decay. First-order decay, which is commonly observed in biodegradation studies, is represented as:

$$C = C_0 e^{-\lambda t} \qquad \text{(Eq. 12.1)}$$

where C is the contaminant concentration of interest at time t, C_0 is the initial concentration, and λ is the decay constant. The decay constant is related to the contaminant concentration half-life ($t_{1/2}$) through:

$$t_{1/2} = \frac{0.693}{\lambda} \qquad \text{(Eq. 12.2)}$$

It is important to point out that using first-order decay to represent biodegradation in a mathematical model is a simplification of what actually occurs. In fact, the assumption of first-order biodegradation kinetics may not apply in all cases, even under controlled laboratory conditions (Simkins and Alexander, 1984), though it often provides a useful approximation of measured degradation rates. Simple first-order decay incorporated into the standard advection–dispersion equation results in a model that can compute decreases in concentration, but cannot explicitly account for processes such as lag, rate of biomass growth, competition, electron donor utilization, and other variables. Nevertheless, the use of analytical and numerical models based on simple first-order decay is prevalent at the present time, and they can provide useful screening-level predictions, even though they are subject to considerable uncertainty.

The most widely used analytical model for chlorinated solvent biodegradation is BIOCHLOR, which has proven useful for many situations. However, BIOCHLOR has significant limitations (West et al., 2008), and recent efforts have focused on improving analytical modeling of chlorinated solvent biodegradation (Burnell et al., 2012). The recent REMChlor model is an analytical model designed to evaluate source treatments, including ISB (Falta et al., 2005).

One source of uncertainty associated with a model based on first-order decay is the adopted contaminant half-life (or decay constant). The half-life is site-specific and can vary considerably with time and location at any particular site. These issues are often overlooked, and their effects are difficult to quantify. Suarez and Rifai (1999) report half-lives ranging from less than a day to many years for chlorinated ethenes in groundwater. Consequently, relying upon literature sources alone to predict ISB performance is not recommended except at the most basic level of screening. Most remedial designs should rely on site-specific evaluations of half-lives, through model calibration, microcosm studies or column studies (Newell et al., 2002).

An alternative to employing a simple analytical or numerical model based on first-order decay to predict ISB performance is to employ a multiphase/multicomponent reactive transport model. The advantage of employing such a model includes the ability to simulate specific processes of interest such as rate of biomass growth, competition, rate of electron donor utilization, etc. The challenge in employing such sophisticated models stems primarily from the need for site-specific values of a number of coefficients and rate constants, many of which are difficult to obtain in a field setting.

In addition to adopting some level of solute transport modeling to predict ISB performance, it is important to perform groundwater flow modeling to properly assess and design aspects of ISB implementation such as flow from injection wells to extraction wells within closed-loop systems, capture associated with extraction wells, and lateral spreading associated with the delivery of amendments from single-well injections. Groundwater flow modeling is typically performed as part of ISB design, but the model should also be continuously updated and employed as ISB remedy implementation takes place.

One recent approach to numerical modeling (a model called DNAPL3D-RX) is highlighted in the following section, to provide greater detail and insight into the key modeling issues. This modeling effort suggests that ISB may not remove all of the contaminant mass from a source zone and that rebound may occur after ISB at some sites, although it may not occur for months to years following the cessation of active treatment. The rebound is primarily attributable to the dissolution of DNAPL remaining at the end of active treatment. Furthermore, although active ISB treatment for 2–3 years can reduce mass discharge downgradient for several years after treatment is stopped, it may not provide a long-term improvement over natural dissolution alone. These conclusions suggest that practitioners should run such predictive models prior to selecting and designing ISB systems, carefully consider the potential for eventual rebound and include contingency plans to ensure long-term protectiveness.

Practitioners also should be aware that the uncertainty when modeling ISB performance can be considerable. Key sources of uncertainty include the following:

- The assumption of first-order kinetics, which can be a useful approximation but often ignores real-world complexities
- The difficulties in characterizing the spatial variability of hydraulic conductivity that will govern groundwater flow directions and rates of substrate delivery
- The potential for treatment-induced effects such as bioclogging that alter conditions during and after treatment
- The complexities of the subsurface microbial ecology, in which numerous interacting populations change over time and space
- The uncertainties regarding contaminant mass distribution, particularly mass in less permeable regions and mass in DNAPL pools
- Variation in degradation rate over time in response to electron donor addition and changing geochemical conditions
- Difficulty in extrapolating from degradation rates observed in laboratory-scale design studies to those actually achieved during field implementation

Nevertheless predicting ISB performance is critical, and the results can be highly useful in designing the ISB system, developing realistic expectations for ISB performance, and deciding when to stop active treatment or adjust the system to optimize performance. Importantly, the models employed in predicting performance must be updated continuously as the system is operated to reduce uncertainty and reflect the new information from operating and monitoring the system.

12.8.1 Modeling ISB at a Field Scale

To evaluate performance metrics and the potential for rebound following ISB treatment at the field scale, the numerical model DNAPL3D-RX (West et al., 2008; West and Kueper, 2012) was employed. This model simulates the complex biodegradation processes that occur during active treatment of TCE DNAPL source zones in porous media. DNAPL3D-RX is a three-dimensional finite-difference numerical model that integrates two-phase flow simulation (Gerhard and Kueper, 2003; Kueper and Frind, 1991), interphase mass transfer processes (Grant et al., 2007), and the reactive transport model RT3D (Clement, 1997, 2003; Clement et al., 1998).

The example simulations presented here were conducted deterministically for seven idealized template sites impacted with TCE DNAPL. Each template site was varied by one of three integral factors (Table 12.5): (1) TCE DNAPL release volume (small, medium or large); (2) geologic heterogeneity (low, medium or high variability in the hydraulic conductivity); or (3) the mean bulk permeability of the aquifer (low, medium or high). The performance metrics of interest were TCE DNAPL mass removal within the source zone and TCE solute mass flux reduction at a downgradient boundary. In addition, an enhancement factor was computed for both metrics that evaluates the potential for improvement from ISB treatment relative to abiotic processes (no treatment, dissolution only).

Given the complex and dynamic nature of ISB at real sites, models must integrate mechanisms, stoichiometry, and reaction kinetics that capture salient processes at the field scale. For the simulations presented herein, DNAPL3D-RX incorporated the following key biodegradation processes:

- Equilibrium dissolution of TCE DNAPL
- ISB by biostimulation with pulsed (1 day/week) lactate injection
- Lactate is converted *in situ* to H_2 by fermenters
- Competitive consumption of H_2 by both dechlorinators and methanogens

Table 12.5. Summary of Template Sites

Template Site	DNAPL	Initial Mass (kg)	Initial Volume (m^3)	Mean k (m^2)	Variance ln k $(\ln (m^2))^2$
Base case	TCE	3,520	2.41	3.03×10^{-12}	1.74
High mean k	TCE	3,496	2.39	3.02×10^{-11}	1.74
Low mean k	TCE	3,535	2.42	3.04×10^{-13}	1.74
Low heterogeneity	TCE	3,355	2.30	1.87×10^{-12}	0.87
High heterogeneity	TCE	3,186	2.18	7.41×10^{-12}	3.48
Small DNAPL volume	TCE	785	0.54	3.03×10^{-12}	1.74
Large DNAPL volume	TCE	7,343	5.03	3.03×10^{-12}	1.74

- Bacteria are immobile species adhering to soil grains
- The degradation of TCE to *cis*-DCE, where *cis*-DCE is the terminal end product
- TCE degradation to *cis*-DCE via dual-Monod kinetics, including the influence of both biomass concentrations and H_2 concentrations
- No toxic inhibition
- Solute transformation only occurs in the aqueous phase
- Reversible isothermal linear sorption
- First-order decay of biomass
- Bioclogging of soil pores due to biomass growth

The stoichiometric equations and kinetic reactions for each of the above processes are provided below. The governing equations for two-phase flow with sources and sinks, and the reactive transport equations for the mobile and immobile species are omitted here for brevity; the interested reader is referred to the aforementioned citations for those details.

As noted earlier, dechlorination can proceed in a step-wise process reducing PCE to ethene. For this work, only the transformation of TCE to *cis*-DCE is modeled:

$$\mathrm{TCE} + \mathrm{H_2} \rightarrow cis - \mathrm{DCE} + \mathrm{Cl^-} + \mathrm{H^+} + \mathrm{biomass} \qquad \text{(Eq. 12.3)}$$

where TCE is the electron acceptor, the electron donor substrate is H_2, and biomass is explicitly synthesized. Following Fennell and Gossett (1998), H_2 is generated through the *in situ* fermentation of injected lactate given by:

$$\frac{\partial[\mathrm{lactate}]}{\partial t} = -q_{lactate}^{MAX}[X_{lactate}]\left(\frac{[\mathrm{lactate}]}{K_{lactate} + [\mathrm{lactate}]}\right) \qquad \text{(Eq. 12.4)}$$

where q is the maximum utilization rate {M $M^{-1}T^{-1}$}, K is the Monod half-saturation constant {M L^{-3}}, and X is the biomass concentration {M L^{-3}}. The subscript *lactate* denotes a parameter related to the fermentation of lactate to H_2 and the brackets ([ ]) denote molar concentration.

Competition for the consumption of H_2 exists between dechlorinators and methanogens. In the case of the latter species, the utilization of H_2 is given by (e.g., Amos et al., 2007a; Fennell and Gossett, 1998):

$$\frac{\partial\left[\mathrm{H_2^{meth}}\right]}{\partial t} = -q_{meth}^{MAX}[X_{meth}]\left(\frac{([\mathrm{H_2}] - H_{meth}^*)}{K_{H_2}^{meth} + ([\mathrm{H_2}] - H_{meth}^*)}\right)I_{toxic} \qquad \text{(Eq. 12.5)}$$

where the subscript *meth* denotes the immobile species methanogens, H^* is the threshold H_2 concentration for subsistence, I_{toxic} is the TCE inhibition coefficient accounting for the influence of TCE toxicity on methanogensis and dechlorination (Amos et al., 2007a). The rate of TCE consumption by dechlorinators is given by (Christ and Abriola, 2007; Chu et al., 2003; Fennell and Gossett, 1998):

$$\frac{\partial[\mathrm{TCE}]}{\partial t} = -\frac{q_{TCE}^{MAX}[X_{CE}]}{R_{TCE}}\left(\frac{[\mathrm{TCE}]}{K_{TCE} + [\mathrm{TCE}]}\right)\left(\frac{([\mathrm{H_2}] - H^*)}{K_{H_2} + ([\mathrm{H_2}] - H^*)}\right)I_{toxic} \qquad \text{(Eq. 12.6)}$$

where the subscripts TCE and CE denotes the mobile solute species and the immobile dechlorinator species, respectively, and R is the retardation factor for TCE. For these simulations

$I_{toxic} = 1$ as a conservative bias that promotes greater degradation rates but also increased competition.

Given the above processes, the overall rate of change in H_2 concentration is given by (Christ and Abriola, 2007):

$$\frac{\partial[\mathrm{H_2}]}{\partial t} = F_{lactate}\frac{\partial[\mathrm{lactate}]}{\partial t} - \left(F_{TCE}\frac{\partial[\mathrm{TCE}]}{\partial t} + F_{meth}\frac{\partial\left[\mathrm{H_2^{meth}}\right]}{\partial t}\right) \quad \text{(Eq. 12.7)}$$

where F is the stoichiometric consumption or production coefficient (Bagley, 1998). Lactate serves as the only source of H_2; hence, when lactate injection ceases, no additional sources of H_2 are available for ISB processes, and all H_2 may eventually become depleted.

The biomass synthesis and decay for the fermenters ($X_{lactate}$), methanogens (X_{meth}), and dechlorinators (X_{CE}) can be described by (Fennell and Gossett, 1998):

$$\frac{\partial[X_{lactate}]}{\partial t} = -Y_{lactate}\frac{\partial[\mathrm{lactate}]}{\partial t} - \lambda_{lactate}[X_{lactate}] \quad \text{(Eq. 12.8)}$$

$$\frac{\partial[X_{meth}]}{\partial t} = -Y_{meth}\frac{\partial\left[\mathrm{H_2^{meth}}\right]}{\partial t} - \lambda_{meth}[X_{meth}] \quad \text{(Eq. 12.9)}$$

$$\frac{\partial[X_{CE}]}{\partial t} = -Y_{TCE}\frac{\partial[\mathrm{TCE}]}{\partial t} - \lambda_{CE}[X_{CE}] \quad \text{(Eq. 12.10)}$$

where Y is the biomass yield coefficient {−} for each species and λ is a first-order decay rate constant (Cupples et al., 2004).

The production of biomass can lead to the onset of bioclogging of soil pores, and on a larger scale, the biofouling of injection wells and geologic media. The mechanisms and kinetics of bioclogging have been studied by many researches yielding various mathematical representations (Chu et al., 2003; Clement et al., 1996; Thullner et al., 2002; Vandevivere and Baveye, 1992; Vandevivere et al., 1995; Zysset et al., 1994). For this work, the analytical approach represented by Clement et al. (1996) is adopted and modified to simulate reductions in permeability (k) as a function of total biomass (X) growth and decay (see West, 2009). Following Criddle et al. (1991), it is assumed that 20% of the maximum biomass is recalcitrant yielding a permanent reduction in permeability during biomass decay.

The performance metrics of interest for ISB of the TCE DNAPL source zone are total DNAPL mass (M_{DNAPL}) and total boundary mass flux (M_f), where the latter is computed by:

$$M_f = \sum_{i=1}^{N} C_i q_i \quad \text{(Eq. 12.11)}$$

where i denotes an individual node, q is the Darcy flux {L T^{-1}}, and C is the concentration of TCE at each node {M L^{-3}}. The metrics M_{DNAPL} and M_f are further utilized to define enhancement factors that directly compare the performance of ISB to abiotic processes only (i.e., no treatment, dissolution only).

The enhancement factor of total DNAPL mass removal (E_m) is given by:

$$E_m = \frac{M_{DNAPL}^0 - M_{DNAPL}^{ISB}(t)}{M_{DNAPL}^0 - M_{DNAPL}^{diss}(t)} \quad \text{(Eq. 12.12)}$$

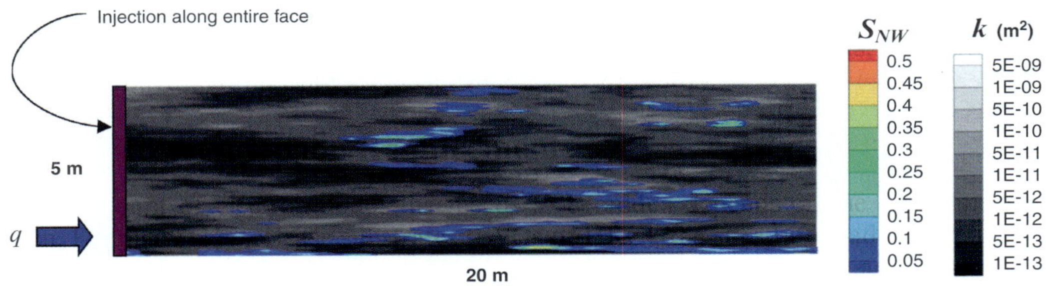

Figure 12.8. Cross-sectional view of model domain utilized for ISB simulations. Groundwater flow is from left to right with amendment injection along the upgradient boundary face. The TCE DNAPL saturations (S_{NW}) and permeability (k) field highlight the influence of geologic heterogeneity on DNAPL distribution in the subsurface.

where M^0_{DNAPL} is the initial DNAPL mass in the domain ($t = 0$), and M^{ISB}_{DNAPL} and M^{diss}_{DNAPL} are the DNAPL mass values for ISB and dissolution only (no treatment), respectively, at time (t). The enhancement factor for total boundary mass flux (E_f) is defined as:

$$E_f = \frac{M_f^{diss}(t)}{M_f^{ISB}(t)} \qquad \text{(Eq. 12.13)}$$

where the superscripts *ISB* and *diss* denote biostimulation and abiotic dissolution (no treatment), respectively. An enhancement factor for either E_m or E_f greater than 1 indicates that ISB performance was greater than dissolution only (no treatment).

Each template site represents a DNAPL source zone measuring 20 m (65.6 ft) long, 10 m (32.8 ft) wide, and 5 m (16.4 ft) thick, comprising both pooled and residual TCE DNAPL. A cross-sectional depiction of the model domain and TCE DNAPL saturations (S_{NW}) at mid-width (5 m) is provided in Figure 12.8. Hydraulic displacement (i.e., water flooding) was conducted on each template site prior to initiating ISB as a preliminary mass removal strategy that reduced the degree of pooled DNAPL. The model was discretized into 0.4 m by 0.4 m by 0.05 m (vertically) blocks for a total of 125,000 nodes. General model input parameters and ISB specific input parameters are provided in West (2009). An average hydraulic gradient of 0.05 was generated across each source zone, which is considered representative of continuous pumping conditions. A constant concentration injection boundary condition was applied to the upgradient face of the domain to simulate an amendment injection trench or line of injection wells positioned further upgradient of the source zone.

Simulations were conducted for each template site for both ISB and dissolution only (no treatment) scenarios. For ISB, lactate was injected at a concentration of 39,130 mg/L at a frequency of 1 day per week for a period of 2.5 years. Subsequently, active injection of lactate was discontinued, while the model was executed for an additional 7.5 years. The total simulation time for all scenarios was 10 years. Both M_{DNAPL} and M_f were computed for each time step, whereas E_m and E_f were selectively computed at 2.5 years (end of injection) and 10 years (end of simulation).

Figure 12.9 presents the DNAPL mass and boundary mass flux for the base case from the initiation of treatment (0 years) to 10 years; similar plots were constructed for all scenarios but are not presented here (see West, 2009). For dissolution only, the initial DNAPL mass of 3,520 kg is reduced to 900 kg over the duration of the simulation. For ISB, the rate of DNAPL mass depletion is less than the dissolution only case for 0–2 years, indicating no enhancement during this initial lag period. Similar occurrences of lag were observed for all template sites

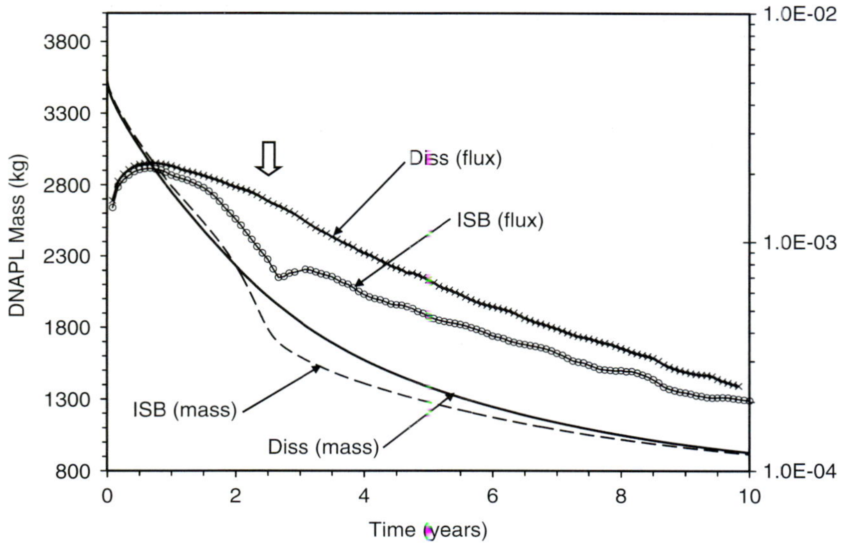

Figure 12.9. Plot of TCE DNAPL mass remaining and TCE boundary mass flux for the base case template site. The *arrow* indicates the time that active injection of lactate for ISB was terminated (2.5 years). Approximately 900 kg of TCE DNAPL mass remained after 10 years for both ISB and dissolution only (Diss).

Table 12.6. Comparison of Enhancement Factors for DNAPL Mass Removal (E_m) and Boundary Mass Flux Reduction (E_f) at the End of Active Injection of ISB and at 10 Years

Template Site	2.5 Years (End of Active Injection for ISB)		10 Years (End of Simulation)	
	E_m	E_f	E_m	E_f
Base case	1.15	1.81	1.00	1.13
High mean *k*	1.12	6.51	1.03	7.48
Low mean *k*	1.53	1.06	1.94	1.34
Low heterogeneity	1.25	1.40	1.01	1.31
High heterogeneity	1.11	7.01	1.01	1.16
Small DNAPL volume	1.15	6.61	1.00	1.75
Large DNAPL volume	1.27	1.58	1.03	1.21

Values greater than 1 indicate that ISB removed more DNAPL mass relative to dissolution or had relatively less boundary mass flux.

with the exception of the high mean *k* scenario. This reduction in ISB performance relative to dissolution only for M_{DNAPL} is due to bioclogging as biomass preferentially develops at higher saturation nodes that generally coincide with higher permeability pathways. However, from 2 years to the termination of injection (2.5 years) there is an accelerated reduction in DNAPL mass for ISB, leading to an E_m of 1.15 (see Table 12.6).

In contrast, a lag period was not observed with respect to boundary mass flux for the base case, and the $E_f > 1$ for the entire simulation. At 2.5 years the $E_f = 1.81$ with a peak $E_f = 2.07$

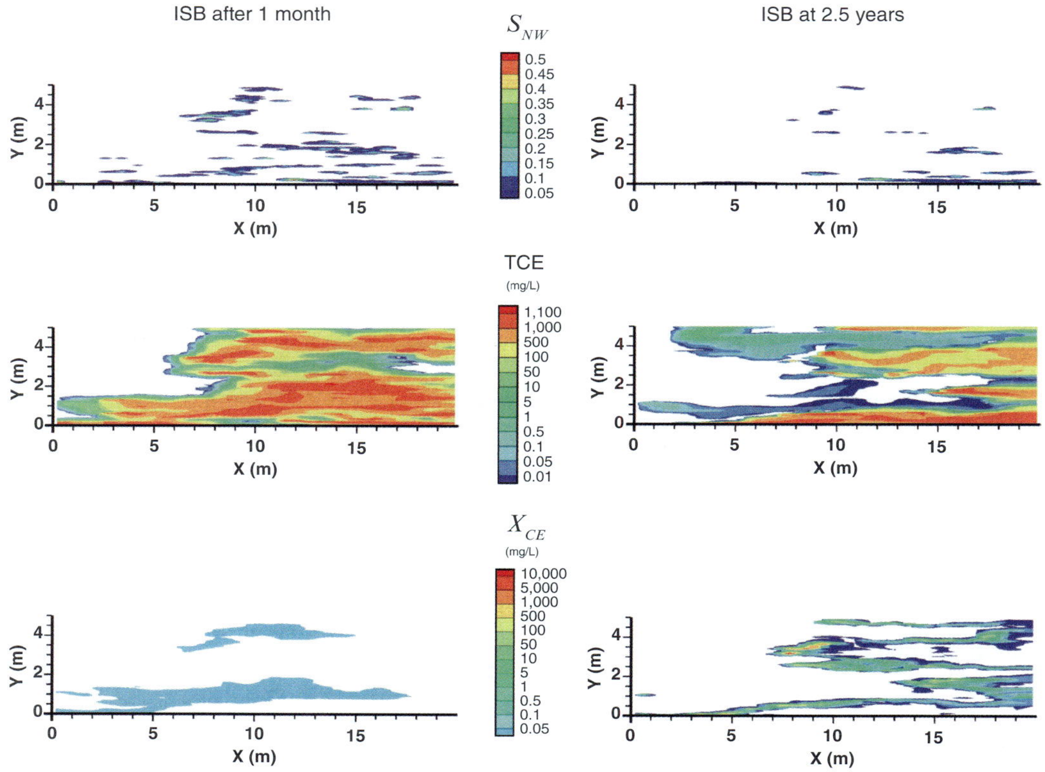

Figure 12.10. Cross-sectional plots of non-wetting saturations (S_{NW}), TCE concentration and dechlorinator biomass (X_{CE}) after 1 month and 2.5 years of ISB. Groundwater flow is from left to right and lactate is injected along the left upgradient boundary face.

occurring shortly after the termination of lactate injection. Following this period, H_2 concentrations cannot be replenished and the rate of dechlorination decreases markedly. In fact, the rebound of mass flux is observed for several months following the termination of injection. Of note is that the numerical model did not allow for accumulated biomass to act as a carbon source and electron donor following the cessation of active treatment. In practice, this biomass may delay the time to which rebound will be observed. We note that further research is required to further understand this process.

For all template sites considered here, the $E_f \geq 1$ throughout the simulation. However, the occurrence of rebound was case dependent: rebound was observed for the base case (medium variance and medium mean k), the large volume scenario, the small volume scenario, the high mean k scenario, and the high heterogeneity scenario. The latter three scenarios exhibited both the greatest E_f at 2.5 years and the most pronounced rebound (see Table 12.6). In these scenarios rebound occurred for 1 year or less following the termination of injection. From Table 12.6 it also can be observed that the high mean k scenario was the only ISB simulation to achieve a greater enhancement in mass flux reduction (M_f) at 10 years than at 2.5 years.

A comparison of DNAPL saturations (S_{NW}), TCE solute concentrations, and biomass growth (X_{CE}) is provided in Figure 12.10 for a longitudinal cross-section through the domain at mid-width (5 m). The first pane of plots is for 1 month after the initiation of ISB, while the alternate pane shows model output at the termination of active lactate injection (2.5 years). The top panes for S_{NW} illustrates mass depletion over the 2.5-year period, where the total number of nodes with $S_{NW} > 0$ is greatly reduced, as is the overall magnitude of S_{NW} at nodes with DNAPL remaining; similar results occur for dissolution only (no treatment), but the extent and magnitude of depletion is not as pronounced. TCE concentrations are presented in the middle pane, which demonstrates several interesting results.

Initially TCE concentrations are widely distributed throughout the domain with the higher saturation nodes yielding the highest concentrations that are selectively transported through the most transmissive pathways. Some of the solute impacted zones are generated from out-of-plane transport. At 2.5 years, the TCE solute preferential pathways become more pronounced due to geologic heterogeneity, variable DNAPL saturations, spatially selective biomass growth and flow bypassing. The installation of short monitoring well screens or intermittent groundwater grab samples along the downgradient boundary could produce markedly different concentrations depending on vertical placement. Thus a sufficiently dense monitoring/sampling network is required to accurately capture representative concentrations and mass flux signatures from DNAPL source zones.

The bottom pane demonstrates the selective nature of dechlorinator (X_{CE}) biomass growth in these simulations. A low concentration biomass initial condition was assumed at the initiation of ISB to represent background; hence the distribution of low concentration biomass in the plot of ISB after 1 month. However, following an additional 2.4 years of lactate injections (1 day per week), biomass concentrations (expressed in mg/L) increased by up to six orders of magnitude at some nodes. The dechlorinator biomass is highly localized around higher saturation nodes and along higher concentration TCE solute pathways. This marked increase in biomass density may not be as significant when toxicity is incorporated. Despite the absence of toxicity, Table 13.6 summarizes that the maximum simulated enhancement for ISB at 2.5 years for E_m and E_f was 1.53 and 7.01, respectively.

Modeling results demonstrate that enhancements of DNAPL mass removal after 2.5 years of ISB ranged between 1.11 and 1.53 for the template sites and that these enhancement factors generally diminished after 7.5 years of post-treatment monitoring due to dissolution tailing. Conceivably, tailing effects might be less pronounced if a residual H_2 concentration was maintained during post-treatment monitoring. Experimental work with ISB and PCE has found enhancement factors of 2–3 (Sleep et al., 2006; Yang and McCarty, 2002), 5 (Yang and McCarty, 2000) and up to 16 (Cope and Hughes, 2001). In the field, enhanced dissolution may be more limited due to heterogeneity, flow bypassing and many other complicating factors. Only the low mean k scenario demonstrated an increase in E_m during post-treatment monitoring, due primarily the low seepage velocity and slow travel time of amendments relative to the length of the source zone and distance to the downgradient boundary face.

Because ISB is a spatially kinetic process, the positioning of the downgradient monitoring point influences the interpretation of enhancement for a given treatment time: moving the compliance point further up- or downgradient can produce different results. It is important to note that both ISB and dissolution only (no treatment) were unable to completely remediate any of the seven template sites; in all scenarios DNAPL mass and boundary mass flux persisted. In addition, dissolution tailing generally reduced E_m to 1 by 10 years, suggesting that terminating ISB prior to complete DNAPL mass removal is an ineffective long-term strategy.

In general, the enhancement of mass flux reduction (E_f) was greater than the enhancement of DNAPL mass removal (E_m) at both 2.5 and 10 years; the low mean k scenario was the only exception. The magnitude of M_f was variable between template site as was the occurrence, duration and magnitude of mass flux rebound following the termination of injection. It is important to note that E_f is not necessarily correlated to E_m, and significant reductions in boundary mass flux do not necessarily imply significant reductions in DNAPL mass.

12.9 OPERATIONS AND MONITORING ISSUES

12.9.1 Operating and Optimizing ISB Systems

Active remediation of source zones using ISB may require frequent monitoring to optimize treatment. Electron donor and amendment concentrations may need adjustment to respond to process monitoring information. Interim results should be evaluated carefully for evidence of stagnant zones requiring more intense treatment and for evidence of biofouling or other operational problems that need to be addressed before they become serious issues. Pulsing or reversals in extraction/injection wells may improve delivery, and it may be necessary to add wells to target specific areas or depth intervals.

In addition, any technology relying on fluid injection into an aquifer entails several operational risks including contaminant migration, preferential flow and therefore poor distribution, and short-circuiting. To control these risks during operations, the following operational issues should be evaluated carefully: (1) injection pressure limits, (2) DNAPL mobilization, (3) hydraulic responses in confined and semi-confined aquifer formations and (4) groundwater displacement (ITRC, 2008).

Injecting fluids into an aquifer can lead to fracturing and uncontrolled spreading of the fluids, along with any amendments and/or contaminants. Injection pressures should be designed to minimize unintentional hydraulic fracturing and avoid contaminant mobilization. Direct-push injections or conventional screened injection wells should be tested to assure that the aquifer can accommodate fluid insertion at the design flow rate. Conservative tracers such as bromide can be used to identify and quantify any such impacts, and if detected, the injection pressures and volumes should be modified to limit the effects.

Finally, injecting fluids near or above pooled NAPL can alter the capillary pressure and potentially mobilize the DNAPL. This may not be detrimental if the mobilized DNAPL enters already-impacted regions of the subsurface, but is likely undesirable if the mobilization of DNAPL leads to an expansion of the footprint or depth of the source zone. Related to this, the risks associated with drilling through a DNAPL source zone should be evaluated and considered prior to installing the injection, extraction, and monitoring wells needed to support ISB implementation.

Because of the uncertainties inherent in source-zone treatment, especially with an injection-based technology such as ISB, optimization is critical. Optimization is particularly important during the early stages of treatment. The injection volumes, pressures, and amendment concentrations should be estimated based on mathematical modeling and the best available knowledge (see Borden et al., 2008). However, practitioners should be prepared to carefully monitor the actual operations and to adjust the conditions as needed during treatment.

12.9.2 Monitoring ISB Systems

Monitoring an ISB system includes both process and performance monitoring (ITRC, 2008). Process monitoring is used to evaluate whether the system is meeting its design objectives, or if optimization is needed. It includes evaluating amendment distribution and longevity, measuring numbers of specific microbes or gene copies, ensuring that environmental conditions are favorable and that DNAPL is not migrating, and determining the extent and duration of any accumulations of potentially harmful daughter products. Performance monitoring is used to evaluate whether the treatment is meeting the remedial objectives and when it can be shut off. Performance monitoring involves evaluating multiple lines of evidence, including: (1) the concentrations of the parent chlorinated ethenes, and all daughter and end products; (2) any geochemical changes, particularly the impacts to secondary water quality parameters; and possibly (3) the mass discharge and flux of the COCs (ITRC, 2010).

The differences and similarities in these monitoring phases are discussed in detail in ITRC (2008). The following sections briefly describe the key monitoring parameters, the reasons they should be monitored, and issues to consider when measuring these parameters and evaluating the results. The key parameters can be classified into four categories: (1) contaminant concentrations, (2) geochemical characteristics, (3) biological indicators and (4) stable isotopes.

12.9.2.1 Contaminant Concentrations and Mass Discharge

Enhancing the dissolution and degradation of the DNAPL constituents produces a characteristic pattern of sequential increases and subsequent decreases in the daughter product concentrations (Figure 12.11). This pattern should occur over time in a well near DNAPL accumulations, as it reflects the increased rate of formation of daughter products closer to the DNAPL sources. However, the pattern may also reflect the greater aqueous solubilities of the daughter products, so that the concentrations of DCE and VC may temporarily be greater (on an absolute or molar basis) than those of the parent compounds (Carr et al., 2000; ITRC, 2008). Also the parent compound may be degrading faster in solution than some of the daughter products. As a result, the patterns actually observed at many sites where reductive dechlorination is working may differ from the somewhat idealized pattern shown in Figure 12.11.

Ethene, and in some cases ethane, are key breakdown products to monitor because they are innocuous compounds, and their presence at reasonable concentrations (>5–10% of the parent concentration on a molar basis) demonstrates that complete conversion through VC is occurring. However, ethene and ethane generally are not present in stoichiometric amounts, and mass balances are notoriously incomplete at sites undergoing ISB. There are at least three reasons for the difficulty in closing mass balances: (1) some of the ethene and ethane produced is further biodegraded to carbon dioxide and/or methane (Bradley, 2003); (2) non-biological reactions, notably reductive dehalogenation by Fe(II) minerals, can contribute to chlorinated ethene destruction by a different pathway (through acetylene), without production of chlorinated ethene daughter products (Ferrey et al., 2004; Cwiertny and Scherer, 2010); and (3) some aerobic biodegradation of VC (and possibly *cis*-DCE) may be occurring at very low oxygen levels, with conversion to carbon dioxide and chloride and no ethene formation (Coleman et al., 2002; Gossett, 2010).

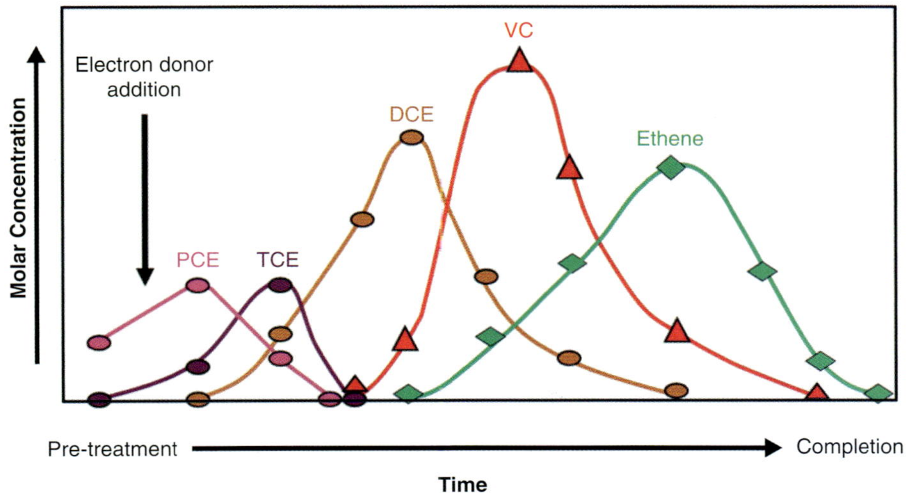

Figure 12.11. Idealized pattern of daughter product appearance and removal during *in situ* bioremediation of a chlorinated solvent source zone (modified from ITRC, 2008).

Contaminant monitoring may be required for several years after treatment. Although rebound in contaminant concentrations is typically not observed following ISB application in source zones (McGuire et al., 2006; Adamson and Newell, 2009), few sites have monitoring records extending more than a few years post-treatment, and potential certainly exists for some rebound after the residual effects of treatment (high biomass, back diffusion of electron donors) have subsided, so ongoing monitoring is prudent. In the event that rebound does not occur, such sustained treatment may be attributed to the slow degradation of the microbial biomass produced during active treatment (Sleep et al., 2005; Adamson and Newell, 2009), as well as to the possible storage of reducing power in iron minerals capable of abiotic reductive dechlorination (Elsner et al., 2004) and/or the possible diffusion of electron donors into lower-permeability materials, and their slow back diffusion into the more transmissive regions (Adamson et al., 2011).

Mass discharge is increasingly recognized as a valuable performance metric for partial source-zone remediation, that combines concentration data and the groundwater velocity to measure the mass leaving a source zone (Wood et al., 2004). However, the uncertainty involved in mass flux and discharge measurements can be large and often is not quantified (Li et al., 2007). ISB in particular has important features that need to be recognized by those measuring and interpreting mass discharge results; it involves partial degradation to daughter products as well as possibly the complete reduction to ethene (or ethane) or even abiotic reduction to products such as acetylene that typically are not measured (Brown et al., 2009). Therefore, the mass fluxes of several compounds must be monitored, and the uncertainties involved must be considered (Cai et al., 2012). In addition, ISB often results in a temporary increase in mass discharge, followed by a large (and hopefully permanent) reduction from pre-treatment levels, so it is important to interpret ISB mass discharge data in the context of the time and location of the measurements taken (ITRC, 2010; Cai et al., 2012). Finally, there is some potential for biomass accumulations to change flow paths during ISB, as biomass growth near DNAPL accumulations or in areas exposed to higher levels of organic carbon can cause groundwater at a local scale to be diverted around these areas (Lebrón et al., 2007).

12.9.2.2 Geochemical Characteristics

The main parameters needed to confirm that the redox conditions are appropriate for reductive dechlorination are the dissolved oxygen (DO) concentration and the oxidation–reduction potential (ORP or Eh). DO should be <0.5 mg/L, and the Eh should be < −100 mV (AFCEE, 2007). However, one should keep in mind that DO measurements from the field are notoriously untrustworthy (Wilkin et al., 2001). Other key indicators of the dominant terminal electron accepting processes include dissolved iron (Fe[II]), sulfate and sulfide concentrations, and methane. At a minimum, there should be evidence of ongoing sulfate reduction, and preferably there should be evidence of methanogenesis (detectable CH_4) (Henry, 2010).

Sufficient electron donor should be available to sustain reducing conditions and provide hydrogen for reductive dechlorination. Although fatty acid concentrations are often measured, the DOC concentrations are typically sufficient to ensure that amendments are being distributed as planned and are at levels sufficient to ensure effective treatment. Existing guidance suggests that maintaining DOC > 50 mg/L in monitoring wells within the treatment zone should be sufficient for soluble substrate systems (Suthersan et al., 2002; AFCEE, 2007). However, lower levels of DOC may be acceptable for slow-release substrates such as mulch and vegetable oil.

The pH is a key factor both for characterizing the site's suitability (as discussed earlier) and for ensuring continued effective treatment. The fermentation process can be highly acidifying, especially in poorly buffered aquifers when large amounts of electron donor sources are added, so ensuring that the pH remains >5.5, or preferably ≥6.0, is important to ensuring that complete dechlorination can continue to occur (Robinson et al., 2009).

Chloride is particularly important to monitor when treating source zones. Chloride is the end product of dechlorination, so the molar volume of chloride produced is a direct measure of the rate of biodegradation occurring (ITRC, 2008). The background chloride concentrations often make it difficult to detect any increase, especially when treating plumes (AFCEE et al., 2004). In source zones, which often have relatively high chlorinated ethene concentrations compared to the background levels of Cl^-, the production of Cl^- often can provide a direct measure of the rate of dechlorination and can be converted back to the original moles of parent compound that were desorbed and degraded during treatment. Estimating the rate of biodegradation on the basis of such data should be carried out using a mathematical solute transport model. Estimating the rate of biodegradation cannot be carried out by simply matching a first-order decay curve to the concentration versus time signature in a monitoring well because that signature is likely influenced by a number of additional processes such as changing rates of upgradient DNAPL dissolution as the DNAPL is depleted, hydrodynamic dispersion and seasonal changes in the direction of groundwater flow.

Other conditions that may need to be monitored include any secondary water quality parameters that may be of concern besides pH, sulfides and DOC (Henry, 2010). Establishing strongly reducing conditions, and lowering the pH, can increase the concentrations of dissolved metals, notably iron, manganese, arsenic and possibly selenium, and generate undesirable fermentation products (ketones and aldehydes, as well as hydrogen sulfide and methane gas).

12.9.2.3 Biological Indicators

Dhc bacteria are very slow-growing and difficult to culture. They are therefore difficult to isolate or enumerate using conventional microbiological methods such as plate counts (Löffler et al., 2013b). Fortunately, molecular biological methods have made it possible to identify and quantify the *Dhc* populations at a site, as well as some of the genes involved in reductive dehalogenation (notably VC RDases). As mentioned previously, the numbers of these biomarkers have been useful for deciding whether bioaugmentation is needed and if ISB systems are operating effectively (Stroo et al., 2012). The technique can also be used to track bioaugmentation cultures or even to discriminate between indigenous and introduced *Dhc* (Holmes et al., 2006).

Quantifying *Dhc* and VC dehalogenase genes (specifically vcrA and bvcA) is done by using the qPCR technique (Cupples, 2008). Although other biomarkers and molecular biology methods may be used, qPCR has proven to be highly sensitive and specific for key gene sequences (SERDP, 2006). Recent work has produced guidance on collecting and analyzing samples by qPCR as well as on interpreting the results of *Dhc* and VC Rdase biomarkers (Lebrón et al., 2008; Ritalahti et al., 2010; ITRC, 2011). One important finding is that onsite biomass collection using filtration techniques is important for accurate analyses, as compared to the more typical practice of shipping the unfiltered groundwater samples to the off-site laboratory (Ritalahti et al., 2010).

12.9.2.4 Stable Isotopes

Compound-specific isotope analysis (CSIA) relies on specialized mass spectrometers to distinguish between stable isotopes of several important atoms, including $^{13}C/^{12}C$ and $^{35}Cl/^{37}Cl$. Biodegradation reactions preferentially deplete the pool of lighter atoms in the parent compound, as these bonds are slightly easier to break. As a result, CSIA of the parent compounds and daughter products can distinguish between biological and non-biological processes affecting the contaminants (Liang et al., 2009) and also can indicate the rate and extent of

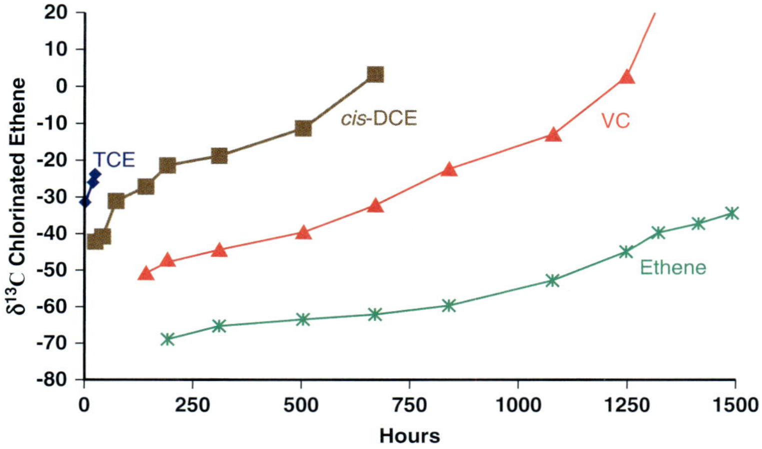

Figure 12.12. Pattern of carbon stable isotope ratios during biodegradation of PCE (modified from Slater et al., 2001). The δ ^{13}C value refers to the normalized $^{13}C/^{12}C$ ratio, expressed as per mil, or 0/00.

biodegradation (Sherwood-Lollar et al., 2001; Song et al., 2002; Aeppli et al., 2010). CSIA has proven valuable, largely because it provides an unequivocal indication of natural biodegradation and sometimes can discriminate between differing sources (Hunkeler et al., 2008; Wilson, 2010). The typical pattern of sequential enrichment in the heavier isotopes of the parent and daughter products is illustrated in Figure 12.12.

CSIA also can indicate that biodegradation and DNAPL depletion are occurring during source-zone bioremediation. In a well near the DNAPL source material, the expected pattern following electron donor addition (and bioaugmentation, if performed) is a reduction in the concentrations of the parent compounds (PCE and TCE), because these are the most rapidly biodegraded under reducing conditions (Vogel et al., 1987). The ratios of $^{13}C/^{12}C$ in the PCE in groundwater near a PCE DNAPL, for example, should increase during biodegradation, with an "enrichment factor" that is characteristic of the compound and reaction mechanism (Slater et al., 2001). The TCE, DCE, VC and ethene produced will have relatively low $^{13}C/^{12}C$ ratios initially, and these ratios will increase as the compounds are further biodegraded. The enrichment factors of parent and daughter compounds can be used to quantify biodegradation rates with greater accuracy than less direct methods (Morrill et al., 2009). However, it is important to realize that, if the well is near DNAPL, a high concentration of PCE in the aqueous phase may represent molecules in equilibrium with the PCE in the DNAPL, and until the DNAPL is depleted extensively the PCE in groundwater will have the same isotope ratio as the DNAPL (Morrill et al., 2009).

12.10 CASE STUDIES

The following case studies present four well-monitored demonstrations of the performance of ISB for source zones. The first two are relatively contained and small-scale. The first is a comparison of biostimulation and bioaugmentation in a controlled reactor (roughly 400 ft^3) containing a known volume of PCE DNAPL. Results indicated roughly a two- to threefold enhancement in flushing due to biological activity. The second is from Dover Air Force Base (AFB), Deleware where a contained test cell was emplaced into the soil, with approximately 1,500 ft^3 of saturated aquifer. Those results indicated a flushing enhancement of between 2 and 4.5, with possibly greater enhancement in the later stages of flushing.

In addition, two field demonstrations are included. In the first, hydraulically isolated test cells were established at Fort Lewis, Washington and mass flux was measured in the cells with or without ISB treatment. The enhancement in mass flux ranged from a factor of 2–8, though some of this enhancement may have been due to the whey used as the electron donor. Finally, the potential for treating an existing TCE source zone by injecting edible oil was tested at Tarheel Army Missile Plant, North Carolina. In this case, the groundwater concentrations of chloroethenes showed a sustained decrease of roughly 90% after the oil injections, with extensive conversion to daughter products.

Together, these case studies suggest that it is reasonable to expect a two- to threefold increase in DNAPL dissolution rate when using active ISB over that achieved by hydraulic flushing alone, during the period of treatment. The results also indicate passive treatment can achieve a 90% reduction in source-zone concentrations, during the period of time that the oil or accumulated biomass persists. None of these case studies address the long-term impacts of ISB, an area where reliable information is still needed.

12.10.1 Rice University Experimental Controlled Release System: Biostimulation and Bioaugmentation

In the first phase of this project (Phase I), two 11.7 m^3 (413 ft^3) experimental controlled release systems (ECRS) were packed with sandy model aquifer material and amended with PCE DNAPL source zones (Figure 12.13). The tanks then were operated in parallel with identical flow regimes and electron donor amendments to measure the impacts of

Figure 12.13. One of the Experimental Controlled Release System tanks being filled prior to the experiment. Photograph provided courtesy of C. Herb Ward, Rice University, Houston, Texas.

bioremediation on the mass discharge of contaminants from the sources. HRC®, and later dissolved lactate, served as electron donors to promote dechlorination. One ECRS was bioaugmented with an anaerobic dechlorinating consortium directly into the source zone, and the other served as a control (biostimulated only) to determine the benefits of bioaugmentation. The presence of halorespiring bacteria in the aquifer matrix prior to bioaugmentation, shown by nested PCR with phylogenetic primers, suggests that dechlorinating catabolic potential may be somewhat widespread. Polymerase chain reaction analyses demonstrated that the bacteria present in the culture used for bioaugmentation in the ECRS prevailed for almost a year. Unfortunately, even with *Dehalococcoides* present, the low concentrations of ethene produced (<1 μM) indicate that complete dechlorination to nontoxic end products was limited.

The results demonstrated that the low concentration of ethene produced in this first phase was not due to washout of the dechlorinating organisms. The experiment also demonstrated that as long as the electron acceptor was not limiting, there was greater energy flow to the dechlorinating populations than to the methanogens. Overall, the results obtained in the Phase I corroborate that:

- Source-zone reductive dechlorination of PCE is possible at near field scale, and
- A system bioaugmented with a competent halorespiring consortium can enhance DNAPL dissolution at significantly greater rates than a system that is biostimulated only.

The Phase II testing compared the fate of a mixed DNAPL source zone under a natural attenuation scenario (no treatment, natural rates of dissolution) with a most probable engineering approach that included biostimulation and bioaugmentation. The same experimental ECRS tanks used in Phase I described above were emptied and repacked with uncontaminated sandy soil. HRC® was continuously added in the influent as a pre-hydrolized (diluted) mixture consisting of 50:50 v/v HRC®: deionized water. The effluent concentration of ethene measured in the biostimulated and bioaugmented tank (~4 μM) was fourfold higher than Phase I. This suggests a more complete dechlorination activity that was most likely the result of the slower groundwater seepage velocity used in this experiment (0.4 m/d) compared to the Phase I experiment (1.6 m/d).

Cumulative mass-balance calculations showed that the total mass removed at the end of the experiment in the biostimulated and bioaugmented tank was nearly 47% of the total mass of PCE added to the tank (Da Silva et al., 2006; see Figure 12.14). Of this removal, 26% was removed by dissolution (as measured by the mass of PCE in the effluent) and 21% by dechlorination to lesser chlorinated products, mainly TCE and *cis*-DCE. In the control tank, 34% of the PCE added to the tank was removed, with 31% being removed by dissolution and 3% by dechlorination. The benefit of biostimulation and bioaugmentation was observed with higher (sevenfold) dechlorination activity compared to the control tank.

These results from a carefully monitored system indicate that biological activity can enhance dissolution from DNAPLs in the subsurface, perhaps by a factor of 2–3. The results also suggest that long periods of time may be required for effective treatment – the 200-day treatment time under these conditions corresponds to roughly 80 pore volumes, but under field conditions it can take several months to years to move a pore volume through a source zone.

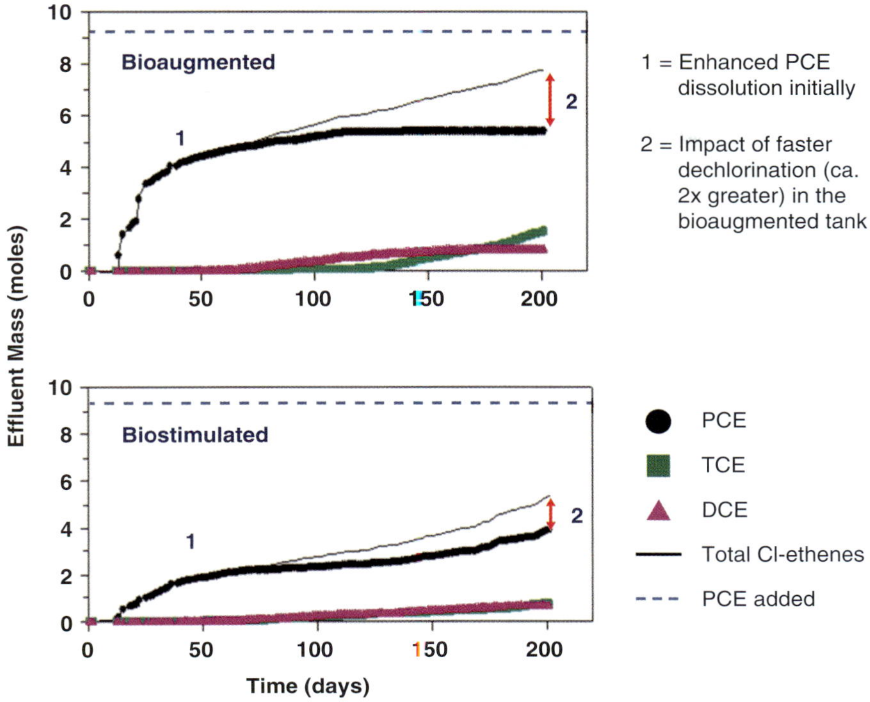

Figure 12.14. Cumulative mass of chloroethenes flushed over time from the bioaugmented and biostimulated only ECRS tanks. Approximately twice as much mass was flushed from the bioaugmented tanks initially than from the biostimulated tank, which showed little evidence of biodegradation until later in the experiment (modified from Da Silva et al., 2006).

12.10.2 Dover AFB: Bioaugmentation and Lactate Recirculation

A known volume of PCE was released in a test cell created by inserting sheet metal walls to a depth of 6 ft (2 m) in place at a shallow site at Dover AFB, Dover, Delaware, USA (Figure 12.15). This demonstration evaluated the use of biological processes to enhance PCE dissolution by establishing a baseline dissolution rate, followed by biostimulation (lactate solution continuously recirculated through the source) and finally by bioaugmentation (addition of KB-1™ with continued recirculation). Initial laboratory tests showed that bioaugmentation was needed for complete dechlorination at this site, and that bioaugmentation could enhance dissolution by a factor of roughly 2–5 compared to the flushing-only baseline (Sleep et al., 2006).

During the field test, effluent from the test cell was analyzed for all chloroethenes as well as chloride. The results confirmed that bioaugmentation was required to promote dechlorination of the PCE to *cis*-1,2-DCE, VC, and ethene and that the dissolution of the DNAPL remaining after the initial water flushing was enhanced by a factor of 2–4.5, depending on the method used to calculate dissolution. The effluent chloroethene concentrations were significantly higher than was predicted based on the kinetics observed prior to bioaugmentation (Figure 12.16).

The field test also demonstrated that bioclogging occurred as a result of electron donor addition, and this clogging eventually caused the flow paths within the test cell to change and prevented delivery of the electron donor to those zones with the greatest amounts of residual PCE. This finding underscores the need for careful performance monitoring during full-scale ISB applications and the need to adequately prevent bioclogging.

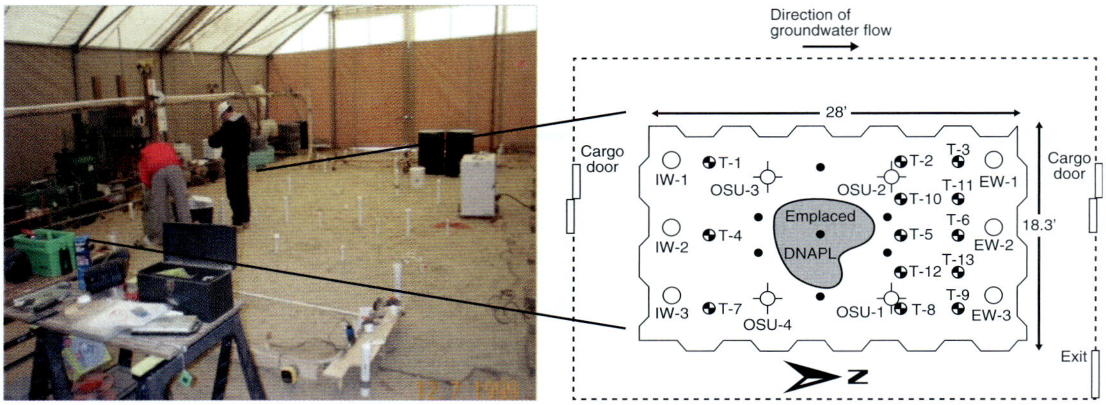

Figure 12.15. Dover AFB test cell during operations, with expanded schematic of the cell showing location of emplaced PCE and injection wells, extraction wells, and all monitoring points.

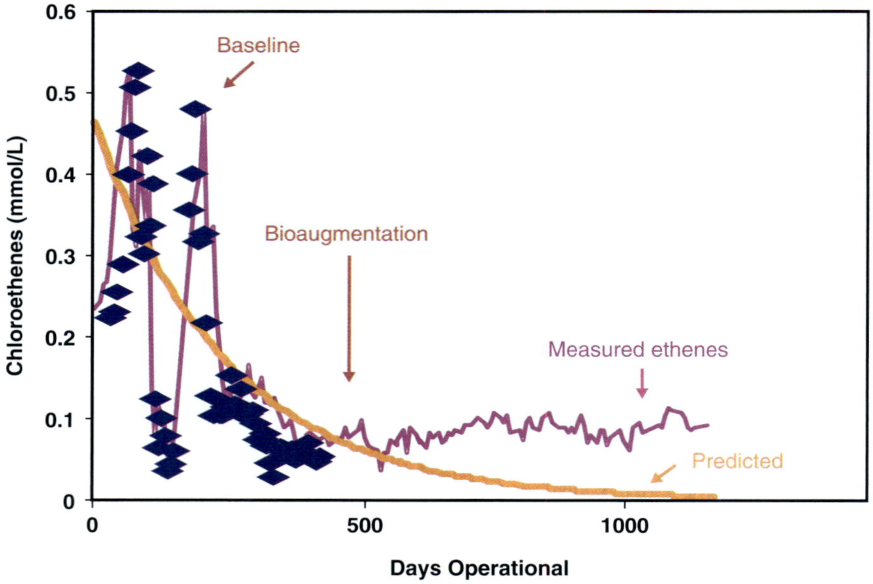

Figure 12.16. Concentrations (molar) of chloroethenes during Dover AFB demonstration. After bioaugmentation, the PCE was largely converted to daughter products, and the total concentration of daughter products increased significantly over the concentrations that were predicted based on the prior kinetics, and over the concentrations measured prior to bioaugmentation. Modified from Lebrón et al. (2007) and Geosyntec (2008).

12.10.3 Fort Lewis East Gate Disposal Yard: Whey Injections

Two hydraulically isolated treatment cells, each consisting of an injection well, an extraction well, and a network of monitoring wells were installed at a DNAPL source zone in Fort Lewis, WA, USA (Figure 12.17; details in Macbeth and Sorenson, 2008). The site had a TCE DNAPL source within a shallow, gravelly aquifer, with rapid groundwater flow (1 ft/day or 0.3 m/day). In Treatment Cell 1, high concentration (10%) whey powder was injected initially, followed by low concentration (1%) whey powder injections. In Treatment Cell 2, the reverse was done: the low concentration whey was injected first, followed by the 10% whey powder solution.

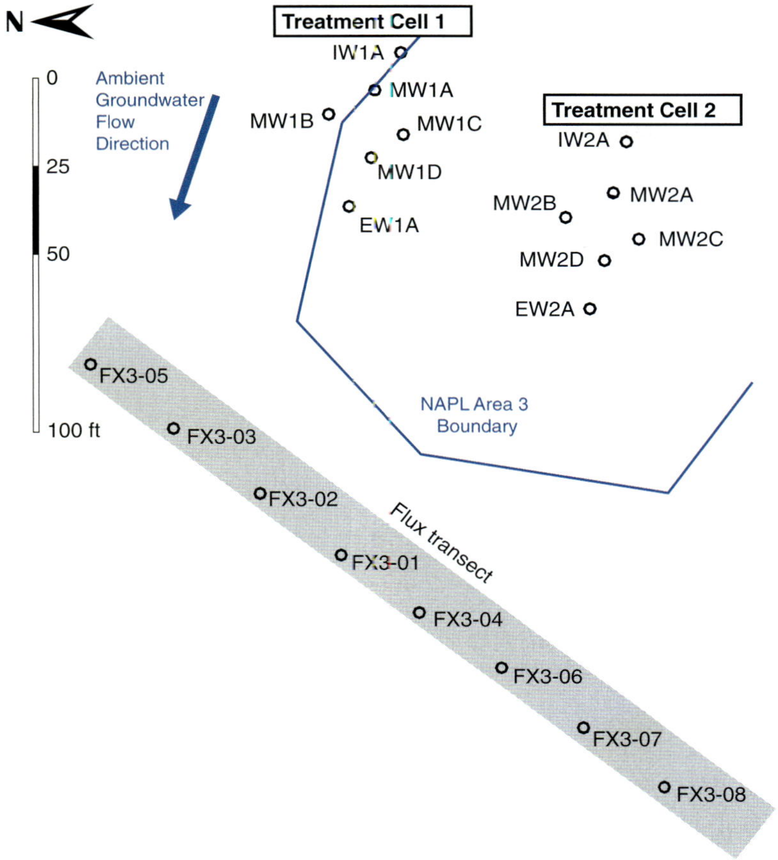

Figure 12.17. Layout of Fort Lewis treatment cells within the TCE source zone and the mass flux monitoring wells located immediately downgradient (from Macbeth and Sorenson, 2008).

Treatment Cell 2, in the core of the source zone, had significantly higher concentrations of TCE and DCE than Treatment Cell 1, located on the fringe of the source. In both cells, whey injection quickly stimulated iron- and sulfate-reducing conditions, and reductive dechlorination of TCE to DCE was rapid and complete. An initial drop in pH due to the whey injections delayed the onset of methanogenic conditions and further dechlorination of the DCE to VC and ethene. Statistical comparisons demonstrated that VOC molar concentrations were significantly increased during treatment (Figure 12.18). The calculated dissolution enhancement factors ranged from approximately 3–8 under optimal conditions (that is, when 10% whey powder was injected). This enhancement apparently resulted from both the increased biodegradation to daughter products and enhanced solubilization of TCE by the organic matter in the whey.

ISB not only increased DNAPL dissolution during treatment, but also the treatment effects persisted after the whey injection stopped. The total VOC concentrations remained well below baseline for at least 4 months after treatment (concentrations were 94–99% lower at the last sampling in May 2006 than in the baseline sampling of July 2005).

The costs of the demonstration were carefully tracked, and a realistic estimate of the full-scale cleanup cost was developed. Based on an assumed 3-year operations period, the total cost for cleanup of the 0.5-acre site was estimated to be \$0.9M (a unit cost of \$56 per cubic yard [yd^3] or \$74/m^3). The actual cost of cleaning up this same source zone using electrical resistance heating (ERH) was \$5M or \$313/yd^3 (\$412/m^3).

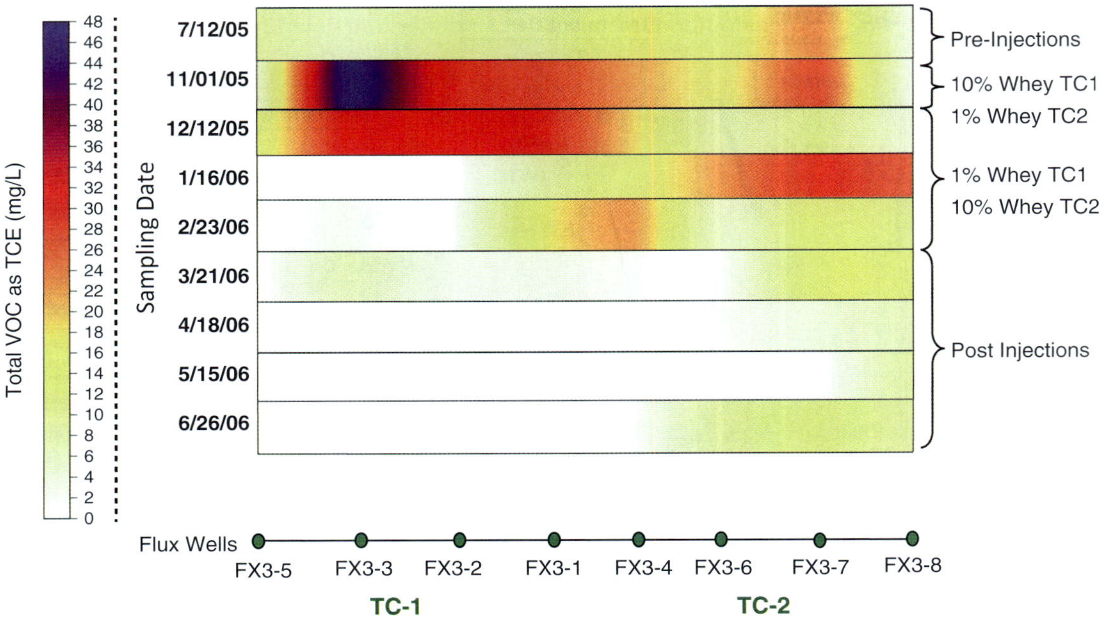

Figure 12.18. Site layout and total VOC concentrations (as mg/L of TCE) in monitoring wells downgradient of the source zone during the Fort Lewis ISB demonstration (from Macbeth and Sorenson, 2008).

12.10.4 Tarheel Army Missile Plant, North Carolina: Edible Oil Injection

In preparation for transfer of ownership of the property, the Army evaluated ISB to treat TCE in groundwater, using commercially available edible oil (EOS®). The pilot test was designed to treat a 100 × 100 ft (approximately 30 m × 30 m) zone that included the primary source zone (Borden et al., 2007). The source and plume were within a more transmissive zone present at the transition from bedrock to weather material where the rock fractures remain open allowing water flow.

The EOS was injected in two separate events, as a dilute emulsion into eight wells (6 in [2 cm]-diameter), spaced approximately 30–35 ft [9–11 m] apart. After each injection, groundwater was recirculated to distribute the emulsion throughout the targeted treatment zone, by extracting and injecting one pore volume from the four well pairs. No bioaugmentation was done. Over the two phases of injections, approximately 18,500 pounds (lb) (8,400 kg) of EOS concentrate were added, and 163,000 gallons (314,000 L) of groundwater were recirculated. The active treatment was completed in two separate injection events over 18 and 32 days.

After injection, the ORP changed from slightly positive to strongly reducing (<-100 mV), DO was no longer detectable, and methane was consistently detected. Volatile organic compounds in groundwater within the source zone were monitored for over a year, and concentrations declined by roughly 90% on a molar basis. Moreover, the contaminants that remained showed evidence of biodegradation. TCE was completely depleted with the production of *cis*-DCE and VC (Figure 12.19).

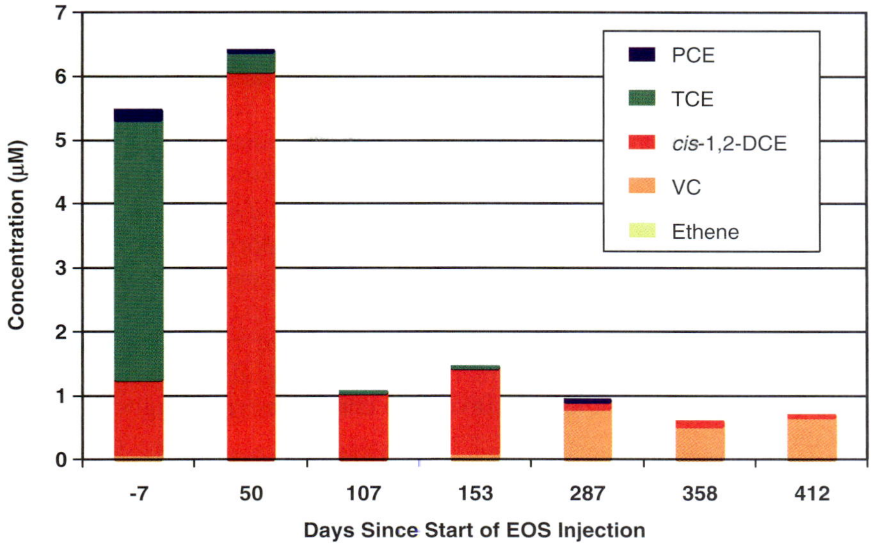

Figure 12.19. Average molar concentrations of chlorinated ethenes in groundwater after source-zone treatment with EOS™ (emulsified edible oil) at Tarheel Army Missile Plant, NC, USA (graphic provided courtesy of Robert C. Borden, North Carolina State University, Raleigh, North Carolina, 2013; updated from ITRC, 2007).

12.11 LESSONS LEARNED

Case studies and performance reviews provide some lessons regarding the use of ISB for remediation of source zones:

- *Set realistic objectives.* Complete cleanup over a few years is unlikely. Reducing mass discharge from a source by 90–99% and achieving similar reductions in contaminant concentrations in the permeable portions of the source zone are reasonable expectations, although lower reductions have been observed in some field-scale treatments (McGuire et al., 2006). However, undiscovered and/or untreated source material, especially in less transmissive zones, generally will require some long-term management.
- *Use adaptive management when possible.* Typically, uncertainty exists in the source characterization and site-specific issues limit performance (stagnant areas, unknown contaminated areas). ISB, like any injection-based technology, is best applied in stages, learning along the way, and optimizing the design and operations over time.
- *The long-term treatment impacts are not fully known.* Some rebound in contaminant concentrations may occur after treatment is completed and the aquifer returns to pre-treatment conditions. Although ISB may remove much of the source zone mass and reduce the short-term mass discharge, model simulations indicate that long-term monitoring in combination with natural attenuation may be required at many sites. Recent research suggests that buildup of bacterial biomass and reduced minerals during ISB may sustain treatment after electron donor addition ends. As a result, contaminant concentrations may not rebound for several years after treatment, providing site managers with time to detect rebound and implement contingency actions, if required.

- *Monitor and address biofouling aggressively.* Adding large amounts of readily degradable materials stimulates more than the dechlorinators. Fouling of wells due to microbial growth or activities is a common problem, and it can be particularly critical for source zones if large amounts of electron donor are needed and water is continuously extracted and injected. Several remedies are available (Geosyntec, 2005), but it is critical to monitor the wells carefully and respond quickly to indications of fouling.
- *Carefully evaluate the concentrations of parent and daughter products.* The distribution and relative concentrations of the different chloroethenes, along with the geochemical parameters that exist prior to treatment, can be useful in designing the eventual ISB system. For example, substantial preexisting dechlorination suggests that less electron donor will be needed to achieve reducing conditions and may demonstrate that a complete dechlorination pathway already exists. Similarly, a lack of VC or ethene given otherwise favorable conditions can suggest a need for bioaugmentation.
- *Effective treatment must overcome significant challenges.* Practitioners considering ISB for a source zone should be aware of several potential difficulties. The electron donor demand can be so great that it is not feasible to supply enough donor and/or pH buffer, or it may cause other problems such as methane production or clogging of the subsurface. Long-term performance of ISB is not clear, and treated sources may still sustain a plume that requires ongoing monitoring or even treatment. These challenges require careful design, monitoring and continuous optimization throughout treatment.

12.12 FUTURE DEVELOPMENTS

ISB for chlorinated ethenes is a rapidly evolving technology, especially when applied to source zones. Source-zone treatment in general has gained acceptance and developed a record of experience only over the last decade. The next decade is likely to see refinement of the approaches taken, applications of the basic principles to more complex sites and more cost-effective monitoring and operation (Suthersan et al., 2011). Moreover, it will probably include deliberate combinations with other technologies, sequentially or simultaneously (ITRC, 2011). This section examines some of the potential future developments in this area.

12.12.1 Improved Understanding and Treatment of Low-Permeability Regions

Understanding if targeted treatment of lower permeability zones is needed, and then designing cost-effective ways to treat these zones, are important issues for ISB. Like any injection-based strategy, most of the amendments move through and remain in the most transmissive regions. In a recent spill, or at the leading edge of a plume, most of the contaminants are in the transmissive regions, and injection-based strategies may be relatively successful. But all too often, source zones already have been exposed for decades, and a large fraction of the remaining mass may be in the lower permeability regions. Targeting amendment delivery to these regions could lead to lower cost and more effective treatment.

One potential approach to improving delivery is to co-inject the amendments with shear-thinning agents (such as gum agar), to temporarily clog the transmissive regions and force amendments into the less permeable areas (Zhong et al., 2008; Newell, 2009). Another possibility is to combine ISB or ISCO or ISCR with electrokinetics (EK) to improve amendment delivery to the low-permeability zones (Norris et al., 1995, Reynolds et al., 2008; Jones et al., 2011; Wu et al., 2012 a, b). The use of EK as the driving force for amendment migration overcomes the traditional limitations of hydraulic injection approaches and is relatively insensitive to the soil type and the hydraulic permeability.

Ongoing research is examining the fundamental aspects of chlorinated solvent storage in, and release from, lower permeability materials, including sorption to clays and organic materials, and clay-DNAPL interactions. The goal of this work is to improve predictions of the magnitude of contaminant storage in low permeability zones, and help managers decide if treatment is needed to control any back diffusion.

12.12.2 Improving Delivery to DNAPL Accumulations

The difficulty of delivering water soluble electron donors to the DNAPL phase is obvious, and makes effective treatment of sites with DNAPL accumulations very challenging. One solution currently being tested is the formation of mixed DNAPLs containing both the contaminant and electron donor. Partitioning electron donors (PEDs) are organic compounds such as *n*-butyl acetate and *n*-hexanol that can be delivered in water but can partition into the DNAPL phase, enhancing biodegradation at and near the DNAPL:water interfaces (Lebrón, 2007). Laboratory-scale testing has indicated that both these PEDs will partition to DNAPLs and persist for several years, reducing the need for subsequent electron donor reinjections to sustain continued reductive dechlorination (Cápiro et al., 2011). Alternatively, vegetable oils can be injected where the chlorinated solvent partitions into the oil droplets, forming a mixed NAPL that will provide both electron donor and electron acceptor for bacterial growth. This strategy has been tested in the laboratory with EVO (Hiortdahl and Borden, 2011) and in the field with neat vegetable oil injections (Henry et al., 2007). Heating the oil prior to injection may increase the distribution of the oil in the subsurface (Williams, 2003).

12.12.3 ISB at Fractured Rock Sites

DNAPL sources located in bedrock are particularly difficult to treat and manage (Steimle, 2002). It is difficult to characterize the DNAPL nature and extent (Mercer et al., 2008), matrix storage can sustain plumes for decades (Mutch et al., 1993; Parker et al., 1994; West and Kueper, 2010), and delivering amendments can be very challenging (Goldstein et al., 2004). Thermal treatment may be effective, but can be difficult and costly (Johnson et al., 2009; Baston et al., 2010; Chen et al., 2010), so there is considerable interest in less costly technologies, including bioremediation.

Laboratory research has indicated that bioaugmentation can be highly effective in enhancing the removal of DNAPL from fractures within bedrock (Schaefer et al., 2009b, 2010). Limited field testing also has suggested that bioaugmentation can be effective in bedrock (Kane et al., 2005; De Flaun et al., 2006). There is a need for well-monitored field-scale demonstrations to understand the costs and performance, and for improvements in the technology to increase the use of ISB at fractured bedrock sites.

12.12.4 Long-Term Performance Predictions and Improvements

As mentioned earlier, ISB treatment in DNAPL source zones must be sustained over several years in order to achieve significant mass reductions. However, because long-term monitoring data sets are not yet available, there is still some uncertainty regarding the long-term performance. For example, what will be the post-treatment equilibrium state, and how long will it take before re-equilibration occurs? The modeling results provided in the text box suggest that ISB may have relatively little effect on the total source mass and the mass discharge from the source several years after treatment is stopped, as compared to natural dissolution of the source zone, and that the greatest long-term enhancement will be in heterogeneous and low-permeability aquifers.

There will almost certainly be some remaining mass within the source zone after ISB is terminated. Untreated DNAPL pools and zones of residual DNAPL may still be present, especially in inaccessible areas such as the less permeable regions in highly heterogeneous porous media and in dead-end fractures. These remaining source materials can sustain plumes for decades above regulatory criteria, but they will be attenuated so long as reducing conditions persist, especially if elevated levels of organic carbon and/or reactive reduced iron minerals are still present. Validated models will help determine the need for monitoring and its frequency, better long-term monitoring methods will reduce the costs and materials may be developed specifically to ensure long-term treatment.

REFERENCES

Acar Y, Gale RJ, Alshawabkeh AN, Marks RE, Puppala S, Bricka M, Parker R. 1995. Electrokinetic remediation: Basics and technology status. J Hazard Mater 40:117–137.

Adamson DT, Newell CJ. 2009. Support of source zone bioremediation through endogenous biomass decay and electron donor recycling. Bioremediaton J 13:29–40.

Adamson DT, McDade JM, Hughes JB. 2003. Inoculation of a DNAPL source zone to initiate reductive dechlorination of PCE. Environ Sci Technol 37:2525–2533.

Adamson DT, McGuire TM, Newell CJ, Stroo H. 2011. Sustained treatment: Implications for treatment timescales associated with source depletion technologies. Remediat J 2:27–50.

Aeppli C, Hofstetter TB, Amaral HIF, Kipfer R, Schwarzenbach RP, Berg M. 2010. Quantifying *in situ* transformation rates of chlorinated ethenes by combining compound-specific stable isotope analysis, groundwater dating, and carbon isotope mass balances. Environ Sci Technol 44:3705–3711.

AFCEE. 2007. Protocol for *In Situ* Bioremediation of Chlorinated Solvents Using Edible Oil. Prepared by Solutions-IES, Inc., Terra Systems, Inc., and Parsons Corp. http://www.clu-in.org/download/remed/Final-Edible-Oil-Protocol-October-2007.pdf. Accessed September 14, 2013.

AFCEE (Air Force Center for Engineering and the Environment), NFESC (Naval Facilities Engineering Service Center), ESTCP (Environmental Security Technology Certification Program). 2004. Principles and Practices of Enhanced Anaerobic Bioremediation of Chlorinated Solvents. Prepared by Parsons Infrastructure & Technology Group, Inc., Denver, CO, USA. http://www.dtic.mil/cgi-bin/GetTRDoc?Location=U2&doc=GetTRDoc.pdf&AD=ADA511850. Accessed September 14, 2013.

Amos BK, Christ JA, Abriola LM, Pennell KD, Löffler FE. 2007a. Experimental evaluation and mathematical modeling of microbially enhanced tetrachloroethene (PCE) dissolution. Environ Sci Technol 41:963–970.

Amos BK, Daprato RC, Hughes JB, Pennell KD, Löffler FE. 2007b. Effects of the nonionic surfactant Tween 80 on microbial reductive dechlorination of chlorinated ethenes. Environ Sci Technol 41:1710–1716.

Amos BK, Suchomel EJ, Pennell KD, Löffler FE. 2008. Microbial activity and distribution during enhanced contaminant dissolution from a NAPL source zone. Water Res 42:2963–2974.

Bagley DM. 1998. Systematic approach for modeling tetrachloroethene biodegradation. J Environ Eng 124:1076–1086.

Baston DP, Falta RW, Kueper BH. 2010. Numerical modeling of thermal conductive heating in fractured bedrock. Ground Water 48:836–843.

Basu NB, Rao PSC, Falta RW, Annable MD, Jawitz JW, Hatfield K. 2008. Temporal evolution of DNAPL source and contaminant flux distribution: Impacts of source mass depletion. J Contam Hydrol 95:93–109.

Battelle. 2004. Demonstration of Biodegradation of Dense, Nonaqueous-phase Liquids (DNAPL) Through Biostimulation and Bioaugmentation at Launch Complex 34 in Cape Canaveral Air Force Station, Florida. Final Report. EPA/540/R-07/007. U.S. Environmental Protection Agency National Risk Management Research Laboratory Superfund Innovative Technology Evaluation Program, Cincinnati, OH, USA. http://nepis.epa.gov/Exe/ZyPURL.cgi?Dockey=P10030M1.txt. Accessed September 14, 2013.

Becker J. 2006. A modeling study and implications of competition between *Dehalococcoides ethenogenes* and other tetrachloroethene-respiring bacteria. Environ Sci Technol 40:4473–4480.

Becvar E, Evans P, Lebrón C, Stroo H, Wilson J, Wymore R. 2008. Workshop on *In Situ* Biogeochemical Transformation of Chlorinated Solvents. Prepared for AFCEE, Brooks City-Base, TX; ESTCP, Arlington, VA; NFESC, Port Hueneme, CA. February 2008. http://www.dtic.mil/cgi-bin/GetTRDoc?AD=ADA501302. Accessed September 14, 2013.

Borden RC. 2003. Anaerobic bioremediation of chlorinated solvent source areas – What can be achieved? AFCEE Annual Technology Transfer Workshop, San Antonio, TX, USA.

Borden RC, Beckwith WJ, Lieberman MT, Akladiss N, Hill SR. 2007. Enhanced anaerobic bioremediation of a TCE source at the Tarheel Army Missile Plant using EOS. Remediat J 17:12–19.

Borden RC, Clayton M, Weispfenning AM, Simpkin T, Lieberman MT. 2008. Guidance Document: Development of a Design Tool for Planning Aqueous Amendment Injections. ESTCP, Arlington, VA, USA. Project ER-200626. http://www.serdp.org. Accessed September 14, 2013.

Bouwer EJ, McCarty PL. 1983. Transformations of 1- and 2-carbon halogenated aliphatic organic compounds under methanogenic conditions. Appl Environ Microbiol 45:1286–1294.

Bradley PM. 2003. History and ecology of chloroethene biodegradation: A review. Bioremediation J 7:81–109.

Bradley PM, Chapelle FH. 1996. Anaerobic mineralization of vinyl chloride in Fe(III)-reducing aquifer sediments. Environ Sci Technol 30:2084–2086.

Bradley PM, Chapelle FH. 2010. Biodegradation of Chlorinated Ethenes. In Stroo HF, Ward CH, eds, *In Situ* Remediation of Chlorinated Solvent Plumes. Springer, New York, NY, USA, pp 39–67.

Bradley PM, Chapelle FH. 2011. Microbial mineralization of dichloroethene and vinyl chloride under hypoxic conditions. Ground Water Monit Remediat 31:39–49.

Bradley PM, Landmeyer JE, Dinicola RS. 1998. Anaerobic oxidation of [1,2-14C] dichloroethene under Mn(IV)-reducing conditions. Appl Environ Microbiol 64:1560–1562.

Brown RA, Mueller JG, Seech AG, Henderson JK, Wilson JT. 2009. Interactions between biological and abiotic pathways in the reduction of chlorinated solvents. Remediat J 20:9–20.

Burnell DK, Mercer JW, Sims LS. 2012. Analytical models of steady-state plumes undergoing sequential first-order degradation. Ground Water 50:394–411.

Butler EC, Hayes KF. 1999. Kinetics of the transformation of trichloroethylene and tetrachloroethylene by iron sulfide. Environ Sci Technol 33:2021–2027.

Cai Z, Wilson RD, Lerner DN. 2012. Assessing TCE source bioremediation by geostatistical analysis of a flux fence. Ground Water 50:908–917.

Cápiro N, Granbery EK, Lebrón CA, Major DW, McMaster ML, Pound MJ, Löffler FE, Pennell KD. 2011. Liquid-liquid mass transfer of partitioning electron donors in chlorinated solvent source zones. Environ Sci Technol 45:1547–1554.

Carr CS, Garg S, Hughes JB. 2000. Effect of dechlorinating bacteria on the longevity and composition of PCE-containing nonaqueous phase liquids under equilibrium dissolution conditions. Environ Sci Technol 34:1088–1094.

Chapman SW, Parker BL. 2005. Plume persistence due to aquitard back diffusion following dense nonaqueous phase liquid source removal or isolation. Water Resour Res 41:W12411, doi:10.1029/2005WR004224.

Chen F, Liu X, Falta RW, Murdoch LC. 2010. Experimental demonstration of contaminant removal from fractured rock by boiling. Environ Sci Technol 44:6437–6442.

Christ JA, Abriola LM. 2007. Modeling metabolic reductive dechlorination in dense non-aqueous phase liquid source-zones. Adv Water Resour 30:1547–1561.

Christ JA, Ramsburg CA, Löffler FE, Pennell KD, Abriola LM. 2005. Coupling aggressive mass removal with microbial reductive dechlorination for remediation of DNAPL source zones – A review and assessment. Environ Health Perspect 113:465–477.

Chu M, Kitanidis PK, McCarty PL. 2003. Effects of biomass accumulation on microbially enhanced dissolution of a PCE pool: A numerical simulation. J Contam Hydrol 65:79–100.

Clement TP. 1997. RT3D – A Modular Computer Code for Simulating Reactive Multispecies Transport in 3-Dimensional Groundwater Systems (Version 1.0). PNNL-11720. Pacific Northwest National Laboratory, Richland, WA, USA. 59 p.

Clement TP. 2003. RT3D v2.5 Update document, February. Battelle Pacific Northwest Division, Columbus, OH, USA. http://bioprocess.pnnl.gov/rt3d.htm. Accessed September 14, 2013.

Clement TP, Hooker BS, Skeen RS. 1996. Macroscopic models for predicting changes in saturated porous media properties caused by microbial growth. Ground Water 34:934–942.

Clement TP, Sun Y, Hooker BS, Petersen JN. 1998. Modeling multispecies reactive transport in ground water. Ground Water Monit Remediat 18:79–92.

Coleman NV, Mattes TM, Gossett JM, Spain JC. 2002. Biodegradation of *cis*-dichloroethene as the sole carbon source by a β-protobacterium. Appl Environ Microbiol 68:2726–2730.

Cope N, Hughes JB. 2001. Biologically-enhanced removal of PCE from NAPL source zones. Environ Sci Technol 35:2014–2021.

Costanza J, Fletcher KE, Löffler FE, Pennell KD. 2009. Fate of TCE in heated Fort Lewis soil. Environ Sci Technol 43:909–914.

Criddle CS, Alvarez LM, McCarty PM. 1991. Microbial Processes in Porous Media. In Bear J, Corapcioglu MY, eds, Transport Processes in Porous Media. Kluwer Academic Publishers, Dordrecht, Boston, London, pp 641–691.

Cupples AM. 2008. Real-time PCR quantification of *Dehalococcoides* populations: Methods and applications. J Microbiol Methods 72:1–11.

Cupples AM, Spormann AM, McCarty PL. 2004. Vinyl chloride and *cis*-dichloroethene dechlorination kinetics and microorganism growth under substrate limiting conditions. Environ Sci Technol 38:1102–1107.

Cwiertny DM, Scherer MM. 2010. Abiotic Processes Affecting the Remediation of Chlroinated Solvents. In Stroo HF, Ward CH, eds, *In Situ* Remediation of Chlorinated Solvent Plumes. Springer, New York, NY, USA, pp 69–108.

Da Silva MLB, Daprato RC, Gomez DE, Hughes JB, Ward CH, Alvarez PJJ. 2006. Comparison of bioaugmentation and biostimulation for the enhancement of DNAPL source zone bioremediation. Water Environ Res 78:2456–2465.

De Flaun MF, Drew S, Dale J, Lacombe P, Schauble P. 2006. Application of bioaugmentation for TCE DNAPL in fractured bedrock. In Sass BM, ed, Proceedings, 5th International Conference on Remediation of Chlorinated and Recalcitrant Compounds, Monterey, CA, USA. May 22–25. Battelle Press, Columbus, OH, USA. Paper E-21.

Duhamel M, Wehr SD, Yu L, Rizvi H, Seepersad D, Dworatzek S, Cox EE, Edwards EA. 2002. Comparison of anaerobic dechlorinating enrichment cultures maintained on tetrachloroethene, trichloroethene, *cis*-dichloroethene and vinyl chloride. Water Res 36:4093–4202.

Duhamel M, Mo K, Edwards E. 2004. Characterization of a highly enriched *Dehalococcoides*-containing culture that grows on vinyl chloride and trichloroethene. Appl Environ Microbiol 70:5538–5545.

Eaddy A. 2008. Scale-up and characterization of an enrichment culture for bioaugmentation of the P-area chlorinated ethene plume at the Savannah River site. MS Thesis. Clemson University, Clemson, SC, USA.

Ellis DE, Lutz EJ, Odom JM, Buchanan Jr RL, Bartlett CL, Lee MD, Harkness MR, Deweerd KA. 2000. Bioaugmentation for accelerated *in situ* anaerobic bioremediation. Environ Sci Technol 34:2254–2260.

Elsner M, Schwarzenbach RP, Haderlein SB. 2004. Reactivity of Fe(II)-bearing minerals toward reductive transformation of organic contaminants. Environ Sci Technol 38:799–807.

ESTCP (Environmental Security Technology Certification Program). 2005. Bioaugmentation for Remediation of Chlorinated Solvents: Technology Development, Status, and Research Needs. ESTCP, Arlington, VA, USA. http://www.serdp.org. Accessed September 14, 2013.

ESTCP. 2006. Protocol for Enhanced *In Situ* Bioremediation Using Emulsified Edible Oil. Project ER-200221. ESTCP, Arlington, VA, USA. Prepared by Solutions-IES. http://www.serdp.org. Accessed September 14, 2013.

Falta RW, Stacy MB, Ahsanuzzaman ANM, Wang M, Earle R. 2005. REMChlor Remediation Evaluation Model for Chlorinated Solvents User's Manual Version 1.0. http://www.epa.gov/ada/csmos/models/remchlor.html. Accessed September 14, 2013.

Fathepure BZ, Nengu JP, Boyd SA. 1987. Anaerobic bacteria that degrade perchlorothene. Appl Environ Microbiol 53:2671–2674.

Feenstra S, Cherry JA, Parker BL. 1996. Conceptual Models for the Behavior of Dense Nonaqueous Phase Liquids (DNAPLs) in the Subsurface. In Pankow JF, Cherry JA, eds, Dense Chlorinated Solvents and Other DNAPLs in Groundwater. Waterloo Press, Toronto, Canada, pp 53–88.

Fennell DE, Gossett JM. 1998. Modeling the production of and competition for hydrogen in a dechlorinating culture. Environ Sci Technol 32:2450–2460.

Fennel DE, Carroll AB, Gossett JM, Zinder SH. 2001. Assessment of indigenous reductive dechlorinating potential at a TCE-contaminated site using microcosms, polymerase chain reaction analysis, and site data. Environ Sci Technol 35:1830–1839.

Ferrey ML, Wilkin RT, Ford RG, Wilson JT. 2004. Nonbiological removal of *cis*-dichloroethylene and 1,1-dichloroethylene in aquifer sediment containing magnetite. Environ Sci Technol 38:1746–1752.

Fletcher KE, Costanza J, Pennell KD, Löffler FE. 2011. Electron donor availability for microbial reductive processes following thermal treatment. Water Res 45:6625–6636.

Fogel S, Smoler D, Findlay M. 2007. Lessons learned as a result of completing 250 microcosm studies. Proceedings, 9th International *In Situ* and On-Site Bioremediation Symposium, Baltimore, MD, USA, May 7–10.

Fogel S, Findlay M, Folsom S, Kozar M. 2009. The Importance of pH in Reductive Dechlorination of Chlorinated Solvents. In Wickramanayake GB, Rectanus HV, eds. Proceedings,

10th International *In Situ* and On-Site Bioremediation Symposium, Baltimore, MD, USA. May 5–8. Battelle Memorial Institute, Columbus, OH, USA. Paper L-47.

Friis AK, Albrechtsen H-J, Heron G, Bjerg PL. 2005. Redox processes and release of organic matter after thermal treatment of a TCE-contaminated aquifer. Environ Sci Technol 39:5787–5795.

Friis AK, Albrechtsen H-J, Cox E, Bjerg PL. 2006. The need for bioaugmentation after thermal treatment of a TCE-contaminated aquifer: Laboratory experiments. J Contam Hydrol 88: 235–248.

Friis AK, Heimann AC, Jakobsen R, Albrechtsen H-J, Cox E, Bjerg PL. 2007. Temperature dependence of anaerobic TCE-dechlorination in a highly enriched *Dehalococcoides*-containing culture. Water Res 41:355–364.

Fure AD, Jawitz JW, Annable MD. 2006. DNAPL Source depletion: Linking architecture and flux response. J Contam Hydrol 85:118–140.

Gantzer CJ, Wackett LP. 1991. Reductive dechlorination catalyzed by bacterial transition-metal coenzymes. Environ Sci Technol 25:715–722.

Gent D, Bricka RM, Truax DD, Zappi ME. 2001. Electrokinetic Movement of Biological Amendments Through Natural Soils to Enhance *In Situ* Bioremediation. In Leeson A, Peyton BM, Means JL, Magar VS, eds, Bioremediation of Inorganic Compounds. Battelle Press, Columbus, OH, USA, pp 241–248.

GeoSyntec. 2004. Assessing the Feasibility of DNAPL Source Zone Remediation: Review of Case Studies. Contract Report CR 04-002-ENV. Naval Facilities Engineering Services Center, Port Hueneme, CA, USA.

Geosyntec. 2005. A Review of Biofouling Controls for Enhanced *In Situ* Bioremediation of Groundwater. ESTCP, Arlington, VA, USA. http://www.serdp.org. Accessed September 14, 2013.

Geosyntec. 2007. Cost and Performance Report: Demonstration of Bioaugmentation at Kelly AFB, TX, USA. Project ER-199914. http://www.serdp.org. Accessed September 14, 2013.

Geosyntec. 2008. Cost and Performance Report: Biodegradation of Dense Non-Aqueous Phase Liquids (DNAPLS) Through Bioaugmentation of Source Areas – Dover National Test Site, Dover, DE, USA. Project ER-200008. http://www.serdp.org. Accessed September 14, 2013.

Gerhard JI, Kueper BH. 2003. Influence of constitutive model parameters on the predicted migration of DNAPL in heterogeneous porous media. Water Resour Res 39:1279, doi:10.1029/2002WR001570.

Giovannoni SJ, Tripp HJ, Givan S, Podar M, Vergin KL, Baptista D, Bibbs L, Eads J, Richardson TH, Noordewier M, Rappe MS, Short JM, Carrington JC, Mathur EJ. 2005. Genome streamlining in a cosmopolitan oceanic bacterium. Sci 309:1242–1245.

Glover KC, Munakata-Marr J, Illangasekare TH. 2007. Biologically enhanced mass transfer of tetrachloroethene from DNAPL in source zones: Experimental evaluation and influence of pool morphology. Environ Sci Technol 41:1384–1389.

Goldstein KJ, Vitolins A, Navon D, Parker BL, Chapman S, Anderson GA. 2004. Characterization and pilot studies of chemical oxidation remediation of fracture shale. Ground Water 14:19–38.

Gossett J. 2010. Sustained aerobic oxidation of vinyl chloride at low oxygen concentrations. Environ Sci Technol 44:1405–1411.

Grant GP, Gerhard JI, Kueper BH. 2007. Multidimensional validation of a numerical model for simulating a DNAPL release in heterogeneous porous media. J Contam Hydrol 92:109–128.

Harkness MR, Bracco AA, Brennan MJ Jr, DeWeerd KA, Spivack JL. 1999. Use of bioaugmentation to stimulate complete reductive dechlorination of trichloroethene in Dover soil columns. Environ Sci Technol 33:1100–1109.

Hartmans S, deBont JAM, Tramper J, Luyben KCAM. 1985. Bacterial degradation of vinyl chloride. Biotechnol Lett 7:383–388.

He J, Ritalahti KM, Aiello MR, Löffler FE. 2003. Complete detoxification of vinyl chloride by an anaerobic enrichment culture and identification of the reductively dechlorinating population as a *Dehalococcoides* species. Appl Environ Microbiol 69:996–1003.

He J, Sung Y, Krajmalnik-Brown R, Ritalahti KM, Löffler FE. 2005. Isolation and characterization of *Dehalococcoides* sp. strain FL2, a trichloroethene (TCE)- and 1,2-dichloroethene-respiring anaerobe. Environ Microbiol 7:1442–1450.

He J, Holmes VF, Lee PKH, Alvarez-Cohen L. 2007. Influence of vitamin B_{12} and cocultures on the growth of *Dehalococcoides* isolates in defined medium. Appl Environ Microbiol 73:2847–2853.

Heimann AC, Friis AK, Jakobsen R. 2005. Effects of sulfate on anaerobic chloroethene degradation by an enriched culture under transient and steady-state hydrogen supply. Water Res 39:3579–3586.

Hendrickson ER, Payne JA, Young RM, Starr MG, Perry MP, Fahnestock S, Ellis DE, Ebersole RC. 2002. Molecular analysis of *Dehalococcoides* 16S ribosomal DNA from chloroethene-contaminated sites throughout North America and Europe. Appl Environ Microbiol 68:485–495.

Henry B. 2010. Biostimulation for Anaerobic Bioremediation of Chlorinated Solvents. In Stroo HF, Ward CH, eds, *In Situ* Remediation of Chlorinated Solvent Plumes, Springer, New York, NY, USA, pp 357–423.

Henry BM, Turner AL, Becvar ESK, Haas PE. 2007. Long-term source reduction using neat vegetable oil at CCAFS, Florida. Proceedings, 9th International *In Situ* and On-Site Bioremediation Symposium, Baltimore, MD, USA, May 7–10. Battelle Press, Columbus OH, USA, Paper K-12.

Hiortdahl KM, Borden RC. 2011. Anaerobic bioremediation of DNAPL in lab columns. Bioremediation and Sustainable Environmental Technologies Symposium, Reno, NV, USA.

Holliger C, Schraa G, Stams AJM, Zehnder AJB. 1993. A highly purified enrichment culture couples the reductive dechlorination of tetrachloroethene to growth. Appl Environ Microbiol 59:2991–2997.

Holliger C, Hahn D, Harmsen H, Ludwig W, Schumacher W, Tindall B, Vazquez F, Weiss N, Zehnder AJB. 1998. *Dehalobacter restrictus* gen. nov. and sp. nov., a strictly anaerobic bacterium that reductively dechlorinates tetra- and trichloroethene in an anaerobic respiration. Arch Microbiol 169:313–321.

Holmes VF, He J, Lee PKH, Alvarez-Cohen L. 2006. Discrimination of multiple *Dehalococcoides* strains in a trichloroethene enrichment by quantification of their reductive dehalogenase genes. Appl Environ Microbiol 72:5877–5883.

Hood E, Major D, Driedger G. 2007. The effect of concentrated electron donors on the solubility of trichloroethene. Ground Water Monit Remediat 27:93–98.

Hood ED, Major DW, Quinn J, Yoon S, Gavaskar A, Edwards EA. 2008. Demonstration of enhanced bioremediation in a TCE source area at Cape Canaveral Air Force Station, Launch Complex 34. Ground Water Monit Remediat 28:98–107.

Hrapovic L, Sleep BE, Major D, Hood ED. 2005. Laboratory study of treatment of trichloroethene by chemical oxidation followed by bioremediation. Environ Sci Technol 39:2888–2897.

Huang D, Becker JG. 2011. Dehalorespiration model that incorporates the self-inhibition and biomass inactivation effects of high tetrachloroethene concentrations. Environ Sci Technol 45:1093–1099.

Hunkeler D, Meckenstock R Sherwood-Lollar B, Schmidt T, Wilson J. 2008. A Guide for Assessing Biodegradation and Source Identification of Organic Ground Water Contaminants Using Compound Specific Isotope Analysis (CSIA). EPA/600/R-08/148. USEPA, Washington, DC, USA.

ITRC (Interstate Technology & Regulatory Council). 2005. Overview of *In Situ* Bioremediation of Chlorinated Ethene DNAPL Source Zones. ITRC, Bioremediation of DNAPLs Team, Washington, DC, USA. http://www.itrcweb.org/Documents/BioDNAPL-1.pdf. Accessed September 14, 2013.

ITRC. 2007. *In Situ* Bioremediation of Chlorinated Ethene DNAPL Source Zones: A Review of Case Studies. ITRC, Bioremediation of DNAPLs Team, Washington, DC, USA. http://www.itrcweb.org/Documents/bioDNPL_ Docs/BioDNAPL-2.pdf. Accessed September 14, 2013.

ITRC. 2008. *In Situ* Bioremediation of Chlorinated Ethene: DNAPL Source Zones. ITRC, Bioremediation of DNAPLs Team, Washington, DC, USA. http://www.itrcweb.org/Documents/bioDNPL_Docs/BioDNAPL3.pdf. Accessed September 14, 2013.

ITRC. 2010. Technology Overview: Use and Measurement of Mass Flux and Mass Discharge. ITRC, Washington, DC, USA. http://www.itrcweb.org/Documents/MASSFLUX1.pdf. Accessed September 14, 2013.

ITRC. 2011. Environmental Molecular Diagnostics Fact Sheets. ITRC, Environmental Molecular Diagnostics Team, Washington, DC, USA. http://www.itrcweb.org/Documents/EMD1.pdf. Accessed September 14, 2013.

Johnson P, Dahlen P, Triplett-Kingston J, Foote E, Williams S. 2009. Critical Evaluation of the State-of-the-Art *In Situ* Thermal Treatment Technologies for DNAPL Source Zone Treatment. ESTCP, Arlington, VA, USA. Project ER-200314. http://www.serdp.org. Accessed September 14, 2013.

Jones EH, Reynolds DA, Wood AL, Thomas DG. 2011. Use of electrophoresis for transporting nano-iron in porous media. Ground Water 49:172–183.

Kane A, Vidumsky J, Major DW, Bauer NR. 2005. In-Situ Bioremediation of a Chlorinated Solvent Residual Source in Unconsolidated Sediments and Bedrock Using Bioaugmentation. In Calabrese EJ, Kostecki PT, Dragun J, eds, Contaminated Soils, Sediments and Water: Science in the Real World, Vol 9. Springer, Boston, MA, USA, pp 45–55.

Kavanaugh MC, Rao PSC, Abriola L, Cherry J, Newell C, Sale T, Destouni G, Falta R, Shoemaker S, Siegrist R, Major D, Mercer J, Teusch G, Udell K. 2003. The DNAPL Remediation Challenge: Is There a Case for Source Depletion? EPA/600/R-03/143. USEPA, Washington DC, USA. http://nepis.epa.gov/Exe/ZyPURL.cgi?Dockey=300061GP.txt. Accessed September 14, 2013.

Krajmalnik-Brown R, Hölscher T, Thomson IN, Saunders FM, Ritalahti KM, Löffler FE. 2004. Genetic identification of a putative vinyl chloride reductase in *Dehalococcoides* sp. strain BAV1. Appl Environ Microbiol 70:6347–6351.

Krembs FJ, Siegrist RL, Crimi ML, Furrer RF, Petri BG. 2010. ISCO for groundwater remediation: Analysis of field applications and performance. Ground Water Monit Remediat 30:42–43.

Krumholtz LR, Sharp R, Fishbain SS. 1996. A freshwater anaerobe coupling acetate oxidation to tetrachloroethylene dehalogenation. Appl Environ Microbiol 62:4108–4113.

Kueper BH, Frind EO. 1991. Two-phase flow in heterogeneous porous media 1. Model development. Water Resour Res 27:1049–1057.

Kueper BH, Wealthall GP, Smith JWN, Leharne SA, Lerner DN. 2003. An Illustrated Handbook of DNAPL Transport and Fate in the Subsurface. R&D Publication 133. United Kingdom Environment Agency, Bristol, UK. 67 p.

Lebrón C. 2007. Final Report: Improving Effectiveness of Bioremediation at DNAPL Source Zone Sites Applying Partitioning Electron Donors (PEDs). Project 200716. http://www.serdp.org. Accessed September 14, 2013.

Lebrón CA, McHale T, Young R, Williams D, Bogaart MG, Major DW, McMaster ML, Tasker I, Akladiss N. 2007. Pilot-scale evaluation using bioaugmentation to enhance PCE dissolution at Dover AFB national test site. Remediat J 17:5–17.

Lebrón CA, Acheson C, Dennis P, Druar X, Wilkinson J, Ney E, Major D, Petrovskis E, Barros N, Yeager C, Löffler F, Ritalahti K, Hatt J, Edwards E, Duhamel M, Chan W. 2008. Standardized Procedures for Use of Nucleic Acid-based Tools. SERDP Project ER-1561. Naval Facilities Engineering Command, Port Hueneme, CA, USA. 81 p. http://www.dtic.mil/cgi-bin/GetTRDoc?AD=ADA534133&Location=U2&doc=GetTRDoc.pdf. Accessed September 14, 2013.

Lee W, Batchelor B. 2002. Abiotic reductive dechlorination of chlorinated ethylenes by iron-bearing soil minerals. 1. Pyrite and magnetite. Environ Sci Technol 36:5147–5154.

Lemke LD, Abriola LM, Lang JR. 2004. Influence of hydraulic property correlation on predicted dense nonaqueous phase liquid source zone architecture, mass recovery and contaminant flux. Water Resour Res 40:W12417, doi:10.1029/2004WR003061.

Li KB, Goovaerts P, Abriola LM. 2007. A geostatistical approach for quantification of contaminant mass discharge uncertainty using multilevel sampler measurements. Water Resour Res 43:W06436, doi:10.1029/2006WR005427.

Liang X, Dong Y, Kuder T, Krumholz LR, Philp RP, Butler EC. 2009. Distinguishing abiotic and biotic transformation of tetrachloroethylene and trichloroethylene by stable carbon isotope fractionation. Environ Sci Technol 41:7094–7100.

Löffler FE, Sun Q, Li J, Tiedje J. 2000. 16S rRNA gene-based detection of tetrachloroethene-dechlorinating *Desulfuromonas* and *Dehalococcoides* species. Appl Environ Microbiol 66:1369–1374.

Löffler FE, Yan J, Ritalahti K, Adrian L, Edwards EA, Konstantinidis KT, Müller JA, Fullerton H, Zinder S, Spormann AM. 2013a. *Dehalococcoides mccartyi* gen. nov., sp. nov., obligate organohalide-respiring anaerobic bacteria, relevant to halogen cycling and bioremediation, belong to a novel bacterial class, *Dehalococcoidetes* classis nov., within the phylum *Chloroflexi*. Int J Syst Evol Microbiol 63(Pt 2):625-635. doi: 10.1099/ijs.0.034926-0

Löffler FE, Ritalahti K, Zinder S. 2013b. *Dehalococcoides* and Reductive Dechlorination of Chlorinated Solvents. In Stroo HF, Leeson A, Ward CH, eds, Bioaugmentation for Groundwater Remediation. Springer, New York, NY, USA, pp 39–88.

Lowe SE, Jain MK, Zeikus JG. 1993. Biology, ecology, and biotechnological applications of anaerobic-bacteria adapted to environmental stresses in temperature, pH, salinity, or substrates. Microbiol Rev 57:451–509.

Lu X, Wilson JT, Kampbell DH. 2006. Relationship between *Dehalococcoides* DNA in ground water and rates of reductive dechlorination at field scale. Water Res 40:3131–3140.

Lu X, Wilson JT, Kampbell DH. 2009. Comparison of an assay for *Dehalococcoides* DNA and a microcosm study in predicting reductive dechlorination of chlorinated ethenes in the field. Environ Pollut 157:809–815.

Luitjen MLGC, Smidt H, Boschker HTS, de Vos WM, Schraa G, Stams AJM. 2003. Description of *Sulfurospirillum halorespirans* sp. nov., an anaerobic, tetrachloroethene-respiring bacterium, and transfer of *Dehalospirillum multivorans* to the genus *Sulfurospirillum* as *Sulfurospirillum multivorans* comb. nov. Int J Syst Evol Microbiol 53:787–793.

Macbeth TW, Sorenson K. 2008. Final Report: *In Situ* Bioremediation of Chlorinated Solvent Source Zones With Enhanced Mass Transfer. Prepared for ESTCP, Arlington, VA, USA. Project ER-200218. http://www.serdp.org. Accessed September 14, 2013.

Macbeth TW, Nelson L, Rothermel JS, Wymore RA, Sorenson KS. 2006. Evaluation of whey for bioremediation of trichloroethene source zones. Bioremediation J 10:115–128.

Maillard J, Regeard C, Holliger C. 2005. Isolation and characterization of *Tn*-Dha1, a transposon containing the tetrachloroethene reductive dehalogenase of *Desulfitobacterium hafniense* strain TCE1. Environ Microbiol 7:107–117.

Maymó-Gatell X, Chien, YT, Gossett JM, Zinder SH. 1997. Isolation of a bacterium that reductively dechlorinates tetrachloroethene to ethene. Sci 276:1568–1571.

Maymó-Gatell X, Nijenhuis I, Zinder SH. 2001. Reductive dechlorination of *cis*-1,2-dichloroethene and vinyl chloride by "*Dehalococcoides ethenogenes*". Environ Sci Technol 35:516–521.

McCarty PL. 1997. Breathing with chlorinated solvents. Sci 276:1521–1522.

McCarty PL, Semprini L. 1994. Ground-water treatment for chlorinated solvents. In Norris RD, Hinchee RE, Brown R, McCarty PL, Semprini L, Wilson JT, Kampbell DH, Reinhard M, Bouwer EJ, Borden RC, Vogel TM, Thomas JM, Ward CH, eds, Handbook of Bioremediation. Lewis Publishers, Boca Raton, FL, USA, pp 87–116.

McCarty PL, Chu M-Y, Kitanidis P. 2007. Electron donor and pH relationships for biologically enhanced dissolution of chlorinated solvent DNAPL in groundwater. Eur J Soil Biol 43:276–282.

McDade JM, McGuire TM, Newell CJ. 2005. Analysis of DNAPL source-depletion costs at 36 field sites. Remediat J 15:9–18.

McGuire TM, McDade JM, Newell CJ. 2006. Performance of DNAPL source depletion technologies at 59 chlorinated solvent-impact sites. Ground Water Monit Remediat 26:73–84.

McMurdie PJ, Behrens SF, Müller JA, Göke J, Ritalahti KM, Wagner RD, Holmes S, Richardson P, Löffler FE, Spormann AM. 2009. Localized plasticity linked to reductive dehalogenation in the streamlined genomes of vinyl chloride respiring Dehalococcoides. PLoS Genetics 5:e1000714.

Mercer JW, Cohen RM, Noel MR. 2008. DNAPL Site Characterization Issues at Chlorinated Solvent Sites. In Stroo HF, Ward CH, eds, *In Situ* Remediation of Chlorinated Solvent Plumes. Springer, New York, NY, USA, pp 217–280.

Mohn WW, Tiedje J. 1992. Microbial reductive dehalogenation. Microbiol Mol Biol Rev 56:482–507.

Morrill PL, Sleep BE, Seepersad DJ, McMaster ML, Hood ED, LeBron C, Major DW, Edwards EA, Sherwood-Lollar B. 2009. Variations in expression of carbon isotope fractionation of chlorinated ethenes during biologically enhanced PCE dissolution close to a source zone. J Contam Hydrol 110:60–71.

Mravik SC, Sillan RK, Wood AL, Sewell GW. 2003. Field evaluation of solvent extraction residual biotreatment technology. Environ Sci Technol 37:5040–5049.

Müller JA, Rosner BM, Von Abendroth G, Meshulam-Simon G, McCarty PL, Spormann AM. 2004. Molecular identification of the catabolic vinyl chloride reductase from *Dehalococcoides* sp. strain VS and its environmental distribution. Appl Environ Microbiol 70:4880–4888.

Munakatta-Marr J, Sorenson KS Jr, Petri BG, Cummings JB. 2011. Principles of Combining ISCO With Other *In Situ* Remedial Approaches. In Siegrist RL, Crimi M, Simpkin TJ, eds, *In Situ* Chemical Oxidation for Groundwater Remediation. Springer, New York, NY, USA, pp 285–317.

Mutch RD, Scott JI, Wilson DJ. 1993. Cleanup of fractured rock aquifers: Implications of matrix diffusion. Environ Monit Assess 24:45–70.

NAVFAC, Geosyntec. 2011. DNAPL Tool. http://projects.geosyntec.com/DNAPL/dnapltest.aspx. Accessed September 14, 2013.

Nielsen RB, Keasling JD. 1999. Reductive dechlorination of chlorinated ethene DNAPLS by a culture enriched from contaminated groundwater. Biotechnol Bioeng 62:160–165.

Newell CJ. 2009. Enhanced Amendment Delivery to Low Permeability Zones for Chlorinated Solvent Source Area Bioremediation. Project ER-200913. http://www.serdp.org/. Accessed September 14, 2013.

Newell CJ, Rifai HS, Wilson JT, Connor JA, Aziz CA, Suarez MP. 2002. Calculation and Use of First-Order Rate Constants for Monitored Natural Attenuation Studies. U.S. EPA Ground Water Issue. EPA/540/S-02/500. U.S. EPA National Risk Management Research Laboratory, Cincinnati, OH, USA.

Norris RD, Lageman R, Pool W, van Vulpen M. 1995. *In situ* electro-bioreclamation in low permeable soils. Proceedings, 3rd International Symposium *In Situ* and On-Site Bioreclamation. San Diego, CA, USA. April 24–27.

NRC (National Research Council). 2005. Contaminants in the Subsurface: Source Zone Assessment and Remediation. National Academies Press, Washington, DC, USA. 333 p.

O'Hara S, Krug T, Quinn J, Clausen C, Geiger C. 2006. Field and laboratory evaluation of the treatment of DNAPL source zones using emulsified zero-valent iron. Remediat J 16:32–56.

Parker BL, Gillham RW, Cherry JA. 1994. Diffusive disappearance of immiscible-phase organic liquids in fractured geologic media. Ground Water 32:805–820.

Peterson LN, Sorenson KS, Starr RC. 2000. Field Demonstration Report TAN Final Groundwater Remediation OU 1-07B. DOE/ID-10718. U.S. Department of Energy, Washington, DC, USA.

Philips J, Springael D, Smolders E. 2011. A three-layer diffusion-cell to examine bio-enhanced dissolution of chloroethene dense non-aqueous phase liquid. Chemosphere 83:991–996.

Quinn J, Geiger C, Clausen C, Brooks K, Coon C, O'Hara S, Krug T, Major D, Yoon W-S, Gavaskar A, Holdsworth T. 2005. Field demonstration of DNAPL dehalogenation using emulsified zero-valent iron. Environ Sci Technol 39:1309–1318.

Ramsburg CA, Abriola LM, Pennell KD, Löffler FE, Gamache M, Amos BK, Petrovskis EA. 2004. Stimulated microbial reductive dechlorination following surfactant treatment at the Bachman Road site. Environ Sci Technol 38:5902–5914.

Ramsburg CA, Thornton CE, Christ JA. 2010. Degradation product partitioning in source zones containing chlorinated ethene dense non-aqueous-phase liquid. Environ Sci Technol 44:9105–9111.

Reynolds DA, Jones EH, Gillen M, Yusoff I, Thomas DG. 2008. Electrokinetic migration of permanganate through low-permeability media. Ground Water 46:629–637.

Richardson R, Bhupathiraju VK, Song DL, Goulet TA, Alvarez-Cohen L. 2002. Phylogenetic characterization of microbial communities that reductively dechlorinate TCE based upon a combination of molecular techniques. Environ Sci Technol 36:2652–2662.

Ritalahti KM, Amos BK, Sung Y, Wu Q, Koenigsberg SS, Löffler FE. 2006. Quantitative PCR targeting 16S rRNA and reductive dehalogenase genes simultaneously monitors multiple *Dehalococcoides* strains. Appl Environ Microbiol 72:2765–2774.

Ritalahti KM, Hatt JK, Lugmayr V, Henn K, Petrovskis EA, Ogles DM, Davis GA, Yeager CM, Lebrón CA, Löffler FE. 2010. Comparing on-site to off-site biomass collection for *Dehalococcoides* biomarker gene quantification to predict *in situ* chlorinated ethene detoxification potential. Environ Sci Technol 44:5127–5133.

Robinson C, Barry DA, McCarty PL, Gerhard, JL, Kouznetsova I. 2009. pH control for enhanced reductive bioremediation of chlorinated solvent source zones. Sci Total Environ 407:4560–4573.

Rowlands, D. 2004. Development of Optimal pH for Degradation of Chlorinated Solvents by the KB-1™ Anaerobic Bacterial Culture. PhD Thesis, University of Guelph, Guelph, Ontario, Canada.

Sabalowsky AR, Semprini L. 2010a. Trichloroethene and *cis*-1,2-dichloroethene concentration-dependent toxicity model simulates anaerobic dechlorination at high concentrations: I. batch-fed reactors. Biotechnol Bioeng 107:529–539.

Sabalowsky AR, Semprini L. 2010b. Trichloroethene and *cis*-1,2-dichloroethene concentration-dependent toxicity model simulates anaerobic dechlorination at high concentrations: II. continuous flow and attached growth reactors. Biotechnol Bioeng 107:540–549.

Sahl J, Munakata-Marr J. 2006. The effects of *in situ* chemical oxidation on microbiological processes: A review. Remediat J 16:57–70.

Sahl J, Munakata-Marr J, Crimi M, Siegrist RL. 2007. Coupling permanganate oxidation with microbial dechlorination of tetrachloroethene. Water Environ Res 79:5–12.

Sale TC, McWhorter DB. 2001. Steady-state mass transfer from single component dense non-aqueous phase liquid in uniform flow fields. Water Resour Res 37:393–404.

Sale T, Newell CJ. 2010. Impacts of Source Management on Chlorinated Solvent Plumes. In Stroo HF, Ward CH, eds, *In Situ* Remediation of Chlorinated Solvent Plumes, Springer, New York, NY, USA, pp 185–216.

Sale T, Newell C, Stroo H, Hinchee R, Johnson P. 2008. Frequently Asked Questions Regarding Management of Chlorinated Solvents in Soils and Groundwater. ESTCP, Arlington, VA, USA. http://www.serdp.org/. Accessed September 14, 2013.

Schaefer CE, Condee CW, Vainberg S, Steffan RJ. 2009a. Bioaugmentation for chlorinated ethenes using Dehalococcoides sp.: Comparison between batch and column experiments. Chemosphere 75:141–148.

Schaefer CE, Callaghan AV, King JD, McCray JE. 2009b. Dense nonaqueous phase liquid architecture and dissolution in discretely fractured sandstone blocks. Environ Sci Technol 43:1877–1883.

Schaefer CE, Towne RM, Vainberg S, McCray JE, Steffan RJ. 2010. Bioaugmentation for treatment of dense non-aqueous phase liquid in fractured sandstone blocks. Environ Sci Technol 44:4958–4964.

Scheutz C, Durant ND, Dennis P, Hansen MH, Jørgensen I T, Jakobsen R, Cox EE, Bjerg PL. 2008. Concurrent ethene generation and growth of *Dehalococcoides* containing vinyl chloride reductive dehalogenase genes during an enhanced reductive dechlorination field demonstration. Environ Sci Technol 42:9302–9309.

Seagren EA, Rittman BE, Valocchi AJ. 1993. Quantitative evaluation of flushing and biodegradation for enhancing *in situ* dissolution of nonaqueous-phase liquids. J Contam Hydrol 12:103–132.

Seagren EA, Rittman BE, Valocchi AJ. 1994. Quantitative evaluation of the enhancement of NAPL-pool dissolution by flushing and biodegradation. Environ Sci Technol 28:833–839.

SERDP (Strategic Environmental Research and Development Program). 2006. Expert Panel Workshop on Research and Development Needs for the Environmental Remediation Application of Molecular Biological Tools. SERDP, Arlington, VA, USA. Available at: http://www.serdp.org/. Accessed September 14, 2013.

Sherwood-Lollar B, Slater GF, Sleep B, Witt M, Kleck GM, Harkness M, Spivack J. 2001. Stable carbon isotope evidence for intrinsic bioremediation of tetrachloroethene and trichloroethene at Area 6, Dover Air Force Base. Environ Sci Technol 35:261–269.

Simkins S, Alexander M. 1984. Models for mineralization kinetics with the variables of substrate concentration and population density. Appl Environ Microbiol 47:1299–1306.

Simpkin TJ, Norris RD. 2010. Engineering and Implementation Challenges for Chlorinated Solvent Remediation. In Stroo HF, Ward CH, eds, *In Situ* Remediation of Chlorinated Solvent Plumes. Springer, New York, NY, USA, pp 109–143.

Slater GF, Sherwood-Lollar B, Sleep BE, Edwards EA. 2001. Variability in carbon isotopic fractionation during biodegradation of chlorinated ethenes: Implications for field applications. Environ Sci Technol 35:901–907.

Sleep BE, Brown AJ, Sherwood-Lollar B. 2005. Long-term tetrachloroethene degradation sustained by endogenous cell decay. J Environ Eng Sci 4:11–17.

Sleep BE, Seepersad DJ, Mo K, Heidorn CM, Hrapovic L, Morrill PL, McMaster ML, Hood ED, Lebrón C, Sherwood-Lollar B, Major DW, Edwards EA. 2006. Biological enhancement of tetrachloroethene dissolution and associated microbial community changes. Environ Sci Technol 40:3623–3633.

Song DL, Conrad ME, Sorenson KS, Alvarez-Cohen L. 2002. Stable carbon isotope fractionation during enhanced *in situ* bioremediation of trichloroethene. Environ Sci Technol 36:2262–2268.

Steimle R. 2002. The state of the practice: Characterizing and remediating contaminated groundwater at fractured rock sites. Remediat J 12:23–33.

Stroo HF. 2010. Bioremediation of Chlorinated Solvent Plumes. In Stroo HF, Ward CH, eds, *In Situ* Remediation of Chlorinated Solvent Plumes. Springer, New York, NY, USA, pp 309–324.

Stroo HF, Norris RD. 2009. Alternatives for *In Situ* Bioremediation of Perchlorate. In Stroo HF, Ward CH, eds, *In Situ* Bioremediation of Perchlorate in Groundwater. Springer, New York, NY, USA, pp 79–90.

Stroo HF, Ward CH (eds), 2010. *In Situ* Remediation of Chlorinated Solvent Plumes. Springer, New York, NY, USA. 725 p.

Stroo HF, Unger M, Ward CH, Kavanaugh MC, Vogel C, Leeson A, Marqusee JA, Smith BP. 2003. Remediating chlorinated solvent source zones. Environ Sci Technol 37:224A–230A.

Stroo HF, Major DW, Gossett JM. 2010. Bioaugmentation for Anaerobic Bioremediation of Chlorinated Solvents. In Stroo HF, Ward CH, eds, *In Situ* Remediation of Chlorinated Solvent Plumes. Springer, New York, NY, USA, pp 425–454.

Stroo HF, Lesson A, Marqusee JA, Johnson PC, Ward CH, Kavanaugh MC, Sale TC, Newell CJ, Pennell KD, Lebron CA, Unger M. 2012. Chlorinated ethene source remediation: Lessons learned. Environ Sci Technol 46:6438–6447.

Stroo HF, Major DW, Steffan RJ, Koenigsberg SS, Ward CH. 2013. Bioaugmentation with *Dehalococcoides*: A Decision Guide. In Stroo HF, Leeson A, Ward CH, eds, Bioaugmentation for Groundwater Remediation. Springer, NewYork, NY, USA, pp 117–140.

Suarez MP, Rifai HS. 1999. Biodegradation rates for fuel hydrocarbons and chlorinated solvents in groundwater. Bioremediation J 3:337–362.

Sung Y, Ritalahti KM, Sanford RA, Urbance JW, Flynn SJ, Tiedje JM, Löffler FE. 2003. Characterization of two tetrachloroethene-reducing, acetate-oxidizing anaerobic bacteria and their description as *Desulfuromonas michiganensis* sp. nov. Appl Environ Microbiol 69: 2964–2974.

Sung Y, Ritalahti KM, Apkarian RP, Löffler F. 2006. Quantitative PCR confirms purity of strain GT, a novel trichloroethene-to-ethene-respiring *Dehalococcoides* isolate. Appl Environ Microbiol 72:1980–1987.

Suthersan SS, Lutes CC, Palmer PL, Lenzo F, Payne FC, Liles DS, Burdick J. 2002. Technical Protocol for Using Soluble Carbohydrates to Enhance Reductive Dechlorination of Chlorinated Aliphatic Hydrocarbons. Project ER-199920. Prepared for ESTCP, Arlington, VA, USA. http://www.serdp.org. Accessed September 14, 2013.

Suthersan S, Horst J, Nelson D, Schnobrich M. 2011. Insights from years of performance that are shaping injection-based remediation systems. Remediat J 21:9–25.

Sutton NB, Grotenhuis JTC, Langenhoff AAM, Rijnaarts HHM. 2010. Efforts to improve coupled *in situ* chemical oxidation with bioremediation: a review of optimization strategies. J Soils Sediments 11:129–140.

Thullner M, Zeyer J, Kinzelbach W. 2002. Influence of microbial growth on hydraulic properties of pore networks. Transp Porous Media 49:99–122.

Vainberg S, Steffan RJ, Rogers R, Ladaa T, Pohlmann D, Leigh D. 2006. Production and application of large-scale cultures for bioaugmentation. Proceedings, 5th International Conference on Remediation of Chlorinated and Recalcitrant Compounds, Monterey, CA, USA. May 22–25. Paper A-50.

Van der Zaan B, Hannes F, Hoekstra N, Rijnaarts H, de Vos WM, Smidt H, Gerritse J. 2010. Correlation of *Dehalococcoides* 16S rRNA and chloroethene reductive dehalogenase genes to different geochemical conditions in chloroethene-contaminated groundwater. Appl Environ Microbiol 76:843–850.

Vandevivere P, Baveye P. 1992. Saturated hydraulic conductivity reduction caused by aerobic bacteria in sand columns. Soil Sci Soc Am J 56:1–13.

Vandevivere P, Baveye P, de Lozada DS, DeLeo P. 1995. Microbial clogging of saturated soils and aquifer materials: Evaluation of mathematical models. Water Resour Res 31:2173–2180.

Vogel TM, McCarty PL. 1985. Biotransformation of tetrachloroethylene to trichloroethylene, dichloroethylene, vinyl chloride, and carbon dioxide under methanogenic conditions. Appl Environ Microbiol 49:1080–1083.

Vogel TM, Criddle CS, McCarty PL. 1987. Transformation of halogenated aliphatic compounds. Environ Sci Technol 21:722–736.

West MR. 2009. Mathematical Modeling of DNAPL Source Zone Remediation. PhD Thesis. Queen's University, Kingston, Ontario, Canada, 431 p.

West MR, Grant GP, Gerhard JI, Kueper BH. 2008. The influence of precipitate formation on the chemical oxidation of TCE DNAPL with potassium permanganate. Adv Water Resour 31:324–338.

West MR, Kueper BH. 2010. Plume detachment and recession times in fractured rock. Ground Water 48:416–426.

West MR, Kueper BH. 2012. Numerical Simulation of DNAPL Source Zone Remediation with *In Situ* Chemical Oxidation (ISCO). Adv Water Resour 44:126–139.

Wilkin RT, McNeil MS, Adair CJ, Wilson JT. 2001. Field measurement of dissolved oxygen: A comparison of methods. Ground Water Monit Remediat 21:124–132.

Williams S. 2003. Sequestration of a DNAPL Source with Vegetable Oil. Project ER-200319. http://www.serdp.org. Accessed March 22, 2011.

Wilson JT. 2010. Monitored natural attenuation of chlorinated solvent plumes. In Stroo HF, Ward CH, eds, *In Situ* Remediation of Chlorinated Solvent Plumes. Springer, New York, NY, USA, pp 325–355.

Wilson JT, Wilson BH. 1985. Biotransformation of trichloroethylene in soil. Appl Environ Microbiol 49:242–243.

Wood AL, Annable MD, Jawitz JW, Enfield CG, Falta RW, Goltz MN, Rao PSC. 2004. Impact of DNAPL source treatment on contaminant mass flux. Proceedings, 4th International Conference on Remediation of Chlorinated and Recalcitrant Compounds (Monterey, CA, USA, May 24–27). Battelle Press, Columbus, OH, USA. Paper 1D-07.

Wright DJ, Birak PS, Pedit JA, Miller CT. 2010. Effectiveness of source-zone remediation of DNAPL-contaminated subsurface systems. J Environ Eng 136:452–465.

Wu X, Gent DB, Davis JL, Alshawabkeh AN. 2012a. Lactate injection by electric currents for bioremediation of tetrachloroethylene in clay. Electrochem Acta 86:157–163.

Wu MZ, Reynolds DA, Fourie A, Prommer H, Thomas DG. 2012b. Electrokinetic *in situ* oxidation remediation: Assessment of parameter sensitivities and the influence of aquifer heterogeneity on remediation efficiency. J Contam Hydrol 136–137:72–85.

Yang Y, McCarty PL. 2000. Biologically enhanced dissolution of tetrachloroethene DNAPL. Environ Sci Technol 34:2979–2984.

Yang Y, McCarty PL. 2002. Comparison between donor substrates for biologically enhanced tetrachloroethene DNAPL dissolution. Environ Sci Technol 36:3400–3404.

Yu S, Dolan ME, Semprini L. 2005. Kinetics and inhibition of reductive dechlorination of chlorinated ethylenes by two different mixed cultures. Environ Sci Technol 39:195–205.

Zhong L, Oostrom M, Wietsma TW, Covert MA. 2008. Enhanced remedial amendment delivery through fluid viscosity modifications: Experiments and numerical simulations. J Contam Hydrol 101:29–41.

Zhuang P, Pavlostathis SG. 1995. Effect of temperature, pH and electron-donor on the microbial reductive dechlorination of chloroalkenes. Chemosphere 31:3537–3548.

Zysset A, Stauffer F, Dracos T. 1994. Modeling of reactive groundwater transport governed by biodegradation. Water Resour Res 30:2423–2434.

CHAPTER 13

NATURAL ATTENUATION OF CHLORINATED SOLVENT SOURCE ZONES

Charles J. Newell,[1] Bernard H. Kueper,[2] John T. Wilson[3] and Paul C. Johnson[4]

[1]GSI Environmental Inc., Houston, TX 77098, USA; [2]Queen's University, Kingston, ON, Canada K7L 3N6; [3]U.S. Environmental Protection Agency, Ada, OK 74820, USA; [4]Arizona State University, Tempe, AZ 85287, USA

13.1 INTRODUCTION

Natural attenuation typically is defined as a variety of physical, chemical, or biological processes that under favorable conditions act without human intervention to reduce the mass, toxicity, mobility, volume, or concentration of contaminants in soil or groundwater (USEPA, 1999). When relied upon as a remedy, the term monitored natural attenuation (MNA) is typically employed. The U.S. Environmental Protection Agency (USEPA) defines MNA as "the reliance on natural processes to achieve site-specific remedial objectives" (USEPA, 1999a). The USEPA (2001) acknowledges that definitions of MNA have changed over time and defines MNA as "a remediation approach based on understanding and quantitatively documenting naturally occurring processes at a contaminated site that protect humans and ecological receptors from unacceptable risks of exposure to hazardous contaminants."

In the context of sites impacted by chlorinated solvents, MNA typically has been considered a potential remedy for dissolved phase plumes emanating from source zones, but not the source itself. The USEPA's MNA Directive (USEPA, 1999), for example, presents the following view that MNA is more applicable to dissolved phase plumes than to source zones:

Sources of contamination are more appropriately addressed by engineered removal, treatment or containment technologies.... Where the sources of contamination have been controlled, dissolved plumes may be amenable to MNA because of the relatively small mass of contaminants present in the plume.

This point of view is also expressed in the USEPA's MNA remedial technology fact sheet, which states explicitly that "natural attenuation is not expected to remediate NAPL" (USEPA, 1999). The National Research Council (NRC) document, *Source Zone Assessment and Remediation* (NRC, 2005), presented the viewpoint that MNA targets dissolved phase plumes and should not "constitute in situ source zone remediation." The authors discuss whether active source removal could be coupled with MNA and how this might be accomplished. The authors state that a commonly expressed goal is to reduce concentrations of source zone contaminants enough to "allow MNA to take over" but conclude that there was "very little scientific evidence" to address this issue.

These and other regulatory agency documents clearly focus on using MNA as a plume remediation technology, not a source remediation technology. What has become apparent since the publication of many of these documents, however, is that chlorinated solvent source zones often do exhibit natural attenuation behavior, as evidenced by observations of long term decreases in concentration from untreated near-source wells (Newell et al., 2006) and historical surveys

B.H. Kueper et al. (eds.), *Chlorinated Solvent Source Zone Remediation*,
doi: 10.1007/978-1-4614-6922-3_13, © Springer Science+Business Media New York 2014

that show MNA being used for some sites for source zone management (McGuire et al., 2003). The role of MNA as a component of source management strategies and treatment trains is now being discussed (ITRC, 2011) and field studies of its significance are underway.

The natural attenuation of a chlorinated solvent source zone (also referred to as source zone natural attenuation, or SZNA) is defined by us as:

A variety of physical, chemical or biological processes that act without human intervention to reduce the mass, toxicity, mobility, volume, mass flux, mass discharge or concentration of contaminants in soil or groundwater within that volume of the subsurface currently or previously occupied by chlorinated solvents in the form of nonaqueous phase liquids.

Monitored natural attenuation (MNA) of a chlorinated solvent source zone (source zone MNA) is defined by us as:

The reliance upon a variety of physical, chemical or biological processes that act without human intervention to achieve site-specific remedial objectives, as verified through ongoing monitoring and data analysis, within and downgradient of that volume of the subsurface currently or previously occupied by chlorinated solvents in the form of nonaqueous phase liquids.

This chapter describes the rationale for using MNA to remediate and manage source zones and summarizes the tools available for evaluating and implementing source zone MNA. After first presenting a conceptual model of the life cycle of a dense nonaqueous phase liquid (DNAPL) source zone, the chapter introduces the key processes involved in natural attenuation of a chlorinated solvent source and then discusses the ability of source zone MNA to meet various remedial objectives for source zone management. The chapter then summarizes the state of the practice and the different approaches available for implementing source zone MNA, including a pending new protocol for SZNA. The chapter then introduces several models available for evaluating attenuation rates and the contributions of different processes and finally summarizes a few case studies of the successful use of source zone MNA for source zone management.

13.1.1 MNA as a Source Zone Remedy

One of the first documents discussing MNA for source zones was a chapter by Wiedemeier et al. (1999) entitled "Attenuation of Source Zones and Formation of Plumes." In this chapter, the authors describe the dissolution of both single-component and multicomponent NAPLs and discuss the time frames associated with the dissolution of a finite amount of source mass.

In 2000, the National Research Council (NRC, 2000) provided a more nuanced discussion of MNA and source zones in their book *Natural Attenuation for Groundwater Remediation*. The NRC made the following observations:

- In many cases, environmental regulators require source removal or containment as part of a natural attenuation remedy
- Expert opinions conflict on whether source removal is advisable when using natural attenuation as a remedy, even when such removal is technically feasible
- More commonly, the source cannot be delineated completely and/or cannot be removed to any significant degree even if located perfectly
- In some cases, source removal efforts may directly and adversely affect natural attenuation

Since 2004, emphasis on MNA as a remedy for source zones has increased with the development of field studies, process information, models, and protocols designed specifically for source zone MNA. For example, Farhat et al. (2004) introduced the software package *SourceDK*, a screening- and planning-level tool that can be applied to a variety of different types of source zones in groundwater for estimating groundwater remediation time frame and the uncertainties associated with the estimated time frame. *SourceDK* was introduced for the explicit purpose of evaluating source zone attenuation. It included empirical data, data analysis tools, and simple source zone attenuation models.

In a document entitled "Frequently Asked Questions Regarding the Management of Chlorinated Solvents in Soils and Groundwater" (Sale et al., 2008a), MNA was cited as a frequently used in situ source treatment technology, but under these conditions:

Often used as polishing step for weathered source zones and/or following active treatment. Requires knowledge of attenuation processes and, in most cases, temporal data demonstrating decreasing trend in source zone concentrations. May require longer time frame than engineered approaches.

Johnson et al. (2006) presented a methodology for SZNA assessment at hydrocarbon sites that for the first time defined a mix of field analyses and mass flux assessments to evaluate and quantify source zone attenuation and demonstrated its application to a large hydrocarbon release site (Lundegard and Johnson, 2006). This approach was adopted by the Interstate Technology & Regulatory Council (ITRC) in 2009 for light nonaqueous phase liquid (LNAPL) sites and the authors are currently extending this technique to chlorinated solvent source zones as part of an Environmental Security Technology Certification Program (ESTCP, 2013) project.

Currently, source zone MNA is being applied as a stand alone remedy (McGuire et al., 2004), or as a polishing step after active remediation, at some sites. The ITRC (2011) stated that active remediation technologies rarely achieve complete remediation of all contaminant mass, and thus, MNA is in effect a near-universal component of chlorinated solvent site remedies. In 2011, the ITRC reiterated this conclusion by stating:

In general, performance data from multiple site studies show that a single application of current-generation remediation technologies is not likely to achieve low part-per-billion drinking water standards. Therefore, active remediation at many or perhaps most sites will be followed by MNA of residual contaminants.

More recently, the NRC (2012) proposed an expanded role for MNA at sites with persistent contamination that would include MNA in source zones. It is presumed here that the persistent contamination in the plume is the result of limitations of source zone treatment:

At many complex sites, contaminant concentrations in the plume remain stalled at levels above cleanup goals despite continued operation of remedial systems. There is no clear path forward to a final end state embodied in the current cleanup programs, such that money continues to be spent, with no concomitant reduction in risks. If the effectiveness of site remediation reaches a point of diminishing returns prior to reaching cleanup goals and optimization has been exhausted, the transition to monitored natural attenuation or some other active or passive management should be considered using a formal evaluation. This transition assessment would determine whether a new remedy is warranted at the site or whether long term management is appropriate.

13.1.2 Factors Affecting Applicability

The set of factors that affect the applicability of source zone MNA are somewhat different from those affecting active remediation technologies. In particular, many active remediation technologies are much more sensitive to the types of contaminants, hydrogeologic conditions, and starting concentrations than source zone MNA, which is sensitive to a combination of technical, risk, and regulatory factors. The USEPA MNA Directive (1999) listed these factors when considering MNA for plumes, and these same factors are also applicable to source zone MNA.

In determining whether MNA is an appropriate remedy for soil or groundwater at a given site, the USEPA or other regulatory authorities should consider the following:

- Whether the contaminants present in soil or groundwater can be effectively remediated by natural attenuation processes
- Whether or not the contaminant plume is stable and the potential for the environmental conditions that influence plume stability to change over time
- Whether human health, drinking water supplies, other groundwaters, surface waters, ecosystems, sediments, air, or other environmental resources could be adversely impacted as a consequence of selecting MNA as the remediation option
- Current and projected demand for the affected resource over the time period that the remedy will remain in effect
- Whether the contamination, either by itself or as an accumulation with other nearby sources (on site or off site), will exert a long term detrimental impact on available water supplies or other environmental resources
- Whether the estimated time frame of remediation is reasonable (see section on "Reasonable Time Frame for Remediation") compared to time frame required for other more active methods, including the anticipated effectiveness of various remedial approaches on different portions of the contaminated soil and/or groundwater
- The nature and distribution of sources of contamination and whether these sources have been, or can be, adequately controlled
- Whether the resulting transformation products present a greater risk, due to increased toxicity and/or mobility, than do the parent contaminants
- The impact of existing and proposed active remediation measures upon the MNA component of the remedy or the impact of remediation measures or other operations/activities (pumping wells) in close proximity to the site
- Whether reliable site-specific mechanisms for implementing institutional controls such as zoning ordinances are available and if an institution responsible for their monitoring and enforcement can be identified

13.1.3 Life Cycle of Chlorinated Solvent Sources

Source zone MNA explicitly recognizes the fact that sources change over time as they progress through a defined life cycle. This is sometimes called "weathering" of source materials and is a term often referred to at hydrocarbon sites. Monitored natural attenuation may be a sole remedy, but often it is used after active treatment, which essentially accelerates the aging by stimulating one or more of these processes. The following sections describe the natural life cycle of a chlorinated solvent source zone, to illustrate the changes that can be expected over time and to provide a framework for the later discussion of the natural attenuation processes that control this life cycle.

A source zone's life cycle progresses from the initial release of DNAPL to the subsurface through complete depletion by natural processes. The life cycle is described here for a single DNAPL release into unconsolidated deposits overlying fractured bedrock. The DNAPL is assumed to be composed of more than one chemical component. In practice, the life cycle of a source zone may be more complex than that described here, with complicating factors including multiple DNAPL releases in time at multiple locations, changing geochemical conditions with time, and the presence of other contaminants such as metals and inorganics. The following five stages of the source zone life cycle are illustrated in Figure 13.1 and are

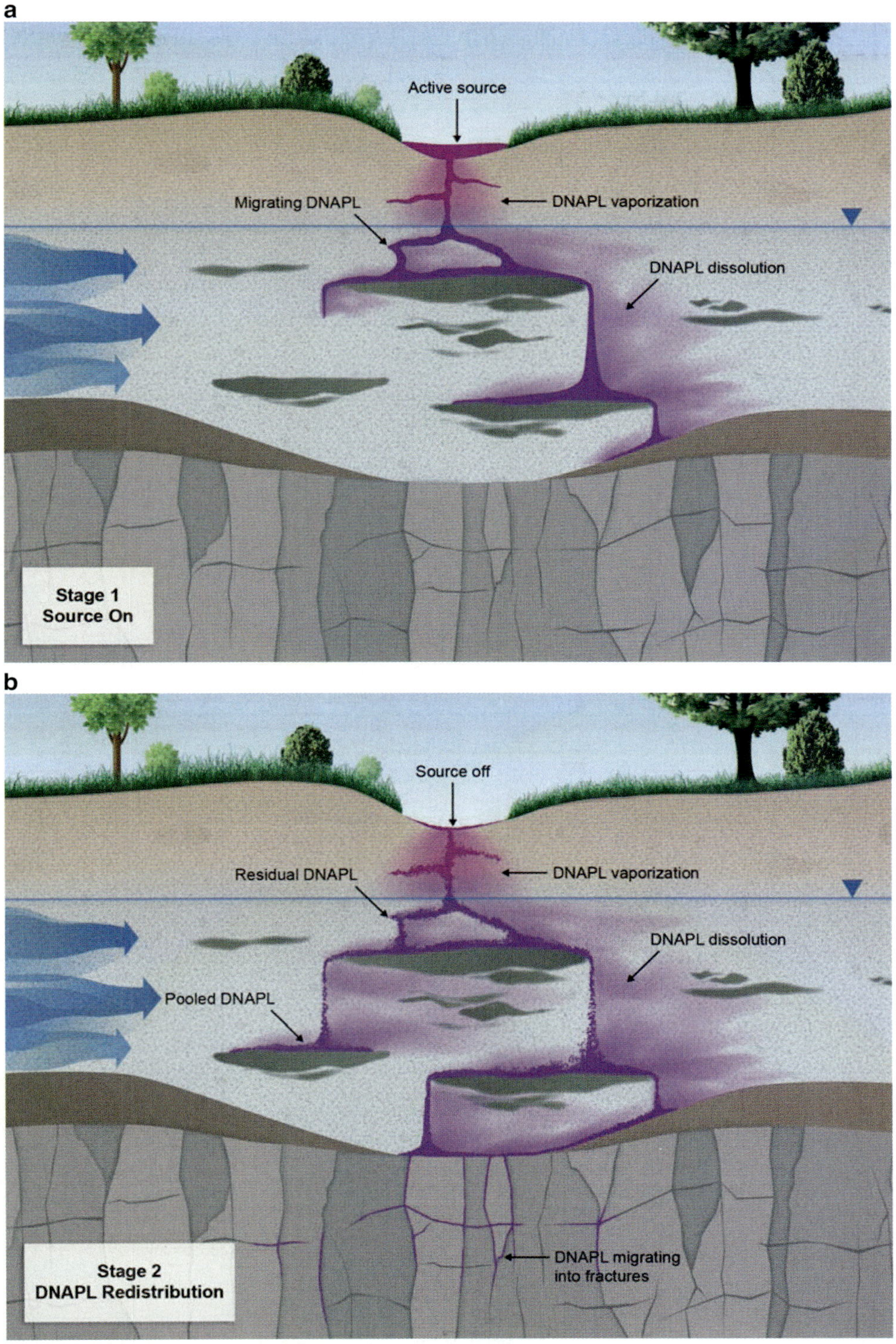

Figure 13.1. Life cycle of a chlorinated solvent source zone. (a) Stage 1, initial DNAPL release; (b) Stage 2, DNAPL redistribution; (c) Stage 3, continued DNAPL dissolution and weathering; (d) complete DNAPL depletion; and (e) desorption and back diffusion (from Kueper et al., 2014).

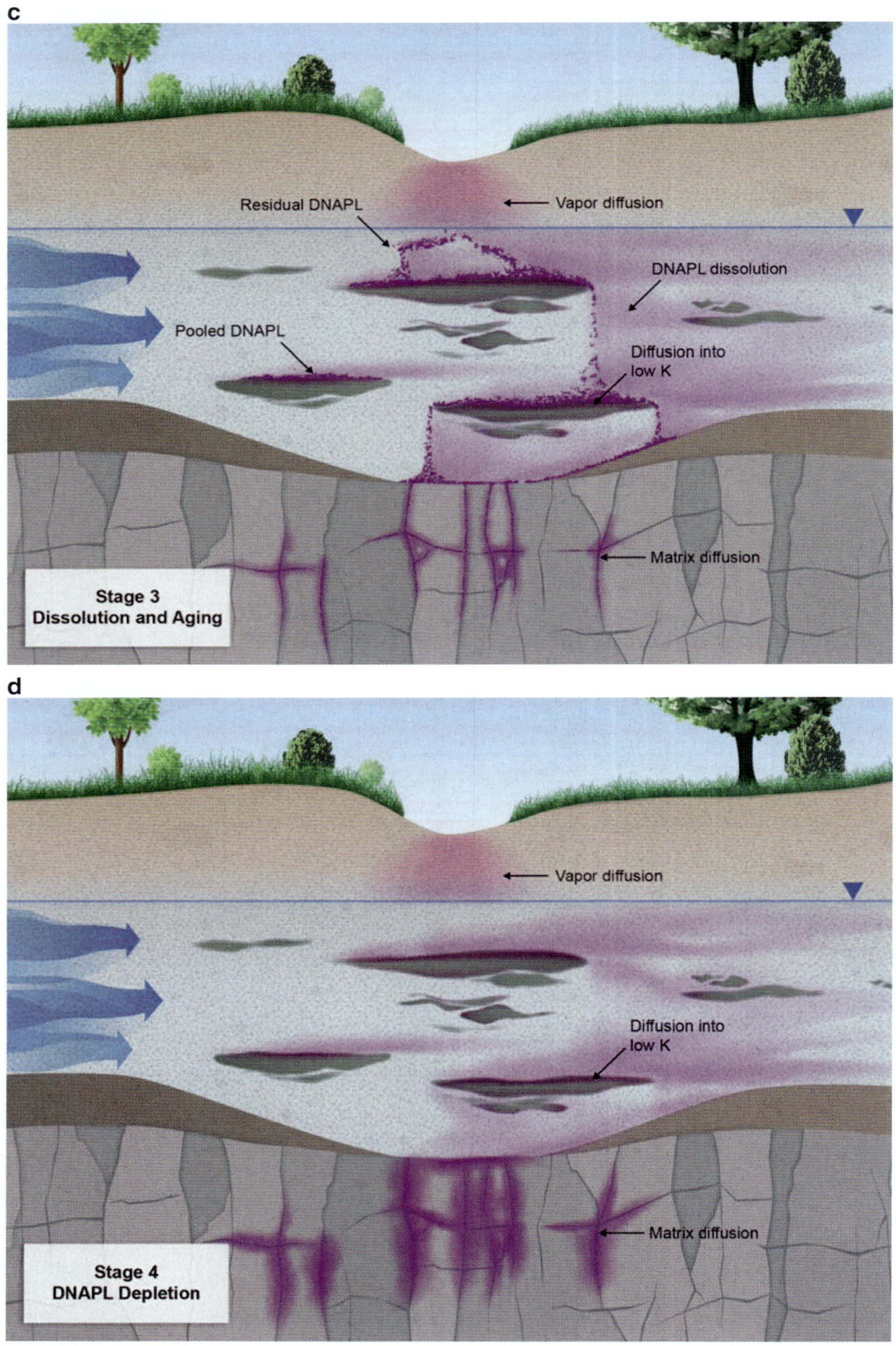

Figure 13.1. (continued)

e

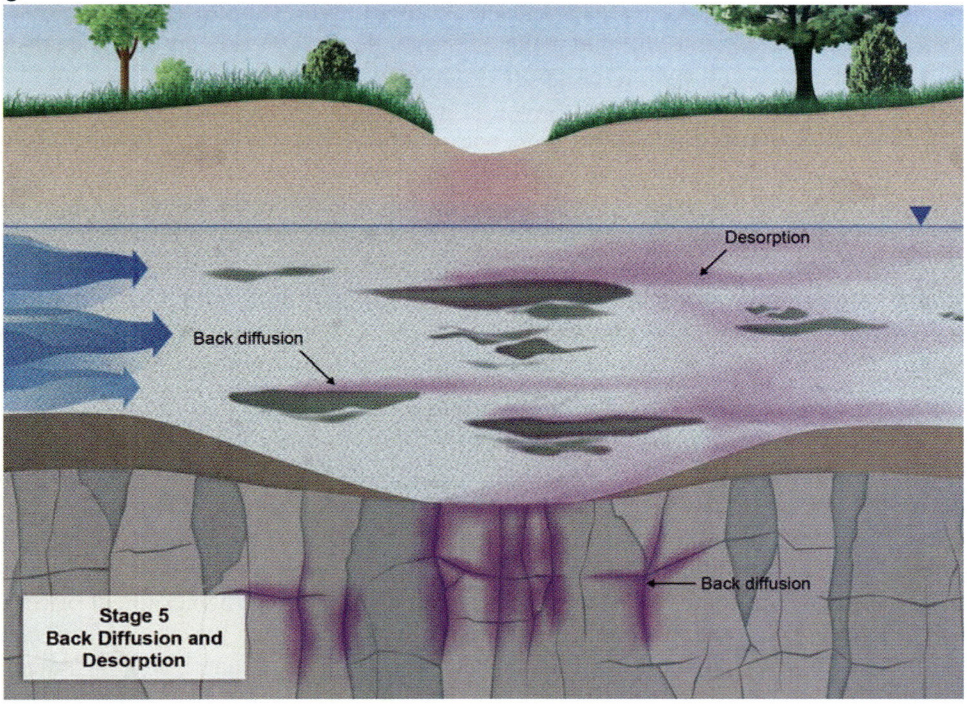

Figure 13.1. (continued)

also presented and discussed in Chapter 1. Similarly, Sale et al. (2008a) describe three life cycle stages (early, middle, late) for a DNAPL source zone. These correspond to the end of Stage 2, Stage 3, and Stage 5 below.

Stage 1: Initial DNAPL Release

The evolution of a chlorinated solvent source zone begins with an initial DNAPL release to the subsurface. Releases can be short in duration, such as a catastrophic failure of a storage vessel, or longer term in nature, such as persistent leaking from a disposal pond or burn pit being operated for many years. The nature and duration of the release influences the subsurface migration pathways of the DNAPL. For example, slow releases at low capillary pressure will tend to result in more lateral spreading of the DNAPL and a sparser distribution of residual and pools compared to fast, high capillary pressure releases (Poulsen and Kueper, 1992; Kueper and Gerhard, 1995). Understanding the nature and duration of the initial DNAPL release is an important component of conceptual model development.

Stage 2: DNAPL Redistribution

When the release of DNAPL to the subsurface has ceased, DNAPL will continue to migrate until it comes to rest in a static distribution of residual DNAPL (disconnected blobs and ganglia) and DNAPL in accumulations referred to as pools. During this process, capillary pressures are generally decreasing and large changes in DNAPL surface area to volume are taking place. The particular configuration of residual DNAPL and pools that arises (referred to as the ganglia to pool [GTP] ratio) has been used by some researchers to define the "DNAPL architecture." The DNAPL architecture is influenced by a variety of factors including the DNAPL viscosity, the DNAPL–water interfacial tension, wettability, the DNAPL density, geologic structure and permeability, the volume of the release, and the nature and duration of the release.

In the unsaturated zone, additional influences include the DNAPL–air surface tension, the DNAPL vapor pressure, and spatial/temporal variations in moisture content. The length of time required for DNAPL redistribution can vary from months (Kueper et al., 1993) to decades (Gerhard et al., 2007) depending on the factors listed above. For example, a low-volume release of low-viscosity, high interfacial tension DNAPL in a highly permeable setting will redistribute for only a short period (perhaps months), while a large volume, moderate to high-viscosity release in lower permeability setting may take much longer (perhaps decades). During the DNAPL redistribution period, DNAPL vaporization takes place above the water table, DNAPL dissolution into groundwater takes place below the water table, and biodegradation may remove contaminants from the aqueous phase thereby increasing the DNAPL dissolution driving force.

Stage 3: Continued DNAPL Dissolution and Aging

As continued dissolution and vaporization of DNAPL takes place, vapor and dissolved phase plumes continue to develop. This results in a continuous reduction of DNAPL mass with time. If present, biodegradation increases the dissolution rate. For a multicomponent DNAPL, the higher effective solubility components are depleted faster than the lower effective solubility components, thereby enriching the DNAPL in the lower effective solubility components. This results in changes to the composition of the dissolved phase plume downgradient of the source zone and a lowering of total concentrations within the plume. Given knowledge of the original DNAPL composition, the current concentrations in the plume immediately downgradient of the source zone, the rate of biodegradation, and the rate of groundwater flow through the source zone, one can in theory estimate how much the DNAPL mass has been depleted. In addition to changes in aqueous-phase concentrations stemming from changes in DNAPL composition, total concentrations will be reduced as a result of complete depletion of DNAPL from certain areas of the source zone. In general, DNAPL in higher permeability pathways will deplete faster than DNAPL in lower permeability pathways. Source zones with biodegradation will deplete faster than sources without biodegradation.

During this stage of the source zone life cycle, significant quantities of dissolved phase mass can diffuse into low permeability zones such as silts and clays and the bedrock matrix. If DNAPL is pooled above a capillary barrier, the forward diffusion process can transfer mass directly from the DNAPL into that zone. DNAPL mass is therefore depleted by this mechanism in addition to dissolution into flowing groundwater.

Stage 4: Complete DNAPL Depletion

Eventually, all DNAPL mass will have been depleted as a result of vaporization, dissolution, and diffusion mechanisms. The length of time required for complete depletion to occur below the water table will be a function of many factors including the volume of DNAPL originally released to the subsurface, the DNAPL solubility, the groundwater velocity, and the degree of biodegradation. For small volume DNAPL releases into high permeability media, the life span of the DNAPL may be as short as a few years. For large volume DNAPL releases into low to moderate permeability media, the life span could be on the order of hundreds of years. In the unsaturated zone, the length of time required for complete depletion to occur will be a function of many factors including the volume of DNAPL released, the DNAPL vapor pressure, and the moisture content. For relatively small volumes of high vapor pressure DNAPLs released to dry soils, the life span may be as short as a few months. Conversely, large volumes of DNAPL released to wet soils can persist for many decades or longer.

Stage 5: Desorption and Back Diffusion

Following complete depletion of DNAPL, all mass below the water table will be present as contaminants sorb to grain surfaces and those present in the aqueous phase. Concentrations in the permeable zones will start to decrease as clean groundwater enters the source zone from upgradient. This lowering of concentrations in the permeable zones triggers back diffusion of contaminant mass from the low permeability zones. This transfer of mass from low- to high-permeability zones continues for a longer period of time than the original forward diffusion process. How relevant back diffusion is with respect to meeting cleanup goals is largely dependent on the ability of groundwater flow in the permeable zones to dilute the back-diffusing concentrations. In general, back diffusion is expected to sustain plumes above cleanup standards for long periods of time in geologic settings where the proportion of low-permeability media exceeds that of the high permeability media. Examples include fractured clay, fractured sedimentary bedrock, and interbedded sand/clay sequences characterized by relatively thin occurrences of sand. Large amounts of organic carbon, low groundwater velocity, and little to no biodegradation also serve to prolong the desorption and back diffusion stage.

Considerable discussion in the groundwater community has centered on the percentage of current sites in any one of the above five stages. Given better disposal, handling, and storage practices, it would be reasonable to conclude that relatively few sites are currently in Stage 1. However, the length of time that a site will remain in Stages 2 and 3 can vary considerably from site to site given the variety of factors that dictate DNAPL redistribution, dissolution, and aging timescales. It is also important to recognize that not all regions within a source zone will reach each life cycle stage at the same time.

Some individuals contend that from a management perspective, understanding how much DNAPL still remains within the source zone is unimportant and that expending significant characterization funds to determine this mass is wasted effort. To others, determining how much DNAPL is present in a source zone is critical from a remediation perspective. Attempting to remediate a source zone with a technology specifically designed for DNAPL removal would be inappropriate, for example, if little or no DNAPL was present, while the design for some remediation technologies such as chemical oxidation or reduction requires a reasonable estimate of the initial mass present.

The fact that source zones may go through a prolonged exhausted stage, where concentrations are low but still exceed cleanup standards, has important implications for site management. Notably, managers should expect that continued application of certain remediation technologies (treatment trains) may have little benefit in such source zones, while others may have a longer lasting effect. For example, injection of relatively short lived oxidation reagents that only remove contaminants from transmissive zones without any other secondary effects will not have a long term beneficial effect on concentrations at sites where transmissive zone concentrations are sustained by back diffusion. On the other hand, injection of electron donors that support biomass growth and subsequent *endogenous decay*, or charge reactive minerals to make them reactive, may result in extended sustained treatment of back diffusion dominated sources that may last several years (Adamson et al., 2011).

13.2 KEY ATTENUATION PROCESSES

Chlorinated solvent source zones are affected by a number of attenuation processes including physical, chemical, and biological processes (Figure 13.2). The degree to which these processes contribute to source zone attenuation is influenced by a number of site specific conditions including the chemical composition and mass of contaminant present,

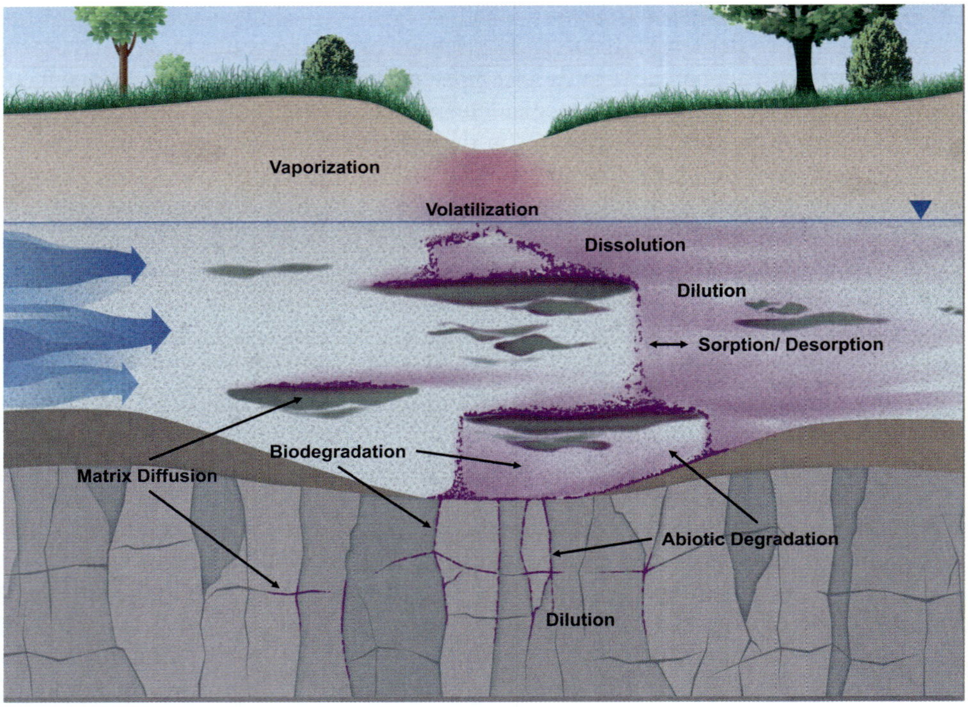

Figure 13.2. Natural attenuation processes within a chlorinated solvent source zone.

source zone architecture, the distribution and magnitude of groundwater flow, ambient geochemical conditions, hydraulic conductivity distribution, and the age of the source. The various processes of interest are briefly described in the following sections.

13.2.1 Vaporization and Volatilization

Vaporization refers to the transfer of contaminant mass from the DNAPL phase to the air phase. Vaporization is particularly relevant to the natural attenuation of volatile DNAPL sources in the unsaturated zone. The rate at which DNAPL will deplete as a result of vaporization is highly dependent upon the vapor pressure of the DNAPL, the volume of DNAPL present, the moisture content of the porous or fractured medium, and the DNAPL architecture. Highly volatile solvents in dry unsaturated media may persist for only a few months to a few years, a much shorter time frame than is associated with those same solvents when submerged below the water table. At such sites, the lack of DNAPL in unsaturated media should not be taken as evidence that DNAPL releases to the saturated zone did not occur.

To illustrate the importance of moisture content, Table 13.1 presents the effective vapor-phase diffusion coefficient for trichloroethene (TCE) for a variety of soil conditions. The diffusive mass flux in unsaturated media, and hence the rate of DNAPL depletion due to vaporization, is directly proportional to the effective diffusion coefficient. Table 13.1 demonstrates that the effective vapor-phase diffusion coefficient varies by a factor of more than 500 between very dry and very moist conditions. Estimating the life span of chlorinated solvent DNAPLs in unsaturated media can be performed using a variety of analytical and numerical modeling approaches. Key input parameters to such models are the moisture content, the DNAPL vapor pressure, and the volume of DNAPL released.

Table 13.1. Effective Vapor Phase Diffusion Coefficient for TCE in Unsaturated Porous Media

Total Porosity	Volumetric Moisture Content	Volumetric Air Content	Effective Vapor Phase Diffusion Coefficient (m^2/day)
0.35	0.05	0.30	0.0511
0.35	0.10	0.25	0.0269
0.35	0.15	0.20	0.0124
0.35	0.20	0.15	0.0046
0.35	0.25	0.10	0.0012
0.35	0.30	0.05	0.0001

Effective vapor-phase diffusion coefficient calculated using Millington and Quirk (1961). TCE dimensionless Henry's Law constant = 0.428; TCE free air diffusion coefficient = 7.9E-02 cm^2/s; TCE free water diffusion coefficient = 9.1E-06 cm^2/s; TCE organic carbon partition coefficient = 126 ml/g; soil fraction organic carbon = 0.003

Volatilization refers to the transfer of contaminant mass dissolved in groundwater to the air phase and is characterized by the Henry's Law constant for the compound of interest. Volatilization is not a significant source zone MNA process while DNAPL is still present (Stages 1–3). However, volatilization may be a significant depletion process during Stages 4 and 5 when DNAPL is absent. An Arizona State University research group (ESTCP, 2014) evaluated vaporization and volatilization at three chlorinated solvent source zones using the SZNA methodology developed by Johnson et al. (2006) for ESTCP. At the sites considered, however, vaporization and volatilization were a small contributor to SZNA because the sources were primarily submerged below the water table.

13.2.2 DNAPL Dissolution

Advection through chlorinated solvent source zones results in the transfer of aqueous-phase contaminant mass from the source zone to the plume. The mass discharge from the source zone can be measured using one of five methods (ITRC, 2010) and is discussed further in Chapter 7. Given that hydraulic conductivity typically varies by a factor of 100 or more at any particular site and the dissolved concentrations can also be highly spatially variable, it follows that the spatial distribution of mass flux can be highly spatially variable and that the quantification of overall mass discharge can be challenging.

Newell et al. (2011) present a plume magnitude classification system that relies upon mass discharge to categorize plumes into one of ten separate categories, where a Magnitude 1 plume is weak with a mass discharge of 0.001 grams per day (g/d) or less and a Magnitude 10 plume is strong with a mass discharge of 100,000 g/d or more. The same classification can be applied to chlorinated solvent source zones, where the mass discharge from the source zone to the plume can be described as ranging from Magnitude 1 through to Magnitude 10 as summarized in Table 13.2.

DNAPL dissolution is an important process in the first three stages of the chlorinated solvent source zone life cycle. Considerable research has focused on the dissolution of DNAPLs, with early studies concentrating on laboratory studies employing homogeneous porous media and the development of mass transfer coefficients (Miller et al., 1990). Subsequent work included laboratory studies and modeling to assess DNAPL dissolution in heterogeneous media (Sale and McWhorter, 2001; Parker and Park, 2004; Fure et al., 2006; Parker et al., 1994), as well as field-scale assessments (Rivett et al., 2001). Key concepts from these works include the following: (1) the importance of DNAPL architecture in controlling dissolution, (2) the value of the Power Law Model in describing the relationship between mass

Table 13.2. Source Zone Mass Discharge Classification (Modified After Newell et al., 2011)

Source Zone Mass Discharge Classification	Mass Discharge (g/d)	Equivalent Number of 55 gal Drums of TCE Leaving Source Zone per Year	Typical Site Conditions
Mag 1	<0.001	<1.2E-06	*Low-transmissivity units with small and weak source zones*
Mag 2	0.001 to <0.01	1.2E-06 to <1.2E-05	
Mag 3	0.01 to <0.1	1.2E-05 to <1.2E-04	
Mag 4	0.1 to <1	1.2E-04 to <1.2E-03	
Mag 5	1 to <10	1.2E-03 to <0.012	
Mag 6	10 to <100	0.012 to <0.12	
Mag 7	100 to <1,000	0.12 to <1.2	*High-transmissivity units with large and strong source zones*
Mag 8	1,000 to <10,000	1.2 to <12	
Mag 9	10,000 to <100,000	12 to <120	
Mag 10	>100,000	>120	

removal and mass discharge from a source, (3) the need to consider the effective solubility since many sources contain mixtures of solvents, and (4) the importance of dissolution into low-permeability zones in controlling longevity and the effects of treatment technologies. These concepts are discussed in the following sections.

13.2.2.1 DNAPL Architecture

A source with little of the DNAPL present as pools (a low fraction of DNAPL in pools compared to ganglia, or a high ganglia to pool ratio) will result in faster overall depletion of DNAPL mass through dissolution than one with a high pool fraction (Suchomel and Pennell, 2006; Christ et al., 2008). Ganglia (residual DNAPL) generally have a higher surface area to volume ratio than pools, which, when combined with a higher relative permeability to water than within pools, are subject to faster rates of dissolution into flowing groundwater (Grant et al., 2007). In addition to the pool fraction, other aspects of DNAPL architecture that influence the rate of DNAPL depletion through dissolution into groundwater include the overall volume of DNAPL present, the continuity of higher permeability pathways and the proportion of DNAPL present in such pathways as compared to lower permeability pathways, and the sparseness of individual zones of residual and pooled DNAPL. Of particular note is the fact that DNAPL present in high permeability pathways may dissolve much quicker than DNAPL present in low permeability pathways, resulting in a significant decline in mass discharge from the source zone with time.

13.2.2.2 The Power Law Model

The Power Law Model is an empirical relationship between mass discharge emanating from a source zone and the amount of mass remaining in that zone (Rao et al., 2001; Rao and Jawitz, 2003; Parker and Park, 2004; Zhu and Sykes, 2004; Falta et al., 2007) (Figure 13.3). While this model was originally developed to express how sources change after source remediation (Chapters 5 and 6), it can also be applied to describe source attenuation via natural processes. The Power Law Model expression is:

$$\frac{M_d}{M_{do}} = \left[\frac{M}{M_o}\right]^{\Gamma} \quad \text{(Eq. 13.1)}$$

where M_d is the mass discharge leaving the source zone at a particular time (mass per time), M is the mass remaining in the source zone at this time (mass), M_{do} is the mass discharge initially leaving the source zone ($t = 0$), M_o is the source mass at the beginning ($t = 0$) of the application of the Power Law analysis (mass), and $\Gamma$is an exponent reflective of source zone architecture, DNAPL characteristics, and geology.

Those who use the Power Law Model assert that $\Gamma < 1$ can represent a source zone dominated by DNAPL pools, or the case in which DNAPL is mostly in high-conductivity zones (Falta et al., 2005a; Newell et al., 2011), and $\Gamma > 1$ can represent a relatively homogeneous DNAPL distribution (Rao and Jawitz, 2003), or DNAPL in mostly low permeability zones in a heterogeneous system (Falta et al., 2005a, b). Newell (2011), the Center for Sustainable Groundwater & Soil Solutions (CSGSS, 2011), and Seyedabbasi et al. (2012) point out that an aged source zone dominated by back diffusion of aqueous and sorbed phase mass from low permeability zones is likely to be represented by $\Gamma > 1$. A value of $\Gamma = 1$ simplifies to an assumption of first-order decay of the source zone mass discharge and mass over time (Falta et al., 2005a). Figure 13.3 shows examples of different types of sources with the Power Law Model. The REMChlor model (Falta et al., 2005a, b; USEPA, 2007; Chapter 6) incorporates the Power Law Model as a source depletion representation that can be used for either evaluating the effects of remediation or for evaluating long term source behavior controlled by natural attenuation.

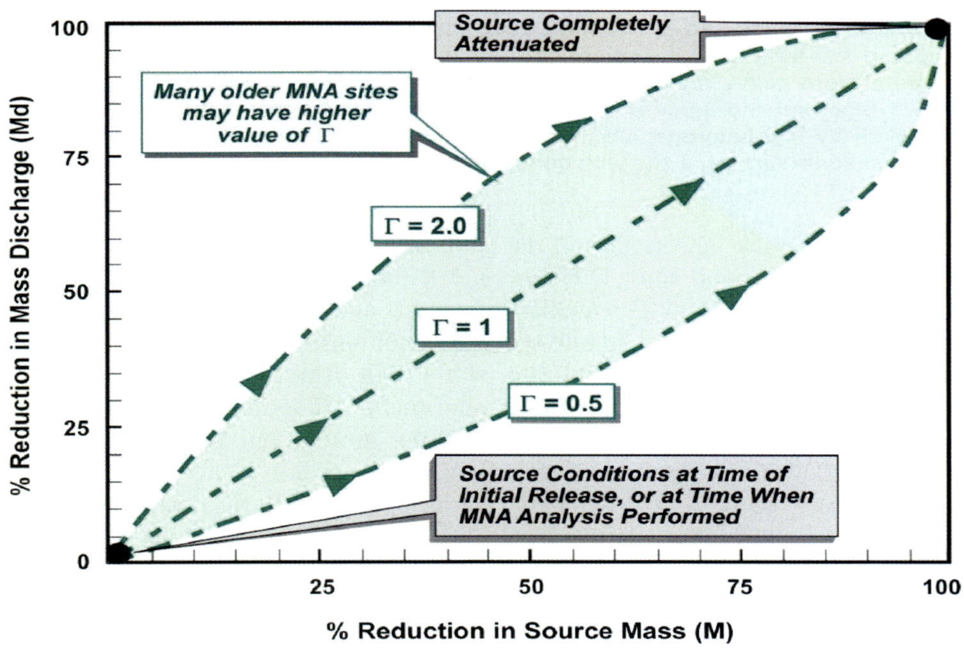

Figure 13.3. Relationship between source zone mass loss and reduction in mass discharge from the source zone when applied to source zone MNA. Power Law relationships are shown for typical range in gamma values ($0.5 < \Gamma < 2$) (Falta et al., 2007). Older sources will likely have higher values of Γ (Falta et al., 2007), while CSGSS (2011) suggests $\Gamma > 1$ will be needed to represent the long concentration versus time tail associated with old sites dominated by desorption and back diffusion from low permeability zones.

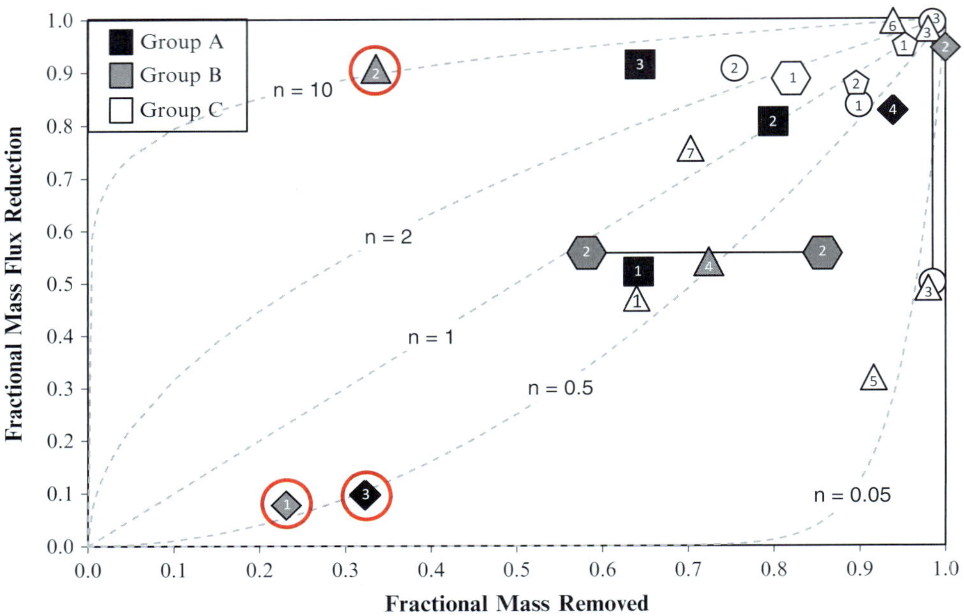

Figure 13.4. Mass flux reductions as a function of source zone mass removal for several field studies (DiFilippo and Brusseau, 2008). Group A (largest symbols with darkest background) comprises the studies with the lowest uncertainty and Group C (smallest symbols with the white background) are those with the highest uncertainty based on the methods used to determine initial mass, mass flux, and mass removal. The dashed lines represent curves generated using the simple Power Law Model. Red circles have been added to the original figure to show three water flushing/pump-and-treat remediation projects that might be similar to source zone MNA (the other projects were more active remediation technologies, such as surfactant flushing or chemical oxidation). 1-Diamond and 3-Diamond were controlled release experiments at the Borden site (a site with relatively low heterogeneity) while 2-Triangle represents a pump-and-treat system at Tucson International Airport, a site with considerable heterogeneity.

There is considerable uncertainty in the application of a simple relationship such as the Power Law Model to actual sites. DiFilippo and Brusseau (2008) evaluated mass discharge versus mass relationships from 21 remediation projects and concluded that "the use of simple tools, especially those not tuned to site-specific conditions, to predict mass-flux-reduction/mass-removal relationships may be fraught with a high degree of uncertainty." Figure 13.4 shows mass flux (mass discharge) reductions as a function of source zone mass removal from several field studies (DiFilippo and Brusseau, 2008). In addition, Seyedabbasi et al. (2012) showed how the Power Law exponent changed over time for a hypothetical source zone, with $\Gamma < 1$ in early years for a DNAPL-dominated source and $\Gamma > 1$ during later years for a source dominated by matrix diffusion.

13.2.2.3 Effective Solubility

For chlorinated solvent DNAPLs with a mixture of components, the effective solubility is the concentration of a particular constituent in water in equilibrium with the DNAPL. This effective solubility is controlled by the mole fraction of the particular constituent in the DNAPL and Raoult's Law (see Equation 3.4 in Chapter 3). The ramification of effective solubility is that high-solubility and high mole fraction constituents will elute first, followed by the lower effective solubility components. Thus, the mole fraction of low effective solubility constituents

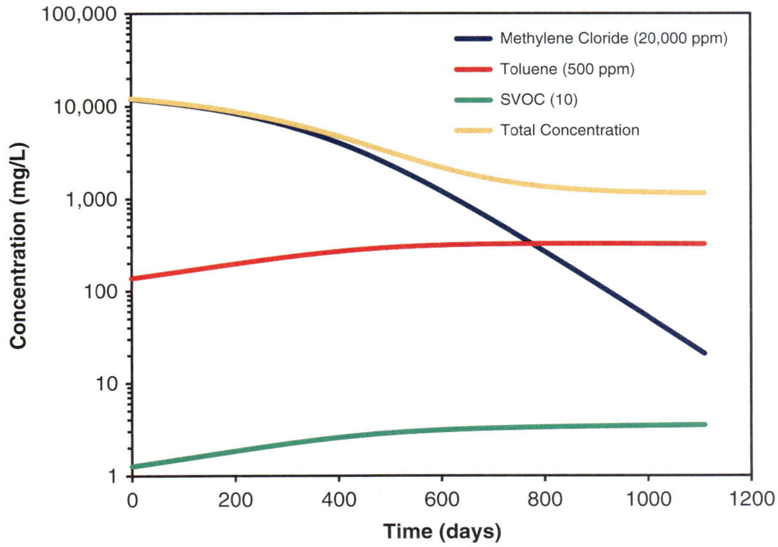

Figure 13.5. Concentration (log scale) versus time downgradient of a 1 m^3 source consisting of fine sand containing a three-component DNAPL at residual saturation. The concentration of the initially highest effective solubility compound (methylene chloride) decreases with time while the concentrations of the other two compounds (toluene and SVOC) increase with time. Total concentration decreases with time exhibiting source attenuation.

will increase over the lifetime of the dissolution event as the DNAPL weathers, with corresponding changes in the downgradient plume chemistry. The overall manifestation of this phenomenon is that the total mass flux from the DNAPL will decrease with time.

Figure 13.5 illustrates the dissolution history of a three-component DNAPL initially composed of 50% methylene chloride, 25% toluene, and 25% semivolatile organic compound (SVOC) by mass. The single-component solubilities of methylene chloride, toluene, and SVOC are taken to be 20,000, 500, and 10 milligrams per liter (mg/L), respectively. The example considers 1 cubic meter (m^3) (1.3 cubic yards) of fine sand having a porosity of 0.3 and containing residual DNAPL at a saturation of 20% of pore space. The 1 m^3 of fine sand is subject to a hydraulic gradient of 0.01 and the residual DNAPL is assumed to dissolve under equilibrium partitioning conditions following Raoult's Law. Figure 13.5 illustrates that the initially high effective solubility component (methylene chloride) decreases in concentration with time while the initially lower effective solubility components (toluene and methylene chloride) increase in concentration with time. The total concentration of organic compounds decreases with time, exhibiting attenuation behavior. The DNAPL becomes increasingly depleted of methylene chloride with time, leading to an enrichment of the toluene and SVOC mole fractions within the remaining DNAPL.

13.2.2.4 Dissolution into Low Permeability Zones

In special cases where a small volume of DNAPL is present adjacent to a low permeability zone acting as a diffusive sink, the DNAPL can dissolve relatively quickly as a result of diffusive mass transfer into the low permeability zone. DNAPL present in small aperture fractures in clay or bedrock exhibiting high matrix porosity, for example, can deplete in relatively short time frames (Parker et al., 1994). Using the calculation procedure outlined in Parker et al. (1994), TCE DNAPL in a 40-micron (μ) fracture in clay having an effective matrix porosity of 25% will persist for approximately 1 year. However, TCE DNAPL in a 200 μ fracture in sandstone having an effective matrix porosity of 5% will persist for over 600 years. Significant dissolution into low permeability zones as a mechanism for DNAPL depletion is most relevant for smaller volumes of DNAPL directly adjacent to significant diffusive sinks as can be the case for chlorinated solvent DNAPL releases into fractured clay. Note that the Power Law is now being used to represent these effects; CSGSS (2011) suggests using $\Gamma > 1$ to account for the long tail associated with back diffusion from low permeability zones.

13.2.3 Desorption

Typically sorption and desorption have been considered to be linear and reversible processes, although considerable hysteresis has been observed in lab and field studies. This hysteresis is also referred to as availability effects, where sorption to different types of organic material corresponds to different release rates during desorption. Several researchers have developed models describing this hysteresis, most focusing on simulating desorption as a kinetics-controlled phenomena (Linz and Nakles, 1997). One other approach is to consider the hysteresis as an equilibrium-dominated process in which two different partition coefficients are active: one during an initial phase with relatively high concentrations (the linear desorption isotherm) and the second during periods with lower concentrations when availability effects are more important (Kan et al., 1998; Chen et al, 2002). The equilibrium model predicts a long extended tail of desorption that may be very important to consider for low concentration sites approaching the groundwater cleanup objectives.

13.2.4 Matrix Diffusion

Matrix diffusion refers to the process of mass transfer between transmissive zones and low permeability zones (Figure 13.6) (Lipson et al., 2005; Chapman and Parker, 2005; Sale et al., 2008b; West and Kueper, 2010). Two forward matrix diffusion processes might occur in a

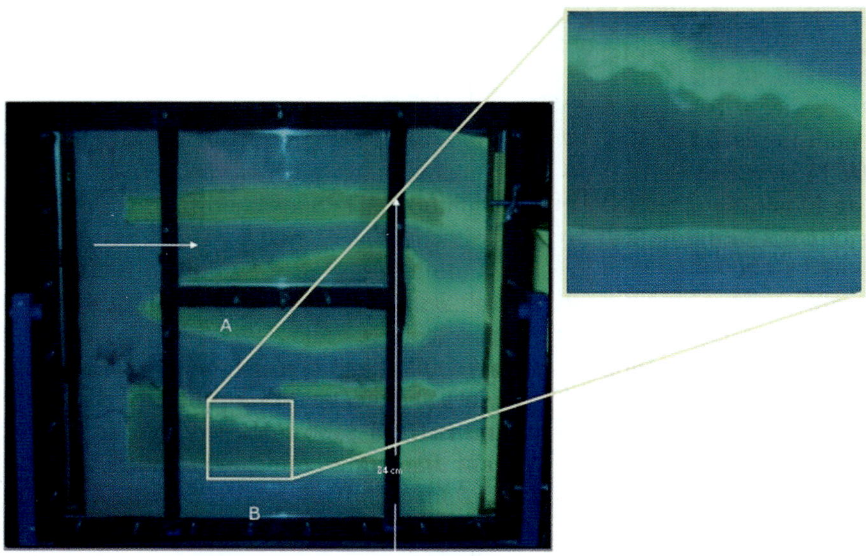

Figure 13.6. Groundwater research tank showing diffusive mass transfer of green dye between sand zones and clay layers. The forward diffusion period for dye was 23 days; this picture shows conditions on day 4 of the back diffusion period (i.e., 27 days after the initial introduction of dye into flowing groundwater). The green dye in the sands stems from back diffusion out of the clay layers (Doner and Sale, 2008).

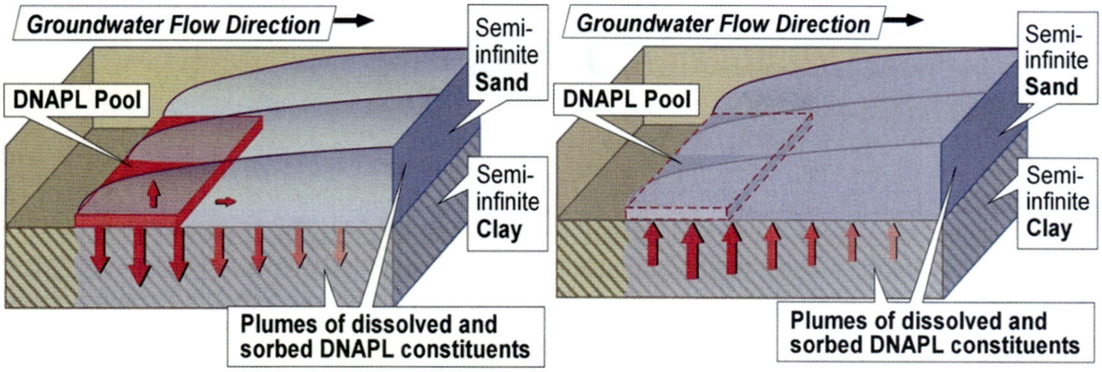

Figure 13.7. Forward diffusion (*left panel*) and back diffusion of contaminants to/from low-permeability media. Forward diffusion can be considered an attenuation process if compliance points are located in the transmissive zones (sand in the figures above). During the back diffusion period, however, attenuation would slow due to mass transfer back into the transmissive zones (Farhat et al., 2012).

source zone during life cycle Stages 1–3 (the loading period for matrix diffusion): (1) direct DNAPL dissolution into low permeability zones adjacent to the DNAPL (Section 13.2.2.4) and (2) DNAPL dissolution into transmissive zones followed by subsequent forward diffusion into low permeability zones. This distinction may be important in making remediation decisions at sites that are in life cycle Stages 1–3, but not for sites in life cycle Stages 4 or 5 where back diffusion may be the dominant mass transfer process.

During life cycle Stages 1–3, matrix diffusion can be considered a nondestructive attenuation process for the transmissive zone. During the back diffusion phase (also called the release period), however, attenuation of the transmissive zone will slow due to the source-like behavior of the low permeability zones (Figure 13.7). The importance of the back diffusion process is largely dependent on the ability of groundwater flux in the transmissive zone to dilute the concentrations contributed by the low permeability zones. These concepts are discussed further with example modeling in Chapter 5.

One key question regarding matrix diffusion is whether contaminants that have diffused into low permeability zones degrade. Even slow attenuation processes (biotic or abiotic) in low permeability zones could be important for source attenuation. The limited work performed to date has demonstrated biological degradation in a clayey till (Scheutz et al., 2010), in a clayey aquitard (Takeuchi et al., 2011), and in a sedimentary rock matrix (Lima et al., 2011). This likely is not universal, however, and sites should be evaluated on a case by case basis regarding the potential for biological degradation in low permeability media. Sale et al. (2008b) employed a two layer porous media analytical model to conclude that both contaminant degradation and contaminant adsorption in the low permeability zone, if they occur, are critical factors governing the timing and magnitude of downgradient improvements in water quality.

Sale et al. (2008a) report that the general conditions where low permeability storage effects may be significant include the following:

- Geologic settings with transmissive zones that are a small fraction of the aquifer's total volume
- Contaminants present at high concentrations
- Contaminants stable in their physical setting (TCE in an aerobic aquifer)
- Systems with relatively slow groundwater flow rates

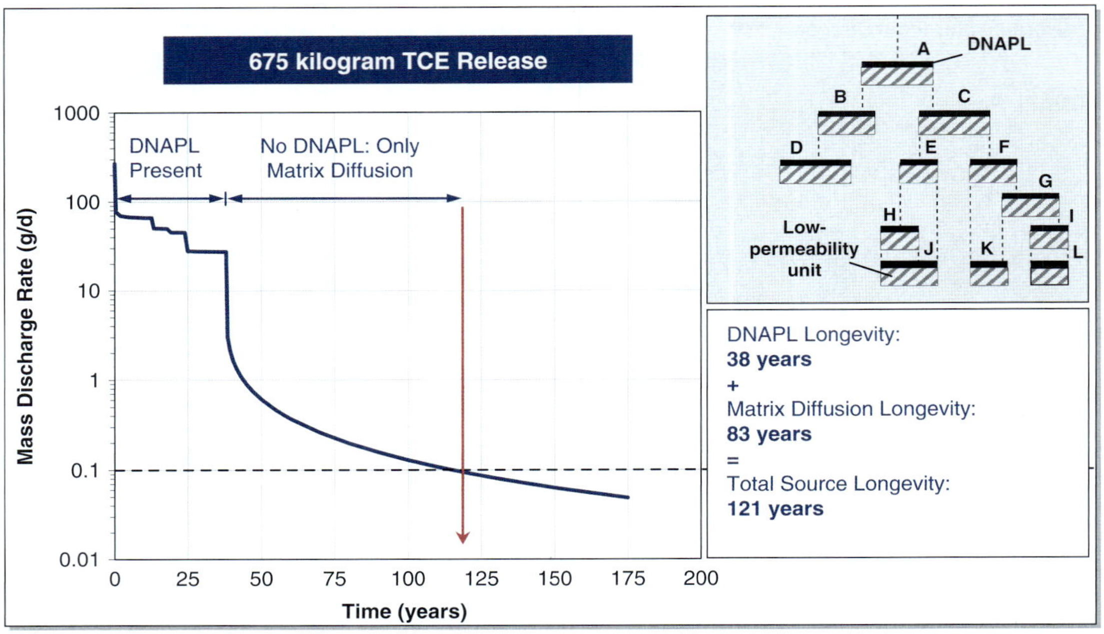

Figure 13.8. Source zone attenuation curve for 675 kg (approximately two 55 gal drums) TCE release distributed into source architecture shown in top right panel. Simple DNAPL dissolution and matrix diffusion models indicate the total source lifetime (defined as >0.1 g/d mass discharge, or an average source zone concentration of >6 μg/L) to be 121 years, with matrix diffusion being responsible for 83 years (approximately 70% of the source lifetime). Adapted from Seyedabbasi et al., 2012.

- Low permeability sediments with high fractions of organic carbon
- Sites where large amounts of contaminant were released
- Older sites with time for large amounts of contaminants to move into low permeability zones

Seyedabbasi et al. (2012) compared the contribution of DNAPL dissolution and matrix diffusion to the longevity of chlorinated solvent source zones. They modified a hypothetical DNAPL source zone architecture consisting of twelve DNAPL pools and thirteen DNAPL fingers originally devised by Anderson et al. (1992) by placing the DNAPL pools on low permeability zones conducive to matrix diffusion. Simple DNAPL dissolution and matrix-diffusion models were used to develop mass discharge versus time curves for three different contaminants, each with a 675 kilogram (kg) release (Figure 13.8). The authors calculated that for the hypothetical TCE release, the dissolution of DNAPL lasted for 38 years, while the back diffusion from low permeability zones maintained the source strength above 0.1 grams per day (g/d) for an additional 83 years. In the context of Figure 13.1, this implies that Stage 4 (DNAPL depletion) was reached at 38 years. The length of time required to reach Stage 4 will vary from site to site. The example presented here considered a relatively small volume of DNAPL released (two drums of TCE); it follows that larger volume releases would likely require longer periods of time to reach Stage 4. Of note, however, is that Stage 5 (back diffusion and desorption) is expected to persist for a longer period of time than Stages 1–4 combined.

13.2.5 Abiotic Degradation

For many years, MNA researchers and practitioners focused on a handful of hydrolysis reactions for certain dissolved chlorinated solvents, such as the transformation of 1,1,1-trichloroethane (1,1,1 TCA) to 1,1-dichloroethene (1,1 DCE) by dehydrochlorination and chloroethane to ethanol by abiotic hydrolysis. These reactions, while important to dissolved plume MNA, were only rarely associated with source zone attenuation processes. One study that did correlate abiotic chemical degradation reactions to source zone processes was by Newell et al. (2006), where long term temporal records of perchloroethene (PCE), TCE, *cis*-DCE, and 1,1,1-TCA concentrations in groundwater were evaluated. The point attenuation rate for 1,1,1-TCA was found to be approximately twice as fast as the point decay rate for TCE, a finding that the authors stated could have been attributed to the abiotic dissolved phase reaction of 1,1,1-TCA that would be increasing the driving force between source materials and the aqueous phase.

More recently, there has been increased attention on the abiotic reactions involving ferrous and mixed-valence iron-bearing minerals. In particular, pyrite, magnetite, and green rust are known to catalyze reductive dechlorination. Reducing power can be stored in these minerals and promote reductive dechlorination (He et al., 2009). At some sites naturally occurring minerals support this abiotic degradation (Ferrey et al., 2004). At other sites, when electron donors are released along with chlorinated solvents or added as part of an active bioremediation project, these electron donors can drive biological reductive dechlorination but can also charge iron and other minerals in the aquifer matrix so that abiotic degradation reactions can occur along with the biological ones (Kennedy et al., 2006; ITRC, 2008; Adamson et al., 2011). For sites where some type of hydrocarbon was released together with chlorinated solvents, these mineral-driven abiotic reactions can serve as an important destructive attenuation process in chlorinated solvent source zones (Brown et al., 2007).

13.2.6 Biotic Degradation

Biodegradation is likely to be a key component of source zone attenuation at many chlorinated solvent sites. This chapter provides only a limited overview of this topic, as the impact of enhanced biodegradation processes on source zones is discussed in detail in Chapter 12. In general, there is strong consensus that biological reactions in source zones, even naturally occurring ones, will result in an increased rate of DNAPL dissolution (ITRC, 2008). Biodegradation of contaminants in the aqueous phase will enhance the local-scale concentration gradient driving DNAPL dissolution compared to conditions of nonaqueous phase biodegradation. At some sites, the necessary bacterial population, electron donors, and geochemical conditions are present to sustain biodegradation without engineered intervention. For example, McGuire et al. (2003) reported anaerobic biodegradation to be active at more than 90% of 45 chlorinated solvent sites employing MNA. Biodegradation within source zones can be considered a means of enhancing the rate of DNAPL mass depletion.

In addition to enhancing the rate of DNAPL mass depletion, biodegradation within a source zone can result in lowered aqueous-phase concentrations and mass flux exiting the source zone. Such a situation could occur if complete biodegradation (dechlorination of all daughter products including DCE and vinyl chloride) occurs within or immediately downgradient of the source. Such a situation could occur in an anaerobic environment with a relatively slow aqueous-phase loading of solvents from the source. This aspect of biodegradation in contributing to source zone attenuation is perhaps of greater significance than DNAPL mass depletion and can be an important aspect of attenuation for aged source zones being sustained by desorption and back diffusion from low permeability zones (Stage 5).

13.3 REMEDIAL GOALS AND METRICS FOR SOURCE ZONE MNA

The objectives for chlorinated solvent source zone MNA, and the metrics used for evaluating success, generally are similar to those used for active remediation. However, it is important to consider that natural attenuation processes are generally slower than those measured during active remediation, so some modifications to the methods may be needed.

13.3.1 Reducing Source Zone Concentrations and Mass Discharge

Monitored natural attenuation projects for plumes have traditionally focused on reduction of concentrations in monitoring wells. However, even as far back as 1998, there was an emerging recognition that mass flux/mass discharge could improve our understanding of MNA for plumes. For example, a USEPA MNA seminar in 1998 made –the following points (USEPA, 1998):

Flux estimate across the boundary to a receptor is the best estimate of loading to a receptor. The reduction in the flux along the flowpath is the best estimate of natural attenuation of the plume as a whole. The flux is the best estimate of the amount of contaminant leaving the source area. This information would be needed to scale an active remedy if necessary.

In 2001, the USEPA's Scientific Advisory Board concluded (USEPA, 2001):

In summary, measurements of mass flux of the contaminants and footprint parameters – not just concentrations – are necessary to document cause-and-effect and to assess long term sustainability/permanence. Site characterization and monitoring plans should be proactively designed to accommodate mass-flux estimates.

In practice, however, many regulatory programs are still focused on meeting concentration based goals, either because of statutory requirements, perceived difficulty in obtaining mass flux/mass discharge data, or inertia. The ITRC's mass flux/mass discharge technology overview (2010) stated that while it was unlikely that mass flux will globally replace point concentrations for regulatory compliance, it is a powerful tool for developing remediation goals.

New scientific initiatives, such as the SZNA methodology (Section 13.4.3.3), make the case that source zone attenuation should be evaluated in a similar fashion to active remediation projects where mass removal is often quantified as the product of flow rate and concentration (e.g., soil vapor extraction and pump-and-treat systems). For SZNA projects, a macroscopic mass balance is developed based on natural groundwater flow controlled mass loss, biodegradation, and vapor transport controlled loss. With this type of analysis, the specific attenuation processes are identified, and the overall SZNA removal rate can be compared to the expected removal rate and performance of other technologies. In other words, a near-term snapshot and quantification of source zone mass loss can, in less than a few months, help provide an apple-to-apples comparison of source zone attenuation to other remedial options in proper context for decision makers.

Natural attenuation will reduce the concentration and mass discharge from chlorinated solvent source zones over time, but at a rate that may be difficult to measure or predict in the short term. Empirical point concentration data from 26 chlorinated solvent source zones without any active remediation suggested that when long term concentration versus time records (Figure 13.9) were fit to a first-order source decay model, the resulting median PCE, TCE, and 1,1,1-TCA point decay rates as defined by the USEPA (2002) (k_{point}) were 0.23, 0.11, and 0.54 per year, respectively. A point decay rate is a first-order decay rate fit to

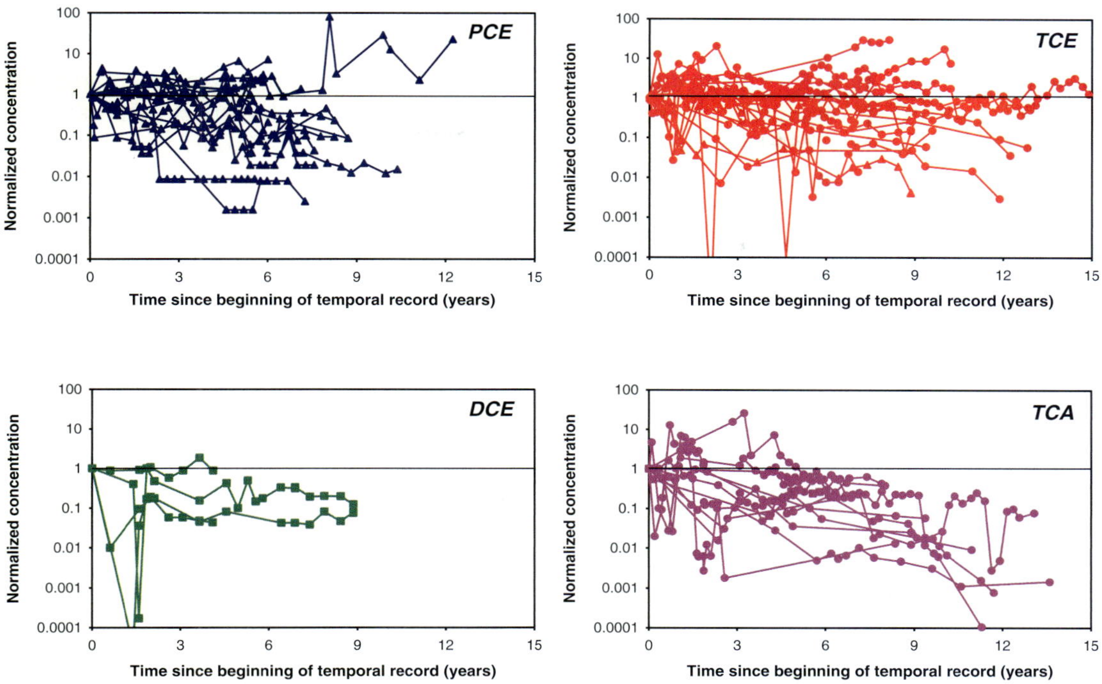

Figure 13.9. Long term temporal records at untreated chlorinated solvent source zones (Newell et al., 2006). Each curve represents a normalized concentration versus time record from a monitoring well in a source zone. For PCE, 17 wells from 9 sites are shown; TCE, 21 wells from 13 sites; DCE, 4 wells from 2 sites; and TCA, 10 wells from 6 sites. Note that vertical axis is log scale. The median percent reductions in concentration for these temporal records are PCE, 71%; TCE, 81%; DCE, 86%; and TCA, 99%. 28 of 52 temporal records had statistically significant first-order decay trends. With permission from ASCE.

concentration versus time data and is different from a bulk decay rate or a biodegradation decay rate in a plume. For these 26 sites (defined as low risk sites because no active remediation had been performed), the median half-lives corresponding to the above point decay rates for PCE, TCE, and 1,1,1-TCA were 3 years, 6 years, and 2 years, respectively. The authors concluded that "if the median point decay rates from these sites are maintained over a 20 year period, the resulting reduction in concentration will be similar to the reported reduction in source zone concentrations achieved by active in situ source remediation technologies (typical project length: 1–2 years)" (Newell et al., 2006). Note that sites with relatively large releases of chlorinated solvents and sites with extensive observed pooled accumulation of DNAPL were not represented in this 26-site database. Because of the inherent variability (noise) in groundwater monitoring data, a one order of magnitude reduction in concentration may be required to obtain a statistically significant source zone MNA attenuation rate at many sites (see Section 13.4.3.1). The time required to observe such a concentration reduction will vary between sites depending on the strength of various source zone MNA processes and fundamental parameters such as groundwater velocity and mass of contaminant present.

Several simple planning-level models can be used to estimate the change in mass discharge over time due to chlorinated solvent source zone MNA. The BIOCHLOR model (Aziz et al., 2000a) incorporates a source decay term, and mass discharge at several locations downgradient of the source can be evaluated. The REMChlor model (Remediation Evaluation Model for

Chlorinated Solvents) provides the same functionality as BIOCHLOR but allows the user to enter a particular chlorinated solvent source zone architecture using the power function described in Section 13.1.3. Chapter 6 provides further discussion and example simulations, evaluating the effect of source zone depletion on plume response. With respect to monitoring, it may be worthwhile to collect groundwater quality samples from within source zones as early as possible to begin establishing concentration versus time records indicative of whether or not source zone natural attenuation is occurring. More frequent monitoring, or special sampling during the beginning phase of an MNA investigation, may be helpful in understanding how seasonal and year to year hydrogeologic variations influence source zone attenuation rates (Pope et al., 2004; Wilson, 2011).

13.3.2 Effect on Source Longevity

One of the arguments used by certain investigators for performing partial source removal is that it will significantly reduce the source longevity compared to using only source zone MNA. However, because of the long time frames involved and the difficulty in establishing a control, it is difficult to demonstrate the quantitative impact of partial source removal on source longevity. Newell and Adamson (2005) used several planning-level source models, including ones based on exponential decay, to estimate the reduction in source longevity, which they referred to as *remediation time frame.* The exponential decay model is the “middle of the road” approach when using the USEPA’s REMChlor model (Chapter 6) where gamma = 1.0 (CSGSS, 2011). With a first-order source decay model, the remediation time frame under different conditions (before and after remediation) can be estimated using the approach presented in Newell and Adamson (2005):

$$RTF = \frac{-M_o\left[\ln\left(\frac{M_{dg}}{M_{do}}\right)\right]}{M_{do}} \qquad \text{(Eq. 13.2)}$$

where RTF is the remediation time frame to achieve a mass discharge goal (time), M_o is the current source mass (mass), M_{dg} is the mass discharge goal (mass per time), and M_{do} is the current mass discharge from the source zone (mass per time). Note that this equation can be applied where source zone MNA is a stand alone remedy with the original mass and mass discharge from the source zone, or after partial source depletion, where the remaining mass and mass discharge are used. When both calculations are performed, the relative change in remediation time frame between active remediation and source zone MNA can be estimated.

As an example (Newell and Adamson, 2005), consider a source zone containing 80 kg of TCE and producing a mass discharge of 2 kg/yr (kilograms per year). The remediation goal is to reduce the mass discharge by 99–0.02 kg/yr. An active remediation application is projected to reduce mass and mass discharge by 70%. Using Equation 13.2 (first-order decay, gamma = 1), planning-level estimates for source longevity (time to reach the mass discharge goal of 0.02 kg/yr) can be developed. Equation 13.2 predicts that the source longevity for the active remediation application followed by source zone MNA is calculated to be 136 years. The source longevity for source zone MNA alone is calculated to be 184 years. Removal of 70% of the contaminant mass by active remediation does not reduce source longevity by 70%, rather only 26%.

With this Power Law Model (gamma = 1), significant mass removal (generally well above 90% removal) is required to realize significant reductions in source longevity compared to source zone MNA alone. This is a crucial point in understanding the net benefit of partial source removal.

13.3.3 Potential for Attaining MCLs

All indications are that source zone MNA will reach stringent groundwater cleanup standards, such as maximum contaminant levels (MCLs), given enough time. The ability of source zone MNA to reach MCLs in a reasonable time frame, however, may not be achievable depending on what the expectation for the length of the time frame is. In all likelihood, source zone MNA, either as a stand alone remedy or as a polishing step, may only be acceptable when long term management over decades is possible, even after partial source removal. In fact, the authors are unaware of any sites where MCLs have been reached at all locations within the source zone as a result of MNA.

However, other than relatively small source zones and those amenable to excavation, the authors are not aware of any chlorinated solvent source zones where groundwater MCLs have been achieved at all locations as a result of any active remediation technology applications. The Interstate Technology & Regulatory Council (2011) addresses this important issue in their "Integrated DNAPL Site Strategy" document, and NRC (2012) concluded that "Significant limitations with currently available remedial technologies persist that make achievement of MCLs throughout the aquifer unlikely at most complex groundwater sites in a time frame of 50–100 years." Attaining MCLs has not been widely demonstrated for either in situ remediation projects or source zone MNA.

Thus, a key question is not the absolute time frame associated with source zone MNA, but a realistic estimate of the relative time frame between source zone MNA and an active remediation technology. This type of analysis is important before any active remediation is performed or if a treatment-train approach of active technologies is being considered. Our experience in both cases is that the simple screening-level models may often suggest that the improvements in remediation time frame will be much less than the expectations of the site stakeholders. While these models have considerable uncertainty, they can provide useful process based information for making site management decisions.

13.3.4 Costs

Source zone MNA is perceived by some within the groundwater community to cost less than active remediation. The Remedial Action Cost Engineering and Requirements (RACER) software tool (AECOM, 2009) will generally indicate lower cost over a 30 year period for source zone MNA compared to almost any active remediation alternative except for very small source zones. The life cycle cost comparison between source zone MNA and active remediation is more complicated than a 30 year evaluation, however, and dependent on several factors:

- Use of net present value to calculate life cycle costs
- Whether MNA monitoring is required after active remediation (Many active remediation projects do not reach MCLs and therefore require some type of post-remediation or MNA monitoring)
- Intensity of long term monitoring, both in terms of sampling frequency and number of locations

McGuire et al. (2003) reported on typical costs for MNA studies performed in the early 2000s. Most of these studies were for plume MNA analysis, as opposed to purely source zone MNA. The authors determined that the average total cost for an initial MNA analysis from over 178 sites was $188,000, with a range between $10,000 and $750,000. Average annual costs for continued monitoring ranged from $3,000 to $150,000, with an average annual cost of $32,000.

Note that because source zone MNA is monitoring intensive, source zone MNA at fractured rock sites may be more expensive than at sites where less expensive site characterization technologies, such as direct push rigs, can be employed.

A rough rule of thumb that has been reported for long term optimization programs is that a one-time sampling event for a single groundwater monitoring well typically costs $750 (Hunter, 2004). This cost includes any needed travel, sampling time, waste disposal, laboratory analyses, and data reduction costs. The Sustainable Remediation Tool (SRT) (AFCEE, 2010) uses these simple planning level cost-estimation formulas based on cost algorithms in the RACER system:

<u>Capital cost of MNA</u>
(Depth of well × no. of wells × $200) + (no. of sampling events for characterization + no. of sampling events in the first year) × (no. of samples per event) × [($2.5 × depth of well) + $2,957]

<u>Operations and maintenance (O&M) cost of MNA</u>
Sampling events after the first year × samples per event × [($2.5 × depth of well) + $2,957]

13.3.5 Sustainability and Intangible Goals

Source zone MNA is perceived to be less intensive and have less adverse impact on the environment in terms of sustainability metrics, such as carbon footprint and energy use, than active treatment. However, if active remediation will close a site quickly so that no further monitoring is required, then active remediation may be more favorable with respect to several sustainability metrics because of the energy use, carbon footprint, and worker risk associated with long term MNA monitoring, particularly if travel to the site for monitoring is required for decades. Programs such as the U.S. Air Force's SRT (AFCEE, 2010) and Battelle's SiteWise tool (Battelle, 2011) can be used to compute sustainability metrics for different site management scenarios.

In terms of intangibles as expressed in the "Relative Merits of Source Depletion" chart shown in Sale et al. (2008a), active remediation is more favorable to managers who desire an active remedy, wish to test new technologies, and seek to reduce the stewardship burden on future generations and for sites with a clear public preference for source treatment. However, source zone MNA would likely be more favorable from an intangibles perspective to those who desire a low-impact remedy, prefer to use proven technologies, and do not wish to expend financial resources for limited risk reduction benefits or where a site has few or no public perception issues.

13.4 STATE OF THE PRACTICE

Source zone MNA is an emerging remediation technology. Despite having been discussed in the technical literature since at least 1999, and having specific tools such as SourceDK and REMChlor introduced in the 2000s, it still lacks several of the developments that marked the emergence of MNA for plumes. For example, source zone MNA has no formal protocol or methodology document, only a few source zone MNA-specific datasets can be found in the technical literature, and good case studies are still rare.

Recently, discussion and development of source zone MNA as a remediation alternative has increased. Its role as a polishing step after partial source depletion was discussed extensively in the ITRC's Integrated DNAPL Site Strategy document (ITRC, 2011). A natural source zone depletion technology overview was recently published for light nonaqueous phase liquid (LNAPL) sites by the ITRC (ITRC, 2009), and the Department of Defense's ESTCP program has funded a source zone natural attenuation project in 2010. The NRC's 2012 "Alternatives for Managing the Nation's Complex Contaminated Groundwater Sites" advocated a "transition assessment step" to determine if a site should go into a long term management mode (NRC, 2012). This transition assessment would include considering risk from residual contamination in subsurface zones, life cycle costs and the incremental costs compared to the level of risk reduction achieved, and the likely reaction of stakeholders. The long term management step would likely include source zone MNA at many sites.

Perhaps the key implementation issue regarding source zone MNA is remediation time frame. It will only be attractive when long term management over decades to centuries is possible, even after partial source removal. But there is a growing recognition that long term management will be required even for intensive active remediation projects as some contaminant mass is inevitably left behind. One of the key messages of the NRC's 2012 report was that although a reasonable remediation time frame for MNA is usually considered to be less than 100 years, the use of longer remediation time frames for passive long term management (which includes MNA) is "critical to avoid inefficient use of limited resources." This NRC report may be an important catalyst for new thinking about the applicability of source zone MNA at many sites.

13.4.1 Regulatory Practices

The USEPA (1999) expected that "...source control measures will be taken at most sites where practicable," where source control measures include removal, treatment, containment, or a combination of these actions. Key definitions provided in the 1999 directive include:

Source material is defined as "material that includes or contains hazardous substances, pollutants or contaminants that act as a reservoir [either stationary or mobile] for migration of contamination to the groundwater, to surface water, to air, [or other environmental media,] or acts as a source for direct exposure. Contaminated groundwater generally is not considered to be a source material although nonaqueous phase liquids (NAPLS [occurring either as residual- or free-phase]) may be viewed as source materials." (USEPA, 1991).

Principal threat wastes are those source materials that are "highly toxic or highly mobile that generally cannot be reliably contained, or would present a significant risk to human health or the environment should exposure occur. They include liquids and other highly mobile materials (e.g., solvents) or materials having high concentrations of toxic compounds." (USEPA, 1991).

Low level threat wastes are "source materials that generally can be reliably contained and that would present only a low risk in the event of release" (USEPA, 1991). Since contaminated groundwater is not source material, it is neither a principal nor a low level threat waste.

With these definitions and associated discussion, the USEPA (1999) directive presumed that MNA was more applicable to dissolved plumes than to source zones and that MNA alone would not be applicable to source materials at most sites.

The USEPA's Technical Protocol for Evaluating Natural Attenuation of Chlorinated Solvents in Ground Water (Wiedemeier et al., 1999) focused largely on biodegradation via reductive dechlorination of dissolved constituents, with a much more limited review of source processes. The protocol presented a conceptual model where *contaminant sources* generally consisted of mobile or residual NAPL that needed to be addressed via active means before MNA could be implemented (Wiedemeier et al., 1999):

Removal, treatment or containment of NAPLs may be necessary for MNA to be a viable remedial option or to decrease the time needed for natural processes to attain site-specific remediation objectives. In cases where removal of mobile NAPL is feasible, it is desirable to remove this source material and decrease the time required to reach cleanup objectives. Where removal or treatment of NAPL is not practical, source containment may be practicable and necessary for MNA to be a viable remedial option.

These key regulatory documents show that, in the 1990s, the USEPA perceived MNA as not being amenable for treating contaminant sources. Rather, active treatment was prescribed for most source zones, with the implication that MNA was more suitable as a primary remediation technology for contaminant plumes or perhaps as a polishing technology for sources following active treatment. The presumption of active source remediation prior to considering source zone MNA remains the preferred restoration strategy, particularly at the federal level.

Some states, however, have accepted source zone MNA as an acceptable primary remediation technology. For example, the Texas Risk Reduction Programs' MNA Guidance (TNRCC, 2001) states that:

For the second and third source mechanisms, there often is a continuing release of COCs to groundwater from source materials. Under Remedy Standard A and B, MNA is an appropriate remedy if natural attenuation processes are strong enough to "decontaminate" both the dissolved phase PCLE zone (defined below) *and any source materials residing in the groundwater, including NAPLs (if the PCL is less than COC solubility). If natural attenuation processes cannot be demonstrated to decontaminate the groundwater source materials in a reasonable time frame then MNA is not appropriate as a sole remedy. Under Remedy Standard B, MNA is an appropriate "control" remedy if natural attenuation processes can control the dissolved phase PCLE zone resulting from source materials. If there are any NAPLs present in the groundwater, the readily recoverable NAPLs must be recovered and the NAPLs must be precluded from spreading horizontally or vertically, or from creating any other hazard, and therefore, MNA may not be solely adequate for addressing NAPLs.*

In the quotation above, PCL is a protective concentration level. In some cases, this is an MCL, while a PCLE is a *PCL exceedence*, or plume boundary as defined by the appropriate PCL.

In other words, in the state of Texas, source attenuation can be used as a remediation technique if a particular cleanup goal is achieved in a reasonable time frame (reasonable time frame is not defined specifically, however). In addition, at some Texas sites where a Plume Management Zone has been established, there is an option in which no remediation time frame is specified as long as MNA controls the dissolved plume being discharged by the source. In this case, no source attenuation is needed as long as the Plume Management Zone is in place.

The California Regional Water Quality Control Board, San Francisco Region, (CRWQCB, 2009) has developed guidance that allows source zone MNA after active remediation. The CRWQCB "embarked on a project to develop criteria for evaluating if and when chlorinated solvent sites that pose little threat to human and ecological health, water quality, and beneficial uses but do not yet meet cleanup standards at all locations, could be closed. This process is referred to as 'low-threat closure.'" Under this guidance, sites that met nine criteria can be closed under the low-threat closure process. In particular, two of the nine criteria deal exclusively with sources: (1) pollutant sources are identified and evaluated and (2) pollutant sources are remediated to the extent feasible. Therefore, under this guidance, chlorinated solvent source zones that have been remediated to the extent feasible, but still exceed cleanup standards, could be closed if "cleanup standards can be met under natural conditions within a reasonable time frame, following completion of cleanup and monitoring."

The U.S. Air Force has developed a *low risk site closure* guidance (LoRSC) manual that employs several lines of evidence for determining if a site should be considered a low risk site requiring no further active remediation (Farhat et al., 2012). The overall low risk concept is based on the assumption that source zone MNA is acceptable at some sites. This guidance uses footprints of source zone attenuation as one of twelve key low risk decision questions.

13.4.2 Changing Source Paradigms and Impacts on Management

The chlorinated solvent site life cycle may have important implications for managing chlorinated solvent source zones using MNA. As discussed earlier, few sites still experience active releases, given that most chlorinated solvent source zones being investigated today were created several years or decades ago. Similarly, sites still experiencing spatial redistribution of the DNAPL are fewer in number than those in the continued dissolution and aging stages. However, sites in these first two stages of the source zone life cycle may be good candidates for source depletion using in situ technologies given that low permeability media are likely not fully loaded with aqueous and sorbed phase contamination. Source zone MNA by itself is probably not a good choice for most such sites unless the release is very small and/or concentrations are close to regulatory limits.

For sites that are further along the aging process, including those at which DNAPL has been completely depleted, a strong argument may exist that little or no principal threat wastes must be remediated, and source zone MNA may be an appropriate response. Contaminants present in aqueous and sorbed phases in low permeability media may be considered to be source materials by some but low level threat wastes by others. This regulatory distinction may determine whether active remediation will be required in late stage chlorinated solvent source zones or if source zone MNA can be used as a sole remedy.

Removal of some or all of the DNAPL at many sites experiencing continued dissolution and aging may reduce the mass in the source zone and result in the reduction of groundwater concentrations and mass discharge by one to two orders of magnitude. At some sites, assuming a reasonably fast plume response time, significant benefits will result such as reducing the size of the plume, which can decrease long term monitoring costs. At other sites, the reductions in risk, plume footprint, and monitoring costs resulting from active remediation may not be significantly advantageous over source zone MNA, particularly if the natural source zone attenuation rate is relatively rapid (see Section 13.2.1).

It is important to understand that natural attenuation of source zones does occur and that it is likely to be the final phase of most source zone restoration programs. Since most source zone concentrations are three to five orders of magnitude higher than the typical cleanup goals, active remediation may not achieve these goals and natural attenuation will be necessary. Thus, in some cases, the goal of active remediation will be to remove or destroy sufficient mass so that MNA can be protective and can be relied upon for long term management. This vision is often a difficult one for some stakeholders to endorse, although it is increasingly being adopted (ITRC, 2011).

13.4.3 Source Zone MNA Evaluation Techniques

Methods presented in MNA protocols for plumes, such as the reduction in concentration in wells and the identification of degradation footprints, have been adapted for sites where source zone MNA has been used as the sole remedy. Pope et al. (2004) describe methods to monitor the performance of MNA remedies in plumes, and many of the same concepts are directly applicable to source zones. However, plumes and source zones have important differences, and recently there has been more focused development on evaluating source zone MNA.

Source zone natural attenuation (SZNA) is a technique in which the mass discharge associated with dissolution, biodegradation, and volatilization/vaporization are measured.

The Environmental Security Technology Certification Program (ESTCP) has funded the testing and development of an SZNA protocol for chlorinated solvent sites. The key advantages for a mass discharge, SZNA type approach is that SZNA can be compared more directly to other remediation technologies and specific attenuation processes can be quantified. One potential disadvantage at this time is regulatory acceptance; many regulators and regulatory programs still focus on point concentrations from monitoring wells for compliance purposes. Overall, however, the mass discharge/mass balance SZNA approach will likely get wider acceptance in the future.

Mass balance tools and models are available to evaluate source zone MNA if both source mass and mass discharge can be estimated, even with significant uncertainty. These include relationships to calculate first-order source attenuation rates (Equation 13.4) and analytical models such as BIOCHLOR and REMChlor (Section 13.5.2), both of which are distributed by the USEPA. These tools and models can help establish a source zone attenuation "glide path" that with proper expectations and understanding of uncertainty can be used for source zone MNA (Section 13.6.1). These models are particularly useful at comparing the relative remediation time frames associated with source zone MNA versus active remediation followed by source zone MNA.

13.4.3.1 Decreasing Source Strength

Changes in source strength have traditionally been evaluated using concentration versus time data, and many of the methods and tools focused on concentration data. More recently, there has been increased interest in using mass discharge as a remediation metric (ITRC, 2010). Mass discharge has the advantage of providing quantitative direct evidence of the amount of mass leaving the source and integrates both flow and concentration. Therefore, in the future, source zone attenuation studies will likely utilize both mass discharge and concentration data. The Interstate Technology & Regulatory Council (ITRC, 2010) outlines five methods to estimate mass discharge that can be used downgradient of source zones. For source zone MNA, mass discharge alone demonstrates that source zone attenuation is occurring, while mass discharge trends with time can be used to predict whether source lifetimes are likely to be years, decades, or centuries.

Temporal records can be analyzed to determine that the concentration and/or mass discharge versus time trend is at a site. Techniques outlined in Aziz et al. (2003) and provided in the Air Force's MAROS program (Aziz et al., 2000b) and a public domain calculator tool (Connor et al., 2012) can be used to statistically define concentration/mass discharge trends as "increasing," "probably increasing," "stable," "probably decreasing," "decreasing," or "no trend." For a source zone MNA evaluation, declining concentration/mass discharge is typically the primary line of evidence used for an MNA demonstration.

The amount of monitoring required to accurately characterize a concentration or mass discharge versus time trend depends on a number of factors, the most important of which are (1) the magnitude of the long term trend (the attenuation rate), (2) the magnitude of short-term variability, and (3) the uncertainty in the raw data and data reduction. When the change attributable to the long term trend is small relative to short-term variability and uncertainty, a longer monitoring period is required to characterize the long term trend. The uncertainty in a mass discharge estimate might be reduced by increased spatial resolution of the sampling. At many sites, more than one half life of monitoring data may be required to establish a source zone attenuation rate with certainty. At many sites, this implies that approximately 10 years of monitoring may be required. An evaluation of 52 chlorinated solvent temporal records showed that 28 (53%) had a statistically significant decreasing trend (Newell et al., 2006). In the USEPA MNA Directive (USEPA, 1999), J.T. Wilson concluded that "analysis of natural attenuation rates from many sites indicates that a measured decrease in contaminant concentrations of at least one order of magnitude is necessary to determine the appropriate rate law to describe the rate of attenuation, and to demonstrate that the estimated rate is statistically different from zero at a 95% level of confidence. Due to variability resulting from sampling and analysis, as well as plume variability over time, smaller apparent reductions are often insufficient to demonstrate (with 95% level of confidence) that attenuation has in fact occurred at all."

Wilson (2011) then developed statistical rules for estimating the time required to reach a concentration cleanup goal, testing whether an observed reduction in concentration is statistically significant, and determining whether a reduction in concentration is adequate to meet site remediation goals. This guidance focused on concentration versus time rates, which closely track the rate of weathering in a source zone rather than the rate of attenuation for dissolved phase contaminants. The simple relationship used by Wilson et al. (2005) for estimating the time required to demonstrate MNA of MTBE source zones is also applicable to chlorinated solvent source zones. In this relationship, the source attenuation rate (k_n) necessary to meet an attenuation goal within a given time (t) is calculated by:

$$k_n = \frac{-\ln\left[\frac{C_g}{C_o}\right]}{t} \qquad \text{(Eq. 13.3)}$$

where C_g is the desired groundwater concentration goal (mass per volume) and C_o is the current groundwater concentration (mass per volume). When this methodology was applied to 20 underground storage sites, statistically significant attenuation of MTBE was found at 11 of the sites (Wilson et al., 2005). At these sites, 4–13 monitoring events were required to identify a statistically significant concentration trend. However, for the cases where a trend was identified using only 4 or 5 monitoring events, this trend was commonly an artifact of short-term variability and the identified trend was not consistent with the longer monitoring record.

Wilson (2011) also provide guidance on how to (1) incorporate MNA progress into 5 year reviews, potentially with a 5 year monitoring interval; (2) deal with data quality issues; (3) evaluate a site as a whole; (4) understand behavior of transformation products; and (5) deal with many other important technical issues regarding analyzing temporal data. Although mass flux/mass discharge are not specifically mentioned in this guidance, almost all the data analysis and statistical methods can be applied to temporal mass discharge data.

One study of groundwater variability used three large groundwater monitoring datasets (approximately 3,900 wells at three sites, all with 10+ years of monitoring data) to suggest that the long term trend in wells (primarily source attenuation) comprised only 30–40% of the monitoring variability (McHugh et al., 2012).

13.4.3.2 Biodegradation "Footprints"

The National Research Council's *Natural Attenuation for Groundwater Remediation* (NRC, 2000) prescribed that "footprints" of natural attenuation be evaluated to confirm that known attenuation processes were in fact occurring. These footprints included observations of degradation daughter products, such as *cis*-DCE, the daughter product of TCE biodegradation, and geochemical conditions indicating that biodegradation was occurring (such as anaerobic conditions that are supportive of reductive dechlorination). These footprints are useful for both dissolved plume and source zone MNA. For example, the presence of *cis*-DCE, a biodegradation product of TCE, that is found 1–2 m (3–6 feet) within a low permeability zone may have significance to understanding site history. At many sites, it may take decades for contaminants to diffuse 1–2 m into a low permeability zone. If *cis*-DCE is present within these low permeability zones, it may mean that *cis*-DCE was being produced in the high permeability zone adjacent to the low permeability zone. Further discussion is provided in Section 13.2.4.

It should be noted that biodegradation, although typically desirable, is not a requirement for natural attenuation of either a source zone or a plume. Abiotic reactions, dispersion, and diffusion can eventually result in a steady-state (naturally attenuated) plume.

One attenuation footprint not mentioned by the NRC, but being used more extensively in recent years, is compound-specific stable isotope analysis (CSIA); CSIA of contaminants has recently been applied to characterize fate and transport processes for a number of groundwater constituents (see Chapter 4; Hunkeler et al., 2008). The primary applications of this technique are for potentially distinguishing specific sources of contamination and quantifying biodegradation versus dilution rates in plumes (Murphy and Morrison, 2002). Application of CSIA to support source zone MNA is often complicated by the lack of a baseline for the original release and because high concentrations of undegraded material dominate the isotope ratio signal. Isotope fractionation is controlled by substrate mass transfer limitations (Thullner et al., 2008; Kampara et al., 2008). Consequently, the bioavailability of source materials to degrading microbes controls the strength of the kinetic isotope effect. Stable isotope signatures of chlorinated solvents in source zones tend to be conservative of the initial release conditions, particularly if DNAPL is present (Lollar et al., 2001; Hunkeler et al., 2004). The utility of CSIA to demonstrate MNA of chlorinated solvent sources appears to be limited in most cases at this time.

13.4.3.3 Source Zone Natural Attenuation Methodology

The SZNA strategy for hydrocarbon sites developed by Johnson et al. (2006) in part considers SZNA as a benchmark for assessing the performance and relative benefits of engineered remediation options and provides a stand alone methodological approach independent of engineered options. An ESTCP guidance document tells how to extend the SZNA concept to chlorinated solvent sites (ESTCP, 2014). Preliminary elements of the DNAPL SZNA approach include the following:

- There has been an increased recognition that SZNA is an implicit component of engineered remediation options for dealing with post-remediation residuals at most sites
- Demonstrating and quantifying natural attenuation of sources requires different techniques and conceptual models than those used for MNA in dissolved plumes
- Source zone natural attenuation methodology requires data gathering and data reduction grouped by three purposes: Group 1 activities address the question of whether SZNA is occurring, Group 2 activities focus on determining the current rate of SZNA, and Group 3 activities evaluate the long term implications for residuals, groundwater quality, and SZNA rates

The SZNA guidance developed by Johnson et al. (2006) is based on a mass balance approach intended to develop estimates of the various natural attenuation processes (Figure 13.10). It relies on vapor monitoring within the vadose zone and upgradient and downgradient monitoring wells to monitor contaminants and indicators of natural attenuation processes. The necessary data include contaminant and daughter product concentrations in vapors and groundwater leaving the source zone, the vapor migration rates and groundwater velocities, and the fluxes of electron donors and acceptors into and out of the source zone. These data are input into a series of equations used to estimate the natural attenuation rates.

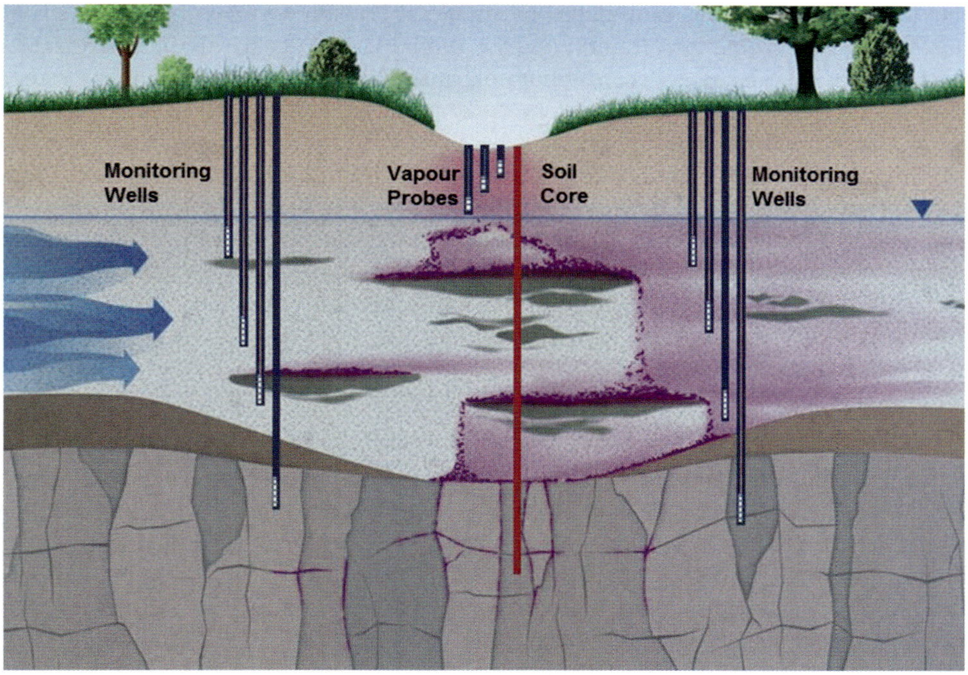

Figure 13.10. Schematic of source zone natural attenuation monitoring methods.

The Group 2 activities are generally the key component of SZNA assessment and involve obtaining detailed data needed for estimating the fluxes of contaminants leaving the source zone via dissolved plume transport, volatilization, and biodegradation.

13.4.3.4 Modeling Tools

A number of mathematical modeling tools can be applied for source zone MNA to help understand key processes and to make predictions about future performance. These are discussed in detail in Chapters 5 and 6, and models specifically useful for source zone MNA are described in Section 13.5. One key point is that groundwater modeling results are subject to uncertainty, and this uncertainty can be particularly important for source zone natural attenuation, which may extend over decades if not centuries. Models are useful for providing insights into specific processes, or demonstrating important details for the current conceptual model, but any one model simulation should not be relied upon alone to provide specific predictions of concentrations far into the future. Uncertainty can be assessed by a variety of techniques such as sensitivity analyses and probabilistic approaches. The probabilistic version of REMChlor (Liang et al., 2010), for example, can provide a depiction of uncertainty in a model's output when appropriate ranges for key input data (source mass, groundwater velocity, source architecture) are incorporated. This uncertainty can be assessed both for remediation alternatives and source zone MNA scenarios.

The one limitation of most of these models is that the user must input a parameter that prescribes the source strength with time; for example, either a decay rate or gamma value. The first-order decay rate for a source zone is either based on mass balance concepts (Section 13.5.1) or from historical observations of source zones (Newell et al., 2006). For gamma, there are physically based rules for determining gamma based on site conditions, such as using a gamma between 1.0 and 2.0 for sites dominated by matrix diffusion sources. However, all of these methods have considerable uncertainty, and the modeling results should be considered indicative of the general style of source attenuation and not a precise prediction of remediation time frames.

One pertinent recommendation on the reliability of future projections was presented in the ITRC's Integrated DNAPL Site Strategy document (ITRC, 2011), which recommended establishing quantitative functional goals (goals that are specific, measurable, attainable, relevant, and time bound) for no more than 20 years in the future. This philosophy means that modeling results may be more useful in establishing a likely glide path of future source behavior, but one that is updated periodically (e.g., every 5–10 years) based on new data. In other words, the model expresses the best knowledge at that time about the site, but with the expectation that new data will change the model.

13.4.3.5 Reconstructing Source History

One emerging source zone MNA technique is using matrix diffusion signals to reconstruct source zone attenuation history. Many MNA protocols prescribe that a primary line of evidence be developed, where declining mass, mass discharge, or concentrations are demonstrated. One limitation to this is that the declining trends can be relatively slow (multiple year half lives) but that the monitoring record is short, only a few years. This makes it difficult to demonstrate conclusively a declining trend.

In theory, but not always applicable at all sites, a source reconstruction method can extend the concentration record back into time to the time of the original release, even if it occurred decades ago (Liu and Ball, 1999). Therefore, if the source has been declining over this entire time period, a much more effective demonstration of the primary line of evidence can be established.

The source reconstruction method was implemented by Chapman and Parker (2005) at a site in Connecticut, USA. At locations where DNAPL was still present, diffusion into a clayey aquitard resulted in relatively smooth soil concentration versus penetration depth profiles that were consistent with diffusion processes from a constant source, where the maximum soil concentration occurred at the interface and then decreased with depth into the aquitard. However, in other locations, the maximum soil concentration was encountered deeper within the aquitard, with lower concentrations measured at the interface. This pattern was presumed to result from historically high concentrations at the interface due to a high strength (DNAPL) source, which was later depleted, such that concentrations at the interface declined during more recent times, and the subsequent back diffusion from the intervals nearest the interface decreased the concentration relative to those within deeper layers in the aquitard. Figure 13.11 shows the resulting reconstruction of source zone history extending back 50 years at this site.

The utility of this method for supporting source zone MNA is outlined in the right-hand panels in Figure 13.12, developed as part of an ongoing research project (ESTCP, 2013; Newell, 2011). The collection of depth-discrete soil samples from a low permeability zone underlying an aquifer potentially allows for the historical reconstruction of the source loading. When the characteristic back diffusion profiles from locations within the source and in the downgradient plume are encountered, they can be used to demonstrate a decreasing concentration trend resulting from source decay that has occurred in the time elapsed following the release (assuming that a reasonable estimate of the time of release can be made). This is done by back-calculating the source loading (concentration history at the low permeability interface) that would have resulted in the measured soil concentrations in the low permeability layers as the result of diffusion-dominated transport in these layers.

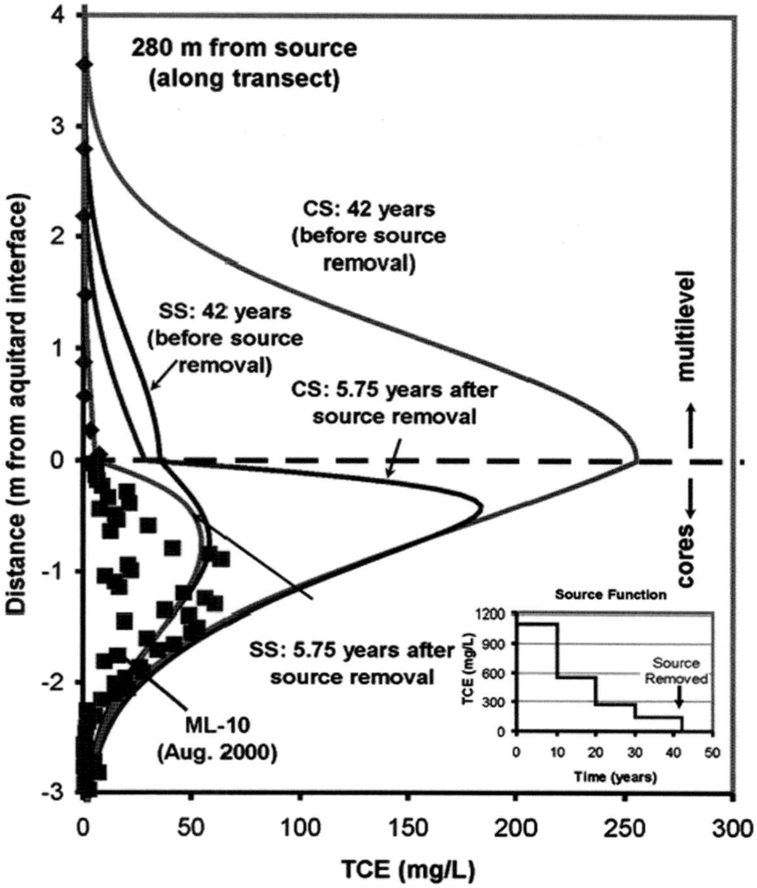

Figure 13.11. Matrix diffusion sampling in plume 280 m downgradient of source zone (low permeability zone is lower part of figure) with 1-D model fits to the data. CS = constant source, SS = step declining source. The best fit was with a first-order decay of source concentrations over time, starting 50 years prior to sampling, and with a 10 year half life (see source function graph above) (Chapman and Parker, 2005). This material is reproduced with permission of John Wiley & Sons, Inc.

This analysis can be compared to the more typical case, in which only a limited temporal record is available (left-hand panels of Figure 13.12), so that the true extent of source attenuation cannot be adequately appreciated because the majority of source decay occurred prior to the collection of monitoring data. This lack of source history data hampers an evaluation of the attenuation potential at a site, both in terms of what has already occurred and what can be expected to occur in the future.

13.4.4 Options for Implementation

Source zone MNA can be used in two basic ways:

1. As a polishing approach after some type of active source remediation has been implemented. Many practitioners, researchers, and regulators expect that source zone MNA will be an almost universal remedy after most active source remediation efforts because active source remediation typically does not restore groundwater to drinking water conditions site-wide.

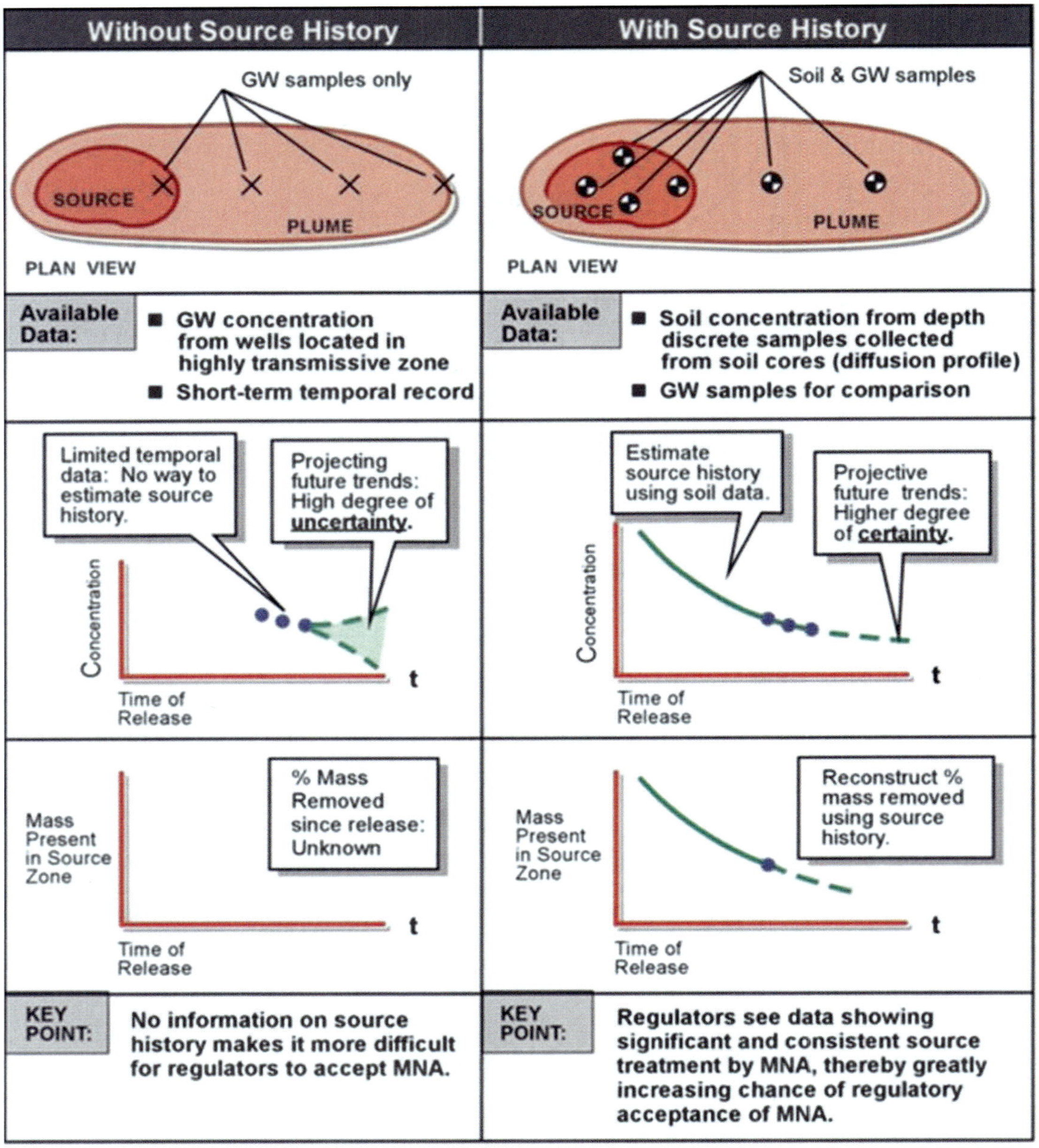

Figure 13.12. Source history derived from low permeability zone soil concentration data and 1-D diffusion models (modified after Newell, 2011).

2. As the sole remedy for source remediation. Source zone MNA may be particularly applicable to source zones far along in the aging process (late Stage 3) and for source zones in which DNAPL has been depleted (Stages 4 and 5).

One emerging concept for implementing source zone MNA is to enhance source zone biodegradation processes by reducing the influx of competing electron acceptors using vertical barriers. Typically, containment is considered to be a static technology, with no destruction of the contained source materials. However, the biodegradation efficiency within chlorinated solvent source zones containing both solvents and anthropogenic electron donors is reduced by competition from the influx of dissolved oxygen, nitrate, and sulfate, thereby reducing the source attenuation rate (Newell et al., 2003). By installing a vertical barrier at the site, this influx is reduced, biodegradation efficiency is increased, and source attenuation rate is increased. Tools such as the BioBalance model (Kamath et al., 2006) develop the mass balance between electron donor and competing electron acceptor and can be used to quantify the benefits of chlorinated source zone isolation to enhance source zone MNA.

13.5 MODELING SOURCE ZONE MONITORED NATURAL ATTENUATION

Models to assist in the evaluation, selection, and design of source zone MNA can be classified as data interpretation/extrapolation tools, analytical models, and numerical models. These approaches are discussed separately in the following sections, with descriptions of specific models where appropriate. While most of these tools and methods were developed to analyze concentration trends, most can also be used to analyze mass discharge data or to simulate how source zone attenuation will reduce mass discharge in the future. When employing modeling tools to assess source zone MNA, it is important to recognize that predictions will have uncertainty. The degree of uncertainty will vary depending on the amount and quality of data available to the modeler. It is also important to update modeling efforts as new data become available throughout the life of a project.

13.5.1 Simple Data Extrapolation Modeling

The USEPA's "Use of First Order Decay for Monitored Natural Attenuation" (USEPA, 2002) describes different types of first-order rate constants. A concentration versus time rate constant, or k_{point}, was defined as:

Concentration versus *Time Rate Constants: A rate constant derived from a concentration* versus *time (C* versus *T) plot at a single monitoring location provides information regarding the potential plume lifetime at that location, but cannot be used to evaluate the distribution of contaminant mass within the ground-water system. The C* versus *T rate constant at a location within the source zone represents the persistence in source strength over time and can be used to estimate the time required to reach a remediation goal at that particular location. To adequately assess an entire plume, monitoring wells must be available that adequately delineate the entire plume, and an adequate record of monitoring data must be available to calculate a C* versus *T plot for each well. At most sites, the rate of attenuation in the source area (due to weathering of residual source materials such as NAPLs) is slower than the rate of attenuation of materials in ground water, and concentration profiles in plumes tend to retreat back toward the source over time. In this circumstance, the lifecycle of the plume is controlled by the rate of attenuation of the source, and can be predicted by the C* versus *T plots in the most contaminated wells. At some sites, the rate of attenuation of the source is rapid compared to the rate of attenuation in ground water. This pattern is most common when contaminants are readily soluble in ground water and when contaminants are not biodegraded in ground water. In this case, the rate of attenuation of the source as predicted by a C* versus *T plot will underestimate the lifetime of the plume.*

The method used to calculate a concentration versus time first-order rate constant is illustrated with this diagram (Figure 13.13).

This USEPA document from 2002 went beyond concentration based rates and included a detailed discussion of how to calculate and apply mass based rate constants, including rates based on mass flux (now called mass discharge) data.

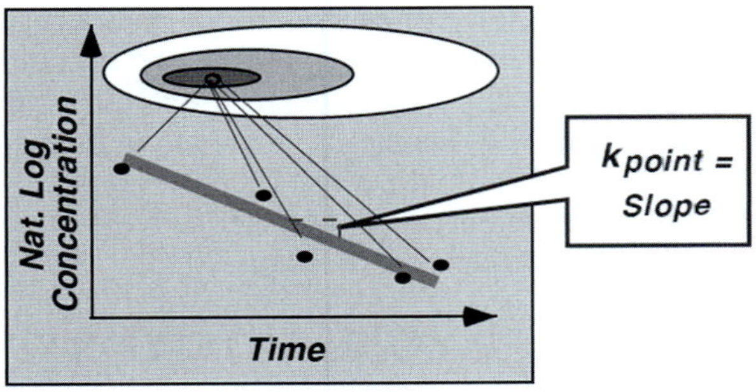

Figure 13.13. Method to determine k_{point} for source attenuation (USEPA, 2002).

Table 13.3. Source Zone Concentration Versus Time Half-Lives from High Concentration Wells at Chlorinated Solvent Sites (Farhat et al., 2004)

Statistics on Sites	Source Half Life from Natural Log Concentration[a] Versus Time Regression (Years)					
	PCE	TCE	*cis*-DCE	VC	1,2-DCA	1,1,1-TCA
75th Percentile of sites	*(Increasing)*	*(Increasing)*	5.0	*(Increasing)*	*(Increasing)*	*(Increasing)*
Median of sites	6.2	4.7	1.1	10.5	4.9	4.5
25th Percentile of sites	1.3	2.5	0.89	1.5	1.4	2.0
Regression r^2	0.28	0.21	0.47	0.33	0.48	0.32
Number of sites	32	37	11	18	13	23

Increasing: concentration versus time trend was increasing
[a]Concentration is from the well with highest historical concentration during temporal record at each site

The Air Force's SourceDK tool (Farhat et al., 2004) has a module where concentration versus time data from a monitoring well, either in plumes or in a source zone, can be entered into a spreadsheet template and then give the user a k_{point} by plotting the natural log of concentration versus time and finding the slope of the resulting regression. In addition, if a concentration goal, such as an MCL, is entered by the user, the time to reach the concentration goal is calculated along with 90 and 95% uncertainty levels. This approach was to calculate k_{point} at 37 chlorinated solvent sites using data compiled by McNabb et al. (1999) and McNabb (Walt W. McNabb, Lawrence Livermore National Lab, Livermore, California) (Table 13.3).

For example, in Table 13.3, 32 sites had at least 3 years of PCE concentration versus time data. At each site, the well with the highest historical concentration was used and the natural log of the PCE concentration was plotted versus time. A regression line was constructed for all 32 of these sites to yield a k_{point} for each site. The median k_{point} for the 32 PCE wells was 0.11 per year; this value was converted to a representative source half life by dividing into 0.693 (i.e., ln 2) to yield a median source half life of 6.2 years. More than 25% of the sites had regression lines with positive slopes (designated as an increasing trend in Table 13.3).

The SourceDK model and the USEPA models BioScreen (Newell et al., 1996), BIOCHLOR (Aziz et al., 2000a), and REMChlor (Falta et al., 2005a) all use a method for estimating source zone first-order decay coefficients that is based on a mass balance method in which source mass discharge is divided by source mass at some time:

$$k_s = \frac{M_d}{M} \qquad \text{(Eq. 13.4)}$$

where k_s is the source decay coefficient assuming first-order decay of source zone concentration and mass discharge, M_d is the mass discharge leaving the source zone at the time when the source zone analysis was performed (typically the year when the source zone data has been collected), and M is the mass in the source zone at the time the source zone analysis was performed. Note that k_s is similar to a k_{point} from a well in the source zone.

The ESTCP's Mass Flux Toolkit (Farhat et al., 2006) is another data interpretation/data analysis tool that can be used to evaluate source zone MNA. Mass discharge data at different times can be entered into this spreadsheet tool, and then these data are plotted against time to show if and to what extent attenuation has occurred over time.

13.5.2 Analytical Models

Several analytical models can be used to evaluate source zone MNA at chlorinated solvent sites. These include SourceDK (Farhat et al., 2004), which is a spreadsheet package providing three different tools for evaluating source zone attenuation including (1) an empirical extrapolation tool for groundwater monitoring data from the source zone, (2) a simple box model (mass balance model of mass within a source zone and mass flux leaving the source zone assuming a Power Law Model gamma = 1.0) for chlorinated solvent sites and a modified model for hydrocarbon sites, and (3) a more complex set of tools based on DNAPL dissolution equations.

BIOCHLOR is a spreadsheet package based on an approximate analytical solution that allows the user to enter in source decay rates. The package calculates and displays the resulting concentration versus time data at various locations within the downgradient plume. The model estimates concentrations and mass discharge immediately downgradient of the source zone, allowing analysis of source zone MNA (Aziz et al., 2000a).

BioBalance (Kamath et al., 2006) simulates changes in mass flux and/or source zone mass on remediation time frame. The model performs a stoichiometric mass balance on electron donors and competing electron acceptors acting on aqueous-phase chlorinated solvents in the source zone with anthropogenic electron donors such as oils or fuels. The model evaluates the sustainability of chlorinated solvent degradation within the source zone given a potentially limiting supply of electron donors.

REMChlor is an analytical solution that allows the user to enter source zone information, such as inputs to a Power Law Model, and subsequently evaluate concentration and/or mass discharge versus time immediately downgradient of the source zone (USEPA, 2007; Falta et al., 2005a, b). A second-generation version, PREMChlor, is now available that allows users to evaluate uncertainty in modeling simulations using a Monte Carlo-type probabilistic approach (Liang et al, 2010). REMChlor is discussed extensively in Chapter 6.

Natural Attenuation Software (NAS) includes the ability to simulate NAPL dissolution over time and therefore evaluate MNA in both the plume and the source zone (Widdowson et al., 2005; Chapelle et al., 2003). Natural Attenuation Software comprises a combination of analytical and numerical solute transport models. The natural attenuation processes considered by NAS include advection, dispersion, sorption, NAPL dissolution, and biodegradation. Natural Attenuation Software estimates and applies separate biodegradation rates from one redox zone to the next.

13.5.3 Numerical Models

Most of the commercially used numerical groundwater flow and contaminant transport models (such as MODFLOW/MT3D or MODFLOW/RT3D) are not designed to explicitly simulate source zone attenuation processes. These models are designed to take source information at designated source cells in the form of constant concentration or constant flux or some user-entered concentration versus time function and then simulate the fate and transport of the resulting plume that is formed. These models do not explicitly simulate the presence of DNAPL, matrix diffusion sources, or attenuation processes acting on source materials.

Several research codes have been used to evaluate source zone processes. These models are capable of simulating the entire life cycle of a chlorinated solvent source zone beginning with the initial release and redistribution of DNAPL, followed by continued DNAPL dissolution and aging, complete DNAPL depletion, and long term back diffusion of solute from low permeability zones. These model codes are discussed in Chapter 5 and include DNAPL3D-RX (West et al., 2008), COMPSIM (O'Carroll and Sleep, 2007), UTCHEM (Pope and Delshad, 2000), TMVOC (Pruess and Battistelli, 2002), MVALOR (Christ et al., 2008), and others. The expertise required to run these models goes beyond that required to run groundwater flow and solute transport models such as MODFLOW/MT3D. In addition, these models require additional input parameters such as the DNAPL composition and fluid properties and constitutive relationships such as capillary pressure and relative permeability functions.

13.6 OPERATIONS, MONITORING AND OPTIMIZATION

Like active remediation projects, source zone MNA often comprises an initial phase and then an operations and monitoring phase. The initial phase, described in Section 13.6, consists of detailed sampling of both contaminant and geochemical parameters and then detailed evaluation of the data using statistics, rate calculations, and frequently the use of computer models as well.

The operations and monitoring phase typically consists of contaminant sampling to confirm and update the MNA conceptual model and track the progress of attenuation. Temporal records (contaminant concentration versus time data from groundwater monitoring wells) or temporal mass discharge data are collected and analyzed and compared to predictions. Numerous existing texts describe standard groundwater monitoring procedures, although recent SERDP studies are presenting insights on methods to reduce variability and noise in long term concentration versus time temporal records. The ITRC (2010) presented an overview of mass discharge/mass flux data acquisition techniques that can be useful in monitoring natural attenuation.

13.6.1 Observational Approach and Comparison to Predictions

The observational approach is a method for managing the uncertainty at chlorinated solvent sites, similar to the concepts of adaptive management (Sale et al., 2008a). The observational approach was originally developed for geotechnical engineering by Terzaghi and Peck (1948) as a technique for dealing with the uncertainty inherent in subsurface environments by constantly refining the conceptual model and modifying the design and operations in response to new data. This approach consists of the following five steps (Peck, 1969; NRC, 2006):

1. Assess probable conditions and develop contingency plans for adverse outcomes
2. Establish key parameters for observation
3. Measure observational parameters and compare to calculations
4. Compare predicted and measured parameters
5. Change the design as needed

When applied to source zone attenuation, a "glide path" is calculated (Step 2) by extrapolating existing trends (such as with Tier 1 of the SourceDK tool) or using a source zone attenuation model (see Section 13.4.3). Typical observation parameters for this glide path are source zone groundwater concentrations and/or mass discharge rates from the source zone (Step 3). In the future, the predicted and measured parameters are compared (Step 4), and if necessary, the design is either maintained, or if necessary, the design is changed (Step 5). For example, if after 10 years the source attenuation rate is considered to be too slow for some required land use or regulatory objective, then active remediation could be applied.

The Texas Risk Reduction Program (TNRCC, 2001) describes a glide path approach to demonstrate decontamination via MNA, which typically involves attenuation of the source zone:

To evaluate the effectiveness of MNA as a method to decontaminate COCs, a routine monitoring program should be established. In most cases, the routine monitoring will focus on the change in COCs over time as they trend toward the required PCLs. This routine monitoring might not require the sampling of all the monitoring wells or all the COCs at a particular location, but uses the most important wells/COCs to ensure that a correct "glide path" is being maintained to achieve the PCLs.

13.7 CASE STUDIES

Several applications of source zone MNA have been found, although this area is still relatively innovative. The following case studies include (1) a survey of chlorinated solvent MNA sites, including several in which the MNA remedy included the source as well as the plume, to examine some of the common features of such sites and the experiences to date; (2) a hypothetical case study used by the ITRC Integrated DNAPL Site Strategy team to illustrate the factors that managers should consider when incorporating MNA into an overall strategy; and (3) an application of MNA for the source zone at a manufacturing facility, illustrating the results from a specific mixed-solvent site.

13.7.1 Survey of Source Zone MNA Applications

McGuire et al. (2003) compiled characteristics of 43 chlorinated solvent sites using MNA. One key result was that MNA was described as the *sole remedy* in about 30% of the sites, meaning MNA was likely being used for both the plume and the source zone.

13.7.2 Hypothetical Case Study

The Interstate Technology & Regulatory Council (ITRC, 2011) developed a hypothetical example of key decision points during the life cycle of a site where active remediation had been performed in a chlorinated solvent source zone, followed by MNA. This hypothetical example described the relationship between remediation objectives, time frame, contaminant reduction, and uncertainty. In this example (Figure 13.14), Panel A represents the intended outcome, Panel B illustrates what really happened after active remediation, and Panel C is an evaluation of the benefits of continued aggressive treatment versus source zone MNA:

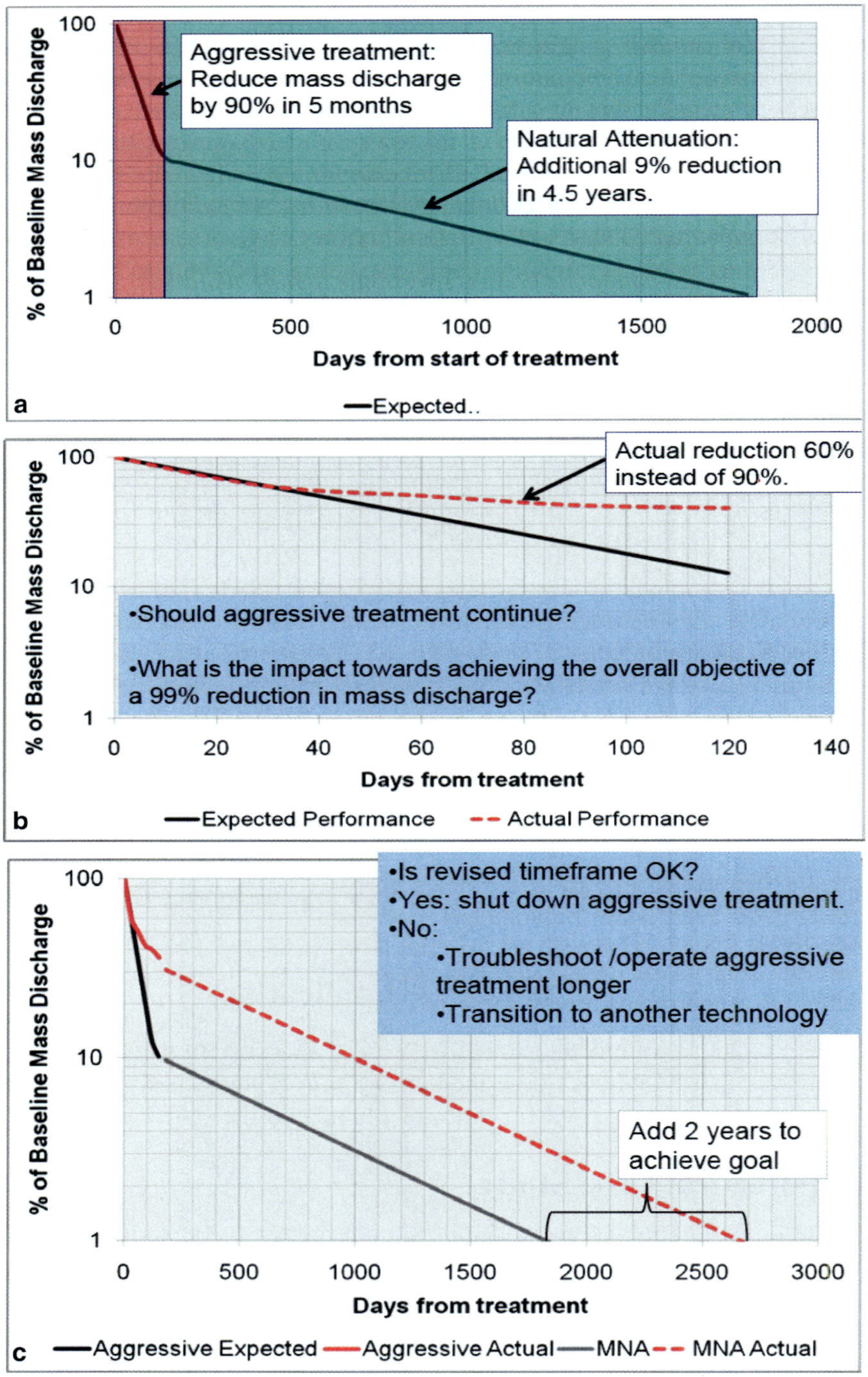

Figure 13.14. **Hypothetical decision point illustrating expected outcome of an aggressive source treatment (a), expected and actual performance (b), and impact to the overall remedy (c) (ITRC, 2011).**

The overall objective of the source zone treatment is to reduce contaminant mass discharge from the source zone by 99% within 5 years. To achieve this objective, it is expected that aggressive treatment will reduce contaminant mass discharge by 90% within 5 months followed by transition to MNA to reduce mass discharge by an additional 9% over the next 4.5 years (see A). After aggressive treatment, performance is evaluated near the time of expected shutdown to decide whether the objectives have been met and whether the aggressive treatment can be discontinued. As shown in B, only a 60% reduction in contaminant mass discharge was achieved within the treatment time period instead of the desired 90%. Should aggressive treatment be halted on schedule or continued until the planned 90% reduction in mass discharged is achieved? Part C of Figure 13.14 *compares the two options' impact on the overall objective of a 99% reduction in mass discharge. Without additional treatment, it will take MNA 6.5 years to achieve the functional objective. If this expanded time frame is acceptable, then the functional objective can be revised to achieve a 99% reduction in 7 years instead of 5, and aggressive treatment can be discontinued. If this revision is not acceptable, then a reevaluation of the remedy treatment options is necessary, including evaluating why the aggressive treatment did not achieve objectives – is it technology specific* (e.g., *it may be more difficult to extract mass stored in low-permeability zones) or due to issues with the CSM* (e.g., *there may be additional source material outside the treatment zone)? This example shows that several reevaluation points may be needed and different types of data may be appropriate for different decisions.*

13.7.3 Midwest Manufacturing Facility Case Study

At a chemical manufacturing facility located in the Midwest, a burn pit was used for fire training from the 1960s to the mid 1970s. After use was discontinued, the burn pit was filled. Based on subsequent site investigations, the burn pit was inferred to be the source of a dissolved groundwater plume, consisting primarily of 1,1,1-TCA. From November 2003 to January 2004, source area remediation was implemented and consisted of excavation of affected soils. A total of nearly 2,000 tons (1.8×10^6 kg) of soil was removed, and an estimated 560 kg of organic chemicals were removed with these soils, including 160 kg 1,1,1-TCA, 3 kg 1,1-DCE, and 21 kg polychlorinated biphenyls (PCBs). It is estimated that the excavation remedy removed 99% of 1,1,1-TCA and >99% of 1,1-DCE contained in the vadose zone soils at concentrations exceeding performance standards.

Only two constituents, 1,1,1-TCA and 1,1-DCE, were consistently detected above groundwater protection standards in the dissolved groundwater plume emanating from the vicinity of the burn pit. As implemented in 2003, the groundwater remedy for the remaining contaminants in groundwater consisted of monitored natural attenuation. From 2003 to 2011, groundwater monitoring for 1,1,1-TCA and 1,1-DCE shows stable or decreasing concentration trends for both constituents at all wells (see Figure 13.15). A summary of Mann-Kendall statistical analyses (Connor et al., 2012) for groundwater data collected from 2003 to 2011 for wells in the monitoring well network is summarized in Table 13.4.

Currently, the facility is managing the remaining groundwater constituents using MNA in the plume and in the former source zone (at location of Well F).

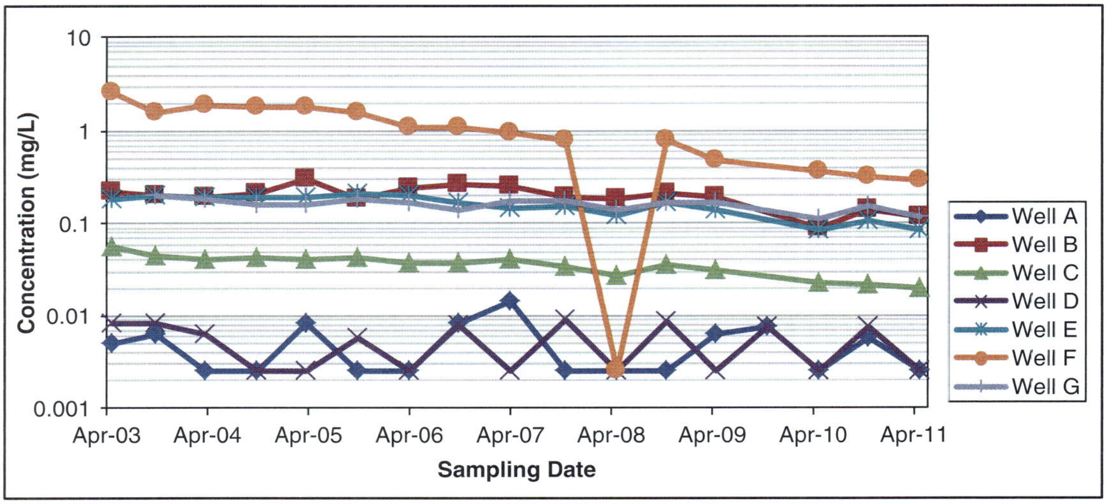

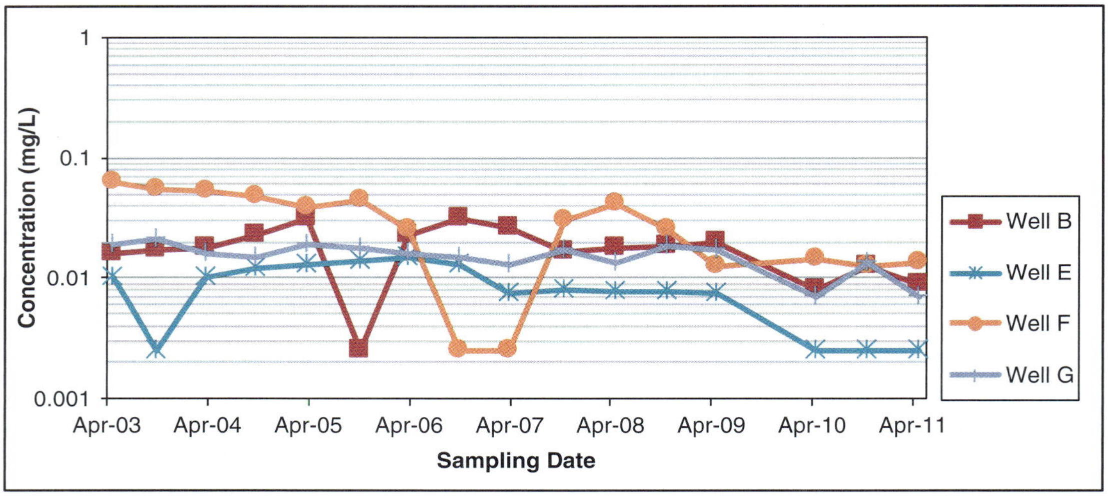

Figure 13.15. Concentration versus time data from 2003 to 2011 for 1,1,1-TCA (*top* graph) and 1,1-DCE (*bottom* graph), Midwest Manufacturing Facility. Well F is the well underlying the former source zone. The sampling results from this well in April 2008 are unexplained.

Table 13.4. Location of Wells Relative to Source Zone and Statistical Concentration versus Time Trend Using Aziz et al. (2003) Methodology

Well ID	Distance from Source (feet)	1,1,1-TCA Trend	1,1-DCE Trend
Well A	700	Stable	–
Well B	650	Decreasing	Stable
Well C	400	Decreasing	–
Well D	1,080	Stable	–
Well E	250	Decreasing	Decreasing
Well F	0	Decreasing	Decreasing
Well G	250	Decreasing	Decreasing

Note: "–" = 1,1-DCE is not historically detected in monitoring well

13.8 SUMMARY

The experience of the last decade has provided several key lessons for practitioners interested in evaluating or using MNA for chlorinated solvent source zones. These lessons are summarized in the following sections and organized by attenuation processes, methods available for practitioners, and lessons from field-scale applications.

13.8.1 Processes

- DNAPL source attenuation results from the processes of advection, dispersion, dissolution, sorption/desorption, matrix diffusion, biodegradation, abiotic degradation, and vaporization/volatilization. Vaporization/volatilization of chlorinated solvent sources can be significant processes in the unsaturated zone but will be less important below the water table.
- The life cycle of a chlorinated solvent source zone begins with an original release of DNAPL to the subsurface. Following this, a period of DNAPL redistribution takes place, typically lasting from months to one or two decades, depending on the amount and type of DNAPL released. The redistribution period is followed by continued DNAPL dissolution and source zone aging. At some point in time, all DNAPL will be depleted, and the source will be sustained by back diffusion from any low permeability zones that may be present. Sites in the late stages of aging (also called weathering), and those being sustained by back diffusion, are likely candidates for source zone MNA because they have low mass discharge rates that persist for long periods of time (the "long tail"), they probably will not have primary threat wastes, and they will be less amenable to remediation via injection based technologies than sites with recent releases.
- Many sites will not have enough data to precisely delineate the source zone architecture, the distribution of contaminants between DNAPL pools, DNAPL ganglia, and the surrounding matrix. Therefore, simplified mass discharge/mass relationships, such as a first-order decay or Power Law Model, are often used to simulate source attenuation. For sites dominated by matrix diffusion, use of larger gamma values in the Power Law Model (gamma $>$ 1.0) are needed to represent the long tail in concentrations.

13.8.2 Methods

- We can with confidence demonstrate that source zone MNA is occurring and can bound mass loss rates within approximately an order of magnitude with good field data. Temporal trend analysis will require several years of quality data at most sites, and even then it will be a retrospective analysis with no guarantee that extrapolations to predict future source behavior will be accurate.
- While source zone MNA has been based on concentration based plume MNA techniques in the past, new mass discharge based techniques such as the SZNA approach are now being developed. With an SZNA-type approach, the specific attenuation processes (vapor phase, dissolved transport, and biodegradation) can be quantified and compared to active remediation technologies. This mass balance type of approach will probably see much broader acceptance in the future.
- First-order source decay coefficients (k_s) can be derived either from temporal concentration records or from a simple mass balance relationship (Equation 13.4).
- Several data analysis tools are available to help process site data to evaluate source zone MNA, including SourceDK and the Mass Flux Toolkit. Several screening-level

analytical models are available including BIOCHLOR, REMChlor, and the Natural Attenuation Software (NAS). Numerical models that incorporate a wider range of processes than analytical models are also available. These models typically require estimates of source mass and mass discharge or an estimated source attenuation rate.

- Matrix diffusion data have been used to reconstruct source strength histories and therefore could be used to better understand multidecade attenuation processes at some sites.

13.8.3 Applications

- A study of chlorinated solvent sites where MNA was being pursued found that about 30% of the sites were relying on source zone MNA as the sole remedy for source zone remediation. The use of MNA as the sole source zone remedy will likely increase in the future.
- Several regulatory agencies and other groups have issued guidance that allows some form of source zone MNA for low risk or low threat chlorinated solvent sites.
- Because it is very difficult to restore groundwater to MCLs, MNA is envisioned to be applied as a polishing step after applying active remediation technologies at early life-cycle stage sites and possibly as the sole remedy at late-stage sites (Stage 5) where DNAPL depletion has occurred or is soon to occur (late Stage 3).
- Source zone MNA will only be attractive when long term management over decades to centuries is possible, even after partial source removal. However, because most active technologies will only remove part of the contaminant mass, decades-to-centuries time frames may also apply to many active remediation projects. Therefore it can be useful to compare the relative remediation time frame of source zone MNA, either as a stand-alone remedy or as a polishing step, versus active remediation using simple predictive models that embody our best understanding of source zone processes. A 2012 National Research Council report (NRC, 2012) stressed that the use of longer remediation time frames (sometimes more than 100 years) for passive long term management approaches such as MNA is "critical to avoid inefficient use of limited resources."
- Our ability to predict future conditions under source zone MNA is limited due to uncertainty, but when used under an adaptive observational approach, model predictions can provide useful projections for site stakeholders. Overall, source zone attenuation models can be useful, but they should be applied with the appropriate expectations with regard to accuracy and reliability.

Disclaimer: The USEPA through its Office of Research and Development conducted a portion of the research described here under in-house task 3674 (MNA of Chlorinated Solvents). It has not been subjected to agency review and therefore does not necessarily reflect the views of the agency, and no official endorsement should be inferred.

REFERENCES

Adamson DT, McGuire TM, Newell CJ, Stroo H. 2011. Sustained treatment: Implications for treatment timescales associated with source depletion technologies. Remediat J 21:27–50.

AECOM. 2009. Remedial Action Cost Engineering and Requirements (RACER™) software, version 10.2. Los Angeles, CA, USA.

AFCEE (Air Force Center for Engineering and the Environment). 2010. Sustainable Remediation Tool (SRT). Version 2.1. http://www.afcee.af.mil/resources/technologytransfer/programsandinitiatives/sustainableremediation/srt/index.asp. Accessed January 2, 2013.

Anderson MR, Johnson RL, Pankow JF. 1992. Dissolution of dense chlorinated solvents into groundwater, 3, Modeling contaminant plumes from fingers and pools of solvent. Environ Sci Technol 26:901–908.

Aziz CE, Newell CJ, Gonzales JR, Haas PE, Clement TP, Sun Y. 2000a. BIOCHLOR Natural Attenuation Decision Support System, User's Manual, Version 1.0. EPA/600/R-00/008. USEPA, Washington DC, USA. www.gsi-net.com. Accessed January 2, 2013.

Aziz JJ, Newell CJ, Rifai HS, Ling M, Gonzales JR. 2000b. Monitoring and Remediation Optimization System (MAROS), Software User's Guide. www.gsi-net.com. Accessed January 2, 2013.

Aziz JJ, Ling M, Rifai HS, Newell CJ, Gonzales JR. 2003. MAROS: A decision support system for optimizing monitoring plans. Ground Water 41:355–367.

Battelle. 2011. SiteWise Version 2, User Guide. Developed for the Naval Facilities Engineering Command, Port Hueneme, CA, USA. https://portal.navfac.navy.mil/portal/page/portal/navfac/navfac_ww_pp/navfac_nfesc_pp/environmental/erb/resourceerb/ug-2092-env_sitewisetm-v2_2011-06.pdf. Accessed February 2 2013.

Brown RA, Wilson JT, Ferrey M. 2007. Monitored natural attenuation forum: The case for abiotic MNA. Remediat J 17:127–137.

Chapelle FH, Widdowson MA, Brauner JS, Mendez E, Casey CC. 2003. Methodology for estimating times of remediation associated with monitored natural attenuation: U.S. Geological Survey Water-Resources Investigations Report 03-4057. http://pubs.usgs.gov/wri/wri034057. Accessed September 30, 2013.

Chapman SW, Parker BL. 2005. Plume persistence due to aquitard back diffusion following dense nonaqueous phase liquid removal or isolation. Water Resour Res 41:W12411.

Chen W, Kan AT, Newell CJ, Moore E, Tomson MB. 2002. More realistic soil cleanup standards with dual-equilibrium desorption. Ground Water 40:153–164.

Christ J, Lemke L, Abriola L. 2008. The influence of dimensionality on simulations of mass recovery from nonuniform dense non-aqueous phase liquid (DNAPL) source zones. Adv Water Resour 32:401–412.

Connor J, Farhat S, Vanderford M, Newell C. 2012. GSI Mann-Kendall Toolkit. GSI Environmental Inc., Houston, TX, USA. http://www.gsi-net.com/en/software/free-software/gsi-mann-kendall-toolkit.html. Accessed March 14, 2013.

CRWQCB (California Regional Water Quality Control Board). 2009. Assessment Tool for Closure of Low-Threat Chlorinated Solvent Sites. San Francisco Bay Region, San Francisco, CA, USA.

CSGSS (Center for Sustainable Groundwater & Soil Solutions). 2011. Practical Models for Supporting Remediation of Chlorinated Solvents. Short course. Aiken, SC, USA. www.gsi-net.com

DiFilippo EL, Brusseau ML. 2008. Relationship between mass-flux reduction and source-zone mass removal: Analysis of field data. J Contam Hydrol 98:22–35.

Doner L, Sale T. 2008. Back Diff Video Movie, Colorado State University Center for Contaminant Hydrology. http://www.serdp-estcp.org/Tools-and-Training/Environmental-Restoration/DNAPL-Source-Zones/Chlorinated-Solvents-On-Demand-Video. Accessed March 14, 2013.

ESTCP (Environmental Security Technology Certification Program). 2013. Determining Source Attenuation History to Support Closure by Natural Attenuation. ER-201032. Project conducted by GSI Environmental, University of Guelph, and Colorado State University. http://www.serdp.org/. Accessed January 30, 2014.

ESTCP. 2014. Assessment of the Natural Attenuation of NAPL Source Zones and Post-Treatment NAPL Source Zone Residuals. Project ER-0705. http://www.serdp.org/. Accessed January 12, 2014.

Falta RW, Rao PSC, Basu N. 2005a. Assessing the impacts of partial mass depletion in DNAPL source zones: I. Analytical modeling of source strength functions and plume response. J Contam Hydrol 78:259–280.

Falta RW, Basu N, Rao PSC. 2005b. Assessing the impacts of partial mass depletion in DNAPL source zones: II. Coupling source strength functions to plume evolution. J Contam Hydrol 79:45–66.

Falta RW, Stacy MB, Ahsanuzzaman ANM, Wang M, Earle RC. 2007. REMChlor: Remediation Evaluation Model for Chlorinated Solvents, User's Manual Version 1.0. www.epa.gov/ada/csmos/models/remchlor.html. Accessed August 9, 2012.

Farhat SK, de Blanc PC, Newell CJ, Gonzales JR, Perez J. 2004. SourceDK Remediation Timeframe Decision Support System, User's Manual, April 2004. www.gsi-net.com. Accessed August 9, 2012.

Farhat SK, Newell CJ, Nichols E. 2006. Mass Flux Toolkit To Evaluate Groundwater Impacts, Attenuation, and Remediation Alternatives. Groundwater Services, Houston, TX, USA. www.gsi-net.com. Accessed August 9, 2012.

Farhat SK, Newell CJ, Vanderford MV, McHugh TE, Mahler NT, Gillespie JL, Jurena PN, Bodour AA. 2012. Low-Risk Site Closure Guidance Manual to Accelerate Closure of Conventional and Performance Based Contract Sites. GSI Environmental, Houston, TX, USA.

Ferrey ML, Wilkin RT, Ford RG, Wilson JT. 2004. Nonbiological removal of cis-dichloroethylene and 1,1-dichloroethylene in aquifer sediment containing magnetite. Environ Sci Technol 38:1746–1752.

Fure AD, Jawitz JW, Annable MD. 2006. DNAPL source depletion: Linking architecture and flux response. J Contam Hydrol 85:118–140.

Gerhard JI, Pang T, Kueper BH. 2007. Time scales of DNAPL migration in sandy aquifers examined via numerical simulation. Ground Water 45:147–157.

Grant GP, Gerhard JI, Kueper BH. 2007. Multidimensional validation of a numerical model for simulating a DNAPL release in heterogeneous porous media. J Contam Hydrol 92:109–128.

He Y, Su C, Wilson J, Wilkin R, Adair C, Lee T, Bradley P, Ferrey M. 2009. Identification and characterization methods for reactive minerals responsible for natural attenuation of chlorinated organic compounds in ground water. EPA 600/R-09/115. USEPA, Washington, DC, USA.

Hunkeler D, Chollet N, Pittet X, Aravena R, Cherry JA, Parker BL. 2004. Effect of source variability and transport processes on carbon isotope ratios of TCE and PCE in two sandy aquifers. J Contam Hydrol 74:265–282.

Hunkeler D, Meckenstock RU, Sherwood-Lollar B, Schmidt TC, Wilson JT. 2008. A guide for assessing biodegradation and source identification of organic ground water contaminants using compound specific isotope analysis (CSIA). EPA 600-R-08-148. USEPA, Ada, OK, USA. 82 p.

Hunter P. 2004. Overview of Air Force Long-Term Monitoring Optimization, Programs and Case Studies. Center for Environmental Excellence, Brooks Air Force Base, TX, USA. www.clu-in.org/siteopt/proceedings_04/track_b/wed/18_hunter_philip.pdf.

ITRC (Interstate Technology & Regulatory Council). 2008. Enhanced Attenuation: Chlorinated Organics. EACO-1. Washington, DC, USA.

ITRC. 2009. Evaluating Natural Source Zone Depletion at Sites with LNAPL. LNAPL-1. Washington, DC, USA.

ITRC. 2010. Use and Measurement of Mass Flux and Mass Discharge. MASSFLUX-.1. Washington, DC, USA.

ITRC. 2011. Integrated DNAPL Site Strategy Technical/Regulatory Guidance. IDSS-1 Washington, DC, USA.

Johnson P, Lundegard P, Liu Z. 2006. Source zone natural attenuation at petroleum hydrocarbon spill sites. I: Site-specific assessment approach. Ground Water Monit Remediat 26:82–92.

Kamath RK, Newell CJ, Looney BB, Vangelas KM, Adamson DT. 2006. BioBalance – A Mass Balance Toolkit, User's Manual. http://www.gsi-net.com/en/software/free-software/biobalance-toolkit.html. Accessed March 14, 2013.

Kampara M, Thullner M, Richnow HH, Harms H, Lukas Y Wick LY. 2008. Impact of bioavailability restrictions on microbially induced stable isotope fractionation. 2. Experimental evidence. Environ Sci Technol 42:6552–6558.

Kan AT, Fu G, Hunter M, Chen W, Ward CH, Tomson MB. 1998. Irreversible sorption of neutral hydrocarbons to sediments: Experimental observations and model predictions. Environ Sci Technol 32:892–902.

Kennedy LG, Everett JW, Gonzales J. 2006. Assessment of biogeochemical natural attenuation and treatment of chlorinated solvents, Altus Air Force Base, Altus, Oklahoma. J Contam Hydrol 83:221–236.

Kueper BH, Gerhard JI. 1995. Variability of point source infiltration rates for two-phase flow in heterogeneous porous media. Water Resour Res 31:2971–2980.

Kueper BH, Redman D, Starr RC, Reitsma S, Mah M. 1993. A field experiment to study the behavior of tetrachloroethylene below the water table: Spatial distribution of residual and pooled DNAPL. Ground Water 31:756–766.

Kueper BH, Stroo HF, Vogel CM, Ward CH. 2014. Source Zone Remediation: The State of the Practice. Chapter 1 in Chlorinated Solvent Source Zone Remediation. In Ward CH, ed, SERDP/ESTCP Remediation Technology Monograph Series. Springer Science + Business Media, LLC, New York, NY, USA.

Liang H, Falta R, Newell C, Farhat S, Rao S, Basu N. 2010. Decision and Management Tools for DNAPL Sites: Optimization of Chlorinated Solvent Source and Plume Remediation Considering Uncertainty (Final Report). ESTCP Project ER-0704. ESTCP, Arlington, VA, USA.

Lima GU, Parker B, Meyer J. 2011. Dechlorinating microorganisms in a sedimentary rock matrix contaminated with a mixture of VOCs. Environ Sci Technol 46:5756–5763.

Linz DG, Nakles DV, eds. 1997. Environmentally Acceptable Endpoints in Soil. American Academy of Environmental Engineering, Annapolis, MD, USA.

Lipson D, Kueper BH, Gefell MJ. 2005. Matrix diffusion-derived plume attenuation in fractured bedrock. Ground Water 43:30–39.

Liu C, Ball WP. 1999. Application of inverse methods to contaminant source identification from aquitard diffusion profiles at Dover AFB, Delaware. Water Resour Res 35:1975–1985.

Lollar BS, Slater GF, Sleep B, Witt M, Klecka GM, Harkness M, Spivack J. 2001. Stable carbon isotope evidence for intrinsic bioremediation of tetrachloroethene and trichloroethene at Area 6, Dover Air Force Base. Environ Sci Technol 35:261–269.

Lundegard PD, Johnson PC. 2006. Source zone natural attenuation at petroleum hydrocarbon spill sites II: Application to a former oil field. Ground Water Monit Remed 26:93–106.

McGuire TM, Newell CJ, Looney, BB Vangelas KM. 2003. Historical and retrospective survey of monitored natural attenuation: A line of inquiry supporting monitored natural attenuation and enhanced passive remediation of chlorinated solvents. WSRC-TR-2003-00333, Westinghouse Savannah River Company, U.S. Department of Energy, Aiken, SC USA.

McGuire TM, Newell CJ, Looney BB, Vangelas KM, Sink CH. 2004. Historical analysis of monitored natural attenuation: A survey of 191 chlorinated solvent sites and 45 solvent plumes. Remediat J 15:99–112.

McHugh T, Beckley L, Liu C, Newell C. 2012. Factors influencing variability in groundwater monitoring data sets. Groundwater Monit Remediat 31:92–101.

McNabb WW, Rice DW, Bear J, Ragaini R, Tuckfield C, Oldenburg C. 1999. Historical Case Analysis of Chlorinated Volatile Organic Compound Plumes. Lawrence Livermore National Laboratory, Livermore, CA, USA.

Miller CT, Poirier-McNeil MM, Mayer AS. 1990. Dissolution of trapped nonaqueous phase liquids: Mass transfer characteristics. Water Resour Res 26:2783–2796.

Millington RJ, Quirk JM. 1961. Permeability of porous solids. Trans Faraday Soc 57:1200–1207.

Murphy B L, Morrison RD. 2002. Environmental Forensics. Academic Press, London, UK.

Newell C. 2011. Overview of Matrix Diffusion Processes and Its Effects on Managing Chlorinated Solvent Sites. Webinar presentation to USEPA, May 5, 2011. http://www.gsi-net.com/en/publications/gsi-environmental-papers.html?year=2011. Accessed March 14, 2013.

Newell CJ, Adamson DT. 2005. Planning-level source decay models to evaluate impact of source depletion on remediation time frame. Remediat J 15:27–47.

Newell CJ, Gonzales J, McLeod R. 1996. BIOSCREEN Natural Attenuation Decision Support System. U.S. Environmental Protection Agency. EPA/600/R-96/087. August.

Newell CJ, Aziz CA, Cox GA. 2003. Enhanced Anaerobic Treatment Zones in Groundwater. U.S. Patent No. 6562235.

Newell CJ, Cowie I, McGuire TM, McNabb W. 2006. Multiyear temporal changes in chlorinated solvent concentrations at 23 MNA sites. J Environ Eng 132:653–663.

Newell CJ, Farhat SK, Adamson DT, Looney BB. 2011. Contaminant plume classification system based on mass discharge. Ground Water 49:914–919.

NRC (National Research Council). 2000. Natural Attenuation for Groundwater Remediation. National Academies Press, Washington, DC, USA.

NRC. 2005. Contaminants in the Subsurface: Source Zone Assessment and Remediation. National Academies Press, Washington, DC, USA.

NRC. 2006. Geological and Geotechnical Engineering in the New Millennium: Opportunities for Research and Technological Innovation. National Academies Press, Washington, DC, USA.

NRC. 2012. Alternatives for Managing the Nation's Complex Contaminated Groundwater Sites. National Academies Press, Washington, DC, USA.

O'Carroll DM, Sleep BE. 2007. Hot water flushing for immiscible displacement of a viscous NAPL. J Contam Hydrol 91:247–266.

Parker JC, Park E. 2004. Modeling field-scale dense nonaqueous phase liquid dissolution kinetics in heterogeneous aquifers Water Resour Res 40:W05109.

Parker BL, Gillham RW, Cherry JA. 1994. Diffusive disappearance of immiscible-phase organic liquids in fractured geologic media. Ground Water 32:805–820.

Peck RB. 1969. Advantages and limitations of the observational method in applied soil mechanics. Geotechnique 19:171–187.

Pope D, Acree S, Levine H, Mangion S, van Ee J, Hurt K, Wilson B. 2004. Performance monitoring of MNA remedies for VOCs in ground water. EPA/600/R-04/027. USEPA, Washington, DC, USA. http://www.epa.gov/ada/gw/mna.html. Accessed January 3, 2013.

Pope G, Delshad M. 2000. Volume I: User's Guide for UTCHEM-9.0, A three-dimensional chemical flood simulator. Reservoir Engineering Research Program, University of Texas at Austin, Austin, TX, USA. 153 p.

Poulsen M, Kueper BH., 1992. A field experiment to study the behavior of tetrachloroethylene in unsaturated porous media. Environ Sci Technol 26:889–895.

Pruess K, Battistelli A. 2002. TMVOC, A numerical simulator for three-phase non-isothermal flows of multicomponent hydrocarbon mixtures in saturated-unsaturated heterogeneous media. LBNL-49375. Lawrence Berkeley National Laboratory, Berkeley, CA, USA.

Rao PSC, Jawitz JW, Enfield CG, Falta RW, Annable MD, Wood AL. 2001. Technology integration for contaminated site remediation: Cleanup goals and performance criteria. In Thornton S, Oswald S, eds. Groundwater Quality 2001: Natural and Enhanced Restoration of Groundwater Pollution. Int Ass Hydrological Sciences. 273:571–578.

Rao PSC, Jawitz JW. 2003. Comment on "Steady-state mass transfer from single-component dense non-aqueous phase liquids in uniform flow fields" by T. C. Sale & D. B. McWhorter. Water Resour Res 39:1068.

Rivett M, Feenstra S, Cherry J. 2001. A controlled field experiment on groundwater contamination by a multicomponent DNAPL: Creation of the emplaced-source and overview of dissolved plume development. J Contam Hydrol 49:111–149.

Sale TC, McWhorter DB. 2001. Steady state mass transfer from single-component dense nonaqueous phase liquids in uniform flow fields. Water Resour Res 37:393–404.

Sale T, Newell C, Stroo H, Hinchee R, Johnson P. 2008a. Frequently Asked Questions Regarding Management of Chlorinated Solvent in Soils and Groundwater. Environmental Security Testing and Certification Program, Arlington, VA, USA. http://www.serdp.org/. Accessed January 30, 2014.

Sale T, Zimbron J, Dandy D. 2008b. Effects of reduced contaminant loading on downgradient water quality in an idealized two layer system. J Contam Hydrol 102:72–85.

Scheutz C, Broholm MM, Durant ND, Begtrup Weeth E, Jorgensen TH, Dennis P, Jacobsen CS, Cox EE, Chambon JC, Bjerg PL. 2010. Field evaluation of biological enhanced reductive dechlorination of chloroethenes in clayey till. Environ Sci Technol 44:5134–5141.

Seyedabbasi MA, Newell CJ, Adamson DT, Sale TC. 2012. Relative contribution of DNAPL dissolution and matrix diffusion to the long-term persistence of chlorinated solvent source zones. J Contam Hydrol 134–135:69–81.

Suchomel E, Pennell K. 2006. Reductions in contaminant mass discharge following partial mass removal from DNAPL source zones. Environ Sci Technol 40:6110–6116.

Takeuchi M, Kawabe Y, Watanabe E, Oiwa T, Takahashi M, Nanba K, Kamagata Y, Hanada S, Ohko Y, Komai T. 2011. Comparative study of microbial dechlorination of chlorinated ethenes in an aquifer and a clayey aquitard. J Contam Hydrol 124:14–24.

Terzaghi K, Peck RB. 1948. Soil Mechanics in Engineering Practice. John Wiley & Sons, New York, NY, USA.

Thullner M, Kampara M, Richnow HH, Harms H, Wick LY. 2008. Impact of bioavailability restrictions on microbially induced stable isotope fractionation: 1. Theoretical calculation. Environ Sci Technol 42:6544–6551.

TNRCC (Texas Natural Resources Conservation Commission). 2001. Monitored natural attenuation demonstrations. RG-366/TRPP-33. Austin, TX, USA. http://www.tceq.texas.gov/publications/rg/rg-366_trrp_33.html/view. Accessed January 3, 2013.

USEPA (U.S. Environmental Protection Agency). 1991. A guide to principal threat and low level threat wastes. Superfund Publication 9380.3-06FS. Fact Sheet, November. Office of Emergency Remedial Response. Washington, DC, USA.

USEPA. 1998. Seminars: Monitored Natural Attenuation for Ground Water. EPA/625/K-98/001. Washington, DC, USA.

USEPA. 1999. Use of Monitored Natural Attenuation at Superfund, RCRA Corrective Action, and Underground Storage Tank Sites. OSWER Directive 9200.4-17P. Washington, DC, USA.

USEPA. 2001. Monitored Natural Attenuation: USEPA Research Program – an EPA Science Advisory Board Review. EPA-SAB-EEC-01-004. Washington, DC, USA.

USEPA. 2002. Calculation and Use of First-Order Rate Constants For Monitored Natural Attenuation Studies. EPA/540/S-02/500. Washington, DC, USA. http://www.epa.gov/ada/pubs/issue.html. Accessed January 4, 2013.

USEPA. 2007. REMChlor Version 1.0 – December 2007. http://www.epa.gov/ada/csmos/models/remchlor.html. Accessed January 4, 2013.

West MR, Kueper BH. 2010. Plume detachment and recession times in fractured rock. Ground Water 48:416–426.

West MR, Grant GP, Gerhard JI, Kueper BH. 2008. The influence of precipitate formation on the chemical oxidation of TCE DNAPL with potassium permanganate. Adv Water Res 31:324–338.

Widdowson MA, Mendez E III, Chapelle FH, Casey CC. 2005. Natural Attenuation Software (NAS) User's Manual Version 2. http://www.nas.cee.vt.edu. Accessed January 4, 2013.

Wiedemeier TH, Rifai HS, Newell CJ, Wilson J. 1999. Natural Attenuation of Fuels and Chlorinated Solvents. John Wiley & Sons, New York, NY, USA.

Wilson JT. 2011. An Approach for Evaluating the Progress of Natural Attenuation in Groundwater. EPA/600/R-11/204. Office of Research and Development, USEPA, Washington, DC, USA.

Wilson JT, Kaiser PM, Adair C. 2005. Monitored Natural Attenuation of MTBE as a Risk Management Option at Leaking Underground Storage Tank Sites EPA600/R-04/179. http://www.epa.gov/ada/gw/mna.html. Accessed January 29, 2014.

Zhu J, Sykes JF. 2004. Simple screening models of NAPL dissolution in the subsurface. J Contam Hydrol 72:245–258.

CHAPTER 14

IN SITU THERMAL TREATMENT OF CHLORINATED SOLVENT SOURCE ZONES

Jennifer L. Triplett Kingston,[1] Paul C. Johnson,[2] Bernard H. Kueper[3] and Kevin G. Mumford[3]

[1]Haley and Aldrich, Overland Park, KS 66210, USA; [2]Arizona State University, Tempe, AZ 85287, USA; [3]Queen's University, Kingston, ON, Canada K7L 3N6

14.1 INTRODUCTION

Combining *in situ* heating with physical recovery, chemical reaction and biodegradation processes has led to a spectrum of *in situ* thermal remediation options for the cleanup of soil, rock and groundwater impacted by dense, nonaqueous phase liquids (DNAPLs). The growth in the application and understanding of these technologies over the past two decades has been significant – to the point that their potential application is considered for many sites having short target cleanup time frames (less than a few years) and for contaminants that are not accessible by other cleanup technologies (such as mass diffused into fine-grained media). This chapter presents an overview of the most practiced thermal remediation technologies, summarizes our understanding of their performance to date and presents a few case studies. From this point forward, *in situ* thermal technologies will simply be referred to as *thermal technologies* throughout this chapter.

The development of thermal technologies stems in part from the use of heating and displacement technologies dating back to the 1930s for enhanced oil recovery but also from knowledge gained since the 1980s on the physical, chemical and biological treatment of subsurface contamination and the dependencies of the underlying processes (such as vaporization, volatilization, dissolution, flow, reaction) on temperature. The initial development and proof-of-concept testing of thermal technologies at contaminated sites occurred principally in the 1980s and 1990s (Udell, 1996), with a deeper understanding emerging more recently as a result of increased application in the past decade and empirical review of thermal technology performance.

Thermal technologies generally employ one of three primary heating options: steam-enhanced extraction (SEE), thermal conductive heating (TCH, sometimes referred to as *in situ* thermal desorption or ISTD) and electrical resistance heating (ERH), a variation of which is called the Electro-Thermal Dynamic Stripping Process (ET-DSP™). The use of radio frequency heating (RFH) has also been explored, and *in situ* soil mixing with large-diameter augers (LDA) combined with steam and hot air injection has been applied at full scale alone and in combination with delivery of reactants, notably iron particles. Depending on operating temperatures and contaminant concentrations, heating may decrease DNAPL viscosity, decrease DNAPL–water interfacial tension, increase biodegradation and other reaction rates, increase the aqueous solubility of the DNAPL and/or increase volatility. The primary differences between these thermal technologies are the methods of heating or energy delivery, the need for and methods used for contaminant recovery and the processes targeted for thermal enhancement.

B.H. Kueper et al. (eds.), *Chlorinated Solvent Source Zone Remediation*,
doi: 10.1007/978-1-4614-6922-3_14,

14.2 FUNDAMENTAL MECHANISMS

Thermal technologies are applied to remove DNAPL by increasing subsurface temperatures. The manner in which this heating takes place varies according to the specific technology used (ERH, TCH or SEE), but all of these technologies are being applied either (1) to increase vapor pressures to induce liquid-to-gas phase change (boiling) and/or increase partitioning to the gas phase (vaporization and volatilization) to improve vapor recovery or (2) to increase the mobility of DNAPL through reductions in viscosity and interfacial tension to enhance liquid recovery.

Certain thermal technologies can offer a significant advantage over fluid migration (injection-based) technologies, such as *in situ* chemical oxidation (ISCO) and cosolvent flushing, because the mechanisms that govern heating and heat transfer (flow of electrical current for ERH and heat conduction for TCH) are much less sensitive to geologic heterogeneity than the mechanisms that govern fluid migration. Because SEE relies on the injection of steam, it is affected by the sensitivity of fluid migration to variations in permeability. As such, this advantage is strongest for ERH and TCH technologies.

This section introduces the relationship of several important physicochemical parameters to temperature, including vapor pressure, partitioning coefficients, density, viscosity and interfacial tension. This section also includes sample calculations for use in preliminary design. It is important to note that other fundamental mechanisms also are important for successful design of a specific thermal technology application, particularly those related to heat transfer, multiphase flow, and electrical current flow. The mechanisms discussed here, however, are those that are common to the various technologies and relate to the behavior of the DNAPL and associated dissolved and sorbed-phase volatile organic compounds (VOCs).

14.2.1 Vapor Formation

14.2.1.1 Boiling

Thermal technologies, when used in water-saturated zones, are most often applied to create a gas phase by boiling water. At the pore scale, gas is created at nucleation sites on the surface of solid grains when, in the absence of DNAPL or VOCs, the temperature is sufficiently high that the vapor pressure of water is equal to the sum of the water pressure and the capillary pressure:

$$P_w^v = P_w + P_c \qquad \text{(Eq. 14.1)}$$

where P_w^v is the vapor pressure of water, P_w is the liquid water pressure and P_c is the capillary pressure. Equation 14.1 must be satisfied at all locations where a gas phase is present. At the initial stages of gas-phase development, P_c is the capillary pressure characteristic of a heterogeneous nucleation site, typically considered to be a small, hydrophobic crevice in the surface of a grain that contains trapped gas (Atchley and Properetti, 1989; Satik and Yortsos, 1996). As vapor formation continues, the pore-scale vapor bubble expands in a series of pore-filling and pressurization steps (Satik and Yortsos, 1996) as it moves through pore throats to adjacent pore bodies, and the capillary pressure is dictated by the evolving curvature of the gas–liquid interface. This expansion continues until the coalescence of adjacent bubbles occurs and a macroscopic, connected gas phase is formed. The creation and subsequent recovery of this connected gas phase are critical to the successful application of thermal technologies.

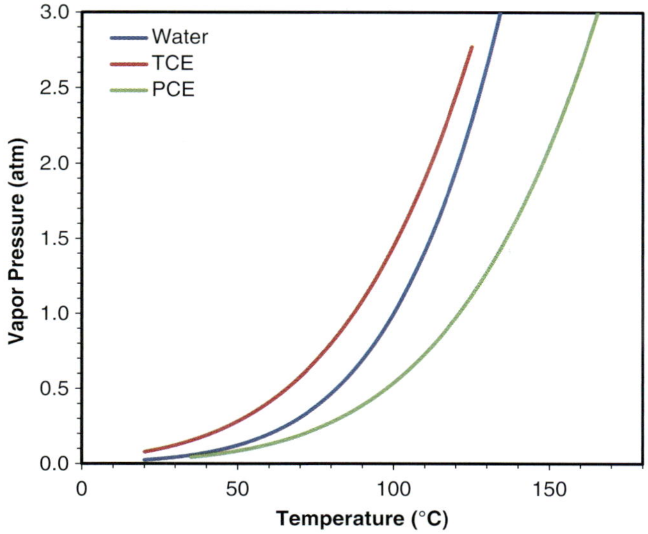

Figure 14.1. Vapor pressure versus temperature for water, TCE, and PCE. Note: °C - degrees Celsius.

Table 14.1. Antoine Equation Parameters[a]

Compound	A	B	C	Valid Temperature Range (°C)
Water	8.07131	1,730.63	233.426	1–100
Water	8.14019	1,810.94	244.485	99–374
TCE	7.0281	1,315.1	230.01	−13–127
PCE	7.02	1,415.49	221.01	34–187

[a]From Yaws and Yang (1989) for Equation 14.2 with P^V in torr and T in °C

The increase of vapor pressure with temperature for water and many DNAPLs can be approximated by the Antoine equation (Lide and Kehiaian, 1994):

$$\log P_i^V = A_i - \frac{B_i}{T + C_i} \quad \text{(Eq. 14.2)}$$

where P_i^V is the vapor pressure of compound; i, A_i, B_i, and C_i are compound-specific empirical constants; and T is the temperature. Tabulated values of A_i, B_i and C_i can be found in engineering and science reference texts such as Lide and Kehiaian (1994) and Perry and Green (1997) and can be presented for different units of pressures (atm, Pa, torr) and temperature (°C, °K). Figure 14.1 shows the vapor pressures of water, trichloroethene (TCE) and tetrachloroethene (also known as perchloroethene or PCE) calculated using Equation 14.2 and the A_i, B_i and C_i values listed in Table 14.1. Additional correlations for estimating vapor pressure can be found in, for example, Reid et al. (1987).

The temperature at which Equation 14.1 is satisfied is the boiling point. While the normal boiling point of water is 100°C (212°F) (for $P_g = 1$ atm), Equation 14.1 shows that below the water table a higher temperature must be reached to increase the vapor pressure above the sum of the liquid water pressure (which increases with depth below the water table) and the capillary pressure created by the curved surface of the gas–liquid interface. Approximating the liquid pressure of water as the hydrostatic pressure (no vertical components of groundwater flow) and assuming negligible capillary pressure, the combination of Equations 14.1 and 14.2 gives

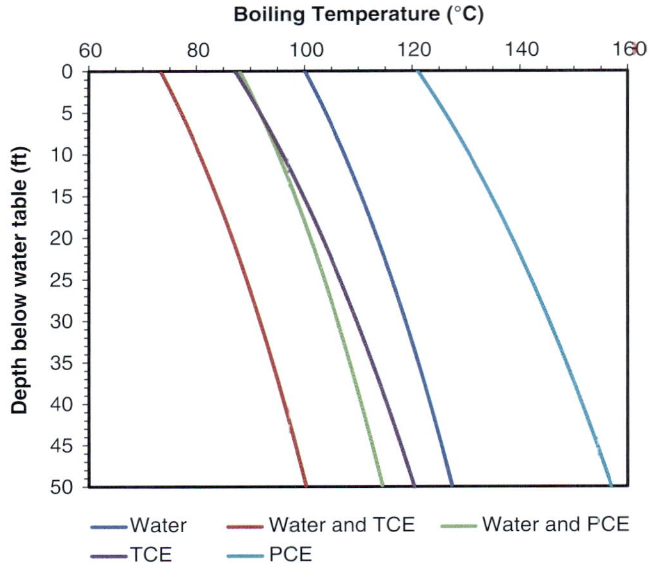

Figure 14.2. Boiling and co-boiling point versus depth for water, TCE and PCE, estimated using parameters in Table 14.1 applied outside of their reported ranges as an approximation.

$$T_{b,i} = \frac{B_i}{A_i - log(\rho_w gz + P_{atm})} - C_i \qquad \text{(Eq. 14.3)}$$

where $T_{b,i}$ is the boiling point of compound i at depth z below the water table, ρ_w is the density of liquid water and P_{atm} is the atmospheric pressure. The boiling temperature of pure water as a function of depth below the water table is shown in Figure 14.2.

It is important to note that the presence of dissolved gases (such as O_2 or N_2) in the groundwater will allow the initial formation of small amounts of vapor below the boiling point of water as exsolved gases and water vapor will both contribute to the total gas pressure. The extent to which these gases contribute to the gas pressure will depend on their aqueous concentrations and partitioning constant (Henry's Law). This initial degassing will not significantly affect the boiling points shown in Figure 14.2.

14.2.1.2 Co-boiling

Under conditions where two immiscible liquids are present, such as at the interface between residual or pooled DNAPL and water, the vapor pressures of both the liquids will contribute to the total gas pressure that must satisfy Equation 14.1. This phenomenon is referred to as co-boiling. For an n component DNAPL whose vapor pressure can be calculated using Raoult's Law, which assumes behavior as an ideal mixture, the onset of co-boiling occurs when

$$P_w^v + \sum_j^n x_j P_j^v = P_w + P_c \qquad \text{(Eq. 14.4)}$$

where P_j^v is the vapor pressure of DNAPL component j, x_j is the mole fraction of component j in the nonaqueous phase liquid (NAPL), and n is the total number of volatile components in the DNAPL. Equation 14.4 assumes that any water dissolved in the NAPL, or VOC dissolved in the water, does not significantly affect the vapor pressures of the components. Because the vapor

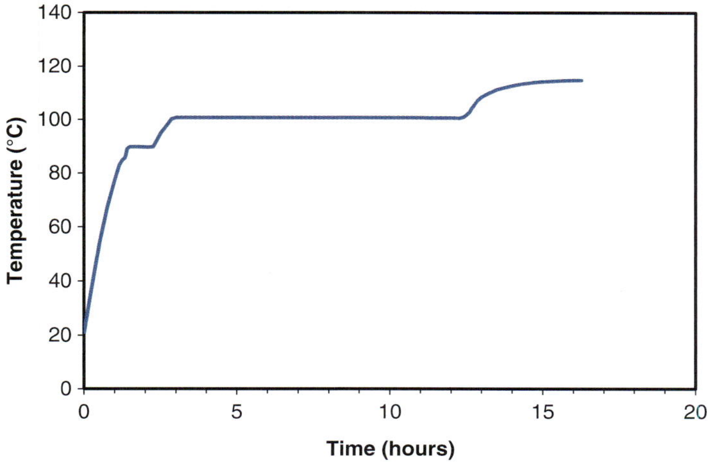

Figure 14.3. Temperature versus time data during heating of PCE DNAPL and water in sand (Burghardt and Kueper, 2008).

pressures of the two immiscible liquids are additive, vapor is formed at a lower temperature (the co-boiling point) than the boiling point of any of the pure components alone. For the simple case of water and a single-component DNAPL ($n = 1$ in Equation 14.4), the increase in co-boiling points with depth for immiscible mixtures of water and TCE and water and PCE is shown in Figure 14.2.

When temperature can be monitored in the immediate area of the NAPL–water interface where co-boiling occurs, as can be done in laboratory experiments, a co-boiling plateau can be observed (DeVoe and Udell, 1998; Burghardt and Kueper, 2008; Martin and Kueper, 2011). Figure 14.3 illustrates the PCE–water co-boiling plateau observed by Burghardt and Kueper (2008). In the first plateau shown (constant temperature with time), energy is being used to co-boil DNAPL and water. When all the DNAPL has been removed, vapor pressure is no longer sufficient to satisfy Equation 14.1, and heating of water continues until the boiling point of water is reached (second plateau). When all water has been removed, the soil temperature increases (superheating period) dependent on the energy input. This superheating period does not occur in ERH applications, which rely on the aqueous phase to conduct electrical current. Furthermore, the DNAPL–water co-boiling plateau is not often observed in data from field applications because of the unlikely positioning of temperature monitoring points close (within centimeters [cm]) to DNAPL–water interfaces.

For a multicomponent DNAPL, the co-boiling plateau may not necessarily exhibit a constant temperature as seen in Figure 14.3. This deviation stems from the fact that the second term of Equation 14.4 (the effective vapor pressure of the DNAPL mixture) changes over time as the more volatile components are removed preferentially. As such, to maintain co-boiling conditions, the temperature increases, increasing the vapor pressures of the remaining, less volatile components. The slope of the temperature versus time curve is less than during the initial heating of the water and DNAPL or the final heating of the water because some of the energy is being used for phase change while some is heating the remaining liquids. A theoretical temperature versus time profile, based on Equation 14.4, as well as the heat capacities and heats of vaporization of water, TCE and PCE, for a multicomponent DNAPL (50% TCE and 50% PCE by volume) is presented in Figure 14.4 to illustrate the temperature profile during co-boiling of a DNAPL mixture.

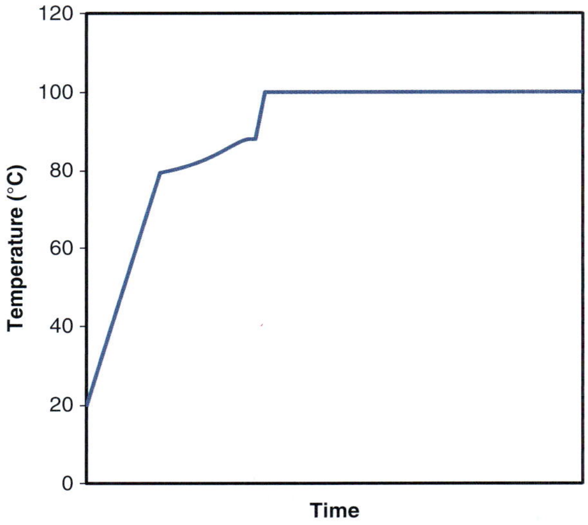

Figure 14.4. Theoretical temperature versus time profile for heating of a multicomponent DNAPL (50% TCE, 50% PCE) in the presence of water.

During co-boiling, the composition of the vapor phase is controlled by the relative vapor pressures of the components at the co-boiling point. Using the left-hand term of Equation 14.4 to define the total gas pressure, Dalton's Law gives the mole fraction in the gas phase:

$$y_j = \frac{P_j^v}{\sum_j^n x_j P_j^v + P_w^v} \qquad \text{(Eq. 14.5)}$$

where y_j is the mole fraction of component j in the gas phase. For example, at a depth of 10 feet (ft), or 3.1 meters (m), below the water table, vapor formed during co-boiling of water and TCE would be 63% TCE by mole fraction, which is equivalent to 93% by mass fraction. Vapor formed during co-boiling of water and PCE would be 35% PCE by mole fraction, which is equivalent to 83% by mass fraction. For a single-component DNAPL, these percentages would remain constant throughout the co-boiling period. For a multicomponent DNAPL, however, the vapor phase composition will vary throughout the co-boiling period as discussed above. It is expected that in full-scale applications, where vapors from many extraction wells are typically combined before sampling for VOC concentrations, measured mole fractions of VOCs will be lower because of dilution from areas where the vapor is pure steam.

14.2.2 Partitioning

The partitioning of VOC mass between the water, gas and solid phases is an important consideration for the successful application of thermal technologies, particularly later in the remediation application when most NAPL has been removed. The equilibrium partitioning, where the chemical potential of the VOC in each phase is equal, will change with temperature as chemical potential changes with temperature. The partitioning between the water phase and other phases is typically described using the aqueous solubility (water–NAPL), Henry's Law constant (water–gas) and the soil–water distribution coefficient (water–solid).

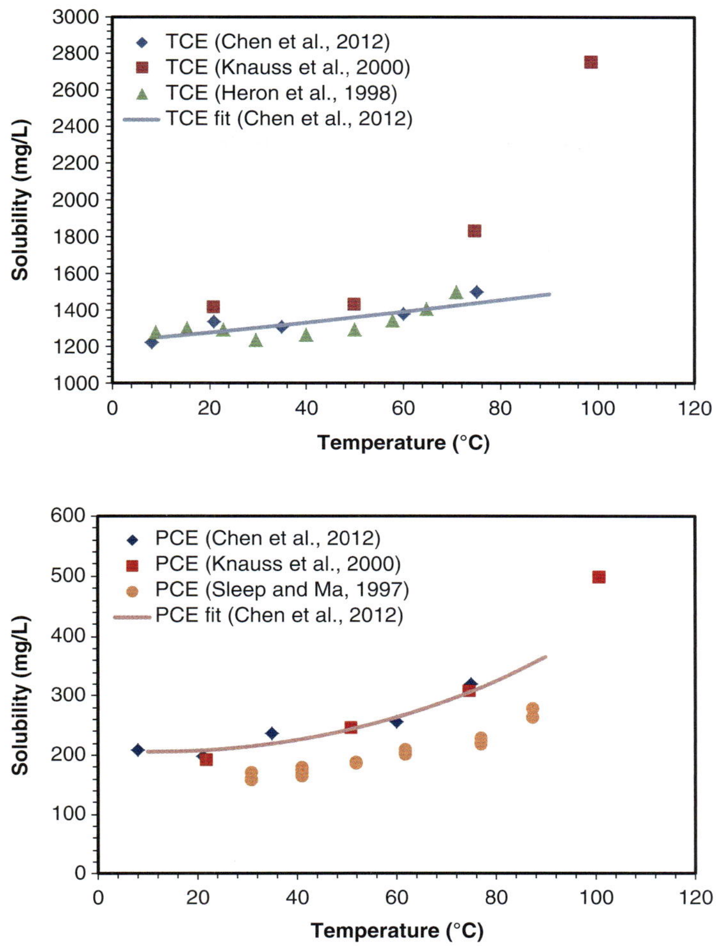

Figure 14.5. Solubility versus temperature for TCE (*upper panel*) and PCE (*lower panel*) from three separate laboratory studies. Values reported by Knauss et al. (2000) and regression equations by Chen et al. (2012) (Equations 14.6 and 14.7) have been converted from mole fraction to mg/L.

14.2.2.1 Aqueous Solubility

Heron et al. (1998) and Knauss et al. (2000) showed that the solubilities of TCE and PCE increase with increasing temperature. In addition, Chen et al. (2012) showed that this relationship is not true for all chlorinated VOCs, as the solubility of some VOCs investigated in their study increased and decreased between 8 and 75°C (1,1,1-trichloroethane [TCA], chloroform and carbon tetrachloride), while the solubility of others did not change substantially (1,1-dichloroethane and *cis*-1,2-dichloroethene), although they reported that hydrolysis reactions may have affected their results for some compounds at higher temperatures. Solubility versus temperature data from these three studies are presented in Figure 14.5 for TCE and PCE. Also shown are the regression equations presented by Chen et al. (2012) for TCE and PCE:

$$R \ln K_{TCE} = -220.293 + \frac{5136}{T} + 23.03885 \ln T \qquad \text{(Eq. 14.6)}$$

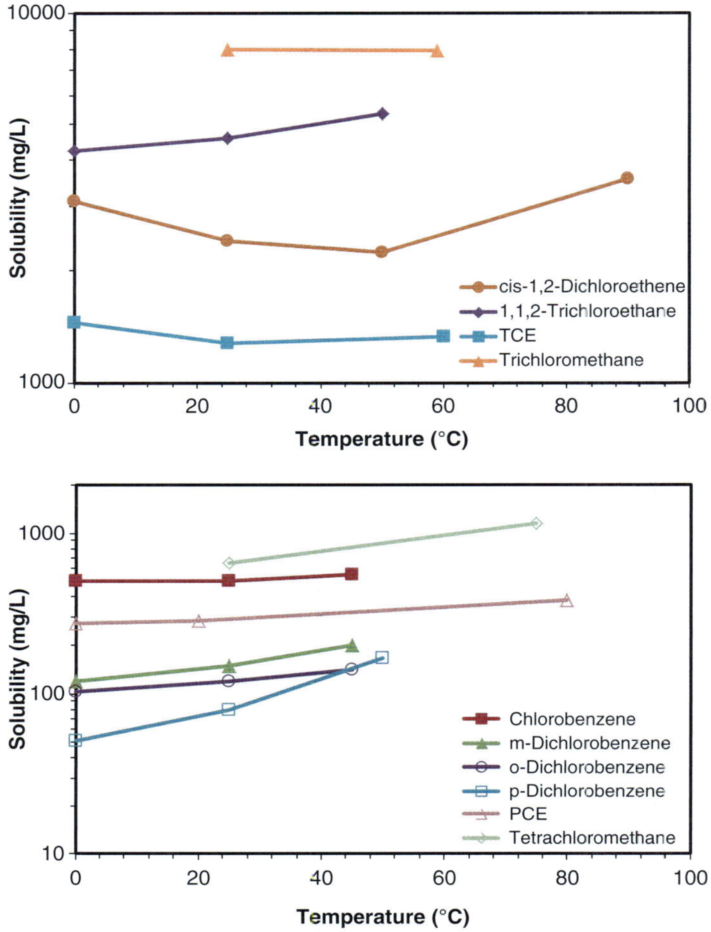

Figure 14.6. Solubility versus temperature for ten chlorinated organics between 1,000 and 10,000 mg/L (*upper panel*) and between 10 and 2,000 mg/L (*lower panel*) (CRC Handbook of Chemistry and Physics, 2012).

$$\ln K_{PCE} = -1313.04 + \frac{52334}{T} + 184.0563 \ln T \qquad \text{(Eq. 14.7)}$$

where K is the mole fraction of VOC (TCE or PCE) in the aqueous phase and T is the temperature (K). The coefficients in these regression equations account for the temperature dependency of the enthalpy and entropy of solution, which is required to estimate solubility over this temperature range (Knauss et al., 2000).

Laboratory studies (Knauss et al., 2000; Heron et al., 1998; Chen et al., 2012; Sleep and Ma, 1997) and data from the CRC Handbook of Chemistry and Physics (2012) (Figure 14.6) show that the increase in the aqueous solubility of many chlorinated compounds is generally less than an order of magnitude. For example, the solubilities of TCE and PCE near the boiling point of water (70–100°C [158–212°F]) are approximately 1.2–2.0 and 1.6–2.7 times greater, respectively, than their solubilities at 20°C. Increased mass transfer rates from DNAPL to the aqueous phase can also result from increased temperature independent of increases in aqueous solubility. For example, Imhoff et al. (1997) reported an increase in the mass transfer rate coefficient for the dissolution of residual PCE by a factor of two between temperatures of 5 and 40°C

(41 and 104°F), with no significant increase in aqueous solubility. Compared to relative increases in vapor pressure and Henry's Law constant for volatile chlorinated compounds, these increases in solubility and mass transfer rate coefficients represent only a moderate increase, and it is unlikely that substantial decreases in remediation times will be realized because of increased dissolution at these temperatures. The increased solubility and mass transfer rate coefficients, however, do present the possibility of increased risk to downgradient groundwater receptors during initial heating phases unless adequate hydraulic control is implemented as part of the remediation application.

14.2.2.2 Henry's Law

Once DNAPL has been removed by co-boiling or in applications where the target heating areas contain only dissolved and sorbed VOC, mass transfer of VOC from the aqueous phase to the gas phase (volatilization) is the primary treatment mechanism. The relative concentrations of VOC in the water and gas phases are typically calculated using Henry's Law, which describes the relative concentrations at equilibrium and also characterizes the magnitude of the driving force for mass transfer. The dimensionless form of Henry's Law is

$$C_{i,g} = H_i C_{i,w} \qquad \text{(Eq. 14.8)}$$

where H_i is the dimensionless Henry's Law constant for compound i, $C_{i,g}$ is the concentration of compound i in the gas phase, and $C_{i,w}$ is the concentration of compound i in the water phase.

Alternatively, Henry's Law can be expressed as

$$P_i = K_{H,i} C_{i,w} \qquad \text{(Eq. 14.9)}$$

where $K_{H,i}$ is Henry's Law constant (atm·m^3/mol) for compound i and P_i is the partial pressure of compound i (atm). Equations 14.8 and 14.9 can be related by the ideal gas law:

$$C_{i,g} = \frac{n_i}{V} = \frac{P_i}{RT} \qquad \text{(Eq. 14.10)}$$

where n_i is the number of moles of component i in the gas phase, R is the ideal gas constant, and T is the temperature (°K). The value of $K_{H,i}$ can be adjusted for temperature over small temperature ranges, where the enthalpy of dissolution is assumed to be constant, using the van't Hoff equation (Heron et al., 1998; Knauss et al., 2000; Chen et al., 2012; Gossett, 1987). However, Heron et al. (1998) showed that the enthalpy of dissolution for TCE varied significantly between 10 and 95°C [50 and 203°F] and proposed the use of a linear relationship between the enthalpy of dissolution and temperature, combined with the van't Hoff equation, to give an expression for Henry's Law constant as a function of temperature:

$$\text{In}\, K_{H,i} = D_i - \frac{E_i}{T} + F_i \,\text{In}\, T \qquad \text{(Eq. 14.11)}$$

where D_i, E_i and F_i are empirical constants.

Laboratory studies (Knauss et al., 2000; Heron et al., 1998; Chen et al., 2012; Sleep and Ma, 1997; Gossett, 1987; Görgényi et al., 2002) and data compiled by Baker and Hieste (2009) show that Henry's Law constant for many chlorinated compounds increases significantly with temperature. Correlations based on Equation 14.11 developed by Chen et al. (2012) predict increases between 4 and 10 times, between 20 and 90°C [68 and 194°F] for 12 different chlorinated organics (Figure 14.7). Data presented by Heron et al. (1998) shows an increase in Henry's Law constant for TCE by a factor of 20 between 10 and 95°C [50 and 203°F]. For a discussion of the thermodynamics that underlie this increase, please see Görgényi et al. (2002). Values of D_i, E_i and F_i for PCE and TCE presented by Chen et al. (2012) are listed in Table 14.2, along with those of Heron et al. (1998).

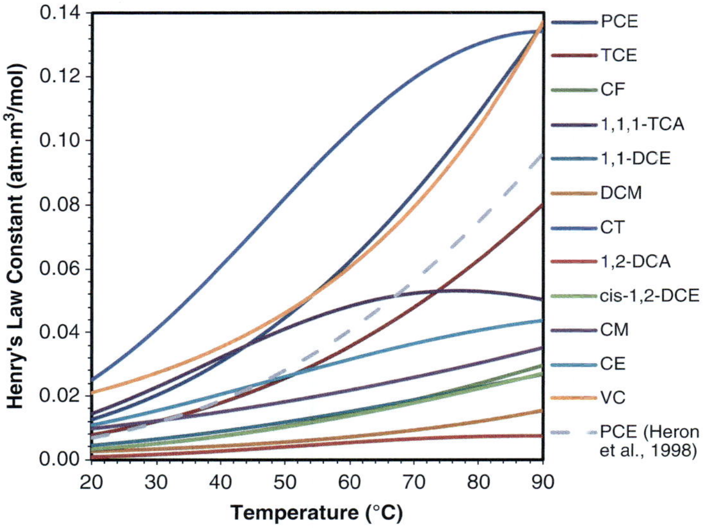

Figure 14.7. Henry's Law constant versus temperature for several chlorinated organic compounds.

Table 14.2. Regression Coefficients for Henry's Law Constant (Equation 14.11)

Compound	A	B	C	Reference
PCE	152.2	10,547	−21.23	Chen et al. (2012)
TCE	186.1	12,540	−26.[illegible]1	Heron et al. (1998)
TCE	98.26	7,936	−13.39	Chen et al. (2012)

14.2.2.3 Distribution Coefficient

Remediation targets at most sites require the achievement of low concentrations in soil and groundwater. At these low concentrations, sorbed-phase VOCs can represent a significant fraction of the mass. The relative concentration of VOC in the water and solid phases is typically described using a distribution coefficient as follows:

$$C_{i,s} = K_{d,i}\, C_{i,w} \qquad \text{(Eq. 14.12)}$$

where $K_{\mathrm{d},i}$ is the soil–water distribution coefficient for compound i and $C_{i,s}$ is the concentration of compound i in the solid phase. Equation 14.12 is a linear isotherm. Nonlinear isotherms, including the Langmuir and Freundlich isotherms, can also be used to describe sorption. In addition, the distribution of sorption sites within particles can lead to rate-limited sorption. A detailed discussion of isotherms and kinetic sorption related to changes in temperature is presented by Werth and Reinhard (1997a, b). When applying Equation 14.12, $K_{d,i}$ is typically assumed to be proportional to the fraction of organic carbon as follows:

$$K_{d,i} = f_{oc} K_{oc,i} \qquad \text{(Eq. 14.13)}$$

where f_{oc} is the fraction of organic carbon in the porous medium and $K_{\mathrm{oc},i}$ is the organic carbon partitioning coefficient for compound i. Equation 14.13 ignores any sorption to mineral surfaces, which may be important in low f_{oc} material. It is expected that the sorption of VOCs will decrease at higher temperatures. For example, Sleep and McClure (2001) reported a decrease in the distribution coefficients for toluene, naphthalene and PCE. Adjusting their K_{d}

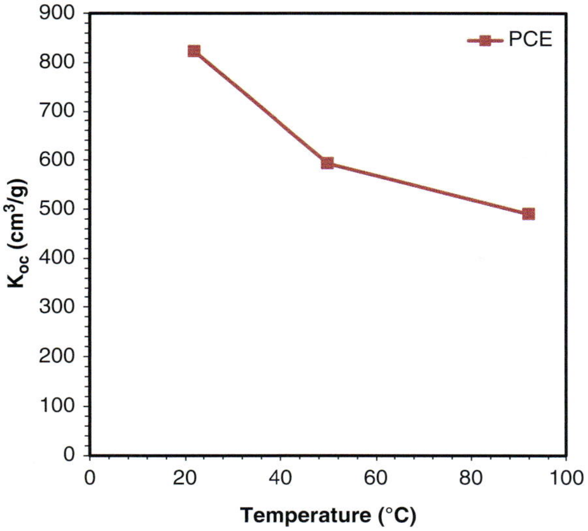

Figure 14.8. Organic carbon partitioning coefficient versus temperature for PCE (data from Sleep and McClure, 2001).

values for their reported f_{oc} of 0.0045 gives K_{oc} values for PCE, using Equation 14.13, that decrease from approximately 820 to 490 cm^3/g between 22 and 92°C. Their data for PCE is shown in Figure 14.8. Due to significant site-to-site variation in sorption behavior, particularly if nonlinear sorption or rate-limited sorption is expected, the experimental determination of site-specific sorption isotherms may be required to accurately predict sorption behavior throughout a thermal remediation application.

It is important to note that Burghardt and Kueper (2008) observed a persistent fraction of PCE that remained following prolonged heating and drying of a well-graded sand with a f_{oc} of approximately 0.0013. They attributed this persistence to PCE sorbed to mineral surfaces, which could have different temperature dependence than sorption to organic matter.

14.2.3 Fluid Properties

Although boiling, vaporization and volatilization are typically the target mechanisms for removing chlorinated organics, increases in temperature also can be used to increase the mobility of DNAPL. This effect is most useful for pumping continuous DNAPL, the rate of which is affected by changes in the density and viscosity of the DNAPL, consistent with Darcy's Law. Even if increased mobility for pumping is not the intention, increased mobility of a continuous DNAPL phase can occur during preliminary heating before the onset of boiling or co-boiling conditions or during the formation of a DNAPL condensation bank in SEE applications.

14.2.3.1 Increased Mobility of Continuous DNAPL

Sleep and Ma (1997) presented data for the temperature dependence of the density and viscosity of PCE and fit their data to the functions:

$$\rho_{PCE} = 1.6294 - 6.6655 \times 10^{-4} T - 4.9643 \times 10^{-6} T^2 \qquad \text{(Eq. 14.14)}$$

$$\mu_{PCE} = exp\left(-4.273 + \frac{1.890 \times 10^3}{T} - \frac{2.035 \times 10^5}{T^2}\right) \qquad \text{(Eq. 14.15)}$$

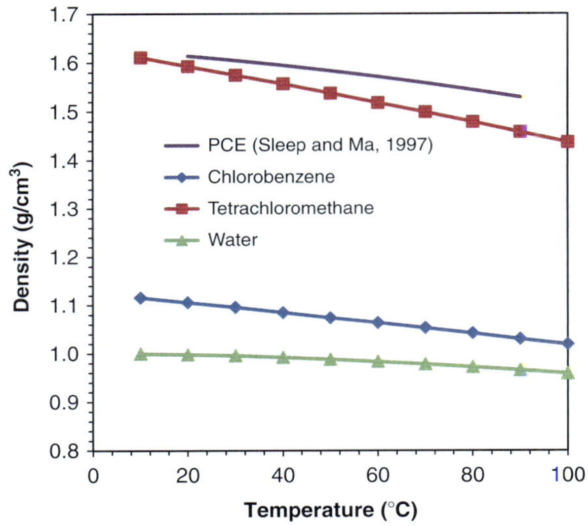

Figure 14.9. Liquid density versus temperature.

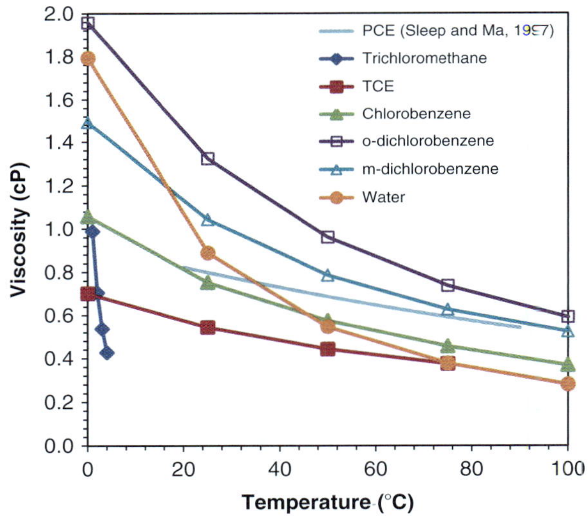

Figure 14.10. Liquid viscosity versus temperature.

where ρ_{PCE} (g/cm^3) and μ_{PCE} (cP) are the density and viscosity of PCE, respectively. Temperature (T) is expressed in degrees Celsius for Equation 14.14 and in Kelvin for Equation 14.15. Note that the first term in Equation 14.15 has been modified from the published value of -4.723 to match the published data (B. Sleep, University of Toronto, Toronto, Canada, Personal communication, 2013). Equations 14.14 and 14.15 are plotted in Figures 14.9 and 14.10, respectively, along with data for water and other chlorinated NAPLs (CRC Handbook of Chemistry and Physics, 2012).

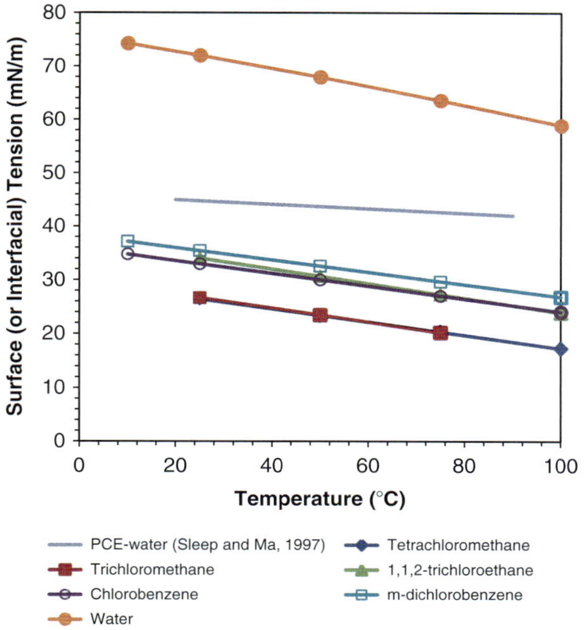

Figure 14.11. Surface tensions of several NAPLs and water as a function of temperature. PCE–water interfacial tension is included for reference.

For chlorinated organics, including the data from Sleep and Ma (1997), there is little decrease in density with temperature between 20 and 90°C [68 and 194°F]. For the NAPLs listed in Figure 14.9, densities at 90°C [194°F] are within 92–95% of the densities at 20°C. Sleep and Ma (1997) maintain that this minimal decrease in density is unlikely to improve recovery due only to swelling of the NAPL. The decrease in viscosity is more substantial, and the viscosities of the NAPLs listed in Figure 14.10 at 100°C [212°F] are 56–69% of the viscosities at 25°C. This decrease in viscosity will aid recovery of continuous DNAPL by pumping. However, the increase in vapor pressure and Henry's Law constant for most chlorinated organics makes recovery of compounds from a gas phase the more favorable remediation mechanism.

14.2.3.2 Mobilization of Trapped or Retained DNAPL

In addition to decreases in the density and viscosity of the DNAPL, increases in temperature also will decrease the density and viscosity of water (Figures 14.9 and 14.10) and decrease the interfacial tension between water and the DNAPL (Figure 14.11). These effects can lead to mobilization of trapped DNAPL, particularly continuous DNAPL retained on capillary barriers. For DNAPL, downward mobilization through capillary barriers is a significant negative outcome of a remediation application. The maximum height of DNAPL that can be retained on a capillary barrier is directly proportional to the DNAPL–water interfacial tension and the cosine of the contact angle between the DNAPL–water interface and the solid grain surface and inversely proportional to the difference in density between the DNAPL and water (Kueper et al., 1993). Sleep and Ma (1997) showed that the interfacial tension of PCE, while expected to decrease with increasing temperature, does not decrease substantially. They present the following regression expression for the interfacial tension between laboratory grade PCE and water:

$$\sigma_{PCE-water} = 45.808 - 0.042T \qquad \text{(Eq. 14.14)}$$

where $\sigma_{PCE-water}$ is the interfacial tension between PCE and water (milliNewton per meter, mN/m) and T is in °C. This equation predicts a decrease in the interfacial tension between PCE and water from 45 to 42 mN/m between 20 and 90°C (Figure 14.11). Similar results were reported by Imhoff et al. (1997) who observed a decrease in the interfacial tension between PCE and water from 43 to 40 mN/m between 5 and 40°C (41 and 104°F). The surface tension (the interfacial tension between a liquid and its vapor) decreases with increasing temperature for most liquids (Adamson, 1990). The effects of temperature on the surface tension of PCE as well as several other chlorinated DNAPLs are shown in Figure 14.11.

The difference between the surface tensions of water and the NAPL can provide a rough estimate of NAPL–water interfacial tension (see Rowlinson and Widom (1982) for a discussion of Antonow's rule and its limitations). Figure 14.11 shows that this difference does not change substantially with increasing temperature, which is consistent with the results reported by Sleep and Ma (1997) for PCE. The effect of decreasing interfacial tension with increasing temperature on the maximum height of DNAPL that can be retained on a capillary barrier is offset by decrease in DNAPL density, especially for those DNAPLs whose densities decrease more substantially than does the density of water with increasing temperature (density difference between the DNAPL and water decreases with increasing temperature; Figure 14.9). In addition, the contact angle at most sites is expected to be small (water-wet conditions), and while the contact angle is likely to decrease with increasing temperature, this decrease is expected to be small (Sleep and Ma, 1997). As such, the maximum height of DNAPL that can be retained on a capillary barrier is likely not a strong function of temperature.

The mobilization of trapped DNAPL blobs depends on the balance of gravity, viscous and capillary forces. The balance between gravity and capillary forces is typically characterized by the bond number, and the balance between viscous and capillary forces by the capillary number. A total trapping number has also been developed, which includes all three forces (Pennell et al., 1996). Sleep and Ma (1997) stated that increases in temperature up to approximately 90°C (194°F) are unlikely to result in viscous mobilization of trapped PCE, as the viscosity of water decreases faster with increasing temperature than the interfacial tension (Figures 14.10 and 14.11), and the critical capillary number required for mobilization will not be reached.

14.3 DESCRIPTIONS OF THERMAL TECHNOLOGIES

The three most commonly evaluated thermal technologies in the context of DNAPL source zone remediation are SEE, TCH and ERH/ET-DSP™. These technologies vary according to the method of heating or energy delivery, the maximum temperature that can be attained, the means of contaminant recovery and the processes that are targeted for thermal enhancement. Other thermal technologies that are sometimes considered include steam injection in conjunction with soil mixing using large-diameter augers (LDA) and radio frequency heating (RFH). Technologies also may be combined in some cases to exploit the advantages of different approaches for different subsurface conditions (Heron et al., 20[illegible]). Vendors are available for implementation of these technologies throughout North America and several other regions around the world, and useful design guidance is available (USACE, 2009).

14.3.1 Steam-Enhanced Extraction

14.3.1.1 Technology Description

Steam injection was used in the petroleum industry for tertiary oil recovery as early as the 1930s (White and Moss, 1983). In that application, steam injection assisted in viscous oil recovery, primarily by reducing oil viscosity and increasing the driving pressure to cause more effective displacement of oil toward recovery wells. By 1979, steam injection was the principal enhanced oil recovery method, accounting for 90% of all oil produced by such methods (Ali and Meldau, 1979). The main improvement to the technology over the years has been a shift to steam flooding, with cyclic steam injection becoming an important operating strategy (Ali and Meldau, 1979).

As in the case of oil production, steam has been used to reduce DNAPL viscosity and alter capillary forces and to increase pressure gradients to effect increased recovery of relatively nonvolatile liquids, such as creosote and heavy oils, in remediation. It also has been applied where the primary goal is to vaporize and recover DNAPL via condensate and vapor capture systems, as is the case with volatile chlorinated solvents (such as PCE and TCE). Mechanisms contributing to contaminant removal during steam injection include vaporization of water and DNAPL, formation of a steam zone, and hydraulic displacement of condensed volatile organic compounds in front of the steam zone (Udell, 1996; NRC, 2005). Variants on steam injection technology are known by a range of names, including dynamic underground stripping (when used in combination with ERH), steam injection preceded by hot water injection, and steam injection combined with hot air injection. These technologies are grouped into a category referred to here as SEE.

Steam-enhanced extraction involves injecting steam under pressure into the subsurface and recovering liquids and vapors through a network of injection and extraction wells (Figure 14.12).

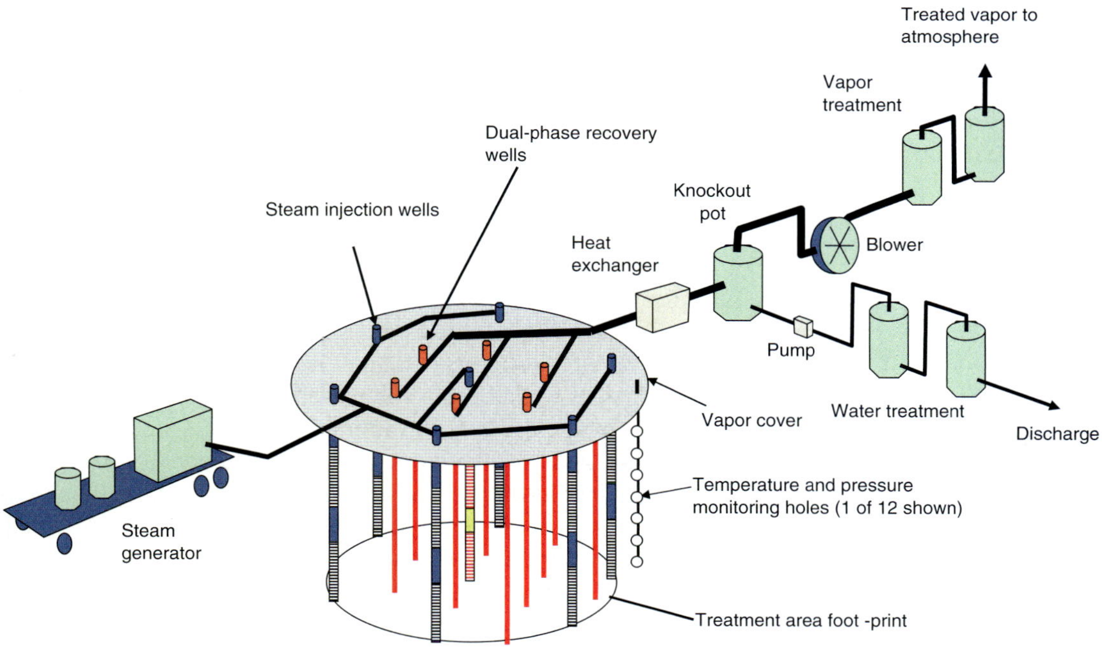

Figure 14.12. Steam-enhanced extraction process schematic (figure by Heron and Crisp as presented in Johnson et al., 2009).

In the early stages of the process, the injected steam displaces groundwater and DNAPL, as well as soil gas if injected above the water table, and heating occurs as energy is transferred from the steam to the formation. This energy transfer results in the formation of a steam and contaminant condensate front that moves out from the injection wells and is followed by a steam zone. The movement and geometry of the steam front are controlled by the steam injection rate, subsurface permeability to vapor and liquid flow, buoyancy effects, and the positioning of the injection and extraction wells. The steam injection pressure and rate may be increased and decreased in a cyclical fashion to promote steam migration into zones not actively swept by the initial application.

One potential risk associated with injecting steam is vertical mobilization of DNAPL into previously unoccupied areas. The accumulation of DNAPL at the steam front, resulting from condensation of vapors and coalescing of displaced DNAPL, is associated with an increase in capillary pressure. This increase in capillary pressure may cause exceedances of entry pressures and consequential vertical mobilization of the DNAPL. This mobilization will be exacerbated if the DNAPL–water interfacial tension is reduced at elevated temperatures because entry pressures are directly proportional to interfacial tension. Methods to reduce the risk of vertical DNAPL mobilization include injecting air along with the steam to reduce the rate of energy application (Kaslusky and Udell, 2002) and injecting steam beneath the target zone to catch any vertically mobilized DNAPL (referred to as creating a *hot floor*). However, full-scale applications of either air injection or hot floors to reduce the potential for vertical DNAPL mobilization have been relatively few.

Steam-enhanced extraction typically is used to heat soils to temperatures up to the boiling point of water, which varies with pressure (and therefore depth) when steam is injected below the water table. Energy delivery follows vapor (steam) and liquid (condensate) flow, so the initial heating occurs along the steam flow path, which is largely dictated by the structure and permeability distribution in the subsurface. Steam will migrate preferentially through the higher permeability portions of the subsurface, potentially bypassing some contaminated regions. Energy transfer into and heating of bypassed regions must then occur by conduction away from the primary steam flow path. Except in simple geology (such as a permeable zone bounded above and below by impermeable layers), migration of the steam front may be difficult to predict. However, steam temperatures are typically still able to be achieved in the target treatment zone.

14.3.1.2 State of the Practice

A generalized schematic of an SEE system is shown in Figure 14.12. Steam injection and vapor/liquid extraction wells are typically configured either in a square pattern (referred to as a 5 spot) or in a triangular pattern (7 spot). The adopted configuration need not be regular, however, since SEE is a fluid-delivery-based process without electrical phasing considerations. Extraction wells can be either vertical wells within the pattern or horizontal or angled wells located in optimized positions to capture the heated vapors and liquids. Steam injection wells are typically screened to the base of the treatment zone or slightly deeper to compensate for steam rise into the target treatment zone. Extraction wells are often screened fully across the target treatment zone, allowing for both liquid and vapor recovery from the same well. Some wells may be screened deeper if it is desirable to dewater the target treatment zone, or it is necessary to reduce groundwater flow to reduce thermal losses.

Steam injection well spacing is dictated by two primary considerations: the need to create steam conditions throughout the target treatment zone and the desire to deliver a target volume of steam. Typical well spacings range from approximately 20 to 50 ft (6 to 16 m), with closer spacing employed where upward steam migration limits the lateral steam distribution (relatively homogenous and permeable settings with shallow target treatment zones) and larger spacings employed where the stratigraphy facilitates lateral steam front migration. Practitioners specify a volume of steam to be delivered to the treatment zone that is typically expressed as equivalent liquid water pore volumes (one steam condensate pore volume is the volume of water it would take to fill the soil pores of the target treatment zone). Typically, between one and three pore volumes of steam are injected, with less volume required for more volatile compounds such as chlorinated solvents. However, for creosote and heavy soils, if remedial goals are stringent, up to 20 pore volumes (typically 8 pore volumes) may need to be injected.

In a typical SEE application, a constant rate of steam is injected into the treatment zone initially. This phase is continued until the desired steam zone is achieved, and then cyclical (increasing and decreasing the delivery rate) or pulsed (repeated on-off) steam injection is initiated. This phase is continued until the desired cumulative pore volumes of steam are delivered to the treatment zone. In the event that groundwater flow into the treatment zone cannot be controlled, steam injection wells upgradient of the treatment zone can be operated continuously to form a steam barrier against the influx of groundwater, while other wells within the treatment zone are cycled or pulsed. Throughout the application, temperature and pressure monitoring are conducted to evaluate steam migration. Temperature- and pressure-monitoring wells are simple vertical borings used to document performance and pneumatic control during treatment. These monitoring wells are located inside and outside the treatment area, typically at different distances from the operational wells to illustrate the progression of the SEE process in the subsurface.

The aboveground equipment varies from site to site depending on treatment area size, volume, nature of contamination, and local regulatory requirements for treating the effluents. Since the extracted fluids can include water, DNAPL, air and steam, at varying temperatures and pressures, the treatment system must be a robust combination of cooling, separation and treatment units previously proven to be effective for their functions. Typical aboveground components are shown in Figure 14.12. Effluent gases are cooled to promote liquid condensation before vapor treatment, and conventional vapor and water treatment technologies are used. The heat exchanger/condenser reduces the temperature of the extracted vapors to remove steam and increase the efficiency of the water and vapor separator. The vapor treatment system might be as simple as a granular activated carbon (GAC) system and a vacuum blower. Other vapor treatment options include catalytic or thermal oxidation. Condensate treatment is typically completed by water–NAPL separation, filtering, liquid phase GAC filtration or air stripping.

14.3.2 Thermal Conductive Heating

14.3.2.1 Technology Description

Thermal conductive heating involves the simultaneous application of heat and vacuum to the subsurface; TCH is currently a patented technology of TerraTherm that was originally developed more than 15 years ago by Shell Exploration and Production (Shell E&P), a division of Royal Dutch Shell (Shell), to accomplish enhanced oil recovery (Johnson et al., 2009). From 1994 to 1996, Shell Technology Ventures, Inc., a wholly owned subsidiary of Shell E&P that held the TCH patents, conducted several TCH demonstrations and established TerraTherm Environmental Services, Inc., to be a stand-alone remediation company based in Houston, TX, USA, offering TCH services to both the public and private sectors. Shell subsequently elected to exit the environmental cleanup business and, in January of 2000, donated the TCH rights within the

Figure 14.13. Proprietary TerraTherm heating element deployed inside each TCH heater boring. The metal rod has a diameter of ca. 0.5 in. The white beads are ceramic isolators. Electrical power flows through the steel rod, causing it to heat resistively. See U.S. Patent Nos 5,190,405 (Vinegar et al., 1993), 5,318,116 (Vinegar et al., 1994), 6,485,232 (Vinegar and Stegmeier, 2002) and 6,632,047 (Vinegar and Stegmeier, 2003). Photograph provided courtesy of TerraTherm Inc.

USA to the University of Texas at Austin (UT). A new company, TerraTherm, Inc. (TerraTherm), secured the exclusive license from UT to commercialize TCH within the USA. In 2002, Shell granted TerraTherm exclusive rights to commercialize the TCH technology outside the USA (Johnson et al., 2009).

During TCH, heat is applied to the subsurface using simple electrical heaters (Figure 14.13) installed inside a casing in contact with the soil, so that radiation and conductive heat transfer are effective near the heater (Stegemeier and Vinegar, 2001; Udell, 1996). As a result, thermal conduction and convection occur in the bulk of the soil volume. As heating progresses, the heater wells are heated to temperatures of approximately 500–800°C (930° to 1,470°F), creating significant temperature gradients in the formation around each well. Since the spatial distribution of thermal conductivity in the subsurface is much less variable than the spatial distribution of permeability, the delivery of heat by TCH is not as susceptible to geologic heterogeneity as the delivery of steam. A unique feature of conductive heating approaches, relative to other thermal technologies, is the potential to achieve subsurface temperatures well above the boiling point of water (100°C [212°F] at 1 atm pressure), provided that all of the moisture can be boiled away from the treatment zone.

As the heat front moves away from the heaters through the soil by thermal conduction and convection, the superposition of heat from the many heaters results in a temperature rise throughout the target treatment zone. As soil temperatures increase, contaminants and water contained in the soil matrix are vaporized. While locations near heaters may achieve temperatures well above the boiling point of water, locations between heaters need only achieve temperatures near the boiling point of water to accomplish steam distillation of common chlorinated VOCs.

The vacuum that is applied to the vapor extraction points from the process system draws the vapors through the soils and into the off-gas piping network for treatment (Figure 14.14). For chlorinated solvents such as TCE and PCE, vaporization is the most important physical removal and remediation mechanism. In addition to the physical removal described above, biological and chemical degradation processes may occur during and after thermal remediation. These processes may include thermal destruction by oxidation and pyrolysis near heating elements at temperatures around 400°C (750°F), microbial mineralization of NAPL components, and hydrolysis at elevated temperature (Baker and Kuhlman, 2002).

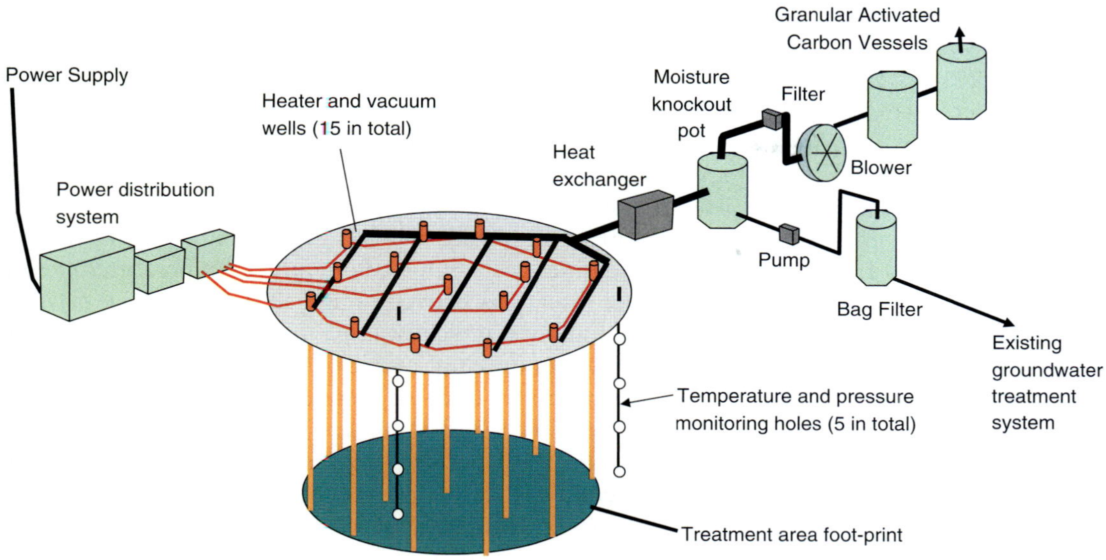

Figure 14.14. Schematic representation of TCH system components (Lebrón et al., 2012).

The advantages of TCH compared to certain other DNAPL remediation technologies include the following:

- Predictable heating due to simplicity of the conductive heating approach, providing relatively uniform heat distribution and treatment
- The ability to achieve low VOC concentrations in soil (single-digit mg/kg levels or lower; Heron et al., 2008)
- Flexibility with respect to the depth of application (Johnson et al., 2009)

The potential limitations of TCH are listed below:

- An energy demand on the order of 120–300 kWh per cubic yard treated, corresponding to an energy cost of approximately \$10–30 per cubic yard, and potentially significant carbon dioxide emissions and other environmental impacts (Lemming et al., 2010)
- The need for intensive invasive drilling and on-site construction activities, which may disrupt site activities temporarily
- Sensitivity to groundwater flow and associated cooling in high-permeability zones that can slow heating and prevent reaching target temperatures (Johnson et al., 2009)

14.3.2.2 State of the Practice

Over the past 5 years, TCH has undergone a number of improvements in technology (Kingston et al., 2010). Heater wells have become simpler, less expensive and more able to resist corrosive conditions. They are now amenable to installation by most available drilling methods, with installation rates in the range of 200–400 ft per day (~60–120 m/d) per rig. Control systems have become simplified, and off-gas treatment can be accomplished by a wider array of components, with the choice depending on project requirements. Thermal conductive heating has been applied over a wide range of time durations, and the energy requirements for a range of subsurface conditions are well understood. Vapor and liquid extraction wells can be either vertical wells within the pattern or horizontal/angled wells located in optimized positions to capture the heated fluids. To avoid overheating of wells and heater materials, the heater element power input is typically limited to below 400 watts (W)/ft of heater (<120 W/m).

Heat losses stem from conduction of heat outside the treatment zone. Accordingly, TCH heaters typically extend between 2 and 5 ft (0.6 and 1.6 m) outside the target treatment zone to ensure heating of the entire volume to the target temperature. These heat losses are inevitable and part of any heating technology, and care must be taken to treat the edges, including top and bottom, of the target volume. The spacing of TCH heater wells varies based on the type of contaminant and formation. For treatment of VOCs, the heater spacing typically ranges between 12 and 20 ft (3.7 and 6 m). For treatment of SVOCs where target temperatures exceed the boiling point of water, a typical heater spacing is between 7 and 12 ft (2.1 and 3.6 m).

Thermal conductive heating applications are typically implemented in a single phase. Temperatures between heaters are typically monitored at centroid locations. Power applied to the heaters can be varied based on the observed rates of temperature rise to facilitate a relatively uniform rate of heating throughout the target volume. Throughout the application, temperature and pressure monitoring must be conducted to track the formation and location of vapors to ensure that adequate containment and extraction are occurring. Process monitoring typically focuses on power delivery to the heaters, verification of heating, verification of contaminant containment, and both instantaneous and cumulative mass removal rate.

14.3.3 Electrical Resistance Heating

14.3.3.1 Technology Description

Electrical resistance heating (ERH) was originally developed as an enhanced oil recovery technique (Harvey and Govier, 1980; Wattenbarger and McDougal, 1988). Electrical heating of reservoir formations employing alternating current (AC) was field tested as early as 1969 for enhanced oil recovery (Pizarro and Trevisan, 1990), and a number of variants of the process were patented in the 1970s (Gill, 1970; Hagedorn, 1976). In the early 1990s, Pacific Northwest National Laboratory (PNNL) developed ERH for soil remediation with funding provided by the US Department of Energy's Office of Science and Technology (Buettner and Daily, 1995). The technology was originally developed as a dewatering tool for PNNL's *in situ* vitrification process, but soon afterward it was proposed as a method for remediation of subsurface contamination (Buettner et al., 1999; Gauglitz et al., 1994; McGee et al., 1998). Electrical resistance heating became commercially available for use as a stand-alone technology in 1997 (Johnson et al., 2009).

As shown schematically in Figure 14.15, ERH heats the soil by passing electrical current between a network of electrodes. The resistance of the soil allows *in situ* heating to occur. The lateral and vertical positioning of the electrodes, the voltage difference applied between electrodes, and the electrical resistance of the soil determine the electrical current strength and path and ultimately the energy delivery and heating pattern in the subsurface. The power source is often low-frequency (60 Hz) AC electrical power (Wattenbarger and McDougal, 1988) with a typical voltage of 150–450 volts (V) (Beyke et al., 2007).

The electrical current is delivered throughout the subsurface using electrodes installed in hexagonal or triangular array patterns. The electrodes can be vertical, angled, or horizontal and can be installed by traditional subsurface drilling and installation techniques. Three- or six-phase electricity is applied. Three-phase heating is usually used in full-scale systems and six-phase in pilot tests (Beyke and Fleming, 2005; USACE, 2006). The subsurface electrical resistance varies by lithology; for example, low-permeability silts and clay lenses have higher electrical conductivity and are thus heated preferentially (Beyke and Fleming, 2005).

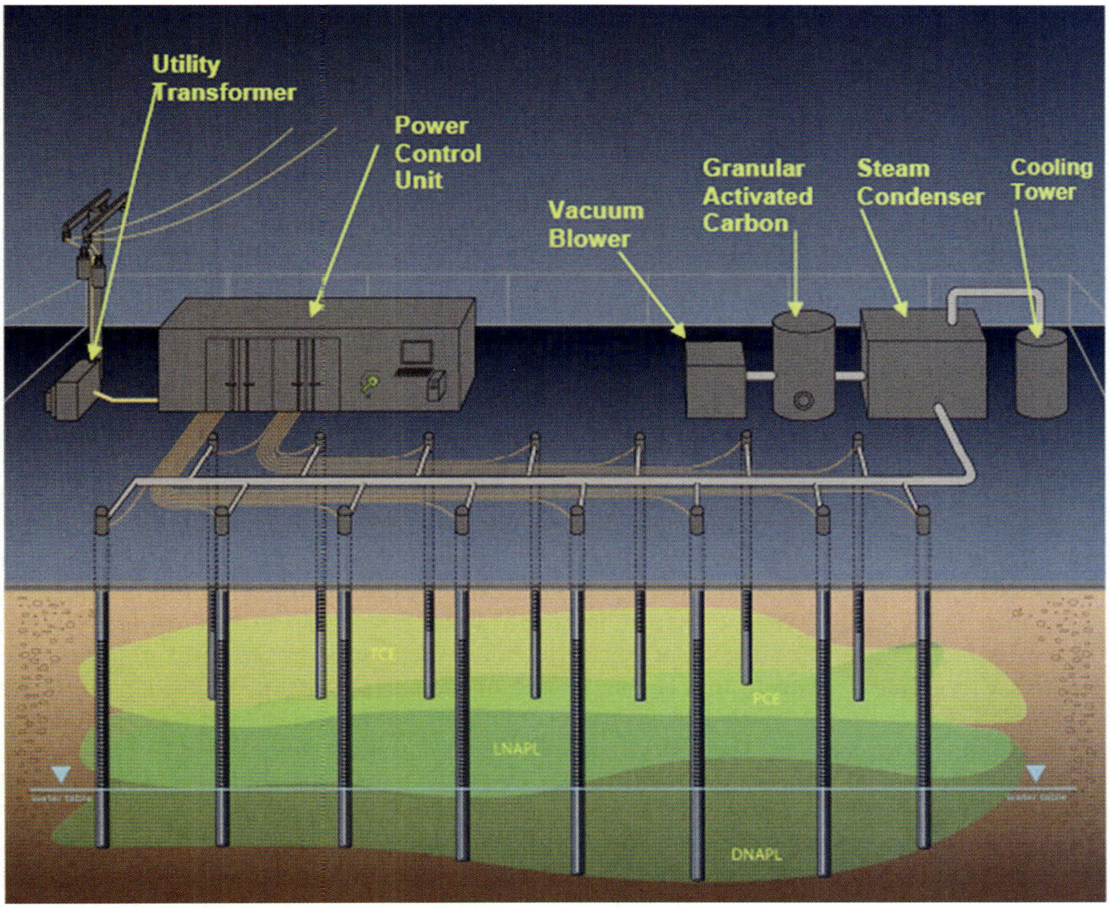

Figure 14.15. Conceptual depiction of an ERH application (from Johnson et al., 2009).

Electrical resistance heating is limited to raising the temperature of the subsurface to the boiling point of water (100°C [212°F] at 1 atm pressure). Soil moisture is converted to steam, and this steam helps drive volatilized chemicals from their point of volatilization to more permeable vapor flow zones, where they can be captured by soil vapor extraction (SVE) systems.

The primary factors that can limit the effectiveness of heating using ERH are moisture loss and the cooling influence of groundwater flow. The former is of concern because electrical resistance increases (electrical conductivity decreases) as soil moisture decreases, thereby resulting in reduced energy delivery to drier portions of the target treatment zone. Electrical resistance heating systems therefore often incorporate wetting systems around the electrodes and in the targeted treatment area to maintain a specified amount of soil moisture.

14.3.3.2 State of the Practice

Technological improvements have been made in equipment and mode of application of ERH over the past 5–8 years. The modifications to the mode of application have incorporated physical, chemical and biological processes that have been observed to occur during ERH. Improvements made to the equipment include simplification of power control units (PCUs), improvements in electrode design, and modification of water drip systems to maintain soil moisture around electrodes. Improvements have been made to the efficiency of operations, in both installation and energy, but also from an operational focus. More maintenance-friendly condensers are now being used to control costs and improve efficiency. Various electrode designs have been developed over the years for a variety of applications. Vapor and water extraction wells either are typically co-located with the electrode or are placed at the centroid of the hexagonal or triangle pattern.

Electrical resistance heating electrodes are typically constructed using either steel pipe or copper plate, such that multiple electrodes can be installed within the same boring and can be designed to allow independent control of the energy input to discrete depth intervals. For some applications, sheet piling has been used as electrodes. In the electrically conductive intervals, the surrounding borehole annulus is packed with a conductive material, such as graphite and/or steel shot, to increase the effective diameter of the electrode. All other portions of the borehole are insulated, and the surrounding annulus is filled with relatively nonelectrically conductive materials such as sand or cement. Based on the current state of the ERH technology and experience, the practical minimum thickness of a target heated zone is 8 ft (2.5 m) considering electrical fanning and thermal conduction (Johnson et al., 2009).

The ET-DSP™, an adaptation of ERH, uses electrodes that are up to 12 in. (30.5 cm) in diameter and up to 10 ft (3.05 m) long. These electrodes are rated for up to 180°C (356°F) at more than 50 kilowatts (kW) and have an imbedded water circulation system (cooling system) within each electrode (McMillan-McGee, 2011). These electrodes are installed in a borehole packed with granular graphite that is compressed to the surface of the electrode. The water circulation system for the ET-DSP™ electrodes is integral to the operation of the electrode via the heat transfer mechanism. The system injects water at the ends of the electrode. The water is then heated to steam temperatures and transported into the target treatment zone. Typical injection rates are 0.1–0.2 gallons per minute (gpm) [0.4–0.8 L/min] per electrode.

Electrodes are typically spaced 15–23 ft (4.6–7 m) apart. The spacing is dependent upon the characteristics of the contaminants to be treated, the desired rate of heating, the expected heat losses, the construction of the electrodes, and the desired final temperature to be achieved. Electrodes typically extend 2–3 ft (0.6–1 m) below the target treatment zone and 2–5 ft (0.6–1.6 m) outside the target treatment volume, similar to TCH, to ensure heating of the entire volume to the target temperature. In cases where treatment is targeted below the water table, groundwater flow carries warm water and energy from the treatment zone and replaces it with cooler water. This energy sink can sometimes be significant and can limit the heating of the target treatment zone. The cooling influence of inflowing cold groundwater sometimes can be reduced by using physical barriers, groundwater extraction and/or groundwater injection wells.

Temperature monitoring points are located throughout the treatment area and typically are located equidistant between groups of electrodes to monitor temperatures at the furthest distance from the energy application point. Each temperature monitoring point consists of a string of thermocouples, typically set at 5 ft (1.6 m) depths. Process monitoring typically focuses on power delivery to electrodes, verification of heating, verification of contaminant containment, and instantaneous and cumulative mass removal rate. Aboveground equipment varies from site to site depending on treatment area size, volume, nature of contamination, and local regulatory requirements for treating the effluents. The aboveground vapor treatment systems are essentially the same for all thermal technologies, as are the monitoring requirements.

14.3.4 Large-Diameter Auger Mixing with Steam Injection

14.3.4.1 Technology Description

In situ soil mixing with steam and hot air injection has been in use since the early 1990s but has changed and been modified throughout the years. The initial version of the technology functioned by using a pair of hollow stem augers to move vertically through the soil, breaking up the soil and facilitating the flow of gases through the subsurface (de Percin, 1991). Steam and compressed air are piped through the augers to nozzles located on the cutter blades. The heat from the injected steam and hot air vaporizes the VOCs, and the gas flow carries the contaminants to ground surface where a shroud covers the treatment area. The shroud is kept under vacuum to assist the flow of gases from the soil and to ensure against leakage to the outside environment (de Percin, 1991). The off-gases are conveyed to a treatment system.

The technology evolved in the early 2000s to the use of a single 6–10 ft (1.8–3 m) large-diameter auger (LDA) which can be used to mix soil with hot air and steam or with other additives such as iron powder and extend to depths of over 70 ft (21 m). Steam and hot air are pumped through the auger and mixed into the soil. The augers act to break up the soil structure and increase permeability. The injected steam and hot air heat the soil to temperatures approaching the boiling point of water (100°C, 212°F). The shroud at the top is under vacuum to extract all the vapors. The process is operated in a batch mode, with movement of the auger across the treatment zone in an overlapping treatment pattern.

14.3.4.2 State of the Practice

The major improvement to the LDA technology over the last 5–10 years has been combining steam/hot air injection with injections of zero valent iron (ZVI) powder in a water/guar slurry. The ZVI is intended to continue the remediation after the thermal treatment has stopped. The technology operates as a batch process with each cell being remediated separately. Once the cell is remediated to preselected real-time vapor sampling criteria as set by the goals of each individual project, the auger is removed from the cell and set up over the next cell. The cells are laid out to overlap to ensure 100% areal coverage. The overlaps are typically on the order of 3–4 ft (roughly 1 m), depending on the size of the auger.

During the initial drilling of each LDA cell, penetration rates of approximately 0.5 ft per minute (0.15 m/min) or less can be achieved. The penetration rate controls the heating rate. After the initial pass, the drilling rates can increase to 1 ft/min (0.3 m/min). The drilling passes will continue until the preselected vapor criterion is achieved. The subsurface target temperature is about 170°F (76°C). This desired target temperature is high enough to increase the vapor pressure of most VOCs enough to ensure high removal rates. Heating a column of soil to this temperature typically occurs in less than 1 h.

Process monitoring typically focuses solely on instantaneous and cumulative mass removal rate of contaminant and water. The aboveground equipment varies from site to site depending on treatment auger diameter and local regulatory requirements for treating the effluents. The aboveground vapor treatment systems are essentially the same for all thermal technologies, but the monitoring requirements for LDA mixing are significantly different. For LDA mixing with steam injection, typical monitoring includes the depth of auger progress, steam flow rates, air flow rates, off-gas flow rates, off-gas temperatures and real-time gas chromatography.

14.3.5 Radio Frequency Heating

14.3.5.1 Technology Description

Radio frequency heating (RFH) has existed since the 1930s. It is commonly divided into dielectric heating and induction heating with applications ranging from medical diathermy of muscles and joints to industrial dielectric heating techniques used to dry wood and cure adhesives. In the mid-1970s, RFH was under development for the recovery of hydrocarbons by heating large volumes of earth (Dev et al., 1984).

Radio frequency (RF) refers to frequencies used in wireless communication with a wide frequency range from 45 hertz (Hz) to 10 gigahertz (GHz), but the frequencies of interest for *in situ* soil remediation are between 2 and 45 megahertz (MHz) (Dev et al., 1987). Radio frequency technologies deliver electromagnetic energy in the RF band to nonconductive materials through carefully controlled RF transmissions. The RF energy is made by converting standard AC electricity by an RF generator.

The RF energy is then conveyed into the soil by one or more antennae (USEPA, 1995) with heat transfer occurring at the molecular level. Radio frequency heating occurs by ohmic and dielectric mechanisms. Ohmic heating results from the resistance a medium offers to the flow of electrical current or from a voltage drop, pushing electrons up into the conduction band and moving them through the soil mass, producing resistance heating. Dielectric heating is the result of distortion of atomic or molecular structure in response to an applied electrical field. The dipole moments of the molecules in the polar substance have random orientation; thus, adding the external electrical field will cause the dipole moments to begin to align (Smith and Hinchee, 1993). This alignment is used to determine the best frequency for each substance. The technology can be used to produce temperatures ranging from 100 to 400°C (Dev et al., 1987).

During RF heating, water and contaminants are volatilized and converted to vapor. This conversion to steam significantly enhances the extent of the RF energy's penetration from the borehole because of the attendant reduction in the material dielectric losses where steam is produced (Kasevich et al., 1996). Steam does not absorb RF energy, while water does, so the diameter of the heating zone expands rapidly. A typical RFH applicator does not produce uniform heating with the treated zone, but a system employing an array of four applicators spaced 3–6 m (9.8–19.7 ft) apart can achieve near uniform heating (Kasevich et al., 1996). The heating of the applicators is elliptical and radially outward.

Radio frequency heating applications have not been widely used over the past 10 years. In fact only three known applications have occurred over the past 10 years; two were European applications. The state of the practice of this technology will not be discussed based on this knowledge.

14.4 SYNTHESIS OF AVAILABLE DATA

14.4.1 Data Summary

The performance of thermal technologies was assessed through compilation and critical review of data available from pilot- and full-scale applications conducted from 1988 to 2007 (Johnson et al., 2009). Particular emphasis was placed on gaining a better understanding of settings in which thermal technologies have been applied, the design and operating conditions that were used, and the performance of the systems. The following information is available in more detail in Johnson et al. (2009). Although the focus of this chapter is on the remediation of chlorinated solvent sources, the summary information presented here also considers petroleum and other types of contaminants in an effort to put chlorinated solvent applications into perspective.

Table 14.3. Summary of Technology Applications by Technology Type (Kingston et al., 2010)

Technology	Number of Applications	Pilot Scale[a]	Full Scale[a]	Number Since Year 2000
Steam-based heating	46	26	19	15
Electrical resistance heating	87	23	56	48
Thermal conductive heating	26	12	14	17
Others (including mixing/heating)	23	14	9	4
Total	182	75	98	84

[a]Some sites have an unknown application size and thus are not included in the pilot- and full-scale count

The *in situ* thermal treatment application data were obtained from a variety of sources including site reports; published literature; US Environmental Protection Agency (USEPA) cost and performance reports; discussions with project managers, vendors and consultants; and unpublished data and observations. Sites for which data were collected encompassed *in situ* thermal technology applications worldwide and included ERH, SEE with and without hot water or hot air injection, TCH, RFH and LDA mixing with steam.

For each technology application studied, emphasis was placed on identifying:

- The setting (geology, depth to groundwater, source zone boundaries, chemicals present)
- System design parameters (number of energy delivery points, area and depth of the treatment zone)
- Operating conditions (temperature achieved, duration of treatment, duration of monitoring)
- Performance data (emphasizing improvement in groundwater quality and reduction in mass discharge of contaminant to the aquifer)

Capture of this data involved data interpretation and the use of professional judgment, especially when comparing pre- and post-treatment groundwater impacts.

A total of 182 *in situ* thermal treatment technology applications at 163 different sites were identified in this study. Table 14.3 presents the number of *in situ* thermal applications by technology. It also indicates how many were full-scale versus pilot-scale applications and how many occurred since 2000. As can be seen, approximately 98 of 182 applications were implemented at full scale, with roughly half (56) of those being ERH systems. Table 14.3 also shows that 84 of 182 applications have been implemented since 2000, more than half (57%) of which were ERH systems.

Table 14.4 summarizes key features of *in situ* thermal technology applications conducted since 2000 to demonstrate the current state of the practice. Note that the number of applications totaled in each column may not total 84 due to the fact that the information might not be available for all 84 applications.

Table 14.4 shows that the majority of the thermal applications were carried out in generalized scenarios B and C. Scenario B (low-permeability with high-permeability lenses) accounts for 36 of 84 of thermal treatments. Most TCH and ERH applications occurred in scenario B settings. Scenario C (high-permeability with low-permeability lenses) settings account for roughly another one-third of all applications. The majority of applications in scenario C settings are ERH, although steam heating had six (of 15) of its applications within

Table 14.4. Summary of the State of the Practice for Thermal Technologies

Generalized Conceptual Scenario		Technology	Total Sites[a]		Chemical(s) Treated[a]				Design Parameters[a]					
			# of Sites		# of Sites Treating These Chemicals (C-Chlorinated Solvents, P-Petroleum Hydrocarbons, W-Wood Treating, O-Others)				# of Sites w/Target Treatment Zones With Sizes in This Range [ft²]			# of Sites w/Density of Energy Delivery Points (Electrodes or Wells) in This Range [per 100ft²]		
			Pilot Tests	Full scale	C	P	W	O	$< 10^4$	$10^4 - 4 \times 10^4$	$> 4 \times 10^4$	< 0.25	$0.25 - 0.50$	> 0.5
Generalized scenario A: relatively homogeneous and permeable unconsolidated sediments (mixtures of sands, gravels, silts, etc.)	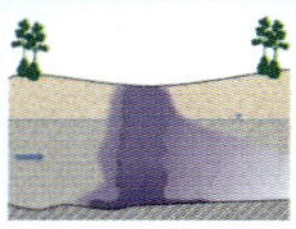	SEE	1	0	0	1	0	0	1	0	0	1	0	0
		ERH	0	1	0	1	0	1	1	1	0	0	1	1
		TCH	1	4	4	3	0	1	[illegible]	1	0	0	0	3
		Others	0	0	0	0	0	0	0	0	0	0	0	0
Generalized scenario B: largely impermeable sediments with interbedded layers of higher-permeable material	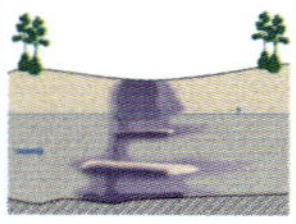	SEE	1	1	2	0	0	0	1	0	0	1	0	0
		ERH	6	18	16	9	1	1	14	7	0	3	8	9
		TCH	3	7	7	1	2	2	7	3	0	0	1	9
		Other	0	0	0	0	0	0	0	0	0	0	0	0
Generalized scenario C: largely permeable sediments with interbedded lenses of low-permeable material	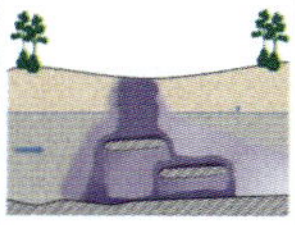	SEE	3	3	3	2	1	1	[illegible]	2	1	4	0	0
		ERH	2	13	14	5	0	0	[illegible]	9	0	1	4	8
		TCH	0	0	0	0	0	0	0	0	0	0	0	0
		Other	1	2	3	3	0	0	1	2	0	0	0	0
Generalized scenario D: competent, but fractured bedrock	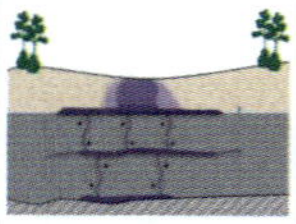	SEE	2	1	3	1	0	0	[illegible]	0	0	1	0	1
		ERH	0	0	0	0	0	0	0	0	0	0	0	0
		TCH	0	0	0	0	0	0	0	0	0	0	0	0
		Other	0	1	1	0	0	0	0	0	0	0	0	0
Generalized scenario E: weathered bedrock	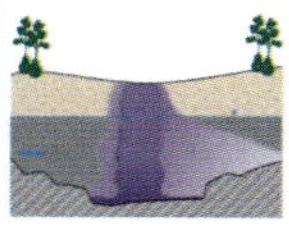	SEE	0	0	0	0	0	0	0	0	0	0	0	0
		ERH	0	0	0	0	0	0	0	0	0	0	0	0
		TCH	1	0	1	1	0	0	1	0	0	0	0	1
		Other	0	0	0	0	0	0	0	0	0	0	0	0
Unknown scenario		SEE	1	2	2	1	1	0	0	0	0	0	0	0
		ERH	4	4	7	2	0	0	1	3	0	1	2	0
		TCH	1	0	0	0	0	0	1	0	0	0	1	0
		Other	0	0	0	0	0	0	0	0	0	0	0	0

Notes: *AFB*, Air Force Base; *AFP*, Air Force Plant; *AFS*, Air Force Station; *LLNL*, Lawrence Livermore National Laboratory

[a]Limited to information from sites with systems installed since 2000 with designs reasonably similar to state-of-the-practice designs

Operating Parameters[a]									Performance Measures[a]												
# of Sites w/Temperatures in Target Treatment Zone in These Ranges [°C]			# of Sites w/Active Heating Durations in These Ranges [yr]			# of Sites w/Post-treatment Monitoring in These Ranges [yr]			Sites w/Final Dissolved Concentrations Generally in This Range [μg/L]			Sites w/Concentration Reduction in This Range [%]			Sites w/Final Mass Discharges Generally in This Range [kg/yr]			Sites w/Mass Discharge Reduction in this Range [%]			
< 80	80 – 110	> 110	< 0.5	0.5 – 1.0	> 1.0	< 0.5	0.5 – 2.0	> 2.0	< 10	10 – 100	100 – 1,000	10×	100×	1,000×	< 1	1.0 – 10	> 10	10×	100×	1,000×	Name(s) of Example Well-Studied Sites
0	0	0	0	1	0	0	0	0	0	0	1	1	0	0	0	0	0	0	0	0	Guadalupe
0	3	0	1	0	0	0	0	1	0	0	1	1	0	0	1	0	0	0	1	0	Hunter Army Airfield
0	2	2	1	2	0	0	1	0	0	0	1	0	0	1	0	0	0	0	0	0	
0	0	0	0	0	0	0	0	0	0	0	0	0	0	0	0	0	0	0	0	0	
0	1	0	1	0	1	1	0	0	0	1	0	1	0	0	0	0	0	0	0	0	
3	10	0	11	7	2	1	0	0	1	1	5	2	2	3	0	0	1	1	0	0	AFP 4
0	7	2	1	5	2	0	0	0	0	0	2	0	2	0	0	0	0	0	0	0	Alhambra Pole Yard
0	0	0	0	0	0	0	0	0	0	0	0	0	0	0	0	0	0	0	0	0	
2	2	0	2	0	3	0	0	0	0	0	1	1	0	0	0	0	0	0	0	0	Visalia, LLNL
1	11	0	8	4	1	0	2	0	0	0	9	7	1	1	2	4	2	6	2	0	NAS Alameda Site 5-1 ERH, Young Rainey Star, Fort Lewis Areas 1, 2, 3
0	0	0	0	0	0	0	0	0	0	0	0	0	0	0	0	0	0	0	0	0	
0	0	0	2	0	0	0	0	0	0	2	0	0	0	2	0	0	0	0	0	0	Cape Canaveral AFS
1	1	0	3	0	0	1	0	0	0	0	2	2	0	0	1	0	0	1	0	0	Edwards AFB, Loring AFB
0	0	0	0	0	0	0	0	0	0	0	0	0	0	0	0	0	0	0	0	0	
0	0	0	0	0	0	0	0	0	0	0	0	0	0	0	0	0	0	0	0	0	
0	0	0	0	0	0	0	0	0	0	0	0	0	0	0	0	0	0	0	0	0	
0	0	0	0	0	0	0	0	0	0	0	0	0	0	0	0	0	0	0	0	0	
0	0	0	0	0	0	0	0	0	0	0	0	0	0	0	0	0	0	0	0	0	
0	1	0	1	0	0	0	0	0	0	0	0	0	0	0	0	0	0	0	0	0	
0	0	0	0	0	0	0	0	0	0	0	0	0	0	0	0	0	0	0	0	0	
1	0	0	0	0	0	0	0	0	0	0	0	0	0	0	0	0	0	0	0	0	
1	0	0	0	0	0	1	0	0	0	0	0	0	0	0	0	0	0	0	0	0	
0	0	0	0	0	0	0	0	0	0	0	0	0	0	0	0	0	0	0	0	0	
0	0	0	0	0	0	0	0	0	0	0	0	0	0	0	0	0	0	0	0	0	

Table 14.5. Aggregate Design Information of Thermal Technologies

Technology	Number of Sites with Target Treatment Zones with Sizes in This Range [ft^2]				Number of Sites with Density of Energy Delivery Points (Electrodes or Wells) in This Range [# per 100 ft^2]			
	$<10^4$	10^4–4×10^4	$>4 \times 10^4$	Unknown	<0.25	0.25–0.50	>0.5	Unknown
Steam-based heating	16	6	4	20	20	2	4	20
Electrical resistance heating	36	24	0	27	10	23	27	27
Thermal conductive heating	19	6	0	1	1	1	23	1
Others (including mixing/heating)[a]	8	2	0	13	2	0	8	13

[a]For three steam auger sites, the density is one energy point per cell. This density does not fit into the number calculation so it is classified as >0.5

this geologic setting. Few applications in generalized scenarios A, D and E were identified (7, 4 and 1 of 84 total documented applications, respectively). This result may reflect the low frequency of occurrence of homogeneous settings in nature (scenario A) as well as the difficulty and risks in dealing with complex fractured and bedrock settings.

Table 14.4 also summarizes information available on the chemicals present at 83 of 84 sites. Of those 83 sites, chlorinated solvents were treated at 63 of the sites. Petroleum hydrocarbons were the other main contaminant category treated by thermal applications and represent about 35% of sites in this study. Wood treating and other chemicals accounted for about 7% of sites.

Table 14.4 categorizes the number of energy delivery points per 100 ft^2 (~per 10 m^2) and indicates that most SEE designs had densities of less than one energy delivery point per 400 ft^2 (about 1 per 40 m^2 or greater than 20 ft [6 m] spacings), while most TCH applications involved densities greater than one energy delivery point per 200 ft^2 (about 1 per 20 m^2 or less than 14 ft [4.2 m] spacings). Electrical resistance heating applications spanned the range of density categories but were weighted more toward higher densities and electrode spacings less than 20 ft (6 m).

Other basic application data was synthesized using all available data which include aggregate design information and basic operating conditions. Aggregate design information for all applications reviewed is summarized in Table 14.5. Of the 121 applications for which data were available, 117 involved treating areas $<4 \times 10^4$ ft^2 ($<3{,}700$ m^2), and roughly two-thirds of those involved treatment zones smaller than 10^4 ft^2 (ca. 900 m^2).

Table 14.6 summarizes the basic operating conditions for all of the applications reviewed. Of the 95 applications for which temperature data were available, 63 were operated at temperatures in the 80–110°C range in the target treatment zone. With respect to technology, 37 of the ERH applications were operated within that 80–110°C range while seven of the steam applications were operated at temperatures less than 80°C, and about half of the conductive heating applications were operated at temperatures greater than 110°C.

Of note in Table 14.6 are the durations of application. Of the applications for which data were available, 81 of 84 were operated for less than 6 months, and this pattern is true for all thermal technologies. Also in Table 14.6 is information on post-treatment groundwater monitoring ranges. Of the 182 sites, only 14 (8%) have known information on post-treatment groundwater monitoring. One-half of the 14 sites were monitored for fewer than 6 months, while only one site was monitored for more than 2 years. With respect to soil sampling, very little data were provided for review.

Table 14.6. Basic Operating Conditions of Thermal Technologies

Technology	Number of Sites with Temperatures in Target Treatment Zone in These Ranges [C]				Number of Sites with Active Heating Durations in These Ranges [y]				Number of Sites with Post-treatment Monitoring in These Ranges [y]			
	<80	80 - 110	>110	Unknown	<0.5	0.5 - 1.0	>1.0	Unknown	<0.5	0.5 - 2.0	>2.0	Unknown
Steam-based heating	7	13	1	25	14	0	3	29	2	0	0	44
Electrical resistance heating	9	37	0	41	38	2	0	47	1	5	1	80
Thermal conductive heating	0	11[a]	12[a]	4	18	3	0	5	1	1	0	24
Others (including mixing/ heating)	2	2	1	18	6	0	0	17	3	0	0	20

[a]One site had two different temperature values. The 80–110 °C temperature was for the saturated zone and the >110 °C temperature for the vadose zone

Table 14.7. Mass Discharge Reduction for State-of-the-Practice Thermal Applications

Site No.	Heating Technology	Generalized Scenario/Site	Dissolved Groundwater Concentration Reduction	Mass Discharge Reduction				
				<10×	10×	100×	1,000×	>1,000×
1	ERH	Scenario A (SDC)	10×			X		
2	ERH	Scenario B[a] (SDC)	<10×	X	X			
3	ERH	Scenario C	10×		X			
4	ERH	Scenario C[b] (SDC)	>10× to <100×		X			
5	ERH	Scenario C[c]	<10×	X				
6	ERH	Scenario C[c]	<10×	X		X		
7	ERH	Scenario C	<10×				X	
8	ERH	Scenario C (SDC)	10×		X			
9	ERH	Scenario C (SDC)	100×			X		
10	ERH	Scenario C	1,000×		X			
11	SEE	Scenario C	100×			X		
12	SEE	Scenario C	10×	X				
13	SEE	Scenario C[c]	10,000×				X	X
14	SEE	Scenario D[b]	<10×	X				

Note: *SDC*, supplemental data collection site for this project

[a]Mass discharge assessment involved two calculations using, first, only the post-treatment field investigation data and then the post-treatment field investigation data supplemented with data from a set of monitoring wells that were directly in line with the field investigation transect

[b]Pilot application appeared to encompass the entire source zone based on documentation reviewed

[c]Site used two different vertical intervals to calculate mass discharge: (1) only shallow geology and (2) shallow and deep geology

Using data from the 14 sites with pre- and post-treatment groundwater monitoring, order-of-magnitude concentration and mass discharge percent reductions were estimated. These data are presented in Table 14.7. In 9 of 14 sites, the groundwater concentration reduction was about one order of magnitude or less, and four sites had concentration reductions equal to or greater than two orders of magnitude. Because mass discharge calculations involve spatially variable hydraulic conductivity data, the mass discharge reduction can differ from the overall concentration reduction.

14.4.2 Lessons Learned

During the state-of-the-practice thermal remediation reviews, a number of lessons learned were noted:

- Site-specific treatment system shutdown criteria should be linked to remediation goals (groundwater and/or soil concentrations, mass discharge reduction). Reviews of site applications suggested that most had used arbitrary time–temperature criteria (e.g., operate for 2 months after a target temperature is reached, regardless of concentration reached) to determine the treatment duration.
- It is important to have characterized adequately the extent of the source area and contaminant distribution within the source zone before designing the treatment system. Reviews of site applications suggested poor pre-treatment source delineation, leading to undersized treatment footprints and incomplete treatment.
- Temperature data in and around the treatment zone are needed to confirm adequate heating, and system revisions may be needed while operating. Past applications have shown that insufficient temperature monitoring can lead to the occurrence of cold spots and in turn lead to suboptimal treatment effectiveness.
- Post-treatment monitoring should be long enough to allow groundwater flow through the treatment zone from upgradient to downgradient and to allow the treatment zone to cool down to near-ambient conditions for sites with groundwater remediation requirements. As was seen through the study, a number of sites had no documented post-treatment groundwater quality data.
- Communication is a key to ensuring a successful thermal treatment. Clients, vendors, regulators, consultants and other key stakeholders must have open communication with clearly defined goals to be assured that the thermal system is adequately designed and monitored to meet the overall project objectives.

14.4.3 Recommendations for Technology Selection, Application and Performance Assessment

Based on empirical analysis of thermal applications, reasonable cleanup objectives include:

- Partial removal of DNAPLs
- One order of magnitude or greater concentration reduction in soil and/or groundwater within the target treatment zone after temperatures return to ambient conditions
- Two orders of magnitude or greater mass flux reduction at the downgradient edge of the target treatment zone
- Rapid cleanup (usually 1 year or less) to expedite redevelopment of properties

Site objectives should include an economic analysis to ensure that cost combined with the expected results of the remediation will meet client and regulatory objectives.

All of the thermal treatment technologies discussed above are applicable to DNAPL removal, with some having more potential for success than others at any given site, depending on contaminant, target treatment zone depth, thickness and area, depth to groundwater, and lithology. When considering the different thermal treatment technologies for possible use, the following points may be useful:

- In general, all methods except TCH are self-limited to achieving temperatures at about the boiling point of water; if higher temperatures are needed, TCH may be the only option at this time.
- With respect to energy delivery rate, in high-permeability settings, steam has the potential to accomplish the most rapid initial heating.
- In settings with high-permeability contrasts and contaminants in the low-permeability layers, technologies like ERH and TCH that create/deliver heat directly to the low-permeability materials may be favored over others.
- With respect to anticipating heating distribution, some technologies are more predictable than others, with steam injection being the least predictable, except in constrained layered settings (permeable layer between two impermeable layers).
- In layered geologic settings, the use of two thermal technologies (steam and ERH or steam and TCH) may provide a more successful remediation by allowing each technology to focus on the geology it can treat with the most success.

After selecting a particular thermal technology, design and monitoring of the well field will be the most important drivers to a successful remediation. The target treatment zone should include the entire source zone and beyond, considering the uncertainty in source zone delineation. Heating elements should be installed using state-of-the-practice applications specified above for each thermal technology. Sufficient multi-depth temperature monitoring points should be installed within the treatment zone to understand temperature distribution. Reviewing the state of the practice (Kingston et al., 2010) indicates that approximately one point should be installed per 250–450 ft^2 (one every 23–42 m^2). Pressure monitoring should follow the same monitoring strategy as temperature monitoring.

Table 14.8 summarizes types of data needs for future applications, including source zone delineation, pre- and post-treatment groundwater sampling and analysis, depth–discrete aquifer characterization, hydraulic gradient and treatment zone temperature monitoring. This table assumes that typical site and source zone characterization data are available.

Table 14.8. Data Needs for Future Thermal Applications

Data Need and Methodology	Purpose and Justification
• Collection of soil cores.	• To confirm site conceptual model and to identify zones of interest for depth–discrete hydraulic conductivity tests and water sampling.
• Pre- and post-treatment groundwater quality data from depth–discrete samples collected along a transect perpendicular to groundwater flow at the downgradient edge of the source zone. • Samples collected at a vertical density that allows sampling within each layer and adequate vertical delineation of the dissolved chemical distribution (typically five to ten samples across the vertical interval of interest).	• To verify source zone width, the location of the downgradient edge, and to facilitate pre-treatment mass flux calculations.

(*continued*)

Table 14.8 (continued)

Data Need and Methodology	Purpose and Justification
• Pre-treatment groundwater quality data from depth–discrete sampling along transects that move upgradient from the initial transect across the entire length of the source. • Samples can be collected at a vertical and horizontal density that is less than the initial transect, as long as it is sufficient to define the lateral and vertical boundaries of the source zone within acceptable design tolerances.	• To verify the up- and downgradient edges of the source zone and to determine the length of the source zone.
• Hydraulic conductivity data at each depth–discrete sampling interval along the initial transect downgradient of the source zone.	• Quantification of the hydraulic conductivity throughout the saturated zone is necessary for the mass flux calculations. It also provides information necessary to anticipate the movement of fluids and vapor during the heating process.
• Hydraulic conductivity data from all existing permanent monitoring wells.	• To provide a greater understanding of the geologic conditions across the site and for use, as necessary, in mass flux calculations.
• Groundwater quality data from permanent, multilevel/nested monitoring wells in the source zone. • Multilevel/nested monitoring wells should be installed within each stratigraphic layer. The wells should be sampled pre-treatment, at the boiling point of the main contaminant of concern and after treatment optimization changes. These changes would include changes to energy applied within a stratigraphic layer and lengthening of treatment duration based on concentration data.	• To provide depth–discrete groundwater quality data from areas within the source zone before and during treatment. • Multilevel/nested permanent monitoring wells will allow continuous monitoring of groundwater quality within differing stratigraphic zones throughout the period of treatment. • The ability to monitor groundwater quality throughout the heating process would provide feedback on groundwater quality changes and allow operators to have real-time data on groundwater concentrations. This data could then be used to adjust energy input to different stratigraphic layers and duration of treatment. • These data would aid in optimizing a thermal treatment technology based on discrete–depth concentrations within the treatment zone during treatment when used in conjunction with energy applied and temperatures achieved in those zones.
• Depth to groundwater in all surveyed wells.	• To calculate groundwater elevations, determine groundwater flow direction, and calculate gradient.
• Temperature data from a network of monitoring locations throughout the treatment zone. • Temperature points should be placed at the centroid between all energy delivery points and on the outer edges of the treatment zone. These should be placed at a vertical density that allows delineation of the temperature within each geologic layer (typically five to ten points across the vertical interval of interest) and within any critical target treatment zone.	• To verify the temperature across the treatment zone and within each geologic layer. Historically, a predetermined average treatment zone temperature has been used to decide when to shut down a system. By using a multilevel temperature monitoring system, temperatures for the entire treatment zone can be monitored real time. With this increased level of monitoring, treatment can be maintained until the set-point temperature is reached for the whole treatment zone. This would ensure a more effective treatment. Temperature data versus time provides insight into the dynamics of the heating process and behavior of the subsurface.

(continued)

Table 14.8 (continued)

Data Need and Methodology	Purpose and Justification
• Vapor stream and liquid stream contaminant concentration data and flow rates of those process streams. • Ideally, online real-time monitoring would be used to guide the timing of sample collection.	• Quantifying mass removal (flow rate × concentration) and monitoring a real-time mass balance will allow insight into the performance versus time that can be used to optimize the system operating conditions.
• Post-treatment groundwater quality data from downgradient transect after near pre-treatment temperatures have been achieved.	• This data used to calculate post-treatment performance in terms of groundwater quality and mass discharge.

14.5 CASE STUDIES

The three case studies presented in the following sections represent state-of-the-practice applications of the major thermal remediation technologies to treat chlorinated ethene source zones. Each case study represents a relatively well-documented and well-monitored field-scale project. The first is an ERH project performed at Naval Air Station (NAS) Alameda Point in California, USA; the second is an ERH test done at Fort Lewis, Washington, USA; and the third is a demonstration of TCH to treat fractured bedrock at the Naval Air Warfare Center in Trenton, New Jersey, USA.

14.5.1 NAS Alameda Point, California

14.5.1.1 Background

The information contained in this case study is based on Kingston et al. (2010). The Naval Air Station (NAS) Alameda Point (site) is located in Alameda, CA, USA. The site consists of more than 18 acres (7.3 hectares [ha]) of land located in the central portion of Alameda Point and includes Building 5, which covers approximately 12.5 acres (5 ha). Building 5 housed specialty shops for aircraft component repair and maintenance from 1942 until the base was closed in April 1997. Building 5 also housed a plating shop and a selective plating shop where small parts were plated by hand. A hazardous waste storage area at Site 5 was closed in 1988. This area was located outside of Building 5 in the southeastern corner of the site. Additional activities at Site 5 included a lead–acid and nickel–cadmium battery service area.

Chemical contaminants from the various industrial processes inside Building 5 are believed to have been released directly to the subsurface beneath certain operational areas. Solvents are believed to have been released as spills and as leakage from a solvent tank in the hazardous waste storage area outside the southeast corner of Building 5. Solvent releases are also believed to have occurred from a solvent tank located on the eastern side of Building 5, and solvents and metals are believed to have been released from the plating shop via floor drains.

The DNAPL source zone was determined to occupy approximately 0.33 acre (0.13 ha) and consisted mainly of TCE and TCA and associated degradation products. A pilot-scale six-phase ERH application was performed in June 2002. Based on the results of the pilot, full-scale ERH applications were performed throughout the entire source zone.

14.5.1.2 Geologic Conditions

The conceptual subsurface model for Site 5 includes five geologic units. The Lower San Antonio unit, or Yerba Buena mud, is a clay that extends from a depth of approximately 125 ft

(38 m) below ground surface (bgs) to 170–200 ft (ca. 52–60 m) bgs. The Upper San Antonio unit overlies the Yerba Buena mud and extends from 100 to about 125 ft (~30–38 m) bgs. It consists of interbedded very fine-grained, silty sand and green-gray silty clay. The Merritt Sand formation overlies the San Antonio formation and extends from 35 to about 100 ft (roughly 10–30 m) bgs and contains 3 sediment types: (1) yellow-brown clayey sand, with approximately 5 percent clay; (2) moist, silty sand; and (3) fine-grained, well-sorted sand with some shell fragments. The Bay Sediment unit (BSU) overlies the Merritt Sand formation and extends from 15 to about 35 ft (4.5–10 m) bgs and is composed of three sediment types: (1) a stiff, moist, dark olive clay; (2) sand and clay with a number of shell fragments; and (3) silty sand with interbedded layers of fine-grained sand.

Artificial fill overlies the BSU and is composed of olive brown, unconsolidated fine- to medium-grained sand with lenses of silty sand, gravelly sand, or sandy gravel. Groundwater is encountered in the artificial fill between 4 and 7 ft (1.2 and 2.1 m) bgs. The BSU separates the first and second water-bearing zones (WBZ) with low-permeability sediments. The first WBZ is located in the artificial fill and upper part of the BSU. The first WBZ general flow direction is to the northeast. The second WBZ is situated within the lower part of the BSU, the Merritt Sand, and the Upper San Antonio unit. The second WBZ general flow direction is to the south.

14.5.1.3 System Installation

Installation for full-scale six-phase heating began in 2004. The system consisted of seven sheet pile electrodes installed to a depth of 19 ft (5.8 m) bgs and 22 sheet pile electrodes installed to a depth of 14 ft (4.3 m) bgs and one sheet pile electrode installed to 15 ft (4.6 m) bgs. Electrodes were spaced approximately 25 ft (7.6 m) apart. The total treatment area was approximately 1/3 acre (0.13 ha) (Figure 14.16). Two monitoring wells were installed inside the treatment area. These two monitoring wells were used along with 12 monitoring wells installed during the pilot-scale ERH application. Soil vapor extraction wells were installed on approximately 15 ft (4.6 m) centers.

The full-scale system was brought online in July 2004 and was operated until November 2004. The remedial system performance was monitored continuously during operation, and an estimated 3,000 pounds (lb) (equivalent to ~1,360 kilograms [kg]) of VOC contamination was removed in recovered vapors and groundwater. Table 14.9 provides a summary of the key geologic, hydrogeologic, and treatment area information.

This application was favorable to the application of ERH:

- Site contaminants were chlorinated VOCs.
- Large contaminant mass and concentrations had a reasonably well-defined lateral and depth extent.
- Medium-permeability formation enabled groundwater influx to be controlled.
- Depth of site contaminants was greater than 10 ft (3.1 m), and the area was easily accessible.

14.5.1.4 Post-treatment Field Investigation

The post-treatment field investigation focused on groundwater sampling across a transect perpendicular to groundwater flow immediately downgradient of the treatment zones. The goal was to collect sufficient groundwater and aquifer characterization data to assess post-treatment groundwater quality and estimate mass discharge.

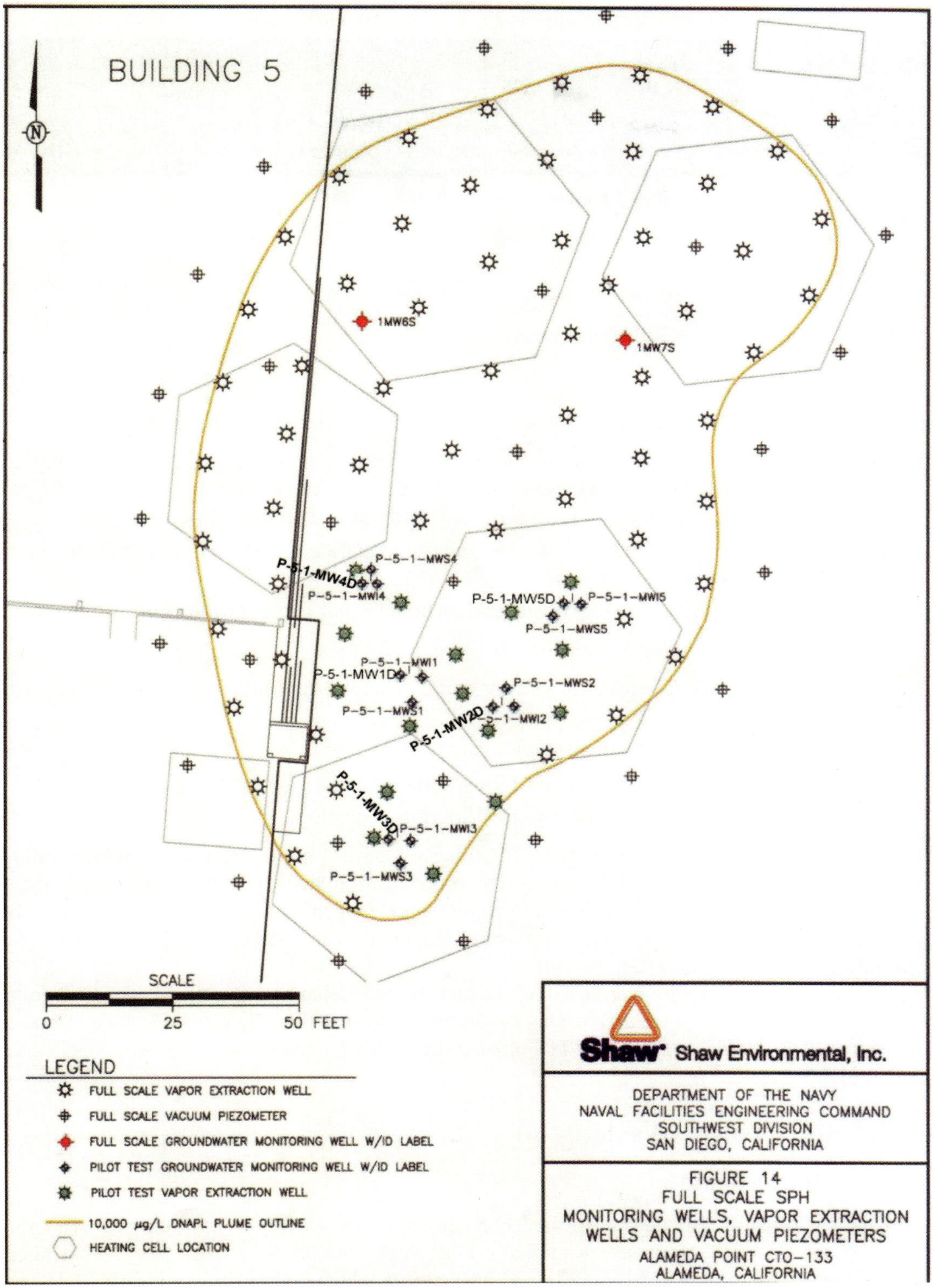

Figure 14.16. NAS Alameda ERH system layout (courtesy of Shaw Environment, Inc. as presented in Kingston et al., 2010).

Table 14.9. Site Geology, Hydrogeology, and Treatment Area Information

Site ID	Technology	Geology at This Site is Most Like This Conceptual Scenario[a]	Number of Electrodes	Number of Permanent Monitoring Wells	Type of Chemicals Treated (C-Chlorinated Solvents)	Size of Target Treatment Area [ft²]	Thickness of Target Treatment Interval [ft]	Depth to water [ft]
NAS Alameda Bldg 5, Site 5-1	ERH	C	30	15	C	14,520	20	6

C, permeable sediments with interbedded lenses of low-permeable material
ERH, electrical resistance heating
[a]Scenario descriptors (for the target treatment zone)

To accomplish this goal, the following field activities were undertaken (Table 14.10):

- Groundwater sampling and aquifer characterization at seven downgradient sampling locations, each with at least five depth–discrete sampling points, along a transect equal in width to the original source zone and downgradient dissolved plume
- Groundwater sampling and aquifer characterization at 11 monitoring wells in or adjacent to the treatment zone
- Analysis of water samples for general chemistry and hydrocarbon concentrations

Aquifer characterization involved the following activities:

- Aquifer-specific capacity tests at 39 depth–discrete sampling points with seven sampling locations
- Slug testing at 11 permanent monitoring wells
- Depth-to-water measurements for flow direction and gradient
- Collection of two soil cores for confirmation of geology

Groundwater samples from the 39 discrete–depth post-treatment sampling points indicated that ERH decreased groundwater concentration by two orders of magnitude, a 99% reduction. Table 14.11 provides the groundwater concentration ranges for the pre- and post-treatment data. Pre-treatment data were obtained from existing site reports and past groundwater concentration data.

These groundwater data were used to calculate mass discharge using the Mass Flux Toolkit software (Farhat and Newell, 2011). This evaluation indicated that the mass discharge through the source zone decreased by two orders of magnitude (99.7% reduction, from 49 to 0.13 kg/yr) (Table 14.12).

14.5.2 Fort Lewis, Washington

14.5.2.1 Background

The 86,000-acre (34,800-ha) former Fort Lewis Army facility is located near Tacoma, Washington, USA, on the southeastern shore of Puget Sound. The main degreasing agent used at the facility from 1946 to 1970 was TCE, after which it was replaced with 1,1,1-TCA. Waste TCE was disposed of with waste oils at several locations including the East Gate Disposal

Table 14.10. Transect Details for Supplemental Site Investigations

Site ID	Number of Transect Sampling Locations	Transect Length (ft)	Vertical Sampling Interval (ft bgs)	Number of Depth-Specific GW Samples	Number of Aquifer-Specific Capacity Tests
NAS Alameda Site 5-1, Bldg. 5	7	115	6.5–21	39	39

Note: *bgs*, below ground surface; *ft*, foot/feet

Table 14.11. Range of Pre- and Post-treatment Concentration Data (μg/L)

Site	Contaminant	Pre-treatment Concentration Ranges from Site Documentation (μg/L)		Post-treatment Concentration Ranges from Supplemental Field Investigations Performed Under This Study (μg/L)	
		High	Low	High	Low
NAS Alameda, Site 5-1, Bldg. 5	Vinyl chloride	8,140	ND <0.5	29	ND <1
	1,1-Dichloroethene	15,100	ND <0.5	2	ND <1
	trans-1,2-Dichloroethene	300	ND <0.5	2	ND <1
	1,1-Dichloroethane	48,800	15	2	ND <1
	cis-1,2-Dichloroethene	13,700	ND <1.3	71	ND <1
	1,2-Dichloroethane	ND[a] <250	ND <0.5	ND <1	ND <1
	1,1,1-Trichloroethane	42,000	ND <0.5	ND <1	ND <1
	Trichloroethene	1,600	ND <0.5	76	1
	1,1,2-Trichloroethane	ND <250	ND <0.5	ND <1	ND <1
	Tetrachloroethene	54	ND <0.5	47	ND <1

[a]ND denotes not detected at the detection level indicated

Table 14.12. Summary of Mass Discharge (Mass Flux) Results Pre- and Post-treatment

Site	Contaminant	Pre-treatment Discharge (kg/yr)[a]	Post-treatment Mass Discharge (kg/yr)[b]	Post-treatment Mass Discharge per Linear Foot (kg/yr/ft)
NAS Alameda Site 5-1, Bldg. 5[c]	Total contaminant flux	4.9×10^{1}	1.3×10^{-1}	9.6×10^{-4}

[a]Mass discharge calculations were based on monitoring well data from the documentation
[b]Mass discharge calculations were based on discrete–depth sampling data or a combination of discrete–depth sampling data and monitoring well data
[c]Mass discharge calculations were based on discrete–depth sampling data only

Yard (EGDY). The EGDY was used between 1946 and 1970 as a waste disposal site storing barrels and vats in trenches around the yard.

To mitigate impacts to groundwater, in 1995 the Army built two groundwater pump-and-treat systems, which are still in operation today. Extraction wells were installed at the leading edge of the groundwater plume and near the source area to prevent off-site migration. Although the extraction/treatment system was successfully removing chlorinated organics from groundwater, monitoring results indicated increased contaminant levels in groundwater near the source and possible expansion of the contaminant plume in some areas.

Between January and September 2001, the Army excavated and removed more than 1,000 buried drums containing TCE wastes from the EGDY. The remedial activities identified DNAPL interspersed throughout the soil matrix mainly in the form of ganglia and globules. Three thermal treatment systems were installed to treat chlorinated solvents and some petroleum hydrocarbons in soils and groundwater. These treatment systems were run in series. This case study is focused on EGDY Area 3. The EGDY Area 3 full-scale ERH system was operated from October 2006 to January 2007.

14.5.2.2 Geologic Conditions

The site geology generally consists of glacial outwash, till, and alluvial deposits from the Wisconsinan glaciation time period. Man-made materials and debris (bricks, wood waste, metal, glass) were encountered at several locations. Nonaqueous phase liquid was generally observed at various locations ranging from near ground surface to approximately 37 ft (11 m) bgs. Site lithology was unpredictable, and groundwater velocity in the highly transmissive aquifer was as high as 10–20 ft (roughly 3–6 m) per day. Depth to groundwater varied slightly in each area but on average was 5 ft (1.5 m) bgs.

14.5.2.3 System Installation

Area 3 is approximately 18,200 square feet (0.4 acre or 0.16 ha) in size and up to 30 ft (9 m) deep, for a total volume of approximately 20,100 cubic yards or approximately 15,000 cubic meters (Figure 14.17). An estimated 140,000 lb (~63,500 kg) of NAPL was present in the subsurface.

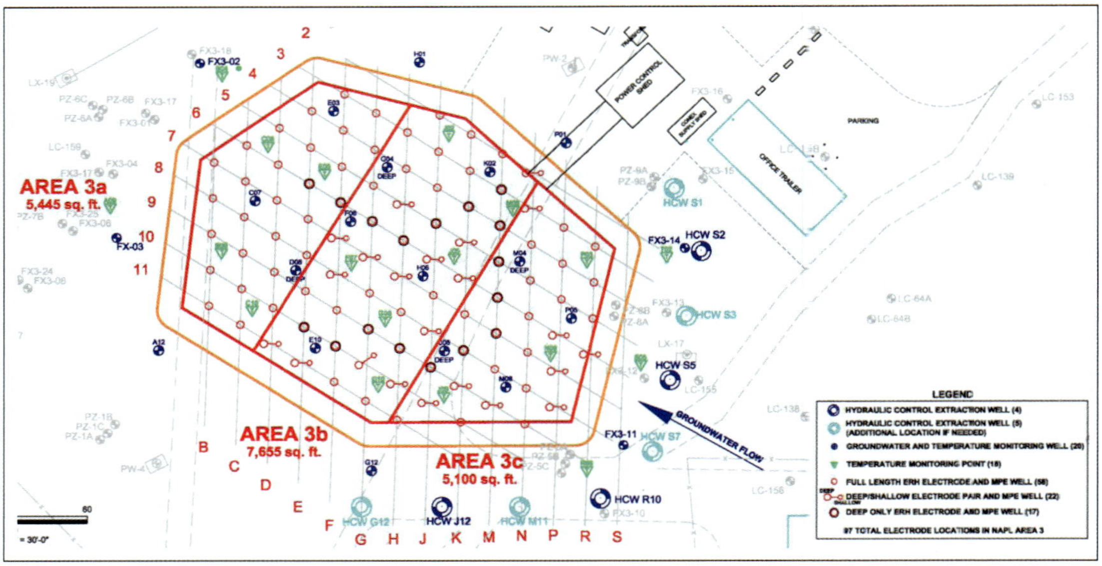

Figure 14.17. Schematic illustration of Fort Lewis ERH application (Kingston et al., 2010).

Table 14.13. Site Geology, Hydrogeology and Treatment Area Information

Site ID	Technology	Geology at This Site Is Most Like This Conceptual Scenario[a]	Number of Electrodes	Number of Permanent Monitoring Wells	Type of Chemicals Treated	Size of Target Treatment Area [ft^2]	Thickness of Target Treatment Interval [ft]	Depth to Water [ft]
Ft. Lewis EDGY Area 3	ERH	C	93	17	Chlorinated Solvents, Petroleum	18,200	30	7

C, permeable sediments with interbedded lenses of low-permeable material
ERH, electrical resistance heating
N/A, not available
[a]Scenario descriptors (for the target treatment zone)

The treatment system consisted of 97 electrodes installed to 30 ft (9.2 m) bgs at a spacing of approximately 15 ft (4.6 m). Each electrode was accompanied by a co-located vapor extraction well. A total of 18 temperature monitoring points and 20 monitoring wells were installed, which equates to approximately one every 400 ft^2 (37 m^2).

The full-scale system was brought online in October 2006 and was operated until January 2007. The remedial system performance was continuously monitored during operation, and an estimated 3,665 lb (1,664 kg) of VOC and petroleum contaminants was removed in recovered vapors and groundwater. Groundwater samples were taken within and at the downgradient edge of the treatment zone to monitor groundwater concentration changes with temperature increases (Truex et al., 2009). Table 14.13 provides a summary of the key geologic, hydrogeologic and treatment area information.

This site was favorable to the application of ERH for several reasons, including:

- Site contaminants were chlorinated VOCs and petroleum hydrocarbons.
- Large contaminant mass and concentrations had a reasonably well-defined lateral and depth extent.
- Depths of site contaminants were greater than 10 ft (3.1 m), and the area was easily accessible.

The contractor installed the electrodes with an approximate 15 ft (4.6 m) spacing encompassing the source area using ERH electrodes with co-located soil vapor extraction wells.

14.5.2.4 Post-treatment Field Investigation

Groundwater samples were collected from existing upgradient, treatment zone and downgradient groundwater monitoring wells before, during, and after treatment. This monitoring approach provided a comprehensive set of data that allowed for pre- versus post-treatment groundwater data analysis, the ability to assess treatment temperature and groundwater concentration versus time and location, and with existing hydraulic conductivity information generated by the contractor, the ability to calculate a mass discharge from the treatment zone immediately following the application. Groundwater samples from five upgradient, eight treatment zone and four downgradient groundwater monitoring wells were analyzed for dissolved chlorinated and petroleum hydrocarbon concentrations.

Hydraulic conductivity measurements made by the US Army Corps of Engineers (USACE) personnel/contractors using monitoring wells around Area 3 ranged from 38 to 120 ft/d

Table 14.14. Ranges of Pre- and Post-treatment Concentration Data (μg/L)

Site	Contaminant	Pre-treatment Concentration Ranges from Site Documentation (μg/L)		Post-treatment Concentration Ranges from Supplemental Field Investigations Performed Under This Study (μg/L)	
		High	Low	High	Low
Ft. Lewis EGDY Area 3	Vinyl chloride	5,800	ND[a] <1	170	ND <1
	1,1-Dichloroethene	N/A	N/A	24	ND <1
	trans-1,2-Dichloroethene	480	ND <1	38	ND <1
	cis-1,2-Dichloroethene	30,000	ND <1	2,200	ND <1
	Trichloroethylene	17,000	2	2,200	ND <1
	Tetrachloroethene	9	ND <1	1	ND <1
	1,3,5-Trimethylbenzene	88	ND <1	19	ND <1
	1,2,4-Trimethylbenzene	22	ND <1	ND <1	ND <1

[a]ND denotes not detected at the detection level indicated

Table 14.15. Summary of Mass Discharge Results Pre- and Post-treatment

Site	Contaminant	Pre-treatment Discharge (kg/yr)[a]	Post-treatment Mass Discharge (kg/yr)[b]	Post-treatment Mass Discharge per Linear Foot (kg/yr/ft)
Ft. Lewis EGDY Area 3[c]	Total contaminant flux	3.2×10^{1}	2.1	1.9×10^{-2}

[a]Mass discharge calculations were based on monitoring well data from the documentation
[b]Mass discharge calculations were based on discrete–depth sampling data or a combination of discrete–depth sampling data and monitoring well data
[c]Mass discharge calculations were based on monitoring well data analyzed by Arizona State University personnel

(11.6–36.6 m/d), with an average of 65 ft/d (20 m/d). Water level monitoring throughout 2006 and 2007 indicated that the site is characterized by variable hydraulic gradients based on season. The range of measurements was used in determining the mass flux to provide a reasonable range of values based on seasonal variations.

Groundwater samples from 16 monitoring events indicated that application of ERH decreased groundwater concentration by one order of magnitude. Table 14.14 provides the groundwater concentration ranges for the pre- and post-treatment data.

These groundwater data were used to calculate a mass discharge using the Mass Flux Toolkit software (Farhat and Newell, 2011). This evaluation indicated that the mass discharge through the source zone decreased by one order of magnitude. Table 14.15 provides the mass discharge data.

14.5.3 Former Naval Air Warfare Center, Trenton, New Jersey

14.5.3.1 Background

The former Naval Air Warfare Center (NAWC) located in West Trenton, New Jersey, USA, operated as a US Navy jet engine testing facility for military aircraft from the mid-1950s to the late 1990s. The NAWC covered 67 acres (27 ha) and contained large jet engine test buildings, associated service buildings, hangars, and scores of smaller support structures all interconnected with a

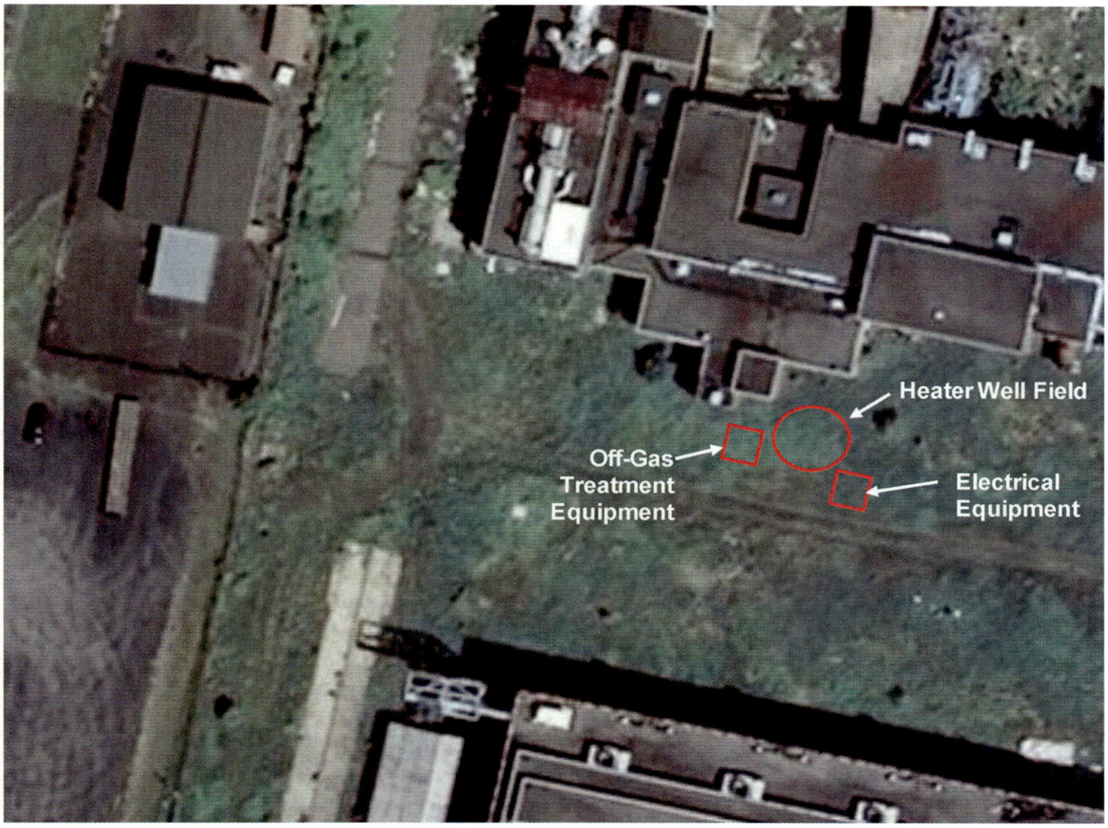

Figure 14.18. Location of the TCH demonstration in fractured rock at the NAWC site.

network of aboveground and underground service lines. The NAWC was decommissioned on October 15, 1998, and has been sub-parceled and sold. As a result of past activities, TCE, jet fuel, and other chemicals leaked into the subsurface. Trichloroethene that intercepted spilled jet fuel began to biodegrade rather rapidly, whereas TCE that flowed deeply into the fractured bedrock biodegraded slowly if at all. Investigations of groundwater contamination at the site began in the late 1980s, and a pump-and-treat system was in operation by the late 1990s.

In 1993, the US Geologic Survey (USGS) began studies at the NAWC site in cooperation with the US Navy. In 2001, the NAWC site became a fractured rock research site under the USGS Toxic Substances Hydrology Program. The cooperative effort that began in 2001 includes scientists from the Navy, New Jersey Department of Environmental Protection, New Jersey Geological Survey, and universities and other research institutions. The specific area designated for the demonstration of TCH in fractured bedrock is depicted in Figure 14.18. Trichloroethene is believed to be both present in bedrock as residual DNAPL in fractures and diffused into the rock matrix. The TCH demonstration was performed by TerraTherm, Inc., with assistance from the USGS, Queen's University (Kingston, Ontario, Canada), and the US Navy.

14.5.3.2 Geologic Conditions

The NAWC site is located within the Newark Basin geologic province and is underlain by mudstones of the Skunk Hallow, Byram, and Ewing Creek members of the Lockatong Formation (Lacombe and Burton, 2010). Folds, faults and joints within the bedrock were developed at great depths by tectonic compression during the Jurassic. All strata in the TCH demonstration area are folded with a similar strike and dip, typically N66°E and 28°NW, respectively. Weathered mudstone from land surface to about 25 to 30 ft (7.6–9.2 m) bgs is visually, chemically, geophysically and hydraulically different from the unweathered mudstone of about 25 to 30 to 55 ft (7.6–9.2 m to 16.8 m) bgs. Weathered mudstone stratum ranges from unconsolidated muds to highly fissile and highly fractured bedrock as a result of differential degradation of the various types of mudstone strata.

Tiedeman et al. (2010) report the transmissivity for the following strata within the overall USGS research site: upper weathered zone (6.5×10^{-4} meters per second [m/s]), lower weathered zone (2.4×10^{-6} m/s), unweathered dipping mudstone with high hydraulic conductivity (3.5×10^{-6} to 3.5×10^{-3} m/s), and unweathered dipping mudstone with low hydraulic conductivity (9.3×10^{-9} m/s). Primary porosity of the shallow highly weathered but indurated bedrock is up to 15%. Fissile rocks likely have a much higher porosity. At depth, the indurated rock has a porosity of 3–5% (Lebrón et al., 2012). The top of the bedrock is encountered at a depth of approximately 6 ft (1.8 m) bgs in the area of the TCH demonstration, and the water table is located at 5 ft (1.5 m) bgs.

14.5.3.3 System Installation

As illustrated in Figure 14.19, the TCH demonstration included installation of 15 heater borings (designated HO-1 through HO-15) installed approximately 5 ft (1.5 m) apart, 15 vapor extraction points installed adjacent to the heater borings, and eight temperature monitoring borings. All borings were completed using sonic drilling, which allowed collection of rock core for logging and testing for both TCE content and physical characteristics. Each heater boring consisted of a 3-in. diameter, nonperforated carbon steel casing with a bottom seal, installed to a depth of approximately 56 ft (17 m) bgs. Each vapor extraction well consisted of 1-in. (2.54-cm) diameter, stainless steel screen with bottom seal, installed in the same borehole as a heater well, to a depth of approximately 45 ft (13.7 m) bgs, and screened from 5 to 45 ft (1.6–13.7 m) bgs. The heater borings with co-located vacuum extraction points are depicted schematically in Figure 14.20. A single heater element was placed inside each stainless steel liner and set inside the heater can. Groups of heater wells were wired in series, to deliver up to approximately 350 watts per foot of heated length to the subsurface at full power.

A silicon-controlled rectifier power controller and remote temperature controllers were used to regulate the power application to the TCH heaters based on temperature input from thermocouples (TCs). The TCs were placed on the outside of the stainless steel sleeve adjacent to selected heaters to allow monitoring and control of the temperature of the heaters. Data from the TCs were used to control the amount of power delivered to each heater electrical circuit to maintain an optimum heater temperature. The amount of power delivered to each circuit was controlled manually by the operator during the initial start-up and ramping period, after which the heaters were set to constant power outputs based on the observed operating temperature of the heater.

The eight formation temperature monitoring points were constructed of 1.5-in. (3.81-cm) diameter, nonperforated carbon steel pipe with bottom cap, installed to a depth of ~50 ft (15 m) bgs in a 4-in. (10.6-cm) cased hole. Each temperature monitoring well had approximately 10 thermocouples located 5, 10, 15, 20, 25, 30, 35, 40, 45 and 50 ft bgs (roughly every 1.5 m to a depth of 15 m bgs). The annular space between the pipe and the borehole wall was filled with a cement–bentonite grout using a tremie pipe to the ground surface. The points were further sealed at the ground surface as the surface cover was installed.

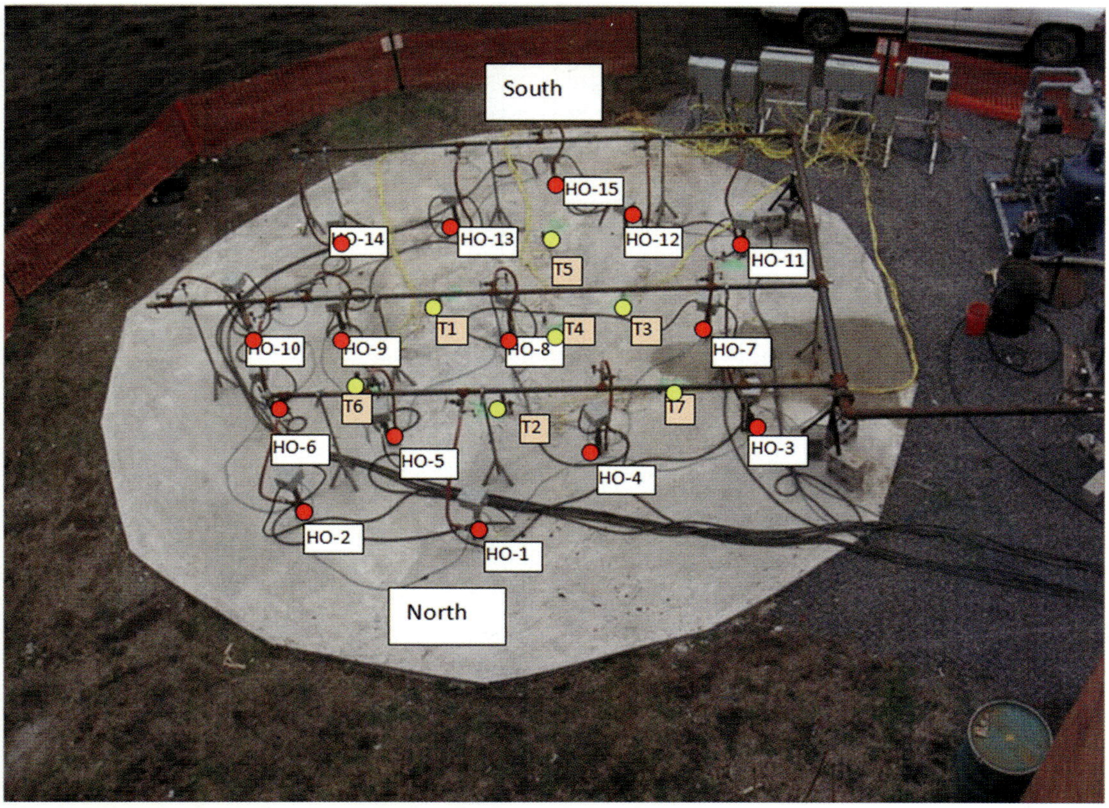

Figure 14.19. Illustration of TCH system components, including 15 heater borings with associated vapor extraction wells (HO-1 through HO-15) and eight temperature monitoring borings (T1 through T7 depicted) installed over an area of approximately 45.6 m^2. Photograph courtesy of USGS.

Thermal conductive heating operations ran continuously for 106 days without any major shutdowns other than small interruptions for scheduled maintenance and minor equipment replacement and GAC changeouts. The heating period lasted a total of 97 days, while the extraction system operated for 106 days. The overall schedule included 6 days of start-up, 97 days of operation and 3 days of cool down.

14.5.3.4 Post-treatment Field Investigation

The primary objective of the TCH demonstration was to evaluate the ability to remove TCE diffused in the rock matrix. Figure 14.21 illustrates a plot of average temperature for a particular depth versus time throughout the duration of the demonstration. Temperatures reached steady-state values after approximately 30 days of heating. Temperatures in the lower 15 ft (4.6 m) of the test zone did not reach target values, presumably because of the influence of inflowing cold groundwater and the more competent nature of bedrock at those depths.

To evaluate the removal of TCE from the rock matrix, three boreholes (BR-1, BR-2 and BR-3) were drilled in the center of the treatment area in which samples were collected both pre- and post-treatment in adjacent boreholes. The samples were placed in a rock-crushing device, preserved in methanol and shaken for 24 h to allow the TCE to partition into the methanol.

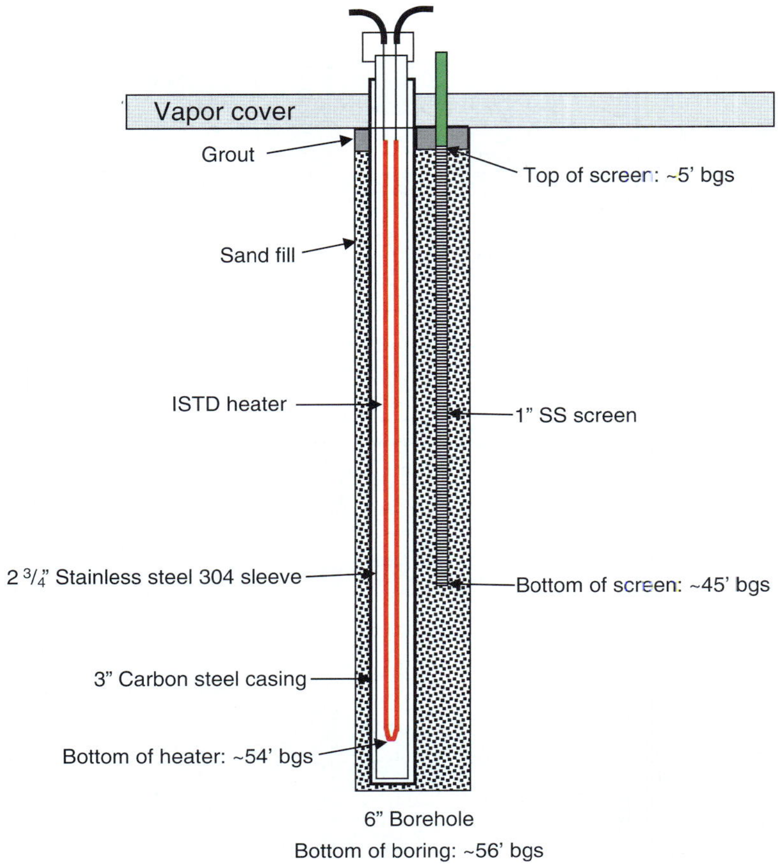

Figure 14.20. Schematic illustration of heater boring with co-located vapor extraction well (Lebrón et al., 2012).

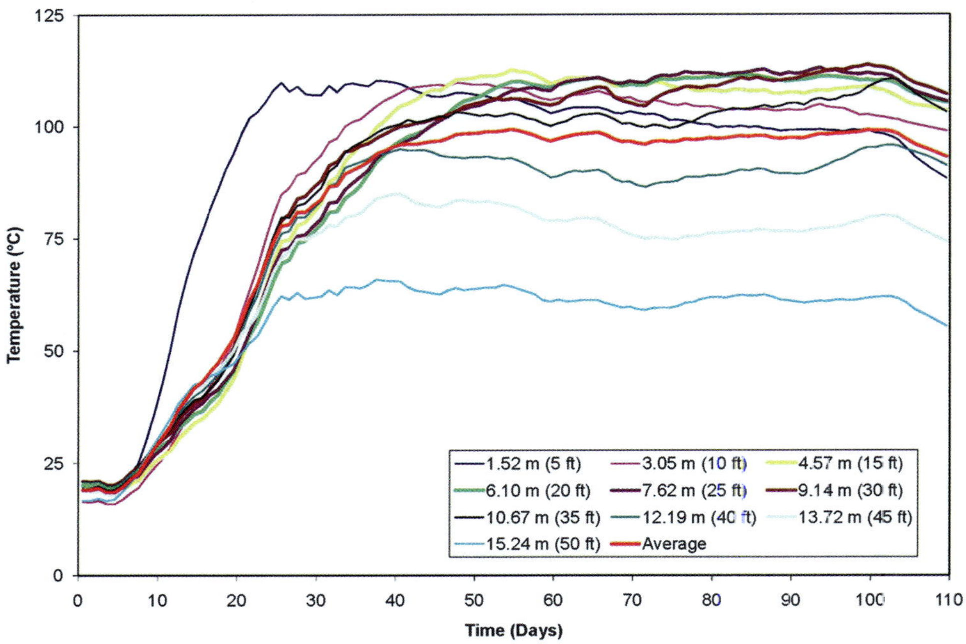

Figure 14.21. Average temperature for eight temperature monitor borings (T1–T8) that had 1.5 m interval temperature points from 1.5 to 15.2 m below ground surface.

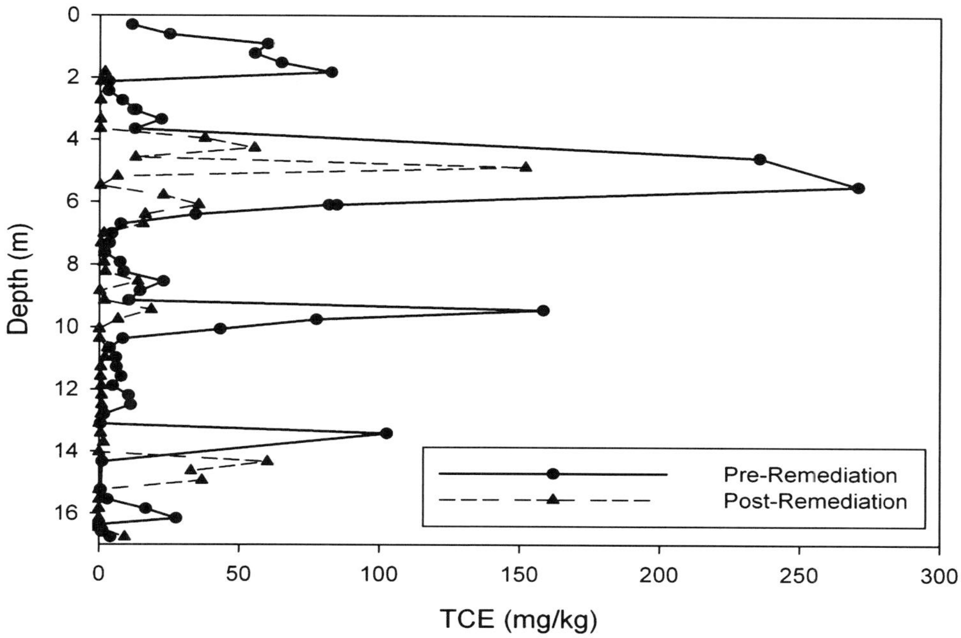

Figure 14.22. TCE concentration in the rock matrix versus depth from location BR-1 pre- and post-treatment (Lebrón et al., 2012).

The methanol solution was then extracted from each rock sample and analyzed for TCE concentration by an external laboratory. Figure 14.22 presents the rock matrix sampling results as a function of depth from the BR-1 pair of coring locations pre- and post-treatment. This location exhibited higher initial concentrations of TCE in the rock matrix than BR-2 and BR-3 and contained a greater number of discrete sample results. Considering results from all three paired locations, approximately 63.5% of the TCE was removed from the rock matrix as a result of TCH application. The concentration peaks depicted in Figure 14.22 correspond to the locations of fractures and demonstrate that TCE had not fully diffused into the rock matrix at the time of the demonstration.

14.6 SUMMARY

In situ thermal treatment has been applied to chlorinated solvent source zones for approximately 15 years. In that time, it has proven to be capable of rapidly removing the vast majority (>99%) of the mass of chlorinated solvents from the targeted volume at some sites and reducing the associated groundwater concentrations by two to four orders of magnitude. Given a thorough understanding of contaminant distribution and reasonable access to the contaminants, thermal treatment can be effective. It may even meet closure criteria, although in most cases studied some residual contamination remains and the groundwater concentrations remain above the MCL values. Thermal treatment can remove significant quantities of contaminant mass, even for less permeable materials, because most of the thermal technologies are less sensitive to subsurface heterogeneities than injection-based technologies such as ISCO and cosolvent flushing.

Thermal treatment is applied either (1) to increase vapor recovery by inducing boiling and/or increasing the vaporization and volatilization of water and contaminants or (2) to increase liquid recovery by increasing DNAPL mobility by reducing the viscosity and interfacial tension.

Given the variety of strategies for implementing the technology and the variety of processes occurring during any form of thermal treatment, it is important to fully understand these processes when selecting or designing a remediation system. Therefore, this chapter includes a thorough discussion of the processes responsible for vaporization (boiling and co-boiling of water and DNAPL), partitioning of contaminants between phases, and fluid transport (focusing on the mobilization of contaminants present as continuous, residual or trapped DNAPL).

Thermal technology has gone through a rapid evolution. Several different commercially successful approaches have been developed. The approaches most used at the present time are electrical resistance heating and thermal conductive heating, although radio frequency heating, steam and heating while mixing the soil with large augurs are being used commercially as well. The technology evolution also has led to combinations of different thermal techniques with each other or with other remediation technologies. Each of these thermal treatment strategies acts through different mechanisms, and each has its own advantages and limitations. However, all approaches require relatively high inputs of energy, compared to the other technologies available. Also, all thermal technologies require vapor extraction and aboveground systems to capture and treat volatilized contaminants.

The state of the practice has also evolved rapidly, to the point that some guidance and standardized protocols are available. A recently released summary of the state of the practice reviewed the design parameters and performance at 182 field sites. Based largely on this review, several practical recommendations have been presented: (1) define reasonable remediation objectives and link the system shutdown criteria to these objectives; (2) adequately characterize the source area, as much of the contamination remaining after thermal treatment is due to insufficient delineation; (3) collect sufficient thermal data in and around the treatment zone; and (4) monitor long enough after treatment to allow groundwater to flow entirely through the treatment zone and the treatment zone to cool down to near-ambient conditions. In addition, the chapter includes a summary of the specific data needs during thermal treatment. Finally, several case studies are presented to provide examples of designs, performance and site-specific issues.

ACKNOWLEDGMENT

The authors of this chapter would like to thank Cindy Zhao and Julia Starostenko, both students at Queen's University, Kingston, ON, Canada, for their support and contributions to this chapter.

REFERENCES

Adamson AW. 1990. Physical Chemistry of Surfaces, 5th ed. John Wiley & Sons, New York, NY, USA.

Ali SMF, Meldau RF. 1979. Current steamflood technology. J Petrol Technol 31:1332–1342.

Atchley AA, Properetti A. 1989. The crevice model of bubble nucleation. J Acoust Soc Am 86:1065–1084.

Baker RS, Hieste U. 2009. Large-Scale Physical Models of Thermal Remediation of DNAPL Source Zones in Aquitards. Project ER-1423 Final Report. SERDP, Arlington, VA, USA.

Baker R, Kuhlman M. 2002. A description of the mechanics of *in situ* thermal destruction (ISTD) reactions. In El-Akabi H, ed, Proceedings, 2nd International Conference, Oxidation and Reduction Technologies for Soil and Groundwater. ORTs-2, San Diego, CA, USA.

Beyke G, Fleming D. 2005. In situ thermal remediation of DNAPL and LNAPL using electrical resistance heating. Remediat J 15:5–22.

Beyke G, Dodson ME, Powell TD, Wolf JL. 2007. Electrode heating with remediation agent. U.S. Patent No. 7,290,959.

Buettner HM, Daily WD. 1995. Cleaning contaminated soil using electrical heating and air stripping. J Environ Eng 121:580–588.

Buettner HM, Daily WD, Aines RD, Newmark RL, Ramirez AL, Siegel WH. 1999. Electrode wells for powerline-frequency electrical heating of soils. U.S. Patent No. 5,907,662.

Burghardt JM, Kueper BH. 2008. Laboratory study evaluating heating of tetrachloroethylene impacted soil. Ground Water Monit Remediat 28:95–106.

Chen F, Freedman DL, Falta RW, Murdoch LC. 2012. Henry's Law constants of chlorinated solvents at elevated temperatures. Chemosphere 86:156–165.

CRC. 2012. Handbook of Chemistry and Physics, 92nd ed. CRC Press, Boca Raton, FL, USA.

de Percin PR. 1991. Demonstration of in situ steam and hot-air stripping technology. J Air Waste Manage Assoc 41:873–877.

Dev H, Bridges J, Sresty G. 1984. Decontamination of hazardous waste substances from spills and uncontrolled waste sites by radio frequency in situ heating. EPA-600/D-84-077. U.S. Environmental Protection Agency, Cincinatti, OH, USA.

Dev H, Condorelli P, Bridges J, Rogers C, Downey D. 1987. In situ radio frequency heating process for decontamination of soil. In Exner J, ed, Solving Hazardous Waste Problems: Learning from Dioxins. American Chemical Society, Washington, DC, USA, pp 332–339.

DeVoe C, Udell KS. 1998. Thermodynamic and hydrodynamic behavior of water and DNAPLs during heating. Proceedings, First Conference on Remediation of Chlorinated and Recalcitrant Compounds, May 18–21, Monterey, CA, USA. Battelle Press, Columbus, OH, USA, pp 61–66.

Farhat S, Newell CJ. 2011. Mass Flux Toolkit. http://www.gsi-net.com/software/free-software/mass-flux-toolkit.html. Accessed January 31, 2013.

Gauglitz PA, Roberts JS, Bergsman TM, Caley SM, Heath WO, Miller MC, Moss RW, Schalla R, Jarosch TR, Eddy-Dilek CA. 1994. Six-phase soil heating accelerates VOC extraction from clay soil. Proceedings, SPECTRUM '94: International Nuclear and Hazardous Waste Management Conference, August 14–18, Atlanta, GA, USA. 10 p.

Gill WG. 1970. Electrical method and apparatus for the recovery of oil. U.S. Patent No. 3,642,066.

Görgényi M, Dewulf J, Van Langenhove H. 2002. Temperature dependence of Henry's Law constant in an extended temperature range. Chemosphere 48:757–762.

Gossett JM. 1987. Measurement of Henry's Law constants for C1 and C2 chlorinated hydrocarbons. Environ Sci Technol 21:202–208.

Hagedorn AR. 1976. Oil recovery by combination steam stimulation and electrical heating. U.S. Patent No. 3,946,809.

Harvey AH, Govier JP. 1980. Petroleum production method. U.S. Patent No. 4,228,853.

Heron G, Christensen TH, Enfield CG. 1998. Henry's Law constant for trichloroethylene between 10 and 95 degrees C. Environ Sci Technol 32:1433–1437.

Heron G, Carroll S, Nielsen SG. 2005. Full-scale removal of DNAPL constituents using steam-enhanced extraction and electrical resistance heating. Ground Water Monit Remediat 25:92–107.

Heron G, Parker K, Galligan J, Holmes TC. 2008. Thermal treatment of eight CVOC source zones to near nondetect concentrations. Ground Water Monit Remediat 29:56–65.

Imhoff PT, Frizzel A, Miller CT. 1997. Evaluation of thermal effects on the dissolution of a nonaqueous phase liquid in porous media. Environ Sci Technol 31:1615–1622.

Johnson P, Dahlen P, Triplett Kingston J, Foote E, Williams S. 2009. State of Practice Overview: Critical Evaluation of State-of-the-Art In Situ Thermal Treatment Technologies for DNAPL Source Zone Treatment. ESTCP Project ER-0314. ESTCP, Arlington, VA, USA. http://www.serdp.org/. Accessed July 20, 2012.

Kasevich RS, Price SL, Marley MC, Faust DL. 1996. Thermal enhancement for site remediation using radio frequency heating (RFH). Proceedings, Annual Meeting Air Waste Management Association, Anaheim, CA, USA.

Kaslusky SF, Udell KS. 2002. A theoretical model of air and steam co-injection to prevent the downward migration of DNAPLs during steam-enhanced extraction. J Contam Hydrol 55:213–232.

Kingston JLT, Dahlen PR, Johnson PC. 2010. State-of-the-practice review of in situ thermal technologies. Ground Water Monit Remediat 30:64–72.

Knauss KG, Dibley MJ, Leif RN, Mew DA, Aines RD. 2000. The aqueous solubility of trichloroethene (TCE) and tetrachloroethene (PCE) as a function of temperature. Appl Geochem 15:501–512.

Kueper BH, Redman JD, Starr RC, Reitsma S, Mah M. 1993. A field experiment to study the behavior of tetrachloroethylene below the watertable: Spatial distribution of residual and pooled DNAPL. Ground Water 31:756–766.

Lacombe PJ, Burton WC. 2010. Hydrogeologic framework of fractured sedimentary rock, Newark Basin, New Jersey. Ground Water Monit Remediat 30:35–45.

Lebrón CA, Phelan D, Heron G, LaChance J, Nielsen SG, Kueper B, Rodriguez D, Wemp A, Baston D, Lacombe P, Chapelle FH. 2012. Final Report: Dense Non Aqueous Phase Liquid (DNAPL) Removal from Fractured Rock Using Thermal Conductive Heating (TCH). Project ER-200715. ESTCP, Arlington, VA, USA.

Lemming G, Hauschild MZ, Chambon J, Binning PJ, Bulle C, Margni M, Bjerg PL. 2010. Environmental impacts of remediation of a trichloroethene-contaminated site: Life cycle assessment of remediation alternatives. Environ Sci Technol 44:9163–9169

Lide DR, Kehiaian HY. 1994. CRC Handbook of Thermophysical and Thermochemical Data. CRC Press, Boca Raton, FL, USA.

Martin EJ, Kueper BH. 2011. Observation of trapped gas during electrical resistance heating of trichloroethylene under passive venting conditions. J Contam Hydrol 126:291–300.

McGee BCW, Vermeulen FE, Vinsome PKW, Buettner MR, Chute FS. 1998. *In situ* decontamination of soil. J Can Petrol Technol 37:15–22

McMillan-McGee Corporation. 2011. Technical Description: ET-DSP™ *In Situ* Thermal Remediation. http://www.mcmillan-mcgee.com/mcmillan-mcgee/technology/etdsp-process-description.php. Accessed January 31, 2013.

NRC (National Research Council). 2005. Contaminants in the Subsurface: Source Zone Assessment and Remediation. National Academies Press, Washington, DC, USA.

Pennell KD, Pope GA, Abriola LM. 1996. Influence of viscous and buoyancy forces on the mobilization of residual tetrachloroethylene during surfactant flushing. Environ Sci Technol 30:1328–1335.

Perry RH, Green DW, eds. 1997. Perry's Chemical Engineers' Handbook, 7th ed. McGraw-Hill, New York, NY, USA.

Pizarro JOS, Trevisan OV. 1990. Electrical heating of oil reservoirs: Numerical simulation and field test results. J Petrol Technol 42:1320–1326.

Reid RC, Prausnitz JM, Poling BE. 1987. The Properties of Liquids and Gases, 4th ed. McGraw-Hill, New York, NY, USA.

Rowlinson JS, Widom B. 1982. Molecular Theory of Capillarity. Dover Publications, Mineola, NY, USA.

Satik C, Yortsos YC. 1996. A pore-network study of bubble growth in porous media driven by heat transfer. J Heat Transfer 118:455–462.

Sleep BE, Ma YF. 1997. Thermal variation of organic fluid properties and impact on thermal remediation feasibility. J Soil Contam 6:281–306.

Sleep BE, McClure PD. 2001. The effect of temperature on adsorption of organic compounds to soils. Can Geotech J 38:46–52.

Smith LA, Hinchee RE. 1993. *In Situ* Thermal Technologies for Site Remediation. Lewis Publishers, Ann Arbor, MI, USA.

Stegemeier G, Vinegar HJ. 2001. Thermal conduction heating for *in situ* thermal desorption of soils. In Chang HO, ed, Hazardous and Radioactive Waste Treatment Technologies Handbook. CRC Press, Boca Raton, FL, USA, pp 1–37.

Tiedeman CR, Lacombe PJ, Goode DJ. 2010. Multiple well-shutdown tests and site-scale flow simulation in fractured rocks. Ground Water 48:401–415.

Truex MJ, Gillie JM, Powers JG, Lynch KP. 2009. Assessment of *in situ* thermal treatment for chlorinated organic source zones. Remediat J 19:7–17.

Udell KS. 1996. Heat and mass transfer in clean-up of underground toxic wastes. In Tien C-L, ed, Advances in Heat Transfer Research. Begell House, Redding, CT, USA.

USACE (U.S. Army Corps of Engineers). 2006. Unified Facilities Criteria (UFC) Design: In Situ Thermal Remediation. Publication FC 3-280-05. USACE, Washington, DC, USA.

USACE. 2009. Design: In Situ Thermal Remediation. Publication EM 1110-1-4015. USACE, Washington, DC, USA.

USEPA (U.S. Environmental Protection Agency). 1995. In situ remediation technology status report: Thermal enhancements. EPA 542-K-94-009. Office of Solid Waste and Emergency Response, Washington, DC, USA.

Vinegar HJ, Stegmeier GL. 2002. Low cost, self regulating heater for use in an in situ thermal desorption soil remediation system. U.S. Patent No. 6,485,232.

Vinegar HJ, Stegmeier GL. 2003. Heater element for use in an in situ thermal desorption soil remediation system. U.S. Patent No. 6,632,047.

Vinegar HJ, Stegmeier GL, deRouffignac EP, Chou CC. 1993. Vacuum method for removing soil contaminants utilizing thermal conduction heating. U.S. Patent No. 5,190,405.

Vinegar HJ, Stegmeier GL, deRouffignac EP, Chou CC. 1994. Vacuum method for removing soil contaminants utilizing thermal conduction heating. U.S. Patent No. 5,318,116.

Wattenbarger RA, McDougal FW. 1988. Oil production response to *in situ* electrical resistance heating (ERH). J Can Petrol Technol 27:45–50.

Werth CJ, Reinhard M. 1997a. Effects of temperature on trichloroethylene desorption from silica gel and natural sediments. 1. Isotherms. Environ Sci Technol 31:689–696.

Werth CJ, Reinhard M. 1997b. Effects of temperature on trichloroethylene desorption from silica gel and natural sediments. 1. Kinetics. Environ Sci Technol 31:697–703.

White PD, Moss JT. 1983. Thermal Recovery Methods. PennWell Books, Tulsa, OK, USA. 384 p.

Yaws CL, Yang HC. 1989. To estimate vapor pressure easily. Hydrocarb Process 68:65–70.

CHAPTER 15

COMBINED REMEDIES

Dean F. Williamson

CH2M Hill, Gainesville, FL 32608, USA

15.1 INTRODUCTION

Contaminated sites containing dense nonaqueous phase liquids (DNAPLs) represent some of the most difficult and complex remediation challenges (USEPA, 2003; NRC, 2004). Achieving stringent remediation goals, such as meeting drinking water maximum contaminant levels (MCLs) across the entire site, may require removing or destroying nearly all of the contaminant mass. Several remedial technologies are capable of removing a high percentage of the mass (more than 90%), although lower removal efficiencies are commonly measured during full-scale cleanups (Stroo et al., 2003; McGuire et al., 2006). Removal of well over 99% of the mass, however, is required to meet typical risk-based closure criteria at most source zones. Even aggressive technologies such as *in situ* thermal treatment (ISTT), which can remove 99% or more of the contaminant mass under favorable conditions (Heron et al., 2005a), may not meet closure criteria, because the initial source concentrations can be three to five orders of magnitude higher than the MCLs (Johnson et al., 2009).

For these reasons, multiple remedial technologies often will be required to achieve site closure at DNAPL sites. In fact, recent guidance has recognized these technical limitations and in response has advocated development of integrated strategies at DNAPL sites (ITRC, 2011). An integrated strategy has many components, including developing appropriate remedial objectives and performance metrics, but the remediation component is specifically envisioned as a combination of "treatment technologies applied in sequence or in parallel, designed to optimize performance and take advantage of potential synergistic effects" (ITRC, 2011). In this chapter, any such technology combination is classified as a combined remedy or a deliberate combination of two or more treatment technologies that is designed to provide a more effective or efficient overall remedy than is possible when using any one technology alone.

Since there are several demonstrated source zone technologies, each with its own strengths and limitations, numerous combinations are possible. Potential synergistic and inhibitory interactions between various technologies have been explored over the past 20 years, and some products and procedures have been developed to exploit such synergies. This chapter summarizes the rationale and relevant experiences for several of these technology combinations and the potential for improving source zone treatment through combined remedies.

15.1.1 Evolution of Combined Remedies

In practice, combined remedies have been implemented at DNAPL sites for as long as DNAPL remediation has been attempted. However, combining technologies often has been a matter of necessity rather than design, due to the difficulties in achieving complete remediation of DNAPL source zones by using individual technologies. Often, site managers had to select, design and implement new technologies at sites that had been previously addressed, because an

B.H. Kueper et al. (eds.), *Chlorinated Solvent Source Zone Remediation*,
doi: 10.1007/978-1-4614-6922-3_15, © Springer Science+Business Media New York 2014

initial technology did not achieve the desired goals. Typically, these *de facto* combined remedies were not preplanned, and the potential for synergistic or antagonistic effects was not evaluated before implementing the initial remedy, because the initial remedy was expected to be the sole remedy required.

During the 1980s and 1990s, extensive research was conducted on DNAPL site remediation technologies, primarily on individual technologies including pump-and-treat, hydraulic displacement, dual phase extraction, air sparging, *in situ* chemical oxidation (ISCO) and *in situ* bioremediation (ISB). Not until the late 1990s, and increasingly in the 2000s, were articles published that addressed the concept of combining remediation technologies. For example, Jackson (2001) suggested that an initial aggressive *in situ* technology should be used to remove DNAPL, while less aggressive technologies may be more appropriate for polishing the remaining contamination. Other authors suggested more specific combinations, involving physical or chemical technologies such as ISTT or surfactant-enhanced flushing, followed by ISB as the polishing technology (Dettmer, 2002; Mravik et al., 2003; Ramsburg et al., 2004).

Researchers have continued to evaluate the benefits of combining an increasing number of technologies. For example, Christ et al. (2005) proposed coupling aggressive mass removal in the source zone with ISB and presented modeling results demonstrating the costs and benefits of this approach for different site conditions. Other research has addressed the potential effects of one treatment technology on subsequent technologies. For example, Azadpour-Keely et al. (2004) evaluated the impacts of field-scale ISCO and thermal treatment on post-treatment bioremediation. Hrapovic et al. (2005) conducted laboratory studies on the impact of permanganate on subsequent biodegradation of trichloroethene (TCE), and Siegrist et al. (2006) have tested the potential for coupling ISCO with surfactant-enhanced aquifer remediation (SEAR) and with ISB.

In practice, it has become more common for site managers to plan to combine technologies to treat sources, often involving aggressive treatment followed by monitored natural attenuation (MNA) (ITRC, 2011). During the last decade, several commercial products have been developed specifically to combine different treatment processes, for example, chemical and biological dechlorination. A wide range of technology combinations have been used commercially. For example, all of the most commonly used oxidants (catalyzed hydrogen peroxide, permanganate, activated persulfate and ozone) have been used in field-scale combined remedies, with or after a variety of other technologies, including pump-and-treat, air sparging/soil vapor extraction, electrical resistive heating and surfactant-based technologies (ITRC, 2005). In addition, all major ISCO oxidants have been used in combinations with subsequent ISB (Siegrist et al., 2013).

A workshop sponsored by the U.S. Environmental Protection Agency (USEPA) and the National Institute of Environmental Health Sciences to explore the potential for further development of combined remedies was held in 2006. Experts on various remediation technologies were brought together to explore the possibilities for selected remedy combinations. Key questions explored at the workshop included:

- What opportunities exist for combining specific technology combinations?
- What type of performance might we expect from combined technologies?
- What site conditions favor applying specific combined technologies?
- What are the most critical research needs for developing combined remedies?
- What are the barriers or limitations to combining remedies?

Although no proceedings from the workshop were published, the workshop sponsors believed that the workshop was successful in increasing the awareness of the potential benefits for combining remedies and in identifying potential areas for further research (J. Cummings, USEPA, Washington, DC, USA, personal communication, 2010).

Today, the concept of combined remedies is generally accepted as a necessary approach for remediating some DNAPL sites and as a potentially more cost-effective remediation approach at many others (ITRC, 2011). Evaluating combinations of remediation technologies is a standard element of both feasibility studies and optimization assessments at most DNAPL sites (Battelle, 2010a). Significant progress has been made in better understanding the synergistic benefits and practical implementation aspects of at least some of the many possible technology combinations.

15.1.2 Reasons to Use Combined Remedies

As noted above, the use of combined remedies may be necessary to achieve closure or effective DNAPL containment at some sites. At other sites, a combined remedy may offer economic or other benefits over any single technology. The key reasons for considering a combined remedy at any site will vary and could include technical, regulatory or business-related factors:

- Relatively large source zone size or large contaminant mass present
- DNAPL distribution throughout geologic units with significantly different hydrogeologic properties or characteristics
- Presence of mixed solvent DNAPLs, such as combinations of chloroethanes, chlorobenzenes and chloromethanes, or the presence of other contaminants (semivolatile organic compounds, metals, radionuclides) with the DNAPL
- Remedial action objectives (RAOs) that require a very high level of remediation performance, such as achieving drinking water MCLs within the source zone
- RAOs that require achieving a significant or rapid reduction in the dissolved contaminant flux migrating from the source zone
- Relatively short desired timeframe for achieving site closure
- Geologic setting or source zone architecture not favorable for optimal remedy implementation
- Presence of infrastructure or other facilities that constrain optimal remedy implementation

15.1.3 Challenges in Combining Technologies

A large number of conceptually feasible remedial technology combinations exist, but few have received significant research attention. Among the most studied to date are combinations including ISCO, notably ISCO/ISB and surfactant-enhanced ISCO, although work on other combinations is ongoing (described in later sections). More broadly, little research has been conducted on systematic design and implementation approaches for combining more than two remediation technologies or in comparing the relative efficiencies of two different combination pairs for remediating a specific site. An overarching technical framework for identifying or selecting optimal technology combinations for specific sites has not been developed. Significant additional work remains to be done to fully develop the principles of optimally combining remediation technologies.

For teams evaluating potential remedies at DNAPL sites, it is usually impracticable to evaluate more than a few of the many potential combinations of technologies that exist.

The limited information available regarding many technology combinations contributes to the difficulties that teams experience in identifying the technology combinations worth evaluating in detail. The remediation community would benefit from additional research and development of tools that allow the most practicable combinations of technologies to be identified for different site conditions.

To illustrate the relatively large number of technology combinations that could be evaluated at a site and the need for further research to improve these evaluations, consider a scenario in which potential remedies are being considered for initial application to treat a relatively small source (<0.25 acres, or <1,010 square meters [m^2]). A common objective for similar sites is to apply an active remedial technology only to the extent that MNA can be used as a polishing technology, to address any remaining contaminants. Figure 15.1 shows possible remedial technologies that might be considered for this representative scenario. For this scenario, the team conducting the remedy evaluation might evaluate in detail some or all of the five remedial alternatives shown.

In cases when none of the five technologies are capable of providing sufficient treatment so that MNA can effectively manage any remaining risk, the use of a sequential combined remedy may allow efficient treatment in stages. This may then eventually achieve the goal of MNA as a final remedial alternative. Figure 15.2 presents potential sequential technology combinations that would result from adding three potential follow-on technologies to the five initial primary technologies.

For this scenario, after eliminating like technology combinations (such as ISCO–ISCO), a total of 12 new sequential combinations exist in addition to the original five alternatives.

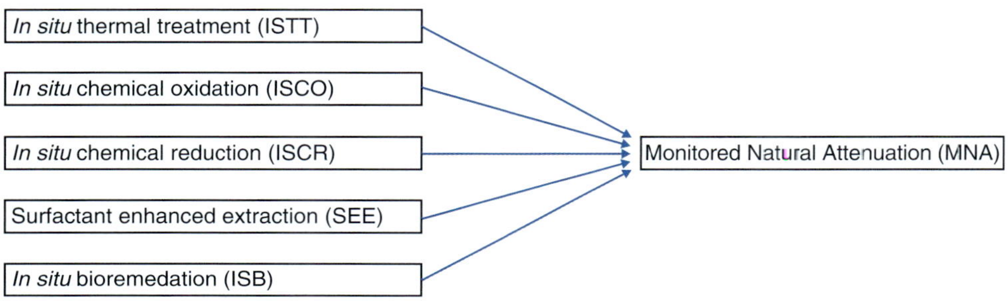

Figure 15.1. Representative remedial alternatives for treating small to medium source zone (primary remedy followed by monitored natural attenuation).

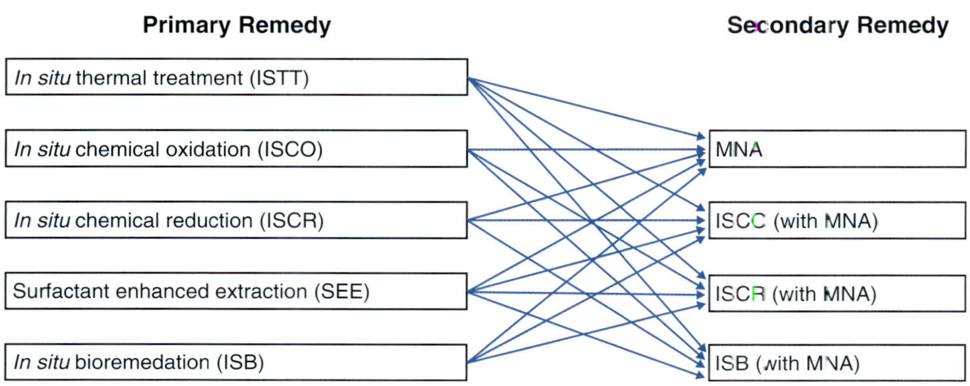

Figure 15.2. Potential sequential combinations for treating small to medium source zone.

As a practical matter, this number of possible combinations is too many for project teams to consider. Faced with so many choices, project teams typically will focus on identifying the primary remedies most likely to achieve the greatest degree of initial remediation and eliminating those that are likely to be less effective. They will then evaluate potential follow-up remedies that would be most compatible with the retained primary remedies.

While this approach can yield effective solutions, what may be missed is an appreciation of the synergistic effects that a technology combination may have, including combinations in which the primary treatment may not perform as well as other technologies initially, but may result in a more efficient or cost-effective solution over the longer term when combined with a highly compatible follow-on technology. For example, ISTT results in a treated zone that retains residual heat for a significant period after the cessation of heating. The application of *in situ* bioremediation after ISTT while the site temperature remains elevated could take advantage of the residual heat to enhance the effectiveness of these other treatment processes. Additional research on the effectiveness and synergies of combining remedies would allow for the most efficient solutions to be identified more easily and ensure that effective combinations are not being overlooked.

15.2 GENERAL TYPES OF COMBINED REMEDIES

The concept of combined remedies covers a broad range of applications and a large number of possible technology combinations. Various definitions of combined remedies are possible. The simplest definition would include any remedial approach that includes more than one remedial technology. Recognizing that the overall goal of research into combined technologies is to identify optimal and effective combinations, the following functional definition of an optimal combined remedy is suggested:

> *The sequential or concurrent application of two or more remediation technologies that use different physical, chemical, or biological treatment processes, targeting the use of individual technologies toward their optimal range while blending and transitioning between the different technologies, in space and time, to achieve remedial action objectives as efficiently as possible.*

A large universe of potential technology combinations exists within this broad definition. Because of the large number of potential combinations conceivable, it is useful to establish some general definitions or terms for referring to different categories of combined remedies to facilitate discussion. Three general categories of combined remedies can be identified:

- ***Sequential technology combinations*** – A sequential technology combination typically targets a specific source zone volume with two or more technologies applied sequentially, offset in time and largely independently of each other, although the beneficial aftereffects of the first technology may be used to enhance the treatment process of the subsequent technology. This combined remedy approach is sometimes referred to as a *treatment train*. Examples of sequential technology combinations include the application of a primary aggressive technology such as ISTT or ISCO, followed by a secondary polishing technology such as ISB or MNA.
- ***Concurrent technology combinations*** – A concurrent technology combination may target different areas of the source zone by applying multiple technologies at the same time. One example of this type of combined remedy is the use of soil mixing with clay and zero-valent iron (ZVI) in a low permeability portion of a source zone concurrently with application of ISCO in a more permeable portion of the source zone. Another example is the application of a source treatment technology such as ISTT to remove contaminant mass from the source, combined with a plume treatment technology such

as a permeable reactive barrier (PRB) applied downgradient to reduce contaminant flux from the source. DNAPL source zone containment strategies may also apply concurrent combinations of technology, with different technologies focused on controlling different potential chemical migration or exposure pathways.

- ***Coupled process combinations*** – Coupled process combinations include combined remedies that apply more than one physical or chemical process together to enhance the overall remediation result. An example of a coupled process combination is the application of a surfactant to an aquifer shortly before application of an oxidant, to increase the amount of dissolved phase contamination available for reaction with the oxidant. Another example is the use of an *in situ* remediation reagent that contains multiple active materials that stimulate different treatment processes, such as products that contain both ZVI and an organic substrate to achieve both abiotic chemical reduction and enhanced bioremediation.

In practice, a combined remedy can be applied in a planned manner, or one component of the remedy may be applied only if needed. These two categories of combined remedies (planned vs. contingent) represent different strategic decisions regarding the management of the site that should be considered when developing the overall remedy. In this chapter, these categories are defined as follows:

- ***Planned combinations*** – Planned combinations involve a commitment to implementing more than one technology, such as applying an initial DNAPL extraction followed by a polishing technology. Planned combinations are deliberately selected to take advantage of potential synergies, such as the post-treatment conditions created by an initial technology.
- ***Contingent combinations*** – A contingent combination involves identifying a secondary remedial technology that may be used in the event that a primary technology applied does not fully meet its treatment objectives. Contingent combinations are beginning to be used in Records of Decisions (RODs) and other remedial decision documents as a strategy to expedite implementation of the second technology if the initial technology does not meet its performance objectives.

The categories above are not mutually exclusive or intended to represent the full range of combinations or categories that may exist. Other classification methods could be used, but these categories are useful for describing the many different types of combined technologies.

15.3 STRATEGIES FOR COMBINING REMEDIAL TECHNOLOGIES

The different categories of combined remedies can result in a wide range of technology combinations. This section presents generalized strategies that can be used for combining technologies for sequential and concurrent combinations.

Managers of chlorinated solvent site remediation efforts must first develop an overall strategy for the site to guide the selection of remedial technologies. Source zone management strategies range from those based primarily on source zone treatment to those based primarily on source zone containment and include a variety of approaches that blend both treatment and containment. The range of management approaches constitute a spectrum of approaches with complete or nearly complete source zone containment on one end of the spectrum and complete or nearly complete source zone treatment or removal on the other end. The source zone management spectrum also includes combined containment and treatment and partial

Figure 15.3. DNAPL source zone management spectrum (photograph courtesy of B.H. Kueper, Queen's University, Kingston, ON, Canada).

contaminant mass removal approaches. Figure 15.3 shows the source zone management spectrum and some of the remedial technologies that may be appropriate for sites applying a particular strategy.

Teams often consider the following factors in developing an overall source zone management strategy:

- Nature and size of DNAPL source zone
- Site geologic and hydrogeologic setting, including aquifer geochemistry
- Proximity, type and number of potential receptors
- Business objectives of site owner, including desired time to achieve all RAOs and the site-specific differences between the containment and treatment costs

The specific management strategy selected will influence the particular remedial technologies that can achieve these RAOs. The management strategy selected will also influence the potential types of technology combinations that may be appropriate for the site. At sites that require a high level of treatment, sequential technology combinations may be applicable. At sites that need a high level of containment, concurrent technology combinations may be applicable. The potential exists, however, to apply nearly any type of technology combination (sequential, concurrent, coupled process or contingent combination) for nearly any source zone management approach selected. The following two sections discuss generalized approaches for applying sequential and concurrent remedial technologies.

15.3.1 Sequential Technology Combinations

Because of the wide range of site conditions and factors that contribute to the technology selection process, it is difficult to develop a single universal strategy applicable to active DNAPL remediation at all sites. However, based on the cumulative knowledge and experience gained from implementing various technologies at sites over the past several decades,

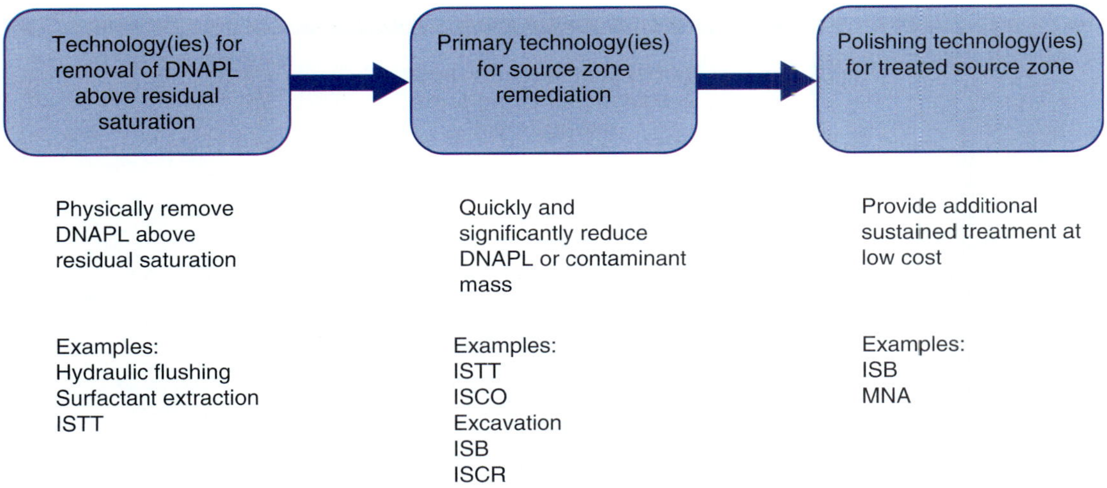

Figure 15.4. Example of an overarching sequential remedy strategy.

a generalized conceptual approach for sequential treatment of a source zone has been developed (Figure 15.4). This generalized approach provides a framework for considering specific sequential combinations and includes the following activities:

- An initial remedial technology designed to extract any DNAPL present above residual saturation to the extent practicable. This activity may not be required or applicable at every site.
- A primary remedial technology designed to remove or treat residual DNAPL to the extent practicable and to address the relatively high contaminant concentrations within the source zone.
- A polishing remedial technology designed to address any remaining soil and groundwater contamination.

At sites with large DNAPL masses or complex geologic or site settings, alternate strategies that include containment options also may be appropriate for consideration.

The general rationale for this conceptual sequential strategy is as follows:

- At sites with DNAPL above residual saturation, physical removal or extraction of this DNAPL is often a practicable and cost-effective step prior to application of methods that rely on *in situ* contaminant destruction via chemical or biological processes such as ISCO, ISCR, or ISB. Technologies such as hydraulic displacement, ISTT, and surfactant-enhanced extraction may be quicker, more cost effective, or more efficient at removing this DNAPL than ISCO, ISCR, or ISB.
- At sites where DNAPL is not present above residual saturation, or it is not practicable to remove the DNAPL, a primary treatment technology is applied within the source zone to rapidly treat the highest-strength areas within the source. Candidate primary treatment technologies are usually identified based on site conditions and selection criteria such as effectiveness, implementability and cost. Once the candidate technologies are identified, comparative criteria such as those prescribed by Superfund or other regulations are used to identify the technologies most applicable to the specific site. Typical performance requirements for primary treatment technologies may include the ability to directly treat DNAPL, to effectively treat high dissolved- and

sorbed-phase volatile organic compound (VOC) concentrations, to effectively reduce the mass discharge from the source zone (the plume loading), and to reduce or control any exposure of receptors to contaminants. Primary technologies typically are intended to yield results within a short timeframe (several months to a year) after implementation.

- At many sites, the available source zone technologies will not result in complete DNAPL removal or achieve closure criteria such as drinking water MCLs, and therefore, a secondary or polishing technology will be required. Secondary technologies often are implemented after the primary technology has reached its point of diminishing returns. Typical performance requirements for polishing remedies may include the ability to enhance or sustain treatment results achieved by the primary technology, to further reduce concentrations of soil and groundwater contaminants, to reduce dissolved phase flux from the source zone, and to reduce or control exposure of receptors to contamination. The polishing technology often is selected because it has lower capital and operations and maintenance costs and is a more passive technology than the primary one. The polishing technology may be targeted at addressing contamination or the flux related to contamination that remains in finer-grained portions of the source zone, from which transport into the main groundwater flow system is controlled primarily by diffusion. The polishing technology may also be applied for a significantly longer period of time than the primary technology.

Table 15.1 summarizes commonly applied remedial technologies for active remediation of chlorinated solvent source zones and identifies the typical role in the conceptualized source zone treatment approach for each technology. A summary of potential combined remedies (a total of 32 combinations) that could be assembled from the technologies in Table 15.1 is presented in Table 15.2. Research on potential synergistic and antagonistic interactions has been completed on only a few of these potential technology combinations.

In practice, it is often possible to reduce the number of combinations to be considered, based on factors such as the RAOs or site conditions. For example, air sparging may not be practicable in some geologic settings, such as where contamination is present within the saturated zone with interbedded higher- and lower-permeability layers. Other factors, such as the contaminant mass distribution between aquifer zones of higher and lower permeability, the degree of contamination present, and the short- and long-term RAOs, also can influence the selection of specific remediation technologies and eliminate some potential technology combinations from detailed consideration.

Table 15.1. Functional Role for Commonly Used Remediation Technologies in Generalized Sequential Treatment Strategy

Mobile DNAPL Extraction Technologies	Source Zone Primary Treatment Technologies	Source Zone Polishing Technologies
Hydraulic displacement	Excavation	*In situ* chemical oxidation
In situ thermal treatment	*In situ* thermal treatment	*In situ* bioremediation
Surfactant-enhanced extraction	Surfactant/cosolvent flushing	*In situ* chemical reduction
	In situ chemical oxidation	*In situ* air sparging
	In situ bioremediation	Natural attenuation
	In situ chemical reduction	
	Soil mixing with ZVI or other reagent	
	In situ air sparging	

Table 15.2. Sequential Combinations That Can be Assembled Based on Technologies from Table 15.1

Primary Treatment Technology	Secondary/Polishing Treatment Technologies				
	In Situ Chemical Oxidation	*In Situ* Bioremediation	*In Situ* Chemical Reduction	*In Situ* Air Sparging	Natural Attenuation
Excavation	X	X	X	X	X
In situ thermal treatment	X	X	X	X	X
In situ chemical oxidation		X	X	X	X
In situ bioremediation	X		X	X	X
In situ chemical reduction	X	X		X	X
Soil mixing w/ZVI or other reagent					X
Surfactant/ cosolvent flushing	X	X	X	X	X
In situ air sparging	X	X	X		X

Conceptually, a combined sequential technology for a site can be assembled by simply selecting a technology from each category in Table 15.1 and applying them sequentially. However, developing an optimal approach generally requires careful consideration of a variety of factors, including the geologic and hydrogeologic setting, specific contaminants present, aquifer geochemistry and the remedial action objectives, as well as any synergistic and antagonistic effects that a technology may have on subsequent technologies. Post-implementation impacts of each of the technologies listed in Table 15.1 are briefly described in the following section and summarized in Table 15.3. The reader is encouraged to review more detailed technology descriptions, such as those in earlier chapters of this book, and other references to better understand the specific potential impacts of each of these technologies at a particular site.

15.3.1.1 Post-treatment Impacts of Primary Source Zone Technologies

In considering any sequential technology combination, it is particularly important to understand the post-treatment impacts of the initial technology used. In most cases, these technologies are relatively aggressive and can be expected to cause significant and lasting changes in the subsurface environment. These changes may affect several features of the system, including the geochemical, microbiological, and hydrogeological conditions, and thus may impact the performance of subsequent technologies in a variety of ways. The following subsections discuss what is known of these post-treatment impacts for individual technologies.

Hydraulic Displacement

Little literature has been published on the post-implementation impacts of hydraulic displacement (see Chapter 8 in this volume) on subsequent remedial technologies. However, based on the manner in which hydraulic displacement is implemented (typically involving the extraction of groundwater and nonaqueous phase liquid (NAPL) and potentially the injection of

Table 15.3. General Summary of Post-application Impacts of Primary Treatment Technologies

Technology	Post-application Impacts to Consider	References
Hydraulic displacement	Minimal post-application impacts expected to the aquifer. Groundwater and DNAPL extraction will typically cause little change in overall groundwater geochemistry. If water injection is implemented with this remedy, some groundwater geochemistry changes may occur, depending on the source of injected water. This technology is compatible with various follow-up remedies.	Richards et al., 2005
In situ thermal treatment	Elevated temperature will remain in thermally treated area for some period at ISTT shutdown, ranging from several months up to a year or longer, depending on site conditions. Contaminant solubility and bioavailability may be elevated during this period. The temperature of downgradient groundwater may increase temporarily. Significant changes to microbial populations and community structure and diversity are likely to occur. Thermophilic and hyperthermophilic bacterial populations may be stimulated during heating. Changes in redox conditions, compared to pre-treatment conditions, may occur, depending on thermal process used. An increase in bioavailable organic carbon may occur, which may stimulate subsequent bioremediation. Bioaugmentation may be appropriate if *in situ* bioremediation is considered as a follow-up remedy. Abiotic and biological degradation rates may be elevated with elevated temperatures.	Beyke and Fleming, 2005; Davis, 1997; Dettmer, 2002; Friis et al., 2005, 2007; Heron et al., 2005b; Johnson et al., 2009; Kingston et al., 2010; Pennell et al., 2009; Powell et al., 2007; Richardson et al., 2002; Truex et al., 2007; USACE, 2009; USEPA, 2004
Surfactant/cosolvent flushing	Some amount of surfactant or cosolvent will remain in the aquifer, which may enhance or hinder microbiological processes, depending on the specific surfactant/cosolvent, resulting in changes in microbial populations or community structure. Residual surfactant or cosolvent may act as organic substrate or electron donor, lowering redox potential and stimulating anaerobic biodegradation processes. Dissolved concentrations of contaminants of concern (COCs) or mass flux from source zone may change. Ganglia-to-pool ratio of DNAPL may decline.	Amos et al., 2007; McGuire and Hughes, 2003; Ramakrishnan et al., 2005; Ramsburg et al., 2004; Suchomel and Pennell, 2006
Soil excavation	Minimal changes are expected to aquifer outside the excavated volume. Excavation is not expected to result in significant changes to overall site geochemistry. The excavated volume is compatible with follow-up remedies and may be adapted for use for reagent delivery for application of a subsequent *in situ* technology.	Parsons, 2000; USEPA, 2009
In situ chemical oxidation	Geochemical changes will vary, depending on oxidant and process used. Many geochemical changes will be temporary and geochemistry will generally tend to revert toward pre-ISCO conditions after oxidant delivery is completed. Hydrogen peroxide is expected to attenuate quickly. Persulfate and permanganate may persist for weeks and months, respectively, keeping the redox potential (Eh) elevated. Some heavy metals, such as chromium, may occur at elevated concentrations in groundwater but will attenuate when Eh declines. Other impacts may include changes in pH	Battelle Memorial Institute, 2010b; Block et al., 2004; Brown et al., 2009a; Chapelle et al., 2005; Huling and Pivetz, 2006; ITRC, 2005; Jones et al., 2009; Krembs et al., 2010;

(*continued*)

Table 15.3. (continued)

Technology	Post-application Impacts to Consider	References
	(an increase or decrease, depending on the process), increase in alternate electron acceptors (dissolved oxygen, Mn^{4+} [in manganese oxides], iron, sulfate, carbon dioxide), increase in bioavailable organic carbon, and decrease in soil sorption sites. If permanganate is used, hydraulic conductivity may decrease in areas with large amounts of manganese dioxide (MnO_2) solids precipitate. Changes in microbial populations may occur but sterilization of aquifer will not. Populations of some microbes may rebound post-ISCO. Bioaugmentation may be required for some sites at which ISB is implemented after ISCO.	Luhrs et al., 2006; Siegrist et al., 2013; Sahl and Munakata-Marr, 2006; Siegrist et al., 2006; Sutton et al., 2010
In situ chemical reduction	Geochemical changes may vary, depending on reductant and ISCR process used. ZVI is most commonly used reagent. Depressed Eh may persist for some period. Other impacts may include an increase in pH, alkalinity, dissolved iron, dissolved manganese, and dissolved hydrogen. Changes in microbial populations are likely to occur, with bacteria that thrive at low redox potentials (such as sulfate reducers and methanogens) likely to proliferate.	Chen et al., 2002; McElroy et al., 2002; Dolfing et al., 2008; Brown et al., 2009b; Shetty et al., 2009; Kirshling et al., 2010
Soil mixing	If clay is used, hydraulic conductivity will be significantly lower after soil mixing. Depending on reagent used (e.g., ZVI versus an oxidant), geochemistry impacts will be similar to those listed for ISCO and ISCR. Lower hydraulic conductivity after mixing will increase difficulty in applying follow-up technologies that rely on liquid-phase injection of reagents.	NRC, 2004; Olson et al., 2012
In situ air sparging	Addition of oxygen may increase Eh of aquifer. Conceptually, geochemical changes could induce some plugging of aquifer, particularly if high concentrations of reduced iron or manganese are present; however, this has not been widely observed in practice. Aquifer expected to return to pre-sparging conditions after *in situ* air sparging (IAS) is completed. Addition of oxygen may negatively impact some anaerobic bacteria, such as *Dehalococcoides*. If *in situ* bioremediation is being considered as a follow-on remedy, bioaugmentation may be appropriate.	Sutherson, 1997; Leeson et al., 2002
In situ bioremediation	Addition of organic substrate as part of ISB will lower oxidation–reduction potential, shift microbial respiration from those using more energetic electron acceptors (aerobic respiration and nitrate reduction) to less energetic electron acceptors (iron reduction, sulfate reduction, methanogenesis). Post-ISB (enhanced reductive dechlorination [ERD]) conditions may also include depletion of alternate electron acceptors, elevated methane, and presence of residual total organic carbon (TOC). Iron sulfides and mixed-valence iron oxides, formed *in situ* due to the effects of geochemical changes due to ISB implementation, may also be present.	USEPA, 2000; ITRC, 2007; 2008

water to assist in creating the desired hydraulic gradients), the post-implementation impacts to the aquifer are not expected to be significant. Little if any change to the aquifer geochemistry would be expected from this technology, and hydraulic displacement should be compatible with all follow-up technologies.

In Situ Thermal Treatment

In situ thermal treatment may cause a variety of impacts to subsurface conditions. The thermally treated volume remains at elevated temperature for a period of time, from several months to over a year, depending on site conditions after thermal treatment is completed (Friis et al., 2005; Chapter 14 this volume). During cooldown, abiotic processes may be more active than biological processes above 70 degrees Celsius (°C), with biological processes predominating below 70°C (Powell et al., 2007). Microbial reductive dechlorination is likely to be negligible above 40°C (Pennell et al., 2009). The heating and cooling processes associated with ISTT may cause significant changes in soil microbial diversity and populations (Dettmer, 2002; Richardson et al., 2002). Some microorganisms have been found to be capable of surviving thermal treatment, and the rate of cooling appears to have an impact on bacterial survival with greater post-thermal bacterial populations associated with slower cooling (Richardson et al., 2002). Total bacterial populations after thermal treatment can rebound to levels similar to those measured before thermal treatment (Azadpour-Keely et al., 2004).

Friis et al. (2007) found that no *Dehalococcoides* (*Dhc*) bacteria survived in soil microcosms heated to 100°C, but reduced levels of other dechlorinating bacteria survived the heating regime. These data suggest that although some bacteria are expected to survive thermal treatment processes, the ability for naturally occurring bacteria to completely dechlorinate solvents may be temporarily lost after thermal treatment. Repopulation of the thermally treated volume by advection of groundwater containing dechlorinating bacteria into the area may allow for recovery of dechlorinating bacteria. Alternatively, bioaugmentation may be necessary to optimize ISB after thermal treatment.

Changes in redox conditions after thermal treatment have not been well documented. Heron et al. (2005b) indicated that steam injection would likely increase the groundwater redox potential (particularly if air is co-injected with steam), that thermal conductive heating would likely have relatively small impact on redox conditions, and that electrical resistance heating (ERH) may induce a slightly lower redox potential after heating. Friis et al. (2005) observed few changes in redox potential in heated microcosms and at a field application of ERH.

Application of ISTT may increase the amount of dissolved organic carbon (DOC) or water-soluble organics in the aquifer (Friis et al., 2005; Beyke and Fleming, 2005; Pennell et al., 2009). The increase in DOC may stimulate or enhance biological activity. Elevated temperature can increase contaminant solubility and decrease the relative amount sorbed to soil (Davis, 1997); these effects can increase the bioavailability of the chemicals. These conditions may enhance the effectiveness of bioremediation after ISTT, provided that appropriate dechlorinating bacteria are present.

Overall, the post-implementation impacts of ISTT are expected to be compatible with most typical follow-up treatment technologies. Residual heat from an ISTT system may provide a useful, temporary post-implementation condition that can be used to enhance the effectiveness of some follow-up technologies. However, for some follow-up technologies, such as ISB, it is important to ensure implementation at a temperature compatible with the growth range of the key microorganisms (Fletcher et al., 2011).

Surfactant/Cosolvent Extraction or Flushing

Surfactant and cosolvent flushing involve the injection and recovery of solutions containing surfactants or cosolvents (Simpkin et al., 1999; Chapter 11 this volume). Some portion of the surfactant or cosolvent injected into the aquifer will remain after application of surfactant or cosolvent flushing, but little published information is available regarding concentrations. In general, the amount of residual surfactant or cosolvent remaining in an aquifer after application of this process will depend on the initial concentrations injected and the thoroughness of the extraction process.

Surfactants have the ability to affect microbiological activity of the endogenous bacteria present within the aquifer at a site. The effects of surfactants on the physiology of microorganisms range from inhibition of growth due to surfactant toxicity to stimulation of growth caused by the use of surfactants as a co-substrate (Volkering et al., 1998). The toxic effects of surfactants on bacteria are typically caused by two factors: disruption of cellular membranes by interaction with lipid components and reactions of surfactant molecules with proteins essential to the functioning of the cell.

Residual surfactants or cosolvents can serve as electron donors, stimulating dechlorinating microorganisms after flushing is stopped. For example, Ramsburg et al. (2004) showed that anaerobic biodegradation of perchloroethene (PCE) was enhanced after flushing with a nonionic surfactant (Tween® 80) at the Bachman Road site. Similarly, post-treatment monitoring at a PCE-impacted site where cosolvent extraction with ethanol was used also found enhanced anaerobic biodegradation of PCE (Mravik et al., 2003).

McGuire and Hughes (2003) found that some surfactants may impede reductive dechlorination of PCE. Tween® 80 inhibited dechlorination of solvents by certain *Dhc* species in laboratory tests (Amos et al., 2007). Because of the range of effects of surfactants on reductive dechlorination activity, the potential impact of the selected surfactant on microorganisms should be considered prior to implementing surfactant flushing.

Post-treatment conditions at surfactant/cosolvent flushing sites may include lower redox potential and increased DOC due to the residual surfactant or cosolvent. Groundwater concentrations and mass flux from the source zone may be significantly reduced compared to pre-treatment conditions (Suchomel and Pennell, 2006).

Residual surfactants in an aquifer also have the potential to interact with subsequently applied oxidants. Dugan et al. (2009) evaluated 72 surfactants for compatibility with permanganate and found a wide range in reactivity between the surfactants and permanganate. Some surfactants, such as Tween® 80, were highly reactive with permanganate, while others, such as Aerosol® OT, were minimally reactive with permanganate. These tests suggest that oxidants applied after surfactant or cosolvent flushing may react to varying degrees with the residual surfactants or cosolvents, potentially causing an increased oxidant demand.

Overall, surfactant/cosolvent flushing is expected to be compatible with a variety of follow-up treatment technologies. However, the impact of residual surfactant or cosolvent on microorganisms or reactions with subsequently applied reagents such as oxidants should be considered prior to implementing follow-up remedies.

Soil Excavation

Soil excavation can be an effective technology for removing DNAPL and other forms of contaminant mass. Where conditions are suitable, it may remove up to 99% of contaminant mass (Parsons, 2000). Excavation has been an element of combined remedy strategies at several DNAPL sites that have achieved regulatory closure (USEPA, 2009). Excavation is easily implemented at relatively shallow depths, up to approximately 40 feet (ft) (12 m) below ground surface (bgs), using conventional excavation equipment. Excavation is not typically applied very deeply into the saturated zone, due to the need for dewatering the excavation. Dewatering may be used to increase the depth of an excavation. However, DNAPL located in saturated portions of the target excavation zone may be mobilized downward if that aquifer volume becomes unsaturated, due to the lower residual saturation capacity of unsaturated soil versus saturated soil. This is of particular concern at sites without a low permeability layer, which may impede such downward DNAPL migration.

Excavation is compatible with many follow-up remedial technologies. Excavation is not expected to directly change geochemical or hydrogeologic conditions outside of the excavation volume. The excavated volume may be backfilled with a permeable material like pea gravel and used for reagent delivery for subsequent treatment technologies, such as ISCO or ISB. For example, if contamination remains below the excavated area, a high-concentration ISCO solution could be delivered into the excavation bottom, from which it may flow downward into deeper zones due to hydraulic or density-driven flow.

Overall, excavation is expected to be compatible with a variety of follow-up remediation technologies.

***In Situ* Chemical Oxidation**

Post-treatment impacts following application of ISCO (see Chapter 9 this volume) will generally vary depending on the oxidant and chemical process applied. The redox potential (Eh) of the groundwater will remain elevated until the oxidants have attenuated. Catalyzed hydrogen peroxide (CHP) and ozone are generally the most short-lived oxidants, with persulfate and permanganate capable of sustaining activity for weeks to months, respectively (ITRC, 2005). Sra et al. (2010) found that persulfate decomposition in uncontaminated media followed first-order rate law for all aquifer materials tested, with the half-lives for persulfate varying widely, from 2 to 600 days.

The pH of the groundwater post-ISCO may vary, depending on the oxidant selected. For example, application of CHP is often conducted at a pH in the range of 3.5–5, and acids may be added prior to CHP application to modify the natural aquifer pH into this range as part of the ISCO process (ITRC, 2005). Alternately, base-activated persulfate may require an increase in the pH to approximately 10.5 or higher as part of the ISCO process (Block et al, 2004). Permanganate is effective over a range of pH conditions, and therefore, pH modification is typically not implemented with permanganate. However, permanganate can increase or decrease pH depending on site conditions (Nelson et al, 2001; Crimi and Siegrist, 2003). Generally, these modifications to pH are expected to be temporary due to buffering capacity of the soil (ITRC, 2005).

Moore (2008) evaluated post-implementation impacts of permanganate at 30 field sites and reported that several geochemical parameters, such as manganese, pH, and oxidation–reduction potential (ORP), began to approach pre-ISCO levels within 2 years regardless of site and design conditions. In addition to pH, post-ISCO impacts will include a change in the amount and type of electron acceptors present. Depending on the ISCO process used, these changes could include increases in dissolved oxygen (DO), manganese dioxides, ferric iron, sulfate, and carbon dioxide. These additional electron acceptors may influence subsequent microbial activity. Microbial populations may decline during ISCO but are expected to rebound after ISCO is completed (Sahl and Munakata-Marr, 2006). *In situ* chemical oxidation also may increase substrate reactivity and bioavailability.

The application of permanganate leads to formation of manganese dioxide minerals, typically represented as MnO_2. Excessive precipitation of MnO_2 solids that may occur near DNAPL–water interfaces may lead to plugging of portions of the aquifer (MacKinnon and Thomson, 2002) or a reduction in relative hydraulic conductivity (Schroth et al., 2001). Recent testing has shown that the addition of polyphosphate with a permanganate solution may mobilize MnO_2 precipitates away from a TCE-permanganate reaction zone and reduce the formation of a mass transfer resistant film over residual DNAPL (Sra, 2010). The presence of MnO_2 solids may also increase subsequent electron donor demand if ISB is subsequently implemented, resulting in hydrogen consumption and inhibitory activity of dechlorinating bacteria (Geosyntec, 2010).

Overall, ISCO is expected to be compatible with a variety of follow-up treatment technologies. However, depending on the specific technology selected for subsequent application, consideration should be given to actions to mitigate residual impacts from oxidants.

In Situ Chemical Reduction

The most common ISCR reagent is the various forms of ZVI, including microscale, nanoscale, and emulsified ZVI (see Chapter 10 this volume). Because the ZVI in each of these consists of particulates, injection methods used for liquid-phase reagents are often not effective for ZVI delivery. Consequentially, ZVI is often delivered via hydraulic or pneumatic fracturing. For this reason, impacts from an ISCR application may include those that result from the delivery method and may include changes in the hydraulic conductivity within the injection zone, creation of preferential flow pathways, and physical displacement of contaminants.

There is relatively little published information regarding post-ISCR geochemical conditions. Residual effects due to the chemical processes of ZVI may include elevated pH due to generation of hydroxyl anions as part of the ZVI reaction process, strongly reducing conditions (which could be a positive effect if the technology is followed with ERD) and an increase in ferrous and ferric iron concentrations (Gavaskar et al., 2005). However, the pH and redox conditions following the ISCR application typically return to baseline conditions within a reasonable timeframe. Overall, ISCR is expected to be compatible with most typical follow-up treatment technologies.

Soil Mixing

Hydrogeologic conditions within an area subjected to soil mixing will be significantly different than premixing conditions. The mixing process will physically disrupt vertical and horizontal heterogeneities within the mixed zone, creating more uniform conditions throughout the mixed zone, although the mixed soil will not necessarily be homogenous throughout the mixed zone (USEPA, 1991). If clay is mixed into the soil during the process, the overall hydraulic conductivity of the treated volume may be reduced by a factor of 300–1,000 (Olson et al., 2012). Mass flux from the mixed volume zone is expected to decline significantly as well. This lower hydraulic conductivity may make application of follow-up remedies that rely on gas- or liquid-phase injection (such as *in situ* air sparging [IAS], ISCO, or ISB) difficult to apply within the soil mixed volume.

In Situ Air Sparging

In situ air sparging involves the injection of air into the saturated zone to effect volatilization of VOCs from the dissolved phase to the gas phase. Little published information regarding post-IAS geochemical impacts is available. The injection of air may raise the ORP of the aquifer, creating aerobic conditions. Precipitation of dissolved reduced metals, such as divalent iron and manganese, may occur during IAS. Introduction of oxygen may also be harmful to certain anaerobes involved in reductive dechlorination, such as *Dhc*. Therefore, bioaugmentation may be appropriate if ISB is implemented after IAS. Overall, IAS is expected to be compatible with most follow-up treatment technologies, although establishing reducing conditions after IAS may require additional effort and costs.

In Situ Bioremediation

In situ bioremediation of many chlorinated solvents can be achieved either anaerobically, typically via enhanced reductive dechlorination (ERD), or aerobically via cometabolic degradation (see Chapter 12 this volume). A few chlorinated solvents, such as vinyl chloride and chlorobenzenes, can also be biodegraded through aerobic direct oxidation. In practice, anaerobic biodegradation methods are most commonly implemented for chlorinated ethanes and ethenes for both source zones and plumes. A variety of guidance documents have been developed that present information regarding implementation of ERD (USEPA, 2000; ITRC, 2007; ITRC, 2008). Complete dechlorination via ERD requires strongly reducing geochemical conditions, the presence of adequate fermentable organic substrate and an appropriate microbial community, proper pH and nutrients, and the absence of inhibitory chemicals, such as chloroform.

Enhanced reductive dechlorination can cause several post-treatment impacts to the aquifer. Key impacts may include the following:

- *Low redox conditions and depletion of alternate electron acceptors.*

  Enhanced reductive dechlorination is typically implemented by injecting a fermentable organic substrate, an electron-rich material that functions as an electron donor for the desired anaerobic reactions. Injection of such a material is often referred to as biostimulation. The addition of the electron donor results in the creation of lower redox conditions as the more energetic electron acceptors, such as oxygen and nitrate, become depleted and microbiological processes shift to lower energetic electron acceptors, such as iron and sulfate. Sulfidogenic and methanogenic conditions are often created by this process.

  After implementing ERD, geochemical conditions may continue to reduce for some period of time, as residual carbon and microbial biomass that accumulated throughout the ISB process continue to provide sources of electron donors to sustain reducing conditions (Adamson et al., 2011). If upgradient groundwater entering the treated zone is aerobic or has a higher redox condition than within the treated zone, the aquifer redox may be expected to rise over time, returning toward pre-ISB conditions. Depletion of alternate electron acceptors, such as oxygen, nitrate, and sulfate, may continue in the aquifer for some time following completion of ERD. Incoming groundwater may replenish these electron acceptors over time.

- *Elevated concentrations of reduced species of some labile metals.*

  Experience at many sites has shown the potential for concentrations of several redox-sensitive metals, including iron, manganese, and arsenic, to increase within an ISB treatment zone. This increase is due primarily to the reduction of naturally occurring iron- and manganese-containing minerals within the ISB treatment zone being used as electron acceptors, through a process referred to as dissimilatory iron reduction (Kappler and Straub, 2005). Ferric iron oxide and oxyhydroxide minerals are often present in aquifers and under appropriate conditions are highly susceptible to this process. Because arsenic is often present within the matrix of these minerals, their dissolution often results in arsenic entering the groundwater system in addition to iron and manganese.

- *Presence of metal sulfides and other mineral precipitates.*

  In situ bioremediation can cause both the creation of sulfidogenic conditions, in which sulfate is converted to sulfide, and an increase in dissolved iron and manganese concentrations due to dissimilatory iron reduction. Together, these processes can lead to the *in situ* formation of iron sulfide and manganese sulfide minerals (Becvar et al., 2008). In addition, mixed-valence iron hydroxide minerals (sometimes referred to as "green rust") may also form within ISB treatment zones. Some of these minerals possess some degree of dechlorinating capability and may provide for residual treatment of VOCs after active maintenance of an ERD system has ceased.

Generally, it should be feasible to implement a wide variety of remedial technologies after beginning ISB, although it is more common to use ISB after a more aggressive technology.

Technologies that rely on injections of oxygen or chemical oxidants may be impacted by the increased oxidant demand due to the residual substrate and microbial biomass, as well as by the presence of reduced metals and minerals.

15.3.2 Concurrent Technology Combinations

Concurrent combinations typically refer to multiple remediation technologies applied in different spatial areas of the contaminated aquifer volume. Categories of concurrent technology combinations include:

- Concurrent remedies for source treatment (e.g., ISTT for high-strength areas and ISB for the lower-strength regions)
- Concurrent remedies for source containment
- Concurrent remedies for separate treatment of the source and plume

The strategies for selecting concurrent remedies for each of these categories are likely to be different, because the RAOs for these different source zone strategies are likely to be different from each other. The first and second types of concurrent strategies are discussed briefly below. The third type, although worthy of discussion, is beyond the scope of this chapter and book, which addresses only source zones and not plumes resulting from source zones. The application of remediation technologies for treating chlorinated solvent plumes is covered in Stroo and Ward (2010).

15.3.2.1 Concurrent Remedies for Source Treatment

A variety of factors may affect the remediation technology selection strategy for applying multiple concurrent technologies within a treatment strategy:

- DNAPL mass distribution between zones of lower and higher permeability
- Presence of mixed waste (such as different types of chlorinated solvents, presence of semi-volatile organics, pesticides, heavy metals, or other contaminants) in some portions of the source zone
- Physical constraints for some portions of the source zone, such as the presence of buildings or other infrastructure, which require use of different technologies for addressing different portions of the source zone

Generally, the same remediation technologies commonly considered or applied for the sequential treatment of source zones, such as those listed in Table 15.1, are also considered for use as concurrent combinations within a treatment strategy. The concurrent combination strategy is different from a sequential combination in that the concurrent combination targets different spatial areas of the source zone, as opposed to treating the same spatial area at different times. However, both types of technology combinations must consider the potential interactions of one technology with the other if the technologies are applied in close enough proximity that they could impact each other. In that event, the technologies considered for combination should be compatible with each other and minimize the potential for deleterious interactions.

15.3.2.2 Concurrent Remedies for Source Containment

Source zone management strategies based on containment must ensure that potential pathways for contaminant migration and receptor exposure are effectively controlled. Key potential migration pathways would include vertical or lateral migration of DNAPL,

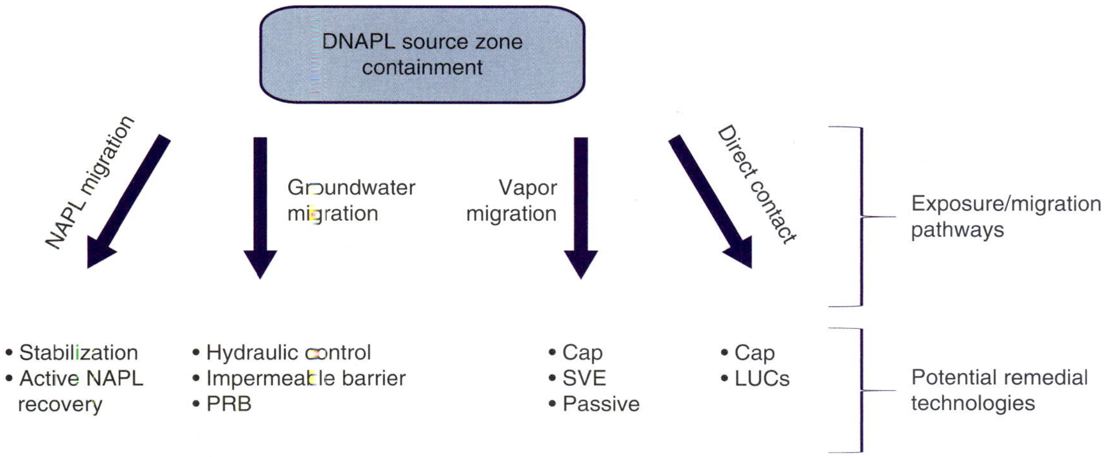

Figure 15.5. Generalized DNAPL containment strategy using concurrent combined remedies. Notes: LUC = land use controls; SVE = soil vapor extraction.

groundwater advection of dissolved contaminants, and vadose zone transport of contaminant vapors. In addition, potential exposure pathways that may allow for physical contact with the contaminants, such as dermal contact, should be controlled so that unacceptable risks to receptors do not occur.

Within this framework, a variety of remedial technologies could be applied in a concurrent manner to provide effective DNAPL containment. Figure 15.5 depicts a generalized containment strategy using concurrent technologies to address various potential exposure pathways, with representative technologies that could be applied for each of these functional roles.

At some sites, some degree of source zone treatment also could be applied within a broader containment strategy. For example, if hydraulic containment is implemented, the extraction and treatment of contaminated groundwater would provide some degree of source mass depletion. A large number of potential technology combinations, including concurrent and sequential combinations, could be developed for source zone containment strategies.

15.4 PRIOR APPLICATIONS OF SELECTED TECHNOLOGY COMBINATIONS

A literature review identified published research and papers related to some of the more frequently considered technology combinations. A summary of these reports is presented below.

15.4.1 ISCO–ISB Sequential Combination

A variety of papers and articles have been published regarding the combination of ISCO followed by ISB of chlorinated solvents. Most of these investigations focused on evaluating the effects of oxidants on microbial populations and the ability of the microorganisms to dechlorinate the solvents.

Sahl and Munakata-Marr (2006) reviewed the effects of ISCO on microbial processes. They identified several effects that may negatively impact microbial processes, including:

- Increases in Eh in groundwater
- Changes in pH, increases or decreases, depending on the oxidant used

- Changes in concentrations of specific electron acceptors in groundwater
- Antibacterial impacts, including cell death, decomposition of deoxyribonucleic acid (DNA) structure, and damage to protein and lipids

The authors reported that while biological processes may be affected in the short term, a rebound in biological activity, some of which may actively contribute to contaminant degradation, can generally be expected following ISCO due to:

- Lower contaminant concentrations after ISCO implementation
- Repopulation of ISCO-treated areas by upgradient groundwater migrating into the area
- Geochemical conditions reverting to pre-oxidation conditions

The authors reported that ISCO may produce more biodegradable substrates, such as by oxidation of fulvic or humic acids, which may enhance bioremediation activity of microbes downgradient from the zone of oxidant application or within the source zone after the geochemical effects of ISCO have attenuated.

Enhancement of *in situ* biodegradation of chlorinated solvents after ISCO implementation has been reported in at least one field application. Westersund et al. (2006) described the implementation of full-scale ISB after treatment of a TCE source zone using permanganate. Reductive dechlorination daughter products, including vinyl chloride (VC), were present in the source zone prior to ISCO treatment, indicating that indigenous bacteria at the site included dechlorinators. After ISCO, evidence of biological reductive dechlorination appeared. Oxidation–reduction potential levels declined, and the levels of daughter compound ratios increased in the lower portion of the treatment zone, with increases in both dichloroethene (DCE) and VC. Ethene also was detected, at up to 2.2 milligrams per liter (mg/L), within 18 months after ISCO injections. Based on these observations, biostimulation was implemented by injecting emulsified vegetable oil (EVO) throughout most of the original treatment zone. The substrate injection resulted in further reductive dechlorination of the VOCs. Dissolved manganese concentrations in groundwater increased after biostimulation, with a maximum concentration of 173 mg/L. The authors suggested that the increase in enhanced reductive dechlorination after ISCO may have been caused by an increase in electron donors or nutrients by the ISCO process, possibly due to partial oxidation of natural soil organic matter or biomass.

Droste et al. (2002) reported evidence of enhanced reductive dechlorination after implementing a pilot test of ISCO, using a combination of persulfate and permanganate. Persulfate was applied to the aquifer first to satisfy nontarget oxidant demand, followed by the application of permanganate for treatment of TCE. Data collected during the post-oxidation monitoring period indicated an unanticipated enhancement of the reductive dechlorination process in and downgradient of the treated area. Evidence of enhanced reductive dechlorination included an increased ratio of daughter to parent compounds. This increased ratio of daughter to parent products occurred for both *cis*-1,2-DCE:TCE and VC:*cis*-1,2-DCE. An evaluation of hydrogen concentrations, phospholipid fatty acid analysis, and other geochemical data suggested that enhancement of sulfate reduction occurred in the zone in which ERD was observed. The authors suggested that the following factors may have contributed to the enhanced reductive dechlorination:

- Enhanced growth of sulfate-reducing bacteria due to the formation of sulfate from persulfate
- ISCO-related degradation of naturally occurring complex organic carbon in the aquifer into simpler, more biodegradable organics
- Increased VOC bioavailability due to the destruction of sorption sites by the ISCO process

Macbeth et al. (2005) evaluated the impacts of permanganate on indigenous microbes in a PCE-impacted aquifer using molecular analytical methods (polymerase chain reaction (PCR), terminal restriction fragment length polymorphism, and quantitative PCR). DNA samples were collected prior to and approximately 1 year after ISCO application to assess overall changes to microbial populations. Initially, high permanganate concentrations reduced overall microbial biomass and diversity. After permanganate declined below a threshold level, a stimulatory effect was observed and biomass of some, but not all, bacterial populations increased. After further decline in permanganate concentrations, additional bacterial populations were stimulated and diversity increased. After 1 year, the overall diversity remained below the original diversity level.

Koch et al. (2007) were able to successfully achieve reductive dechlorination in a TCE source zone after previous application of permanganate was unsuccessful in meeting treatment objectives. Post-ISCO conditions included low pH and elevated hexavalent chromium in groundwater. Lactate, EVO, micronutrients, and pH buffer were injected to support biostimulation. Bioaugmentation was not implemented. Within 6 months after injection of the substrate, evidence of reductive dechlorination was observed, including an increase in the ratio of *cis*-1,2-DCE to TCE. The paper did not report whether VC or ethene were observed. A decline in ORP and DO and an increase in concentrations of acetate and propionate (indicative of lactate fermentation), signified conditions more favorable for reductive dechlorination, were created compared to initial post-ISCO conditions. Dissolved manganese concentrations increased from 5–10 micrograms per liter (μg/L) prior to biostimulation to 100–1,000 μg/L after biostimulation. Hexavalent chromium concentrations also declined after biostimulation. *Dhc* concentrations were <10 cells/milliliter, indicating relatively low *Dhc* populations.

Hrapovic et al. (2005) conducted laboratory studies to evaluate the feasibility of ISCO using permanganate as an initial source zone treatment technology followed by ERD for polishing of the source zone. Trichloroethene DNAPL source zones were created in laboratory columns packed with soil from a contaminated site. A permanganate solution was flushed through the columns, resulting in rapid treatment of dissolved phase TCE and partial DNAPL mass depletion. A post-oxidation flush with distilled water removed residual permanganate. A biostimulation phase using ethanol and acetate was then implemented, with a halorespiring culture, KB-1®, later added to several columns. The biostimulation phase was initially conducted with distilled water, followed by groundwater from a contaminated site.

No microbial activity was detected in either the biostimulated or bioaugmented columns during the post-oxidation period using distilled water for acetate and ethanol delivery. After switching to contaminated site groundwater, biodegradation of electron donors began within 20 days. After the onset of biodegradation of the electron donors, *cis*-1,2-DCE was observed in the bioaugmented columns but not in the columns not bioaugmented. Trichloroethene dechlorination occurred in the bioaugmented columns concurrently with an increase in dissolved manganese, indicating that manganese reduction was occurring. Dechlorination of *cis*-1,2-DCE to VC and ethene was observed only in one column after a third bioaugmentation event.

The results of this study indicate that implementation of bioremediation after permanganate flushing is a viable sequential treatment combination. However, permanganate flushing may significantly alter the indigenous microbial population if close contact with organisms is achieved. Establishing an effective *in situ* bioremediation system after permanganate flushing may require a period of biostimulation to first create conditions conducive to reductive dechlorination. If the indigenous groundwater bacteria at the site contain *Dhc*, biostimulation alone may be adequate to develop an effective ERD system. If *Dhc* are not naturally present at the site or if excessive delay occurs before the onset of complete dechlorination of TCE, bioaugmentation with a halorespiring culture containing *Dhc* may be required to establish a robust ERD system. Elevated dissolved manganese concentrations indicated the potential for formation of a significant dissolved manganese plume concurrent with the biodegradation of chlorinated solvents.

Sahl et al. (2007) conducted laboratory studies on the impact of permanganate flooding on a commercially available halorespiring culture (KB-1®). Test columns containing PCE DNAPL were inoculated with KB-1®, followed by application of permanganate solution at different concentrations and hydraulic loading rates. The tests were completed without reinoculation after applying ISCO to determine whether the original culture was sterilized by the oxidant or could resume dechlorination activity when pre-oxidation conditions were reestablished. During application of the oxidant, PCE reductive dechlorination daughter products were not detected. Post-oxidant application, after an initial period of inactivity, TCE and *cis*-1,2-DCE were detected in all bioaugmented test columns. Vinyl chloride and ethene were detected eventually in some but not all of the columns. The post-oxidation ratio of *cis*-1,2-DCE to PCE was greater than pre-oxidation ratio.

The results indicate that some dechlorinating microbes survived the permanganate flushing. While *Dhc* cells were detected throughout the column, using PCR methods, the test method was not capable of determining whether the *Dhc* cells were living or dead. The absence of VC and ethene in some columns suggests that *Dhc* activity may have been suppressed by the permanganate. Thus, the potential impacts to microbial communities and changes in the relative percentages of different microbial species must be considered when evaluating the potential for applying a combined ISCO (permanganate) and bioremediation strategy.

The tests revealed differences in the rate at which dechlorinating activity resumed after completing the oxidation period, based on the rate of oxidant solution application and its concentration. The most rapid rebound in dechlorinating activity occurred using a low concentration of oxidant applied as a large volume of solution at a high application rate. Thus, the design of the oxidant delivery system may be an important consideration in optimizing a combined ISCO–ISB remedy.

Elevated manganese concentrations were measured in the column effluent during the post-ISCO period, indicating that manganese reduction of manganese oxide solids that precipitated during permanganate application may be reduced post-oxidation. The results indicate that reductive dechlorination of PCE may occur concurrently with manganese reduction. Although the manganese mineral product that forms as a result of permanganate reactions is commonly considered to be a form of manganese dioxide (MnO_2), Loomer et al. (2010a) recently demonstrated that the valence of the Mn oxyhydroxides (MnO_x) formed through permanganate oxidation of TCE is between Mn(II) and Mn(IV), and that the Mn minerals formed are comprised of mixed-valence Mn. Loomer et al. (2010b) completed additional testing that showed that when reducing conditions were established, the MnO_x could be used as an electron acceptor and engage in reductive dissolution, as well as cation exchange. The authors concluded that after an initial equilibrium period during which a permanganate-treated aquifer readjusts to natural groundwater conditions, large amounts of MnO_x may buffer the redox state of the aquifer. If reducing conditions develop, subsequent release of dissolved Mn may occur along with trace metals that may be associated with the MnO_x solids. These metals may attenuate through the formation of rhodochrosite form of $MnCO_3$ and other minerals.

15.4.1.1 Conclusions Regarding the ISCO–ISB Sequential Combination

The following conclusions regarding implementation of ISCO followed by ISB can be reached based on these studies:

- ISB of chlorinated solvents via reductive dechlorination is generally compatible as a follow-up technology where ISCO has been previously implemented, although microbial populations may be temporarily adversely impacted by ISCO processes. Populations typically recover after ISCO although the community structure may be different than prior to ISCO.

- ISCO may enhance subsequent ISB through several mechanisms, such as increases in the DOC, improved bioavailability of contaminants through a decrease in sorption sites, or by enhancing the biodegradability of natural organic compounds through partial degradation.
- Available data suggest that the geochemical effects of ISCO on an aquifer, such as changes in pH and Eh, are generally temporary, and over time, the aquifer geochemistry tends to revert to pre-ISCO conditions.
- If application of bioremediation is desired before groundwater conditions have reverted to pre-ISCO conditions, or if conditions are not within the optimal range for bioremediation, groundwater conditions can be adjusted by adding electron donors, nutrients or pH buffers.
- Excess electron acceptors that may be present in the aquifer after ISCO, such as dissolved oxygen and sulfate, depending on the ISCO process used, will not interfere with establishing conditions conducive for ISB if an effective reagent amendment approach is implemented.
- If native bacterial populations have been impacted such that dechlorinating bacteria are not sufficiently active for effective bioremediation, a bioaugmentation culture can be added to the aquifer to facilitate ISB. It is important to establish appropriate reducing conditions in the aquifer prior to adding a bioaugmentation culture.
- Where permanganate is used as the oxidant preceding ISB, a dissolved manganese plume may be generated, and this possibility should be anticipated in the overall strategy for this combination.

15.4.2 Surfactant/Cosolvent–ISCO Combinations

The use of surfactants and cosolvents can be combined with ISCO processes in several different ways. Surfactants and cosolvents can be added before or concurrently with an oxidant to increase the dissolved phase contaminant concentrations, thereby increasing the efficiency of the oxidant. Alternately, surfactant/cosolvent flushing can be implemented first as a separate (primary) technology, with ISCO then being applied as a secondary or polishing technology. For both types of approaches, the surfactants, cosolvents and oxidants must be selected in a manner that optimizes treatment process chemistry and is compatible with the treatment objectives.

15.4.2.1 Surfactant/Cosolvent Flushing with ISCO: Sequential Combinations

One potentially effective sequential technology combination is to initially flush an aquifer with a surfactant or cosolvent to remove DNAPL, prior to using ISCO as a polishing step. Conrad et al. (2002) tested this combination in a laboratory-scale experiment, using the nonionic surfactant Tween® 80 before adding potassium permanganate. While the surfactant extraction process successfully removed approximately 90% of the TCE, subsequent application of permanganate caused significant formation of manganese dioxide minerals in close proximity to the remaining DNAPL, which limited the contact between the permanganate and remaining TCE. Additional details, such as the amount of surfactant remaining in solution when permanganate was applied and the overall compatibility of Tween® 80 and permanganate, were not reported.

Shiau (2008) described one-dimensional column testing that combined an initial surfactant flush to remove TCE DNAPL, using a variety of surfactants including sodium dioctylsulfosuccinate (Aerosol® OT-100) and linear alkyl diphenyl oxide disulfonate (Calfax®), followed by permanganate treatment. Trichloroethene concentrations in the column effluent during the surfactant flooding portion of the test reached as high as 100,000 mg/L. Subsequent ISCO treatment of the column using permanganate reduced the TCE in the column effluent to approximately 0.001 mg/L.

While no field-scale studies were found that have confirmed the effectiveness of this sequential technology combination, surfactant/cosolvent flushing followed by ISCO appears to be a viable combination that warrants further evaluation. For optimal implementation of this technology combination, the residual surfactant/cosolvent should have minimal impact on subsequent oxidant demand during ISCO application.

15.4.2.2 Surfactant-Enhanced ISCO: Coupled Process Combination

Surfactant-enhanced ISCO represents a coupled process combination that is conceptually similar to the sequential combination of surfactant/cosolvent flushing followed by ISCO. However, because the surfactant or cosolvent and oxidant are applied together, a high degree of compatibility between the reagents is required. Conceptually, the advantage of this combination over ISCO alone is that the amount of contaminant in the aqueous phase with which oxidant may react is greatly increased, leading to more efficient oxidant use. However, if the oxidant is overly reactive with the surfactant or cosolvent, the overall oxidant demand will increase, which may counteract any benefit achieved by the higher contaminant concentration. Thus, a key to success for this process combination is finding a compatible surfactant–oxidant combination that allows for effective oxidation of the contaminant with minimal reaction of the oxidant with the surfactant.

Li and Hanlie (2008) evaluated combinations of surfactants with permanganate oxidation of TCE, using batch, flow-through columns, and three-dimensional tank tests. Surfactants tested included Ninate®, sodium dodecyl sulfonate (SDS), Triton™ X-100, and Witconol™, at concentrations ranging from 0.1% to 1%. The results of the testing indicated that the observed pseudo-first-order rate constants for permanganate reaction with the contaminant increased with increasing surfactant dose. The observed pseudo-first-order rate constant increased as much as 3–5 times in the presence of SDS at a concentration above its critical micelle concentration. The researchers also noted that the fast reaction rate between TCE and permanganate may lead to the formation of manganese oxide precipitates at the interface of the TCE DNAPL, which may hinder overall treatment performance.

Dugan et al. (2009) tested the compatibility of 72 surfactants and 7 cosolvents with potassium permanganate for treating PCE. The results identified several surfactants that were effective in causing increased solubilization of PCE while maintaining relatively low reactivity with permanganate. The most compatible surfactants included sodium dioctylsulfosuccinate (Aerosol® OT-100), sodium hexadecyl diphenyl oxide disulfonate (Dowfax™ 8390), (C16 linear) sodium diphenyl oxide sulfonate (Calfax® 16L-35), and SDS. Of the cosolvents tested, *tert*-butyl alcohol was the most compatible.

These tests also revealed that combinations of these surfactants could increase the solubility of PCE more than the use of a single surfactant. Surfactant combinations resulted in an increase in PCE solubility from 200 mg/L without surfactant to as high as 62,700–107,000 mg/L with surfactants and sodium bromide. The comparative efficiency of permanganate oxidation of PCE under these conditions was not reported. This work resulted in a patented process combining surfactants, cosolvents, and chemical oxidation (Crimi et al., 2009).

Hoag et al. (2010) applied plant-derived surfactants, such as ethoxylated vegetable oils, and biodegradable citrus-based solvents such as citrus terpenes with and without chemical oxidants for treatment of light nonaqueous phase liquid and DNAPL. When combined with oxidants, the process is referred to as surfactant-enhanced ISCO. Solubility enhancement factors for PCE as high as 250 were reported using plant-based surfactants and cosolvents. Oxidants evaluated with these surfactants and cosolvents included activated persulfate and CHP. Significant increases in the mass of contaminants oxidized per mole of oxidant were reported.

Dugan et al. (2010) presented a critical review of literature related to the use of coupling surfactants and cosolvents with oxidants for enhanced DNAPL removal. They concluded that surfactant application with standard ISCO technology may enhance and improve removal effectiveness and therefore could markedly improve project life cycle costs. They presented the following rationale for coupling surfactant-enhanced extraction with ISCO:

- Surfactant-enhanced aquifer remediation (SEAR) only is useful for DNAPL phase remediation and not for dissolved phase COCs.
- In general, ISCO is more effective in treating the dissolved phase COCs.
- The combination of surfactants and ISCO could lead to significant savings in chemical use and project costs.
- Reuse and recycling of recovered surfactant should be minimized and/or completely eliminated.
- The combination has been developed and used commercially by at least one vendor, under the trade name S-ISCO® (http://www.verutek.com/s-isco-whitepaper-summary-0, accessed January 16, 2014).

15.4.2.3 Cosolvent-Enhanced ISCO: Coupled Process Combination

Like surfactants, cosolvents can enhance dissolution and desorption of contaminants, thus making the contaminant more available for treatment with ISCO. However, a cosolvent must be selected that will not significantly increase the oxidant demand. Otherwise, the efficiencies gained by treating higher contaminant concentrations may be offset by the increased oxidant demand.

Using permanganate, Zhai et al. (2006) evaluated the feasibility of coupling cosolvents with ISCO for enhancing PCE oxidation. A total of 17 cosolvents were first tested for their recalcitrance to permanganate oxidation. Of those tested, acetone and tertiary butyl alcohol (TBA or tert-butyl alcohol) were found to be the most recalcitrant and selected for further study. The effects of these cosolvents on enhancing PCE solubility were then evaluated. The effects of these cosolvents on interfacial tension also were evaluated to minimize the potential for DNAPL mobilization.

The results of the study indicated that cosolvent-enhanced ISCO with permanganate appears to be a viable treatment process. Both cosolvents significantly increased the solubility of PCE. The PCE solubility measured in water with 30% TBA was 4,076 mg/L, which is significantly greater than solubility values of 150–200 mg/L typically reported for PCE in water. In the absence of PCE DNAPL, the reaction rate constant for oxidation of PCE by permanganate was reduced by approximately 25% when 20% TBA or acetone were used, compared to reaction rates without cosolvents. Although the reaction rate with cosolvents is lower than without them, the much higher dissolved concentrations of PCE when using cosolvents provided an increase in the overall oxidation efficiency. In addition, the study found that acetone exhibited a much lower tendency than TBA to mobilize PCE DNAPL. This study indicates that efficiencies in solvent oxidation can be achieved with cosolvents. Subsequent measures to address the presence of residual cosolvents in groundwater may be required.

Conclusions Regarding Surfactant- and Cosolvent-Enhanced ISCO

The following conclusions regarding surfactant–/cosolvent–ISCO coupled process combination can be reached based on these studies:

- The amount of dissolved phase contaminant available for reaction with an oxidant can be increased by one or more orders of magnitude using surfactants and cosolvents.
- Oxidant–contaminant reaction rates can be increased by using suitable surfactants and cosolvents. In addition, the amount of contaminant oxidized per mole of oxidant can be increased using surfactants and cosolvents.
- Identification of a compatible surfactant/cosolvent and oxidant combination is critical to the successful implementation of this coupled process combination. If the surfactant or cosolvent is too reactive with the oxidant, efficiencies gained by higher contaminant concentrations can be lost due to oxidant consumption by the surfactant or cosolvent.
- When applying surfactants to remediation of DNAPL, consideration should be given to the possibility of inadvertent DNAPL mobilization. Application of surfactant should be implemented in a manner that eliminates or minimizes the potential for inadvertent DNAPL mobilization.
- Most studies of this coupled process combination have been conducted under laboratory conditions. The degree to which the success of this process can be achieved under field conditions has not yet been determined. However, based on the results of laboratory testing, this process combination appears to have potential for accelerating remediation at sites suitable for its application.

15.4.2.4 Polymer-Enhanced Surfactants and ISCO: Coupled Process Combination

Ongoing research is addressing the potential for injecting oxidants simultaneously with shear-thinning polymers to improve the delivery of the ISCO reagents, especially into the lower-permeability regions that are difficult to treat with injection-based technologies. Shear-thinning polymer solutions are non-Newtonian fluids that undergo viscosity reductions under conditions of shear stress. The use of polymers for improving delivery of liquid-phase reagents, such as surfactant-containing solutions, for environmental remediation is an idea adapted from enhanced oil recovery methods applied in the petroleum industry.

Martel et al. (1998) evaluated several polymers for use with surfactants to increase overall sweep efficiency and achieve more uniform treatment of an aquifer. Of the polymers tested, xanthan gum was found to be most suitable. The results of the laboratory testing concluded that injection of a xanthan solution after surfactant injection decreased fluid velocity in more permeable layers while increasing fluid velocity in less permeable layers and eliminated viscous fingering at the polymer/surfactant front. A xanthan solution pre-flush also helped to limit surfactant solution mobility and adsorption to solids.

Smith et al. (2008) evaluated the compatibility of polymers and chemical oxidants. Combinations of nontoxic polymers (xanthan and hydrolyzed polyacrylamide [HPAM]) and several chemical oxidants (potassium permanganate and sodium persulfate) were investigated to determine the suitability of these mixtures for polymer-enhanced ISCO applications. After 72 hours of exposure, only one of the polymer–oxidant combinations tested was able to maintain more than 10% of its initial viscosity. Combinations with sodium persulfate decreased solution viscosity more strongly than those with permanganate. Xanthan solutions retained a much larger portion of its original viscosity after contact with permanganate than did HPAM. The xanthan-permanganate combination retained its viscosity to a greater degree than other combinations tested. In the xanthan-permanganate experiments, oxidant demand ranged from 5% to 20% and was correlated to initial polymer concentration. Subsequent testing of the xanthan-permanganate combination on oxidation rate constants of PCE showed that the presence of xanthan did not hinder the oxidation rate of PCE by permanganate.

The successful identification of a compatible polymer–oxidant combination suggests the possibility of applying this technology in the field, although to date this combination has not been used commercially. Initial results are promising, however, and indicate it may be possible to achieve more uniform delivery of permanganate into aquifers with zones of varying hydraulic conductivity (McCray et al., 2010). However, testing also has shown that it probably does not make sense for highly heterogeneous sites (Crimi et al., 2013).

15.4.3 ISCR–ISB Combinations

Over the last few years, ISCR has been more frequently paired with bioremediation and MNA to address DNAPL zones. As discussed by Brown et al. (2009a), the biotic–abiotic interactions are either causative, when biological reactions are used to create reducing minerals, such as iron sulfide or reduced-iron oxides such as magnetite, or synergistic, when one process is used to enhance the reactivity or efficacy of the other.

The causative interactions occur when the metabolism of a carbon substrate results in reduction of ferric iron to ferrous iron and subsequent formation of magnetite, green rust, or other reduced-iron-bearing materials capable of mediating abiotic dechlorination reactions. This approach is exemplified by the Biogeochemical Reductive Dechlorination (BiRD) process (Kennedy et al., 2006). The sequential process includes the (1) application of a substrate (mulch in this case), (2) creation of iron- and sulfate-reducing conditions, (3) formation of iron sulfides, and (4) dechlorination of the chlorinated solvents via abiotic and biotic mechanisms. Shen and Wilson (2007) attributed as much as one half of the TCE removal to abiotic reactions with iron monosulfides and the remainder to biotic processes. The content of iron monosulfide in geological material can be estimated from a determination of acid volatile sulfide (AVS). Analyses of AVS are simple, affordable, and commercially available Brown et al. (2009b).

In the synergistic applications the abiotic and biotic reactions occur concurrently. Such applications have commonly included the following amendments:

- EHC® – A commercial ISCR product from FMC Corporation that envelops ZVI with hydrophilic carbon
- EZVI – A product developed in the early 2000s at the National Aeronautics and Space Administration's Launch Complex 34 (LC34) at Cape Canaveral Air Station, Florida that includes nanoscale ZVI particles contained within an oil emulsion droplet
- ZVI with a soluble carbon substrate (e.g., molasses or lactate)

The overarching objective of the ZVI-substrate combinations is to use the ZVI to quickly create strongly reducing conditions that are suitable for abiotic and biotic processes and the carbon to extend the duration and reach of treatment effectiveness. The benefits or objectives of each of the components are as follows:

ZVI

- Rapidly creates strongly reducing conditions (e.g., Eh <-550 millivolts (mV)), when many chlorinated solvents are highly unstable (Dolfing et al., 2008). The highly reducing environment supports abiotic (Gavaskar et al., 2005) and biotic dechlorination.
- Rapidly reduces chlorinated solvents via β-elimination and hydrogenolysis pathways.
- Generates hydrogen, which is used by the bacteria as an electron donor (Dolfing et al., 2008).

- Enhances the dissolution of the DNAPL, making more mass accessible for both microbial dechlorination and reaction with the ZVI (Martin et al., 2006).
- Minimizes the residual concentrations the chlorinated ethene daughter products and improves the long-term effectiveness of the biological component of the remedy.

Carbon

- Eliminates common competing electron acceptors by facilitating the biologically mediated reduction of oxygen, nitrate and sulfate (Brown et al., 2009a; Martin et al., 2006).
- Produces organic acids, which counters the generation of hydroxyl ions during the metal corrosion process (Brown et al., 2009b).
- Facilitates metal precipitation or immobilization with the reduction of sulfate (Dolfing et al., 2008).
- Enhances the long-term biological sequential dechlorination process.
- Enhances the solubility of the DNAPL, making more mass accessible for both microbial dechlorination and reaction with the ZVI (Martin et al., 2006).

A variety of papers and articles have been published regarding the effectiveness of various ISCR-ISB applications. These are briefly reviewed in the following sections.

15.4.3.1 Carbon Plus Zero-Valent Iron (EHC®)

The EHC® products combine a carbon source to stimulate biological activity with ZVI to rapidly destroy accessible contaminants. This combination can lead to rapid mass removal as well as long-term control of any residual contamination. For example, Shetty et al. (2009) observed rapid dechlorination of 1,1,2,2-tetrachloroethane (TeCA), TCE, and chloroform at locations where EHC® was present following the injections. Unfortunately, the product did not reach some locations because of the spatial heterogeneity at the site (a saprolite and weathered bedrock aquifer in North Carolina). As a result, contaminants remained after the initial injections, in areas outside the highly reducing influence of the ZVI.

Over time, however, the monitoring results showed that the residual contaminant concentrations were gradually decreasing, as the organic carbon stimulated biological reductive dechlorination over a 2-year period. Evidence of biodegradation included parent compound decreases coupled with daughter product concentration increases, increasing and persistent TOC concentrations, and low ORP and DOC levels throughout the observed treatment zones. Molin et al. (2010) found a similar synergistic effect over time and concluded that over time, diffusion, advection, and dispersion increase the effective distribution of EHC® constituents in the subsurface and thereby provide treatment well beyond the initial fractures.

15.4.3.2 EHC® with Bioaugmentation

EHC® can also be combined with bioaugmentation to increase the initial removal of DNAPL constituents. For example, Peale et al. (2010) describe the successful application of EHC® and a halorespiring bioaugmentation culture (KB-1®, Sirem Labs) to remediate a TCE DNAPL source zone (nearly 600 mg/L) in Portland, Oregon. EHC® and KB-1® were directly injected into the subsurface from 40 to 112 ft bgs within the source area and as a PRB just downgradient of the source (KB-1® was injected 7–14 days after the EHC®). Five months after the application, the TCE concentrations decreased by at least 98%, ethene concentrations increased above baseline levels, the *Dhc* population increased by several orders of magnitude, and downgradient monitoring data suggested that areas that were inaccessible to injection were being treated due to migration of dissolved organic carbon.

15.4.3.3 Emulsified Zero-Valent Iron

Alone, nanoscale ZVI (nZVI) is highly reactive, due to its high surface area per unit mass, and it can readily be injected into the subsurface using conventional injection technologies instead of pneumatic fracturing. However, nZVI is rapidly depleted because it is so reactive, and the particles have a tendency to agglomerate in many groundwater conditions, thereby counteracting the conventional delivery strategy. As described in O'Hara et al. (2006), the addition of nZVI with emulsified vegetable oil in a commercial product called EZVI provides the following potential enhancements to nZVI alone:

- Increases the contact between the DNAPL constituents and the ZVI by drawing the target compounds into the oil emulsion (sequestration) where the abiotic reaction occurs within a water phase surrounding the nZVI
- Stimulates biological reductive dechlorination to treat contaminants that are not chemically degraded
- Provides a protective layer around the iron that minimizes contact with inorganic compounds such as nitrate that would ordinarily accelerate the passivation of the nZVI

Bench-scale tests conducted by O'Hara et al. (2006) indicated that the EZVI approach was more effective in addressing DNAPL than using nZVI or emulsified oil independently. However, uniform distribution of the EZVI during field-scale testing was elusive despite using pneumatic fracturing, hydraulic fracturing, pressure pulse, and direct push injections.

15.4.3.4 ZVI and Soluble Substrates

Martin et al. (2006) provide a discussion on the use of an effective coupled ISCR–ISB strategy to remediate a PCE DNAPL source area in bedrock. Molasses was injected along the property boundary to create an anaerobic treatment zone that controlled off-site plume migration. It also was injected into the source area, where strongly reducing conditions supported the complete dechlorination of PCE. In addition, ZVI was added to the injection fluid (800 pounds [approximately 360 kilograms] in three locations using pneumatic fracturing) to help accelerate degradation and shorten the duration of the overall remediation. After declines of 77–96% in groundwater concentrations in the injection and monitoring wells, additional molasses was injected as a contingency measure to address the residual chlorinated volatile organic compounds.

Conclusions Regarding ISCR–ISB Combinations

The following conclusions regarding the ISCR–ISB process combination can be reached based on these studies:

- The combined use of a chemical reductant, such as ZVI, and organic materials to stimulate biodegradation is a compatible coupled process combination.
- ZVI may facilitate a rapid decline in dissolved phase VOC concentrations, while the organic materials facilitate a longer-term biological treatment of VOCs.

15.4.4 ISCR–ISCO Combinations

Little published information was found on combinations of chemical reductants and oxidants within the context of DNAPL remediation. Generally, applying chemical reductants and oxidants concurrently would not be expected to be a highly effective treatment process due to the potential for the reductant and oxidant to react with each other rather than with the target contaminant.

One study on the treatment of TCE in the combined presence of a chemical reductant and a chemical oxidant was found in the literature. Okwi et al. (2005) evaluated the impact of permanganate on the ability of granular ZVI to degrade PCE. The tests were designed to evaluate the impact of injecting permanganate upgradient of a PRB comprised of granular ZVI on the effectiveness of the ZVI for treating contaminants. The study found that permanganate rapidly oxidized the ZVI, forming insoluble precipitates of both manganese and iron. The precipitates formed as films on the surface of the iron, thus inhibiting transport of electrons from the iron surface. The study demonstrated that permanganate was not compatible with ZVI. Although no other studies on combining reductants and oxidants have been found, it seems unlikely that ISCR–ISCO combinations would provide significant treatment benefits over other available remedial technologies.

15.4.5 *In Situ* Thermal Treatment: ISCO/ISCR/ISB Combinations

The addition of heat to chemical or biological processes would be expected to have direct effects on the rates of many degradation reactions, because many chemical and biological process rates vary with temperature, often exhibiting an Arrhenius type relationship (Laidler, 1978). Research into the effects of combining thermal treatment processes with ISCR, ISCO and ISB is ongoing, and the amount of information regarding the degree to which heat can enhance other technologies is expected to increase significantly over the next few years.

Heat is an effective activation method for persulfate (Huang et al., 2005). The effectiveness of heat-activated persulfate for treating 59 VOCs was evaluated at temperatures ranging from 20°C to 40°C. Contaminant degradation rates were found to increase with increasing reaction temperature. Waldemer et al. (2007) evaluated the oxidation rates of chlorinated ethenes using heat-activated persulfate at temperatures ranging from 30°C to 70°C and found that contaminant degradation rate constants fit the Arrhenius relationship.

Johnson et al. (2008) evaluated the persistence of persulfate under thermal activation conditions. The researchers found that the rate of persulfate decomposition varied directly with temperature according to an Arrhenius type relationship. They also noted that persulfate activated via thermal methods may have limited ability to access lower-permeability aquifer zones due to accelerated rates of persulfate decomposition at elevated temperatures, with persulfate half-lives on the order of hours to days at high to moderate temperatures. These results are consistent with those of Costanza et al. (2010), who found that the half-life of persulfate declined with higher temperatures, from 31°C to 70°C.

Huang et al. (1999) and Dai and Reitsma (2005) evaluated the kinetics of permanganate oxidation of chlorinated ethenes at temperatures ranging from 10°C to 20°C, which is typical for ambient conditions. Additional research into the effect of higher temperatures on permanganate oxidation would be useful in further assessing any advantage this process may offer over ambient temperature oxidation.

Fletcher et al. (2011) evaluated the performance, viability and biomarker (DNA and ribonucleic acid [RNA]) abundances of *Dhc* during exposure to elevated temperatures. They found that several PCE-dechlorinating cultures produced ethane when incubated at temperatures of 30°C. However, VC accumulated when cultures were incubated at 35°C or 40°C. Cultures incubated at 40°C for less than 49 days resumed VC dechlorination following cooling, but cultures incubated at 45°C completely lost the ability to dechlorinate VC. The researchers concluded that complete dechlorination of chlorinated ethenes should not be expected at sites where ISTT is applied until temperatures cool to below 35°C. These conclusions are consistent with the research of Costanza et al. (2009), who found that dechlorination of TCE in soil and groundwater within sealed ampoules ceased when heated to either 50°C or 95°C.

Truex et al. (2011) conducted a field test to assess the potential for treating sources by combining ZVI injections with moderate temperature ISTT using electrical resistance heating. They reported that the overall rate of TCE transformation by ZVI was 3.6–4.8 times higher at temperatures above 30°C compared to the rates at ambient temperatures (~10°C).

15.4.5.1 Conclusions Regarding ISTT: ISCO/ISCR/ISB Combinations

The research to date leads to the following conclusions regarding the combinations of ISTT with other *in situ* technologies (ISCO, ISCR or ISB):

- Research into potential synergistic combinations between the application of heat and other remedial technologies is increasing. Additional research into the cost competitiveness of thermally enhanced remediation technologies would be helpful.
- For combinations of ISTT and ISB, an increase in temperature to the range of 30–35°C may enhance the rate of reductive dechlorination compared to rates at ambient temperature. However, temperatures of 40°C or greater are inhibitory to reductive dechlorination. Temperatures significantly over 40°C are expected to be biocidal to key dechlorinating bacteria.
- For the combination of ISTT and ISCO using persulfate, increases in temperature are expected to increase the rate at which persulfate degrades to sulfate. Higher temperatures may decrease the longevity of persulfate within an aquifer, significantly reducing its ability to diffuse into and provide treatment for lower-permeability zones.
- For the combination of ISTT and ISCR, an increase in temperature to 50°C increases considerably the rate at which ZVI degrades TCE.

15.4.6 Surfactant–ISB Combinations

Conceptually, surfactants could be used with ISB to improve delivery of the electron donors and thereby improve efficacy. In this process, a surfactant would be added to the aquifer to increase the aqueous solubility of the target organic contaminants, so that biodegradation can proceed more quickly or efficiently. A separate organic substrate could be added to provide an electron donor or the surfactant itself may function as the electron donor.

Little published research into this process combination was found, perhaps because of concerns that the surfactants can be highly toxic to bacteria. Surfactants can be toxic to bacteria by either disrupting their cellular membranes or by reacting with proteins that are essential to cell functions (Volkering et al., 1998). However, the use of biosurfactants, which are generally less toxic than chemical surfactants, may be suitable for combination with ISB. Further research into this potential combination would be useful.

15.4.7 Contingent Combined Remedies

As described earlier, a contingent combined remedy includes selecting and planning for implementing a secondary remedial technology in the event that the primary technology applied did not fully meet its treatment objectives. The general intent of a contingent combined remedy is to expedite implementation of the second technology if the initial technology does not meet its performance objectives.

The most significant benefits of specifying a contingent combined remedy may be realized in regulatory programs that have significant procedural requirements involved in changing a remedy. By specifying a contingent remedy in the key decision documents that prescribe the required remedial approach, it may be possible to avoid or minimize the regulatory transaction costs required to change the originally selected remedy if it is less successful than expected.

The 2011 ROD for the Cabot Carbon/Koppers Superfund site in Gainesville, Florida, provides a practical example of a contingent combined remedy. The Cabot Carbon/Koppers site includes a former wood-treating site that operated for nearly 100 years, beginning in 1916. Creosote DNAPL is present in the subsurface in several source zones at this site. For two of the source zones, the ROD has specified the application of *in situ* solidification/stabilization, using appropriate reagents, such as cement, applied with deep soil mixing augers.

However, for two other source zones with data suggesting that DNAPL may not have penetrated to as great a depth, the ROD provides for a contingent remedy. In this case, pilot testing of an innovative chemical stabilization method will be performed, using permanganate delivered through temporary or permanent injection points. Provided that the pilot testing demonstrates that this innovative treatment will be effective and meet the RAOs, chemical stabilization will be implemented in the two source zones on a full-scale basis. However, if the pilot test finds that the chemical stabilization does not achieve the target performance requirements, the contingent remedy for these two source zones is *in situ* solidification/stabilization, as implemented at the other two source zones.

Incorporating the contingent remedy into the ROD circumvents the need for a ROD amendment if the remedy needs to be changed from chemical stabilization to *in situ* solidification/stabilization. Avoiding a ROD amendment will allow faster remediation of the site, with fewer administrative impacts.

15.5 SUMMARY

The use of combined remedies for treating source zones has become an established and increasingly popular remediation strategy over the last decade. The increasing emphasis on deliberately combining remedies has resulted from an increased understanding of both the limitations of individual remediation technologies and the economic and performance benefits of combining remediation technologies.

The combined remedies concept covers a broad range of potential technology combinations and application strategies. To date, research has been conducted on a relatively limited number of technology combinations. While progress has been made in better understanding optimal methods for implementing single remedies, additional research into combining various remedies would be beneficial.

A variety of technology combinations have been used to treat DNAPL source zones, including sequential combinations, concurrent combinations, and coupled process combinations. A generalized strategy for implementing combined remedies is presented in this chapter, although to date little systematic research into optimal strategies for combining technologies has been carried out. Cost optimization models that consider uncertainty in site and remediation characteristics and allow for comparisons of different technologies may be valuable tools for identifying optimal technology combinations (Parker et al., 2011). However, because several technical, regulatory and business-related factors may impact remediation planning and technology selection, developing an overall remediation strategy applicable to all DNAPL sites is not feasible.

Of the available sequential process combinations, ISCO followed by ISB has been perhaps the most widely evaluated combination, both in the laboratory and in the field. A significant number of published articles on this topic are available, covering both laboratory and field applications for this combination. *In situ* bioremediation is the most frequently suggested active follow-up technology after implementing a primary source treatment method, such as ISTT, ISCO, or SEAR. A key aspect to consider when planning any sequential combination is the post-implementation impacts of the primary treatment technology on the biological and geochemical conditions. These impacts vary widely, depending on the specific primary technology applied, but they can be profound and can strongly impact the effectiveness and costs of a subsequent technology.

Of the available coupled process combinations, the surfactant-ISCO coupled process has received the greatest amount of research interest. Significant progress has been made in understanding the general compatibility of many commonly available surfactants with several commonly used oxidants. Several patents have been granted for combining surfactants/cosolvents with ISCO. However, little information on the field effectiveness or cost-effectiveness of these technologies compared to ISCO alone is yet available. Further data on field applications of these processes would be valuable.

Another coupled process that has been successfully demonstrated is the ISCR–ISB combination. Commercial products such as EHC® and EZVI have been developed specifically to combine a chemical reductant (ZVI) with an organic substrate to stimulate anaerobic biodegradation. To date, little research has been published on the cost competitiveness of these combined products compared to ISB or ISCR alone.

Research also has focused on the coupled process combinations of ISTT with other technologies, such as ISCO, ISCR and ISB. The results to date indicate that while some important reaction rates can be accelerated through the application of heat, each technology combined with thermal treatment may have a unique temperature range within which optimal process enhancements occur and beyond which, competing reactions or detrimental effects outweigh the benefits of the higher reaction rates.

Overall, significant progress has been made in understanding several combined remedies, although many have not yet been carefully researched. Additional research into the following areas would be helpful to advance the state of the practice related to source zone remediation:

- *What type of performance can be expected when combining technologies?*

  While consistently achieving a high degree of treatment performance within a laboratory-scale setting is often possible for remedial technologies, the performance of remedial technologies in the field is often highly variable and less successful than in the laboratory. This is often due to factors such as greater geologic heterogeneity and less uniformity in reagent delivery in field settings compared to laboratory conditions. Further research into methods that enable more consistent performance of both individual and combined technologies would be beneficial.

- *What site conditions favor applying specific technology combinations?*

  Significant progress has been made in understanding many conditions that affect the performance of individual technologies. However, little research has been published on the effect of various site conditions, such as heterogeneous geology, hydrogeology, aquifer geochemistry, and contaminant mass distribution on the selection of technology combinations. Additional research into this area would allow easier identification of combinations with potentially greater performance and elimination from further consideration of combinations with less potential.

- *What potential technology combinations have the greatest synergistic benefits and are the most cost-effective?*

  Application of some technology combinations, such as the sequential combination of ISCO followed by ISB, has been conducted on a field- or full-scale basis at enough sites that the general compatibility of these two technologies is known to be acceptable. The widespread use of ISB over the past 10 years has demonstrated that it is often a cost-effective remedy as a follow-up or polishing technology at sites where conditions for its use are favorable. However, available data for many other combinations is much more limited, particularly regarding the cost-effectiveness of various combinations. Combinations involving the use of thermal treatment with ISCO, ISCR and ISB, for example, are beginning to be explored, and information regarding the potential effectiveness of these combinations is beginning to be developed. However, the cost-effectiveness of these combinations at full-scale sites is not yet clearly known. Similarly, the effectiveness of several coupled process combinations, such as the use of surfactants or cosolvents with ISCO, has been effectively demonstrated in the laboratory, but the degree of performance achievable in the field and the cost-effectiveness of this combination for full-scale sites have not yet been demonstrated. Greater research to determine the combinations that provide the most cost-effective treatment performance would be helpful.

Further research into these and other questions could identify remedial technology combinations that offer significant benefits over current remediation methods. It should be recognized that the development of remediation technologies is a dynamic field and that as remediation technologies change and evolve and new information becomes available, the answers to the questions above will also change and evolve. Thus, the evaluation of combined remedy concepts is expected to be an ongoing process that offers the potential for many new discoveries and opportunities for greater efficiency in achieving the challenging goal of DNAPL source zone remediation.

REFERENCES

Adamson DT, McGuire TM, Newell CJ, Stroo H. 2011. Sustained treatment: Implications for treatment timescales associated with source-depletion technologies. Remediat J 20:27–50.

Amos BK, Deprato RC, Hughes JB, Pennell KD, Löffler FE. 2007. Effects of the non-ionic surfactant Tween 80 on microbial reductive dechlorination of chlorinated ethenes. Environ Sci Technol 41:1710–1716.

Azadpour-Keely A, Wood AL, Lee TR, Mravik S. 2004. Microbial responses to *in situ* chemical oxidation, six-phase heating, and steam injection remediation technologies in groundwater. Remediat J 14:5–17.

Battelle Memorial Institute. 2010a. Guidance for Optimizing Remedy Evaluation, Selection, and Design. Naval Facilities Engineering Service Center (NFESC), Port Hueneme, CA, USA

Battelle Memorial Institute. 2010b. Cost and Performance Report for Persulfate Treatability Studies. NFESC, Port Hueneme, CA, USA.

Becvar E, Evans P, Lebrón C, Stroo H, Wilson J, Wymore R. 2008. Workshop Report: *In Situ* Biogeochemical Transformation of Chlorinated Solvents. Prepared for AFCEE (Air Force Center for Environmental Excellence), NFESC, Environmental Security Technology Certification Program (ESTCP). http://www.dtic.mil/cgi-bin/GetTRDoc?AD=ADA501302. Accessed January 17, 2014.

Beyke G, Fleming D. 2005. *In situ* thermal remediation of DNAPL and LNAPL using electrical resistance heating. Remediat J 15:5–22.

Block PA, Brown RA, Robinson D. 2004. Novel activation technologies for sodium persulfate in situ chemical oxidation. Proceedings, 4th International Conference on Remediation of Chlorinated and Recalcitrant Compounds, Monterey, CA, USA, May 24-27, Paper 2A-05.

Brown RA, Lewis R, Leahy MC, Luhrs R. 2009a. Is there life after ISCO? The effect of oxidants on *in situ* bioremediation. Proceedings, 10th International Symposium on *In Situ* and On-site Bioremediation, Baltimore, MD, USA, May 5–8. Paper No. K-12.

Brown RA, Mueller JG, Seech AG, Henderson JK, Wilson JT. 2009b. Interactions between biological and abiotic pathways in the reduction of chlorinated solvents. Remediat J 20:9–10.

Chapelle FH, Bradley PM, Casey CC. 2005. Behavior of a chlorinated ethene plume following source-area treatment with Fenton's Reagent. Ground Water Monit Remediat 25:131–141.

Chen S, Markesic SJ, Abrams SH. 2002. Injection of zero-valent iron into a bedrock formation for the reduction of TCE. Proceedings, Third International Conference on Remediation of Chlorinated and Recalcitrant Compounds, Monterey, CA, USA, May 20–23.

Christ JA, Ramsburg CA, Abriola LM, Pennell KD, Löffler FE. 2005. Coupling aggressive mass removal with microbial reductive dechlorination for remediation of DNAPL source zones: A review and assessment. Environ Health Perspect 113:465–477.

Conrad SH, Glass RJ, Peplinski WJ. 2002. Bench-scale visualization of DNAPL remediation processes in analog heterogeneous aquifers: Surfactant floods and *in situ* oxidation using permanganate. J Contam Hydrol 58:13–49.

Costanza J, Fletcher KE, Löffler FE, Pennell KD. 2009. Fate of TCE heated in Fort Lewis soil. Environ Sci Technol 43:909–914.

Costanza J, Otano G, Callaghan J, Pennell KD. 2010. PCE oxidation by sodium persulfate in the presence of solids. Environ Sci Technol 44:9445–9450.

Crimi ML, Siegrist RL. 2003. Geochemical effects on metals following permanganate oxidation of DNAPLs. Ground Water 41:458–469.

Crimi ML, Dugan PJ, Siegrist RL. 2009. Method and compositions for treatment of subsurface contaminants. U.S. Patent 7,553,105 B1.

Crimi M, Silva JAK, Palaia T. 2013. Final Report: Cooperative Technology Demonstration: Polymer-Enhanced Subsurface Delivery and Distribution of Permanganate. ESTCP Project ER-200912. Prepared for Environmental Security Technology Certification Program, Alexandria, VA, USA. http://www.serdp.org/. Accessed January 17, 2014.

Dai Q, Reitsma S. 2005. Kinetic study of permanganate oxidation of tetrachloroethylene at a high pH under acidic conditions. Remediat J 14:67–79.

Davis EL. 1997. How Heat Can Enhance *In Situ* Soil and Aquifer Remediation: Important Chemical Properties and Guidance on Choosing the Appropriate Technique. EPA/540/S-97/502. U.S. Environmental Protection Agency (USEPA), Ada, OK, USA.

Dettmer K. 2002. A Discussion of the Effects of Thermal Remediation Treatments on Microbial Degradation Processes. USEPA, Office of Solid Waste and Emergency Response, Washington, DC, USA.

Dolfing J, Van Eekert M, Seech A, Vogan J, Mueller J. 2008. *In situ* chemical reduction (ISCR) technologies: Significance of low Eh reactions. Soil Sediment Contam 17:63–74.

Droste EX, Marley MC, Parikh JM, Lee AM, Dinardo PM, Woody BA, Hoag GE, Chheda PV. 2002. Observed enhanced reductive dechlorination after in situ chemical oxidation pilot test. Proceedings, 3rd International Conference of Remediation of Chlorinated and Recalcitrant Compounds, Monterey, CA, USA. Paper No. 2C-01.

Dugan PJ, Siegrist RL, Crimi ML. 2009. Method and compositions for treatment of subsurface contaminants. U.S. Patent 2009/7553105.

Dugan PJ, Siegrist RL, Crimi ML. 2010. Coupling surfactants/cosolvents with oxidants for enhanced DNAPL removal: A review. Remediat J 20:27–50.

Fletcher KE, Costanza J, Cruz-Garcia C, Ramaswamy NS, Pennell KD, Löffler FE. 2011. Effects of elevated temperature on *Dehalococcoides* dechlorination performance and DNA and RNA biomarker abundance. Environ Sci Technol 45:712–718.

Friis AK, Albrechtsen HJ, Heron G, Bjerg PL. 2005. Redox processes and release of organic matter after thermal treatment of a TCE-contaminated aquifer. Environ Sci Technol 39:5787–5795.

Friis AK, Edwards EA, Albrechtsen HJ, Udell KS, Duhamel M, Bjerg PL. 2007. Dechlorination after thermal treatment of a TCE-contaminated aquifer: Laboratory experiments. Chemosphere 67:816–825.

Gavaskar A, Tatar L, Condit W. 2005. Cost and performance report for nano-scale zero valent iron technologies for source remediation. Prepared for NFESC, Port Hueneme, CA, USA. http://www.dtic.mil/cgi-bin/GetTRDoc?Location=U2&doc=GetTRDoc.pdf& AD=ADA446916. Accessed January 17, 2014.

Geosyntec. 2010. Cost and Performance Report: Remediation of DNAPL Through Sequential In Situ Chemical Oxidation and Bioaugmentation. ESTCP Project ER-200116. Prepared for ESTCP, Alexandria, VA, USA. http://www.serdp.org/. Accessed January 17, 2014.

Heron G, Carroll S, Nielsen SG. 2005a. Full-scale removal of DNAPL constituents using steam-enhanced extraction and electrical resistance heating. Ground Water Monit Remediat 25:92–107.

Heron G, Baker R, LaChance J, Friis AK. 2005b. Using heat to accelerate remediation. Proceedings, 8th International Symposium In-Situ and On Site Bioremediation, Baltimore, MD, USA, June 6–9. Paper No. D-23.

Hoag GE, Collins JB, Anderson DK. 2010. Method for Extraction and Surfactant-Enhanced Subsurface Contaminant Recovery. U.S. Patent 2010/0185039 A1.

Hrapovic L, Sleep BE, Major DJ, Hood ED. 2005. Laboratory study of treatment of trichloroethene by chemical oxidation followed by bioremediation. Environ Sci Technol 39:2888–2897.

Huang K, Hoag GE, Chheda P, Woody BA, Dobb GM. 1999. Kinetic study of oxidation of trichloroethene by potassium permanganate. Environ Eng Sci 16:265–274.

Huang K, Zhao Z, Hoag GE, Dahmani A, Block PA. 2005. Degradation of volatile organic compounds with thermally activated persulfate. Chemosphere 61:551–560.

Huling SG, Pivetz BE. 2006. In-Situ Chemical Oxidation: Engineering Issue. EPA/600/R-06/072. Office of Research and Development, USEPA, Cincinnati, OH, USA.

ITRC (Interstate Technical and Regulatory Council). 2005. Technical and Regulatory Guidance for In-Situ Chemical Oxidation of Contaminated Soil and Groundwater, 2nd ed. Washington, DC, USA. http://www.itrcweb.org/Guidance/ListDocuments?TopicID=10&SubTopicID=17. Accessed January 17, 2014.

ITRC. 2007. *In Situ* Bioremediation of Chlorinated Ethene DNAPL Source Zones: Case Studies. ITRC, Washington, DC, USA. http://www.itrcweb.org/Guidance/ListDoc uments?TopicID=3&SubTopicID=3. Accessed January 17, 2014.

ITRC. 2008. *In Situ* Bioremediation of Chlorinated Ethene DNAPL Source Zones. ITRC, Washington, DC, USA. http://www.itrcweb.org/Documents/bioDNPL_Docs/BioDNAPL 3.pdf. Accessed January 17, 2014.

ITRC. 2011. Technical and Regulatory Guidance: Integrated DNAPL Site Strategy. ITRC, Washington, DC, USA. http://www.itrcweb.org/Guidance/ListDocuments?TopicID=5&SubTopicID=10. Accessed January 17, 2014.

Jackson RE. 2001. DNAPL remediation: Which "new paradigm" will prevail? Ground Water Monit Remediat 21:54–58.

Johnson P, Dahlen P, Kingston J, Foote E, Williams S. 2009. State-of-the-Practice Overview: Critical Evaluation of State-of-the-Art *In Situ* Thermal Treatment Technologies for DNAPL Source Zone Treatment. ESTCP Project ER-0314. Prepared for ESTCP, Alexandria, VA, USA. http://www.serdp.org/. Accessed January 17, 2014.

Johnson RL, Tratnyek PG, Johnson RO. 2008. Persulfate persistence under thermal activation conditions. Environ Sci Technol 42:9350–9356.

Jones A, Escobar M, Serlin C, Sercu B, Holden P, Stollar R, Murphy P. 2009. Microbial community composition assessment during in-situ chemical oxidation with permanganate: The effect of oxidants on in-situ bioremediation. Proceedings, Tenth International In Situ and On-site Bioremediation Symposium, Baltimore, MD, May 5–8.

Kappler A, Straub K. 2005. Geomicrobiological cycling of iron. Rev Mineral Geochem 59:85–108.

Kennedy LG, Everett JW, Becvar E, DeFeo D. 2006. Field-scale demonstration of induced biogeochemical reductive dechlorination at Dover Air Force Base, Dover, Delaware. J Contam Hydrol 88:119–136.

Kingston JL, Dahlen PR, Johnson PC. 2010. State-of-the-practice review of *in situ* thermal technologies. Ground Water Monit Remediat 30:64–72.

Kirshling TL, Gregory KB, Minkley EG Jr, Lowry GV, Tilton RD. 2010. Impact of nano-scale zero valent iron on geochemistry and microbial populations in trichloroethylene contaminated aquifer materials. Environ Sci Technol 44:3474–3480.

Koch SA, Rice MJ, Stolzenburg TR, Baker KB, Haselow J. 2007. Reversing full-scale ISCO. Proceedings, 9th International Symposium, In Situ and On-Site Bioremediation, Baltimore, MD, USA, May 7–10. Paper No. G-06.

Krembs FJ, Siegrist RL, Crimi ML, Furrer RF, Petri BG. 2010. ISCO for groundwater remediation: Analysis of field applications and performance. Ground Water Monit Remediat 30:42–53.

Laidler KJ. 1978. Physical Chemistry with Biological Applications. The Benjamin/Cummings Publishing Co, San Francisco, CA, USA.

Leeson A, Johnson PC, Bruce CL, Johnson RL, Amerson IL, Vogel CM, Hinchee RE, Marley M, Peargin T, Coonfare CT, Gillespie RD. 2002. Design Paradigm: Air Sparging Technology Transfer and Multi-Site Evaluation. ESTCP Project ER-9808. Prepared for ESTCP, Alexandria, VA, USA. http://www.serdp.org/. Accessed January 17, 2014.

Li Z, Hanlie H. 2008. Combination of surfactant solubilization with permanganate oxidation for DNAPL remediation. Water Res 42:605–614.

Loomer DB, Al TA, Banks VJ, Parker BL, Mayer KU. 2010a. Manganese valence in oxides formed from in situ chemical oxidation of TCE by $KMnO_4$. Environ Sci Technol 44:5934–5939

Loomer DB, Al TA, Banks VJ, Parker BL, Mayer KU. 2010b. Manganese and trace metal mobility under reducing conditions following in situ chemical oxidation of TCE by $KMnO_4$: A laboratory column experiment. J Contam Hydrol 119:13–24.

Luhrs RC, Lewis RW, Huling SG. 2006. ISCO's long-term impact on aquifer conditions and microbial activity. Proceedings, 5th International Conference on Remediation of Chlorinated and Recalcitrant Compounds, Monterey, CA, USA. Paper No. D-48.

Macbeth TW, Starr RC, Sorenson KS, Goehlert R, Moor KS. 2005. ISCO Impacts on indigenous microbes in a PCE/DNAPL-contaminated aquifer. Proceedings, 8th International In Situ and On-Site Bioremediation Symposium, Baltimore, MD, USA, June 6–9. Paper No. D-22.

MacKinnon LK, Thomson NR. 2002. Laboratory-scale *in situ* chemical oxidation or a perchloroethylene pool using permanganate. J Contam Hydrol 56:49–74.

Martel KE, Martel R, Lefebvre R, Gelinas PJ. 1998. Laboratory study of polymer solutions used for mobility control during in situ NAPL recovery. Ground Water Monit Remediat 18:103–113.

Martin JP, Horst JF, Lenzo F. 2006. Full-scale ZVI and molasses treatment of PCE DNAPL in fractured bedrock. Proceedings, 5th International Conference on Remediation of Chlorinated and Recalcitrant Compounds, Monterey, CA, USA, May. Paper No. C-50.

McCray JE, Munakata-Marr J, Silva JAK, Davenport S, Smith MM. 2010. Final Report: Multi-Scale Experiments to Evaluate Mobility Control Methods for Enhancing the Sweep Efficiency of Injected Subsurface Remediation Amendments. SERDP Project ER-1486. Prepared for SERDP, Alexandria, VA, USA. http://www.serdp.org/. Accessed January 17, 2014.

McElroy B, Keith A, Glasgow J, Dasappa S. 2002. In situ chemical reduction pilot tests – Marshall Space Flight Center, Huntsville, Alabama. Proceedings, 3rd International Conference on Remediation of Chlorinated and Recalcitrant Compounds, Monterey, CA, USA. Paper No. 2A-17.

McGuire T, Hughes JB. 2003. Effects of surfactants on the dechlorination of chlorinated ethenes. Environ Toxicol Chem 22:2630–2638.

McGuire TM, McDade JM, Newell CJ. 2006. Performance of DNAPL source depletion technologies at 59 chlorinated solvent-impact sites. Ground Water Monit Remediat 26:73–84.

Molin J, Mueller J, Hanson D, Fowler T, Skrotzki T. 2010. In situ remediation of PCE at a site with clayey lithology and significant smear zone. Remediat J 20:51–62.

Moore K. 2008. Geochemical Impacts from Permanganate Oxidation Based on Field Scale Assessments. MS Thesis. East Tennessee State University, Johnson City, TN, USA. http://etd-submit.etsu.edu/etd/theses/available/etd-0910108-191936/unrestricted/MooreK100808f.pdf. Accessed January 17, 2014.

Mravik SC, Sillan RK, Wood AL, Sewell GW. 2003. Field evaluation of the solvent extraction residual biotreatment technology. Environ Sci Technol 37:5040–5049.

Nelson MD, Parker BL, Al TA, Cherry JA, Loomer D. 2001. Geochemical reactions resulting from in situ chemical oxidation of PCE-DNAPL by KMnO4 in a sandy aquifer. Environ Sci Technol 35:1266–1275.

NRC (National Research Council). 2004. Contaminants in the Subsurface, Source Zone Assessment and Remediation. The National Academies Press, Washington, DC, USA.

O'Hara S, Krug T, Quinn J, Clausen C, Geiger C. 2006. Field and laboratory evaluation of the treatment of DNAPL source zones using emulsified zero-valent iron. Remediat J 16:35–56.

Okwi GJ, Thomson NR, Gillham RW. 2005. The impact of permanganate on the ability of granular iron to degrade TCE. Ground Water Monit Remediat 25:123–128.

Olson MR, Sale TC, Shackelford CD, Bozzini C, Skeean J. 2012. Chlorinated solvent source zone remediation via ZVI-clay soil mixing: 1-year results. Ground Water Monit Remediat 32:63–74.

Parker J, Kim U, Kitanidis P, Cardiff M, Liu X, Lee J. 2011. Final Report: Practical Cost-Optimization of Characterization and Remediation Decisions at DNAPL Sites with Consideration of Prediction Uncertainty. SERDP Project ER-1611. Prepared for SERDP, Alexandria, VA, USA. http://www.serdp.org/. Accessed January 17, 2014.

Parsons Engineering Science, Inc. 2000. Remediation of Chlorinated Solvent Contamination on Industrial and Airfield Sites. Prepared for AFCEE, San Antonio, TX, USA. http://www.clu-in.org/download/contaminantfocus/dnapl/Treatment_Technologies/remediation_of_ chlorinated_DNAPL_AFCEEAF_Chlor.pdf. Accessed January 17, 2014.

Peale JGD, Mueller J, Molin J. 2010. Successful ISCR-enhanced bioremediation of TCE DNAPL source utilizing EHC® and KB-1®. Remediat J 20:63–81.

Pennell KD, Löffler FE, Costanza J, Fletcher KE, Ramaswamy NA, Otano G, Callaghan J. 2009. Final Report: Investigation of Chemical Reactivity, Mass Recovery and Biological Activity During Thermal Treatment of DNAPL Source Zones. SERDP Project ER-1419. Prepared for SERDP, Alexandria, VA, USA. http://www.serdp.org/. Accessed January 17, 2014.

Powell T, Smith G, Sturza J, Lynch K, Truex M. 2007. New advances for in-situ treatment using electrical resistance heating. Remediat J 17:51–70.

Ramakrishnan V, Ogram AV, Lindner AS. 2005. Impacts of co-solvent flushing on microbial populations capable of degrading trichloroethylene. Environ Health Perspect 113:55–61.

Ramsburg CA, Abriola LM, Pennell KD, Löffler FE, Gamache M, Amos BK, Petrovskis EA. 2004. Stimulated microbial reductive dechlorination following surfactant treatment at the Bachman road site. Environ Sci Technol 38:5902–5914.

Richards A, Kueper BH, Gerhard JI. 2005. Hydraulic displacement of DNAPL source zones. First International Conference on Challenges in Site Remediation, Chicago, IL, USA, October 23–27.

Richardson RE, James CA, Bhupathiraju VK, Alvarez-Cohen L. 2002. Microbial activity in soils following steam treatment. Biodegradation 13:285–295.

Sahl J, Munakata-Marr J. 2006. The effects of *in situ* chemical oxidation on microbiological processes: A review. Remediat J 16:57–70.

Sahl JW, Munakata-Marr J, Crimi ML, Siegrist RL. 2007. Coupling permanganate oxidation with microbial dechlorination of tetrachloroethene. Water Environ Res 79:5–12.

Schroth MH, Oostrom M, Wietsma TW, Istok JD. 2001. *In situ* oxidation of trichloroethylene by permanganate: Effects on porous medium hydraulic properties. J Contam Hydrol 50:79–98.

Siegrist RL, Crimi M, Munakata-Marr J, Illangasekare T, Lowe K, Van Cuyk S, Dugan P, Heiderscheidt J, Jackson S, Petri B, Sahl J, Seitz S. 2006. Final Report: Reaction and Transport Processes Controlling In Situ Chemical Oxidation of DNAPLs. Project ER-1290. Prepared for SERDP, Alexandria, VA, USA. http://www.serdp.org/. Accessed January 17, 2014.

Siegrist RL, Crimi M, Simpkin TJ, eds. 2013. *In Situ* Chemical Oxidation for Groundwater Remediation. SERDP/ESTCP Remediation Technology Monograph Series, Springer Science + Business Media, LLC, New York, NY, USA. p. 678

Shen H, Wilson JT. 2007. Trichloroethylene removal from groundwater in flow-through columns simulating a permeable reactive barrier constructed with plant mulch. Environ Sci Technol 41:4077–4083.

Shetty N, Zenker M, McKeon J, Moreno J. 2009. Remediation of chlorinated ethenes, ethanes, and methanes in groundwater using carbon- and iron-based electron donor. Remediat 19:35–48.

Shiau B. 2008. *In situ* surfactant and chemical oxidation flushing for complete remediation of contaminants and methods of using same. U.S. Patent 2008/7364386 B2.

Simpkin T, Sale T, Kueper B, Pitts M, Wyatt K. 1999. Surfactants and Cosolvents for NAPL Remediation: A Technology Practices Manual. Lewis Publishers, Boca Raton, FL, USA.

Smith MM, Silva JAK, Munakata-Marr J, McCray JE. 2008. Compatibility of polymers and chemical oxidants for enhanced groundwater remediation. Environ Sci Technol 42:9296–9301.

Sra KS. 2010. Mobility enhancement of Mn oxide during permanganate oxidation of TCE. Ground Water Monit Remediat 30:99–106.

Sra KS, Thomson NR, Barker J. 2010. Persistence of persulfate in uncontaminated aquifer materials. Environ Sci Technol 44:3098–3104.

Stroo HF, Unger M, Ward CH, Kavanaugh MC, Vogel C, Leeson A, Marqusee JA, Smith BP. 2003. Remediating chlorinated solvent source zones. Environ Sci Technol 37:224A–230A.

Stroo HF, Ward CH. 2010. *In Situ* Remediation of Chlorinated Solvent Plumes. SERDP/ESTCP Remediation Technology Monograph Series, Springer Science + Business Media LLC, New York, NY, USA.

Suchomel EJ, Pennell KD. 2006. Reductions in contaminant mass discharge following partial mass removal from DNAPL source zones. Environ Sci Technol 40:6110–6116.

Sutherson SS. 1997. Remediation Engineering: Design Concepts. CRC Press, Inc., Boca Raton, FL, USA.

Sutton NB, Grotenhuis JTC, Langenhoff AAM, Rijnaarts HHM. 2010. Efforts to improve coupled in situ chemical oxidation with bioremediation: A review of optimization strategies. J Soils Sediments 11:129–140.

Truex M, Powell T, Lynch K. 2007. *In situ* dechlorination of TCE during aquifer heating. Ground Water Monit Remediat 27:96–105.

Truex MJ, Macbeth TW, Vermuel VR, Fritz BG, Mendoza DP, Mackley RD, Wietsma TW, Sandberg G, Powell T, Powers J, Pitre F, Michaelsen M, Ballock-Dixon SJ Zhong L, Oostrom M. 2011. Demonstration of combined zero-valent iron and electrical resistance heating for *in situ* remediation. Environ Sci Technol 45:5346–5351.

USACE (U.S. Army Corp of Engineers). 2009. Design: *In Situ* Thermal Remediation. Engineering Manual. EM 1110-1-4015. USACE, Washington, DC, USA.

USEPA (U.S. Environmental Protection Agency). 1991. Toxic Treatments, In Situ Steam/Hot-Air Stripping Technology, Applications Analysis Report. EPA/540/A5-90/008. Risk Reduction Engineering Laboratory, Office of Research and Development, USEPA, Cincinnati, OH, USA. March.

USEPA. 2000. Engineered Approaches to *In Situ* Bioremediation of Chlorinated Solvents: Fundamentals and Field Applications. EPA 542-R-00-008. USEPA, Washington, DC, USA.

USEPA. 2003. The DNAPL Remediation Challenge: Is There a Case for Source Depletion? EPA/600/R-03/143. USEPA, Cincinnati, OH, USA.

USEPA. 2004. *In Situ* Thermal Treatment of Chlorinated Solvents. Fundamentals and Field Applications. EPA 542-R-04-010. USEPA, Washington, DC, USA.

USEPA. 2009. DNAPL Remediation: Selected Projects Where Regulatory Closure Goals Have Been Achieved. EPA/542/R-09/008. USEPA, Washington, DC, USA. www.clu-in.org. Accessed January 17, 2014.

Volkering F, Breure AM, Rulkens WH. 1998. Microbiological aspects of surfactant use for biological soil remediation. Biodegradation 8:401–417.

Waldemer RH, Tratnyek PG, Johnson RL, Nurmi JT. 2007. Oxidation of chlorinated ethenes by heat-activated persulfate: Kinetics and products. Environ Sci Technol 41:1010–1015.

Westersund J, Fernandes L, Jones S, Clough H. 2006. Stimulating anaerobic reductive dechlorination following chemical oxidation treatment. Proceedings, 5th International Conference on Remediation of Chlorinated and Recalcitrant Compounds, Monterey, CA, USA, May 22–25. Paper No. D-56.

Zhai X, Hua I, Rao PS, Lee LS. 2006. Cosolvent-enhanced chemical oxidation of perchloroethylene by potassium permanganate. J Contam Hydrol 82:61–74.

CHAPTER 16

COST ANALYSES FOR REMEDIAL OPTIONS

Mark Harkness[1] and Julie Konzuk[2]

[1]GE Global Research, Niskayuna, NY 12309, USA; [2]Geosyntec Consultants, Guelph, ON, Canada N1G 5G3

16.1 INTRODUCTION

The economic analysis of remedial options is a key activity in any remedial selection process. It is typically employed during the feasibility study phase of remedy selection to aid in selecting the preferred remedial option for a site, in concert with a number of other criteria. Typically, the capital and recurring costs of various alternatives are compared using a net present value (NPV) calculation, which allows the costs to be summed and compared on an equivalent (life cycle) basis. In addition to the ability to compare remedial options, a detailed economic analysis also allows the design engineer to determine which cost elements of a specific remedy drive the overall cost of the remedy. Understanding the primary cost drivers improves the potential for cost optimization and therefore results in more cost-effective remedial designs.

This chapter provides a general approach to developing cost information for several remedial technologies often used to treat dense nonaqueous phase liquid (DNAPL) source areas containing chlorinated aliphatic compounds. Cost information for each technology is derived for two template site situations to illustrate the process. Costs are broken down into design elements, capital expenditures, operation and maintenance (O&M), and monitoring costs to help understand the primary cost drivers for each technology.

All design engineering firms will use a version of this approach. The basis for estimating the components of cost will vary based on past experience with similar projects, specific site characteristics, local economic conditions and company costing methodologies, which are often proprietary. The specific costs presented during this exercise were derived either from solicited vendor quotes or from the combined internal expertise of General Electric and Geosyntec in performing economic analyses. The presentation of all underlying assumptions, detailed calculations and source spreadsheets is beyond the scope of this chapter.

This chapter does not discuss the applicability of individual remedies to DNAPL source areas in any detail. The reader is directed to other chapters in this book for this information. The intent of this chapter is also not to provide definitive cost information for each remedial technology, since these are highly site-specific. The examples given are meant to illustrate cost categories that should be included in the economic analysis of remedial options and to provide the reader with some general observations about cost drivers for various DNAPL remediation technologies and how they might influence remedy selection and implementation.

B.H. Kueper et al. (eds.), *Chlorinated Solvent Source Zone Remediation*,
doi: 10.1007/978-1-4614-6922-3_16,

16.2 COST ANALYSIS PROCESS

The economic analysis presented here is based upon a method first used by DuPont to compare *in situ* remediation technologies on a consistent economic basis (Quinton et al., 1997). This approach, involving detailed costing for hypothetical template sites, is used in this chapter to cost remedial technologies for two source areas containing trichloroethene (TCE) DNAPL.

16.2.1 Template Site Descriptions

The source areas of the two template sites differ in basic stratigraphy – one representing a relatively permeable aquifer containing predominantly sands and a second representing a less permeable aquifer containing sands, silts and clays. This choice is designed to capture some of the variety in unconsolidated aquifer conditions typically encountered at field sites.

The first template source area (Case 1) consists of a permeable sandy aquifer having an area of 1,500 square meters (m^2; 16,140 square feet [ft^2]) and depth of 4.5 m (14.8 ft). This source area nominally contains 15,000 kilograms (kg; 33,000 pounds [lbs]) of TCE that is either in the aqueous (dissolved) phase or sorbed to the soil or present as DNAPL in the aquifer matrix. This situation is representative of a site where a TCE release has occurred over a period of time prior to remediation. In the course of the cost analysis, the TCE mass will be varied from 1,500 kg (3,300 lbs) to 60,000 kg (132,000 lbs) to understand how total contaminant mass may impact the cost of various remedial options.

The depth to groundwater is 1.5 m (4.9 ft), and it is 4.5 m (14.8 ft) to a clay aquitard that underlies the saturated zone, providing a saturated thickness of 3 m (9.8 ft). It is assumed that most of the TCE mass exists in the saturated thickness or sits on the aquitard surface. The hydraulic conductivity is 5×10^{-3} centimeters per second (cm/s; 14.4 ft/day). Groundwater velocity through the source area is 32 m (74.5 ft) per year (yr). The groundwater in this area contains 500 milligrams per liter (mg/L) of TCE and 20 mg/L of *cis*-1,2-dichloroethene (*cis*-DCE), along with 5 mg/L dissolved iron and 100 mg/L sulfate. These and other characteristics of the subsurface matrix and groundwater are summarized in Table 16.1.

The second template source area scenario (Case 2) consists of a significantly less permeable aquifer containing sands, silts and clays of similar dimensions to the first. The nominal mass of TCE is the same as in Case 1. The depth to groundwater is 1.5 m (4.9 ft) and it is 4.5 m (14.8 ft) to a clay aquitard that underlies the saturated zone. In Case 2, the hydraulic conductivity ranges between 10^{-5} and 10^{-4} cm/s (0.3–3.0 ft/day). In this case, the groundwater velocity through the source area is less than 1 m (3.2 ft) per year. As before, the groundwater in this area contains 500 mg/L of TCE and 20 mg/L *cis*-DCE, along with 5 mg/L dissolved iron and 100 mg/L sulfate.

The site for both template cases is assumed to be an active industrial site in the northeastern part of the United States. It is assumed the remediation will take place on an inactive portion of the property away from any buildings or other structures. Utilities and water are available. A wastewater treatment facility (WWTF) onsite is capable of accepting treated water. However, untreated waste liquids and soils generated during remedial activities will require either pretreatment prior to discharge or drummed disposal as hazardous waste.

When the sensitivity analysis is performed on TCE source strength, total TCE mass in the saturated zone will be allowed to vary from 1,500 to 60,000 kg. In each case, the TCE will be distributed between the aqueous phase (dissolved TCE), soil matrix (sorbed TCE) or DNAPL phase. It was assumed that negligible TCE would be present as vapor phase within the saturated interval. Distribution of TCE between the aqueous phase and soil was calculated assuming equilibrium partitioning. The remaining TCE was assumed to reside in the DNAPL phase.

The results of this calculation appear in Table 16.2. This calculation does not attempt to account for TCE that may have diffused into the clay aquitard. Although not quantified, some attempt will be made to consider the impact of TCE in this low permeability zone in the remedial designs and costing that follow.

Table 16.1. Description of Template Site Cases

Parameter	Case 1	Case 2
Location	Source area	Source area
Area (m^2)	1,500	1,500
Depth to groundwater (m)	1.5	1.5
Depth to aquitard (m)	4.5	4.5
Saturated thickness (m)	3.0	3.0
Hydraulic conductivity (cm/s)	5×10^{-3}	10^{-5} to 10^{-4}
Groundwater gradient (m/m)	3×10^{-3}	3×10^{-3}
Groundwater velocity (m/yr)	32	<1
Porosity	0.3	0.3
Effective porosity	0.15	0.2
Fraction organic carbon (kg/kg)	0.002	0.002
Injection rate (L/min)	18	3.8
NOD (permanganate; g/kg solids)	2	4
EVO adsorption (g/kg solids)	1	2
Residual TCE – base case (kg)	15,000	15,000
TCE (aqueous phase, mg/L)	500	500
cis-DCE (aqueous phase, mg/L)	20	20
VC (aqueous phase, mg/L)	0	0
pH (standard units)	7.0	7.0
Dissolved oxygen (mg/L)	0	0
Nitrate (mg/L)	0.1	0.1
Dissolved iron (mg/L)	5	5
Sulfate (mg/L)	100	100

Note: EVO – emulsified vegetable oil; g – gram(s); NOD – natural oxidant demand; VC – vinyl chloride

Table 16.2. Distribution of TCE in Template Source Area

Template Scenario	Total TCE (kg)	Dissolved TCE (mg/L)	Dissolved TCE (kg)	Sorbed TCE (mg/kg)	Sorbed TCE (kg)	DNAPL TCE (kg)
Low DNAPL	1,500	100	135 (9%)	26	210 (14%)	1,155 (77%)
Base case	15,000	500	675 (4.5%)	130	1,050 (7.0%)	13,275 (88.5%)
High DNAPL	60,000	700	945 (1.6%)	182	1,475 (2.4%)	57,580 (96.0%)

Note: () indicates percentage distribution of TCE

16.2.2 Cost Categories and Components

Selected remedies applied at these template sites include enhanced *in situ* bioremediation (EISB), *in situ* chemical oxidation (ISCO), *in situ* chemical reduction (ISCR) using micro- or nanoscale zero-valent iron (nZVI), surfactant flooding, thermal treatment using steam and direct resistance heating, pump-and-treat and excavation. Methods used to aid in mixing and amendment emplacement in the low permeability case include *in situ* soil mixing and hydraulic fracturing. Each remedy has been applied to the template site cases where it would be appropriate for the site conditions, although it is recognized that other combinations of technologies are possible. For each remedy, the costs were divided into general categories including design, capital, O&M and monitoring. It is assumed that basic site data derived from remedial site investigation activities are available prior to the design phase. Specific cost components in each cost category are listed in Table 16.3.

The design activities considered in this cost analysis include pre-design sampling or other assessments that are specific to the remedy applied. For example, some technologies may require laboratory studies, groundwater modeling or onsite hydraulic or amendment injection testing prior to implementation. Detailed design, permitting, reporting and procurement costs are also included in this category.

The capital activities include site preparation and all implementation costs incurred during the initial application of a technology in the designated area. These may include mobilization and demobilization of equipment, installation and development of amendment injection wells, installation of a trench, installation of a vapor collection system, acquisition of aboveground or

Table 16.3. Components of Cost Analysis

Design Phase	Capital
• Laboratory studies • Groundwater modeling • In-field hydraulic or injection testing • Detailed design and reports • Procurement • Permitting • Project management and home office support	• Site preparation • Mobilization and demobilization of equipment • Injection and monitoring well installation and development • Utilities and trench installation • Amendment delivery equipment • Aboveground treatment system and housing • Materials and labor for installation and first year of operation • Waste management and disposal • Royalty and licensing fees • Contractor oversight • Project management and home office support • Reports
O&M (per event or year)	**Monitoring (per year)**
• Labor • Electricity • Amendment reinjection • Replacement parts and materials • Waste management and disposal • Contractor oversight • Reports • Project management and home office support	• Labor • Analytical • Data and project management • Air monitoring • Waste management and disposal • Reports

belowground amendment delivery equipment, and acquisition of supporting utilities or installation and housing for an aboveground treatment system. These components may be used in various combinations, depending upon the technology applied. Other capital costs include labor, materials (amendments), waste management and disposal, contractor oversight, reporting, and project management and home office support directly related to implementation of the remedy. Finally, there are licensing and royalty fees associated with some technologies.

Operation and maintenance activities include labor, materials, replacement parts, waste management and disposal, contractor oversight, electricity usage, reporting, and project management and home office support required to keep the remedy operating as designed. Some technologies (ISCO and EISB) may require multiple amendment applications to be effective. Applications required beyond the first year of implementation have been included as O&M costs when appropriate. It has been assumed that the contractors performing periodic maintenance are located near the site, so that travel costs are not significant relative to the time spent onsite.

Monitoring costs include the labor, equipment, analysis, waste management and disposal, and reporting associated with monitoring the performance of the remedy. Although it is likely that downgradient monitoring will be required for most source area cleanups, only monitoring wells in the source area proper will be considered in this cost analysis. The monitoring network used in each case will consist of five 5-cm (2-inch) diameter monitoring wells. The costs provided include the installation of these wells for each option. Monitoring requirements for the various technologies will be as follows:

- For excavation and thermal technologies, it is assumed that removal of >99% of the TCE mass will be achieved in the first year of implementation and that the monitoring requirements will consist of pretreatment and post-treatment soil sampling and 2–3 years of post-treatment quarterly groundwater monitoring to confirm technology effectiveness. Contaminant destruction is unlikely to be as high for *in situ* chemical reduction with soil mixing, but the addition of clay should isolate the source from the surrounding groundwater. Therefore, a similar monitoring program will be utilized. Air monitoring will occur as necessary during active treatment for excavation and *in situ* soil mixing.
- For pump-and-treat, where only minimal mass removal is anticipated on an annual basis, it is assumed that groundwater reductions will tail off to asymptotic levels over time. Therefore, it is assumed that monitoring requirements will consist of quarterly monitoring for the first 3 years and semiannual monitoring for the remainder of the treatment.
- For the remainder of the technologies, it is assumed that groundwater sampling will occur quarterly for the period of time that the remedy is active and for 2 years thereafter. Air monitoring for aboveground treatment systems will occur as necessary only during active treatment. The monitoring program will transition into a monitored natural attenuation (MNA) program with semiannual monitoring of MNA parameters 2 years after active remediation is complete. The sampling frequency will be reduced to annual monitoring 12 years after termination of active treatment.

All economic comparisons have been made on a NPV basis, assuming that the total lifetime of the remediation project is a maximum of 30 years and using a discount rate of 3% for annual costs. This discount rate is derived from Department of Defense guidance on using economic analyses in decision making (DoD, 1995) and is based on current Office of Management and Budget data. The 3% discount rate assumes a 5% nominal return is available on U.S. Treasury notes and bonds and an annual inflation rate of roughly 2%. This real discount rate is appropriate for government entities, while a higher discount rate may be more appropriate for private entities, which can obtain higher returns on investments. A higher discount rate will have the effect of reducing the impact of future expenditures on the net present cost of the remedy.

16.3 REMEDIATION OPTIONS EVALUATED

16.3.1 Case 1: Permeable Source Area

The remedial technologies applied to the permeable source area included EISB, ISCO using permanganate, ISCR using nZVI, surfactant flooding and thermal treatment using steam. While it is recognized that other technologies or combinations of technologies may be applicable for sites with similar characteristics, these technologies were selected for purposes of this cost demonstration to attempt to represent a range of options available. The objective in each case was to reduce the source TCE mass by at least 90% and/or reduce the TCE flux coming from the source area by 90–95%. The timing required to meet the objective is dependent on the technology used and therefore is variable.

Remedial costs for the *in situ* technologies applied to the permeable source area will be compared to pump-and-treat, which is a technology that has been broadly applied to many remedial sites and has traditionally been a benchmark against which other remedies are compared. It should be noted that pump-and-treat only treats the groundwater emanating from the source area without treating the source itself. Thus, it provides containment of the resulting plume rather than remediation of the source area.

16.3.1.1 Enhanced *In Situ* Bioremediation

For the purposes of this analysis, emulsified vegetable oil (EVO) will be used as the electron donor to promote reductive dechlorination of TCE to ethene in the residual source area. The advantages of this electron donor include longevity and ease of distribution in the subsurface relative to other slow-release donors. A laboratory biotreatability study will be used, as it is typically needed to verify that EISB will be effective at a site, and that EVO is a suitable donor.

The EVO will be applied through a series of 50 injection wells spaced on 5.4 m (17.7 ft) centers distributed across the 1,500 m^2 (16,146 ft^2) source area. The 2-inch diameter injection wells will be screened across the saturated zone and developed prior to EVO injection. Pre-design injection testing is included in the cost analysis to establish the injection rate, which is assumed to be 18 liters per minute (L/min; 4.7 gallons per minute [gpm]) and to confirm the effective radius of injection achievable under site conditions (assumed to be 3 m [9.8 ft]). The costs include the addition of 349 kg (768 lbs) of a commercial EVO solution (50% EVO) to each injection point, along with 25,090 L (6,630 gallons [gal]) of groundwater to ensure complete distribution of the EVO. The EVO will be metered through a dosimeter pump connected to a manifold linked to up to 10 groundwater extraction wells to provide a water source for injection, and a second manifold will be used to allow injection into as many as ten injection wells simultaneously. Each injection line will contain a pressure gauge and flow totalizer, to allow accurate measurements of the fluid injections into each well. The injections will be performed by a two-person crew requiring 26 days of labor including mobilization, setup and breakdown. Bioaugmentation with a commercially available bioaugmentation culture will take place during the initial EVO injection program. Three liters (0.8 gal) of culture will be injected into each injection well.

It is assumed that it will take 1.5, 3 and 6 years of active remediation prior to transition to MNA for the low, base case and high DNAPL loadings, respectively. These timeframes are based on the stoichiometric EVO demand for each case, as determined by the electron acceptor (TCE, sulfate) mass present in the source area as DNAPL, dissolved or sorbed phases and assuming that a safety factor of five is required to account for electron donor consumption for nondechlorination activity. It is anticipated that each application of EVO will support reductive dechlorination for 18 months and that a minimum EVO loading is required as determined by the EVO adsorption capacity of the aquifer (totaling 17,500 kg [38,500 lbs] EVO per injection event, or 0.7% of the pore space).

The total stoichiometric EVO demand for the low, base case and high DNAPL scenarios are 2,410 kg (5,302 lbs), 20,408 kg (44,898 lbs) and 80,402 kg (176,884 lbs), respectively. For the low DNAPL scenario, the EVO demand is dictated by the EVO adsorption capacity of the aquifer, and the EVO mass injected to satisfy the adsorption capacity of the aquifer (17,500 kg [38,500 lbs]) is 7.3 times greater than the stoichiometric requirements. Only one injection will be required for the low DNAPL scenario. For the base case scenario, two injections of 0.7% EVO solution will be required to meet the stoichiometric demand, the second of which will occur in year 2. For the high DNAPL case, four injections of 1.0% EVO solution will deliver sufficient EVO to meet the stoichiometric demand, with follow-up injections occurring in years 2, 4 and 5. All additional applications will deliver 100% of the initial EVO loading utilizing the same injection well network. No other operation and maintenance is required with this remedial option.

Quarterly groundwater monitoring will be performed throughout the active phase, followed by quarterly (2 years), semiannual (10 years) and annual MNA monitoring thereafter. All monitoring will include parameters specific to the remedy, including relevant MNA analytes (field parameters [pH, dissolved oxygen, oxidation–reduction potential (ORP)], volatile organic compounds [VOCs], alkalinity, total organic carbon [TOC], anions [nitrate, sulfate, sulfide, chloride], metals [iron, manganese] and gases [ethene, ethane, methane]).

16.3.1.2 *In Situ* Chemical Oxidation

Sodium permanganate will be used as the oxidant to promote degradation of TCE and its daughter products in the source area. The advantages of this oxidant include greater longevity than other oxidants and ease of distribution in the subsurface. Laboratory natural oxidant demand (NOD) testing will be used to estimate the background NOD of the soil. In this case, it is assumed that the NOD of the soil is 2 grams (g) permanganate per kg of soil, which is a representative value for a partially reduced sandy soil with moderate levels of total organic carbon.

The permanganate will be applied through the same network of 50 injection wells spaced on 5.4 m (17.7 ft) centers used in the EISB cost analysis. The 2-inch diameter injection wells will be screened across the saturated zone and developed prior to permanganate injection. Pre-design injection testing is included in the cost analysis to establish the injection rate, which is assumed to be 18 L/min (4.7 gpm), and to confirm the effective radius of injection achievable under site conditions, assumed to be 3 m (9.8 ft). Sodium permanganate will be injected at concentrations ranging from 15 to 19 grams per liter (g/L) for the three DNAPL loading scenarios (optimized to meet the total permanganate demand in the minimum number of injection events), corresponding to between 954 kg (2,099 lbs) to 1,209 kg (2,660 lbs) sodium permanganate injected into each well in each injection event. The sodium permanganate will be mixed with 24,500 L (6,470 gal) of groundwater to ensure complete distribution of the oxidant around each injection point. The sodium permanganate solution will be metered through a pump connected to a manifold linked to up to 10 groundwater extraction wells to provide a water source for injection, and a second manifold will be used to allow injection into as many as ten injection wells simultaneously. Each injection line will contain a pressure gauge and flow totalizer, to allow accurate measurements of the fluid injections into each well. The injections will be performed by a two-person crew and will require 26 days, including setup and breakdown.

It is estimated that it will take 0.5, 1.5 and 3.5 years of active remediation prior to transition to MNA for the low, base case and high DNAPL loadings, respectively. These timeframes are based on the stoichiometric permanganate demand for each case, as determined by the electron donor (for example, TCE) mass present in the source area as DNAPL, dissolved or sorbed phases and also accounting for the permanganate demand exerted by the soil (2 g/kg). A safety factor of 1.25 was also applied to account for the potential for some permanganate mass to advectively migrate out of the source area with groundwater flow prior to reaction with contaminant phase. The total sodium permanganate demand for the low, base case and high DNAPL loadings are 60,436 kg (132,960 lbs), 143,140 kg (314,900 lbs) and 400,790 kg (692,790 lbs), respectively. The injection concentration for sodium permanganate ranges from 15 to 19 g/L to optimize the necessary permanganate mass delivery with a minimum number of injection events. A total of one, three and seven injection events for the low, base case and high DNAPL loading scenarios, respectively, will be required to deliver these amounts of sodium permanganate into the source area. It is anticipated that the permanganate will persist for approximately 6 months after each injection event, and therefore two injections will be completed per year. All additional applications will deliver 100% of the initial permanganate loading utilizing the same injection well network. No other operation and maintenance is required with this remedial option.

Quarterly groundwater monitoring will be performed throughout the active phase, followed by quarterly (2 years), semiannual (10 years) and annual MNA monitoring thereafter. All monitoring will include parameters specific to the remedy, including field parameters, VOCs, chloride, metals and permanganate. Post-treatment monitoring during the MNA phase will include all relevant MNA analyses.

16.3.1.3 *In Situ* Chemical Reduction

The ISCR application will consist of applying nZVI to the DNAPL zone. The nZVI will be injected in a slurry form, created by mixing the nZVI with water containing a dispersant (soy protein) in a 5:1 ratio (by mass). The dispersant is required to improve the transport of the nZVI away from the injection point, as nZVI alone tends to agglomerate into larger particles in the absence of dispersant and filter out near the injection point. To enhance the reactivity of the nZVI, a small amount of palladium (1% by weight) will be added to the slurry to act as a catalyst, and the slurry will be batch-mixed in the field to minimize water contact with the nZVI prior to injection. The soy protein will also act as an electron donor and will enhance biodegradation of the TCE in conjunction with the chemical reduction induced by the nZVI/palladium.

Since the cost of palladium and nZVI is high, nZVI is more cost effective when combined in a treatment train with another follow-up technology such as EISB, particularly for higher DNAPL loadings. One of the strengths of ISCR treatment is that it induces strongly negative ORPs in the groundwater that are conducive to biological activity, and, therefore, provides optimal conditions for follow-up EISB polishing. The degree of treatment completed using ISCR versus EISB is primarily a cost issue, and a cost–benefit analysis is recommended to aid in this decision.

The nZVI slurry will be applied through a series of 50 injection wells spaced on 5.4 m (17.7 ft) centers distributed across the 1,500 m^2 (16,146 ft^2) source area, similar to the EISB and

ISCO approaches. The nZVI injection concentrations were constrained to be within the range typically applied, assuming a minimum concentration of 4 g/L and maximum concentration of 20 g/L. The nZVI slurry typically has a 3–4.5 m (10–15 ft) maximum radius of distribution around the injection point, and a 5.4 m well spacing allows for some overlap between injection points. The 2-inch diameter injection wells will be screened across the saturated zone and developed prior to nZVI injection. Pre-design injection testing is included in the cost analysis to establish the injection rate, which is assumed to be 18 L/min (4.7 gpm), and to confirm the effective radius of injection achievable under site conditions, which is assumed to be 3 m (9.8 ft).

The costs include the addition of 255 kg (561 lbs) of nZVI, combined with 51 kg (112 lbs) of soy protein and 1% palladium, to each injection point, along with 25,090 L (6,630 gal) of groundwater to ensure adequate distribution of the nZVI. The nZVI slurry will be mixed onsite using an onsite batch processor capable of mixing 75,000 L of slurry a day. The slurry will be metered into five injection wells at a time through a manifold linked to up to five groundwater extraction wells to provide a water source for injection, and a second manifold will be used to allow injection into as many as five injection wells simultaneously. The injections will be performed by a seven-person crew requiring 20 days of labor, including mobilization, setup and breakdown. Bioaugmentation with a commercially available bioaugmentation culture will take place after the initial injection program is complete. Three liters (0.8 gal) of culture will be injected into each injection well.

It was assumed for the base case scenario that sufficient nZVI and soy protein would be injected in the first injection event to provide treatment of 40% of the DNAPL utilizing an nZVI concentration of 10 g/L. One follow-up EISB treatment (EVO injection) will be completed to treat the remaining 60% TCE mass. For the high DNAPL scenario, a single application of nZVI at a concentration of 10 g/L will result in treatment of 10% of the DNAPL, with follow-on EISB treatment again required. Substantial cost savings over complete treatment using nZVI are realized for both scenarios through completing only partial treatment of the DNAPL using nZVI. For the low DNAPL scenario, the minimum nZVI concentration of 4 g/L results in treatment of 100% of the DNAPL (assuming ideal performance) and costs were estimated assuming no follow-on treatment. Follow-up EISB treatments will be completed as described for the EISB technology.

The mass of nZVI and EVO required was estimated by calculating the stoichiometric demand as dictated by the mass of electron acceptor (TCE, sulfate) in the source area at the beginning of treatment, increased by a safety factor of five to account for consumption of reactant for nondechlorinating activity. Estimated requirements for nZVI and EVO for the low, base case and high DNAPL loading scenarios are detailed in Table 16.4, along with the corresponding

Table 16.4. Summary of Amendment Requirements for the ISCR Case 1 Remedy

	Low DNAPL	Base Case	High DNAPL
DNAPL treatment using nZVI	100%	40%	10%
Total nZVI mass injected	5,089 kg (11,196 lbs)	12,723 kg (27,991 lbs)	12,723 kg (27,991 lbs)
Total EVO mass injected	0 kg (0 lbs)	19,950 kg (43,890 lbs)	74,814 kg (164,591 lbs)
Total number of injections	1 nZVI + 0 EVO	1 nZVI + 1 EVO	1 nZVI + 2 EVO
nZVI concentration	4 g/L nZVI	10 g/L nZVI	10 g/L nZVI
EVO concentration	0% EVO	0.8% EVO	1.5% EVO

injection concentrations for both the nZVI and EVO. All additional applications will deliver 100% of the initial amendment loading utilizing the same injection well network. Outside of periodic injections, no other operation and maintenance is required with this remedial option.

It is estimated that it will take 0.7, 2.2 and 3.7 years of active remediation prior to transition to MNA for the low, base case and high DNAPL loadings, respectively. These timeframes assume the nZVI and soy protein are consumed within 8 months of injection and that the EVO will support reductive dechlorination for 18 months.

Quarterly groundwater monitoring will be performed throughout the active phase, followed by quarterly (2 years), semiannual (10 years) and annual MNA monitoring thereafter. All monitoring will include parameters specific to the remedy, including relevant MNA analytes (field parameters [pH, DO, ORP], VOCs, alkalinity, TOC, anions [nitrate, sulfate, sulfide, chloride], metals [iron, manganese] and gases [ethene, ethane, methane]).

16.3.1.4 Surfactant Flooding

Surfactant enhanced aquifer remediation (SEAR) will be implemented using Tween® 80, which is a non-ionic food-grade surfactant that enhances DNAPL removal by forming micelles that solubilize the DNAPL phase, effectively increasing solvent solubility by 100–1,000 times. For purposes of this analysis, it is also assumed that SEAR will be successful at removing approximately 90% of the DNAPL mass, with EISB used to treat the remaining mass. Pre-design testing will include laboratory column studies to confirm the sweep efficiency and field injection and hydraulic testing to evaluate potential surfactant delivery issues and confirm well spacings and injection/extraction rates.

During implementation of the surfactant flood, a 6% Tween® 80 solution will be injected into eleven 4-inch injection wells spaced throughout the source area and screened across the 3 m (15 ft) aquifer. Groundwater containing dissolved-phase TCE and surfactant will be extracted from four 4-inch wells, also screened across the aquifer, and then passed through an *ex situ* treatment system before discharge to a WWTF. For the low and base case DNAPL scenarios, it is assumed that *ex situ* treatment of extracted groundwater consists of the addition of a defoamer and air stripping, to remove volatile organic compounds (TCE) to reduce the organic content in the groundwater prior to discharge to the WWTF, and vapor-phase granular activated carbon to capture emissions. The potential exists for the surfactant to interfere with the ability of the air stripper to remove the TCE, resulting in the need to add additional stages or increase air flow rates. This potential issue is not captured in this cost estimate. The discharged groundwater is likely to have a high biochemical oxygen demand (BOD) content, and therefore a surcharge is assumed to apply to WWTF discharge costs for the required BOD pretreatment. For the high DNAPL scenario, additional treatment steps will be required, including DNAPL recovery through decanting, evaporating and condensing units. Recovered DNAPL will then be disposed at a hazardous waste treatment facility, assumed at a cost of $500 per 208 L drum (55 gal).

Surfactant will be mixed with potable water in large mixing tanks prior to conveyance to injection wells via a piping network constructed with appropriate flow control instrumentation. Extracted groundwater will be conveyed to a treatment building containing the previously described treatment units. System operation will be monitored and recorded using a programmable logic controller, and will be overseen by one operator who will be onsite on a daily basis during operations. It is assumed that sanitary sewer and potable water connections are readily available.

As mentioned above, it is assumed that 90% of the DNAPL mass will be removed with the SEAR portion of the treatment train. For the low and base case DNAPL scenarios, it is assumed that one and two pore volume flushes of 6% Tween® 80, respectively, will be required to achieve this: injection of a total of 1,350,000 L of potable water containing 81,000 kg Tween® 80 for the low DNAPL scenario and 2,700,000 L of potable water containing 162,000 kg Tween® 80 for the base case DNAPL scenario. For the high DNAPL scenario, four pore volume flushes will be required (5,400,000 L of potable water mixed with 324,000 kg Tween® 80). At a total injection rate of 22,500 L/day, it is estimated that 2 months will be required to complete each pore volume flush. A follow-up pore volume flush will then be completed using only potable water to flush most of the remaining surfactant and dissolved DNAPL out of the aquifer.

Consistent with observations at field sites where Tween® 80 has been used, it is assumed that some surfactant remains in the aquifer post-treatment and biodegrades (5% retention was observed at the Bachman Road site; Ramsburg et al., 2004). For costing purposes, it is assumed that 5% of the Tween® 80 sorbs to the soil and/or diffuses into the underlying lower permeability clay. As the surfactant desorbs and back-diffuses from the low-permeability layers, it will ferment and support biodegradation of the TCE mass that was not effectively removed with the SEAR phase. For the low and base case DNAPL scenarios, sufficient surfactant is assumed to remain within the aquifer to effectively biodegrade the remaining mass. For the high DNAPL case, however, additional electron donor will be required to fully biodegrade the remaining mass. For all scenarios, the system will be shut down for 1 year post-SEAR to allow for biodegradation of the residual TCE mass (referred to as the passive EISB phase). For the high DNAPL scenario, the system will then be turned back on after the 1-year passive EISB phase, and an additional year of lactate amendment will be performed to biodegrade the TCE mass that remains (referred to as the active EISB phase). The lactate amendment will occur using the existing infrastructure, with groundwater extracted from four wells at 5,625 L/day (1,486 gal/day) per well, mixed with 330 mg/L lactate and then reinjected into eleven wells at 2,045 L/day (540 gal/day) per well. An operator will be present for 8 hours a day on a daily basis during SEAR injection. To enhance the bioactivity, bioaugmentation with a commercially available bioaugmentation culture will take place after the SEAR injection program is complete. Fourteen liters (3.7 gal) of culture will be injected into each injection well.

It is assumed that it will take 1.3, 1.5 and 2.8 years of active remediation prior to transition to MNA for the low, base case and high DNAPL loadings, respectively. These timeframes are based on a 4- and 6-month SEAR phase followed by a 1-year passive EISB phase for the low and base-case DNAPL scenarios, and a 10-month SEAR phase followed by a 1-year passive EISB phase and 1-year active EISB phase for the high DNAPL scenario.

Quarterly groundwater monitoring will be performed throughout the SEAR and EISB phases, followed by quarterly (2 years), semiannual (10 years) and annual MNA monitoring thereafter. All monitoring will include parameters specific to the remedy, including relevant MNA analytes (field parameters [pH, DO, ORP], VOCs, alkalinity, TOC, anions [nitrate, sulfate, sulfide, chloride], metals [iron, manganese] and gases [ethene, ethane, methane]).

16.3.1.5 Thermal Treatment

Thermal treatment in the case of the permeable source area will consist of steam-enhanced extraction combined with multiphase extraction. The remedial objective is to heat the treatment zone to near the boiling point of water (100 degrees Celsius [°C]) using steam. This causes the chlorinated ethenes to volatilize or otherwise be stripped from the soil by the steam, where they can be captured and treated by an aboveground treatment system. The steam generation capacity of the system is assumed to be 1,635 kilograms per hour [kg/hr] (3,600 lbs/hr). The system will consist of 36 steam-injection wells located on 7.6 m (25 ft) centers and 25 multiphase extraction wells on similar spacing. The subsurface temperature and pressure will be measured using 18 temperature-monitoring and 7 pressure-monitoring wells.

The aboveground treatment system will consist of a heat exchanger, liquid–vapor separator, gravity settling tank (to separate DNAPL from water phase) and liquid and vapor phase activated carbon units. The system will handle groundwater extraction rates up to 60 L/min (15.5 gpm) and vapor extraction rates up to 10.8 m^3/min (380 standard cubic feet per minute [scfm]). The maximum TCE removal rate during peak operation is estimated to be 286 kg/day (630 lbs/day).

It is estimated that 4–5 months will be required for mobilization, setup, drilling and system installation. The time for active treatment for the base-case DNAPL loading is estimated to be 120 days. This includes 48 days to establish hydraulic control over the source area and heat the target volume to temperature, 47 days of treatment at temperature and 24 days for sampling and cool down. It is estimated that one operator will be required for 45 hr/week during the operational phase to operate the system and monitor system performance. One month will be required for demobilization at the end of the project.

TCE removal via the multiphase extraction system will be monitored periodically during the active phase of remediation. Soil samples obtained before and after remediation will be used to evaluate remedial effectiveness in the subsurface. Five groundwater monitoring wells will be installed in the source area after the remediation is complete and will be used to evaluate system rebound. It is anticipated that these wells will be sampled quarterly for VOCs only for 2 years post-treatment.

16.3.1.6 Pump-and-Treat

With pump-and-treat, contaminant removal is achieved through extraction of groundwater along the downgradient site boundary and *ex situ* removal of the VOCs from the extracted groundwater. The rate of mass removal is governed by the concentrations in the extracted groundwater and the rate of groundwater extraction. With ongoing operation of pump-and-treat systems, asymptotic reductions in groundwater concentrations are often observed, with the groundwater concentrations typically being greater than cleanup criteria such as maximum contaminant limits. This behavior results in long-term operation of pump-and-treat systems, typically for decades or longer. Due to the uncertainty involved in predicting the life span of operation of these systems, pump-and-treat remediation costs are typically calculated assuming a 30-year operation as a minimum.

For Case 1, it is assumed that groundwater extraction will be performed using two 4-inch diameter wells screened across the saturated zone and developed prior to system start-up. Groundwater extraction rates will correspond to estimated maximum sustainable well yields (5.1 L/min; 1.3 gpm). It is assumed that the extracted groundwater will be treated *ex situ* using granular activated carbon and then discharged to the WWTF onsite.

Pre-design hydraulic testing is included in the cost analysis to establish the sustainable extraction rate and to confirm the width of the hydraulic capture zone created with each extraction well. Numerical modeling is also assumed to be required as a pre-design step to assess capture in a heterogeneous environment. Each extraction well line will contain flow control valves and a flow totalizer. System operation will be automated using a programmable logic controller, to minimize labor requirements for system oversight.

Quarterly groundwater monitoring will be performed for the first 3 years of operation, followed by semiannual monitoring for the remainder of the 30-year duration. All monitoring will include parameters specific to the remedy, including field parameters and VOCs.

16.3.2 Case 2: Low Permeability Source Area

The remedial technologies applied to the low permeability source area included EISB, ISCO using permanganate, *in situ* soil mixing using ZVI and clay and thermal treatment using direct resistance heating. Hydraulic fracturing was used in concert with the first two technologies to increase the permeability of the formation and allow the injection of amendments. Again, while it is recognized that other technologies or combinations of technologies may be applicable for sites with similar characteristics, these technologies were selected for purposes of this cost demonstration to attempt to represent a range of options available. As before, the objective in each case was to reduce the source TCE mass by at least 90% and/or to reduce the TCE flux coming from the source area by 90–95%. The timing required to meet the objective is dependent on the technology used and, therefore, is variable.

In this case, remedial costs for the *in situ* technologies applied to the low permeability source area will be compared to excavation. Pump-and-treat was not considered viable in this situation because of the low permeability of the soils. Like pump-and-treat, excavation has also been a broadly employed technology for remediation of DNAPL sources and, thus, provides a good benchmark for the *in situ* remedies.

16.3.2.1 Hydraulic Fracturing

Hydraulic fracturing will be used with the EISB and ISCO remedial options to enhance the permeability of the formation and allow easier fluid injection. The hydraulic fracturing will consist of pressure injection of sand mixed with guar and will result in propagation of sand-filled fractures within the tighter silt and clay aquifer. The sand will act as a proppant and will create permanent fractures. Sixteen fracture locations will be used on 10 m (32 ft) centers. Three fractures vertically spaced 1.0 m (3.3 ft) apart will be initiated in each fracture location so that 48 fractures will be initiated all together. Each fracture will contain 900 kg (2,000 lbs) of 20/40 Frac sand mixed with guar, creating a radius of influence of 6–8 m (20–25 ft). Thus, individual fractures will overlap. It is estimated that it will take a crew of three people 15 days to perform the hydraulic fracturing. After fracturing is complete, a cluster of three injection wells will be installed at each fracture location, with each well screened to correspond to an individual fracture.

16.3.2.2 Enhanced *In Situ* Bioremediation

For the purposes of this analysis, EVO will be used as the electron donor to promote reductive dechlorination of TCE to ethene in the source area. Similar to Case 1, a laboratory biotreatability study will be used to verify that EISB will be effective at the site.

The EVO will be applied through injection wells screened across the 16 hydraulic fracture locations in the source area. A nested set of three 5-cm (2-inch) injection wells will be installed at each location, with each injection well screened over a designated fracture depth so that each injection will follow a single sand fracture. Thus, 48 injection wells will be required for this design. Injection testing is included in the cost analysis to establish the injection rate, which is assumed to be 3.8 L/min (1.0 gpm) into each injection well. The costs include the addition of 430 kg (950 lbs) of a commercial EVO solution of 45% soybean oil to each injection point, along with 9,400 L (2,400 gal) of fluid (dilute EVO and chase water) to ensure complete distribution of the EVO. The EVO will be metered through a dosimeter pump connected to a nearby potable water source, and a manifold will be used to allow injection into as many as six injection wells simultaneously. Each injection line will contain a pressure gauge and flow totalizer to allow accurate measurements of the fluid injections into each well. The injection program will be performed by a two-person crew and will require 35 days of labor, including setup and breakdown. Bioaugmentation with a commercially available bioaugmentation culture will take place during the initial injection program. Three liters (0.8 gal) of culture will be injected into each injection well.

It was assumed that it would take 3, 6 and 12 years of active remediation prior to transition to MNA for the low, base case and high DNAPL loadings, respectively. The timeframes are longer than for the permeable Case 1 to account for anticipated difficulties in distributing electron donor in the tighter soils and for the slow back diffusion of TCE from the low permeability zones. It is anticipated that each application of EVO will support reductive dechlorination for 2–3 years, so that the base case will require a second application of EVO in year 3. In the same way, the low DNAPL loading will not require reapplication, whereas the high DNAPL loading will require reapplication in years 3, 6 and 9. All additional applications will deliver 100% of the initial EVO loading utilizing the same injection well network. No other operation and maintenance is required with this remedial option.

Quarterly groundwater monitoring will be performed throughout the active phase, followed by quarterly (2 years), semiannual (10 years) and annual MNA monitoring thereafter. All monitoring will include parameters specific to the remedy, similar to those analyzed for the Case 1 template site.

16.3.2.3 *In Situ* Chemical Oxidation

Sodium permanganate will be used as the oxidant to promote degradation of TCE and its daughter products in the source area. Laboratory NOD testing will be used again to estimate the background NOD of the soil. In this case it is assumed that the NOD of the soil is 4 g sodium permanganate per kg of soil, which is a representative value for a partially reduced sandy silt and clay soil with higher levels of total organic carbon.

The permanganate will be applied through the same network of 48 injection wells used in the EISB cost analysis, with each injection following a single sand fracture. Injection testing is included in the cost analysis to establish the injection rate, which is assumed to be 3.8 L/min (1.0 gpm) into each injection well. The costs include the addition of 935 kg (2,057 lbs) of 40% sodium permanganate solution to each injection point, along with 9,400 L (2,400 gal) of potable water to ensure complete distribution of the oxidant. The permanganate solution will be mixed in aboveground tanks prior to injection and injected at a sodium permanganate concentration of 40 g/L. A manifold will be used so that injection can take place in up to six injection well-s simultaneously. Each injection line will contain a pressure gauge and flow totalizer, to allow accurate measurements of the fluid injections into each well. The injection program will be performed by a two-person crew and will require 35 days of labor, including setup and breakdown.

It was assumed that it would take 3, 6 and 12 years of active remediation prior to transition to MNA for the low, base case and high DNAPL loadings, respectively. The timeframes are longer than for permeable Case 1 and similar to those for EISB to account for anticipated difficulties in distributing oxidant in the tighter soils and for the slow back diffusion of TCE from the low permeability zones. It is anticipated that each application of oxidant will support oxidation for 1 year, so that the base case will require four additional applications of oxidant over the following 5 years. In the same way, the low DNAPL loading will require one additional application of oxidant after 1 year, whereas the high DNAPL loading will require eight applications of oxidant spread out over 10 years. All additional applications will deliver 100% of the initial oxidant loading utilizing the same injection well network. No other operation and maintenance is required with this remedial option.

Quarterly groundwater monitoring will be performed throughout the active phase, followed by quarterly (2 years), semiannual (10 years) and annual MNA monitoring thereafter. Monitoring during active treatment will include VOCs, chloride and select metals. Post-treatment monitoring during the MNA phase will include all relevant MNA analytes.

16.3.2.4 ISCR with Soil Mixing

In situ soil mixing will use large diameter augurs to mix clay and microscale zero-valent iron (ZVI) into the source area. The purpose of the clay is to stabilize the source area and limit infiltration of groundwater, while the ZVI reacts with and destroys the TCE. For soil mixing at this relatively shallow depth, a 3-m (10-ft) diameter augur is expected to provide efficient mixing. The crane-mounted auger will consist of three blades, referred to as flights, extending from the center shaft. Injection ports are located over the entire length of the flights to ensure the iron-containing grout is delivered throughout the entire area of the mixed column. A total of 250 overlapping columns will be required to completely cover the source area. Four mixing passes will be completed in each column to provide further homogenization. Quality control soil samples will be collected during mixing to verify iron content and soil uniformity.

50D quality microscale iron and unmodified sodium bentonite clay will each be added to the subsurface soil at 2% by weight for the base case DNAPL loading. This will require a total of 195 metric tons (213 U.S. tons) of iron and clay. The iron costs are estimated to be $860 per metric ton ($780 per U.S. ton) including delivery. Delivered cost for the bentonite clay is estimated to be $275 metric ton ($250 U.S. ton). A grout plant, similar to a pug mill, will be used to combine iron, clay and water into a grout. Components of the plant include material storage silos, grout pumps and a 5 m^3 capacity colloidal tank mixer. The grout will be pumped through the hollow kelly and delivered into the subsurface through the mixing tool. A rotary table, also mounted to the crane, will provide rotational torque (on the order of 369,000 N-m or 500,000 ft-lb), and will also turn the kelly. It will require 2 weeks to mobilize and set up the mixing equipment and 25–30 days to perform the soil mixing. The soil mixing activity will require eight to ten workers onsite during this activity.

The soil will expand during *in situ* mixing. To create space for expansion, the top 1 m (3 ft) of soil will be excavated prior to treatment. Iron and bentonite clay will be mixed into the soil and the soil will be temporarily stockpiled onsite. After the *in situ* treatment is complete, the excavated soil will be moved back on top of the source area to accelerate drainage and compression of the underlying soils.

Air monitoring will be used to protect worker safety during the excavation. It is assumed that the equipment operators will be in Level C personal protective equipment (PPE) when mixing the soil at the high DNAPL loading. Soil samples obtained before and after remediation will be used to evaluate remedial effectiveness in the subsurface. Five groundwater monitoring wells will be installed in the source area after the remediation is complete and will be used to evaluate post-treatment groundwater concentration trends. It is anticipated that these wells will be sampled quarterly for 3 years.

16.3.2.5 Thermal Treatment

Thermal treatment in the case of the low permeability source area will consist of thermal conduction heating (TCH) combined with soil vapor extraction (SVE). As previously described, the remedial objective is to heat the treatment zone to near the boiling point of water (100°C) using thermal conduction, causing the chlorinated ethenes to volatilize where they can be captured by the SVE system and treated by an aboveground treatment system. The system will consist of 110 heater borings located on 4.6 m (15 ft) centers and 25 SVE wells located on 7.6 m (25 ft) centers. The subsurface temperature and pressure will be measured using 21 temperature monitoring and seven pressure monitoring wells.

The aboveground treatment system will consist of a heat exchanger, liquid–vapor separator, gravity settling tank (to separate DNAPL from water phase), and liquid- and vapor-phase activated carbon units. The system is designed to handle vapor extraction rates up to 8.0 m^3/min (280 scfm). The estimated transformer size required for this system is 1,000 kilovolt-ampere (kVA).

It is estimated that 4–5 months will be required for mobilization, setup, drilling and system installation. The active treatment time for the base case DNAPL loading is estimated to be 128 days. This includes 74 days for start-up and to heat the target volume to temperature, 30 days for treatment at temperature, and 24 days for sampling and post-treatment vapor extraction. The maximum TCE removal rate during peak operation is estimated to be 263 kg/day (580 lbs/day). It is estimated that one operator will be required for 45 hr/week during the operational phase to operate the system and monitor system performance. One month will be required for demobilization at the end of the project.

TCE removal via the multiphase extraction system will be monitored periodically during the active phase of remediation. Soil samples obtained before and after remediation will be used to evaluate remedial effectiveness in the subsurface. Five groundwater monitoring wells will be installed in the source area after the remediation is complete and will be used to evaluate TCE concentration rebound. It is anticipated that these wells will be sampled quarterly for 2 years.

16.3.2.6 Excavation

The excavation remedy will utilize sheet piling to isolate the source area and prevent infiltration of water during the excavation work. The excavation will be performed in three sections. Approximately 276 m (900 ft) of 5 m (16.3 ft) long sheet piling will be driven into the clay aquitard and then removed at the end of the project. A total of 6,750 m^3 (8,830 cubic yards [yd^3]) of contaminated soil will be removed. Approximately 2% by weight of kiln dust (lime) will be mixed with the soil during the excavation to adsorb excess water and improve soil-handling properties prior to loading the soil onto trucks for transportation to a secure offsite landfill. The rate of excavation is assumed to be 382 m^3/day (500 yd^3/day). The landfill costs are estimated to be $165 per metric ton ($150 per U.S. ton) and include disposal, transportation, fuel surcharges and taxes. It is assumed that all the soil will meet the alternative soil treatment standards suitable for direct landfill. If the soil were to fail to meet alternative soil treatment standards, additional post-excavation stabilization would be required prior to landfill or the soil would require incineration. The latter option would increase disposal costs by at least five-fold. The excavation will be backfilled with clean soil brought onto the site for this purpose. It will require 25–30 days to install and remove the sheet piling and 25–30 days to perform the excavation. Four to five workers will be onsite during both activities, and oversight will be provided by the primary contractor.

Groundwater will be pumped from the excavation pit simultaneously with soil removal and treated using a rented activated carbon treatment system. The treated water will be pumped to the onsite water treatment facility. The volume of spent carbon is estimated to be 500–1,500 kg (1,000–3,200 lbs) depending on the TCE concentration in the groundwater. The spent carbon will be mixed with the soil and sent to the offsite disposal facility.

Air monitoring will be used to protect worker safety during the excavation. It is assumed that the equipment operators will be in Level C PPE when excavating the soil at the high DNAPL loading. Soil samples obtained before and after remediation will be used to evaluate remedial effectiveness in the subsurface. Five groundwater monitoring wells will be installed in the source area after the remediation is complete and will be used to evaluate system rebound. It is anticipated that these wells will be sampled quarterly for 2 years.

16.4 COST ANALYSES

16.4.1 Case 1: Permeable Source Area

Table 16.5 provides a detailed breakdown of costs for each technology considered for Case 1 for the base case DNAPL scenario. Major cost components include design (pre-design testing, engineering design and procurement), capital costs (purchase and installation of equipment and infrastructure, and the first year of technology implementation), O&M beyond year 1, and groundwater, air and soil monitoring.

Table 16.6 provides a summary and comparison of the breakdown of total costs for each technology described above broken down by major category. Total costs for all categories are calculated on an NPV basis using a 3% discount rate.

For the base case scenario, EISB is the least expensive option, while pump-and-treat is the most expensive. Thermal remediation, ISCO, and ISCR and SEAR followed by EISB are nearly 2–2.5 times more expensive than EISB alone.

As seen in Table 16.6, the breakdown of costs for EISB is fairly evenly distributed between the categories of design, capital, O&M and monitoring. Major components of the capital cost include well installation, equipment and controls (37%), materials and bioaugmentation (20%), and labor and oversight (22%). The EISB approach that was evaluated for Case 1 (passive EVO injections) requires little infrastructure outside of the initial capital investment of 50 shallow wells, and annual O&M costs are limited due to the passive nature of the technology implementation and the low cost of the electron donor. The estimated remedial timeframe for EISB is the longest of the various technologies evaluated and is the primary cost driver for the O&M cost. The timeframe required to meet remedial goals represents the greatest uncertainty with this technology and the driver for increasing costs with increasing DNAPL saturations. Where cost is the primary decision in technology selection, EISB would be an optimal choice for Case 1; however, EISB may not be an appropriate choice where remedial timeframes are restricted.

The breakdown of costs for ISCO is predominantly found in capital and O&M for the base case scenario. In this case, the permanganate costs represent the major contribution to capital costs (33%), followed by well installation, equipment and controls (31%) and labor and oversight (19%). The ISCO approach utilizes passive sodium permanganate injections and is similar to the EISB approach in that little infrastructure is required outside of the initial capital investment of 50 shallow wells, and costs for equipment and labor to complete each injection are limited due to the passive nature of the technology implementation. The primary driver for the O&M cost is the material cost for the oxidant. The higher material costs for ISCO relative to EISB result from a higher unit cost for sodium permanganate compared to EVO and the soil oxidant demand that requires a significant amount of sodium permanganate above the stoichiometric requirements for degradation of the TCE mass.

For the ISCR plus EISB treatment train, the high capital costs shown in the breakdown in Table 16.6 are due predominantly to the cost of the nZVI/palladium mixture, which is significantly more expensive than other amendments on a unit cost basis. Nanoscale zero-valent iron/palladium costs constitute ~50% of the capital costs for this option, followed by well installation, equipment and controls (25%) and labor and oversight (11%). *In situ* chemical reduction costs were substantially reduced by incorporating EISB for treatment of a portion of the TCE mass. For example, 100% reduction of TCE for the base case DNAPL scenario using nZVI/palladium/dispersant alone without follow-up EVO injections is estimated to be $5,444K. In comparison, treating 100% of the TCE using nZVI/palladium/dispersant for the high DNAPL scenario is estimated to cost $15,545K, representing a 185% increase in costs for a 300% increase in DNAPL loading. By incorporating EISB into the treatment train, total remediation costs and

Table 16.5. Detailed Cost Breakdown for Case 1 (Base Case DNAPL Scenario)

Cost Element	EISB	ISCO	ISCR + EISB	SEAR + EISB	Thermal	Pump-and-Treat
Design						
Laboratory studies	$25,000	$15,000	–	$50,000	–	–
Groundwater modeling	–	–	–	–	–	$40,000
In-field hydraulic and injection testing	$19,000	$19,000	$19,000	$19,000	–	$13,000
Detailed design, permitting, and reports	$88,000	$88,000	$162,000	$136,000	$191,000	$178,000
Procurement	$12,000	$12,000	$28,000	$17,000	$57,000	$23,000
Total design	**$144,000**	**$134,000**	**$209,000**	**$222,000**	**$248,000**	**$254,000**
Capital						
Site preparation	–	–	–	–	$10,000	$5,000
Mobilization/ demobilization	$4,000	$4,000	$4,000	$4,000	$210,000	$4,000
Well surveying	$4,000	$4,000	$4,000	$3,000	–	$2,000
Drilling and well installation	$106,000	$106,000	$106,000	$150,000	$241,000	$20,000
Flow control equipment, instrumentation, controls	$114,000	$114,000	$128,000	$559,000	–	$235,000
Ex situ treatment equipment and installation	–	–	–	$373,000	$136,000	
Batching system (nZVI)	–	–	$379,000	–	–	–
Steam generation system – rental (thermal)	–	–	–	–	$48,000	–
Heater system installation (thermal)	–	–	–	–	$675,000	
Start-up costs	$7,000	$7,000	$17,000	$23,000	$24,000	$12,000
Materials (including amendments, shipping, utilities)	$61,000	$231,000	$1,378,000	$938,000	$163,000	$15,000
Implementation labor, travel, per diem	$65,000	$65,000	$173,000	$146,000	$197,000	$42,000
Process monitoring	–	–	–	$16,000	$12,000	$4,000
Bioaugmentation	$57,000	–	$57,000	$57,000	–	–
Waste management and disposal	$24,000	$24,000	$24,000	$90,000	$93,000	$24,000

(*continued*)

Table 16.5. (continued)

Cost Element	EISB	ISCO	ISCR + EISB	SEAR + EISB	Thermal	Pump-and-Treat
Field and home office support	$56,000	$56,000	$78,000	$159,000	$149,000	$53,000
Contractor oversight	$67,000	$67,000	$113,000	$150,000	$75,000	$36,000
Reports	$27,000	$27,000	$35,000	$41,000	$47,000	$18,000
Total capital	**$592,000**	**$705,000**	**$2,496,000**	**$2,709,000**	**$2,080,000**	**$471,000**
O&M (per event/yr)						
Equipment rental	$9,000	$9,000	$9,000	–	–	–
Operation – materials (including shipping and electrical)	$61,000	$231,000	$59,000	–	–	$16,200
Operation – labor, travel, per diem	$65,000	$65,000	$65,000	–	–	$42,000
Operation – oversight	$13,000	$13,000	$13,000	–	–	$17,000
Waste discharge permit renewal (every 5 years)	–	–	–	–	–	$7,000
Process monitoring	–	–	–	–	–	$4,000
Replacement parts and materials, well rehab	$3,000	$3,000	$3,000	–	–	$16,000
Field and home office support	$15,000	$15,000	$15,000	$15,000	–	$46,000
Reports	$18,000	$18,000	$18,000	$18,000	–	$12,000
Total O&M (per injection/yr operations)	**$184,000**	**$354,000**	**$182,000**	**$33,000**	**$0**	**$153,200**
Monitoring Costs (during/post-treatment)						
Soil sampling (first year only)	–	–	–	–	$12,000	–
Monitoring well installation (first year only)	$10,600	$10,600	$10,600	$10,600	$10,600	$10,600
Labor (quarterly monitoring)	$7,200	$7,200	$7,200	$7,200	$7,200	$7,200
Analytical, groundwater (quarterly monitoring)	$8,000	$5,200	$8,000	$8,000	$3,000	$3,000
Waste management and disposal	$1,400	$1,400	$1,400	$1,400	$1,400	$1,400
Reports (annual)	$10,000	$10,000	$10,000	$10,000	$5,000	$5,000
Total monitoring (recurring after year 1)	**$26,600**	**$23,800**	**$26,600**	**$26,600**	**$16,600**	**$16,600**

Table 16.6. NPV Cost Comparison for Case 1 (Base Case Only)

Cost Component	EISB	ISCO	ISCR + EISB	SEAR + EISB	Thermal	Pump-and-Treat
Design	$144K	$134K	$209K	$222K	$248K	$254K
Capital	$592K	$705K	$2,496K	$2,709K	S2,080K	$465K
O&M	$210K	$990K	$261K	$33K	$0	$2,967K
Monitoring	$395K	$373K	$395K	$385K	$60K	$256K
Total	**$1,341K**	**$2,502K**	**$3,361K**	**$3,349K**	**$2,388K**	**$3,924**

sensitivity of the costs to DNAPL loading can be reduced substantially. Although costs remain higher with ISCR than using EISB alone, this approach offers the advantage of reducing remedial timeframes by ~50% below that of EISB alone by initiating treatment with ISCR injections.

Unlike the technologies described above, implementation of SEAR requires a more active remedial approach, with ongoing groundwater extraction and surfactant solution injection, as well as *ex situ* treatment of extracted groundwater, surfactant and TCE mass. This requires significantly more infrastructure and higher waste handling costs than the other treatment options and accounts for the high proportion of the costs being capital costs (Table 16.6). Well installation, treatment equipment and flow controls comprise 41% of the capital cost, with surfactant and operation costs representing the other significant component of the capital cost (40%). As the majority of the treatment is assumed to occur in the first year for the base case scenario, O&M costs are minimal and consist solely of monitoring the passive EISB that occurs as a result of residual surfactant remaining in the subsurface. SEAR combined with EISB as a polishing step may be an attractive approach where shorter remedial timeframes are desired and DNAPL loadings are lower.

Similar to SEAR, thermal remediation requires significant infrastructure for implementation, which accounts for the high capital costs shown in Table 16.6. In this case, well installation and treatment equipment and controls represent 54% of the capital cost (cost driver), with lesser contributions from mobilization/demobilization (10%), utilities (8%) and labor and oversight (13%). Due to the high temperatures that are induced during thermal treatment, TCE removal is highly efficient, and typically >99% mass removal is achieved when the entire DNAPL zone is targeted. The treatment is typically complete in less than 1 year after system start-up resulting in no long-term O&M. Since residual TCE concentrations remaining are very low, there is minimal need for post-treatment monitoring. Due to the relatively fixed capital costs, thermal treatment is typically more cost effective when used to treat higher DNAPL loadings. Thermal remediation may also be the optimal choice where rapid removal and limited post-treatment monitoring is desired.

As shown in Table 16.6, the primary cost driver for pump-and-treat is the long-term O&M, due to the extended treatment duration (assumed to be 30 years). Well installation, equipment and controls constitute about 55% of the capital cost for this option, while labor and oversight are 17% of capital. Operations and maintenance costs are driven primarily by labor and oversight costs (39%), home office support (30%), and materials and electrical costs, including replacement parts (21%). Given the high O&M costs, pump-and-treat is the most expensive option considered in this exercise.

Table 16.7 illustrates the sensitivity of each of these technologies to the amount of DNAPL present within a similar volume of soil for Case 1. As was true for the base case DNAPL scenario, EISB is also the least expensive option for the low and high DNAPL scenarios.

Table 16.7. Total NPV Cost Comparison as a Function of DNAPL Loading

DNAPL Loading	EISB	ISCO	ISCR + EISB	SEAR + EISB	Thermal	Pump-and-Treat
Low DNAPL	$1,133K	$1,357K	$2,311K	$2,741K	$2,181K	$3,901K
Base case	$1,320K	$2,502K	$3,361K	$3,349K	$2,398K	$3,924K
High DNAPL	$1,796K	$3,444K	$3,692K	$4,960K	$2,668K	$3,933K

The most expensive option is pump-and-treat for the lower two DNAPL scenarios, but pump-and-treat is surpassed by SEAR at the highest DNAPL scenario due to the requirement to complete multiple pore volume flushes and the resulting surfactant cost.

The sensitivity of the cost range due to variability in the DNAPL loading varies significantly depending on the technology chosen. *In situ* chemical oxidation, SEAR and ISCR costs are the most sensitive to DNAPL loading while the cost for thermal treatment and pump-and-treat is the least sensitive. Enhanced *in situ* bioremediation costs are also less sensitive to DNAPL loading due to the low cost of the amendment and O&M; however, as noted above, EISB also typically requires longer treatment times than all but pump-and-treat. Because of their sensitivity to DNAPL loading, ISCO, SEAR or ISCR approaches may, therefore, be more attractive when the certainty in the DNAPL loading is high, remedial timeframes are more restricted, and DNAPL mass requiring treatment is low.

The sensitivity of ISCO, ISCR and SEAR to the DNAPL loading is primarily due to the cost of permanganate, the nZVI/palladium slurry and surfactant, which increase proportionally with the higher DNAPL mass. This is also true for EISB; however, the cost of electron donor is significantly less than the other amendments. For SEAR, there are also substantial increases in TCE concentrations in the waste stream at the higher DNAPL loading, resulting in the need for additional treatment infrastructure to decant, evaporate and condense DNAPL prior to discharge to the WWTF. The recovered DNAPL needs to be disposed of as hazardous waste, which is costly ($500 per 208 L/55 gal drum).

The cost sensitivity to DNAPL loading is minimal for thermal treatment and pump-and-treat. The cost driver for thermal treatment is primarily the heating and treatment infrastructure, which is similar in all three scenarios. While this analysis shows that pump-and-treat is also relatively insensitive to DNAPL mass loading, it was assumed that the pump-and-treat system would operate for only 30 years regardless of the mass in place. However, experience has shown that 30 years is not likely to be sufficient where DNAPL is present, and operation may in fact be decades longer, particularly for the high DNAPL scenario. If operation of the pump-and-treat system continues for an additional 25 years (55 years in total), the NPV costs for pump-and-treat would increase to $5,096K from $3,924K for the base case scenario at the 3% discount rate (lower at higher discount rates). Therefore, the insensitivity to DNAPL loading for pump-and-treat may be somewhat artificial given the assumption of 30 years operation for each scenario.

16.4.2 Case 2: Low Permeability Source Area

Detailed cost breakdowns for EISB, ISCO, *in situ* soil mixing, thermal treatment and excavation for Case 2 are provided in Table 16.8. As with Case 1, the major cost components include design; capital costs (including the first year of technology implementation); O&M beyond year 1; and groundwater, air and soil monitoring.

Table 16.8. Detailed Cost Breakdown for Case 2 (Base Case DNAPL Scenario)

Cost Element	EISB	ISCO	ISCR	Thermal	Excavation
Design					
Laboratory studies	$25,000	$15,000	$25,000	–	$5,000
Groundwater modeling	–	–	–	–	–
In-field hydraulic and injection testing	$19,000	$19,000	–	–	–
Detailed design, permitting, and reports	$78,000	$78,000	$75,000	$191,000	$50,000
Procurement	$12,000	$12,000	$11,000	$64,000	$15,000
Total design	**$134,000**	**$134,000**	**$111,000**	**$255,000**	**$70,000**
Capital					
Site preparation	–	–	$20,000	$10,000	$10,000
Hydraulic fracturing	$110,000	$110,000	–	–	–
Mobilization/demobilization	$4,000	$4,000	$150,000	$211,000	$16,000
Sheet piling – rental and installation	–	–	–	–	$203,000
Excavation of soil and backfill (excluding labor)	–	–	$43,000	–	$50,000
Soil mixing (excluding labor)	–	–	$240,000	–	–
Drilling and well installation	$78,000	$78,000	–	$311,000	–
Well surveying	$4,000	$4,000	–	–	–
Flow control equipment, instrumentation, controls	$79,000	$79,000	–	–	–
Ex situ treatment equipment and installation	–	–	–	$62,000	$50,000
ISTD power equipment – rental (thermal)	–	–	–	$71,000	–
ISTD system installation (thermal)	–	–	–	$642,000	–
Start-up costs	$6,000	$6,000	$5,000	$29,000	$5,000
Materials (incl. amendments, shipping, utilities)	$68,000	$236,000	$297,000	$264,000	$199,000
Implementation labor, travel, per diem	$65,000	$65,000	$300,000	$213,000	$150,000
Process monitoring	–	–	–	$12,000	–
Bioaugmentation	$57,000	–	–	–	–
Waste management and disposal	$24,000	$24,000	$25,000	$93,000	$2,066,000
Field and home office support	$51,000	$51,000	$90,000	$161,000	$50,000
Royalty/licensing fees	–	–	$63,000	$70,000	–
Contractor oversight	$67,000	$67,000	$45,000	$75,000	$57,000
Reports	$27,000	$27,000	$50,000	$47,000	$20,000
Total capital	**$640,000**	**$850,000**	**$1,328,000**	**$2,271,000**	**$2,875,000**

(*continued*)

Table 16.8. (continued)

Cost Element	EISB	ISCO	ISCR	Thermal	Excavation
O&M (per event/yr)					
Equipment rental	$9,000	$9,000	–	–	–
Operation – materials (including shipping and electrical)	$68,000	$236,000	–	–	–
Operation – labor, travel, per diem	$65,000	$65,000	–	–	–
Operation – oversight	$14,000	$14,000	–	–	–
Waste discharge permit renewal (every 5 years)	–	–	–	–	–
Process monitoring	–	–	–	–	–
Replacement parts and materials	$2,000	$2,000	–	–	–
Field and home office support	$10,000	$10,000	–	–	–
Reports	$18,000	$18,000	–	–	–
Total O&M (per injection/yr operations)	**$186,000**	**$354,000**	**$0**	**$0**	**$0**
Monitoring Costs (during/post-treatment)					
Soil sampling (first year only)	–	–	$12,000	$12,000	$12,000
Air monitoring (first year only)	–	–	$10,000	–	$10,000
Monitoring well installation (first year only)	$10,600	$10,600	$10,600	$10,600	$10,600
Labor (quarterly monitoring)	$7,200	$7,200	$7,200	$7,200	$7,200
Analytical, groundwater (quarterly monitoring)	$8,000	$5,200	$3,000	$3,000	$3,000
Waste management and disposal	$1,400	$1,400	$1,400	$1,400	$1,400
Reports (annual)	$10,000	$10,000	$5,000	$5,000	$5,000
Total monitoring (recurring after year 1)	**$26,600**	**$23,800**	**$16,600**	**$16,600**	**$16,600**

Note: ISTD – *in situ* thermal desorption

A summary of total costs for each major category throughout the life span of the treatment is provided in Table 16.9 for the base case DNAPL scenario for Case 2. Total costs are calculated on a NPV basis using a 3% discount rate.

As seen in Table 16.9, EISB and *in situ* soil mixing are the lowest cost *in situ* options on an NPV basis for the base case analysis, while ISCO is the most expensive. All of the *in situ* options are less expensive than excavation.

The major costs for EISB are fairly evenly distributed between capital and O&M/monitoring, where capital costs include the first round of donor injection including bioaugmentation. Hydraulic fracturing, well installation and injected materials (including bioaugmentation culture) were each about 20% of the capital cost. Labor for the injection process (including contractor oversight) was about 15% of capital. Operations and maintenance costs assume one additional amendment event in the third year after implementation. An extensive period of MNA monitoring is anticipated once the active phase of the remedy is complete.

Table 16.9. NPV Cost Comparison for Case 2 (Base Case Only)

Cost Component	EISB	ISCO	ISCR	Thermal	Excavation
Design	$134,000	$134,000	$111,000	$255,000	$70,000
Capital	$640,000	$850,000	$1,328,000	$2,260,00	$2,875,000
O&M	$252,000	$1,298,000	$0	$0	$0
Monitoring	$423,000	$388,000	$75,000	$60,000	$59,000
Total	**$1,449,000**	**$2,670,000**	**$1,514,000**	**$2,575,000**	**$3,004,000**

In contrast to EISB, ISCR with *in situ* soil mixing is expected to be complete in 1 year and will encapsulate or destroy most of the TCE in the source area within a relatively short period thereafter, eliminating or significantly reducing O&M and monitoring. In this case, the labor and equipment involved in the soil mixing process makes up about 40% of the capital cost, while materials (iron and clay) make up about 22%. Significant mobilization costs are related to getting the mixing equipment to the site and a royalty fee is associated with using this technology.

The template site likely represents a highly favorable case for *in situ* soil mixing. The relatively shallow contamination permits the use of large-diameter augers, which are less expensive to operate than smaller versions required for deeper mixing. A significant cost savings is realized by treating the upper meter of surface soil onsite instead of sending it to a landfill. Sending the soil offsite would add up to 30% to the capital cost of the remedy or 28% to the total cost. Finally, the present analysis assumes the treated area will not be redeveloped to support building structures. If redevelopment is envisioned, an additional step would be required to add a stabilizing amendment, such as cement, to the upper layers of soil after the monitoring program is complete to improve the load-bearing capacity of the area, adding cost to the O&M part of the project.

In situ chemical oxidation is similar to EISB in terms of design, capital and monitoring costs, but differs in that the oxidant cost is higher than the electron donor cost for EISB, making up 28% of the total capital costs. Operation and maintenance costs are higher because of material costs, requirements for more frequent amendment events, and due to consumption of the oxidant through reaction with natural organic matter in the soil. For the base case DNAPL scenario, the oxidant demand is spread over five amendment events. The treatment frequency for ISCO is dictated by the relatively limited persistence of the oxidant in the subsurface balanced against the need to wait long enough between treatments for TCE to diffuse out of the low permeability regions of the formation.

Thermal remediation is expected to be complete in 1 year and will remove virtually all TCE from the source area. Therefore, monitoring requirements are quite limited for this option. More than half of the capital cost is associated with mobilization and setup of the system at the site, including drilling and installation of the thermal wells. The operational phase is less expensive. For example, electricity, operation of the aboveground treatment system, labor to operate the remedy during the active phase, and waste disposal together account for about 25% of the capital cost or 20% of the total cost. In this light, the template site treatment volume is at the low end of the range where thermal technology becomes cost competitive. Because mobilization and installation costs are the predominant cost drivers for this option, this option will become much more cost competitive at higher treatment volumes.

While excavation was the least expensive to design and had minimal O&M and monitoring costs, it had the highest capital costs of all the remedies considered. The capital costs are predominantly driven by landfill disposal costs, which make up 71% of capital or 69% of total costs. Other significant capital costs include sheet piling rental and installation, excavation and purchase of kiln dust and clean backfill. Each of these contributes about 7% to the capital cost.

Table 16.10 illustrates the sensitivity of each treatment technology with DNAPL loading. On a percentage basis, the NPV cost for ISCO is most sensitive to DNAPL loading, while ISCR with *in situ* soil mixing, thermal treatment and excavation are least sensitive. The high sensitivity of ISCO is due to increased oxidant demand and longer treatment times required by higher levels of DNAPL. These issues are also present for EISB, but are mitigated by lower materials costs and the greater longevity of the electron donor, which permits longer intervals between treatments. Cost increases for thermal treatment are primarily related to increases in aboveground treatment costs at higher DNAPL loadings. Cost increases for ISCR with *in situ* soil mixing are driven by higher iron demand with increased DNAPL levels and some reduction in process efficiency at the highest DNAPL loading due to the need for equipment operators to be in Level C PPE when mixing the soil. Sensitivity to DNAPL mass is least for excavation, where costs rise only marginally at higher DNAPL loadings due to a drop in process efficiency because equipment operators will be in Level C (PPE) during excavation activities. However, the excavation costs do not take into account the possibility that the excavated soil may fail to meet alternative soil treatment standards. This is more likely to occur for the base case and high DNAPL scenarios. Such an outcome would be identified during remedy selection and would dramatically increase disposal costs for these soils (potentially by as much as five times) and could make excavation the most sensitive to DNAPL loading.

Because costs for EISB and ISCO were estimated for both Case 1 and 2, it is possible to compare the numbers in Tables 16.7 and 16.10 to understand how site stratigraphy may impact remedial cost. It is somewhat surprising that subsurface stratigraphy did not appear to have a large impact on overall remedial costs for these remedial options. Enhanced *in situ* bioremediation and ISCO costs for Cases 1 and 2 in most instances differed by less than 10% for all three DNAPL scenarios. In each case the costs were lower for the higher permeability Case 1, largely due to the assumption of better distribution of amendments, greater access to the DNAPL and subsequently shorter cleanup times in Case 1. The differences in cost could have been larger, but were somewhat mitigated by the assumption that both the donor and the oxidant would persist longer in the low permeability stratigraphy due to more limited mass transfer. It should be noted that assumptions of cleanup times are most uncertain for the low permeability soil case, which also increases the uncertainty around these costs relative to Case 1.

Table 16.10. Total NPV Cost Comparison for Case 2 as a Function of DNAPL Loading

DNAPL Loading	EISB	ISCO	ISCR	Thermal	Excavation
Low DNAPL	$1,226,000	$1,613,000	$1,410,000	$2,391,000	$3,000,000
Base case	$1,449,000	$2,670,000	$1,514,000	$2,585,000	$3,004,000
High DNAPL	$1,867,000	$3,708,000	$1,700,000	$2,874,000	$3,046,000

16.5 SUMMARY

This chapter presents a general approach to performing cost analyses for a number of remedial options typically applied to chlorinated solvent source areas and plumes. The approach employs many assumptions, based on the experience of the authors and the advice received from professionals in the field (see Acknowledgements), which may not apply for all sites and are too numerous to fully document. The approach also employs a NPV calculation to allow the total cost of various remedial options to be compared on a total life cycle basis and significant cost drivers for each option identified. In this analysis, two template sites are used to demonstrate the approach for costing of DNAPL source treatment and identification of cost drivers.

In the cases presented here, EISB is consistently found to be the lowest cost DNAPL remediation technology, even with higher DNAPL loading scenarios and lower permeability soils. The major drawback for EISB is the extended timeframe required for treatment (years), which may make this technology less attractive in situations where cleanup time is limited. *In situ* soil mixing applying ISCR (micro-scale ZVI) was cost competitive with EISB for the low permeability soils in Case 2. The shallow source area in the template site was well suited to this technology, which treated the contamination aggressively, resulting in minimal O&M and monitoring costs; it is anticipated, however, that the cost of soil mixing will increase substantially as the depth of the treatment zone increases. Thermal remedies, which also treat source areas aggressively and typically result in near complete mass removal in less than 1 year of active treatment, become more cost effective at larger sites with higher DNAPL loadings. *In situ* chemical oxidation, SEAR, and ISCR using nZVI are more cost effective at lower DNAPL loadings, due to the elevated cost of permanganate, surfactant and nZVI. For ISCR using nZVI remedial costs can be lowered substantially by combining ISCR with EISB. Surfactant-enhanced aquifer remediation is also conducive to having EISB in a treatment train, and costs may be reduced for higher DNAPL loadings by using SEAR to treat a smaller proportion of the DNAPL mass (less than the 90% assumed here) and including an active EISB treatment for the rest. Pump-and-treat and excavation, which have been commonly employed for containment/treatment of DNAPL sources in the past, are the least cost effective of the treatment options for the case studies evaluated herein.

As a final note, the discount rate employed in the NPV analysis impacts the importance of present versus future costs. Specifically, using a higher discount rate will diminish the importance of future O&M and monitoring costs relative to present capital costs. The evaluation completed used a lower discount rate, and thus presents a scenario in which the future O&M and monitoring costs have a larger influence on total life span costs. However, even under these assumptions, technologies that have more significant O&M and long-term monitoring costs, such as EISB, still present a more favorable cost perspective. It is anticipated that a higher discount rate will further favor technologies such as EISB over shorter duration but higher capital cost remedial approaches such as excavation.

ACKNOWLEDGEMENTS

The authors gratefully acknowledge the valuable assistance and input from the following colleagues in putting together these costs estimates. Dr. Erik Petrovskis from Geosyntec provided the majority of input for the cost analysis of the surfactant flooding remedy evaluated in Case 1. Dr. Michael Borda from Golder Associates and Dr. David Reynolds from Geosyntec provided guidance and review for the costing of the nZVI remedy. Dr. Mitchell Olson from Colorado State University provided similar guidance for costing the *in situ* soil mixing remedy used in Case 2. Dr. Gorm Heron from Terratherm provided the majority of the cost inputs for the thermal remedies used in both Case 1 and 2. Mr. Michael LeMoine from Moran Environmental Recovery, LLC provided the cost data for excavation. Mr. Steven Meier and Mr. Damian Foti from the General Electric Company provided more general cost information related to both template sites.

REFERENCES

DoD (Department of Defense). 1995. DoD Instruction 7041.3: Economic Analysis and Program Evaluation for Resource Management. November 7. Washington, DC, USA.

Quinton GE, Buchanan Jr RJ, Ellis DE, Shoemaker SH. 1997. A method to compare groundwater cleanup technologies. Remediat (Autumn):7–16.

Ramsburg CA, Abriola LM, Pennell KD, Löffler FE, Gamache M, Amos BK, and Petrovskis EA. 2004. Stimulated microbial reductive dechlorination following surfactant treatment at the Bachman Road Site. Environ Sci Technol 38:5902–5914.

CHAPTER 17

GROUNDWATER REMEDIATION AND THE USE OF ALTERNATIVE ENDPOINTS AT HIGHLY COMPLEX SITES

Elisabeth L. Hawley[1] Rula A. Deeb[2] and Herbert Levine[3]

[1]ARCADIS/Malcolm Pirnie, Emeryville, CA 94608, USA; [2]Geosyntec, Oakland, CA 94607, USA; [3]U.S. Environmental Protection Agency Region IX, San Francisco, CA 94105, USA

17.1 INTRODUCTION

Sites with contaminated groundwater can be highly complex due to the nature and extent of contamination, hydrogeologic setting and other site circumstances. For example, multiple source zones may be present over a large area in heterogeneous porous media or in a fractured rock environment, contaminating groundwater to depths of several hundred feet (hundreds of meters) below ground surface (ft bgs). There is no definition of *highly complex* sites; however, some of the features of highly complex sites and several examples of cleanup challenges faced at complex sites are illustrated in Figure 17.1, including the presence of a heterogeneous hydrogeologic setting, dense nonaqueous phase liquid (DNAPL), and/or secondary "sources" of contaminants that are stored in the subsurface. Table 17.1 provides several examples of features and issues that may be present at highly complex sites.

Several examples of highly complex sites are described as case studies in this chapter. Site managers may have already implemented or evaluated multiple state of the art remediation technologies for groundwater, such as those described in Chapters 8, 9, 10, 11, 12, 13, 14 and 15. Despite these efforts, groundwater contamination may be expected to remain above long term cleanup goals for many years – in some cases, hundreds or even thousands of years. Although active treatment may still be a component of the remedy to reduce risk and ultimately remediate the site, complex sites must also manage groundwater contamination through containment, monitoring, institutional controls and/or periodic reviews until long term goals are met. Examples of how complex sites have responded to this challenge are described in this chapter.

One of the key questions facing managers and other decision-makers at highly complex sites is how to evaluate the benefits of conducting active remediation (soil vapor extraction, pump-and-treat, chemical oxidation, *in situ* bioremediation, thermal, hydraulic displacement) and/or passive remediation (capping, barrier technologies, monitored natural attenuation [MNA]) in addition to managing groundwater contamination and controlling exposure. Factors limiting the pace of remediation, such as DNAPL dissolution rates, diffusion rates and permeability, may continue to limit the success of active remediation technologies. Some sites must weigh the benefits of gaining only minimal reductions in remedial time frames. The question of whether or not to conduct partial source treatment at complex sites and the associated costs and potential benefits is an area of ongoing discussion and research. These questions are beyond the scope of this chapter; however, this chapter will provide case studies illustrating the variety of remedial approaches that have been taken at several highly complex sites.

B.H. Kueper et al. (eds.), *Chlorinated Solvent Source Zone Remediation*,
doi: 10.1007/978-1-4614-6922-3_17,

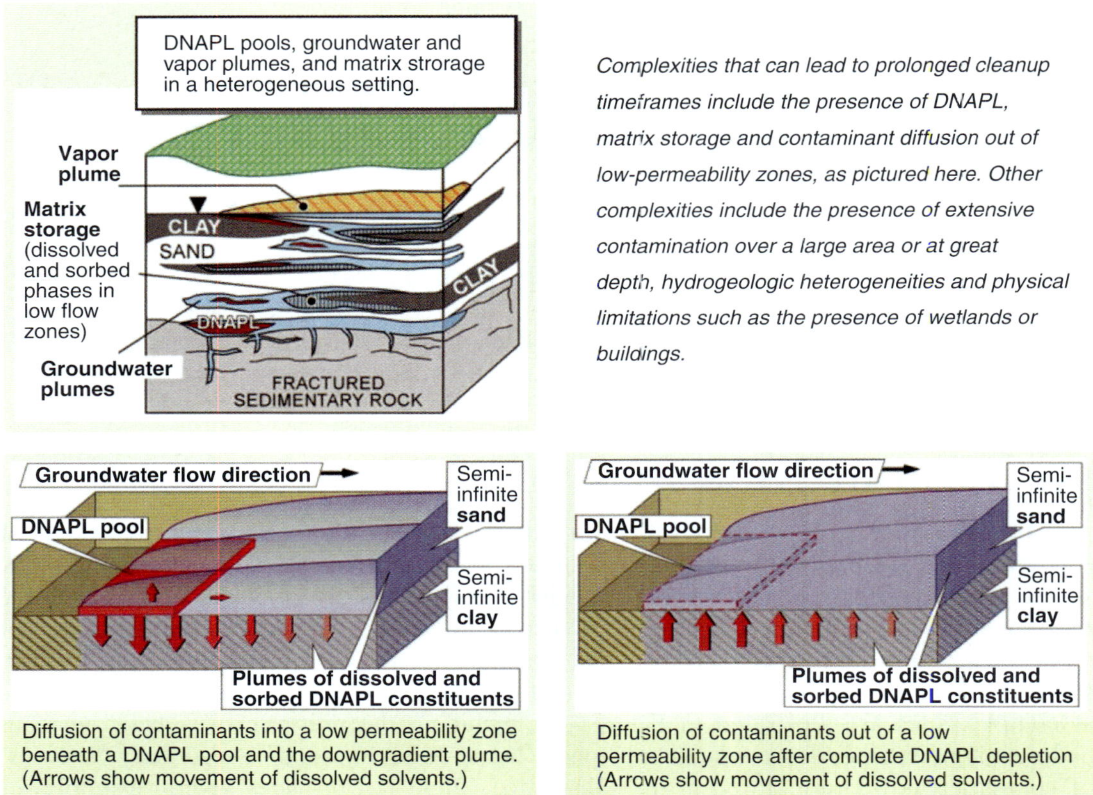

Figure 17.1. Complexities encountered at groundwater cleanup sites (adapted from Sale et al., 2008).

Another key issue that has been addressed through a spectrum of different approaches is setting appropriate long term – ultimate – cleanup goals at highly complex sites. Under current regulations, drinking water requirements and/or background concentrations are typical long-term cleanup goals. Is it possible to remediate to drinking water standards within a reasonable time frame at highly complex sites using today's technology? Current regulations allow site decision-makers to waive or substitute for final cleanup goals under certain circumstances. For example, under the Comprehensive Environmental Response, Compensation, and Liability Act (CERCLA), final numerical cleanup goals such as maximum contaminant levels (MCLs) and other applicable or appropriate and relevant requirements (ARARs) may be waived due to technical impracticability of meeting these goals within a reasonable time frame (USEPA, 1993a). Technical impracticability (TI) waivers and other waivers or substitutions for conventional final cleanup standards are collectively referred to as *alternative endpoints* (Malcolm Pirnie, Inc., 2011a). Through case studies, this chapter provides context for assessing the use and appropriate role of alternative endpoints in final remedies.

Complex sites typically establish near term (5 or 10 year) goals and objectives. Examples of near term goals include eliminating human exposure to contamination, preventing plume expansion and reducing mass in the source area. These objectives can be quantitative, such as decreasing contaminant concentrations to within an order of magnitude of final cleanup goals or achieving a specified percent (%) reduction in mass discharge from a source area. At CERCLA sites, near and long term goals and objectives may be expressed as remedial action objectives (RAOs). Near term objectives can help guide selection of an appropriate final

Table 17.1. Example Features and Issues that May Be Present at "Highly Complex" Sites (modified from Kueper et al., Chapter 18)

Feature/Issue	Description
Highly heterogeneous geology	Wide range of permeability among geologic units, particularly at the small scale. Highly complex sites often exhibit five or six orders of magnitude range in permeability.
Contaminants sequestered in low permeability units	Contaminants that have diffused into low-permeability units (clays and silts) may ultimately back-diffuse at a slow rate that dictates the minimum time frame for remediation.
Contamination in fractured bedrock	Back diffusion from the rock matrix that will likely dictate the time frame for remediation. Delivery of fluids is challenging given the discrete nature of flow through fractures.
Large volumes of DNAPL	Typically require large volumes of amendments, increased time, and increased budget to treat.
Multicomponent DNAPL	Can be challenging to treat because many technologies cannot address all of the components.
Low ganglia to pool ratio	A large amount of pooled DNAPL presents a risk of mobilization, leading to technical challenges with drilling. Pools also have limited DNAPL-water mass transfer rates, extending remediation time frames. Pools are more likely to form in highly heterogeneous geologic settings.
Widespread regional contamination	Extensive regional contamination (acid mine drainage, naturally occurring arsenic) can elevate background concentrations, be inordinately costly to remediate and require regional coordination to be effective.
Long lived inorganic contaminants, such as metals and radioactive compounds	May require long remediation time frames. Many technologies have limited capability for treating inorganic contaminants both *in situ* and *ex situ*.
Presence of high-volume roadways and active use areas overlying contamination	Difficult to treat because of limited access; contamination present beneath areas of human use and active manufacturing.
Multiple stakeholders	Large numbers of stakeholders (regulators, public action groups, press, politicians) can complicate decision making processes.

remedy and determination of appropriate technology specific objectives. The Interstate Technology & Regulatory Council (ITRC) has prepared a technical regulatory guidance document regarding the use of integrated site strategies at chlorinated solvent sites where DNAPL is present (ITRC, 2011a). The document describes the importance of setting near term objectives that are specific, measureable, achievable, realistic and time-bound (SMART) and provides several examples of SMART near term objectives that can be used to guide remedial progress. The Interstate Technology & Regulatory Council has played an important role in developing the acceptance and use of technically sound innovative environmental approaches. As a leading organization of regulators from 50 states, the District of Columbia and US Environmental

Protection Agency (USEPA), as well as multiple federal partners from the Department of Defense, Department of Energy, industry, consultants and other stakeholders, ITRC captures input from many of these agencies in their guidance documents.

This chapter describes different approaches that have been used at highly complex sites including long term cleanup goals, the use of alternative endpoints, near term technology-specific cleanup goals, groundwater management, and/or active remediation. Because there is no well defined population of highly complex sites to systematically assess remedial decisions, case studies have been identified as examples and will be used throughout this chapter to illustrate potential remedial approaches for groundwater at highly complex sites. Some sites have used multiple approaches throughout their cleanup lifetime or in different areas of the site, as illustrated by the case studies. This chapter provides an integrated perspective on how technologies are used in combination with each other to manage groundwater contamination and meet appropriate long term and near term goals.

17.2 OVERVIEW OF REMEDIAL APPROACHES AT COMPLEX SITES

Complex chlorinated solvent sites have used a continuum of different remediation and management strategies to address contaminated groundwater. Table 17.2 illustrates one possible combination of remediation and management approaches used to meet near term and long-term goals at a hypothetical complex site. At this hypothetical site, *in situ* chemical oxidation (ISCO) is being used to actively remove contaminant mass from the source area. After meeting technology specific goals (concentrations decline below 500 micrograms per liter (μg/L)), ISCO will be discontinued. Pump-and-treat will continue to be used to provide hydraulic containment in high concentration areas, while reducing concentrations further towards the ultimate long term goal of meeting MCLs within the source zone. MNA will be used for contaminated areas outside of the containment zone. Institutional controls are in place to prevent groundwater use and restrict site access.

Other combinations of remediation and management could be used to meet near term and long term goals. The following Sections (17.3, 17.4, 17.5, and 17.6) highlight active remediation, passive remediation, and containment aspects of final remedies at complex sites, as well as the use of alternative endpoints. The case studies provided in each section illustrate how a mix of these approaches is typically used at any given site.

17.3 ACTIVE GROUNDWATER REMEDIATION

17.3.1 Description

Most highly complex sites have implemented active remediation technologies at some time, either at full scale or pilot scale. Due to the high uncertainty associated with site conditions at highly complex sites and the long time frames needed to meet ultimate cleanup objectives, near-term goals and objectives must be established to guide active remediation. Successful applications of active remediation systems at highly complex sites often have the following characteristics:

- Active remediation is used to achieve appropriate, achievable and clearly defined near term goals and objectives.
- The design and operation of the system is continually adapted and refined over time in response to empirical data.

Table 17.2. Example of a Remedial Approach for Groundwater at a Complex Hypothetical Site

Goals	Approach		
	Remediation		Management
	Active	Passive	
Near term (examples):			
Reduce exposure	–	–	Institutional controls
Meet technology specific goals (reduce concentrations to below 500 μg/L)	ISCO	–	–
Remove mass to the extent practicable	ISCO	–	–
Reduce mass flux	ISCO	–	–
Prevent off-site migration	Pump-and-treat	–	–
Long term (examples):			
Meet MCLs, action levels, background, other ARARs	ISCO, pump-and-treat	MNA	–
Meet risk-based cleanup values	–	–	–
Use alternative endpoints	–	–	–
ARAR waiver based on technical impracticability, greater risk, other	–	–	–
Alternate concentration limit (ACL)	–	–	–
Other (management/containment zone designations)	–	–	–

- Information gained through active remediation is used to update the conceptual site model and revisit the overall strategy for groundwater remediation and management to meet long term cleanup goals.
- Achievement of near term cleanup goals and objectives is a milestone used to transition from one remedial phase to the next. Active remediation systems can be shut down or used as a component of the next phase of remediation, per the revised remedial strategy, once this milestone is achieved. For example, the goal of active remediation may be to reduce contaminant concentrations to within 100 times the MCL. The ITRC DNAPL strategies document provides more information on developing near term functional goals (ITRC, 2011a).

This performance-based, adaptive and iterative approach allows decisions to be refined as data are evaluated. This is a highly efficient method for making remedial progress at sites with high uncertainty. The approach has been referred to as *adaptive site management* (Malcolm Pirnie, Inc., 2011a) and is similar to the *observational approach* described by Terzaghi and Peck (1948) for geotechnical practices. There are also many parallels between the adaptive site management approach and the philosophy of the Air Force Center for Engineering and Environment (AFCEE)'s Environment Restoration Program Optimization (ERP-O) process, as well as the thought process described by the ITRC's Remediation Risk Management team (ITRC, 2011b) and the Triad approach to site characterization (ITRC, 2007). All these efforts focus on: building, validating and improving the conceptual site model; establishing metrics for evaluating the success of remedial technologies; and continuously evaluating and optimizing model predictions, remedial performance and long term monitoring.

Adaptive site management is often used at complex sites where there are high uncertainty and no clearly defined pathway to clean closure. The approach focuses on remedial progress at the site, making continual corrections to avoid straying far off course from meeting remedial objectives, documenting remedial progress towards metrics, adapting, being innovative and furthering the technology. There is a growing recognition that an iterative approach is often needed at complex sites and that each iteration is not a failure if it effectively generates knowledge that can be used to improve site decision making and overall remedial performance (National Research Council, 2003).

One of the main challenges of using adaptive site management is making real-time decisions while still involving stakeholders with site decision making, informing them of site progress and reaching consensus on the evolving conceptual site model and appropriate near term goals and objectives over time. Flexible language can be used in decision documents to allow for adaptations. Contingency language or milestones for transitioning from one technology to another can be built into the final remedy; technology performance objectives and metrics are selected ahead of time and are approved by the stakeholder group.

Active remediation using an adaptive site management approach can be implemented without needing any type of alternative endpoint. Stakeholders can continue to set and achieve near term goals and objectives with the eventual goal of meeting long term cleanup requirements. As the cleanup progresses and the conceptual site model is refined over time, the site may transition from active to passive remediation or containment (MNA, barrier technologies). Other sites have implemented active remedies in conjunction with an alternative endpoint. For example, a review of case studies with alternative endpoints (Malcolm Pirnie, Inc., 2011a) identified the use of excavation, free product recovery, thermal, ISCO, bioremediation and air sparging/soil vapor extraction in conjunction with TI waivers. Examples are presented in the following section.

17.3.2 Case Studies: Active Remediation

In the area of Building 40 at Watervliet Arsenal, located in the state of New York, USA, active remediation was used to meet a near term remedial objective, build the conceptual site model and provide a technical basis for the next phase of remediation. At this site, high concentrations of perchloroethene (PCE) were present in groundwater, indicating the historical presence of nonaqueous phase liquid (NAPL) down to 150 ft (46 m) bgs (Malcolm Pirnie, Inc., 2011b; Goldstein et al., 2004). Due to the hydrogeologic setting (fractured shale), a significant mass of subsurface PCE had likely diffused into the shale matrix over the past 50 years and would diffuse from the matrix into groundwater slowly over time as PCE concentrations in fractures decreased, prolonging remedial time frames despite active remediation to address dissolved-phase PCE present in the fractures (Goldstein et al., 2004).

Stakeholders wanted to remove PCE from the subsurface to the extent practicable. To combat the effects of PCE back diffusion from the shale matrix, managers decided to deliver large quantities of a strong oxidant to the source zone. The hypothesis was that oxidant could potentially diffuse into the rock matrix and react with PCE, reducing the effects of back diffusion and shortening remedial time frames (Malcolm Pirnie, Inc., 2011b). To test this hypothesis, full scale ISCO injections with sodium permanganate were conducted over a 5 year period. Results were used to refine the conceptual site model and achieve or better define *removal to the extent practicable*. Extensive site characterization was performed including borehole geophysical and hydrological logging and interflow testing, innovative diagnostic tools including mass flux measurement with multilevel sampling systems, and rock core

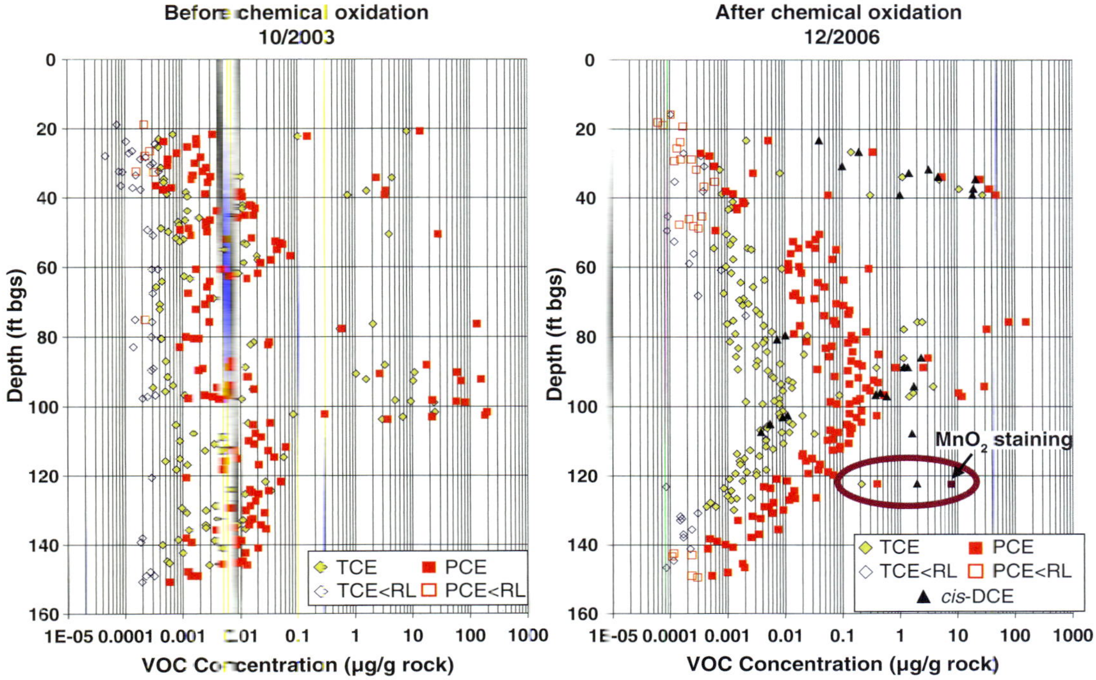

Figure 17.2. Minimal impact of chemical oxidation on PCE and TCE measured in rock core samples at Watervliet Arsenal (adapted from Malcolm Pirnie, 2011b).

crushing to monitor PCE and oxidant concentrations in the rock matrix. Results indicated no measureable benefits of conducting ISCO in terms of PCE concentrations in rock cores or reduction of mass flux along the property boundary (Figure 17.2) (Malcolm Pirnie, Inc., 2011b; Goldstein et al., 2004). The iterative field test design and responsiveness to test data provided a technical basis for a revised final remedy. The ultimate long term goal at the site was to meet MCLs; however, the Army and other stakeholders are now considering an alternative endpoint in conjunction with long term management of contaminated groundwater.

At a second site, The Union Pacific Railroad Company Tie-Treating Plant (The Dalles, Oregon, USA), active remediation is being conducted to achieve the near term goal of removing DNAPL to the extent practicable. The site is contaminated with wood-treatment compounds and fuel oil from site operations since 1923. The groundwater remedy includes DNAPL removal to the extent practicable using extraction wells and water flooding to mobilize DNAPL towards recovery wells and hydraulically contain groundwater in the DNAPL areas (USEPA, 1996). In addition, plume monitoring, hydraulic containment of the plume (if needed), monitoring, and institutional controls are in place. In 2007, the second 5 year review (Oregon Department of Environmental Quality, 2007) reported that more than 81,000 gallons (gal; 307,000 L) of DNAPL had been recovered. The 5 year review reiterated decision-document language stating that long term numeric cleanup goals are not likely achievable at the site. Presumably this site will require an ARAR waiver in the future in order to fully comply with the National Contingency Plan (NCP). Union Pacific plans to prepare a TI waiver evaluation after DNAPL removal modules are shut down (Oregon Department of Environmental Quality, 2007).

17.4 PASSIVE GROUNDWATER REMEDIATION

17.4.1 Description

Other highly complex sites have selected or transitioned to a passive remedy to slowly remediate contamination over a long time frame. Passive remedies generally rely on MNA, although capping, permeable reactive barriers and infrequent injections of long lived remedial agents such as edible vegetable oils may also be considered passive technologies. In addition, containment, monitoring, periodic reviews and/or institutional controls may be put in place to protect human health and the environment.

Monitored natural attenuation is a fairly well accepted component of groundwater remedies, as evidenced by the USEPA report Use of Monitored Natural Attenuation at Superfund, RCRA Corrective Action, and Underground Storage Tank Sites (USEPA, 1999) and several technical reports published in the late 2000s on MNA of inorganic contaminants in groundwater (USEPA, 2007a, b, 2010a; ITRC, 2010). According to the USEPA, implementation of source control measures in conjunction with MNA is almost always necessary (USEPA, 1999). Monitored natural attenuation can be used downgradient of the source area, as a supplement to source control measures or, in some cases, as the primary component of a limited action remedy in the source area. When compared to active technologies, MNA may provide several benefits, including reduced ecological disturbance of sensitive areas, reduced energy consumption, less waste generation and lower remediation costs. Remedial time frames for MNA and active remediation technologies may be similar, due to rate-limiting processes such as diffusion and dissolution.

Institutional controls are often used in conjunction with MNA over long time frames. Examples of institutional controls include zoning restrictions, building/excavation permits, prohibition of well drilling, fencing, and other methods to control exposure pathways. The appropriate role of institutional controls and limited action alternatives was described in a June 2009 memorandum published by USEPA. The memorandum restated the NCP expectations: Institutional controls are generally not substitutes for active remediation but are intended as supplementary protective measures during the implementation of groundwater remedies (USEPA, 2009). Detailed guidance on institutional controls, including full life cycle planning recommendations, effective implementation, maintenance recommendations and enforcement tools, is summarized in recent interim guidance (USEPA, 2010b).

Monitored natural attenuation and other limited action remedies over long time frames at complex sites can be used without an alternative endpoint. NCP Section 300.430(a)(iii)(F) states that USEPA expects to "return usable groundwater to beneficial uses wherever practicable, within a time frame that is reasonable given the particular circumstances of the site." As long as these expectations are achieved, an alternative endpoint is not needed. There is no standard definition of *reasonable time frame*[1]; instead, the definition of reasonable is assessed for each site. The lack of a definition for reasonable time frame has increased the flexibility of site stakeholders to accept longer time frames to reach cleanup requirements and allows for the use of MNA rather than considering remediation to be technically impracticable. Therefore, if longer time frames are warranted for remediation and are accepted by stakeholders, no alternative endpoint is needed. In interviews conducted in 2003, several USEPA and state

[1] The use of a 100 year time frame as *reasonable* is an order-of-magnitude number used as an example by USEPA in its TI waiver guidance document (USEPA, 1993a). In that same document, USEPA states that "no single timeframe can be specified during which restoration must be achieved to be considered technically practicable" (USEPA, 1993a). Some stakeholders have interpreted timescales on the order of 600 years to be technically practicable whereas others have viewed timescales of at least 20–50 years to be technically impracticable (see, e.g., USEPA, 2001b).

regulators referred to MNA as another option at complex sites that are considering TI waivers (Malcolm Pirnie, Inc., 2004).

17.4.2 Case Study: Passive Remediation

At site SA-17 of the former Naval Training Center (NTC) in Orlando, Florida, USA, trichloroethylene (TCE) was present in the source area at concentrations indicative of DNAPL (up to 577,000 µg/L). Trichloroethene in the source area has already been addressed by several remedial technologies, including ISCO with Fenton's reagent followed by enhanced bioremediation. In situ chemical oxidation was unable to treat some portions of the aquifer due to a lack of hydraulic connection, preferential flow paths, and back diffusion. Enhanced bioremediation has been operating using a recirculation well field design (Favara et al., 2006). The site is now in the process of transitioning to MNA (Figure 17.3) (CH2M Hill, 2010). Multiple lines of evidence support a transition to MNA, including the following:

- Favorable geochemical conditions
- The presence of functional genes for dehalogenation (as measured using molecular biological tools)
- The presence of daughter products *cis*-1,2-dichloroethene and vinyl chloride down-gradient of the source zone

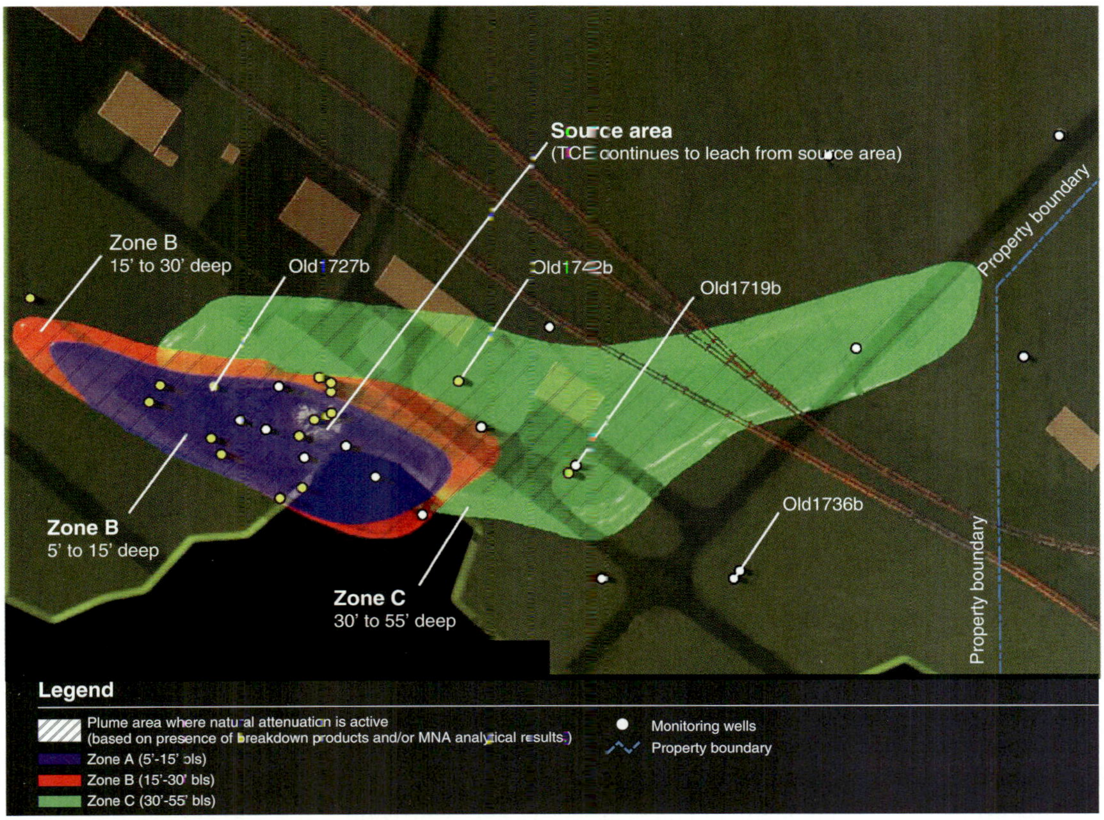

Figure 17.3. Chlorinated solvent plume at NTC Orlando site (adapted from CH2M Hill, 2010).

- Total concentrations of chlorinated volatile organic compounds (VOCs) that are within one to two orders of magnitude above MCLs, approaching Florida's default criteria for natural attenuation
- Natural Attenuation Software estimates indicating that remedial time frames for the downgradient plume would not be further reduced through active remediation in the source zone. Remedial time frames for MNA range from 60 to 70 years for the downgradient plume area

An alternative endpoint was not necessary at the former NTC site. However, passive approaches also can be used in conjunction with alternative endpoints. At McKin Co. (Gray, Maine, USA), chlorinated solvents were present at concentrations indicating DNAPL in bedrock. A 2001 amendment to the final remedy approved transitioning from pump-and-treat to MNA and institutional controls, in conjunction with a TI waiver. Pump-and-treat remained a contingency response to protect surface water. The transition to MNA was supported by the conceptual site model, which indicated that residual DNAPL was likely present in bedrock, but was extremely difficult to locate. Finally, an analysis of remedial time frames indicated that pump-and-treat was not helping to achieve ARARs faster than natural attenuation.

17.5 LONG TERM MANAGEMENT

17.5.1 Description

In addition to active and passive remediation, remedial approaches at highly complex sites also may include some type of long term management, primarily through containment or management in place. These approaches can remove some contaminant mass, but that is not their primary function. For example, a permeable reactive barrier will destroy contaminants as groundwater passes through this zone; however, the primary purpose is to prevent contaminants from migrating beyond the permeable barrier. Containment can be achieved through pump-and-treat, permeable reactive barriers, other barrier technologies, and/or natural attenuation.

Long term management/containment approaches may need to be accompanied with an alternative endpoint to satisfy regulatory requirements. CERCLA sites are expected to meet ARARs everywhere throughout the plume. Therefore, if the remedy is not expected to meet long term cleanup requirements within a reasonable time frame at a highly complex site, an ARAR waiver is required (40 Code of Federal Regulations [CFR] 300.430(f)(1)(ii)(C)). In contrast, at sites regulated under the Resource Conservation and Recovery Act (RCRA), on-site containment and long term management is a widely accepted remedial approach. The idea of a containment or management zone with a defined point of compliance is inherent in RCRA corrective action regulations and in the RCRA approach to managing landfills and other solid waste management units (40 CFR 264.95). Waste management units can be designated to manage wastes in place and meet cleanup requirements at an agreed-upon downgradient point of compliance (40 CFR 264.95). Because of this flexibility, alternative endpoints are rarely needed at RCRA sites that are using long term management/containment approaches.

Several state cleanup programs require an approval process for containment zone designations for areas in which groundwater contamination will be managed over long time frames. Some of these designations help states implement and track land use controls and other institutional controls until long term cleanup goals are reached within the zone. Examples include groundwater management zones in Illinois and New Hampshire. Other designations are alternative endpoints as they waive or replace the final cleanup standard. For example, the term *groundwater management zone* is also used in Delaware's Voluntary Cleanup Program and Hazardous Substances Control Act sites. This state designation prevents the use of groundwater and restricts drilling any new potable water supply wells. This designation also may be used to describe an area where technical impracticability exists for groundwater remediation and contamination will be present for a long time (Delaware, 2008). This designation has been used at 105 sites throughout the state of Delaware as of January 2009, including CERCLA sites (Delaware, 2009).

The Texas Commission on Environmental Quality (TCEQ) has a *plume management zone* designation that modifies groundwater cleanup requirements within any area where cleanup is not technically or economically feasible. This designation is needed at Texas sites using a long-term groundwater management approach. Any DNAPL present at the site must be removed to the extent practicable before approving a plume management zone. In California, sites under the jurisdiction of the State Water Resources Control Board need to designate a *containment zone* if groundwater will be managed over the long term without meeting long term cleanup goals. Containment zones are appropriate at sites where residual contamination is not expected to degrade significantly over time and will remain indefinitely, i.e., where cleanup to water quality objectives is technologically and/or economically infeasible, per Resolution No. 92-49. Several other states (e.g., Georgia, New Jersey, and Wyoming) require *technical impracticability* zone designations at highly complex sites along with any type of long term management/containment zone approach. Examples of state designations and the underlying regulatory citations are provided in Table 17.3. To determine whether a state cleanup program has a comparable designation, state-specific policies and regulations should be reviewed.

17.5.2 Case Studies: Long Term Management

Several sites in Texas have designated groundwater plume management zones, including the Naval Weapons Industrial Reserve Plant (NWIRP) in Dallas, Texas, USA (Navy, 2008). NWIRP Dallas is a RCRA facility with chlorinated solvents in groundwater. Contamination was being addressed by three boundary pump-and-treat systems for more than 10 years (Navy, 2008). The Navy proposed installing two permeable reactive barriers, designating a plume management zone, and conducting, monitoring, and maintaining institutional controls. The remedy was supported by the partnering team and implemented. Long term monitoring is being conducted to ensure that the plume remains within plume management zone boundaries. Other sites with designated plume management zone include the Red River Army Depot Ordnance Training Center Landfill, Spector Salvage Yard, Pioneer Oil Refining Company, Mountain Creek Industrial Center, and State Highway 123 PCE plume. More details on these sites are presented in original site-specific documents (Texas Register, 2007; TCEQ, 2004, 2007; Reed and James, 2010).

The California State Water Resources Board (SWRB) has a website (California SWRB, 2011) listing all sites with containment zones: J.H. Baxter site in Weed, Edwards Air Force Base South Air Force Research Laboratory in Kern County (both CERCLA sites that also have TI waivers), Edwards Air Force Base Arroyos Air Force Research Laboratory in Kern County, and Georgia-Pacific (former Peterbilt Motor Co.) site in Newark, California, USA. More information on these sites is provided in the links posted on the California website (California SWRB, 2011).

Two other facilities in California, Intel Fab 1 (in Santa Clara) and Norge Cleaners (in Napa) had containment zones at one time, but these were rescinded (California SWRB, 2011). The Intel Fab 1 site received an order from the California Regional Water Quality Control

Table 17.3. Examples of State Designations for Groundwater Management/Containment

State Program	Designation	Reference
California, State Water Resources Control Board	Containment zone	State Water Resources Control Board Resolution No. 92-49
Delaware Department of Natural Resources and Environmental Control	Groundwater management zones	Remediation Standards Guidance under the Delaware Hazardous Substance Cleanup Act
Georgia Voluntary Remediation Program Act	Technical impracticability	Senate Bill 78 (Amended Article 3 of Chapter 8 of Title 12 of the Official Code of Georgia Annotated)
Illinois Environmental Protection Agency RCRA Facilities	Groundwater management zone	35 Illinois Administrative Code Part 620.250
Illinois Environmental Protection Agency	Technical impracticability consideration through risk-based cleanup and tiered remedial objectives	35 Illinois Administrative Code Part 742
New Hampshire Department of Environmental Services	Groundwater management zone	New Hampshire Code of Administrative Rules, Chapter Env-Or 600
New Jersey Department of Environmental Protection	Technical impracticability	New Jersey Administrative Code 7:26E-6.1(d)
Texas Commission on Environmental Quality	Plume management zone	30 Texas Administrative Code § 350.33(f)(3)(A)-(E); 30 Texas Administrative Code § 350.37(1)(4)
Wyoming Department of Environmental Quality Voluntary Remediation Program	Technical impracticability determination and establishment of alternative cleanup levels	Wyoming Statutes § 35-11-1605(d)

Board (RWQCB) establishing the containment zone in 1999 for chlorinated solvents in groundwater that had reached asymptotic concentrations but were still above MCLs after years of pump-and-treat. In 2005, the RWQCB determined that the site met criteria for low-risk closure and the order was rescinded. Further groundwater monitoring is no longer needed (California RWQCB San Francisco Bay Region, 2005).

The distinction between containment zones and low-risk closure, and between MNA remedies and low-risk closure, was recently summarized by the San Francisco Regional Board (California RWQCB San Francisco Bay Region, 2009). This document described low-threat closures as potentially applicable before groundwater has been fully restored to beneficial uses, as long as stakeholders have concluded that the site will reach cleanup standards under natural conditions within a reasonable time frame. Other states have adopted similar low-threat closure guidelines including the Colorado Department of Public Health and Environment (Colorado DPHE, 2010). Low threat closure is not typically considered at highly complex sites; however, this is an alternative approach that can be used at sites that are approaching cleanup standards.

Some of the difficulties associated with approving containment zones are illustrated by the Fairchild Semiconductor Corporation South San Jose site (San Jose, California, USA).

The site had been operating a pump-and-treat system (as well as maintaining a slurry wall) to address chlorinated solvent contamination in groundwater. When contaminant removal by the pump-and-treat system approached asymptotic limits, stakeholders considered a containment zone designation. However the site is located in a sensitive hydrogeologic area classified as a recharge zone for groundwater by the local Santa Clara Valley Water District. To avoid potential conflicts with local groundwater management policies, stakeholders decided not to implement any official containment zone policy; however, the approach taken is, in fact, a containment zone system. According to a recent state 5 year review report, the slurry wall present at the site is containing contamination above MCLs. Institutional controls are preventing exposure to contamination. Overall, the remedy is protective, despite several new developments including the detection of 1,4-dioxane inside the slurry wall at concentrations up to 850 μg/L and the evaluation of potential vapor intrusion risks. This is an example of an informal groundwater management/containment approach without a formal containment zone designation. Based on conversations with NAVFAC Southwest representatives, the Navy has considered containment zones and does not accept the procedural requirements outlined in Resolution 92-49. Referencing this resolution and using containment zone language would give the state an expanded role in the remedial decision making process. The Navy has therefore used other alternative endpoints and approaches at the site.

Approaches similar to groundwater management/containment have also been used at CERCLA sites. At the Solvents Recovery Service of New England site, a CERCLA site located in Southington, Connecticut, USA, hydraulic containment with pump-and-treat is being used in conjunction with MNA over a long time frame. Millions of gallons of waste solvents and oils were handled, stored, and processed at the site for more than 30 years. The site has multiple historical potential release areas, including two unlined lagoons, drum storage areas, and truck loading/unloading areas. Approximately 84% of the subsurface contaminant mass is thought to be present as NAPL in the overburden. Nonaqueous phase liquid is also known to be present in the bedrock. The Record of Decision (ROD) (USEPA, 2005a) describes the selected remedy for groundwater in the overburden and bedrock. The overburden remedy includes *in situ* thermal treatment in the NAPL source zone, excavation and capping of soils and wetland soils, pump-and-treat for dissolved plume containment, MNA for areas outside of the pump-and-treat system containment zone, and institutional controls to prevent human exposure. The remedy for the bedrock consists of hydraulic containment using pump-and-treat and MNA in the NAPL area. The pump-and-treat system will be modified as appropriate based on expected reductions in dissolved VOC concentrations due to MNA. The time frame for restoration of the bedrock plume was estimated to be approximately 225 years to reach ARARs in the bedrock aquifer. This time frame was considered reasonable relative to the time frame of other remedial alternatives at the site

17.5.3 Alternative Endpoints

As described previously, the term *alternative endpoints* has been used to refer collectively to various waivers or substitutions for long term cleanup standards (Malcolm Pirnie, Inc., 2011a). Examples of types of alternative endpoints are listed in Table 17.4.

More details on identification of ARARs, ARAR waivers, and ACLs are provided below with case studies. Examples of groundwater management/containment zones were described previously in Section 17.5.

Table 17.4. Types of Alternative Endpoints

Alternative Endpoint	CERCLA	RCRA	State(s)[a]
ARAR waivers	X		
TI waivers	X		
Greater risk waivers	X		
Other waivers (interim remedy, inconsistent application of state standards, fund balancing, equivalent performance)	X		
Alternate concentration limits (ACLs)	X	X	X
Groundwater management/ containment zones	X	X	X
Groundwater reclassification	X	X	X

[a]Various terminology is used under different state cleanup programs

17.5.4 Identification of ARARs

The lead Federal agency is responsible for identifying Federal ARARs; state ARARs are identified by the state regulator. Depending on the site conditions, there may be latitude in defining whether a regulation is an ARAR. For example, if site conditions permit groundwater to be reclassified so that it is no longer a current or potential source of drinking water, drinking water standards such as MCLs will no longer be ARARs. Several states have designations that permit site-specific groundwater reclassification. Examples include the Tennessee site-specific impaired groundwater classification, TCEQ Municipal Setting Designation, and the Ohio Urban Setting Designation. Other states do not allow groundwater to be reclassified at specific sites but do permit entire aquifers to be reclassified according to groundwater use, value, or vulnerability. The reasons for permitting groundwater reclassification are typically not related to site complexity. Most groundwater classification schemes are based on groundwater use, production capacity, value, or vulnerability. Therefore, not all of the sites receiving these designations are highly complex. Examples of several state cleanup program designations and the sites receiving them are provided in the following section.

17.5.5 Case Studies: Identification of ARARs

Two sites in Tennessee have been classified as site-specific impaired groundwater, as indicated in the June 2008 revision of the Rules of the Tennessee Department of Environment and Conservation, Chapter 1200-4-3. This designation is essentially a TI zone, referring to "groundwater that has been contaminated by human activity and the board finds that either it is not technologically feasible to remediate the groundwater. . . or it is not reasonable to remediate to that criteria. . ." One of these sites is the Porter Cable/Rockwell site, where a slowly moving solvent plume will naturally attenuate before it can leave the property boundary. The second site is the Isabella-Eureka mine site, abandoned in bankruptcy court. Additional details about these two sites were not provided in Chapter 1200-4-3, and no other site-specific documents could be found online. Based on the description in Chapter 1200-4-3, these sites do not seem to have particularly unusual characteristics, suggesting that the designation may be appropriate for other sites in Tennessee (Tennessee Water Quality Control Board, 2007).

Texas sites receiving the Municipal Setting Designation under TCEQ are tracked by the state. This designation prevents the current and future use of shallow contaminated groundwater for potable purposes. Approximately 132 properties have received this designation since this legislation was passed in 2003. Eleven more sites are pending approval, one request was denied, one request was withdrawn, and one site was determined to be ineligible (TCEQ, 2010). One example is the Hardy Street Rail Yard site, located in Houston, Texas, USA, under the Texas Voluntary Cleanup Program. At this site, more than 80,000 gallons of diesel fuel NAPL were recovered. The site was thoroughly characterized using a Triad approach. Ten years of monitoring data indicated that chlorinated solvent plumes in groundwater are stable or decreasing in size. Shallow groundwater was not suitable for drinking even if it had not been impacted. Both groundwater and residual NAPL are being addressed by MNA, following mass removal to the extent practicable. In 2003 TCEQ issued a Conditional Certificate of Completion (conditional site closure) approving conditional residential land use at the site (Chapin, 2009).

Similarly, sites receiving the Ohio Urban Setting Designation under the Ohio Voluntary Action Program are tracked by the state. As of December 2010, 60 sites had received this designation (Ohio EPA, 2011). At least 1,879 sites in New Jersey have received Classification Exemption Area designations, totaling more than 32,500 acres (51 square miles; 132 square kilometers) (New Jersey DEP, 2011).

The Navy has used the RCRA Waste Management Unit designation as an ARAR at CERCLA sites to establish a similar concept, a Waste Management Area (WMA), where waste is managed in place and cleanup requirements will be met at a downgradient point of compliance (USEPA, 1998b). An example of a site with designated WMAs is the Barstow Marine Corps Logistics Base Yermo Annex plume, located in Barstow, California, USA. This site has several plumes of PCE, TCE, and other VOCs in an alluvial setting (USEPA, 1998b). The Navy proposed additional areas as WMAs but the EPA did not agree. The Navy agreed to comply with groundwater cleanup standards throughout the plume as a conservative means of demonstrating attainment at the point of compliance, but reserved the right to propose the use of designated points of compliance for these areas in the future (USEPA, 1998b).

17.5.6 Description of ARAR Waivers

Objectives and expectations for remedial actions at CERCLA sites, as described in the NCP, translate into two threshold criteria that all final remedies must meet: overall protection of human health and the environment and compliance with applicable or appropriate and relevant requirements (ARARs) over the long term. Seven other criteria (five primary balancing criteria and two modifying criteria) are used in comparing and selecting the final remedy: long term effectiveness and permanence; reduction in toxicity, mobility, or volume through treatment; short-term effectiveness; implementability; cost; state acceptance; and community acceptance.

Compliance with ARARs includes Federal ARARs and any state ARARs that are more stringent than Federal ARARs. The only grounds for waiving an ARAR are the following six options (40 Code of Federal Regulations (CFR) 300.430(f)(1)(ii)(C)):

- Technical impracticability – if compliance with ARARs is technically impracticable from an engineering standpoint, within a reasonable time frame.
- Greater risk – if compliance with the ARAR would result in greater risk to human health and the environment compared with an alternative which does not comply with ARARs.
- Interim measure – if the remedy is an interim action; final remedy still must achieve ARARs.
- Equivalent standard of performance – if the selected remedy will attain a standard of performance that is equivalent to the ARAR; typically used for action-specific or location-specific ARARs.
- Inconsistent application of state standards – if the ARAR is a state standard that has not been consistently applied to other remedial actions within the state.
- Fund balancing – if compliance with the ARAR would threaten the ability of the fund to respond to and achieve protectiveness at other sites.

Sites incorporating ARAR waivers into final decision documents must ensure that the final remedy is protective of human health and the environment.

After the temporary ARAR waiver for interim remedies, TI waivers are the second most widely used of the six types of ARAR waivers. Guidance from the EPA requires a written evaluation of technical impracticability (TI evaluation report) that must be prepared prior to TI waiver approval to identify ARARs for which the TI decision is being sought, the three-dimensional volume subject to the TI waiver (the TI zone), the conceptual site model, evaluation for the potential for restoration, cost estimate and other parts of the final remedial strategy (USEPA, 1993a). Outside of the TI zone, traditional cleanup objectives will still remain as the final cleanup goal. Details on the TI evaluation process at CERCLA sites are described in USEPA guidance (USEPA, 1993a).

A greater risk ARAR waiver may apply if activities undertaken to meet an ARAR would result in greater risk or harm to human health or the environment than waiving that ARAR and choosing another alternative. The nature of the potential greater risks may vary with the site circumstances. Some examples include the following:

- Greater risk to drinking water aquifer(s) due to potential contaminant mobilization during remedial activity. This line of reasoning might be particularly applicable at a site with DNAPL.
- Greater risk to nearby wetlands, agriculture, and/or ecosystems of implementing pump-and-treat remedies that lead to dewatering or land subsidence.
- Greater risk to sensitive ecosystems in areas where remediation activities would be a disturbance.
- Greater risk posed by explosive hazards or other health and safety hazards associated with particular remedial technologies. If the only technologies suitable for meeting ARARs were determined to pose a greater risk than other technologies, this waiver would be applicable.
- Greater risk to ecosystem of sediment disturbance during dredging or excavation. This waiver would more likely be applicable to sediments or surface waters than groundwater.
- Potential delay for groundwater to reach ARARs, resulting in greater risk because of liner or capping requirements that affect the amount of natural flushing that occurs.

In addition to ARAR waivers based on technical impracticability and greater risk, four other types of waivers are identified in CERCLA regulations. These include ARAR waivers for (1) interim measures, (2) equivalent standard of performance, (3) inconsistent application of state standards and (4) fund balancing. Interim RODs do not have to specify ARARs. However, all interim remedies must be replaced by final remedies. Therefore, this type of ARAR waiver does not provide an alternative endpoint. Applicable or appropriate and relevant requirement waivers for interim measures may be used in conjunction with, or be replaced by, another type of alternative endpoint and approach described in this document. At complex sites, interim remedies may be in place for many years as remedial investigation, pilot testing, and remedy selection proceeds. As stated in the NCP preamble, however, interim measures should be followed within a reasonable time by complete measures that attain ARARs. Interim RODs are reserved for short-term actions that require an immediate response. Long term management strategies are accomplished with a final ROD which specifies RAOs.

An ARAR waiver based on equivalent standard of performance would be considered if the proposed alternative to complying with the ARAR provided equivalent performance and protection of human health and the environment. The intent of this ARAR waiver was clarified in the NCP preamble: "... the purpose of the waiver is to allow alternative technologies that provide a degree of protection as great or greater as the specified technology. ... EPA believes that the... degree of protection, level of performance, and future reliability, should at least be equaled for an alternative to be considered equivalent. While it is possible that there may be redundancy among the three, a lesser level in any of these criteria would compromise equivalency with the original standard" (55 Federal Register 8749-8750, March 8, 1990). No case studies of application of this waiver to groundwater were identified.

Inconsistent application of state requirements refers to a waiver of state ARARs if the state has not consistently applied it to other sites. No specific procedure was established in the NCP for tracking ARARs and demonstrating that they had been applied consistently at sites. The NCP preamble stated that a standard would be presumed to be consistently applied unless there was a reason to believe otherwise. In other words, the burden of proof lies with the individual site rather than with the state.

Fund balancing is another reason for waiving an ARAR. This waiver first appeared in the 1985 NCP and was codified by the Superfund Amendments and Reauthorization Act (SARA) in 1986. This waiver may apply when the costs needed to meet ARARs would be so high as to threaten the superfund's ability to address other sites. The purpose of the waiver was to ensure that USEPA's ability to carry out a comprehensive national response program was not compromised by a single disproportionately high expenditure at a site (USEPA, 1991). Some state cleanup programs may have parallel concerns. This waiver is not applicable at the Department of Defense and Department of Energy sites, which do not use Superfund dollars.

17.5.7 Case Studies: ARAR Waivers

Research efforts have identified 77 CERCLA sites that have received TI waivers for groundwater as of November 2010 (Figure 17.4) (Malcolm Pirnie, Inc., 2011a). Recently, a TI waiver was approved at Edwards Air Force Base South Air Force Research Laboratory, California, USA (ROD dated 9/24/2007). This site has a large TI zone, extending over 16 square miles. Solvent releases occurred during historical rocket motor/fuel testing and subsequent cleaning and disposal into sumps and dry wells. Contaminants include PCE and TCE. There are suspected DNAPL source zones in fractured rock. The geology at the site consists of a thin zone of unconsolidated soil (silty sand) overlying granitic bedrock. The depth to first groundwater

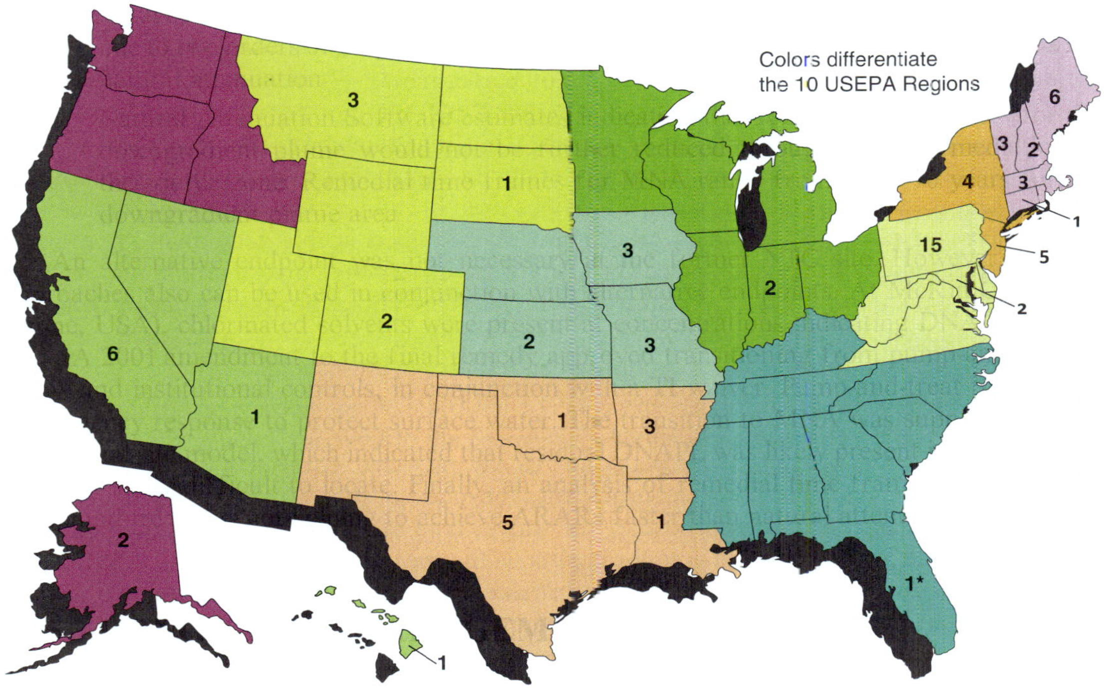

Total is 77* sites as of November 2010
*One additional TI waiver was used and later revoked.

Figure 17.4. Map of CERCLA sites with TI waivers for groundwater (adapted from Malcolm Pirnie 2011a, b).

ranges from 20 to 200 ft (6.1 to 61 m), averaging 120 feet (36 m), flowing through a network of fractures. Hydraulic conductivity has a similarly broad range, from 10^{-7} to 10^{-1} centimeters (cm) per second (3.3×10^{-9} to 10^{-3} ft per second). Wells generally pump at less than 0.5 gal per minute (1.9 L per minute). The Air Force made significant effort over the past decade to characterize the site and evaluate remedial technologies through treatability studies. Site characterization techniques included installing monitoring wells, preparing boring logs, measuring water levels quarterly, mapping surface fractures and lineaments, completing a high-resolution three-dimensional seismic reflection survey, performing aquifer tests and tracer studies, and rock coring. Treatability studies evaluated dual-phase extraction, pump-and-treat, soil vapor extraction, blast fracturing, *in situ* bioremediation, thermal treatment and steam injection. A model was developed to simulate contaminant transport over a large area (the three plumes cover approximately 7.7 square miles). The closest town of Boron, California, USA, is approximately two miles from the base, and the contaminated groundwater is not expected to reach it. Other potential future receptors include production wells located on- and off-base. Based on the conceptual site model substantiated by the significant level of effort and documented field studies, stakeholders supported a ROD that included a TI decision.

The E.I. du Pont de Nemours & Co. site, located in Newport, Delaware, USA, received two types of ARAR waivers – an ARAR waiver for groundwater based on greater risk and a TI waiver for surface water ARARs. This site includes a paint pigment production facility, a chromium dioxide production facility, and two industrial landfills separated by a river. The selected remedy for groundwater consisted of long term monitoring, installation of a public water supply line and establishment of a groundwater management zone. Based on data collected during the remedial investigation and feasibility study, attempts to remediate chlorinated solvents in the lower (Potomac) aquifer would draw more contamination into this aquifer from the more contaminated upper (Columbia) aquifer. Remedial attempts in the upper aquifer would adversely affect wetland areas. More details on the site background and selected remedy can be found in the original 1993 ROD (USEPA, 1993b) and the second 5 year review report (USEPA, 2005c).

No other case studies of greater risk ARAR waivers for chlorinated solvents have been identified. An ARAR waiver for groundwater based on greater risk was approved for the Onondaga Lake site in Syracuse, New York, USA. This site was a manufacturing facility for sodium hydroxide and liquid chlorine using a mercury cell process, followed by subsequent manufacturing of hydrochloric acid and bleach. The primary groundwater contaminant is elemental mercury DNAPL. The groundwater remedy selected in 2000 consisted of a barrier wall installed in the top 55 ft (17 m) down to glacial till, hydraulic containment within the barrier using pump-and-treat, long term monitoring, and deed restrictions. A time frame of 30,000 years to reach ARARs was estimated, indicating that complete restoration of groundwater was also technically impracticable. However, groundwater ARARs were waived on the basis of greater risk, citing losses of wetlands from dewatering if a more aggressive pump-and-treat system were installed. Greater risks from on-site soil excavation and treatment included heavy truck traffic, fugitive dusts and air pollution, lack of community acceptance, and increased traffic accidents.

At the Moss-American Superfund site, located in Milwaukee, Wisconsin, USA, a greater risk waiver was initially applied to RCRA Subtitle C capping requirements and state requirements for a double liner/leachate collection system as part of the original 1990 ROD (USEPA, 1990). Installing an impermeable cap and liner would have reduced the natural flushing and prolonged the treatment time for remediating groundwater contaminated by wood-treatment operations. This greater risk waiver was later revoked by a 1998 ROD amendment (USEPA, 1998a), based on new information that indicated a greater presence of DNAPL. Source control measures were then taken for soils in the area.

Waivers for ARAR based on interim measures have been used at many sites. At some complex sites operating under interim remedies an alternative endpoint or approach is likely to be a component of the final remedy. For example, the Hastings Groundwater Contamination site, OU 19 (located in Hastings, Nebraska, USA), selected an interim remedy for groundwater that consists primarily of institutional controls and groundwater monitoring. Contaminants include chloroethenes (PCE, TCE and daughter products), associated contaminants such as 1,4-dioxane and carbon tetrachloride, and fuel constituents such as benzene and PAHs. Various sources have been discovered, including a grain storage facility, vapor degreasing process, manufactured gas plant, municipal/industrial waste landfill and grain fumigant operations. Nearly all of the soils are deep and are formed in calcareous loess, eolian sands and mixed silty/sandy alluvium (USEPA, 2001a). As stated in the 2001 ROD (USEPA, 2001a), none of the alternatives that were evaluated could achieve ARARs and therefore could not be selected as a final remedy. The USEPA stated that monitoring would be conducted to determine if it would be technically impracticable to meet ARARs. In response to a public comment, the ROD stated that alternative endpoints (TI waiver or ACLs) might be considered as part of the final remedy (USEPA, 2001a): "However, EPA would not consider an application for a TI waiver or ACLs appropriate until response actions have indicated that contaminant concentrations have leveled off after a period of time, or further improvement in ground water quality using available technologies is shown to be impractical." At Hastings, data were not sufficient to support a final ROD incorporating a TI decision; therefore, an interim remedy was proposed instead (USEPA, 2001a). The USEPA's recommendation to consider TI early in the CERCLA cleanup process and to continually refine the conceptual site model during the RI phase may be useful for avoiding further delay of final remedies (USEPA, 1993a).

Site SS-01 of Brandywine Defense Reutilization and Marketing Office, Andrews AFB, Maryland, USA, is another complex site operating under an interim remedy (U.S. Air Force and USEPA Region 3, 2006). The interim ROD stated that it is impractical to treat groundwater within the source zone area to MCLs due to the presence of DNAPL incomplete characterization of the DNAPL source area, and the heterogeneity of the shallow groundwater aquifer (U.S. Air Force and USEPA Region 3, 2006). Data from the interim remedy (hydraulic containment, institutional controls, enhanced bioremediation and bioaugmentation) will be used to evaluate the remediation potential of the final remedy.

Significant discussion of *inconsistent application of state standards* occurred prior to finalizing the off-post ROD dated 12/19/1995 at Rocky Mountain Arsenal, Adams County, Colorado, USA (USEPA, 1995). At issue were Colorado Basic Standards for Groundwater that the state had recently promulgated for the contaminant diisopropyl methylphosphonate (DIMP), a byproduct of manufacturing nerve agent Sarin. Diisopropyl methylphosphonate is found primarily at Rocky Mountain Arsenal. However, the Colorado Basic Standards for Groundwater were accepted as ARARs in the final ROD, and the ARAR waiver does not appear to have been used (USEPA, 1995). Based on a search of the RODs database, ARAR waivers based on equivalent performance, inconsistent application of state standards, and fund balancing have not been used for groundwater at any CERCLA site.

17.5.8 Description of Alternate Concentration Limits

An ACL may be established for groundwater in accordance with CERCLA Section 121(d)(2)(B)(ii).[2] Alternate concentration limits are risk-based concentrations that will not pose a substantial hazard to human health or environmental receptors, given exposure pathways, and other factors. An ACL replaces an ARAR as the new regulatory approved cleanup concentration, rather than waiving the ARAR entirely. In general, ACLs may be considered as part of response actions provided that the following conditions are met (Section 121(d)(2)(B)(ii)):

- Groundwater discharges into surface water (known and projected points of entry to surface water).
- Groundwater discharge does not lead to a statistically significant increase of contaminants in the surface water or any accumulation of contaminants downstream.
- Institutional controls prevent human exposure to contaminated groundwater between the facility boundary and the discharge point(s) of groundwater into surface water.

A recent USEPA policy memorandum (USEPA, 2005b) specified several additional factors to consider prior to establishing ACLs, including the following:

- Are all plumes of contaminated groundwater discharging to surface water? Are contaminants present in a deeper aquifer that does not discharge to surface water?
- Will significant degradation of the aquifer occur prior to discharge to surface water? Could the plume spread to uncontaminated portions of the aquifer?

[2] Note: Alternate concentration limit (ACL) has a second meaning as a term used to describe the allowable limits of hazardous constituents such as metals and pesticides detected in groundwater at landfill units within defined waste management areas (40 CFR 264.94(b)). This second term is not particularly relevant to complex sites and is therefore not discussed in this chapter.

- Have known and projected points of entry of the plume(s) into surface water been, or can they be, specifically identified?
- Are contaminants in sediments or below points of entry accumulating into surface waters?
- Can groundwater be restored?
- What is the potential for degradation by-products within the zone between the source and point(s) of entry to surface waters and the potential for statistically significant increase in degradation products in surface water and corresponding risks?
- Can institutional controls and other enforceable measures preclude human exposure to groundwater contaminants above health-based levels?
- Have total maximum daily loads been established for surface waters and could ACL result in a total maximum daily load exceedance?

The criteria for determining whether ACLs are appropriate at RCRA sites are not as prescriptive. Per RCRA regulations (40 CFR 264.94), ACLs can be established as long as the concentration level does not pose a substantial risk to human health or the environment. This determination is made after considering the potential adverse effects on groundwater quality and the potential adverse effects on hydraulically connected surface water quality. The potential adverse effects of an ACL decision should be considered, including waste characteristics and mobility, hydrogeologic setting, groundwater flow, groundwater and surface water usage (current and future), surface water quality standards, existing groundwater and surface water quality and quantity, rainfall patterns, proximity of source zone to surface waters, potential for human exposure and related health risks, potential for other risks, and permanence of potential adverse effects (40 CFR 264.94).

The basis for the ACL value varies. Some sites estimate an acceptable groundwater concentration based on models or mixing zone analyses of groundwater and surface water, back-calculated from published surface water quality criteria. At another site, ACLs were established as Federal MCLs along the edge of a former landfill at the site (USEPA, 1985).

17.5.9 Case Studies: ACLs

Alternate concentration limits were recently approved at the former Long Beach Naval Complex (Long Beach, California, USA), where groundwater beneath a peninsula is contaminated with VOCs. In addition to excavation and air sparging/soil vapor extraction, ACLs were developed at the land's edge based on California Ocean Plan criteria (Navy, 2011). The site reached response complete in 2007 and has ceased groundwater monitoring. The Navy is still maintaining institutional controls and conducting 5 year reviews. At another site, Jacksonville Naval Air Station OU 3 located in Jacksonville, Florida, USA, contaminant concentrations indicate the potential for residual DNAPL. The site hydrogeology is a complex stratigraphy of interbedded layers of sand, clayey sand, sandy clay, and clays. The original ROD included air sparging/soil vapor extraction, pump-and-treat, enhanced biodegradation and MNA. As part of remedy optimization studies, the CSM was recently updated. The Navy is in the process of conducting fate and transport modeling and a mixing zone analysis to develop ACLs for groundwater that discharges into the St. Johns River (NAVFAC, 2008).

The USEPA (2005b) memorandum appears to make the use of ACLs more difficult at CERCLA sites. The Waterloo Coal Gasification Plant is a CERCLA site in Iowa that recently rescinded ACLs and approved a TI waiver instead. At this site, contaminants of concern were associated with a former manufactured gas plant. Maximum contaminant levels and/or health advisory levels were not exceeded in the nearby Cedar River; there were also no other points of exposure such as domestic water wells. The decision to use ACLs was documented in an Explanation of Significant Differences for OU 1 in 2006 (USEPA, 2006). Although a 2004 ROD had approved ACLs (USEPA, 2004), the actual ACL values had not been approved at the time of the ROD. According to the Explanation of Significant Differences, USEPA (the lead agency) decided as a policy matter not to use the ACL approach to address groundwater at the site. This is consistent with the timing of the EPA memorandum titled "Use of Alternate Concentration Limits in Superfund Cleanups" (USEPA, 2005b).

At the Winthrop Landfill OU 1 site (located in Winthrop, Maine, USA), an Explanation of Significant Differences dated 2/14/2007 upheld ACLs that had been approved in a 1985 ROD (USEPA, 1985, 2007c). More than 3 million gallons (11.4 million L) of chemical wastes were disposed of at this landfill over several decades. Contaminants have been detected in bedrock beneath the site and in lake sediments south of the landfill. Alternate concentration limits were set at Federal MCLs along the edge of the solid waste disposal area (USEPA, 1985). A groundwater pump-and-treat system had been operating from 1995 to 2002 and was determined to be no longer needed to meet ACLs (USEPA, 2007c).

17.6 SUMMARY AND CONCLUSIONS

Complex sites face a number of uncertainties when using today's technology to meet long-term cleanup requirements. Different approaches have been used at complex chlorinated solvent sites, including active remediation guided by near term goals and objectives, use of MNA over long time frames, long term management/containment, or some combination of these approaches. Complex sites have also used a variety of different alternative endpoints, which either waive or substitute for long term cleanup standards.

Integrated strategies often are essential to manage these sites efficiently. The case studies described in this chapter illustrate potential "mix and match" approaches that have been used to meet near term and long term goals at DNAPL sites. In these case studies, progress has been made not by choosing between remediation and management, or between remediation and alternative endpoints, but by combining these approaches into a strategy that works for each site. Finally, the case studies illustrate how different states and cleanup programs have addressed this issue through policy and regulation.

This book has focused primarily on technical issues governing environmental remediation at chlorinated solvent sites. However, at highly complex sites, technical challenges may frequently lead to discussions of policy, management, stakeholder priorities and perceptions, and regulatory interpretation. It is hoped that this discussion of a variety of different approaches to remediation and management to meet near term and long term cleanup goals will broaden the collective thinking about remedies at complex sites and help stakeholders move forward in selecting appropriate cleanup goals and remedial approaches.

REFERENCES

California RWQCB San Francisco Bay Region. 2005. Executive Officer's Report. September 14. www.swrcb.ca.gov/rwqcb2/board_info/agendas/2005/september/09-21-05-4eosr.doc. Accessed November 2012.

California RWQCB San Francisco Bay Region. 2009. Assessment tool for closure of low-threat chlorinated solvent sites. Draft Final. July 31. http://www.waterboards.ca.gov/sanfranciscobay/water_issues/programs/sitecleanup/Low_Threat_Closure_Assessment_Tool.pdf. Accessed July 2013.

California State Water Resources Control Board. 2011. GAMA – Groundwater Ambient Monitoring Program. Sites with Containment Zones. www.swrcb.ca.gov/water_issues/programs/gama/containment_zones.shtml. Accessed November 2012.

Chapin R. 2009. Municipal Setting Designations (MSDs): A new tool for Houston. http://documents.publicworks.houstontx.gov/documents/divisions/planning/msd/public_meeting_presentation_2009_018_hardy_yards.pdf. Accessed November 2012.

Colorado Department of Public Health and Environment. 2010. Draft guidance for the closure of low-threat sites with residual ground water contamination. First edition. August 13. http://www.cobar.org/repository/Inside_Bar/Enviro/DRAFTGWNFAPOLICY-GUIDANCE.pdf.

Delaware Department of Natural Resources and Environmental Control. 2008. Coordinated state agency response policy to detection of volatile synthetic organic contaminants in ground-water and/or drinking water. February. www.awm.delaware.gov/SIRB/Documents/State%20Response.pdf. Accessed November 2012.

Delaware Department of Natural Resources. 2009. HSCA, VCP, and NPL Sites with Groundwater Management Zones (GMZs) Requirements. January. www.awm.delaware.gov/SIRB/Documents/GMZ_sites.pdf. Accessed November 2012.

Favara PJ, Singletary MA, Nwokike B, Tsangaris S. 2006. Using a treatment train to optimize DNAPL source zone remediation. Proceedings, Fifth International Conference on Remediation of Chlorinated and Recalcitrant Compounds, Monterey, CA, USA. Battelle Press, Columbus, OH, USA.

Goldstein KJ, Vitolins AR, Navon D, Parker B, Chapman S, Anderson GA. 2004. Characterization and pilot-scale studies for chemical oxidation remediation of fractured shale. Remed J 14:19-37.

ITRC (Interstate Technology and Regulatory Council). 2007. TRIAD implementation guide. Overview document. May. ITRC Sampling Characterization, and Monitoring Team. www.itrcweb.org/Documents/SCM-3.pdf. Accessed October 2011.

ITRC. 2010. A decision framework for applying monitored natural attenuation processes to metals and radionuclides in groundwater APMR-1. December. www.itrcweb.org/Documents/APMR1.pdf. Accessed November 2012.

ITRC. 2011a. Integrated DNAPL site strategy IDSS-1. Washington, D.C. Interstate Technology & Regulatory Council. Integrated DNAPL Site Strategy Team. www.itrcweb.org/Documents/IntegratedDNAPLStrategy_IDSSDoc/IDSS-1.pdf. Accessed November 2012.

ITRC. 2011b. Project risk management for site remediation. RRM-1. Washington, D.C. Interstate Technology & Regulatory Council. Remediation Risk Management Team. www.itrcweb.org/Documents/RRM-1.pdf. Accessed November 2012.

Malcolm Pirnie, Inc. 2004. Technical impracticability assessments: Guidelines for site applicability and implementation. Phase II report. Prepared for the Army Environmental Center. March. http://aec.army.mil/usaec/cleanup/techimprac.pdf. Accessed November 2012.

Malcolm Pirnie, Inc. 2011a. Alternative endpoints and approaches selected for the remediation of contaminated groundwater. Final Report – Environmental Security Technology Certification Program (ESTCP) Project ER-200832. www.serdp.org. Accessed November 2012.

Malcolm Pirnie, Inc. 2011b. Diagnostic Tools for Performance Evaluation of Innovative In-Situ Remediation Technologies at Chlorinated Solvent-Contaminated Sites. Final Report – Watervliet Arsenal, Watervliet, New York, USA. ESTCP Project ER-200318. July. http://www.serdp.org. Accessed November 2012.

National Research Council. 2003. Environmental Cleanup at Navy Facilities: Adaptive Site Management. National Academies Press, Washington, DC, USA.

NAVFAC (Naval Facilities Engineering Command). 2008. Groundwater risk management handbook. Port Hueneme, CA, USA.

New Jersey Department of Environmental Protection. 2011. Groundwater Contamination Areas (CEA). http://www.state.nj.us/dep/gis/stateshp.html#CEA. Accessed November 2012.

Ohio EPA. 2011. Table listing sites that have received the Urban Setting Designation in Ohio. www.epa.ohio.gov/portals/30/vap/docs/Urban%20Setting%20Designations.pdf. Accessed November 2012.

Oregon Department of Environmental Quality. 2007. Five-Year Record of Decision Review. Second five-year review report, Tie Treating Plant, The Dalles, Oregon. December 20. www.epa.gov/superfund/sites/fiveyear/f2008100002390.pdf. Accessed November 2012.

Reed S, James J. 2010. Environmental Restoration Overview, Mountain Creek Industrial Center. July 16. http://mountaincreekindustrialcenter.com/downloads/MCIC_Industry_Day_Environmental_Presentation.pdf. Accessed November 2012.

Sale T, Newell C, Stroo H, Hinchee R, Johnson P. 2008. Frequently asked questions regarding management of chlorinated solvents in soils and groundwater. Environmental Security Technology Certification Program (ESTCP), Arlington, VA, USA. www.serdp.org. Accessed November 2012.

TCEQ (Texas Commission on Environmental Quality). 2004. Proposed Remedial Action document for State Hwy 123 PCE plume, Proposed state Superfund site, San Marcos, Hays County, Texas. December 9: http://www.tceq.texas.gov/assets/public/remediation/superfund/statehwy123/pdf0303.pdf. Accessed November 2012.

TCEQ. 2007. Proposed Remedial Action document for Spector Salvage Yard, Proposed state Superfund site, Orange, Orange County, Texas. January. http://www.tceq.texas.gov/assets/public/remediation/superfund/spector/pdf0369.pdf. Accessed November 2012.

TCEQ. 2010. Municipal Setting Designations. www.tceq.state.tx.us/remediation/msd.html. Accessed November 2012.

Tennessee Water Quality Control Board, 2007. Rules of Tennessee Department of Environment and Conservation, Tennessee Water Quality Control Board, Division Of Water Pollution Control Chapter 1200-4-3. General water quality criteria. www.tn.gov/sos/rules/1200/1200-04/1200-04-03.pdf. Accessed November 2012.

Terzaghi K, Peck RB. 1948. Soil Mechanics in Engineering Practice. John Wiley & Sons, New York, NY, USA.

Texas Register. 2007. Notice of intent to designate a groundwater plume management zone, Spector Salvage Yard, Orange, Texas. 32 Texas Register 966–967. Accessed November 2012.

U.S. Air Force and USEPA Region 3. 2006. Interim Record of Decision for Site SS-01, Brandywine DRMO, Andrews AFB, Maryland. www.epa.gov/superfund/sites/rods/fulltext/r2006030001338.pdf. Accessed November 2012.

U.S. Navy. 2008. Plume management zone to manage chlorinated solvents, Naval Weapons Industrial Reserve Plant, Dallas, Texas, USA. Department of the Navy Environmental Restoration Program 2008 progress report, p. 32. www.navfac.navy.mil. Accessed November 2012.

U.S. Navy. 2011. Alternative endpoints for groundwater remediation. Draft white paper. Alternative Restoration Technology Team (ARTT). https://portal.navfac.navy.mil/portal/page/portal/NAVFAC/NAVFAC_WW_PP/NAVFAC_NFESC_PP/ENVIRONMENTAL/ERB/WG-ARTT. November 2012.

USEPA (U.S. Environmental Protection Agency). 1985. Record of decision, Winthrop Landfill, OU 1, Winthrop, Maine. EPA ID MED980504435. http://www.epa.gov/superfund/sites/rods/fulltext/r0186500.pdf. Accessed November 2012.

USEPA. 1990. Record of Decision for the Moss-American Site, Milwaukee, Wisconsin, www.epa.gov/superfund/sites/rods/fulltext/r0590112.pdf. Accessed November 2012.

USEPA. 1991. ARARs Q's and A's: The fund-balancing waiver. Office of Solid Waste and Emergency Response (OSWER) Quick Reference Fact Sheet. www.epa.gov/superfund/policy/remedy/pdfs/92-34213fs-s.pdf. Accessed November 2012.

USEPA. 1993a. Guidance for evaluating the technical impracticability of ground-water restoration. OSWER Directive 9234.2-25. EPA/540-R-93-080. www.epa.gov/superfund/health/conmedia/gwdocs/techimp.htm. Accessed November 2012.

USEPA. 1993b. Record of Decision. E.I. du Pont de Nemours & Co., Inc. (Newport Pigment Plant Landfill), OU 01. DED980555122. Newport, Delaware. EPA/ROD/R03-93/170. www.epa.gov/superfund/sites/rods/fulltext/r0393170.pdf. Accessed November 2012.

USEPA. 1995. Record of Decision for Rocky Mountain Arsenal (US Army), OU 4, Adams County, Colorado. CO5210020769. www.epa.gov/superfund/sites/rods/fulltext/r0896128.pdf. Accessed October 2011.

USEPA. 1996. Record of Decision, Union Pacific Railroad Co. Tie-Treating Plant OU 1, The Dalles, Oregon. ORD009049412. www.epa.gov/superfund/sites/rods/fulltext/r1096138.pdf. Accessed October 2011.

USEPA. 1998a. Record of Decision Amendment for the Moss-American Co., Inc. (Kerr-McGee Oil Co.). WID039052626, OU 1, Milwaukee, Wisconsin. www.epa.gov/superfund/sites/rods/fulltext/a0598157.pdf. Accessed October 2011.

USEPA. 1998b. Record of Decision for Barstow Marine Corps Logistics Base, OUs 1 and 2, Barstow, California. CA8170024261. www.epa.gov/superfund/sites/rods/fulltext/r0998031.pdf. Accessed October 2011.

USEPA. 1999. Use of monitored natural attenuation at Superfund, RCRA Corrective Action, and Underground Storage Tank Sites. OSWER Directive 9200.4-17P. www.cluin.org/download/reg/d9200417.pdf. Accessed November 2012.

USEPA. 2001a. McKin Co. Superfund Record of Decision Amendment, OU 2, Gray, Maine. EPA/AMD/R01-01/015. www.epa.gov/superfund/sites/rods/fulltext/a0101015.pdf. Accessed November 2012.

USEPA. 2001b. Record of Decision for Hastings Ground Water Contamination Superfund Site OU 19. NED980862668. EPA/ROD/R07-01/511. Hastings, Nebraska. www.epa.gov/superfund/sites/rods/fulltext/r0701511.pdf. Accessed November 2012.

USEPA. 2004. Record of Decision, Waterloo Coal Gasification Plant, OU 1, Waterloo, IA. IAD980566256. www.epa.gov/superfund/sites/rods/fulltext/r0704653.pdf. Accessed November 2012.

USEPA. 2005a. Record of Decision for the Solvents Recovery Service of New England, OU 3, Southington, Connecticut. CTD009717604. www.epa.gov/superfund/sites/rods/fulltext/r0105008.pdf. Accessed November 2012.

USEPA. 2005b. Use of Alternate Concentration Limits in Superfund cleanups. OSWER 9200.4-39. Technical Memorandum. July 19. www.epa.gov/superfund/health/conmedia/gwdocs/pdfs/aclmemo.pdf. Accessed November 2012.

USEPA. 2005c. Second five-year review report. E.I. du Pont de Nemours & Co., Inc. (Newport Pigment Plant Landfill) Superfund site (a.k.a. DuPont-Newport Site), Newport, Delaware. March 3. www.epa.gov/superfund/sites/fiveyear/f0503006.pdf. Accessed November 2012.

USEPA. 2006. Explanation of significant differences, Waterloo Coal Gasification Plant. IAD980566356. OU 1, Waterloo, IA. www.epa.gov/superfund/sites/rods/fulltext/e0706002.pdf. Accessed November 2012.

USEPA. 2007a. Monitored natural attenuation of inorganic contaminants in ground water. Volume 1. Technical Basis for Assessment. EPA/600/R-07/139. http://epa.gov/ada/gw/mna.html. Accessed November 2012.

USEPA. 2007b. Monitored natural attenuation of inorganic contaminants in ground water. Volume 2. Assessment of non-radionuclides including arsenic, cadmium, chromium, copper, lead, nickel, nitrate, perchlorate and selenium. EPA/600/R-07/140. http://epa.gov/ada/gw/mna.html. Accessed November 2012.

USEPA. 2007c. Explanation of significant differences, Winthrop Landfill, OU 1. Maine. MED980504435. www.epa.gov/superfund/sites/rods/fulltext/e20070100001530.pdf. Accessed November 2012.

USEPA. 2009. Summary of key existing EPA CERCLA policies for groundwater restoration. EPA OSWER Directive 9283.1-33. www.epa.gov/superfund/health/conmedia/gwdocs/pdfs/9283_1-33.pdf. Accessed November 2012.

USEPA. 2010a. Monitored natural attenuation of inorganic contaminants in ground water. Volume 3. Assessment of radionuclides including tritium, radon, strontium, technetium, uranium, iodine, radium, thorium, cesium, and plutonium-Americium. EPA/600/R-10/093.

USEPA. 2010b. Institutional controls: A guide to planning, implementing, maintaining, and enforcing institutional controls at contaminated sites. Interim final. EPA-540-R-09-001. http://www.epa.gov/superfund/policy/ic/guide/icgdraft.pdf. Accessed November 2012.

CHAPTER 18

FUTURE DIRECTIONS AND RESEARCH NEEDS FOR SOURCE ZONE REMEDIATION

Bernard H. Kueper,[1] Alicia J. Shepard,[2] Hans F. Stroo,[3] Catherine M. Vogel[4] and C. Herb Ward[5]

[1]Queen's University, Kingston, ON, Canada K7L 3N6; [2]HydroGeoLogic, Inc., Reston, VA 20190, USA; [3]Stroo Consulting LLC., Ashland, OR 97520, USA; [4]Noblis Inc., Atlanta, GA 30318, USA; [5]Rice University, Houston, TX 77005, USA

18.1 INTRODUCTION

Remediation of sites impacted by chlorinated solvents remains technically challenging even though significant advances in technology development have occurred during the past 20 years. The contamination within the source zones at these sites typically comprises dense, nonaqueous phase liquids (DNAPLs), and dissolved and sorbed phase impacts. If low-permeability media are present within the source zone, these can act as diffusive sinks for aqueous and sorbed phase constituents. Chlorinated solvent source zones are difficult to remediate entirely, and even after treatment, these sources can persist for many decades or even hundreds of years in some cases.

Society, industry and regulatory agencies will likely continue to strive to remediate chlorinated solvent source zones for many years to come. These remediation efforts can have several remedial goals:

- Protection of public health and the environment from exposure to contaminants
- Compliance with applicable and relevant and appropriate requirements (ARARs)
- Prevention of off-site migration of groundwater and vapor phase contamination
- Restoration of off-site and/or onsite groundwater, soil, and soil vapor quality
- Reduction of mass flux of contaminants from source zones
- Reductions in the volume, toxicity and mobility of contaminants within a reasonable time frame

Achieving these goals in a cost-effective manner will require improvements to existing technologies and probably will require development of entirely new technologies. It also will require improvements in the current understanding of contaminant fate and transport in the subsurface. This research likely will require collaboration among experts from many diverse fields and new applications of techniques and insights developed in other disciplines.

Although chlorinated solvents have been present in soil and groundwater since at least the 1940s, research pertaining to the fate and transport of these chemicals in soil and groundwater was not carried out until the mid 1970s. Pioneering work focusing on the fate and transport of

---

Many of the specific research needs identified in this chapter stem from discussions during Research and Development Needs meetings held in Washington, D.C., in 2010 and 2011, and subsequent written contributions. A list of researchers and practitioners that participated in these meetings is provided in the Acknowledgement section at the end of this chapter.

B.H. Kueper et al. (eds.), *Chlorinated Solvent Source Zone Remediation*,
doi: 10.1007/978-1-4614-6922-3_18, © Springer Science+Business Media New York 2014

chlorinated solvent DNAPLs was initiated by F. Schwille and colleagues in West Germany in 1971 (Schwille, 1981, 1988). Following the discovery of Schwille's work in the early 1980s, coincident with the emerging recognition of the significance of chlorinated solvent DNAPLs in soil and groundwater by industry and practitioners in North America, research programs were initiated in Canada and the United States (U.S.) in the mid-1980s. Further initiatives were launched in Germany and other countries. Recognizing that pump-and-treat systems would need to operate indefinitely, given the persistence of chlorinated solvents in the subsurface (NRC, 1994), research quickly focused on developing and testing *in situ* technologies to either remove or destroy contaminant mass in the source zone so that the longevity of such source zones would be reduced.

Remediation of chlorinated solvent source zones remains extraordinarily challenging today, and further advances will require the coordinated efforts of experts from a wide range of scientific disciplines. Several different physical, chemical and biological processes have been investigated for use in source zone remediation, and several treatment technologies have been developed based on these different processes. Moreover, a wide range of physical, chemical, hydrogeological and biological properties of the subsurface environment can impact the fate and transport of DNAPLs and their constituent compounds.

Research in this area is highly interdisciplinary and requires expertise in several areas including mathematics, chemistry, geology, hydraulics, geochemistry, microbiology and engineering. Furthermore, the development of remediation technologies within the contaminant hydrogeology community has built on several decades of research and development in related areas including petroleum engineering, chemical engineering, material sciences and molecular biology; it is likely that much of the progress in the future also will build on results from other disciplines.

Research and development has played a vital role in improving site remediation efforts over the past 30 years, and it will continue to be essential in the future. Future research will be driven by several technical uncertainties and societal trends, which are summarized below and discussed more thoroughly in an earlier volume in this series (Stroo and Ward, 2010). These technical uncertainties and societal trends may conflict in some cases, emphasizing the need for a diverse set of technologies, but research should be linked to these key objectives for remediation in the future.

The overriding objectives for future remediation are to be more *surgical*, more *sustainable*, and more *certain*. The need for more surgical remediation is driven by the costs for remediation, which can be daunting when large volumes of the subsurface have been impacted or when contaminants have moved into challenging matrices such as fractured bedrock or low-permeability materials. Locating, delineating and treating contaminants efficiently will require further improvements in current characterization strategies and technologies and further improvements in treatment technologies.

The sustainability implications of remediation efforts have become increasingly important (U.S. Sustainable Remediation Forum, 2009). Sustainability can be difficult to evaluate, as it involves balancing several environmental, social and economic factors, and depending on the circumstances, sustainability can be viewed as either an impetus to cleanup sites as thoroughly and quickly as possible or as a rationale for use of passive, long-term remediation technologies. There is a growing need to consider other environmental factors when developing and implementing remediation strategies, including energy use, greenhouse gas emissions and life cycle impacts. These trends will drive development of more efficient remediation methods. However, one particularly important environmental factor affecting sustainability is water availability. As the U.S. and world populations continue to increase and demand more water, water scarcity will become an increasing concern (Vörösmarty et al., 2000), and pressure to clean up groundwater (either *in situ* or at the wellhead) will increase.

Source zone remediation is inherently uncertain and that uncertainty must be reduced. The uncertainty begins with the difficulties in locating, delineating and characterizing source zones, which stems largely from the fact that DNAPL migration is influenced by even small variations in geologic structure, can occur over large distances, and can occur in directions not coincident with groundwater flow. The difficulties in delivering reagents effectively and efficiently to impacted areas lead to additional uncertainties in remediating even well-delineated contaminated regions. The uncertainty also includes the difficulties in predictive modeling of remediation performance or natural attenuation. Furthermore, there is uncertainty in actually measuring the performance of remediation systems and in determining when or whether treatment has been sufficient.

Before discussing specific research and development needs, the next section discusses remediation research in a broader context. The goal is to examine the current environment for funding and performing such research and to suggest approaches to better accomplish the needed research in an environment of declining financial support.

18.2 REMEDIATION RESEARCH IN CONTEXT

Applying remediation technologies to source zones has progressed rapidly over the last decade, but many sites still need restoration, and many of these are among the most technically challenging and expensive. For example, the Department of Defense (DoD) has achieved response complete (RC) status at over 80% of the sites in its Installation Restoration Program (IRP), and it has a goal of achieving RC at virtually all of its contaminated sites by the end of fiscal year (FY) 2021 (Leeson and Stroo, 2011). The cumulative costs for treating DoD's remaining IRP sites alone were estimated at $12.8 billion in FY 2010. The total cost to remediate all remaining DNAPL source zones worldwide is highly uncertain and difficult to estimate, but clearly it will be a significant issue for decades.

Similarly, it is difficult to calculate the current and historical research funding in this area, but it is clear that it is declining. In the United States, the federal and state funding for research has decreased as overall government funding for research has declined. Furthermore, many new priorities have emerged, while the issues of contaminated groundwater have been studied over the past 30 years. The major U.S. funding agencies for this type of research are now DoD (largely through the Strategic Environmental Research and Development Program [SERDP] and Environmental Security Technology Certification Program [ESTCP]), the Department of Energy (DOE), and the U.S. Environmental Protection Agency (USEPA) (largely through the National Institute of Environmental and Health Sciences (NIEHS). However, all of these agencies have other growing needs for research funding, and in many cases their total research and development allocations are declining. Industry also has been supporting targeted research in this area for many years in response to regulatory, legal and public concerns and the potential liabilities involved. However, this source of funding also has decreased over time, as implementation of remediation systems has progressed at many of their sites.

The obvious conclusion is that the research community may need to change in response to the decreases in funding. There is a clear need to develop and participate in research partnerships to leverage funds and capabilities. Several potential models are evolving, including research consortia and the National Science Foundation (NSF) research center partnerships. It is important to consider now how researchers and funding agencies can best incentivize and encourage collaboration. Effective collaboration within research teams will be increasingly important, including collaborations between investigators, across technical disciplines, and between different agencies. One model worth considering is public–private partnerships in which public agencies match industry funds, similar to some of the current models used in Canada such as the Natural Sciences and Engineering Research Council (NSERC) of Canada Partnerships Programs.

In developing future research partnerships, it will be important to remember that one of the goals of research is to train highly qualified personnel who will be needed to work in the field of environmental protection. For example, NSERC currently weights 33% of the merit of a Discovery Grant on training of highly qualified personnel, equal to the weightings for the technical merits of the proposal and for the track record of the applicant.

One final general topic to consider in future research efforts is the communication of results. The traditional model is to rely on publication in journal articles and in conference presentations. These avenues will remain essential, but the research community must realize that these may not be the best ways to disseminate the results of research to important audiences, including practitioners and concerned citizens. Also, technology advances have changed communication dramatically and made it important to use new media effectively. Deliberate efforts to use innovative and focused methods to reach a range of different stakeholders are encouraged.

The remainder of this chapter discusses specific research and development needs for different phases of restoration, including characterization, diagnosis and monitoring, and remediation. Although some overlap is inevitable, these phases do have important distinctions.

18.3 RESEARCH NEEDS FOR SITE CHARACTERIZATION

The following sections focus on the need for more effective and efficient characterization of source zones. Characterization of source zones is often inadequate given the difficulties inherent in working in the subsurface and the complex distribution of contaminants between the DNAPL, aqueous, vapor and sorbed phases (NRC, 2005). In many cases, the success of source remediation is directly related to the quality and thoroughness of site characterization. The lack of key information on the spatial extent, mass, and architecture of the contamination from initial characterization efforts can lead to design, implementation, and performance monitoring strategies that are inappropriate or inadequate, leading to increased costs and extended schedules. The same holds for the lack of key information regarding aqueous geochemistry and microbiology, as well as features such as groundwater velocity and the presence of both high- and low-permeability media. It is also important to remember that characterization efforts should continue throughout the technology implementation and remedy optimization phases.

18.3.1 Source Zone Delineation and Mass Estimation

DNAPLs are difficult to locate, yet the performance of many treatment technologies depends on precise delivery of remedial agents to the contaminants. It is now well recognized that the migration of remedial agents will occur primarily through the more permeable regions of the subsurface and that these pathways may not coincide with all occurrences of DNAPL or diffused mass. Thermal treatment using either electrical resistance heating (ERH) or thermal conductive heating (TCH) is not as susceptible to preferential flow pathways as fluid injection technologies, but even thermal treatment will leave some portions of the source untreated if contaminant mass was missed during the initial source characterization efforts (Chapter 14 of this volume).

It is also important to recognize that the source/plume boundary is difficult to define, and over time a significant fraction of the originally released contaminant mass may diffuse into lower permeability regions. These secondary sources can be difficult to detect, yet they can sustain a dissolved-phase plume long after any upgradient sources of contaminants have been removed.

Uncertainty in the spatial extent of the source zone needs to be reduced to allow better informed decisions when selecting and designing remedial technologies and to improve the success of these often expensive technologies. Improved methods and tools are needed for delineating the source zone, DNAPL accumulations, and any contaminant mass that has diffused into less permeable regions. In addition to better delineating the spatial extent of the DNAPL source zone, accurately quantifying contaminant mass is critical to selecting and designing certain technologies, as well as to optimizing their implementation. Several remediation technologies are highly sensitive to the contaminant mass estimates, particularly those requiring chemical amendments (*in situ* chemical oxidation, *in situ* chemical reduction and *in situ* bioremediation). Improved methods are needed for estimating the mass of contaminants present, both the total and the distribution between phases (DNAPL, sorbed, aqueous and vapor) and between more and less transmissive regions.

18.3.2 Source Zone Architecture and Depletion

The architecture of a DNAPL source zone strongly affects its longevity and its response to treatment (Lemke et al., 2004; Fure et al., 2006). For example, the amount of DNAPL present in pools versus ganglia largely determines the performance of mass transfer-based technologies such as chemical oxidation, chemical reduction, and bioremediation (Christ et al., 2005; Park and Parker, 2008). However, adequately characterizing the source zone architecture remains difficult, and techniques to describe the architecture meaningfully could greatly improve the ability to predict and evaluate performance and to predict the long-term depletion of the source (Basu et al., 2008; Christ et al., 2010).

Many sources have aged considerably since the original release of the DNAPL. Aging refers to dissolution of some or all of the DNAPL into groundwater, vaporization of some or all of the DNAPL into soil gas, the transfer of contaminants into low-permeability zones as a result of slow advection and molecular diffusion, and changes in the composition of multicomponent DNAPLs toward the higher molecular weight, lower solubility components. Considerable mass can diffuse into the surrounding matrix and into other inaccessible regions over decades of aging (Chapter 1). Later release of this mass (back diffusion) from low-permeability zones within an aged source can be a serious limitation for any technology where treatment is generally limited to the high-permeability zones. Research is needed to better understand any biotic or abiotic reactions that can be sustained long term within an aged source zone following the removal of contaminant mass from the higher permeability zones through either active treatment or natural processes, so that back diffusion does not continue to sustain downgradient impacts.

The mass of contaminants within the source zone slowly decreases over time due to dissolution, degradation, volatilization, and vaporization. The mass discharge from the source to the plume (the source strength) also decreases over time (Newell et al., 2006). This source zone natural attenuation (SZNA) can be critical when making management decisions, particularly at complex sites. Decision makers need to know how rapidly SZNA is occurring, what natural processes are contributing to SZNA, and if these processes and rates are sustainable. The ESTCP is currently developing guidance for assessing SZNA. A related need is for better methods to quantify the assimilative capacity of an aquifer, which would facilitate better definition of mass discharge/flux targets in support of remedial action objectives for partial source treatment.

Better tools, protocols and decision guides are needed to distinguish between plumes primarily sustained by DNAPL dissolution and those primarily sustained by back diffusion. These tools, protocols, and decision guides are required to select an appropriate remedial response, determine regulatory status, and develop appropriate site management strategies.

18.3.3 Increased Resolution and Fine-Scale Mapping

The overall spatial extent and local-scale distribution of contamination heavily impact the selection, design and implementation of remedial technologies. However, much of our characterization efforts are conducted at a relatively large scale, primarily because of the costs of current site investigation methods. Relatively low-resolution investigations can result in a poor understanding of key characteristics of a site (Li and Abriola, 2009). Higher-resolution mapping of the distribution of contaminants within a source zone would lead to more accurate site conceptual models and a better ability to target subsequent remediation efforts, thereby yielding cost savings and improved cleanup performance through more efficient treatment.

Furthermore, the biological and geochemical conditions affecting contaminant fate often vary significantly over space and time. This variability can lead to inaccurate estimates of degradation rates, for example, and inefficient remediation because poorly treated areas are not identified. However, characterization data typically do not address the variability in biological and geochemical conditions. Better strategies and tools are required to improve understanding of the variability of biological and geochemical conditions and to decide for which sites such efforts should be a priority.

In addition to the above, improved methods of quantifying the local-scale distribution of hydraulic conductivity and the continuity and orientation of geologic structures are needed. This is particularly relevant to the application of fluid injection-based technologies and to properly designing thermal applications where excessive groundwater influx can limit the ability to achieve boiling. The need to better characterize hydraulic conductivity and geologic structure applies to both unconsolidated deposits and fractured bedrock.

The level of resolution needed in any of the above characterization efforts to support management decisions remains a key question. Future research should address this question for a range of different parameters and subsequently seek to improve fine-scale measurements where needed.

18.4 RESEARCH NEEDS FOR REMEDIATION

Numerous technologies have been developed to remove mass from DNAPL source zones, yet extensive research has shown clearly that there are technical limits to their success (Stroo et al., 2012). With this understanding, efforts in recent years have shifted to improving treatment through better delivery techniques, better formulations and deliberate combinations of technologies and to improving the models and predictions of plume responses to source treatment. Many feel that the current suite of remediation technologies will not change significantly and that the greatest need is to design, deploy, diagnose and optimize these existing technologies as much as possible.

A different perspective is that new technologies are still needed, and it is important to ensure that funding is available to develop, test and demonstrate innovative approaches to source zone remediation. It is difficult to predict what areas of research will lead to innovation, particularly the big leaps forward, but such advances are likely given the long-term nature of some contamination problems. The lifespan of contamination at many sites impacted by chlorinated solvents will be well over a hundred years, even with some treatment, so longer-term research goals are certainly appropriate for dealing with these issues. An ongoing effort should be maintained to identify and explore novel applications from other fields and disciplines. Such efforts should stress the need for objective evaluations of new technologies and ideas, including quantitative demonstrations of technologies supported by high levels of experimental design, modeling and analysis.

More effective remediation is needed for the most challenging sites and the most problematic portions of these sites. As many of the relatively simple sites have been addressed, remediation of the most difficult ones has become increasingly important. These include sites with sources in fractured rock, sources in highly heterogeneous porous media, sources containing large amounts of DNAPL, and sources containing many different types of contaminants. Similarly, as it has become clear that contaminants stored in less permeable regions can represent long-term sources, continued research efforts are needed to determine when further remediation of these regions will be needed, how such remediation should be carried out, and how effective it is likely to be. In the future, containment and *ex situ* treatment may play greater roles in addressing contamination remaining after primary efforts at remediation, especially considering the potential for concentration-based cleanup goals (maximum contaminant levels [MCLs]) to be lowered. Although not a focus area in this book, vapor intrusion can be the most important pathway of concern at some sites. Efforts are ongoing to improve understanding of the complexities involved in the groundwater-to-indoor air pathway and to develop more accurate assessment tools (Johnson et al., 2012).

Addressing these challenges will require improvements to existing technologies, better integration of different technologies, and likely entirely new technologies. The following sections identify and discuss the key research and development needed to improve the treatment of chlorinated solvent sources.

18.4.1 Long-Term Impacts of Source Zone Remediation

Some amount of contamination often remains after active source zone treatment, even with aggressive technologies (Kingston et al., 2010). Plumes also may not respond quickly to source treatment depending on the degree of sorption, groundwater velocity and the influence of diffusive sinks. Source zone remediation technologies may exhibit treatment effects that continue for several years after treatment ends (Adamson et al., 2010), and re-equilibration to a new posttreatment steady state may require many more years (West and Kueper, 2010; Falta et al., 2005). It is therefore critical to make reliable predictions of the effect of source treatment on the overall duration of remediation. Useful models have been developed (Chapters 5 and 6 of this volume), but a better fundamental understanding of the long-term impact of applying source zone remediation technologies still is needed to make site-specific projections from current data. For example, what are the relationships between mass remaining or post-remedy concentrations and the long-term plume response? This understanding is needed to set realistic objectives and to determine if further treatment is warranted (Soga et al., 2004).

Efforts should focus on both sources containing primarily DNAPL and sources containing primarily diffused mass, as these may differ in important ways. Also, cost-effective technologies to address any low-level contamination are needed. Preferably, such technologies would be passive with little need for future labor or materials to justify the costs needed to achieve relatively small reductions in risk.

18.4.2 Remedial Fluid Delivery

The experience gained over the past 20 years has demonstrated that one primary challenge associated with fluid injection technologies is incomplete delivery of reagents or incomplete sweep of the source zone with flushing agents. The limitations on delivery remain critical considerations when designing treatment systems and largely control their eventual performance. Research to date has identified many improvements, but better methods of delivering remedial fluids in heterogeneous media will continue to be a goal (Berge and Ramsburg, 2009; Phenrat et al., 2009). The future may hold promising opportunities to leverage innovations from other disciplines and fields to improve delivery of remedial fluids. Delivery and mixing in the subsurface are discussed in detail in a separate volume in this monograph series (Kitanidis and McCarty, 2013).

18.4.3 Combined Remedies

With the understanding that many source zone remediation technologies will not achieve complete removal of contamination, interest in combining technologies has increased (Peale et al., 2008; La Mori et al., 2010; Truex et al., 2011). Technologies can be combined in time or space to develop efficient overall strategies, and such combinations can include a wide variety of processes (Chapter 15). A common approach is to employ an initial technology to remove the majority of the contaminant mass, followed by a subsequent technology to remove any remaining contamination. For example, *in situ* bioremediation has been examined for use after chemical oxidation or thermal treatment (Hrapovic et al., 2005; Friis et al., 2005). Alternatively, two technologies may have a synergistic effect, each increasing the effectiveness of the other (e.g., deliberately combining zero-valent iron and vegetable oil to stimulate biological and chemical reduction or thermally activated persulfate oxidation [Quinn et al., 2005; Johnson et al., 2009]).

Further work is needed to assess the benefits of adopting a combined remedy approach for source zone remediation. From a practical standpoint, how can technologies best be combined, and what are the advantages and drawbacks of different combinations? How are the optimal technology combinations selected and implemented, and can better combinations be developed? When should the transition from one technology to another occur? To address these questions, site managers, field practitioners and regulators need tools.

18.4.4 Diagnostic and Performance Assessment Tools

Progress monitoring during remediation can help managers recognize problems early, optimize technology performance in near real time, determine when to transition to another technology or implement contingency actions, and when to stop technology application. Alternative performance metrics, notably mass flux analyses, have received considerable attention for source zone treatment, largely because of the difficulty in achieving concentration-based cleanup goals (Jawitz et al., 2005; ITRC, 2010). Other techniques also may provide more useful information than the typical data collected. Diagnostic methods, tools, and strategies for monitoring and optimizing remediation are needed.

Better methods, tools, and strategies for assessing performance after active remediation also are needed. Monitoring slow processes such as natural abiotic or biological degradation can be costly and uncertain. Deciding whether additional treatment or some other contingency action is needed can require ongoing monitoring, and the most cost-effective approaches to such monitoring may be reliable and robust automatic sensors capable of identifying key environmental changes that could trigger contingencies. Summaries of long-term performance experience could also be very helpful. Relevant questions should include the appropriate monitoring time period and frequency of data collection, the types of data that should be collected, and the timing and magnitude of any slow rebound or sustained treatment that may occur. To support development of assessment tools, postmortem monitoring could be undertaken at a number of well-characterized sites where full-scale source zone remediation has been implemented.

18.4.5 Improved Containment and *Ex Situ* Treatment

In situ treatment has been the priority for technology development for several years, but there are reasons to develop better containment and *ex situ* treatment options. Two major reasons are the potential lowering of concentration-based regulatory standards and the experience of the past decade showing that some contamination is likely to remain even after aggressive source treatment (Stroo et al. 2012). Decreased standards (MCLs) are anticipated for some frequently encountered contaminants including trichloroethene (TCE). Lowering MCLs could expand the size of dissolved plumes that will require treatment, and it would likely result in more stringent goals for source zone remediation. Remediating large sites to even lower criteria will be difficult and costly and could make wellhead treatment more attractive as an alternative or as a polishing step in a treatment train. Better and more efficient wellhead treatment technologies will be needed for these applications. These technologies also should be capable of removing or destroying common co-contaminants, such as 1,4-dioxane, to increase overall efficiency.

The limitations of source treatment, especially if lower MCLs are required, also may make containment a more attractive strategy for some sites. Some form of hydraulic and/or physical containment will likely be a component in managing challenging sites for the foreseeable future. Research is needed to develop more reliable and robust containment technologies. In addition, better methods and procedures for long-term assessment monitoring are needed.

18.4.6 Economic Analyses

Given the significant expenses involved in remediating DNAPL source zones, economic analyses on several fronts are warranted. First, from a technical perspective, does additional expenditure on characterization up front lead to decreased remediation costs over the long term? Additional research is needed to fully address this question on a technology-specific basis. Second, characterization and remediation efforts have the potential to damage natural resources and worsen the spatial extent of contamination. Efforts should be undertaken to assess the economic and environmental impacts of this damage in relation to the risk reduction achieved. Third, economic analysis techniques should be developed to assist responsible parties in calculating accurate life cycle costs for managing contaminated sites. Finally, in consideration of the current push to develop sustainable approaches to remediation, efforts should focus on how to reduce the energy requirements of remedial technologies.

18.4.7 Technology Performance Models

Despite technology advances, a number of uncertainties remain regarding technology performance. These uncertainties can lead to overdesigning systems or unrealistic expectations. Technology performance hinges on site-specific subsurface characteristics, all of which cannot be quantified without uncertainty. Each site is unique, resulting in technologies that perform beyond expectations under certain conditions, but poorly under others. Improved technology performance models are needed to predict how a given technology or combination of technologies will perform at a particular site. Additional development of deterministic models is needed, as well as more development of simpler models useful for screening technologies and probabilistic models. Further development and application of such tools would likely improve communication and the setting of expectations among site managers, practitioners and regulators.

18.4.8 Impacts of Management Strategies

Management of contaminated sites has undergone several shifts, generally characterized by increasing predictability and cost control over time. Recently, performance-based contracting has been widely adopted for remediation support services at the DoD and other sites. This type of contracting mechanism can have significant impacts on decisions related to technology selection and operation. Although it is intended to foster innovation, it could encourage decisions to use more conventional technologies. A survey of remediation sites that have employed performance-based contracting should be undertaken to capture the lessons learned. Questions include whether and how innovative technology has been used, how these decisions have impacted cleanup, and how research and development have been affected.

18.4.9 Challenging Site Conditions

Challenging sites represent an increasing proportion of the source zones needing remediation, and several site conditions pose particular challenges. The conditions that pose the greatest technical and management challenges are sites with sources in fractured media, sources in highly heterogeneous porous media, sources containing large amounts of DNAPL, sources containing many different types of contaminants, and sites where vapor intrusion is a concern. These conditions are discussed separately in the following sections.

Fractured Media: Contamination within fractured clay and bedrock represents a particularly difficult and costly problem for remediation, given the discrete nature of groundwater flow through fracture networks and the presence of large-volume diffusive sinks in many rock types (Parker et al., 1994; Reynolds and Kueper, 2001). Characterizing and treating contaminants in fractured media can be very challenging (Goldstein et al., 2004), and as many of the less challenging sites have been addressed, these sites have become an increasing fraction of the remaining liabilities for the DoD and other organizations. Further efforts are needed to develop remedial technologies applicable in fractured clay and bedrock.

Highly Heterogeneous Porous Media: Source zones in highly heterogeneous porous media are problematic given the difficulty of delivering injected fluids to all impacted areas. Highly heterogeneous geology, particularly sharp permeability contrasts over small vertical distances, provides an optimal environment for lateral DNAPL migration and pooling.

A small ganglia-to-pool ratio leads to long DNAPL lifespan because of the relatively small DNAPL to water surface area available for mass transfer, which also limits the effectiveness of any mass transfer-based technology. Highly heterogeneous porous media can be challenging for thermal technologies where the produced gases must be removed prior to allowing lateral gas migration out of the treatment zone. Drilling can also be risky in source zones with significant quantities of pooled DNAPL. Further efforts are needed to safely and effectively treat source zones in highly heterogeneous porous media.

Large Amounts of DNAPL: Experience shows that larger volume DNAPL releases typically result in larger source zone footprints and greater depths of impact, increased source zone longevity and longer timescales of migration. Although not representative of the majority of source zones, those that contain large amounts of DNAPL can be problematic to treat because of the large volume of chemical agents that must be employed (such as oxidants and electron donor) and the large number of borings that must be completed to accommodate the application of thermal technologies. Further efforts are needed to develop methods of assessing whether such source zones should receive active treatment at this time or whether management of such source zones should be focused on containment and monitoring.

Different Types of Contaminants: Many, perhaps most, chlorinated solvent sources represent mixtures of compounds or chemical products. Multicomponent DNAPLs can be challenging to treat because only some of the components may be amenable to biodegradation, oxidation, reduction, or dissolution. Treating such source zones may result in leaving behind an altered form of nonaqueous phase liquid (NAPL) and little progress toward achieving site closure. In addition to multicomponent DNAPLs, some sites are characterized by complex inorganic geochemistry in addition to DNAPL presence. The presence of high metals or inorganic constituent concentrations can limit the effectiveness of technologies relying upon biodegradation, oxidation or reduction. Relatively little research has focused on the cost-effective treatment of source zones containing different types of contaminants.

Vapor Intrusion: In recent years, concerns have grown over the migration of contaminated vapors from groundwater plumes and source zones into residential homes. Vapor intrusion is now often the risk driver for many actions at cleanup sites across the DoD (Noblis, 2008) and other organizations. The potential for vapor intrusion increases the costs for monitoring and remediation at chlorinated solvent sites. The risk posed by vapors near a building is a complex issue with many interacting factors (Abreu and Johnson, 2005). An improved understanding of vapor intrusion mechanisms is needed to ensure that resources are directed at real risk reduction. In addition, cost-effective remedies must be developed and demonstrated to manage these risks.

18.5 TECHNOLOGY TRANSFER NEEDS

Technology transfer generally involves transmitting the knowledge gained from researchers and technology developers to technology users (Bozeman, 2000). In this context, the technology users include practitioners, regulators, managers, and the public. The importance of effective transfer of information from the restoration research community to these user groups has been stressed repeatedly (Leeson and Stroo, 2011). The goals of technology transfer are to improve the state of the practice and to foster the adoption of promising technologies.

In the case of chlorinated solvent source zone remediation, many believe that advances in the state of the practice have been slowed by the reluctance of the regulatory community to accept new tools and guidance. Specifically, the experience to date with source zone remediation suggests that some contamination will remain at many sites, despite aggressive remediation efforts. Identifying and managing such sites efficiently could yield substantial savings over current practices, with little risk to human health or the environment. Although regulatory acceptance has been slow, recent guidance has acknowledged these findings (ITRC, 2011), and some regulatory agencies are developing low-threat closure guidelines.

Further development of approved and credible methods to identify sites suitable for low-risk management could lead to more efficient site management without compromising environmental protection. More broadly, strengthening the links between the regulatory and research communities could ensure that the state of the science and technology is communicated effectively and that regulatory objections are addressed early. Stronger links could also help researchers and developers of technology focus on the most important questions.

In addition to the regulatory community, technology transfer efforts should target the practitioners. However, the extent of the information available can be overwhelming, and busy individuals need information available in a highly focused manner. Furthermore, the target audience consists of people who have become accustomed to getting information electronically. Methods of transferring knowledge and technology should be evaluated to inform future efforts and increase their effectiveness. Model systems to compare remedial approaches are a frequently sought tool.

Continuing education requirements are increasing in some states and may represent an opportunity for technology transfer. Public–private consortia represent another potential model for improving communication and fostering innovation and greater success in the field. Such consortia also can lead to additional sources of research and development funding.

18.6 SUMMARY

Society, industry and regulatory agencies likely will continue to strive to remediate chlorinated solvent source zones for many years to come. Research and development has played a vital role in improving site remediation efforts over the past 30 years, and it will continue to be essential in the future. It is apparent, however, that the research community will need to change in response to decreases in funding. Specifically, there is a clear need to develop and participate in research partnerships to leverage funds and capabilities. Research in this area is highly interdisciplinary and requires expertise in several subdisciplines including mathematics, chemistry, geology, hydrogeology, hydrology, geochemistry, microbiology and engineering.

In developing future research partnerships, it will be important to remember that one of the goals of research is to train highly qualified personnel who will be needed to work in the field of environmental protection. It will also be important to consider how the results of research will be communicated. The traditional model is to rely on publication in journal articles and in conference presentations. These avenues will remain essential, but the research community must also realize that these may not be the best ways to disseminate the results of research to important audiences, including practitioners and concerned citizens.

Many of the specific research needs identified in this chapter stem from discussions during Research and Development Needs meetings held in Washington, D.C., in 2010 and 2011, and subsequent written contributions. These meetings were chaired by the authors of this chapter and attended by leading researchers and practitioners in the field of chlorinated solvent source zone remediation. Specific research needs were identified in the general areas of (1) effective and efficient characterization of source zones, (2) remediation technologies and strategies and (3) technology transfer needs.

Within the area of effective and efficient characterization of source zones, it is now recognized that the characterization of source zones is often inadequate given the difficulties inherent in working in the subsurface and the complex distribution of contaminants between the DNAPL, aqueous, vapor, and sorbed phases. In many cases, the success of source zone remediation is directly related to the quality and thoroughness of site characterization. Specific research needs within the general area of effective and efficient characterization of source zones include (1) source zone delineation and mass estimation, (2) source zone architecture and depletion and (3) increased resolution and fine-scale mapping of geologic properties and contaminant distribution.

Within the area of remediation technologies and strategies, it is recognized that many technologies have been developed to remove mass from DNAPL source zones, yet extensive research has clearly shown technical limits to their success. With this understanding, efforts in recent years have shifted to improving treatment through better delivery techniques, to improving formulations or deliberate combinations of technologies, and to improving models and predictions of plume responses to source treatment. Specific research needs within the general area of remediation technologies and strategies include (1) long-term impacts of source zone remediation, (2) remedial fluid delivery, (3) combined remedies, (4) diagnostic and performance assessment tools, (5) improved containment and *ex situ* treatment, (6) economic analyses, (7) technology performance models, (8) impacts of management strategies and (9) development of technologies for challenging site conditions.

With regard to technology transfer needs, the goals of technology transfer are to improve the state of the practice and to foster the adoption of promising technologies. In this context, the technology users include practitioners, regulators, managers, and the public. Communication needs to occur along a number of avenues, particularly communication and open discussions among stakeholders on a frequent basis. Technology transfer should emphasize development of web-based tools that allow users to quickly find focused information specific to their issues. The need for highly qualified personnel is paramount, especially within the field of contaminant hydrogeology which lacks a universal standard for qualification.

ACKNOWLEDGEMENT

The authors of this chapter gratefully acknowledge the discussion, advice, counsel and recommendations of the researchers and practitioners who attended and contributed to meetings held in Washington, D.C., in December 2010 and December 2011. This chapter incorporates the discussions the authors had with the following experts: Linda Abriola, Tufts University; Michael Annable, University of Florida; Michael Basel, Haley & Aldrich; Richard Brown, ERM, Inc.; Natalie Cápiro, Tufts University; Michelle Crimi, Clarkson University; Rula Deeb, Geosyntec; Ron Falta, Clemson University; Rachel Farnum, General Electric (GE) Global Research; Mark Harkness, GE Global Research; Elisabeth Hawley, ARCADIS; Richard Johnson, Oregon Health and Science University; Michael Kavanaugh, Geosyntec; Brian Looney, U.S. Department of Energy, Savannah River National Laboratory; Gregory Lowry, Carnegie Mellon University; David Major, Geosyntec; James Mercer, Tetra Tech GEO; Charles Newell, GSI Environmental; Fred Payne, ARCADIS; Kurt Pennell, Tufts University; Thomas Sale, Colorado State University; Robert Siegrist, Colorado School of Mines; Thomas Simpkin, CH2M Hill; Paul Tratnyek, Oregon Health and Science University; and A. Lynn Wood, U.S. Environmental Protection Agency.

REFERENCES

Abreu LDV, Johnson PC. 2005. Effect of vapor source – building separation and building construction on soil vapor intrusion as studied with a three-dimensional numerical model. Environ Sci Technol 39:4550–4561.

Adamson DT, McGuire TM, Newell CJ, Stroo H. 2010. Sustained treatment: Implications for treatment timescales associated with source-depletion technologies. Remediat J 21:27–50.

Basu NB, Fure AD, Jawitz JW. 2008. Predicting dense nonaqueous phase liquid dissolution using a simplified source depletion model parameterized with partitioning tracers. Water Resour Res 44:W07414, doi:10.1029/2007WR006008.

Berge ND, Ramsburg CA. 2009. Oil-in-water emulsions for encapsulated delivery of reactive iron particles. Environ Sci Technol 43:5060–5066.

Bozeman B. 2000. Technology transfer and public policy: A review of research and theory. Res Policy 29:627–655.

Christ JA, Ramsburg CA, Abriola LM, Pennell KD, Löffler FE. 2005. Coupling aggressive mass removal with microbial reductive dechlorination for remediation of DNAPL source zones – A review and assessment. Environ Health Perspect 113:465-477.

Christ JA, Ramsburg CA, Pennell KD, Abriola LM. 2010. Predicting DNAPL mass discharge from pool-dominated source zones. J Contam Hydrol 114:18–34.

Falta RW, Rao PSC, Basu N. 2005. Assessing the impacts of partial mass depletion in DNAPL source zones: II. Coupling source strength functions to plume evolution. J Contam Hydrol 79:45–66.

Friis AK, Albrechtsen H-J, Heron G, Berg PL. 2005. Redox processes and release of organic matter after thermal treatment of a TCE-contaminated aquifer. Environ Sci Technol 39:5787–5795.

Fure AD, Jawitz JW, Annable MD. 2006. DNAPL Source depletion: Linking architecture and flux response. J Contam Hydrol 85:118–140.

Goldstein KR, Vitolins AR, Navon D, Parker BL, Chapman S, Anderson GA. 2004. Characterization and pilot-scale studies for chemical oxidation remediation of fractured shale. Remediat J 14:19–37.

Hrapovic L, Sleep BE, Major DJ, Hood ED. 2005. Laboratory study of treatment of trichloroethene by chemical oxidation followed by bioremediation. Environ Sci Technol 39:2888–2897.

ITRC (Interstate Technology & Regulatory Council). 2010. Use and Measurement of Mass Flux and Mass Discharge. ITRC, Washington, DC, USA. http://www.itrcweb.org/Guidance/GetDocument?documentID=49. Accessed January 20, 2014.

ITRC. 2011. Technical/Regulatory Guidance: Integrated DNAPL Site Strategy. ITRC, Washington, DC, USA. http://www.itrcweb.org/Guidance/GetDocument?documentID=35. Accessed January 20, 2014.

Jawitz JW, Fure AD, Demmy GG, Berglund S, Rao PSC. 2005. Groundwater contaminant flux reduction resulting from nonaqueous phase liquid mass reduction. Water Resour Res 41: W10408; doi:10.1029/2004WR003825.

Johnson PC, Luo H, Holton C, Dahlen P, Guo Y. 2012. Interim Report: Integrated Field-Scale, Lab-Scale, and Modeling Studies for Improving the Ability to Assess the Groundwater to Indoor Air Pathway at Chlorinated Solvent-Impacted Groundwater Sites. SERDP ER-1686. SERDP, Alexandria, VA, USA. http://www.serdp.org/. Accessed January 20, 2014.

Johnson RL, Tratnyek PG, Johnson RO. 2009. Persulfate persistence under thermal activation conditions. Environ Sci Technol 42:9350–9356.

Kingston JL, Dahlen PR, Johnson PC. 2010. State-of-the-practice review of in situ thermal technologies. Ground Water Monit Remediat 30:64–72.

Kitanidis PK, McCarty PL, eds. 2013. Delivery and Mixing in the Subsurface: Processes and Design Principles for In Situ Remediation. In Ward CH, ed, SERDP/ESTCP Remediation Technology Monograph Series. Springer Science + Business Media, LLC, New York, NY, USA. 325 p.

La Mori P, Kirkland E, Faircloth H, Bogert R, Kershner M. 2010. Combined thermal and zero-valent iron *in situ* soil mixing remediation technology. Remediat J 20:9–25.

Leeson A, Stroo H. 2011. Workshop Report: Investment Strategies to Optimize Research and Demonstration Impacts in Support of DoD Restoration Goals. SERDP, Alexandria, VA, USA. http://www.serdp.org/. Accessed January 20, 2014.

Lemke LD, Abriola LM, Lang JR. 2004. Influence of hydraulic property correlation on predicted dense nonaqueous phase liquid source zone architecture, mass recovery and contaminant flux. Water Resour Res 40:W12417, doi: 10.1029/2004WR003061.

Li KB, Abriola LM. 2009. A multistage multicriteria spatial sampling strategy for estimating contaminant mass discharge and its uncertainty. Water Resour Res 45:W06407, doi: 10.1029/2008WR007362.

Newell CJ, Cowie I, McGuire TM, McNab W. 2006. Multi-year temporal changes in chlorinated solvent concentrations at 23 MNA sites. J Environ Eng 132:653–663.

Noblis. 2008. Tri-Services Handbook for the Assessment of the Vapor Intrusion Pathway. Falls Church, VA, USA. http://www.denix.osd.mil/references/upload/Tri-Serv_VI_Handbook_Final.pdf. Accessed January 20, 2014.

NRC (National Research Council). 1994. Alternatives for Ground Water Cleanup. National Academies Press, Washington, DC, USA. 336 p.

NRC. 2005. Contaminants in the Subsurface: Source Zone Assessment and Remediation. National Academies Press, Washington, DC, USA. 372 p.

Park E, Parker JC. 2008. Effects of mass reduction, flow reduction and enhanced biodecay of DNAPL source zones. Transp Porous Media 73:95–108.

Parker BL, Gillham RW, Cherry JA. 1994. Diffusive disappearance of immiscible-phase organic liquids in fractured geologic media. Ground Water 32:805–820.

Peale JGD, Bakkom E, Lakhwala F, Mueller J, Molin J. 2008. TCE plume remediation via ISCR-enhanced bioremediation utilizing EHC and KB-1. Remediat J 18:19–31.

Phenrat T, Kim H-J, Fagerlund F, Illangasekare T, Tilton RD, Lowry GV. 2009. Particle size distribution, concentration, and magnetic attraction affect transport of polymer modified Fe0 nanoparticles in sand columns. Environ Sci Technol 43:5079–5085.

Quinn J, Geiger C, Clausen C, Brooks K, Coon C, O'Hara S, Krug T, Major D, Yoon W-S, Gavaskar A, Holdsworth T. 2005. Field demonstration of DNAPL dehalogenation using emulsified zero-valent iron. Environ Sci Technol 39:1309–1318.

Reynolds DA, Kueper BH. 2001. Multiphase flow and transport in fractured clay/sand sequences. J Contam Hydrol 51:41–62.

Schwille F. 1981. Groundwater pollution in porous media by fluids immiscible with water. Sci Total Environ 21:173–185.

Schwille F. 1988. Dense Chlorinated Solvents in Porous and Fractured Media. Translated by Pankow JF. Lewis Publishers, Chelsea, MI, USA.

Soga K, Page JWE, Illangasekare TH. 2004. A review of NAPL source zone remediation efficiency and the mass flux approach. J Hazard Mater 110:13–27.

Stroo H, Ward CH, eds. 2010. *In Situ* Remediation of Chlorinated Solvent Plumes. Springer Science + Business Media, New York, NY, USA. 786 p.

Stroo HF, Leeson A, Marqusee JA, Johnson PC, Ward CH, Kavanaugh MC, Sale TC, Newell CJ, Pennell KD, Lebrón CA, Unger M. 2012. Chlorinated ethene source remediation: Lessons learned. Environ Sci Technol 46:6438–6447.

Truex MJ. Macbeth TW, Vermeul VR, Fritz BG, Mendoza DP, Mackley RD, Wietsma TW, Sandberg G, Powers T, Pitre E, Michalsen M, Ballock-Dixon SJ, Zhong L, Oostrom M. 2011. Demonstration of combined zero-valent iron and electrical resistance heating for in situ trichloroethene remediation. Environ Sci Technol 45:5346–5351.

U.S. Sustainable Remediation Forum (SURF). 2009. Sustainable remediation white paper: Integrating sustainable principles, practices, and metrics into remediation projects. Remediat J 19:5–114.

Vörösmarty CJ, Green P, Salisbury J, Lammers RB. 2000. Global water resources: Vulnerability from climate change and population growth. Sci 289:284–288.

West MR, Kueper BH. 2010. Plume detachment and recession times in fractured rock. Ground Water 48:416–426.

APPENDIX A
LIST OF ACRONYMS, ABBREVIATIONS AND SYMBOLS

°C	Degrees Celsius
°F	Degrees Fahrenheit
1,1,1-TCA	1,1,1-Trichloroethane
1,1,2,2-TeCA	1,1,2,2-Tetrachloroethane
1,1,2-TCA	1,1,2-Trichloroethane
1,1-DCA	1,1-Dichloroethane
1,1-DCE	1,1-Dichloroethene
1,2-DCA	1,2-Dichloroethane
1-D	One dimensional
2,4-DCP	2,4-Dichlorophenol
2-D	Two dimensional
3-D	Three dimensional
A/m^2	Amps per square meter
AAP	Alkaline-activated persulfate
AATDF	Advanced Applied (Environmental) Technology Development Facility
ac	Acre
AC	Alternating current
ACL	Alternative concentration limit(s)
ACS	American Chemical Society
ADE	Advection-dispersion equation
AFB	Air Force Base
AFCEE	Air Force Center for Environmental Excellence (renamed the Air Force Center for Engineering and the Environment)
AGU	American Geophysical Union
API	American Petroleum Institute
ARAR	Applicable or relevant and appropriate requirement
ASTM	American Society for Testing and Materials
ASU	Arizona State University
atm	Atmosphere
atm/M	Atmosphere/mole fraction
ATSDR	Agency for Toxic Substances and Disease Registry
AVS	Acid-volatile sulfides
BDL	Below detection limit
bgs	Below ground surface
BiRD	Biogeochemical reductive dechlorination
BOD	Biochemical oxygen demand
BSU	Bay sediment unit
BTC	Breakthrough curves
BTEX	Benzene, toluene, ethylbenzene, and total xylenes
CA	Chloroethane
CAH	Chlorinated aliphatic hydrocarbon
CCAFS	Cape Canaveral Air Force Station
CDI	Chronic daily intake
CDISCO	Conceptual design for ISCO
CERCLA	Comprehensive Environmental Response, Compensation, and Liability Act
CF	Chloroform
CFB	Canadian Forces Base
cfm	Cubic feet per minute
CFR	Code of Federal Regulations
CHP	Catalyzed hydrogen peroxide (H_2O_2) propagations
C-ISCO	Cosolvent *in situ* chemical oxidation
***cis*-DCE**	*Cis*-1,2-dichloroethene

B.H. Kueper et al. (eds.), *Chlorinated Solvent Source Zone Remediation*,
doi: 10.1007/978-1-4614-6922-3, © Springer Science+Business Media New York 2014

cm Centimeter(s)
CM Chloromethane (also termed methyl chloride)
cm/sec Centimeter(s) per second
CMC Carboxymethylcellulose
CMC Critical micelle concentration
CMT Continuous multichannel tubing
COC Contaminant of concern
CORT3D Chemical oxidation reactive transport in three dimensions
cP Centipoise
CPT Cone penetrometer testing
CRD Catalytic reductive dechlorination
CRWQCB California Regional Water Quality Control Board
CSIA Compound-specific isotope analysis
CSM Conceptual site model
CT Carbon tetrachloride
CVOC Chlorinated volatile organic compound
cy Cubic yard

DC Direct current
DCA Dichloroethane
DCE Dichloroethene
DCM Dichloromethane
DERP Defense Environmental Restoration Program
DGGE Denaturing gradient gel electrophoresis
DHC *Dehalococcoides*
DHE Department of Health and Environment
DIMP Diisomethylpropyl phosphonate
DMD Density-modified displacement
DMP Dimethylpentanol
DNA Deoxyribonucleic acid
DNAPL Dense nonaqueous phase liquid
DNAPL3D-RX Three-dimensional DNAPL remediation model
DNT Dinitrotoluene
DO Dissolved oxygen
DOC Dissolved organic carbon
DoD U.S. Department of Defense
DOE U.S. Department of Energy
DP Direct push
DT Dithionite
DTSC Department for Toxic Substances Control
DVI Dual-valent iron

Ea Activation energy
EA Electron acceptor
EC Electrical conductivity
ECD Electron capture detector
ECRS Experimental control release system
ED Electron donor
EDB Ethylene dibromide
EDDS Ethylenediaminedisuccinate
EDTA Ethylenediaminetetraacetic acid
EE/CA Engineering evaluation/cost analysis
EGDY East Gate Disposal Yard
Eh Redox potential
***e*-Hex** 2-Ethyl-1-hexanol
EISB Enhanced *in situ* bioremediation
EK Electrokinetics
EM Electromagnetic
EMD Environmental Molecular Diagnostics
EO Ethylene oxide
EOR Enhanced oil recovery
EOS® Edible Oil Substrate
ER Extraction and reinjection
ERD Enhanced reductive dechlorination
ERH Electrical resistance heating
ESM Equilibrium streamtube model
ESTCP Environmental Security Technology Certification Program
ET-DSP™ Electro thermal-dynamic stripping process
ETH Ethene
EU European Union
EVO Emulsified vegetable oil
EZVI Emulsified oil/nanoscale zero-valent iron

FISH	Fluorescent *in situ* hybridization
f_{OC}	Fraction of organic carbon
FRTR	Federal Remediation Technology Roundtable
ft	Foot/feet
FTA2	Fire Training Area 2
FY	Fiscal year
g	Gram(s)
GAC	Granular activated carbon
gal	Gallon(s)
GC	Gas chromatography
GE	General Electric
gpm	Gallon(s) per minute
GTP	Ganglia to pool (ratio)
GUI	Graphical user interface
GWERD	Ground Water and Ecosystem Restoration Division
ha	Hectare(s)
HD	Hydraulic displacement
HEIDA	Hydroxyethyliminodiacetic acid
HFCS	High-fructose corn syrup
HLB	Hydrophile–lipophile balance
HOC	Hydrophobic organic compounds
HPAM	Hydrolyzed polyacrylamide
hr	Hour(s)
HRC®	Hydrogen release compound
HRT	Hydraulic retention time
HSRC	Hazardous Substance Research Center
IAS	*In situ* air sparging
IC	Institutional control
ID	Inner diameter
IDW	Investigation-derived waste
IFA	Interfacial area
IFT	Interfacial tension
in	Inch(es)
IP	Injection point
IP	Ionization potential
IPT	Integral pump(ing) test
IRP	Installation Restoration Program
ISB	*In situ* bioremediation
ISCO	*In situ* chemical oxidation
ISCR	*In situ* chemical reduction
ISRM	*In situ* redox manipulation
ISSM	*In situ* soil mixing
ISTD	*In situ* thermal desorption
ISTT	*In situ* thermal treatment
ITRC	Interstate Technology & Regulatory Council
IUPAC	International Union of Pure and Applied Chemistry
IW	Injection wells
kg	Kilogram(s)
K_H	Henry's Law constant (H_i also used)
kJ/mole	Kilojoule(s) per mole
kL	Kiloliters
km	Kilometer(s)
K_{oc}	Soil organic carbon/water partition coefficient
K_{ow}	Octanol–water partition coefficient
kPa	Kilopascal(s)
krn-S	Relative permeability–saturation function (organic or non-wetting phase)
kr-S	Relative permeability–saturation relationship
krw-S	Relative permeability–saturation function (wetting or groundwater phase)
kVA	Kilovolt-ampere
L	Liter(s)
L/min	Liter(s) per minute
lb	Pound(s)
LCA	Life cycle analysis
LDA	Large-diameter augers
LEL	Lower explosive limit
LF3	Landfill Number 3
LIF	Laser-induced fluorescence
LLE	Liquid–liquid equilibrium
LLNL	Lawrence Livermore National Laboratory
LNAPL	Light nonaqueous phase liquid
LoRSC	Low-risk site closure
LTM	Long-term monitoring
LUC	Land use control
LUST	Leaking underground storage tank
m	Meter(s)
M	Molar/molarity

MBT Molecular biological tool
MC Methylene chloride
MCL Maximum contaminant level
M_D Mass discharge
MDEQ Michigan Department of Environmental Quality
MDL Minimum detection limit
mg Milligram(s)
mg/kg Milligram(s) per kilogram
mg/L Milligram(s) per liter
MGP Manufactured gas plant
mi Mile(s)
MIBK Methyl isobutyl ketone
min Minute(s)
MIP Membrane interface probe
mL Milliliter(s)
MLM Multilevel monitoring
MLS Multilevel sampler
mM Millimolar
mm Millimeter(s)
mmol Millimole
MNA Monitored natural attenuation
mol Mole(s)
MPPE Macroporous polymer extraction
mRNA Messenger ribonucleic acid
MS Mass spectrometry
MSR Molar solubilization ratio
MTBE Methyl tertiary butyl ether
mV Millivolt(s)
MW Monitoring well

NAPL Nonaqueous phase liquid
NAS Natural Attenuation Software
NAS Naval Air Station
NASA U.S. National Aeronautics and Space Administration
NATO North Atlantic Treaty Organization
NAVFAC Naval Facilities Engineering Command
NAWC Naval Air Warfare Center
NCP National Contingency Plan
ND Non-detect
NFA No further action
NFESC Naval Facilities Engineering Service Center
NGWA National Ground Water Association
NHE Normal hydrogen electrode
NIEHS National Institute of Environmental and Health Sciences
nM Nanomolar
nmol Nanomole(s)
NOD Natural oxidant demand
NOM Natural organic matter
NPE Nonylphenyl ethoxylates
NPL National Priorities List
NPV Net present value
NRC National Research Council
NRD Natural reductant demand
NSERC Natural Sciences and Engineering Research Council
NSF National Science Foundation
NTC Naval Training Center
NTA Nitrilotriacetic acid
NWIRP Naval Weapons Industrial Reserve Plant
nZVI Nanoscale zero-valent iron

O&M Operations and maintenance
OAM Organic aquifer material
OC Organic carbon
OD Outer diameter
OHSU Oregon Health & Science University
OM&M Operations, monitoring, and maintenance
OMB Office of Management and Budget
ORP Oxidation–reduction potential
OSWER Office of Solid Waste and Emergency Response
OTV Optical televiewer
OU Operable unit
OU1 Hill AFB Operable Unit 1
OU2 Hill AFB Operable Unit 2
OVA Organic vapor analyzer
OVM Organic vapor monitor

P&T Pump-and-treat
PAH Polycyclic aromatic hydrocarbon
PCB Polychlorinated biphenyl
PCE Perchloroethene (also termed tetrachloroethene)
PCL Protective concentration level
PCP Pentachlorophenol
PCR Polymerase chain reaction

Pc-S Capillary pressure–saturation relationship
PCU Power control units
PDB Polyethylene diffusion bag
PED Partitioning electron donors
PEL Permissible exposure limit
PF Pool fraction
PFM Passive flux meter
pg/L Picogram(s) per liter
PHS U.S. Public Health Service
PI Principal investigator
PITT Partitioning interwell tracer test
PLFA Phospholipid fatty acid analysis
PLM Power Law source depletion model
PNNL Pacific Northwest National Laboratory
ppb Part(s) per billion
PPE Personal protective equipment
ppm Part(s) per million
ppmv Part(s) per million by volume
PRB Permeable reactive barrier
psi Pounds per square inch
psig Pounds per square inch gauge
PTA Pilot test area
PV Pore volumes
PVC Polyvinyl chloride

QA/QC Quality assurance/quality control
qPCR Quantitative polymerase chain reaction
QSAR Quantitative structure–activity relationship

R&D Research and development
RAO Remedial action objective
RC Response complete
RCRA Resource Conservation and Recovery Act
RDase Reductive dehalogenase
RDX hexahydro-1,3,5-trinitro-1,3,5-triazine (also Royal Demolition eXplosive)
REMChlor Remediation Evaluation Model for Chlorinated Solvents
RF Radio frequency
RFH Radio frequency heating
RFM Recirculation flux measurement
RI Remedial Investigation
RIP Remedy in place
RNA Ribonucleic acid
ROD Record of Decision
ROI Radius of influence
rRNA Ribosomal RNA
RSK Risk-Based Standards for Kansas
RT3D Reactive transport in three dimensions
RTF Remediation time frame
RTZ Reactive treatment zones
RW Recovery well
RWQCB Regional Water Quality Control Board

s/sec Second(s)
SARA Superfund Amendments and Reauthorization Act
SC Soil conductivity
scfm Standard cubic feet per minute
SCR Silicon-controlled rectifier
SDS Sodium dodecyl sulfonate
Shell E&P Shell Exploration and Production
SEAR Surfactant-enhanced aquifer remediation
SEE Steam-enhanced extraction
SER Steam-enhanced remediation
SERDP Strategic Environmental Research and Development Program
SF Cancer risk slope factor
SHMP Sodium hexametaphosphate
S-ISCO Surfactant *in situ* chemical oxidation
SMART Specific, measureable, achievable, realistic, and time bound
SOD Soil oxidant demand
SOM Soil organic matter
SP Spontaneous potential
SRB Sulfate-reducing bacteria
SRSNE Solvents Recovery Service of New England

SRT Sustainable Remediation Tool
STAR Self-Sustaining Treatment for Active Remediation
STOMP Subsurface Transport Over Multiple Phases
SURF Sustainable Remediation Forum
SVE Soil vapor extraction
SVOC Semivolatile organic compound
SWMU Solid waste management unit
SWRB California State Water Resources Board
SZNA Source zone natural attenuation
SZTI Source zone targeted injection

TBA Tertiary butyl alcohol or tert-butyl alcohol
TC Thermocouples
TCA Trichloroethane
TCAAP Twin Cities Army Ammunition Plant
TCDD 2,3,7,8-Tetrachlorodibenzo-p-dioxin
TCE Trichloroethene
TCH Thermal conductive heating
TDEM Time domain electromagnetic
TDP Thermal dissipation probe
TDS Total dissolved solids
TEAP Terminal electron accepting process
TeCA Tetrachloroethane
TCEQ Texas Commission on Environmental Quality
TER Technology Evaluation Report
TI Technical impracticability
TOC Total organic carbon
TOR Time of remediation
TPH Total petroleum hydrocarbons
***trans*-DCE** *Trans*-1,2-dichloroethene
T-RFLP Terminal restriction fragment length polymorphism
TTZ Target treatment zone

U.S. The United States
UF Upper Floridian
UHG Upper Hawthorn Group
UNIQUAC Universal quasi-chemical theory
USACE U.S. Army Corps of Engineers
USCG U.S. Coast Guard
USEPA U.S. Environmental Protection Agency
USGS U.S. Geological Survey
UST Underground storage tank(s)
UT University of Texas at Austin
UV Ultraviolet

V Volt(s)
VC Vinyl chloride
VC RDase Vinyl chloride reductive dehalogenase
VFA Volatile fatty acid
VLF Very low frequency
VOC Volatile organic compound(s)
VOI Value of information

W/m^2 Watts per square meter
WBZ Water-bearing zone
WCGR Waterloo Center for Groundwater Research
WERC Waste Management Education & Research Consortium
WMA Waste Management Area
WSR Weight solubilization ratio
WWTF Wastewater treatment facility

yd Yard(s)
yr Year

ZVI Zero-valent iron
ZVM Zero-valent metal

μ Micron(s)
μg/kg Microgram(s) per kilogram
μg/L Microgram(s) per liter
μm Micrometer
μM Micromolar
μmoles Micromoles

APPENDIX B
UNIT CONVERSION TABLE

Multiply	By	To Obtain
Acres	0.405	Hectares
Acres	1.56 E-3	Square miles (statute)
Centimeters	0.394	Inches
Cubic feet	0.028	Cubic meters
Cubic feet	7.48	Gallons (U.S. liquid)
Cubic feet	28.3	Liters
Cubic meters	35.3	Cubic feet
Cubic yards	0.76	Cubic meters
Feet	0.305	Meters
Feet per year	9.66 E-7	Centimeters per second
Gallons (U.S. liquid)	3.79	Liters
Hectares	2.47	Acres
Inches	2.54	Centimeters
Kilograms	2.20	Pounds (avoir)
Kilograms	35.3	Ounces (avoir)
Kilometers	0.62	Miles (statue)
Liters	0.035	Cubic feet
Liters	0.26	Gallons (U.S. liquid)
Meters	3.28	Feet
Miles (statue)	1.61	Kilometers
Ounces (avoir)	0.028	Kilograms
Ounces (fluid)	29.6	Milliliters
Pounds (avoir)	0.45	Kilograms
Square feet	0.093	Square meters
Square miles	640	Acres

B.H. Kueper et al. (eds.), *Chlorinated Solvent Source Zone Remediation*,
doi: 10.1007/978-1-4614-6922-3, © Springer Science+Business Media New York 2014

APPENDIX C
GLOSSARY[1]

Abiotic - Occurring without the direct involvement of biologic organisms.

Absorption - The uptake of water, other fluids, or dissolved chemicals by a cell or an organism (as tree roots absorb dissolved nutrients in soil).

Acclimation - Physiological adjustment by an organism to environmental change.

Acetogen - A bacterium that generates acetate as a product of anaerobic respiration.

Activated carbon - A highly adsorbent form of carbon used to remove organic chemicals from liquid or gaseous streams.

Active treatment - Application of engineered remediation technologies such as *in situ* bioremediation, *in situ* chemical oxidation and reduction, permeable reactive barrier walls, air sparging, thermal treatment, excavation, and phytoremediation to meet cleanup goals.

Adaptive site management - A management approach that is deliberately iterative, with decisions modified and optimized over time as performance data are obtained and evaluated. It is often used at complex sites where there is high uncertainty and no clearly defined pathway to clean closure.

Adsorption - A process that occurs when a gas or a liquid solute accumulates on the surface of a solid or a liquid (adsorbent), forming a film of molecules or atoms (the adsorbate).

Advection - Transport of molecules dissolved in water along the groundwater flow path at an average expected velocity.

Aerobic - Environmental conditions where oxygen is present.

---

[1]This glossary is a compilation of definitions of terms synthesized by the volume editors and chapter authors from a variety of published and unpublished sources, including previous volumes in the SERDP/ESTCP Remediation Technology Monograph Series.

B.H. Kueper et al. (eds.), *Chlorinated Solvent Source Zone Remediation*,
doi: 10.1007/978-1-4614-6922-3, © Springer Science+Business Media New York 2014

Aerobic cometabolism - An aerobic process in which an organism grows on one compound while fortuitously degrading another compound (see cometabolism). For example, some aerobic bacteria can degrade trichloroethene (TCE) while growing on volatile alkanes (e.g., methane or propane) because the monooxygenase enzymes induced by the alkanes also degrade TCE.

Aerobic oxidation - Process by which some chlorinated solvents (notably dichloroethene [DCE] and vinyl chloride [VC]) are completely biodegraded (with carbon dioxide and chloride as end products) by aerobic bacteria that are capable of using the solvent compounds as growth substrates. Also referred to as *oxic mineralization.*

Aerobic respiration - Process whereby microorganisms use oxygen as an electron acceptor to generate energy.

Air sparging - Technology in which air or oxygen is injected into an aquifer to assist in the volatilization or the biodegradation of contaminants.

Aliphatic compounds - Any chemical compound belonging to the organic class in which the atoms are not linked together to form a ring.

Alkalinity - A measure of the ability of a solution to neutralize acids, equal to the stoichiometric sum of the bases in the solution. An expression of the buffering capacity of the solution.

Alternate concentration limit (ACL) - A risk-based concentration that will not pose a substantial hazard to human health or environmental receptors, given exposure pathways and other factors. Used for both Comprehensive Environmental Response, Compensation, and Liability Act (CERCLA) and Resource Conservation and Recovery Act (RCRA) sites.

Anaerobic - "Without air." Generally refers to occurring or living without oxygen present. Thus, in an anaerobic groundwater system, the chemistry is characterized by reductive conditions. Sometimes anaerobic is used (e.g., in wastewater treatment) to indicate a lack of any electron acceptors (including nitrate and sulfate). In groundwater, a dissolved oxygen concentration below 1.0 milligrams per liter (mg/L) is generally considered anaerobic.

Anaerobic respiration - Process whereby microorganisms use an electron donor such as hydrogen and a chemical other than oxygen as an electron acceptor. Common substitutes for oxygen are nitrate, sulfate, iron, carbon dioxide, and other organic compounds (fermentation).

Analytical model - A mathematical model that has a closed-form solution (the solution to the equations used to describe changes in a system can be expressed as a mathematical analytic function). Analytical solutions can be more exact than numerical models but are typically limited to linear problems, simplified geometry, and homogeneous properties.

Anion - A negatively charged ion.

Anisotropy - In hydrogeology, the condition under which the magnitude of a parameter, such as hydraulic conductivity, varies with direction.

Anoxic - "Without oxygen." For example, anoxic groundwater is groundwater that contains no dissolved oxygen.

Aquifer - An underground geologic unit that is able to produce groundwater at an economical rate or some defined volumetric rate. A confined aquifer lies beneath a confining unit of lower hydraulic conductivity. An unconfined aquifer does not have a confining unit, and its upper surface is defined by the water table.

Aquitard - An underground geological formation of low permeability that does not readily transmit groundwater.

Assimilative capacity - The capacity of a natural body of water to receive and degrade wastewaters or toxic materials, ideally without occurrence of a contaminant plume.

Attenuation - Reduction of contaminant concentrations over space or time. Includes both destructive (e.g., biodegradation, hydrolysis) and nondestructive (e.g., volatilization, dispersion) processes that lower concentrations.

Attenuation rate - The rate of contaminant concentration reduction over time.

Autotrophic - Self-sustaining or self-nourishing. Organisms that have the ability to synthesize their own food from inorganic materials (e.g., carbon dioxide and ammonium).

Bacterium - A single-celled organism of microscopic size (generally 0.3–2.0 microns [μm] in diameter). As opposed to fungi and higher order plants and animals (*eukaryotes*), bacteria are *prokaryotes* (characterized by the absence of a distinct, membrane-bound nucleus or membrane-bound organelles and by deoxyribonucleic acid (DNA) that is not organized into chromosomes).

Baseline - A set of data representing ambient conditions that are collected before remediation is implemented. Compared with post-treatment data to evaluate the effectiveness of remediation.

Bedrock - The solid or the fractured rock, either underlying unconsolidated material or forming ground surface if unconsolidated material is absent.

Bentonite - A colloidal clay, expansible when moist, formed by chemical alteration of volcanic ash.

Bioaugmentation - Addition of microbes to the subsurface to improve the biodegradation of target contaminants. Microbes may be "seeded" from populations already present at a site or from specially cultivated strains of bacteria.

Bioavailability - The degree or the ability to be absorbed and ready to interact in an organism.

Biobarrier - A remediation technology designed to intercept and biologically treat a contaminant plume as it passes through a permeable subsurface barrier. Biobarriers are created by installing wells or trenches across the width of a plume to deliver substrate to the microorganisms in the aquifer as groundwater flows through the barrier.

Biochemical - Produced by or involving chemical reactions of living organisms.

Biodegradation - Biologically mediated conversion of one organic compound to another.

Biofouling - Impairment of the functioning of wells or other equipment as a result of excessive growth of biomass or activity of microorganisms.

Biomarker - A biochemical within an organism that has a particular molecular feature that makes it useful for identifying a specific biological activity.

Biomass - Total mass of microorganisms present in a given amount of water or soil.

Bioremediation - Use of microorganisms to control and destroy contaminants.

Biostimulation - Addition of nutrients and/or other amendments designed to stimulate the desired activities of native organisms.

Biotransformation - Biologically catalyzed transformation of a chemical to some other product.

Buffering capacity - A measure of a solution's ability to resist changes in pH upon addition of acid or base.

Bulk retention capacity - The total volume of dense nonaqueous phase liquid (DNAPL) that has been retained as residual and pooled DNAPL in a unit volume of the subsurface. The bulk retention capacity accounts for the fact that not all lenses, laminations, and geological units within a source zone contain DNAPL, and it is a function of the release history, geology, and DNAPL properties. In unconsolidated media, the bulk retention capacity can be in the range from 0.005 to 0.03, and in fractured media, it can range from 0.0002 to 0.002.

Capillary barriers - Fine-grained lenses, layers, and laminations upon which lateral spreading and pooling of DNAPL can occur. Even if the capillary barrier is penetrated by the DNAPL, it is likely that lateral spreading will have occurred along the top surface of the barrier prior to the capillary pressure having exceeded the entry pressure of the barrier. The finer grained the capillary barrier, the higher the pool height of DNAPL that it can support.

Capillary pressure - The pressure difference between two immiscible liquids that arises because of interfacial tension. It is calculated as the non-wetting-phase pressure minus the wetting-phase pressure. If the DNAPL is the non-wetting phase and water is the wetting phase, for example, the capillary pressure would be the DNAPL pressure minus the water pressure.

Capture zone - The three-dimensional region that contributes the groundwater extracted by one or more wells or drains.

Catalyst - A substance that promotes a chemical reaction but does not itself enter into the reaction.

Catalytic reductive dechlorination - Remediation technology involving addition of a noble metal such as palladium (Pd) capable of catalyzing hydrogenolysis.

Catalyzed hydrogen peroxide (CHP) - Formulations of hydrogen peroxide that include a catalyst, generally iron. Addition of catalysts under appropriate conditions (e.g., iron salts under acidic conditions) "activates" the hydrogen peroxide, causing generation of free radicals, which can degrade contaminants that are not susceptible to oxidation by peroxide only or increase the rate of degradation of more easily oxidized contaminants.

Cation - A positively charged ion.

Chelating agent - A chemical that forms multiple bonds with a single metal ion to produce a soluble, complexed metal–chelant molecule. Used to enhance solubility and uptake of metals or to inhibit production of precipitates or scale.

Chlorinated solvent - A hydrocarbon in which chlorine atoms substitute for one or more hydrogen atoms in the compound's structure. Chlorinated solvents commonly are used for oil and grease removal in manufacturing, dry cleaning, and other operations. Examples include trichloroethene, perchloroethene, and trichloroethane.

Cocontaminant - A contaminant that is present but is not considered the contaminant of concern (COC) as a primary driver of remediation due to relatively lower concentration or level of risk. May or may not be targeted by *in situ* chemical oxidation (ISCO) (or other technologies) when treating the primary COC.

Combined remedy - The deliberate use of two or more technologies, often but not necessarily separated in time or space, to achieve the site management objectives most effectively or efficiently.

Cometabolism - The simultaneous metabolism of two compounds, in which the degradation of the second compound (the secondary substrate) depends on the presence of the first compound (the primary substrate). For example, in the process of degrading methane, some bacteria can degrade hazardous chlorinated solvents that they would otherwise be unable to degrade.

Compound-specific isotope analysis (CSIA) - Analytical method using specialized mass spectrometers to distinguish between stable isotopes of several important atoms, including $^{13}C/^{12}C$ and $^{35}Cl/^{37}Cl$.

Conceptual site model (CSM) - A description of relevant history, features, and subsurface conditions across the site to understand the nature, extent, and fate of contaminants and any potential risks they may represent. The CSM also provides a foundation for remedy evaluation and selection. The CSM describes the history, geology, hydrogeology, surface water features, groundwater-to-surface water interactions, contaminant types, contaminant sources, pathways of contaminant migration, and locations of potential receptors.

Cosolvent flushing - Remediation technology involving injection of a solution containing a cosolvent (a separate water-miscible organic compound) into a subsurface containing NAPL. The cosolvent increases the effective solubility of the NAPL constituents and can thereby increase recovery from extraction wells.

Damkohler number (Da) - Dimensionless number used to relate the chemical reaction time scale to other phenomena occurring in a system.

Darcy flux (*q*) - The volume of groundwater flowing through a unit area of the subsurface per unit time. The Darcy flux is typically calculated using Darcy's Law.

Darcy's Law - The relationship discovered by Henry Darcy that states that the volumetric groundwater flux is equal to the product of the hydraulic conductivity and the negative hydraulic gradient.

Dechlorination - A type of dehalogenation reaction involving replacement of one or more chlorine atoms with hydrogen.

Degradation - The transformation of a compound through biological or abiotic reactions.

Dehalogenation - Replacement of one or more halogens (e.g., chlorine, fluorine, or bromine) with hydrogen atoms.

Dehalorespiration - Energy-yielding respiratory metabolism that encompasses the reductive metabolism of halogenated compounds, such as chlorinated and brominated ethenes.

Dehydrohalogenation - Base-catalyzed chemical reaction occurring under reducing conditions that involves release of a hydrogen and halogen from adjacent carbons, with creation of a new carbon–carbon double bond.

Dense nonaqueous phase liquid (DNAPL) - An organic liquid that is denser than water and does not dissolve or mix easily in water (it is immiscible). In the presence of water, it forms a separate phase from the water. Many chlorinated solvents are DNAPLs. A **single-component DNAPL** is composed of only one chemical. A **multicomponent DNAPL** is composed of two or more chemical components.

Desorption - Opposite of sorption; the release of chemicals from solid surfaces.

Diffusion - Movement of molecules in response to a concentration gradient. Molecules move from areas of high concentration to areas of low concentration.

Dilution - The combined processes of advection, dispersion, and diffusion resulting in a net lowering of concentrations in groundwater.

Discretization - The process of converting a partial differential equation into discrete components (i.e., defining the time steps and spatial intervals over which the equation applies). This process is usually carried out as a first step toward making partial differential equations suitable for numerical evaluation and implementation on digital computers.

Dispersion - The spreading of molecules along and perpendicular to the direction of groundwater flow as a result of pore-scale velocity variations.

Dispersivity - An intrinsic property of a porous or a fractured medium dictating how strong dispersion will be.

Displacement pressure (P_d) - The threshold capillary pressure required for the initial entry of DNAPL into a water-saturated porous or fractured medium. More generally, the threshold capillary pressure required for the initial entry of a non-wetting fluid- into a wetting fluid-saturated porous or fractured medium.

DNAPL architecture - The range of DNAPL saturations and the lateral continuity and thickness of DNAPL pools and zones of residual DNAPL within source zones containing DNAPL.

DNAPL dissolution - Transfer of chemical components from the nonaqueous phase liquid (NAPL) to the aqueous phase.

DNAPL saturation (S_D) - The percentage of the pore space occupied by DNAPL. The sum of the water and DNAPL saturations is always 100% in a DNAPL–water two-phase system ($S_W + S_D = 1$).

DNAPL source zone - The overall volume of the subsurface containing residual and/or pooled DNAPL. Not all portions (e.g., lenses, laminations, or fractures) of the source zone will contain residual and/or pooled DNAPL. The **confirmed DNAPL source zone** is the part of the source zone within which it is known or highly likely that DNAPL exists. The **potential DNAPL source zone** is the part of the source zone within which it is possible that DNAPL exists, but the lines of evidence indicating DNAPL presence are either fewer or are not as strong as those associated with the confirmed DNAPL source zone.

Dual-porosity model - A mathematical model that conceptualizes flow or solute transport as occurring in two overlapping domains, one in which advection dominates (mobile zone) and one in which diffusion dominates (immobile zone). Mass transfer can occur between the mobile and immobile zones.

Effective diffusion coefficient - Product of the free solution diffusion coefficient and the tortuosity of the porous or the fractured medium. An input parameter to Fick's Law to quantify diffusive flux.

Effective porosity - The porosity of porous or fractured media available to a specific process of interest. For example, the effective porosity for groundwater flow accounts only for the pore spaces through which groundwater can flow (does not include occluded pores).

Effective solubility - The equilibrium solubility in water of any component of a multicomponent DNAPL. In general, the various components of a DNAPL suppress each other's aqueous solubility implying that effective solubilities are typically less than single-component (handbook) solubilities. For structurally similar compounds, the effective solubility can be estimated using Raoult's Law.

Electrical resistance heating (ERH) - Form of *in situ* thermal treatment in which the soil is heated by passing electrical current within a network of electrodes placed into the subsurface. The resistance of the soil to the electrical current causes the temperature to increase. Water and contaminants that boil, vaporize, or volatilize are captured using a soil vapor extraction system and treated aboveground. A variation of ERH is *electro thermal-dynamic stripping process* (ET-DSP).

Electron acceptor - Compound that receives electrons (and therefore is reduced) in the oxidation–reduction reactions that are essential for the growth of microorganisms and for bioremediation. Common electron acceptors in the subsurface are oxygen, nitrate, sulfate, iron, and carbon dioxide. Chlorinated solvents (e.g., trichloroethene) can serve as electron acceptors under anaerobic conditions.

Electron donor - Compound that donates electrons (and therefore is oxidized) in the oxidation–reduction reactions that are essential for the growth of microorganisms and bioremediation. Organic compounds (e.g., lactate) generally serve as an electron donor during anaerobic bioremediation. Less chlorinated solvents (e.g., vinyl chloride) can also serve as electron donors. Hydrogen generated in fermentation reactions also can serve as an electron donor.

Electron shuttle - A chemical compound that stimulates the biodegradation of contaminants by facilitating the transfer of electrons to and from bacteria. An electron shuttle compound, like a quinine, can be either reduced or oxidized and thus can shuttle electrons between the ultimate electron donor (e.g., organic carbon) and the ultimate electron acceptor (e.g., ferric iron).

Emergence saturation (S_e) - The DNAPL saturation at which a connected pathway of DNAPL is first formed in a porous or a fractured medium of interest.

Emulsified edible oil - A formulation in which an edible oil (such as soybean oil) is dispersed into water (e.g., through stirring or use of homogenizers) to form a mixture of oil droplets in water. Emulsifying the oil improves the distribution of the oil in the subsurface.

Emulsion - A suspension of small globules of one liquid in a second liquid with which the first will not mix (e.g., oil and water).

Enhanced *in situ* bioremediation (EISB) - See "*In situ* bioremediation."

Entry pressure (P_e) - The threshold capillary pressure required for a connected pathway of DNAPL to form in a water-saturated porous or fractured medium. The entry pressure is higher than the displacement pressure.

Enzyme - A protein created by living organisms to use in transforming a specific compound. The protein serves as a catalyst in the compound's biochemical transformation.

Ex situ - Latin term referring to the removal of a substance from its natural or original position, e.g., treatment of contaminated groundwater aboveground.

Fenton's reagent - A solution consisting of hydrogen peroxide and an iron catalyst used to oxidize contaminants.

Fermentation - Oxidation of organic compounds occurring in the absence of any external electron acceptor.

Ferric iron - Iron in the trivalent state (Fe^{3+}).

Ferrous iron - Iron in the bivalent state (Fe^{2+}), sometimes referred to as dual-valent iron.

First-order decay - Kinetic model based on exponential decay. First-order decay rates are observed commonly during biodegradation, as the biodegradation rate is a function of the concentration. The exponential decay rate is commonly expressed as the half-life for a contaminant.

Ganglia - A disconnected blob of DNAPL that is left behind a migrating DNAPL body. See Residual DNAPL.

Ganglia–pool ratio - The ratio of DNAPL ganglia to DNAPL pools within a DNAPL source zone. Some remediation technologies, such as hydraulic displacement, lead to an increase in the ganglia-to-pool ratio.

Gene probe - A fragment of DNA or ribonucleic acid (RNA) of variable length (usually 100–1,000 bases long) which is used to detect the presence of nucleotide sequences (i.e., a DNA target) that are complementary to the sequence in the probe. Gene probes can be used to determine, and sometimes quantify, the presence of specific genes within a sample.

Glaciofluvial - Pertaining to streams fed by melting glaciers or to deposits and landforms produced by such streams.

Green rust - A stable phase produced upon oxidation of ferrous hydroxides that contains both trivalent and divalent iron, with a formula of $Fe_3(OH)_8$. Commonly produced in the absence of sulfur. Capable of reducing chlorinated ethenes.

Growth substrate - An organic compound upon which bacteria can grow, usually as a sole carbon and energy source.

Half-life - Time required to reduce the concentration of a constituent to half of its initial value.

Hematite - Iron mineral (Fe_2O_3). Capable of reducing chlorinated ethenes.

Heterogeneity - The occurrence of variability of properties in space. For example, hydraulic conductivity is typically heterogeneous within an aquifer.

Hydraulic conductivity - A measure of the rate at which water moves through a unit area of the subsurface under a unit hydraulic gradient.

Hydraulic displacement - A method for recovering pooled DNAPL from the subsurface based on recovering DNAPL from either vertical wells or horizontal drains. Hydraulic displacement can be implemented in a variety of ways, ranging from pumping DNAPL only from a single extraction point to pumping both DNAPL and water from numerous wells or drains while injecting groundwater upgradient to accelerate DNAPL recovery. Also called waterflooding or dual-phase extraction.

Hydraulic fracturing - Method used to create fractures that extend from a borehole into rock-and-clay formations. Fractures are typically maintained by a proppant, a material such as grains of sand or other material that prevent the fractures from closing. Used to increase or restore the ability to transmit fluids.

Hydraulic gradient - The maximum rate of change of hydraulic head with distance and the direction of that maximum rate of change (a vector).

Hydraulic head - The energy (sum of the gravity and pressure components) possessed by a unit weight of water at any particular point in the subsurface. Measured by the elevation of water in a manometer at the laboratory scale or by the elevation of water in a well, borehole, or piezometer in the field. Groundwater flows from points of larger hydraulic head to points of lower hydraulic head.

Hydrocarbons - Chemical compounds that consist entirely of carbon and hydrogen.

Hydrogen bonding - Attractive force between the hydrogen attached to an electronegative atom of one molecule and an electronegative atom of a different molecule. Usually the electronegative atom is oxygen, nitrogen, or fluorine, which have partial negative charges, and the hydrogen then has a partial positive charge.

Hydrogen peroxide (H_2O_2) - An unstable compound used especially as an oxidizing and bleaching agent, antiseptic, and a propellant.

Hydrogenolysis - Chemical reaction in which a carbon–carbon or a carbon–heteroatom single bond is cleaved (or "lysed") by hydrogen. The heteroatom may vary, but it often is O, N, or S. Usually conducted catalytically using hydrogen gas.

Hydrolysis - The decomposition of organic compounds by interaction with water.

Hydrophilic - Having a strong affinity for water. Hydrophilic compounds tend to be found in the aqueous phase.

Hydrophobic - "Water-fearing." Hydrophobic compounds, such as oils and chlorinated solvents, have low solubilities in water and tend to form a separate nonaqueous phase.

Hypoxic - A condition of "low" or "deficient" oxygen content.

Hysteresis - A change in the relationship between two variables depending on history. For example, the relationship between capillary pressure and DNAPL saturation depends on whether capillary pressure is increasing or decreasing.

Immiscibility - The inability of two or more substances or liquids to readily dissolve into one another, such as oil and water.

In situ - Latin term meaning "in place" – in the natural or the original position, e.g., treatment of groundwater in the subsurface.

***In situ* bioremediation** - The use of microorganisms to degrade contaminants in place with the goal of obtaining harmless chemicals as end products. Generally, *in situ* bioremediation is applied to the degradation of contaminants in saturated soils and groundwater, although bioremediation in the unsaturated zone can occur.

***In Situ* chemical oxidation (ISCO)** - Technology that oxidizes contaminants in place by adding oxidants such as permanganate or peroxide, resulting in chemical destruction of the target contaminants.

***In situ* chemical reduction (ISCR)** - Technology that reduces contaminants in place through chemical reductants, with either naturally occurring reductants or addition of reagents such as zero-valent iron, resulting in chemical destruction of the target contaminants.

***In situ* redox manipulation (ISRM)** - Remediation technology involving injection of dithionite, a soluble chemical reductant, to reduce native ferric iron to adsorbed and structural ferrous iron, which can in turn reduce contaminants such as TCE.

***In situ* soil mixing (ISSM)** - A source zone remediation technology in which large-diameter augurs or backhoes are used to move and mix soil in place, leading to homogenization of the impacted subsurface and improved delivery of treatment reagents. The equipment may be used to deliver and mix in a variety of treatment chemicals, such as zero-valent iron, oxidants, electron donors, and/or stabilizing and solidifying agents such as bentonite.

***In situ* thermal treatment (ISTT)** - Treatment system that generates high temperatures to remove and destroy contaminants in place. In practice, three types of technologies have been used – steam injection, electrical resistance heating (generating heat by applying an electrical current), and thermal conductive heating (using electrical subsurface heaters to conduct heat through the solid matrix).

Inorganic compound - A chemical that is not based on covalent carbon bonds. Perchlorate is an inorganic compound, as are metals, nutrients such as nitrogen and phosphorus, minerals, and carbon dioxide.

Integral pump test (IPT) - A method developed to evaluate mass discharge across an entire plume, as opposed to relying on a series of point measurements from one or more individual monitoring wells. In the IPT method, pumping wells positioned along control planes perpendicular to the groundwater flow direction are operated for a time period on the order of days and sampled for contaminants. The concentration time series of the contaminants measured during operation of the pumping wells can then be used to determine the contaminant mass flow rates, mean concentrations, and plume shapes and positions at the control planes.

Interfacial tension - A tensile force within the interface between two immiscible liquids (such as a chlorinated solvent DNAPL and water) that results from the attractive forces between the molecules in the different fluids.

Intrinsic bioremediation - A type of *in situ* bioremediation that uses the innate capabilities of naturally occurring microbes to degrade contaminants without requiring engineering steps to enhance the process.

Intrinsic remediation - *In situ* remediation that uses naturally occurring processes to degrade or remove contaminants without using engineering steps to enhance the process.

Investigation-derived waste (IDW) - Waste generated in the process of investigating or examining an actual or a potentially contaminated site; includes solid and hazardous waste, media (including groundwater, surface water, soils, and sediments), and debris.

Ionization - The physical process of converting an atom or a molecule into an ion by adding or removing charged particles, such as electrons or other ions.

Ionization potential - Work required to remove (to infinity) the topmost electron in an atom or a molecule when the gas atom or molecule is isolated in free space and is in its ground electronic state.

Isotope - Any of two or more species of an element in the periodic table with the same number of protons. Isotopes have nearly identical chemical properties but different atomic masses and physical properties. For example, the isotopes chlorine 37 (^{37}Cl) and chlorine 35 (^{35}Cl) both have 17 protons, but ^{37}C has two extra neutrons and thus a greater mass.

Isotope fractionation - Selective degradation of one isotopic form of a compound over another isotopic form. For example, microorganisms preferentially degrade the ^{12}C isotope of PCE relative to the ^{13}C isotope.

Karst - Geologic formation of irregular limestone deposits with sinks, underground streams, and caverns.

Lactate - A salt or an ester of lactic acid.

Leachate - Solution formed when water percolates through a permeable medium. When passing through contaminated media, the leachate may contain contaminants in solution or in suspension.

Leaking underground storage tank (LUST) trust fund - Created by Congress in 1986 by amending Subtitle I of the RCRA. The LUST Trust Fund has two purposes. First, it provides money for overseeing and enforcing corrective action taken by a responsible party, who is the owner or the operator of the leaking underground storage tank (UST). Second, the Trust Fund provides money for cleanups at LUST sites where the owner or the operator is unknown, unwilling, or unable to respond or which require emergency action. Applicable in the United States.

Life cycle cost - The overall estimated cost for a particular remedial alternative over the time period corresponding to the life of the program including direct and indirect initial costs plus any periodic or continuing costs of operation and maintenance.

Light nonaqueous phase liquid (LNAPL) - A nonaqueous phase liquid less dense than water. Most LNAPLs do not penetrate to significant distances below the water table. Most common petroleum hydrocarbon fuels and lubricating oils are LNAPLs.

Liquid chromatography - A chemical separation technique in which the mobile phase (a liquid) passes over or through a stationary phase.

Log K_{ow} - Logarithmic expression of the octanol–water partition coefficient (K_{ow}), a measure of the equilibrium concentration of a compound between octanol and water.

Long-term monitoring (LTM) - Monitoring conducted after a remedial measure achieves its objectives to evaluate continued protection and performance.

Macroscopic - A length scale above microscopic; large enough to be seen by the unaided eye.

Magnetite - Common mineral of black iron oxide (Fe_3O_4), strongly attracted to magnetic fields. An important mineral capable of reducing chlorinated solvents in groundwater.

Mass balance - An accounting of the total mass inputs and mass outputs to a system.

Mass discharge - The rate of mass flow across an entire plume at a given location. Also referred to as integrated mass flux, it constitutes the entire mass of any solute passing a plane of reference perpendicular to the plume over a given time. Depending on the location, it may represent the source strength (the total mass emanating from a source zone) or the mass loading to a downgradient receptor such as a surface water body.

Mass flux - The amount of mass flowing through a unit cross-sectional area per unit time. Typically calculated as the product of the solute concentration and the Darcy flux. Often incorrectly used interchangeably with mass discharge.

Mass spectrometer - Instrument used to identify the chemical structure of a compound by measuring the masses and relative concentrations of atoms and molecules.

Mass transfer - A general term for the physical processes involving transport of atoms and molecules within physical or chemical systems. For example, DNAPL dissolution is a mass transfer process involving the transport of chemical constituents from DNAPL into groundwater.

Matrix diffusion - Diffusion of compounds in groundwater into, and back out from, low-permeability media such as clay or the primary porosity of rock. Often cited as the time-limiting factor in remediation of chlorinated solvent source zones where low-permeability media are present.

Matrix storage - Storage of compounds in low-permeability media such as clay or the primary porosity of rock. Significant fractions of the total contaminant mass may be stored in the low-permeability or inaccessible areas, complicating subsurface remediation.

Maximum contaminant level (MCL) - Standards set by the U.S. Environmental Protection Agency (USEPA) for drinking water quality that provides a legal threshold limit on the amount of a hazardous substance that is allowed in drinking water under the Safe Drinking Water Act. The limit is usually expressed as a concentration in milligrams or micrograms per liter of water.

Media - Specific environments (air, water, soil) that are the subject of regulatory concern and activities.

Membrane interface probe (MIP) - A probe with a permeable membrane on its side that is heated at desired intervals to volatize nearby organic compounds. The volatile organic compounds (VOCs) permeate the membrane and are delivered to the surface for analysis using one or more detectors, such as a photo-ionization detector, flame ionization detector, or an electron capture detector. Useful as a semiquantitative tool to locate VOC contamination.

Metabolism - The chemical reactions in living cells that convert food sources to energy and new cell mass.

Metabolite - The intermediates and products of metabolism.

Methanogenesis - Process of producing methane gas during biological metabolism.

Micelle - An aggregate of surfactant molecules dispersed in a liquid colloid. A typical micelle in aqueous solution forms an aggregate with the hydrophilic "head" regions in contact with surrounding solvent, sequestering the hydrophobic single-tail regions in the micelle center.

Microcosm - A laboratory vessel established to resemble the conditions of a natural environment.

Microemulsion - Clear, stable, isotropic liquid mixtures of oil, water, and surfactant, frequently in combination with a cosurfactant. The aqueous phase may contain salt(s) or other ingredients; the "oil" may actually be a complex mixture of different hydrocarbons and olefins. Microemulsions form upon simple mixing of the components and do not require the high shear conditions generally used in the formation of ordinary emulsions. The two basic types of microemulsions are direct (oil dispersed in water) and reversed (water dispersed in oil).

Microorganism (microbe) - An organism of microscopic or submicroscopic size. Bacteria are microorganisms.

Mineralization - The complete degradation of an organic chemical or organism to carbon dioxide, water, and possibly other inorganic compounds or elements.

Miscible - Two or more liquids that can be mixed and will remain mixed under normal conditions.

Modified Fenton's reagent - Hydrogen peroxide modified by addition of ferrous (II) or ferric (III) iron to dramatically increase its oxidative strength. This increase is attributed to the production of hydroxyl radicals ($OH^{\bullet}$), and initiation of a chain reaction, involving formation of new radicals. The reaction of iron-catalyzed peroxide oxidation at pH 3–5 is called a "Fenton's Chemistry" after its discoverer H.J.H. Fenton.

Mole fraction - The number of moles of a component of a solution divided by the total number of moles of all components. Mass fractions, as provided by laboratory analysis, can be converted to mole fractions using the molecular weight of each component.

Molecular biological tool (MBT) - Laboratory tests that can measure the presence and activity of microbes at a site. They can be used to assess the potential for and performance of monitored natural attenuation (MNA) and bioremediation strategies for remediation of environmental contaminants.

Monitored natural attenuation (MNA) - Refers to the reliance on natural attenuation processes (within the context of a carefully controlled and monitored site cleanup approach) to achieve site-specific remediation objectives within a time frame that is reasonable compared to that offered by other more active methods.

Monod kinetics - Equation based on the Michaelis–Menten equation for enzyme kinetics that relates a bacterial culture's specific growth rate (μ) to the substrate concentration(s). Requires empirically derived parameters for the maximum growth rate (μ_{max}) with excess substrate available and the half-maximal saturation constant (Ks) – the substrate concentration at which the growth rate is half of μ_{max}. The fundamental equation is $\mu = \mu_{max}$ (s/[Ks + s]).

Monte Carlo simulation - A problem-solving technique used to approximate the probability of certain outcomes by running multiple trial runs, called realizations.

Nanoscale - Generally deals with structures of the size 100 nm or smaller. For example, reactive iron produced in this size range is referred to as nanoscale iron.

Natural attenuation - Reduction in the mass, toxicity, mobility, volume, or concentration of contaminants in soil and/or groundwater caused by natural processes that act without human intervention. These *in situ* processes include biodegradation, dispersion, dilution, volatilization, radioactive decay, and chemical or biological stabilization, transformation, or destruction of contaminants.

Natural organic matter (NOM) - A form of naturally occurring organic matter (originating from organisms) that has been broken down to some base-level compounds (such as cellulose, chitin, protein, lipids). NOM provides nutrients to insects, bacteria, fungi, fish, and other organisms at the base of the food chain.

Natural oxidant demand (NOD) - Refers to one or more chemical reactions that can occur between an oxidant and naturally occurring substances in the subsurface (e.g., NOM, reduced metals, minerals). The oxidant consumed during these reactions is unavailable for reaction with the target contaminants.

Net present value (NPV) - The sum of the present values of future cash inflows and outflows. NPV is used in capital budgeting to analyze the profitability of an investment or a project. NPV analysis is sensitive to the reliability of future cash inflows that an investment or a project will yield. In essence, NPV compares the value of a dollar today to the value of that same dollar in the future, taking inflation and returns into account. Sometimes referred to as *net present worth.*

Nonaqueous phase liquid (NAPL) - An organic liquid that is sparingly soluble in water and remains as a separate fluid phase in the presence of water.

Non-wetting DNAPL - A DNAPL that is non-wetting on a solid surface in the presence of one or more other fluids. The contact angle is measured to be greater than 90° through the non-wetting liquid. Chlorinated solvent DNAPLs are typically non-wetting on solid surfaces in the presence of water. A non-wetting DNAPL will preferentially migrate through coarser grained layers and lenses of porous media and larger aperture fractures.

Numerical model - A mathematical model that uses a numerical time-stepping procedure and spatial discretization to estimate the behavior of a system over time and space (as opposed to an analytical model). Numerical models typically require greater computing power than analytical models, but they can allow more realistic simulations of complex systems.

Octanol–water partition coefficient (K_{ow}) - Ratio of the concentration of a chemical in octanol and in water at equilibrium and at a specified temperature. Octanol is an organic solvent used as a surrogate for NOM. This parameter is used in many environmental studies to help determine the fate of chemicals in the environment. Inversely related to aqueous solubility a high K_{ow} indicates that a compound will preferentially partition into an organic phase rather than into water.

Operation and maintenance (O&M) - Activities conducted at a site to ensure that a technology or an approach is effective and operating properly. The term O&M covers a wide range of activities, from overseeing the proper functioning of a system to conducting monitoring to evaluate the effectiveness of an action. Sometimes referred to in combination with monitoring (OM&M).

Organic - Referring to or derived from living organisms. In chemistry, any compound containing carbon.

Organohalide - Organic compound containing a halogen (such as chloride) linked to carbon.

Organohalide respiration - Process responsible for reductive dehalogenation of chlorinated ethenes, as well as a wide variety of related halogenated compounds, in which bacteria gain energy from reduction of the organohalide. This process has been referred to by several other terms, notably chlororespiration, dechlororespiration, halorespiration, and dehalorespiration.

Oxic - Containing oxygen or oxygenated. Often used to describe an environment, a condition, or a habitat in which oxygen is present.

Oxidation - Transfer (loss) of electrons from a compound, such as an organic contaminant. In the context of biodegradation, the oxidation can supply energy that microorganisms use for growth and reproduction.

Oxidation–reduction potential (ORP) - The tendency of a solution to either gain or lose electrons when it is subject to change by introduction of a new chemical species. A solution with a higher (more positive) reduction potential than the new species will have a tendency to gain electrons from the new species (to be reduced by oxidizing the new species); a solution with a lower (more negative) reduction potential will have a tendency to lose electrons to the new species (to be oxidized by reducing the new species). A positive ORP indicates that the solution is aerobic, while a negative ORP indicates that reducing conditions are dominant.

Oxygenase - An enzyme that introduces oxygen into an organic molecule.

Ozone (O_3) - A simple triatomic molecule, consisting of three oxygen atoms. An allotrope of oxygen that is much less stable than the diatomic oxygen (O_2). A powerful oxidizing agent. Unstable at high concentrations, decaying to ordinary diatomic oxygen.

Partition coefficient - Ratio of the concentrations in one phase to those in another. For example, the soil–water partition coefficient (K_d) is the ratio of the concentration on a solid phase to that in a liquid phase and thus represents a measure of the sorption potential, whereby a contaminant is divided between the soil and water phases.

Partitioning interwell tracer testing (PITT) - Method to estimate the volume of NAPL in a source zone by injecting and recovering a tracer that will partition into the NAPL phase. PITT has the potential to provide information about the NAPL volume distribution in a relatively large-scale area.

Passive injection (passive treatment) - Remediation approach involving additions of amendments to the subsurface on a one-time or a very infrequent basis.

Passive treatment - *In situ* bioremediation approach in which amendments are added to the subsurface on a one-time or an infrequent basis. Passive treatment relies on the use of slow-release electron donors, which can be injected into the subsurface or placed in trenches or wells.

Pathogen - Microorganisms (e.g., bacteria, viruses, or parasites) that can cause disease in humans, animals, and plants.

Percarbonate - Any of a family of perhydrates of carbonate compounds, such as sodium percarbonate ($2Na_2CO_3 \cdot 3H_2O$). Strong oxidant.

Perchloroethene (PCE, perchloroethylene, tetrachloroethene, tetrachloroethylene) - A colorless, nonflammable organic solvent, $Cl_2C{:}CCl_2$, used in dry-cleaning solutions, as an industrial solvent, and a variety of other purposes.

Permanganate - General name for a chemical compound containing the manganate (VII) ion (MnO_4^-). Because manganese is in the +7 oxidation state, the manganate (VII) ion is a strong oxidizing agent.

Permeability - An intrinsic property of a porous or a fractured medium that quantifies the ease with which water or any other fluid can migrate through that medium. Not synonymous with hydraulic conductivity, which accounts for fluid properties as well as permeability.

Permeable reactive barrier (PRB) - A permeable zone containing or creating a reactive treatment area oriented to intercepting and remediating a contaminant plume.

Persulfate - Ions or compounds with more oxygen than normal sulfates, such as potassium persulfate ($KHSO_5$). Strong oxidant.

pH - An expression of the intensity of the basic or the acid condition of a liquid; may range from 0 to 14, where 0 is the most acid and 7 is neutral. Natural groundwater usually has a pH between 6.5 and 8.5.

Pilot scale - A scale of demonstration, testing, or evaluation under laboratory or field conditions that can incorporate certain features and processes that are representative of a full-scale system. A pilot-scale study is often used to investigate the design and performance of a full-scale system.

Pilot test - A trial run of a remediation technology implemented at the field scale. Performed to assess the feasibility of the remediation technology and/or to collect field-scale data on which to base full-scale design. Generally conducted at smaller scale than full-scale treatment.

Plume - A zone of dissolved contaminants. A plume usually originates from a contaminant source zone and extends for some distance in the direction of groundwater flow.

Polychlorinated biphenyls (PCBs) - A group of stable chemicals (often DNAPLs) used in electrical transformers and capacitors for insulating purposes and in gas pipeline systems as lubricants. The sale and new use of these chemicals were banned by law in 1979.

Polymerase chain reaction (PCR) - Technique to amplify a single or a few copies of a specific DNA sequence by several orders of magnitude. Allows detection of a target gene or parts of a gene, even when present at low concentrations in soils or groundwaters for example. PCR relies on thermal cycling, consisting of cycles of repeated heating and cooling of the reaction for DNA melting and enzymatic replication.

Pool fraction (PF) - The fraction of the DNAPL mass present in the subsurface as pooled DNAPL, as opposed to residual DNAPL. Inversely proportional to the ganglia-to-pool (GTP) ratio.

Pooled DNAPL - A local, continuous distribution of DNAPL that accumulates above a capillary barrier. Capillary barriers are typically lower permeability horizons that can occur at any elevation in the subsurface. Within the pool, the DNAPL saturation is typically between 30 and 80% of pore space in both porous media and fractures. Because DNAPL in pools is continuous through the pore structure, pools are potentially mobile and can migrate into monitoring wells and can be mobilized by increases in the hydraulic gradient or lowering of interfacial tension.

Pore volume (PV) - The volume of void space within an overall volume of a porous medium. Used to determine a design metric: the number of pore volumes that is the ratio of the volume of injected reagents to the volume of pore space in a target treatment zone.

Porosity - The fraction of the subsurface volume consisting of pores, fractures, or cavities in which water, air, or DNAPL can exist.

Precipitate - A substance separated from a solution or a suspension by chemical or physical change.

Primary substrates - The electron donors and electron acceptors that are essential to ensure the growth of microorganisms. These compounds can be viewed as analogous to the food and oxygen that are required for human growth and reproduction.

Propagation reaction - Chemical reactions involving free radicals in which the total number of free radicals remains constant.

Protein - Complex nitrogenous organic compounds of high molecular weight made of amino acids. Essential for growth and tissue repair. Many, but not all, proteins are enzymes.

Pyrite - An iron sulfide mineral with the formula FeS_2. The most common of the sulfide minerals.

Radical - An atom, molecule, or ion with unpaired electrons, which are highly reactive. Chemical oxidants like H_2O_2 can be activated during use in ISCO and yield one or more types of radicals (often called free radicals), which serve as the primary oxidizing agents.

Radius of influence (ROI) - The radial distance from an injection/withdrawal point or well to the point where there is no significant impact from activities in the injection/withdrawal point or well.

Raoult's Law - Relates the equilibrium vapor pressure or solubility of components to the composition of the solution. For a multicomponent DNAPL, can be used to calculate the effective vapor pressure or effective solubility of individual components.

Record(s) of Decision - A public document used in the United States that explains which cleanup alternative(s) will be used at National Priorities List sites where, under CERCLA, Trust Funds pay for the cleanup.

Redox (redox potential) - Reduction/oxidation reactions – those in which atoms have their oxidation number changed. For example, carbon may be oxidized by oxygen to yield carbon dioxide or reduced by hydrogen to yield methane. The redox potential (ORP) reflects the tendency of a chemical species to acquire electrons and thereby be reduced. In a redox reaction, one chemical species, the reductant or the reducing agent, loses electrons and is oxidized and the other, the oxidant or the oxidizing agent, gains electrons and is reduced.

Reducing - Environmental conditions that favor a decrease in the oxidation state of reactive chemical species (e.g., reduction of sulfates to sulfides).

Reductase - An enzyme catalyzing a reducing reaction.

Reduction - Transfer of electrons to a compound such as oxygen; occurs when another compound is oxidized.

Reductive dechlorination (hydrogenolysis) - Reaction involving removal of one or more chlorine atoms from an organic compound and their replacement with hydrogen atoms. A subset of reductive dehalogenation. Key reaction for anaerobic degradation of chlorinated solvents.

Reductive dehalogenase - Enzyme responsible for reductive dehalogenation. Also called dechlorinase if specific to chlorine removal. Dehalogenases can provide useful biomarkers for evaluating biodegradation potential at a contaminated site.

Reductive dehalogenation - The process by which a halogen atom (e.g., chlorine or bromine) is replaced on an organic compound with a hydrogen atom. The reactions result in the net addition of two electrons to the organic compound.

Relative permeability - A dimensionless term that accounts for the fact that the overall permeability available to a particular fluid phase in a two-phase flow system is dependent on the saturation of that fluid phase. If the saturation of the phase is low, it will have a low relative permeability because it only occupies a low number of the pore spaces. If the saturation of the phase is high, it will have a high relative permeability, reflecting the fact it occupies a large number of the pore spaces.

Remediation - Implementation of natural or engineered processes that act to meet regulatory based, or self-imposed, cleanup goals in soil and groundwater.

Remediation objective - Remediation objectives can be established to state the purpose for which remediation is intended. They often tend to be high-level outcomes that are desired but, in and of themselves, are often not directly measurable.

Residence time (retention time) - The average amount of time that a particle (molecule) spends in a particular system.

Residual DNAPL - Disconnected blobs (ganglia) of a DNAPL, trapped by capillary forces in the pore space of both porous media and fractures. The blobs and ganglia are typically from 1 to 10 grain diameters in size in unconsolidated deposits and are left behind in the pathways that DNAPL has migrated through.

Residual saturation - The volume of residual DNAPL present in a unit volume of pore space. Residual DNAPL saturations typically vary between 5 and 30% of pore space in both porous media and fractures.

Retardation - Slowing of the movement of compounds in an aquifer relative to the groundwater velocity. Quantified by the retardation factor (R), which is the ratio of the groundwater velocity to the contaminant velocity.

Reverse osmosis - A treatment process used in water systems by adding pressure to force water through a semipermeable membrane. Reverse osmosis removes most drinking-water contaminants. Also used in wastewater treatment. Large-scale reverse osmosis plants are being developed.

Saturated zone - Part of the subsurface that is beneath the water table and in which the pores are filled with water.

Saturation - Refers to the fraction of the pore space that contains a particular fluid phase such as water, air, or NAPL. If no fluid is specified, it is generally taken to refer to water saturation.

Scavenger - Refers to a substance that can react with a free radical to inhibit the free radical from participating in oxidation reactions with COCs. Scavengers include organic compounds like formate and ethanol and inorganic compounds like bicarbonate and carbonate.

Seepage velocity - The average pore water velocity. Since groundwater flow occurs only through interconnected pores and not through the entire subsurface volume the seepage velocity (Vs) is equal to the Darcy flux divided by the effective porosity.

Site characterization - The collection of environmental data that are used to describe the conditions at a property and delineate the nature and extent of a site's contamination.

Soil organic matter (SOM) - Organic constituents in the soil, including undecayed plant and animal tissues, their partial decomposition products, and soil biomass. SOM includes high-molecular-weight organic materials (such as polysaccharides and proteins), simpler substances (such as sugars, amino acids, and other small molecules), and humic substances.

Soil vapor extraction (SVE, soil venting) - A technology for the *in situ* remediation of VOCs in unsaturated soils. The process removes soil vapor contaminated with VOCs and enhances the mass transfer of VOCs from the soil pores to the vapor phase by applying a vacuum to extract soil contaminants and gases.

Solute - A substance dissolved in another substance. A relevant example is an oxidant dissolved in groundwater: oxidant is the solute, and groundwater is the solvent.

Solvent - A substance, usually a liquid, capable of dissolving another substance.

Sorb - To take up and hold by either adsorption or absorption.

Sorption - Partitioning of a substance onto the surface of a solid by physical or chemical attraction. Can refer to either absorption (in which one substance permeates another) or adsorption (surface retention of solid, liquid, or gas molecules, atoms, or ions).

Source strength - The mass discharge from a source zone. Represents the mass loading to the plume per unit time. Can also represent the concentration of contaminants in the source zone.

Source zone - That region of the subsurface that contains, or formerly contained, chlorinated solvents in DNAPL form and contains a sufficient mass of chlorinated solvents to sustain concentrations in soil or groundwater above regulatory limits. This definition recognizes that DNAPL may not be present throughout the entire volume of the subsurface where it was once present and that it may be completely depleted as a result of natural processes or application of engineered remediation technologies.

Source zone architecture - Distribution of contaminants within a source zone. Consists of several features, including (1) the overall shape and dimensions of the source zone, (2) the ratio of residual to pooled DNAPL (also referred to as the pool fraction or the GTP ratio), (3) the lateral continuity of zones of residual DNAPL and DNAPL pools, (4) the thickness of zones of residual DNAPL and DNAPL pools, and (5) the portion of lenses and layers containing DNAPL versus those void of DNAPL. The source zone architecture influences the downgradient dissolved-phase plume concentrations and mass flux distribution.

Source zone natural attenuation (SZNA) - A variety of physical, chemical, or biological processes that act without human intervention to reduce the mass, toxicity, mobility, volume, mass flux, mass discharge, or concentration of contaminants in soil or groundwater within that volume of the subsurface currently or previously occupied by chlorinated solvents in the form of NAPLs.

Specific conductance (electrical conductivity) - Rapid method of estimating the dissolved solid content (total dissolved solids) of a water supply by testing its capacity to carry an electrical current.

Stabilization/solidification - Remediation technique in which contaminants are physically bound or enclosed within a stabilized mass (solidification) or their mobility is reduced due to chemical reactions induced between a stabilizing agent and the contaminants (stabilization).

Stabilizer - Term used to describe a substance that can reduce the rate of reaction of a chemical oxidant during transport in the subsurface.

Stakeholder - A person or a group that has a legitimate interest in a contaminated site.

Steady state - A condition of a physical or a chemical system whereby it does not change over time or in which any one change is continually balanced by another, such as the stable condition of a system in equilibrium.

Stoichiometry - The calculation of quantitative (measurable) relationships of the reactants and products in a balanced chemical reaction.

Substrate - A compound that microorganisms can use in the chemical reactions catalyzed by their enzymes.

Sulfate-reducing bacteria (SRB, sulfate reducer) - Bacteria that convert sulfate to hydrogen sulfide. SRB often play important roles in the oxygen-limited subsurface.

Superoxide radical anion - Superoxide is an anion with the chemical formula O_2^-. It is important as the product of the one-electron reduction of dioxygen O_2, which occurs widely in nature. With one unpaired electron, the superoxide ion is a free radical and, like dioxygen, is paramagnetic. Superoxide is biologically quite toxic and is deployed by the immune system to kill invading microorganisms. Because superoxide is toxic, nearly all organisms living in the presence of oxygen contain isoforms of the superoxide scavenging enzyme, superoxide dismutase, which is an extremely efficient enzyme; it catalyzes the neutralization of superoxide nearly as fast as the two can diffuse together spontaneously in solution.

Surfactant - A surface active agent that serves to lower surface tension, lower interfacial tension, and increase the total aqueous solubility of DNAPLs in groundwater. Generally classified as anionic, cationic, or nonionic.

Surfactant-enhanced *in situ* chemical oxidation (S-ISCO) - Remediation technology involving injection of chemical oxidants along with a surfactant compound into the subsurface. The surfactant increases the aqueous solubility of the NAPL, thereby providing increased contaminant mass available for oxidation.

Surfactant flushing - Remediation technology involving injection of a solution of surfactants into the subsurface containing NAPLs. Surfactants increase the effective aqueous solubility of the NAPL constituents, enhancing NAPL removal during flushing. Also known as surfactant-enhanced aquifer remediation (**SEAR**).

Sustainable remediation - The selection and operation of remediation strategies designed to maximize the net benefit on human health and the environment through the judicious use of limited resources.

Target treatment zone (TTZ) - The portion of the subsurface that the remediation technology or approach is intended to treat.

Technical impracticability (TI) waiver - A formal finding that it is not currently practicable to meet applicable or appropriate relevant requirements (ARARs) in a reasonable time frame and allows for alternative endpoints to be defined.

Terminal pressure (P_t) - The capillary pressure at which DNAPL no longer drains from a unit and residual globules and ganglia are formed.

Thermal conductive heating (TCH) - An *in situ* thermal technology in which electrical heater wells are installed in the subsurface and operated to generate heat that conducts through the porous or the fractured medium. Boiled and volatilized contaminants are recovered using a soil vacuum extraction system.

Tortuosity - The actual length of a solute transport path, which is sinuous in form, divided by the straight-line distance between the ends of the transport path.

Total dissolved solids (TDS) - Combined content of all inorganic and organic substances in a liquid that are present in a molecular, ionized, or microgranular (colloidal sol) suspended form.

Total organic carbon (TOC) - Mass of carbon bound in organic compounds in soils, sediments, and water. Often used as a nonspecific indicator of water quality.

Toxicity - The degree to which a substance or a mixture of substances can cause harm to organisms. Acute toxicity involves harmful effects in an organism through a single or a short-term exposure. Chronic toxicity is the ability of a substance or a mixture of substances to cause harmful effects over an extended period.

Tracer test - Used to "trace" the path of flowing water. Tracer tests are conducted by dissolving a tracer chemical into groundwater at concentrations that do not significantly change the aqueous density. Tracer chemicals must behave conservatively, meaning that no mass is lost through reaction or partitioning into differing phases. Typical tracer chemicals are chloride or bromide.

Transverse dispersion - Dispersion in a direction perpendicular to the overall direction of groundwater flow.

Trichloroethane (TCA) - An industrial solvent (CH_3CCl_3). Other names for it include methyl chloroform and chloroethane. Occurs in two isomers: 1,1,1-TCA and 1,1,2-TCA.

Trichloroethene (TCE) - A stable, low-boiling-point colorless liquid ($CH_3Cl:CHCl_2$). Used as a solvent or a metal-degreasing agent and in other industrial applications. Also called trichloroethylene.

Upscaling - The process of averaging over the local scale in order to determine parameters at the larger scale of interest that are consistent with the important processes at the local scale. Mathematical upscaling methods include volume averaging, homogenization theory, and moment methods.

Validation - Corroboration or the process of establishing the soundness of a technology or a model. In computer modeling and simulation, validation refers to the process of determining the degree to which the model or the simulation is an accurate representation of the real world from the perspective of its intended uses.

Vapor intrusion - Migration of volatile chemicals from the subsurface into overlying buildings and structures.

Vapor pressure - A measure of a substance's propensity to evaporate. The force per unit area exerted by vapor in an equilibrium state with surroundings at a given pressure. It increases exponentially with an increase in temperature. A relative measure of chemical volatility, vapor pressure is used to calculate water partition coefficients and volatilization rate constants.

Vaporization - Conversion of a substance from the liquid phase to the gaseous (vapor) phase. There are two types of vaporization: evaporation and boiling. As opposed to volatilization (transfer of contaminants between the gaseous and aqueous phases), the term vaporization is used to refer to transfer of contaminants from the NAPL directly to the gaseous phase.

Verification - The act of reviewing, inspecting, or testing a technology to establish and document that it meets regulatory or technical standards. In computer modeling, verification refers to the process of determining that a model implementation accurately represents the developer's conceptual description of the model and the mathematical solution to the model.

Vinyl chloride - A chemical compound ($CH_2:CHCl$) that is toxic and believed to be oncogenic. A colorless compound and an important industrial chemical chiefly used to produce the polymer polyvinyl chloride (PVC).

Volatile - Evaporating readily at normal temperatures and pressures.

Volatile organic compound (VOC) - Any organic compound that has a high enough vapor pressure under ambient conditions to significantly vaporize and enter the atmosphere.

Volatilization - Transfer of a chemical from the liquid to the gas phase. As opposed to vaporization (transfer of contaminants to the air from the NAPL phase), the term volatilization is used to refer to the transfer of contaminants dissolved in groundwater to the gaseous phase.

Water saturation (S_W) - The percentage of the pore space occupied by water.

Wettability - The relative degree to which a fluid will spread on or coat a solid surface in the presence of other immiscible fluids. Many DNAPLs are non-wetting, implying that they will preferentially occupy the pore spaces within coarser grained lenses and laminations and larger aperture fractures. Some DNAPLs are wetting with respect to water, however, implying that they will preferentially coat the aquifer materials and thereby occupy the pore spaces of the finer grained media.

Zero-order reaction - Chemical reaction in which the rate is independent of the concentrations of the reactants.

INDEX

D

E

F

G

H

I

Q

R

Printed in the United States of America